Atlas of Clinical Anatomy

By the same author

Gross Anatomy Dissector: A Companion for Atlas of Clinical Anatomy, 1978
An Atlas of Normal Radiographic Anatomy (with Alvin C. Wyman, M.D.), 1976
Clinical Anatomy for Medical Students, Third Edition, 1986
Clinical and Functional Histology, 1984
Clinical Embryology for Medical Students, Third Edition, 1983
Clinical Neuroanatomy for Medical Students, Second Edition, 1986

Atlas of Clinical Anatomy

Richard S. Snell, M.D., Ph.D.
Professor and Chairman, Department of Anatomy,
George Washington University School of Medicine
and Health Sciences, Washington, D.C.

Artist: Terry Dolan

Little, Brown and Company Boston

Third Printing

Library of Congress Catalog Card No. 78-50616

ISBN 0-316-80196-8

Printed in the United States of America

To My Students —
Past, Present, and Future

Contents

Preface

When a physician examines a patient, he first inspects the surface anatomy and relates what he sees, feels, and hears to what he knows exists beneath the skin. In each section of this atlas there are photographs displaying the more important surface landmarks of normal human anatomy.

Medical students are students of medicine and not students of anatomy. They learn the anatomy of those areas that are commonly diseased so that they can easily recognize the abnormal and be able to distinguish it from the normal. They should be continually reminded of the practical application of their knowledge, of how it can be useful in a clinical environment. To this end, a series of clinical notes is placed next to each picture, from which the reader can acquire applied information at a glance.

Most of the illustrations have been based on dissections carried out by freshman medical students. A few have been made from special dissections done by fourth-year medical students and surgical residents.

Each illustration has been carefully planned to show clearly the main structures present in a particular region. The illustrations have been arranged so that superficial structures are examined first, followed by progressively deeper structures. The anatomical terminology is that which is commonly used in clinical practice, and both Latin and English terms are used.

During the preparation of this atlas I have received both encouragement and helpful criticism from clinical and anatomical colleagues and from medical students, for which I am most grateful.

To Mrs. Terry Dolan, my artist, I extend a very special thanks for her patience and determination to produce faithful reproductions of the different regions shown. My thanks also go to Debra Kornbluh, Annette Kinsman, and Kathleen Young for their skill in typing the text and the many hundreds of labels. To James Kendrick, R.B.P., the photographer in the Audiovisual Department of George Washington University Medical Center, my thanks are due for his expert skill in taking the photographs of surface anatomy.

Finally, to the staff of Little, Brown and Company, I wish to express my gratitude and thanks for all their assistance throughout the preparation of this book.

R. S. S.

Figure P-1
Anatomical position, anterior view. A. 27-year-old male. B. 29-year-old female. C. 10-year-old male. D. 10-year-old female.

1. A knowledge of anatomy is of no value to the physician unless he can apply it to the living body. It is strongly recommended that while a student is learning gross anatomy, every opportunity should be taken to confirm the information by first looking at a dissected cadaver and then examining a living body.
2. In order to diagnose disease, the physician must take an accurate medical history, followed by a physical examination. A sound knowledge of the normal anatomy permits a physician to recognize abnormal anatomy. Having obtained an accurate anatomical localization of the disease and an understanding of the pathological process, the prognosis and treatment can be instituted.
3. A physician must not only be familiar with the normal structures seen on the surface of the body but must have a mental image of the normal position of structures that lie beneath the skin. Furthermore, he must be able to relate these unseen structures to identifiable surface landmarks.

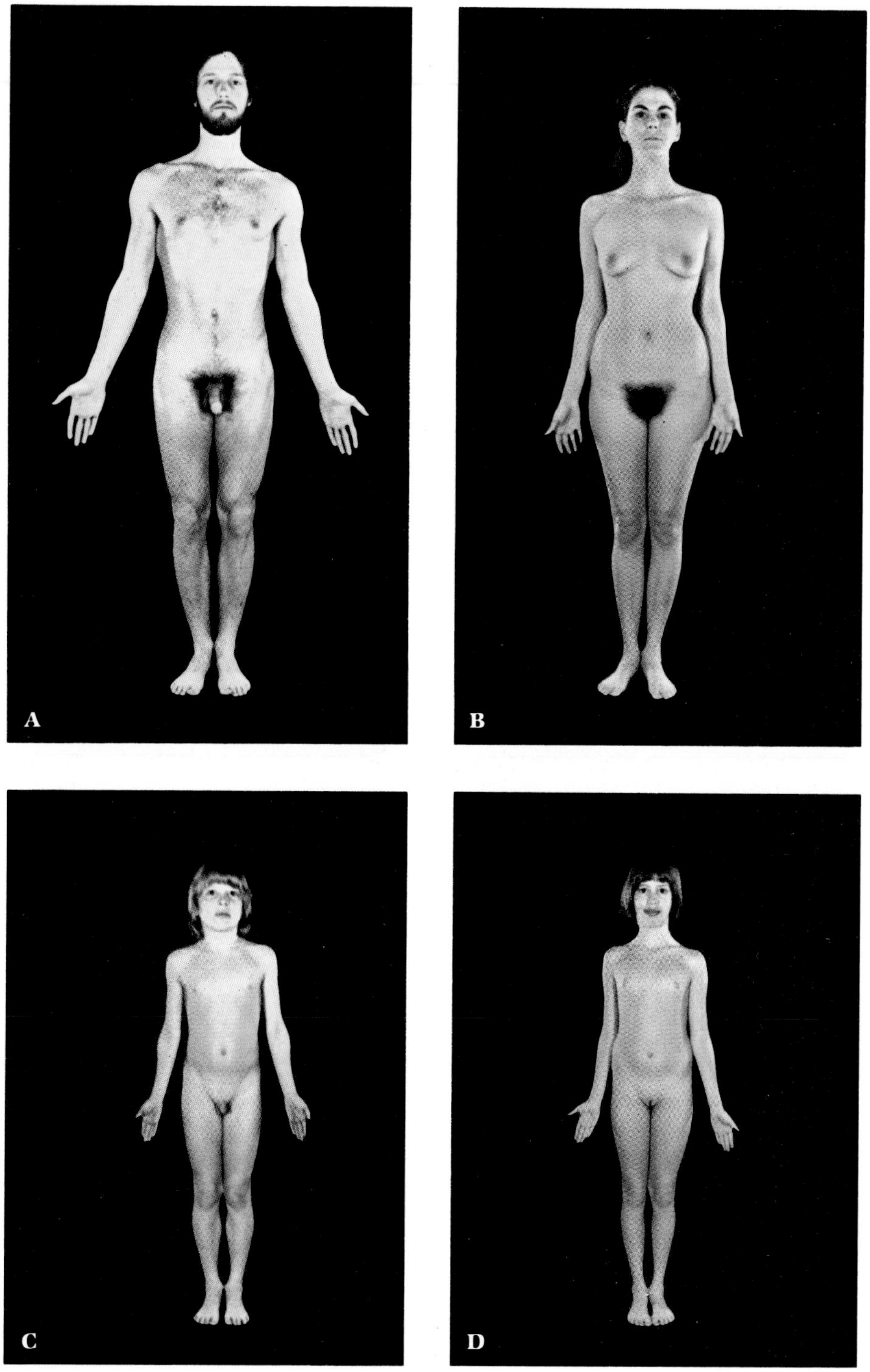
A
B
C
D

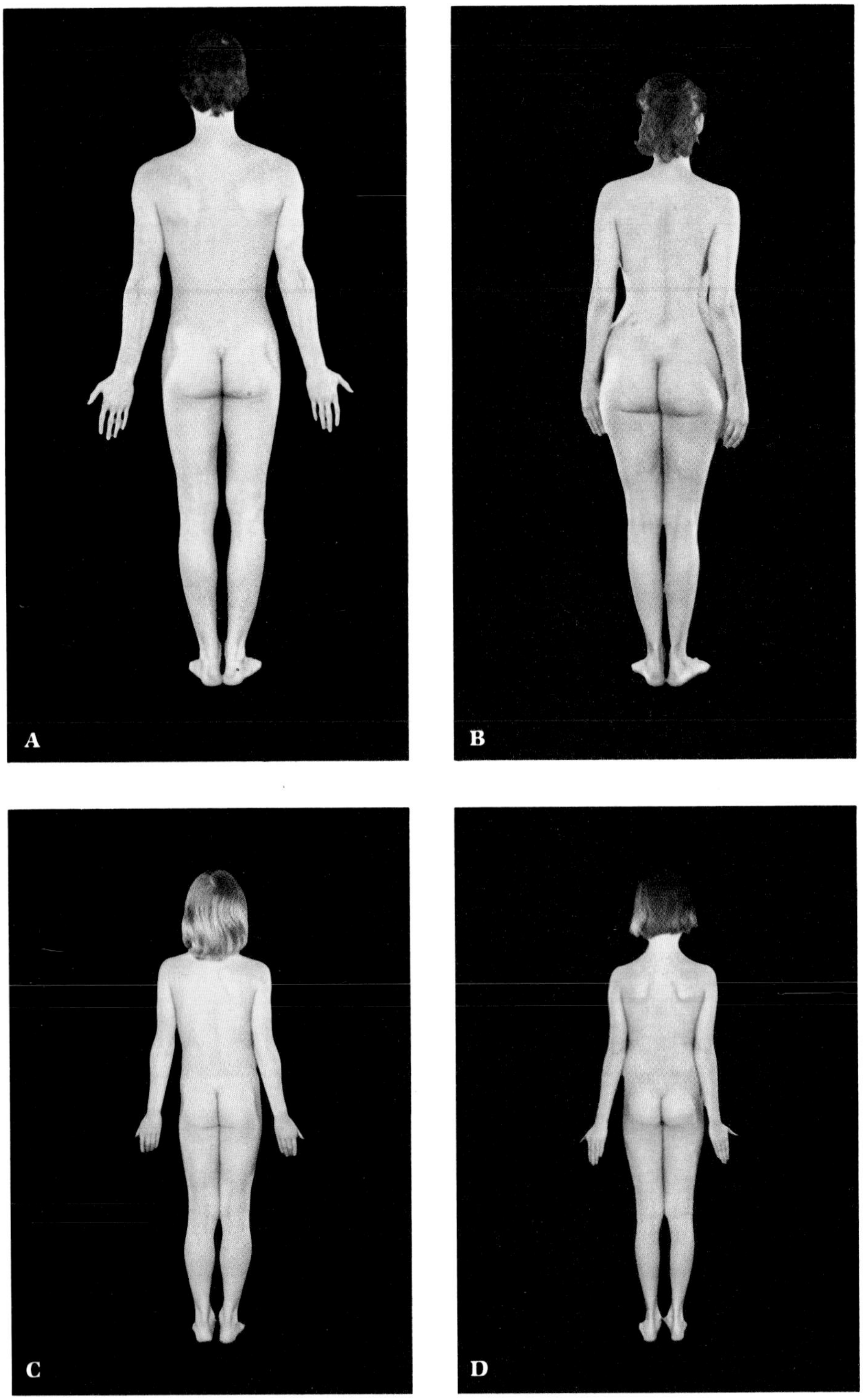

Figure P-2
Anatomical position, posterior view. A. 27-year-old male. B. 29-year-old female. C. 10-year-old male. D. 10-year-old female.

1. Descriptive gross anatomy tends to concentrate on a fixed descriptive form. A physician must always remember that there are sexual and racial differences and that the body structure changes as a person grows and ages.
2. The adult male tends to be taller than the adult female and has longer legs; his bones are bigger and heavier and his muscles are larger. He has less subcutaneous fat, which makes his appearance more angular. His larynx is larger and his vocal cords are longer, so that his voice is deeper. He has a beard and coarse body hair. He possesses axillary and pubic hair, with the latter extending up to the region of the umbilicus.
3. The adult female tends to be shorter than the adult male and has smaller bones, and the muscles are less bulky. She has more subcutaneous fat, accumulations in the breasts, buttocks, and thighs giving her a more rounded appearance. Her head hair is finer and her skin is smoother in appearance. She has axillary and pubic hair, but the latter does not extend above the mons pubis. Women have large breasts and a wider pelvis. They have a wider carrying angle at the elbow, which results in a greater lateral deviation of the forearm on the arm.
4. Up until the age of about 10 years boys and girls grow at about the same rate. Around 12 years the boys start to grow faster. At about 14 years of age the girls overtake the boys for a short period. After the age of 14 the boys grow faster than the girls, so that the majority reach adulthood taller than the female.
5. Puberty begins between ages 10 and 14 in girls and between 12 and 15 in boys. In the girl at puberty the breasts enlarge and the pelvis broadens. At the same time a boy's penis, testes, and scrotum enlarge, and in both sexes axillary and pubic hair appears.
6. Racial differences may be seen in the color of the skin, hair, and eyes, and in the shape and size of the eyes, nose, and lips. Africans and Scandinavians tend to be taller, due to longer legs, whereas Orientals are shorter, with shorter legs. The heads of central Europeans and Orientals also tend to be rounder and broader.

Figure P-3
Anatomical position, lateral view. A. 27-year-old male. B. 29-year-old female. C. 10-year-old male. D. 10-year-old female.

1. All descriptions of the human body are based on the assumption that the person is standing erect, with the arms by the sides and the face and palms of the hands directed forward. This is the so-called *anatomical position.*
2. The *median sagittal plane* is a vertical plane that passes through the center of the body, dividing it into equal right and left halves. Planes situated to one or the other side of the median plane and parallel to it are termed *paramedian planes.*
3. A structure situated nearer to the median plane of the body than another is said to be *medial* to the other. Similarly, a structure that lies farther away from the median plane than another is said to be *lateral* to the other.
4. *Coronal planes* are imaginary vertical planes at right angles to the median plane.
5. *Horizontal* or *transverse planes* are at right angles to both the median and coronal planes.
6. The terms *anterior* and *posterior* are used to indicate the front or back of the body, respectively; thus, when describing the relationship of two structures, one is said to be anterior or posterior to the other insofar as it is closer to the anterior or posterior body surface.
7. The terms *palmar* and *dorsal surfaces* are used instead of anterior and posterior in describing the hand.
8. In reference to the foot the terms *plantar* and *dorsal surfaces* are used instead of lower and upper surfaces.
9. The terms *proximal* and *distal* describe the relative distances from the roots of the limbs; for example, the arm is proximal to the forearm, and the hand is distal to the forearm.
10. The terms *superficial* and *deep* denote the relative distances of structures from the surface of the body.
11. The terms *superior* and *inferior* denote levels relatively high or low with reference to the upper and lower ends of the body.
12. The terms *internal* and *external* are used to describe the relative distance of a structure from the center of an organ or cavity; for example, the internal carotid artery is found inside the cranial cavity and the external carotid artery is found outside the cranial cavity.
13. A person in the *supine position* is lying on his back. A person lying face downward is in the *prone position.*

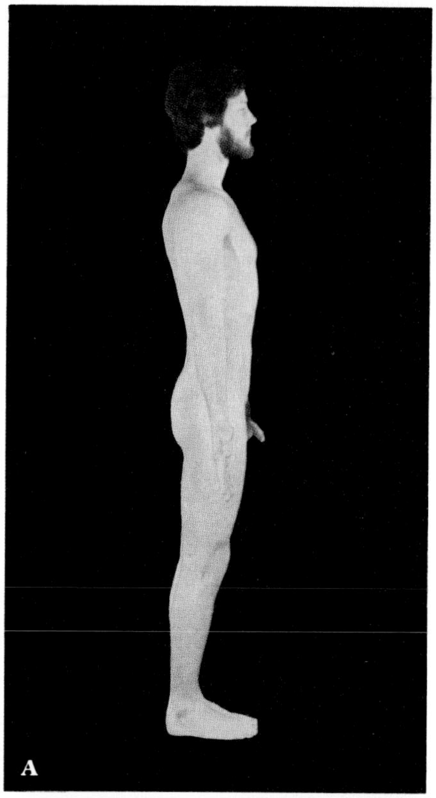

A

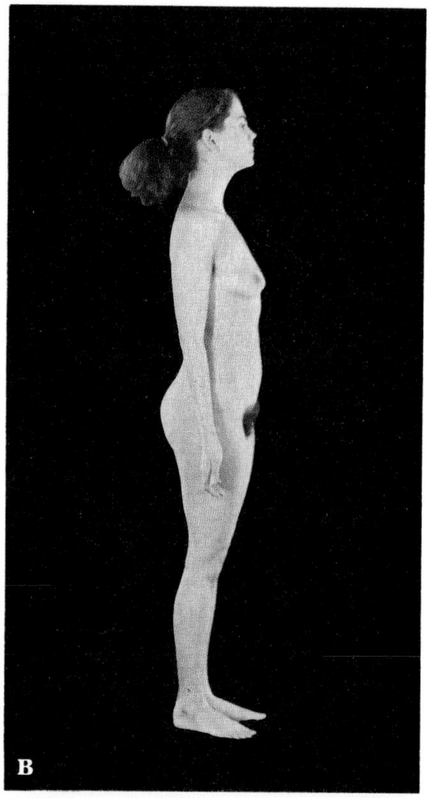

B

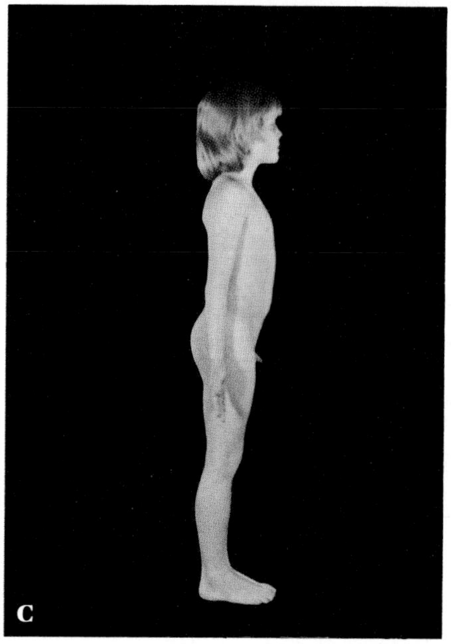

C

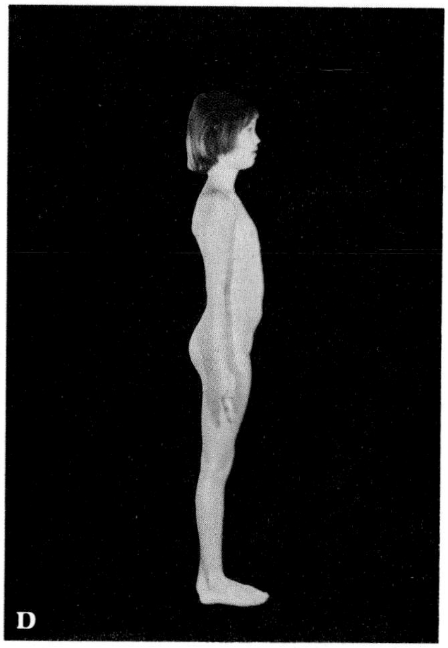

D

Atlas of Clinical Anatomy

1. The Thorax

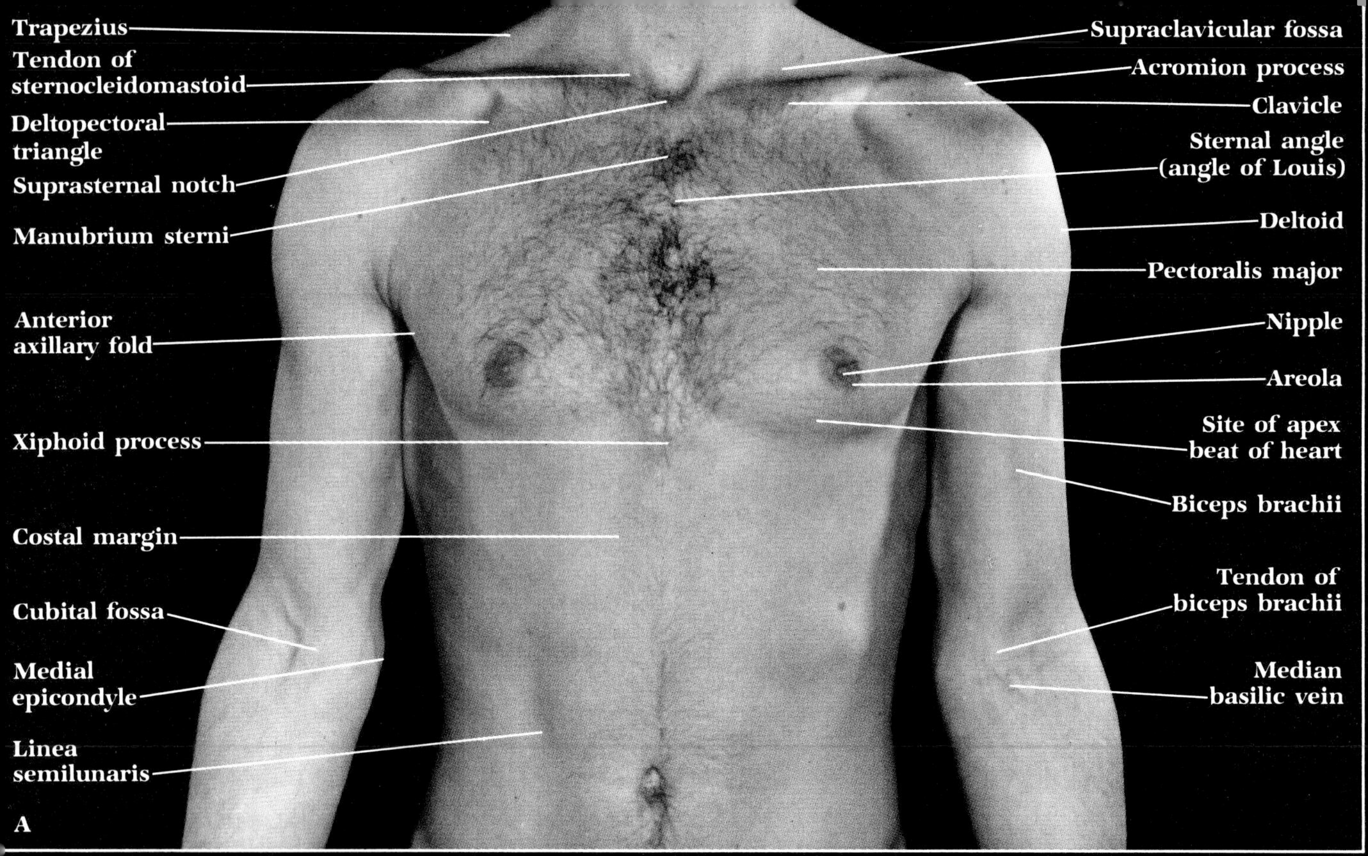
Trapezius
Tendon of sternocleidomastoid
Deltopectoral triangle
Suprasternal notch
Manubrium sterni
Anterior axillary fold
Xiphoid process
Costal margin
Cubital fossa
Medial epicondyle
Linea semilunaris
Supraclavicular fossa
Acromion process
Clavicle
Sternal angle (angle of Louis)
Deltoid
Pectoralis major
Nipple
Areola
Site of apex beat of heart
Biceps brachii
Tendon of biceps brachii
Median basilic vein
A

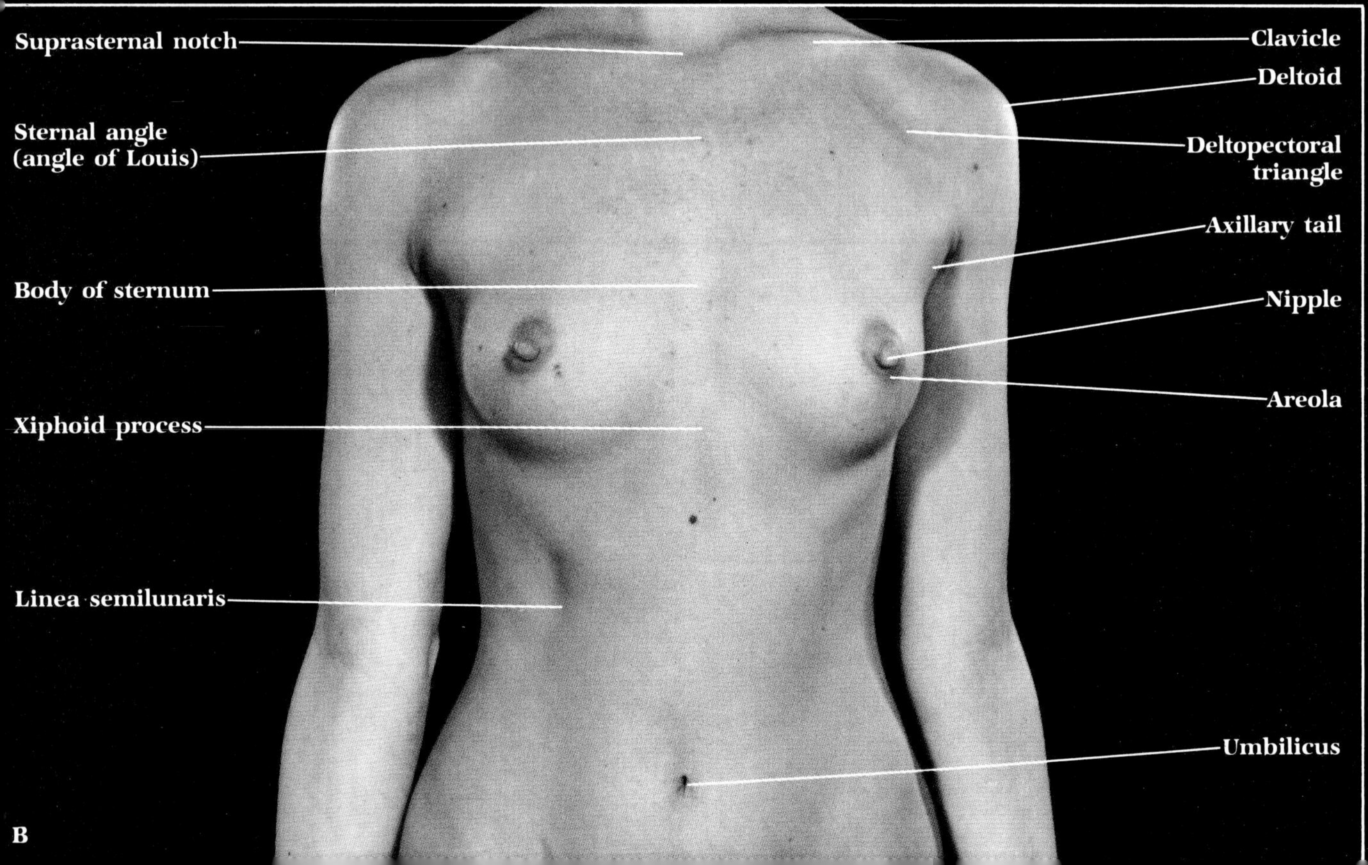
Suprasternal notch
Sternal angle (angle of Louis)
Body of sternum
Xiphoid process
Linea semilunaris
Clavicle
Deltoid
Deltopectoral triangle
Axillary tail
Nipple
Areola
Umbilicus
B

Figure 1-1
Thorax, anterior view.
A. 27-year-old male. B. 29-year-old female.

1. The suprasternal notch is the superior margin of the manubrium sterni and is easily felt between the medial ends of the clavicles in the midline. It lies opposite the lower border of the body of the second thoracic vertebra.
2. The sternal angle (angle of Louis) is the angle made between the manubrium and body of the sternum; at this level the second costal cartilage joins the lateral margin of the sternum. The sternal angle lies opposite the intervertebral disc between the fourth and fifth thoracic vertebrae.
3. The xiphisternal joint is the joint between the xiphoid process of the sternum and the body of the sternum. It lies opposite the body of the ninth thoracic vertebra.
4. The subcostal angle is at the inferior end of the sternum between the sternal attachments of the seventh costal cartilages.
5. The costal margin is the lower boundary of the thorax and is formed by the cartilages of the seventh, eighth, ninth, and tenth ribs, and the ends of the eleventh and twelfth cartilages. The lowest part of the costal margin is formed by the tenth rib and lies at the level of the third lumbar vertebra.
6. The clavicle is subcutaneous throughout its entire length and can be easily palpated. It articulates at its lateral extremity with the acromion process of the scapula.
7. The first rib lies deep to the clavicle and cannot be palpated. The lateral surfaces of the remaining ribs can be felt by pressing the fingers upward into the axilla and drawing them downward over the lateral surface of the chest wall. The twelfth rib, if short, may be difficult to palpate. To identify a particular rib, always first identify the second costal cartilage at the sternal angle and then count the cartilages and ribs downward from this point.
8. The nipple in the male usually lies in the fourth intercostal space about 4 inches (10 cm) from the midline. Its position is not constant in the female.
9. The trachea may be palpated at the root of the neck in the suprasternal notch in the midline.
10. The apex of the lung can be mapped out on the anterior surface of the neck by drawing a curved line, convex upward, from the sternoclavicular joint to a point 1 inch (2.5 cm) above the junction of the medial and intermediate thirds of the clavicle.
11. The anterior border of the right lung begins behind the sternoclavicular joint and runs inferiorly, almost reaching the midline behind the sternal angle. It then continues downward until it reaches the xiphisternal joint.
12. The anterior border of the left lung has a similar course as the right lung, but at the level of the fourth costal cartilage it deviates laterally and extends for a variable distance beyond the lateral margin of the sternum to form the cardiac notch. It then turns sharply inferiorly to the level of the xiphisternal joint.
13. The apex of the heart, formed by the left ventricle, corresponds to the apex beat and is found in the fifth left intercostal space 3½ inches (9 cm) from the midline.
14. The superior border of the heart extends from a point on the second left costal cartilage ½ inch (1.3 cm) from the edge of the sternum to a point on the third right costal cartilage ½ inch (1.3 cm) from the edge of the sternum.
15. The right border of the heart, formed by the right atrium, extends from a point on the third right costal cartilage ½ inch (1.3 cm) from the edge of the sternum downward to a point on the sixth right costal cartilage ½ inch (1.3 cm) from the edge of the sternum.
16. The left border of the heart, formed by the left ventricle, extends from a point on the second left costal cartilage ½ inch (1.3 cm) from the edge of the sternum to the apex beat of the heart.
17. The inferior border of the heart, formed by the right ventricle and the apical part of the left ventricle, extends from the sixth right costal cartilage ½ inch (1.3 cm) from the sternum to the apex beat.
18. The arch of the aorta and the roots of the brachiocephalic and left common carotid arteries lie behind the manubrium sterni.

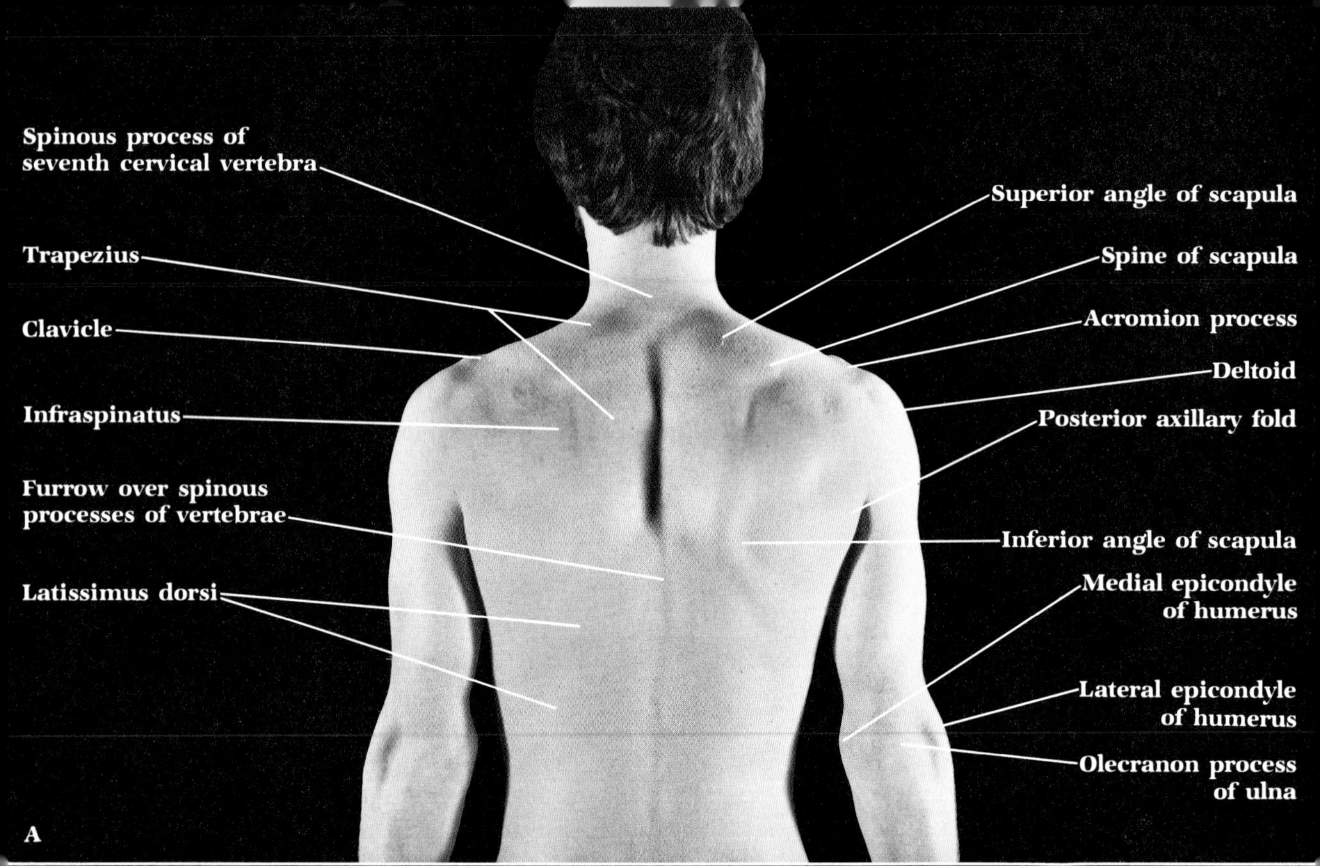
Spinous process of seventh cervical vertebra
Trapezius
Clavicle
Infraspinatus
Furrow over spinous processes of vertebrae
Latissimus dorsi
Superior angle of scapula
Spine of scapula
Acromion process
Deltoid
Posterior axillary fold
Inferior angle of scapula
Medial epicondyle of humerus
Lateral epicondyle of humerus
Olecranon process of ulna
A

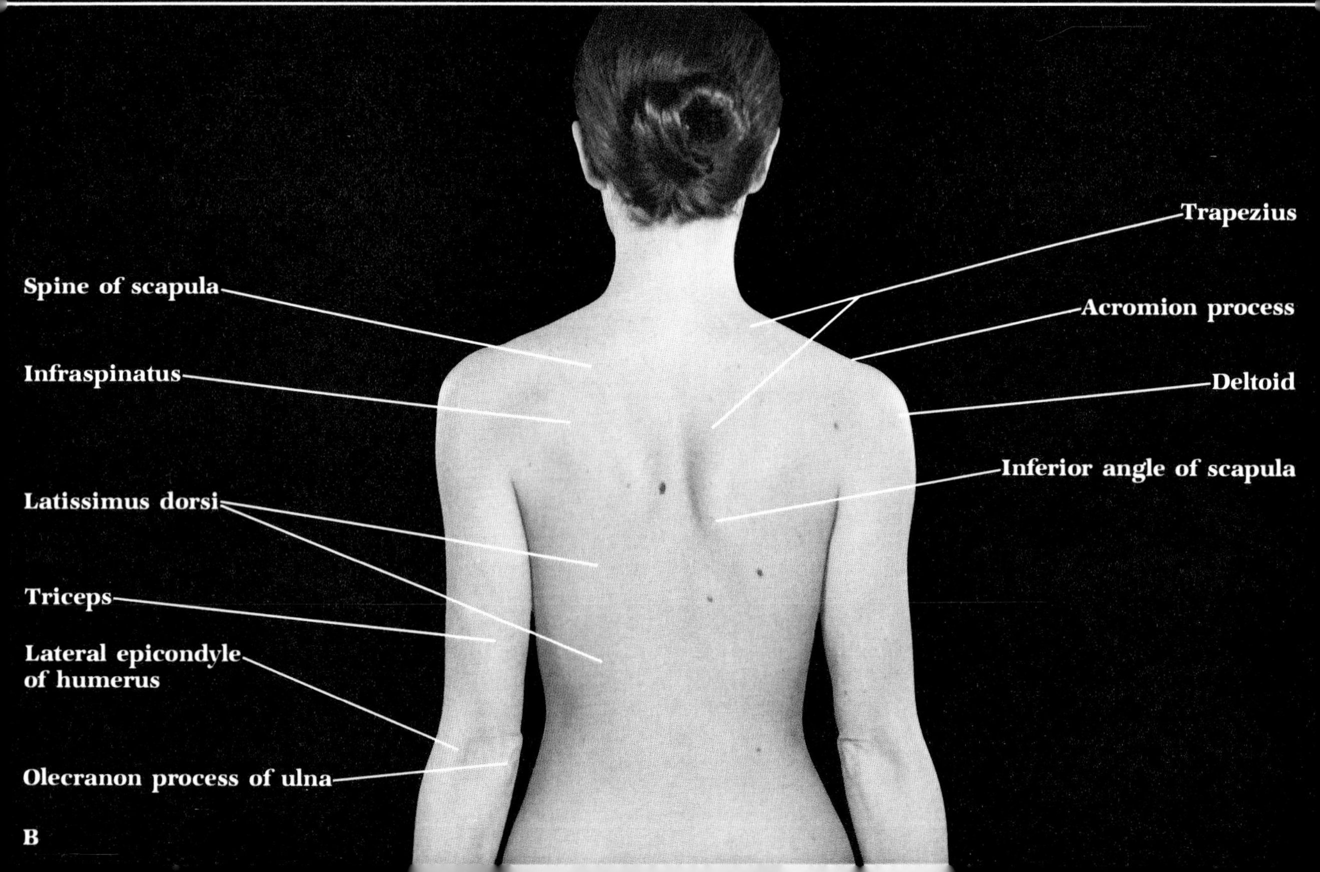
Trapezius
Spine of scapula
Acromion process
Infraspinatus
Deltoid
Inferior angle of scapula
Latissimus dorsi
Triceps
Lateral epicondyle of humerus
Olecranon process of ulna
B

Figure 1-2
Thorax, posterior view.
A. 27-year-old male. B. 29-year-old female.

1. There are anterior and posterior *axillary folds*. The anterior fold is formed by the lower border of the pectoralis major muscle. This may be made to stand out by asking the patient to press his hand hard against his hip. The posterior fold is formed by the tendon of the latissimus dorsi muscle as it passes around the lower border of the teres major muscle.

2. The spinous processes of the thoracic vertebrae can be palpated in the midline posteriorly. The index finger should be placed on the skin in the midline on the posterior surface of the neck and drawn downward in the nuchal groove. The first spinous process to be felt is that of the seventh cervical vertebra (vertebra prominens). Below this level are the spines of the thoracic vertebrae. The tip of the spinous process of a thoracic vertebra lies posterior to the body of the next vertebra below.

3. The superior angle of the scapula lies opposite the spine of the second thoracic vertebra.

4. The spine of the scapula is subcutaneous, and the root of the spine lies on a level with the spine of the third thoracic vertebra.

5. The inferior angle of the scapula lies on a level with the spine of the seventh thoracic vertebra.

6. The posterior border of each lung extends inferiorly from the spinous process of the seventh cervical vertebra to the level of the tenth thoracic vertebra and lies about 1½ inches (4 cm) from the midline.

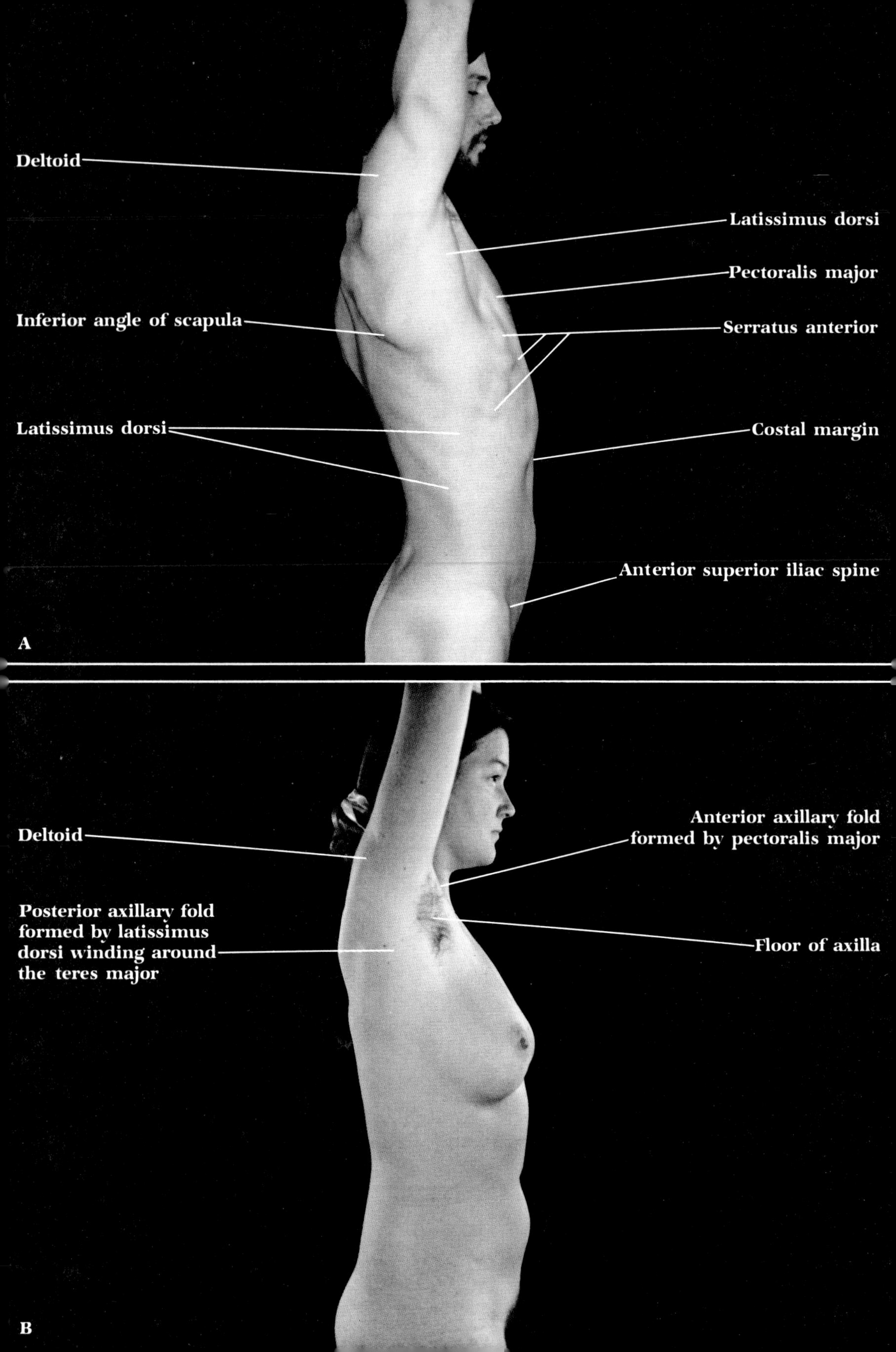
Deltoid
Latissimus dorsi
Pectoralis major
Inferior angle of scapula
Serratus anterior
Latissimus dorsi
Costal margin
Anterior superior iliac spine
A
Deltoid
Anterior axillary fold formed by pectoralis major
Posterior axillary fold formed by latissimus dorsi winding around the teres major
Floor of axilla
B

Figure 1-3
Thorax, lateral view.
A. 27-year-old male. B. 25-year-old female.

1. The lines of orientation in the thorax are as follows:
a. The midsternal line lies in the median plane over the sternum.
b. The midclavicular line runs vertically downward from the midpoint of the clavicle.
c. The anterior axillary line runs vertically downward from the anterior axillary fold.
d. The posterior axillary line runs vertically downward from the posterior axillary fold.
e. The midaxillary line runs vertically downward from a point situated midway between the anterior and posterior axillary folds.
f. The scapular line runs vertically downward on the posterior wall of the thorax, passing through the inferior angle of the scapula (when the arms are at the sides).
2. The lower border of each lung in mid-inspiration follows a curving line, which crosses the sixth rib in the midclavicular line, the eighth rib in the midaxillary line, and reaches the tenth rib adjacent to the vertebral column posteriorly.
3. The oblique fissure of each lung can be indicated on the surface by a line drawn from the root of the spine of the scapula obliquely downward, laterally and anteriorly, following the course of the sixth rib to the sixth costochondral junction. In the left lung, the upper lobe lies above and anterior to this line; the lower lobe lies below and posterior to it.
4. The horizontal fissure in the right lung may be represented by a line drawn horizontally along the fourth costal cartilage to meet the oblique fissure in the midaxillary line. The upper lobe lies above the horizontal fissure and below it, the middle lobe; the lower lobe lies below and posterior to the oblique fissure.

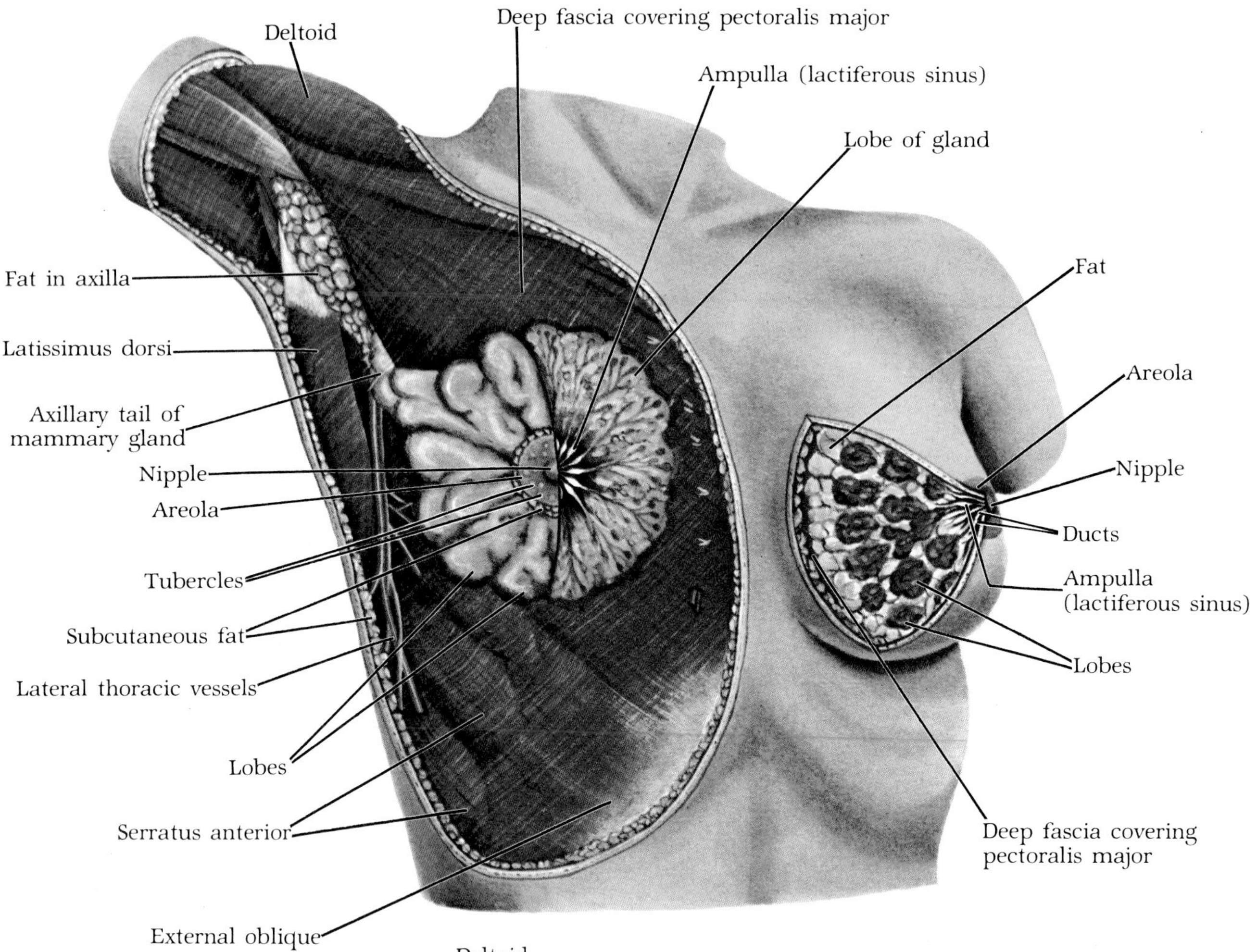

Deltoid
Deep fascia covering pectoralis major
Ampulla (lactiferous sinus)
Lobe of gland
Fat in axilla
Latissimus dorsi
Axillary tail of mammary gland
Nipple
Areola
Tubercles
Subcutaneous fat
Lateral thoracic vessels
Lobes
Serratus anterior
External oblique
Fat
Areola
Nipple
Ducts
Ampulla (lactiferous sinus)
Lobes
Deep fascia covering pectoralis major

A

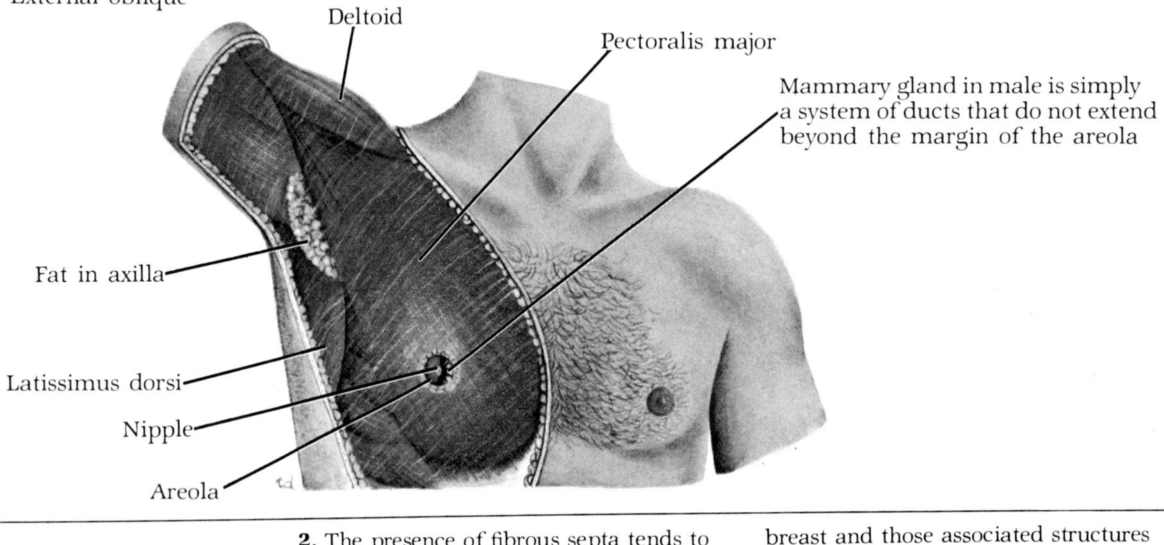

Deltoid
Pectoralis major
Mammary gland in male is simply a system of ducts that do not extend beyond the margin of the areola
Fat in axilla
Latissimus dorsi
Nipple
Areola

B

Figure 1-4
Mature mammary glands.
A. In female. B. In male.

1. The mammary gland is divided into fifteen to twenty compartments by fibrous septa that radiate out from the nipple. Involvement of the ducts of the gland and the fibrous septa by a scirrhous carcinoma or breast abscess will cause dimpling of the skin.

2. The presence of fibrous septa tends to localize infection to one compartment. An abscess should be drained through a radial type of incision to avoid spreading infection into a neighboring compartment; such an incision also minimizes damage to the radially arranged ducts.

3. In the male the glandular tissue is confined to an area beneath the areola. Malignant tumors of the male breast are rare but are usually more invasive than in women.

4. Radical mastectomy for cancer of the mammary gland involves the removal of the breast and those associated structures containing the lymph vessels and nodes. The following are removed en bloc: (a) skin overlying the tumor, including the nipple; (b) pectoralis major and pectoralis minor muscles and associated fascia; (c) fat, fascia, and lymph nodes in the axilla; (d) fascia covering the upper part of the rectus sheath, and the serratus anterior, subscapularis, and latissimus dorsi muscles.

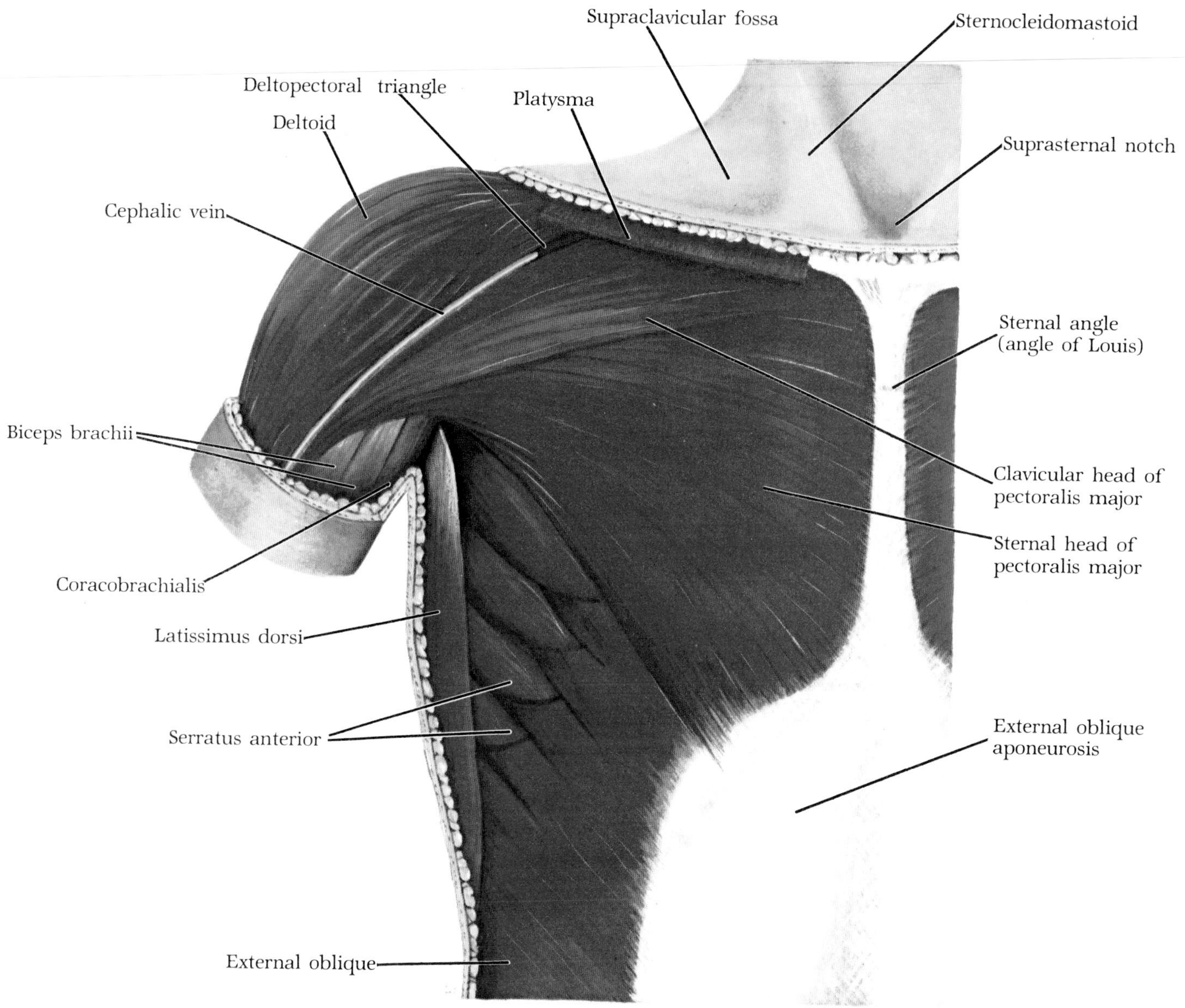
Supraclavicular fossa
Sternocleidomastoid
Deltopectoral triangle
Platysma
Deltoid
Suprasternal notch
Cephalic vein
Sternal angle
(angle of Louis)
Biceps brachii
Clavicular head of
pectoralis major
Sternal head of
pectoralis major
Coracobrachialis
Latissimus dorsi
External oblique
aponeurosis
Serratus anterior
External oblique

Figure 1-5
Pectoral region.

1. The upper part of the thoracic cage is protected by the clavicle and the strong muscles of the shoulder girdle.
2. Because the pectoralis major muscle covers the upper six costochondral junctions lateral to the sternum, the costochondral joints have to be palpated through this muscle.
3. The pectoralis major is frequently congenitally absent, either wholly or in part.
4. The cephalic vein is commonly used as the vein for the insertion of a catheter when performing cardiac catheterization.

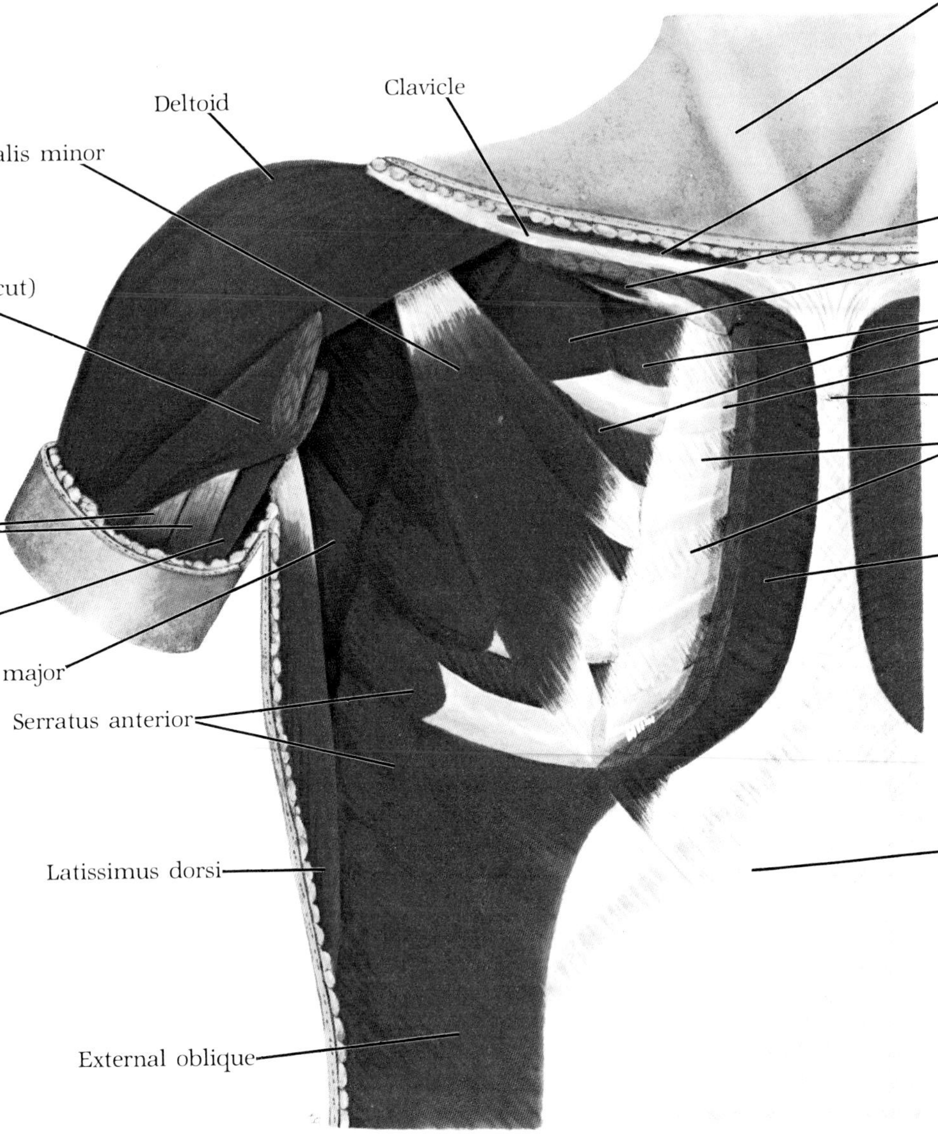
Sternocleidomastoid
Clavicle
Deltoid
Platysma
Pectoralis minor
Subclavius
First digitation of serratus anterior
Pectoralis major (cut)
External intercostal
Second costal cartilage
Sternal angle
Anterior intercostal membrane
Biceps brachii
Sternocostal part of pectoralis major
Coracobrachialis
Teres major
Serratus anterior
External oblique aponeurosis
Latissimus dorsi
External oblique

Figure 1-6
Pectoral region. The pectoralis major muscle has been removed.

1. Note the position of the pectoralis minor: deep to the pectoralis major. Because the lymph vessels draining the mammary gland pierce both muscles to enter the axillary lymph nodes, both muscles are removed during a radical mastectomy.
2. The subclavius muscle assists in stabilizing the sternoclavicular joint. The nerve to the subclavius may contribute fibers to the phrenic nerve and thus form the accessory phrenic nerve.
3. The sternal angle (angle of Louis) formed by the manubrium and body of the sternum lies at the level of the second costal cartilage and is therefore an easy reference point when counting the costal cartilages and ribs.

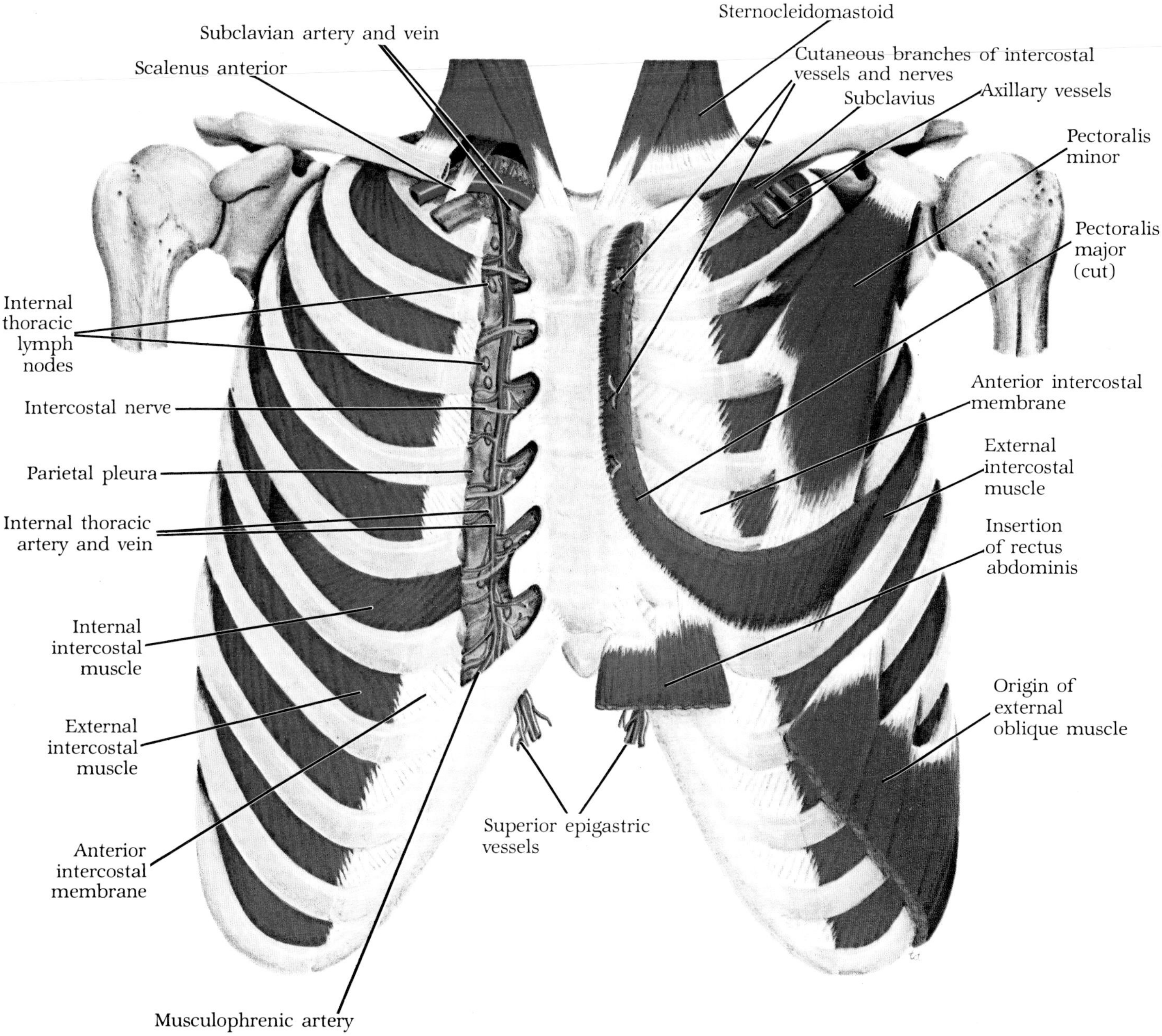
Sternocleidomastoid
Subclavian artery and vein
Cutaneous branches of intercostal vessels and nerves
Scalenus anterior
Subclavius
Axillary vessels
Pectoralis minor
Pectoralis major (cut)
Internal thoracic lymph nodes
Intercostal nerve
Anterior intercostal membrane
Parietal pleura
External intercostal muscle
Internal thoracic artery and vein
Insertion of rectus abdominis
Internal intercostal muscle
Origin of external oblique muscle
External intercostal muscle
Superior epigastric vessels
Anterior intercostal membrane
Musculophrenic artery

Figure 1-7
Anterior thoracic wall. On the right side, portions of the upper six costal cartilages have been removed; on the left side, the pectoralis minor has been left in position.

1. In the adult the middle ribs are the most commonly fractured; they are the longest and least protected and are strongly attached at both ends. The first two ribs are protected by the clavicle, and the last two ribs are unattached to the rib cage anteriorly and are freely movable. Fracture of the ribs in children is rare due to the elasticity of the thoracic wall at that age.

2. When the subclavian artery passes over the first rib, it becomes the axillary artery, lying posterior to the clavicle, and is then susceptible to compression. It may be pressed upon by the scalenus anterior, the first rib, a cervical rib, or even the clavicle. Symptoms of vascular insufficiency to the upper limb will result.

3. The internal thoracic artery and vein lie about a finger's breadth lateral to the sternum. The accompanying lymph nodes drain the medial quadrants of the mammary gland as well as the mediastinum and the intercostal spaces. It follows that a carcinoma occurring in the medial quadrants of the mammary gland tends to metastasize to these nodes within the thorax. If this takes place, eradication of the disease is difficult, if not impossible.

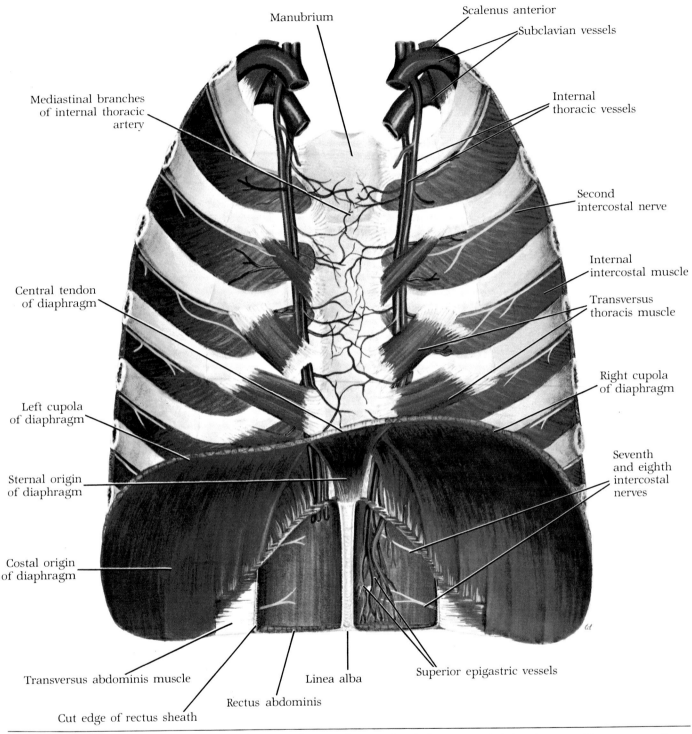

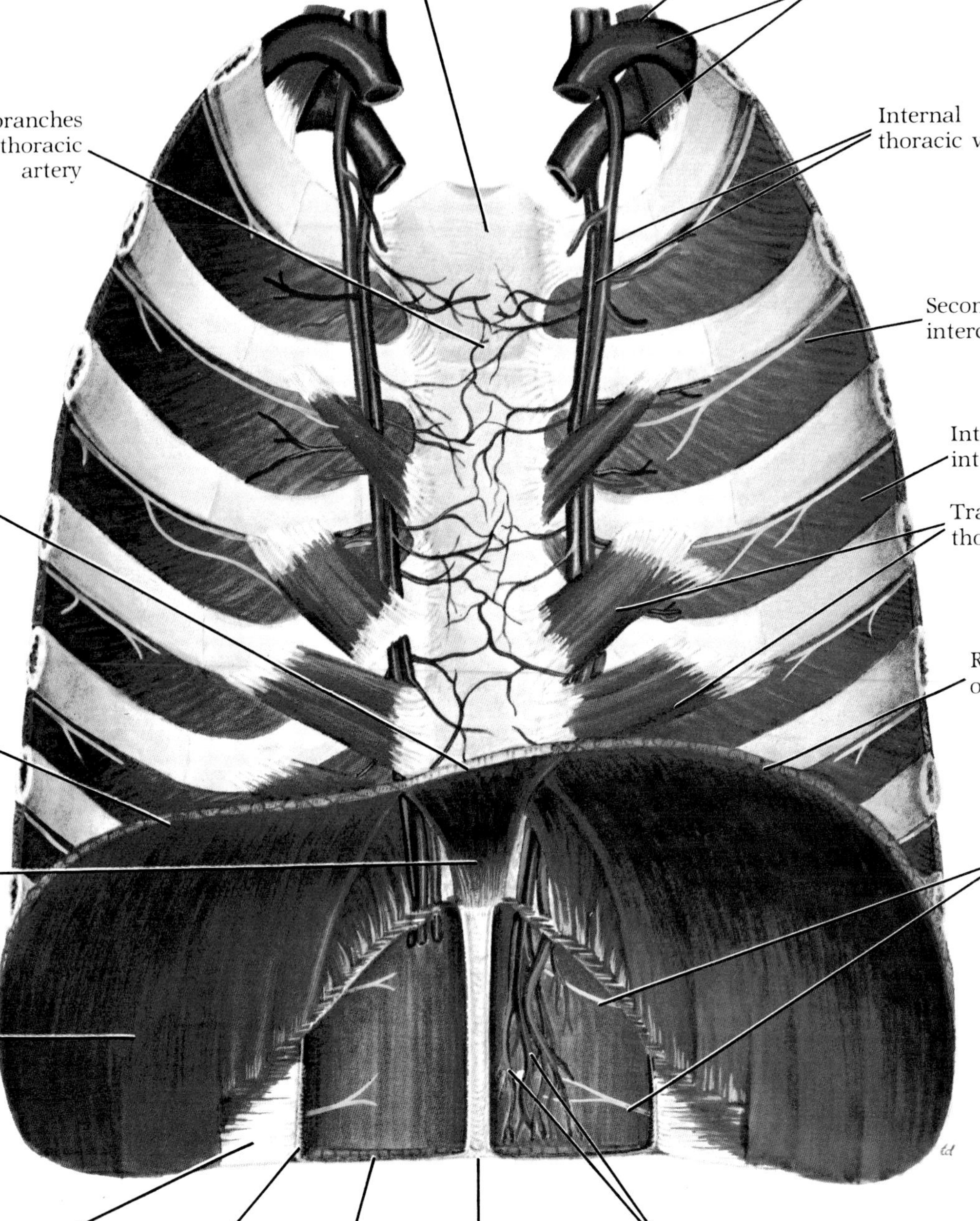
Manubrium
Scalenus anterior
Subclavian vessels
Mediastinal branches of internal thoracic artery
Internal thoracic vessels
Second intercostal nerve
Internal intercostal muscle
Central tendon of diaphragm
Transversus thoracis muscle
Right cupola of diaphragm
Left cupola of diaphragm
Seventh and eighth intercostal nerves
Sternal origin of diaphragm
Costal origin of diaphragm
Transversus abdominis muscle
Linea alba
Superior epigastric vessels
Rectus abdominis
Cut edge of rectus sheath

Figure 1-8
Anterior thoracic wall, posterior view.

1. Congenital absence of ribs, often associated with hemivertebrae, may result in impaired respiratory function. Sternal anomalies also sometimes occur.
2. Crush injuries may cause fractures of several ribs. If limited to one side, the fractures may occur near the rib angles and also anteriorly, near the costochondral junctions. This causes flail chest, during which the segment is sucked in during inspiration and driven out during expiration.
3. A rib fracture may result in a rib fragment being driven into the lung or through the skin.
4. A rib is innervated by its corresponding intercostal nerve. An intercostal nerve block will reduce the pain of rib fracture and thus permit painless inspiration.
5. As a result of the close association of a rib to the intercostal vessels, overlying skin, and underlying pleurae and lung, pneumothorax and/or hemothorax are common complications of rib fractures.
6. Note the route taken by the superior epigastric artery as it passes from the thorax to enter the sheath of the rectus abdominis muscle.

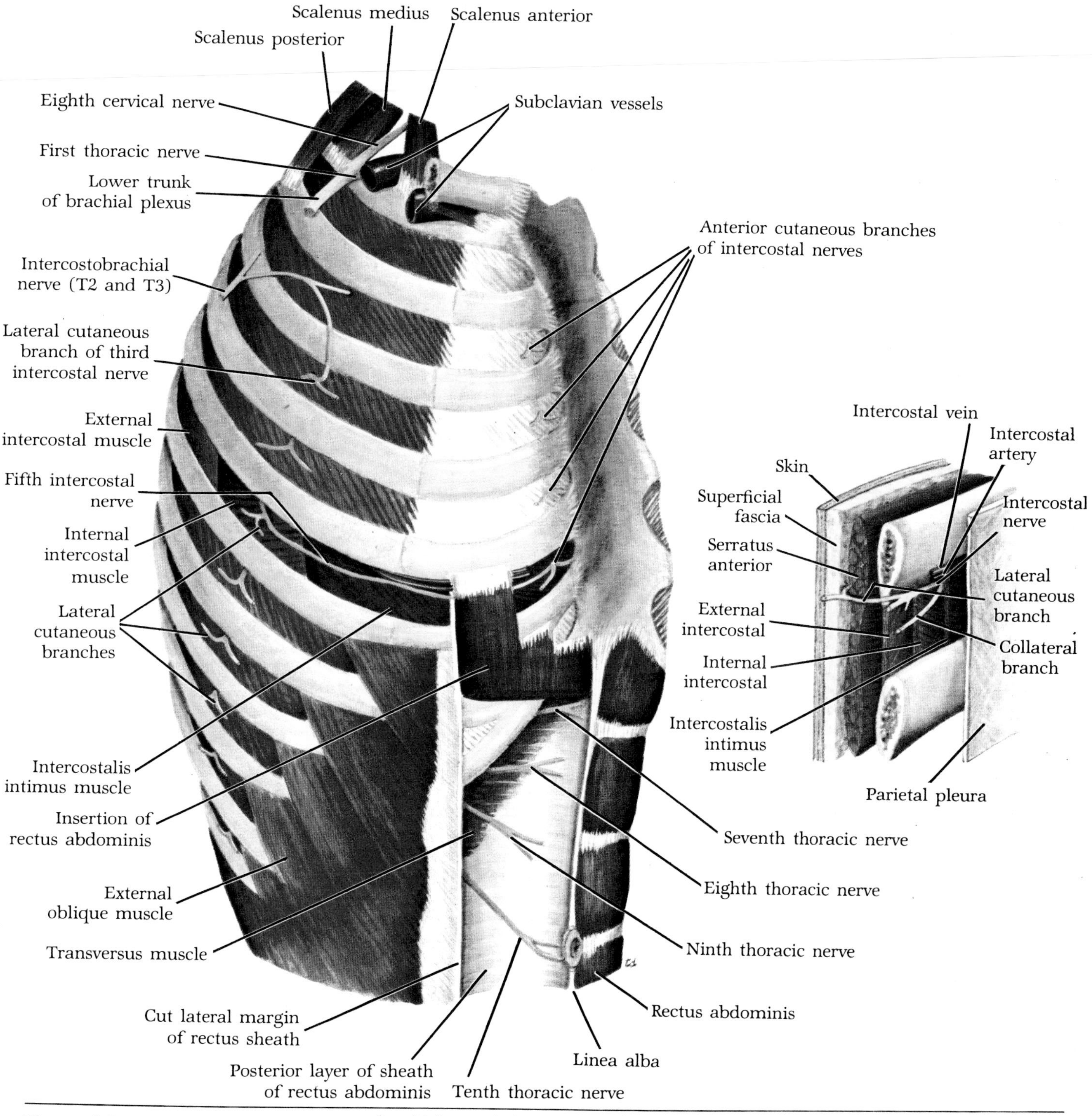
Scalenus medius
Scalenus anterior
Scalenus posterior
Eighth cervical nerve
Subclavian vessels
First thoracic nerve
Lower trunk of brachial plexus
Anterior cutaneous branches of intercostal nerves
Intercostobrachial nerve (T2 and T3)
Lateral cutaneous branch of third intercostal nerve
External intercostal muscle
Fifth intercostal nerve
Internal intercostal muscle
Lateral cutaneous branches
Intercostalis intimus muscle
Insertion of rectus abdominis
External oblique muscle
Transversus muscle
Cut lateral margin of rectus sheath
Posterior layer of sheath of rectus abdominis
Tenth thoracic nerve
Linea alba
Rectus abdominis
Ninth thoracic nerve
Eighth thoracic nerve
Seventh thoracic nerve
Intercostal vein
Intercostal artery
Skin
Superficial fascia
Intercostal nerve
Serratus anterior
Lateral cutaneous branch
External intercostal
Collateral branch
Internal intercostal
Intercostalis intimus muscle
Parietal pleura

Figure 1-9
Anterolateral view of the thoracic and abdominal walls. A section through a typical intercostal space is also shown.

1. A typical intercostal nerve supplies the skin and parietal pleura covering the outer and inner surfaces of each intercostal space, respectively, and the intercostal muscles of each space, the levatores costarum, and the serratus posterior muscles.
2. The first intercostal nerve is joined to the brachial plexus; the second and often the third intercostal nerves are joined to the intercostobrachial nerve. The intercostobrachial nerve is important in that it is associated with referred pain from the heart, especially on the left side in patients with myocardial ischemia.
3. The seventh to the eleventh intercostal nerves supply the skin and parietal pleura and the muscles of their respective intercostal spaces; they also supply the skin and parietal peritoneum covering the outer and inner surfaces of the abdominal wall, and the external oblique, internal oblique, transversus abdominis, and rectus abdominis muscles. Diseases of the lower areas of the parietal pleura may give rise to pain referred to the abdomen, producing increased tone of the abdominal muscles (the muscles feel rigid on palpation of the abdominal wall).
4. The lower part of the thoracic cage protects the upper abdominal organs, especially the liver, spleen, kidneys, and stomach. If the lower ribs are damaged or if there are penetrating wounds of the lower chest wall, however, the physician must ascertain whether these organs are also damaged.
5. The intercostal vessels and nerves are closely related to the lower margins of the ribs. Knowledge of their exact positions is important when an intercostal nerve block is desired or when withdrawing fluid through a needle.

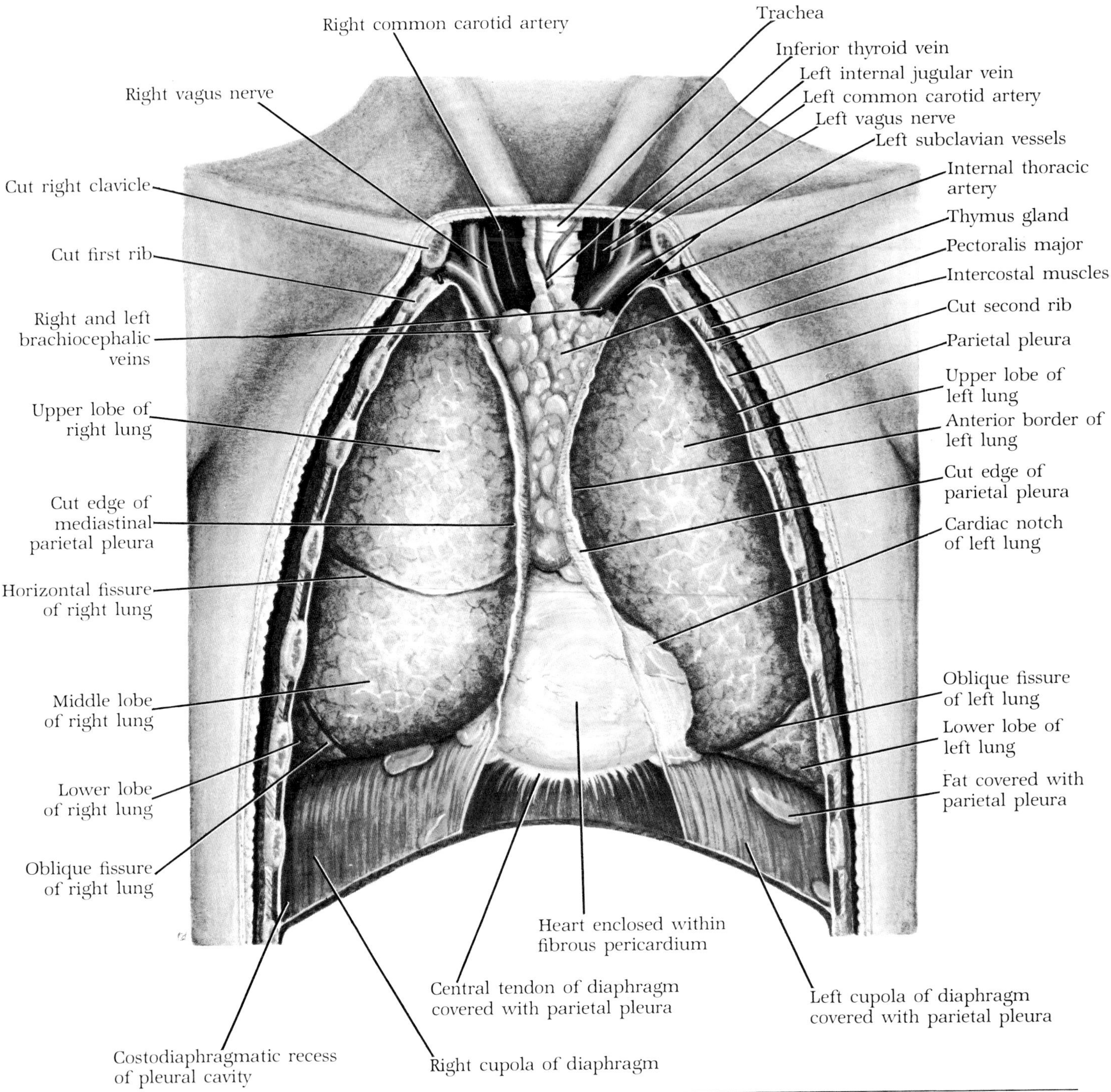

Right common carotid artery
Trachea
Inferior thyroid vein
Left internal jugular vein
Left common carotid artery
Left vagus nerve
Left subclavian vessels
Right vagus nerve
Internal thoracic artery
Cut right clavicle
Thymus gland
Pectoralis major
Cut first rib
Intercostal muscles
Cut second rib
Right and left brachiocephalic veins
Parietal pleura
Upper lobe of left lung
Upper lobe of right lung
Anterior border of left lung
Cut edge of parietal pleura
Cut edge of mediastinal parietal pleura
Cardiac notch of left lung
Horizontal fissure of right lung
Oblique fissure of left lung
Middle lobe of right lung
Lower lobe of left lung
Fat covered with parietal pleura
Lower lobe of right lung
Oblique fissure of right lung
Heart enclosed within fibrous pericardium
Central tendon of diaphragm covered with parietal pleura
Left cupola of diaphragm covered with parietal pleura
Costodiaphragmatic recess of pleural cavity
Right cupola of diaphragm

Figure 1-10
Anterior view of thoracic contents. The sternum, costal cartilages, and parts of the clavicles and ribs have been removed.

1. Closely observe the position of the different parts of the lungs and the heart relative to the anterior thoracic wall. This is particularly important to know when giving a patient a physical examination.

2. The cardiac notch in the left lung and pleura allows a needle to be introduced into the pericardial cavity for the purpose of withdrawing fluid in cases of pericarditis without damaging the lung. The needle is inserted to the left of the xiphoid process in an upward and backward direction at an angle of 45 degrees to the skin.

3. The thymus is a relatively large structure in a child, and its presence is clearly seen in radiographs of the thorax.

4. Disease or injury may result in air (pneumothorax), blood (hemothorax), or pus (empyema) entering the pleural cavity. The lung on the same side collapses, and the flexible mediastinum is displaced to the opposite side.

5. Fluid may be withdrawn from the pleural cavity through a wide-bore needle inserted through a lower intercostal space. When inserting the needle below the seventh intercostal space, one must be extremely careful not to pierce the diaphragm.

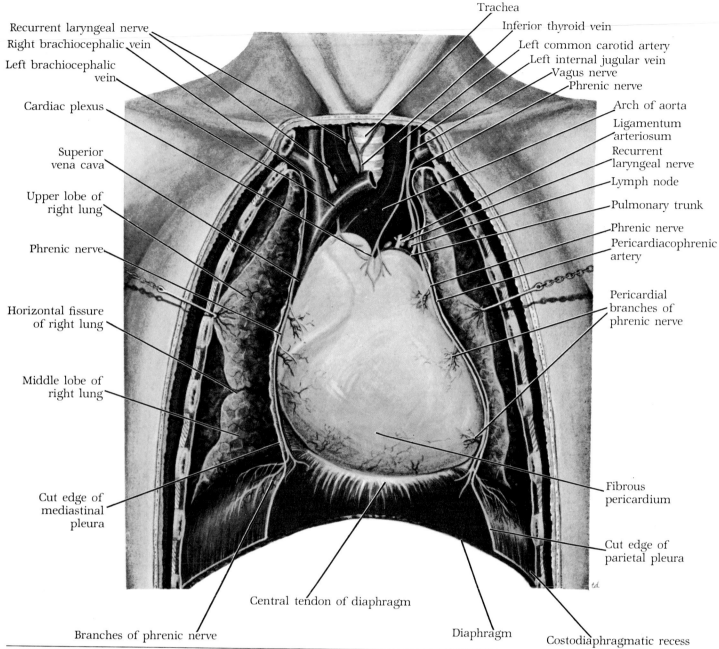

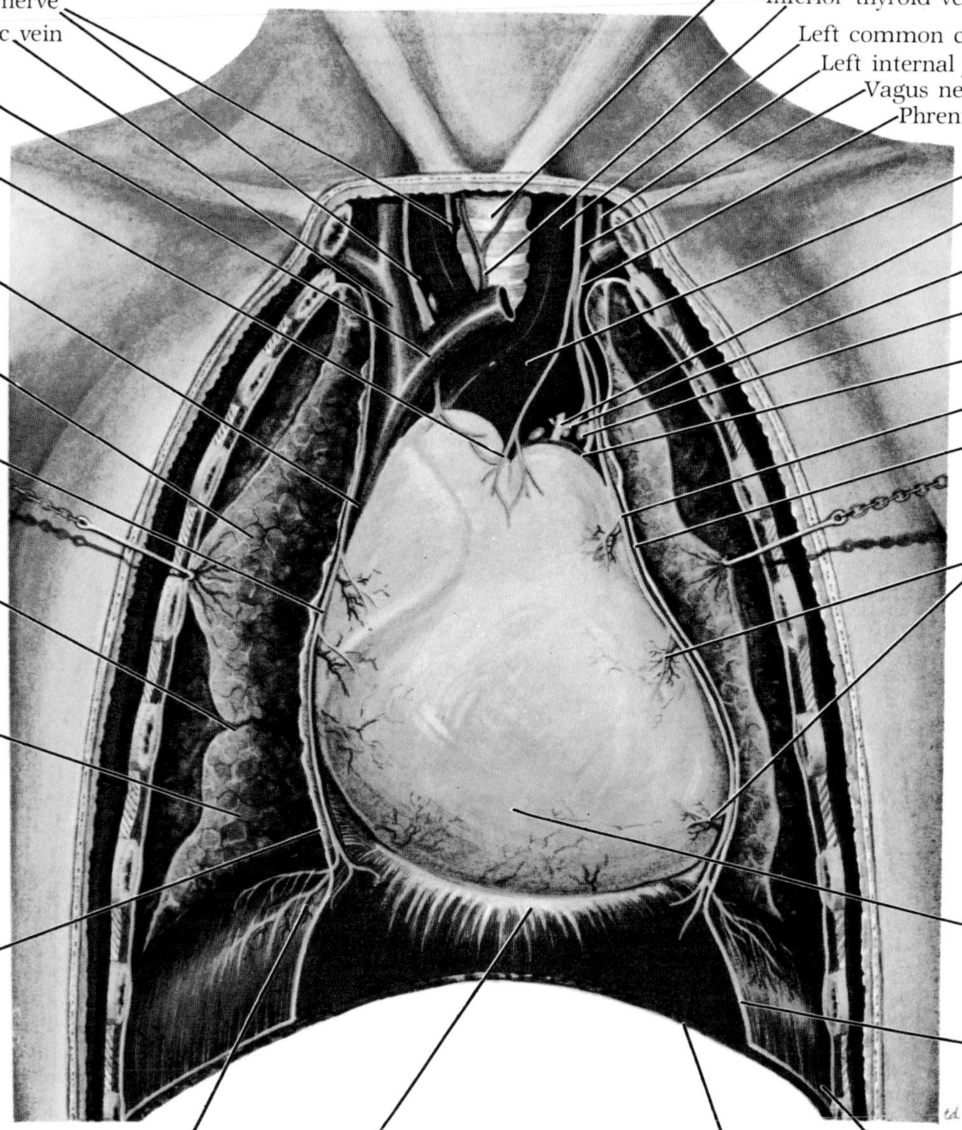
Trachea
Recurrent laryngeal nerve
Inferior thyroid vein
Right brachiocephalic vein
Left common carotid artery
Left brachiocephalic vein
Left internal jugular vein
Vagus nerve
Phrenic nerve
Cardiac plexus
Arch of aorta
Ligamentum arteriosum
Superior vena cava
Recurrent laryngeal nerve
Lymph node
Upper lobe of right lung
Pulmonary trunk
Phrenic nerve
Phrenic nerve
Pericardiacophrenic artery
Pericardial branches of phrenic nerve
Horizontal fissure of right lung
Middle lobe of right lung
Fibrous pericardium
Cut edge of mediastinal pleura
Cut edge of parietal pleura
Central tendon of diaphragm
Branches of phrenic nerve
Diaphragm
Costodiaphragmatic recess

Figure 1-11
Anterior view of the thoracic contents. The pleural cavities have been opened and the lungs pulled laterally.

1. The normally expanded lung with its covering of visceral pleura projects completely into all parts of the pleural cavity so that the cavity is reduced to only a *potential* space. In disease states, however, the pleural cavity may contain gases, fluids, or tumor tissue.
2. The visceral pleura is an integral part of the lung, but the parietal pleura can be easily separated from the thoracic wall.
3. Normally the visceral and parietal layers of pleura are separated by a thin film of fluid, the pleural fluid. The volume of the fluid is determined by a careful balance between absorption and transudation. Disease can upset this equilibrium.
4. Lung tissue and visceral pleura are devoid of pain-sensitive nerve endings. The costal parietal pleura and the peripheral diaphragmatic parietal pleura are innervated by the intercostal nerves. Inflammation of the parietal pleura results in an acute pain referred to the cutaneous distribution of these nerves in the thorax, or in the anterior abdominal wall in the case of the lower nerves. Disease of the central part of the diaphragmatic parietal pleura (innervated by the phrenic nerve, C3, 4, and 5) may result in referred pain to the skin over the shoulder (supraclávicular nerves C3 and 4).

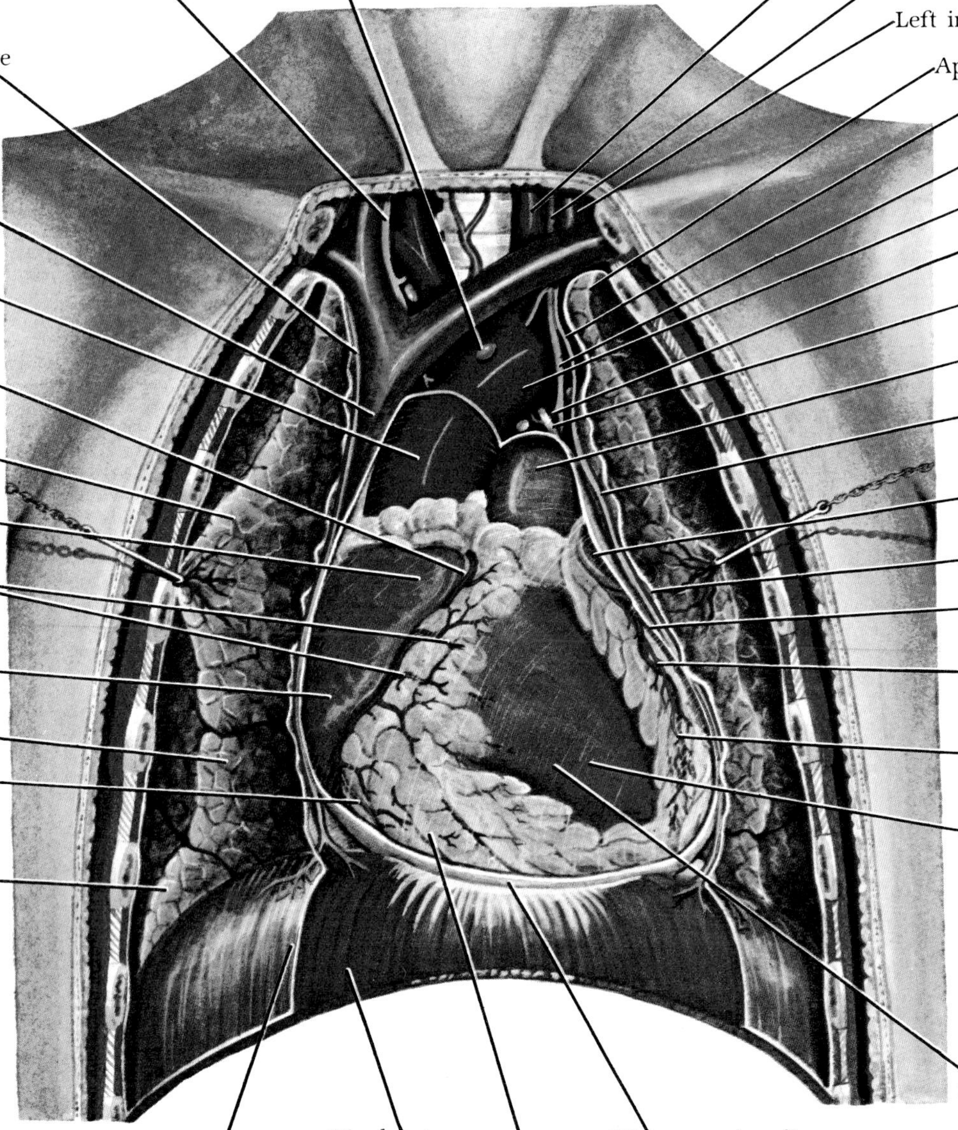
Lymph node
Left common carotid artery
Vagus nerve
Vagus nerve
Left internal jugular vein
Phrenic nerve
Apex of left lung
Phrenic nerve
Superior vena cava
Vagus nerve
Arch of aorta
Ascending aorta
Recurrent laryngeal nerve
Ligamentum arteriosum
Right coronary artery
Pulmonary trunk
Phrenic nerve and pericardiacophrenic artery
Upper lobe of right lung
Right auricle
Left auricle
Cut edge of parietal pleura
Anterior cardiac veins
Cut edge of pericardium
Right atrium
Anterior interventricular branch of left coronary artery
Middle lobe of right lung
Small cardiac vein
Great cardiac vein
Visceral layer of serous pericardium
Lower lobe of right lung
Right ventricle
Diaphragm
Fibrous pericardium
Cut edge of pleura
Fat beneath serous pericardium

Figure 1-12
Anterior view of the thoracic contents. The pleural cavities and the pericardial cavity have been opened. The heart and coronary vessels are seen through the thin visceral layer of serous pericardium.

1. The fibrous pericardium is inelastic; consequently, the sudden accumulation of fluid within the pericardial cavity will impair cardiac filling during diastole. This may occur in pericarditis and is referred to as cardiac tamponade.
2. The ascending aorta and the pulmonary trunk are enclosed by the fibrous pericardium. Rupture of an aneurysm of the ascending aorta would also produce cardiac tamponade.
3. The innervation of the parietal layer of serous pericardium by the phrenic nerve would explain the referred pain to the shoulders (supraclavicular nerves C3 and 4) in patients with pericarditis.
4. In surgical ligation of a patent ductus arteriosus, the left recurrent laryngeal nerve must be identified and carefully preserved.

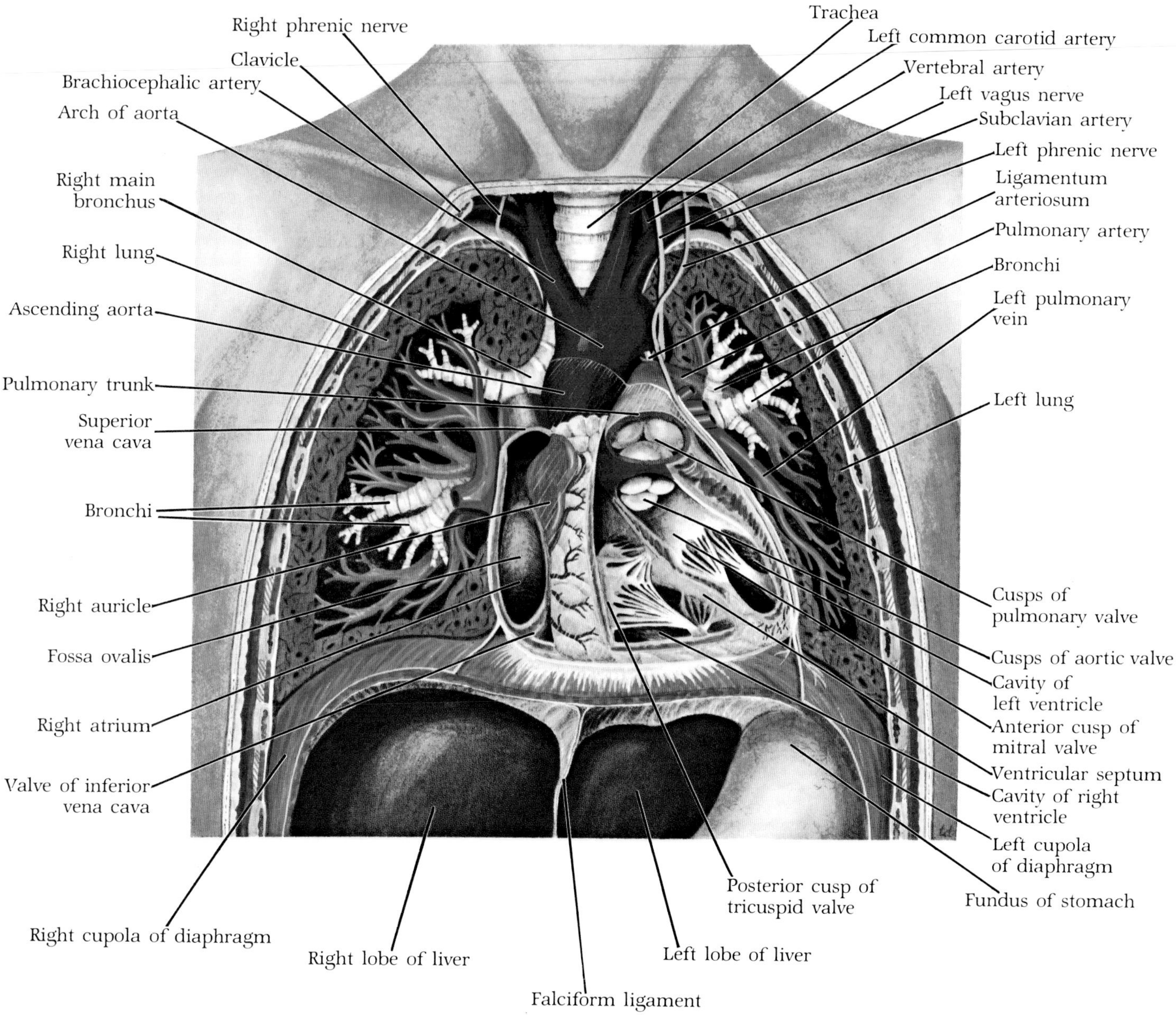

Right phrenic nerve
Clavicle
Brachiocephalic artery
Arch of aorta
Right main bronchus
Right lung
Ascending aorta
Pulmonary trunk
Superior vena cava
Bronchi
Right auricle
Fossa ovalis
Right atrium
Valve of inferior vena cava
Right cupola of diaphragm
Right lobe of liver
Falciform ligament
Left lobe of liver
Posterior cusp of tricuspid valve
Fundus of stomach
Trachea
Left common carotid artery
Vertebral artery
Left vagus nerve
Subclavian artery
Left phrenic nerve
Ligamentum arteriosum
Pulmonary artery
Bronchi
Left pulmonary vein
Left lung
Cusps of pulmonary valve
Cusps of aortic valve
Cavity of left ventricle
Anterior cusp of mitral valve
Ventricular septum
Cavity of right ventricle
Left cupola of diaphragm

Figure 1-13
Contents of the thorax and upper abdomen, anterior view. The lungs have been dissected to show the main blood vessels and air passages, and the heart has been opened.

1. A dilatation of the aortic arch (aneurysm) may compress the trachea. The pulsating aneurysm may tug at the trachea and left bronchus, a clinical sign that may be felt by palpating the trachea in the suprasternal notch.

2. Because of the close proximity of the heart, lungs, diaphragm, liver, and stomach, several of these organs may be involved in penetrating wounds of the thorax or abdomen.

3. Localized infections of the peritoneal cavity immediately below the diaphragm (subdiaphragmatic abscess) may extend through the diaphragm to involve the pleura. Amebic abscesses of the liver may penetrate the diaphragm and involve the pleura and lung.

4. Note the relationship of the pulmonary arteries and veins to the bronchi. This knowledge is essential for the interpretation of pulmonary radiographs, especially pulmonary angiograms and venograms.

5. The position of the heart valves relative to the chest wall should be noted. This must not be confused with the sites where the valves are most easily and separately heard with the stethoscope.

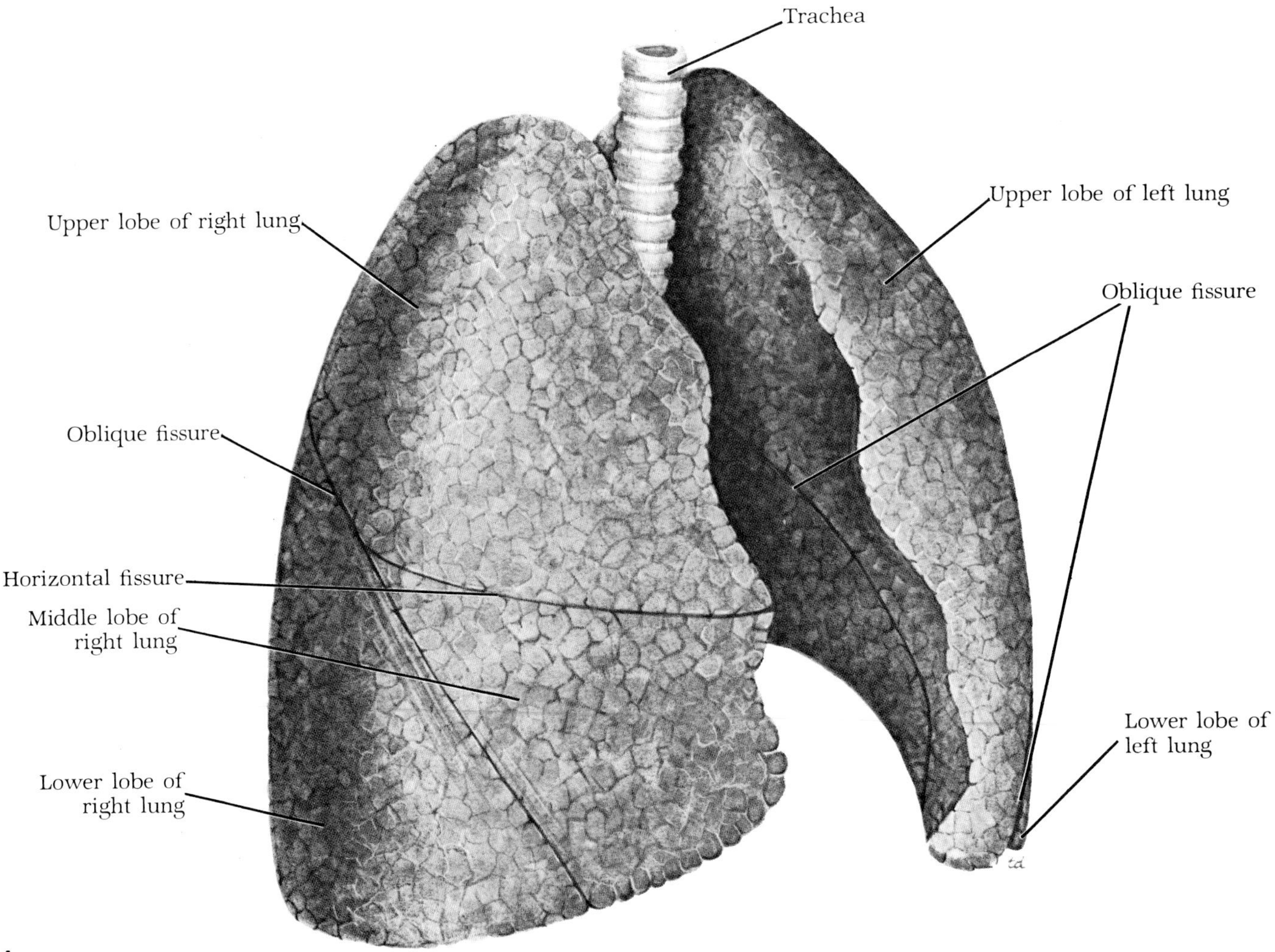
Trachea
Upper lobe of right lung
Upper lobe of left lung
Oblique fissure
Oblique fissure
Horizontal fissure
Middle lobe of right lung
Lower lobe of left lung
Lower lobe of right lung

A

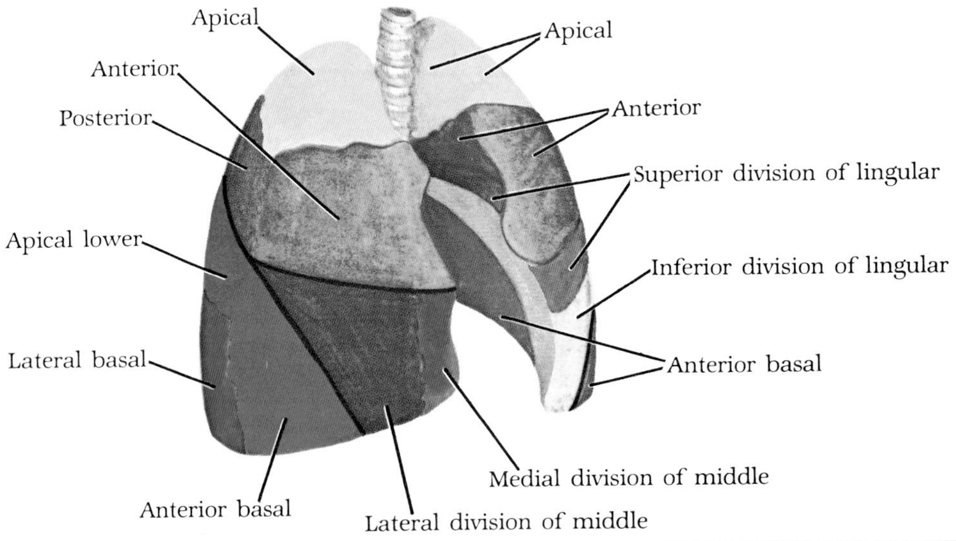
Apical
Apical
Anterior
Anterior
Posterior
Superior division of lingular
Apical lower
Inferior division of lingular
Lateral basal
Anterior basal
Medial division of middle
Anterior basal
Lateral division of middle

B

Figure 1-14
The lungs. A. Lobes.
B. Bronchopulmonary segments.

1. An understanding of the segmental anatomy of the lungs allows the surgeon to dissect and remove a particular segment, leaving the surrounding lung intact (segmental resection). This procedure is particularly useful in patients with localized tuberculosis or a small benign neoplasm.
2. The bronchi and their accompanying pulmonary arteries are segmentally distributed. The branches of the pulmonary veins are located in the intersegmental plane.
3. If the disease process involves several segments of a lung, or if it may be difficult for the surgeon to remove a segment intact with a minimum of trauma to the lung tissue and with little air leakage, a lobectomy should be performed.

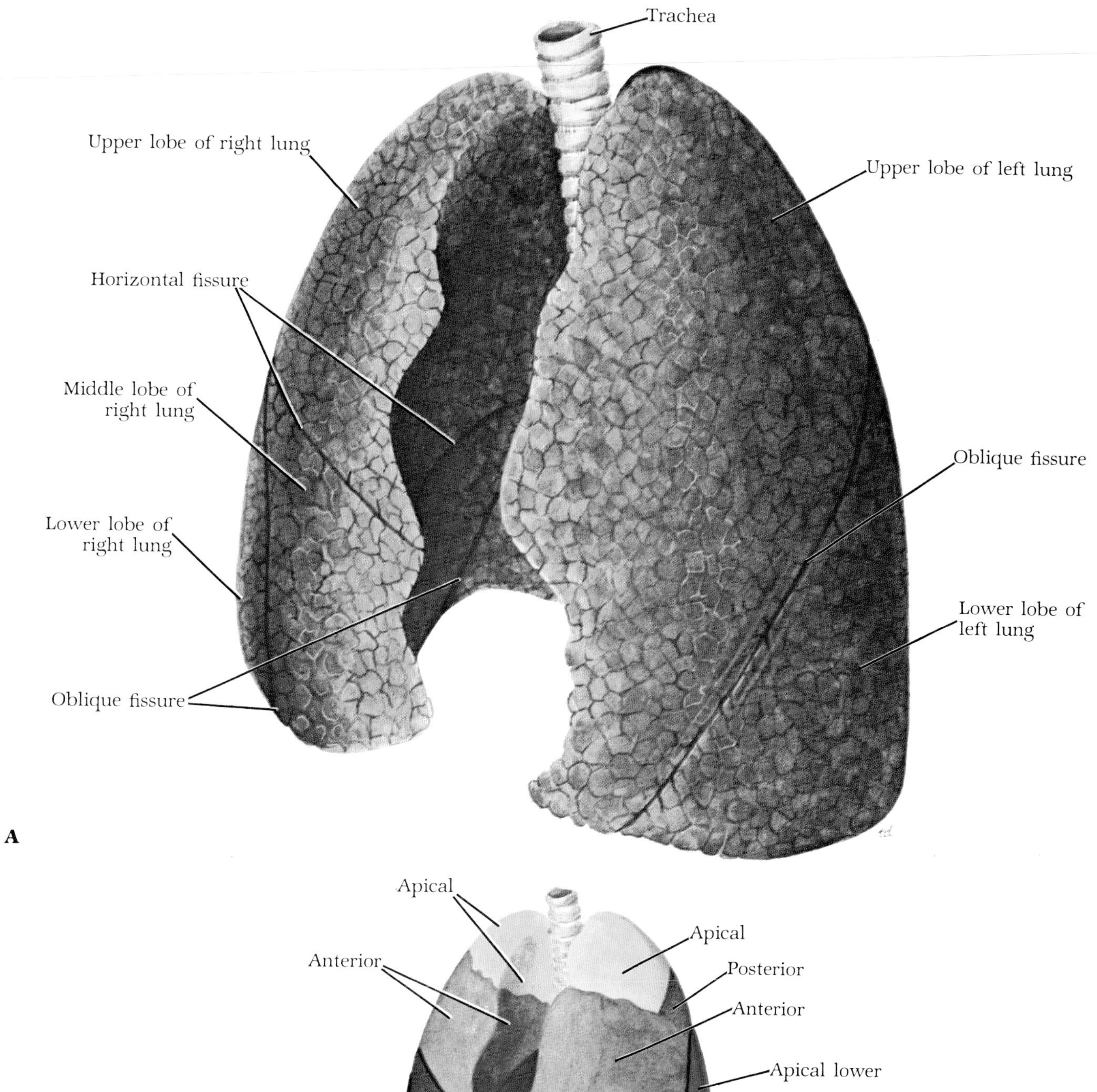
Trachea
Upper lobe of right lung
Upper lobe of left lung
Horizontal fissure
Middle lobe of
right lung
Oblique fissure
Lower lobe of
right lung
Lower lobe of
left lung
Oblique fissure
A

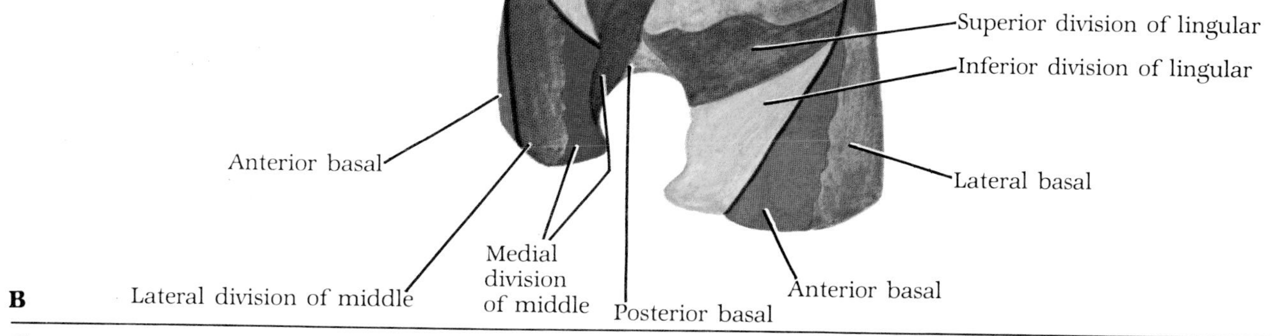
Apical
Apical
Anterior
Posterior
Anterior
Apical lower
Superior division of lingular
Inferior division of lingular
Anterior basal
Lateral basal
B
Lateral division of middle
Medial
division
of middle
Posterior basal
Anterior basal

Figure 1-15
The lungs. A. Lobes.
B. Bronchopulmonary segments.

1. Each segment of a lung is pyramidal in shape, having its apex toward the root of the lung and its base toward the lung surface.

2. Deoxygenated blood is passed into the lungs by the pulmonary arteries, while oxygenated blood is returned to the left atrium by the pulmonary veins. The bronchi and their branches receive their blood supply from the bronchial arteries, branches of the aorta. The bronchial veins drain into the azygos and hemiazygos veins.

3. Each lung is innervated by the pulmonary plexus formed from the branches of the sympathetic trunk and the vagus nerve.

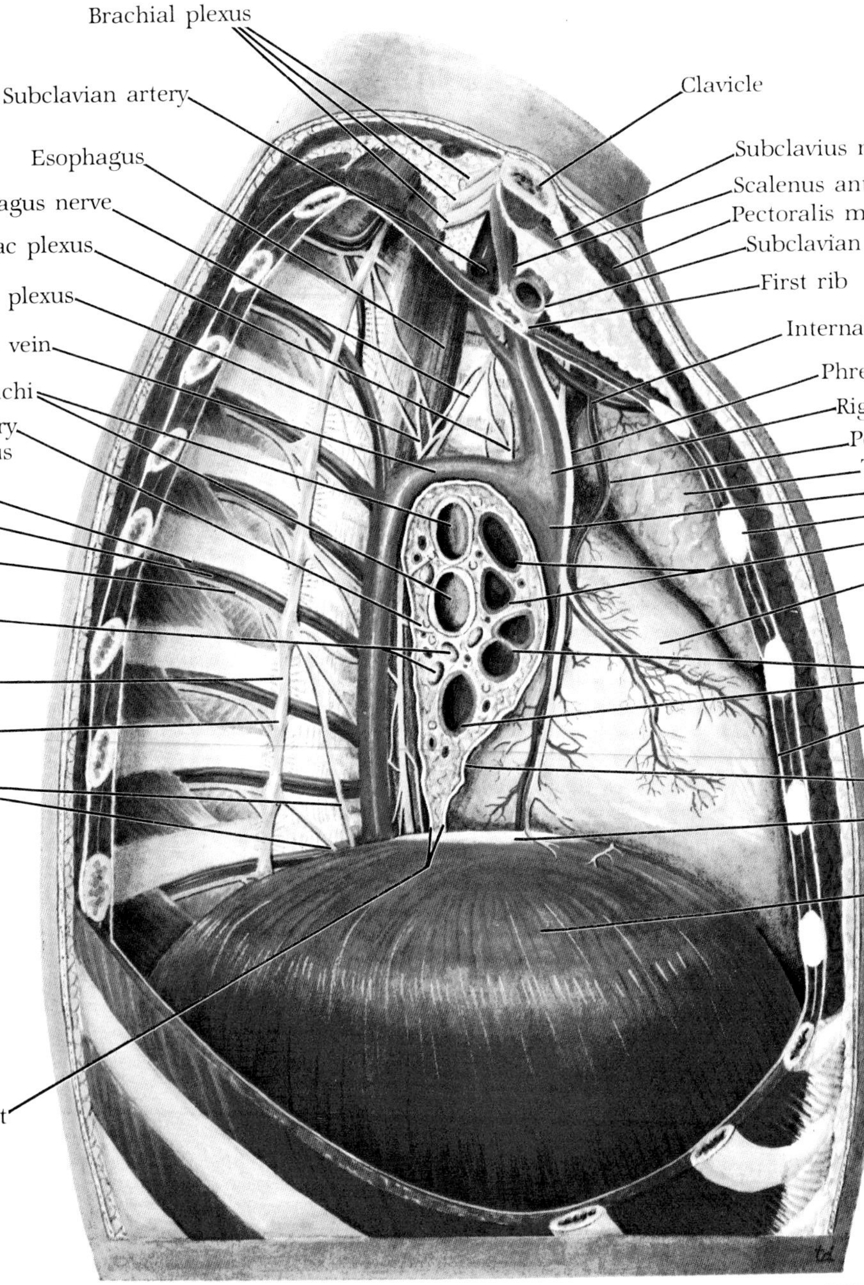
Brachial plexus
Subclavian artery
Esophagus
Vagus nerve
Cardiac plexus
Esophageal plexus
Azygos vein
Bronchi
Posterior pulmonary plexus
Intercostal vein
Intercostal artery
Intercostal nerve
Bronchopulmonary lymph nodes
Sympathetic trunk
Sympathetic ganglion
Roots of greater splanchnic nerve
Pulmonary ligament
Clavicle
Subclavius muscle
Scalenus anterior
Pectoralis major
Subclavian vein
First rib
Internal thoracic artery
Phrenic nerve
Right brachiocephalic vein
Pericardiacophrenic vessels
Thymus gland
Superior vena cava
Second costal cartilage
Pulmonary arteries
Fibrous pericardium
Pulmonary veins
Costal pleura
Cut edge of parietal pleura
Central tendon of diaphragm
Right cupola of diaphragm

Figure 1-16
Right side of mediastinum.

1. The mediastinum is a partition lying between the two pleural cavities. Although solid it is movable and may be displaced laterally by increased pressure in one of the pleural cavities. Such displacement is determined clinically by the lateral deviation of the trachea in the suprasternal notch and by the alteration of the position of the apex beat of the heart; it is easily seen in radiographs of the chest.
2. Continuity of the fascial spaces of the mediastinum with those in the neck may permit the movement of air or fluid from one region to another.
3. Although the mediastinum is largely composed of the heart and great blood vessels, it is made up of many other different tissues. Consequently a large variety of tumors occur in this region, which, as they enlarge, cause widening of the mediastinum and encroach on the pleurae and lungs.
4. Bronchogenic carcinoma with metastases in the lymph nodes at the root of the lung may extend widely in the mediastinum to involve the esophagus, pericardium, heart, and superior vena cava.
5. Note that the pleura extends into the neck for a short distance and may be damaged by penetrating knife wounds in the root of the neck. Also note the large extent of the costodiaphragmatic recess which may be accidently damaged during a surgical procedure below the diaphragm or by penetrating wounds of the abdomen.
6. When inserting a needle into the pleural cavity, it is essential not to damage the intercostal vessels and nerve. This is accomplished by inserting the needle close to the upper border of the rib. To block an intercostal nerve, the anesthetic is infiltrated around the nerve trunk as it lies in the subcostal groove.

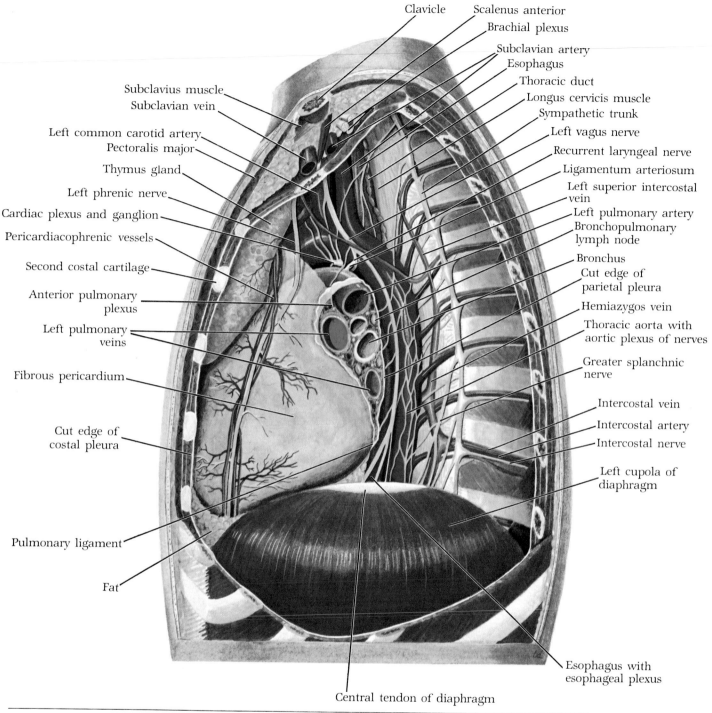

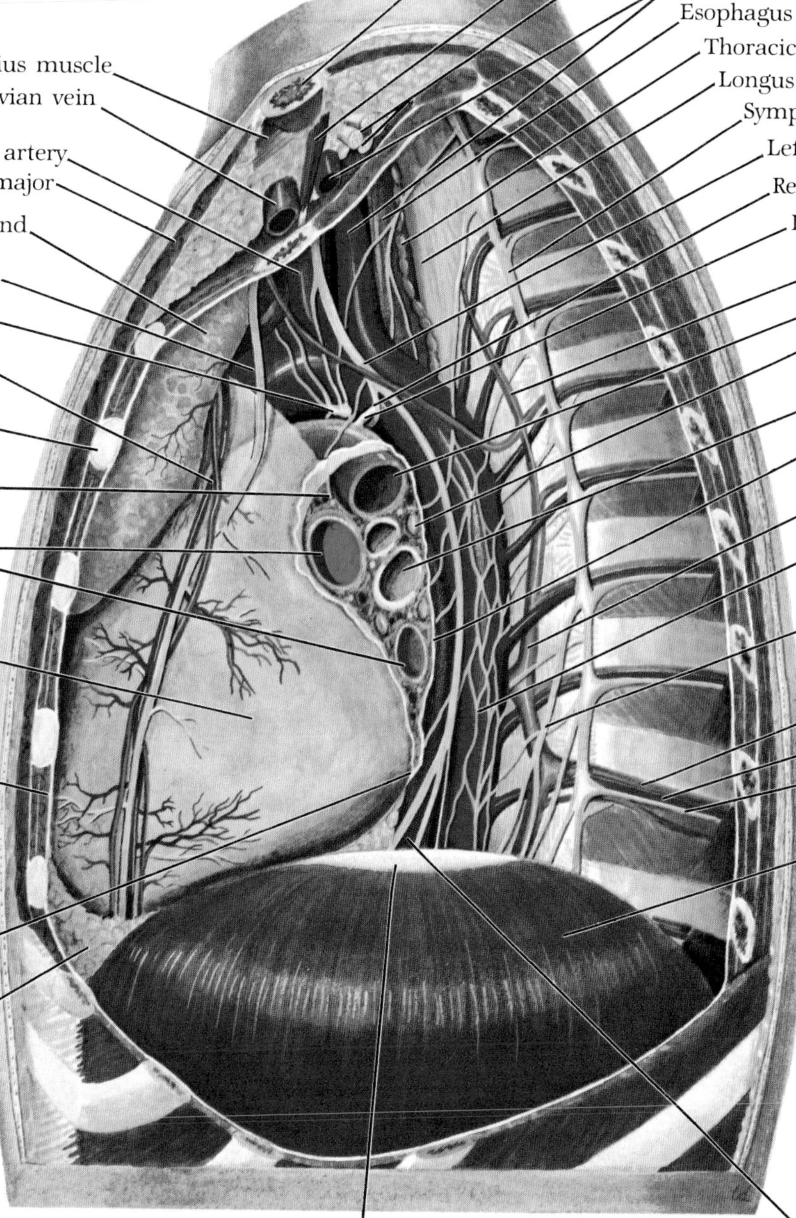
Clavicle
Scalenus anterior
Brachial plexus
Subclavian artery
Esophagus
Thoracic duct
Longus cervicis muscle
Sympathetic trunk
Left vagus nerve
Recurrent laryngeal nerve
Ligamentum arteriosum
Left superior intercostal vein
Left pulmonary artery
Bronchopulmonary lymph node
Bronchus
Cut edge of parietal pleura
Hemiazygos vein
Thoracic aorta with aortic plexus of nerves
Greater splanchnic nerve
Intercostal vein
Intercostal artery
Intercostal nerve
Left cupola of diaphragm
Esophagus with esophageal plexus
Central tendon of diaphragm
Subclavius muscle
Subclavian vein
Left common carotid artery
Pectoralis major
Thymus gland
Left phrenic nerve
Cardiac plexus and ganglion
Pericardiacophrenic vessels
Second costal cartilage
Anterior pulmonary plexus
Left pulmonary veins
Fibrous pericardium
Cut edge of costal pleura
Pulmonary ligament
Fat

Figure 1-17
Left side of mediastinum.

1. The arch of the aorta lies behind the manubrium sterni. A dilatation of the aorta (aneurysm) may show itself as a pulsatile swelling in the suprasternal notch.
2. Coarctation of the aorta, a congenital narrowing just beyond the origin of the left subclavian artery, commonly results in extreme dilatation of the posterior intercostal arteries with erosion of the lower borders of the ribs; this produces the characteristic notching seen on radiographic examination. The dilatation of the arteries is an attempt to bypass the narrowing.
3. Left-sided advanced bronchogenic carcinoma may metastasize to the lymph nodes at the lung root and involve the left recurrent laryngeal nerve, producing hoarseness of the voice.
4. The position of the sympathetic trunk allows the surgeon to perform a sympathectomy for the upper limb by resecting a short segment of the second or third rib through the back or through the axilla. Preganglionic sympathectomy of the second and third thoracic ganglia may be performed to increase the blood flow to the fingers for such conditions as Raynaud's disease. The sympathectomy causes vasodilation of the arterioles in the upper limb.
5. A single dome of the diaphragm may be paralyzed by crushing or sectioning the phrenic nerve in the neck. This may be necessary when the physician wishes to rest the lower lobe of the lung on one side.
6. In posterolateral thoracotomy for lung resection, an incision is made along the fifth, sixth, or seventh rib. The periosteum over the rib is incised and elevated and a segment of the rib removed, after which an incision is made through the rib bed. The opening may be greatly widened by inserting suitable retractors.

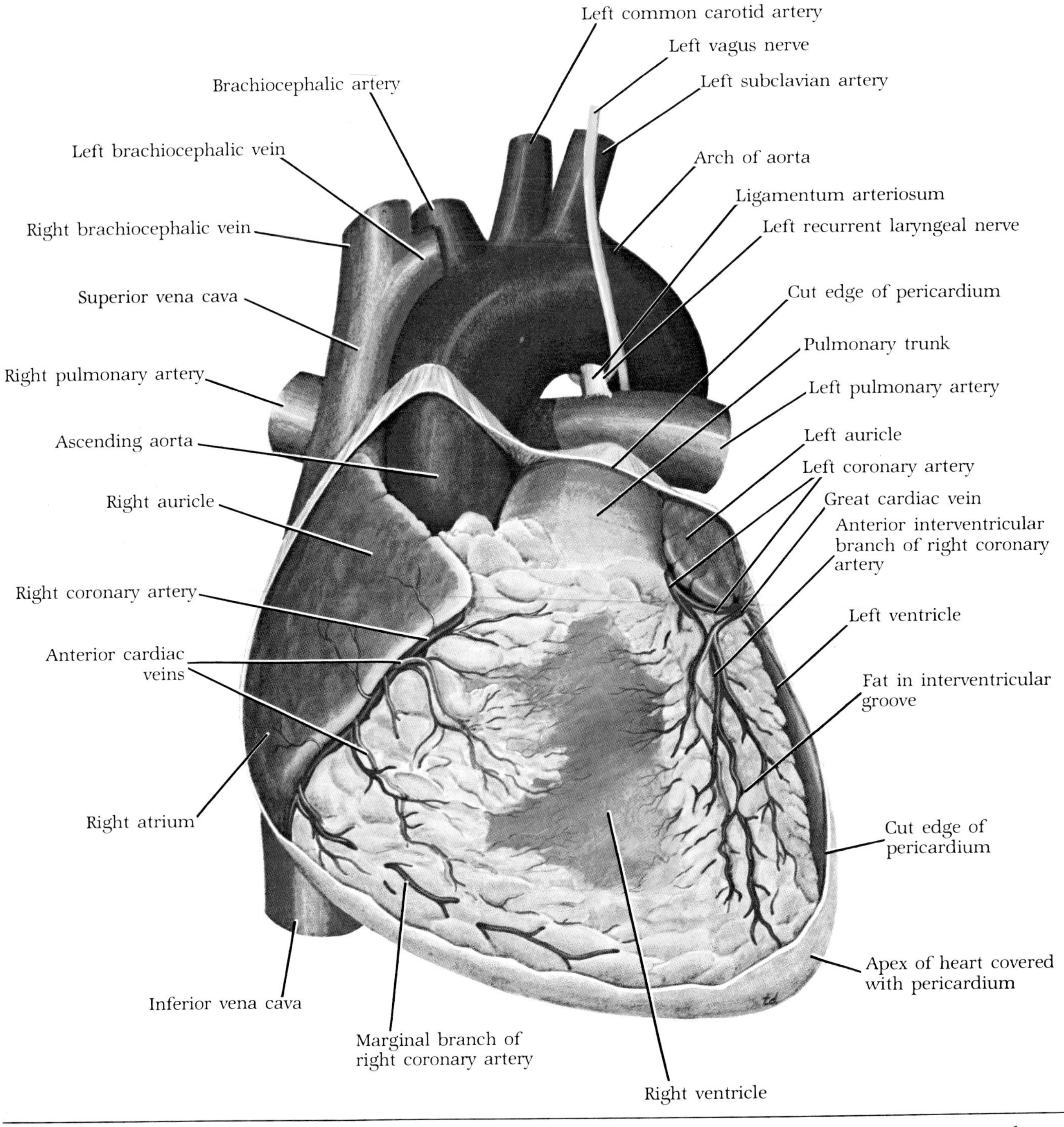
Left common carotid artery
Left vagus nerve
Left subclavian artery
Brachiocephalic artery
Left brachiocephalic vein
Arch of aorta
Ligamentum arteriosum
Right brachiocephalic vein
Left recurrent laryngeal nerve
Superior vena cava
Cut edge of pericardium
Pulmonary trunk
Right pulmonary artery
Left pulmonary artery
Ascending aorta
Left auricle
Left coronary artery
Right auricle
Great cardiac vein
Anterior interventricular branch of right coronary artery
Right coronary artery
Left ventricle
Anterior cardiac veins
Fat in interventricular groove
Right atrium
Cut edge of pericardium
Apex of heart covered with pericardium
Inferior vena cava
Marginal branch of right coronary artery
Right ventricle

Figure 1-18
Anterior surface of the heart and great blood vessels. The anterior wall of the pericardium has been removed.

1. The apex of the heart is formed by the left ventricle.

2. Myocardial infarction (death of heart muscle) is caused by a sudden blocking of a coronary artery by an embolus, arterial disease, or thrombosis. It is one of the most important causes of death in persons past middle age.

3. Angina pectoris is a clinical syndrome in which there is a viselike pain over the sternum that may radiate down the medial aspect of the left or both arms. It is caused by myocardial ischemia due to gradual narrowing of the coronary arteries by atherosclerosis.

4. The fibrous pericardium not only encloses the heart but also the ascending aorta and the pulmonary trunk. An aneurysm of the ascending aorta should it rupture, will open into the pericardial sac, causing cardiac tamponade and immediate death.

5. Effusion into the pericardial cavity in pericarditis may restrict the filling of the heart during diastole.

6. Shrinkage of the fibrous pericardium as the result of disease (constrictive pericarditis) will seriously interfere with filling of the heart.

7. Before a patent ductus arteriosus is surgically ligated, the recurrent laryngeal nerve must be identified and carefully preserved.

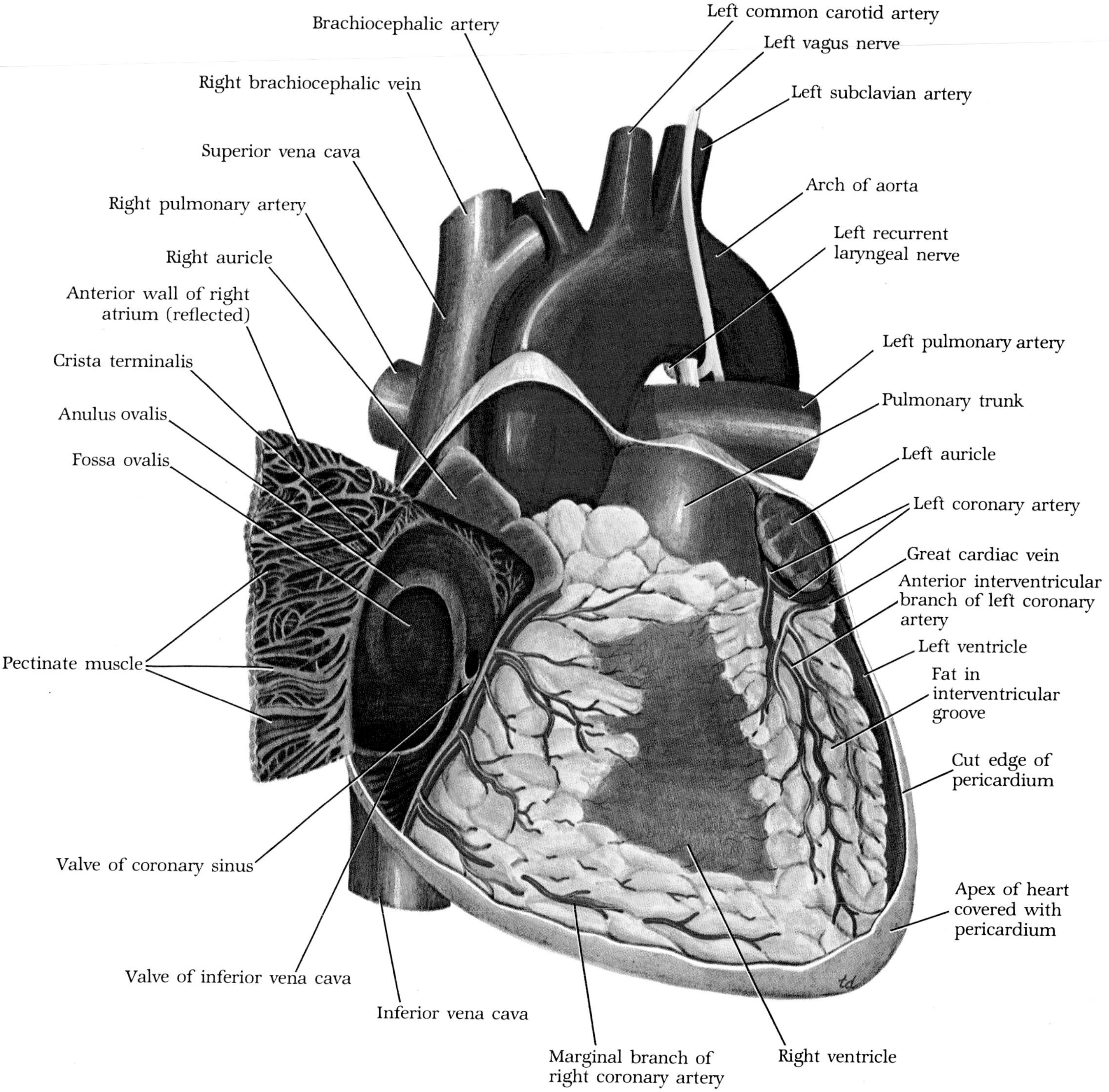
Brachiocephalic artery
Left common carotid artery
Left vagus nerve
Right brachiocephalic vein
Left subclavian artery
Superior vena cava
Arch of aorta
Right pulmonary artery
Left recurrent laryngeal nerve
Right auricle
Anterior wall of right atrium (reflected)
Left pulmonary artery
Crista terminalis
Pulmonary trunk
Anulus ovalis
Left auricle
Fossa ovalis
Left coronary artery
Great cardiac vein
Anterior interventricular branch of left coronary artery
Pectinate muscle
Left ventricle
Fat in interventricular groove
Cut edge of pericardium
Valve of coronary sinus
Apex of heart covered with pericardium
Valve of inferior vena cava
Inferior vena cava
Marginal branch of right coronary artery
Right ventricle

Figure 1-19
Anterior view of heart. The anterior wall of the right atrium has been reflected to show the interior.

1. The anulus ovalis on the septal wall of the right atrium is what remains of the lower edge of the embryological septum secundum. The floor of the fossa ovalis is formed from the septum primum. Atrial septal defects due to a persistence of the foramen secundum (ovale) are the most common congenital cardiac malformations and result from failure of the septum secundum to adequately develop. When a large amount of oxygenated blood from the left atrium passes into the right atrium, the right side of the heart is overworked.

2. In right-sided congestive heart failure due to disease of the tricuspid or pulmonary valves, there is a backup of venous blood in the right atrium and the superior and inferior vena cavae. The right atrium consequently often becomes enlarged due to dilation or hypertrophy, and the right margin of the heart, formed by the right atrium, enlarges to the right as detected by percussion of the chest wall or by chest radiographs.

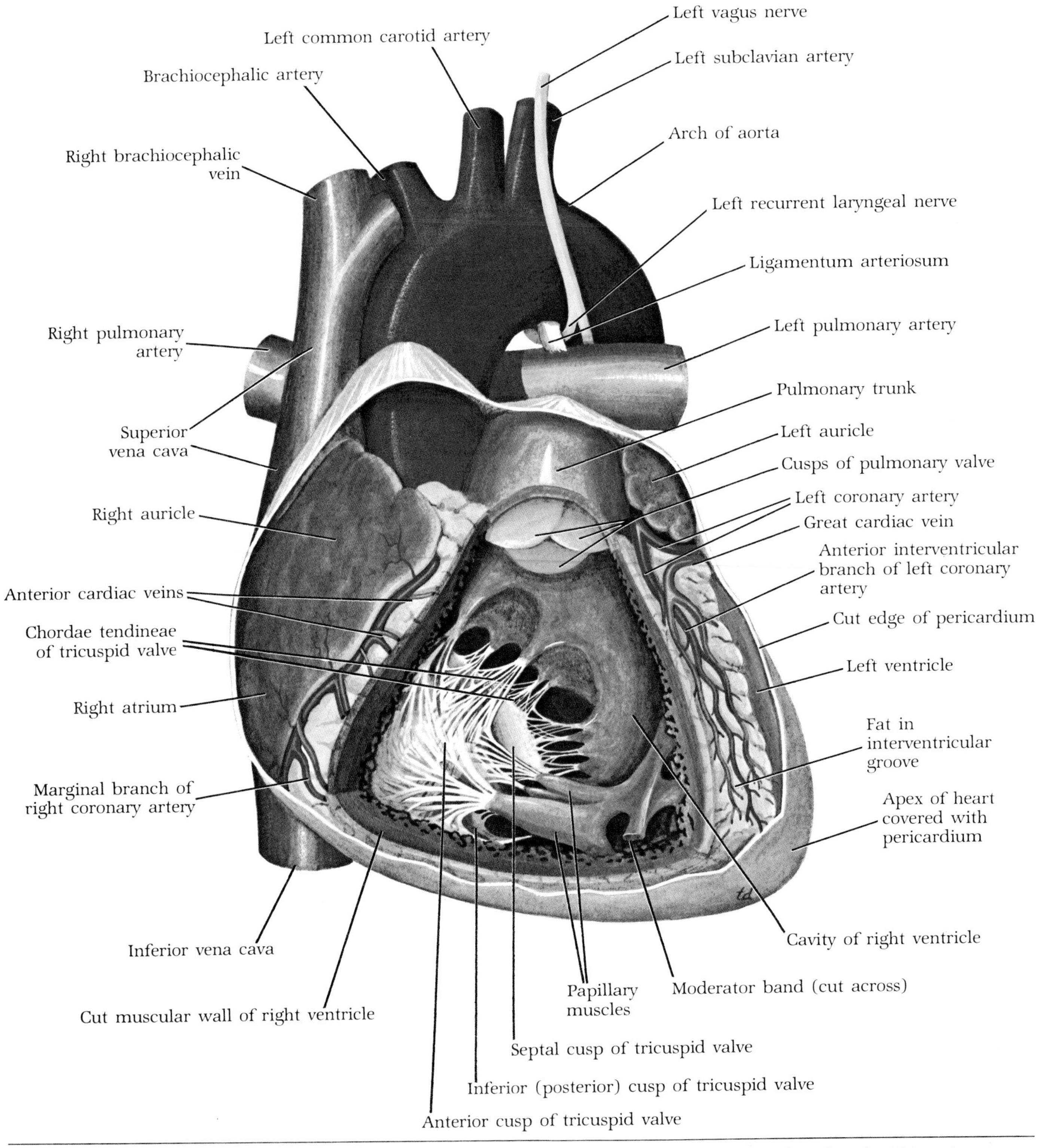

Left vagus nerve
Left common carotid artery
Left subclavian artery
Brachiocephalic artery
Arch of aorta
Right brachiocephalic vein
Left recurrent laryngeal nerve
Ligamentum arteriosum
Left pulmonary artery
Right pulmonary artery
Pulmonary trunk
Superior vena cava
Left auricle
Cusps of pulmonary valve
Right auricle
Left coronary artery
Great cardiac vein
Anterior interventricular branch of left coronary artery
Anterior cardiac veins
Cut edge of pericardium
Chordae tendineae of tricuspid valve
Left ventricle
Right atrium
Fat in interventricular groove
Marginal branch of right coronary artery
Apex of heart covered with pericardium
Inferior vena cava
Cavity of right ventricle
Papillary muscles
Moderator band (cut across)
Cut muscular wall of right ventricle
Septal cusp of tricuspid valve
Inferior (posterior) cusp of tricuspid valve
Anterior cusp of tricuspid valve

Figure 1-20
Anterior view of the heart. The anterior wall of the right ventricle has been removed.

1. Congenital ventricular septal defects, though less frequent than atrial septal defects, are quite common. Normally the ventricular septum is formed from two sources. The inferior muscular part grows up from the floor of the primitive ventricle, and the membranous part is formed by fusion of the lower ends of the bulbar ridges and the septum intermedium (endocardial cushions). The complete ventricular septum is formed when the membranous part fuses with the muscular part. Ventricular septal defects are almost invariably found in the membranous part of the septum.
2. The tricuspid valve is best heard with the stethoscope over the right half of the lower end of the body of the sternum. The pulmonary valve is heard with least interference over the medial end of the second left intercostal space.
3. Disease of the tricuspid valve is rare and almost always accompanies involvement of other cardiac valves by rheumatic disease.
4. Pulmonary valve stenosis may occur as a congenital malformation. Pulmonary valve regurgitation is rare and may accompany severe pulmonary hypertension.

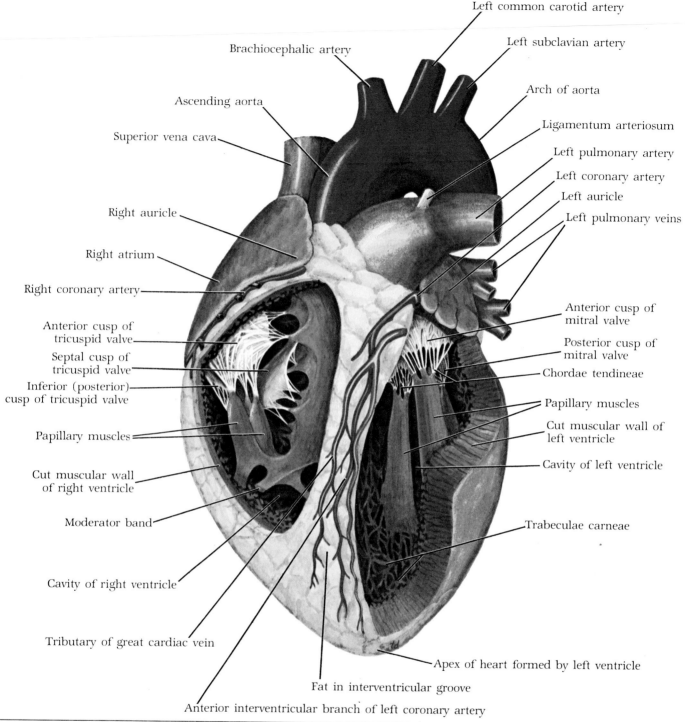

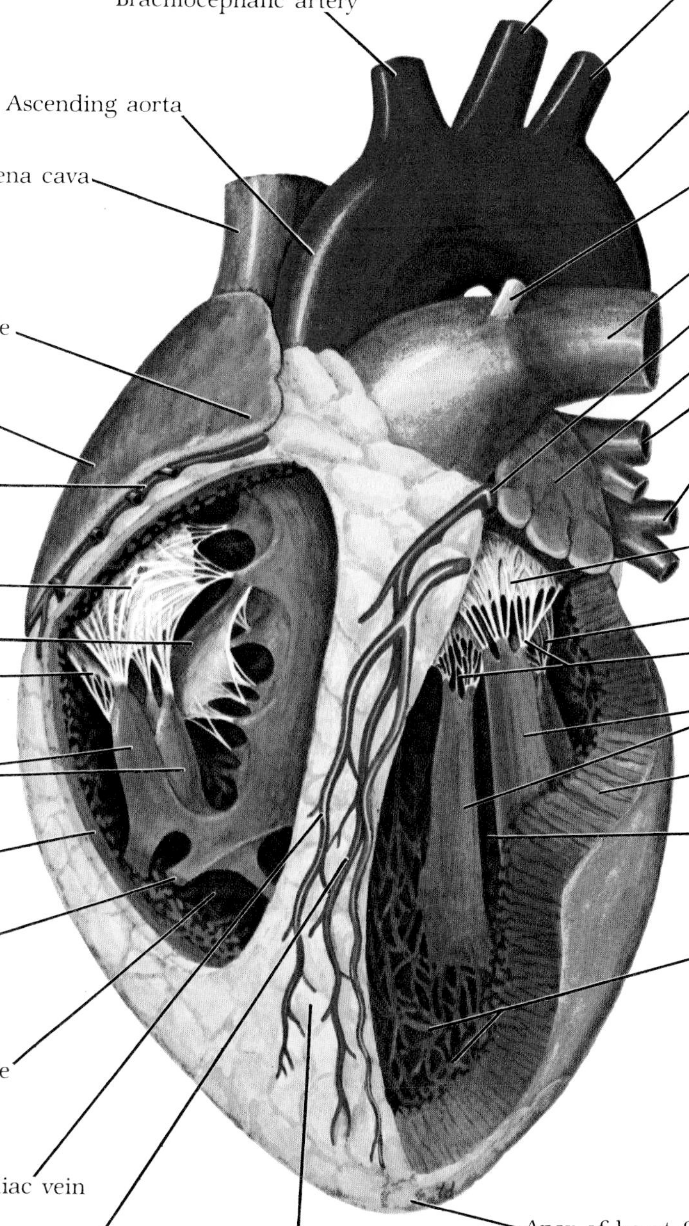
Left common carotid artery
Left subclavian artery
Brachiocephalic artery
Arch of aorta
Ascending aorta
Ligamentum arteriosum
Superior vena cava
Left pulmonary artery
Left coronary artery
Left auricle
Left pulmonary veins
Right auricle
Right atrium
Right coronary artery
Anterior cusp of mitral valve
Anterior cusp of tricuspid valve
Posterior cusp of mitral valve
Septal cusp of tricuspid valve
Chordae tendineae
Inferior (posterior) cusp of tricuspid valve
Papillary muscles
Cut muscular wall of left ventricle
Papillary muscles
Cavity of left ventricle
Cut muscular wall of right ventricle
Moderator band
Trabeculae carneae
Cavity of right ventricle
Tributary of great cardiac vein
Apex of heart formed by left ventricle
Fat in interventricular groove
Anterior interventricular branch of left coronary artery

Figure 1-21
The heart viewed from the left side. The anterior wall of the right ventricle and posterior wall of the left ventricle have been removed.

1. The mitral valve is best heard with the stethoscope over the apex beat (see 4 below).
2. Mitral stenosis and mitral regurgitation usually develop as a sequel to rheumatic fever. The delicate valve cusps fuse, leaving only a reduced central opening for the passage of blood. Later the cusps undergo fibrosis and shrink, after which the free margins no longer come into apposition during systole. The chordae tendineae also shorten, preventing closure of the cusps. The tricuspid valve may also be involved in the disease process.
3. The anterior interventricular branch of the left coronary artery supplies both ventricles and the anterior two-thirds of the ventricular septum. Atherosclerosis occurs in the proximal segments of coronary arteries, especially at the sites of branching, although it does not involve the penetrating muscular branches. It is important to remember that there is no adequate anastomosis between coronary arteries or their branches so that an effective collateral circulation does not take place if a large branch of a coronary artery is occluded.
4. The apex of the heart is formed by the left ventricle. The apex beat, which is due to the apex thrusting forward against the chest wall, may be seen and felt in the fifth left intercostal space about 3½ inches (9 cm) from the midline.

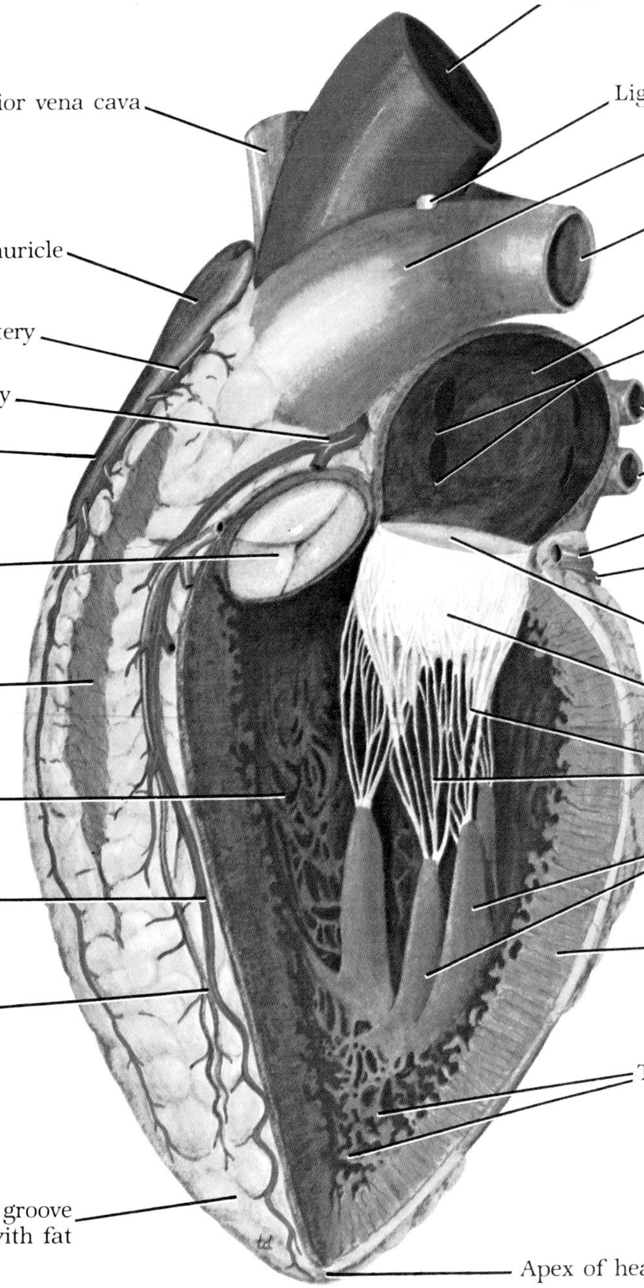
Arch of aorta
Superior vena cava
Ligamentum arteriosum
Pulmonary trunk
Left pulmonary artery
Right auricle
Cavity of left atrium
Right pulmonary veins
Right coronary artery
Left coronary artery
Left pulmonary veins
Anterior wall of right atrium
Great cardiac vein
Left coronary artery
Cusps of aortic valve
Posterior cusp of mitral valve
Anterior wall of right ventricle
Anterior cusp of mitral valve
Chordae tendineae
Interventricular septum
Anterior papillary muscle
Anterior interventricular branch of left coronary artery
Cut muscular wall of left ventricle
Great cardiac vein
Trabeculae carneae
Interventricular groove filled with fat
Apex of heart formed by left ventricle

Figure 1-22
The heart viewed from the left posterior side. The wall of the left ventricle has been removed.

1. The aortic valve is best heard with a stethoscope over the medial end of the second right intercostal space.
2. Congenital aortic valve stenosis may occur at the supravalvular, valvular, and subvalvular levels. Rheumatic fever and atherosclerosis may produce acquired aortic stenosis.
3. Left-sided congestive heart failure is due to overload or damage to the left ventricle, e.g., hypertension or coronary atherosclerosis. This results in a backup of blood in the pulmonary circulation with pulmonary edema; shortness of breath is the common symptom.
4. The ligamentum arteriosum is the fibrous remnant of the ductus arteriosus. Patent ductus arteriosus is a common congenital anomaly in which aortic blood passes into the pulmonary artery, raising the pressure in the pulmonary circulation and causing hypertrophy of the right ventricle. A patent ductus arteriosus endangers life and thus should be ligated and divided surgically.

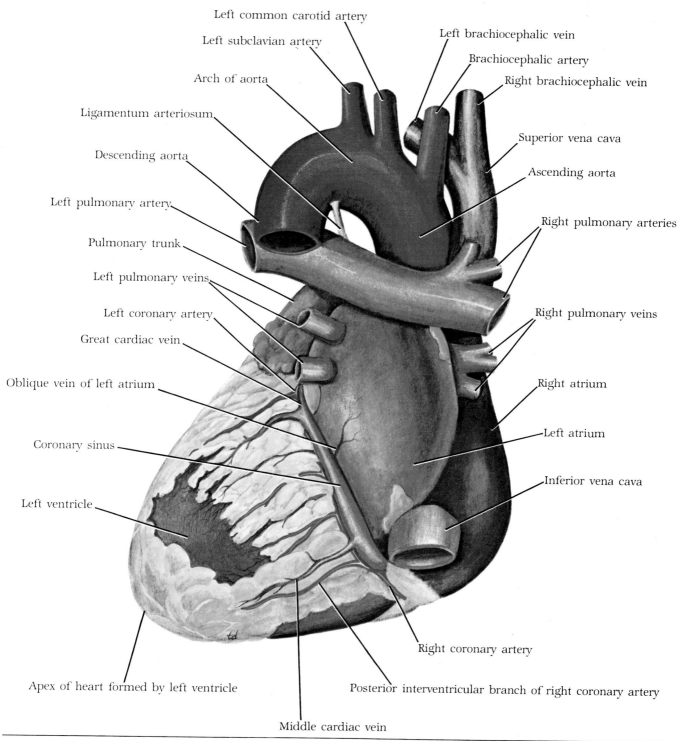

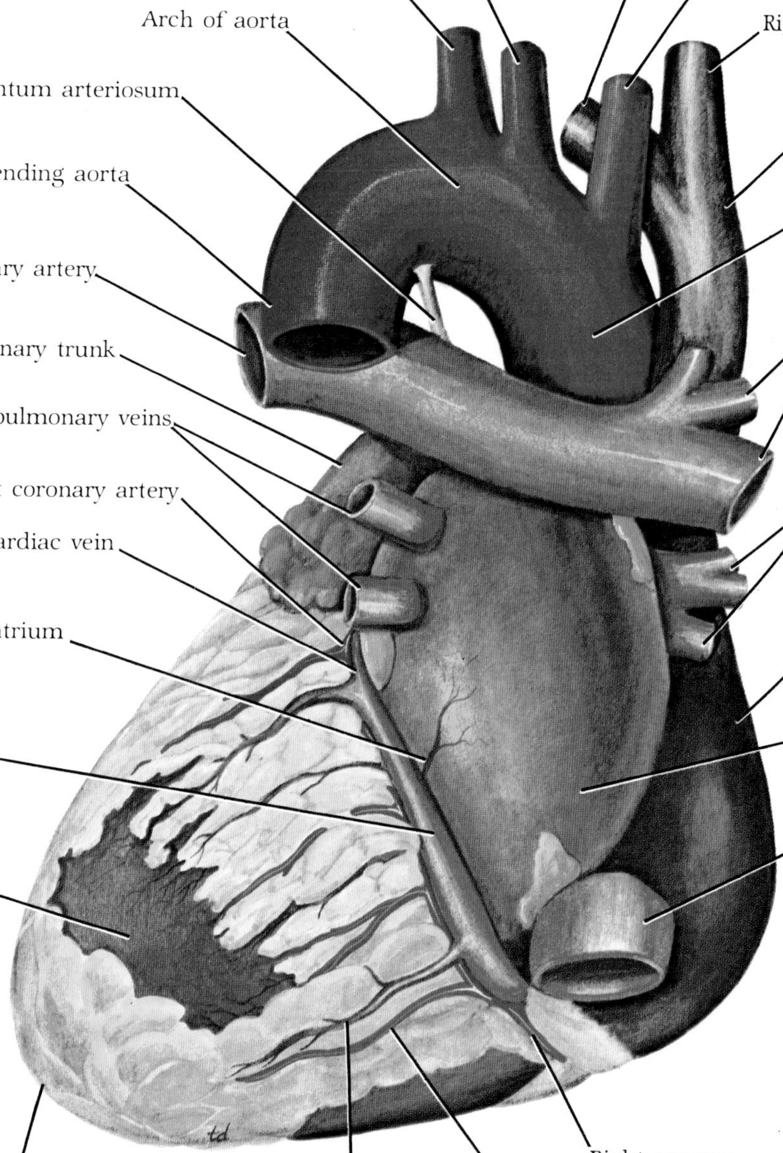
Left common carotid artery
Left subclavian artery
Arch of aorta
Ligamentum arteriosum
Descending aorta
Left pulmonary artery
Pulmonary trunk
Left pulmonary veins
Left coronary artery
Great cardiac vein
Oblique vein of left atrium
Coronary sinus
Left ventricle
Apex of heart formed by left ventricle
Middle cardiac vein
Posterior interventricular branch of right coronary artery
Right coronary artery
Inferior vena cava
Left atrium
Right atrium
Right pulmonary veins
Right pulmonary arteries
Ascending aorta
Superior vena cava
Right brachiocephalic vein
Brachiocephalic artery
Left brachiocephalic vein

Figure 1-23
The posterior surface or base of the heart.

1. Coarctation of the aorta is a congenital anomaly in which there is a narrowing of the aorta just distal to the origin of the left subclavian artery. It is believed to be due to an unusual quantity of ductus arteriosus muscle tissue becoming incorporated into the wall of the aorta during embryonic development. Clinically, the cardinal sign of aortic coarctation is absent or diminished femoral pulses. To compensate for the diminished volume of blood reaching the lower part of the body, an enormous collateral circulation opens up, involving the internal thoracic, subclavian, and posterior intercostal arteries. Operative treatment is required.

2. The posterior surface of the left atrium is directly related to the esophagus. Enlargement of the left atrium, which may occur in patients with mitral stenosis, may be detected in a lateral radiograph of the esophagus following a barium swallow.

3. The left margin and the apex of the heart are formed by the left ventricle. Enlargement of this ventricle may be detected clinically by the fact that the apex beat is further over to the left than normal; also, percussion of the chest wall will show displacement of the left margin of the heart to the left. Radiographic and electrocardiographic examination will also detect or confirm cardiac enlargement.

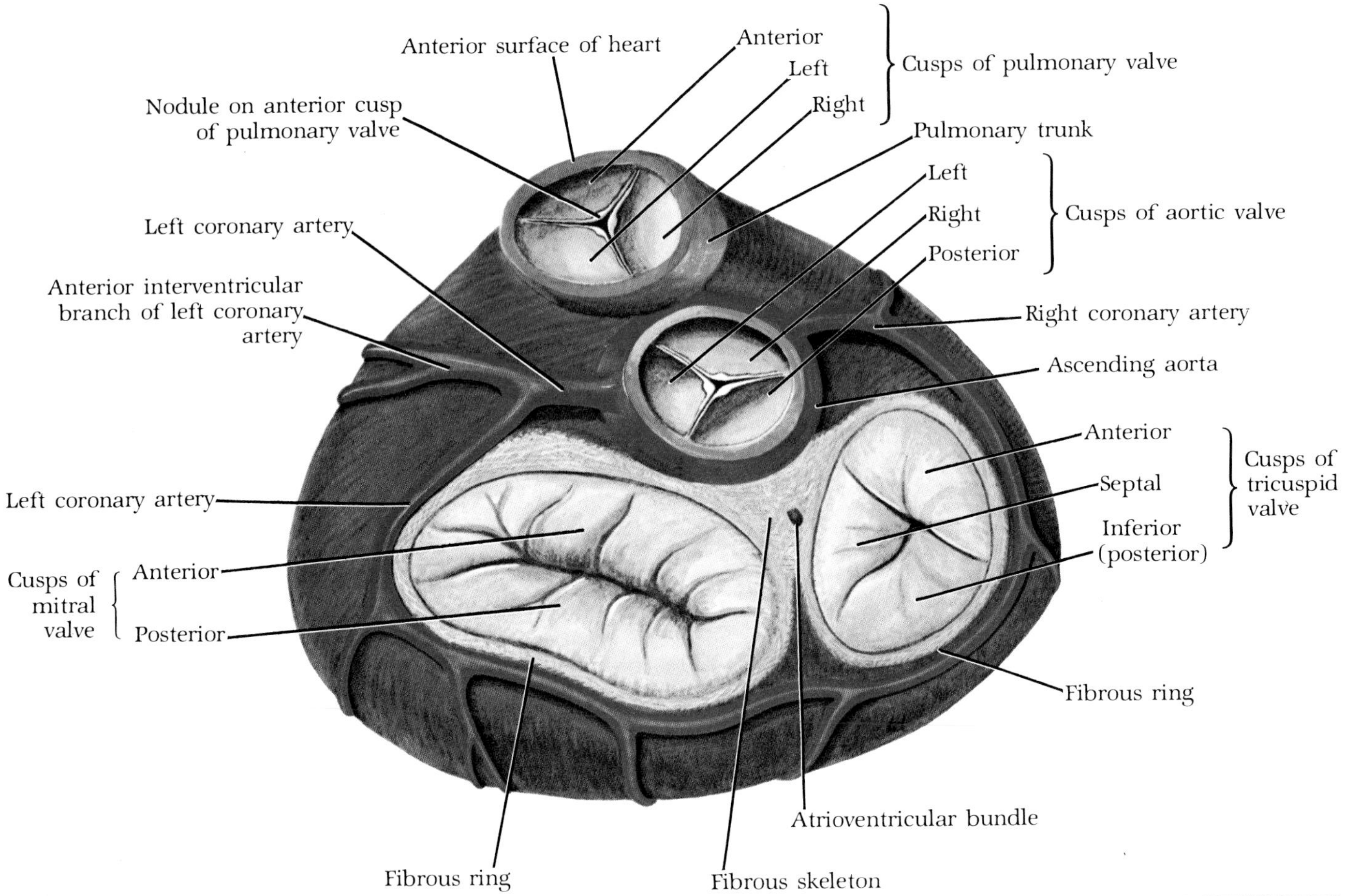
Anterior surface of heart
Anterior
Left
Right
Cusps of pulmonary valve
Nodule on anterior cusp of pulmonary valve
Pulmonary trunk
Left
Right
Posterior
Cusps of aortic valve
Left coronary artery
Anterior interventricular branch of left coronary artery
Right coronary artery
Ascending aorta
Anterior
Septal
Inferior (posterior)
Cusps of tricuspid valve
Left coronary artery
Cusps of mitral valve
Anterior
Posterior
Fibrous ring
Atrioventricular bundle
Fibrous ring
Fibrous skeleton

Figure 1-24
The valves of the heart and the origin of the coronary arteries, superior view. The atria and the great vessels have been removed.

1. Normal heart valves allow one-way flow of blood from the atria into the ventricles and from the ventricles to the aorta and pulmonary trunk. Even though a considerable pressure gradient may exist on the two sides of a valve, the valve must allow passage of the complete stroke output of the heart during one phase of the cycle and yet allow no backflow. To satisfy these criteria the valve cusps must remain separate, thin, and flexible, and must be able to come into close apposition without difficulty.

2. The first heart sound (lūb) is produced by the closure of the atrioventricular valves, and the second sound (dŭp) by closure of the aortic and pulmonary valves. Both closures are accompanied by vibrations through the heart structure that contribute to the sounds heard with the stethoscope.

3. Rheumatic fever is the most common cause of acquired valvular disease. Although the initial inflammatory response may be minimal, it may cause the edges of the cusps to stick together. Later (possibly many years) fibrous thickening occurs followed by loss of flexibility, shrinkage, and deposits of calcium. Narrowing (stenosis) and valvular incompetence (regurgitation) result.

4. Note the precise origin of the coronary arteries.

5. The atrioventricular bundle is the only route by which the wave of conduction can spread from the atria to the ventricles. Failure of the bundle to conduct the normal impulses will result in alterations in the rhythmic contraction of the ventricles (arrhythmias), or, if there is a complete bundle block, there is complete dissociation between the atrial and ventricular rates of contraction.

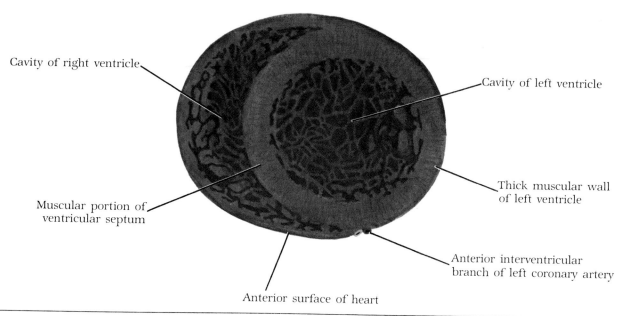

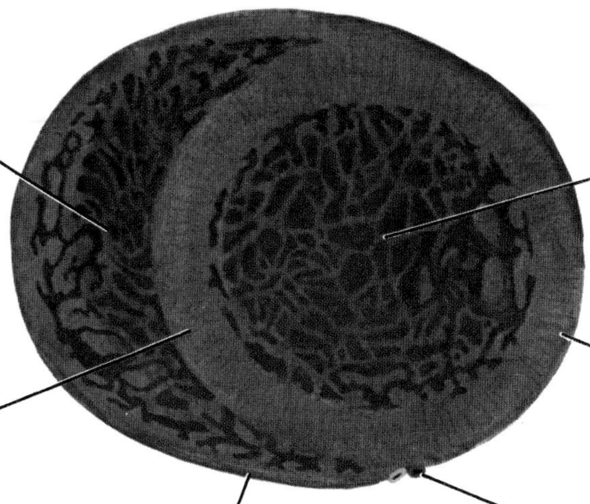
Cavity of right ventricle
Cavity of left ventricle
Thick muscular wall of left ventricle
Muscular portion of ventricular septum
Anterior interventricular branch of left coronary artery
Anterior surface of heart

Figure 1-25
Transverse section of the ventricles of the heart.

1. The thick myocardium of the left ventricular wall permits the left ventricle to pump a volume of blood through the high-resistance systemic circulation, while the thinner myocardium of the right ventricle pumps the same volume of blood through the low-resistance pulmonary circulation.
2. The myocardium depends on adequate coronary blood flow for its nutrition. A sudden coronary occlusion due to thrombosis may result in a fatal myocardial infarction.

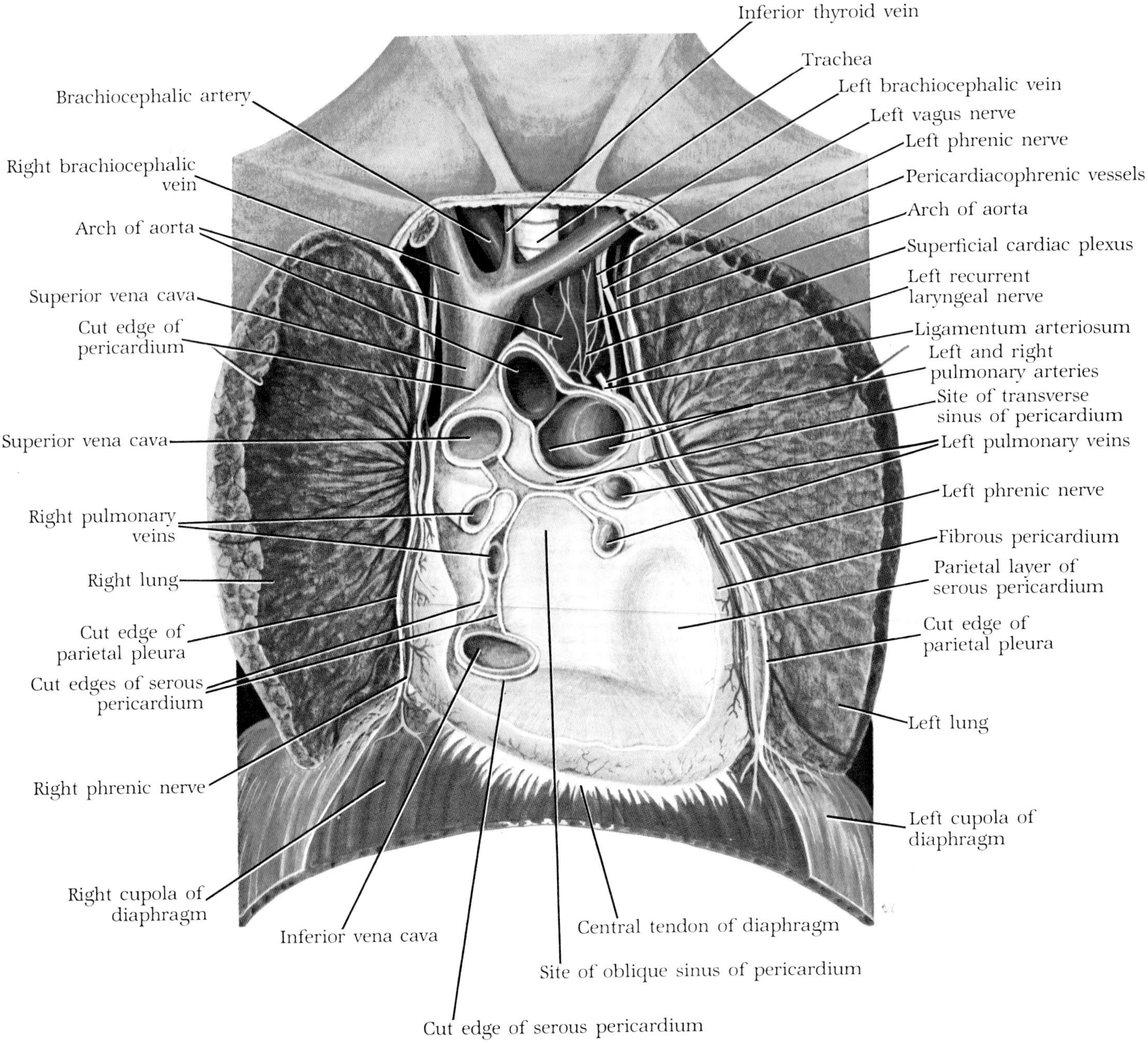
Inferior thyroid vein
Trachea
Left brachiocephalic vein
Left vagus nerve
Left phrenic nerve
Pericardiacophrenic vessels
Arch of aorta
Superficial cardiac plexus
Left recurrent laryngeal nerve
Ligamentum arteriosum
Left and right pulmonary arteries
Site of transverse sinus of pericardium
Left pulmonary veins
Left phrenic nerve
Fibrous pericardium
Parietal layer of serous pericardium
Cut edge of parietal pleura
Left lung
Left cupola of diaphragm
Brachiocephalic artery
Right brachiocephalic vein
Arch of aorta
Superior vena cava
Cut edge of pericardium
Superior vena cava
Right pulmonary veins
Right lung
Cut edge of parietal pleura
Cut edges of serous pericardium
Right phrenic nerve
Right cupola of diaphragm
Inferior vena cava
Central tendon of diaphragm
Site of oblique sinus of pericardium
Cut edge of serous pericardium

Figure 1-26
Interior of the pericardial cavity, anterior view. The heart has been removed.

1. Note that the pericardium is made up of two parts, the fibrous pericardium and the serous pericardium. The fibrous pericardium serves to anchor the heart, while the serous pericardium with its parietal and visceral layers allows the heart to beat without friction between it and the surrounding structures. The pericardial cavity is the slitlike space between the parietal and visceral layers; it normally contains a small amount of fluid that acts as a lubricant to facilitate movements of the heart.
2. The ascending aorta and the pulmonary trunk lie within the fibrous pericardium. Rupture of these vessels would produce immediate cardiac tamponade.
3. Benign tumors of the mediastinum may compress or distort the trachea (producing dyspnea) or the esophagus (producing dysphagia). They rarely compress the large blood vessels to a degree sufficient to produce symptoms, however. Malignant tumors that compress and invade the vessel walls, on the other hand, commonly produce venous congestion due to obstruction of the superior vena cava.

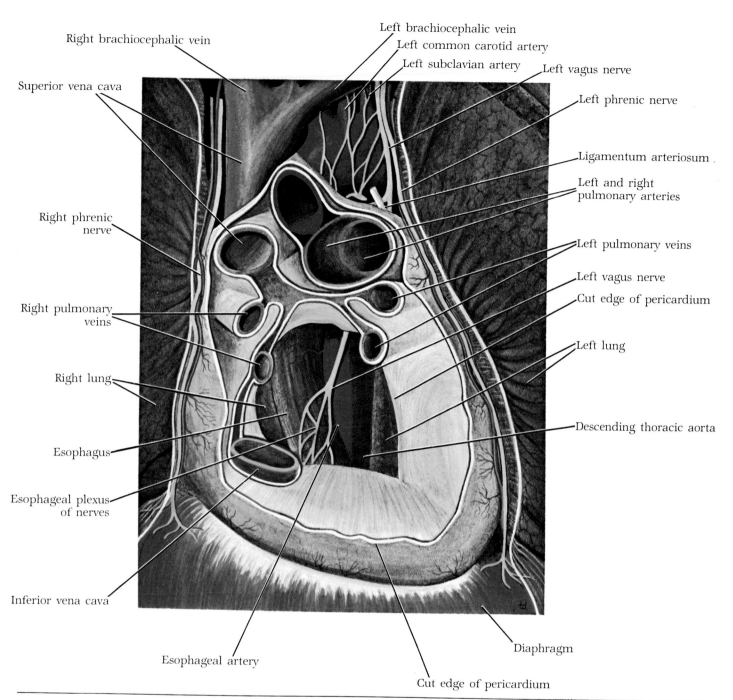

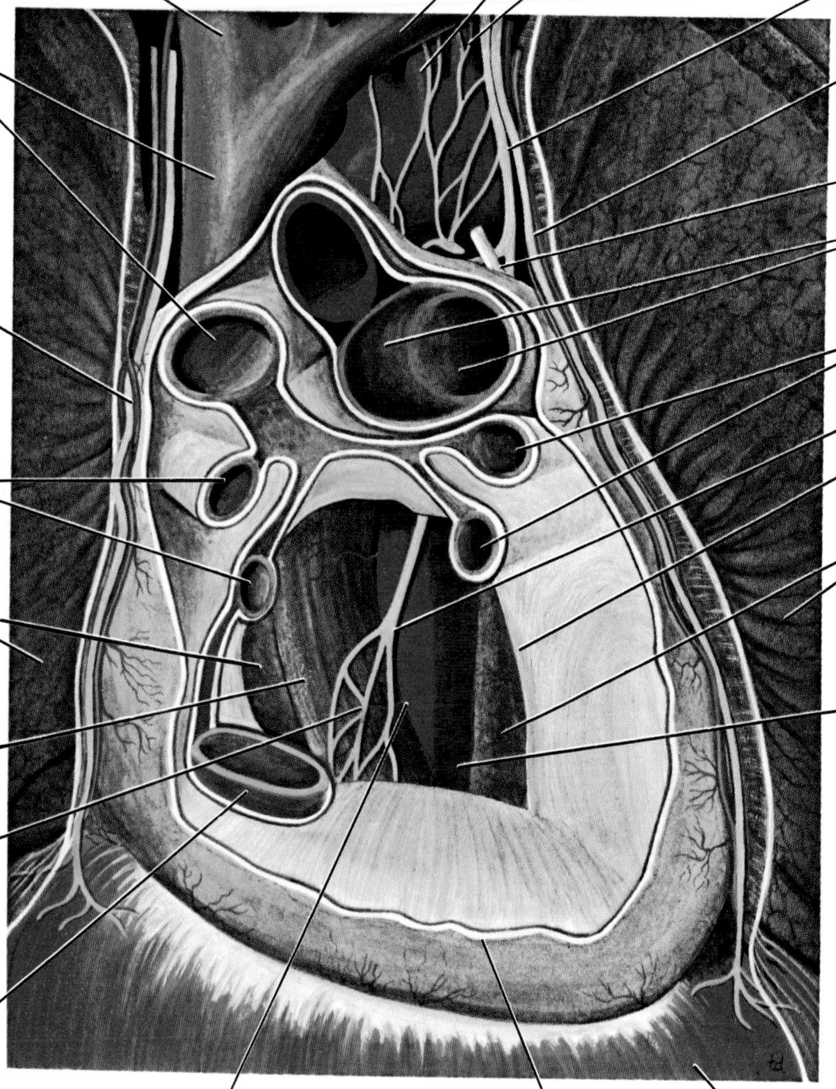
Right brachiocephalic vein
Left brachiocephalic vein
Left common carotid artery
Left subclavian artery
Left vagus nerve
Superior vena cava
Left phrenic nerve
Ligamentum arteriosum
Left and right pulmonary arteries
Right phrenic nerve
Left pulmonary veins
Left vagus nerve
Right pulmonary veins
Cut edge of pericardium
Left lung
Right lung
Descending thoracic aorta
Esophagus
Esophageal plexus of nerves
Inferior vena cava
Diaphragm
Esophageal artery
Cut edge of pericardium

Figure 1-27
Interior of the pericardial cavity, anterior view. Part of the posterior wall of the pericardial cavity has been removed.

1. In the posterior mediastinum the esophagus lies posterior to the left atrium and the aorta lies posterior to the left atrium and left ventricle. Enlargement of the left atrium may be detected clinically by examining a lateral radiograph of the esophagus following a barium swallow.
2. Carcinoma of the esophagus spreads by local invasion and lymphatic metastases. Invasion of the pericardium and heart occurs when the lower, thoracic part of the esophagus is the site of the tumor.

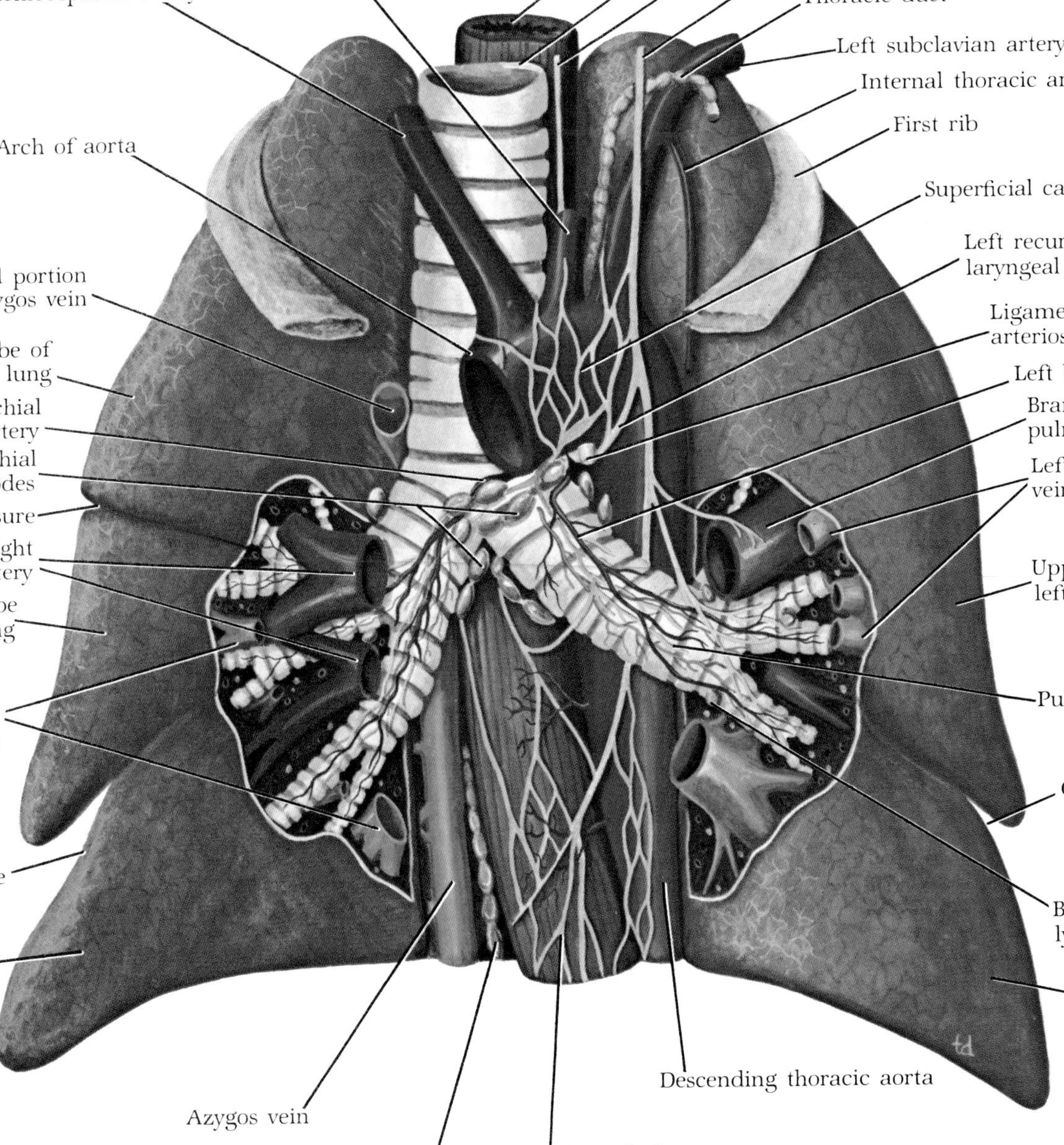
Esophagus
Trachea
Left recurrent laryngeal nerve
Left common carotid artery
Left vagus nerve
Brachiocephalic artery
Thoracic duct
Left subclavian artery
Internal thoracic artery
First rib
Arch of aorta
Superficial cardiac plexus
Left recurrent
laryngeal nerve
Terminal portion
of azygos vein
Ligamentum
arteriosum
Upper lobe of
right lung
Left bronchial artery
Right bronchial
artery
Branch of left
pulmonary artery
Tracheobronchial
lymph nodes
Left pulmonary
veins
Horizontal fissure
Branches of right
pulmonary artery
Upper lobe of
left lung
Middle lobe
of right lung
Right
pulmonary
veins
Pulmonary plexus
Oblique fissure
Oblique fissure
Bronchopulmonary
lymph node
Lower lobe
of right
lung
Lower lobe
of left lung
td
Descending thoracic aorta
Azygos vein
Thoracic duct
Esophageal plexus

Figure 1-28
Trachea, main bronchi, and roots of the lungs, anterior view.

1. Bronchoscopy enables a physician to examine the interior of the trachea, carina, and main bronchi. It is possible with this instrument to obtain biopsy specimens of mucous membrane and remove inhaled foreign bodies.

2. The adult trachea is large in size; an infant's trachea, however, is very small and minor degrees of lumen narrowing caused by disease may be of grave clinical significance.

3. Aspirated foreign bodies may lodge in the trachea. Since the right bronchus is wider than the left and is a more direct continuation of the trachea, foreign bodies tend to enter the right rather than the left bronchus.

4. Congenital tracheoesophageal fistula without esophageal atresia may cause respiratory infection due to inhalation of esophageal or gastric contents.

5. The flexible trachea may be distorted by neighboring structures such as retrosternal goiter or aneurysm of the aorta.

6. Bronchogenic carcinoma not only produces bronchial obstruction but metastasizes to the lymph nodes within the lung and the mediastinum. Carcinoma of the left lung with mediastinal lymph node involvement may compress or invade the left recurrent laryngeal nerve, causing hoarseness of the voice.

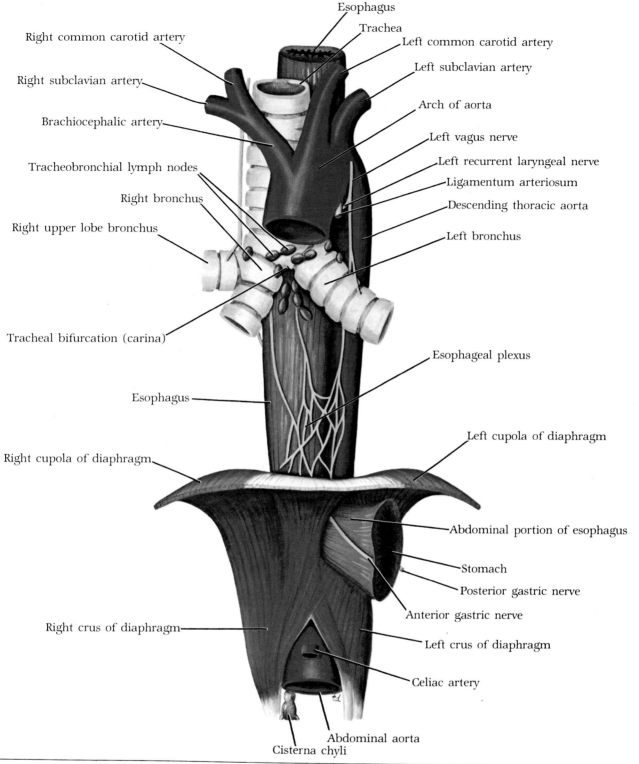

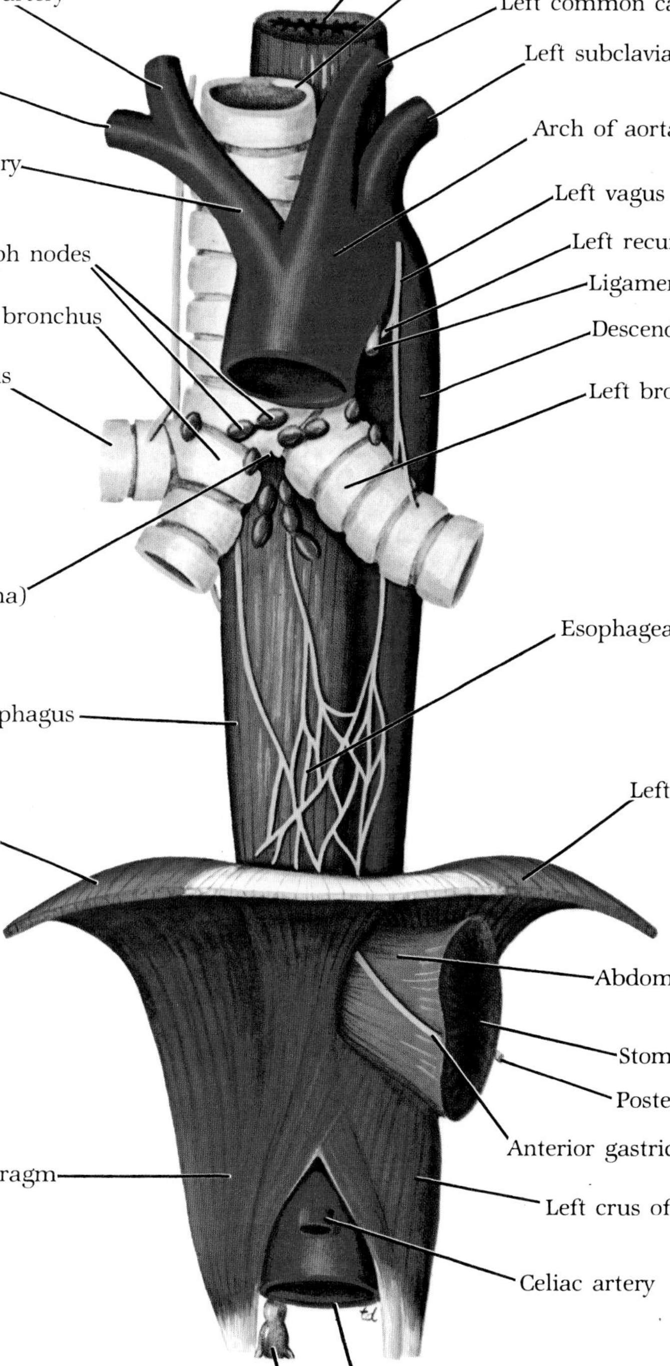
Esophagus
Trachea
Right common carotid artery
Left common carotid artery
Left subclavian artery
Right subclavian artery
Brachiocephalic artery
Arch of aorta
Left vagus nerve
Tracheobronchial lymph nodes
Left recurrent laryngeal nerve
Ligamentum arteriosum
Right bronchus
Descending thoracic aorta
Right upper lobe bronchus
Left bronchus
Tracheal bifurcation (carina)
Esophageal plexus
Esophagus
Left cupola of diaphragm
Right cupola of diaphragm
Abdominal portion of esophagus
Stomach
Posterior gastric nerve
Anterior gastric nerve
Right crus of diaphragm
Left crus of diaphragm
Celiac artery
Abdominal aorta
Cisterna chyli

Figure 1-29
The esophagus, in the thorax and abdomen, anterior view.

1. The esophagus has three anatomical and physiological constrictions: (a) where the pharynx joins the upper end; (b) where the left bronchus and aortic arch cross its anterior surface; and (c) where the esophagus passes through the diaphragm into the stomach. These constrictions are of clinical importance for the following reasons: (a) they are common sites of carcinoma of the esophagus; (b) they may hold up swallowed foreign bodies; (c) strictures develop at them following the drinking of caustic fluids; and (d) they may impede the passage of an esophagoscope.
2. Sliding esophageal hiatal hernia and paraesophageal hiatal hernia are the most common herniae of the diaphragm. In the sliding form the esophagogastric junction and the proximal part of the stomach are pushed superiorly into the posterior mediastinum through the esophageal hiatus. In the paraesophageal type the esophagogastric junction remains stationary and a portion of the stomach passes through the esophageal hiatus alongside the esophagus.
3. The vagus nerves, on emerging below the diaphragm, lie on the anterior and posterior surfaces of the esophagus and are here referred to as anterior and posterior gastric nerves. Operative section of these gastric nerves (vagotomy) combined with gastroenterostomy, pyloroplasty, or an antrectomy is a recognized form of treatment for duodenal ulcer.

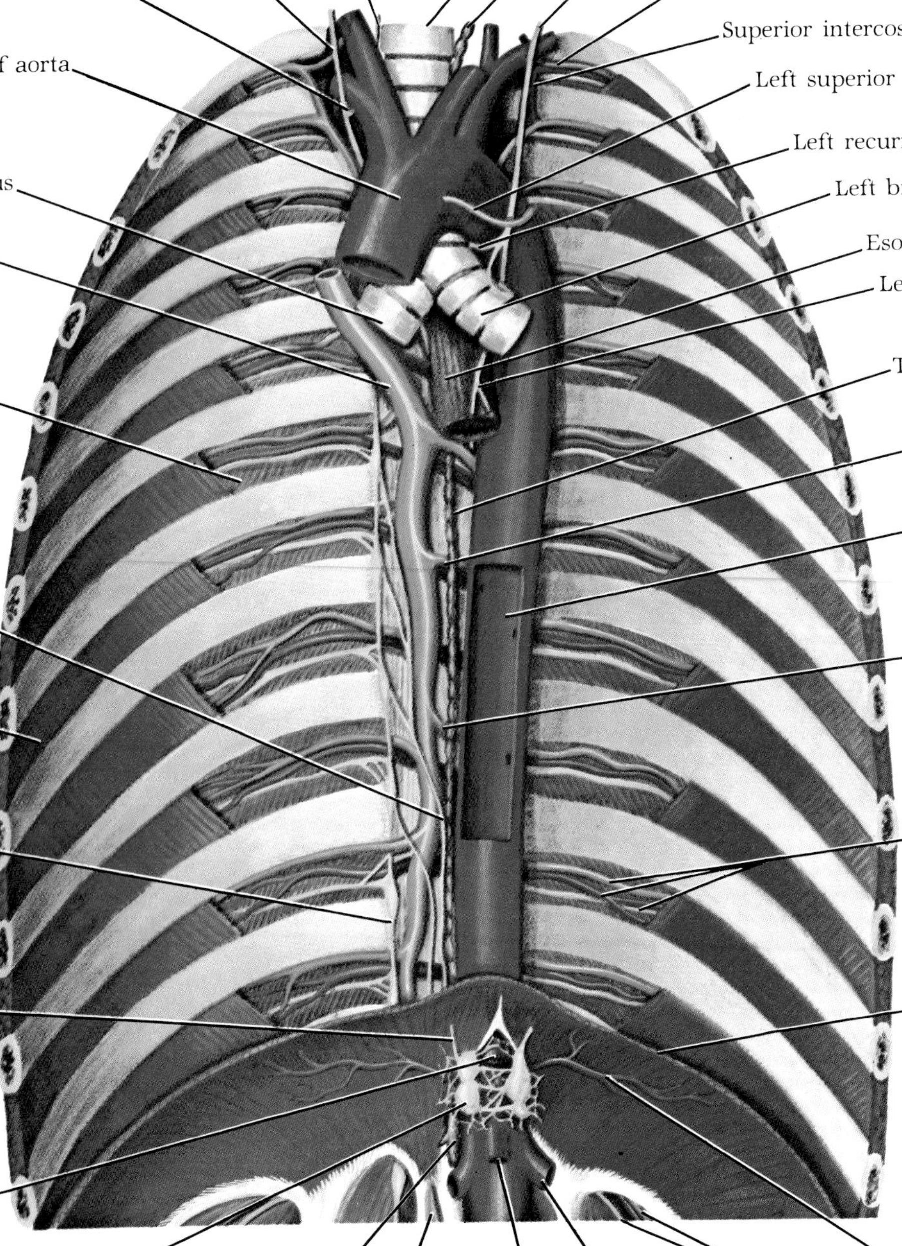
Right recurrent laryngeal nerve
Right vagus nerve
Right recurrent laryngeal nerve
Arch of aorta
Right bronchus
Azygos vein
External intercostal muscle
Greater splanchnic nerve
Innermost intercostal muscle
Sympathetic trunk
Greater splanchnic nerve
Celiac artery
Celiac plexus
Cisterna chyli
Ascending lumbar vein
Trachea
Thoracic duct
Left vagus nerve
First intercostal vein
Superior intercostal artery
Left superior intercostal vein
Left recurrent laryngeal nerve
Left bronchus
Esophagus
Left vagus nerve
Thoracic duct
Terminal portion of superior hemiazygos vein
Aorta opened to show orifices of intercostal arteries
Terminal portion of inferior hemiazygos vein
Intercostal vessels and nerve in tenth intercostal space
Diaphragm
Inferior phrenic artery
Subcostal vessels and nerve
Left renal artery
Superior mesenteric artery

Figure 1-30
Thoracic aorta, trachea, esophagus, azygos vein, and thoracic duct, anterior view.

1. Dilatations of the aorta (aneurysms), which are not uncommon, may be caused by atherosclerosis or syphilitic destruction of the tunica media. Often asymptomatic, they may exert pressure on surrounding structures such as the trachea or superior vena cava.

2. Rupture of the aorta or damage to its wall may occur in severe automobile or plane accidents. The most common site of rupture is at the attachment of the ligamentum arteriosum.

3. Because of the protected location of the thoracic duct and azygos vein in the posterior mediastinum adjacent to the vertebral column, they are rarely damaged. Excessive hyperextension of the vertebral column has been reported to cause rupture of the thoracic duct.

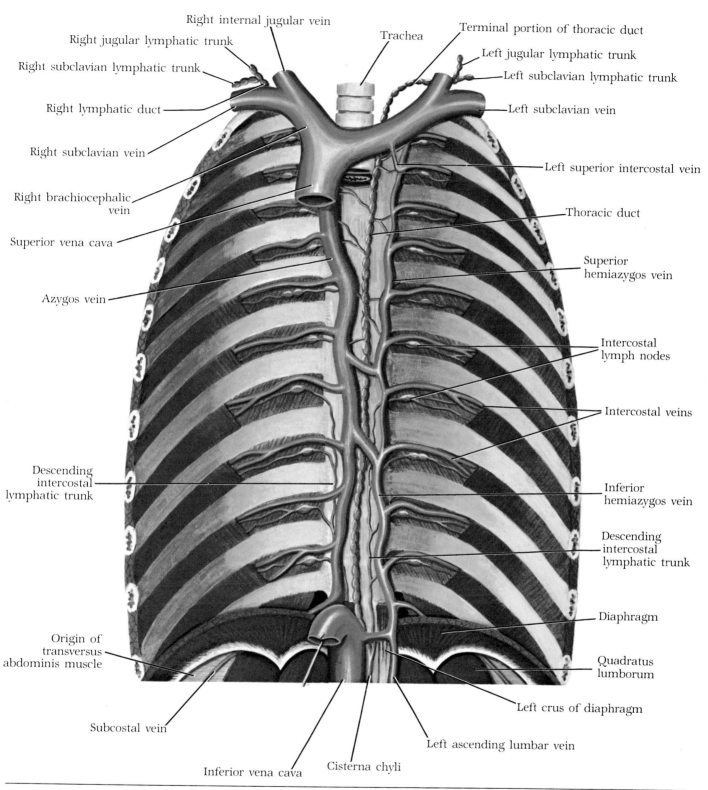

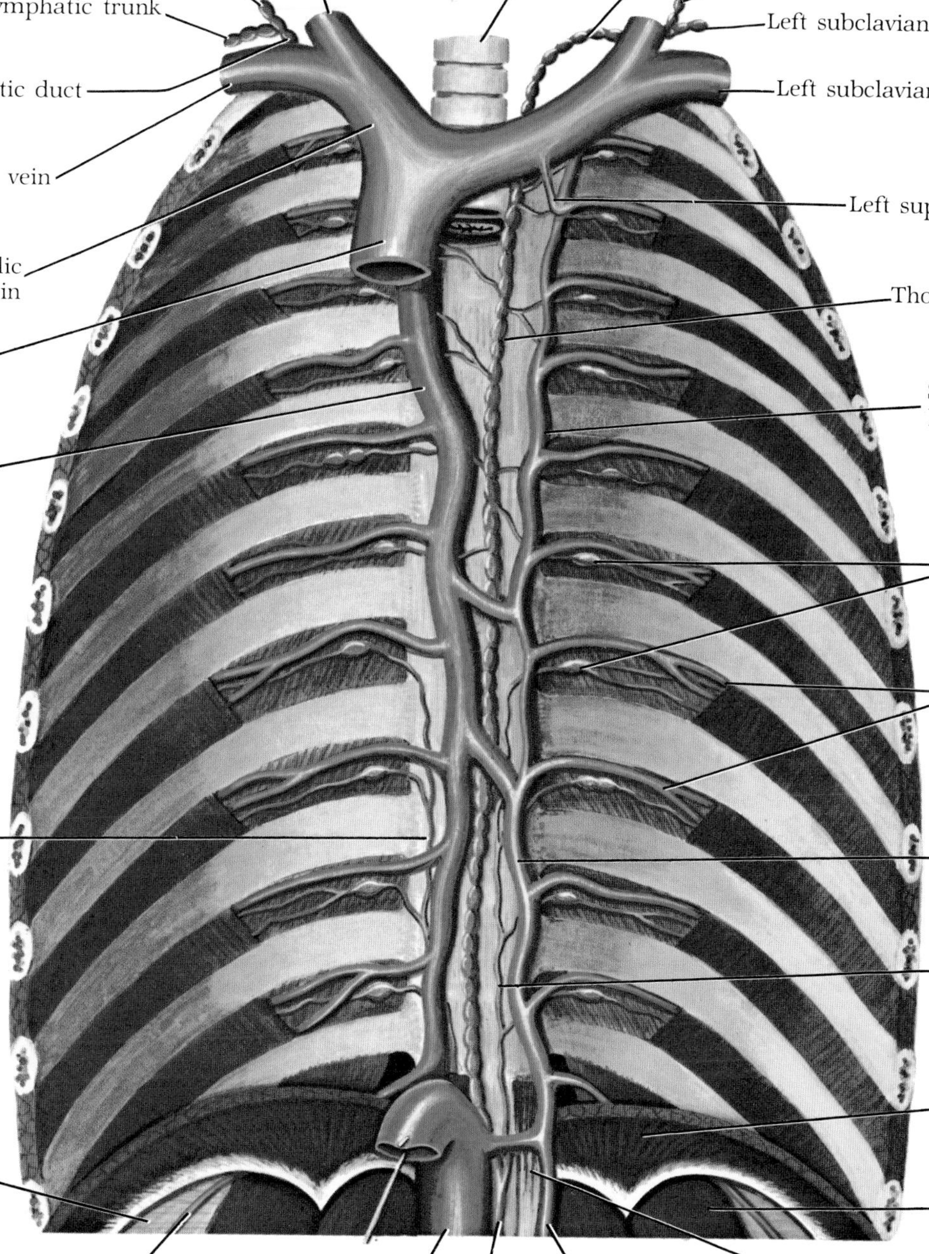
Right internal jugular vein
Right jugular lymphatic trunk
Right subclavian lymphatic trunk
Right lymphatic duct
Right subclavian vein
Right brachiocephalic vein
Superior vena cava
Azygos vein
Descending intercostal lymphatic trunk
Origin of transversus abdominis muscle
Subcostal vein
Inferior vena cava
Cisterna chyli
Trachea
Terminal portion of thoracic duct
Left jugular lymphatic trunk
Left subclavian lymphatic trunk
Left subclavian vein
Left superior intercostal vein
Thoracic duct
Superior hemiazygos vein
Intercostal lymph nodes
Intercostal veins
Inferior hemiazygos vein
Descending intercostal lymphatic trunk
Diaphragm
Quadratus lumborum
Left crus of diaphragm
Left ascending lumbar vein

Figure 1-31
Azygos veins, thoracic duct, and posterior intercostal lymphatics, anterior view.

1. The superior vena cava may become obstructed; the most common cause is malignant neoplasm of the mediastinum, usually bronchogenic carcinoma of the right lung with invasion of the mediastinum. The symptoms of venous congestion of the head and neck and upper extremity are headaches, puffiness of the face, and enlargement of the neck. In advanced cases drowsiness and blurring of vision also occur.

2. The thoracic duct is the largest lymph vessel in the body. Its beaded appearance is due to the presence of numerous valves. The presence of the valves and the negative pressure that exists in the thorax aid the transport of lymph through the thorax to the large veins in the root of the neck. This duct conveys to the blood all lymph from the lower limbs, pelvic cavity, abdominal cavity, left side of the thorax, left side of the head and neck, and left arm.

3. Because of their protected position, the azygos veins are rarely damaged and then usually only during a surgical procedure within the thorax.

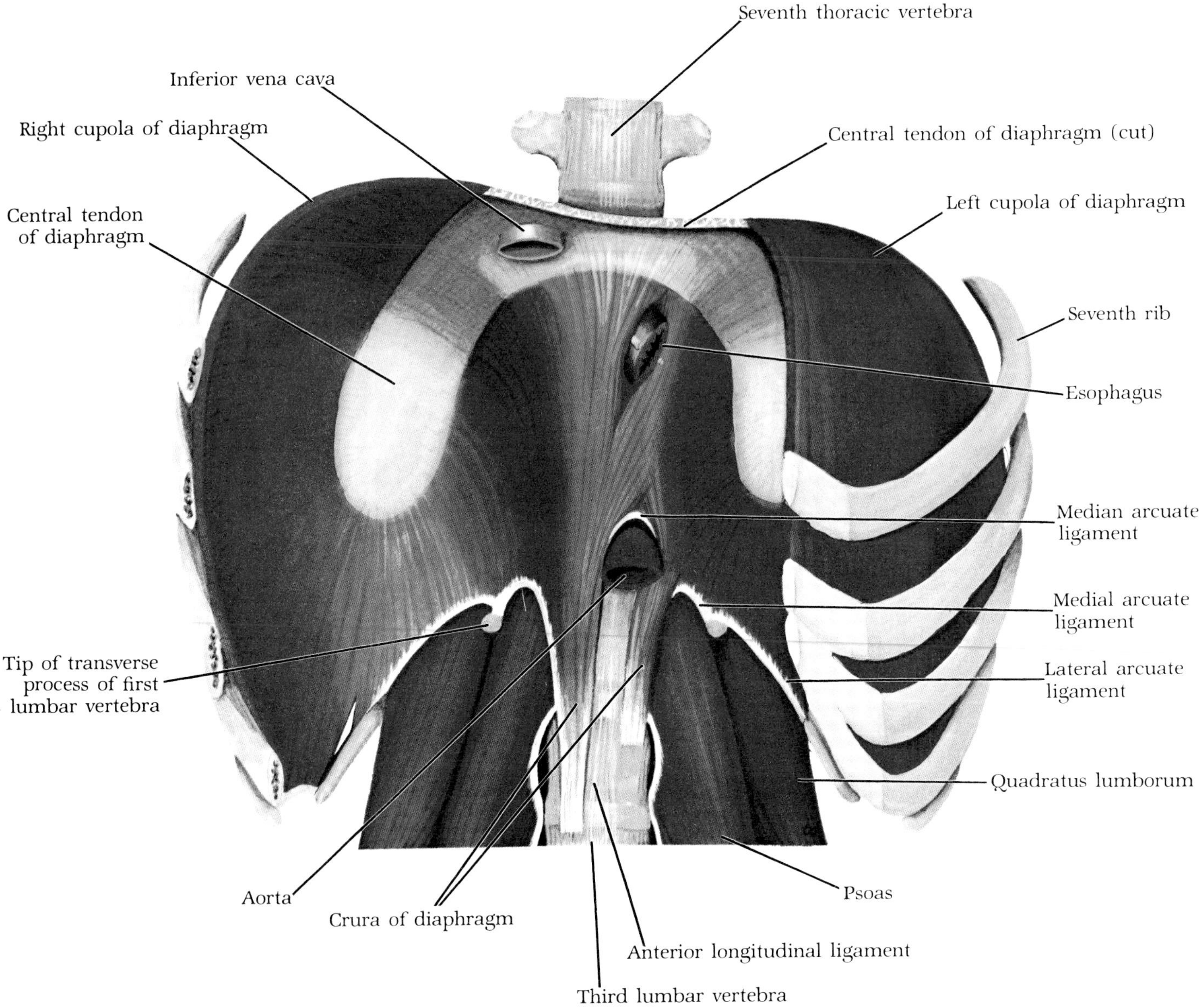
Seventh thoracic vertebra
Inferior vena cava
Right cupola of diaphragm
Central tendon of diaphragm (cut)
Left cupola of diaphragm
Central tendon of diaphragm
Seventh rib
Esophagus
Median arcuate ligament
Medial arcuate ligament
Tip of transverse process of first lumbar vertebra
Lateral arcuate ligament
Quadratus lumborum
Aorta
Crura of diaphragm
Psoas
Anterior longitudinal ligament
Third lumbar vertebra

Figure 1-32
The diaphragm. The anterior portion of the right side has been removed.

1. The diaphragm is the principal muscle of inspiration; consequently its paralysis will result in severe respiratory embarrassment. Poliomyelitis, injury to the cervical region of the spinal cord, or damage to or tumor involvement of the phrenic nerve (C3, 4, and 5) can cause paralysis of the diaphragm.
2. Pain from irritation of the pleura or peritoneum covering the central portion of the diaphragm may be referred to the region of the shoulder. The sensory innervation of the central portion of the diaphragm is the phrenic nerve (C3, 4, and 5), and the sensory nerve supply to the skin over the point of the shoulder is the supraclavicular nerves (C3 and 4).
3. The diaphragm serves as a partition between the cavities of the thorax and abdomen. Imperfect development, weakness, or injury to the diaphragm may permit abdominal viscera to protrude into the thoracic cavity (diaphragmatic hernia).
4. The lymphatic vessels of the diaphragm may allow infection to spread from the peritoneal cavity to the pleural cavity, or vice versa.
5. Gas or fluid (pus, blood, escaped visceral contents) may accumulate under the diaphragm in the subphrenic peritoneal spaces.
6. Esophageal hiatus hernia due to a congenital short esophagus, a sliding esophageal hernia, or a paraesophageal hernia may occur through the esophageal opening in the diaphragm.
7. Hiccup is caused by clonic spasms of the diaphragm.
8. Stitch (side ache) is thought to be due to anoxemia of the diaphragm.

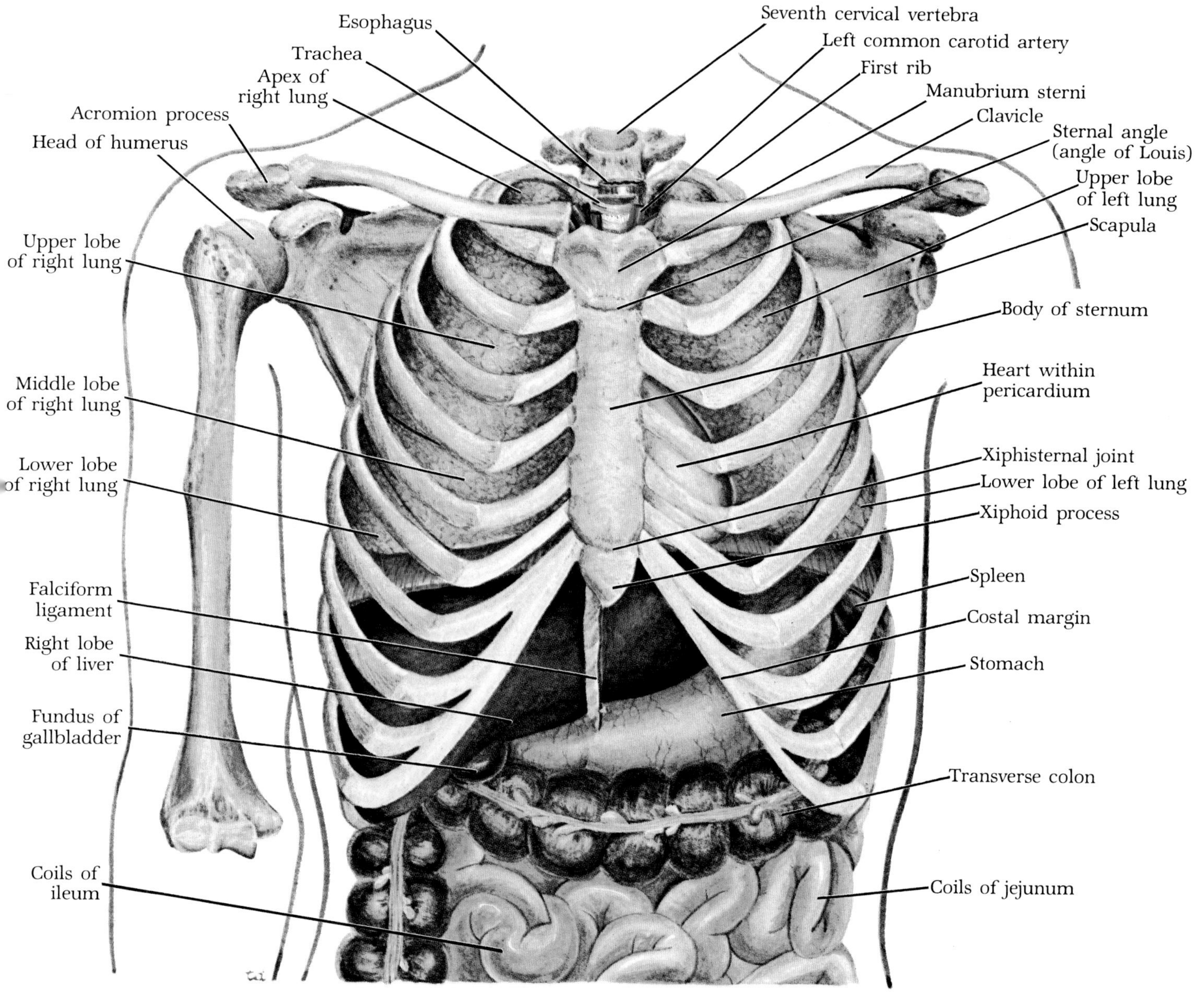
Esophagus
Trachea
Apex of right lung
Acromion process
Head of humerus
Upper lobe of right lung
Middle lobe of right lung
Lower lobe of right lung
Falciform ligament
Right lobe of liver
Fundus of gallbladder
Coils of ileum
Seventh cervical vertebra
Left common carotid artery
First rib
Manubrium sterni
Clavicle
Sternal angle (angle of Louis)
Upper lobe of left lung
Scapula
Body of sternum
Heart within pericardium
Xiphisternal joint
Lower lobe of left lung
Xiphoid process
Spleen
Costal margin
Stomach
Transverse colon
Coils of jejunum

Figure 1-33
Skeleton of the thorax, showing its relationship to thoracic and abdominal viscera, anterior view.

1. A physician must have a mental image of the normal position of the lungs, heart, and upper abdominal viscera in relation to identifiable bony landmarks of the thorax.
2. The sternal angle is an important landmark. It can be felt as a transverse ridge and lies at the level of the second costal cartilage and the second rib. All other ribs may be counted from this point.
3. The apex of the lungs extend up into the neck; because of this they are vulnerable to stab wounds in the root of the neck.
4. A knowledge of the precise surface marking of the lungs enables the physician to accurately listen to the different lobes of the lungs with a stethoscope.
5. When examining a patient remember that the upper lobes of the lungs are most easily examined from in front of the chest and the lower lobes from the back.
6. The surface marking of the heart and the position of the apex beat enables one to determine whether the heart has shifted its position or is enlarged by disease. The apex beat can often be seen and almost always felt. The position of the margins of the heart can be determined by percussion.
7. The lower ribs and costal margins overlap and protect the upper abdominal contents. Severe injuries to the lower part of the thoracic cage may well produce unsuspected injury to the liver, stomach, and spleen.

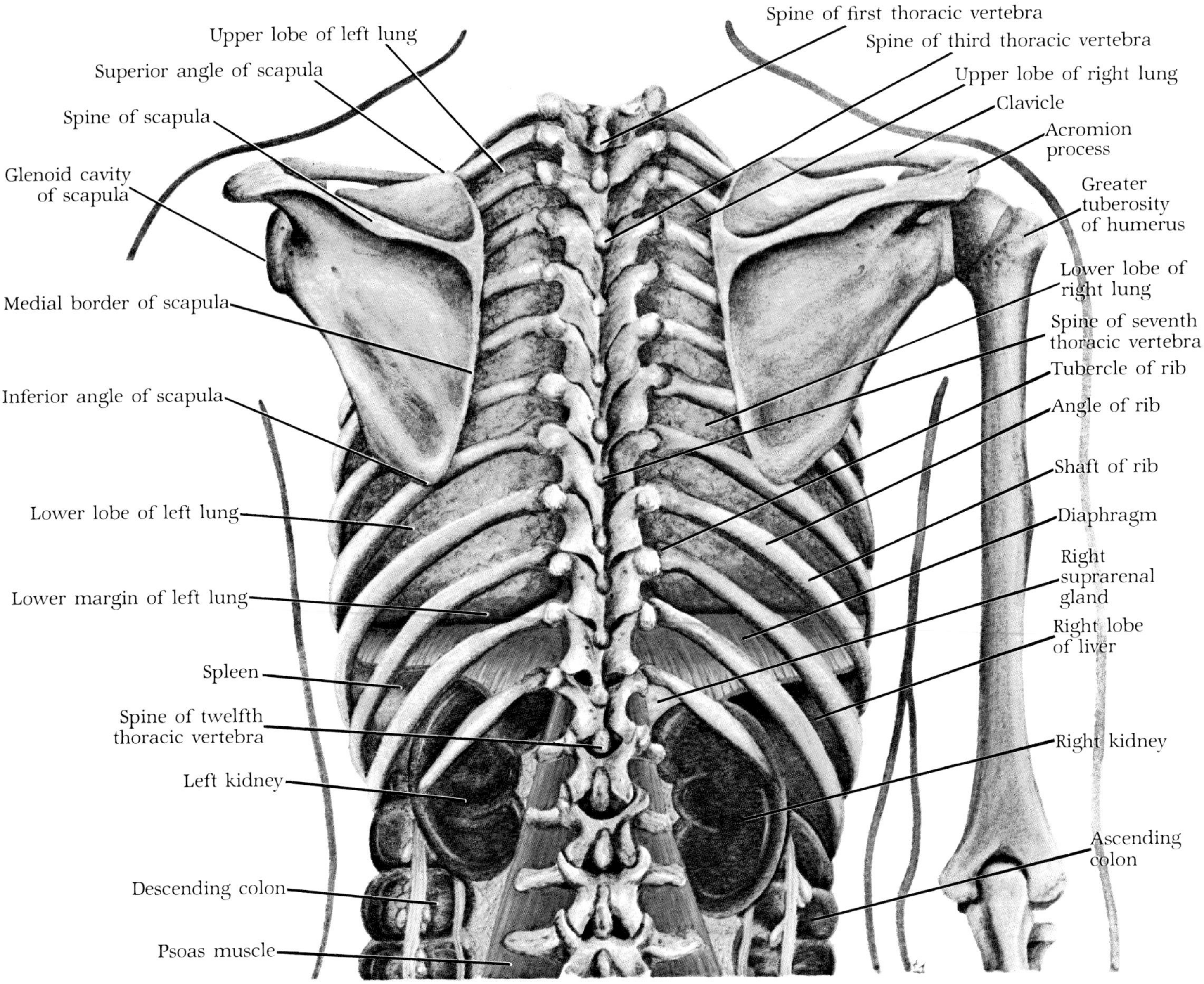

Upper lobe of left lung
Superior angle of scapula
Spine of scapula
Glenoid cavity of scapula
Medial border of scapula
Inferior angle of scapula
Lower lobe of left lung
Lower margin of left lung
Spleen
Spine of twelfth thoracic vertebra
Left kidney
Descending colon
Psoas muscle
Spine of first thoracic vertebra
Spine of third thoracic vertebra
Upper lobe of right lung
Clavicle
Acromion process
Greater tuberosity of humerus
Lower lobe of right lung
Spine of seventh thoracic vertebra
Tubercle of rib
Angle of rib
Shaft of rib
Diaphragm
Right suprarenal gland
Right lobe of liver
Right kidney
Ascending colon

Figure 1-34
Skeleton of the thorax, showing its relationship to thoracic and abdominal viscera, posterior view.

1. As noted earlier it is clinically important to know the surface markings of the pleural reflections and the lobes of the lungs. When listening to breath sounds of the respiratory tract, it should be possible mentally to picture the structures that lie beneath the stethoscope.
2. The lower limit of the pleural reflection may be damaged during nephrectomy operations.
3. The shape of the thorax can be distorted by congenital anomalies of the vertebral column or by the ribs. Destructive disease of the vertebral column producing lateral flexion or scoliosis results in marked distortion of the thoracic cage.

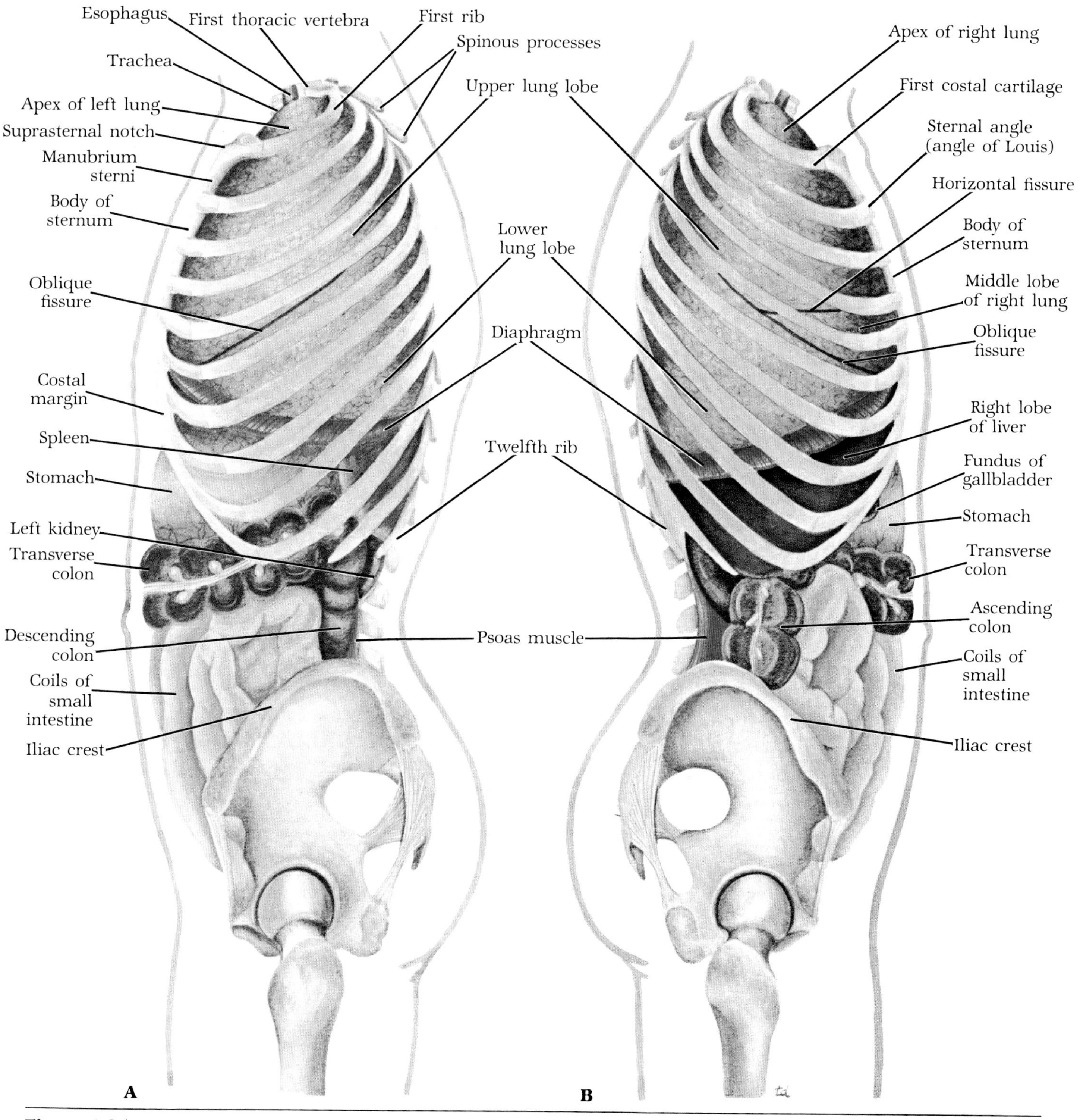
Esophagus
First thoracic vertebra
First rib
Spinous processes
Trachea
Apex of left lung
Suprasternal notch
Manubrium sterni
Body of sternum
Upper lung lobe
Oblique fissure
Lower lung lobe
Diaphragm
Costal margin
Spleen
Stomach
Twelfth rib
Left kidney
Transverse colon
Descending colon
Psoas muscle
Coils of small intestine
Iliac crest
A
Apex of right lung
First costal cartilage
Sternal angle (angle of Louis)
Horizontal fissure
Body of sternum
Middle lobe of right lung
Oblique fissure
Right lobe of liver
Fundus of gallbladder
Stomach
Transverse colon
Ascending colon
Coils of small intestine
Iliac crest
B

Figure 1-35
Skeleton of the thorax, showing its relationship to thoracic and abdominal viscera. A. Left lateral view. B. Right lateral view.

1. Fractures of ribs are common in adults; the ribs tend to break at their weakest part, the region of their angles. One of the great dangers of a fractured rib is that it may damage the underlying lung.
2. Costal cartilages sometimes become ossified, especially in old age, and this changes their radiographic appearance.
3. If one listens to the breath sounds in the respiratory tract in the axilla, areas of all lobes of the lung may be heard.
4. Note again the close relationship of the ninth, tenth, and eleventh ribs to the spleen on the left side and the relationship of the lower ribs to the liver on the right side.

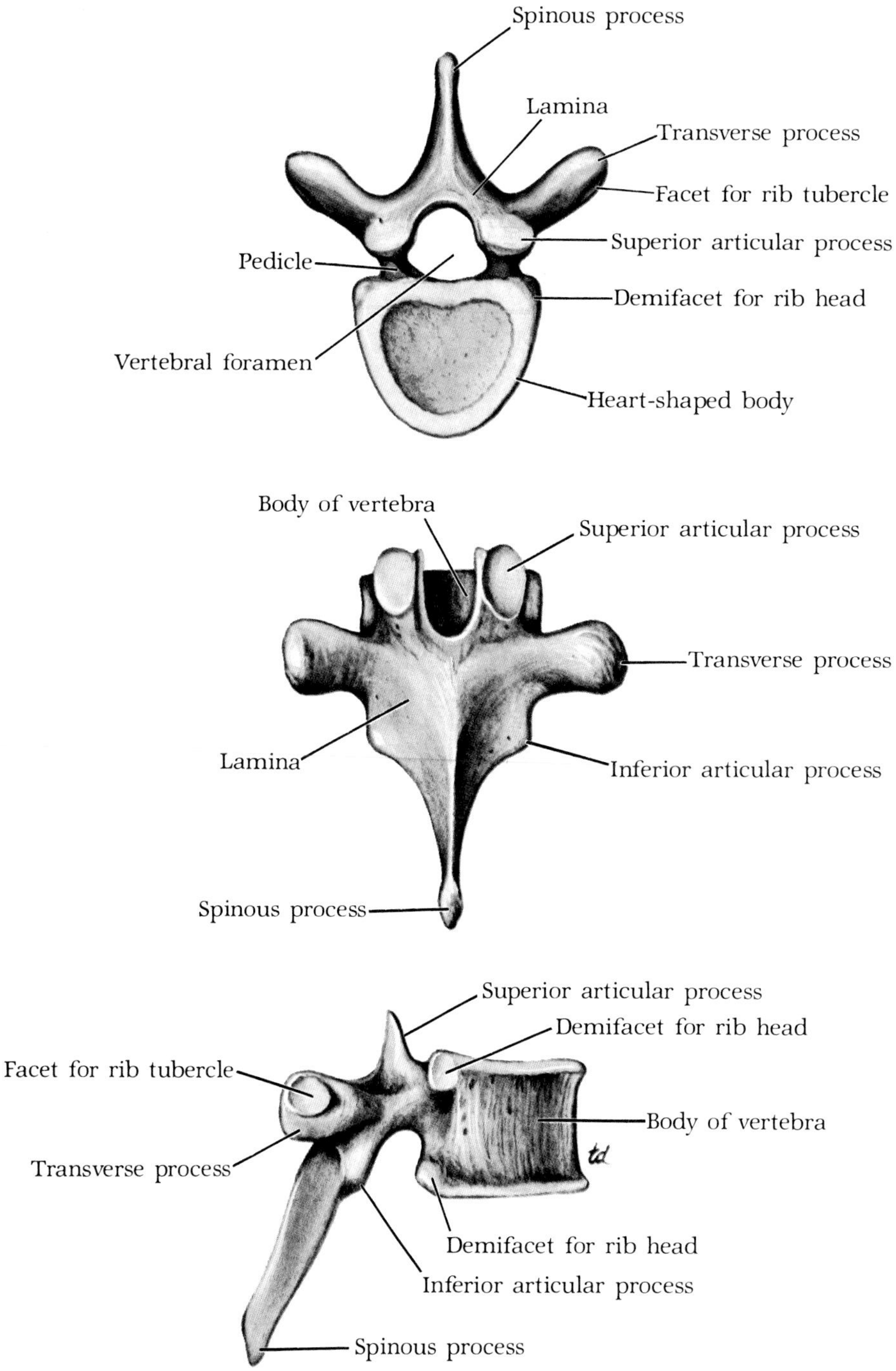
Spinous process
Lamina
Transverse process
Facet for rib tubercle
Superior articular process
Pedicle
Demifacet for rib head
Vertebral foramen
Heart-shaped body
A
Body of vertebra
Superior articular process
Transverse process
Lamina
Inferior articular process
Spinous process
B
Superior articular process
Demifacet for rib head
Facet for rib tubercle
Body of vertebra
Transverse process
td
Demifacet for rib head
Inferior articular process
Spinous process
C

Figure 1-36
Typical thoracic vertebra (sixth).
A. Superior view. B. Posterior view.
C. Lateral view.

1. Because the rib cage provides stability, fractures and dislocations of the upper and middle thoracic vertebrae are uncommon.

2. Compression fractures of the thoracic vertebral bodies are usually caused by an excessive flexion-compression type of injury and occur at the sites of maximum mobility, i.e., the lower thoracic region. With this type of fracture the body of the vertebra is crushed although the strong posterior longitudinal ligament remains intact. Because the vertebral arches remain unbroken and the intervertebral ligaments are intact, vertebral displacement and spinal cord injury do not occur.

3. Fracture-dislocations are caused by an excessive flexion-compression type of injury. Because the articular processes are fractured and the ligaments torn, the vertebrae involved are unstable and the spinal cord is usually severely damaged or severed, with accompanying paraplegia. Note that the vertebral foramen is relatively small in the thoracic region; consequently, the spinal cord is easily damaged in this area.

4. Fractures of the spinous processes, transverse processes, or laminae are caused by direct injury or, in rare cases, by severe muscular activity.

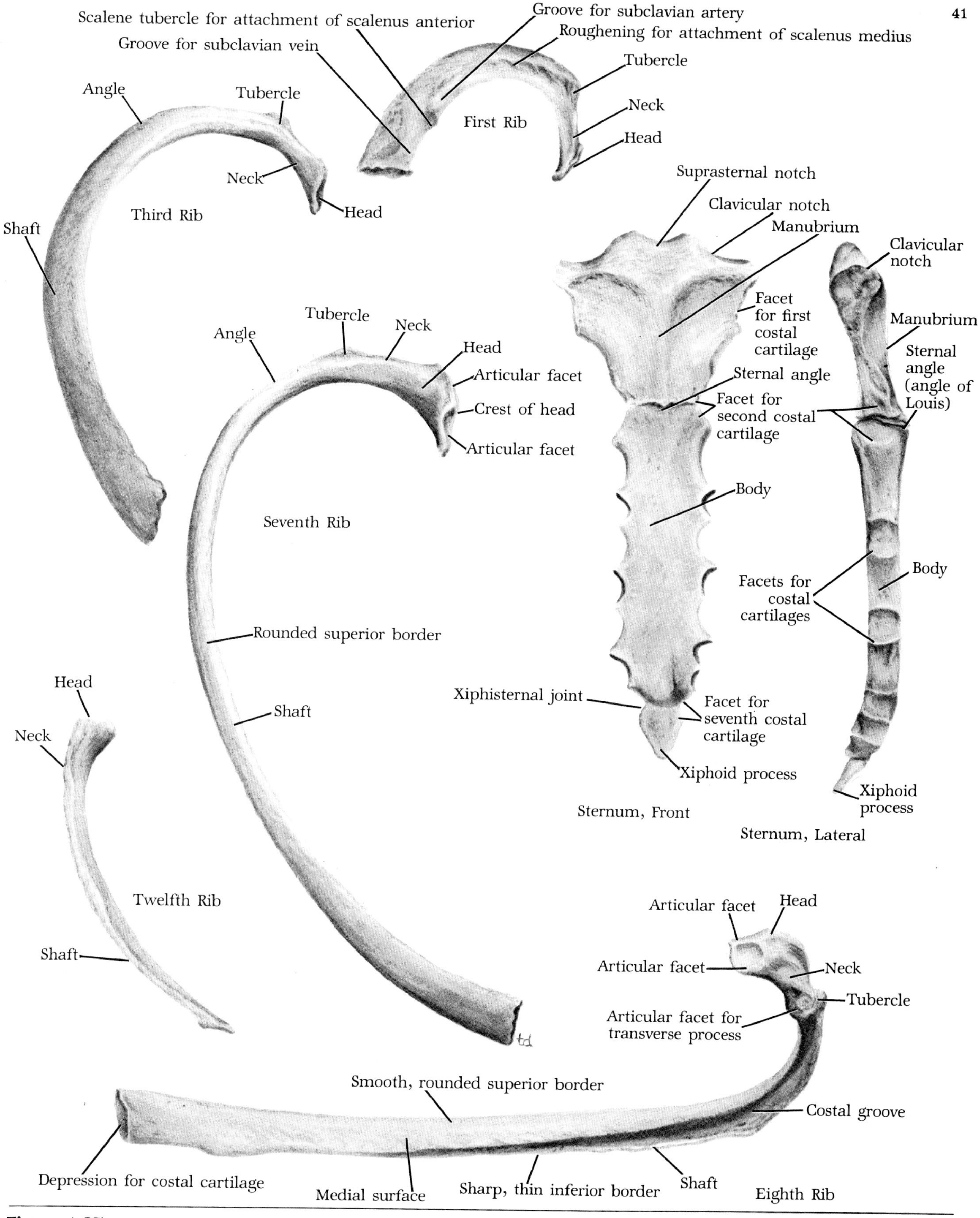
Scalene tubercle for attachment of scalenus anterior
Groove for subclavian artery
Roughening for attachment of scalenus medius
Groove for subclavian vein
Tubercle
First Rib
Neck
Head
Angle
Tubercle
Neck
Head
Third Rib
Shaft
Suprasternal notch
Clavicular notch
Manubrium
Clavicular notch
Facet for first costal cartilage
Manubrium
Sternal angle (angle of Louis)
Tubercle
Neck
Angle
Head
Articular facet
Crest of head
Articular facet
Sternal angle
Facet for second costal cartilage
Body
Seventh Rib
Body
Facets for costal cartilages
Rounded superior border
Head
Neck
Shaft
Xiphisternal joint
Facet for seventh costal cartilage
Xiphoid process
Xiphoid process
Sternum, Front
Sternum, Lateral
Twelfth Rib
Articular facet
Head
Shaft
Articular facet
Neck
Tubercle
Articular facet for transverse process
Smooth, rounded superior border
Costal groove
Depression for costal cartilage
Medial surface
Sharp, thin inferior border
Shaft
Eighth Rib

Figure 1-37
First, third, seventh, and twelfth ribs, viewed from above; eighth rib viewed from below; and sternum viewed from in front and laterally.

1. Congenital anomalies of the first rib or the existence of a cervical rib may cause compression of the roots of the brachial plexus or the subclavian artery. This will cause pain and ischemic symptoms in the upper limb. There may also be evidence of muscular atrophy.

2. As mentioned earlier, the long fixed ribs of the thoracic cage are the most commonly damaged, and they usually break at their weakest part, the region of their angles.

3. Cleft sternum and perforated sternum are rare congenital anomalies that can be corrected by surgery.

4. The sternum possesses red hematopoietic marrow throughout life and is therefore a common site for marrow biopsy. A wide-bore needle is introduced into the marrow cavity through the anterior surface of the sternum under a local anesthetic.

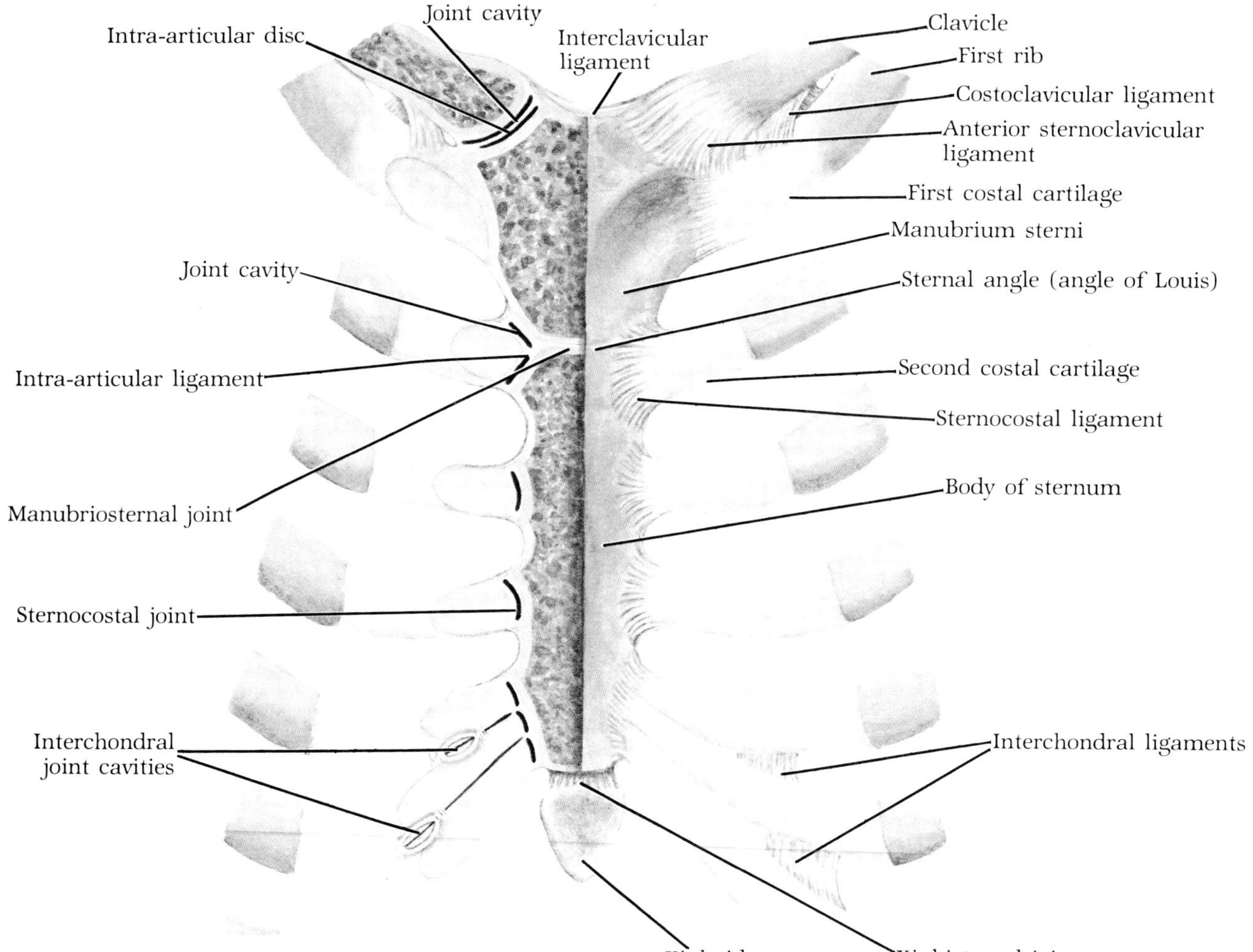

Joint cavity
Intra-articular disc
Interclavicular ligament
Clavicle
First rib
Costoclavicular ligament
Anterior sternoclavicular ligament
First costal cartilage
Manubrium sterni
Joint cavity
Sternal angle (angle of Louis)
Intra-articular ligament
Second costal cartilage
Sternocostal ligament
Manubriosternal joint
Body of sternum
Sternocostal joint
Interchondral joint cavities
Interchondral ligaments
Xiphoid process
Xiphisternal joint

A

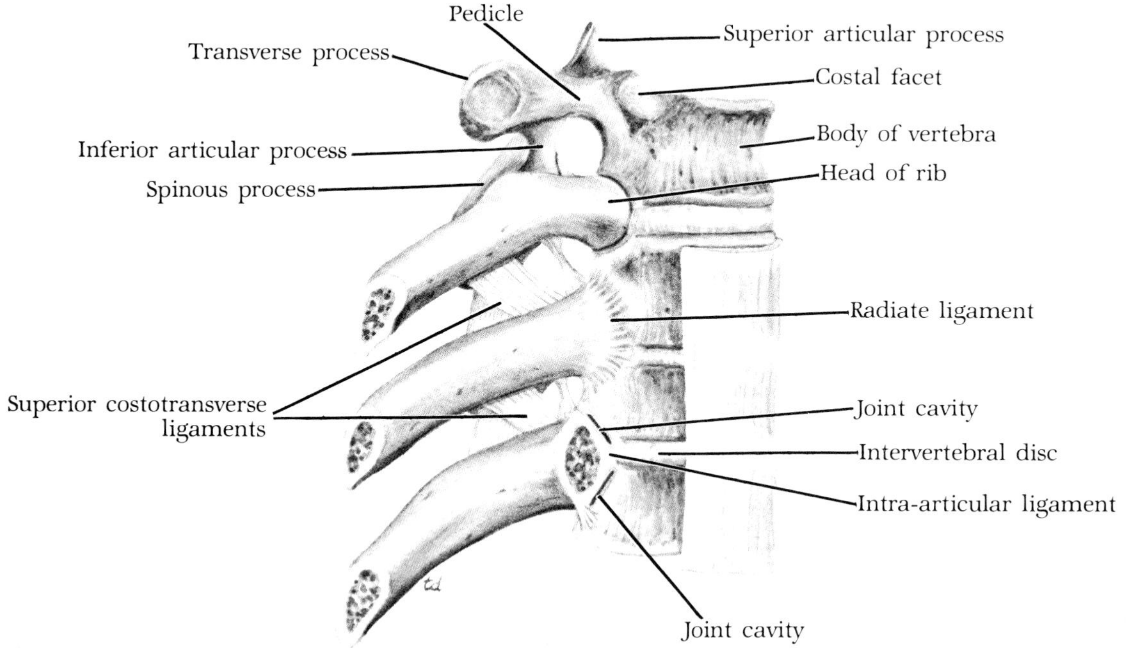

Pedicle
Superior articular process
Transverse process
Costal facet
Body of vertebra
Inferior articular process
Head of rib
Spinous process
Radiate ligament
Superior costotransverse ligaments
Joint cavity
Intervertebral disc
Intra-articular ligament
Joint cavity

B

Figure 1-38
A. Sternoclavicular, sternocostal, and interchondral joints, anterior view. B. Costovertebral joints, right anterolateral view.

1. Excessive force directed along the long axis of the clavicle usually results in fracture of that bone. Dislocation of the sternoclavicular joint does take place occasionally, however, and the clavicle is then displaced forward and downward. Rarely posterior dislocations occur. The strength of the sternoclavicular joint is largely dependent on the very strong costoclavicular ligament and the articular disc.

2. Costochondral or sternochondral injuries occur as the result of a direct blow over the costal cartilage with separation of the cartilage from the rib or sternum. Such injuries are painful and the discomfort may be prolonged.

3. Ankylosing spondylitis, a disease involving the joints of the lumbar and thoracic parts of the vertebral column and the joints of the thoracic cage, results in fixation of the thoracic cage in an inspiratory position.

4. Inspection of the chest during the different phases of respiration will determine its configuration and the range of respiratory movement. It is important to detect any inequalities on the two sides. The impressions gained by inspection can be confirmed by palpation.

2. The Abdomen

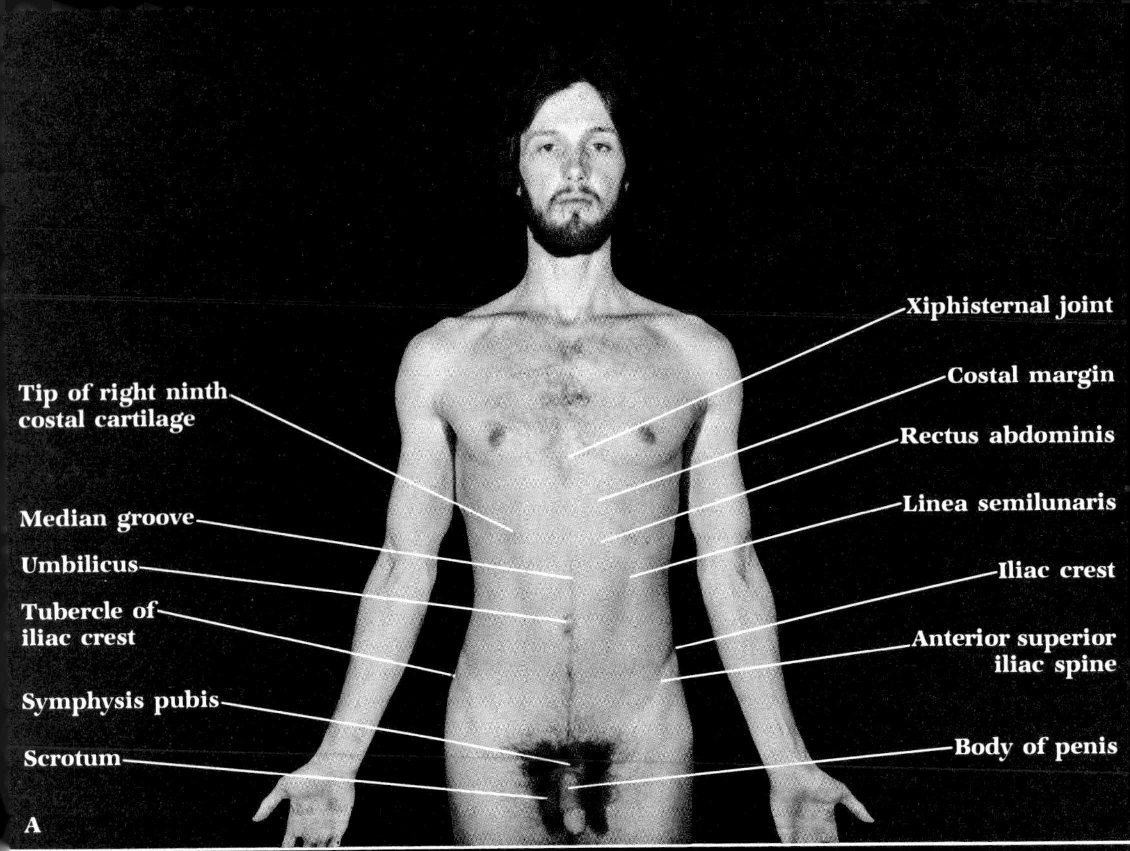

Tip of right ninth costal cartilage
Median groove
Umbilicus
Tubercle of iliac crest
Symphysis pubis
Scrotum
Xiphisternal joint
Costal margin
Rectus abdominis
Linea semilunaris
Iliac crest
Anterior superior iliac spine
Body of penis
A

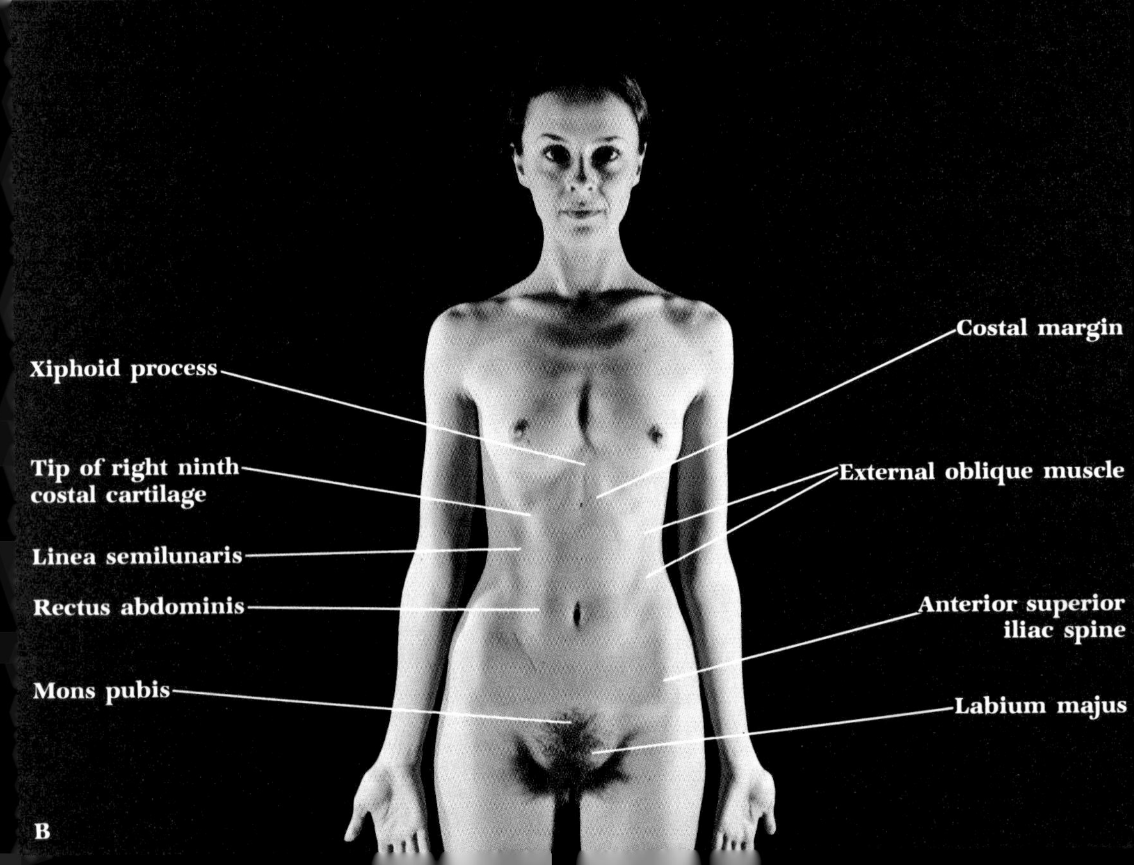

Xiphoid process
Tip of right ninth costal cartilage
Linea semilunaris
Rectus abdominis
Mons pubis
Costal margin
External oblique muscle
Anterior superior iliac spine
Labium majus
B

Figure 2-1
The anterior abdominal wall.
A. 27-year-old male. B. 29-year-old female.

1. The xiphoid process may be palpated in the depression where the costal margins meet at the infrasternal angle.
2. The xiphisternal junction can be felt at the inferior edge of the body of the sternum and lies opposite the body of the ninth thoracic vertebra.
3. The costal margin is formed by the costal cartilages of the seventh, eighth, ninth, tenth, eleventh, and twelfth ribs. Its lowest level at the tenth costal cartilage lies opposite the body of the third lumbar vertebra.
4. The iliac crest can be felt along its entire length. It ends anteriorly at the anterior superior iliac spine and posteriorly at the posterior superior iliac spine. Its highest point lies opposite the body of the fourth lumbar vertebra.
5. The tubercle of the iliac crest is situated on the lateral margin of the crest about 2 inches (5 cm) posterior to the anterior superior iliac spine. It lies at the level of the body of the fifth lumbar vertebra.
6. The symphysis pubis is the cartilaginous joint that lies in the midline between the bodies of the pubic bones.
7. The pubic tubercle can be felt as a small protuberance along the superior surface of the pubis.
8. The pubic crest is the ridge on the superior surface of the pubic bones medial to the pubic tubercle.

Xiphoid process
Tendinous intersections of rectus abdominis
Tip of right ninth costal cartilage
Costal margin
Linea semilunaris
Rectus abdominis
External oblique
Anterior superior iliac spine
Inguinal ligament
Inguinal canal
Male distribution of pubic hair
Symphysis pubis
A

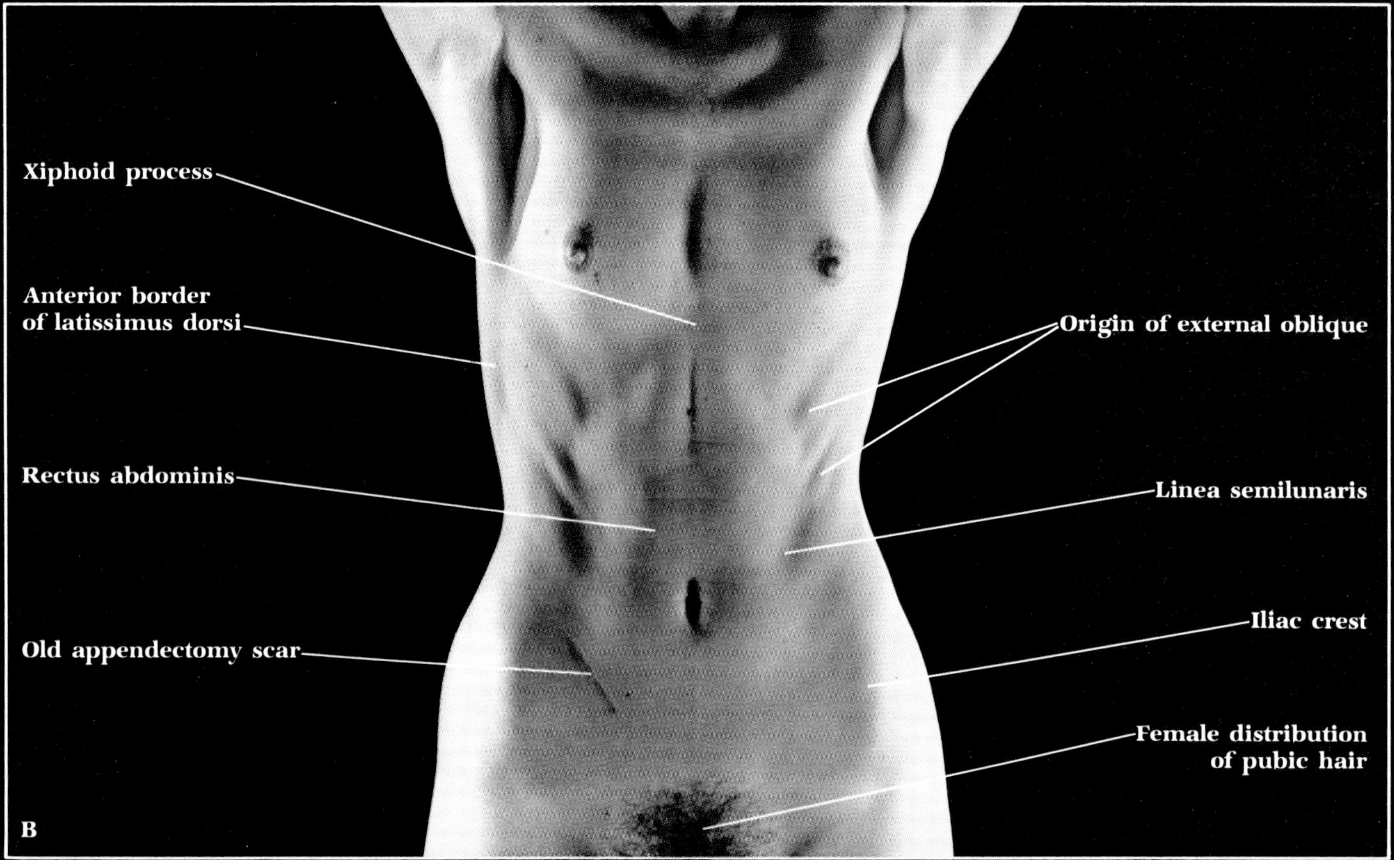
Xiphoid process
Anterior border of latissimus dorsi
Origin of external oblique
Rectus abdominis
Linea semilunaris
Iliac crest
Old appendectomy scar
Female distribution of pubic hair
B

Figure 2-2
The anterior abdominal wall. A. 27-year-old male. B. 29-year-old female.

1. The inguinal ligament can be felt as the rolled inferior margin of the aponeurosis of the external oblique muscle. It is attached laterally to the anterior superior iliac spine and curves downward and medially to become attached to the pubic tubercle.
2. The midinguinal point lies on the inguinal ligament halfway between the symphysis pubis and the anterior superior iliac spine. When this point is palpated the pulsations of the external iliac artery can be felt as it passes under the ligament to become the femoral artery.
3. The linea alba is a fibrous band that extends from the symphysis pubis to the xiphoid process and lies in the midline beneath a slight median groove.
4. The umbilicus lies in the linea alba and is inconstant in position.
5. The linea semilunaris is the lateral edge of the rectus abdominis muscle and crosses the costal margin at the tip of the ninth costal cartilage.
6. In clinical practice it is customary to use two vertical and two horizontal planes to divide the abdomen up into regions.
a. The vertical plane on each side passes through the midinguinal point.
b. The subcostal plane runs through the lowest point of the costal margin on each side. This is the inferior margin of the tenth costal cartilage and lies opposite the third lumbar vertebra.
c. The intertubercular plane joins the tubercles on the iliac crests. This plane lies at the level of the body of the fifth lumbar vertebra.
7. The transpyloric plane is also used clinically. It passes through the tips of the ninth costal cartilages on the two sides, i.e., the point where the lateral margin of the rectus abdominis muscle (the linea semilunaris) crosses the costal margin. To identify these points ask the patient while he is in the supine position to sit up without using his arms. To accomplish this, he contracts the rectus abdominis muscles on both sides, causing the lateral margins of these muscles to then stand out. This plane passes through the pylorus, the duodenojejunal junction, the neck of the pancreas, and the hili of the kidneys.

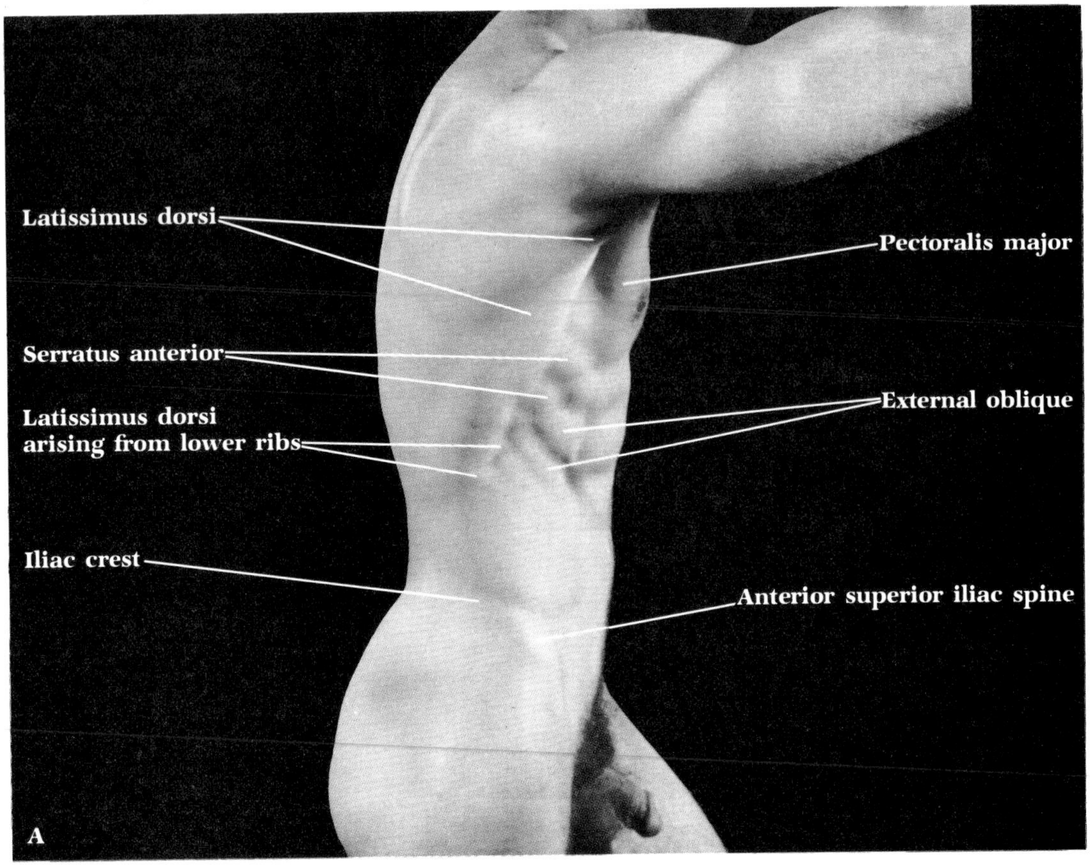
Latissimus dorsi
Pectoralis major
Serratus anterior
External oblique
Latissimus dorsi arising from lower ribs
Iliac crest
Anterior superior iliac spine
A

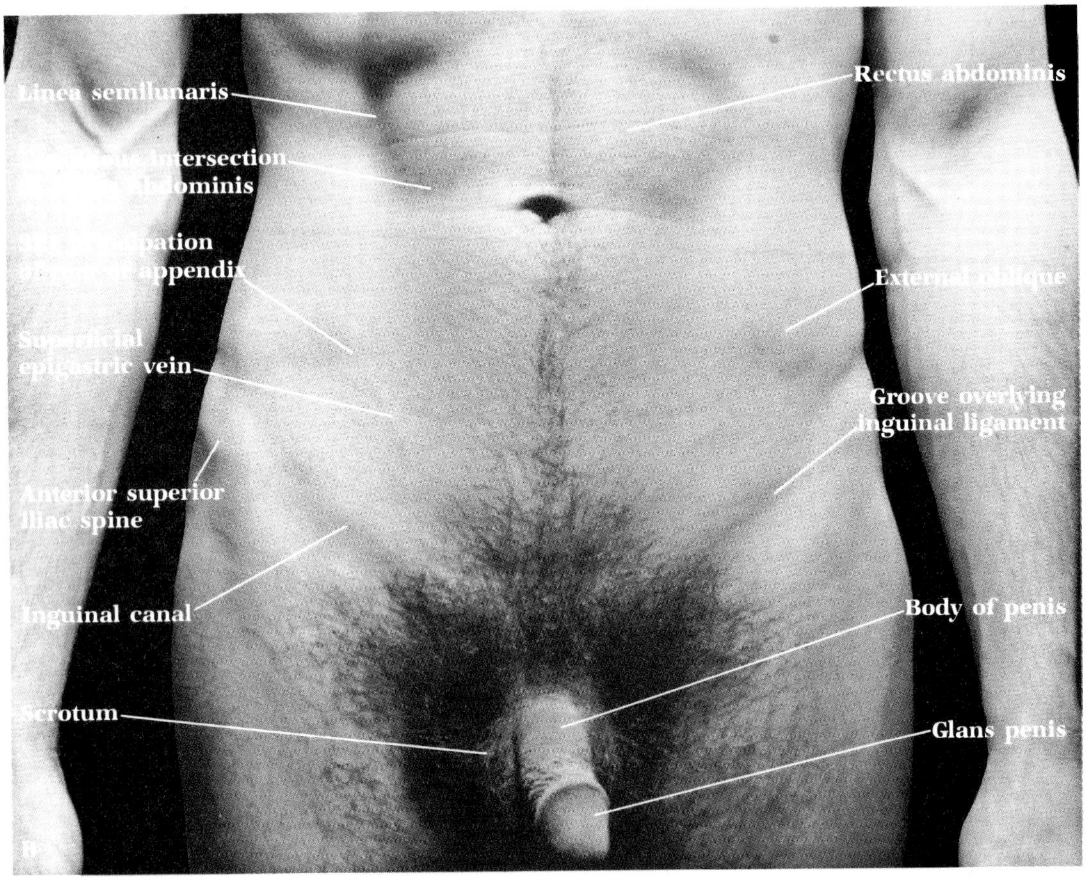
Linea semilunaris
Rectus abdominis
External oblique
Groove overlying inguinal ligament
Anterior superior iliac spine
Inguinal canal
Body of penis
Scrotum
Glans penis

Figure 2-3
The anterior abdominal wall in a 27-year-old male. A. Lateral view, right side. B. Anterior view.

1. The liver lies under cover of the lower ribs with most of its bulk lying in the right hypochondrium and epigastrium. In infants—until about the end of the third year—the lower margin of the liver extends 1 or 2 fingerbreadths below the costal margin. The liver is impalpable in the adult who is obese or who has a well-developed right rectus abdominis muscle. In a thin adult the lower edge of the liver may be felt a fingersbreadth below the costal margin. It is most easily felt when the patient inspires deeply and the diaphragm contracts and pushes down the liver.
2. The fundus of the gallbladder lies opposite the tip of the right ninth costal cartilage, i.e., where the lateral lobe of the rectus abdominis muscle crosses the costal margin.
3. The spleen is situated in the left hypochondrium and lies under cover of the ninth, tenth, and eleventh ribs. Its long axis corresponds to that of the tenth rib, and in the adult it does not normally project forward in front of the midaxillary line. In infants the lower pole of the spleen may just be felt.
4. The pancreas lies across the transpyloric plane. The head lies below and to the right, the neck lies on the plane, and the body and tail lie above and to the left.
5. The right kidney lies at a slightly lower level than the left kidney. Its lower pole can be palpated in the right lumbar region at the end of a deep inspiration in a person with poorly developed abdominal musculature. The normal left kidney is impalpable. Each kidney moves about 1 inch (2.5 cm) in a vertical direction during full respiratory movement of the diaphragm.
6. The aorta lies in the midline of the abdomen and bifurcates below into the right and left common iliac arteries opposite the fourth lumbar vertebra.
7. The superficial inguinal ring is a triangular aperture in the aponeurosis of the external oblique muscle situated above and medial to the pubic tubercle. In the male, the margins of the ring can be felt by invaginating the skin of the upper part of the scrotum with the tip of the little finger. In the female it may be felt with difficulty through the lateral part of the labia majora.
8. The spermatic cord can be felt as a soft tubular structure emerging from the superficial inguinal ring and descending over or medial to the pubic tubercle into the scrotum.
9. The vas deferens can be felt between the finger and thumb in the posterior part of the spermatic cord.
10. The testis can be felt as a firm ovoid body lying free within the scrotum on each side.
11. The epididymis can be felt as an elongated structure posterior to the testis. The enlarged upper end is called the head; below this is the body, and the tail is the narrow lower end. The vas deferens emerges from the tail and ascends medial to the epididymis.

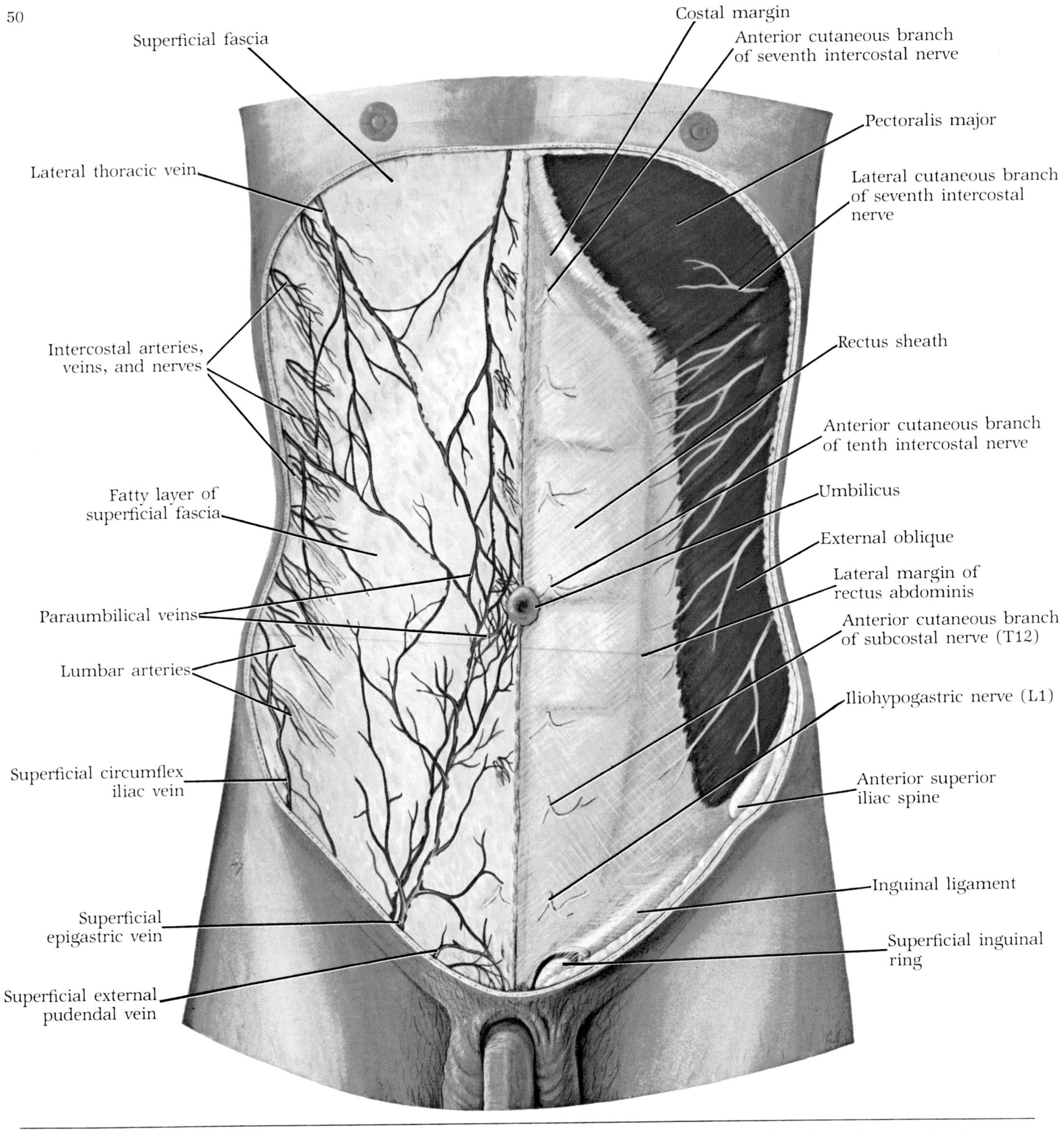
Superficial fascia
Costal margin
Anterior cutaneous branch of seventh intercostal nerve
Pectoralis major
Lateral thoracic vein
Lateral cutaneous branch of seventh intercostal nerve
Intercostal arteries, veins, and nerves
Rectus sheath
Anterior cutaneous branch of tenth intercostal nerve
Fatty layer of superficial fascia
Umbilicus
External oblique
Lateral margin of rectus abdominis
Paraumbilical veins
Anterior cutaneous branch of subcostal nerve (T12)
Lumbar arteries
Iliohypogastric nerve (L1)
Superficial circumflex iliac vein
Anterior superior iliac spine
Inguinal ligament
Superficial epigastric vein
Superficial inguinal ring
Superficial external pudendal vein

Figure 2-4
Anterior abdominal wall. The superficial fascia, the superficial veins, and the nerves shown on the left. On the right, the superficial fascia has been removed to reveal the underlying muscles.

1. The superficial veins around the umbilicus and the paraumbilical veins connecting them to the portal vein may become distended in cases of portal vein obstruction. Obstruction of the superior or inferior vena cava causes distention of the veins running from the anterior thoracic wall to the thigh.
2. The nerves of the anterior and lateral abdominal walls supply the skin, muscles, and parietal peritoneum. Inflammation of the parietal peritoneum will cause not only pain in the overlying skin but also increase the tone of the abdominal musculature in the same area.
3. Because the lower intercostal nerves supply the costal parietal pleura as well as the anterior abdominal wall, the pain of pleurisy may radiate down into the abdomen.
4. The umbilicus is a consolidated scar representing the site of attachment of the umbilical cord in the fetus. Its embryological remains may give rise to clinical problems such as a fistula of the vitello-intestinal duct or a patent urachus. An exomphalos occurs when there is herniation of some of the abdominal contents through the open umbilicus.
5. Acquired infantile umbilical hernia occurs in children and is due to weakness of the umbilicus scar in the linea alba. The majority of these hernias disappear without treatment as the abdominal cavity enlarges.
6. Paraumbilical hernia is a hernia through the linea alba in the region of the umbilicus. It is most common in women and should be treated surgically.

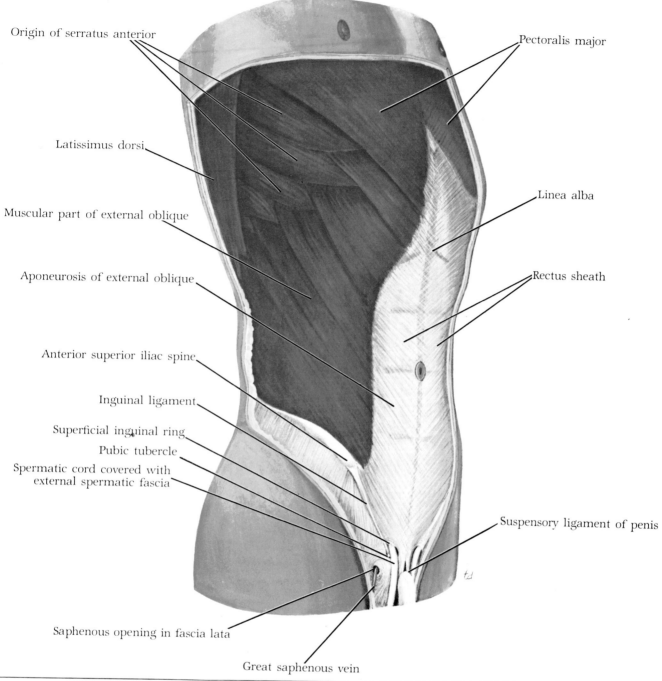

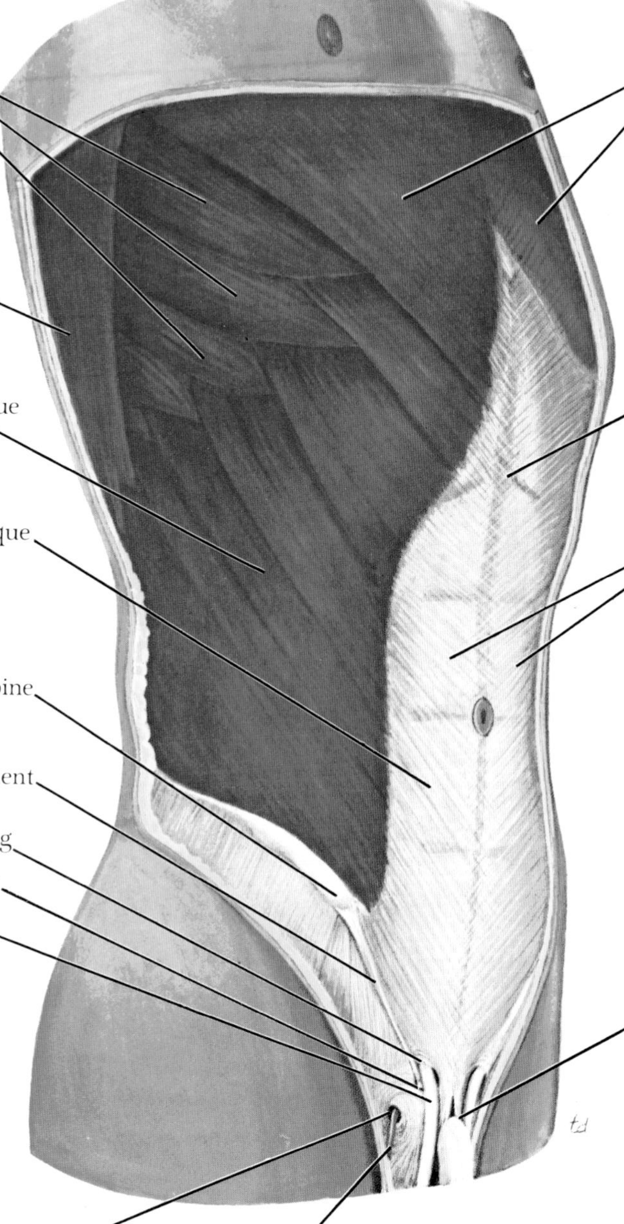
Origin of serratus anterior
Pectoralis major
Latissimus dorsi
Linea alba
Muscular part of external oblique
Aponeurosis of external oblique
Rectus sheath
Anterior superior iliac spine
Inguinal ligament
Superficial inguinal ring
Pubic tubercle
Spermatic cord covered with external spermatic fascia
Suspensory ligament of penis
Saphenous opening in fascia lata
Great saphenous vein

Figure 2-5
Anterior abdominal wall, showing the external oblique muscle, right anterior view.

1. Normally, during inspiration, the abdominal muscles move forward as the diaphragm descends. If the anterior abdominal wall remains stationary or contracts inward during inspiration, it is probable that the parietal peritoneum is inflamed and is responsible for a reflex contraction of the abdominal muscles.

2. Stimulation of the skin of the anterior abdominal wall with a sharp instrument will cause reflex contraction of the underlying abdominal musculature. Failure of this reflex to occur indicates a lesion in the central nervous system involving the upper motor neuron.

3. Note the position of the important inguinal ligament and its bony attachments. Inguinal herniae tend to emerge through the anterior abdominal wall through the superficial inguinal ring. An inguinal hernia can be distinguished from a femoral hernia by the fact that the sac, as it emerges through the superficial inguinal ring, lies above and medial to the pubic tubercle; that of a femoral hernia lies below and lateral to the tubercle.

4. The aponeurosis of the external oblique muscle forms part of the anterior wall of the rectus sheath. Note the attachment of the tendinous intersections of the rectus muscle to the anterior wall of the sheath.

5. The direction of the fibers of the external oblique muscle and its aponeurosis are downward and forward. If surgical incisions can be made in the line of the fibers, the fibers will fall back into position and function normally when the incision is closed.

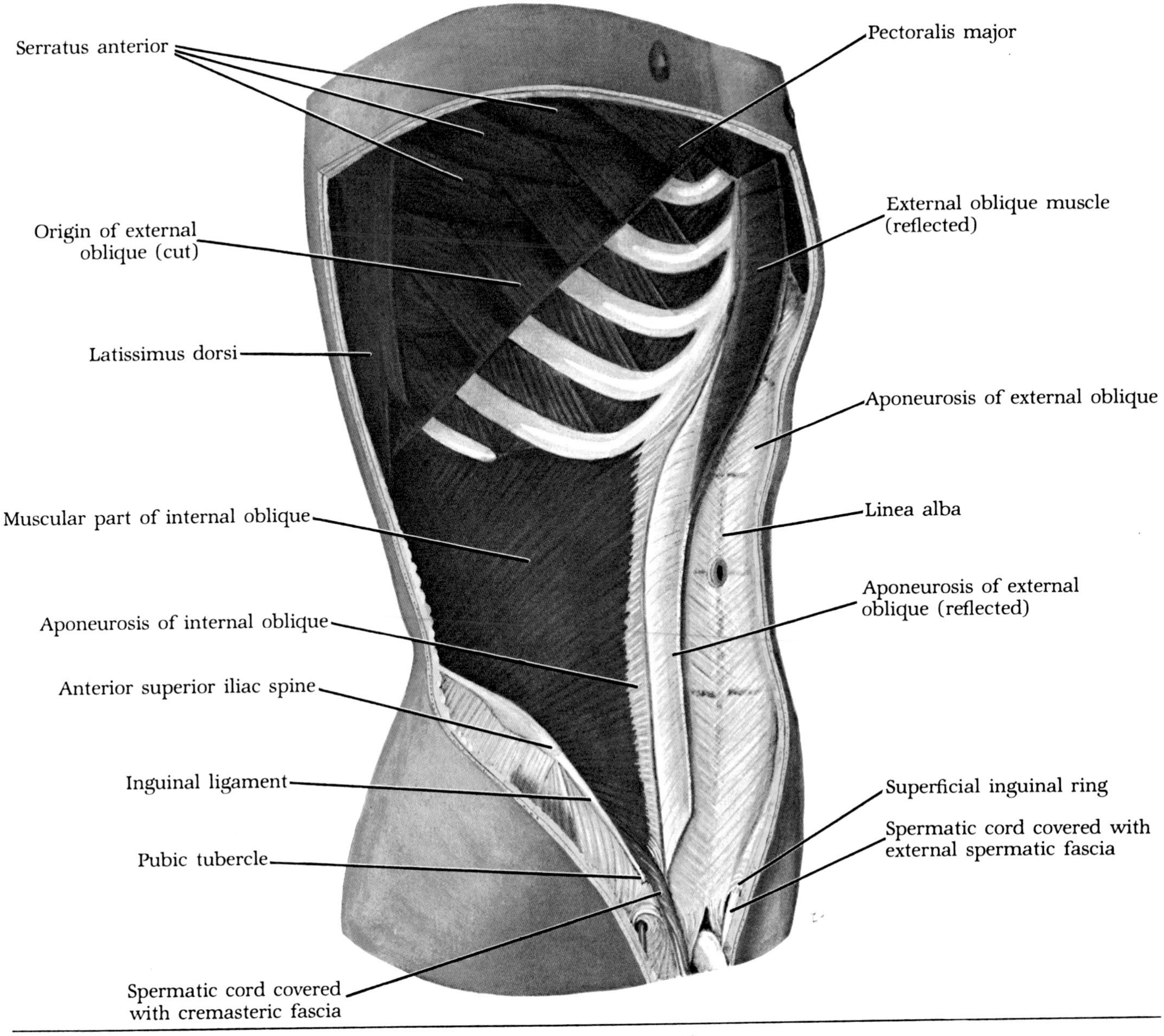
Serratus anterior
Pectoralis major
Origin of external oblique (cut)
External oblique muscle (reflected)
Latissimus dorsi
Aponeurosis of external oblique
Muscular part of internal oblique
Linea alba
Aponeurosis of internal oblique
Aponeurosis of external oblique (reflected)
Anterior superior iliac spine
Inguinal ligament
Superficial inguinal ring
Spermatic cord covered with external spermatic fascia
Pubic tubercle
Spermatic cord covered with cremasteric fascia

Figure 2-6
Anterior abdominal wall, showing the internal oblique muscle, right anterior view. The external oblique muscle has been partly removed and the aponeurosis reflected medially.

1. A hernia is the protrusion of part of the abdominal contents beyond the normal confines of the abdominal wall. Abdominal herniae are of the following common types:
A. Inguinal: indirect or direct
B. Femoral
C. Umbilical: congenital or acquired
D. Epigastric
E. Divarication of the recti abdominis muscles

2. As the spermatic cord (or round ligament of the uterus in the female) passes under the lower border of the internal oblique muscle, it carries with it some muscle fibers called the cremaster muscle. *Cremasteric fascia* is the term used to describe the cremaster muscle and its fascia.

3. The aponeurosis of the internal oblique muscle takes part in the formation of the walls of the rectus sheath. The conjoint tendon, formed by the lowest fibers of the internal oblique and transversus abdominis muscles, greatly strengthens the posterior wall of the inguinal canal.

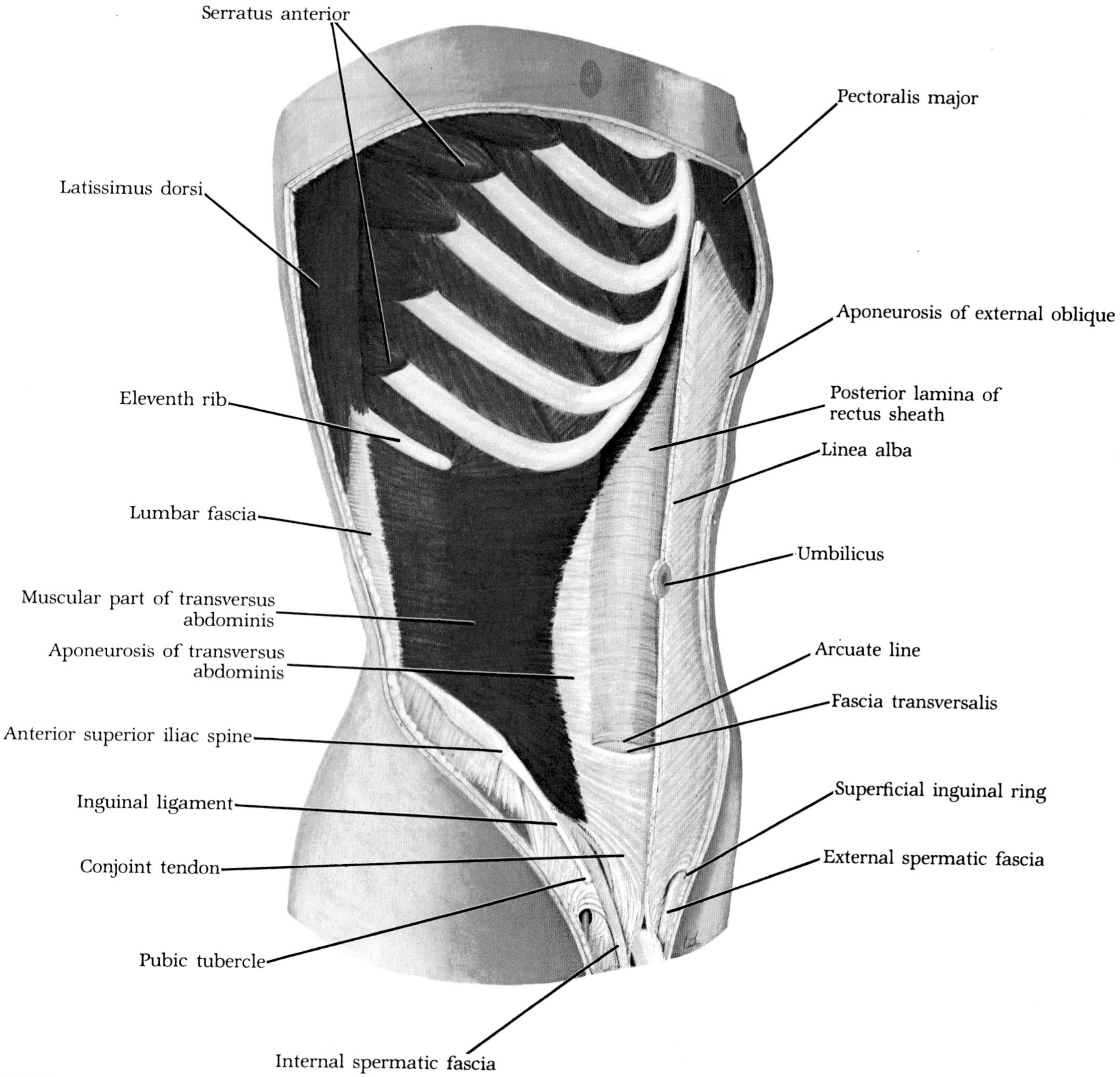
Serratus anterior
Pectoralis major
Latissimus dorsi
Aponeurosis of external oblique
Eleventh rib
Posterior lamina of rectus sheath
Linea alba
Lumbar fascia
Umbilicus
Muscular part of transversus abdominis
Aponeurosis of transversus abdominis
Arcuate line
Fascia transversalis
Anterior superior iliac spine
Inguinal ligament
Superficial inguinal ring
Conjoint tendon
External spermatic fascia
Pubic tubercle
Internal spermatic fascia

Figure 2-7
Anterior abdominal wall, showing the transversus abdominis muscle, right anterior view.

1. The transversus abdominis muscle does not contribute to the coverings of the spermatic cord (round ligament of the uterus in the female).
2. The transversus abdominis muscle and aponeurosis take part in the formation of the rectus sheath. Note that the muscular contribution is made to the posterior wall of the rectus sheath just below the costal margin.
3. The linea alba, a fibrous band extending from the symphysis pubis to the xiphoid process and lying in the midline, is formed from the aponeuroses of the muscles of the anterior abdominal wall. Herniation of small pieces of extraperitoneal fat through the widest part of the linea alba above the umbilicus produces an epigastric hernia (a form of paraumbilical hernia). Making a midline surgical incision through the linea alba is a method of rapidly gaining entrance to the abdomen without causing damage to muscle fibers or their nerve and blood supply.

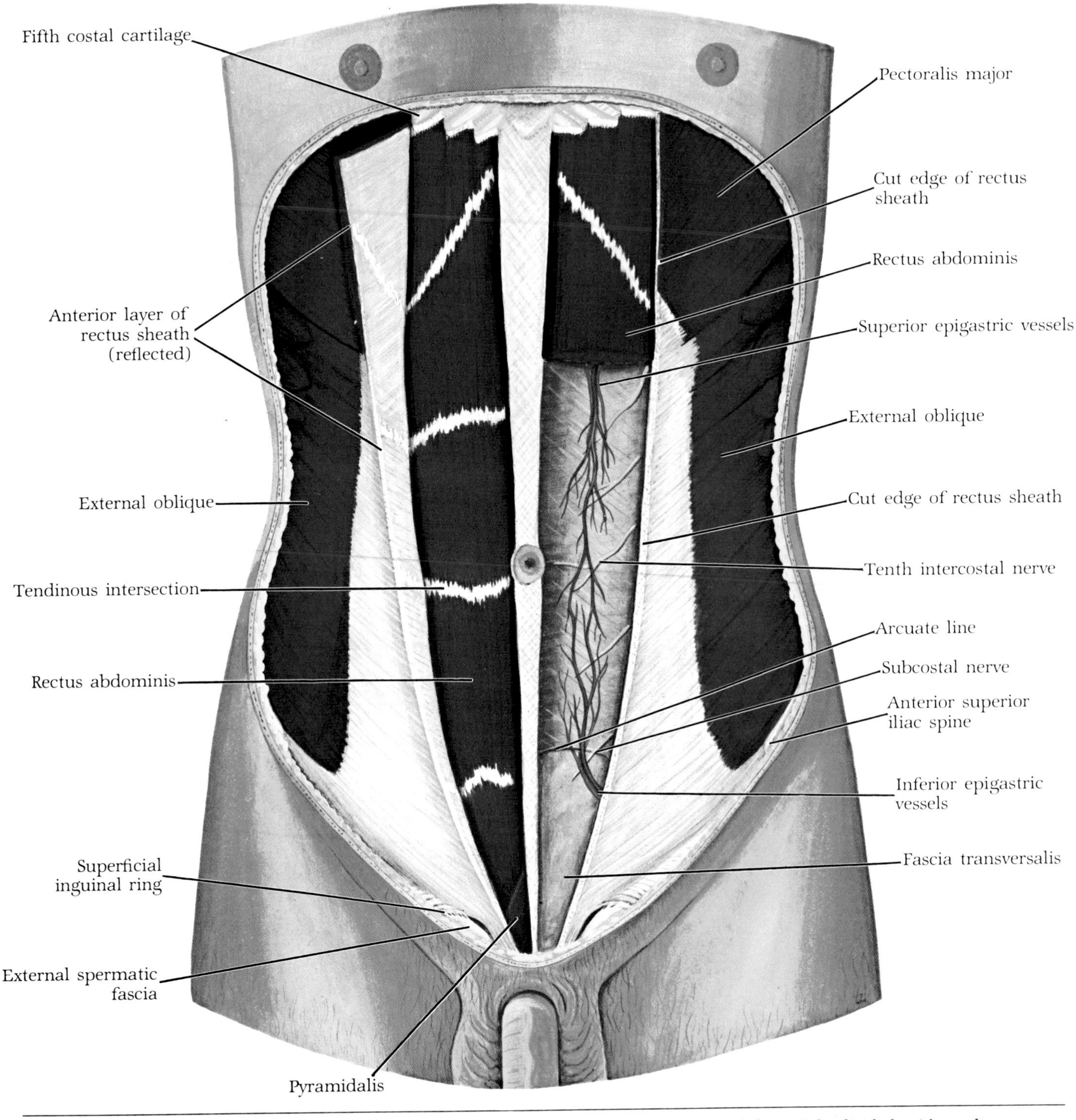
Fifth costal cartilage
Pectoralis major
Cut edge of rectus sheath
Rectus abdominis
Anterior layer of rectus sheath (reflected)
Superior epigastric vessels
External oblique
External oblique
Cut edge of rectus sheath
Tendinous intersection
Tenth intercostal nerve
Arcuate line
Rectus abdominis
Subcostal nerve
Anterior superior iliac spine
Inferior epigastric vessels
Fascia transversalis
Superficial inguinal ring
External spermatic fascia
Pyramidalis

Figure 2-8
Rectus abdominis muscle and rectus sheath, anterior view. On the left, the anterior wall of the rectus sheath has been reflected laterally. On the right, the lower part of the rectus abdominis muscle has been removed.

1. The tendinous intersections of the rectus abdominis muscle are strongly attached to the anterior wall of the rectus sheath. A surgeon making a paramedian incision through the rectus sheath retracts the rectus abdominis laterally, keeping its nerve supply intact, and makes an incision through the posterior wall of the rectus sheath.

2. In pararectal incisions the incision is made through the anterior wall of the rectus sheath parallel to the lateral margin of the rectus muscle. The rectus is freed and retracted medially, exposing the segmental nerves entering its posterior surface; the posterior wall of the sheath is then incised between the segmental nerves. The disadvantage of this incision is that the opening into the abdominal cavity is small, and any longitudinal extension requires that one or more segmental nerves to the rectus abdominis be divided, with resultant postoperative rectus muscle weakness.

3. Divarication of the recti abdominis occurs in elderly multiparous women with weak abdominal muscles. The rectus sheath becomes excessively stretched, the recti separate widely, and the abdominal viscera bulge forward between them.

4. The linea semilunaris, the lateral edge of the rectus abdominis muscle, can be made to stand out by asking the patient to pull in his abdominal muscles. The linea semilunaris crosses the costal margin at the tip of the ninth costal cartilage, an important landmark.

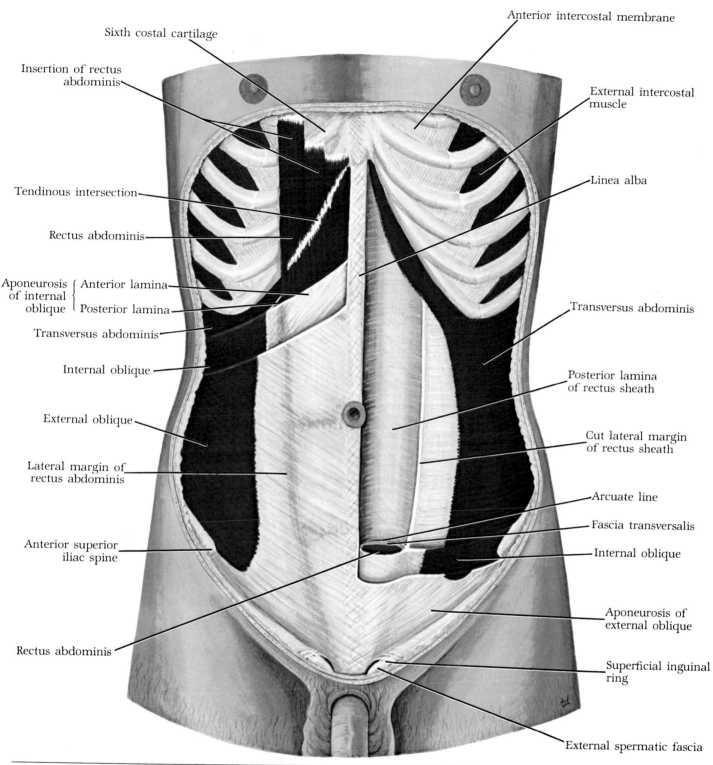

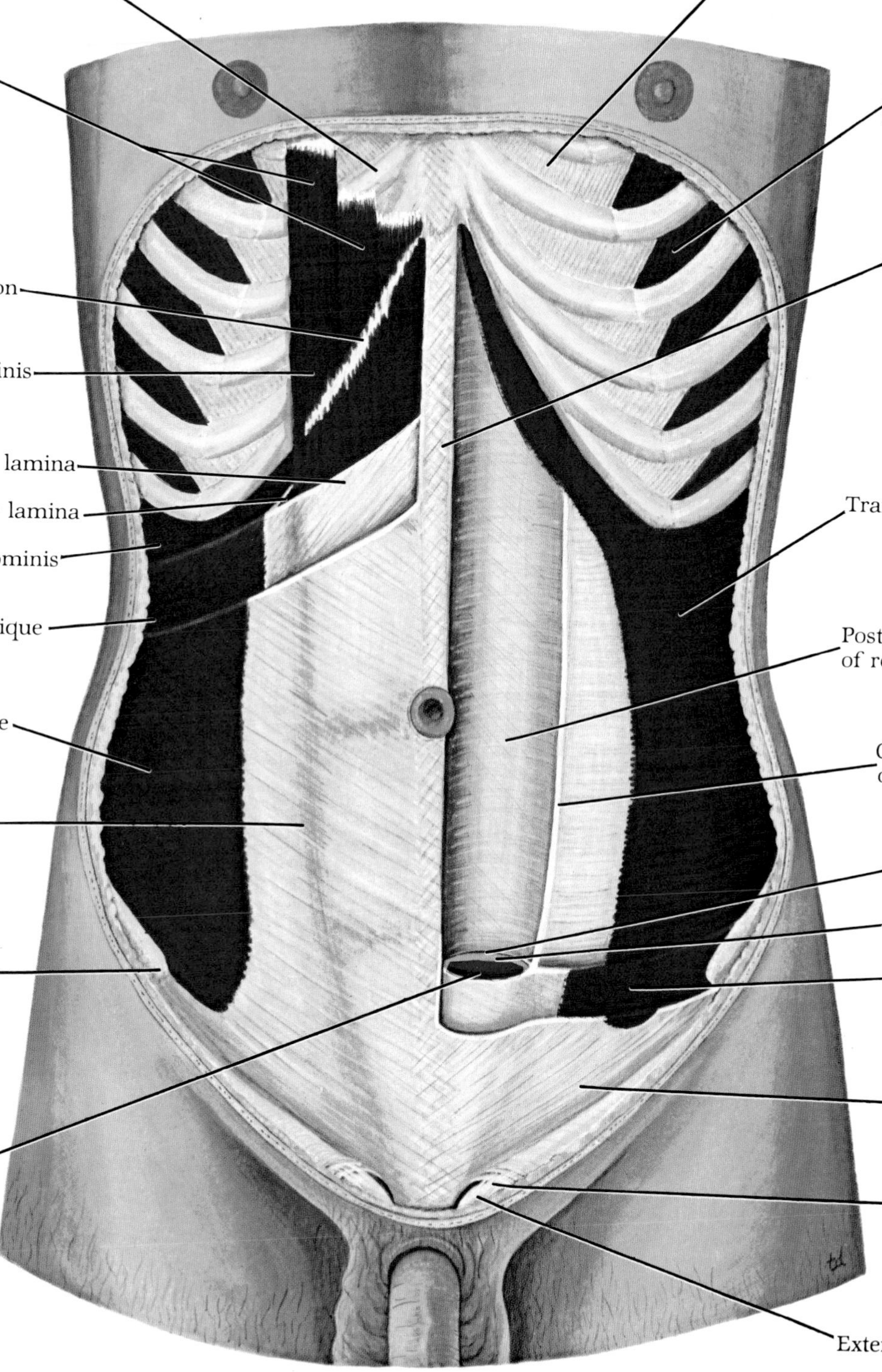
Sixth costal cartilage
Insertion of rectus abdominis
Tendinous intersection
Rectus abdominis
Aponeurosis of internal oblique
Anterior lamina
Posterior lamina
Transversus abdominis
Internal oblique
External oblique
Lateral margin of rectus abdominis
Anterior superior iliac spine
Rectus abdominis
Anterior intercostal membrane
External intercostal muscle
Linea alba
Transversus abdominis
Posterior lamina of rectus sheath
Cut lateral margin of rectus sheath
Arcuate line
Fascia transversalis
Internal oblique
Aponeurosis of external oblique
Superficial inguinal ring
External spermatic fascia

Figure 2-9
Rectus abdominis muscle and rectus sheath, anterior view. On the left, the anterior wall of the rectus sheath has been dissected in layers to show its constituent parts. On the right, the rectus muscle has been removed to show the posterior wall of the sheath.

1. The rectus sheath clinically is a very important part of the anterior abdominal wall. Surgical incisions through the rectus sheath are popular, provided the rectus abdominis muscle and its nerve supply are kept intact. The anterior and posterior walls of the sheath are sutured separately when the incision is closed, allowing the rectus muscle to spring back into position between the suture lines. This provides a very strong repair with minimal interference of function.

2. Between the level of the anterior superior iliac spine and the symphysis pubis, the aponeuroses of all three muscles of the anterior abdominal wall form the anterior wall of the rectus sheath. Thus there is no posterior wall, and the rectus muscle lies in contact with the fascia transversalis.

3. The rare spigelian hernia occurs where the vertical linea semilunaris along the lateral margin of the rectus muscle joins the arcuate line, a potential weak area.

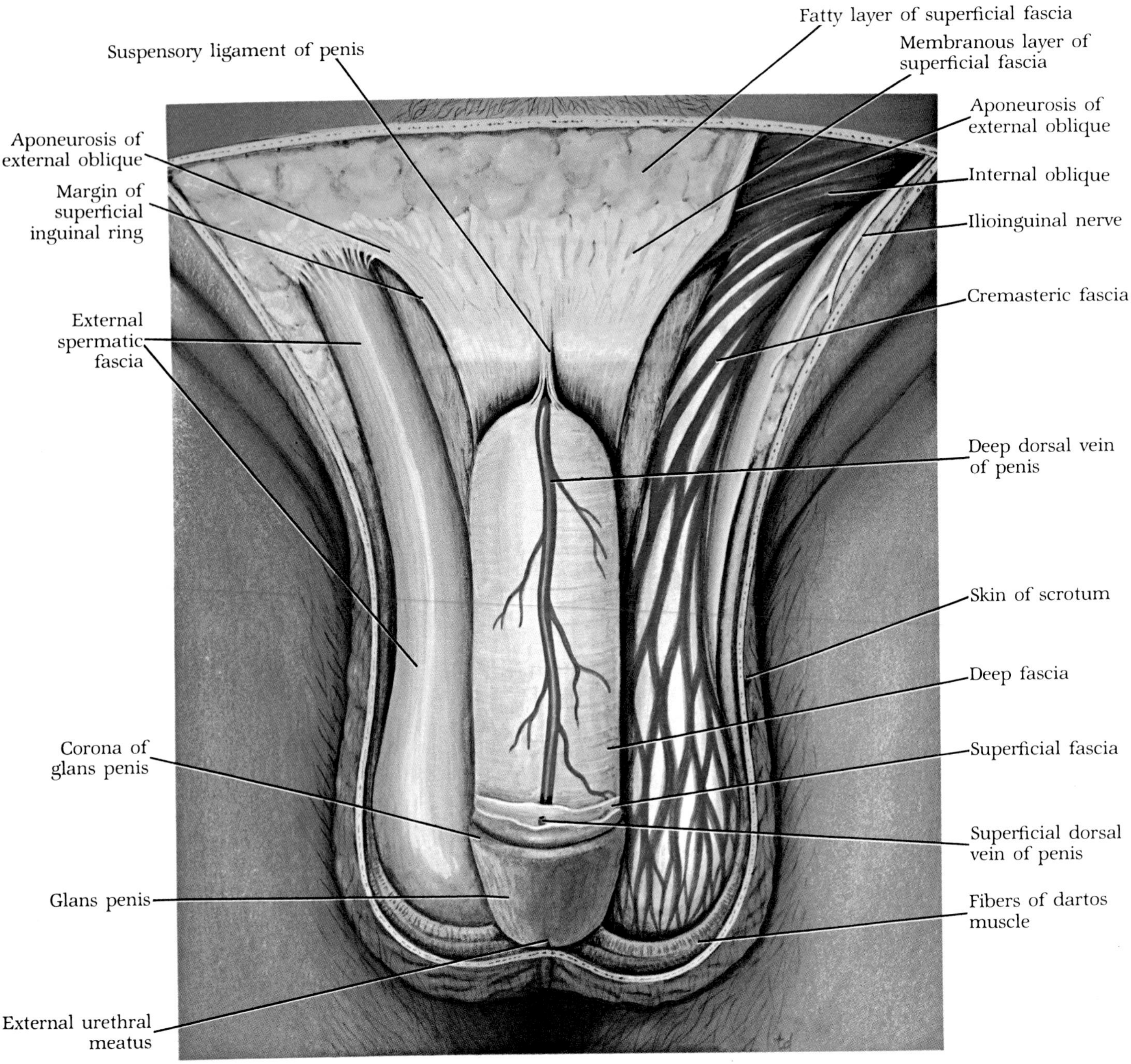
Fatty layer of superficial fascia
Suspensory ligament of penis
Membranous layer of superficial fascia
Aponeurosis of external oblique
Aponeurosis of external oblique
Internal oblique
Margin of superficial inguinal ring
Ilioinguinal nerve
Cremasteric fascia
External spermatic fascia
Deep dorsal vein of penis
Skin of scrotum
Deep fascia
Corona of glans penis
Superficial fascia
Superficial dorsal vein of penis
Glans penis
Fibers of dartos muscle
External urethral meatus

Figure 2-10
Inguinal, pubic, and scrotal regions, anterior view. On the right, the aponeurosis of the external oblique muscle and the external spermatic fascia have been removed to reveal the internal oblique and cremaster muscles.

1. The membranous layer of the superficial fascia (Scarpa's fascia) is continuous with the membranous layer of superficial fascia in the perineum (Colles' fascia). A potential closed space that does not open into the thigh lies beneath this fascia and is continuous with the superficial perineal pouch by way of the penis and scrotum. Rupture of the penile urethra may be followed by extravasation of urine into the scrotum, perineum, and penis, and then up into the lower part of the anterior abdominal wall deep to the membranous layer of fascia.

2. Because inguinal herniae are so common, the superficial inguinal ring is an important structure. Note that the lower part of it lies immediately above and medial to the pubic tubercle. An indirect inguinal hernia leaves the inguinal canal through the superficial ring and must lie within the external spermatic fascia, the cremasteric fascia, and the internal spermatic fascia; it may pass down within these structures to reach the scrotum.

3. A direct inguinal hernia occurs through the posterior wall of the inguinal canal.
It does not pass through the superficial inguinal ring and therefore does not reach the scrotum.

4. The cremaster muscle is supplied by the genitofemoral nerve (L1 and 2) and contracts in response to stimulation of the skin on the medial side of the thigh (cremasteric reflex).

5. The dartos muscle is composed of smooth muscle that normally corrugates the scrotal skin. In ill young boys the dartos muscle loses its tone and the scrotum sags.

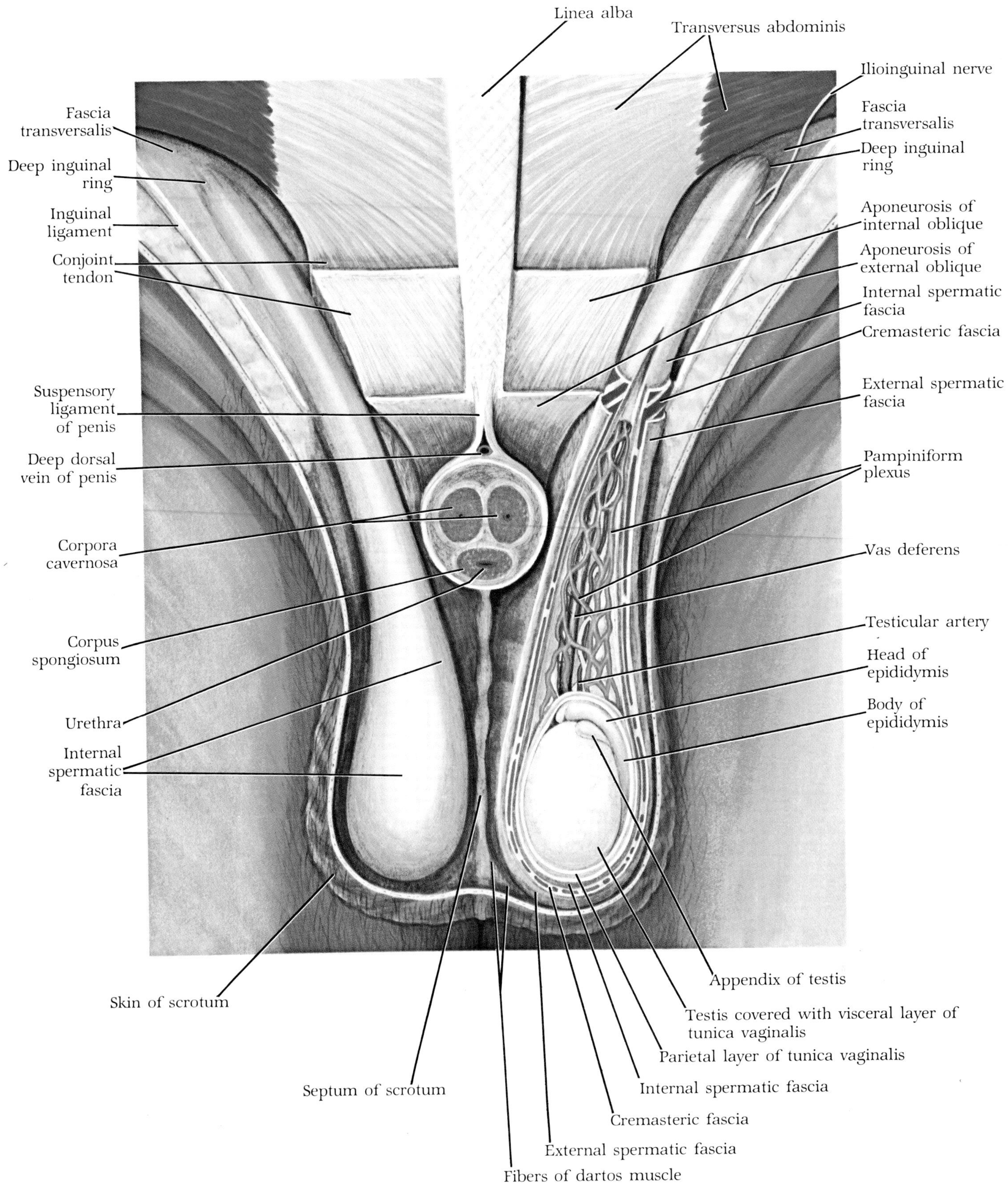
Linea alba
Transversus abdominis
Ilioinguinal nerve
Fascia transversalis
Deep inguinal ring
Aponeurosis of internal oblique
Aponeurosis of external oblique
Internal spermatic fascia
Cremasteric fascia
External spermatic fascia
Pampiniform plexus
Vas deferens
Testicular artery
Head of epididymis
Body of epididymis
Fascia transversalis
Deep inguinal ring
Inguinal ligament
Conjoint tendon
Suspensory ligament of penis
Deep dorsal vein of penis
Corpora cavernosa
Corpus spongiosum
Urethra
Internal spermatic fascia
Skin of scrotum
Septum of scrotum
Fibers of dartos muscle
External spermatic fascia
Cremasteric fascia
Internal spermatic fascia
Parietal layer of tunica vaginalis
Testis covered with visceral layer of tunica vaginalis
Appendix of testis

Figure 2-11
Inguinal and pubic regions, anterior view. The aponeuroses of the external and internal oblique muscles have been removed to show the lower margin of the transversus abdominis muscle and the region of the deep inguinal ring on both sides. On the right, the coverings of the spermatic cord have been removed in layers.

1. The deep inguinal ring is an opening in the fascia transversalis that lies about ½ inch (1.3 cm) above the midinguinal point. The internal spermatic fascia originates at the margin of the ring. An indirect inguinal hernia leaves the abdominal cavity at the deep inguinal ring and passes a variable distance down the canal within the internal spermatic fascia or it may pass all the way through the superficial inguinal ring to enter the scrotum.

2. The posterior wall of the inguinal canal is formed by the fascia transversalis and is strengthened medially by the conjoint tendon. Because of the strength of the conjoint tendon, a direct inguinal hernia is usually nothing more than a generalized bulge of the posterior wall which therefore causes the neck of the hernial sac to be wide.
3. Note that the spermatic cord is a collection of structures that traverse the inguinal canal and pass to and from the testis. It is covered with three concentric layers of fascia derived from the layers of the anterior abdominal wall. It begins at the deep inguinal ring and ends at the testis.
4. A varicocele is a condition in which there is elongation and dilatation of the veins of the pampiniform plexus.
5. The tunica vaginalis is derived embryologically from the distal part of the processus vaginalis. If the proximal part of the processus persists, it may lead to a preformed hernial sac for an indirect inguinal hernia, a congenital hydrocele, or an encysted hydrocele of the cord.

6. The testis may be subject to the following congenital anomalies:
A. Anterior inversion: the epididymis lies anteriorly and the testis and tunica vaginalis posteriorly
B. Polar inversion: the testis and epididymis are completely inverted
C. Imperfect descent: (1) incomplete descent (in which the testis, although travelling down its normal path, fails to reach the floor of the scrotum), and (2) maldescent (in which the testis travels down an abnormal path and fails to reach the scrotum). A testis found in the abdominal cavity or the inguinal canal should be removed surgically and placed in the scrotum before puberty, because the higher temperature in the abdomen retards the normal process of spermatogenesis.

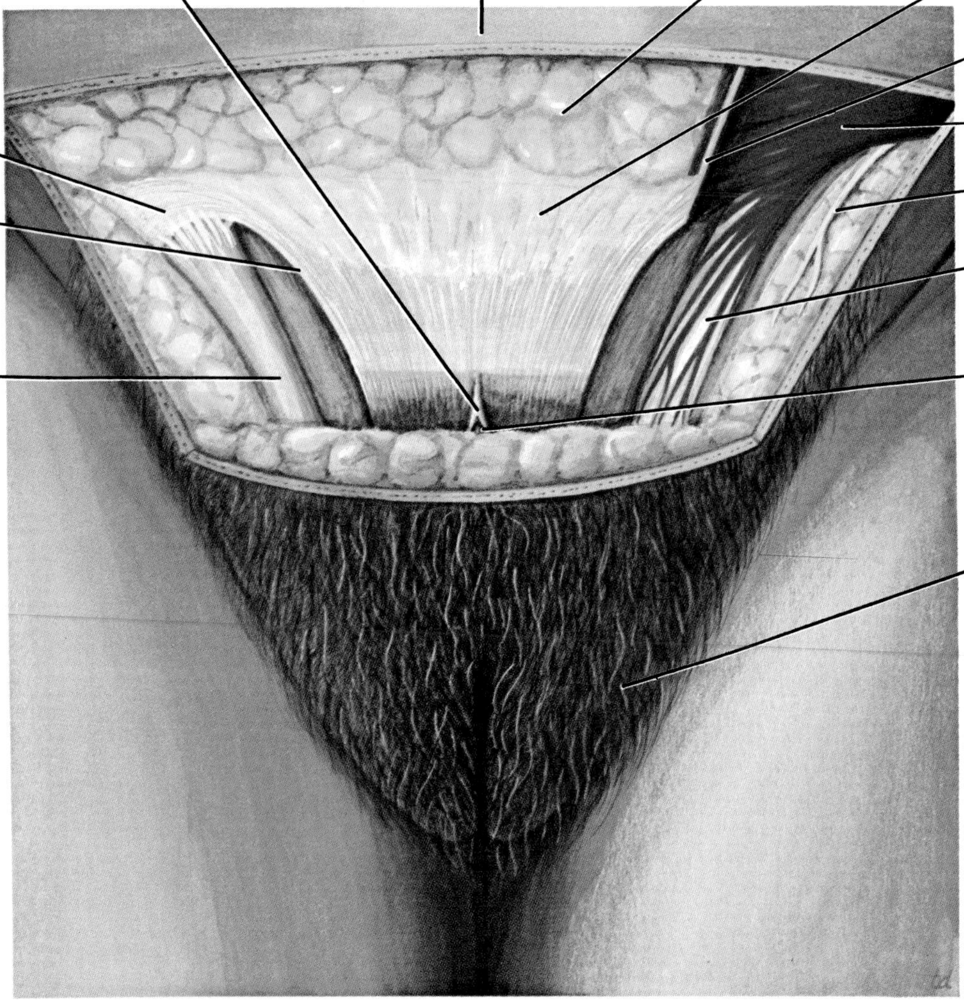
Mons pubis
Fatty layer of superficial fascia
Suspensory ligament of clitoris
Membranous layer of superficial fascia
Aponeurosis of external oblique
Aponeurosis of external oblique
Internal oblique
Ilioinguinal nerve
Margin of superficial inguinal ring
Cremasteric fascia
External covering of round ligament of uterus (external spermatic fascia)
Deep dorsal vein of clitoris
Skin of labium majus

Figure 2-12
Inguinal and pubic regions in the female, anterior view. On the right, the aponeurosis of the external oblique muscle and the external spermatic fascia have been removed to reveal the internal oblique and cremaster muscles.

1. Although an inguinal canal is fully formed in the female, the ovary does not normally descend through it but remains in the pelvis.

2. In the region of the groin, a femoral hernia is more common in women than in men (possibly due to a wider pelvis and femoral canal). An indirect inguinal hernia is much more common in males than females, and nearly one-third are bilateral. Direct inguinal herniae are rare in women, and the majority of those that occur in either sex are bilateral.

3. In the female the only structures that pass through the inguinal canal from the abdominal cavity are the round ligament of the uterus and a few lymph vessels. The lymph vessels convey a small amount of lymph from the body of the uterus to the superficial inguinal nodes.

4. The round ligament of the uterus has the same three coverings as the spermatic cord in the male but they are vestigial.

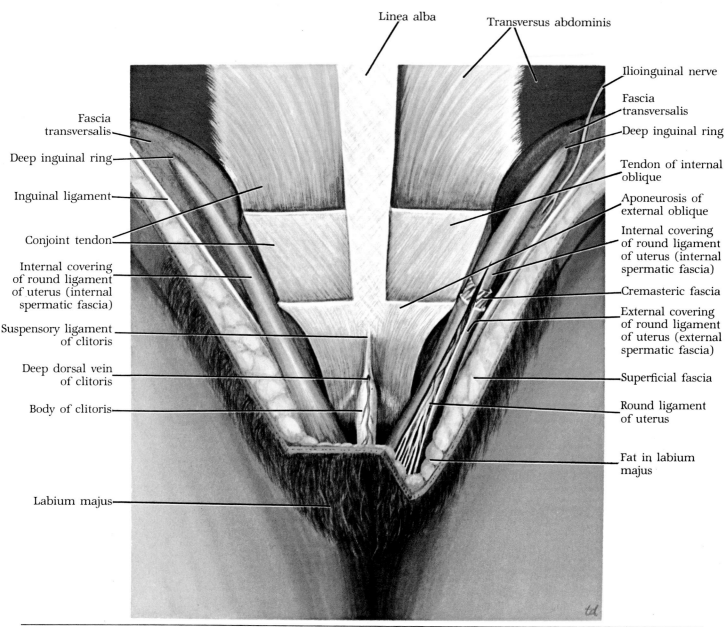

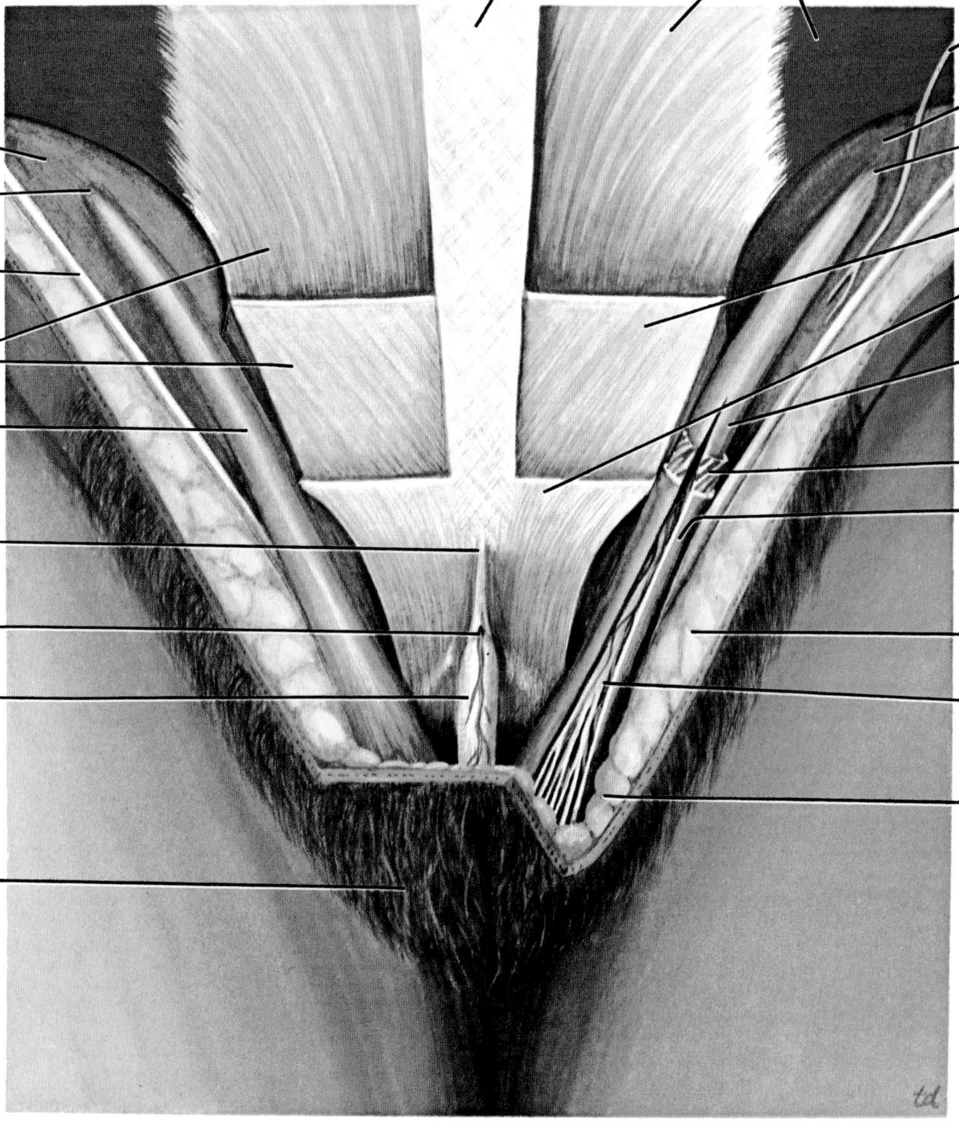
Linea alba
Transversus abdominis
Ilioinguinal nerve
Fascia transversalis
Deep inguinal ring
Tendon of internal oblique
Aponeurosis of external oblique
Internal covering of round ligament of uterus (internal spermatic fascia)
Cremasteric fascia
External covering of round ligament of uterus (external spermatic fascia)
Superficial fascia
Round ligament of uterus
Fat in labium majus
Fascia transversalis
Deep inguinal ring
Inguinal ligament
Conjoint tendon
Internal covering of round ligament of uterus (internal spermatic fascia)
Suspensory ligament of clitoris
Deep dorsal vein of clitoris
Body of clitoris
Labium majus

Figure 2-13
Inguinal canal and pubic regions in the female, anterior view. The aponeuroses of the external and internal oblique muscles have been removed to show the lower margin of the transversus abdominis muscle and the region of the deep inguinal ring on both sides. On the right, the vestigial coverings of the round ligament of the uterus have been removed in layers.

1. An indirect inguinal hernia leaves the abdominal cavity at the deep inguinal ring and passes a variable distance down the canal within the internal spermatic fascia; it may reach the base of the labium majus.

2. A hernia consists of three parts: the sac, the contents of the sac, and the coverings of the sac. The hernial sac is a diverticulum of peritoneum and has a neck and a body. The contents can consist of any structure found within the abdominal cavity and may vary from a small piece of omentum to a large viscus such as the kidney. The hernial coverings are formed from the layers of the abdominal wall through which the hernial sac passes. Sometimes a retroperitoneal viscus becomes incorporated into the wall of the sac and thus slides down into the inguinal canal; such a hernia is referred to as a sliding hernia. A right-sided sliding hernia, for example, may have the cecum in its wall.

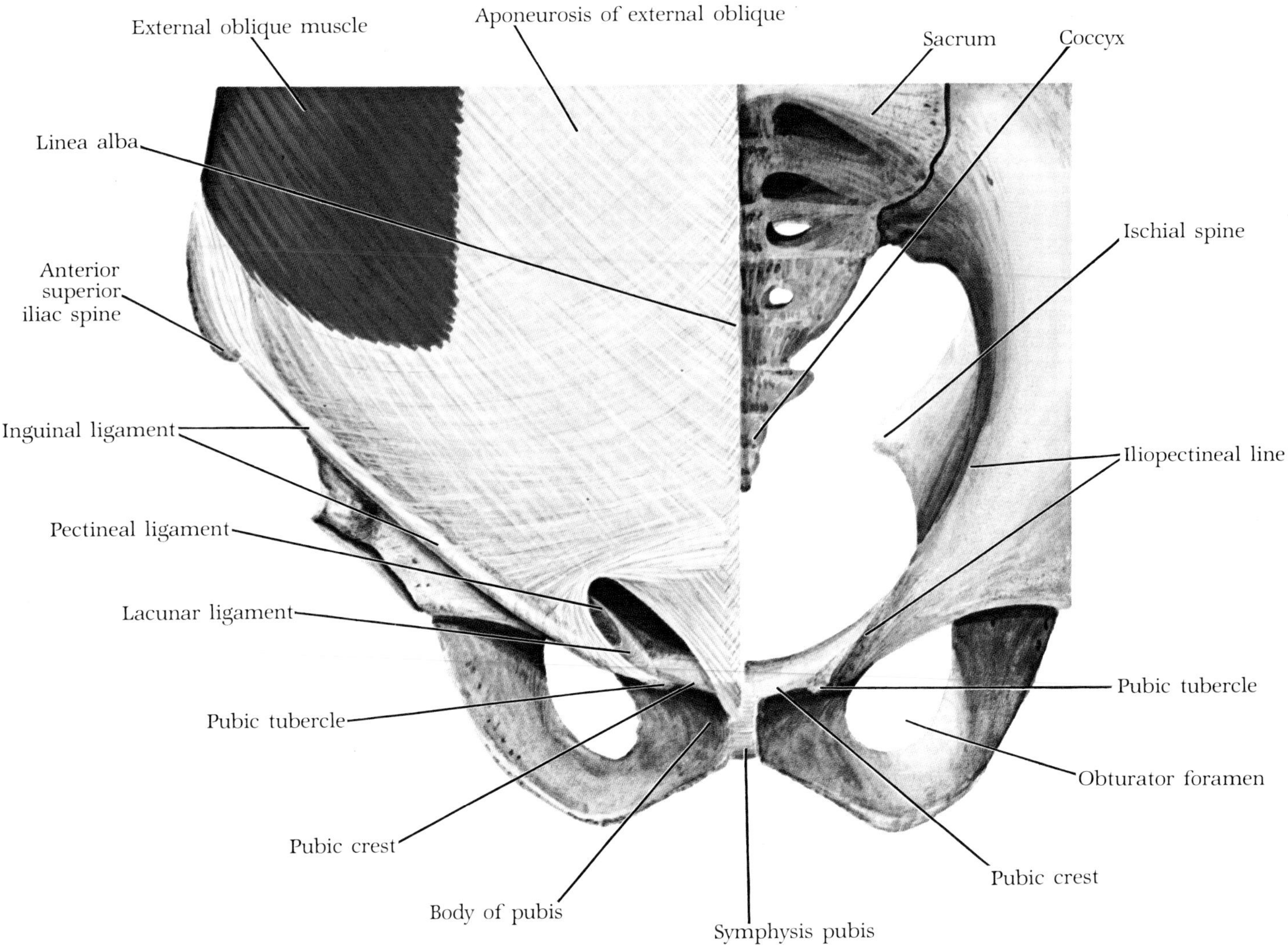
External oblique muscle
Aponeurosis of external oblique
Sacrum
Coccyx
Linea alba
Ischial spine
Anterior superior iliac spine
Inguinal ligament
Iliopectineal line
Pectineal ligament
Lacunar ligament
Pubic tubercle
Pubic tubercle
Obturator foramen
Pubic crest
Pubic crest
Body of pubis
Symphysis pubis

Figure 2-14
Aponeurosis of the external oblique muscle of the abdomen, the inguinal, lacunar, and pectineal ligaments, and the superficial inguinal ring.

1. The inguinal, lacunar, and pectineal ligaments are of great importance to the surgeon operating on an inguinal or femoral hernia. With a neglected indirect inguinal hernia or a direct inguinal hernia the posterior wall of the inguinal canal may have to be strengthened with fascia sutured to the inguinal or pectineal ligaments. In the repair of a femoral hernia, the inguinal ligament is often sutured to the pectineal ligament, thus obliterating the femoral ring.
2. A femoral hernia extends into the thigh beneath the inguinal ligament. It is related medially to the sharp edge of the lacunar ligament and laterally to the femoral vein. The small neck of its sac lies below and lateral to the pubic tubercle, whereas the sac of an indirect inguinal hernia emerges through the superficial inguinal ring above and medial to the pubic tubercle.
3. Because the neck of a femoral hernial sac cannot expand, once an abdominal viscus has passed through the neck into the body of the sac it may be difficult to push it back into the abdominal cavity (irreducible hernia). Also, in a femoral hernia the blood supply to the hernial contents may be compressed at the neck of the sac, and the compression may seriously impair the circulation of the contents (strangulated hernia).

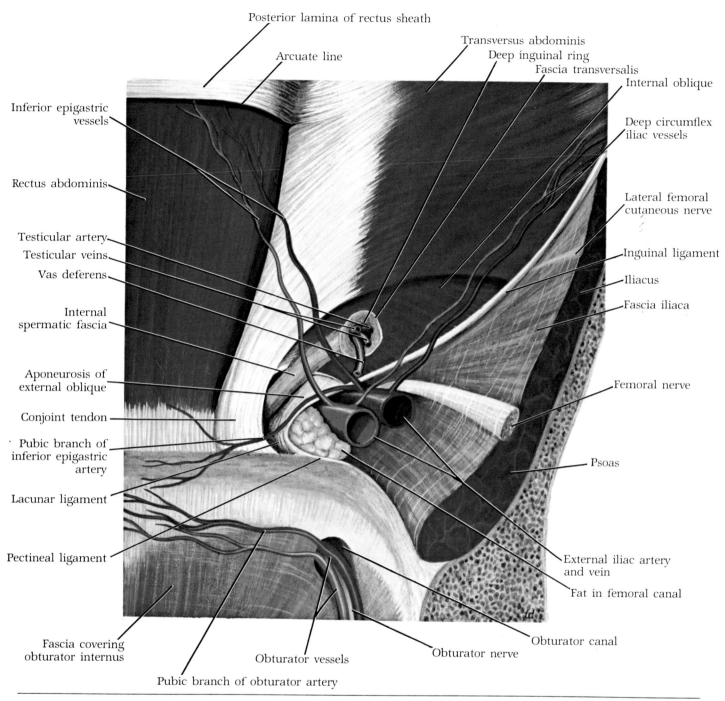

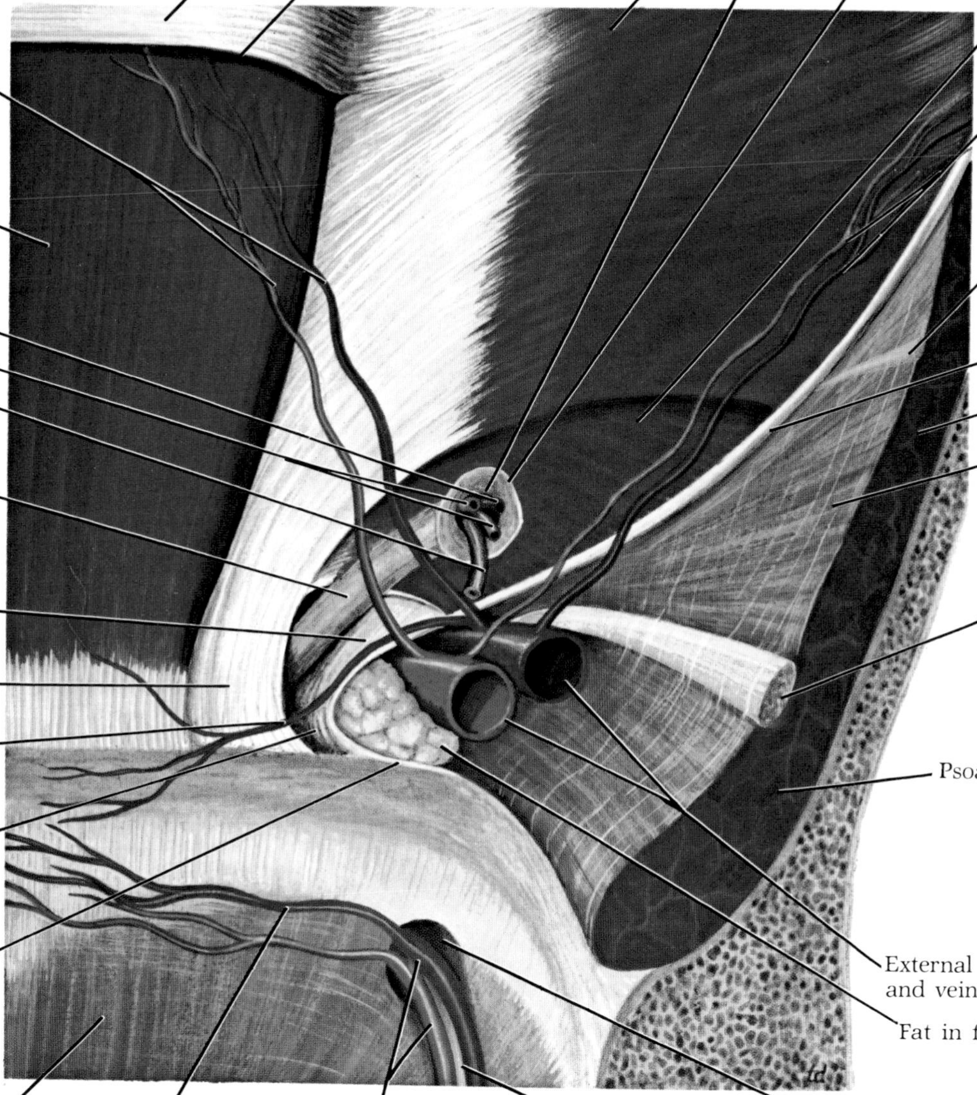
Posterior lamina of rectus sheath
Arcuate line
Transversus abdominis
Deep inguinal ring
Fascia transversalis
Internal oblique
Inferior epigastric vessels
Deep circumflex iliac vessels
Rectus abdominis
Lateral femoral cutaneous nerve
Testicular artery
Testicular veins
Vas deferens
Inguinal ligament
Iliacus
Internal spermatic fascia
Fascia iliaca
Aponeurosis of external oblique
Femoral nerve
Conjoint tendon
Pubic branch of inferior epigastric artery
Psoas
Lacunar ligament
Pectineal ligament
External iliac artery and vein
Fat in femoral canal
Fascia covering obturator internus
Obturator canal
Obturator nerve
Obturator vessels
Pubic branch of obturator artery

Figure 2-15
Deep inguinal ring and femoral ring in the male, posterior view.

1. The inguinal canal in the lower part of the anterior abdominal wall in both sexes constitutes a site of potential weakness. Normally, on coughing and straining, the arching lowest fibers of the internal oblique and transversus abdominis muscles contract, flattening out the arched roof so that it is lowered toward the floor. With severe or prolonged coughing or straining, however, the roof may be approximated to the floor so that the canal is virtually closed.

2. A direct inguinal hernia protrudes forward through the posterior wall of the inguinal canal, thus causing the fascia transversalis and the conjoint tendon to bulge forward. This takes place in an area bounded laterally by the inferior epigastric artery, inferiorly by the inguinal ligament, and medially by the lateral margin of the rectus sheath (Hesselbach's triangle).

3. An indirect inguinal hernia leaves the abdominal cavity through the deep inguinal ring, which is a rounded hole in the fascia transversalis lateral to the inferior epigastric artery. The hernial sac is formed of peritoneum.

4. Note the pubic branch of the inferior epigastric artery. Occasionally this artery is large and forms the obturator artery. If the lacunar ligament was incised surgically to enlarge the neck of a femoral hernia, this artery might be damaged, causing a severe intra-abdominal hemorrhage.

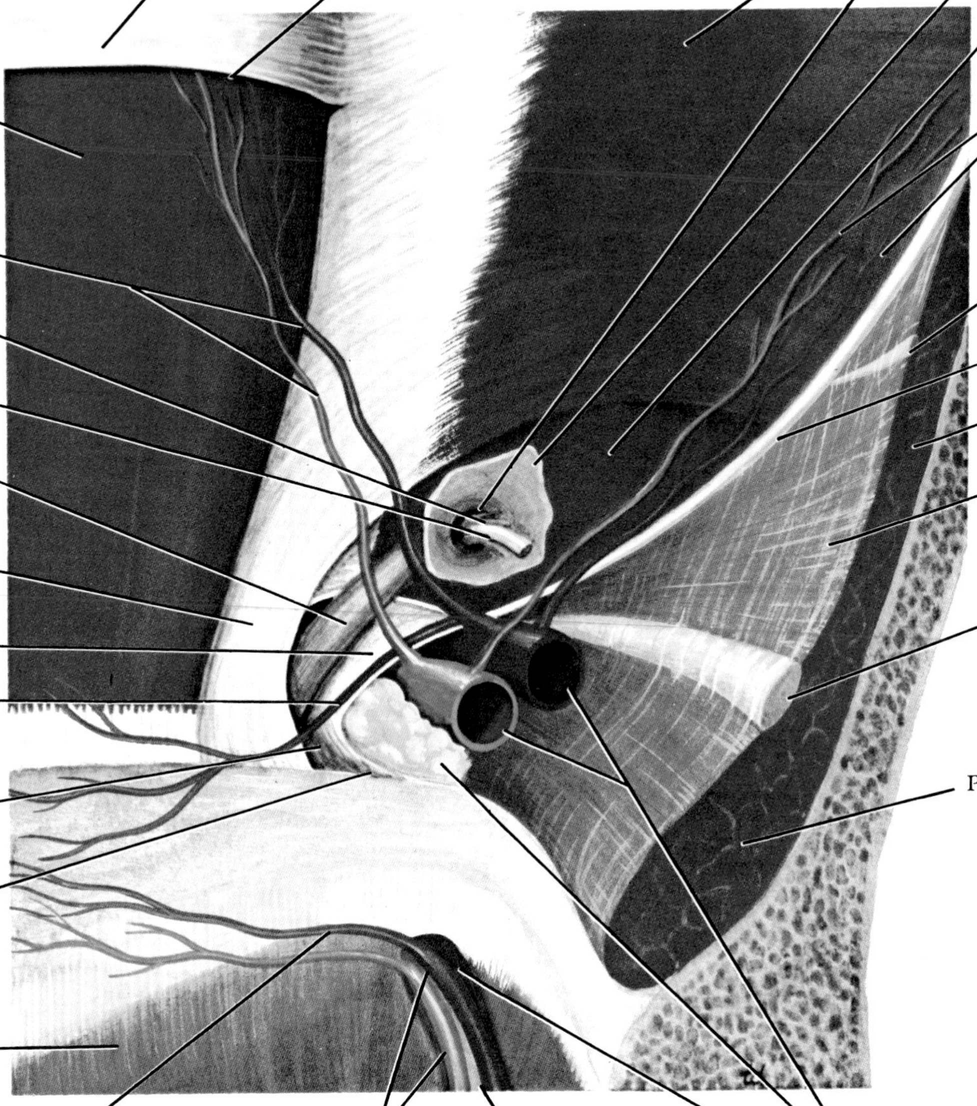
Posterior lamina of rectus sheath
Arcuate line
Transversus abdominis
Deep inguinal ring
Fascia transversalis
Internal oblique
Deep circumflex iliac vessels
Rectus abdominis
Inferior epigastric vessels
Lateral femoral cutaneous nerve
Lymphatic vessel
Inguinal ligament
Round ligament of uterus
Iliacus
Internal covering of round ligament of uterus (internal spermatic fascia)
Fascia iliaca
Conjoint tendon
Femoral nerve
Aponeurosis of external oblique
Pubic branch of inferior epigastric artery
Psoas
Lacunar ligament
Pectineal ligament
Fascia covering obturator internus
External iliac artery and vein
Fat in femoral canal
Pubic branch of obturator artery
Obturator vessels
Obturator nerve
Obturator canal

Figure 2-16
Deep inguinal ring and femoral ring in the female, posterior view.

1. The femoral ring, the name given to the upper end of the femoral canal, is normally filled with a plug of fat called the femoral septum. The femoral ring is related medially to the lacunar ligament, laterally to the femoral vein, superiorly to the inguinal ligament, and inferiorly to the pectineal ligament and the superior ramus of the pubis.

2. In a femoral hernia the peritoneal sac pushes the femoral septum before it as it descends through the femoral canal. After it emerges into the upper part of the thigh, the sac expands and may turn anteriorly through the saphenous opening in the fascia lata and then superiorly to cross the anterior surface of the inguinal ligament.

3. The obturator canal is very rarely the site of a hernia. When it is, the hernial sac passes inferiorly into the adductor region of the thigh.

4. The psoas fascia covers the anterior surface of the psoas muscle. In tuberculous disease of the thoracolumbar region of the vertebral column, pus may extend downward and laterally beneath the psoas fascia into the thigh. A fluctuant swelling then appears above and below the inguinal ligament; the lower swelling may be mistaken for a femoral hernia.

5. Note the round ligament of the uterus and the lymph vessels as they pass through the inguinal canal by way of the deep inguinal ring.

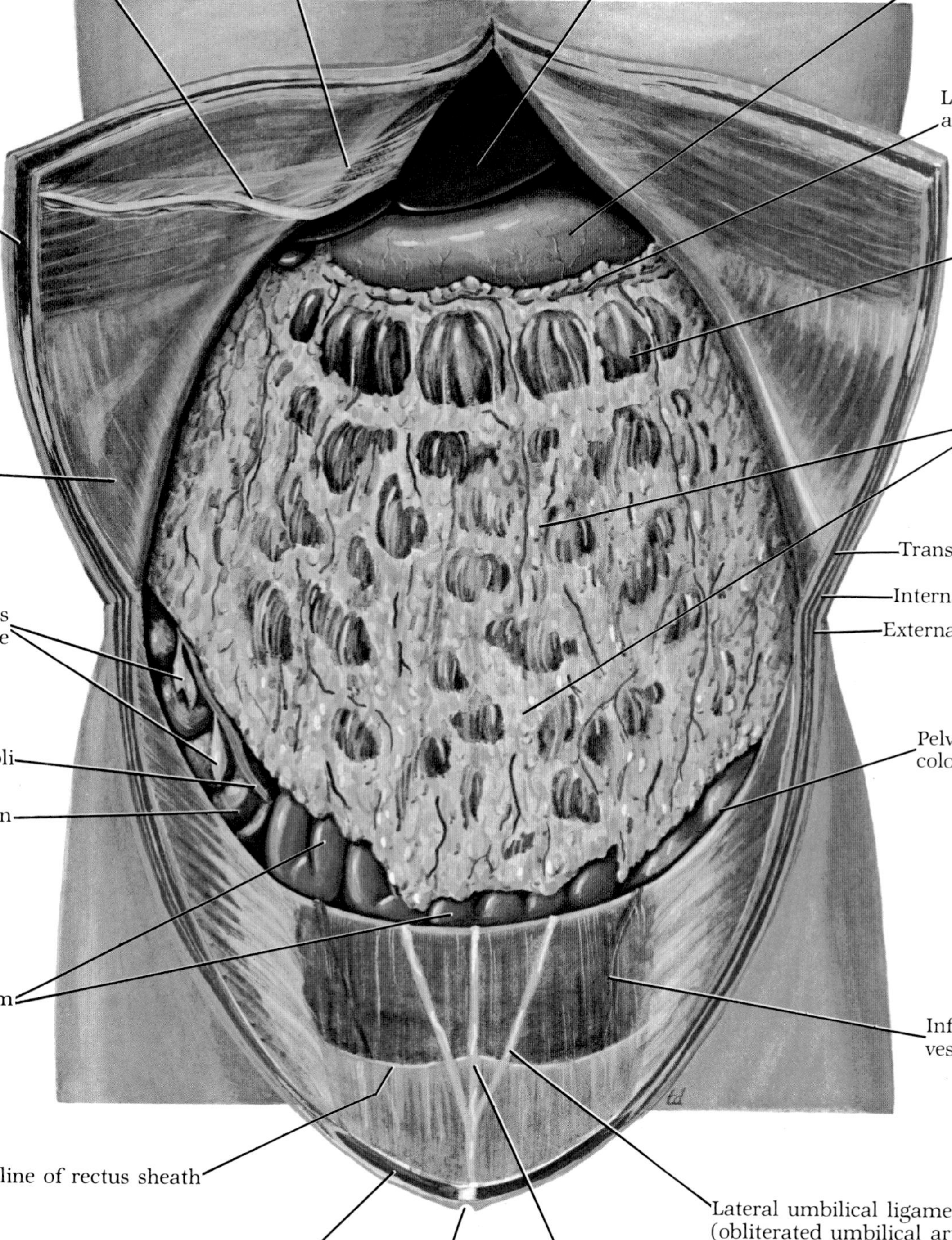
Ligamentum teres
Falciform ligament
Left lobe of liver
Body of stomach
Left gastroepiploic artery
Rectus abdominis
Transverse colon showing through greater omentum
Peritoneum (through which can be seen the transversus abdominis muscle)
Greater omentum
Transversus abdominis
Internal oblique
External oblique
Appendices epiploicae
Teniae coli
Pelvic (sigmoid) colon
Ascending colon
Coils of ileum
Inferior epigastric vessels
td
Arcuate line of rectus sheath
Lateral umbilical ligament (obliterated umbilical artery)
Rectus abdominis muscle
Umbilicus
Median umbilical ligament (urachus)

Figure 2-17
The abdominal cavity. The area has been opened by incising the anterior abdominal wall. The abdominal organs have been left in situ.

1. Most of the liver lies in the right hypochondrium and is protected by the thoracic cage. The lower margin of the liver crosses the epigastrium and is most easily palpated when the patient inspires deeply and the diaphragm contracts and pushes down the liver. Because of the relatively large size of the liver in the child, its lower margin is found at a lower level in children.

2. Although fixed at both ends, the stomach shows considerable variation in size, shape, and position, depending on the individual, degree of filling, the subject's posture, and respiration. A portion of the stomach may be palpated in the epigastric region.

3. The position of the transverse colon is extremely variable but usually crosses the abdomen in the region of the umbilicus.

4. The greater omentum moves about the peritoneal cavity in response to the peristaltic movements of the neighboring gut. At times of acute inflammation, as, for example, in acute appendicitis, the inflammatory exudate causes the omentum to adhere to the appendix. By this means the infection is often localized, thus preventing a serious diffuse peritonitis. Omental adhesions are often encountered by a surgeon when operating on patients who have undergone previous abdominal surgery or peritoneal dialysis.

5. A portion of the omentum is often found in a hernial sac. It may also undergo torsion or even infarction following torsion. Surgeons sometimes use the omentum to buttress an intestinal anastomosis or in the closure of a perforated gastric or duodenal ulcer.

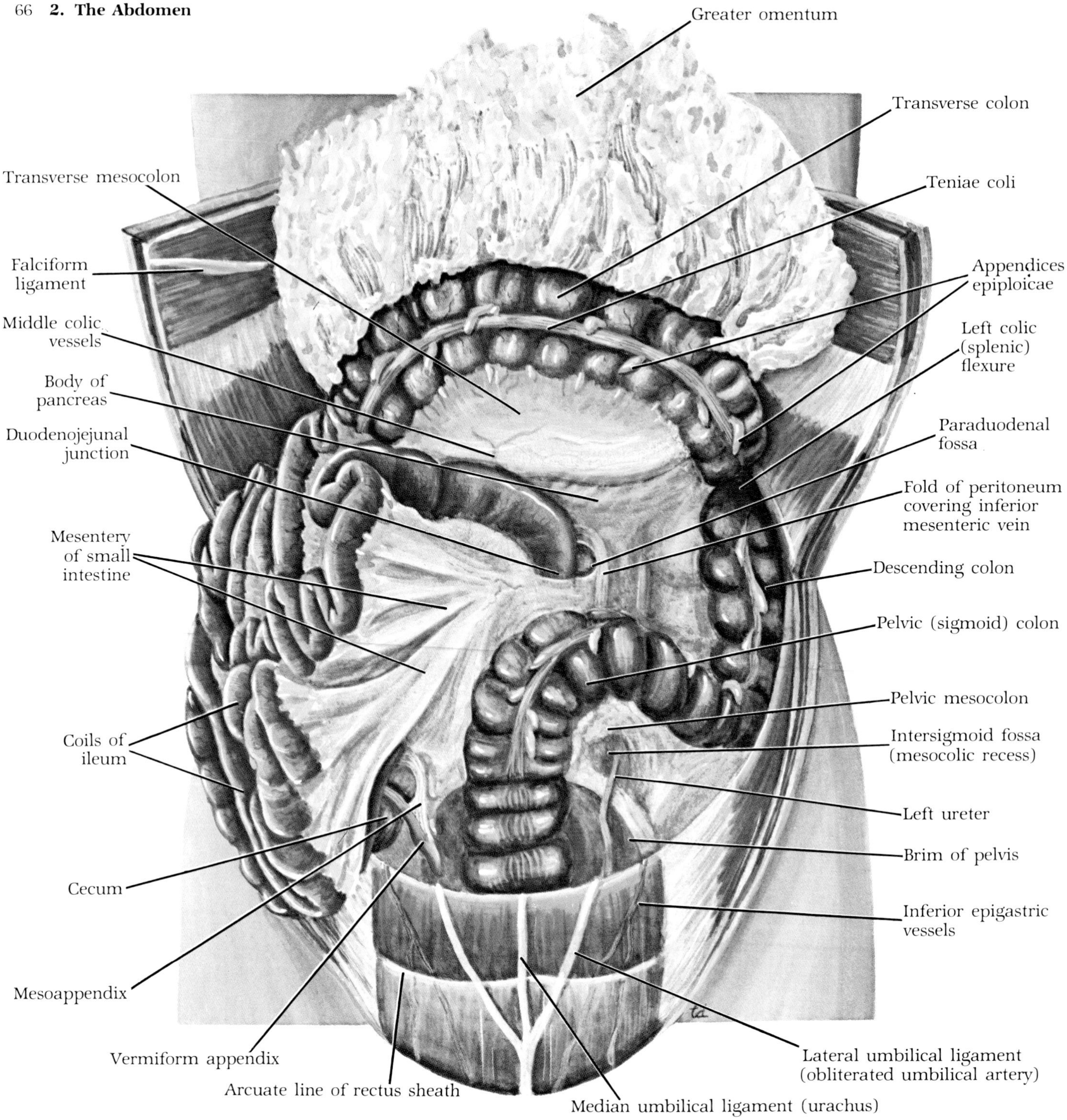
Greater omentum
Transverse colon
Transverse mesocolon
Teniae coli
Falciform ligament
Appendices epiploicae
Middle colic vessels
Left colic (splenic) flexure
Body of pancreas
Paraduodenal fossa
Duodenojejunal junction
Fold of peritoneum covering inferior mesenteric vein
Mesentery of small intestine
Descending colon
Pelvic (sigmoid) colon
Pelvic mesocolon
Coils of ileum
Intersigmoid fossa (mesocolic recess)
Left ureter
Brim of pelvis
Cecum
Inferior epigastric vessels
Mesoappendix
Vermiform appendix
Lateral umbilical ligament (obliterated umbilical artery)
Arcuate line of rectus sheath
Median umbilical ligament (urachus)

Figure 2-18
Abdominal contents after the greater omentum has been reflected superiorly.

1. The peritoneal cavity is the potential space that exists between the parietal and visceral layers of peritoneum. The mesothelial cells of the peritoneum secrete a small amount of serous fluid that moistens the surface and allows free movement between the viscera. The peritoneum is a semipermeable membrane and allows the rapid bidirectional transfer of substances across itself. Because the surface area of the peritoneum is enormous, this transfer property has been made use of in peritoneal dialysis for acute renal insufficiency. Unfortunately, death can easily occur from absorption of toxic bowel contents that have leaked into the peritoneal cavity through a perforation of the bowel wall.
2. An abnormal accumulation of peritoneal fluid is called ascites. This may occur in such conditions as heart failure or malignant invasion of the peritoneum.
3. In the male the peritoneal cavity is closed. In the female infection of the reproductive organs can spread to the cavity through the uterine tubes.
4. The attachment of the transverse mesocolon and the mesentery of the small intestine to the posterior abdominal wall provides barriers that may restrict the movement of infected peritoneal fluid from the upper to the lower part of the peritoneal cavity.
5. Abdominal pain arising from the parietal peritoneum can be precisely localized to the site of origin. The parietal peritoneum lining the anterior abdominal wall is innervated by the anterior rami of the lower six thoracic and the first lumbar nerves. The visceral peritoneum has no pain-afferent fibers. The root of the mesentery is very sensitive to stretch. Pain arising from an abdominal viscus is dull and poorly localized and is referred to the midline.

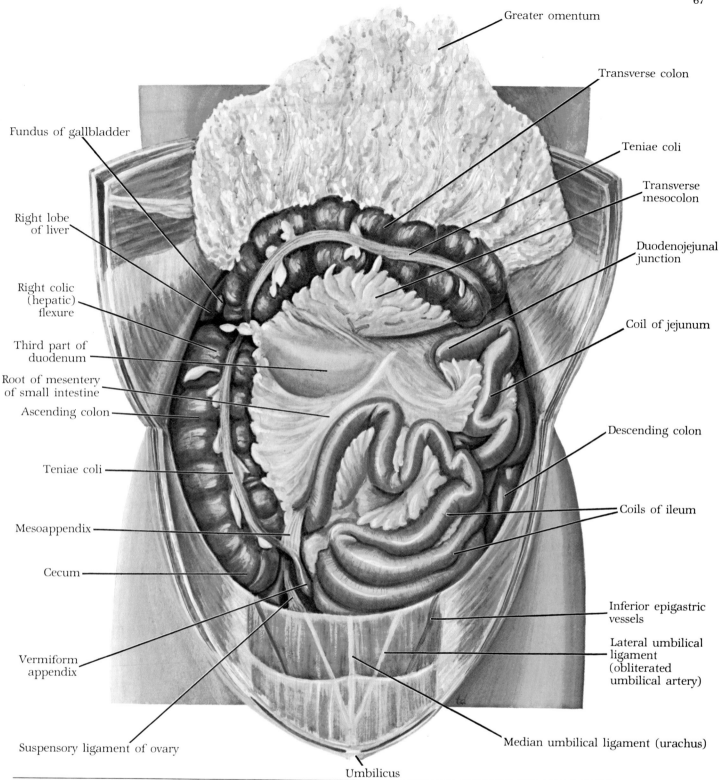

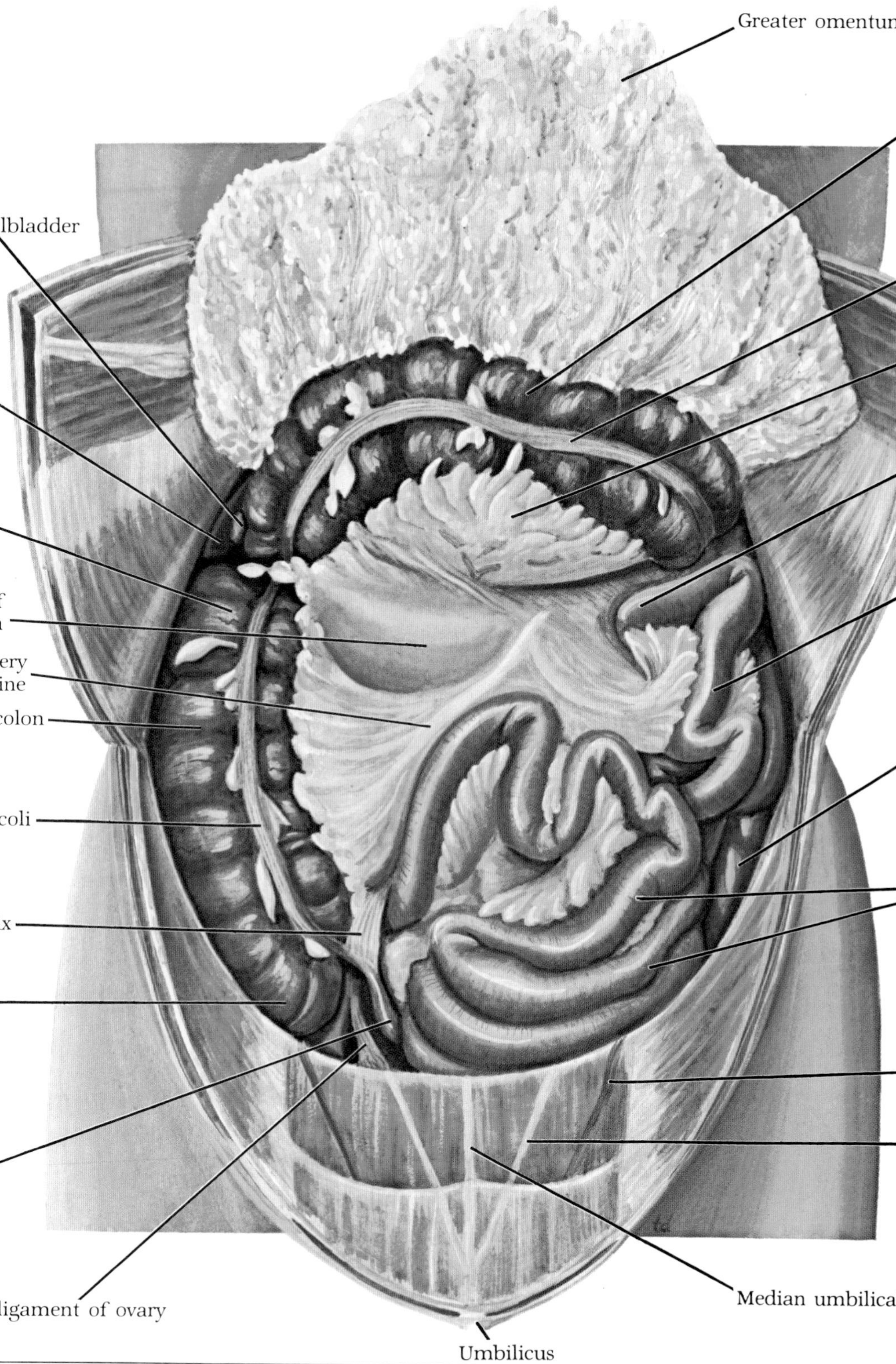
Greater omentum
Transverse colon
Fundus of gallbladder
Teniae coli
Transverse mesocolon
Right lobe of liver
Duodenojejunal junction
Right colic (hepatic) flexure
Coil of jejunum
Third part of duodenum
Root of mesentery of small intestine
Ascending colon
Descending colon
Teniae coli
Coils of ileum
Mesoappendix
Cecum
Inferior epigastric vessels
Lateral umbilical ligament (obliterated umbilical artery)
Vermiform appendix
Suspensory ligament of ovary
Median umbilical ligament (urachus)
Umbilicus

Figure 2-19
Abdominal contents after the greater omentum has been reflected superiorly. Coils of small intestine have been pulled over to the right.

1. It is important to be able to distinguish between the large and small intestines by external examination: (a) the small intestine (with the exception of the duodenum) is mobile, while the ascending and descending parts of the colon are fixed; (b) the small intestine (again with the exception of the duodenum) has a mesentery the attachment of which passes downward across the midline into the right iliac fossa; (c) the longitudinal muscle of the small intestine forms a continuous layer around the gut; in the large intestine (with the exception of the appendix), the longitudinal muscle is collected into three bands, the teniae coli; (d) the small intestine has no fatty tags attached to its wall; the large intestine does have fatty tags, called the appendices epiploicae; (e) the wall of the small intestine is smooth, whereas that of the large intestine is sacculated.

2. A tumor or cyst of the mesentery of the small intestine, when palpated through the anterior abdominal wall, will be more mobile in a direction at right angles to the line of attachment than along the line of attachment.

3. A loop of small intestine may enter a peritoneal recess or fossa (e.g., lesser sac or duodenal fossae) and become strangulated at the edge of the fossa. The inferior mesenteric vein often lies in the anterior wall of the paraduodenal fossa, so it is important that this wall not be blindly incised surgically.

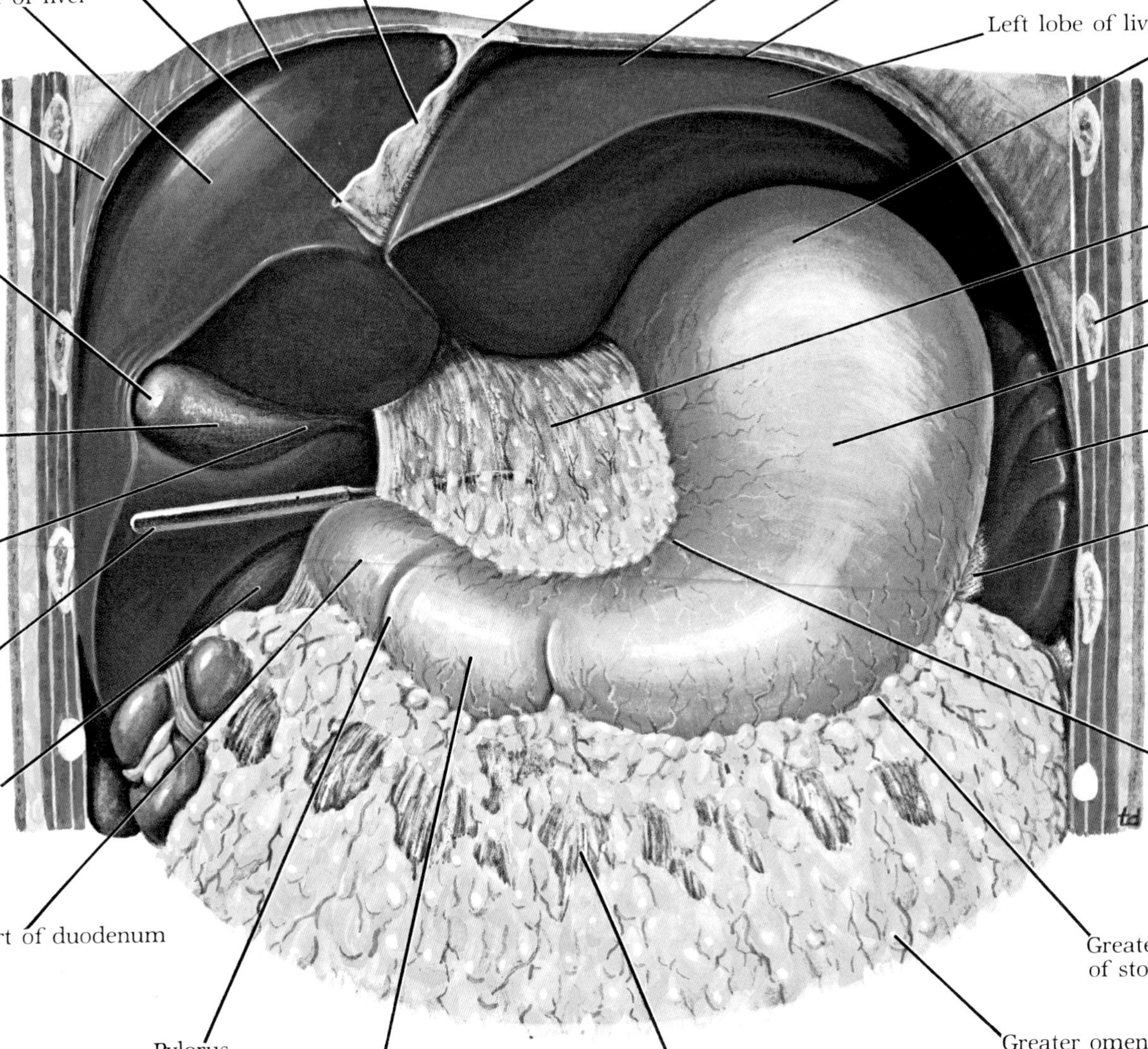
Falciform ligament
Right anterior subphrenic space
Central tendon of diaphragm
Anterior left subphrenic space
Ligamentum teres
Left cupola of diaphragm
Right lobe of liver
Left lobe of liver
Fundus of stomach
Right cupola of diaphragm
Lesser omentum
Fundus of gallbladder
Ninth rib
Body of stomach
Body of gallbladder
Spleen
Gastrosplen omentum (ligament)
Neck of gallbladder
Probe in entrance into lesser sac
Lesser curvature of stomach
Right kidney
First part of duodenum
Greater curvature of stomach
Pylorus
Greater omentum
Antrum of stomach
Transverse colon seen through anterior layer of greater omentum

Figure 2-20
Liver and stomach, anterior view. The liver has been pulled superiorly, and a probe has been inserted into the entrance to the lesser sac.

1. The liver is a soft, friable structure enclosed in a fibrous capsule. Its close relationship to the diaphragm and the lower ribs must be emphasized. Fractures of the lower ribs or penetrating wounds of the lower thorax or upper abdomen are common causes of liver injury. Blunt traumatic injuries during automobile accidents are also common. Severe hemorrhage accompanies tears of this organ.

2. The liver is held in position by the attachment of the hepatic veins to the inferior vena cava; abdominal muscle tone and peritoneal ligaments are of minor importance.

3. The fundus of the gallbladder lies opposite the tip of the right ninth costal cartilage at the point where the lateral edge of the right rectus abdominis muscle crosses the costal margin. This should be remembered when performing a physical examination of the gallbladder.

4. Perforation of the stomach or duodenum is a relatively common complication of peptic ulcer. Gastric ulcers frequently occur along the lesser curvature of the stomach; if they perforate anteriorly, the gastric contents enter the greater sac, while at a posterior perforation the contents will enter the lesser sac. The majority of duodenal perforations occur anteriorly into the greater sac. A chronic anterior gastric ulcer may cause the stomach to adhere to the liver, after which the ulcer may penetrate the liver substance.

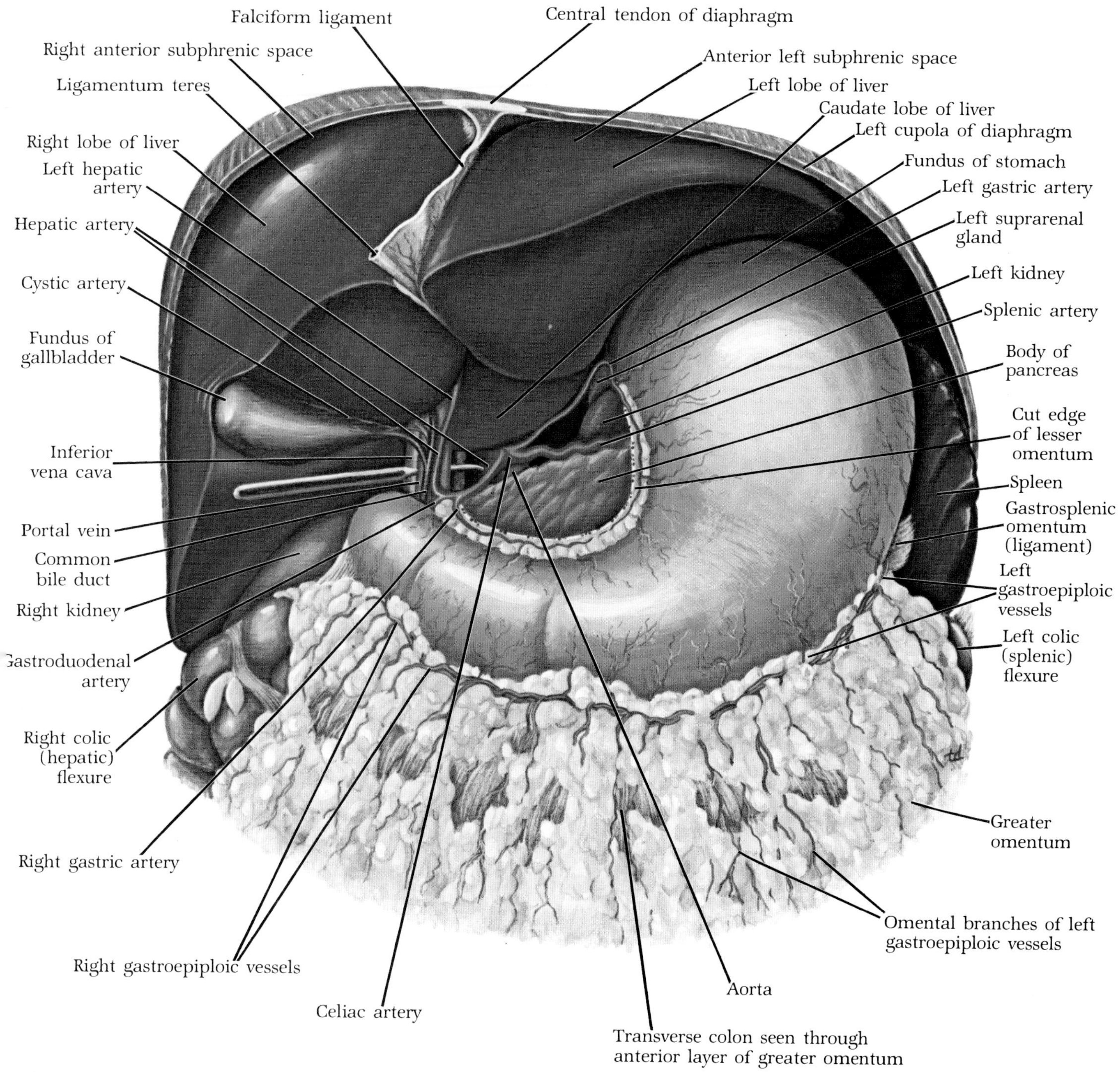
Falciform ligament
Central tendon of diaphragm
Right anterior subphrenic space
Anterior left subphrenic space
Ligamentum teres
Left lobe of liver
Caudate lobe of liver
Left cupola of diaphragm
Right lobe of liver
Left hepatic artery
Fundus of stomach
Left gastric artery
Hepatic artery
Left suprarenal gland
Cystic artery
Left kidney
Splenic artery
Fundus of gallbladder
Body of pancreas
Cut edge of lesser omentum
Inferior vena cava
Spleen
Gastrosplenic omentum (ligament)
Portal vein
Common bile duct
Left gastroepiploic vessels
Right kidney
Left colic (splenic) flexure
Gastroduodenal artery
Right colic (hepatic) flexure
Greater omentum
Right gastric artery
Omental branches of left gastroepiploic vessels
Right gastroepiploic vessels
Aorta
Celiac artery
Transverse colon seen through anterior layer of greater omentum

Figure 2-21
Contents of the lesser omentum and the entrance into the lesser sac, anterior view. The peritoneum of the lesser omentum has been largely removed.

1. The description of the gallbladder and the biliary passages and their arteries in most textbooks applies to only about one-third of patients. A number of congenital anomalies occur in the gallbladder and biliary system and there are numerous variations in the blood supply. The medical student should be aware of such anomalies and accompanying variations, though the details need not be committed to memory.

2. Because of the close proximity of the gallbladder to the duodenum and transverse colon, gallstones have been known to ulcerate through the gallbladder wall into these structures. If they ulcerate into the duodenum they may be held up at the ileocolic junction, producing intestinal obstruction; if into the transverse colon they are passed naturally, per rectum.

3. Cholecystitis may cause irritation of the subdiaphragmatic parietal peritoneum, which is supplied in part by the phrenic nerve (C3, 4, and 5). This in turn may give rise to referred pain over the shoulder since the skin in this area is supplied by the supraclavicular nerves (C3 and 4).

4. The upper part of the common bile duct lies within the free edge of the lesser omentum, to the right of the hepatic artery, and anterior to the portal vein. It is easily accessible to the surgeon.

5. Portacaval shunts for the treatment of portal hypertension may involve the anastomosis of the portal vein, as it lies within the lesser omentum, to the anterior wall of the inferior vena cava behind the entrance into the lesser sac.

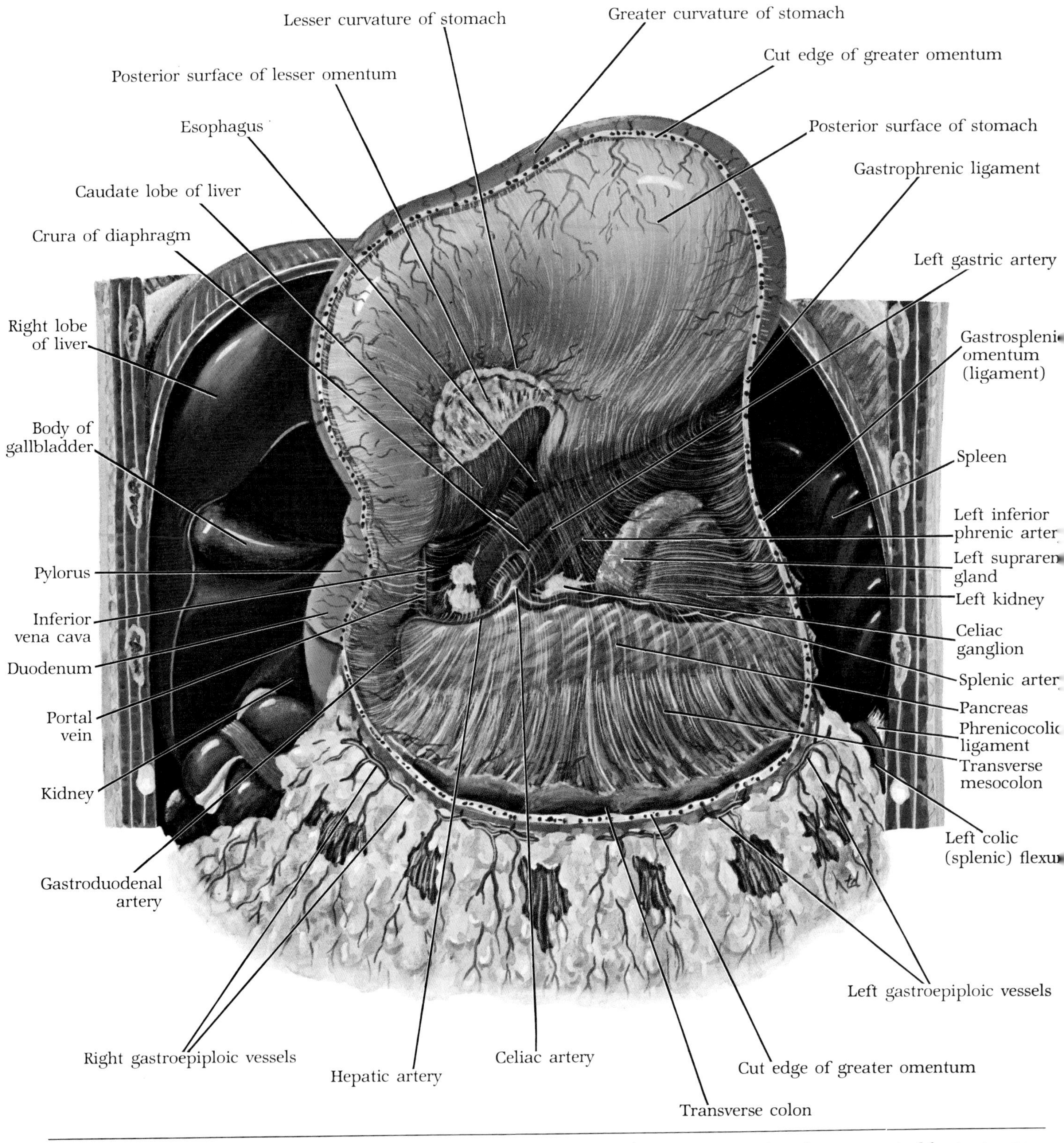
Lesser curvature of stomach
Greater curvature of stomach
Cut edge of greater omentum
Posterior surface of lesser omentum
Esophagus
Posterior surface of stomach
Gastrophrenic ligament
Caudate lobe of liver
Crura of diaphragm
Left gastric artery
Right lobe of liver
Gastrosplenic omentum (ligament)
Body of gallbladder
Spleen
Left inferior phrenic artery
Pylorus
Left suprarenal gland
Left kidney
Inferior vena cava
Celiac ganglion
Duodenum
Splenic artery
Pancreas
Portal vein
Phrenicocolic ligament
Transverse mesocolon
Kidney
Left colic (splenic) flexure
Gastroduodenal artery
Left gastroepiploic vessels
Right gastroepiploic vessels
Hepatic artery
Celiac artery
Cut edge of greater omentum
Transverse colon

Figure 2-22
Stomach bed. The greater omentum has been incised along the greater curvature of the stomach to open the lesser sac, and the stomach has been pulled upward.

1. A peptic ulcer situated on the posterior wall of the stomach may perforate into the lesser sac or become adherent to the pancreas. Erosion of the pancreas produces pain referred to the back. The splenic artery runs along the superior border of the pancreas and erosion of this artery may produce a fatal hemorrhage.
2. A peptic ulcer on the posterior wall of the first part of the duodenum may erode the relatively large gastroduodenal artery, causing a very severe hemorrhage.
3. A perforating peptic ulcer of the anterior wall of the first part of the duodenum may open into the upper part of the greater sac above the transverse colon. The transverse colon then directs the escaping gastric contents into the right lateral paracolic gutter and thence down into the right iliac fossa.
4. Pain in the stomach is caused by the stretching or spasmodic contraction of the smooth muscle in the stomach walls and is referred to the epigastrium.

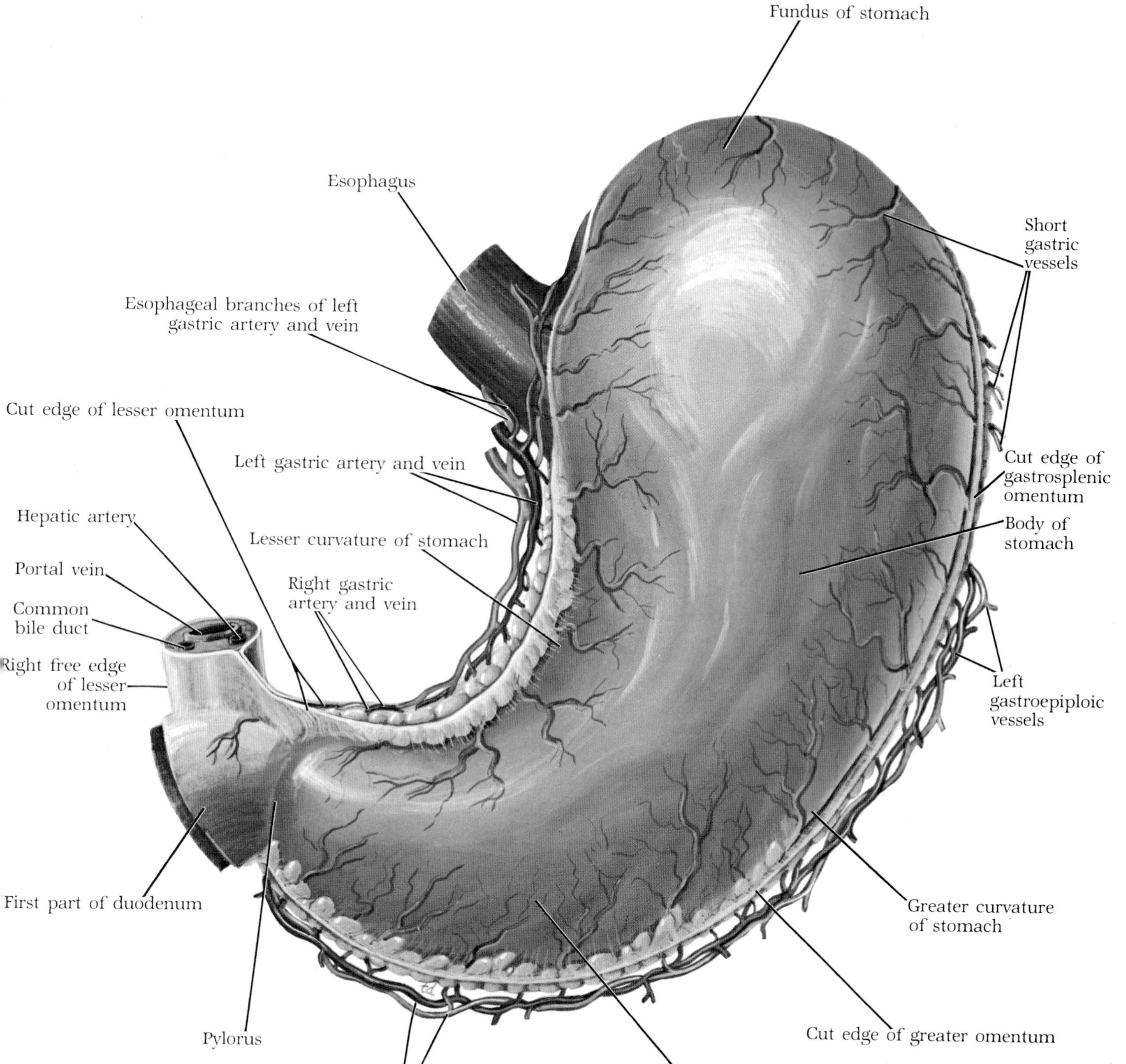
Fundus of stomach
Esophagus
Short gastric vessels
Esophageal branches of left gastric artery and vein
Cut edge of lesser omentum
Left gastric artery and vein
Cut edge of gastrosplenic omentum
Hepatic artery
Body of stomach
Lesser curvature of stomach
Portal vein
Right gastric artery and vein
Common bile duct
Right free edge of lesser omentum
Left gastroepiploic vessels
First part of duodenum
Greater curvature of stomach
Pylorus
Cut edge of greater omentum
Antrum of stomach
Right gastroepiploic vessels

Figure 2-23
Stomach, showing its profuse blood supply, anterior view.

1. The stomach may be examined by palpation of the epigastric region of the anterior abdominal wall, by radiology following a barium meal, and by gastroscopy. Gastric secretions can be analyzed biochemically by examining a sample of gastric contents obtained through a swallowed nasal tube.

2. The stomach is relatively fixed at both ends but very mobile in between. It tends to be high and transversely arranged in the short, obese person and elongated vertically in the tall, thin person. Considerable variations in size and position occur with alteration in the volume of contents, position of the body, and phase of respiration.

3. At the lower third of the esophagus, the esophageal branches of the left gastric vein (portal tributary) anastomose with the esophageal veins draining the middle third of the esophagus into the azygos veins (systemic tributary). Hypertension in the portal vein due to liver disease or blockage of the extrahepatic part of the portal vein leads to congestion of collateral pathways, which may reveal itself by severe bleeding from esophagogastric varices.

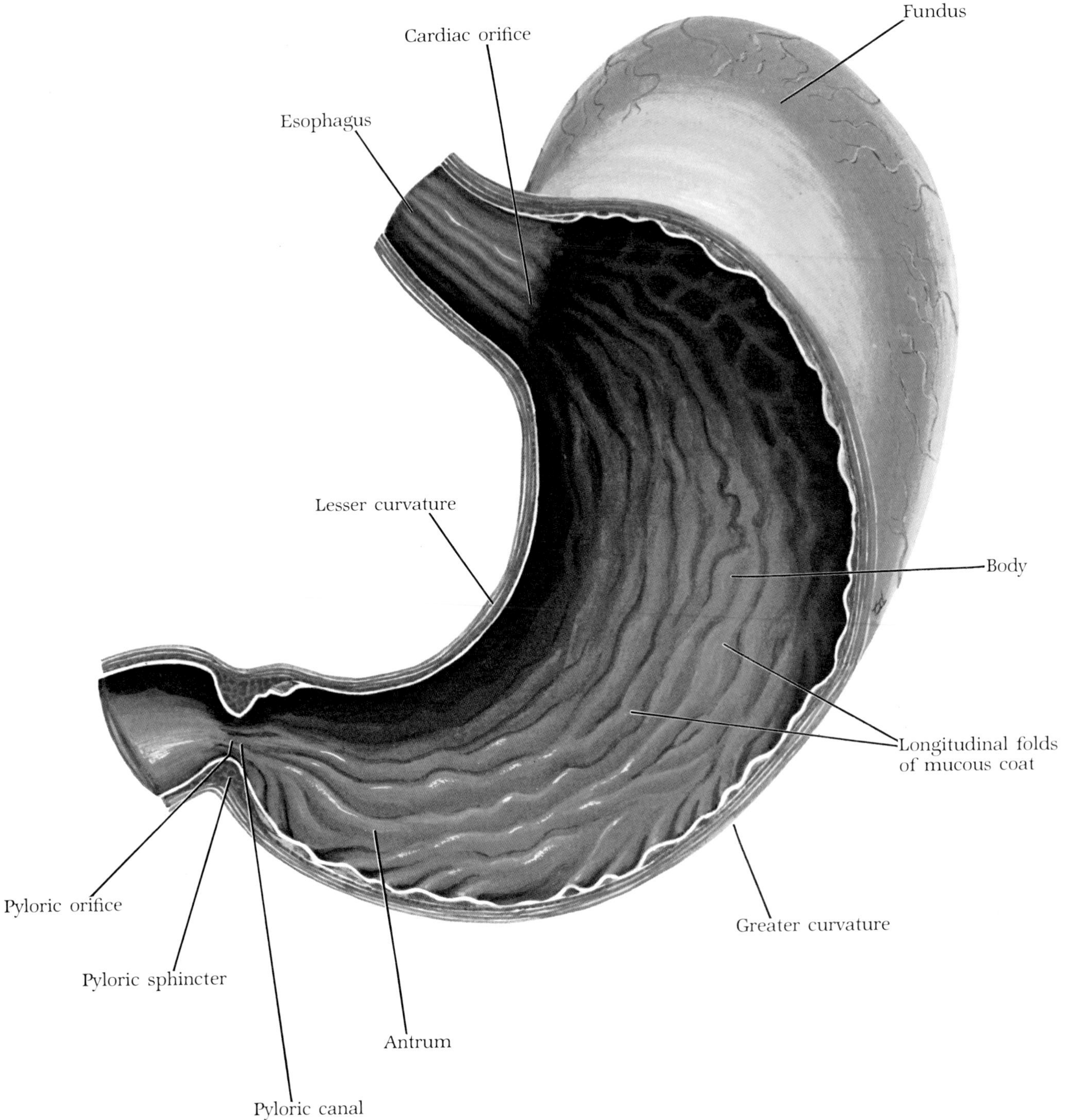
Fundus
Cardiac orifice
Esophagus
Lesser curvature
Body
Longitudinal folds of mucous coat
Pyloric orifice
Greater curvature
Pyloric sphincter
Antrum
Pyloric canal

Figure 2-24
Stomach, showing the muscular coats and mucosal lining.

1. There is no anatomical sphincter at the cardiac orifice of the stomach. The following have been suggested as the mechanisms that prevent regurgitation of stomach contents into the esophagus. The looping fibers of the right crus of the diaphragm which encircle the esophagus may compress it. The circular muscle of the lower end of the esophagus may contract independently of the remaining muscle and thus serve as a physiological sphincter. Achalasia of the cardia is a clinical condition, producing symptoms of obstruction to swallowing; it is thought to be due to a degeneration or absence of the nerve cells in Auerbach's plexus in the esophagus.
2. The pyloric sphincter is a thickening of the circular muscle coat of the stomach. Congenital hypertrophic pyloric stenosis is a condition in which there is an increase in the number and size of the smooth muscle fibers of the pylorus, producing obstruction to the distal portion of the stomach. The treatment is a longitudinal incision through the hypertrophied muscle to permit herniation of the submucosa and mucosa through the incision.
3. Gastroscopy, using modern fiberoptic instrumentation, permits direct visualization of the gastric mucosa, brushing of the mucosa cells, or performance of a mucosal biopsy.
4. Chronic gastric ulcers usually occur on or close to the lower part of the lesser curvature of the stomach. The ulcer invades the muscular and peritoneal coats in such a way that the stomach aheres to neighboring structures.

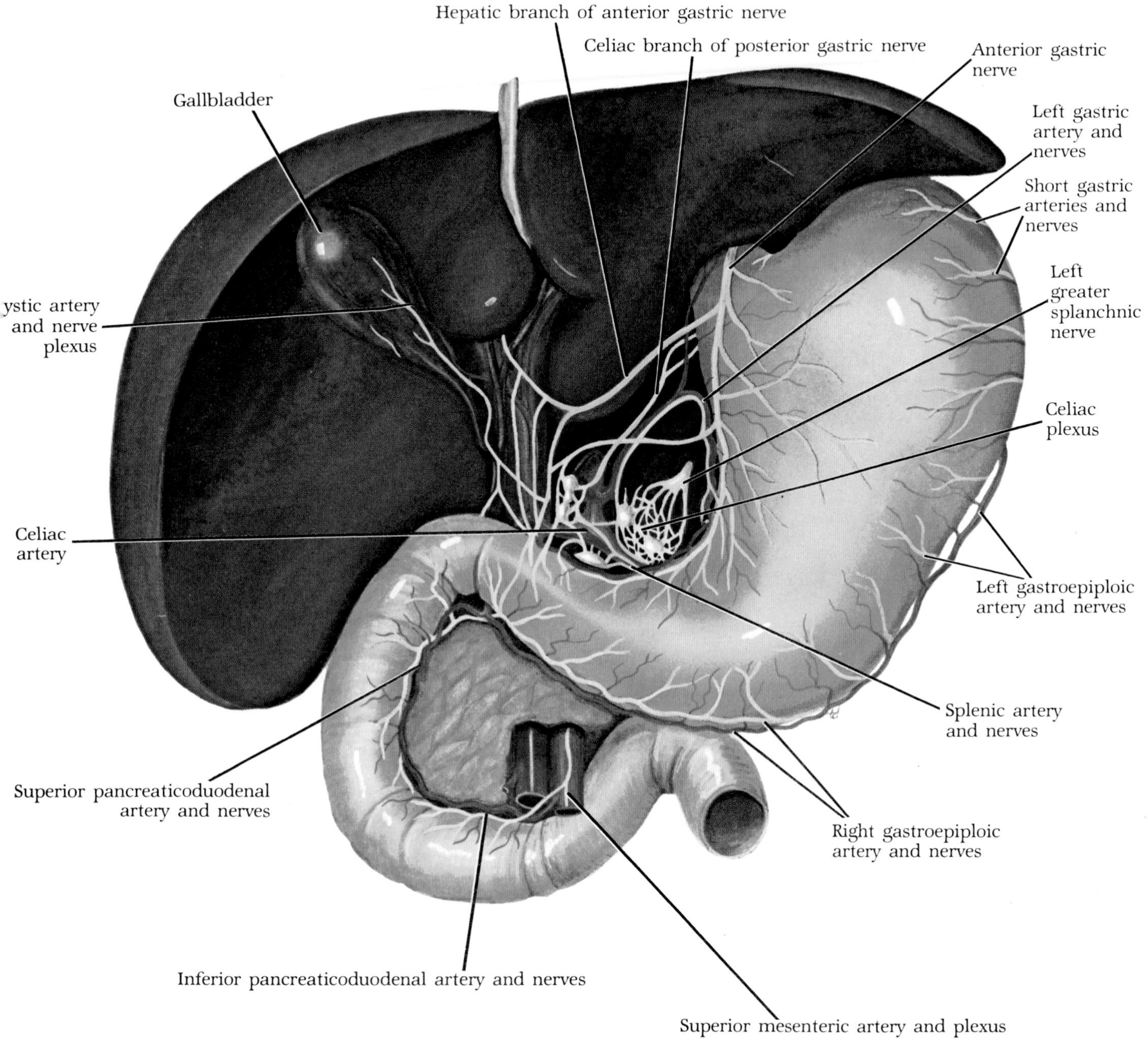
Hepatic branch of anterior gastric nerve
Celiac branch of posterior gastric nerve
Anterior gastric nerve
Gallbladder
Left gastric artery and nerves
Short gastric arteries and nerves
ystic artery and nerve plexus
Left greater splanchnic nerve
Celiac plexus
Celiac artery
Left gastroepiploic artery and nerves
Splenic artery and nerves
Superior pancreaticoduodenal artery and nerves
Right gastroepiploic artery and nerves
Inferior pancreaticoduodenal artery and nerves
Superior mesenteric artery and plexus

Figure 2-25
Innervation of the stomach and duodenum, anterior view.

1. The sensation of pain in the stomach is caused by the stretching or spasmodic contraction of the smooth muscle in its walls and is referred to the epigastrium. The pain-transmitting fibers leave the stomach with the sympathetic nerves, pass through the celiac ganglia, and reach the spinal cord by way of the greater splanchnic nerves. The dermatomes involved on the anterior abdominal wall are those of the seventh, eighth, and ninth thoracic nerves. Pain arising from the stomach, like that from all abdominal viscera, is dull, poorly localized, and referred to the midline.

2. Gastric peristaltic movement depends in part upon a reflex arc. The sensory receptors are stimulated by distention and the afferent impulses pass up in the vagus nerves, which also carry the efferent nerves fibers.

3. Gastric secretion depends on a vagal-antral phase and an intestinal phase. The thought, sight, smell, and taste of food give rise to nervous impulses that are mediated to the stomach by the vagus nerves. The gastric secretion is stimulated by the direct effect of the vagus nerves and indirectly by the vagally induced release of the hormone gastrin by the antrum of the stomach. The direct vagal stimulation produces a juice rich in pepsin; gastrin stimulation produces a juice rich in hydrochloric acid. Between meals the stomach continues to produce secretion as the result of the presence of food in the intestine, although this mechanism is not fully understood.

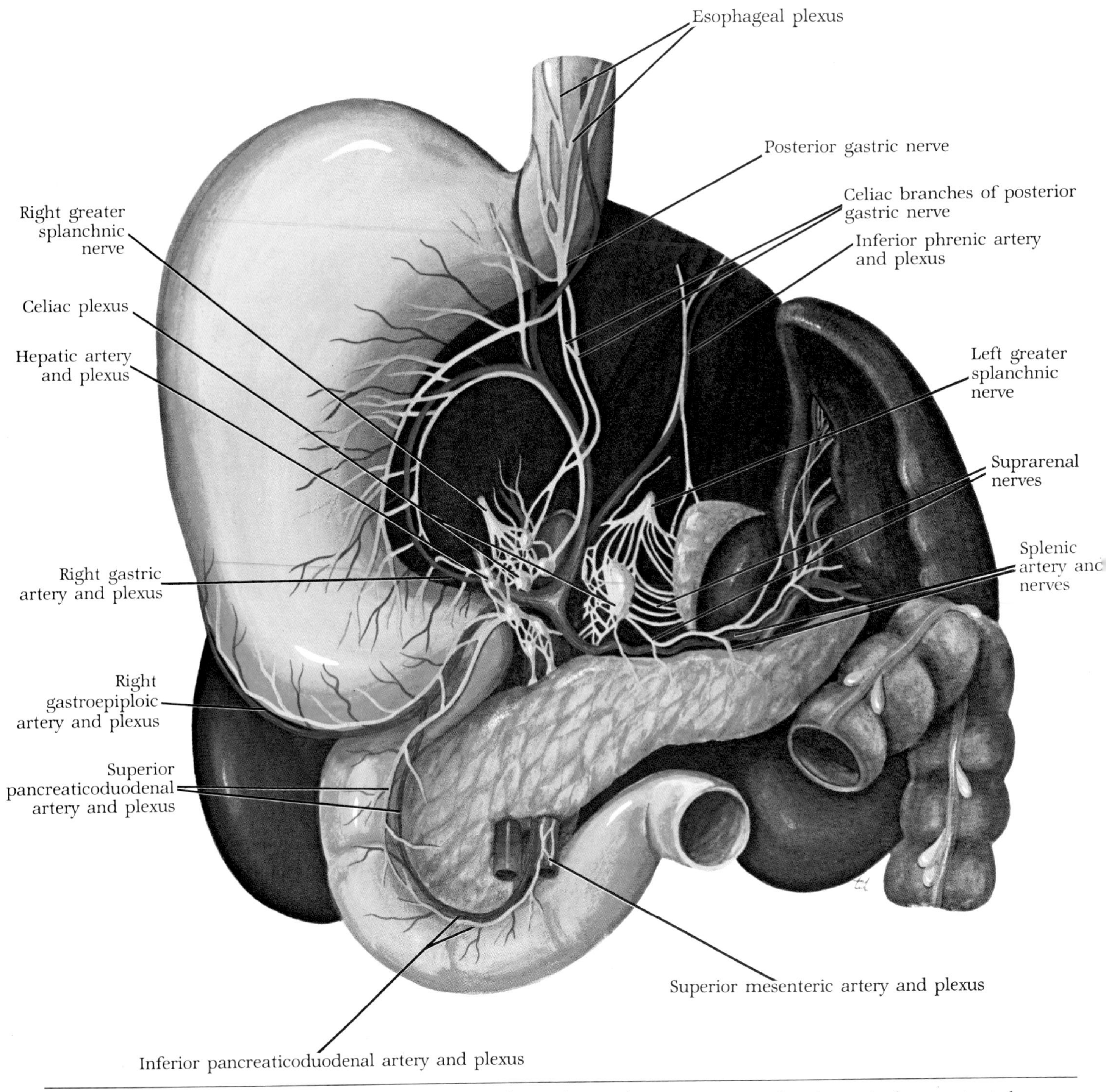
Esophageal plexus
Posterior gastric nerve
Celiac branches of posterior gastric nerve
Inferior phrenic artery and plexus
Right greater splanchnic nerve
Celiac plexus
Hepatic artery and plexus
Left greater splanchnic nerve
Suprarenal nerves
Splenic artery and nerves
Right gastric artery and plexus
Right gastroepiploic artery and plexus
Superior pancreaticoduodenal artery and plexus
Superior mesenteric artery and plexus
Inferior pancreaticoduodenal artery and plexus

Figure 2-26
Innervation of the stomach and duodenum, posterior view.

1. The surgical treatment of duodenal ulcers is directed toward reducing the gastric acidity. Subtotal gastric resection removes 75 percent of the distal part of the stomach, thus effectively removing the gastrin-producing area but leaving the acid-producing parietal cells in the body and fundus that are still innervated by the vagus nerves. Vagotomy, the operation in which the anterior and posterior gastric nerves are sectioned just beneath the diaphragm, removes direct vagal stimulation of the parietal cells and the vagally induced release of gastrin. Unfortunately this operation inhibits antral motility and the normal opening of the pyloric sphincter so that gastric stasis and distention occur. Vagotomy coupled with pyloroplasty, antrectomy, or gastroenterostomy removes the main stimulatory mechanism of gastric secretion and also solves the problem of gastric stasis that occurs following vagotomy alone. Because the operative risk is small, this form of surgical treatment of chronic peptic ulcer is popular.

2. Chronic gastric ulcers are treated surgically by partial gastrectomy and, in patients with a high rate of acid gastric secretion, by vagotomy with pyloroplasty or antrectomy.

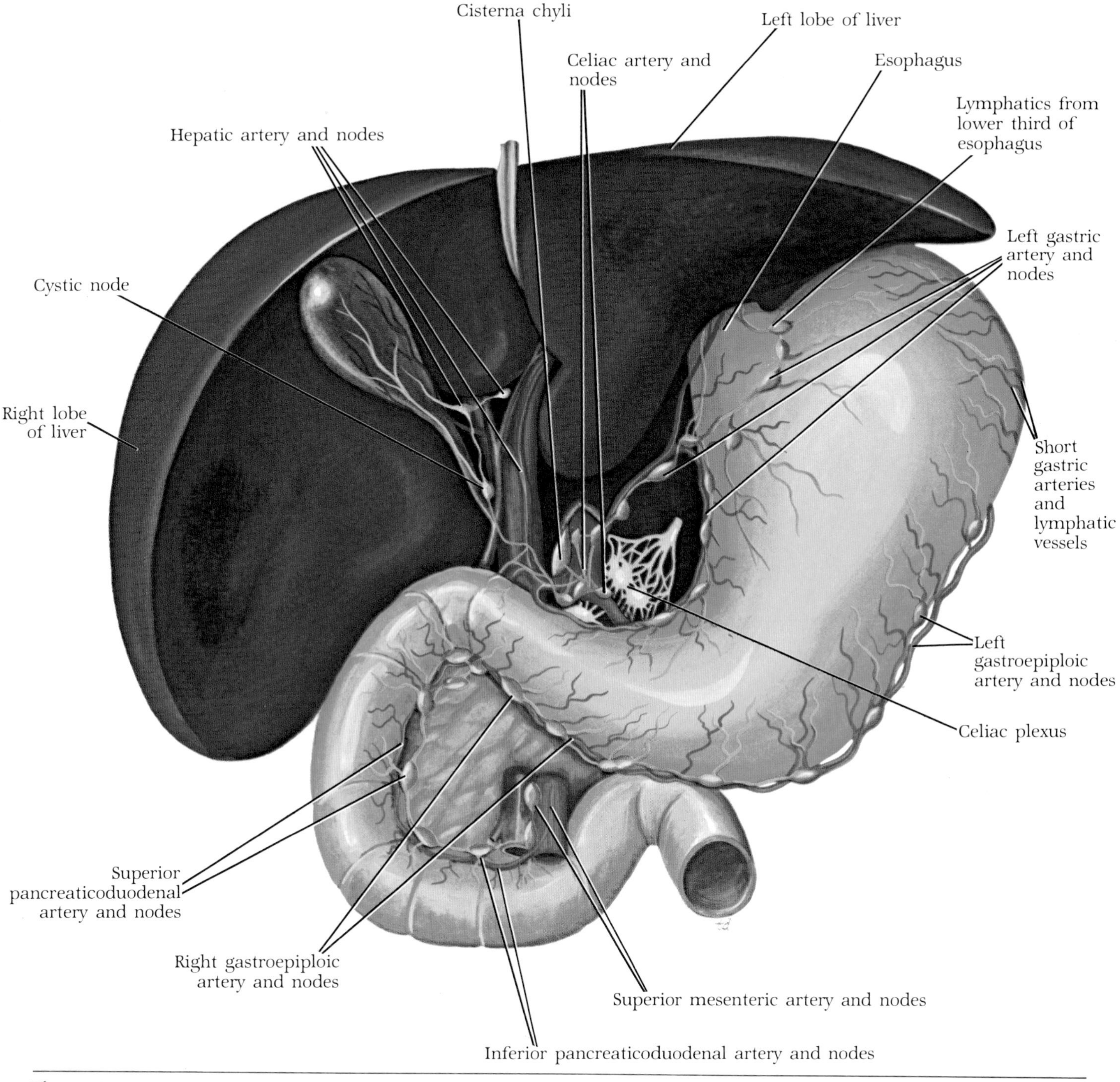
Cisterna chyli
Left lobe of liver
Celiac artery and nodes
Esophagus
Lymphatics from lower third of esophagus
Hepatic artery and nodes
Left gastric artery and nodes
Cystic node
Right lobe of liver
Short gastric arteries and lymphatic vessels
Left gastroepiploic artery and nodes
Celiac plexus
Superior pancreaticoduodenal artery and nodes
Right gastroepiploic artery and nodes
Superior mesenteric artery and nodes
Inferior pancreaticoduodenal artery and nodes

Figure 2-27
Lymphatic drainage of the stomach and duodenum.

1. The gastric mucosal lymph vessels are in continuity, and it is thus possible for cancer cells to travel to different parts of the stomach via this route. These cells often pass through local lymph nodes but are held up in the regional nodes. For this reason malignant disease of the stomach is treated by total gastrectomy, which includes the removal of: (a) the distal end of the esophagus and the first part of the duodenum; (b) the spleen and the gastrosplenic and lienorenal ligaments and their associated lymph nodes; (c) the splenic vessels; (d) the tail and body of the pancreas and their associated nodes; (e) the nodes along the lesser curvature of the stomach; and (f) the nodes along the greater curvature, along with the greater omentum. The continuity of the gut is restored by anastomosing the esophagus with the jejunum. Some surgeons prefer a less radical procedure and perform a subtotal gastrectomy, removing the structures listed above but leaving the pancreas in situ.

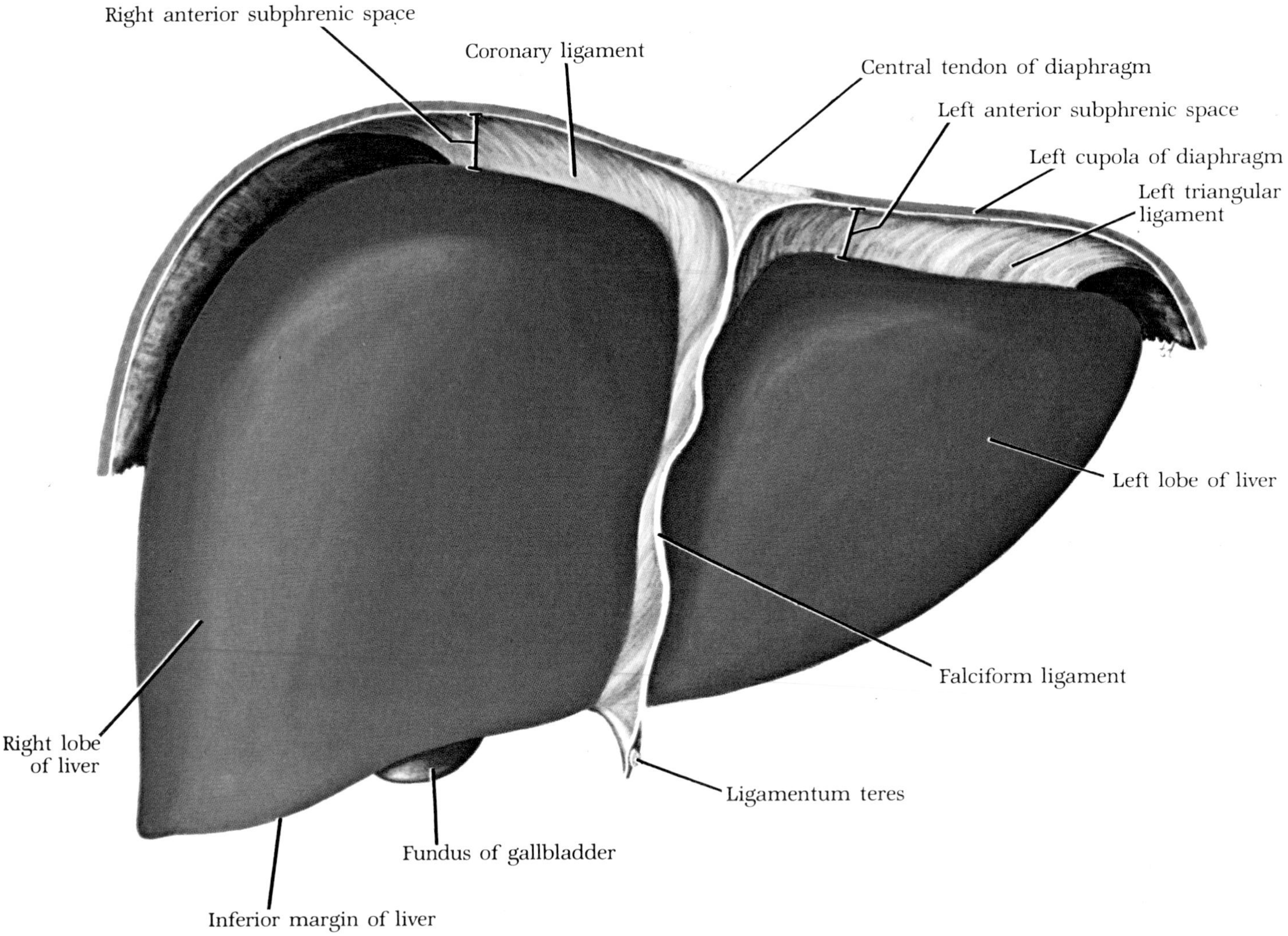
Right anterior subphrenic space
Coronary ligament
Central tendon of diaphragm
Left anterior subphrenic space
Left cupola of diaphragm
Left triangular ligament
Left lobe of liver
Falciform ligament
Ligamentum teres
Right lobe of liver
Fundus of gallbladder
Inferior margin of liver

Figure 2-28
Liver, anterior view.

1. The liver is a segmental organ with definite vascular and biliary cleavage planes. The precise modern anatomical division into lobes differs from the classic description shown in the diagram. The true dividing line of the liver into right and left lobes is to the right of the falciform ligament in line with the inferior vena cava and the gallbladder fossa. Furthermore, a right segmental fissure divides the right lobe into anterior and posterior segments, while the falciform ligament divides the left lobe into medial and lateral segments.

2. Because anatomical research has shown that the bile ducts, hepatic arteries, and portal vein are distributed in a segmental manner, appropriate ligation of these structures allows the surgeon to remove large portions of the liver in patients with severe traumatic lacerations of the liver or with a liver tumor, for example. Even large, localized carcinomatous metastatic tumors have been successfully removed.

3. The majority of amebic abscesses in the liver are solitary and are located in the upper part of the right lobe. Diaphragmatic irritation may cause referred pain over the shoulder, the nervous impulses having ascended in the phrenic nerve (C3, 4, and 5); the supraclavicular nerves (C3 and 4) supply the skin over the shoulder.

4. Infected peritoneal fluid may localize in one of the subphrenic spaces to form an abscess. The subphrenic region lies between the diaphragm above and transverse colon below; the liver divides it into a number of clinically important peritoneal spaces. The important relationship of these spaces to the pleura must be emphasized.

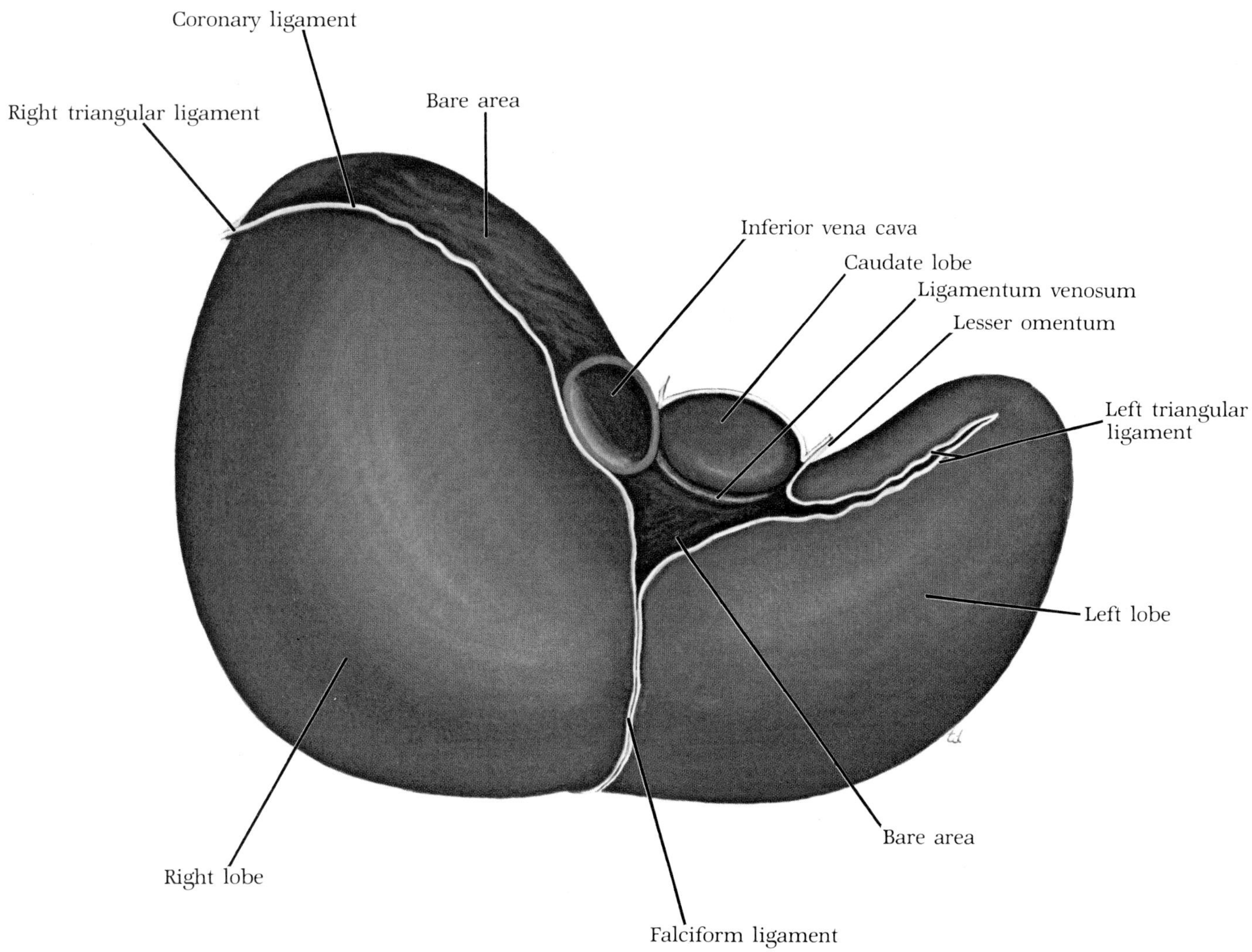
Coronary ligament
Right triangular ligament
Bare area
Inferior vena cava
Caudate lobe
Ligamentum venosum
Lesser omentum
Left triangular ligament
Left lobe
Bare area
Right lobe
Falciform ligament

Figure 2-29
Liver, superior view.

1. The liver is held in position in the upper part of the abdominal cavity by the attachment of the hepatic veins to the inferior vena cava. The peritoneal ligaments play only a minor role in its support. This fact is important to the surgeon because even if the peritoneal ligaments are cut, the liver can be only slightly rotated.

2. The afferent blood supply to the liver is provided by: (a) the hepatic artery that carries oxygenated blood, contributing about 30 percent of the hepatic blood flow; and (b) the portal vein, which contributes about 70 percent. When there is severe trauma of the liver region and the hepatic artery is torn, it is possible to ligate the hepatic artery provided the patient has a normally functioning liver. Tearing of the portal vein is usually fatal although ligation or repair has resulted in survival of the patient.

3. Evaluation of the position, size, and shape of the liver can be determined by radiographic examination of the thorax and abdomen, noting deformities in the domes of the diaphragm, scintillation scanning, and angiography.

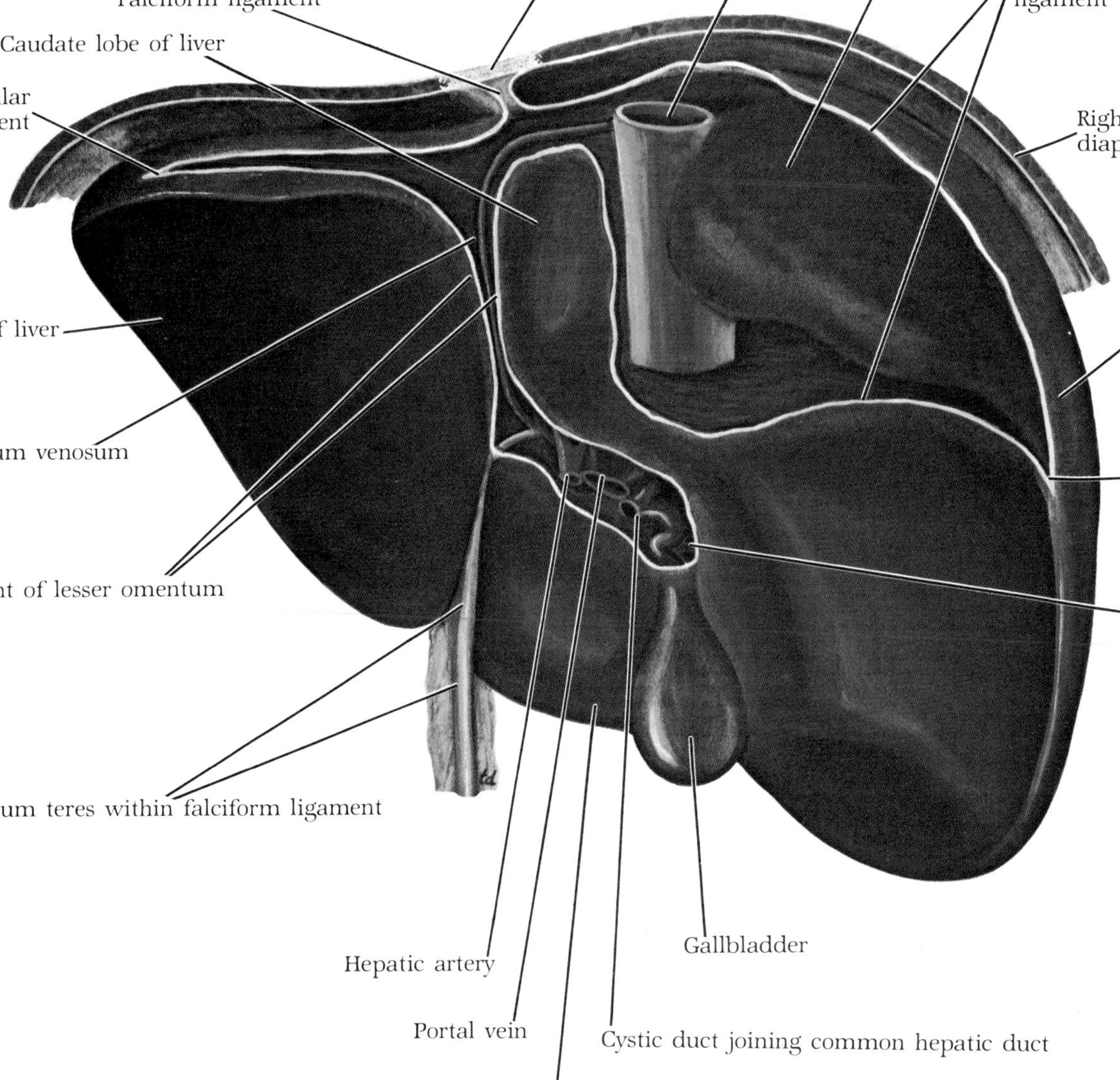
Central tendon of diaphragm
Inferior vena cava
Bare area
Superior and inferior layers of coronary ligament
Falciform ligament
Caudate lobe of liver
Left triangular ligament
Right cupola of diaphragm
Left lobe of liver
Right lobe of liver
Ligamentum venosum
Right triangular ligament
Attachment of lesser omentum
Porta hepatis
Ligamentum teres within falciform ligament
Gallbladder
Hepatic artery
Portal vein
Cystic duct joining common hepatic duct
Quadrate lobe

Figure 2-30
Liver, posterior view.

1. As emphasized previously the peritoneal ligaments provide only minor support for the liver, the main support being from the attachment of the hepatic veins to the inferior vena cava. The peritoneal ligaments are important, however, in that they subdivide the region of the abdominal cavity below the diaphragm into subphrenic spaces. A perforated part of the gut—for example, a perforated appendix or a peptic ulcer—may result in infected fluid collecting in one of the subphrenic spaces and forming a subphrenic abscess.
2. The wide angle of union of the splenic vein with the superior mesenteric veins to form the portal vein leads to the streaming of blood in the portal vein. The right lobe of the liver receives blood mainly from the intestine, whereas the left lobe, quadrate lobe, and caudate lobe receive blood from the stomach and spleen. This blood distribution may explain the distribution of secondary malignant deposits in the liver.
3. Spasm of the smooth muscle of the wall of the gallbladder in an attempt to expel a gallstone gives rise to referred pain in the epigastrium.
4. Cholecystitis or inflammation of the gallbladder may cause irritation of the subdiaphragmatic parietal peritoneum, which is supplied in part by the phrenic nerve (C3, 4, and 5). This may in turn give rise to referred pain over the shoulder because the skin in this area is supplied by the supraclavicular nerves (C3 and 4).

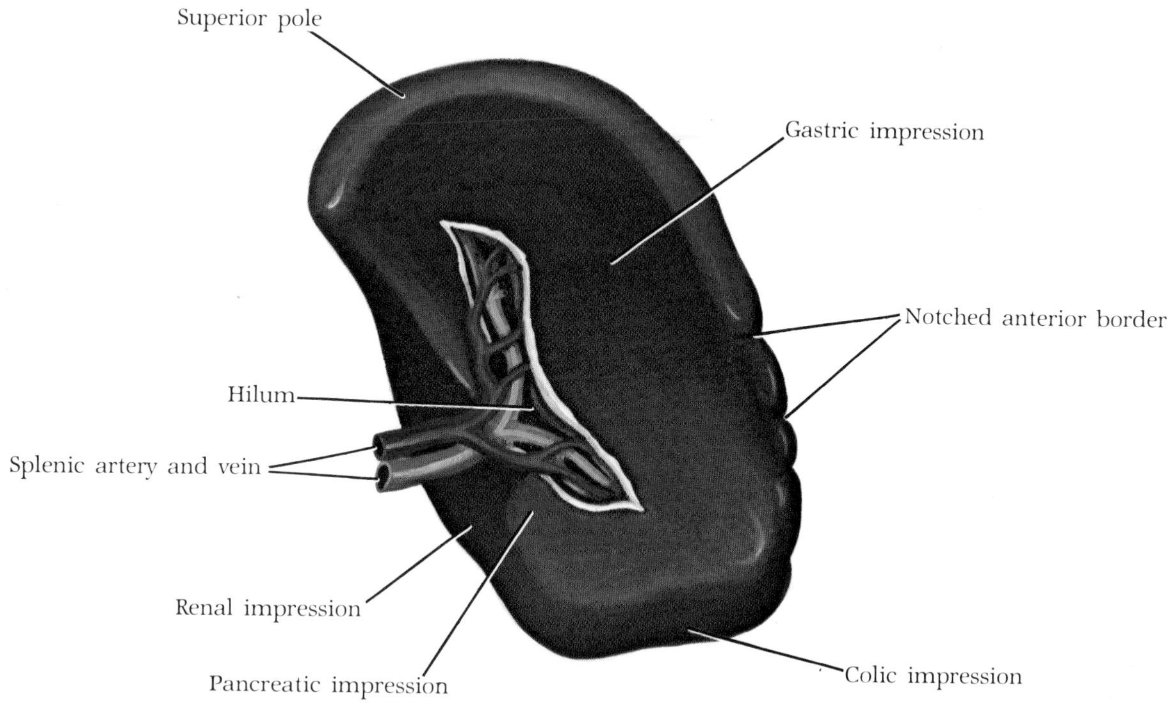

Superior pole
Gastric impression
Notched anterior border
Hilum
Splenic artery and vein
Renal impression
Pancreatic impression
Colic impression

A

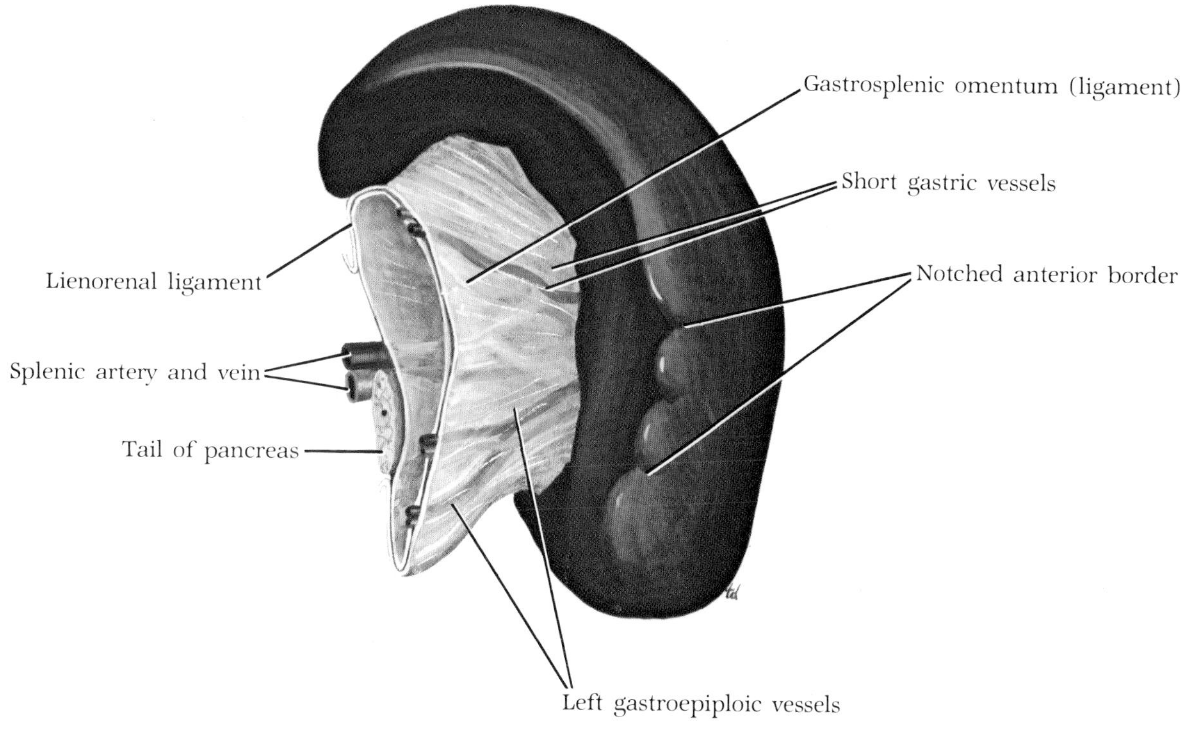

Gastrosplenic omentum (ligament)
Short gastric vessels
Notched anterior border
Lienorenal ligament
Splenic artery and vein
Tail of pancreas
Left gastroepiploic vessels

B

Figure 2-31
Spleen. A. Anterior view. B. With peritoneal attachments, anterior view.

1. Normally the spleen cannot be palpated on abdominal examination. It lies in the left hypochondrium and is related to the ninth, tenth, and eleventh ribs. It does not project forward in front of the midaxillary line. In healthy individuals percussion of the area does not produce significant dullness. As the spleen enlarges, however, dullness may be detected along the line of the ninth rib, and a radiograph of the abdomen may reveal an increase in size.
2. As a diseased spleen enlarges, it extends downward and medially. Once it escapes from below the left costal margin, its notched anterior border can be recognized by palpation through the anterior abdominal wall.
3. Accessory spleens are common (occurring in 10 to 15 percent of the population). These accessory organs usually receive their blood supply from the splenic artery and are found at the hilum of the spleen, the gastrosplenic omentum, the lienorenal ligament, the greater omentum, and, on occasion, in the pelvis. In hematophilic disorders in which splenectomy is therapeutic, the accessory spleens must also be removed.
4. The spleen is the most common organ to be injured from blunt trauma to the abdomen. Injury during automobile accidents and while playing contact sports are the most common occasions for such trauma. Penetrating wounds of the lower left thorax may also damage the spleen.
5. The tail of the pancreas lies in the lienorenal ligament and is in contact with the hilus of the spleen. The pancreas can be damaged during splenectomy.

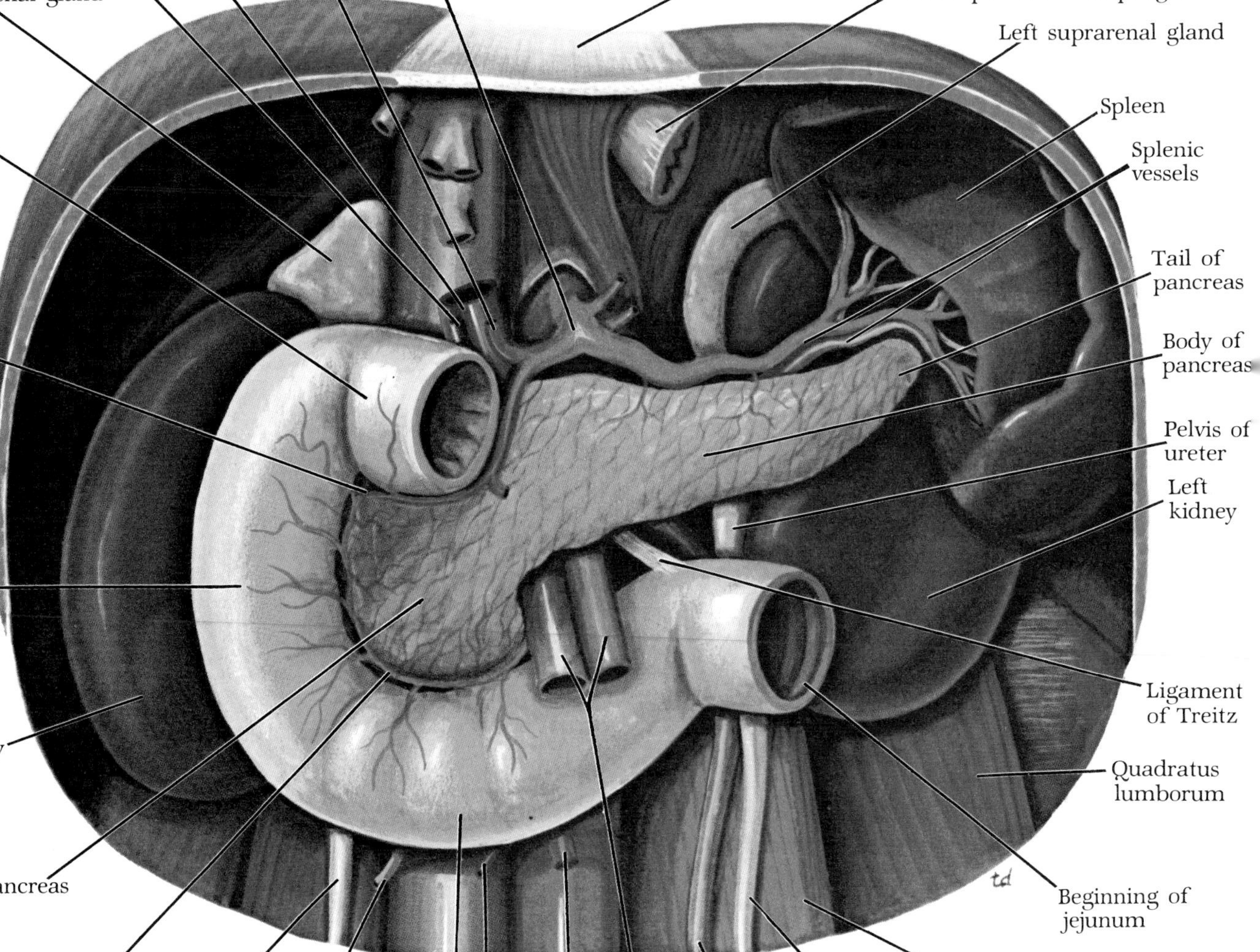
Hepatic artery
Celiac artery
Portal vein
Common bile duct
Central tendon of diaphragm
Abdominal portion of esophagus
Right suprarenal gland
Left suprarenal gland
Spleen
First part of duodenum
Splenic vessels
Tail of pancreas
Superior pancreatico-duodenal artery
Body of pancreas
Pelvis of ureter
Left kidney
Second part of duodenum
Ligament of Treitz
Right kidney
Quadratus lumborum
Head of pancreas
Beginning of jejunum
Psoas
Inferior pancreaticoduodenal artery
Ureter
Right ureter
Inferior mesenteric vein
Superior mesenteric vessels
Right testicular (ovarian) vein
Inferior mesenteric artery
Inferior vena cava
Right testicular (ovarian) artery
Third part of duodenum

Figure 2-32
Duodenum and pancreas, Anterior view.

1. The first part of the duodenum is fairly mobile, the second and third parts are fixed and retroperitoneal, and the fourth part joins the fixed duodenum to the very mobile jejunum. The duodenojejunal junction is suspended by the ligament of Treitz. Injury to the duodenum as the result of blunt injury to the abdomen, such as from an automobile steering wheel, is commonly retroperitoneal.

2. The first part of the duodenum is a common site for peptic ulcer. It is also the first part to receive the gastric acid chyme, which is believed to be an important factor in the production of a duodenal ulcer at this site.

3. A peptic ulcer of the anterior wall of the duodenum may penetrate into the liver or more commonly perforate into the upper part of the greater sac. An ulcer on the posterior wall of the first part of the duodenum may erode the gastroduodenal artery, causing a very severe hemorrhage.

4. Radiographic examination of the duodenum following a barium meal is the most commonly used technique for diagnosing a peptic ulcer. In the first part of the normal duodenum the barium forms a triangular homogenous shadow, the duodenal cap. In the second, third, and fourth parts the circular mucosal folds break up the barium emulsion, giving it a floccular appearance.

5. Because of the close relationship of the right colic flexure and the right kidney to the duodenum, great care has to be exercised during right hemicolectomy and right nephrectomy to avoid damage to the duodenum.

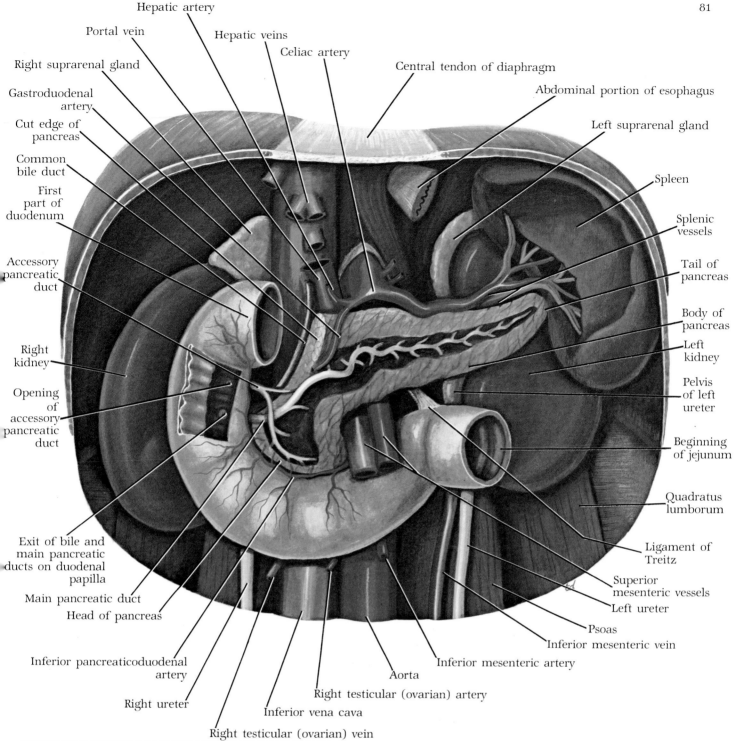

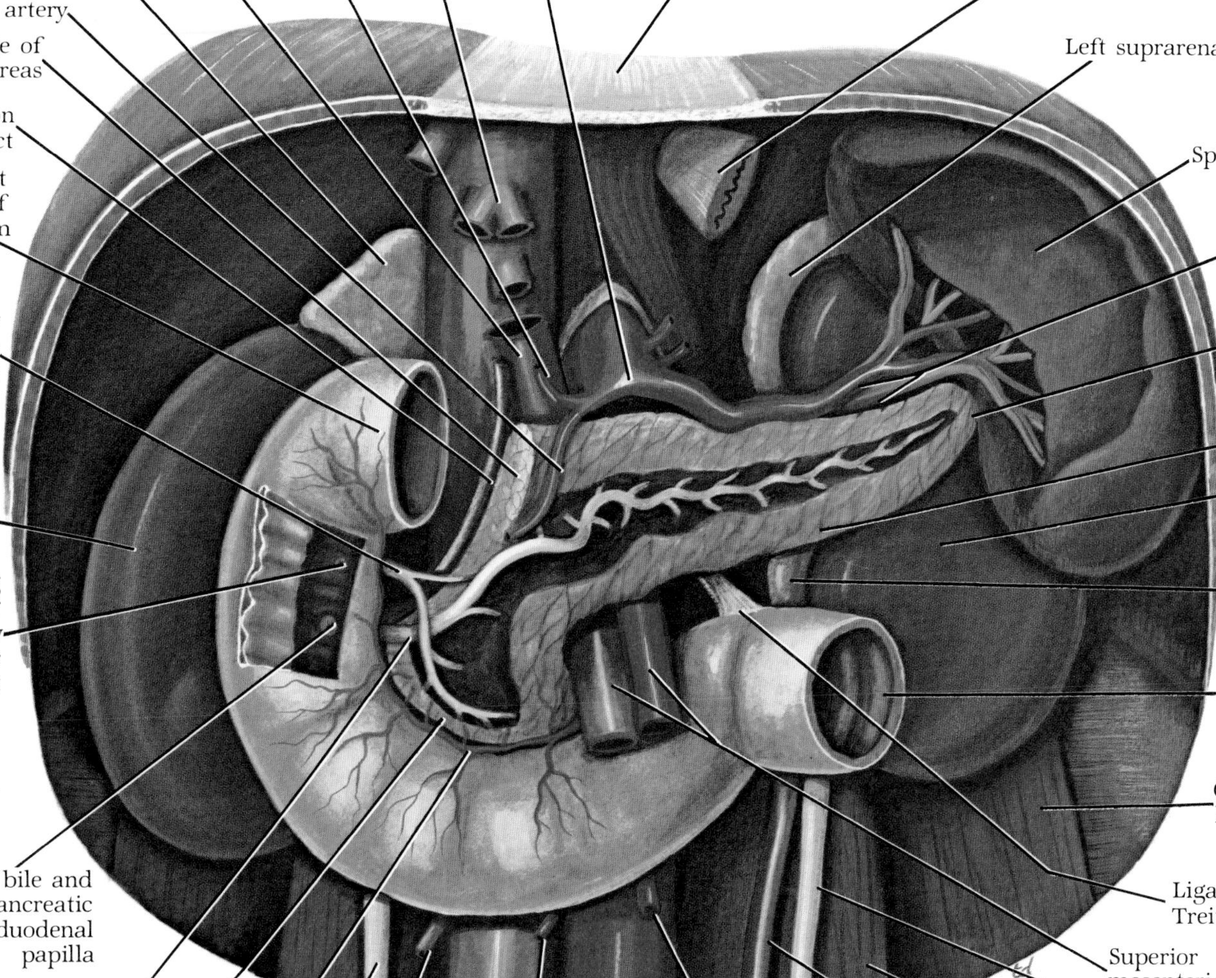
Hepatic artery
Portal vein
Hepatic veins
Celiac artery
Right suprarenal gland
Central tendon of diaphragm
Gastroduodenal artery
Abdominal portion of esophagus
Cut edge of pancreas
Left suprarenal gland
Common bile duct
Spleen
First part of duodenum
Splenic vessels
Accessory pancreatic duct
Tail of pancreas
Body of pancreas
Right kidney
Left kidney
Opening of accessory pancreatic duct
Pelvis of left ureter
Beginning of jejunum
Quadratus lumborum
Exit of bile and main pancreatic ducts on duodenal papilla
Ligament of Treitz
Main pancreatic duct
Superior mesenteric vessels
Head of pancreas
Left ureter
Psoas
Inferior mesenteric vein
Inferior pancreaticoduodenal artery
Inferior mesenteric artery
Aorta
Right testicular (ovarian) artery
Right ureter
Inferior vena cava
Right testicular (ovarian) vein

Figure 2-33
Duodenum and pancreas, anterior view. The anterior wall of the second part of the duodenum has been incised and reflected, revealing the duodenal papillae; the common bile duct and the pancreatic ducts have been dissected out.

1. The close relationship of the pancreas to the stomach may result in this organ's becoming involved in a deep penetrating ulcer of the stomach. In this situation the pain tends to be referred to the back.

2. Inflammation of or damage to the pancreas may result in effusion of peritoneal fluid into the lesser sac. Pseudocysts of the pancreas are due to cystic accumulations of fluid in the lesser sac.

3. Carcinoma of the head of the pancreas often causes obstructive jaundice by pressing on the common bile duct within its substance.

4. The close relationship of the tail of the pancreas to the hilum of the spleen sometimes results in damage to the pancreas during splenectomy.

5. Annular pancreas is a rare congenital anomaly in which the ventral bud of the pancreas becomes tethered to the posterior abdominal wall and fails to fuse correctly with the dorsal bud. This condition may cause duodenal obstruction.

6. The immobility of the pancreas on the posterior abdominal wall makes it vulnerable to blunt trauma. It commonly breaks across the rigid vertebral column.

7. Gallstones in the common bile duct are usually removed through a longitudinal incision in the supraduodenal part of the duct. If a stone is impacted at the lower end of the duct or in the ampulla of Vater, the anterior wall of the duodenum is incised.

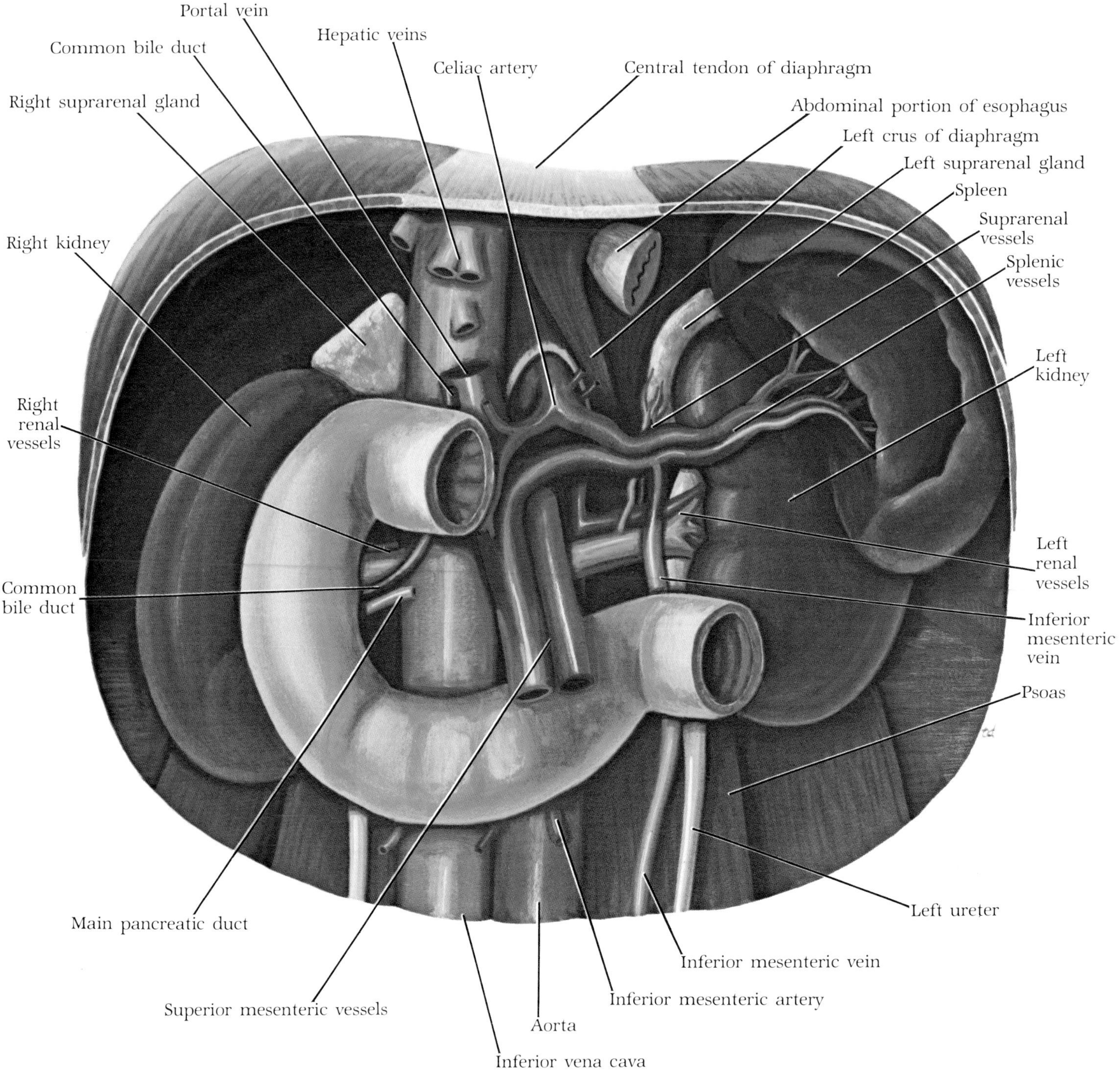
Portal vein
Common bile duct
Hepatic veins
Celiac artery
Central tendon of diaphragm
Right suprarenal gland
Abdominal portion of esophagus
Left crus of diaphragm
Left suprarenal gland
Spleen
Suprarenal vessels
Splenic vessels
Right kidney
Left kidney
Right renal vessels
Left renal vessels
Common bile duct
Inferior mesenteric vein
Psoas
Left ureter
Main pancreatic duct
Inferior mesenteric vein
Inferior mesenteric artery
Superior mesenteric vessels
Aorta
Inferior vena cava

Figure 2-34
Posterior relations of the duodenum and pancreas.

1. Portal hypertension may be caused by hepatic disease or blockage of the extrahepatic portion of the portal vein. As a result of this elevated pressure, collateral pathways open up, the spleen enlarges, and ascites may occur.

2. The following important communications exist between the portal and systemic veins, both of which become enlarged should the direct route through the liver become blocked:
A. At the lower third of the esophagus, the esophageal branches of the left gastric vein (portal tributary) anastomose with the esophageal veins draining the middle third of the esophagus into the azygos veins (systemic tributary).
B. Halfway down the anal canal, the superior rectal veins (portal tributary) draining the upper half of the anal canal anastomose with the middle and inferior rectal veins (systemic tributaries).
C. The paraumbilical veins connect the left branch of the portal vein with the superficial veins of the anterior abdominal wall (systemic tributaries).
D. The veins of the ascending colon, descending colon, duodenum, pancreas, and liver (portal tributaries) anastomose with the renal, lumbar, and phrenic veins (systemic tributaries).

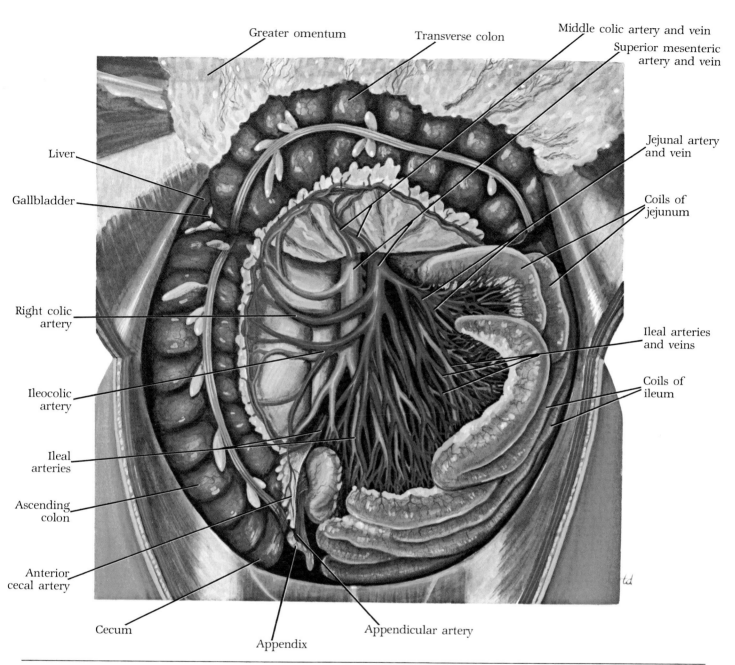

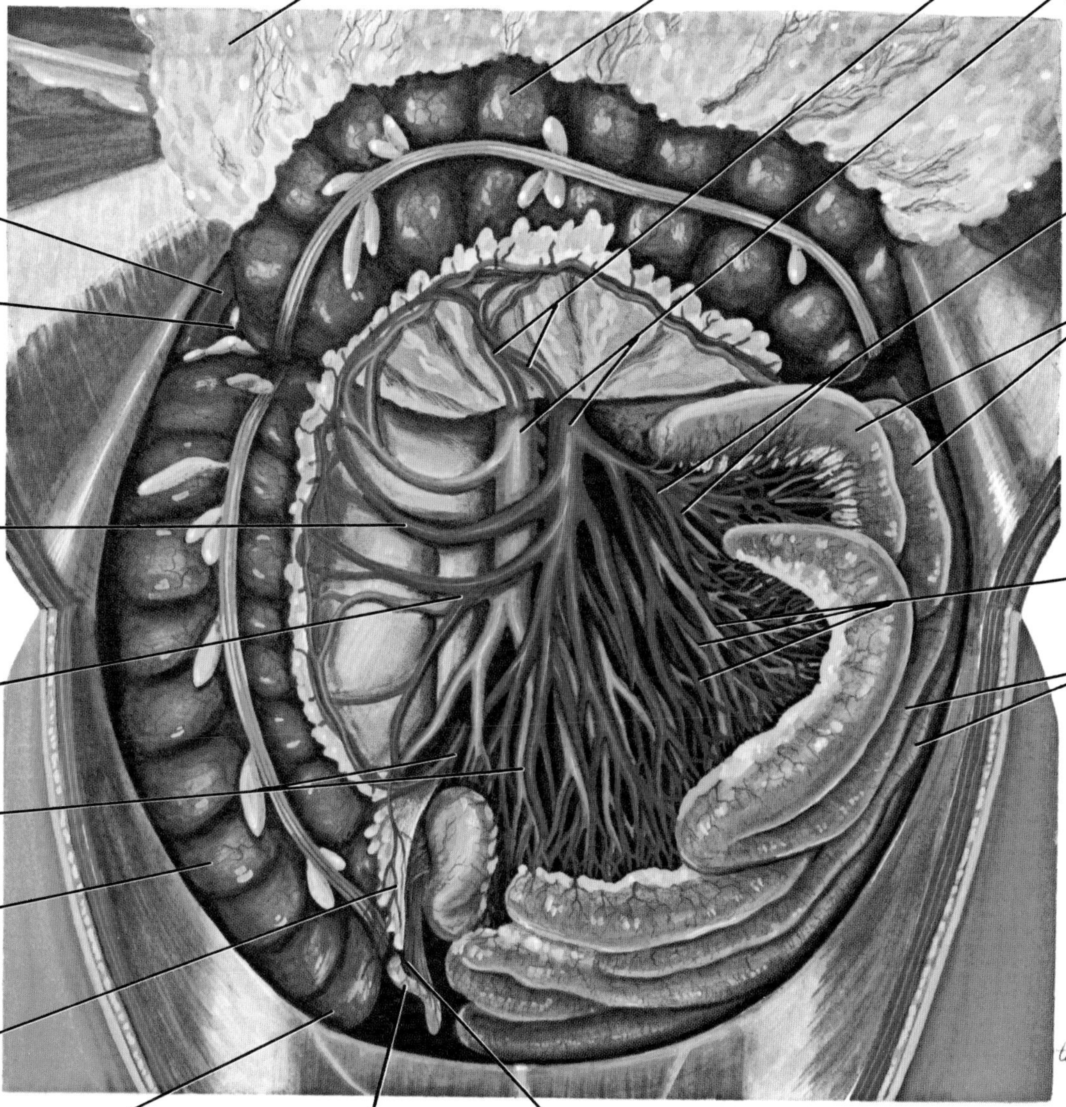
Greater omentum
Transverse colon
Middle colic artery and vein
Superior mesenteric artery and vein
Liver
Gallbladder
Jejunal artery and vein
Coils of jejunum
Right colic artery
Ileal arteries and veins
Coils of ileum
Ileocolic artery
Ileal arteries
Ascending colon
Anterior cecal artery
Cecum
Appendix
Appendicular artery

Figure 2-35
Superior mesenteric artery and vein.

1. Sudden complete occlusion of the main trunk of the superior mesenteric artery is most frequently due to an embolus. If thrombosis should occur, it nearly always occurs in an artery in which there is atherosclerosis. Extreme abdominal pain followed by vomiting is the earliest clinical feature. Paralysis and death of the affected segment of the intestine result.

2. Should occlusion of arterial segments to the jejunum or ileum occur, the anastomotic arcades within the mesentery provide alternative pathways for blood to reach the intestinal wall.

3. Each terminal straight branch from an arcade passes alternately to opposite sides of the jejunum and ileum and supplies 1 to 2 cm of bowel length. They enter the bowel wall at right angles to its length and do not anastomose until they form the submucous plexus. When performing a segmental resection of the jejunum or ileum, the bowel should be cut obliquely to ensure an adequate blood supply to the antimesenteric border.

4. The superior mesenteric artery supplies the gastrointestinal tract from the middle of the second part of the duodenum to as far as the distal one-third of the transverse colon.

5. Superior mesenteric vein occlusion is usually due to thrombosis, and the symptoms are similar to those of sudden arterial occlusion. Because the thrombosis tends to occur in the peripheral tributaries, shorter segments of the bowel have to be resected as compared with what has to be done after arterial thrombosis, which usually occurs in the main arterial stem.

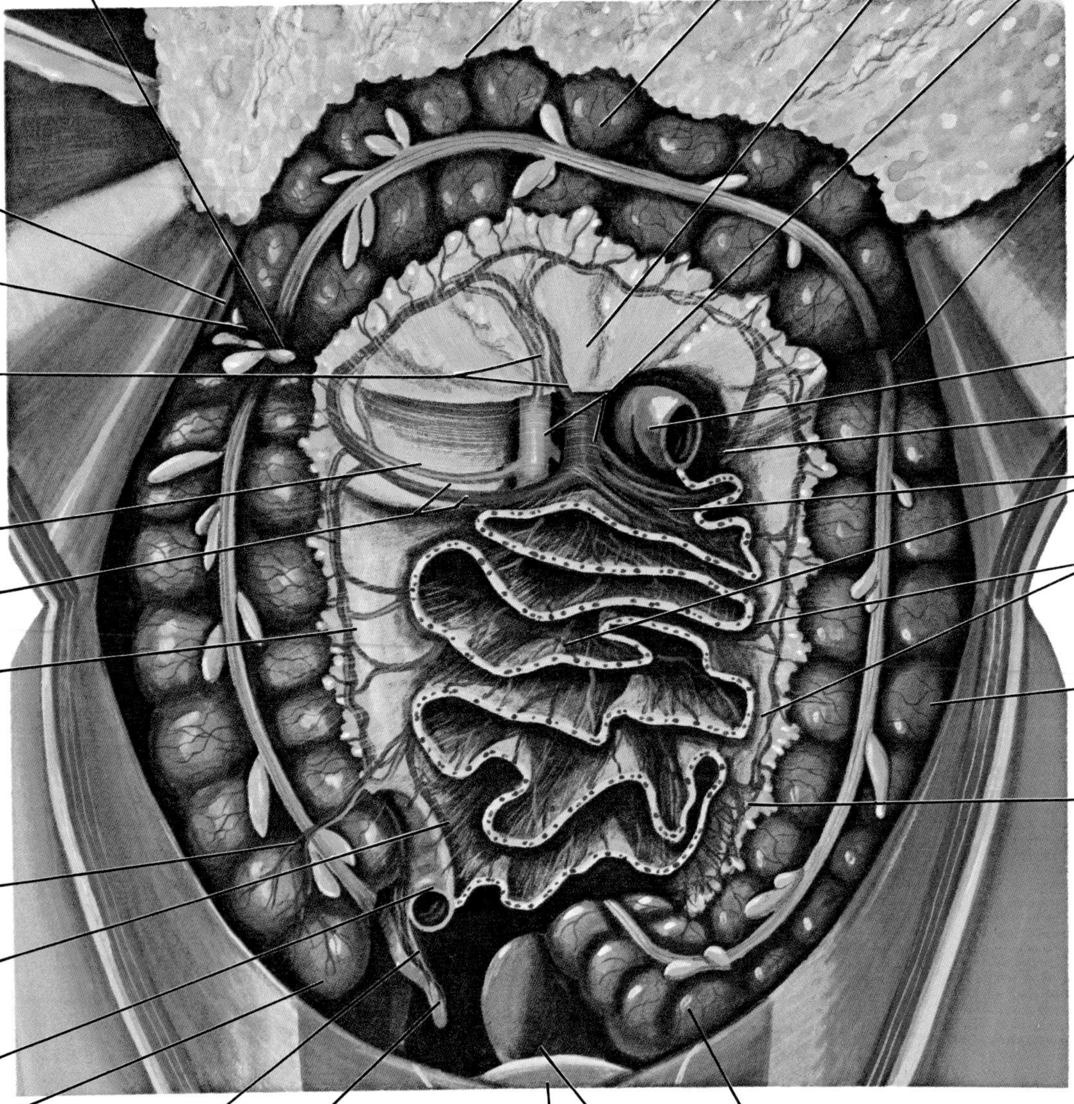
Greater omentum
Superior mesenteric vessels
Right colic (hepatic) flexure
Transverse colon
Transverse mesocolon
Left colic (splenic) flexure
Liver
Gallbladder
Middle colic artery and vein
Duodenojejunal flexure
Branch of superior left colic artery
Jejunal and ileal arteries
Duodenum
Folds of mesentery of small intestine (cut)
Right colic vessels
Marginal artery and vein
Descending colon
Inferior left colic artery
Anterior cecal artery
Ileal artery
Ileum
Cecum
Appendicular vessels
Appendix
Bladder
Rectum
Pelvic colon

Figure 2-36
Mesentery of the small intestine.

1. The mesentery serves as a suspensory ligament of the jejunum and ileum. Within its fused layers of peritoneum run the superior mesenteric artery and its branches and accompanying vein, lymph vessels, lymph nodes, nerve fibers, and a variable amount of adipose tissue. The mesentery itself has no pain-afferent fibers, although the root of the mesentery is very sensitive to stretch. The mesentery, like the omentum, has absorptive and bactericidal properties. During inflammations it can form adhesions in an attempt to localize intraperitoneal infections.

2. Congenital defects or holes in the mesentery rarely occur; the majority are produced during the course of intra-abdominal surgery. Such defects form potential sites for internal herniation of loops of bowel.

3. The superior mesenteric artery supplies the gastrointestinal tract from the middle of the second part of the duodenum to as far as the distal one-third of the transverse colon. The inferior mesenteric artery supplies the large intestine from the distal one-third of the transverse colon to halfway down the anal canal. Anastomosis between the superior pancreaticoduodenal artery from the gastroduodenal branch of the celiac and the inferior pancreaticoduodenal artery, a branch of the superior mesenteric, enables the third and fourth parts of the duodenum and the first few inches of jejunum to survive when the superior mesenteric artery is occluded.

4. The colic arteries around the concave margin of the large intestine from the ileocolic junction to the rectum anastomose to form a single important arterial trunk called the marginal artery. This anastomosis can supply the bowel even though one of the major arteries is ligated.

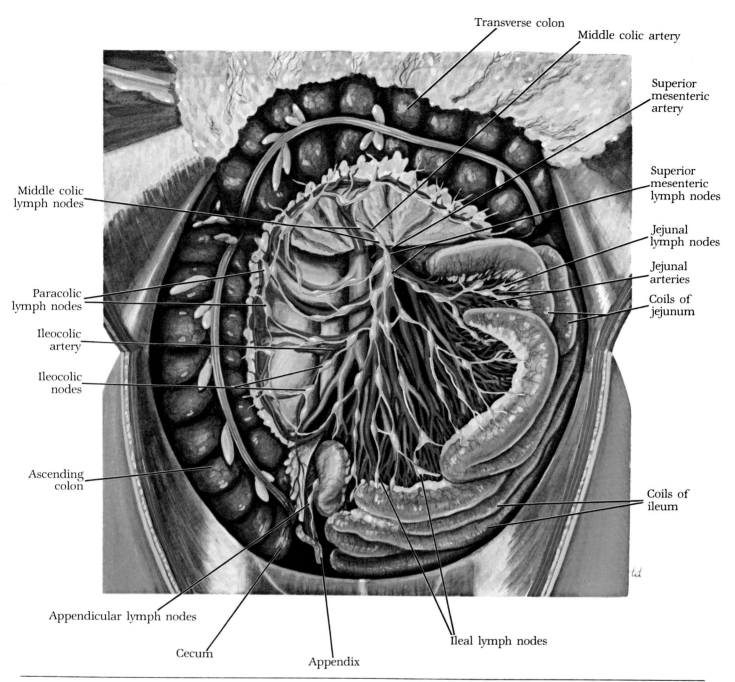

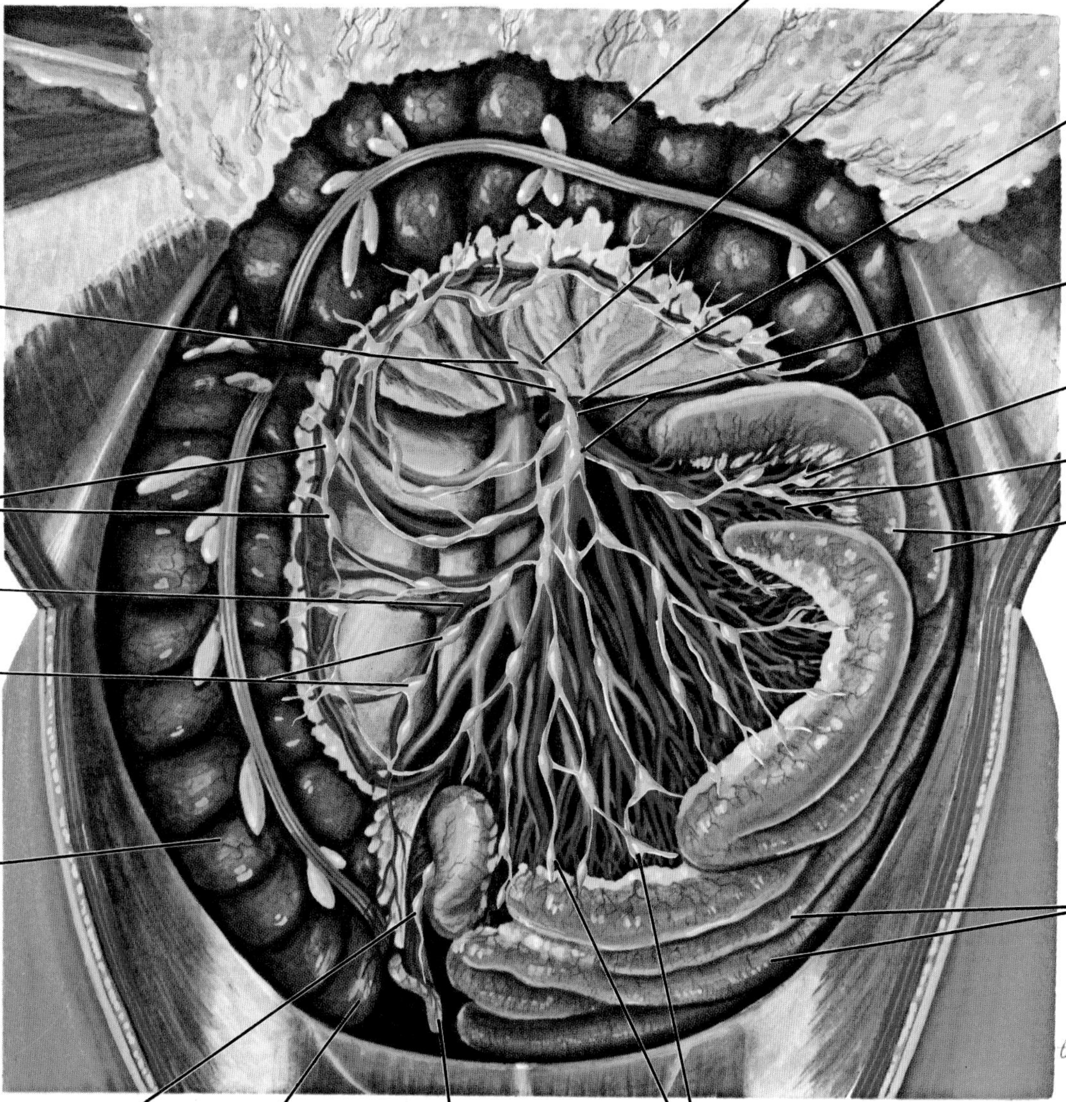
Transverse colon
Middle colic artery
Superior mesenteric artery
Superior mesenteric lymph nodes
Jejunal lymph nodes
Jejunal arteries
Coils of jejunum
Middle colic lymph nodes
Paracolic lymph nodes
Ileocolic artery
Ileocolic nodes
Ascending colon
Coils of ileum
Appendicular lymph nodes
Cecum
Appendix
Ileal lymph nodes

Figure 2-37
Superior mesenteric lymph vessels and nodes.

1. Carcinoma of the colon and rectum is a relatively common disease. About 20 percent occur in the segment formed by the cecum, ascending colon, hepatic flexure, and transverse colon. If a diagnosis is made early and a partial colectomy and removal of the lymph vessels and lymph nodes draining the area are performed, a cure can be anticipated.

2. For a cancer occurring in the cecum, ascending colon, or hepatic flexure, the resection should include a short segment of the terminal ileum, cecum, ascending colon, and right half of the transverse colon. The ileocolic, right colic, and right branch of the middle colic vessels with their accompanying lymphatic vessels and nodes, mesentery, and overlying peritoneum should also be removed en bloc. The continuity of the intestinal tract is restored by joining the ileum to the left half of the transverse colon.

3. A cancer occurring in the middle of the transverse colon should be treated by removal of the middle segment of the colon together with the middle colic blood vessels and lymphatics and associated part of the transverse mesocolon. The continuity of the bowel is restored by joining the hepatic to the splenic flexures.

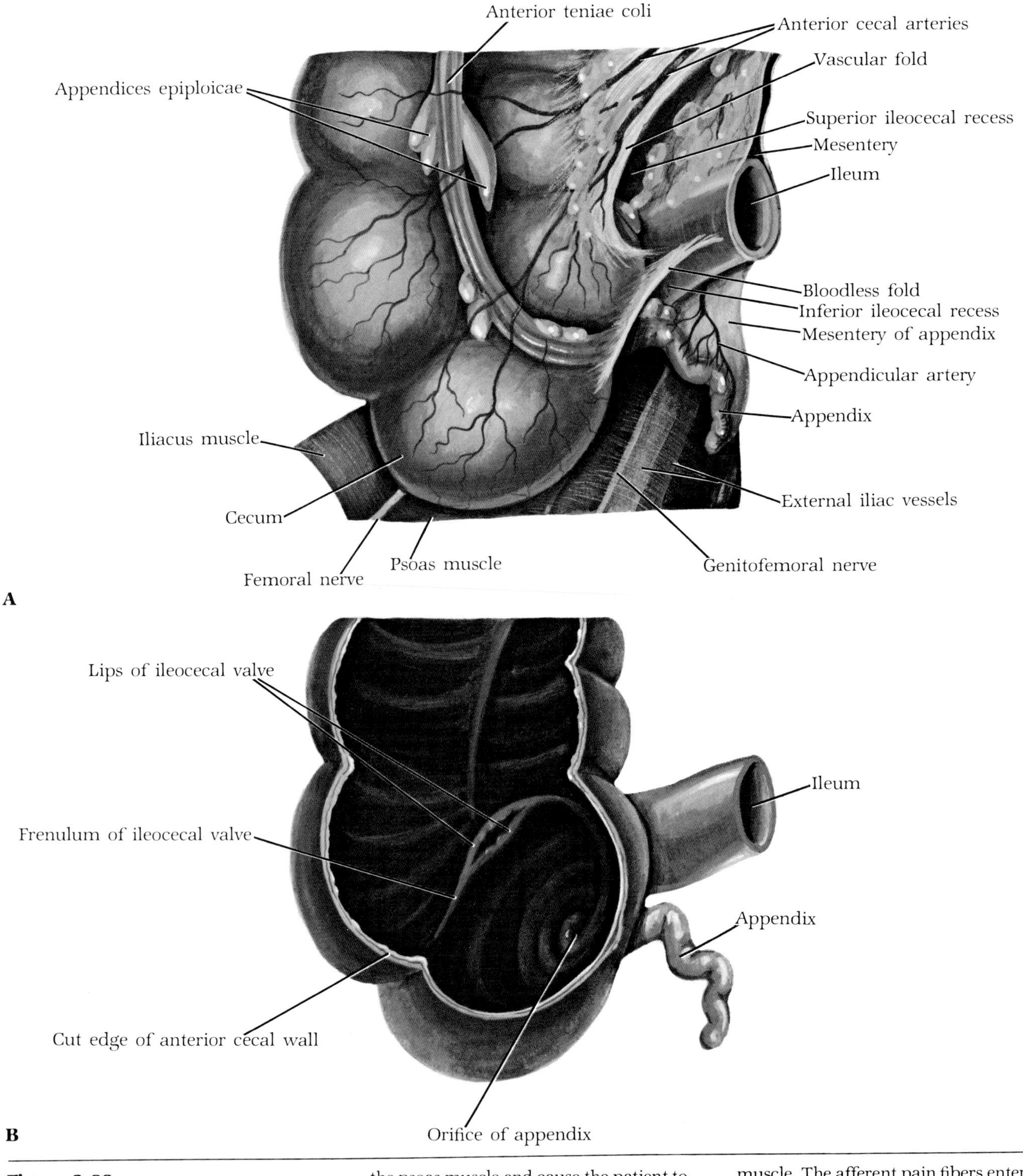
Anterior teniae coli
Anterior cecal arteries
Vascular fold
Appendices epiploicae
Superior ileocecal recess
Mesentery
Ileum
Bloodless fold
Inferior ileocecal recess
Mesentery of appendix
Appendicular artery
Appendix
Iliacus muscle
External iliac vessels
Cecum
Psoas muscle
Genitofemoral nerve
Femoral nerve
A
Lips of ileocecal valve
Ileum
Frenulum of ileocecal valve
Appendix
Cut edge of anterior cecal wall
B
Orifice of appendix

Figure 2-38
Terminal part of ileum, cecum, and appendix, anterior views. A. Intact. B. Anterior wall of cecum has been removed to show the ileocecal valve and the orifice of the appendix.

1. The variability in position of the free end of the appendix must be remembered when attempting to diagnose appendicitis. The free end may be pelvic, retrocecal, or retroileal. A retrocecal appendix may irritate the psoas muscle and cause the patient to keep his right hip joint flexed.
2. Because the three teniae coli meet at the junction of the cecum with the appendix, the base of the appendix can be located at operation by tracing the anterior teniae downward over the cecal wall. This may be particularly helpful when the appendix is buried in omental adhesions.
3. The lumen of the normal appendix is small, and obstruction by a fecalith is the most common cause of acute appendicitis.
4. Visceral pain in the appendix is produced by distention of its lumen or spasm of its muscle. The afferent pain fibers enter the spinal cord at the level of the tenth thoracic segment, and a vague referred pain is felt in the region of the umbilicus. Later, when the pain shifts to where the inflamed appendix irritates the parietal peritoneum, it becomes precise, severe, and localized.

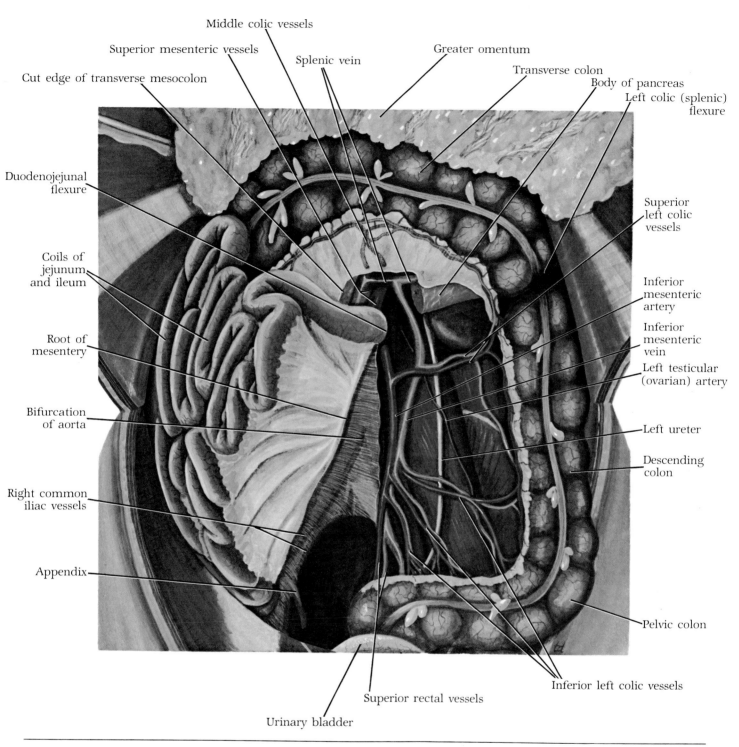

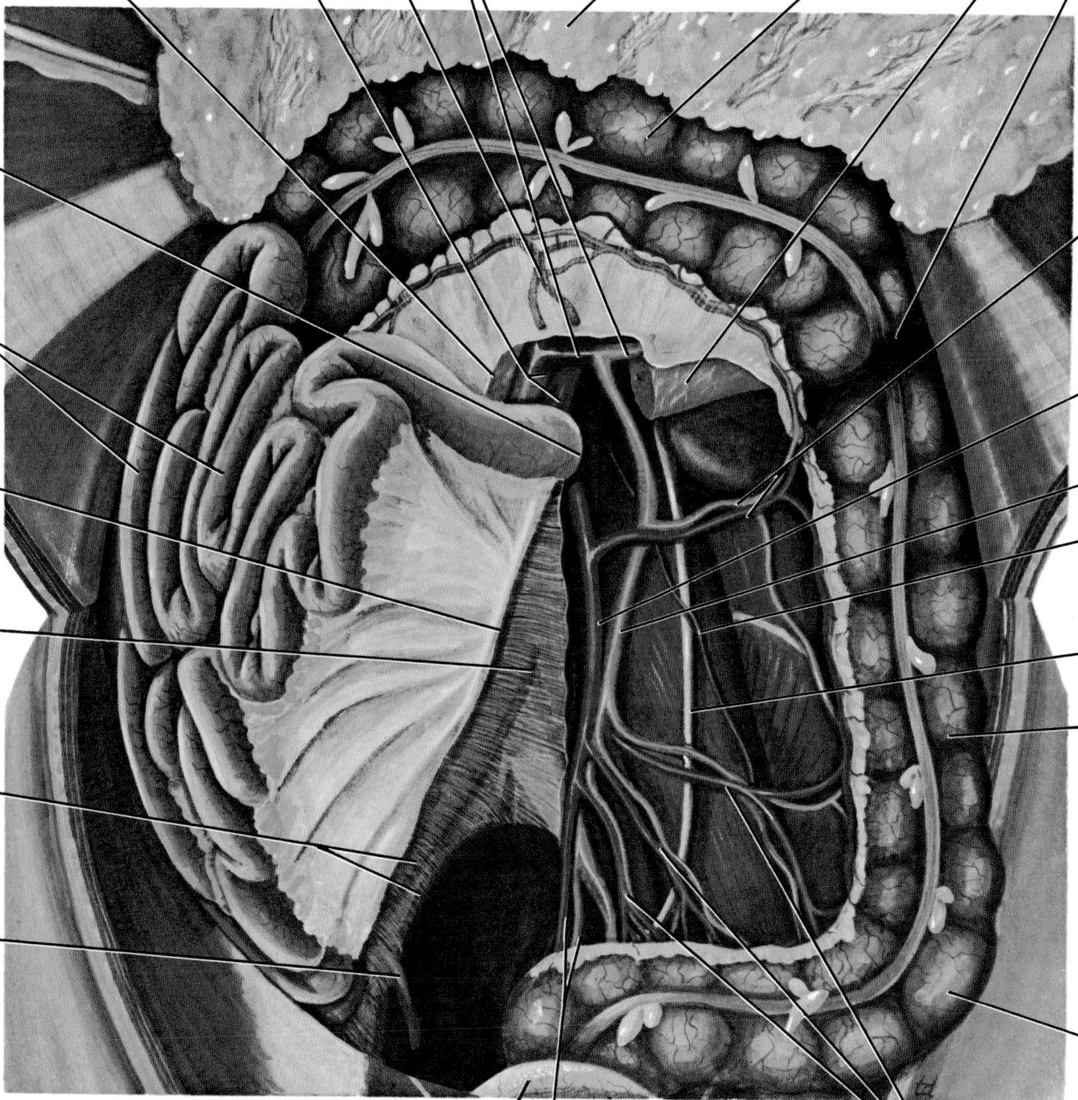
Middle colic vessels
Superior mesenteric vessels
Splenic vein
Greater omentum
Transverse colon
Cut edge of transverse mesocolon
Body of pancreas
Left colic (splenic) flexure
Duodenojejunal flexure
Superior left colic vessels
Coils of jejunum and ileum
Inferior mesenteric artery
Inferior mesenteric vein
Root of mesentery
Left testicular (ovarian) artery
Bifurcation of aorta
Left ureter
Descending colon
Right common iliac vessels
Appendix
Pelvic colon
Inferior left colic vessels
Superior rectal vessels
Urinary bladder

Figure 2-39
Inferior mesenteric artery and vein. The peritoneum covering these vessels on the posterior abdominal wall has been removed.

1. The inferior mesenteric artery supplies the large intestine from the distal one-third of the transverse colon to halfway down the anal canal. The importance of the marginal artery has already been emphasized. This anastomosing trunk is capable of supplying the bowel even though one of the major arteries is ligated.

2. The inferior mesenteric vein is a continuation of the superior hemorrhoidal vein. The main trunk does not follow the artery but passes to the left of the duodenojejunal flexure where it may be related to the paraduodenal fossa. It finally passes posterior to the pancreas to join the splenic vein.

3. Sudden occlusion of the inferior mesenteric artery may occur as the result of a block by thrombosis or an embolism. The inferior mesenteric vein may be blocked by thrombosis. Low abdominal pain accompanied by bloody stools and tenderness over the descending colon is the usual clinical picture. Prompt surgical exploration is essential.

4. Volvulus of the sigmoid colon is a common cause of colonic obstruction. Chronic constipation, an extra-long loop of sigmoid colon, and a narrow root of the mesentery predispose to the condition. Strangulation of the blood supply is a complication and requires emergency surgery.

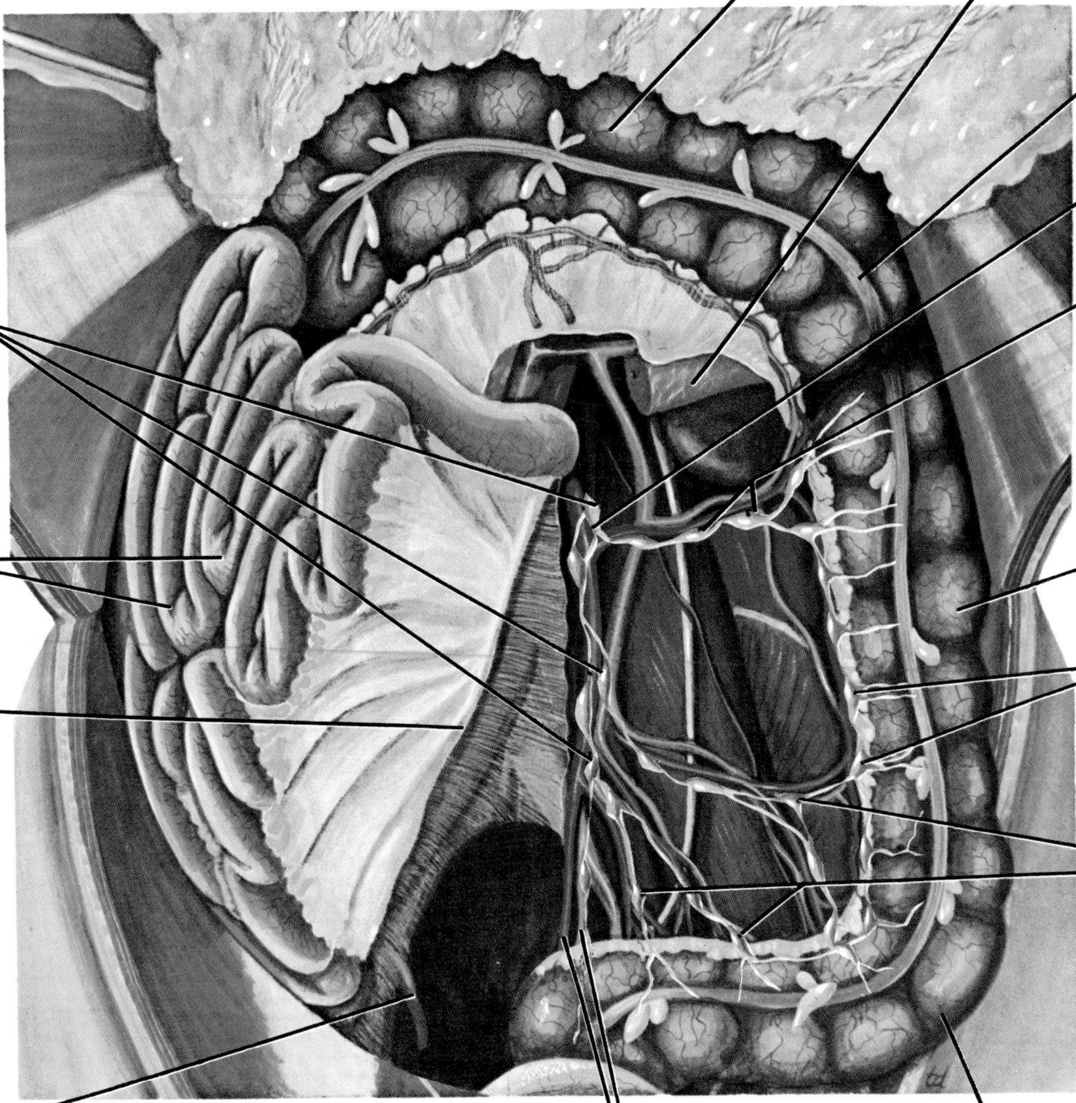

Transverse colon
Pancreas
Left colic (splenic) flexure
Inferior mesenteric artery
Superior left colic lymph nodes
Inferior mesenteric lymph nodes
Coils of small intestine
Root of mesentery
Descending colon
Paracolic lymph nodes
Inferior left colic lymph nodes
Appendix
Superior rectal vessels
Pelvic colon

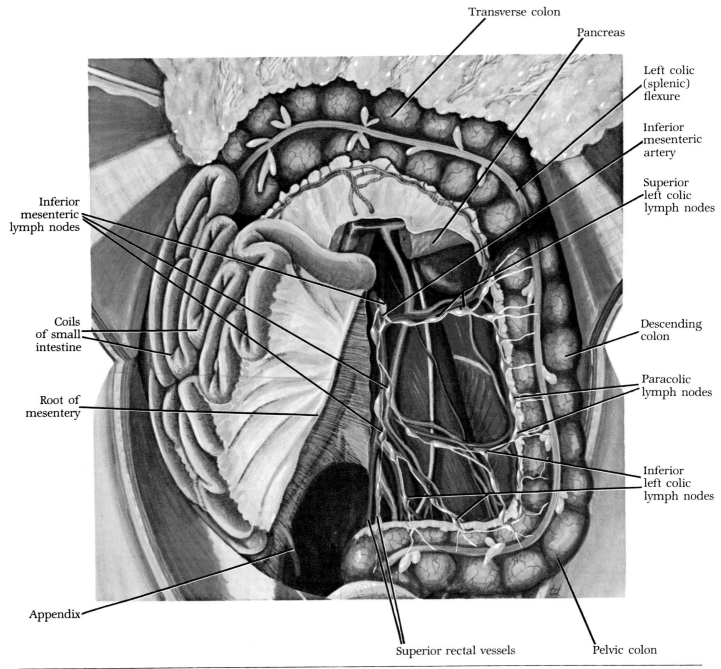

Figure 2-40
Inferior mesenteric lymph vessels and nodes.

1. Carcinoma of the colon and rectum is a relatively common disease. About 80 percent occur in the descending colon, sigmoid, and rectum. The treatment is a partial colectomy accompanied by removal of the lymph vessels and lymph nodes draining the area. For cancers of the left colic flexure, resection includes the transverse and descending colons along with the middle colic and left colic vessels and lymphatics. Continuity of the bowel is reestablished by anastomosing the ascending colon to the sigmoid colon.
2. For cancers of the descending colon, the colon from the left colic flexure to the proximal part of the sigmoid and the left colic vessels and lymphatics are removed. Continuity of the bowel is restored by anastomosing the transverse colon to the distal sigmoid colon.
3. When the cancer involves the sigmoid or rectosigmoid junction, the distal descending colon as far down as the proximal rectum is removed, along with the lower left colic and superior rectal vessels and lymphatic vessels. Continuity of the bowel is reestablished (if possible) by anastomosing the descending colon to the rectum. Cancers of the rectum are best treated by an abdominoperineal resection of the rectum and anal canal followed by a permanent colostomy.

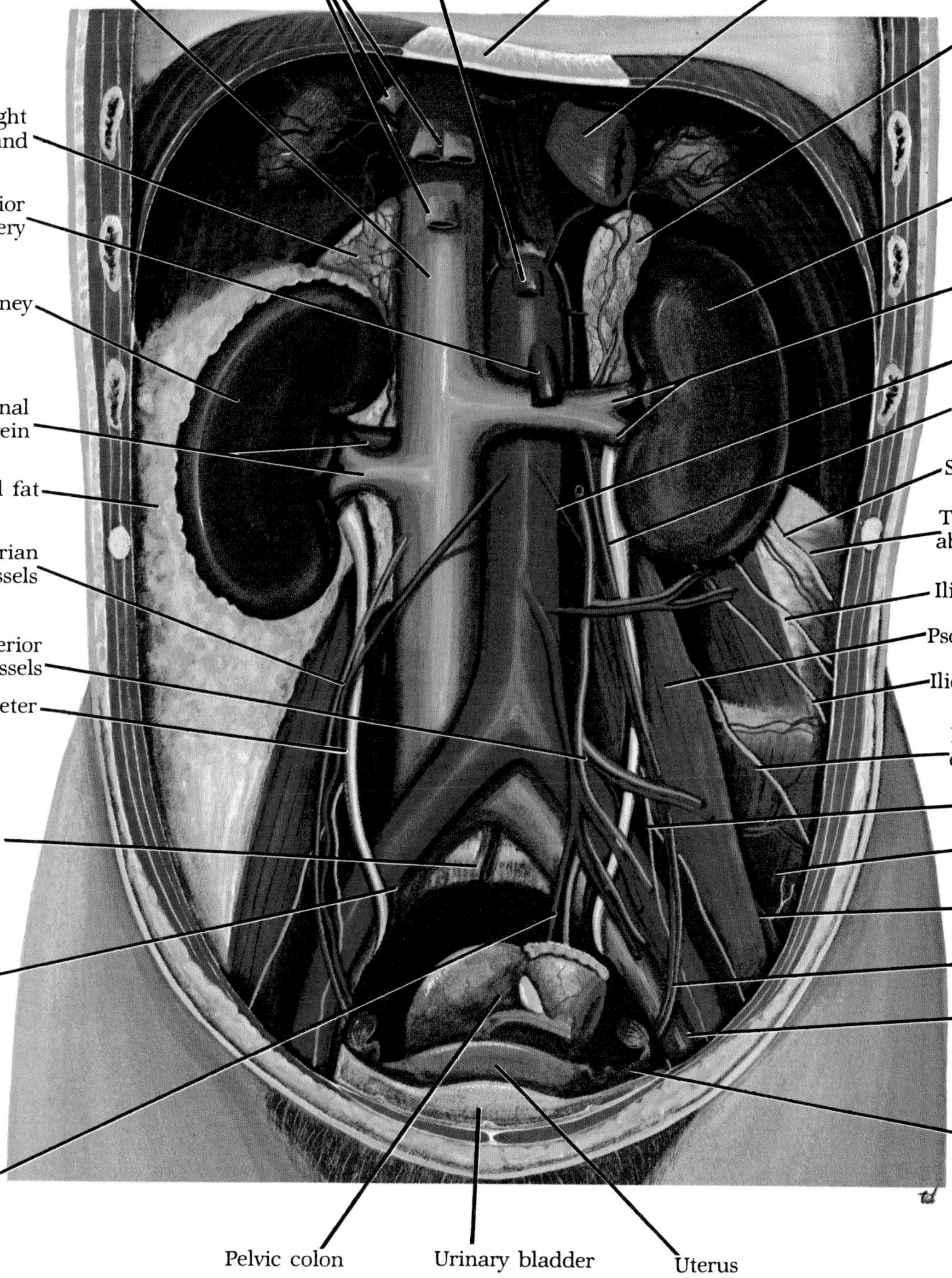

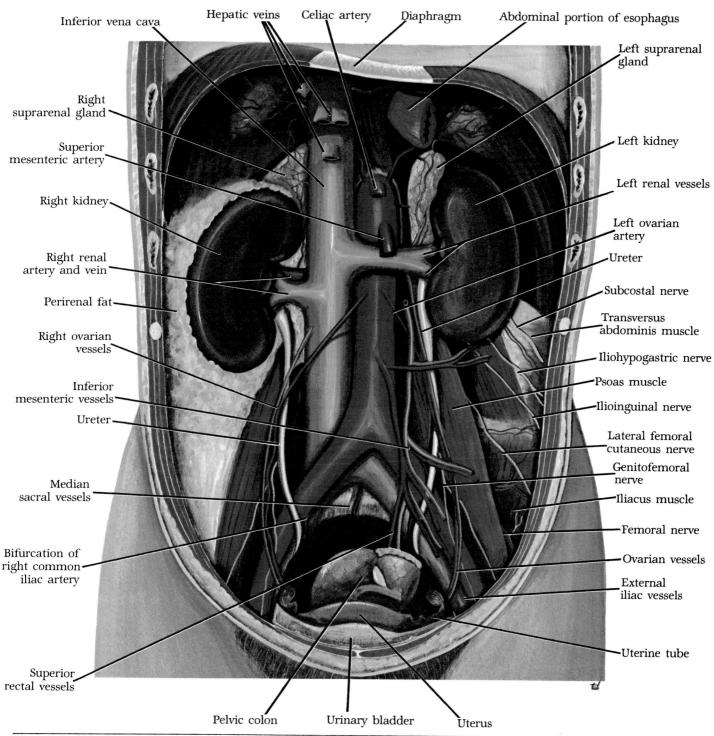

Inferior vena cava
Hepatic veins
Celiac artery
Diaphragm
Abdominal portion of esophagus
Left suprarenal gland
Right suprarenal gland
Superior mesenteric artery
Left kidney
Left renal vessels
Right kidney
Left ovarian artery
Right renal artery and vein
Ureter
Perirenal fat
Subcostal nerve
Transversus abdominis muscle
Right ovarian vessels
Iliohypogastric nerve
Inferior mesenteric vessels
Psoas muscle
Ilioinguinal nerve
Ureter
Lateral femoral cutaneous nerve
Genitofemoral nerve
Median sacral vessels
Iliacus muscle
Femoral nerve
Bifurcation of right common iliac artery
Ovarian vessels
External iliac vessels
Uterine tube
Superior rectal vessels
Pelvic colon
Urinary bladder
Uterus

Figure 2-41
Posterior abdominal wall showing the kidneys and ureters in situ.

1. The kidneys are retroperitoneal structures surrounded by perinephric fat and fascia. A small abscess in the kidney may burst through the renal capsule to form a perinephric abscess.
2. The right kidney lies at a lower level than the left kidney; the lower pole of the right kidney may be palpated in the right lumbar region at the end of deep inspiration in a thin person and in one with poorly developed abdominal musculature.
3. Supernumerary renal arteries, which are relatively common, are clinically important since they may cross the pelviureteral junction and obstruct the outflow of urine.
4. Malignant tumors of the kidney commonly spread by direct invasion of the renal vein. Venous occlusion may result from spread of the tumor into the inferior vena cava.
5. The ureter has three sites of anatomical narrowing where urinary calculi may be arrested: at the pelviureteral junction, at the pelvic brim, and where the ureter enters the bladder.
6. When examining a radiograph of the ureter, notice that the ureter crosses the anterior surface of the psoas muscle, lies close to the tips of the transverse processes of the lumbar vertebrae, crosses the region of the sacroiliac joint, swings laterally close to the ischial spine, and then turns medially to enter the bladder.
7. Pain fibers from the ureter enter the spinal cord at segments T11 and T12 and L1 and 2. In ureteric colic the agonizing pain is referred to the skin areas supplied by the nerves of these segments of the spinal cord, i. e., the pain extends from the loin to the groin.

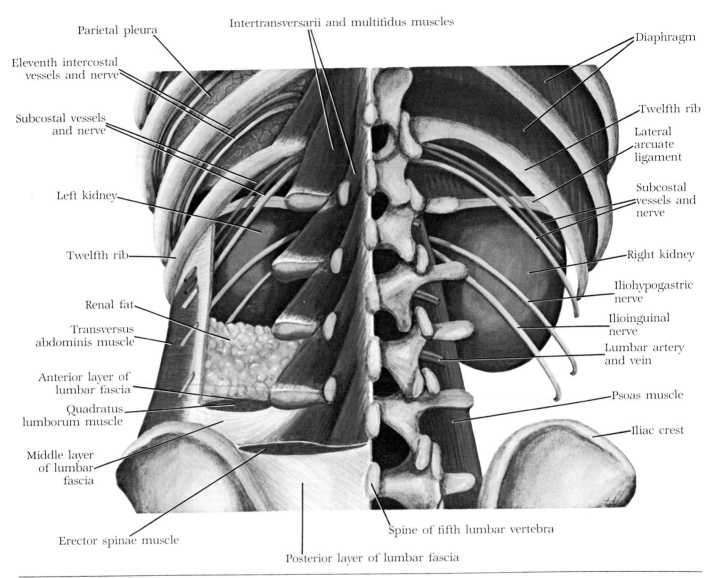

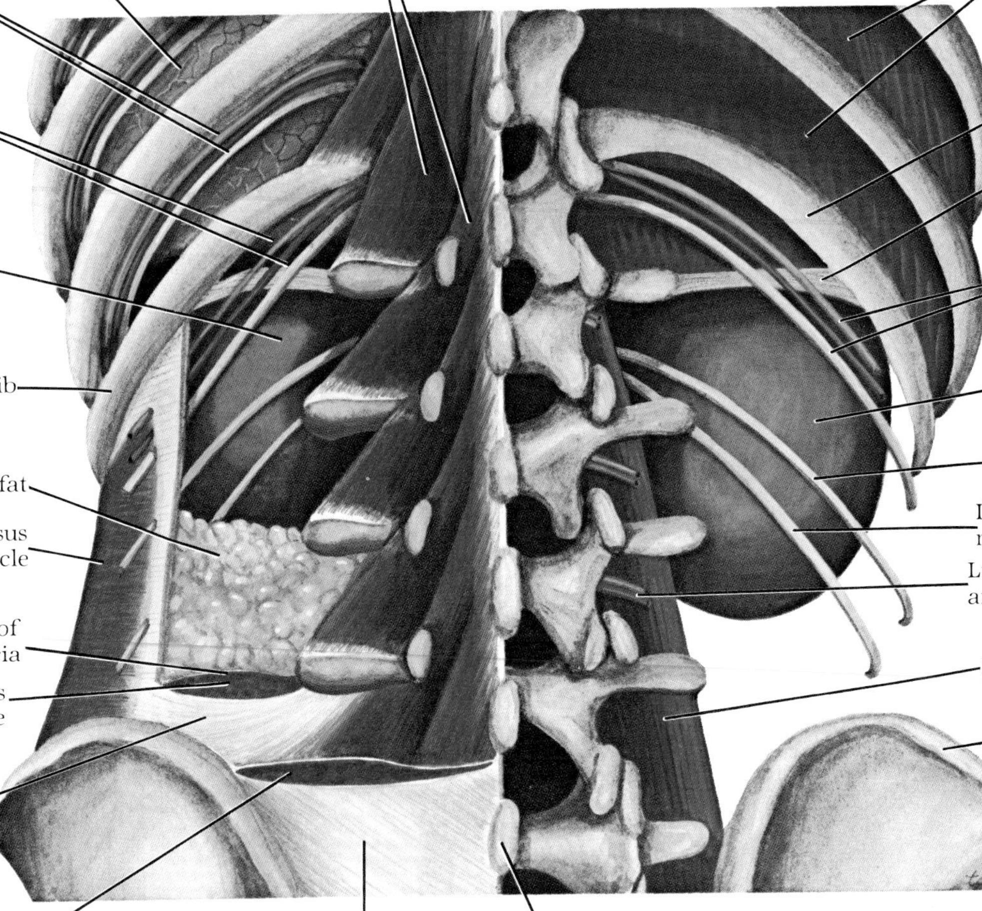
Parietal pleura
Intertransversarii and multifidus muscles
Diaphragm
Eleventh intercostal vessels and nerve
Twelfth rib
Subcostal vessels and nerve
Lateral arcuate ligament
Subcostal vessels and nerve
Left kidney
Twelfth rib
Right kidney
Iliohypogastric nerve
Renal fat
Ilioinguinal nerve
Transversus abdominis muscle
Lumbar artery and vein
Anterior layer of lumbar fascia
Psoas muscle
Quadratus lumborum muscle
Iliac crest
Middle layer of lumbar fascia
Spine of fifth lumbar vertebra
Erector spinae muscle
Posterior layer of lumbar fascia

Figure 2-42
Posterior relations of the kidneys.

1. Although both kidneys are well protected by the lower part of the thoracic cage and the powerful back muscles, they may be damaged by blunt trauma to the loin as from a kick or an automobile or plane accident.

2. Lumbar nephrectomy is a popular method of removing a kidney, since the approach is retroperitoneal and avoids contamination of the peritoneal cavity. The disadvantages are a limited exposure, and the fact that the renal blood vessels can only be tied after the kidney has been manipulated.

3. Bimanual examination of a kidney in this region is carried out with the patient relaxed in the supine position. The examiner places one hand on the back behind and below the twelfth rib and the other hand below the rib cage anteriorly. When, as requested, the patient inspires deeply, the diaphragm descends, pushing the kidneys inferiorly. In slender individuals, the lower pole of the right kidney can usually be felt between the examiner's hands. An inflamed kidney will produce tenderness in the costovertebral angle. A perinephric abscess will also produce tenderness and rigidity of the muscles in the costovertebral angle. As the abscess enlarges, a swelling will develop in this area.

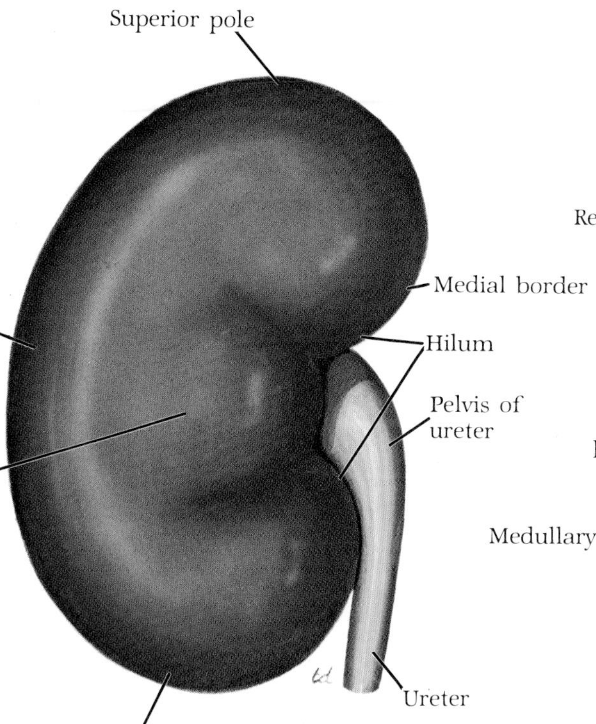
Superior pole
Medial border
Lateral border
Hilum
Pelvis of ureter
Anterior surface
Ureter
Inferior pole

A

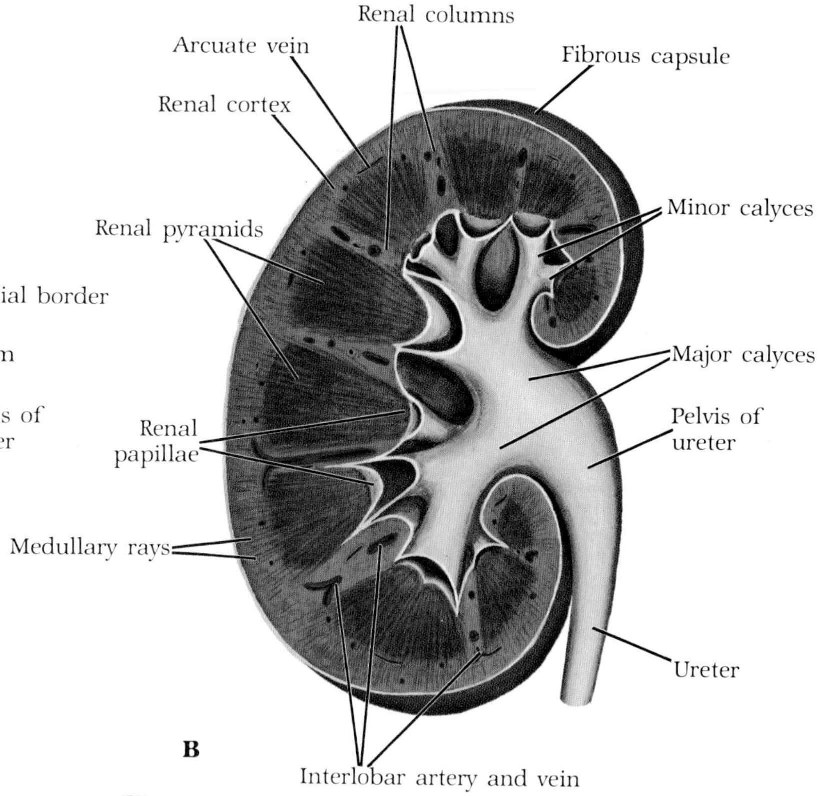
Renal columns
Arcuate vein
Fibrous capsule
Renal cortex
Minor calyces
Renal pyramids
Major calyces
Pelvis of ureter
Renal papillae
Medullary rays
Ureter
Interlobar artery and vein

B

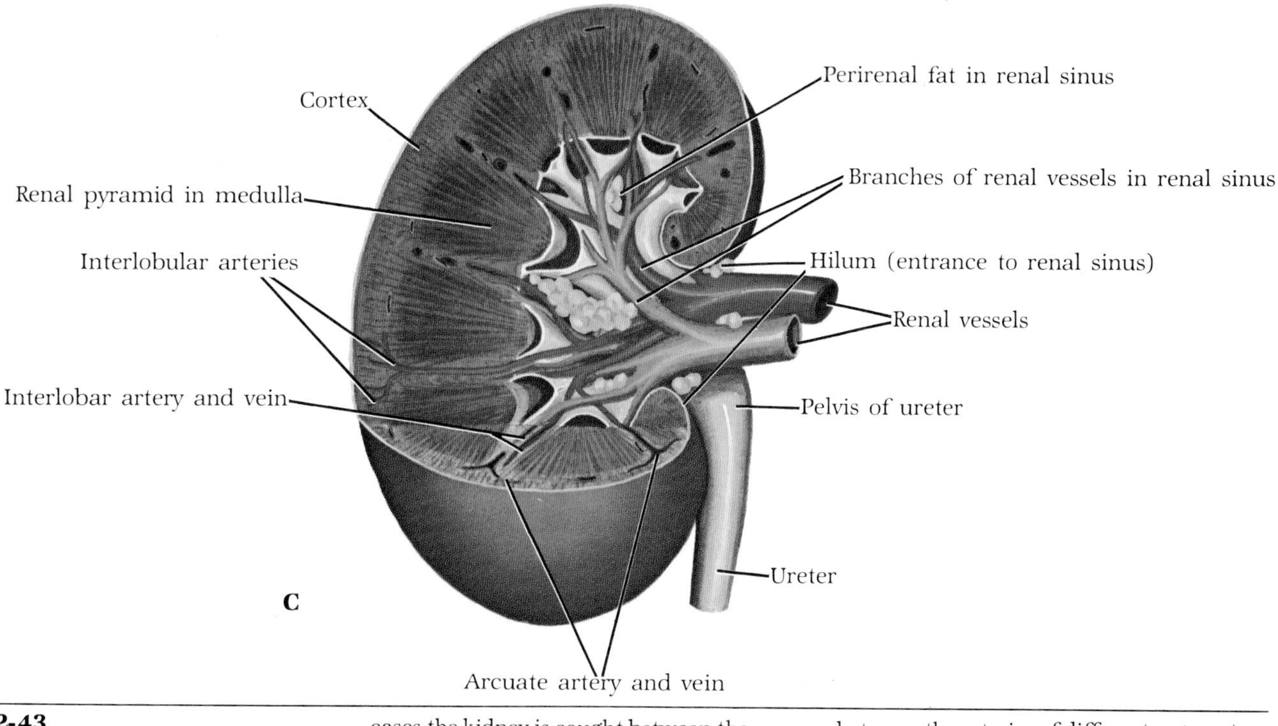
Perirenal fat in renal sinus
Cortex
Branches of renal vessels in renal sinus
Renal pyramid in medulla
Interlobular arteries
Hilum (entrance to renal sinus)
Renal vessels
Interlobar artery and vein
Pelvis of ureter
Ureter
Arcuate artery and vein

C

Figure 2-43
A. Right kidney, anterior views. B. Coronal section showing cortex, medulla, pyramids, renal papillae, and calyces. C. Combined coronal and horizontal sections showing renal vessels and pelvis of ureter.

1. Blunt trauma to the loin, such as occurs in automobile accidents or in contact sports, results in injury to the kidney more frequently than to any other abdominal organ. In some cases the kidney is caught between the twelfth rib and the vertebral column in extreme lateral flexion.
2. The kidneys are normally held in position by the renal vessels, the surrounding perinephric fat, and the perinephric fascia.
3. When the renal artery enters the hilum of the kidney, it subdivides into several branches that pass between the renal papillae to be distributed to the kidney. For practical purposes the kidney is divided into five segments, each of which is supplied by its own artery. There is no anastomosis between the arteries of different segments. The intrarenal veins freely anastomose and do not follow a segmental pattern.
4. Accessory renal arteries arise either from the renal artery or directly from the aorta and usually enter the inferior pole of the kidney. An accessory artery may press on the ureter and obstruct the outflow of urine.
5. The arrangement of the perinepric fascia limits the spread of blood or urine from a ruptured kidney and the spread of pus from an infected kidney.

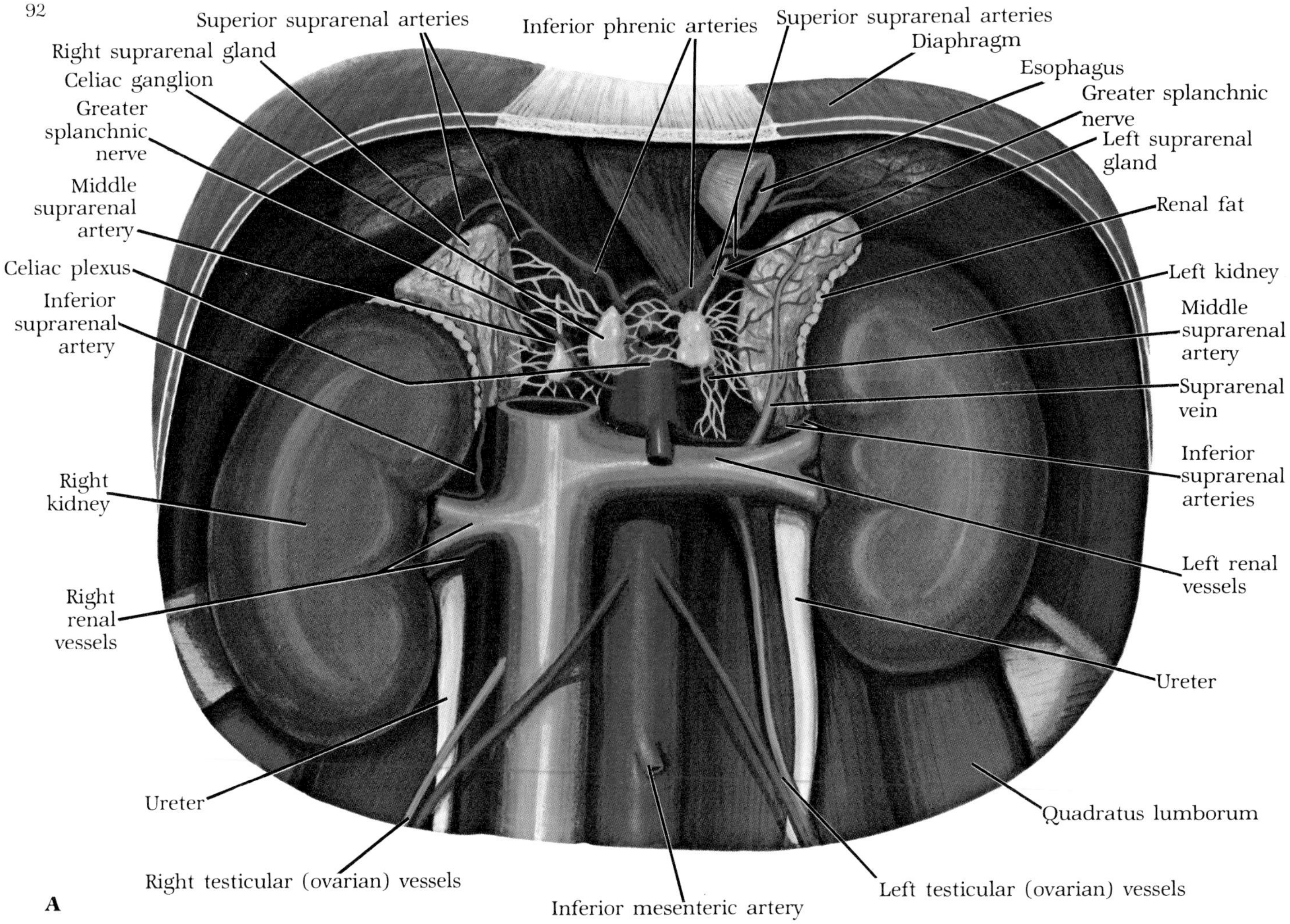
Superior suprarenal arteries
Inferior phrenic arteries
Superior suprarenal arteries
Right suprarenal gland
Diaphragm
Celiac ganglion
Esophagus
Greater splanchnic nerve
Greater splanchnic nerve
Left suprarenal gland
Middle suprarenal artery
Renal fat
Celiac plexus
Left kidney
Inferior suprarenal artery
Middle suprarenal artery
Suprarenal vein
Right kidney
Inferior suprarenal arteries
Left renal vessels
Right renal vessels
Ureter
Ureter
Quadratus lumborum
Right testicular (ovarian) vessels
Left testicular (ovarian) vessels
Inferior mesenteric artery

A

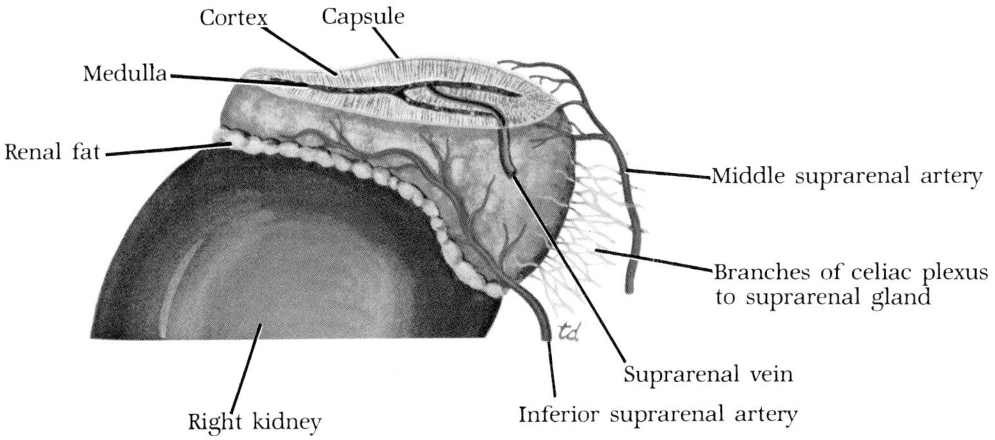
Cortex
Capsule
Medulla
Renal fat
Middle suprarenal artery
Branches of celiac plexus to suprarenal gland
td
Suprarenal vein
Right kidney
Inferior suprarenal artery

B

Figure 2-44
Suprarenal glands. A. In situ, anterior view. B. Horizontal section through right suprarenal gland showing internal structure.

1. The suprarenal gland consists of two distinct endocrine organs: the suprarenal cortex and the suprarenal medulla. Of the two, the suprarenal cortex is the one essential to life. It produces many steroid hormones that have, depending on their structure, electrolytic, glycogenic, progestational, androgenic, or estrogenic activities. The suprarenal medulla, which is not essential to life, is stimulated through the celiac plexus and responds by releasing epinephrine into the bloodstream, thus producing a general sympathetic effect on all cells of the body.

2. Suprarenal cortical hyperplasia is the commonest cause of Cushing's syndrome, the clinical manifestations of which include moon-shaped face, truncal obesity, hirsutism, and hypertension; if the syndrome occurs later in life, it may result from an adenoma or carcinoma of the cortex. Adrenocortical insufficiency (Addison's disease), which is characterized clinically by increased pigmentation, muscular weakness, weight loss, and hypotension, may be caused by tuberculous destruction or bilateral atrophy of both cortices.

3. Pheochromocytoma, a tumor of the medulla, produces a paroxysmal or sustained hypertension. The symptoms and signs are due to the production of a large amount of catecholamines which are then poured into the bloodstream. Neuroblastoma is a malignant tumor of the medulla that occurs in young children. These tumors are metabolically inactive.

4. Accessory chromaffin tissue normally is found outside the suprarenal medulla, on the posterior abdominal wall along the anterior surface of the aorta. Small pieces may also be found in relation to the sympathetic trunk in the abdomen or pelvis. If these small pieces do not disappear after childhood a pheochromocytoma may develop from them.

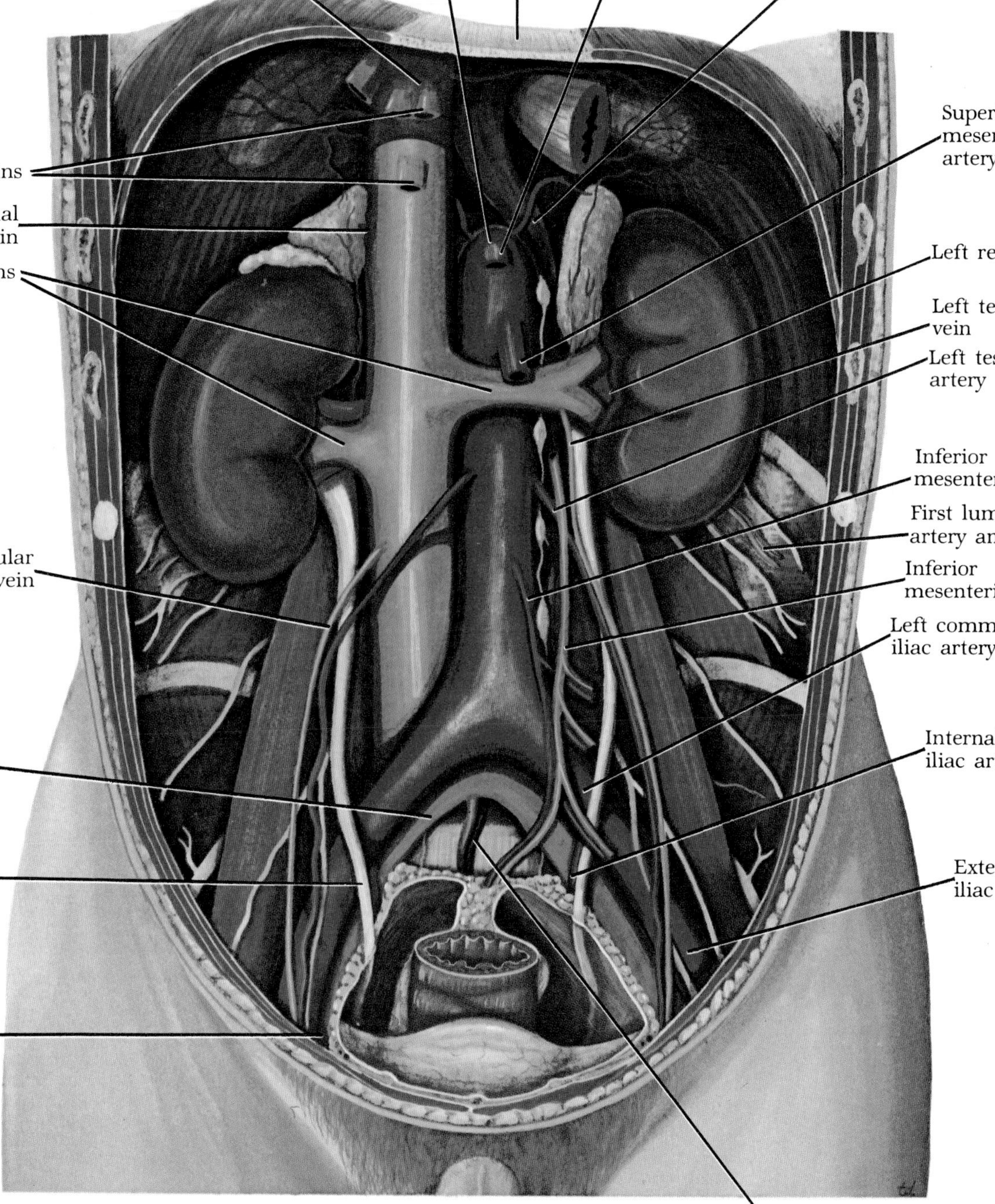

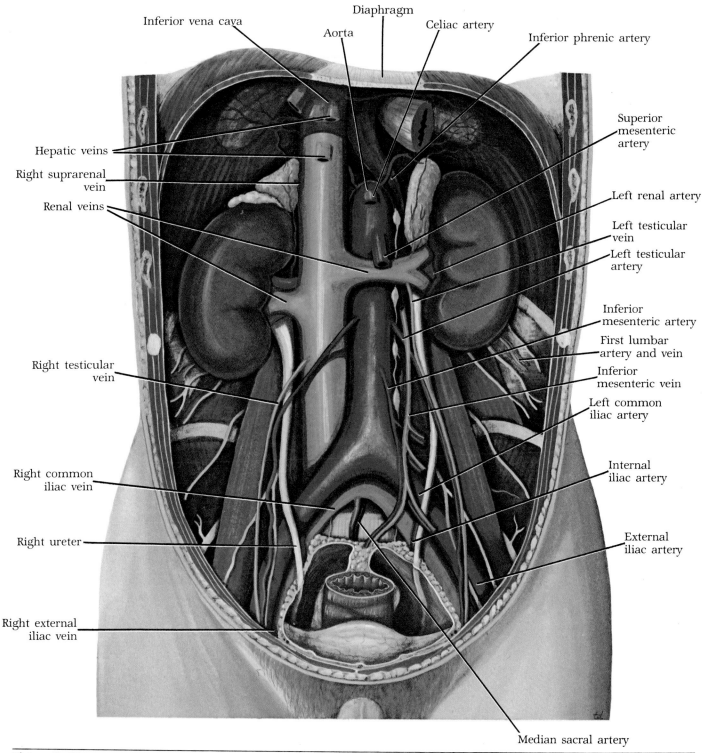

Diaphragm
Inferior vena cava
Aorta
Celiac artery
Inferior phrenic artery
Superior mesenteric artery
Hepatic veins
Right suprarenal vein
Left renal artery
Renal veins
Left testicular vein
Left testicular artery
Inferior mesenteric artery
First lumbar artery and vein
Right testicular vein
Inferior mesenteric vein
Left common iliac artery
Internal iliac artery
Right common iliac vein
Right ureter
External iliac artery
Right external iliac vein
Median sacral artery

Figure 2-45
Aorta and inferior vena cava.

1. Localized or diffuse dilatations of the abdominal part of the aorta (aneurysms) usually occur below the origin of the renal arteries. The majority result from atherosclerosis and occur most commonly in elderly men. The aneurysm characteristically starts below the renal arteries and extends down to involve the aortic bifurcation. Large aneurysms should be surgically excised and replaced with a prosthetic graft.

2. Gradual, progressive occlusion of the bifurcation of the abdominal aorta produced by atherosclerosis results in the characteristic clinical symptoms of pain in the legs on walking (claudication) and impotence, the latter being due to lack of blood in the internal iliac arteries. In otherwise healthy individuals, surgical treatment by thromboendarterectomy or a bypass graft should be considered.

3. The bifurcation of the abdominal aorta where the lumen suddenly narrows may be a lodging site for an embolus discharged from the heart. Severe ischemia of the lower limbs then results.

4. Because of the loose retroperitoneal connective tissue, a malignant retroperitoneal tumor often becomes very large before producing symptoms. It may eventually compress the inferior vena cava, producing swelling and varicose veins of a lower extremity.

5. During the later stages of pregnancy, the enlarged uterus often compresses the inferior vena cava, producing edema of the ankles and feet and temporary varicose veins.

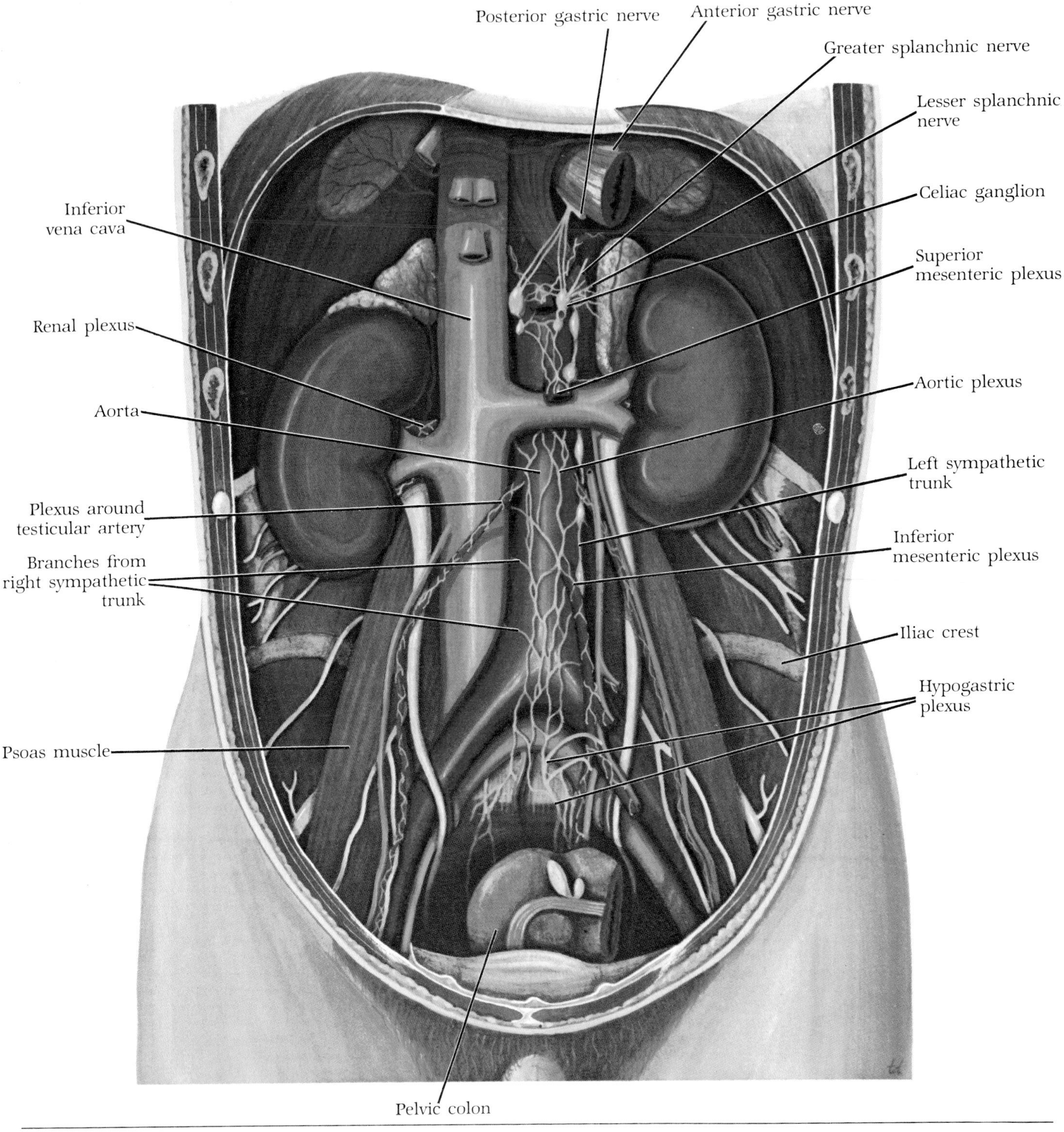
Posterior gastric nerve
Anterior gastric nerve
Greater splanchnic nerve
Lesser splanchnic nerve
Celiac ganglion
Superior mesenteric plexus
Aortic plexus
Left sympathetic trunk
Inferior mesenteric plexus
Iliac crest
Hypogastric plexus
Inferior vena cava
Renal plexus
Aorta
Plexus around testicular artery
Branches from right sympathetic trunk
Psoas muscle
Pelvic colon

Figure 2-46
Aorta and related sympathetic plexuses.

1. The retroperitoneal area is bounded anteriorly by the peritoneum, posteriorly by the vertebral column and psoas and quadratus lumborum muscles, superiorly by the diaphragm, and inferiorly by the brim of the pelvis. It contains many important organs and large blood vessels; the spaces between these structures are filled with fatty connective tissue. The retroperitoneal area is important clinically since there are no boundaries, and disease processes can extend widely on both sides of the body. Moreover since the area is large any pathological condition is usually advanced before there are any signs or symptoms. Radiological examination combined with pyelography, arteriography, and presacral air insufflation all assist in making a diagnosis.

2. Lumbar sympathectomy to produce vasodilatation of the arteries of the leg is a common treatment for patients with vasospastic disorders. The preganglionic sympathetic fibers that supply the vessels of the lower limb leave the spinal cord from segments T11 to L2 and synapse in the lumbar and sacral ganglia of the sympathetic trunks. The postganglionic fibers join the lumbar and sacral nerves and are distributed to the vessels of the limb as branches of these nerves. Additional postganglionic fibers pass directly from the lumbar ganglia to the common and external iliac arteries. In the male a bilateral lumbar sympathectomy may be followed by loss of ejaculatory power although erection is not impaired.

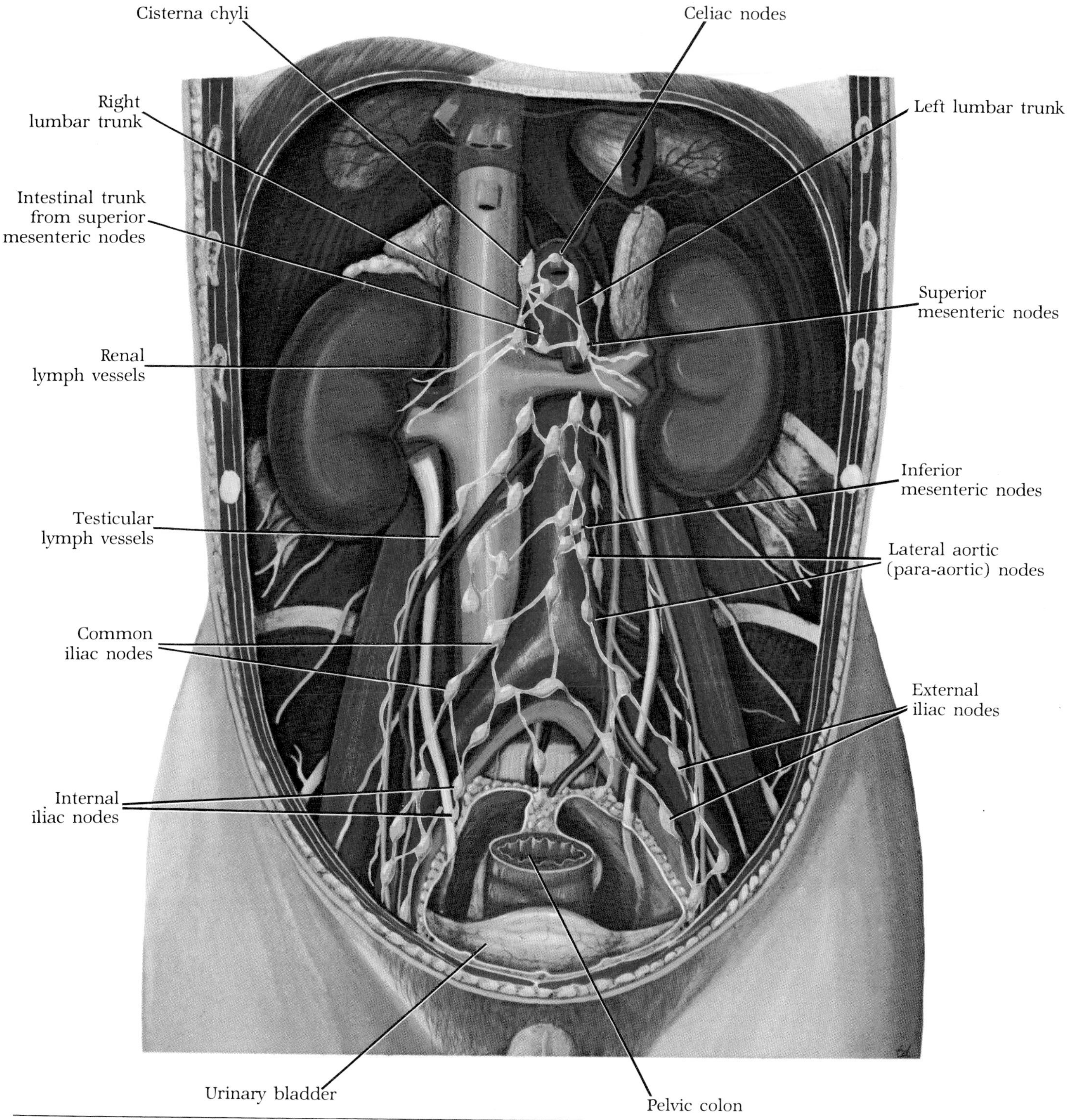
Cisterna chyli
Celiac nodes
Right lumbar trunk
Left lumbar trunk
Intestinal trunk from superior mesenteric nodes
Superior mesenteric nodes
Renal lymph vessels
Inferior mesenteric nodes
Testicular lymph vessels
Lateral aortic (para-aortic) nodes
Common iliac nodes
External iliac nodes
Internal iliac nodes
Urinary bladder
Pelvic colon

Figure 2-47
Lymph vessels and lymph nodes on the posterior abdominal wall.

1. The cisterna chyli is a dilatation of the lower end of the thoracic duct. The lumbar trunks are efferent vessels of the lumbar (lateral aortic) lymph nodes that receive lymph from the lower limbs, kidneys, suprarenals and the testes or ovaries, and the deep lymph vessels from the greater part of the abdominal wall, as well as from the pelvic walls and pelvic viscera. The intestinal trunks receive lymph from the gastrointestinal tract, pancreas, spleen, and from part of the liver.

2. The deep lymph nodes should be examined in patients with generalized enlargement of the lymph nodes in the cervical, axillary, and inguinal regions. Enlargement of the mediastinal nodes is detected by chest radiograph. The deep inguinal, iliac, and lumbar (lateral aortic) nodes may be examined by lymphangiography. A radiopaque dye is injected into the lymph vessels of the feet. From there it drains into the inguinal, external iliac, common iliac, and lumbar nodes; enlargement of these nodes can then be visualized on a radiograph.

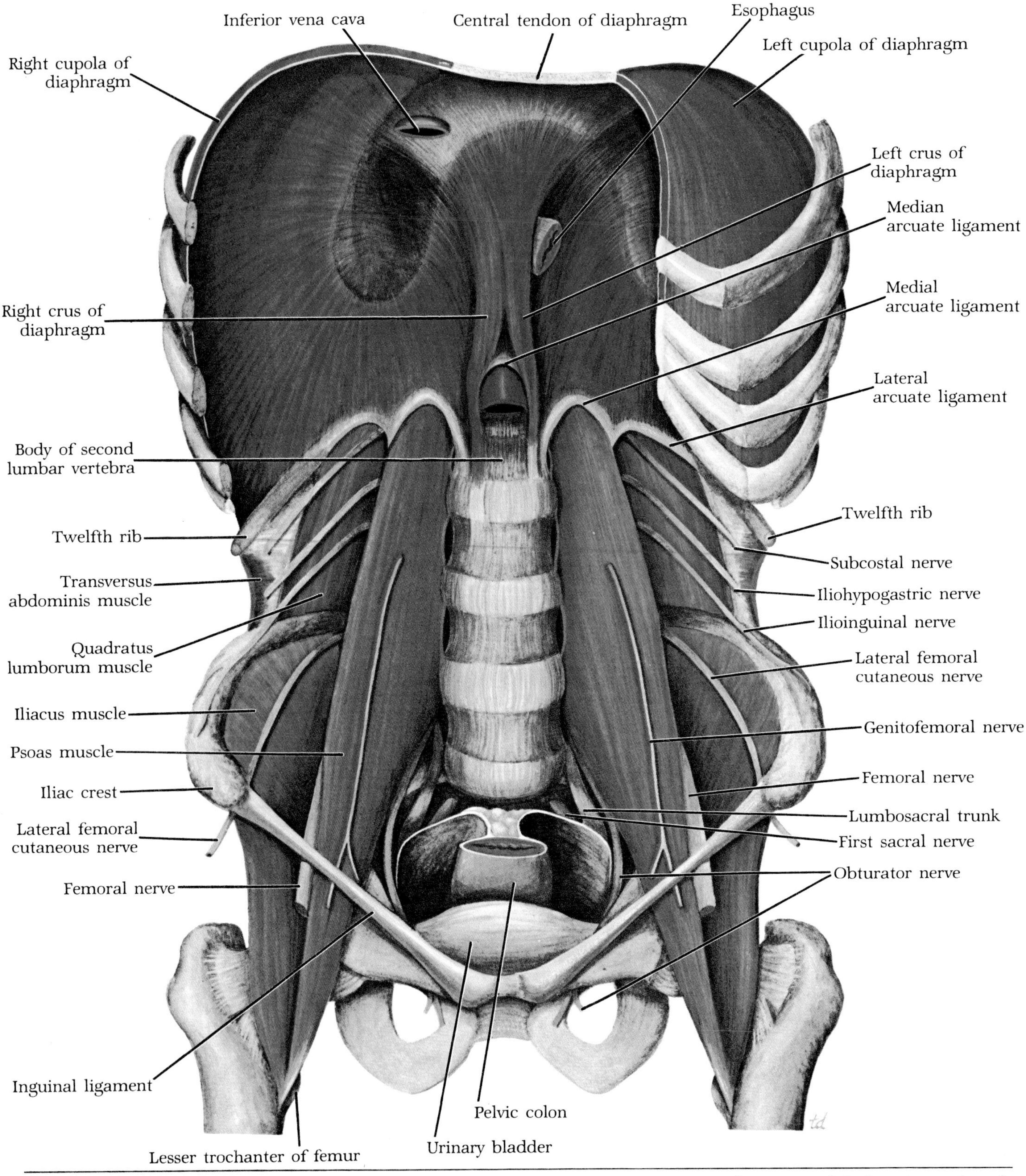
Inferior vena cava
Central tendon of diaphragm
Esophagus
Left cupola of diaphragm
Right cupola of diaphragm
Left crus of diaphragm
Median arcuate ligament
Right crus of diaphragm
Medial arcuate ligament
Lateral arcuate ligament
Body of second lumbar vertebra
Twelfth rib
Twelfth rib
Subcostal nerve
Transversus abdominis muscle
Iliohypogastric nerve
Ilioinguinal nerve
Quadratus lumborum muscle
Lateral femoral cutaneous nerve
Iliacus muscle
Genitofemoral nerve
Psoas muscle
Femoral nerve
Iliac crest
Lumbosacral trunk
Lateral femoral cutaneous nerve
First sacral nerve
Femoral nerve
Obturator nerve
Inguinal ligament
Pelvic colon
Urinary bladder
Lesser trochanter of femur
td

Figure 2-48
Posterior abdominal wall.

1. The extensive origin of the diaphragm from the thoracic outlet is clearly shown in this illustration. It must be remembered that the lower margins of the lungs and pleura extend inferiorly behind the diaphragm and are closely related to the upper abdominal organs and the subphrenic peritoneal spaces. Peritoneal infection can spread through the diaphragm to involve the pleura. Penetrating wounds of the upper abdomen can also involve the thoracic contents.

2. Note the placement of the psoas sheath here. Tuberculous infection of a lumbar vertebra may result in the extravasation of pus down the psoas sheath into the thigh. The presence of a swelling above and below the inguinal ligament, together with clinical signs and symptoms referred to the vertebral column, should make the diagnosis obvious.

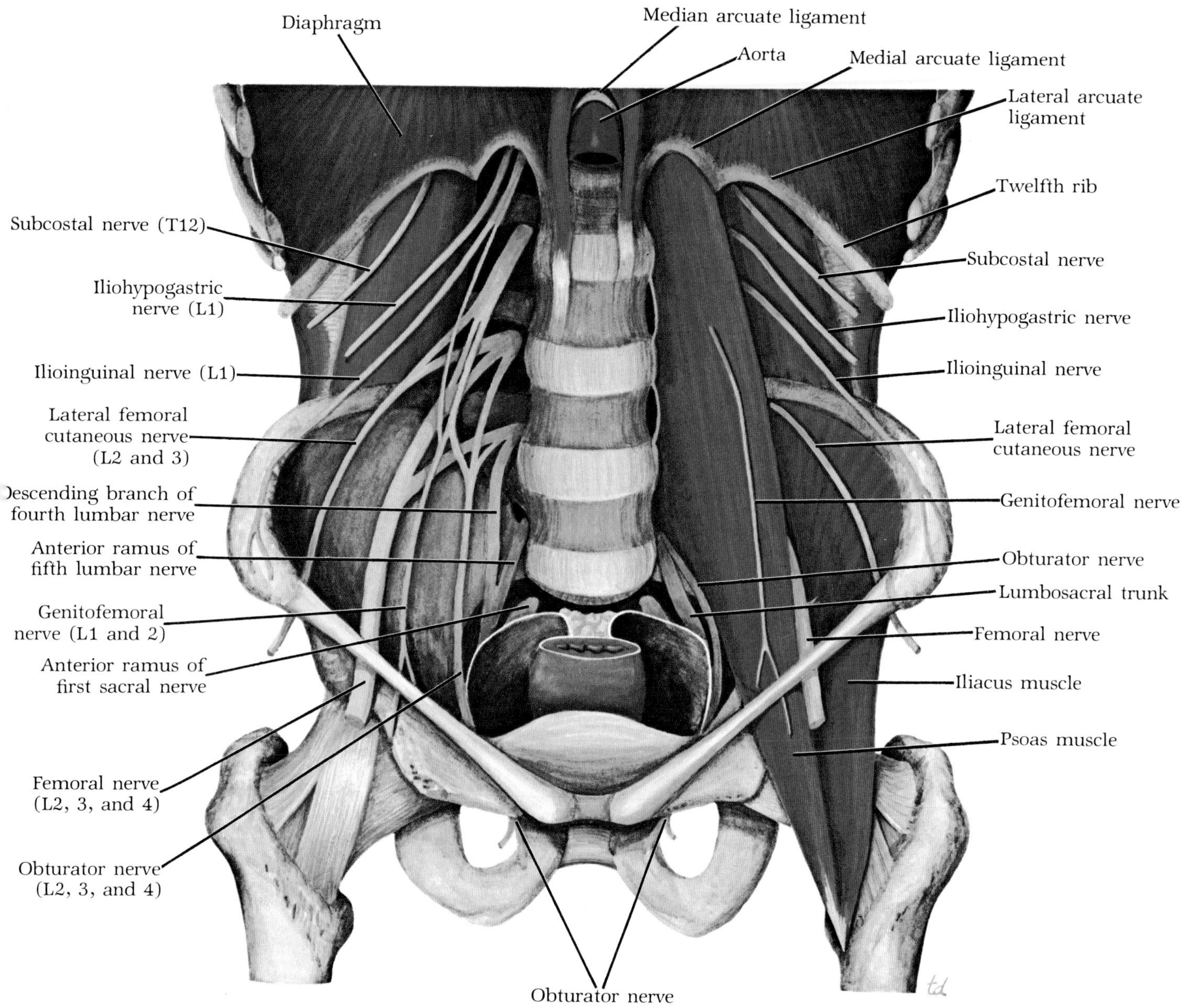
Diaphragm
Median arcuate ligament
Aorta
Medial arcuate ligament
Lateral arcuate ligament
Twelfth rib
Subcostal nerve (T12)
Subcostal nerve
Iliohypogastric nerve (L1)
Iliohypogastric nerve
Ilioinguinal nerve (L1)
Ilioinguinal nerve
Lateral femoral cutaneous nerve (L2 and 3)
Lateral femoral cutaneous nerve
Descending branch of fourth lumbar nerve
Genitofemoral nerve
Anterior ramus of fifth lumbar nerve
Obturator nerve
Lumbosacral trunk
Genitofemoral nerve (L1 and 2)
Femoral nerve
Anterior ramus of first sacral nerve
Iliacus muscle
Psoas muscle
Femoral nerve (L2, 3, and 4)
Obturator nerve (L2, 3, and 4)
Obturator nerve

Figure 2-49
Lumbar plexus, anterior view. On the left, the psoas muscle has been removed to reveal the main parts of the lumbar plexus.

1. The lumbar plexus lies within the substance of the psoas muscle. It is formed from the anterior rami of the first four lumbar nerves. The main branches are the femoral and obturator nerves that arise from the second, third, and fourth lumbar nerves. Except for the obturator nerve, which emerges at the medial border of the psoas, and the genitofemoral nerve, which emerges on the anterior aspect, all remaining branches of the plexus emerge on the lateral border.

2. Because of its protected position, the lumbar plexus is rarely damaged and then only by penetrating injuries. The roots of the plexus, however, are commonly involved in diseases of the vertebral column such as herniation of the intervertebral discs, osteoarthritis, etc.

3. The lateral femoral cutaneous nerve sometimes pierces the inguinal ligament; if the ligament then presses on the nerve, as occasionally happens, there may be pain, then anesthesia, down the lateral side of the thigh.

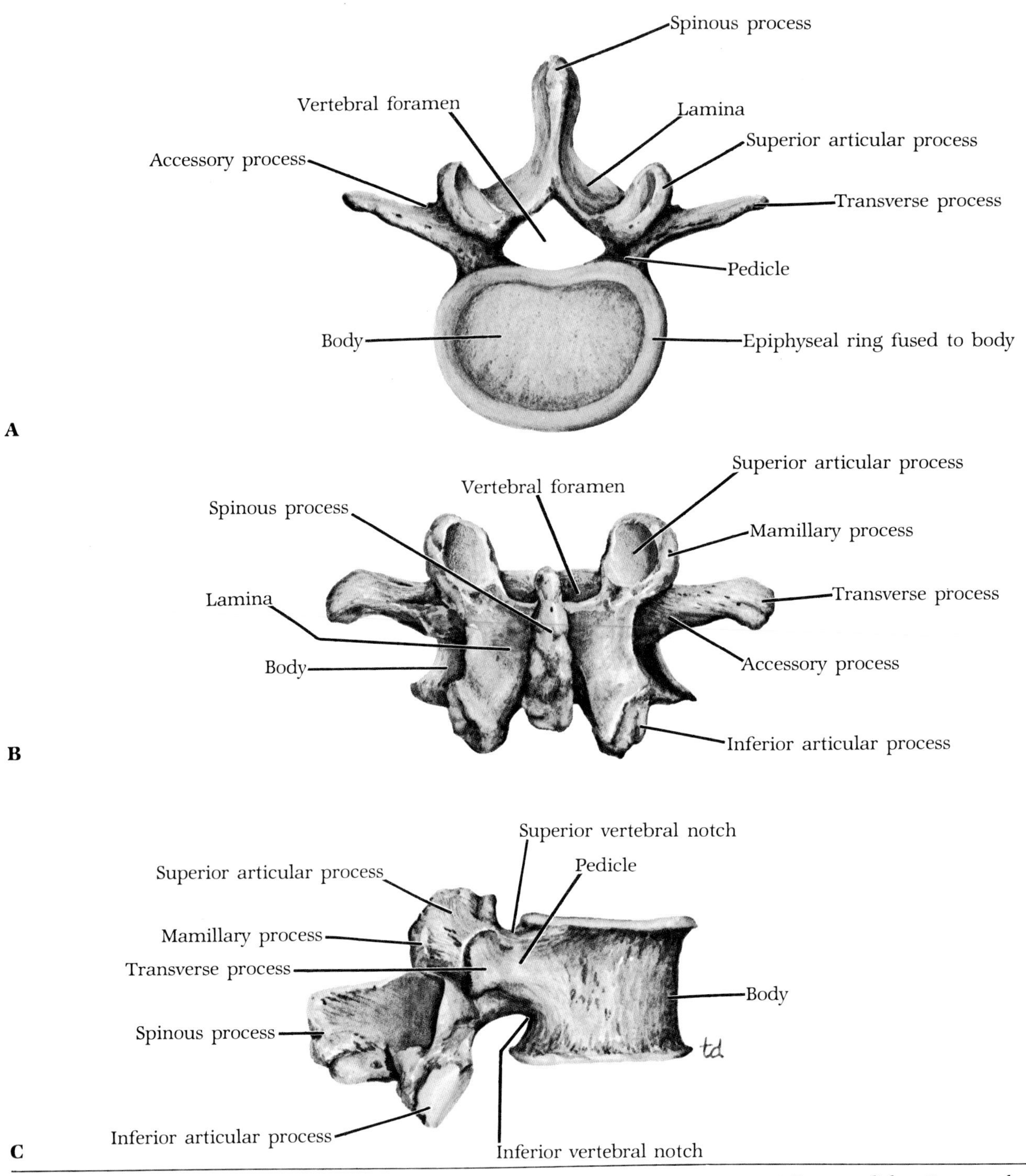
Spinous process
Vertebral foramen
Lamina
Superior articular process
Accessory process
Transverse process
Pedicle
Body
Epiphyseal ring fused to body
A
Superior articular process
Vertebral foramen
Spinous process
Mamillary process
Transverse process
Lamina
Accessory process
Body
Inferior articular process
B
Superior vertebral notch
Pedicle
Superior articular process
Mamillary process
Transverse process
Body
Spinous process
td
Inferior articular process
C
Inferior vertebral notch

Figure 2-50
Typical lumbar vertebra (third). A. Superior view. B. Posterior view. C. Lateral view.

1. Each intervertebral foramen is bounded above and below by the pedicles of adjacent vertebrae, anteriorly by the lower part of the vertebral body and by the intervertebral disc, and posteriorly by the articular processes and the joint between them. The spinal nerve thus is very vulnerable and may be pressed on or irritated by disease of the surrounding structures. Herniation of the intervertebral disc, fractures of the vertebral bodies, and osteoarthritis involving the joints of the articular processes or the joints between the vertebral bodies, may all result in pressure, stretching, or edema of the emerging spinal nerve.
2. Fractures of the spinous processes, transverse processes, or laminae are caused by direct injury or, in rare cases, by severe muscular activity.
3. Compression fractures of the vertebral bodies are usually caused by an excessive flexion-compression type of injury.
4. Fracture-dislocations are also caused by an excessive flexion-compression type of injury. Because the articular processes are fractured and the ligaments torn, the vertebrae are unstable and the spinal cord or cauda equina is usually severely damaged.
5. In spondylolisthesis, the body of a lower lumbar vertebra, usually the fifth, moves forward on the body of the vertebra below and carries with it the whole of the upper portion of the vertebral column. The condition is due to a congenital defect in the pedicles of the migrating vertebra.

3. The Pelvis

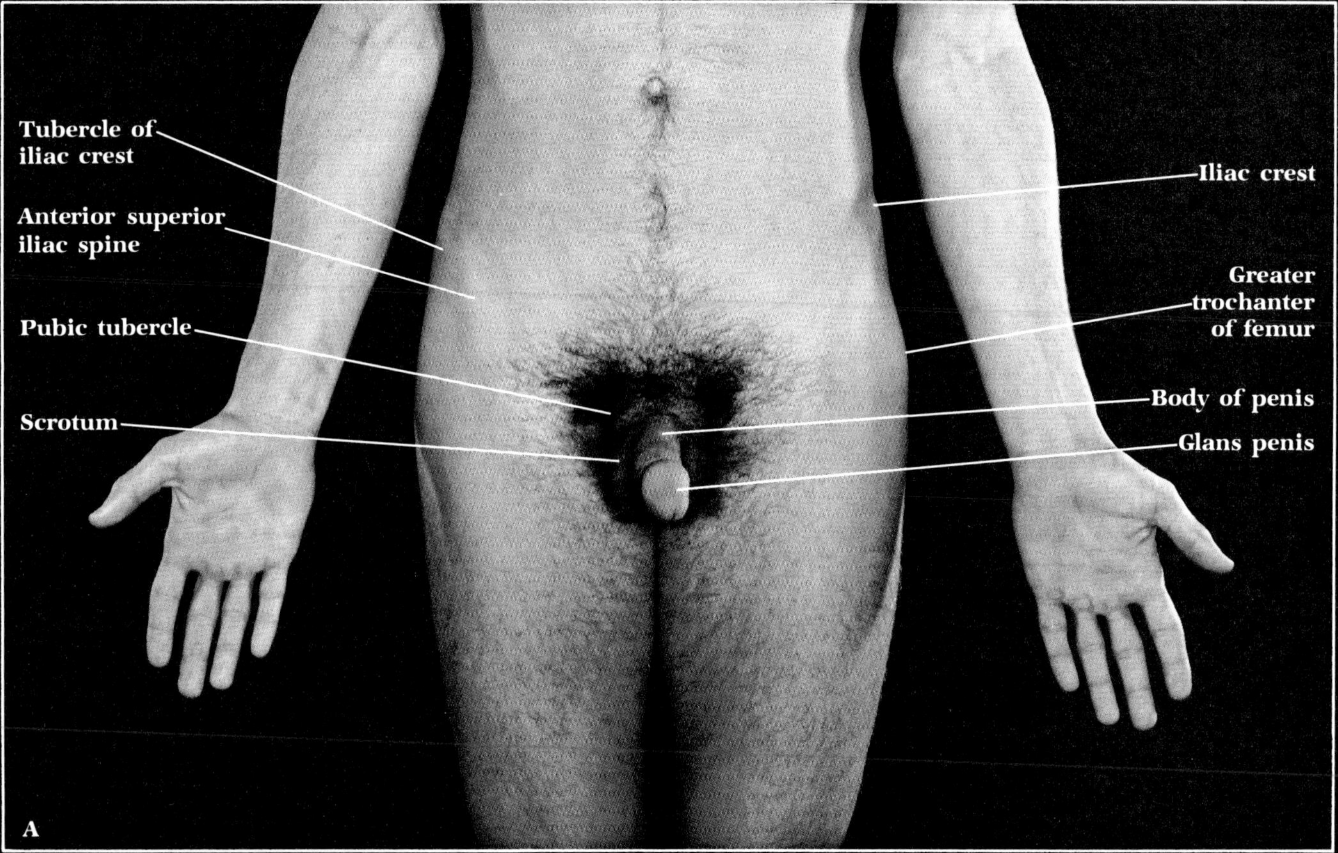
Tubercle of iliac crest
Anterior superior iliac spine
Pubic tubercle
Scrotum
Iliac crest
Greater trochanter of femur
Body of penis
Glans penis
A

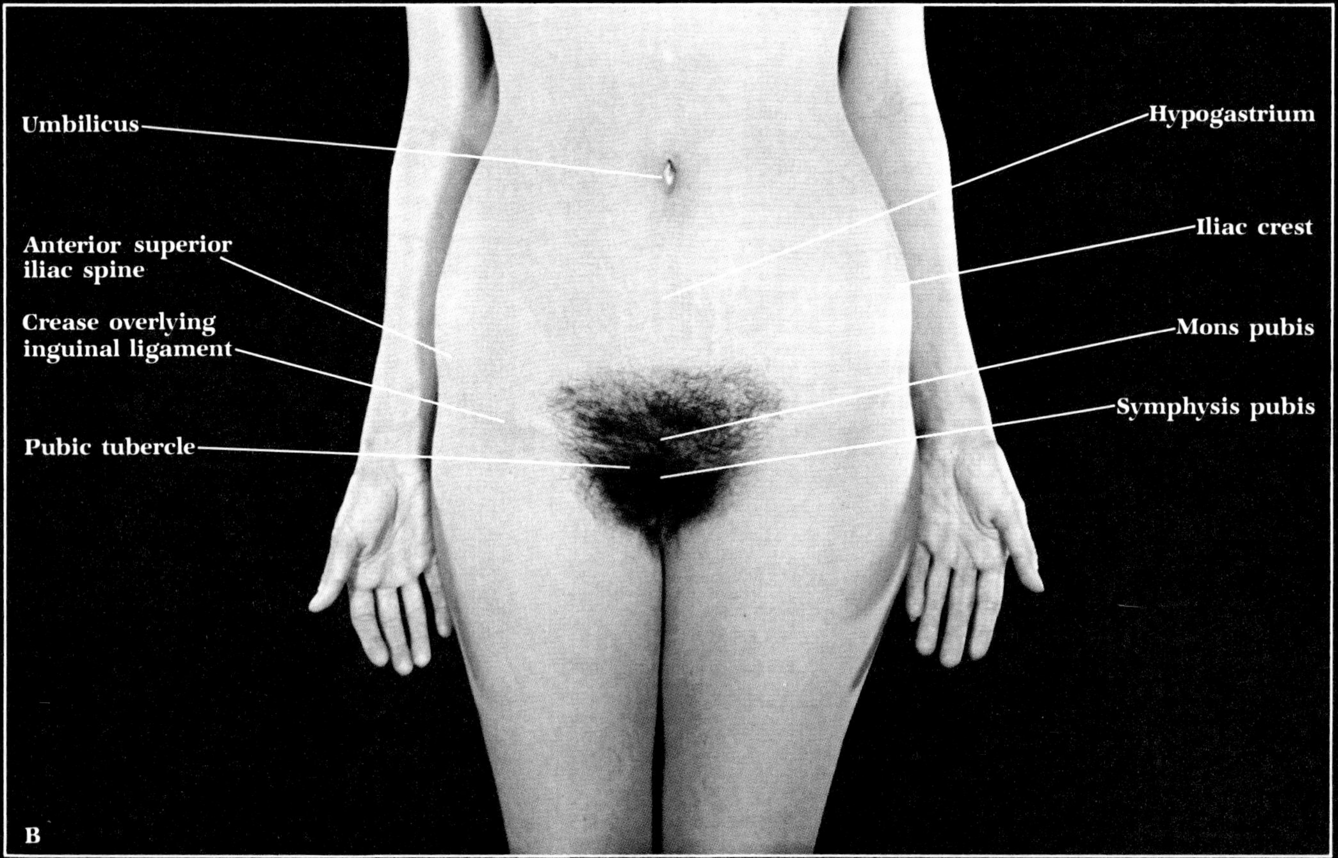
Umbilicus
Anterior superior iliac spine
Crease overlying inguinal ligament
Pubic tubercle
Hypogastrium
Iliac crest
Mons pubis
Symphysis pubis
B

Figure 3-1
The pelvis, anterior view.
A. 27-year-old male. B. 29-year-old female.

1. The iliac crest can be felt through the skin along its entire length.
2. The anterior superior iliac spine is situated at the anterior end of the iliac crest and lies at the upper lateral end of the fold of the groin.
3. The posterior superior iliac spine is situated at the posterior end of the iliac crest.
4. The tubercle of the iliac crest is about 2 inches (5 cm) posterior to the anterior superior iliac spine and projects from the outer edge of the iliac crest. The tubercle lies at the level of the body of the fifth lumbar vertebra.
5. The pubic crest is the ridge of bone on the superior surface of the pubic bone, medial to the pubic tubercle.
6. The symphysis pubis is the cartilaginous joint that lies in the midline between the bodies of the pubic bones.
7. The urinary bladder in the young child is an abdominopelvic organ, and, when filled, it can be palpated through the anterior abdominal wall above the symphysis pubis. As the child grows, the pelvic cavity enlarges and the bladder becomes a pelvic organ. It is important to remember that, even in the adult, when the bladder becomes distended with urine its superior wall rises out of the pelvis and may be palpated through the anterior abdominal wall in the hypogastrium.

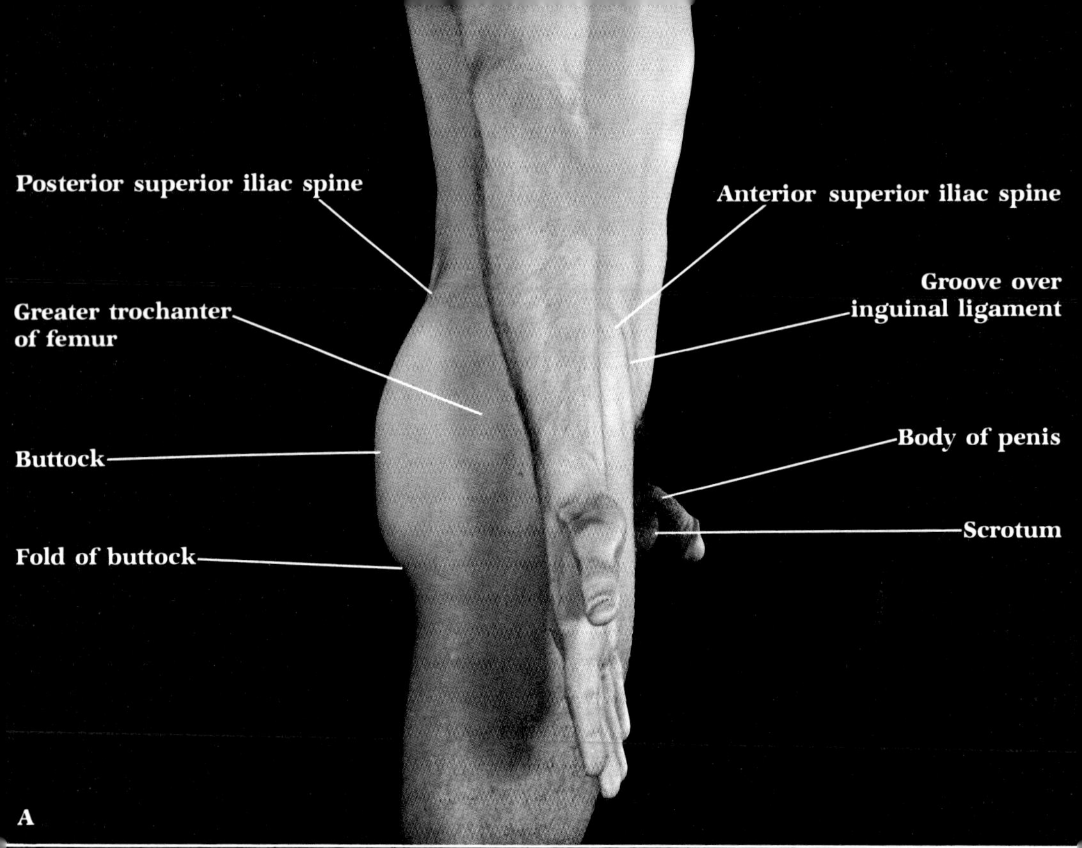

Posterior superior iliac spine
Anterior superior iliac spine
Groove over inguinal ligament
Greater trochanter of femur
Buttock
Body of penis
Scrotum
Fold of buttock

A

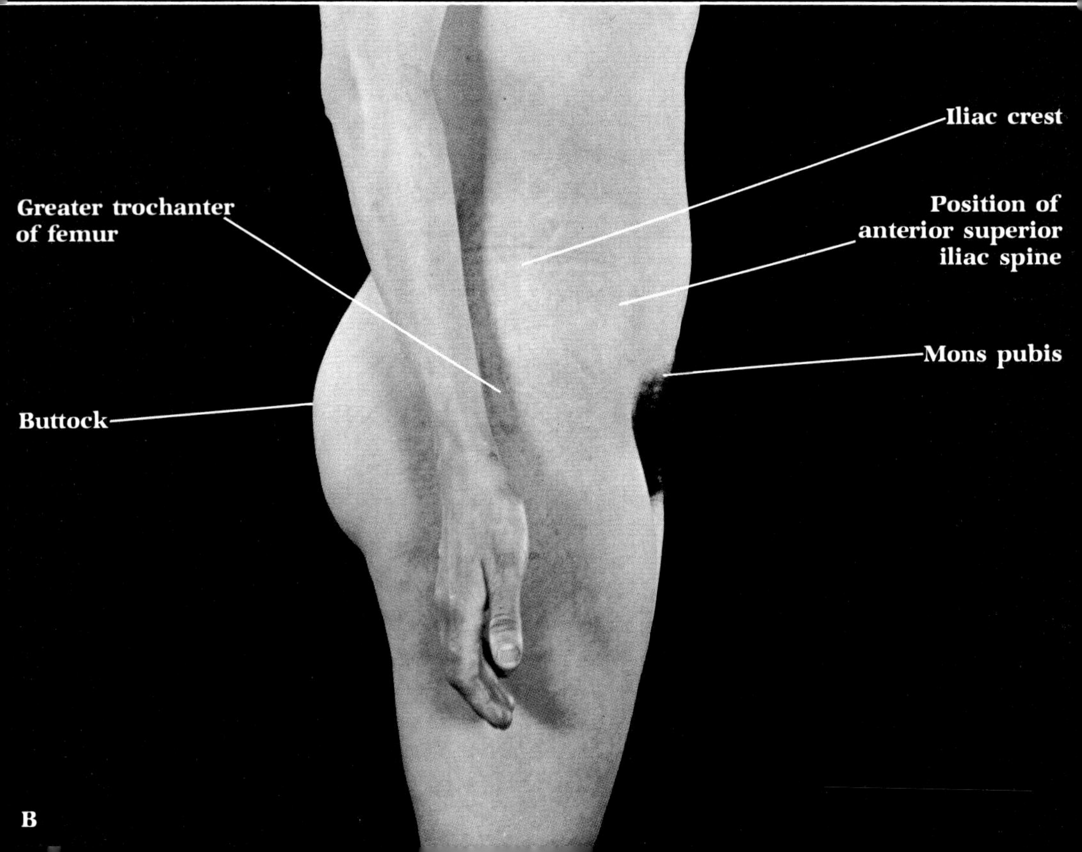

Iliac crest
Greater trochanter of femur
Position of anterior superior iliac spine
Mons pubis
Buttock

B

Figure 3-2
The pelvis, lateral view.
A. 27-year-old male. B. 29-year-old female.

1. The iliac crest can be felt along its entire length. It ends anteriorly at the anterior superior iliac spine and posteriorly at the posterior superior iliac spine. Its highest point lies opposite the body of the fourth lumbar vertebra.

2. The tubercle of the iliac crest is situated on the lateral margin of the crest about 2 inches (5 cm) posterior to the anterior superior iliac spine. It lies on the level of the body of the fifth lumbar vertebra.

3. The symphysis pubis is the cartilaginous joint that lies in the midline between the bodies of the pubic bones.

4. As mentioned previously, the urinary bladder in the young child is an abdominopelvic organ and when filled can be palpated through the anterior abdominal wall. In a similar manner, the overdistended adult bladder can also be palpated through the anterior abdominal wall.

5. The fundus of the pregnant uterus can be palpated through the lower part of the anterior abdominal wall toward the end of the second month of pregnancy.

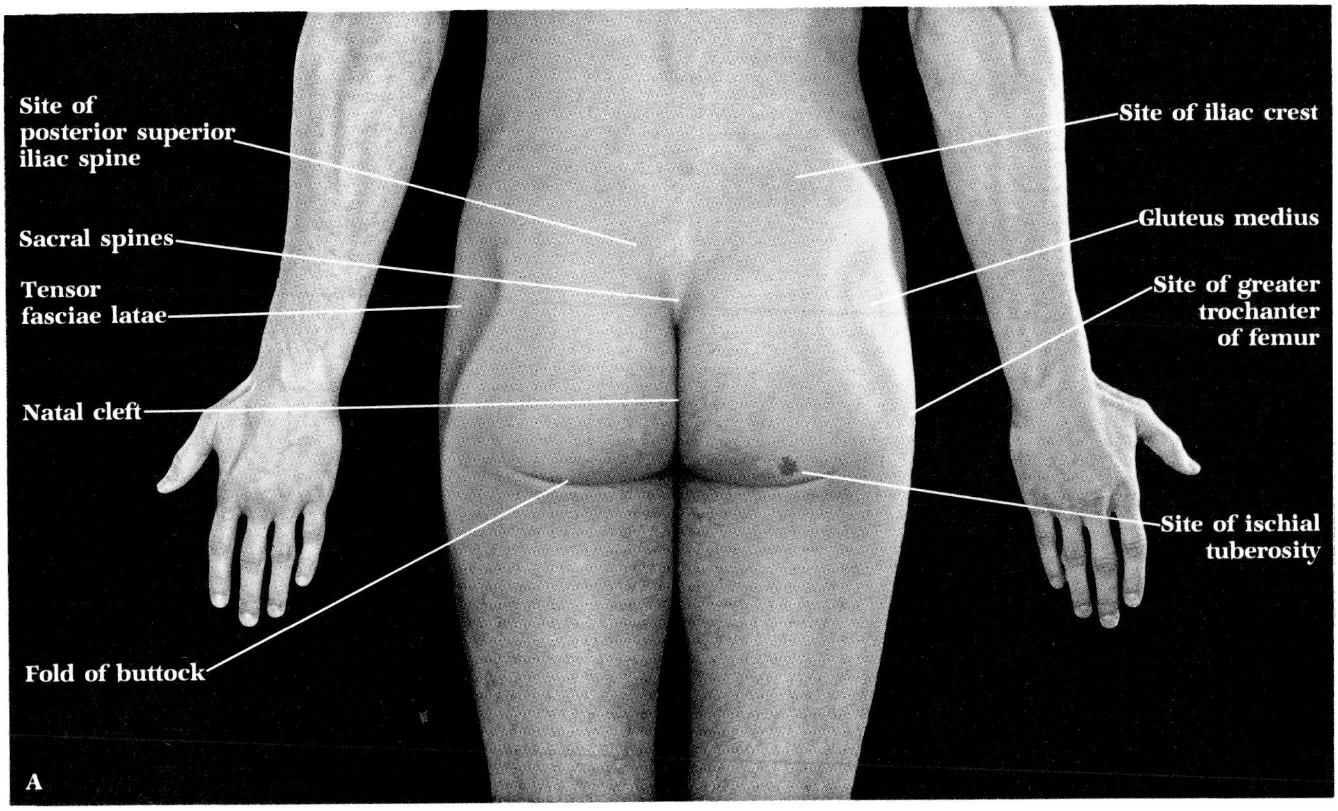
Site of posterior superior iliac spine
Sacral spines
Tensor fasciae latae
Natal cleft
Fold of buttock
Site of iliac crest
Gluteus medius
Site of greater trochanter of femur
Site of ischial tuberosity
A

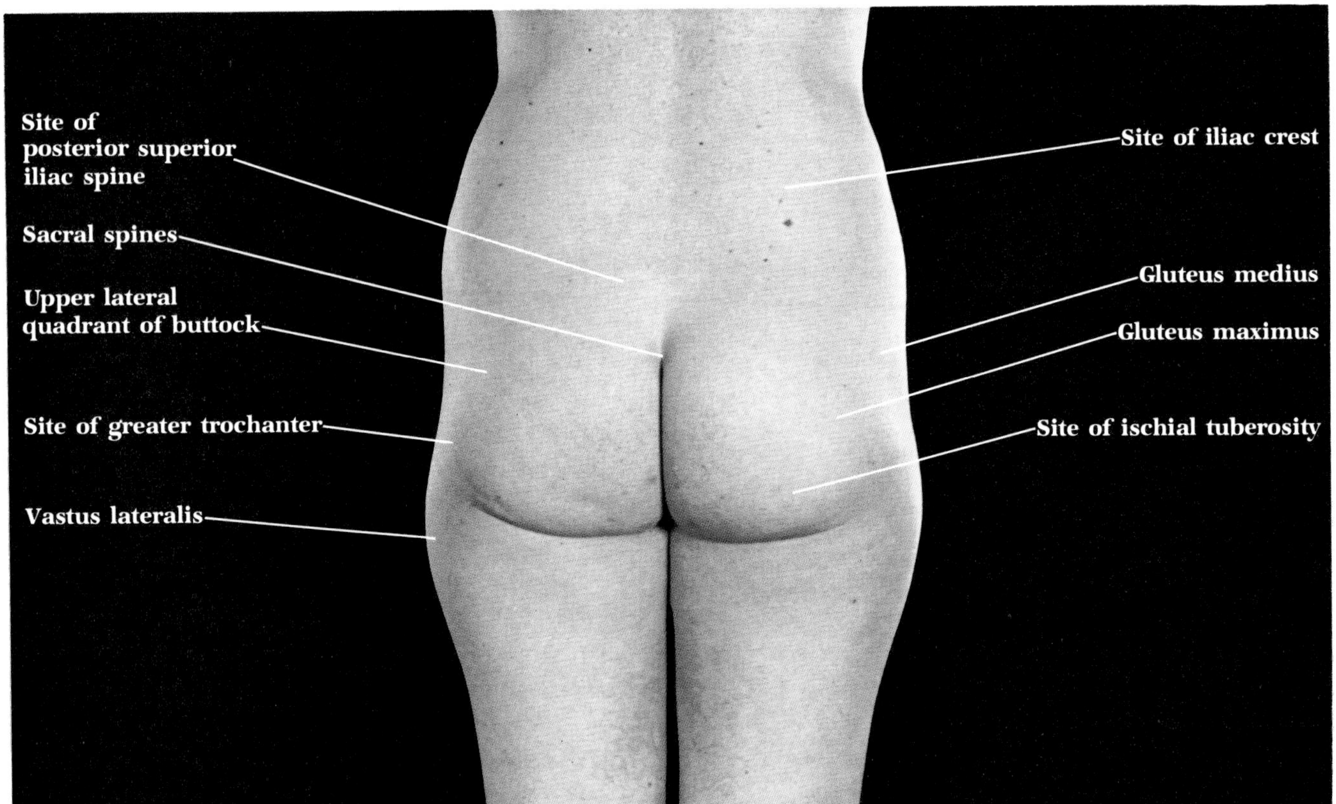
Site of posterior superior iliac spine
Sacral spines
Upper lateral quadrant of buttock
Site of greater trochanter
Vastus lateralis
Site of iliac crest
Gluteus medius
Gluteus maximus
Site of ischial tuberosity
B

Figure 3-3
The pelvis, posterior view.
A. 27-year-old male. B. 25-year-old female.

1. The posterior superior iliac spine is situated at the posterior end of the iliac crest. It lies at the bottom of a small skin dimple and on a level with the second sacral spine, which coincides with the lower limit of the subarachnoid space.

2. The spinous processes of the sacrum are fused with each other in the midline to form the median sacral crest. The crest can be felt beneath the skin in the uppermost part of the natal cleft between the buttocks.

3. The sacral hiatus is situated on the posterior aspect of the lower end of the sacrum, and it is here that the extradural space terminates. The hiatus lies about 2 inches (5 cm) above the tip of the coccyx and beneath the skin of the natal cleft.

4. The tip of the coccyx can be palpated in the natal cleft about 1 inch (2.5 cm) behind the anus. The anterior surface of the coccyx may be palpated with a gloved finger in the anal canal.

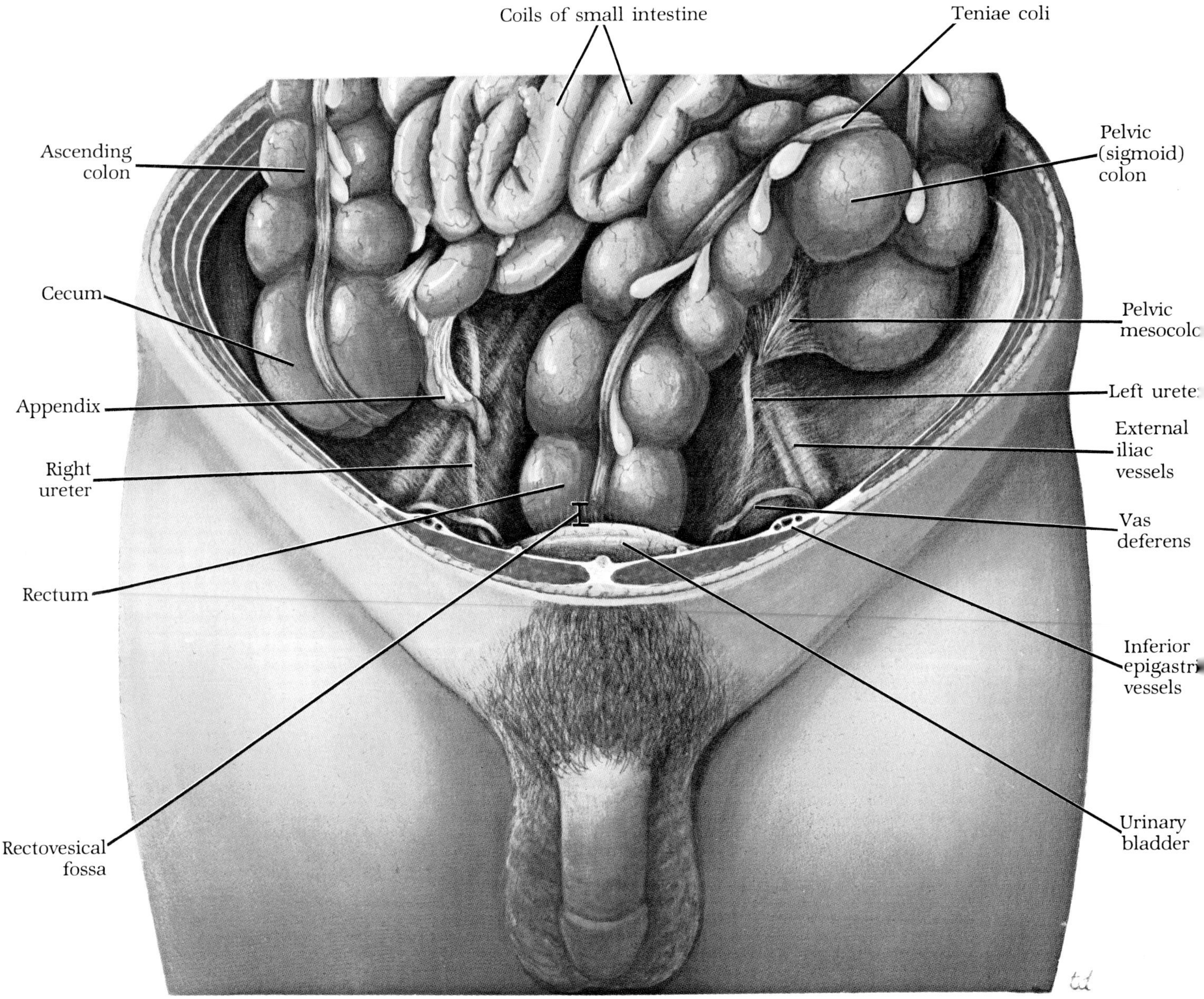
Coils of small intestine
Teniae coli
Ascending colon
Pelvic (sigmoid) colon
Cecum
Pelvic mesocolo
Appendix
Left urete
External iliac vessels
Right ureter
Vas deferens
Rectum
Inferior epigastri
vessels
Urinary bladder
Rectovesical fossa

Figure 3-4
Lower part of the abdominal cavity in the male, oblique anterior view.

1. Coils of small and large intestine normally hang down into the pelvis where they are in close proximity to the bladder and the rectum. In diseases of the pelvic (sigmoid) colon such as ulcerative colitis or diverticulitis, the bowel may become adherent to the bladder, rectum, ileum, or ureter and produce an internal fistula.

2. In the child the bladder is in a higher position than in an adult due to the relatively smaller size of the pelvis. A full bladder in the young child is easily palpated through the lower part of the anterior abdominal wall because the internal urethral orifice lies at the level of the superior border of the symphysis pubis. When a low abdominal incision into the abdomen is made, the surgeon must make sure that the infant's bladder is empty.

3. If an inflamed appendix is hanging down into the pelvis there may be no abdominal tenderness in the right iliac region, but deep tenderness may be experienced in the hypogastric region. Rectal examination (or vaginal examination in the female) may reveal tenderness of the peritoneum in the pelvis on the right side.

4. Perforation of an inflamed appendix may produce a localized pelvic peritonitis.

5. The mucous membrane of the pelvic sigmoid colon can be examined with a sigmoidoscope. Such examination is important because carcinoma of the sigmoid colon is relatively common.

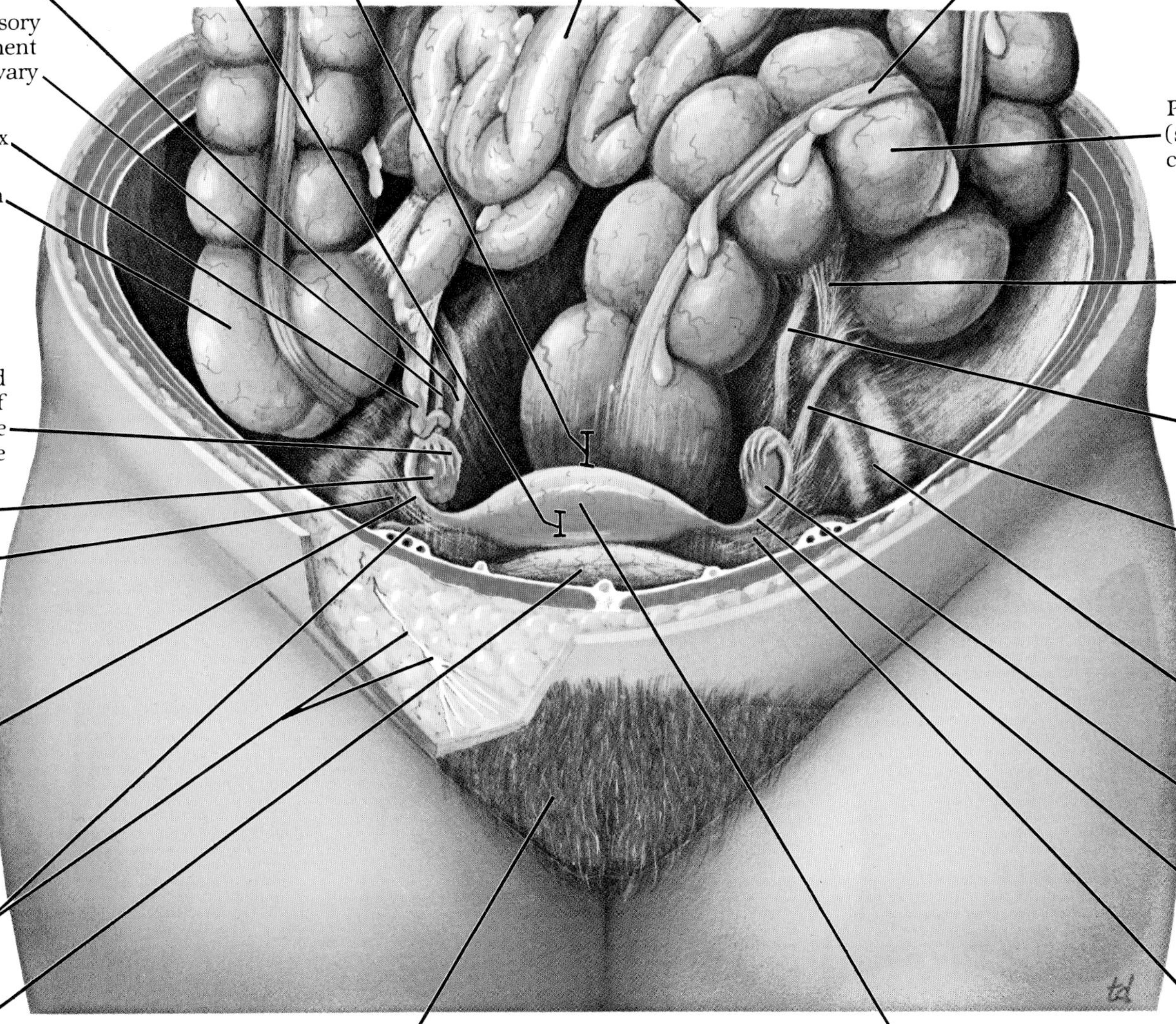

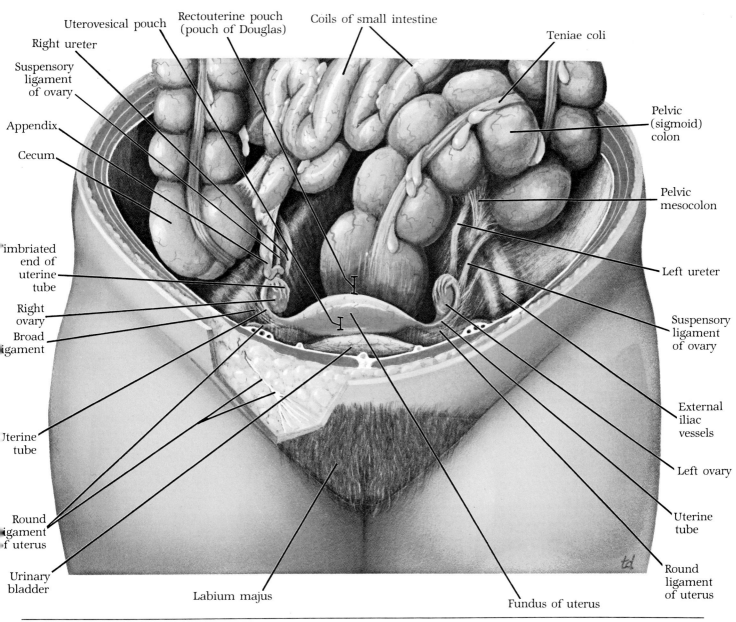

Uterovesical pouch
Rectouterine pouch (pouch of Douglas)
Coils of small intestine
Teniae coli
Right ureter
Suspensory ligament of ovary
Appendix
Cecum
Pelvic (sigmoid) colon
Pelvic mesocolon
imbriated end of uterine tube
Left ureter
Right ovary
Broad igament
Suspensory ligament of ovary
External iliac vessels
Jterine tube
Left ovary
Round igament f uterus
Uterine tube
Round ligament of uterus
Urinary bladder
Labium majus
Fundus of uterus

Figure 3-5
Lower part of the abdominal cavity in the female, oblique anterior view.

1. The peritoneal cavity in the female is open down to the vulva by means of the uterine tubes, uterus, and vagina. Ascending infection, for example gonorrhea, can consequently produce a pelvic peritonitis.
2. A bimanual examination of the uterus may be made if the bladder is empty. The index and middle fingers of the gloved right hand are inserted into the anterior fornix of the vagina, and the left hand is placed on the anterior abdominal wall in the hypogastrium; the fundus and body of the uterus may then be palpated between the abdominal and vaginal fingers. This physical examination is often of great help in diagnosing pregnancy and detecting tumors of the uterus.
3. The position of the ovary is very variable and largely depends on the position of the uterus to which it is attached by the round ligament.
4. Retroversion (a bending posteriorly) of the uterus will result in the ovary prolapsing into the rectouterine pouch (pouch of Douglas). When in this position, the ovary may be palpated through the posterior vaginal wall. Such a position also may cause pain during intercourse (dyspareunia).
5. Note the course of the ureter in the pelvis. In advanced cases of carcinoma of the cervix, the ureter may be invaded or pressed upon so that dilatation of the ureter proximal to this site may occur.
6. Acute inflammation of the uterine tube and ovary (salpingo-oophoritis) produces lower abdominal pain that is usually bilateral. The pain in acute appendicitis is in the center of the abdomen, subsequently localizing to the right iliac fossa.
7. The round ligament of the uterus consists of smooth muscle and connective tissue and hypertrophies during pregnancy. It is larger in multiparae than in nulliparae. The lymph vessels that pass along it drain the fundus of the uterus to the superficial inguinal lymph nodes. Advanced carcinoma of the fundus may result in metastases to the inguinal nodes.

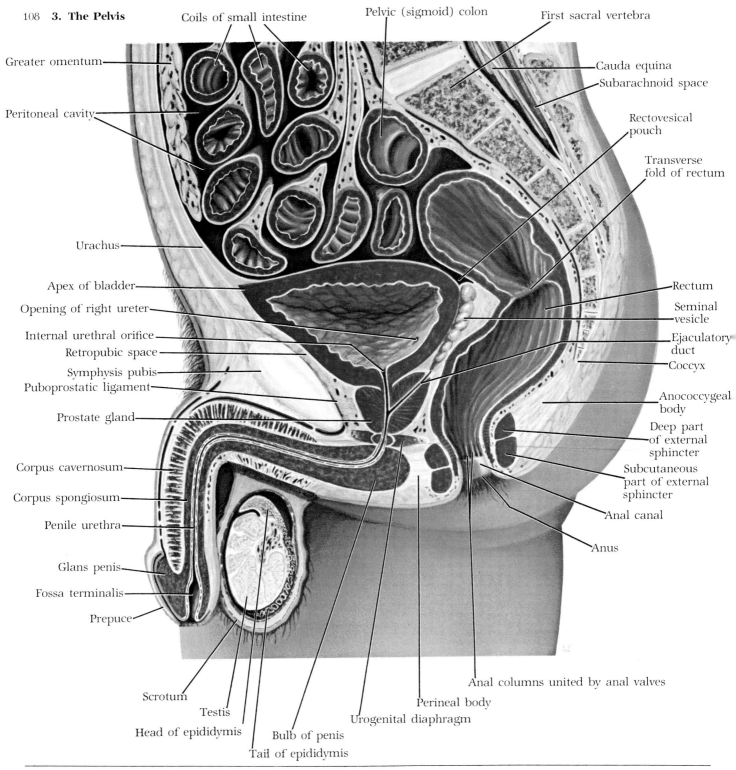

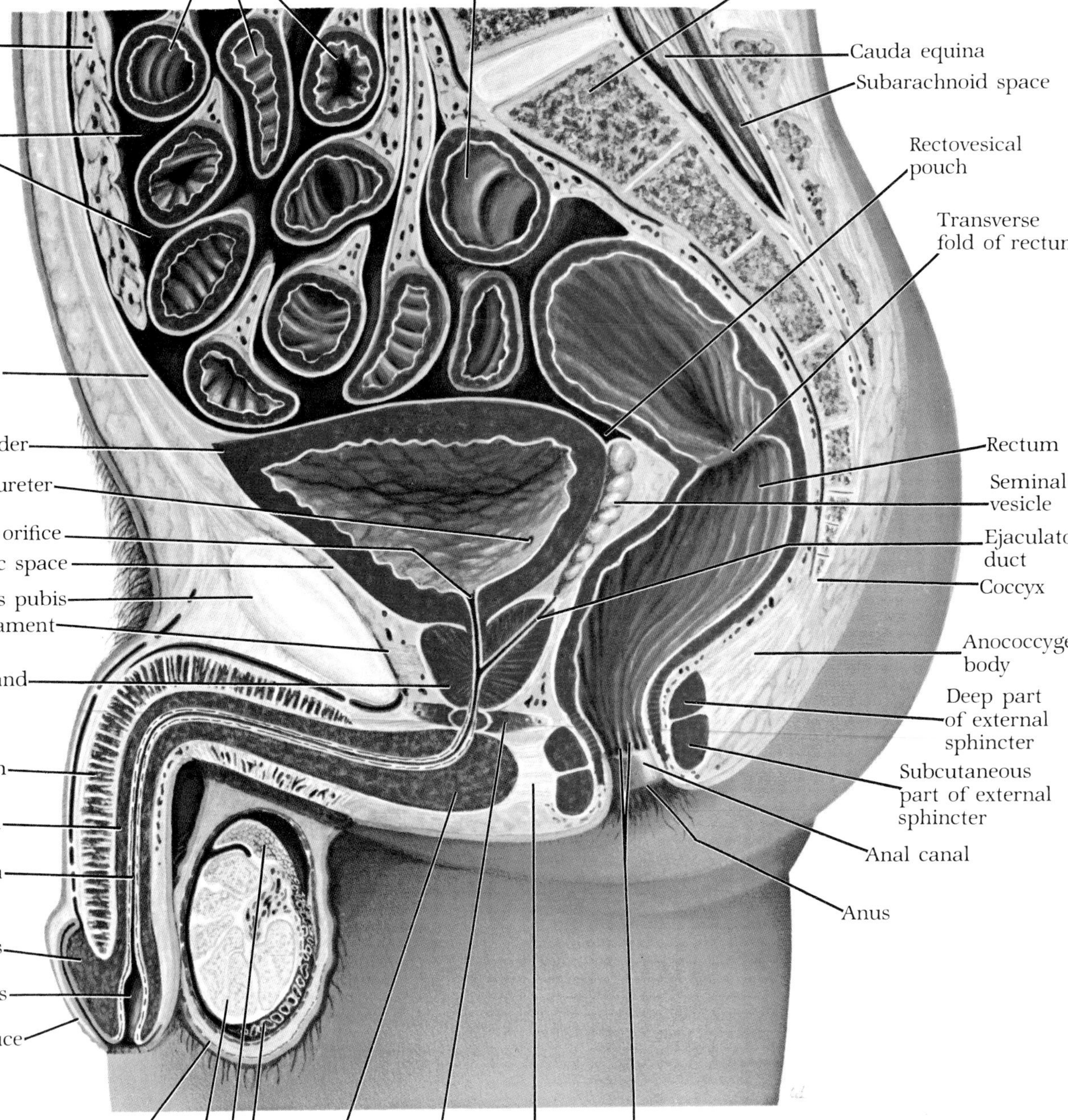
Coils of small intestine
Pelvic (sigmoid) colon
First sacral vertebra
Greater omentum
Cauda equina
Subarachnoid space
Peritoneal cavity
Rectovesical pouch
Transverse fold of rectum
Urachus
Apex of bladder
Rectum
Opening of right ureter
Seminal vesicle
Internal urethral orifice
Ejaculatory duct
Retropubic space
Symphysis pubis
Coccyx
Puboprostatic ligament
Anococcygeal body
Prostate gland
Deep part of external sphincter
Corpus cavernosum
Subcutaneous part of external sphincter
Corpus spongiosum
Penile urethra
Anal canal
Anus
Glans penis
Fossa terminalis
Prepuce
Anal columns united by anal valves
Scrotum
Perineal body
Testis
Urogenital diaphragm
Head of epididymis
Bulb of penis
Tail of epididymis

Figure 3-6
Sagittal section of male pelvis.

1. The tributaries of the superior rectal vein drain venous blood from the mucous membrane of the rectum and the upper half of the anal canal. Abnormal dilatations of these tributaries form internal hemorrhoids.
2. Partial and complete prolapse of the rectum through the anus occurs relatively frequently.
3. Carcinoma of the rectum is relatively common. In advanced cases the malignant neoplasm may extend posteriorly to invade the sacrum and the sacral plexus, laterally to involve the ureter, and anteriorly to involve the prostate, seminal vesicle, or bladder.
4. The mucous membrane of the lower rectum can be examined by inserting a gloved finger through the anal canal. The great majority of cancers of the rectum can be diagnosed by this means.
5. Proctoscopy, the introduction of an internally illuminated tubular instrument through the anus, enables the physician to examine the greater part of the rectal mucosa.
6. The full bladder in the adult projects up into the abdomen and, in extreme cases, may even rise above the umbilicus. A severe blow to the lower part of the anterior abdominal wall in a person with a full bladder may result in intraperitoneal rupture of the bladder.
7. As the bladder fills, the superior wall rises out of the pelvis and peels the peritoneum off the posterior surface of the anterior abdominal wall. It is thus possible, as an emergency procedure, to pass a needle into a full bladder through the lower part of the anterior abdominal wall without entering the peritoneal cavity.
8. Benign enlargement of the prostate is common in men over 50 years of age. The resulting compression of the urethra and interference in function with the sphincter vesicae cause difficulties in micturition and possibly adverse back-pressure effects on the ureters and both kidneys.
9. The prostate and seminal vesicles may be palpated by means of a gloved finger inserted into the rectum.

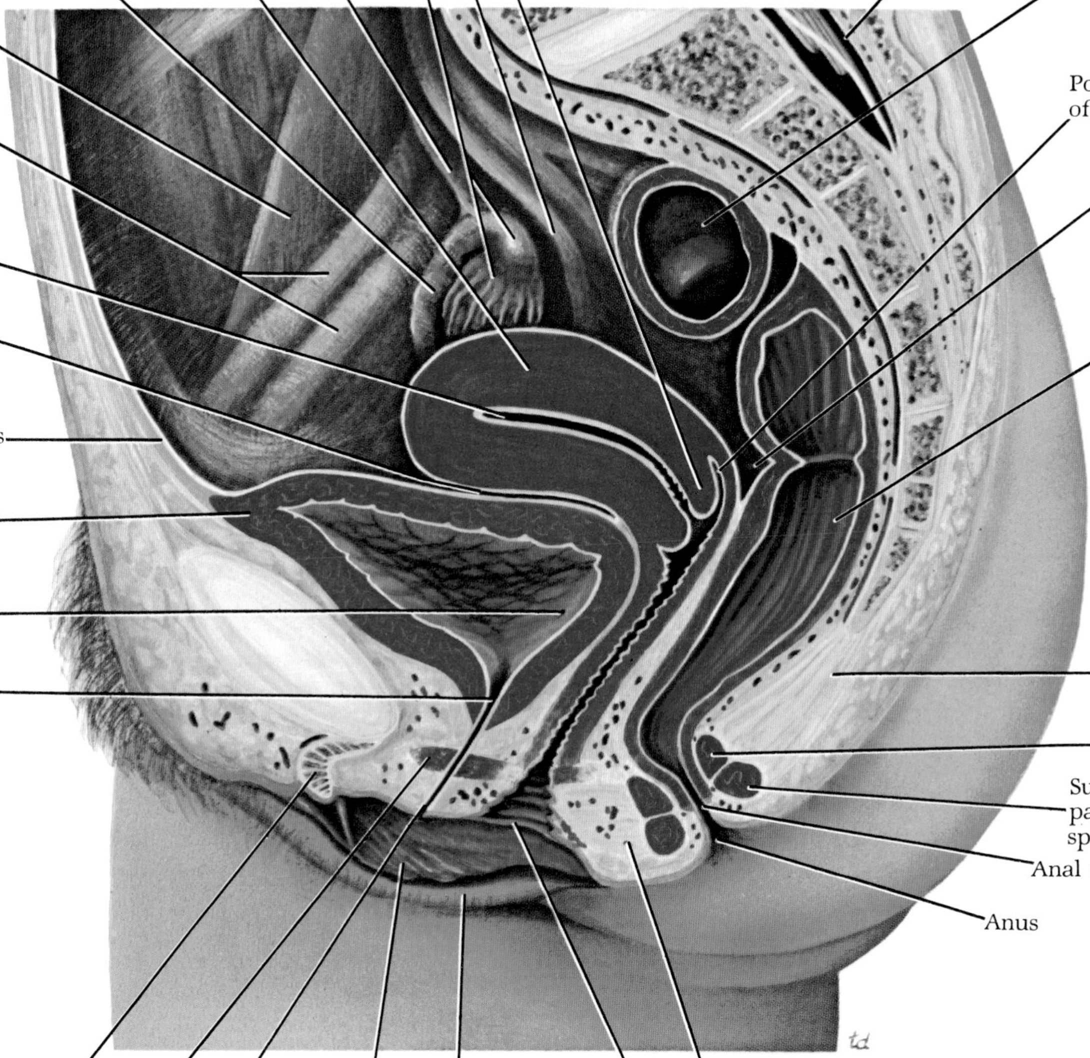

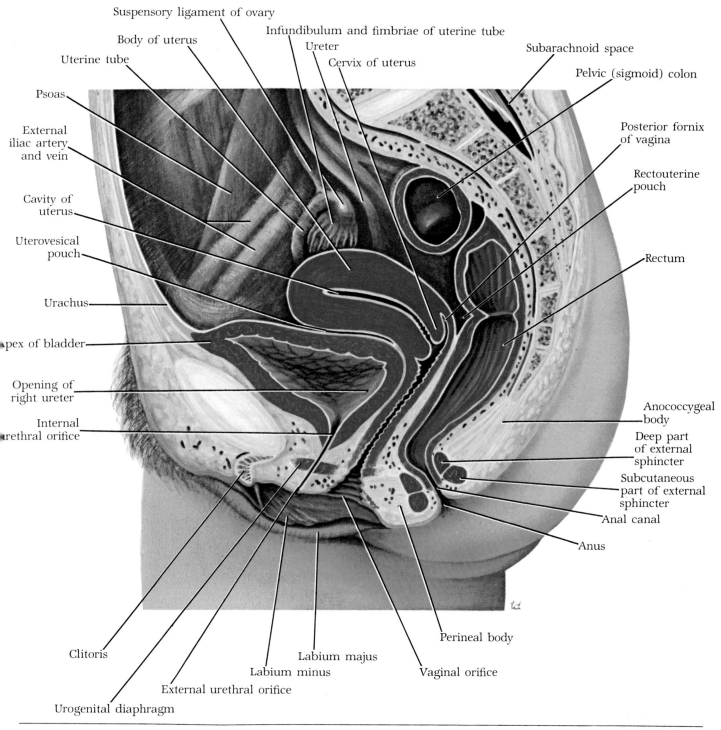

Suspensory ligament of ovary
Infundibulum and fimbriae of uterine tube
Body of uterus
Ureter
Uterine tube
Cervix of uterus
Subarachnoid space
Pelvic (sigmoid) colon
Psoas
External iliac artery and vein
Posterior fornix of vagina
Rectouterine pouch
Cavity of uterus
Uterovesical pouch
Rectum
Urachus
pex of bladder
Opening of right ureter
Anococcygeal body
Internal rethral orifice
Deep part of external sphincter
Subcutaneous part of external sphincter
Anal canal
Anus
Perineal body
Clitoris
Labium majus
Labium minus
Vaginal orifice
External urethral orifice
Urogenital diaphragm

Figure 3-7
Sagittal section of female pelvis.

1. The rectouterine pouch (pouch of Douglas) is the most inferior part of the peritoneal cavity in the female. The peritoneum passes from the anterior wall of the rectum to cover the upper part of the posterior wall of the vagina and the posterior wall of the uterus. Collections of pus or blood in the pouch can be easily evacuated by incising the vaginal wall in the posterior fornix.

2. Fractures of the pelvis commonly occur in automobile accidents. Splinters of bone are driven into the pelvic cavity and frequently penetrate the bladder, vagina, and rectum. The pelvic veins that lie beneath the parietal peritoneum, if damaged, are a source of severe hemorrhage under these circumstances.

3. Many disease conditions occurring in the female pelvis may be diagnosed by a vaginal examination. The following structures may be palpated through the vaginal walls from above downward: *anteriorly*, (a) the bladder and (b) urethra; *posteriorly*, (a) loops of ileum and pelvic colon in the rectouterine pouch (pouch of Douglas), (b) the rectal ampulla, and (c) the perineal body; *laterally*, (a) the ureters, (b) the pelvic fascia and the anterior fibers of the levatores ani muscles, and (c) the urogenital diaphragm.

4. The normal uterus is a mobile organ, and its position is altered by the state of the bladder. In the majority of normal women the uterus is anteverted and anteflexed. A retroverted, retroflexed uterus may cause dyspareunia and congestive dysmenorrhea.

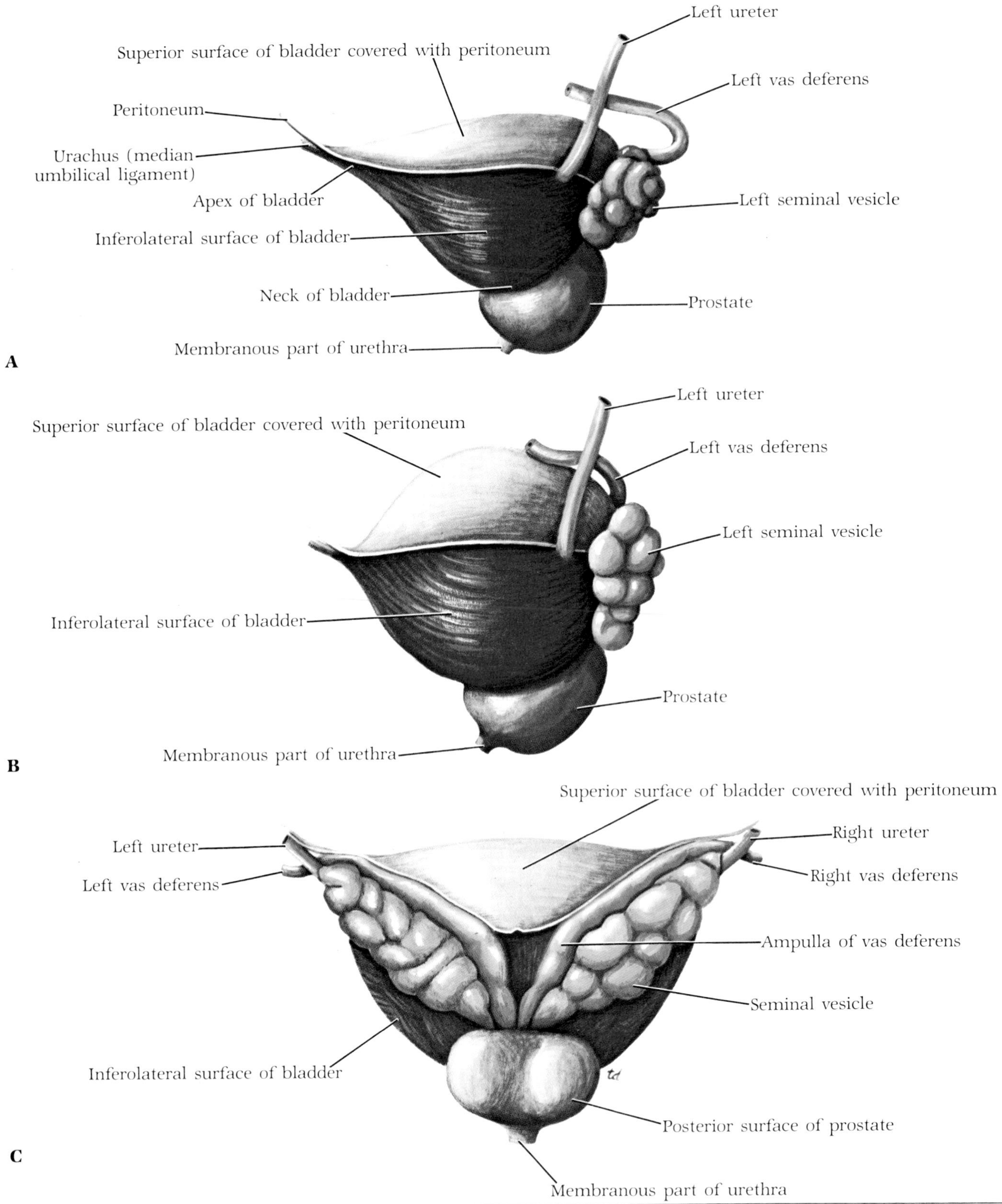
Left ureter
Superior surface of bladder covered with peritoneum
Left vas deferens
Peritoneum
Urachus (median umbilical ligament)
Apex of bladder
Left seminal vesicle
Inferolateral surface of bladder
Neck of bladder
Prostate
Membranous part of urethra
A
Left ureter
Superior surface of bladder covered with peritoneum
Left vas deferens
Left seminal vesicle
Inferolateral surface of bladder
Prostate
Membranous part of urethra
B
Superior surface of bladder covered with peritoneum
Left ureter
Right ureter
Left vas deferens
Right vas deferens
Ampulla of vas deferens
Seminal vesicle
Inferolateral surface of bladder
Posterior surface of prostate
C
Membranous part of urethra

Figure 3-8
Urinary bladder. A. Empty, showing prostate and seminal vesicle, lateral view. B. Full, showing prostate and seminal vesicle, lateral view. C. Empty, showing prostate and seminal vesicle, posterior view.

1. The urinary bladder becomes ovoid in shape as it fills with urine. This alteration in shape is largely due to the stretching of the inferolateral and superior surfaces.
2. The terminal parts of the vas deferens, the seminal vesicles, and the posterior surface of the prostate can be palpated by inserting a gloved index finger into the anal canal and rectum. It is possible to diagnose disease of these organs by this method.
3. In benign hypertrophy of the prostate the median lobe of the gland enlarges upward and encroaches within the sphincter vesicae, located at the neck of the bladder. The leakage of urine into the prostatic urethra causes an intense reflex desire to micturate. The enlargement of the median and lateral lobes of the gland produces elongation and lateral compression and distortion of the urethra, so that the patient experiences difficulty in passing urine and the stream is weak.

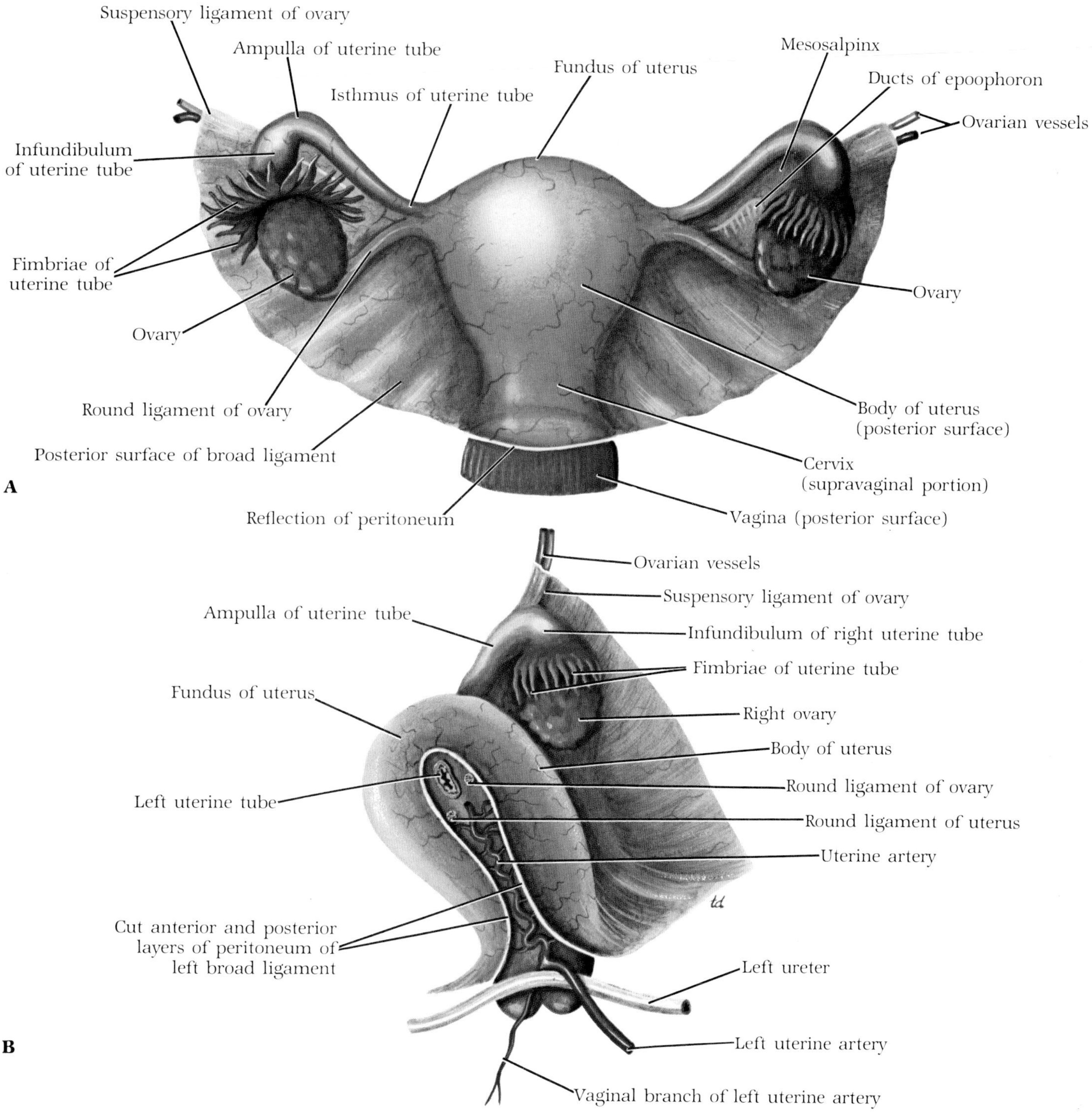
Suspensory ligament of ovary
Ampulla of uterine tube
Isthmus of uterine tube
Fundus of uterus
Mesosalpinx
Ducts of epoophoron
Ovarian vessels
Infundibulum of uterine tube
Fimbriae of uterine tube
Ovary
Ovary
Round ligament of ovary
Posterior surface of broad ligament
Body of uterus (posterior surface)
Cervix (supravaginal portion)
Vagina (posterior surface)
Reflection of peritoneum
A
Ovarian vessels
Suspensory ligament of ovary
Ampulla of uterine tube
Infundibulum of right uterine tube
Fimbriae of uterine tube
Fundus of uterus
Right ovary
Body of uterus
Left uterine tube
Round ligament of ovary
Round ligament of uterus
Uterine artery
Cut anterior and posterior layers of peritoneum of left broad ligament
Left ureter
Left uterine artery
Vaginal branch of left uterine artery
B

Figure 3-9
Uterus. A. Showing broad ligaments and ovaries, posterior view. B. Lateral view. Note structures that lie within broad ligament.

1. The ovary is smooth before puberty but becomes increasingly puckered by repeated ovulation. After the menopause the ovary becomes shriveled and shrinks in size.
2. The ovary is kept in position by the broad ligament and the mesovarium. Following pregnancy the broad ligament is lax, and the ovaries may prolapse into the rectouterine pouch (pouch of Douglas). In these circumstances the ovary may be tender and cause dyspareunia; it may be palpated through the posterior fornix of the vagina.
3. The ovarian artery supplies the ovary and the uterine tube and ends by anastomosing with the uterine artery. The uterine tube is thus supplied by two arteries: the ovarian artery and the anastomosing branch of uterine artery which supplies its medial end. Because it receives two sources of blood, the uterine tube does not become gangrenous when acutely inflamed.
4. Because of the attachment of the cervix to the vaginal vault, it follows that prolapse of the uterus is always accompanied by some prolapse of the vagina.
5. The round ligaments of the uterus may be shortened surgically or attached to the anterior abdominal wall to produce anteversion in a retroverted uterus.
6. A variety of ectopic pregnancy occurs when implantation of a fertilized ovum occurs in the wall of the uterine tube. The trophoblast quickly erodes through the wall of the tube and a large quantity of blood effuses into the peritoneal cavity.
7. The ureter is crossed by the uterine artery as the former passes beneath the broad ligament. A badly placed ligature around the uterine artery during the course of a hysterectomy operation could easily include the ureter.

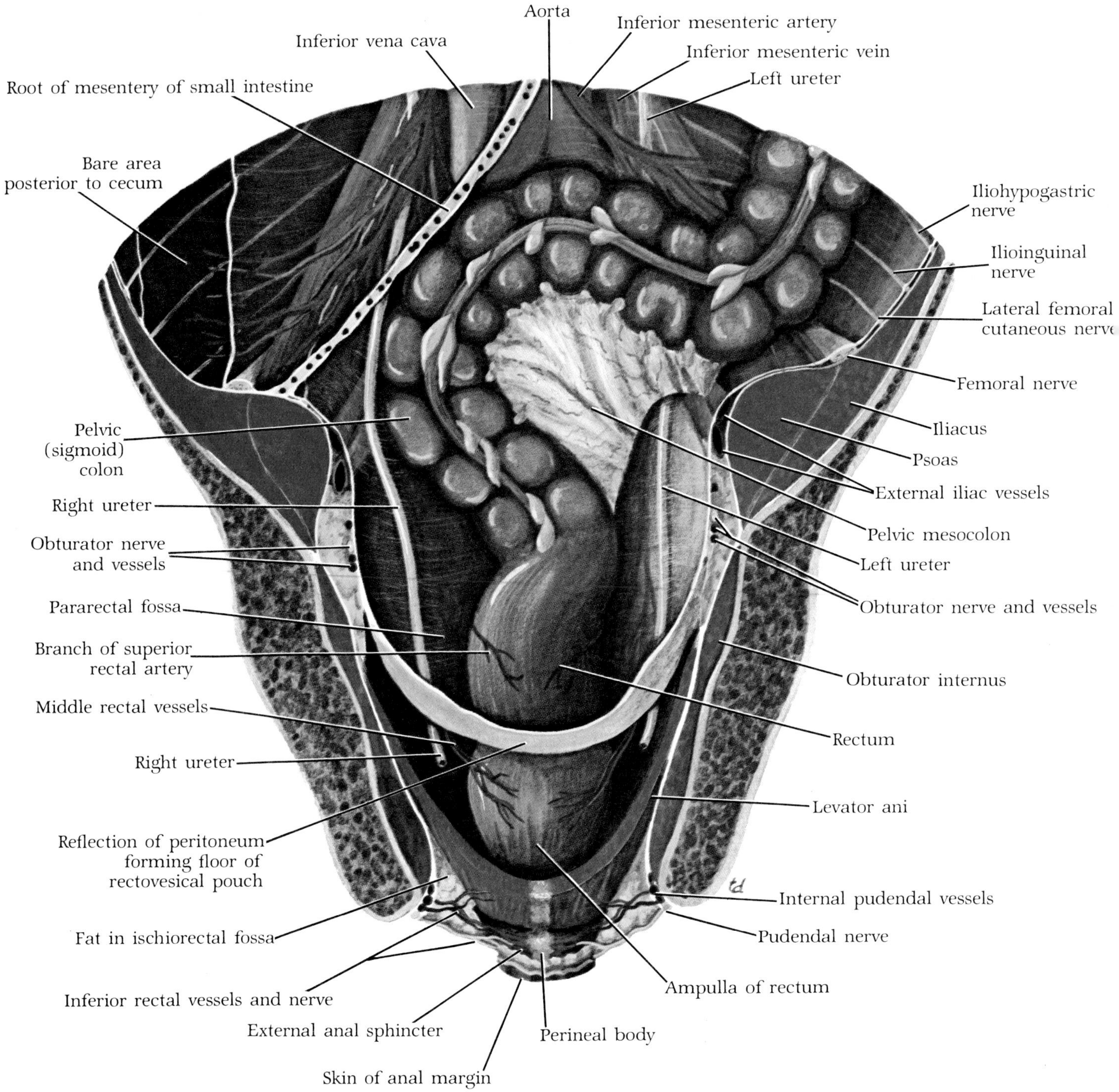

Aorta
Inferior mesenteric artery
Inferior vena cava
Inferior mesenteric vein
Left ureter
Root of mesentery of small intestine
Bare area posterior to cecum
Iliohypogastric nerve
Ilioinguinal nerve
Lateral femoral cutaneous nerve
Femoral nerve
Pelvic (sigmoid) colon
Iliacus
Psoas
External iliac vessels
Right ureter
Pelvic mesocolon
Obturator nerve and vessels
Left ureter
Pararectal fossa
Obturator nerve and vessels
Branch of superior rectal artery
Obturator internus
Middle rectal vessels
Rectum
Right ureter
Levator ani
Reflection of peritoneum forming floor of rectovesical pouch
Internal pudendal vessels
Fat in ischiorectal fossa
Pudendal nerve
Ampulla of rectum
Inferior rectal vessels and nerve
External anal sphincter
Perineal body
Skin of anal margin

Figure 3-10
Coronal section of pelvis, showing rectum, anterior view.

1. The rectum begins as a continuation of the pelvic colon anterior to the third piece of the sacrum. It ends inferiorly by passing through the pelvic floor to become the anal canal. It measures about 5 inches (13 cm) long.
2. The sites of the anteroposterior flexure of the rectum (as it follows the curvature of the sacrum and coccyx) and the three lateral flexures must be remembered when passing a sigmoidoscope.
3. Partial and complete prolapse of the rectum may follow damage to the levatores ani muscles during childbirth or because of poor muscle tone in the aged.
4. The rectum and rectosigmoid junction are frequent sites for malignant disease. There are three routes of spread of carcinoma of the rectum: (a) slow, gradual, direct spread; (b) by venous spread, leading to hepatic metastases; and (c) by lymphatic spread. The general direction of lymphatic spread is superior and follows the lymph vessels along the course of the superior rectal artery. A deeply ulcerated growth situated about the middle of the rectum may also spread along the lymph vessels accompanying the middle rectal vessels.
5. The transverse folds of the rectum are permanent folds formed of the mucous membrane, submucosa, and the circular muscle. They are semicircular in shape with two placed on the left and one on the right rectal wall. Their presence should be remembered when passing a proctoscope or sigmoidoscope.

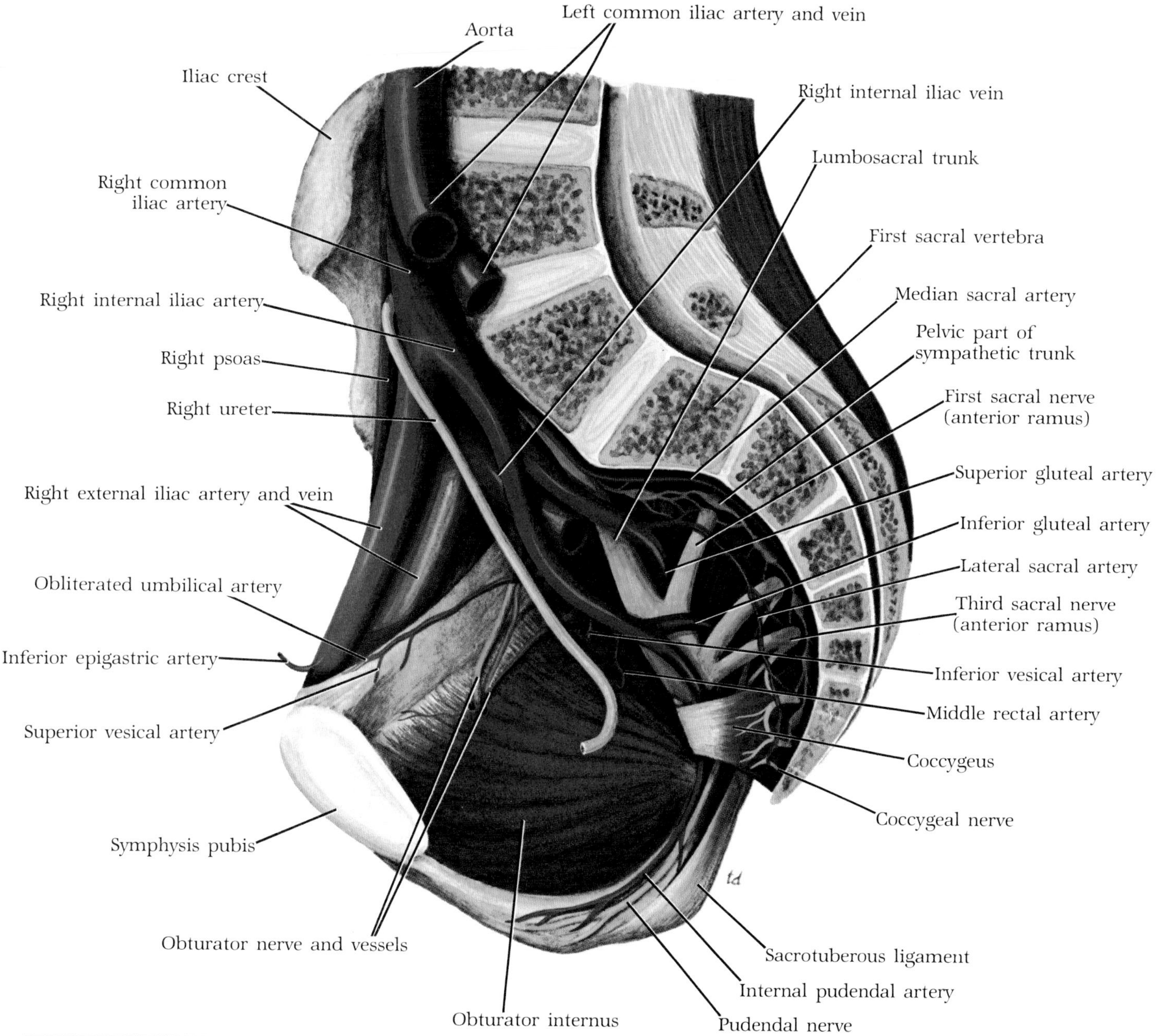
Left common iliac artery and vein
Aorta
Iliac crest
Right internal iliac vein
Lumbosacral trunk
Right common iliac artery
First sacral vertebra
Right internal iliac artery
Median sacral artery
Pelvic part of sympathetic trunk
Right psoas
Right ureter
First sacral nerve (anterior ramus)
Superior gluteal artery
Right external iliac artery and vein
Inferior gluteal artery
Lateral sacral artery
Obliterated umbilical artery
Third sacral nerve (anterior ramus)
Inferior epigastric artery
Inferior vesical artery
Middle rectal artery
Superior vesical artery
Coccygeus
Coccygeal nerve
Symphysis pubis
td
Obturator nerve and vessels
Sacrotuberous ligament
Internal pudendal artery
Obturator internus
Pudendal nerve

Figure 3-11
Lateral wall of pelvis.

1. Pelvic veins, unlike the arteries, are thin-walled and form extensive retroperitoneal plexuses. Fractures of the pelvis often result in the tearing of these veins, causing severe local retroperitoneal hemorrhage. Such hemorrhages are a common cause of death in patients with pelvic fractures.

2. The sacral plexus may be involved in a direct extension of a carcinoma of the rectum. This will cause severe, intractable pain that follows distribution of the sciatic nerve down the leg.

3. The ureter crosses the pelvic inlet anterior to the bifurcation of the common iliac artery and in front of the sacroiliac joint. It runs downward and backward anterior to the internal iliac artery to the region of the ischial spine. Here it turns forward and medially to enter the upper lateral angle of the bladder. This fact is important when interpreting a pyelogram of the course of the ureter in the pelvis.

4. The obturator canal is rarely the site of an internal hernia.

5. The pudendal nerve leaves the pelvis through the greater sciatic foramen, enters the gluteal region for a short distance, then enters the perineum through the lesser sciatic foramen. As it runs anteriorly on the medial side of the ischial tuberosity, it may be blocked by an anesthetic to produce analgesia of the perineum in patients in whom a forceps delivery at childbirth is required.

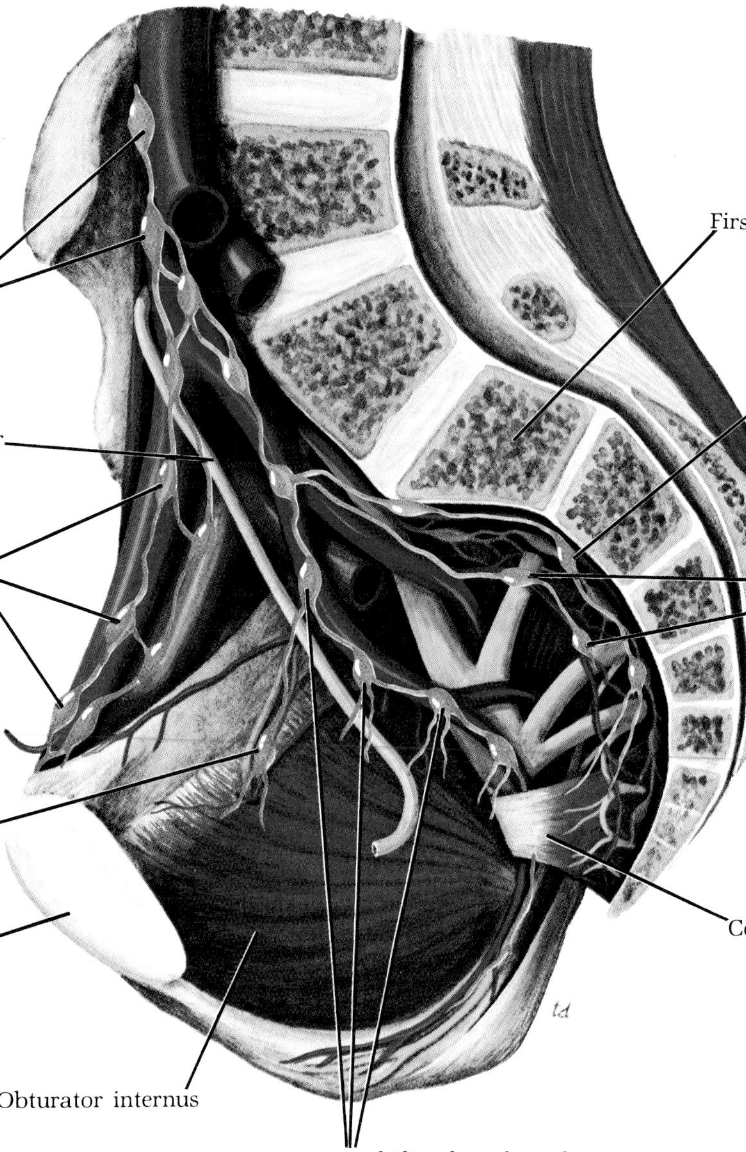
Common iliac lymph nodes
Right ureter
External iliac lymph nodes
Obturator node
Symphysis pubis
Obturator internus
Internal iliac lymph nodes
First sacral vertebra
Median sacral node
Lateral sacral nodes
Coccygeus

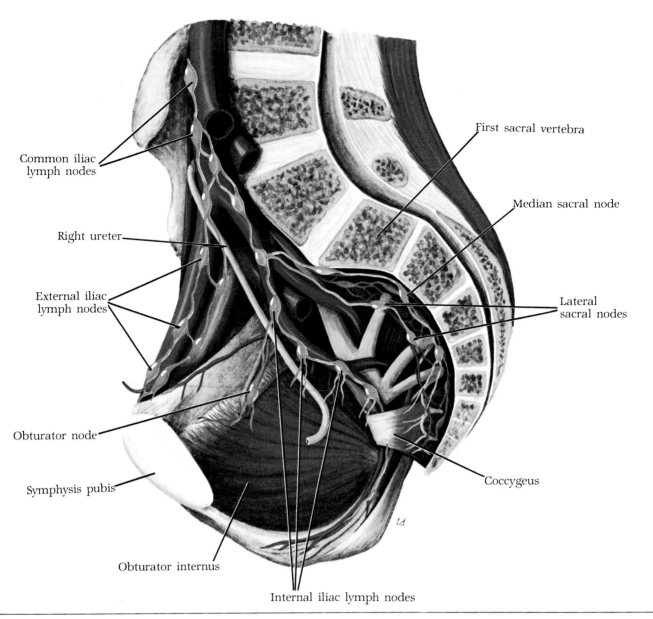

Figure 3-12
Lymphatic drainage of the pelvis.

1. The lymphatic vessels and nodes are of great importance when considering the treatment and prognosis of malignant disease of the pelvic viscera. In patients with carcinoma of the cervix, for example—whether they be treated by surgery or irradiation—it is essential that the metastases in the internal or external iliac nodes be adequately dealt with. Because it is impossible to know whether or not the lymph nodes are involved by clinical testing, bilateral lymph node removal or adequate high-voltage radiotherapy should be routine treatment.

2. It must be remembered that the ovaries developed high up on the posterior abdominal wall and later descended into the pelvis, bringing their lymphatic drainage and blood vessels with them. The lymph from the ovaries drains into the para-aortic (lumbar) nodes at the level of the first lumbar vertebra. Malignant tumors of the ovaries tend to spread locally early rather than metastasize to the para-aortic nodes.

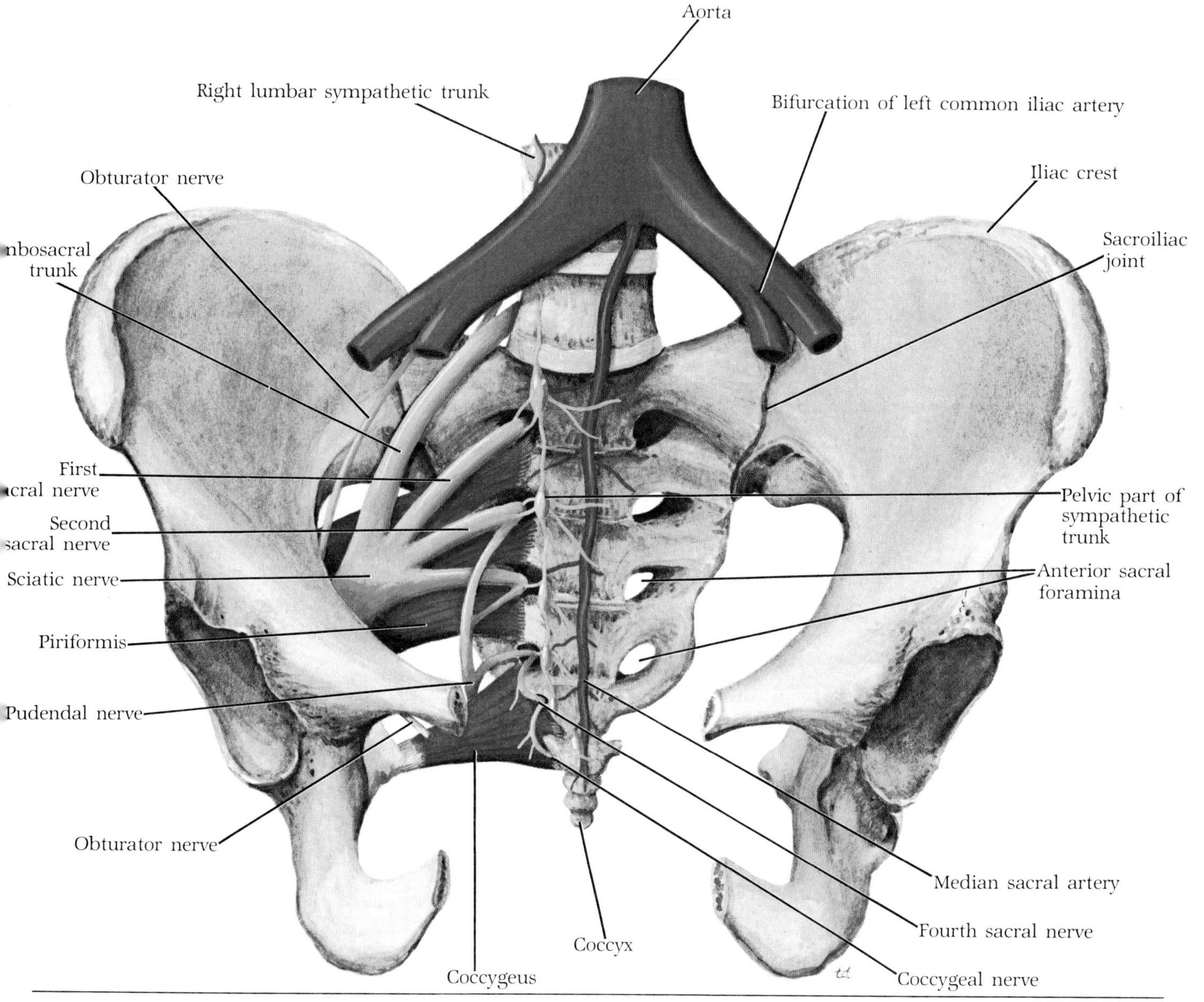
Aorta
Right lumbar sympathetic trunk
Bifurcation of left common iliac artery
Obturator nerve
Iliac crest
nbosacral trunk
Sacroiliac joint
First
cral nerve
Second
sacral nerve
Pelvic part of sympathetic trunk
Sciatic nerve
Anterior sacral foramina
Piriformis
Pudendal nerve
Obturator nerve
Median sacral artery
Fourth sacral nerve
Coccyx
Coccygeus
Coccygeal nerve

Figure 3-13
Posterior pelvic wall showing the right sacral plexus and the pelvic part of the right sympathetic trunk.

1. A pelvic appendicular abscess or a pyosalpinx, by irritating the obturator nerve, can give rise to referred pain down the medial side of the thigh; this may be mistaken for disorders of the hip joint. Moreover, pelvic inflammation can cause irritative spasm of the psoas, iliacus, piriformis, or obturator internus muscles and cause restriction of hip movements because of pain if the movement is forced.

2. A tumor, an inflammatory mass, or a fetal head in pregnancy may press upon or invade the sacral plexus or its branches, causing pain which is then distributed along the course of the sciatic nerve. A pelvic tumor or mass can usually be palpated on rectal or vaginal examination.

3. Occlusion of the aortic bifurcation or the iliac arteries produces pain in the legs on walking. The pain may be located in the lower back or buttocks and may be confused with the pain of prolapsed intervertebral disc.

4. The sciatic nerve may be damaged in fractures of the pelvis.

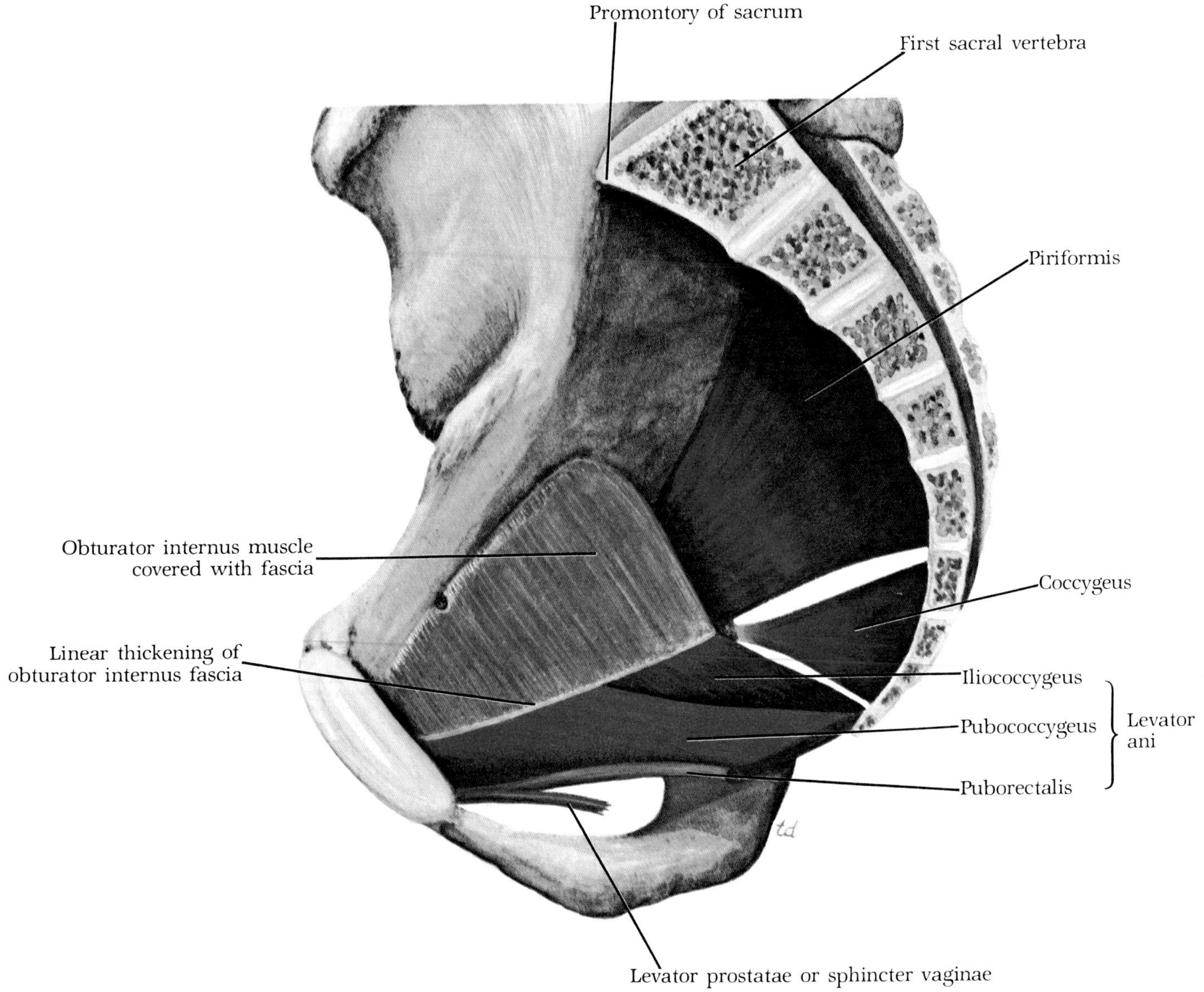
Promontory of sacrum
First sacral vertebra
Piriformis
Obturator internus muscle covered with fascia
Coccygeus
Linear thickening of obturator internus fascia
Iliococcygeus
Pubococcygeus
Puborectalis
Levator ani
td
Levator prostatae or sphincter vaginae

Figure 3-14
Right half of pelvis showing pelvic walls.

1. The anterior pelvic wall is the shallowest of the pelvic walls and is formed by the posterior surfaces of the bodies of the pubic bones and the symphysis pubis. The posterior pelvic wall is extensive and is formed by the sacrum and coccyx and by the piriformis muscles and their covering of parietal pelvic fascia. The lateral pelvic wall is formed by part of the innominate bone below the pelvic inlet and by the obturator internus muscle; the obturator membrane and the sacrotuberous and sacrospinous ligaments also assist in forming this wall. The floor of the pelvis supports the pelvic viscera and is formed by the pelvic diaphragm.
2. The pelvic diaphragm is formed by the important levatores ani muscles and the small coccygeus muscles and their covering fasciae. It is incomplete anteriorly to allow for the passage of the urethra in the male and the urethra and vagina in the female.

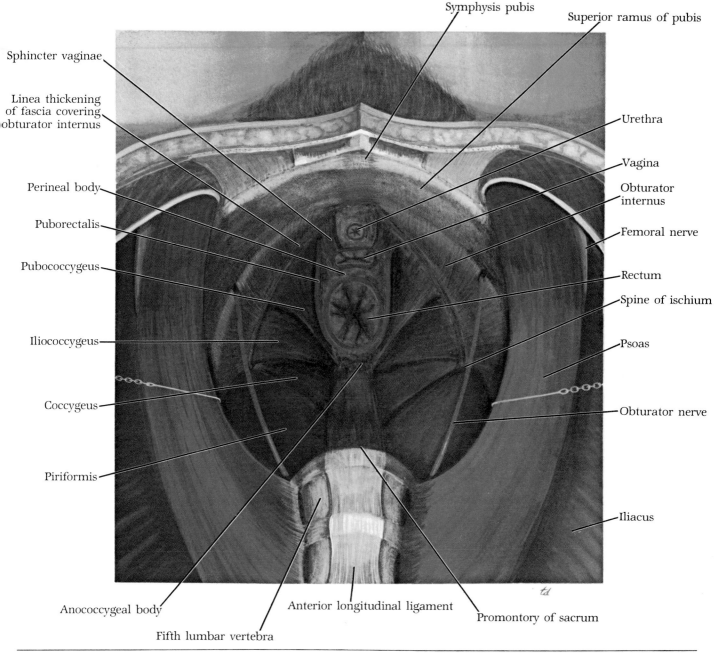

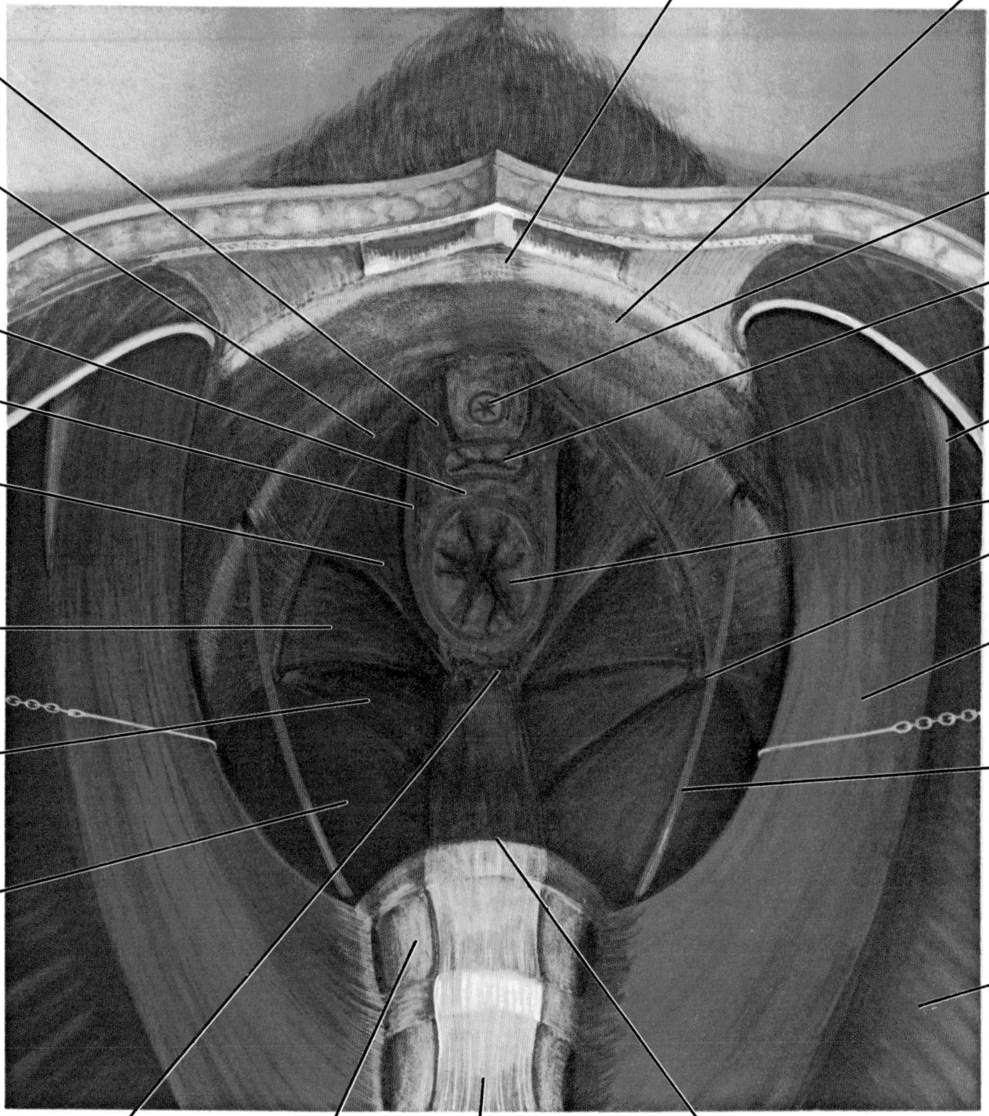
Symphysis pubis
Superior ramus of pubis
Sphincter vaginae
Linea thickening of fascia covering obturator internus
Urethra
Vagina
Obturator internus
Perineal body
Puborectalis
Femoral nerve
Pubococcygeus
Rectum
Spine of ischium
Iliococcygeus
Psoas
Coccygeus
Obturator nerve
Piriformis
Iliacus
Anococcygeal body
Anterior longitudinal ligament
Promontory of sacrum
Fifth lumbar vertebra

Figure 3-15
Pelvic floor in the female.

1. The pelvic diaphragm is a gutter-shaped sheet of muscle formed by the levatores ani and coccygeus muscles and their covering fasciae. The tone of these muscles supports the pelvic viscera; an increase in the tone counteracts the rise in intra-abdominal pressure that occurs during coughing, micturition, or defecation.

2. During the second stage of labor the gutter shape of the floor plays a part in the rotation of the baby's head.

3. The most medial fibers of the pubococcygeus are extremely important in maintaining the urethra, the vagina, and the rectum in their normal positions. They fuse with the fascia and smooth muscle of the vagina and decussate between the vagina and the rectum. If these muscle fibers tear during a difficult childbirth, there is an increased tendency for the vagina and uterus to prolapse through the pelvic floor. One of the most important surgical procedures for treating uterine prolapse is the suturing together of the pubococcygeus muscles between the vagina and the rectum.

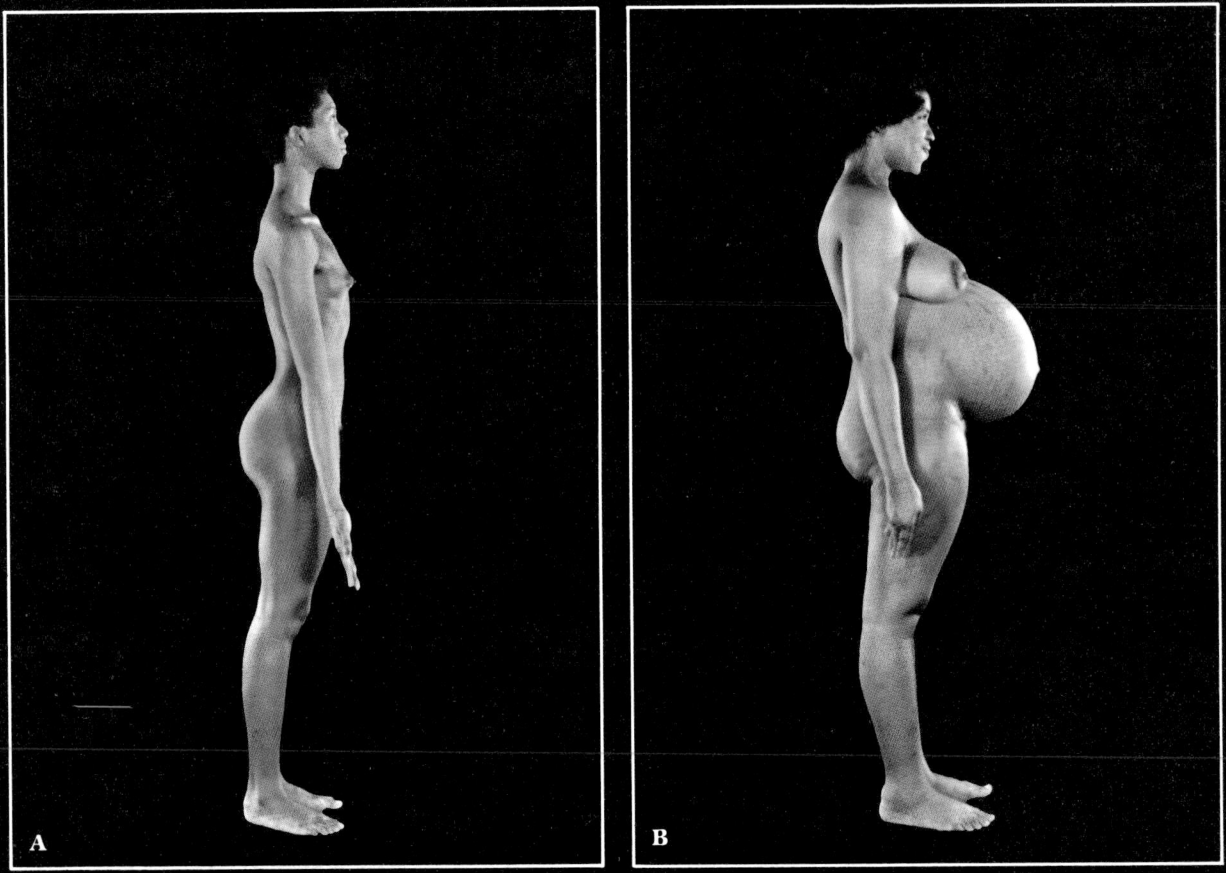

Figure 3-16
Lateral views of adult females. A. Nonpregnant. B. Pregnant, expected date of delivery approximately ten days (courtesy of Dr. S. Fabro).

The pregnant woman undergoes many profound physiological and anatomical changes during pregnancy to accommodate the growth and development of the fetus.

1. Amenorrhea. Cessation of periods strongly suggests pregnancy. It should be remembered, however, that this symptom is not completely reliable, since a few women bleed a little at irregular intervals during pregnancy, while others may have amenorrhea for other reasons, such as sudden climatic or environmental changes or hormonal imbalances.

2. Morning sickness. More than half of all pregnant women are troubled by morning nausea and even vomiting during the first three months of pregnancy. It is believed to be due to the increased production of progesterone, which relaxes the smooth muscle of the stomach wall.

3. Heartburn. A retrosternal burning pain due to regurgitation of acid gastric contents into the esophagus, heartburn is due to the poor tone of the lower esophageal musculature and the raised intra-abdominal pressure caused by the expanding uterus.

4. Constipation is initially due to lack of smooth muscle tone in the gastrointestinal tract but later may be caused by pressure on the rectum by the enlarging uterus.

5. Breast enlargement, tingling sensations in and tenderness of the breasts occur as the result of hormonal stimulation of the glandular structures in the breast. In primigravidae the nipples and areolae become more pigmented and the tubercles of the areola become enlarged.

6. Bladder changes. Pressure on the bladder by the enlarging uterus causes increased frequency of micturition.

7. Body posture. As the uterus enlarges and the weight of the fetus increases, the mother, in an effort to balance her body, compensates for this by increasing the normal lordosis present in the lumbar part of the vertebral column. To compensate for the lumbar lordosis she increases the kyphosis in the thoracic region and flexes the cervical part of her vertebral column, thus protruding the head forward. Lumbosacral backache is a common complaint as the result of these structural changes.

8. The ligaments of the sacroiliac joints and the symphysis pubis undergo softening and relaxation during pregnancy to help in the passage of the fetal head at birth. In some women the separation of the bones is so great that difficulty in walking is experienced.

9. Swelling of the feet, varicose veins, and hemorrhoids are all common in pregnancy and are due to tissue edema, loss of tone of the smooth muscle in the walls of veins, and the pressure of the enlarging uterus on the iliac veins and the inferior vena cava.

10. Shortness of breath is sometimes troublesome toward the end of pregnancy and is due to the expanding uterus pressing upward and restricting the movements of the diaphragm.

11. Cramps in the legs often occur in the latter part of pregnancy and are due to the pressure of the uterus on the sacral plexus and its branches. This condition is probably enhanced by pelvic venous congestion caused by pressure of the uterus on the iliac veins and inferior vena cava.

12. Brown pigmentation of the skin of the face, vulva, perineum, and linea alba is due to an increase in hormonal stimulation of the epidermal melanocytes.

13. White streaks in the skin, striae gravidarum, are seen over the anterior abdominal wall, breasts, thighs, and buttocks. This is due to the rupture of elastic fibers in the dermis following excessive stretching of the skin. It is present in some women and absent in others.

14. The vulval and vaginal walls show venous congestion and are bluish in color. This is one of the early signs of pregnancy and is present within a few weeks of conception.

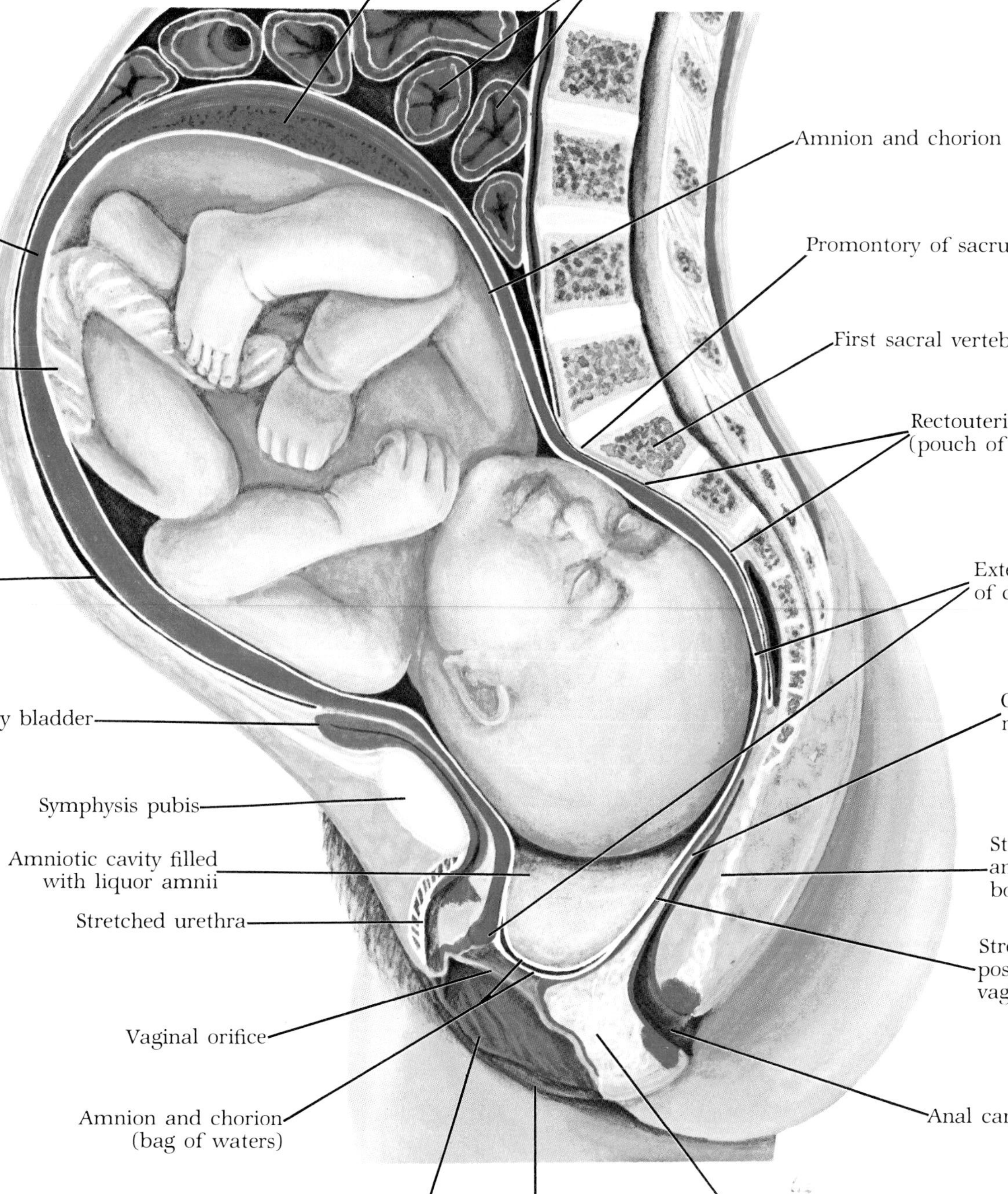
Placenta
Coils of small intestine
Amnion and chorion
Wall of uterus
Promontory of sacrum
First sacral vertebra
Umbilical cord
Rectouterine pouch (pouch of Douglas)
Peritoneal cavity
External os of cervix
Compressed rectum
Urinary bladder
Symphysis pubis
Amniotic cavity filled with liquor amnii
Stretched anococcygeal body
Stretched urethra
Stretched posterior vaginal wall
Vaginal orifice
Anal canal
Amnion and chorion (bag of waters)
Labium minus
Labium majus
Perineal body

Figure 3-17
Sagittal section of the maternal abdomen and pelvis during the second stage of labor. The fetal membranes have pouched downward into the cervix and vagina, and the external os of the cervix is dilated.

1. The child shows a vertex presentation with the occiput of the skull in the right anterior quadrant of the pelvis [right occipitoanterior presentation (R.O.A.)].

2. The rectum and bladder have been emptied to provide maximum room in the pelvis for the fetus. Removal of fecal material from the rectum by enema prevents later contamination of the genital tract from this source.

3. A rectal or vaginal examination permits the obstetrician to determine whether the fetal membranes have ruptured, the degree of dilatation of the cervix, and the nature and position of the presenting part.

4. When the uterus contracts during labor, it exerts a hydraulic pressure on its contents. Because the cervical canal is the weak spot in the uterine wall, the contents are forced downward toward the internal os of the cervix. The uterine contractions bring about the protrusion of the fetal membranes and amniotic fluid (bag of waters) through the cervix, the dilatation of the lower part of the uterine body and cervix, and the expulsion of the baby and placenta, in that order.

5. To be born, the child must be passed through the cervix, the opening in the pelvic diaphragm, the urogenital diaphragm, and the perineum. The bladder is pushed forward and upward and the rectum is flattened against the sacrum. The perineal body becomes greatly stretched.

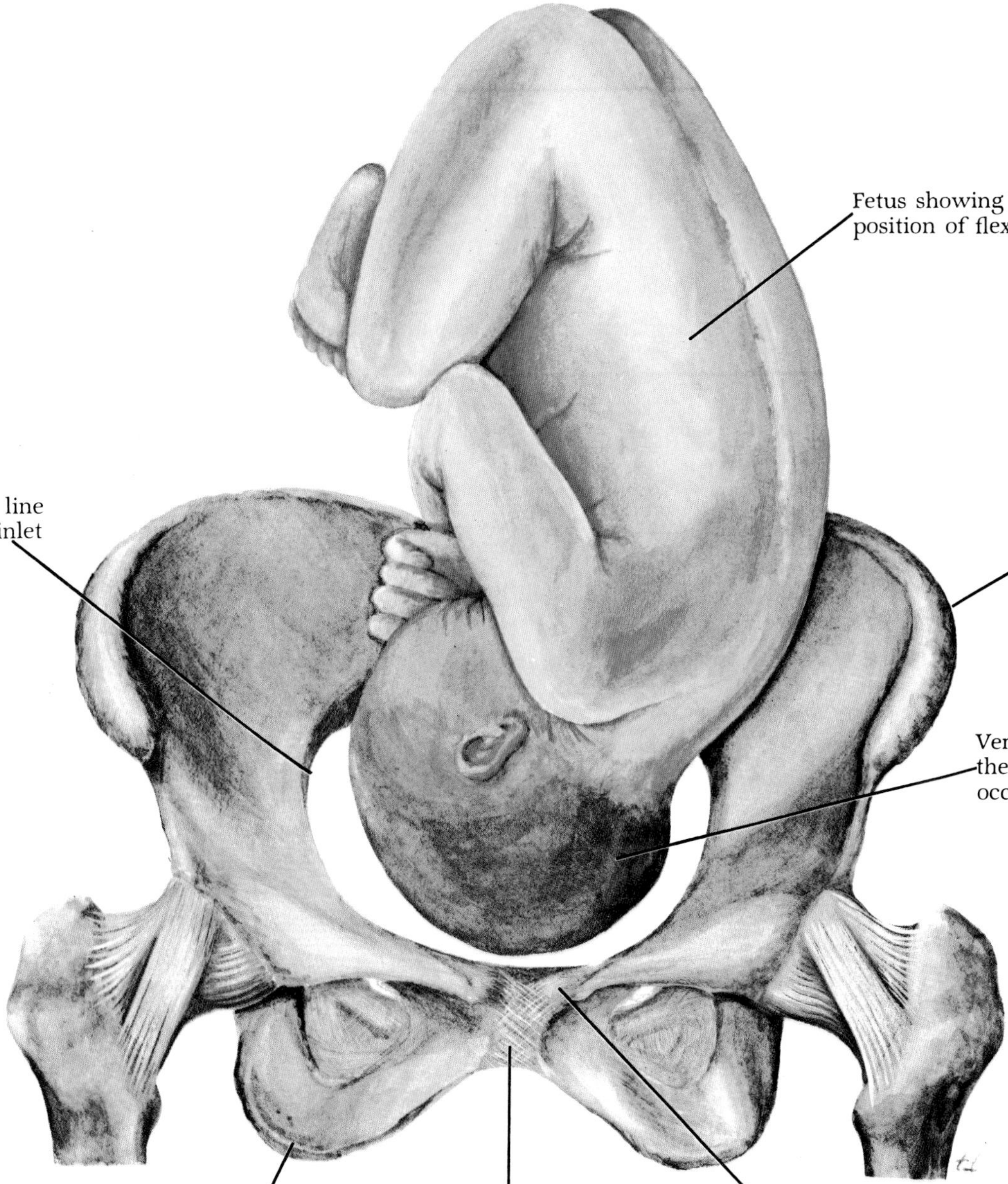
Fetus showing the normal position of flexion
Iliopectineal line of pelvic inlet
Iliac crest
Vertex presentation with the head in the left occipitoanterior position
Ischial tuberosity
Symphysis pubis
Pubic tubercle

Figure 3-18
The fetal head about to enter the true pelvis — a vertex presentation.

1. The upper part of the bony pelvis, the false pelvis, is of little obstetrical importance. Its funnel shape does provide some support for the pregnant uterus, and during the latter part of pregnancy it helps to guide the fetus into the true pelvis.

2. The inlet of the true pelvis is bounded posteriorly by the sacral promontory, laterally by the iliopectineal lines, and anteriorly by the symphysis pubis.

3. Because the widest diameter at the pelvic inlet is transverse, the longest axis of the baby's head takes up the transverse position. Descent into the pelvis begins about two weeks before the onset of labor in the primigravida. Further descent during the first stage of labor is brought about by the force of uterine contractions.

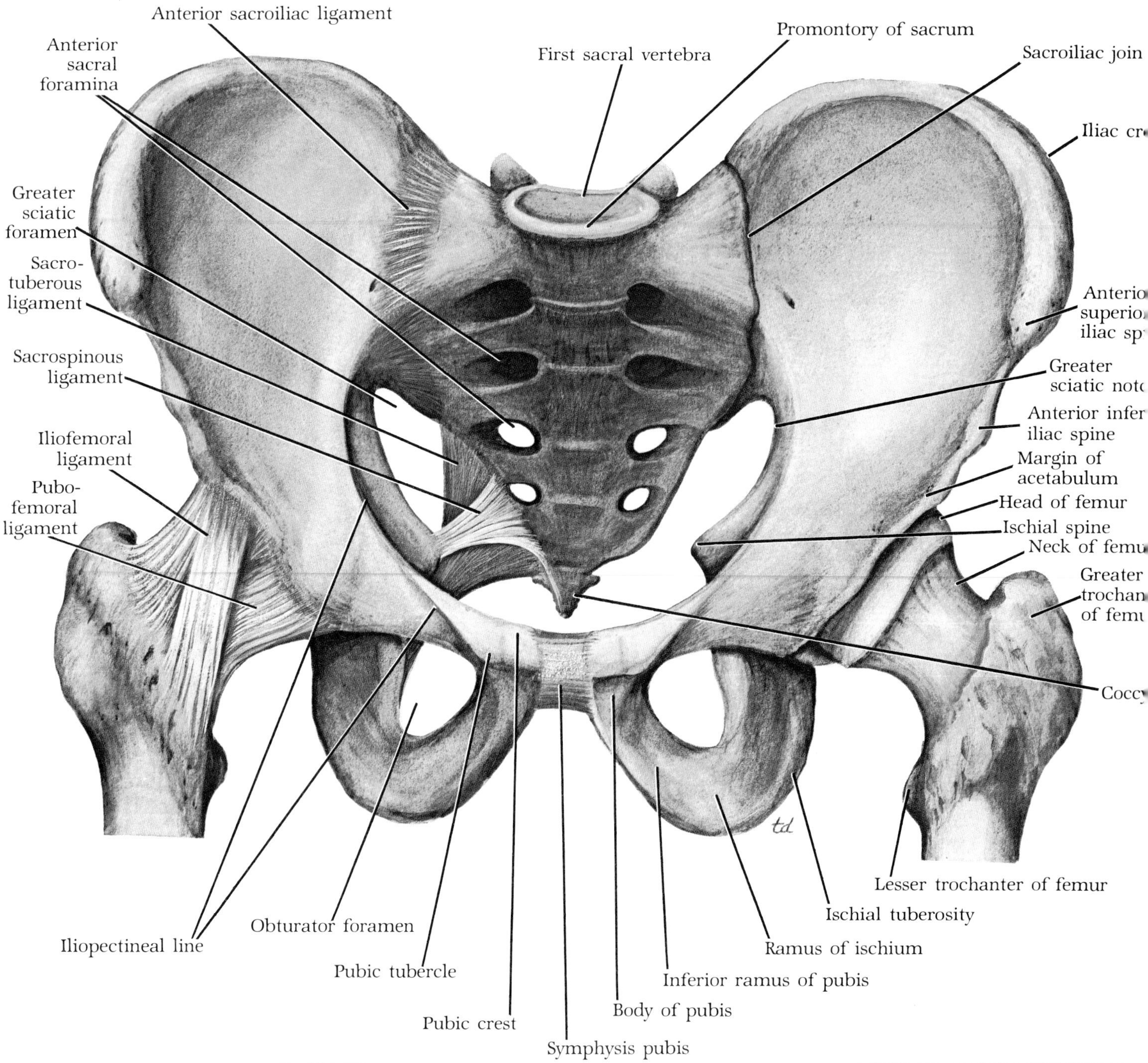
Anterior sacroiliac ligament
Anterior sacral foramina
First sacral vertebra
Promontory of sacrum
Sacroiliac join
Iliac cr
Greater sciatic foramen
Sacro-tuberous ligament
Sacrospinous ligament
Iliofemoral ligament
Pubo-femoral ligament
Anterio superio iliac sp
Greater sciatic note
Anterior infer iliac spine
Margin of acetabulum
Head of femur
Ischial spine
Neck of femu
Greater trochan of femu
Coccy
td
Lesser trochanter of femur
Ischial tuberosity
Ramus of ischium
Inferior ramus of pubis
Body of pubis
Symphysis pubis
Pubic crest
Pubic tubercle
Obturator foramen
Iliopectineal line

Figure 3-19
Male pelvis, anterior view.

1. The pelvis, composed of the two innominate bones, sacrum and coccyx, provides: (a) protection for the pelvic viscera; (b) attachments of muscles; and (c) the transmission of weight from the trunk to the lower extremities.
2. Fractures of the false pelvis due to direct trauma occasionally occur. Bone fragments are seldom displaced since the iliacus is attached over a wide area on the inside and the gluteal muscles are attached on the outside.
3. Isolated fractures of the true pelvis sometimes occur as the result of direct trauma in elderly patients with osteoporosis. The pubic rami may be fractured in this manner.
4. Fractures of the true pelvis are commonly caused by automobile accidents. It is useful to regard the bony pelvis as a rigid ring which, if broken at one point on its circumference, usually breaks at a second place. For example, anteroposterior compression may produce fractures through the pubic rami and fracture of the lateral part of the sacrum.
5. The head of the femur may be driven through the floor of the acetabulum during a heavy fall on the greater trochanter of the femur.
6. Remember that fractures of the true pelvis result in a high mortality, the majority of patients dying from severe internal hemorrhage resulting from the tearing of the large intrapelvic veins. Fragments of bone may be thrust into the pelvic cavity, penetrating the urethra, bladder, or vagina.

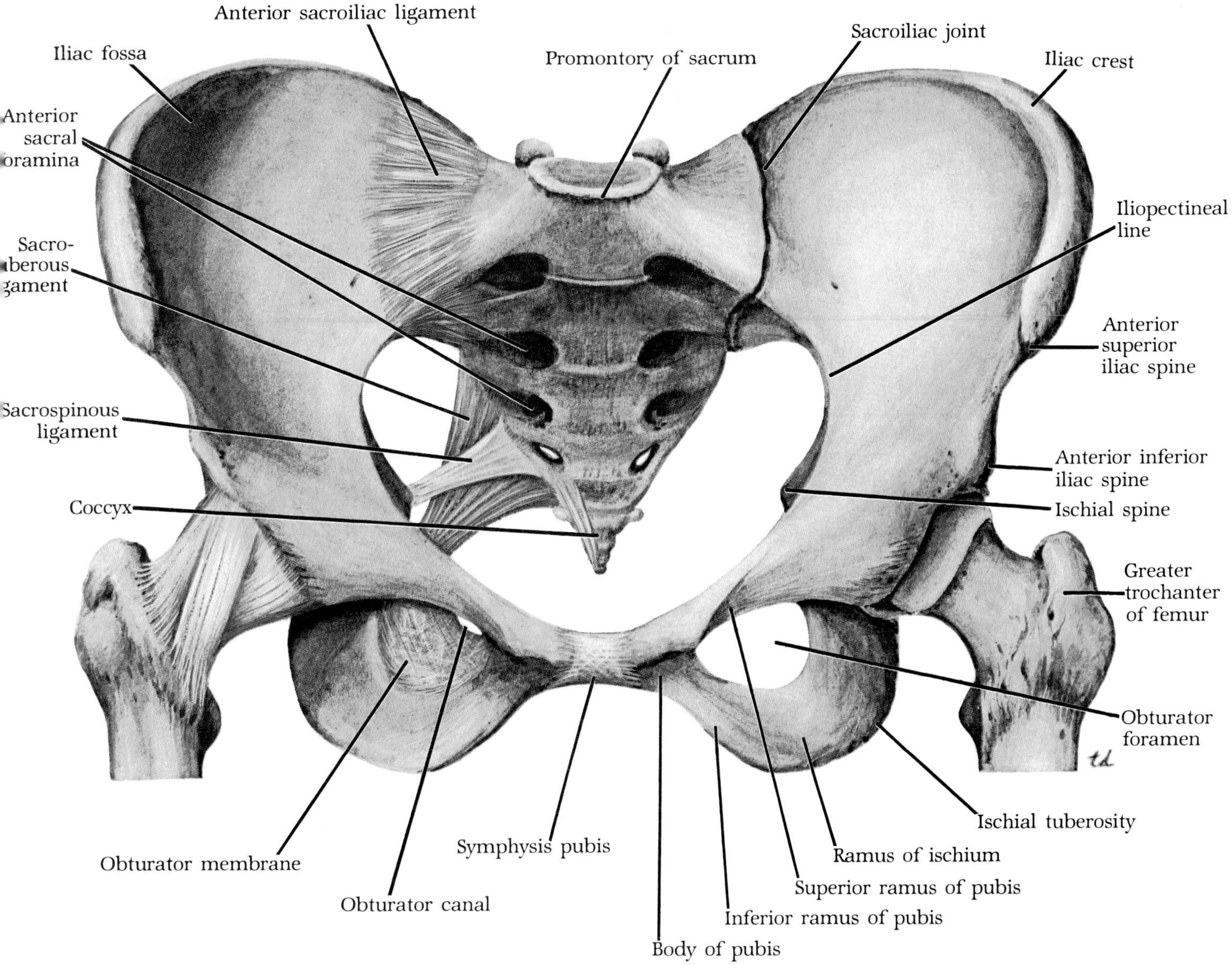
Anterior sacroiliac ligament
Iliac fossa
Promontory of sacrum
Sacroiliac joint
Iliac crest
Anterior
sacral
oramina
Iliopectineal
line
Sacro-
berous
gament
Anterior
superior
iliac spine
Sacrospinous
ligament
Anterior inferior
iliac spine
Coccyx
Ischial spine
Greater
trochanter
of femur
Obturator
foramen
td
Ischial tuberosity
Symphysis pubis
Ramus of ischium
Obturator membrane
Superior ramus of pubis
Obturator canal
Inferior ramus of pubis
Body of pubis

Figure 3-20
Female pelvis, anterior view.

1. In the female the pelvis is shallower, the cavity is larger and more cylindrical, and the outlet is wider in anteroposterior and transverse diameters than in the male.
2. There are four pelvic joints: two sacroiliac, the symphysis pubis, and the sacrococcygeal. In the nonpregnant woman there is little movement possible at these joints; during pregnancy, however, the ligaments are softened and stretch in response to hormonal changes, and slight separation of the bones is possible. This slight relaxation of the joints provides additional space for the passage of the fetal head during labor.
3. During the later stages of pregnancy the mother is carrying an additional weight of approximately 25 pounds (11.2 kg) and has an altered center of gravity when walking which subjects the sacroiliac joints to added strain. For this reason women frequently complain of backache during the last months of their pregnancy.
4. The following areas are of great importance in obstetrical practice: (a) the pelvic inlet or brim of the true pelvis is bounded anteriorly by the symphysis pubis, laterally by the iliopectineal lines, and posteriorly by sacral promontory; (b) the pelvic outlet of the true pelvis is bounded anteriorly by the pubic arch, laterally by the ischial tuberosities, and posteriorly by the coccyx; the sacrotuberous ligaments also form part of the margin of the outlet; (c) the pelvic cavity is the space between the inlet and the outlet; (d) the axis of the pelvis is an imaginary line joining the central points of the anteroposterior diameters from the inlet to the outlet and is the curved course taken by the baby's head as it descends through the pelvis during parturition.

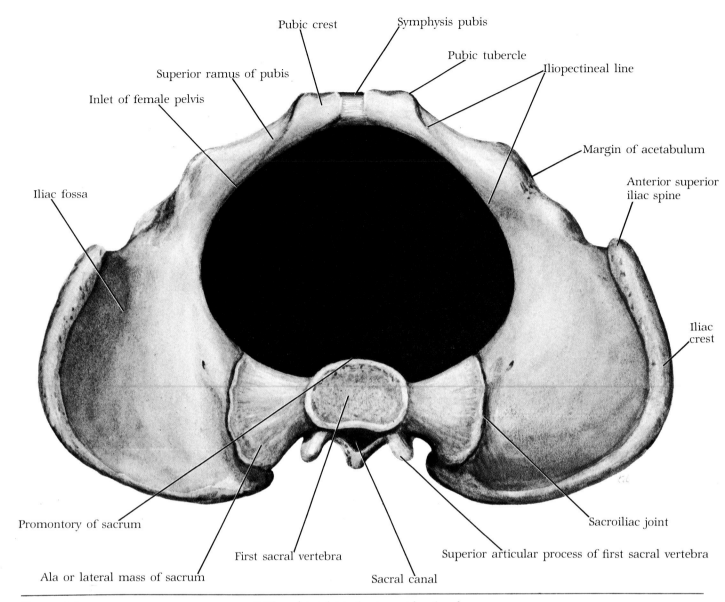

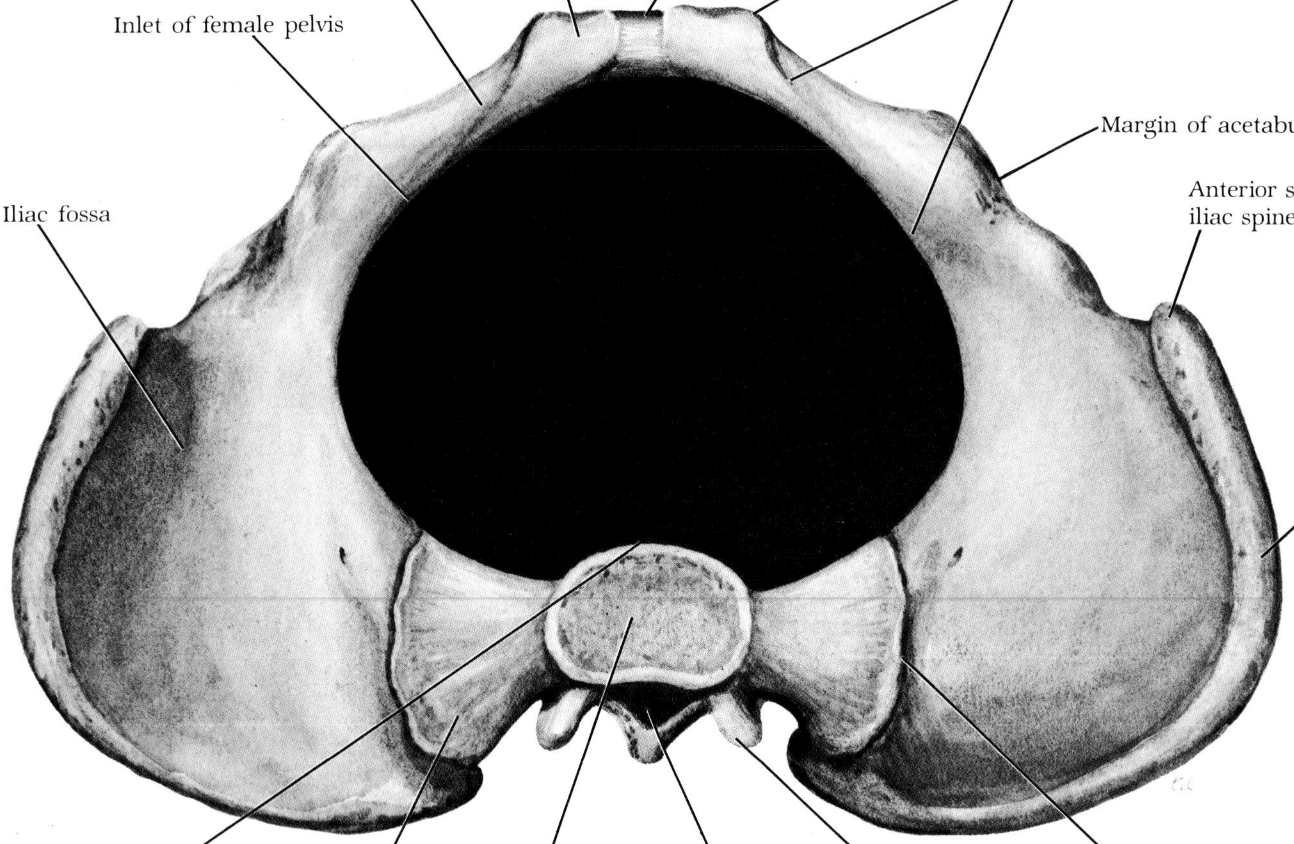
Pubic crest
Symphysis pubis
Pubic tubercle
Superior ramus of pubis
Iliopectineal line
Inlet of female pelvis
Margin of acetabulum
Anterior superior iliac spine
Iliac fossa
Iliac crest
Promontory of sacrum
Sacroiliac joint
First sacral vertebra
Superior articular process of first sacral vertebra
Ala or lateral mass of sacrum
Sacral canal

Figure 3-21
Female pelvis as seen from above — pelvic inlet or brim.

1. The false pelvis is that part above the brim and consists of the flared-out iliac bones. It has very little obstetrical importance.
2. The pelvic inlet or brim of the true pelvis is bounded anteriorly by the symphysis pubis, laterally by the iliopectineal lines, and posteriorly by the sacral promontory.
3. The pelvic inlet in the gynecoid or female pelvis is round, except where the promontory of the sacrum projects into it.
The transverse diameter of the inlet is the widest part so that as the fetal head enters the pelvis the longest axis of the head tends to take up the transverse position.
4. A mother whose bony pelvis is of normal shape and size will give birth without difficulty to a baby of normal size. Pelves, however, vary in size and shape and may be deformed by disease, e.g., rickets. It is imperative, therefore, that the obstetrician detect deviations from the normal so that the woman does not go into an obstructed labor that may result in injury to or death of the child and mother.

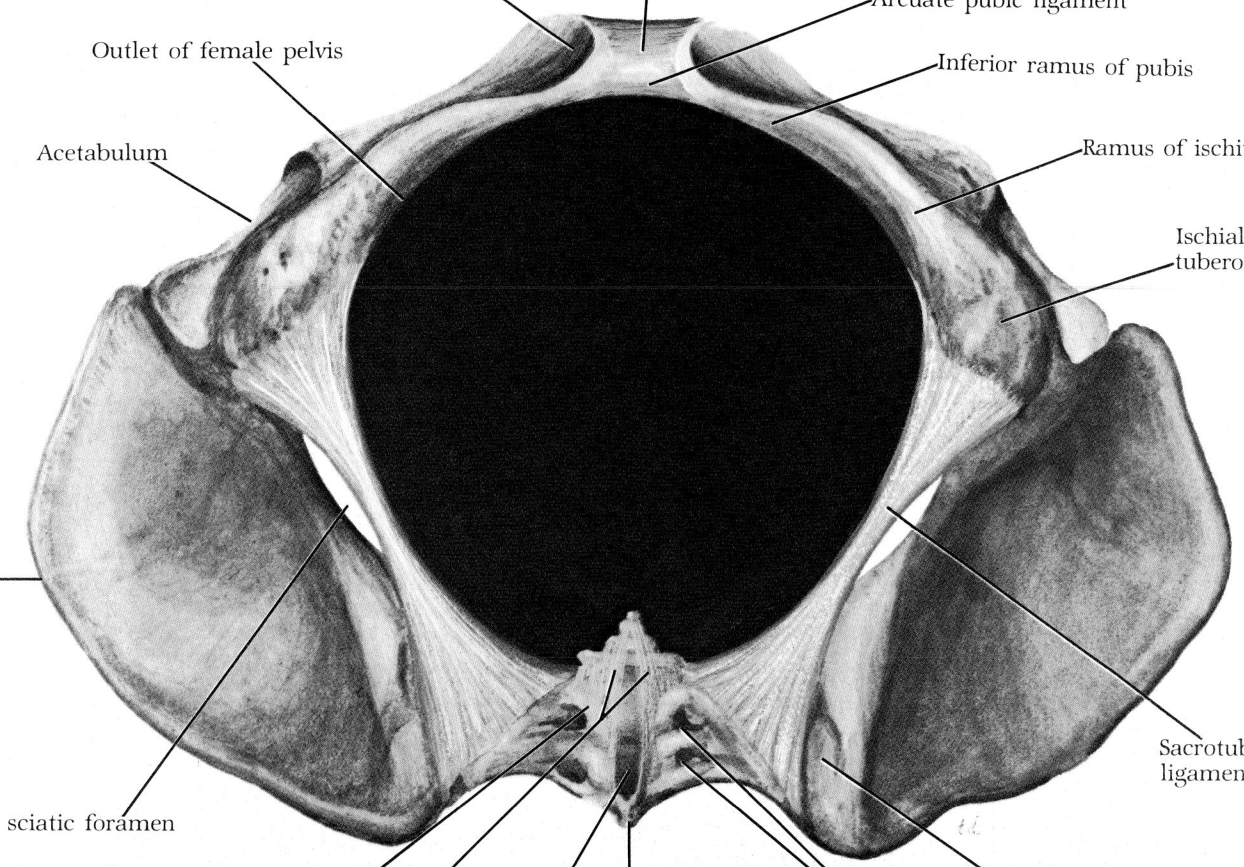
Symphysis pubis
Body of pubis
Arcuate pubic ligament
Outlet of female pelvis
Inferior ramus of pubis
Acetabulum
Ramus of ischium
Ischial tuberosity
Iliac crest
Sacrotuberous ligament
Greater sciatic foramen
Sacral cornu
Posterior sacrococcygeal ligament
Sacral hiatus
Median crest of sacrum
Posterior superior iliac spine
Posterior sacral foramina

Figure 3-22
Female pelvis as seen from below — pelvic outlet.

1. The pelvic outlet of the true pelvis is bounded anteriorly by the pubic arch, which, in the gynecoid pelvis, forms an angle of ninety degrees. Laterally it is bounded by the ischial tuberosities and posteriorly by the coccyx. The sacrotuberous ligaments also form part of the outlet.

2. During a pelvic examination on a pregnant woman the characteristics of the pubic arch and the distance between the ischial tuberosities should be determined. Spread the fingers under the pubic arch and examine its shape. Is it broad or angular? Four of the examiner's fingers should be able to rest comfortably in the angle below the symphysis pubis. The distance between the ischial tuberosities may be estimated by making a fist. Because it normally measures about 4 inches (10 cm), the closed fist should be able to fit between the ischial tuberosities in the perineum.

3. As the result of the softening of the ligaments of the sacrococcygeal joint during pregnancy, the coccyx is able to move posteriorly during the passage of the baby's head.

4. The Perineum

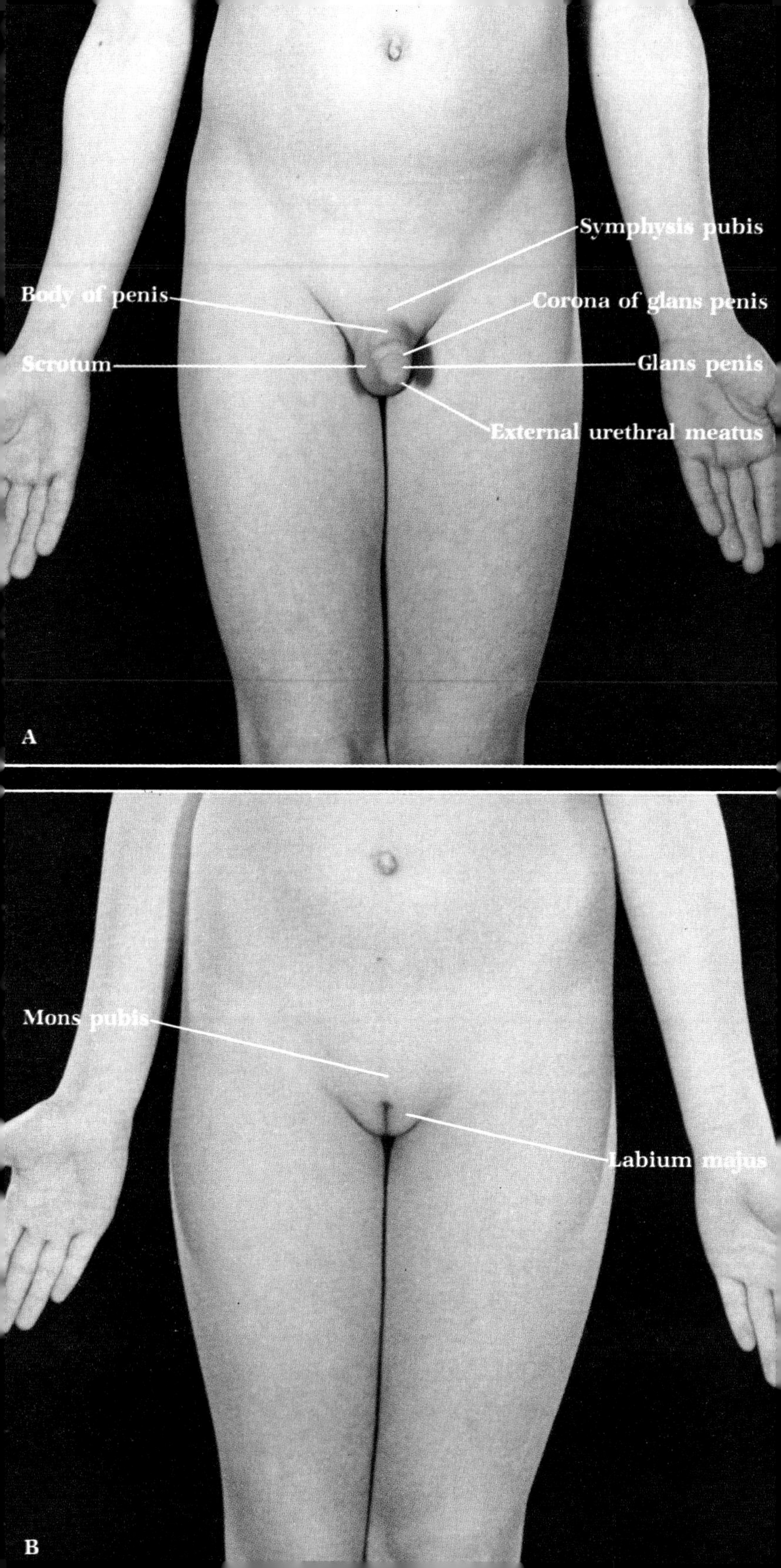

Symphysis pubis
Body of penis
Corona of glans penis
Scrotum
Glans penis
External urethral meatus
A
Mons pubis
Labium majus
B

Figure 4-1
The pelvis and perineum, anterior view. A. 10-year-old circumcised male. B. 10-year-old female.

1. The pelvic cavity, or cavity of the true pelvis, may be defined as the area situated between the pelvic inlet and the pelvic outlet. The pelvic diaphragm subdivides the cavity into the main pelvic cavity above and the perineum below.
2. The perineum, when seen from below, is diamond-shaped and is bounded anteriorly by the symphysis pubis, posteriorly by the tip of the coccyx, and laterally by the ischial tuberosities. An imaginary line joining the ischial tuberosities divides the region into an anterior urogenital triangle and a posterior anal triangle.
3. In the male child the scrotum is small and both testes should be present within it. The size and consistency of each testis should be determined, since frequently they are not the same size.
4. The size of the penis in the child varies. It often becomes erect when the bladder is full. The glans forms the extremity of the body of the penis. At the summit of the glans is the external urethral meatus. Extending from the lower margin of the external meatus is a fold of membrane, called the frenulum, that connects the glans to the prepuce.
5. The edge of the base of the glans is called the corona. The prepuce is formed by a fold of skin attached to the neck of the penis. In the newborn infant the prepuce cannot be retracted. Circumcision is a minor operation in which the prepuce is removed.
6. At puberty in the male (12 to 15 years) the testes, scrotum, and penis start to enlarge and pubic hair makes its appearance.
7. In the female child the labia are small but the clitoris is prominent. Up until puberty the vaginal wall has a thin mucosa and has a redder appearance than in the adult female. The hymen is large and the opening remains fairly constant in size from the young child until puberty, permitting adequate instrumental examination of the vagina and cervix.
8. At puberty in the female (10-14 years) the pelvis widens and pubic hair appears. The Doderlein's bacillus appears in the vaginal lumen and starts breaking down the glycogen in the desquamated surface cells to form lactic acid, thus changing the vaginal secretions from alkaline to acid. This is a protective phenomenon that reduces the incidence of vaginitis. The labia, vagina, and uterus all increase in size.

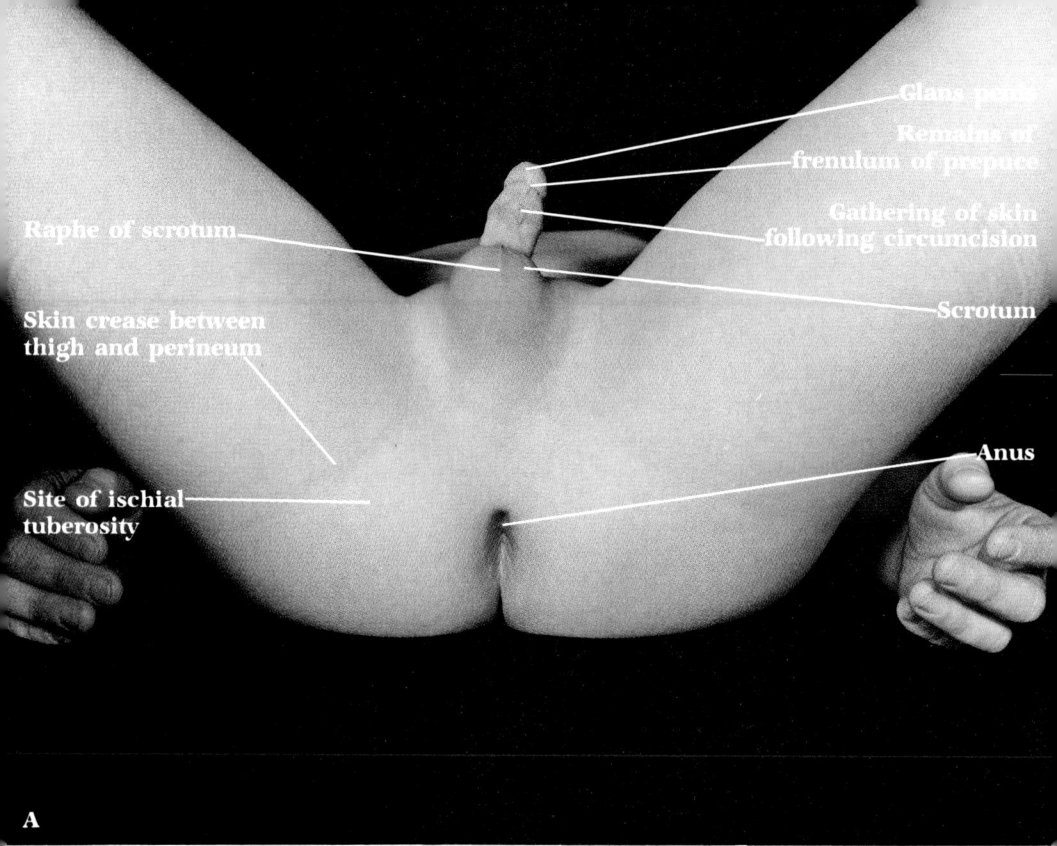

Glans penis
Remains of frenulum of prepuce
Gathering of skin following circumcision
Raphe of scrotum
Scrotum
Skin crease between thigh and perineum
Anus
Site of ischial tuberosity

A

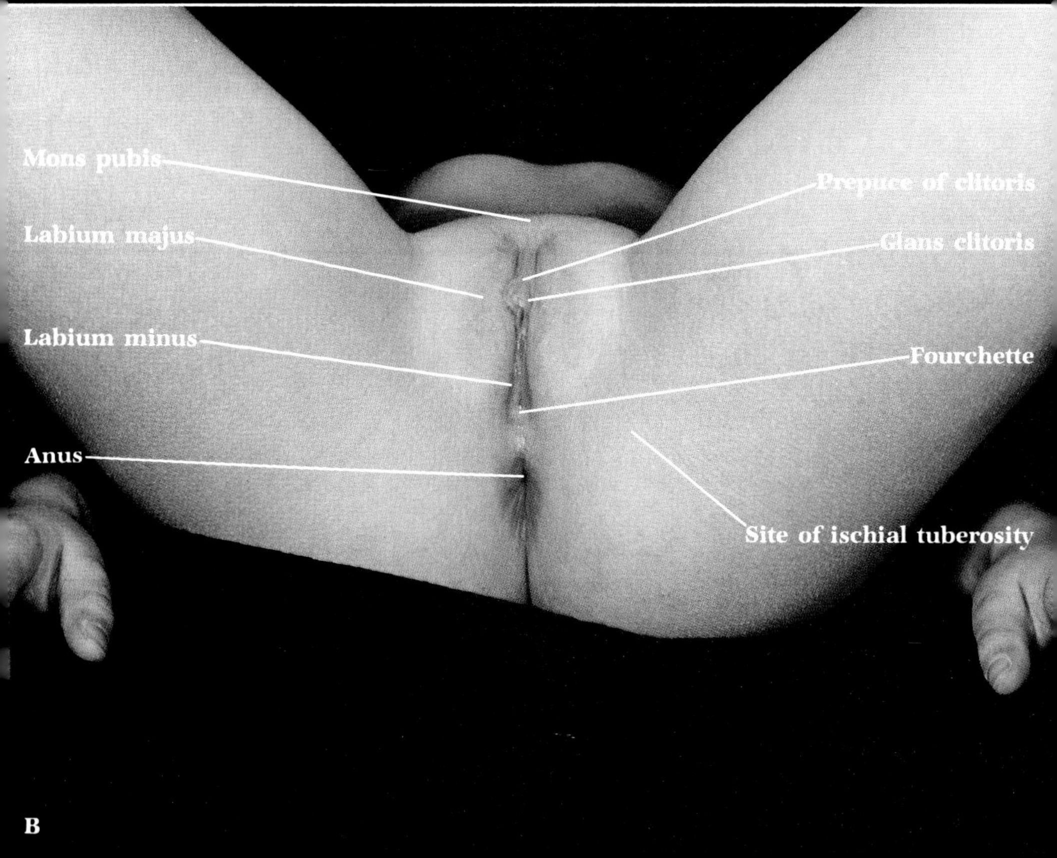

Mons pubis
Prepuce of clitoris
Labium majus
Glans clitoris
Labium minus
Fourchette
Anus
Site of ischial tuberosity

B

Figure 4-2
The perineum, inferior view.
A. 10-year-old circumcised male.
B. 10-year-old female.

1. The urogenital triangle is bounded anteriorly by the pubic arch and laterally by the ischial tuberosities. The anal triangle is bounded posteriorly by the tip of the coccyx, and on each side by the ischial tuberosity and the sacrotuberous ligament, overlapped by the border of the gluteus maximus muscle.

2. In young boys the testes are often retractile and when examined, especially with a cold hand, the testis may be pulled up to the superficial inguinal ring. For this reason the misdiagnosis of undescended testes is frequently made. Failure to find a testis in the scrotum should be followed by a careful search along the inguinal canal for an incompletely descended testis, or on the inner side of the thigh or in the perineum posterior to the scrotum for an ectopic or imperfectly descended testis.

3. The external urethral meatus should be examined in the young boy by first retracting the prepuce (remember that the prepuce is normally adherent to the glans in young infants). If the urethra opens onto the surface at some point on the undersurface of the penis or in the perineum the child has a congenital anomaly known as hypospadias.

4. In the infant female the labia minora are relatively large but they soon regress.

5. In the adrenogenital syndrome the clitoris is hypertrophied and there is complete or incomplete fusion of the labia.

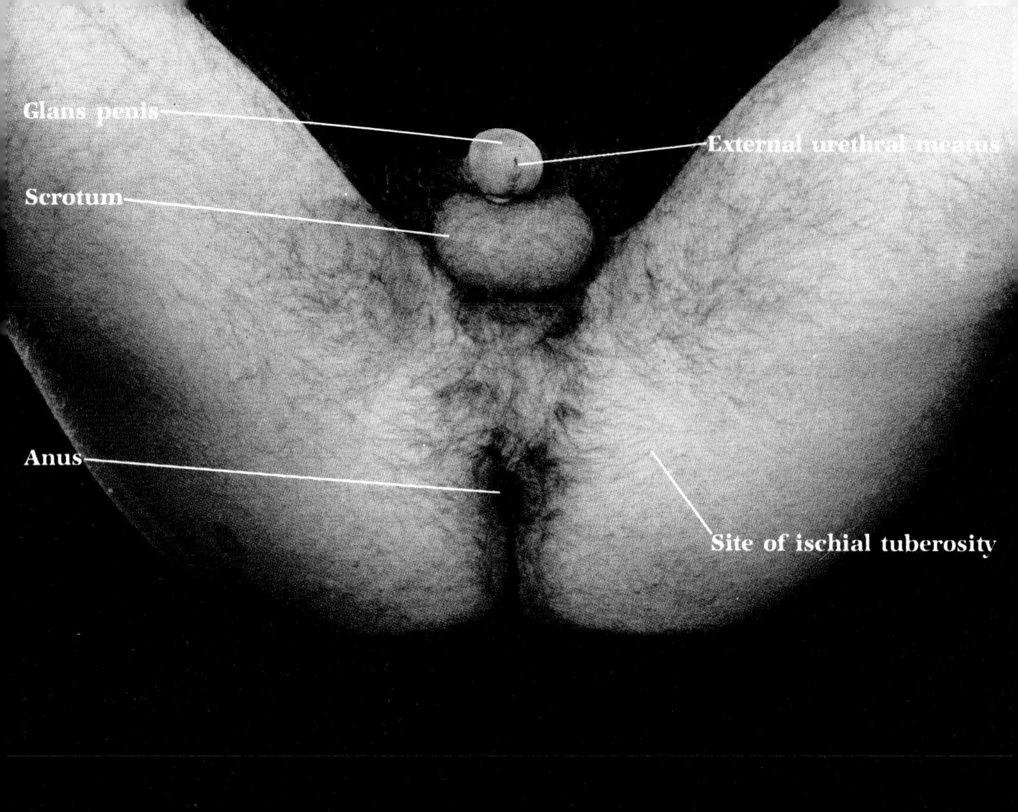
Glans penis
External urethral meatus
Scrotum
Anus
Site of ischial tuberosity

A

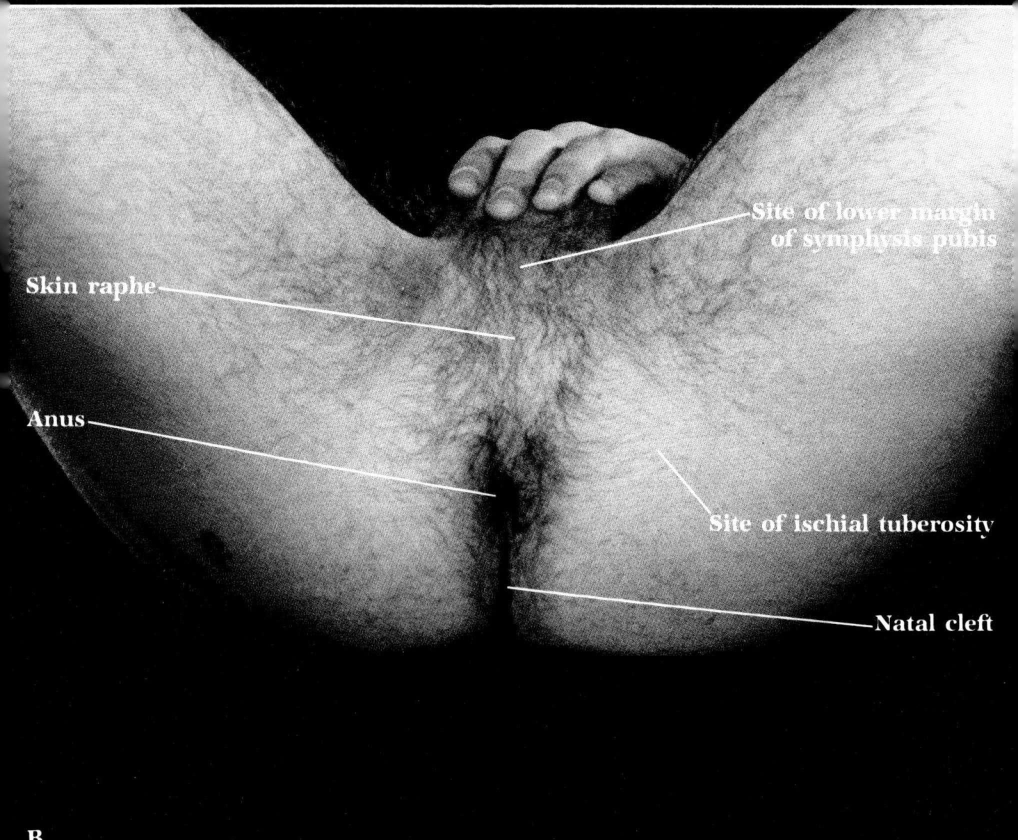
Site of lower margin
of symphysis pubis
Skin raphe
Anus
Site of ischial tuberosity
Natal cleft

B

Figure 4-3
The perineum in a 27-year-old circumcised male, inferior view. A. With the scrotum and penis in the anatomical position. B. With the scrotum and penis displaced superiorly.

1. When examining the penis note that it consists of a root, body, and glans. The root of the penis, which should be palpated, consists of three masses of erectile tissue: the bulb of the penis and the right and left crura of the penis. The bulb may be felt on deep palpation in the midline of the perineum posterior to the scrotum. Since it is traversed by the urethra, a thickening of the urethral wall due, for example, to a fibrous stricture can be felt. Each crus is attached to the side of the pubic arch.
2. The glans and prepuce must be carefully examined. The prepuce should be drawn back and signs of infection or ulcers should be noted.
3. The opening of the external urethral meatus should be spread wide open so that the distal end of the urethral mucous membrane can be inspected. Ask the patient to "strip" the penis by milking any discharge or pus through the external meatus.
4. Inspect the scrotum and note that the skin is rugose and covered with sparse hair. Edema is often seen here in patients with chronic nephritis and those with cardiac failure because of the laxity of the subcutaneous tissues. Carcinoma of the scrotum, sebaceous cysts, cellulitis, and intertrigo all occur in this structure. Remember that the lymphatic drainage of the tissues of the scrotum is into the horizontal group of superficial inguinal nodes.
5. The testes when palpated should move freely within the scrotum. They should be oval in shape and have a firm consistency. The surface is smooth. A normal testis when gently squeezed gives rise to a sickening pain.
6. The epididymis should be palpated posterior to the testis. It is a long, narrow, firm structure having the vas deferens on its medial side. The epididymis has an expanded upper end—the head, a body, and a pointed tail inferiorly.
7. Palpate the spermatic cord in the upper part of the scrotum between the finger and thumb and note the presence of a firm, smooth cordlike structure in its posterior part called the vas deferens.
8. Always examine the part of the urogenital triangle posterior to the scrotum. A periurethral abscess associated with a urethral stricture or an abscess of the bulbourethral gland may be identified in this area.
9. The anus should be inspected by separating the buttocks. Normally the anal margin has a reddish brown mucosa that is puckered by the contraction of the external anal sphincter. There should be no evidence of perianal redness or other signs of inflammation. Look for fissures, ulcers, and protruding veins or hemorrhoids. Observe whether there are small skin tags at the orifice; these are the remnants of thrombosed external hemorrhoids.
10. Ask the patient to bear down and observe the relaxation of the currugator cutis ani and the external sphincter. Normally the anal mucosa should protrude only slightly.

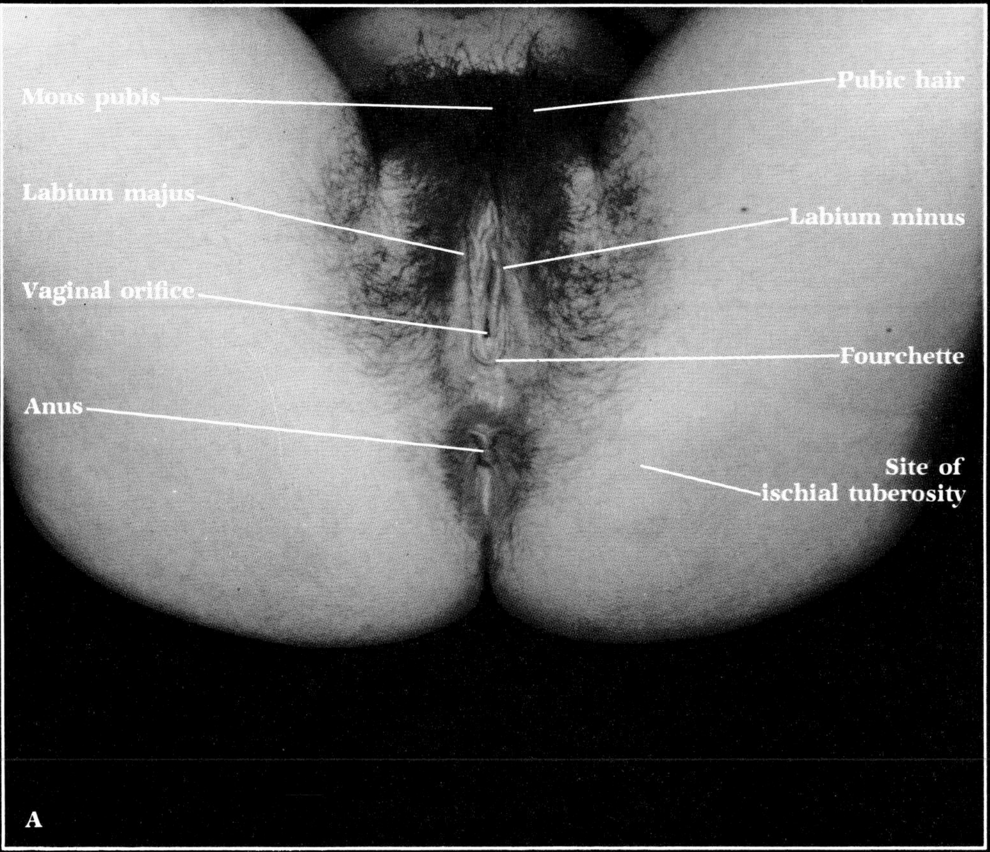
Mons pubis
Pubic hair
Labium majus
Labium minus
Vaginal orifice
Fourchette
Anus
Site of
ischial tuberosity
A

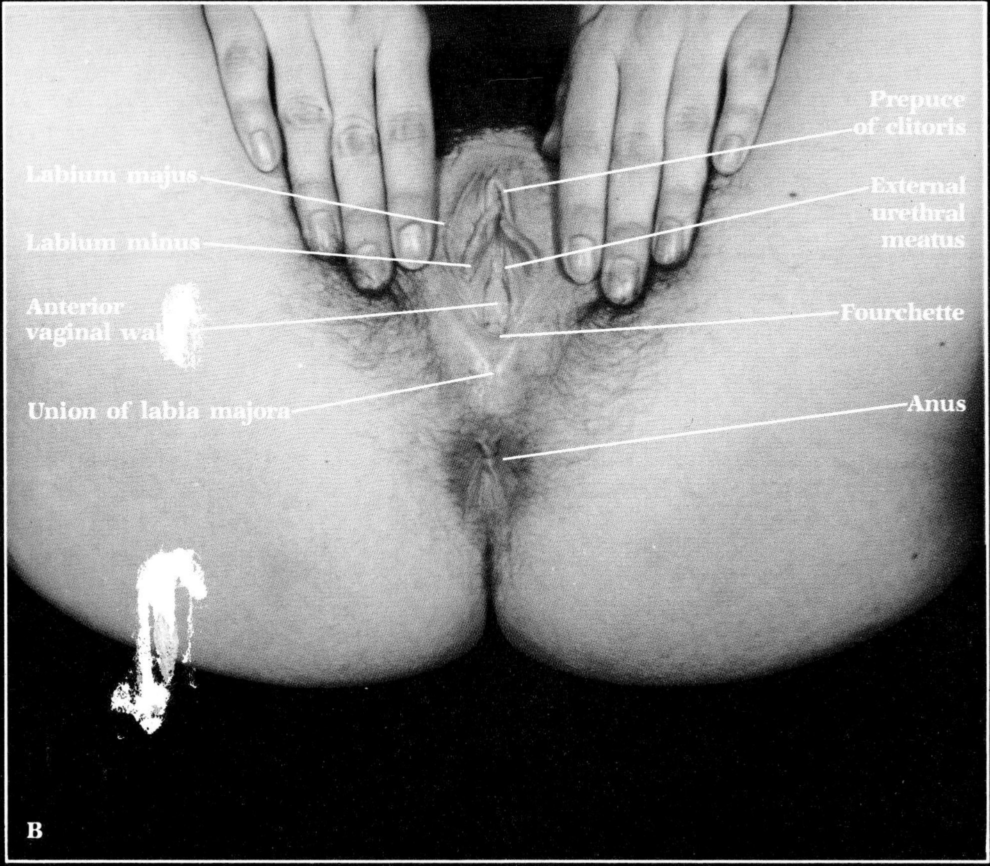
Prepuce
of clitoris
Labium majus
External
urethral
meatus
Labium minus
Anterior
vaginal wal
Fourchette
Union of labia majora
Anus
B

Figure 4-4
The perineum in a 25-year-old female, inferior view. A. With the labia together. B. With the labia separated.

1. The *vulva* is the name applied to the female external genitalia.
2. The mons pubis is a rounded, hair-bearing elevation of skin found anterior to the pubis. It is produced by an underlying pad of fat.
3. The labia majora are prominent, hair-bearing folds of skin extending posteriorly from the mons pubis to unite posteriorly in the midline.
4. The labia minora are two smaller folds of soft skin devoid of hair that lie between the labia majora. Their posterior ends are united to form a sharp fold, the fourchette. Anteriorly they split to enclose the clitoris, forming an anterior prepuce and a posterior frenulum.
5. The vestibule is a smooth triangular area bounded laterally by the labia minora; it has the clitoris at its apex and the fourchette at its base. It is perforated by the urethra (which lies immediately behind the clitoris) and the vagina.
6. The vaginal orifice is protected in the virgin by a thin mucosal fold called the hymen, which is perforated at its center. At the first coitus the hymen tears, usually posteriorly or posterolaterally, and after childbirth only a few tags of the hymen remain.
7. The greater vestibular glands, a pair of small mucus-secreting glands, drain their secretion on each side into the groove between the hymen and the posterior part of the labium minus.
8. The clitoris is situated at the apex of the vestibule anteriorly. It has a structure similar to the penis, and the glans of the clitoris is partly hidden by the prepuce.
9. On clinical examination, inspect the vulva systematically, starting with the mons pubis, then the lateral and medial surfaces of the labia. These are common sites for such lesions as syphilitic chancre, herpes, lymphogranuloma inguinale, and carcinoma. Examine the urethral orifice; its margins are usually slightly everted. A urethral caruncle (an extremely tender nodule), or prolapse of the urethra may be noted.
10. The greater vestibular glands are common sites for infection. They are examined by palpating the posterior part of the labia majora between the finger and thumb. The gland is deeply placed.
11. Inspect the vaginal orifice and note the presence of blood, pus, or discharge. A bluish coloration of the vaginal mucosa is one of the early signs of pregnancy.
12. Ask the patient to bear down and note the adequacy of the vaginal supports. If a cystocele (prolapse of the anterior vaginal wall) is present, the anterior vaginal wall will bulge downward. A rectocele (prolapse of the posterior vaginal wall) will reveal itself by a bulging downward of the posterior vaginal wall.
13. The anal margin has a reddish brown mucosa puckered by contraction of the external anal sphincter. Look for fissures, ulcers, and protruding hemorrhoids or skin tags. Ask the patient to bear down. Normally the anal mucosa should protrude only slightly.

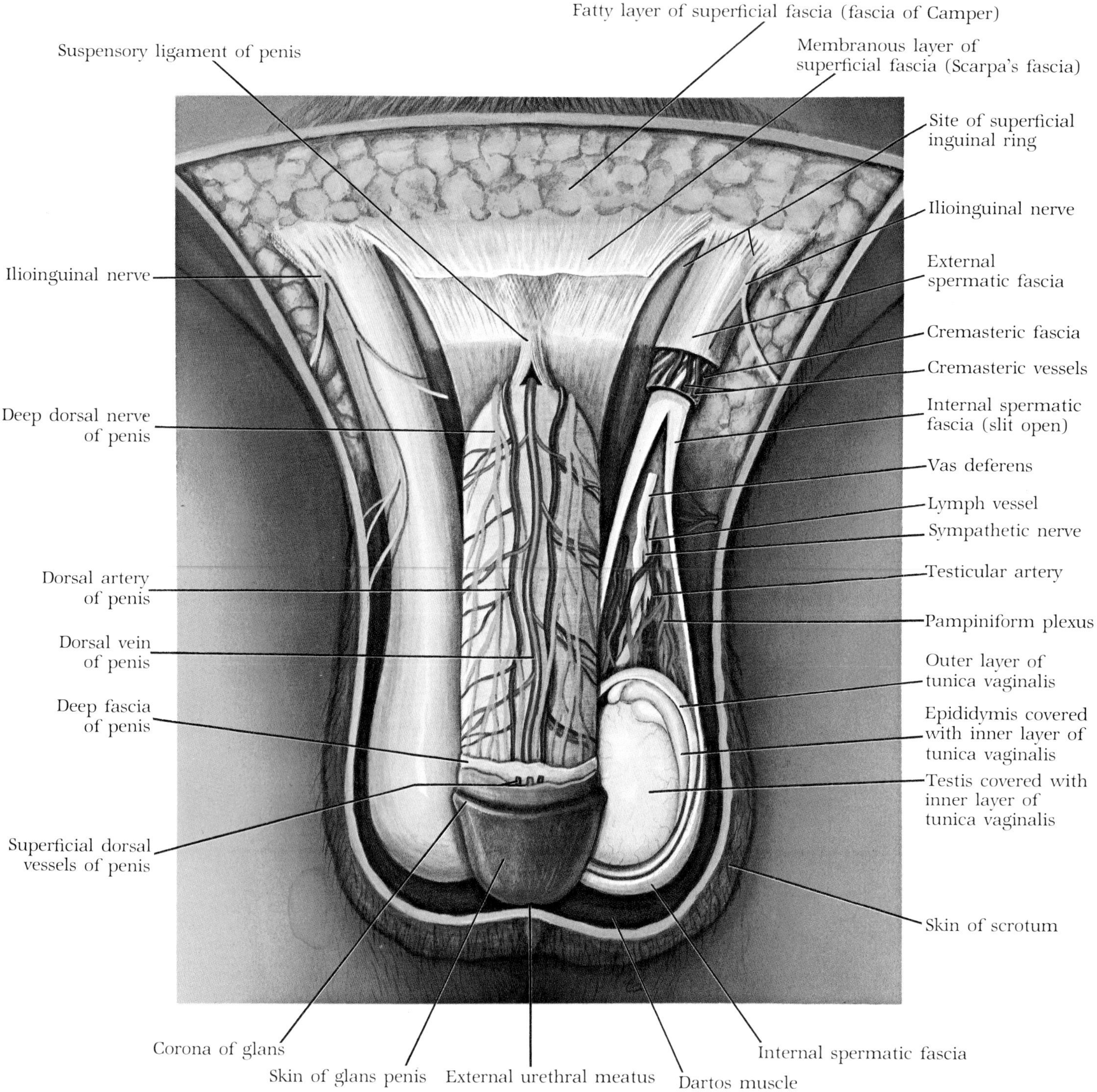
Fatty layer of superficial fascia (fascia of Camper)
Suspensory ligament of penis
Membranous layer of superficial fascia (Scarpa's fascia)
Site of superficial inguinal ring
Ilioinguinal nerve
Ilioinguinal nerve
External spermatic fascia
Cremasteric fascia
Cremasteric vessels
Internal spermatic fascia (slit open)
Deep dorsal nerve of penis
Vas deferens
Lymph vessel
Sympathetic nerve
Testicular artery
Dorsal artery of penis
Pampiniform plexus
Dorsal vein of penis
Outer layer of tunica vaginalis
Deep fascia of penis
Epididymis covered with inner layer of tunica vaginalis
Testis covered with inner layer of tunica vaginalis
Superficial dorsal vessels of penis
Skin of scrotum
Corona of glans
Skin of glans penis
External urethral meatus
Dartos muscle
Internal spermatic fascia

Figure 4-5
Penis and contents of scrotum, anterior view.

1. The penis is composed of three bodies of erectile tissue: the two corpora cavernosa and the corpus spongiosum. All three bodies are bound together by deep fascia (Buck's fascia).

2. Normally the glans penis is covered by the prepuce, a hoodlike fold of skin. Congenital or inflammatory narrowing of the preputial opening is known as phimosis. Carcinoma of the glans penis is common. Circumcision in infancy prevents the development of this carcinoma in later life.

3. Lymphatic drainage of the penis is important in relation to spread of carcinoma of the penis. The skin of the penis is drained into the medial group of superficial inguinal lymph nodes. The deeper structures of the penis drain into the internal iliac nodes.

4. Incomplete descent and maldescent are two types of imperfect descent of the testis. In incomplete descent the testis, although travelling down its normal path, fails to reach the floor of the scrotum. It may be found within the abdomen or inguinal canal, at the superficial inguinal ring, or high up in the scrotum. In maldescent the testis travels down an abnormal path and fails to reach the scrotum. It may be found in the superficial fascia of the anterior abdominal wall above the inguinal ligament, anterior to the pubis, in the perineum, or in the thigh.

5. Normal spermatogenesis will not take place if the testes are situated within the abdomen or in the inguinal canal because the higher temperatures there retard the normal process.

6. A malignant tumor of the testis spreads superiorly via the lymph vessels to the lumbar (para-aortic) lymph nodes at the level of the first lumbar vertebra. Later, when the tumor invades the skin of the scrotum, the superficial inguinal nodes become involved.

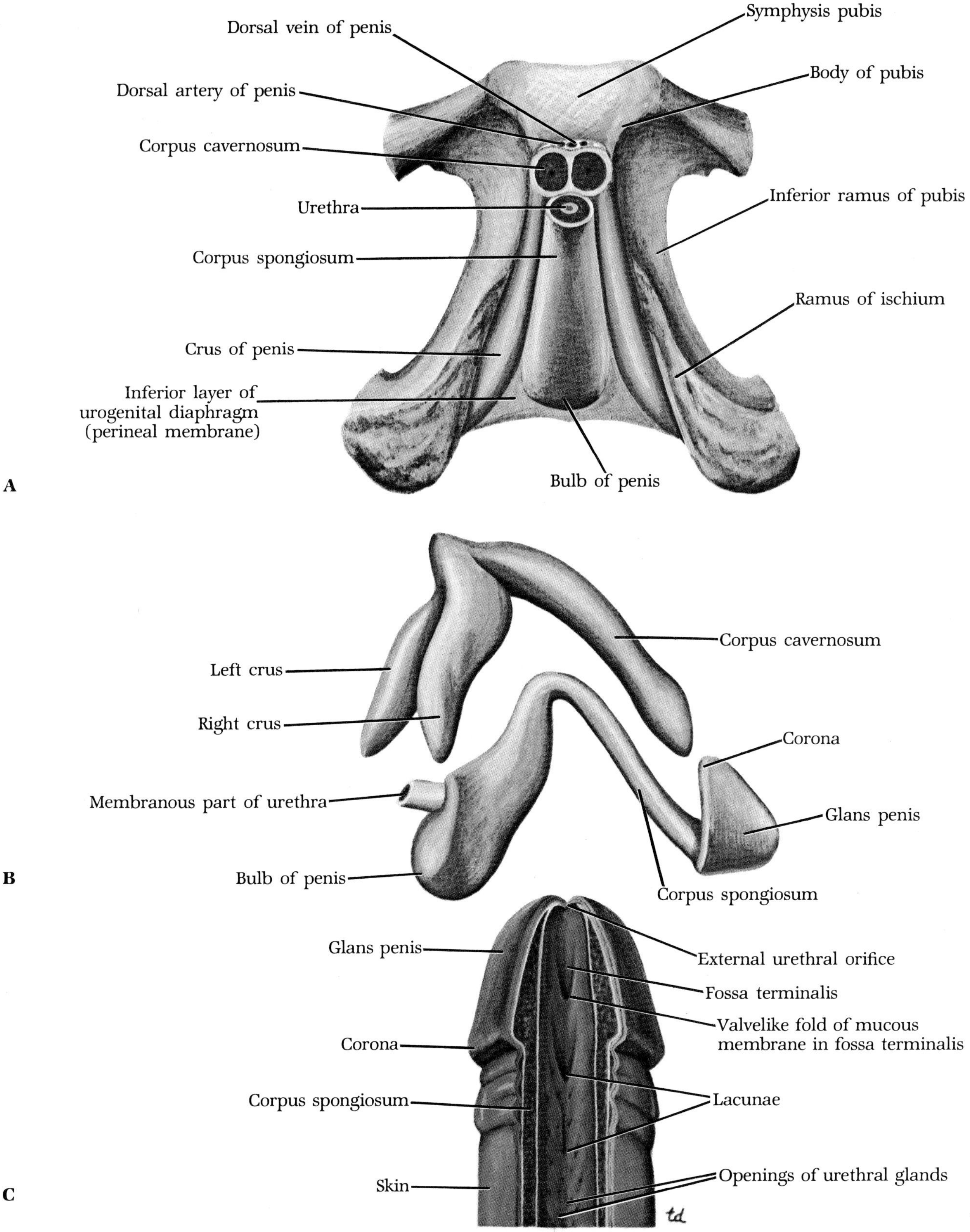
Dorsal vein of penis
Symphysis pubis
Dorsal artery of penis
Body of pubis
Corpus cavernosum
Urethra
Inferior ramus of pubis
Corpus spongiosum
Ramus of ischium
Crus of penis
Inferior layer of urogenital diaphragm (perineal membrane)
Bulb of penis
A
Corpus cavernosum
Left crus
Right crus
Corona
Membranous part of urethra
Glans penis
Bulb of penis
Corpus spongiosum
B
Glans penis
External urethral orifice
Fossa terminalis
Valvelike fold of mucous membrane in fossa terminalis
Corona
Corpus spongiosum
Lacunae
Openings of urethral glands
Skin
C
td

Figure 4-6
The penis. A. Root. B. The three bodies of erectile tissue. C. The penile urethra slit open to show the folds of mucous membrane and glandular orifices in the roof of the urethra.

1. Chronic mild inflammation of the urethra may lead to inflammation of the fibrous sheath of the corpora cavernosa. This produces swelling and induration of the penis (Peyronie's disease).

2. Note that in the urethra there are depressions in the mucous membrane of the roof and the openings of the ducts of the urethral glands (Littre's glands). In gonorrheal urethritis the gonococci invade these areas; their persistence in the ducts and glands may maintain the inflammation and be a source of chronic infection.

3. Chronic gonococcal infection of the penile urethra results in the formation of irregular scar tissue, which then contracts and leads to a stricture of the urethra.

4. The following facts must be kept in mind before inserting a catheter in a male: (a) the external urethral orifice at the glans penis is the narrowest part of the entire urethra; (b) within the glans, the urethra dilates slightly to form the fossa terminalis (fossa navicularis); (c) near the posterior end of the fossa, a fold of mucous membrane projects into the lumen from the roof; if the point of the catheter passes through the external orifice and is then directed toward the urethral floor until it has passed the mucosal fold, it should easily pass along a normal urethra into the bladder.

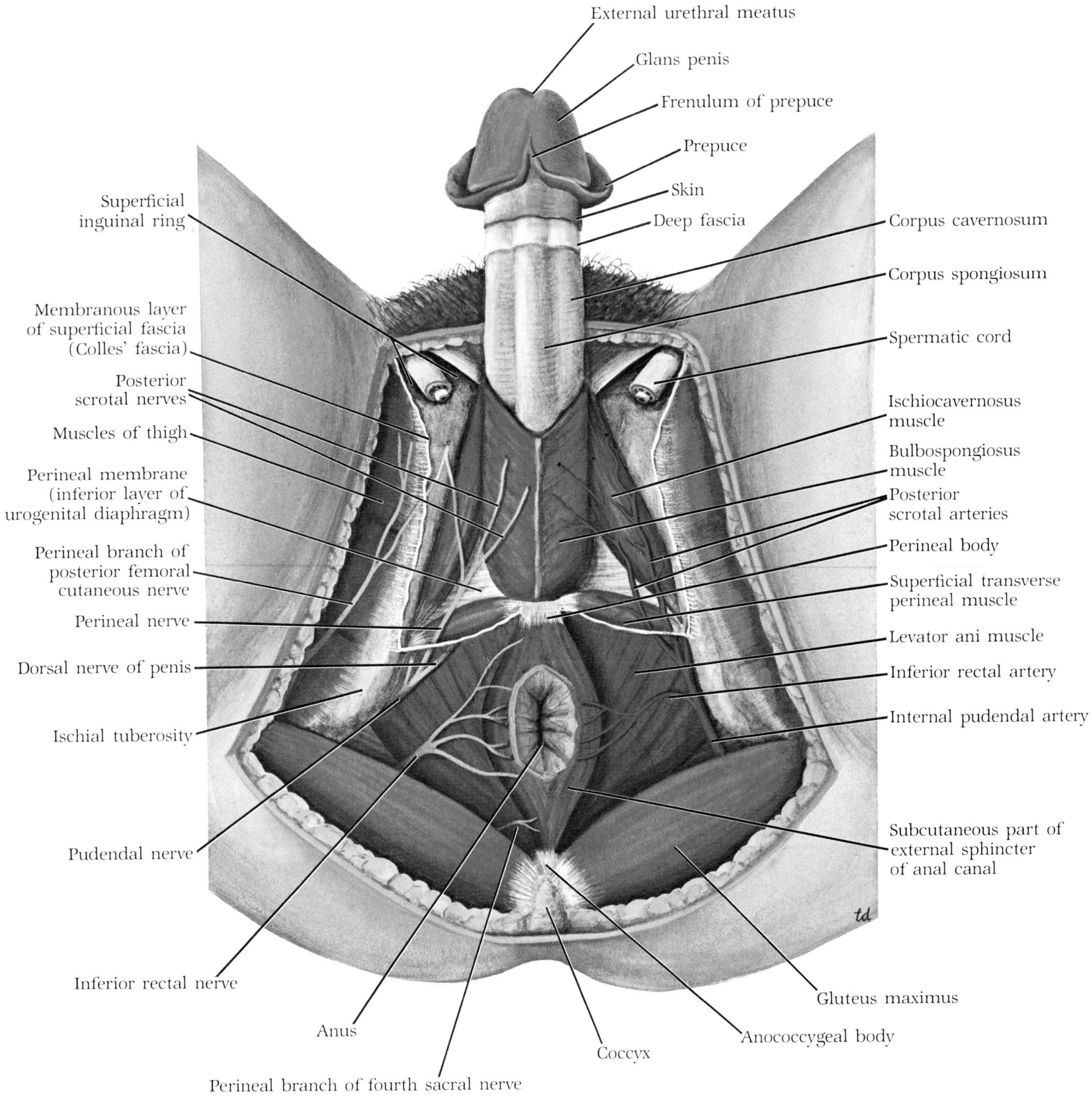
External urethral meatus
Glans penis
Frenulum of prepuce
Prepuce
Skin
Deep fascia
Superficial inguinal ring
Corpus cavernosum
Corpus spongiosum
Membranous layer of superficial fascia (Colles' fascia)
Spermatic cord
Posterior scrotal nerves
Ischiocavernosus muscle
Muscles of thigh
Bulbospongiosus muscle
Perineal membrane (inferior layer of urogenital diaphragm)
Posterior scrotal arteries
Perineal body
Perineal branch of posterior femoral cutaneous nerve
Superficial transverse perineal muscle
Perineal nerve
Levator ani muscle
Dorsal nerve of penis
Inferior rectal artery
Internal pudendal artery
Ischial tuberosity
Subcutaneous part of external sphincter of anal canal
Pudendal nerve
td
Inferior rectal nerve
Gluteus maximus
Anus
Anococcygeal body
Coccyx
Perineal branch of fourth sacral nerve

Figure 4-7
Male urogenital triangle and anal triangle as seen from below. The membranous layer of superficial fascia (Colles' fascia) has been cut away to expose the contents of the superficial perineal pouch, and the scrotum has been removed.

1. The prepuce, a hoodlike covering of the glans penis, may have to be surgically removed (circumcision) for such conditions as phimosis, paraphimosis, and chronic balanitis. The frenulum of the prepuce contains a small artery that can cause severe bleeding during circumcision unless it is secured.

2. The penis sustains an erection by a rapid filling of the blood spaces in the corpora cavernosa and corpus spongiosum and a compressing of the draining veins. An excessively prolonged erection, known as priapism, can be caused by clotting of blood in the vascular spaces, sickle cell disease, and malignant neoplasia.

3. Rupture of the penile urethra may result in extravasation of urine into the superficial perineal pouch; the urine then extends over the scrotum, beneath the membranous layer of superficial fascia.

4. The most dependent part of the male urethra lies within the bulb where it is subject to chronic inflammation and stricture formation.

5. The lymphatic drainage of the mucous membrane of the lower half of the anal canal and the perianal skin is into the medial group of superficial inguinal lymph nodes. These areas should always be examined when the superficial inguinal nodes are enlarged.

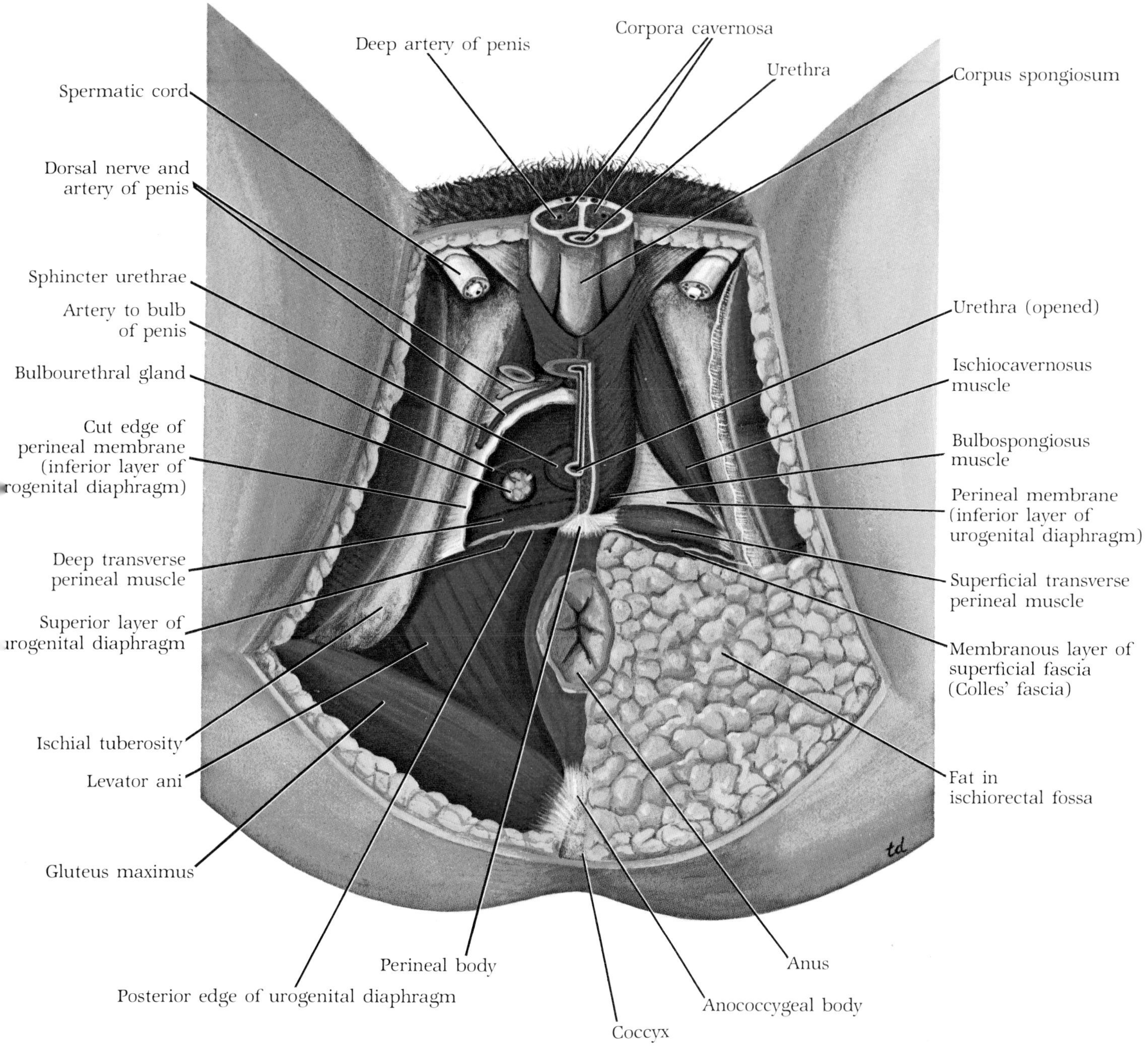
Corpora cavernosa
Deep artery of penis
Urethra
Corpus spongiosum
Spermatic cord
Dorsal nerve and artery of penis
Sphincter urethrae
Artery to bulb of penis
Bulbourethral gland
Cut edge of perineal membrane (inferior layer of urogenital diaphragm)
Deep transverse perineal muscle
Superior layer of urogenital diaphragm
Ischial tuberosity
Levator ani
Gluteus maximus
Urethra (opened)
Ischiocavernosus muscle
Bulbospongiosus muscle
Perineal membrane (inferior layer of urogenital diaphragm)
Superficial transverse perineal muscle
Membranous layer of superficial fascia (Colles' fascia)
Fat in ischiorectal fossa
td
Perineal body
Anus
Posterior edge of urogenital diaphragm
Anococcygeal body
Coccyx

Figure 4-8
Male urogenital triangle and anal triangle as seen from below. The bulb of the penis has been sectioned to show the urethra; the inferior layer of the urogenital diaphragm has been cut away on the left to show the bulbourethral gland within the deep perineal pouch.

1. Rupture of the urethra may result from a severe blow on the perineum, the commonest site being within the bulb of the penis. The urine extravasates into the superficial perineal pouch as described previously. When the membranous part of the urethra ruptures, the urine escapes into the deep perineal pouch and may extravasate superiorly around the prostate and bladder.

2. A gonococcal urethritis may infect the bulbourethral glands, giving rise to an abscess that can be palpated through the anterior rectal wall.

3. The ischiorectal fossae are filled with fat. Their close proximity to the anal canal makes them particularly vulnerable to infection. Infection commonly tracks laterally from the anal mucosa through the external anal sphincter. Infection of the perianal hair follicles or sweat glands may also cause infection of the fossae. An ischiorectal abscess may involve the opposite fossa if the infection spreads across the midline posterior to the anal canal.

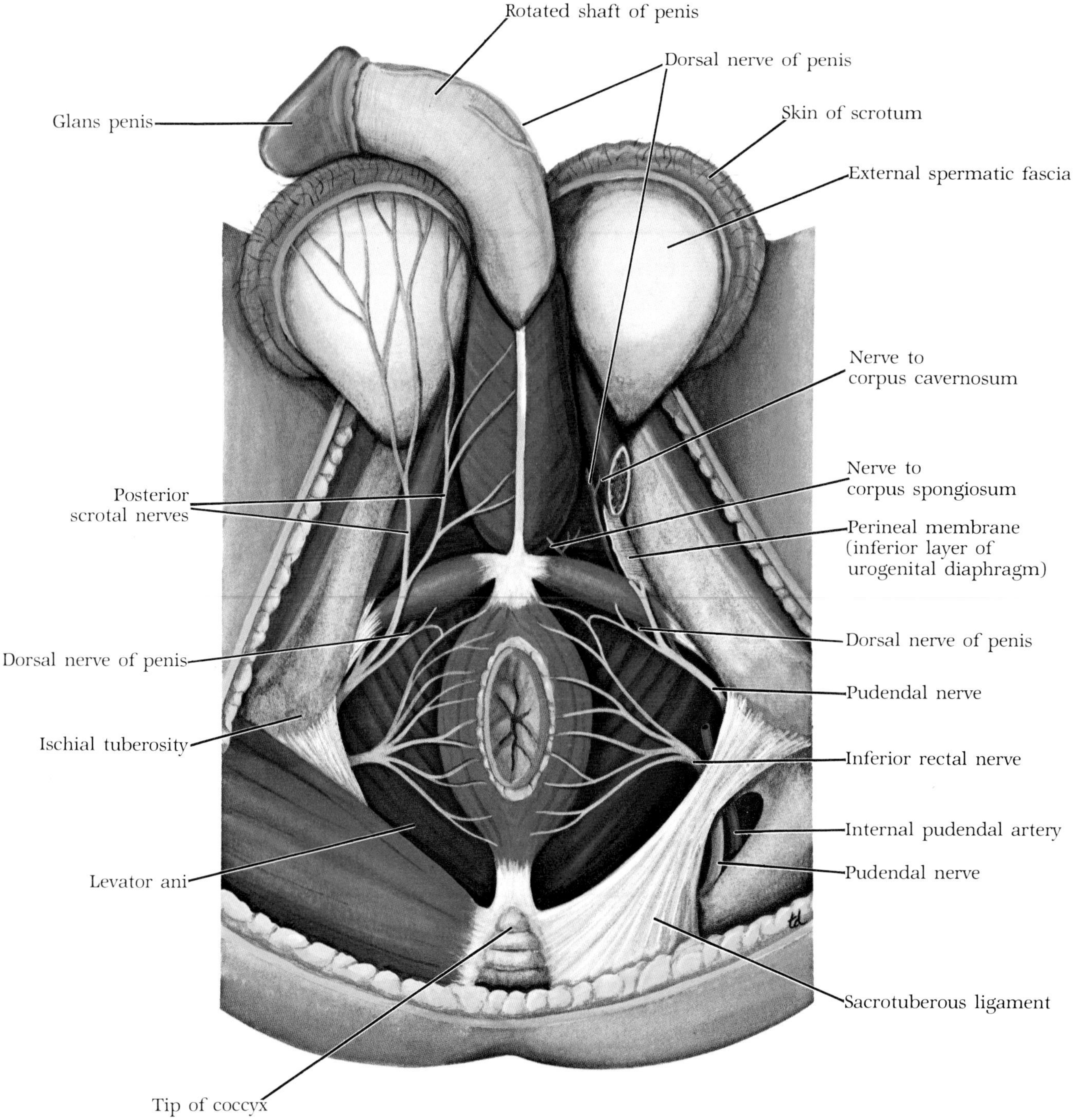
Rotated shaft of penis
Dorsal nerve of penis
Glans penis
Skin of scrotum
External spermatic fascia
Nerve to
corpus cavernosum
Posterior
scrotal nerves
Nerve to
corpus spongiosum
Perineal membrane
(inferior layer of
urogenital diaphragm)
Dorsal nerve of penis
Dorsal nerve of penis
Pudendal nerve
Ischial tuberosity
Inferior rectal nerve
Internal pudendal artery
Levator ani
Pudendal nerve
Sacrotuberous ligament
Tip of coccyx

Figure 4-9
Male urogenital triangle and anal triangle as seen from below. The scrotum has been divided, and the two halves have been pulled forward. The left crus of the penis has also been divided.

1. The lymphatic drainage of the wall of the scrotum — i.e., from the skin and fasciae, including the tunica vaginalis — is into the superficial inguinal lymph nodes. A malignant tumor of the testis spreads superiorly via the lymph vessels to the lumbar (para-aortic) lymph nodes at the level of the first lumbar vertebra. The superficial inguinal lymph nodes do not become involved until the tumor spreads locally to involve the tissues and skin of the scrotum, a late event.

2. Perianal abscesses are produced by fecal trauma to the anal mucosa. The abscess may be localized to the submucosa (submucous abscess), may occur beneath the perianal skin (subcutaneous abscess), or may occupy the ischiorectal fossa (ischiorectal abscess).

3. Anal fistulae result from spread or inadequate treatment of anal abscesses. One end of the fistula is at the lumen of the anal canal or lower rectum and the other end appears on the skin surface close to the anus.

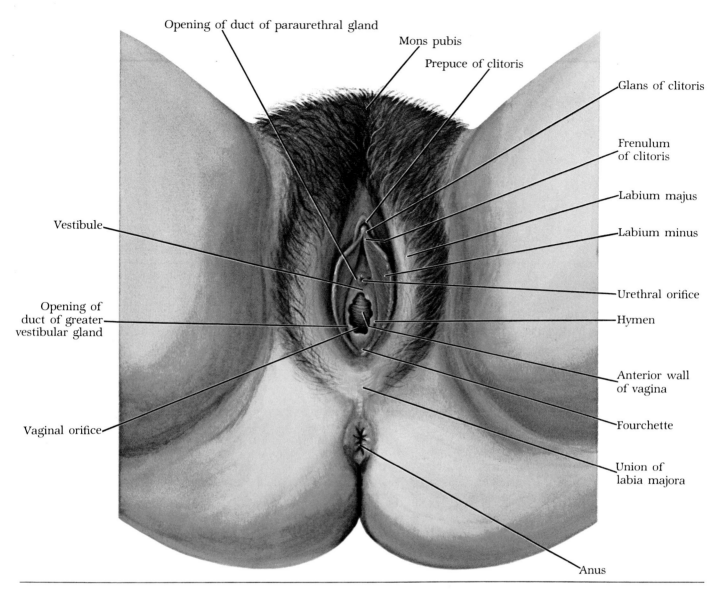

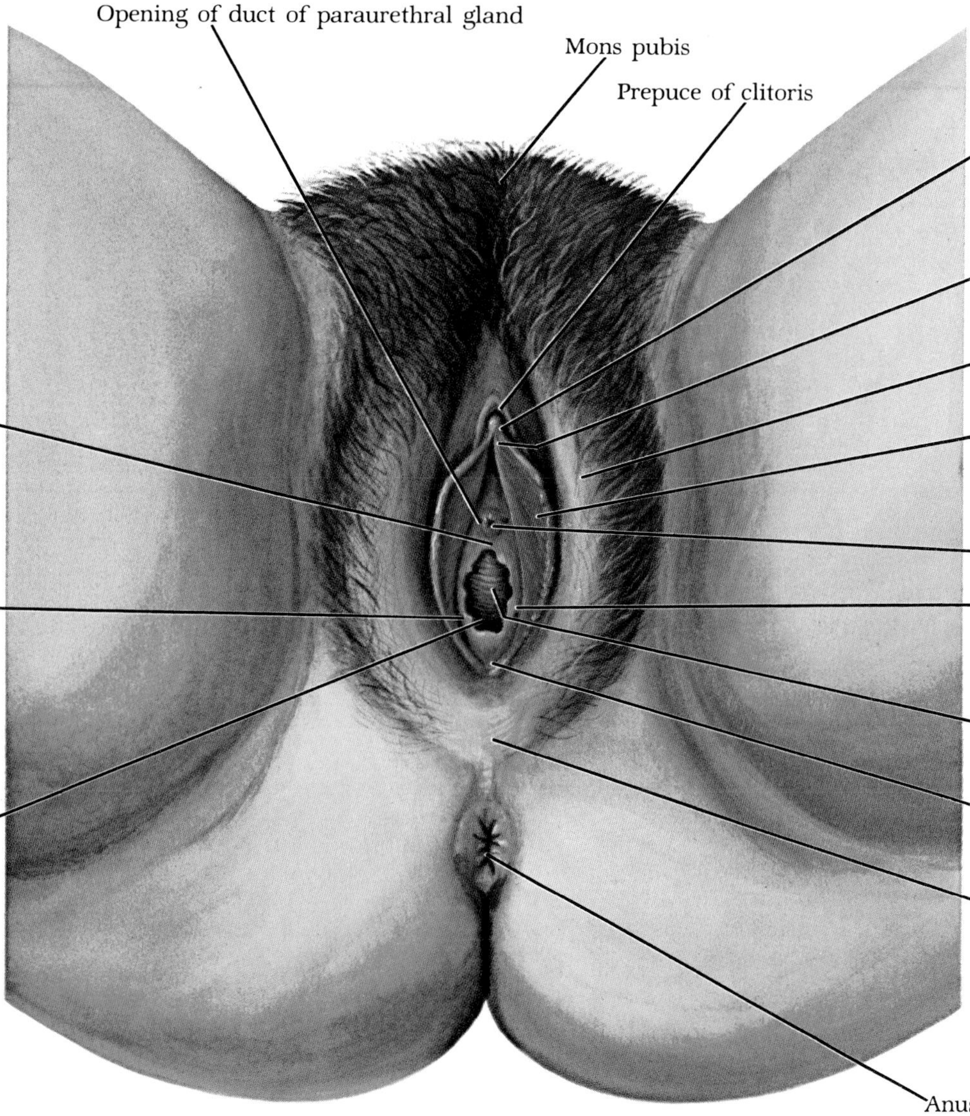
Opening of duct of paraurethral gland
Mons pubis
Prepuce of clitoris
Glans of clitoris
Frenulum of clitoris
Labium majus
Labium minus
Vestibule
Urethral orifice
Opening of duct of greater vestibular gland
Hymen
Anterior wall of vagina
Fourchette
Vaginal orifice
Union of labia majora
Anus

Figure 4-10
Vulva of a woman who has had sexual intercourse.

1. A bluish discoloration due to local venous congestion of the vulva, vestibule, and vagina is one of the objective signs of pregnancy. It appears at about the eighth to the twelfth week and increases as the pregnancy progresses. It is most obvious in multigravidas. Vulvar varices of the labia majora are also common in pregnancy, regressing completely after delivery.

2. Infections of the greater vestibular gland (Bartholin's gland) due to gonococci, E. coli, and pyogenic organisms are common. The posterior part of the labium majus becomes swollen and tender as the gland expands and pus emerges from the duct. The paraurethral glands may become infected secondary to a chronic urethritis.

3. Cysts of the greater vestibular gland are a common complication of a preceding infection. The duct becomes blocked as the result of fibrous tissue formation; the posterior part of the labium majus then becomes enlarged.

4. Carcinoma of the vulva is a rare form of carcinoma of the female genital tract, but when it does occur lymph node metastases are common. The lymph passes first to the medial group of superficial inguinal nodes, then into the deep inguinal nodes, and finally drains into the external and common iliac nodes. Surgical excision of the vulva and careful dissection of all the lymph nodes —whether actually or potentially affected— is the best method of treatment.

5. Cystocele (prolapse of the anterior vaginal wall), urethrocele, and prolapse of the uterus may also produce swellings that can protrude outside the vaginal orifice.

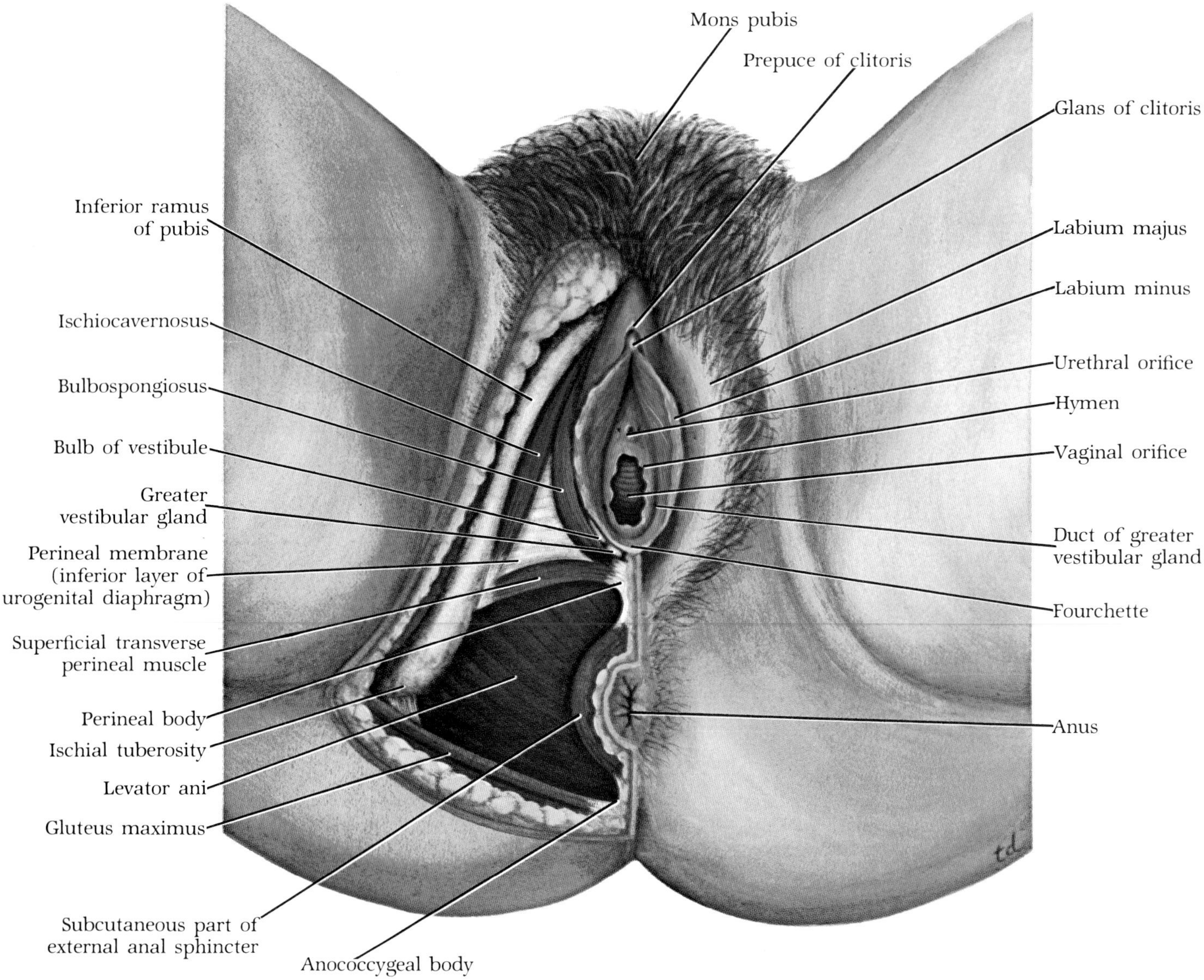
Mons pubis
Prepuce of clitoris
Glans of clitoris
Inferior ramus of pubis
Labium majus
Labium minus
Ischiocavernosus
Urethral orifice
Bulbospongiosus
Hymen
Bulb of vestibule
Vaginal orifice
Greater vestibular gland
Duct of greater vestibular gland
Perineal membrane (inferior layer of urogenital diaphragm)
Fourchette
Superficial transverse perineal muscle
Perineal body
Anus
Ischial tuberosity
Levator ani
Gluteus maximus
Subcutaneous part of external anal sphincter
Anococcygeal body

Figure 4-11
Female perineum. On the left, in the anal triangle, the fatty tissue has been removed from the ischiorectal fossa to show the inferior surface of the pelvic diaphragm and the external anal sphincter; in the urogenital triangle, the membranous layer of superficial fascia (Colles' fascia) has been removed to show the muscles of the superficial perineal pouch.

1. Because of numerous glands and ducts opening onto the surface in the region of the vulva, this area is prone to infection. The sebaceous glands of the labia majora, the ducts of the greater vestibular glands, the vagina, the urethra, and the paraurethral glands can all become infected. Provided that the pH of the interior of the vagina is kept low, it is capable of resisting infection to a remarkable degree.

2. The external urethral orifice lies in the vestibule below the clitoris. Before introduction of a catheter, the anterior ends of the labia minora should be parted by the fingers in order to reveal the urethral opening.

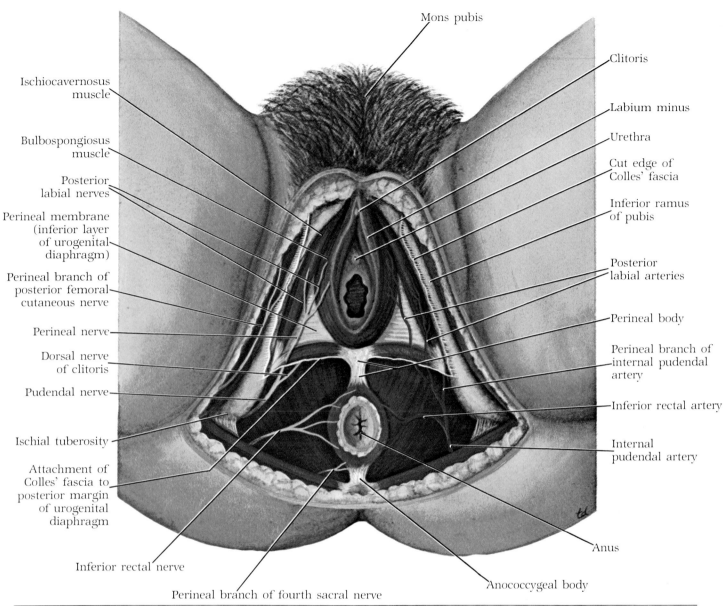

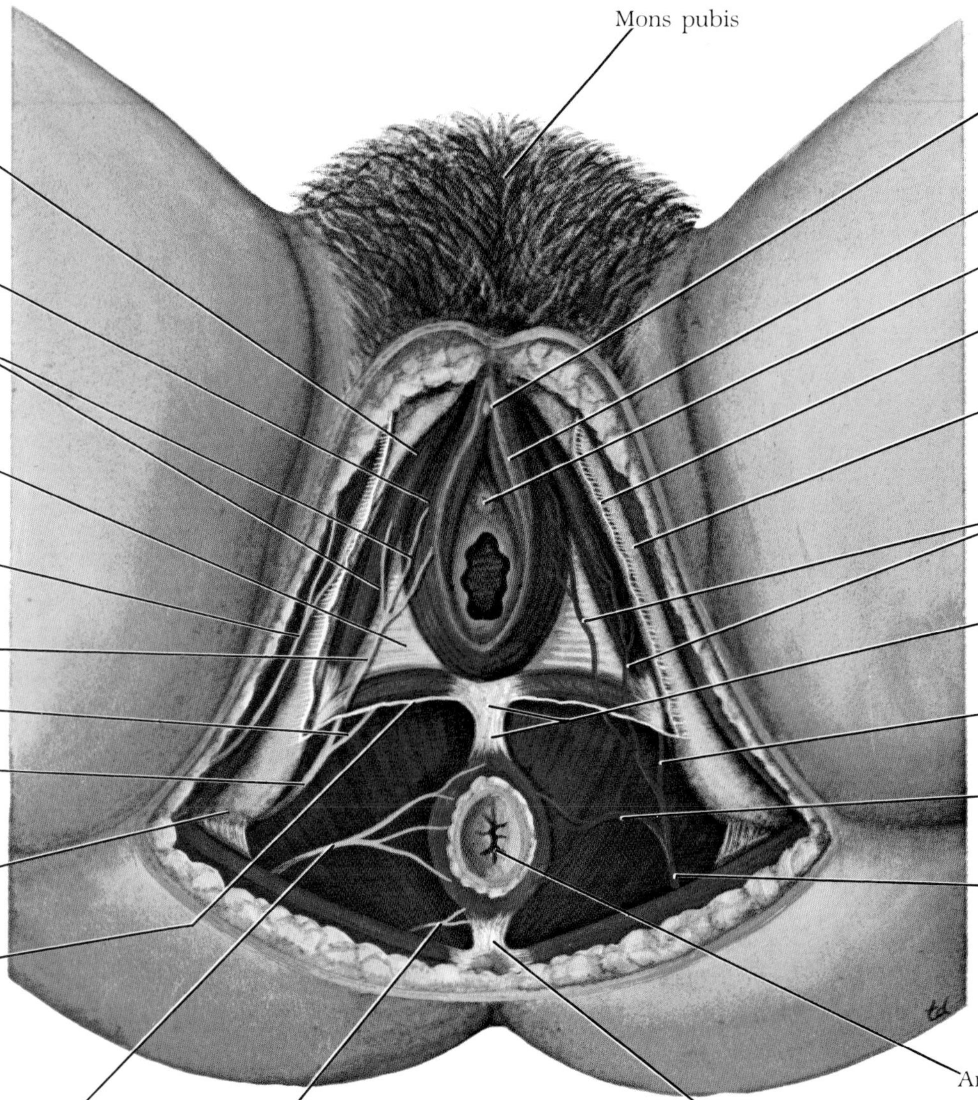
Mons pubis
Clitoris
Ischiocavernosus muscle
Labium minus
Urethra
Bulbospongiosus muscle
Cut edge of Colles' fascia
Posterior labial nerves
Inferior ramus of pubis
Perineal membrane (inferior layer of urogenital diaphragm)
Posterior labial arteries
Perineal branch of posterior femoral cutaneous nerve
Perineal body
Perineal nerve
Perineal branch of internal pudendal artery
Dorsal nerve of clitoris
Pudendal nerve
Inferior rectal artery
Ischial tuberosity
Internal pudendal artery
Attachment of Colles' fascia to posterior margin of urogenital diaphragm
Anus
Inferior rectal nerve
Anococcygeal body
Perineal branch of fourth sacral nerve

Figure 4-12
Female perineum. In the anal triangle, the fatty tissue has been removed from the ischiorectal fossae to show the inferior surface of the pelvic diaphragm and the external anal sphincter; in the urogenital triangle, the superficial fascia (Colles' fascia) has been removed to show the muscles, nerves, and arteries in the superficial perineal pouch.

1. The perineal body is a wedge of fibromuscular tissue that lies between the lower part of the vagina and the anal canal. It is held in position by the insertion of the perineal muscles and by the attachment of the levatores ani muscles. It supports the posterior wall of the vagina and may be damaged by laceration during childbirth, leading to permanent weakness of the pelvic floor.

2. The lymphatic drainage of the vulva and the mucous membrane of the lower half of the anal caal and the perianal skin is into the medial group of the superficial inguinal lymph nodes. These areas should always be examined when this group of nodes is enlarged.

3. The part of the pudendal nerve that runs forward in the fascial pudendal canal on the lateral wall of the ischiorectal fossa may be blocked by an anesthetic to produce analgesia of the perineum in women who are about to have a forceps delivery.

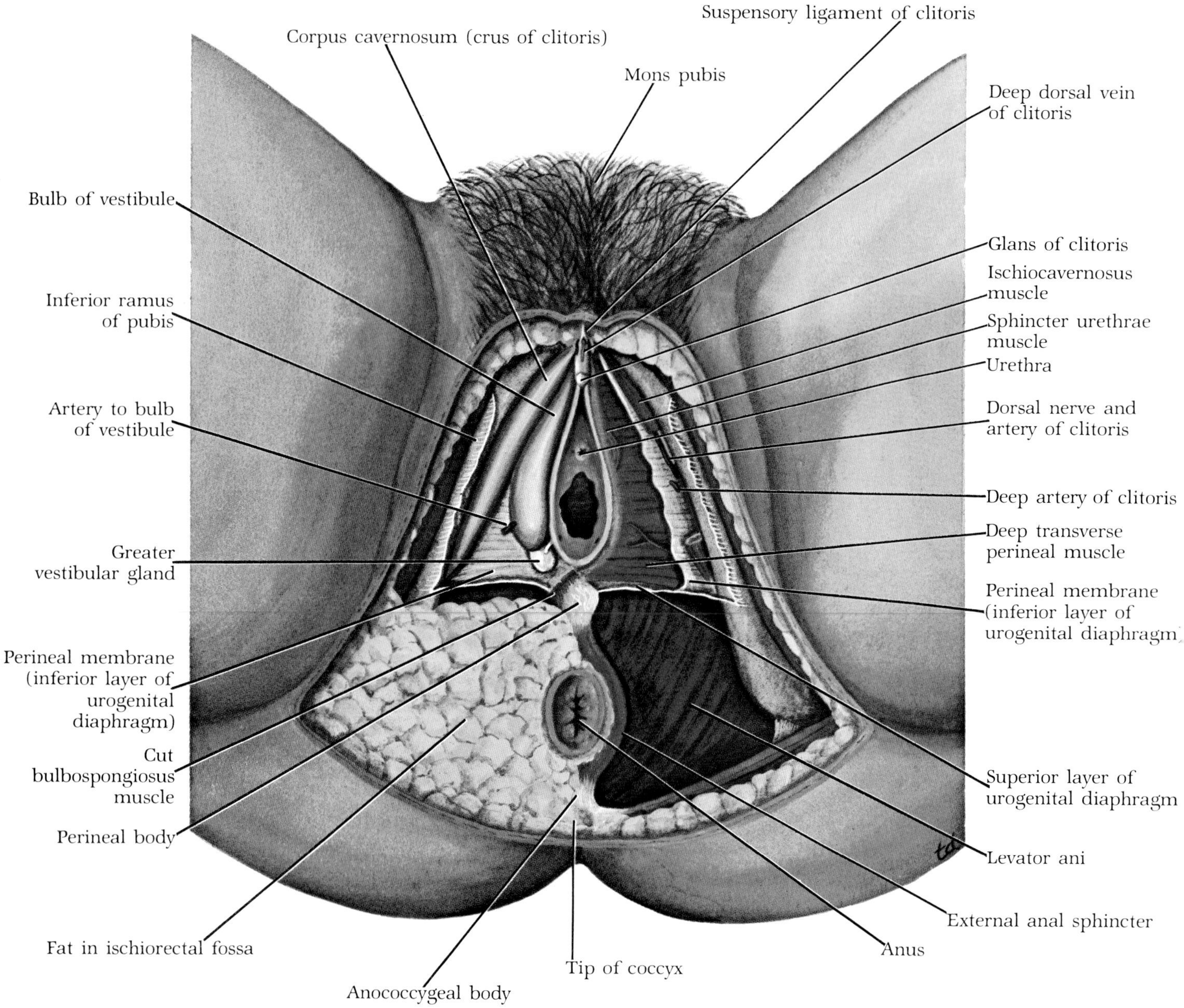
Suspensory ligament of clitoris
Corpus cavernosum (crus of clitoris)
Mons pubis
Deep dorsal vein of clitoris
Bulb of vestibule
Glans of clitoris
Ischiocavernosus muscle
Inferior ramus of pubis
Sphincter urethrae muscle
Urethra
Artery to bulb of vestibule
Dorsal nerve and artery of clitoris
Deep artery of clitoris
Greater vestibular gland
Deep transverse perineal muscle
Perineal membrane (inferior layer of urogenital diaphragm)
Perineal membrane (inferior layer of urogenital diaphragm)
Cut bulbospongiosus muscle
Superior layer of urogenital diaphragm
Perineal body
Levator ani
External anal sphincter
Fat in ischiorectal fossa
Anus
Tip of coccyx
Anococcygeal body

Figure 4-13
Female perineum. In the anal triangle, on the right, the fatty tissue has been removed from the ischiorectal fossa to show the inferior surface of the pelvic diaphragm and the external anal sphincter. In the urogenital triangle, the superficial perineal muscles on the left have been removed to expose the bulb of the vestibule and the crus of the clitoris; on the right, the inferior fascial layer of the urogenital diaphragm (perineal membrane) has been cut away to show the sphincter urethrae and the deep transverse perineal muscle. A segment of the crus of the clitoris has also been removed.

1. In the majority of women there is little more than an abrasion of the posterior vaginal wall during childbirth. Disproportion between the baby and the maternal soft tissues — either the head is too large or the birth canal too small — is the commonest cause of perineal lacerations. A severe tear of the lower third of the posterior wall of the vagina extending through the perineal body and overlying skin may occur, and in some cases the lacerations may extend further posteriorly into the anal canal and damage the external sphincter. In these circumstances accurate repair of the walls of the anal canal, vagina, and perineal body should be undertaken as soon as possible.

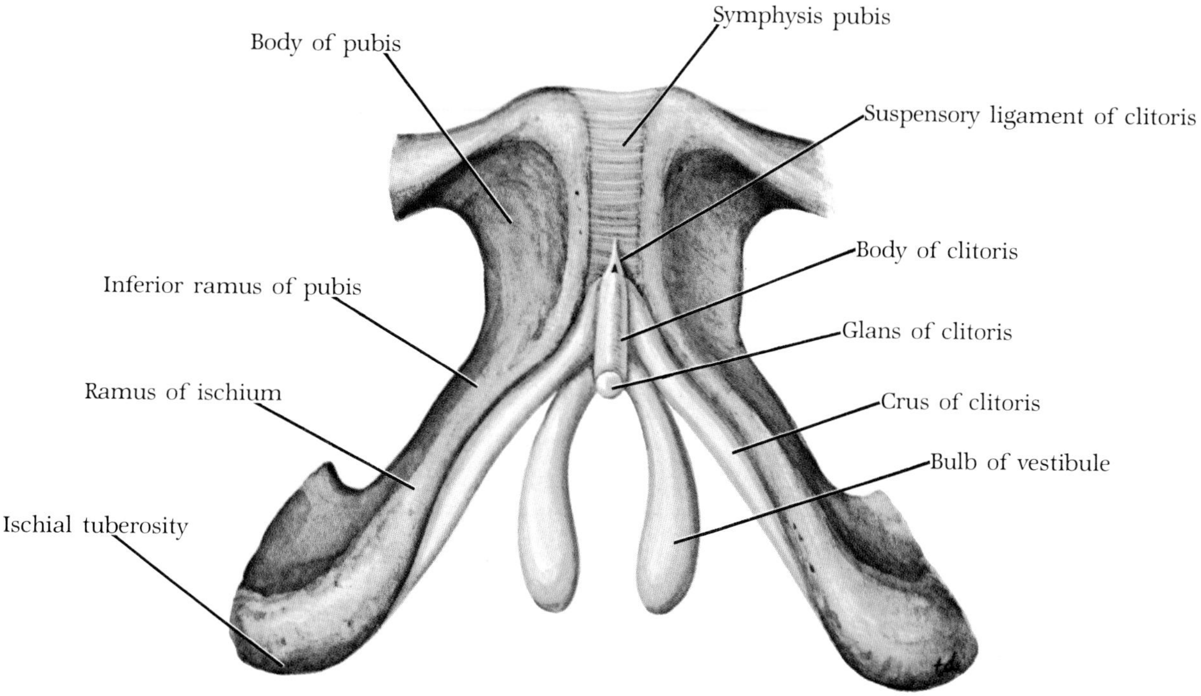
Symphysis pubis
Body of pubis
Suspensory ligament of clitoris
Body of clitoris
Inferior ramus of pubis
Glans of clitoris
Ramus of ischium
Crus of clitoris
Bulb of vestibule
Ischial tuberosity

Figure 4-14
Root of clitoris.

1. Erection of the clitoris occurs when there is a rapid filling of the blood spaces in the crura of the clitoris and bulb of the vestibule and a compression of the draining veins. Note that the crura of the clitoris are homologous with the crura of the penis and that the bulb of the vestibule corresponds with the bulb of the penis.

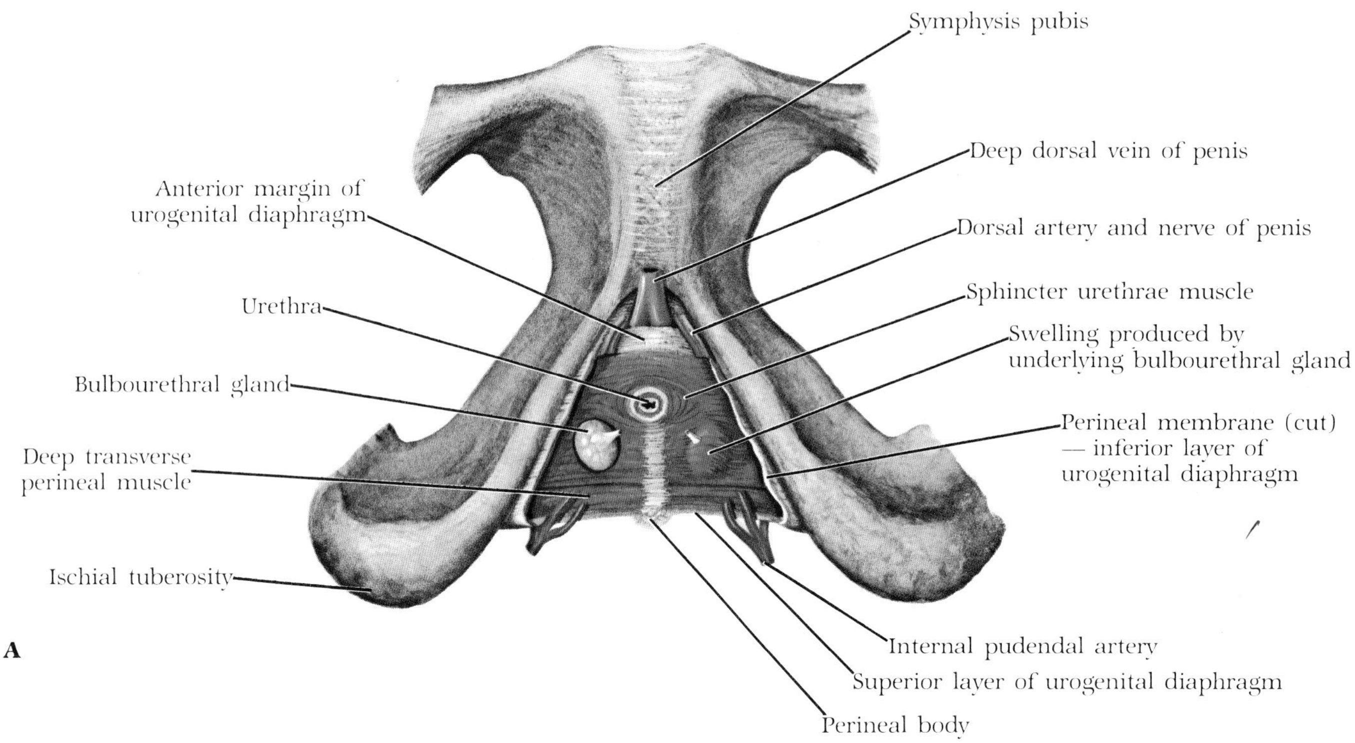
Symphysis pubis
Deep dorsal vein of penis
Anterior margin of urogenital diaphragm
Dorsal artery and nerve of penis
Urethra
Sphincter urethrae muscle
Swelling produced by underlying bulbourethral gland
Bulbourethral gland
Perineal membrane (cut) — inferior layer of urogenital diaphragm
Deep transverse perineal muscle
Ischial tuberosity
A
Internal pudendal artery
Superior layer of urogenital diaphragm
Perineal body

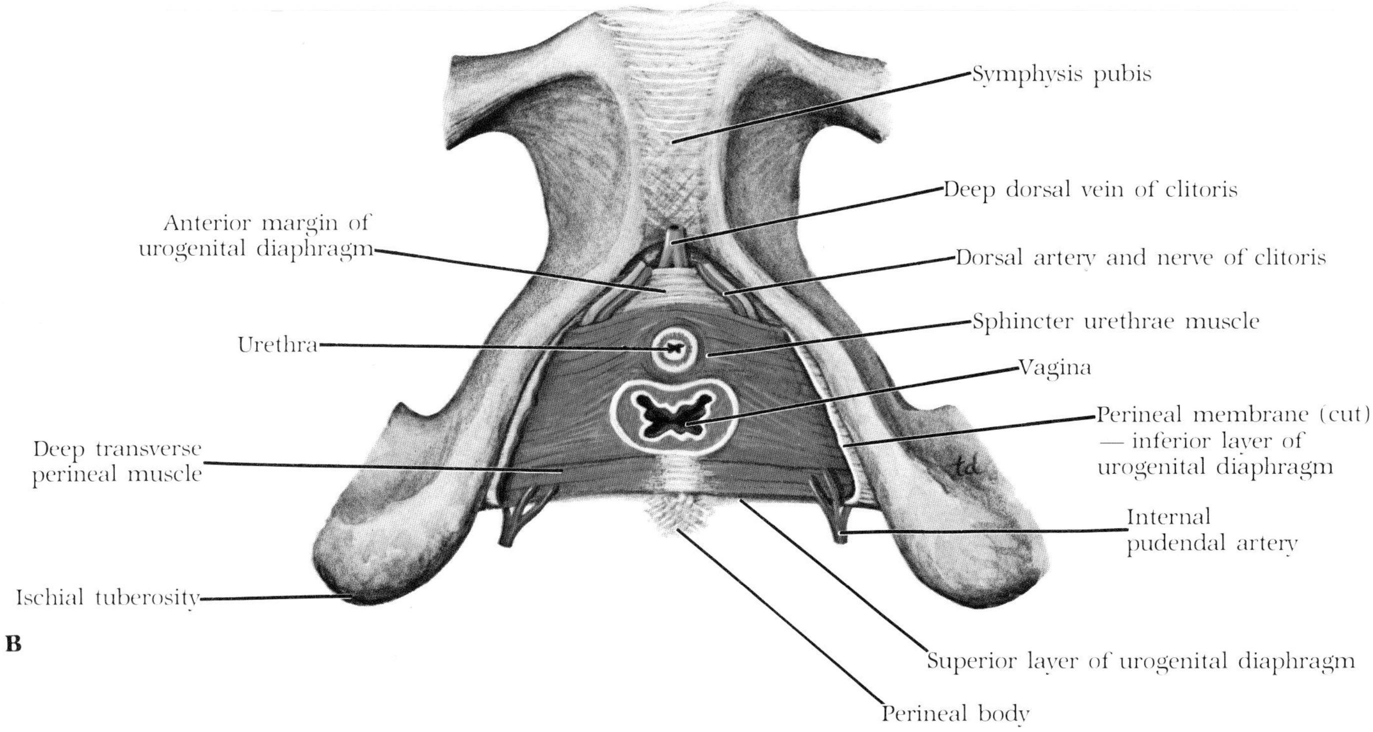
Symphysis pubis
Deep dorsal vein of clitoris
Anterior margin of urogenital diaphragm
Dorsal artery and nerve of clitoris
Urethra
Sphincter urethrae muscle
Vagina
Perineal membrane (cut) — inferior layer of urogenital diaphragm
Deep transverse perineal muscle
Internal pudendal artery
Ischial tuberosity
B
Superior layer of urogenital diaphragm
Perineal body

Figure 4-15
Urogenital diaphragm. A. In male. B. In female. In both parts of the figure the inferior layer of fascia (perineal membrane) has been cut and removed.

1. The fact that the membranous part of the urethra is the narrowest, most fixed, and least dilatable part of the entire length of the urethra should be remembered when passing a catheter.
2. If the membranous part of the urethra in a male is ruptured, urine escapes into the deep perineal pouch within the urogenital diaphragm and may extravasate superiorly around the prostate and bladder, or inferiorly into the superficial perineal pouch.
3. The bulbourethral glands commonly become infected from a gonococcal urethritis, which may in turn give rise to an abscess that can be palpated through the anterior rectal wall.
4. The urethral sphincter is composed of voluntary muscle and is innervated by branches of the pudendal nerve (S2,3, and 4). Although it assists in the arrest of the passage of urine and helps to empty the urethra at the end of micturition, its function is small compared to that of the sphincter vesicae.
5. The urogenital diaphragm in the female is not as strongly constructed as in the male. It does assist in the support and positioning of the vagina and is attached posteriorly to the important perineal body.

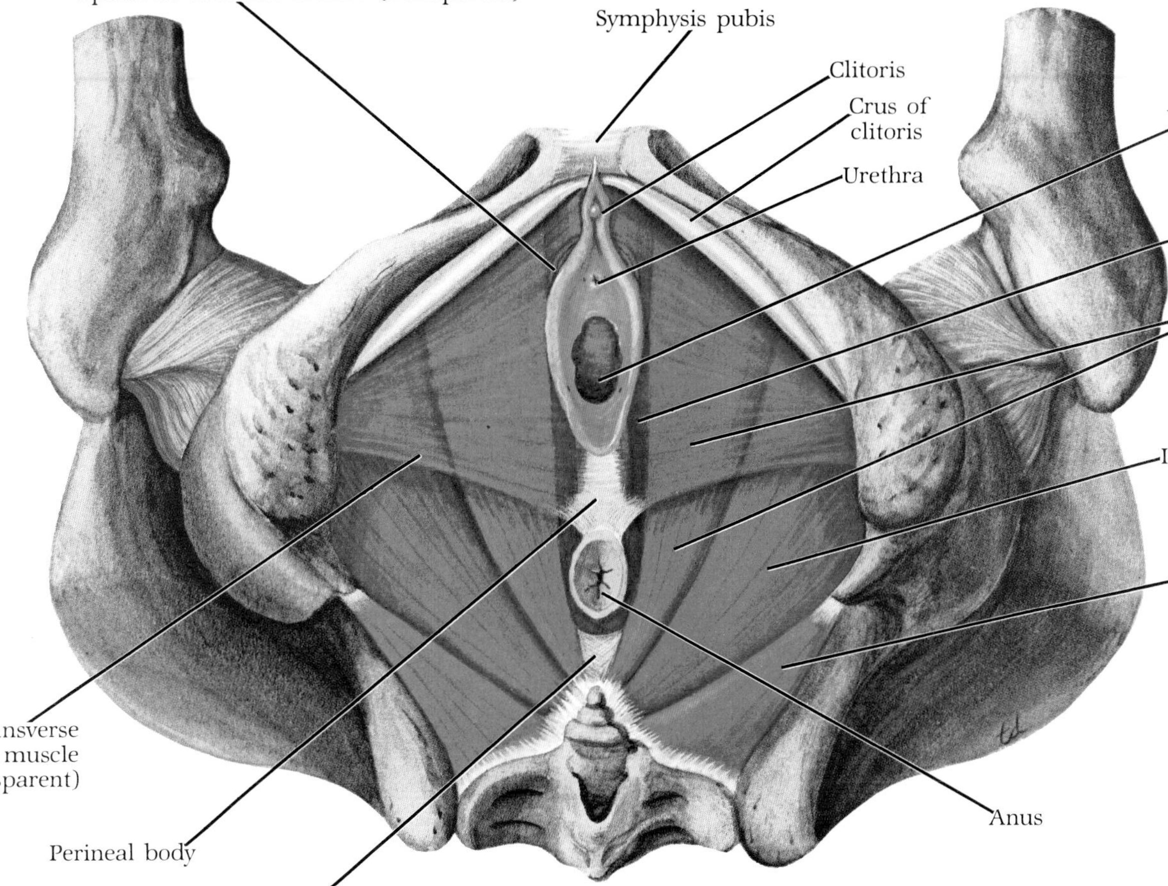
Sphincter urethrae muscle (transparent)
Symphysis pubis
Clitoris
Crus of clitoris
Urethra
Vaginal orifice
Puborectalis
Pubococcygeus
Iliococcygeus
Coccygeus
Anus
Deep transverse perineal muscle (transparent)
Perineal body
Anococcygeal body

Figure 4-16
Female perineal muscles, inferior view.

1. The pelvic diaphragm, which supports the pelvic viscera, is formed by the important levatores ani muscles and the small coccygeus muscles and their covering fasciae. It is incomplete anteriorly to allow for the passage of the urethra and the vagina. Some of the muscle fibers from the two sides of the pelvic diaphragm interlace between the anal canal and the vagina in the perineal body and many form a strong slinglike band (puborectalis muscle) from the pubis to around the lower part of the rectum. It is important to realize that the muscles of the two sides slope together in the midline to form a V-shaped gutter that is directed forward under the pubic arch. Vaginal examination permits the puborectalis portion of the levator ani to be felt as two round pillars lateral to the vagina.

2. The urogenital diaphragm is a musculofascial diaphragm situated in the anterior part of the perineum and filling in the gap of the pubic arch. It is formed by the sphincter urethrae and the deep transverse perineal muscles, all of which are enclosed between a superior and an inferior layer of fascia of the urogenital diaphragm. The inferior layer of fascia is often referred to as the perineal membrane.

3. The perineal body is of great importance in obstetrics, since its laceration produces weakness of the pelvic floor. It is pyramidal in shape, with its apex in the pelvic diaphragm and its base at the skin between the vagina and the anus.

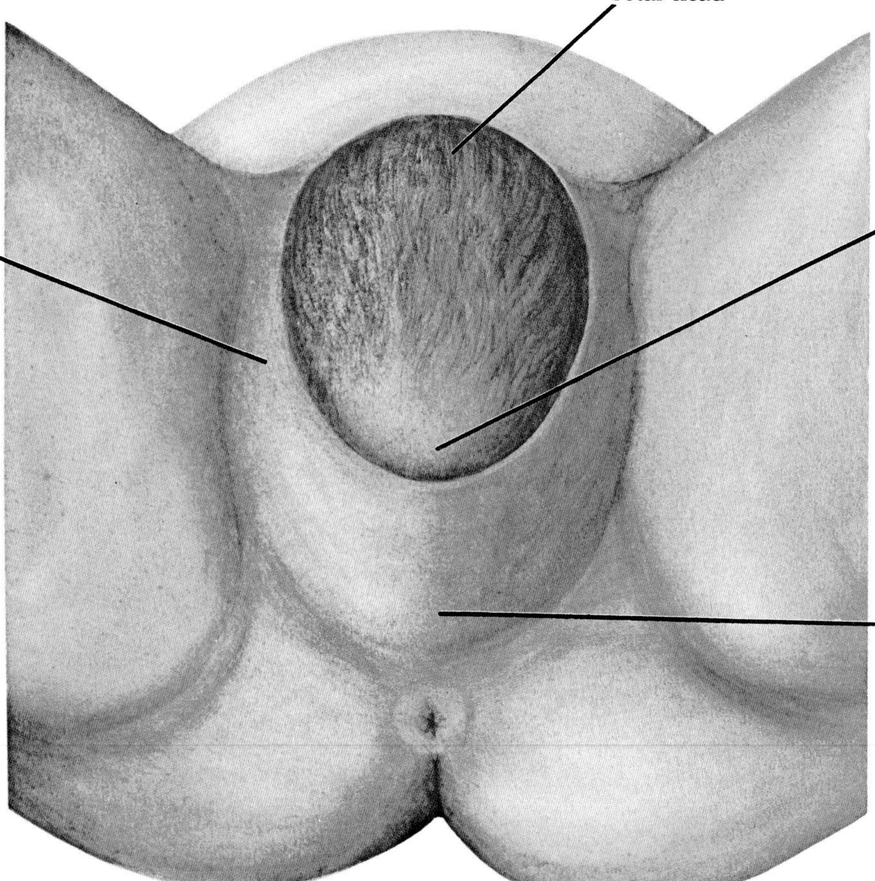
Fetal head
Distended maternal perineum
Forehead of fetus
Region of perineal body greatly stretched and thinned

A

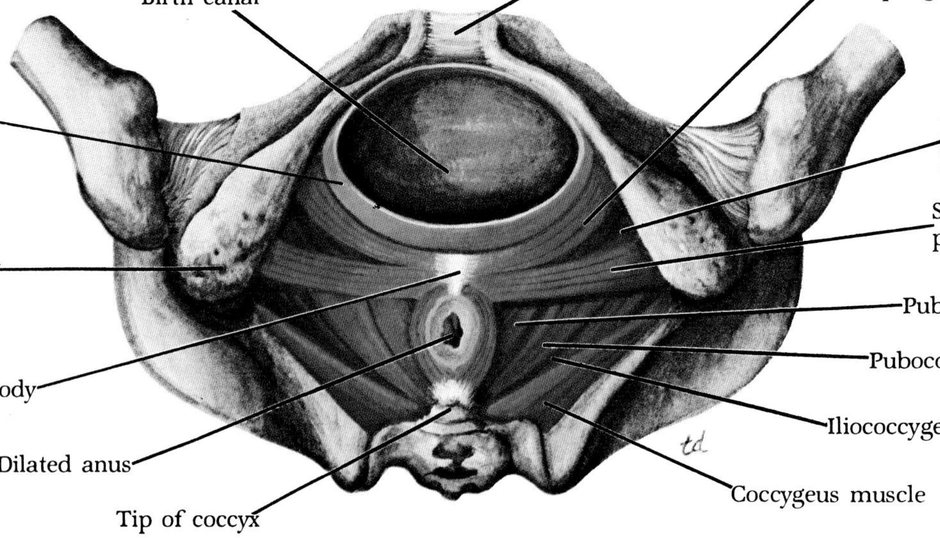
Symphysis pubis
Birth canal
Bulbospongiosus muscle
Margin of vaginal wall
Deep transverse perineal muscle
Superficial transverse perineal muscle
Ischial tuberosity
Puborectalis muscle
Pubococcygeus muscle
Perineal body
Iliococcygeus muscle
Dilated anus
Coccygeus muscle
Tip of coccyx
td

B

Figure 4-17
A. Perineum becoming stretched and thinned by the child's head.
B. Distended urogenital and pelvic diaphragms.

1. Considerable relaxation and softening of the urogenital and pelvic diaphragms occur during the later months of pregnancy. The dilation of the urogenital diaphragm and the pelvic diaphragm during the second stage of labor allows them, together, to form a fibromuscular birth canal that is attached to the walls and outlet of the bony pelvis.
2. The levatores ani muscles of the two sides of the body have been separated and drawn apart so that with the pubic arch they form a canal equal to the circumference of the presenting part of the child, which is usually the head, having a circumference of about 33 to 35 cm.
3. The perineal body is flattened, and the rectum and anal canal are flattened against the sacrum and coccyx during this stage of labor.
4. When there is a disproportion between the baby and the soft parts of the perineum, or the delivery occurs so rapidly that there is insufficient time for the soft tissues to dilate, severe lacerations of the urogenital and pelvic diaphragms and perineal body may occur. As a result, the vaginal walls, the urethra, and the neck of the bladder sag. This leads to vaginal prolapse and stress incontinence. Severe lacerations may involve the sphincters of the anal canal and even the rectal wall, producing fecal incontinence.

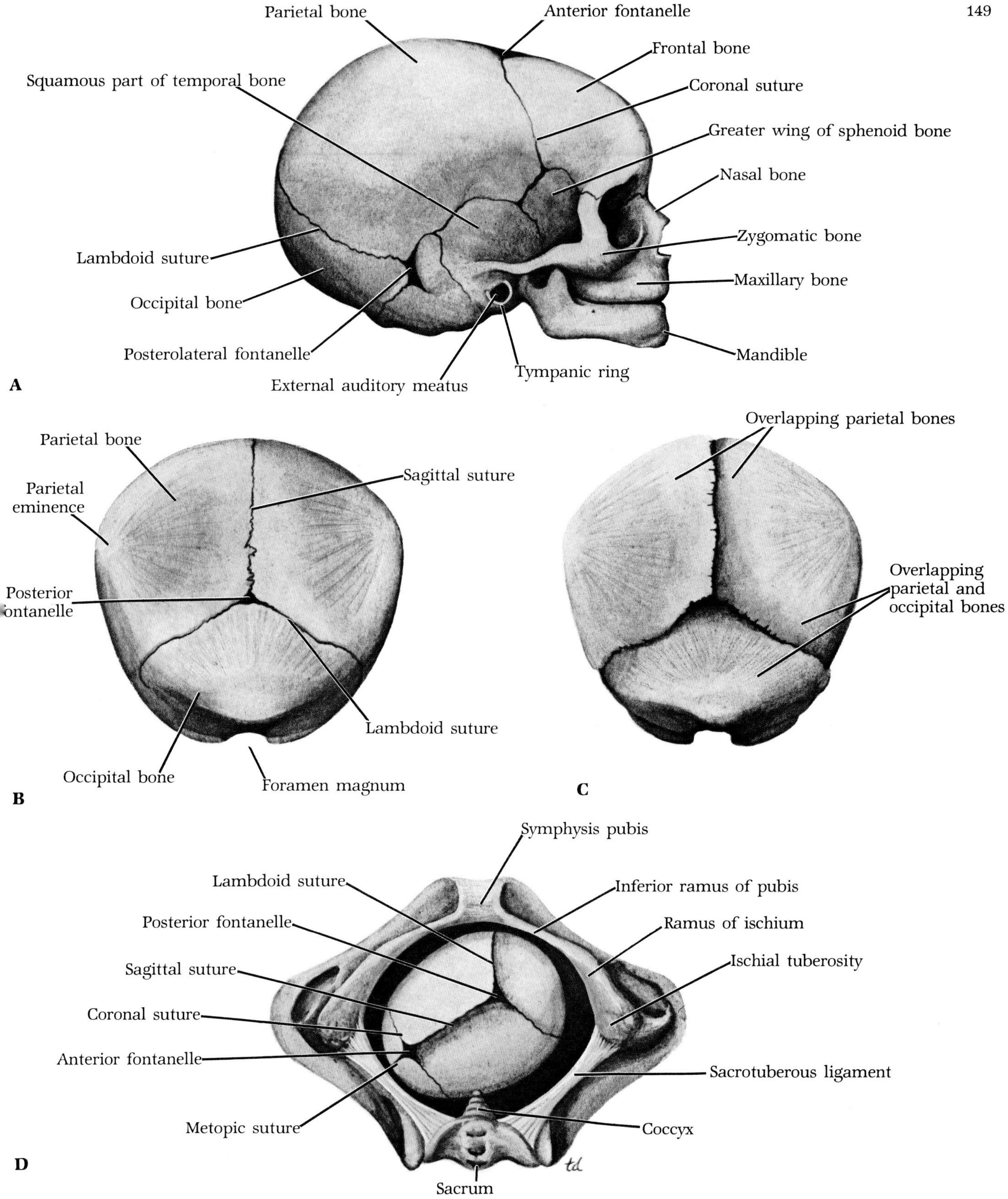
Parietal bone
Anterior fontanelle
Frontal bone
Squamous part of temporal bone
Coronal suture
Greater wing of sphenoid bone
Nasal bone
Zygomatic bone
Lambdoid suture
Maxillary bone
Occipital bone
Posterolateral fontanelle
Mandible
Tympanic ring
External auditory meatus
A
Parietal bone
Sagittal suture
Parietal eminence
Posterior fontanelle
Lambdoid suture
Occipital bone
Foramen magnum
B
Overlapping parietal bones
Overlapping parietal and occipital bones
C
Symphysis pubis
Lambdoid suture
Inferior ramus of pubis
Posterior fontanelle
Ramus of ischium
Sagittal suture
Ischial tuberosity
Coronal suture
Anterior fontanelle
Sacrotuberous ligament
Metopic suture
Coccyx
D
td
Sacrum

Figure 4-18
Fetal head. A. Lateral view. B. Posterior view. C. Posterior view, showing molding. D. Within maternal bony pelvis during labor.

1. Most of the bones of the skull are ossified by the time of birth, but they are mobile on each other, with their mobility most marked in the vault. The ability to overlap provides the molding of the cranium that is so important during the process of childbirth. The bones of the vault are separated by fibrous tissue (the sutures); at their corners they are separated by large areas known as fontanelles.

2. As the child's head is pushed down the birth canal, the occipital bone is usually pressed under the two parietal bones; one parietal overlaps the other, with the depressed one against the promontory of the maternal sacrum. When molding is absent, as in a fetus whose delivery is long overdue, making it postmature, labor is more difficult.

3. Prior to the onset of labor, the ligaments of the sacroiliac, the sacrococcygeal joints, and the symphysis pubis soften and become slack. Clearly this increases the potential capacity of the pelvic cavity and its inlet and outlet.

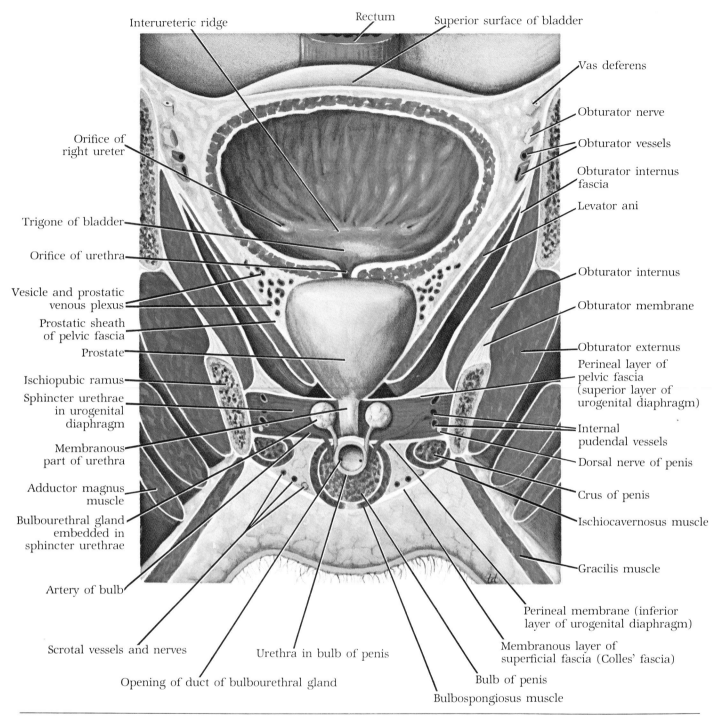

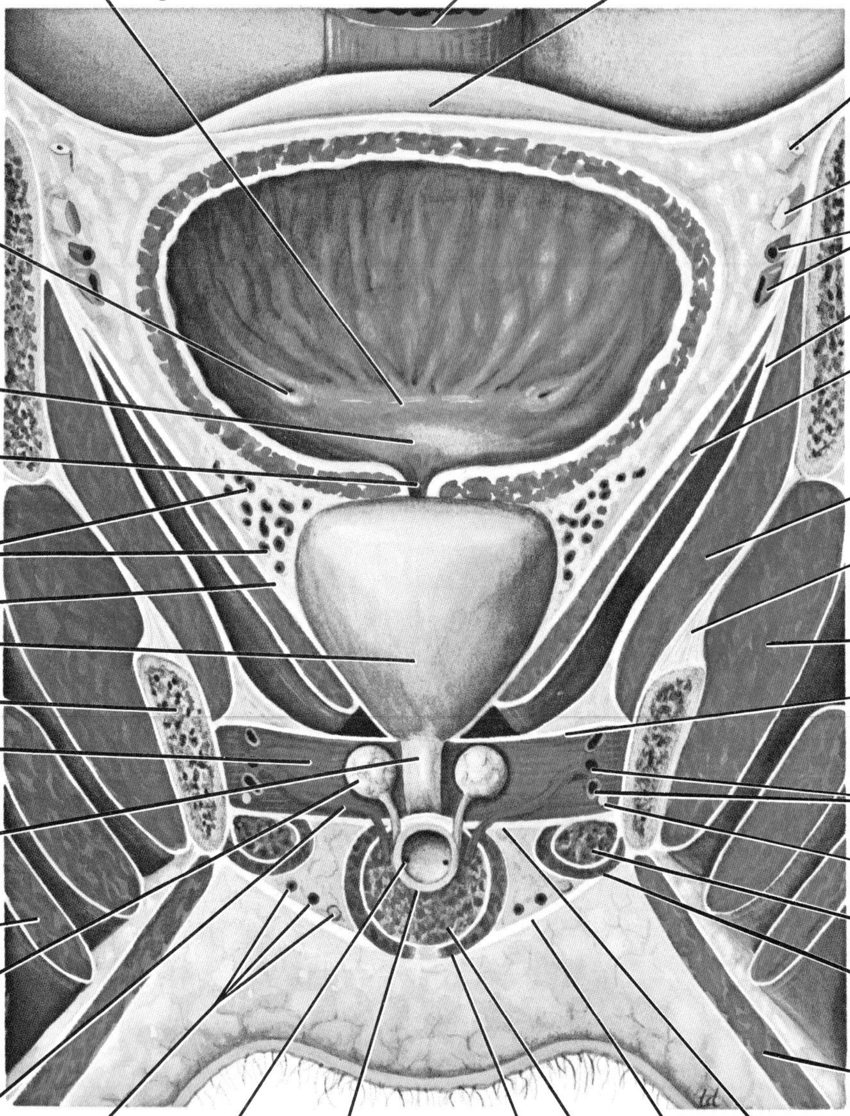
Interureteric ridge
Rectum
Superior surface of bladder
Vas deferens
Obturator nerve
Obturator vessels
Obturator internus fascia
Levator ani
Orifice of right ureter
Trigone of bladder
Orifice of urethra
Vesicle and prostatic venous plexus
Prostatic sheath of pelvic fascia
Prostate
Ischiopubic ramus
Sphincter urethrae in urogenital diaphragm
Membranous part of urethra
Adductor magnus muscle
Bulbourethral gland embedded in sphincter urethrae
Artery of bulb
Obturator internus
Obturator membrane
Obturator externus
Perineal layer of pelvic fascia (superior layer of urogenital diaphragm)
Internal pudendal vessels
Dorsal nerve of penis
Crus of penis
Ischiocavernosus muscle
Gracilis muscle
Perineal membrane (inferior layer of urogenital diaphragm)
Membranous layer of superficial fascia (Colles' fascia)
Scrotal vessels and nerves
Urethra in bulb of penis
Opening of duct of bulbourethral gland
Bulb of penis
Bulbospongiosus muscle

Figure 4-19
Coronal section of male pelvis, showing bladder, prostate, urogenital diaphragm, and contents of superficial perineal pouch, anterior view.

1. The mucous membrane of the bladder, the two ureteric orifices, and the urethral meatus can easily be observed by means of a cystoscope. An illuminated tube fitted with lenses is introduced into the bladder through the urethra. When the bladder is distended with fluid, the mucous membrane is pink and smooth over the trigone. If the bladder is partially emptied, the mucous membrane over the trigone remains smooth but is thrown into folds elsewhere. The ureteric orifices are slitlike, and the interureteric ridge and the uvula vesicae may be recognized easily.

2. The nervous control of micturition may be disrupted due to spinal cord injuries, resulting in the following conditions.

A. *Atonic Bladder.* This occurs during spinal shock immediately following injury; the bladder wall is relaxed, the sphincter vesicae are tightly contracted, and the sphincter urethrae are relaxed.

B. *Automatic Reflex Bladder.* This condition occurs when the cord lesion lies above the level of the parasympathetic outflow (S2, 3, and 4). The bladder fills and empties reflexly.

C. *Autonomous Bladder.* If the sacral segments of the spinal cord are destroyed, the bladder merely fills to capacity and overflows; continual dribbling results.

3. Benign enlargement of the prostate is common in men over 50 years of age. The median lobe enlarges upward and encroaches within the sphincter vesicae. The leakage of urine causes an intense desire to micturate, but enlargement of the median and lateral lobes produces elongation, lateral compression, and distortion of the urethra so that the patient experiences difficulty in passing urine, and the stream is weak.

4. The bulbourethral glands are common sites of chronic gonoccocal infection.

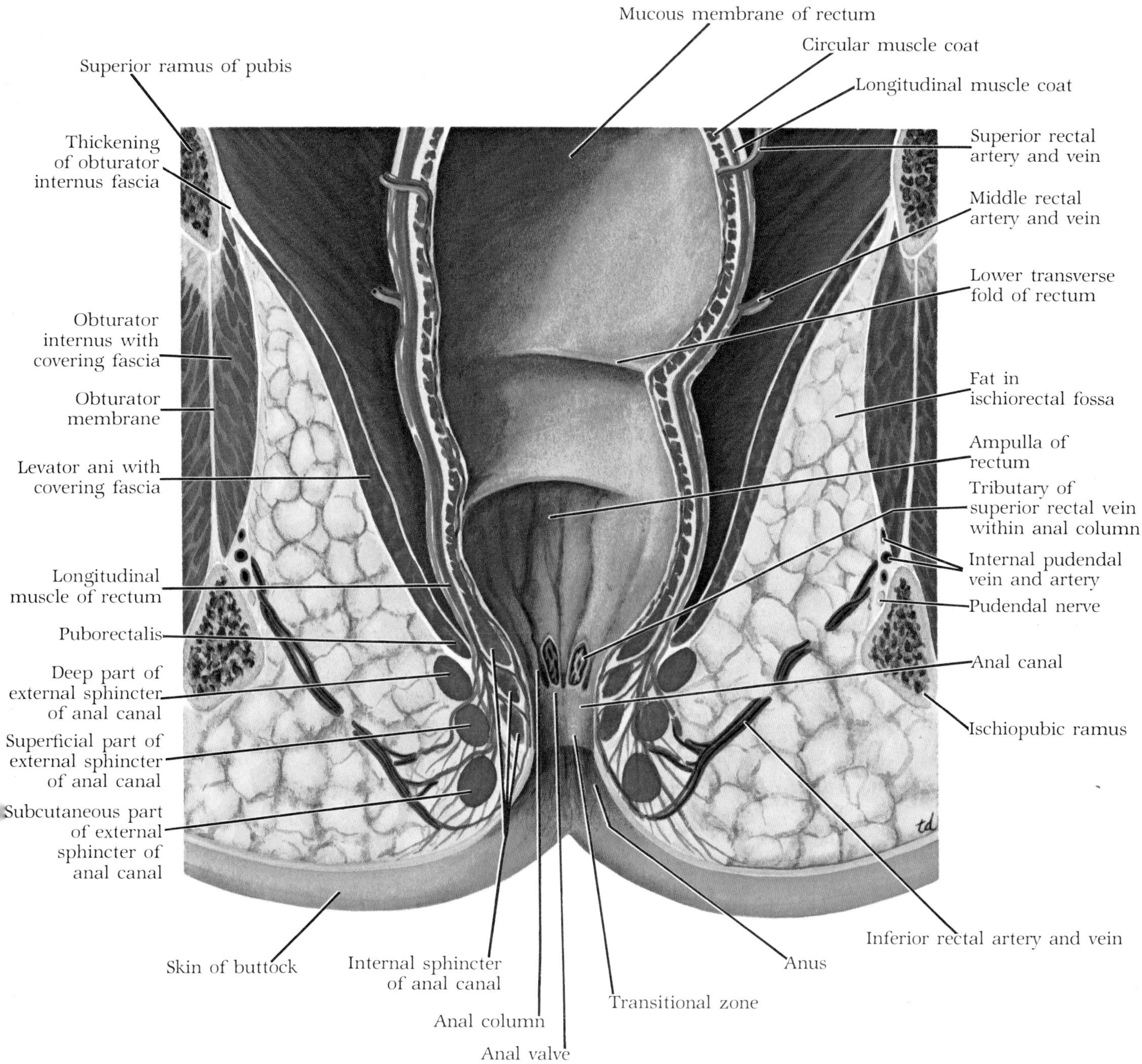
Mucous membrane of rectum
Circular muscle coat
Longitudinal muscle coat
Superior ramus of pubis
Thickening of obturator internus fascia
Superior rectal artery and vein
Middle rectal artery and vein
Lower transverse fold of rectum
Obturator internus with covering fascia
Obturator membrane
Fat in ischiorectal fossa
Levator ani with covering fascia
Ampulla of rectum
Tributary of superior rectal vein within anal column
Internal pudendal vein and artery
Longitudinal muscle of rectum
Pudendal nerve
Puborectalis
Anal canal
Deep part of external sphincter of anal canal
Ischiopubic ramus
Superficial part of external sphincter of anal canal
Subcutaneous part of external sphincter of anal canal
td
Inferior rectal artery and vein
Skin of buttock
Internal sphincter of anal canal
Anus
Transitional zone
Anal column
Anal valve

Figure 4-20
Coronal section of pelvis and perineum, showing rectum and anal canal.

1. Internal hemorrhoids (piles) are varicosities of the tributaries of the superior rectal (hemorrhoidal) vein and are covered by mucous membrane. The tributaries of the vein, which lie in the anal columns at 3, 7, and 11 o'clock positions when the patient is viewed in the lithotomy position, are particularly liable to become varicosed.
2. External hemorrhoids are varicosities of the tributaries of the inferior rectal (hemorrhoidal) vein as they run laterally from the anal margins.
3. An anal fissure is an elongated ulcer in the anal canal caused by a hard fecal mass catching on an anal valve and tearing it down to the anal margin. It occurs most often in persons suffering from chronic constipation. The site of the fissure, which is supplied by the inferior rectal nerve, is extremely sensitive. The pain causes reflex spasm of the external anal sphincter.
4. Perianal abscesses are produced by fecal trauma to the anal mucosa.
5. Anal fistulae develop as the result of spread or inadequate treatment of anal abscesses. The fistula, at one end, opens at the lumen of the anal canal or lower rectum and on the skin surface at the other end.
6. The ischiorectal fossae are common sites of abscess. Infection commonly tracks laterally from the anal mucosa through the external anal sphincter.
7. The anorectal ring, consisting of the deep part of the external sphincter, the internal sphincter, and the puborectalis part of the levator ani, is the most important part of the sphincteric mechanism of the anal canal. Surgical damage to the anorectal ring will produce fecal incontinence.
8. Lymphatic drainage of the mucous membrane of the upper half of the anal canal is superiorly to the lymph nodes along the course of the superior rectal artery. The lower half is drained inferiorly to the medial group of superficial inguinal nodes.

5. The Upper Limb

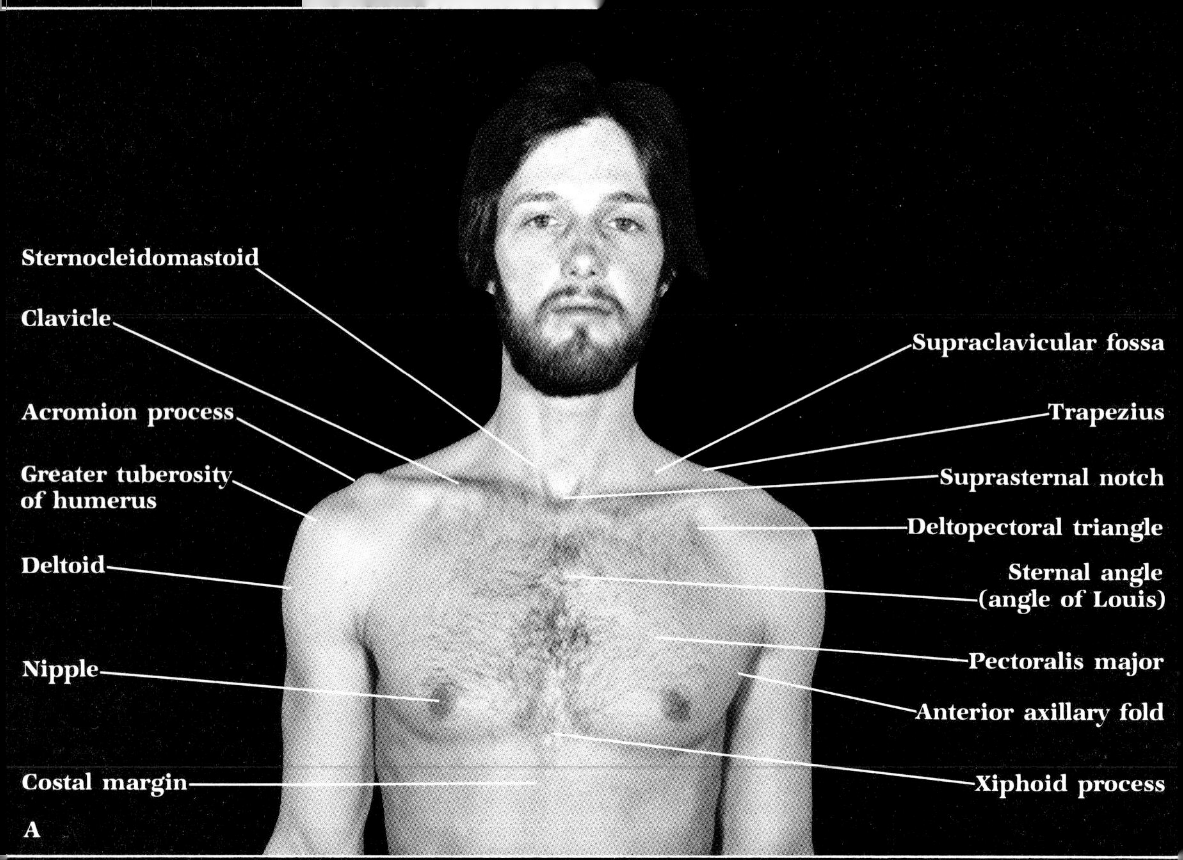
Sternocleidomastoid
Clavicle
Acromion process
Greater tuberosity of humerus
Deltoid
Nipple
Costal margin
A
Supraclavicular fossa
Trapezius
Suprasternal notch
Deltopectoral triangle
Sternal angle (angle of Louis)
Pectoralis major
Anterior axillary fold
Xiphoid process

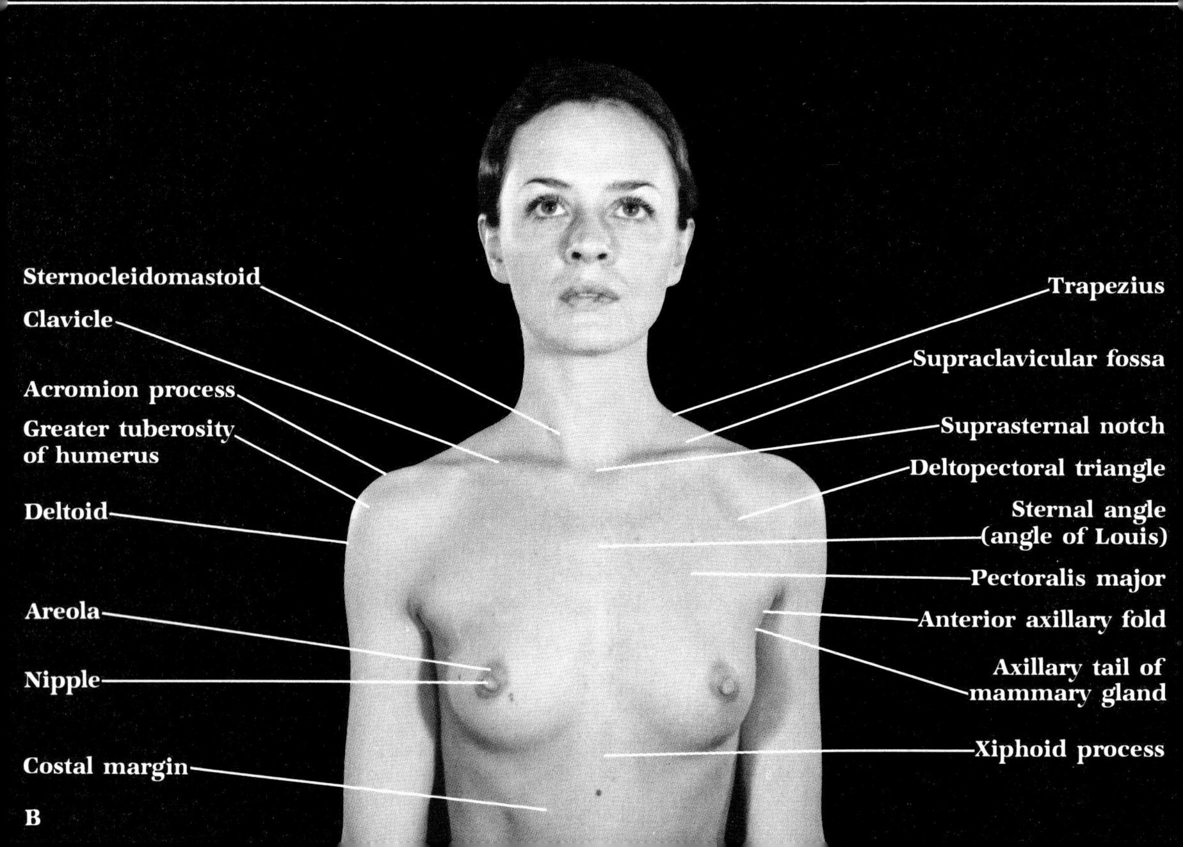
Sternocleidomastoid
Clavicle
Acromion process
Greater tuberosity of humerus
Deltoid
Areola
Nipple
Costal margin
B
Trapezius
Supraclavicular fossa
Suprasternal notch
Deltopectoral triangle
Sternal angle (angle of Louis)
Pectoralis major
Anterior axillary fold
Axillary tail of mammary gland
Xiphoid process

Figure 5-1
Chest and shoulder, anterior view. A. 27-year-old male. B. 29-year-old female.

1. The suprasternal notch is the superior margin of the manubrium sterni and is felt between the medial ends of the clavicles in the midline.
2. The sternal angle (angle of Louis) is the angle made between the manubrium and body of the sternum; at this level the second costal cartilage joins the lateral margin of the sternum.
3. The xiphisternal joint is between the xiphoid process of the sternum and the body of the sternum.
4. The subcostal angle is situated at the inferior end of the sternum between the sternal attachments of the seventh costal cartilages.
5. The costal margin is formed by the cartilages of the seventh, eighth, ninth, and tenth ribs and the ends of the eleventh and twelfth cartilages.
6. The clavicle is subcutaneous throughout its entire length and can be easily palpated. It articulates at its medial extremity with the sternum and first costal cartilage and at its lateral end with the acromion process of the scapula.
7. The first rib lies deep to the clavicle and cannot be palpated. The lateral surfaces of the remaining ribs can be felt by pressing the fingers upward into the axilla and drawing them downward over the lateral surface of the chest wall. To identify a particular rib, always first identify the second costal cartilage at the sternal angle and then count the cartilages and ribs downward from this point.
8. The deltopectoral triangle is a small triangular depression situated below the outer third of the clavicle; it is bounded on either side by the pectoralis major and deltoid muscles.
9. The tip of the coracoid process of the scapula can be felt on deep palpation in the lateral part of the triangle; it is covered by the anterior fibers of the deltoid.
10. The acromion process of the scapula forms the lateral extremity of the spine of the scapula. It is subcutaneous and easily located.
11. The rounded curve of the shoulder is produced by the deltoid muscle covering the greater tuberosity of the humerus.

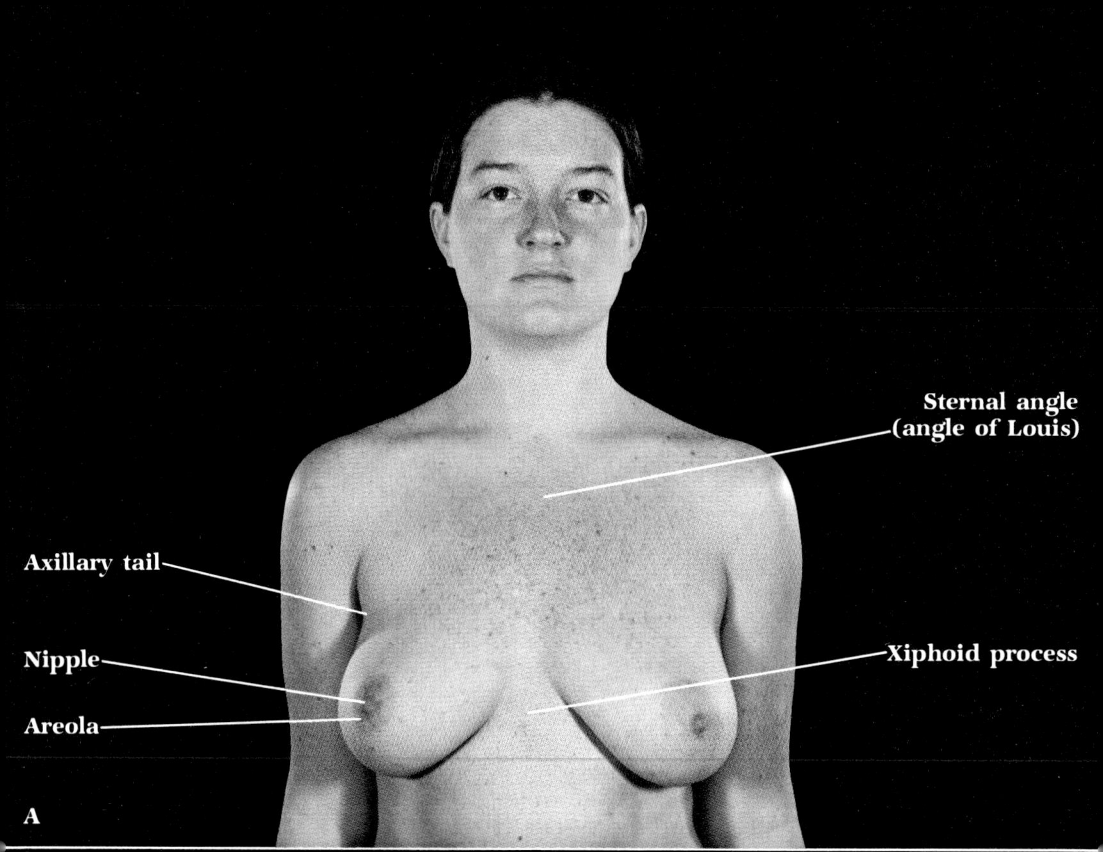
Sternal angle
(angle of Louis)
Axillary tail
Nipple
Xiphoid process
Areola
A

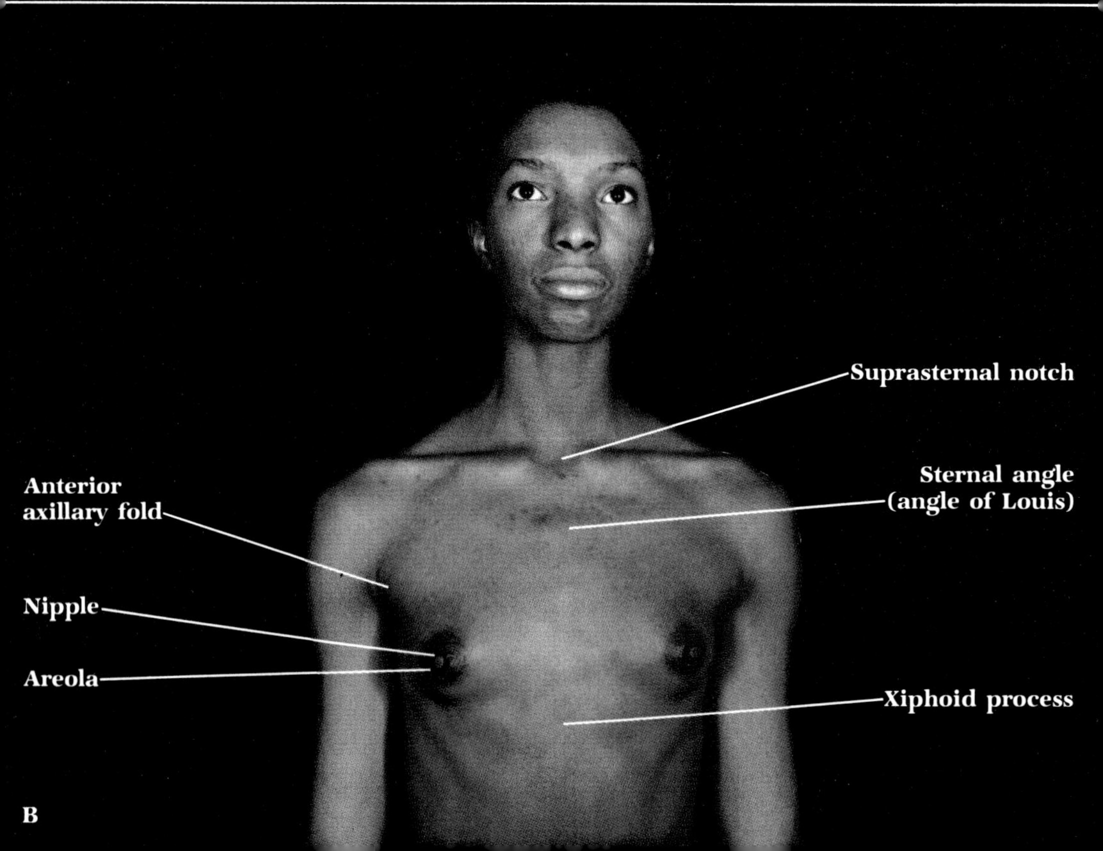
Suprasternal notch
Anterior
axillary fold
Sternal angle
(angle of Louis)
Nipple
Areola
Xiphoid process
B

Figure 5-2
Mammary glands of adult female. A. 25-year-old female. B. 22-year-old female.

1. The mammary glands are examined with the patient undressed to the waist and sitting upright. The patient is then asked to lie down so that the breasts can be palpated against the underlying thoracic wall. Finally the patient is asked to sit up again and raise both arms above her head. With this maneuver a carcinoma tethered to the skin will produce dimpling of the skin or retraction of the nipple.

2. The mammary glands in the post-pubertal female are hemispherical in shape and slightly pendulous; they extend from the second to the sixth rib and from the lateral margin of the sternum to the midaxillary line. The greater part of the mammary gland lies in the superficial fascia and can be moved freely in all directions. A small part of the breast known as the axillary tail extends upward and laterally, pierces the deep fascia, and enters the axilla.
3. The nipple and areola and the breast tissue immediately deep to this area are palpated between finger and thumb for swellings, tenderness, or discharge from the nipple.

4. With the examiner's open hand held flat against the thoracic wall, the mammary gland is then carefully palpated in its entirety. The breast is soft because the fat contained within it is fluid. The glandular tissue gives the breast a firm, overall lobulated consistency.

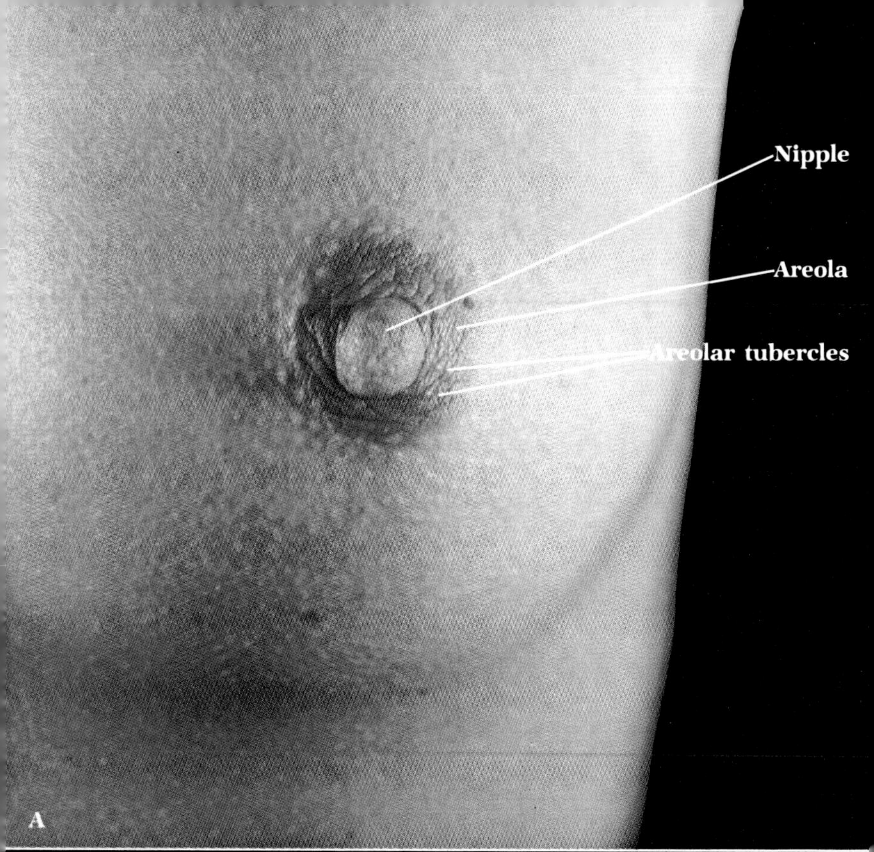
Nipple
Areola
Areolar tubercles
A

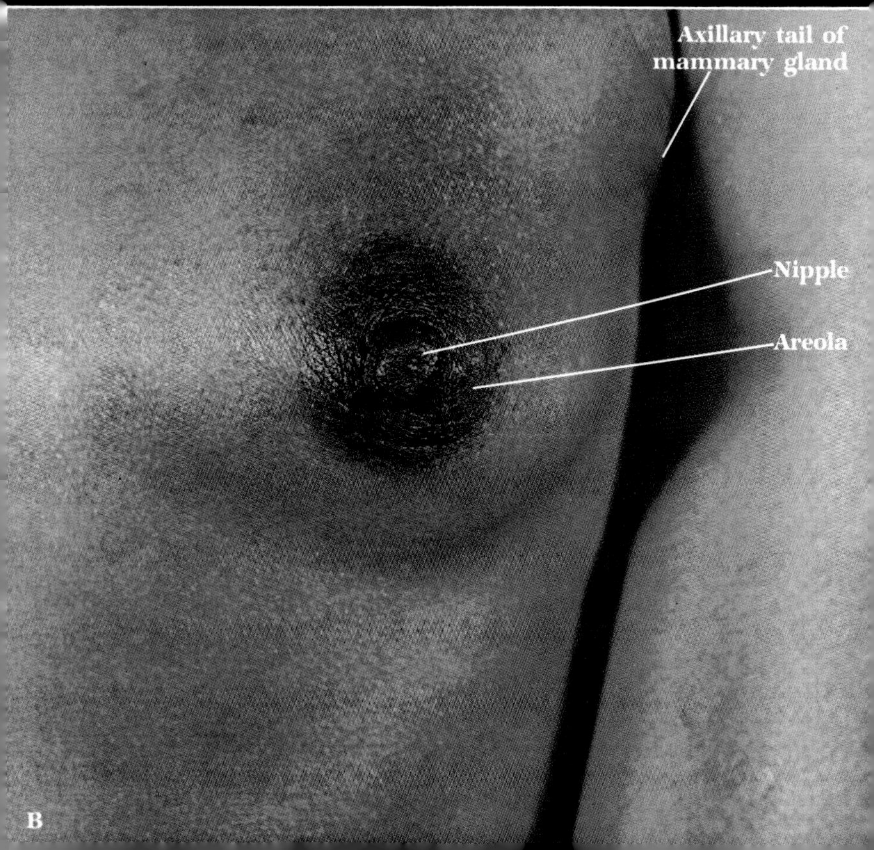
Axillary tail of
mammary gland
Nipple
Areola
B

Figure 5-3
Left female mammary gland.
A. 29-year-old female. B. 22-year-old female.

1. The mammary glands in the post-pubertal female are hemispherical in shape and lie for the most part in the superficial fascia.
2. The nipple projects from the lower half of the gland but its position in relation to the chest wall varies greatly and depends on the degree of development of the gland.
3. The areola is a circular area of pigmented skin that surrounds the base of the nipple. Pink in color in the young girl, the areola becomes darker in color in the second month of the first pregnancy and never regains its former tint. Tiny tubercles of the areola are produced by the underlying areolar glands.
4. In the immature female and in the male, the mammary glands are rudimentary. The nipples are small and usually lie over the fourth intercostal space.
5. Retracted nipple or inverted nipple is a congenital anomaly due to a failure in the later stages of development. In about one-quarter of the cases it is bilateral. It is important clinically, since normal suckling of an infant cannot take place and the nipple is prone to infection. This condition must not be confused with a nipple that has recently become retracted from an underlying scirrhous carcinoma.

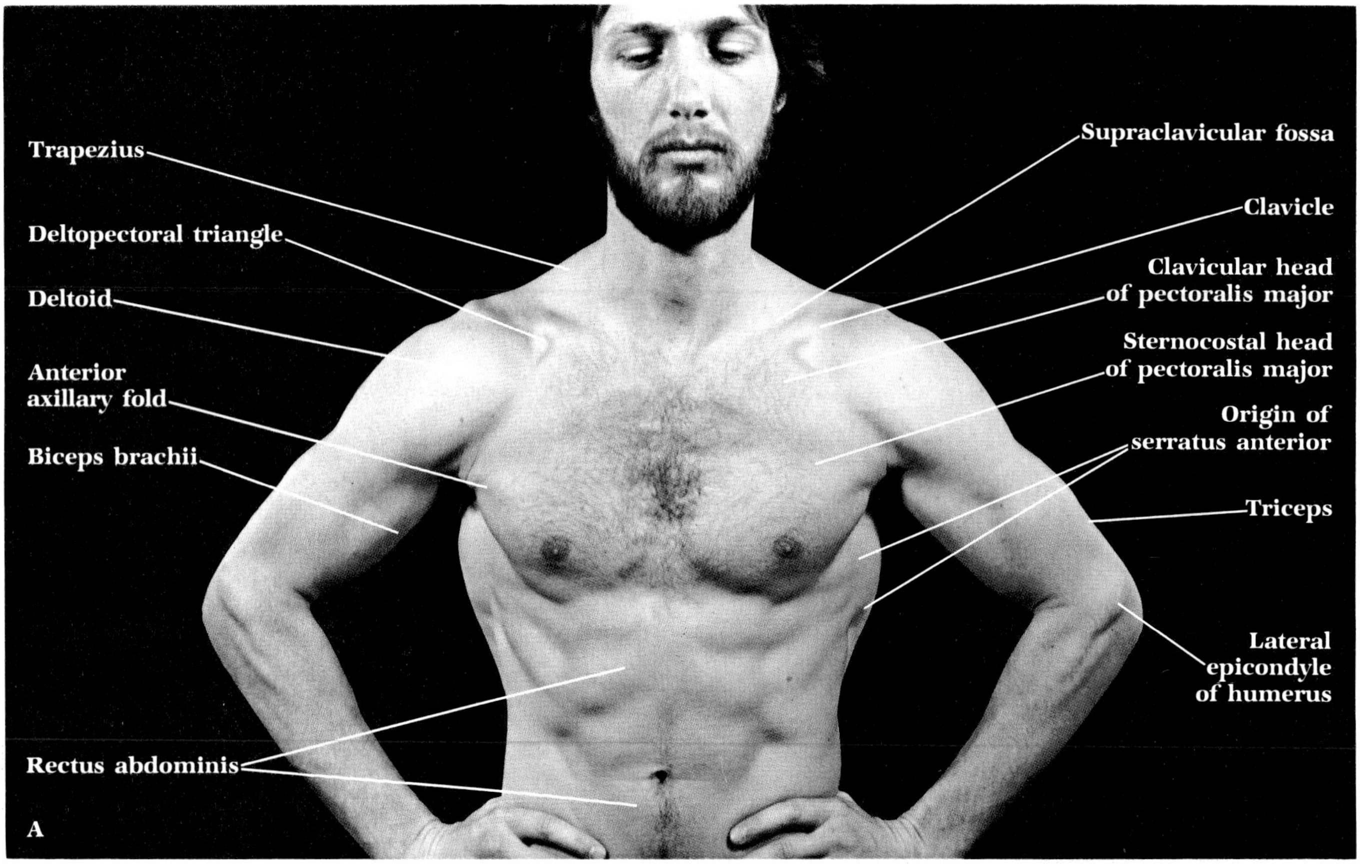
Trapezius
Deltopectoral triangle
Deltoid
Anterior axillary fold
Biceps brachii
Rectus abdominis
Supraclavicular fossa
Clavicle
Clavicular head of pectoralis major
Sternocostal head of pectoralis major
Origin of serratus anterior
Triceps
Lateral epicondyle of humerus
A

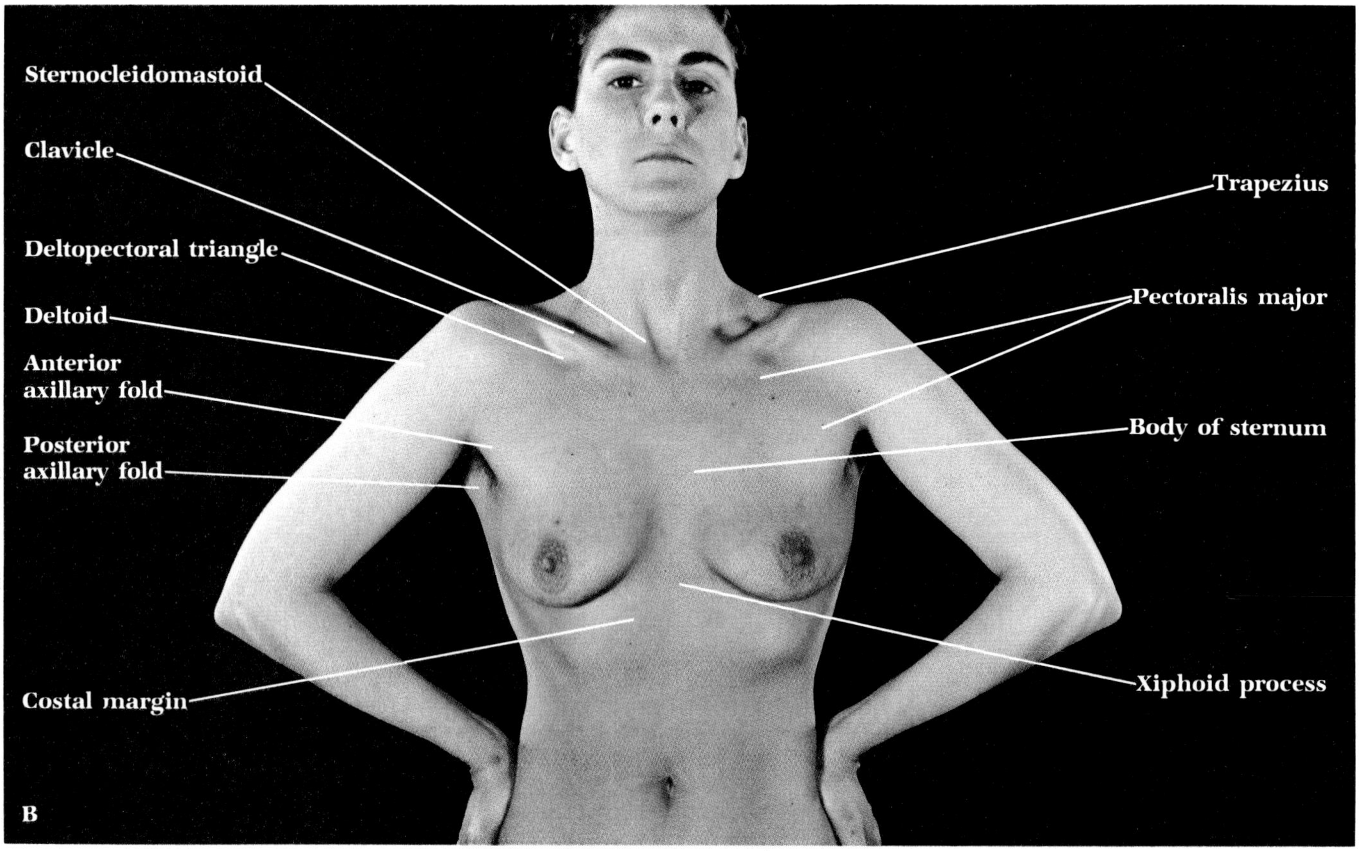
Sternocleidomastoid
Clavicle
Deltopectoral triangle
Deltoid
Anterior axillary fold
Posterior axillary fold
Costal margin
Trapezius
Pectoralis major
Body of sternum
Xiphoid process
B

Figure 5-4
The pectoral region. A. 27-year-old male. B. 29-year-old female.

1. The clavicle can be palpated throughout its length. The positions of the sternoclavicular and acromioclavicular joints can be easily identified.
2. The supraclavicular fossa can be seen and felt as a depression between the sternocleidomastoid muscle and the trapezius muscle just above the clavicle.
3. The deltopectoral triangle is a depression situated below the outer third of the clavicle; it is bounded on either side by the pectoralis major and deltoid muscles.
4. The anterior axillary fold is formed by the lower margin of the pectoralis major muscle and can be palpated between the finger and thumb.
5. The posterior axillary fold is formed by the tendon of the latissimus dorsi muscle winding around the lower border of the teres major muscle; it can also be palpated between the finger and thumb.
6. The axilla should be examined when the forearm is supported and the pectoral muscles relaxed. With the arm by the side, the inferior part of the head of the humerus can be easily palpated through the floor of the axilla.
7. The pulsations of the axillary artery can be felt high up in the axilla. The cords of the brachial plexus may be palpated around the axillary artery.
8. The medial wall of the axilla is formed by the upper ribs covered by the serratus anterior muscle; in a muscular subject the serrations of this muscle can be seen and felt interdigitating with those of the external oblique muscle of the anterior abdominal wall. The lateral wall is formed by the coracobrachialis and biceps brachii muscles.

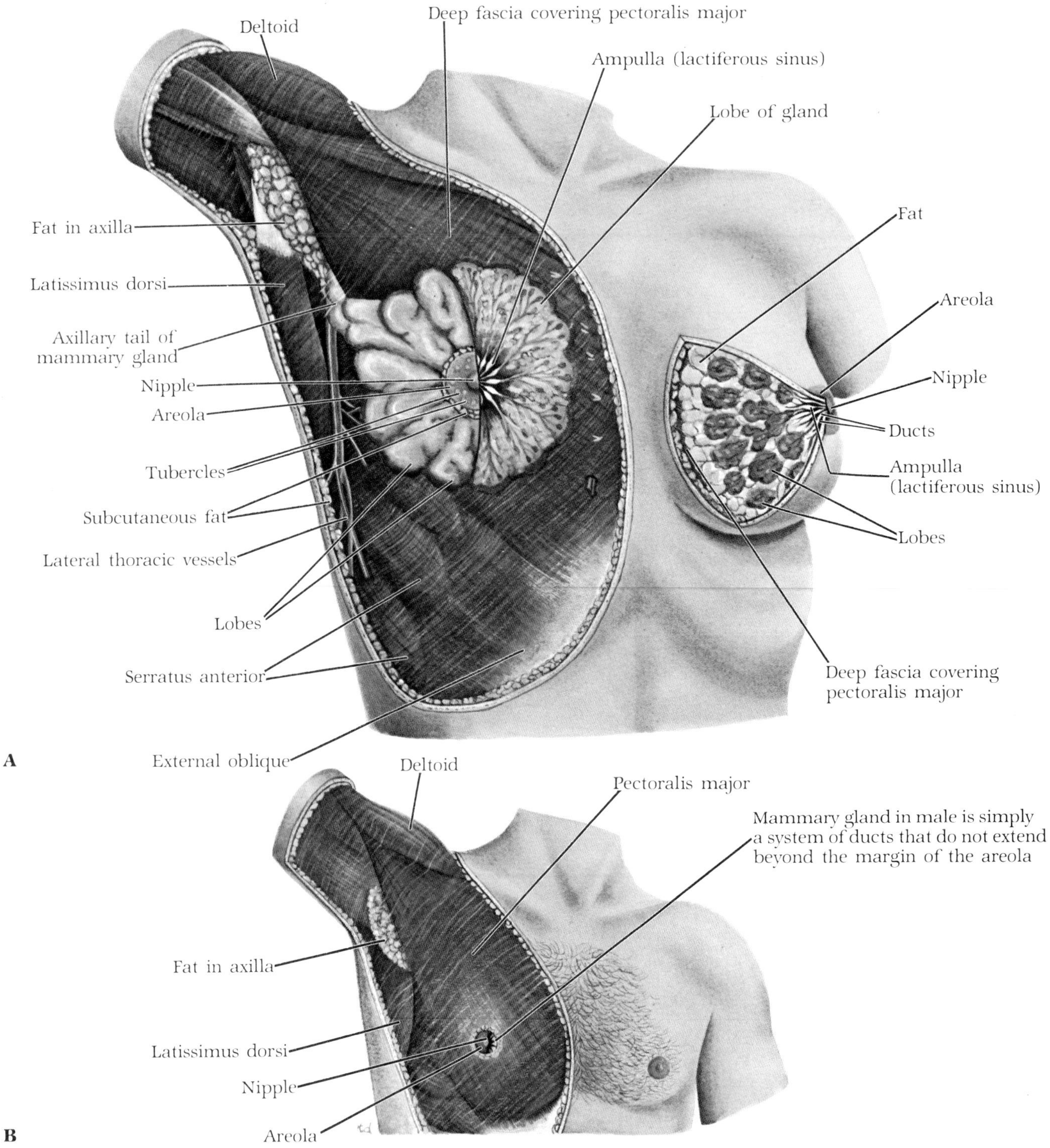
Deltoid
Deep fascia covering pectoralis major
Ampulla (lactiferous sinus)
Lobe of gland
Fat in axilla
Latissimus dorsi
Axillary tail of mammary gland
Nipple
Areola
Tubercles
Subcutaneous fat
Lateral thoracic vessels
Lobes
Serratus anterior
External oblique
Fat
Areola
Nipple
Ducts
Ampulla (lactiferous sinus)
Lobes
Deep fascia covering pectoralis major
A
Deltoid
Pectoralis major
Mammary gland in male is simply a system of ducts that do not extend beyond the margin of the areola
Fat in axilla
Latissimus dorsi
Nipple
Areola
B

Figure 5-5
Mature mammary glands.
A. In female. B. In male.

1. The mammary gland is divided into fifteen to twenty compartments by fibrous septa that radiate out from the nipple. Involvement of the ducts of the gland and the fibrous septa by a scirrhous carcinoma or breast abscess will cause dimpling of the skin.

2. The presence of fibrous septa tends to localize infection to one compartment. An abscess should be drained through a radial type of incision to avoid spreading infection into a neighboring compartment; such an incision also minimizes damage to the radially arranged ducts.

3. In the male the glandular tissue is confined to an area beneath the areola. Malignant tumors of the male breast are rare but are usually more invasive than in women.

4. Radical mastectomy for cancer of the mammary gland involves the removal of the breast and those associated structures containing the lymph vessels and nodes. The following are removed en bloc: (a) skin overlying the tumor, and including the nipple; (b) pectoralis major and pectoralis minor muscles and associated fascia; (c) fat, fascia, and lymph nodes in the axilla; (d) fascia covering the upper part of the rectus sheath, and the serratus anterior, subscapularis, and latissimus dorsi muscles.

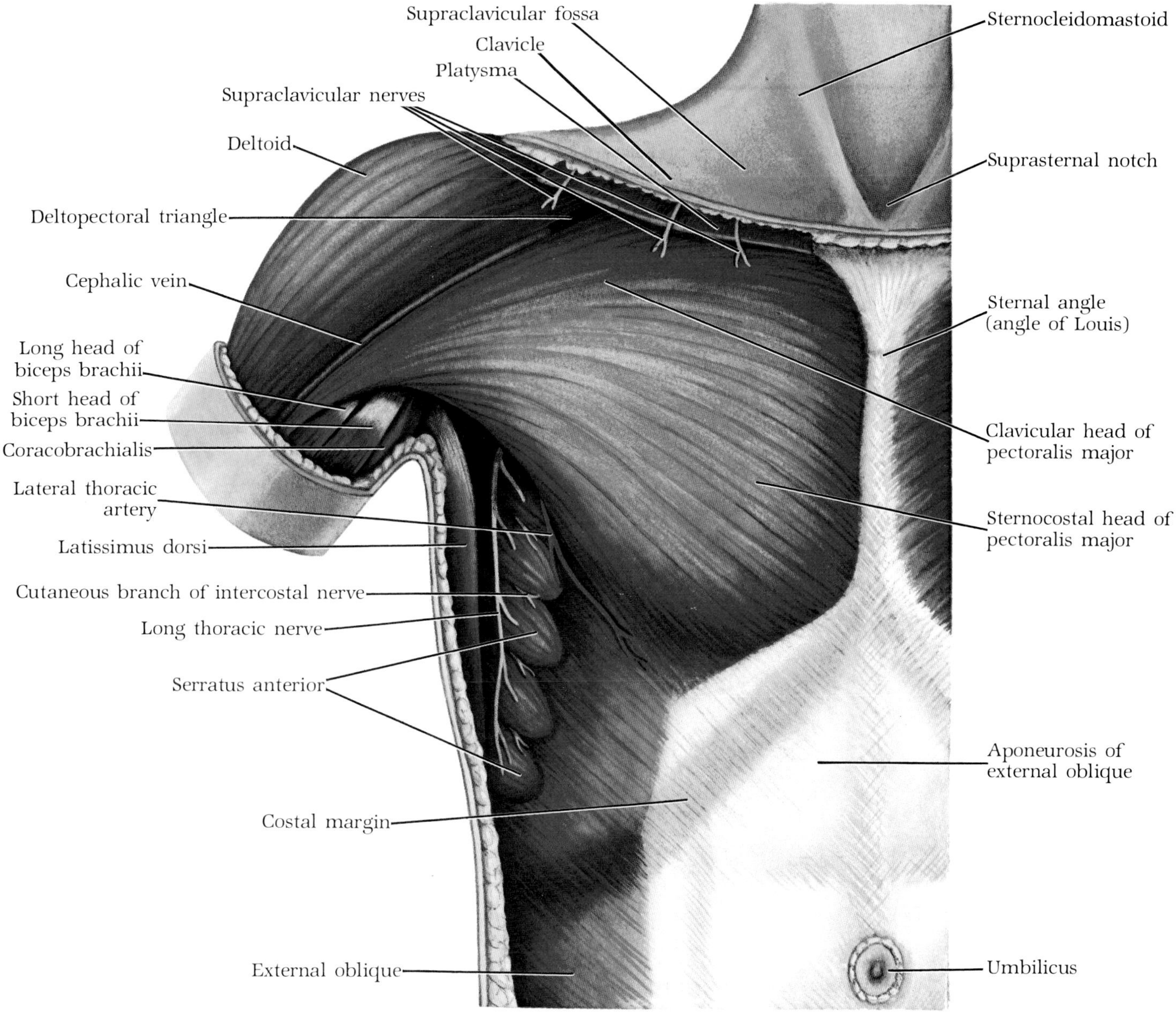
Supraclavicular fossa
Clavicle
Platysma
Supraclavicular nerves
Deltoid
Deltopectoral triangle
Cephalic vein
Long head of biceps brachii
Short head of biceps brachii
Coracobrachialis
Lateral thoracic artery
Latissimus dorsi
Cutaneous branch of intercostal nerve
Long thoracic nerve
Serratus anterior
Costal margin
External oblique
Sternocleidomastoid
Suprasternal notch
Sternal angle (angle of Louis)
Clavicular head of pectoralis major
Sternocostal head of pectoralis major
Aponeurosis of external oblique
Umbilicus

Figure 5-6
Pectoral region and axilla.

1. The natural curve of the shoulder is produced by the deltoid muscle's being pushed laterally by the underlying greater tuberosity of the humerus. A dislocation of the shoulder joint results in a displacement of the greater tuberosity, causing a loss of the curve of the shoulder.

2. The clavicle, because of its position, is exposed to trauma and transmits forces from the upper limb to the trunk. It is the most commonly fractured bone in the body. A fracture usually follows a fall on the shoulder or outstretched hand, with the bone commonly breaking at the junction of the middle and outer thirds, its weakest point. Following the fracture, the lateral fragment is depressed by the weight of the arm and is pulled medially by the pectoralis major. The medial end is tilted superiorly by the sternocleidomastoid.

3. The supraclavicular nerves (C3 and 4) that supply the skin over the shoulder are important because pain caused by irritation of that part of the peritoneum beneath the diaphragm that is supplied by the phrenic nerve (C3, 4, and 5) may be referred to the skin of the shoulder region. This may occur, for example, in inflammation of the gallbladder.

4. The supraclavicular nerves may be involved in callus formation following fracture of the clavicle, resulting in persistent pain at the side of the neck and over the shoulder.

5. Partial or complete congenital absence of the pectoralis major occasionally occurs.

6. The long thoracic nerve may be damaged during a radical mastectomy operation. This would result in paralysis of the serratus anterior muscle, and the patient would have difficulty in raising her arm above her head.

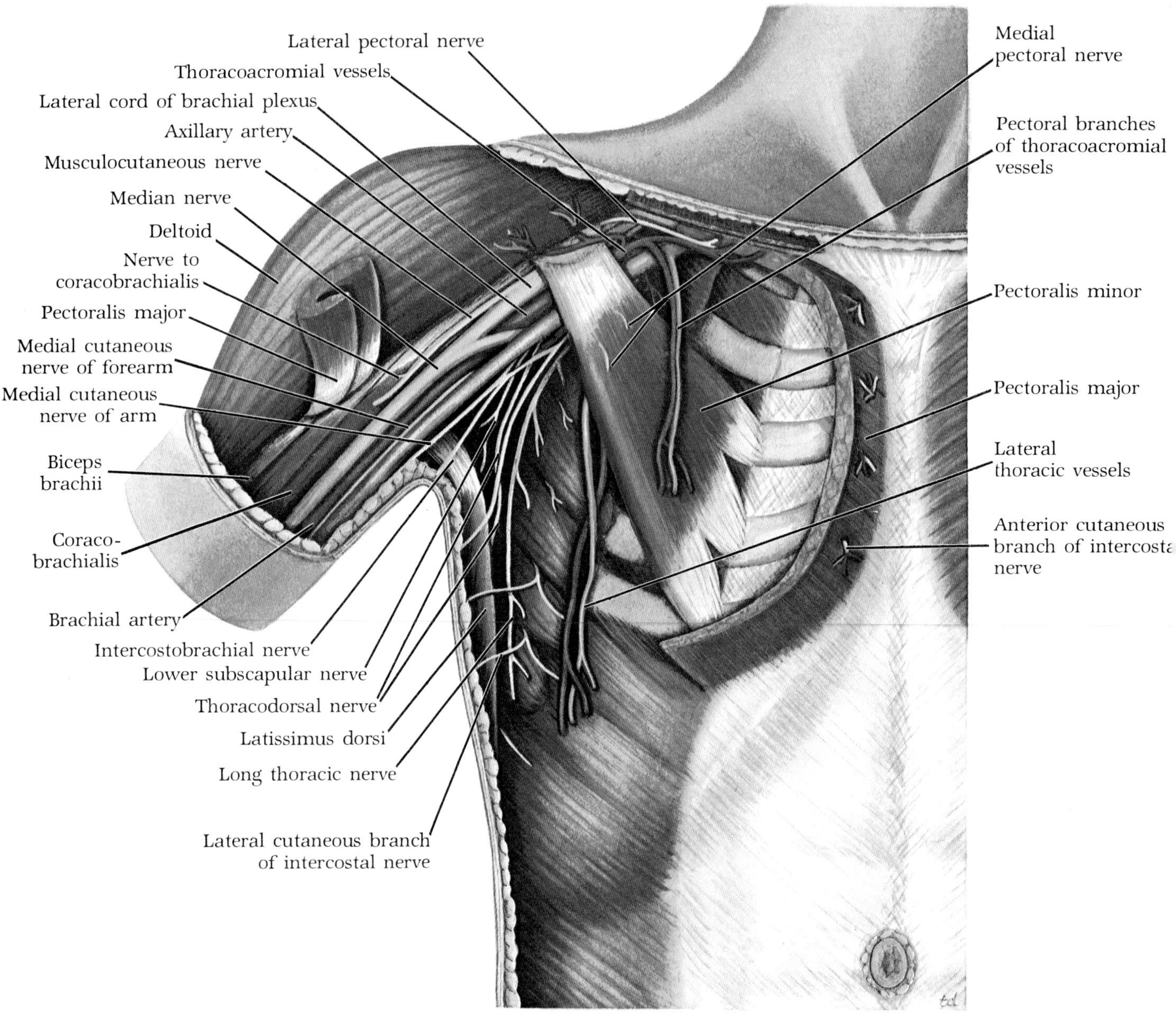
Lateral pectoral nerve
Thoracoacromial vessels
Lateral cord of brachial plexus
Axillary artery
Musculocutaneous nerve
Median nerve
Deltoid
Nerve to coracobrachialis
Pectoralis major
Medial cutaneous nerve of forearm
Medial cutaneous nerve of arm
Biceps brachii
Coraco-brachialis
Brachial artery
Intercostobrachial nerve
Lower subscapular nerve
Thoracodorsal nerve
Latissimus dorsi
Long thoracic nerve
Lateral cutaneous branch of intercostal nerve
Medial pectoral nerve
Pectoral branches of thoracoacromial vessels
Pectoralis minor
Pectoralis major
Lateral thoracic vessels
Anterior cutaneous branch of intercosta
nerve

Figure 5-7
Pectoral region and axilla. The pectoralis major muscle has been removed to display underlying structures.

1. A clinician must know where the arteries of the upper limb can be palpated or compressed in an emergency. The third part of the axillary artery can be felt in the axilla as it lies anterior to the teres major muscle. The brachial artery can be palpated in the arm as it lies on the brachialis and is overlapped from the lateral side by the biceps brachii.

2. Spontaneous thrombosis of the axillary vein occasionally occurs following excessive and unaccustomed movements of the arm at the shoulder joint.

3. *Brachial plexus.* The roots, trunks, and divisions of the brachial plexus are found in the inferior part of the posterior triangle of the neck, with the cords and most of the branches of the plexus lying in the axilla. Because of this protected position within the axilla, the cords are rarely damaged by direct injury, although individual cords may be divided by gunshot or stab wounds. Incomplete injuries to the plexus are common and are usually caused by traction or pressure: a dislocated shoulder joint can produce traction, and the armpit rest of a poorly adjusted crutch can produce pressure.

4. The intercostobrachial nerve is formed by the union of the lateral branch of the second intercostal nerve with the medial cutaneous nerve of the arm. It is important to know this because this is the route along which the pain from myocardial ischemia is referred to the armpit and medial side of the left arm.

5. The long thoracic nerve may be damaged during a radical mastectomy, as mentioned previously. Not only would the patient have difficulty in raising her arm above her head because of the paralysis of the serratus anterior muscle, but the inferior angle of the scapula would protrude posteriorly, a condition known as winged scapula.

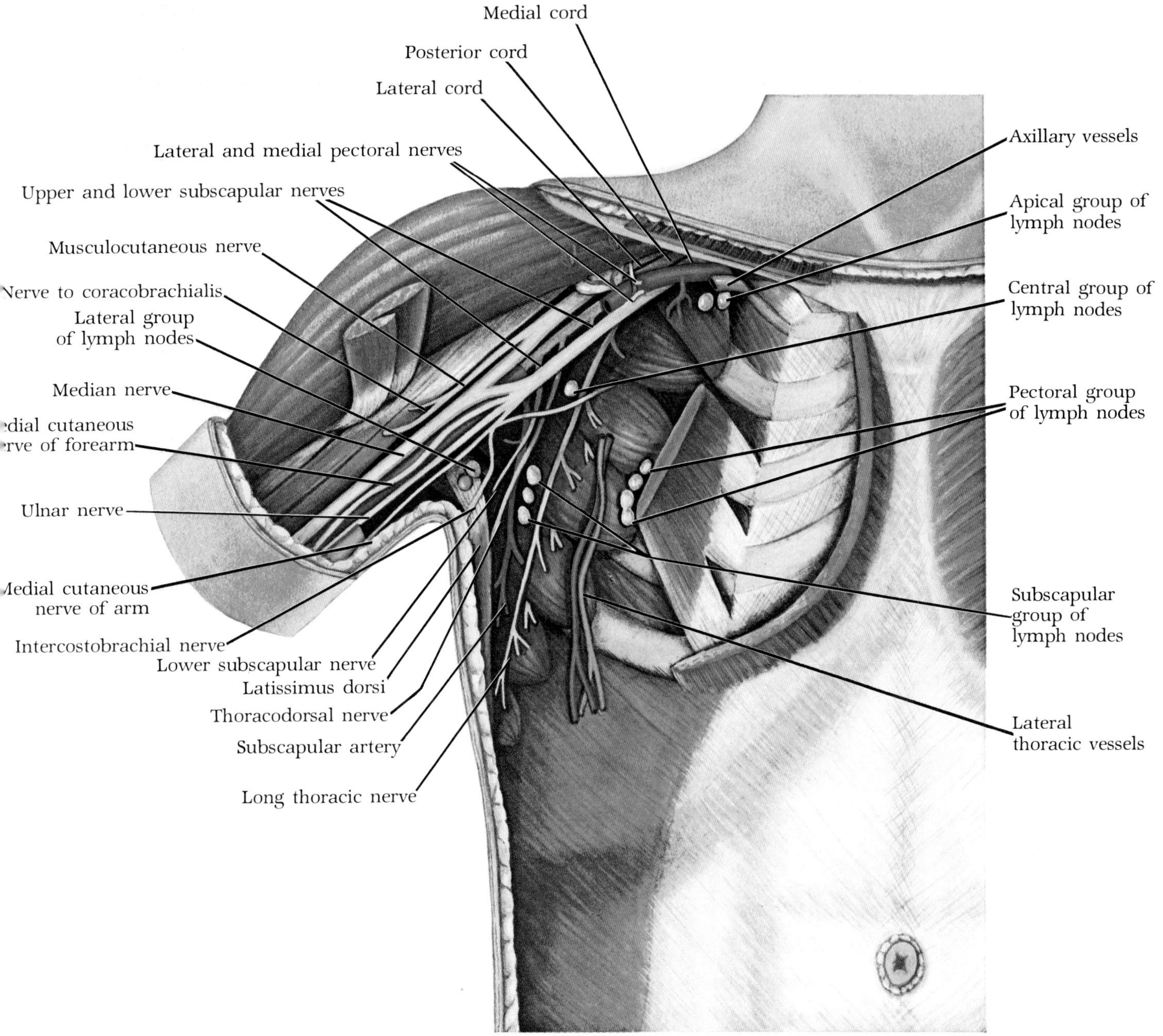
Medial cord
Posterior cord
Lateral cord
Lateral and medial pectoral nerves
Upper and lower subscapular nerves
Musculocutaneous nerve
Nerve to coracobrachialis
Lateral group of lymph nodes
Median nerve
Medial cutaneous nerve of forearm
Ulnar nerve
Medial cutaneous nerve of arm
Intercostobrachial nerve
Lower subscapular nerve
Latissimus dorsi
Thoracodorsal nerve
Subscapular artery
Long thoracic nerve
Axillary vessels
Apical group of lymph nodes
Central group of lymph nodes
Pectoral group of lymph nodes
Subscapular group of lymph nodes
Lateral thoracic vessels

Figure 5-8
Pectoral region and axilla. The pectoralis major and minor muscles and clavipectoral fascia have been removed to display underlying structures.

1. The axillary nerve, a terminal branch of the posterior cord of the brachial plexus, may be injured by the pressure of a crutch pressing upward into the armpit. The axillary nerve is also vulnerable as it passes posteriorly through the quadrilateral space. Here it may be damaged by downward displacement of the humeral head in a shoulder dislocation or fracture of the surgical neck of the humerus with resultant paralysis of the deltoid and teres minor muscles and a loss of skin sensation over the lower half of the deltoid muscle.

2. The radial nerve, also a terminal branch of the posterior cord of the brachial plexus, can be damaged by the pressure of the upper end of a crutch pressing into the armpit, or by a drunkard's falling asleep with his arm over the back of a chair. Fractures and dislocations of the upper end of the humerus can injure the radial nerve; when the humerus is displaced inferiorly, the radial nerve is pulled on, stretching the nerve excessively. The patient is then unable to extend the elbow joint and has wrist drop. There is a small loss of skin sensation down the posterior surface of the lower part of the arm and down a narrow strip on the back of the forearm. There is also a variable area of sensory loss on the lateral side of the dorsum of the hand and base of the thumb.

3. Cancer of the breast, a common disease, tends to spread by way of the lymph vessels. The lymphatic vessels from the medial half of the breast pierce the second, third, and fourth intercostal spaces and drain into the internal thoracic nodes, while the lymphatic vessels from the lateral half of the breast drain into the anterior or pectoral group of axillary nodes. A cancer occurring in the lateral half of the breast would therefore tend to spread to the axillary nodes. Thoracic metastases are difficult or impossible to treat, but the lymph nodes of the axilla can be removed surgically.

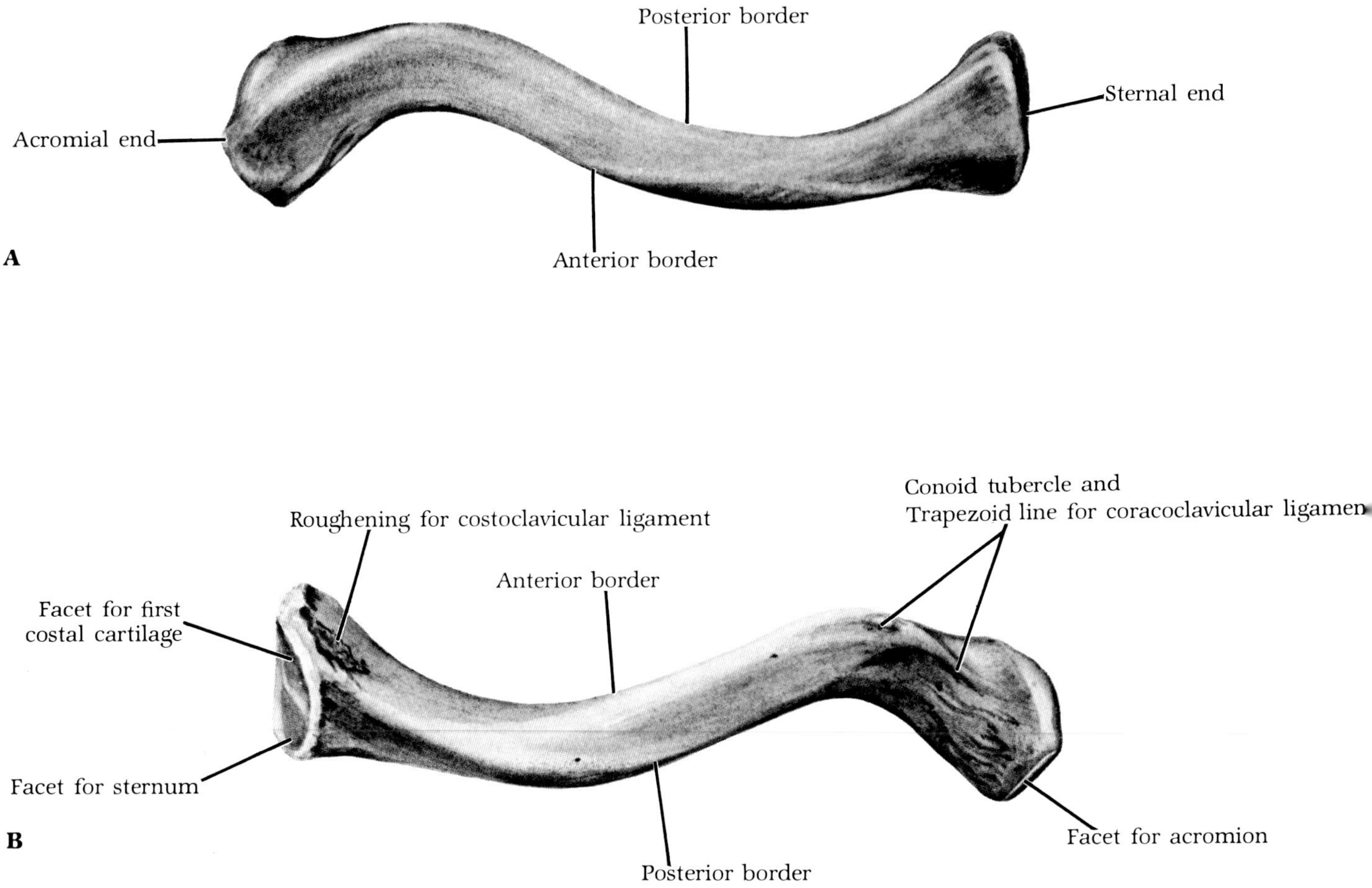
Posterior border
Sternal end
Acromial end
A
Anterior border
Conoid tubercle and
Trapezoid line for coracoclavicular ligament
Roughening for costoclavicular ligament
Anterior border
Facet for first
costal cartilage
Facet for sternum
B
Facet for acromion
Posterior border

Figure 5-9
Right clavicle. A. Superior surface. B. Inferior surface.

1. The clavicle is a strut that holds the arm laterally to allow it to move freely on the trunk.

2. The clavicle is the most commonly fractured bone in the body. As the result of a fall on the shoulder or outstretched hand, the force is transmitted along the clavicle, which then breaks at the junction of the middle and outer thirds, its weakest point.

3. Because the clavicle is subcutaneous along its entire length, fractures can easily be palpated.

4. The close relationship of the supraclavicular nerves to the clavicle can result in their involvement in callus formation following fracture of the bone.

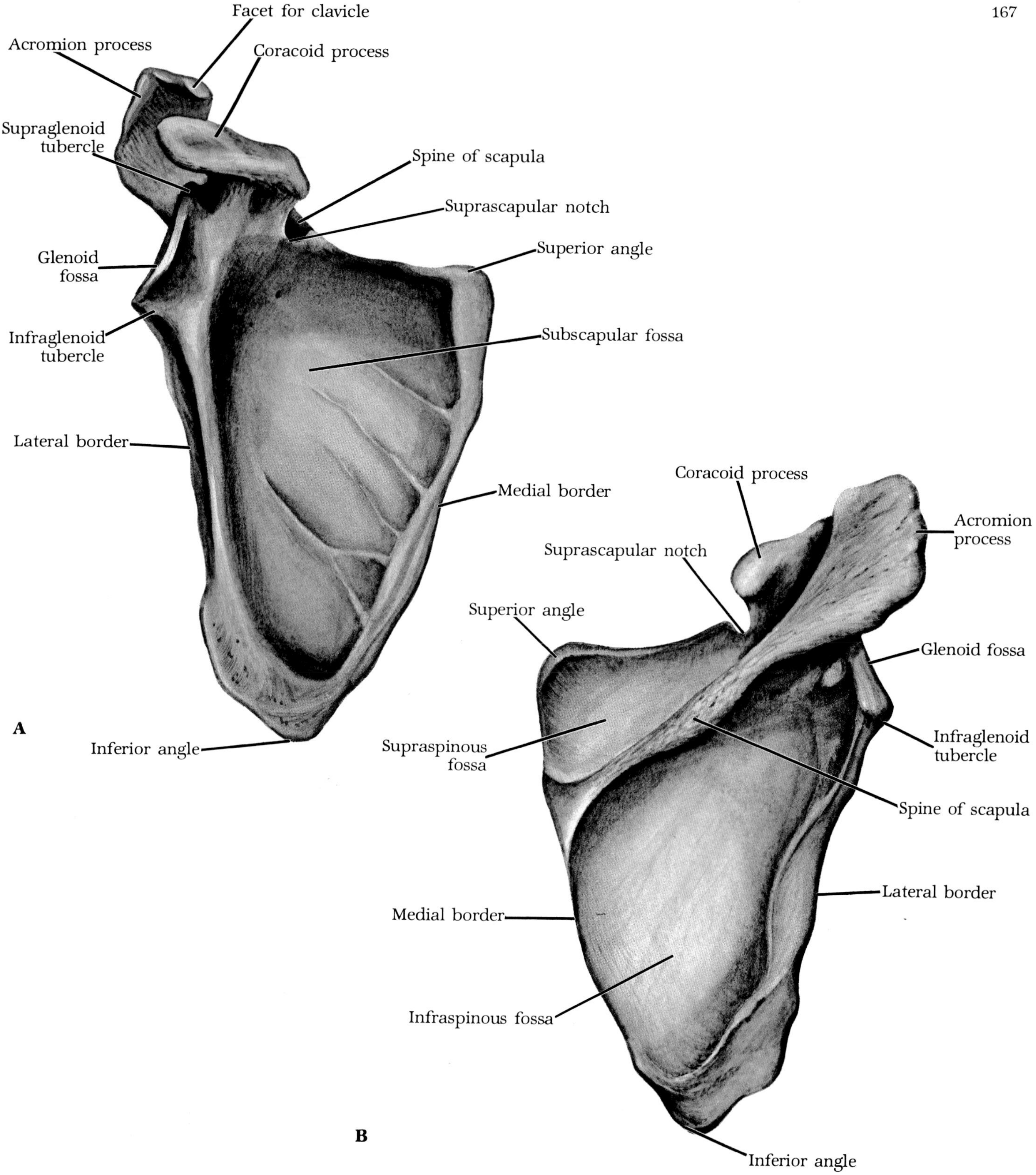
Facet for clavicle
Acromion process
Coracoid process
Supraglenoid tubercle
Spine of scapula
Suprascapular notch
Superior angle
Glenoid fossa
Infraglenoid tubercle
Subscapular fossa
Lateral border
Medial border
A
Inferior angle
Coracoid process
Acromion process
Suprascapular notch
Superior angle
Glenoid fossa
Supraspinous fossa
Infraglenoid tubercle
Spine of scapula
Lateral border
Medial border
Infraspinous fossa
B
Inferior angle

Figure 5-10
Right scapula. A. Anterior view. B. Posterior view.

1. Congenital elevation of the scapula (Sprengel's shoulder) is a condition in which one or both scapulae are smaller than normal and situated at a higher level. There may be a fibrous connection with the vertebral column. The shoulder joint can be abducted to a right angle, but the arm cannot be raised further because of fixation of the scapula.

2. Fractures of the scapula are usually the result of severe direct trauma as occurs in run-over accidents or in occupants of automobiles involved in crashes. Injuries are usually associated with fractured ribs and cardiopulmonary trauma. Fortunately most fractures of the scapula require little treatment because the muscles on the anterior and posterior surfaces adequately splint the fragments.

3. The position of the scapula on the posterior wall of the thorax opposite the second to the seventh ribs is maintained by the tone and balance of the muscles attached to it. If one of these muscles is paralyzed, this balance is upset—for example, dropped shoulder, which occurs with paralysis of the trapezius, or winged scapula, which occurs with paralysis of the serratus anterior. Such imbalance can be detected by careful physical examination.

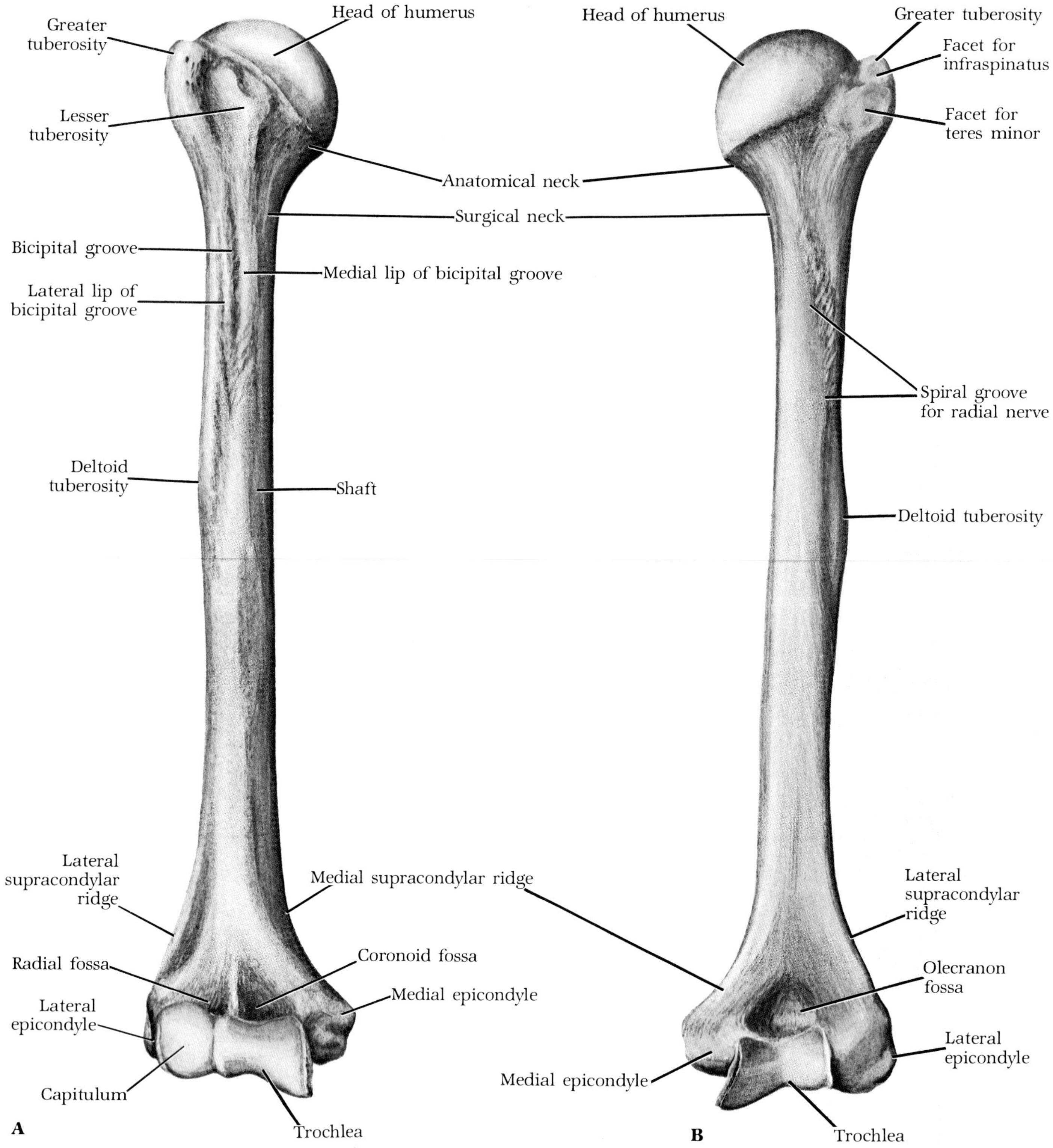
Head of humerus
Greater tuberosity
Lesser tuberosity
Anatomical neck
Surgical neck
Bicipital groove
Medial lip of bicipital groove
Lateral lip of bicipital groove
Deltoid tuberosity
Shaft
Lateral supracondylar ridge
Medial supracondylar ridge
Radial fossa
Coronoid fossa
Medial epicondyle
Lateral epicondyle
Capitulum
Trochlea
A
Head of humerus
Greater tuberosity
Facet for infraspinatus
Facet for teres minor
Spiral groove for radial nerve
Deltoid tuberosity
Lateral supracondylar ridge
Olecranon fossa
Lateral epicondyle
Medial epicondyle
Trochlea
B

Figure 5-11
Right humerus. A. Anterior view. B. Posterior view.

1. *Fractures of the Proximal End of the Humerus.* The greater tuberosity may be fractured by direct trauma or avulsed by violent contractions of the supraspinatus muscle. The surgical neck of the humerus may be fractured by a direct blow on the lateral aspect of the shoulder or in an indirect manner by the person falling on the outstretched hand. In children injury to the proximal epiphyseal cartilage may be followed by retardation of growth in the length of the humerus.

2. *Fractures of the Shaft of the Humerus.* These fractures are common, with the displacement of the fragments dependent on the relation of the site of the fracture to the insertion of the deltoid. When the fracture line is proximal to the deltoid insertion, the proximal fragment is adducted by the pectoralis major, latissimus dorsi, and teres major muscles; the distal fragment is pulled proximally by the deltoid, biceps, and triceps. When the fracture is distal to the deltoid insertion, the proximal fragment is abducted by the deltoid, and the distal fragment is pulled proximally by the biceps and triceps. The radial nerve may be damaged in shaft fractures.

3. *Fractures of the Distal End of the Humerus.* Supracondylar fractures are common in children and occur when the child falls on the outstretched hand with the elbow joint partially flexed. The median nerve and the brachial artery are occasionally injured in this type of fracture.
The medial epicondyle can be avulsed by the medial collateral ligament of the elbow joint if the forearm is forcibly abducted. The ulnar nerve may be injured at the time of a fracture, may become involved in the callus formation, or may undergo irritation on the irregular bony surface after the bone fragments are reunited.
Fracture of the trochlea can occur when a person falls on the outstretched hand with the elbow joint slightly flexed and the forearm pronated.

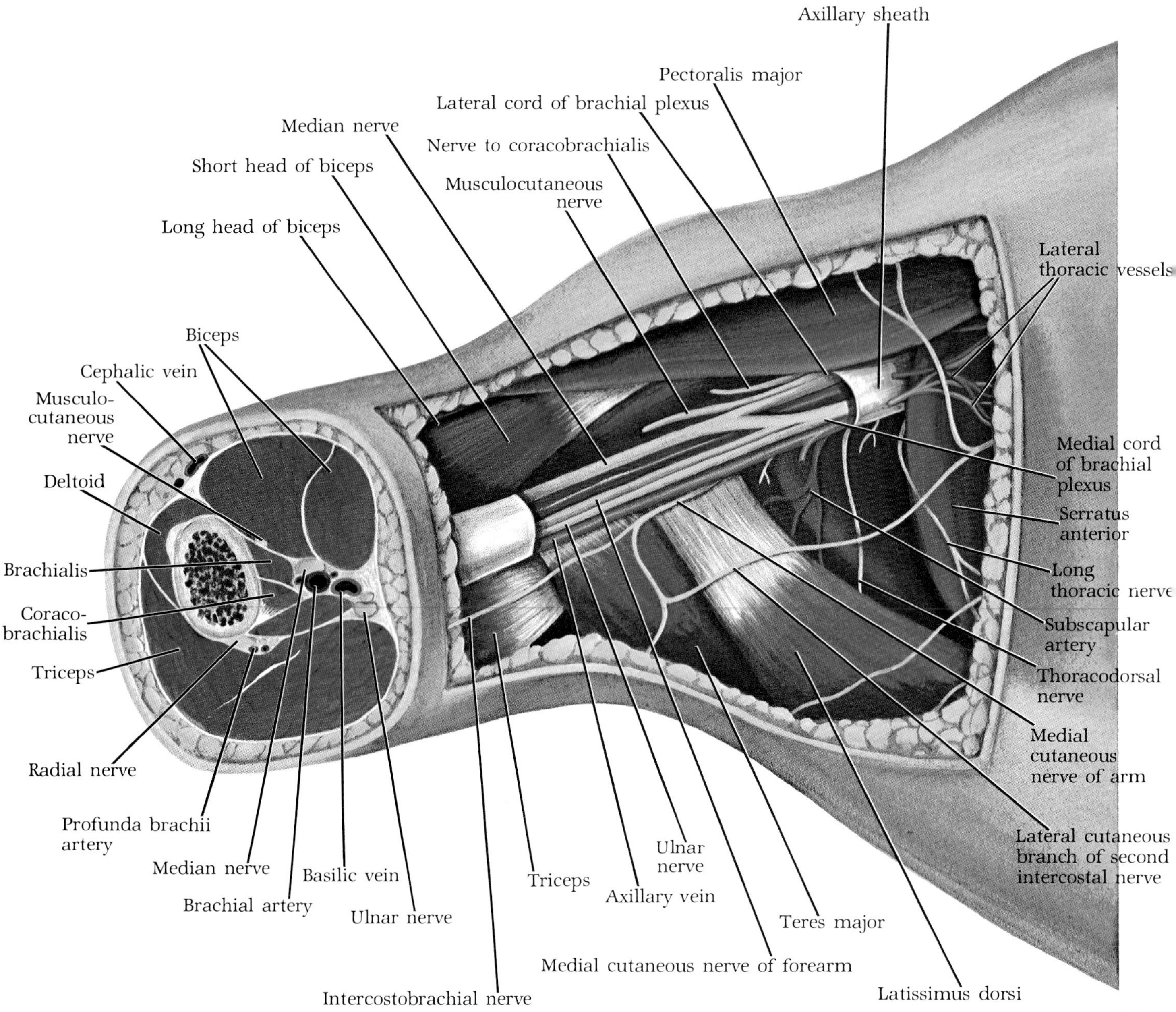

Axillary sheath
Pectoralis major
Lateral cord of brachial plexus
Median nerve
Nerve to coracobrachialis
Short head of biceps
Musculocutaneous nerve
Long head of biceps
Lateral thoracic vessels
Biceps
Cephalic vein
Musculo-cutaneous nerve
Deltoid
Medial cord of brachial plexus
Serratus anterior
Brachialis
Coraco-brachialis
Long thoracic nerve
Subscapular artery
Triceps
Thoracodorsal nerve
Medial cutaneous nerve of arm
Radial nerve
Profunda brachii artery
Lateral cutaneous branch of second intercostal nerve
Median nerve
Basilic vein
Ulnar nerve
Triceps
Brachial artery
Ulnar nerve
Axillary vein
Teres major
Medial cutaneous nerve of forearm
Intercostobrachial nerve
Latissimus dorsi

Figure 5-12
Axilla, inferior view.

1. The anterior skin fold of the axilla is formed by the rounded lower margin of the pectoralis major. The posterior skin fold is formed by the latissimus dorsi winding around the lower margin of the teres major muscle.

2. The axillary sheath, formed of deep fascia derived from the prevertebral layer of deep fascia in the neck, encloses the axillary artery and the brachial plexus; the axillary vein is on the outside of the sheath. A brachial plexus nerve block can easily be obtained by closing the distal part of the sheath with finger pressure, inserting a syringe needle into the proximal part of the sheath, and then injecting a local anesthetic. The anesthetic solution is massaged along the sheath, producing a nerve block. The position of the sheath can be verified by feeling the pulsations of the third part of the axillary artery.

3. Note the close relationship of the shaft of the humerus to the radial nerve. Fractures of the humerus at this site can easily involve the radial nerve, causing the patient to have wrist drop.

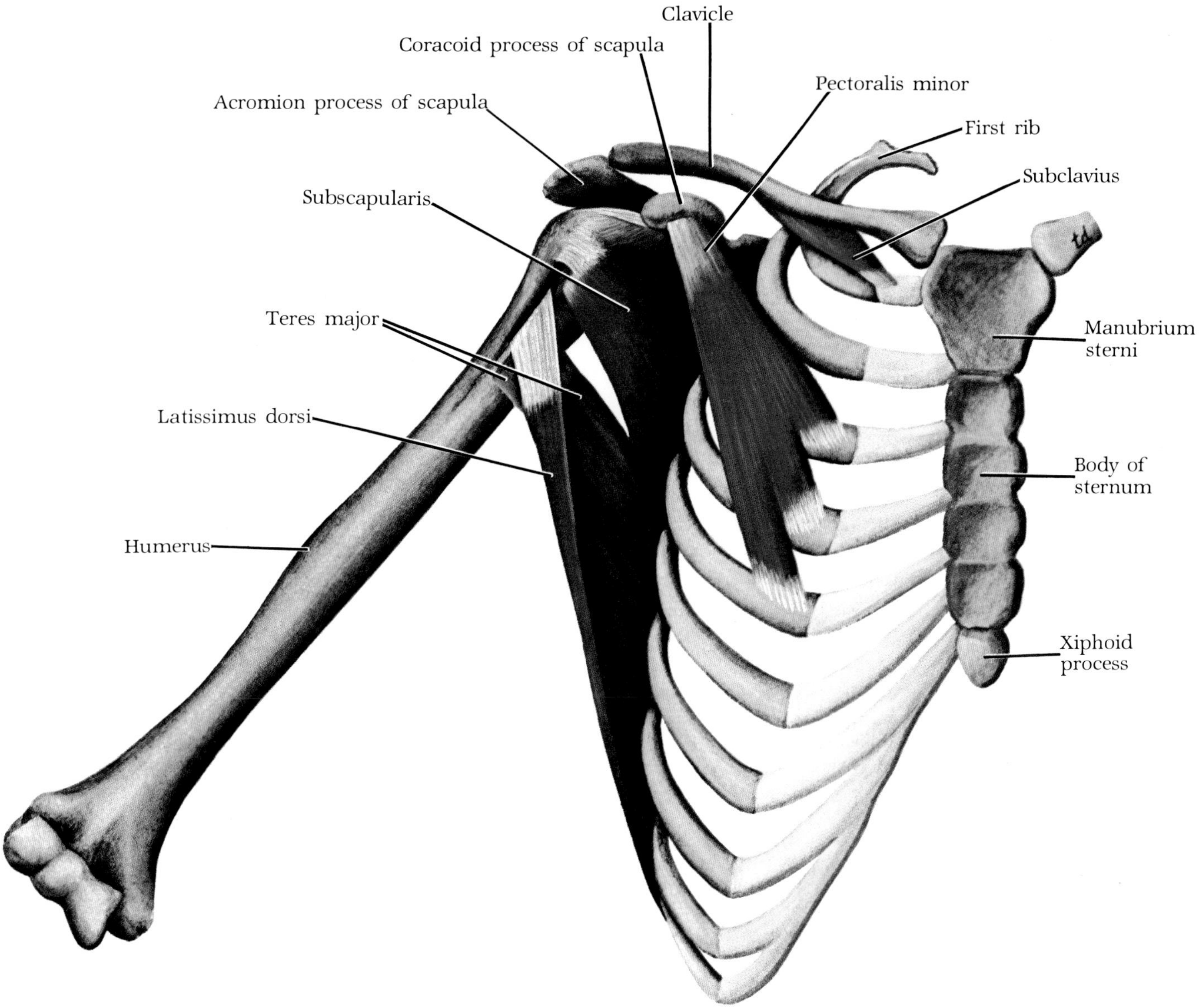
Clavicle
Coracoid process of scapula
Pectoralis minor
Acromion process of scapula
First rib
Subclavius
Subscapularis
Manubrium sterni
Teres major
Latissimus dorsi
Body of sternum
Humerus
Xiphoid process

Figure 5-13
Subclavius, pectoralis minor, subscapularis, teres major, and latissimus dorsi muscles, anterior view.

1. The subclavius muscle assists in stabilizing the sternoclavicular joint. The nerve to the subclavius (C5 and 6) is important clinically, since it may give a contribution (C5) to the phrenic nerve; when present, this contributory branch is referred to as the accessory phrenic nerve.

2. The pectoralis minor muscle, on contraction, depresses the point of the shoulder and the lateral end of the clavicle. Following a fracture of the clavicle the pectoralis minor takes part in displacing the lateral fragment inferiorly. The pectoralis minor muscle is removed surgically in radical mastectomies (along with the pectoralis major) since the lymphatic vessels draining the breast pass through this muscle en route to the pectoral lymph nodes.

3. The tone of the subscapularis muscle greatly strengthens the shoulder joint and this muscle, together with the other short muscles that surround this joint, helps to keep the head of the humerus in the glenoid fossa.

4. The latissimus dorsi tendon and the teres major muscle form the rounded posterior fold of the axilla. When a fracture of the shaft of the humerus is proximal to the insertion of the deltoid, these muscles will cause adduction of the proximal fragment of the shaft.

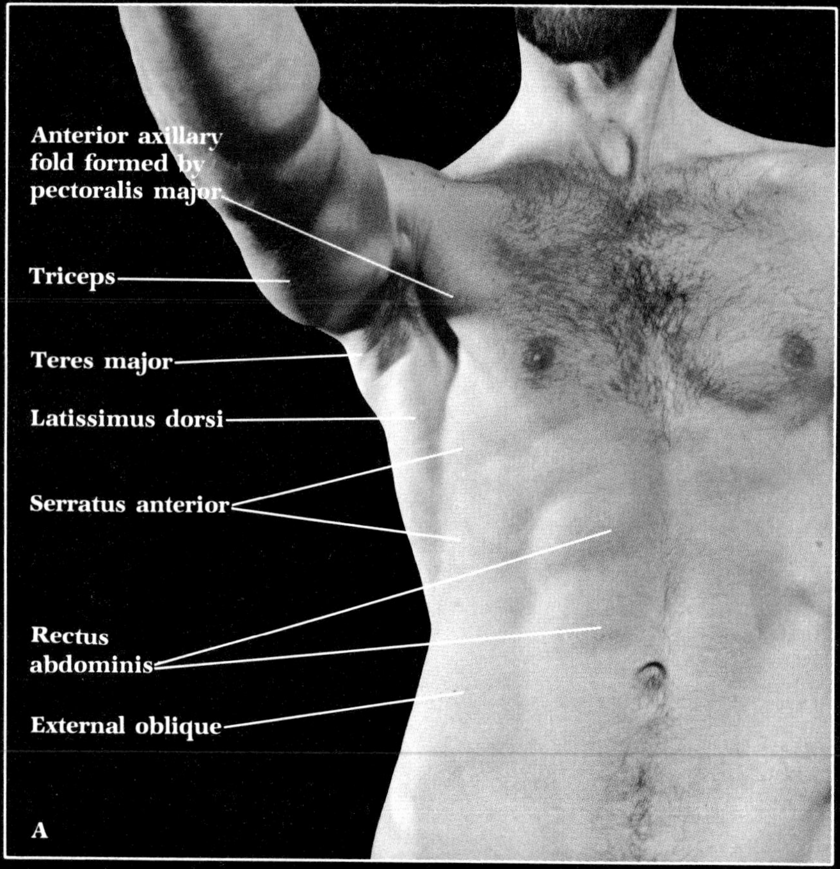

Anterior axillary fold formed by pectoralis major
Triceps
Teres major
Latissimus dorsi
Serratus anterior
Rectus abdominis
External oblique
A

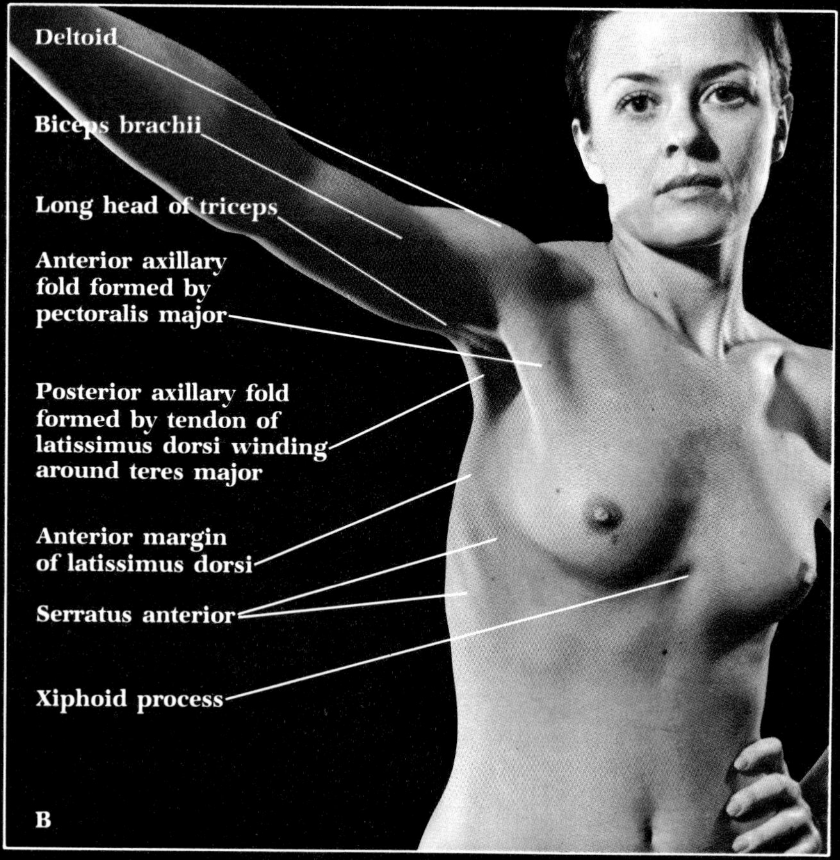

Deltoid
Biceps brachii
Long head of triceps
Anterior axillary fold formed by pectoralis major
Posterior axillary fold formed by tendon of latissimus dorsi winding around teres major
Anterior margin of latissimus dorsi
Serratus anterior
Xiphoid process
B

Figure 5-14
The axilla and pectoral region.
A. 27-year-old male. B. 29-year-old female.

1. The axilla should be examined when the forearm is supported and the pectoral muscles are relaxed. With the arm by the side, the inferior part of the head of the humerus can be palpated through the floor of the axilla.
2. The pulsations of the axillary artery can be felt high up in the axilla, and the cords of the brachial plexus may be palpated around the artery.
3. The anterior axillary fold is formed by the lower margin of the pectoralis major muscle. The posterior axillary fold is formed by the tendon of the latissimus dorsi muscle winding around the lower border of the teres major muscle.
4. The medial wall of the axilla is formed by the upper ribs covered by the serratus anterior muscle. The lateral wall is formed by the coracobrachialis and biceps brachii muscles.

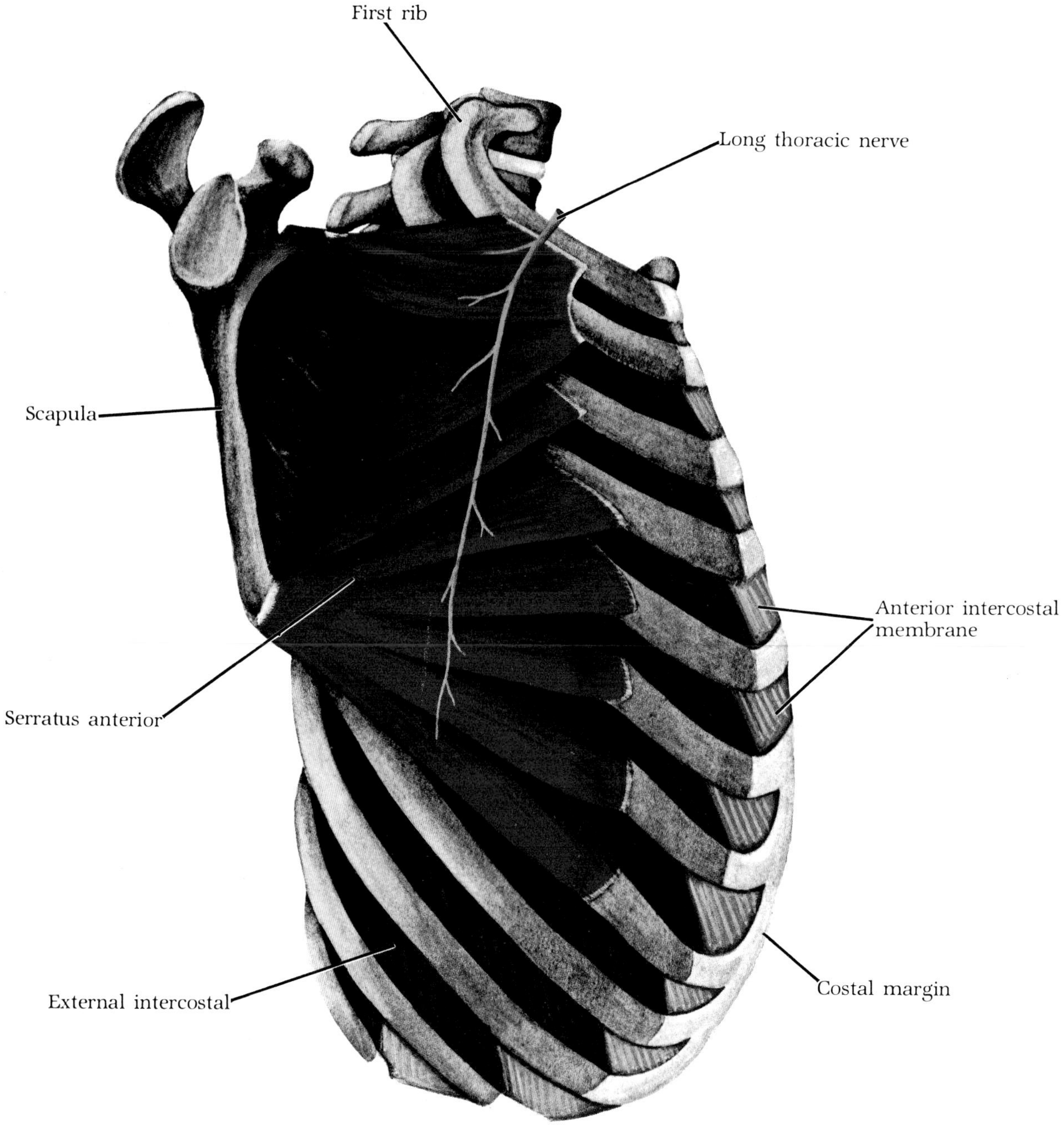
First rib
Long thoracic nerve
Scapula
Anterior intercostal membrane
Serratus anterior
Costal margin
External intercostal

Figure 5-15
Serratus anterior muscle, lateral view.

1. The serratus anterior muscle draws the scapula forward around the thoracic wall and, because of the greater pull exerted on the inferior angle, rotates the scapula so that its inferior angle passes laterally and forward; in this action, the serratus anterior is assisted by the trapezius. This rotation of the scapula takes place when the arm is raised from the horizontal abducted position to a vertical position above the head. This muscle is also used when the arm is pushed forward in the horizontal position, as in a forward punch.

2. The serratus anterior muscle is supplied by the long thoracic nerve (C5, 6, and 7). This nerve may be injured by blows to or pressure on the posterior triangle in the neck or during the surgical procedure of radical mastectomy. Paralysis of the serratus anterior results in the inability to rotate the scapula as described above, and the patient experiences difficulty in raising her arm above the head. The medial border and inferior angle of the scapula can no longer be kept closely applied to the chest wall and will protrude posteriorly, a condition known as winged scapula.

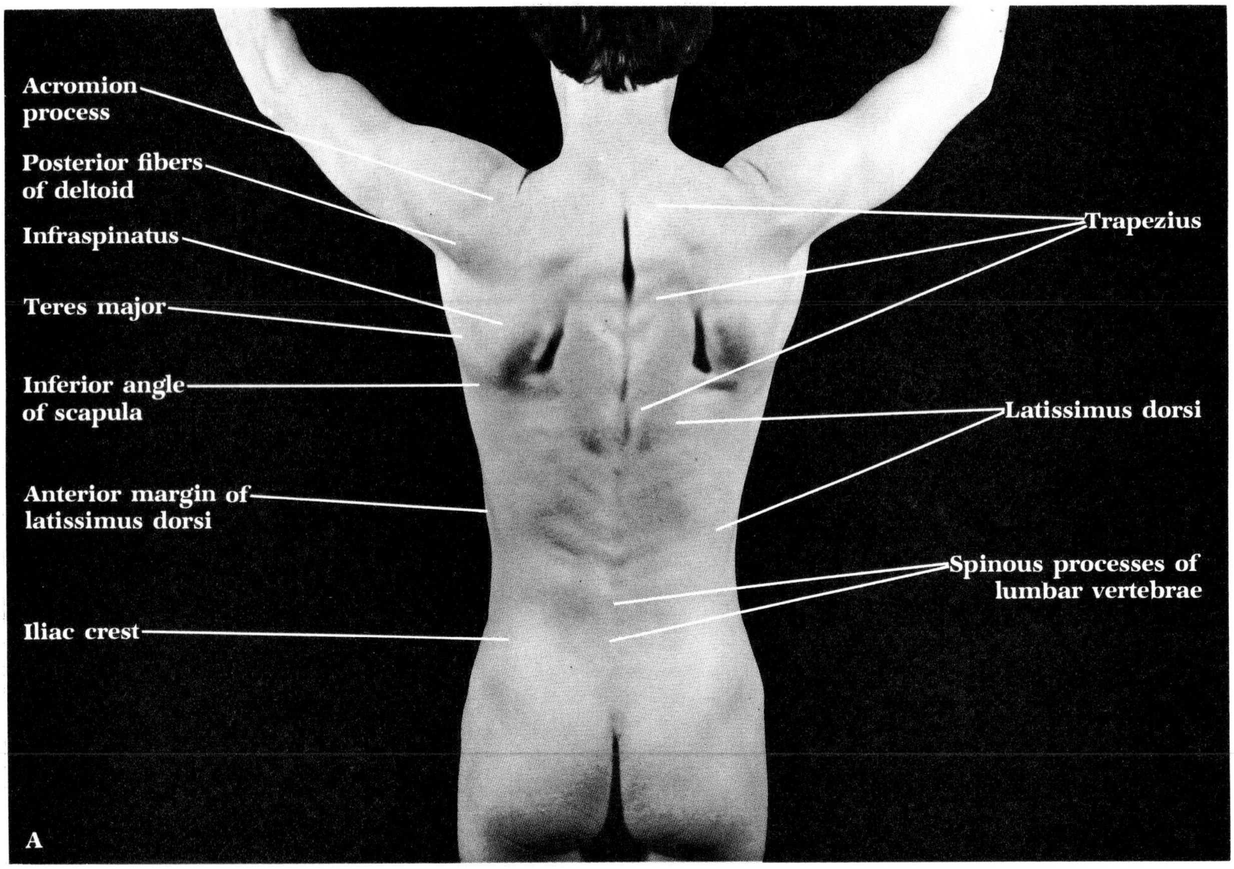
Acromion process
Posterior fibers of deltoid
Infraspinatus
Teres major
Inferior angle of scapula
Anterior margin of latissimus dorsi
Iliac crest
Trapezius
Latissimus dorsi
Spinous processes of lumbar vertebrae
A

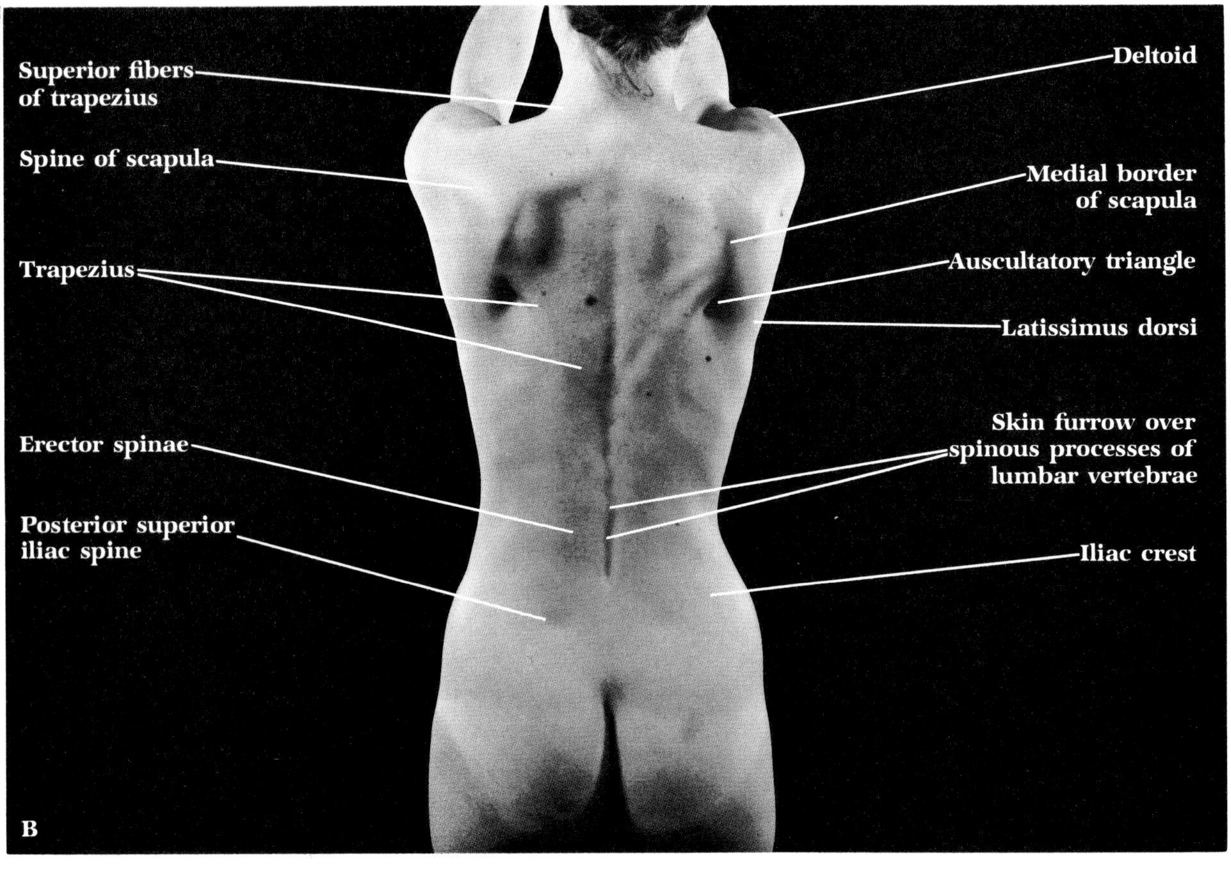
Superior fibers of trapezius
Spine of scapula
Trapezius
Erector spinae
Posterior superior iliac spine
Deltoid
Medial border of scapula
Auscultatory triangle
Latissimus dorsi
Skin furrow over spinous processes of lumbar vertebrae
Iliac crest
B

Figure 5-16
The back. A. 27-year-old male. B. 29-year-old female.

1. The external occipital protuberance lies at the junction of the head and neck. If the index finger of the examiner is placed on the skin in the midline, it can be drawn downward from the protuberance in the nuchal groove. The first spinous process to be felt is that of the seventh cervical vertebra (vertebra prominens). The first to sixth spines are covered by the ligamentum nuchae.
2. The nuchal groove is continuous below with a furrow that runs down the middle of the back over the tips of the spines of the thoracic and lumbar vertebrae. The most prominent spine is that of the first thoracic vertebra.
3. The acromion process of the scapula forms the lateral extremity of the spine of the scapula. Below the lateral edge of the acromion process the smooth rounded curve of the shoulder is produced by the deltoid muscle, which covers the greater tuberosity of the humerus.
4. The crest of the spine of the scapula can be palpated and traced medially to the medial border of the scapula, which it joins at the level of the third thoracic spine.
5. The superior angle of the scapula can be palpated opposite the first thoracic spine.
6. The inferior angle of the scapula can be palpated opposite the seventh thoracic spine.
7. The erector spinae muscles are large and lie on either side of the spines of the lumbar and thoracic vertebrae. They should be examined with the flat of the hand.
8. The iliac crests are easily palpated along their entire length. They lie at the level of the fourth lumbar spine. Each crest ends posteriorly at the posterior superior iliac spine, which lies beneath a skin dimple at the level of the second sacral vertebra.

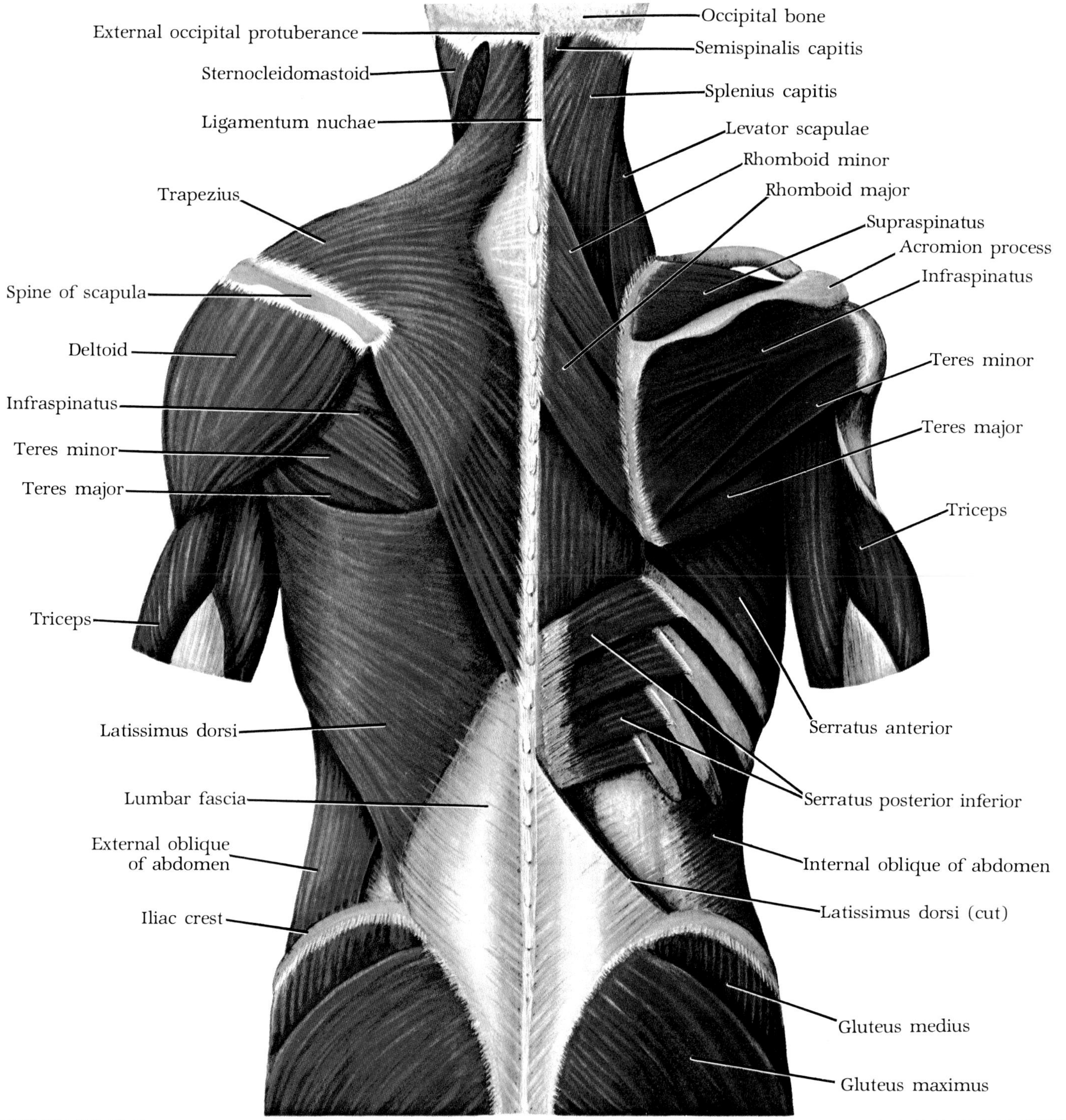
External occipital protuberance
Occipital bone
Semispinalis capitis
Sternocleidomastoid
Splenius capitis
Ligamentum nuchae
Levator scapulae
Rhomboid minor
Trapezius
Rhomboid major
Supraspinatus
Acromion process
Infraspinatus
Spine of scapula
Deltoid
Teres minor
Infraspinatus
Teres minor
Teres major
Teres major
Triceps
Triceps
Serratus anterior
Latissimus dorsi
Serratus posterior inferior
Lumbar fascia
External oblique of abdomen
Internal oblique of abdomen
Iliac crest
Latissimus dorsi (cut)
Gluteus medius
Gluteus maximus

Figure 5-17
Superficial and deep muscles of the back.

1. The two scapulae afford protection against trauma to the upper part of the rib cage. Good tone in the muscles attached to the scapulae allow them to maintain their normal postural position.
2. Paralysis of the trapezius results in a dropped shoulder. Paralysis of the levator scapulae and rhomboids produces a sagging of the medial border of the scapula, and the patient then experiences difficulty in "bracing the shoulders backward." Paralysis of the serratus anterior results in winged scapula.
3. The so-called auscultatory triangle lies between the superior border of the latissimus dorsi, the lateral border of the trapezius, and the medial border of the rhomboid major. Because there are fewer superficial back muscles in this area, a stethoscope placed here will be closer to the lungs, allowing the breath sounds to be heard more clearly than elsewhere.
4. The lumbar triangle is bounded laterally by the posterior border of the external oblique, medially by the lateral border of latissimus dorsi, and inferiorly by the iliac crest; the floor of the triangle is formed by the internal oblique. If infection of the vertebral column results in abscess formation, pus may track laterally and inferiorly and emerge from beneath the latissimus dorsi in the lumbar triangle.
5. Again note that the curve of the shoulder is produced by the greater tuberosity of the humerus displacing the deltoid laterally.

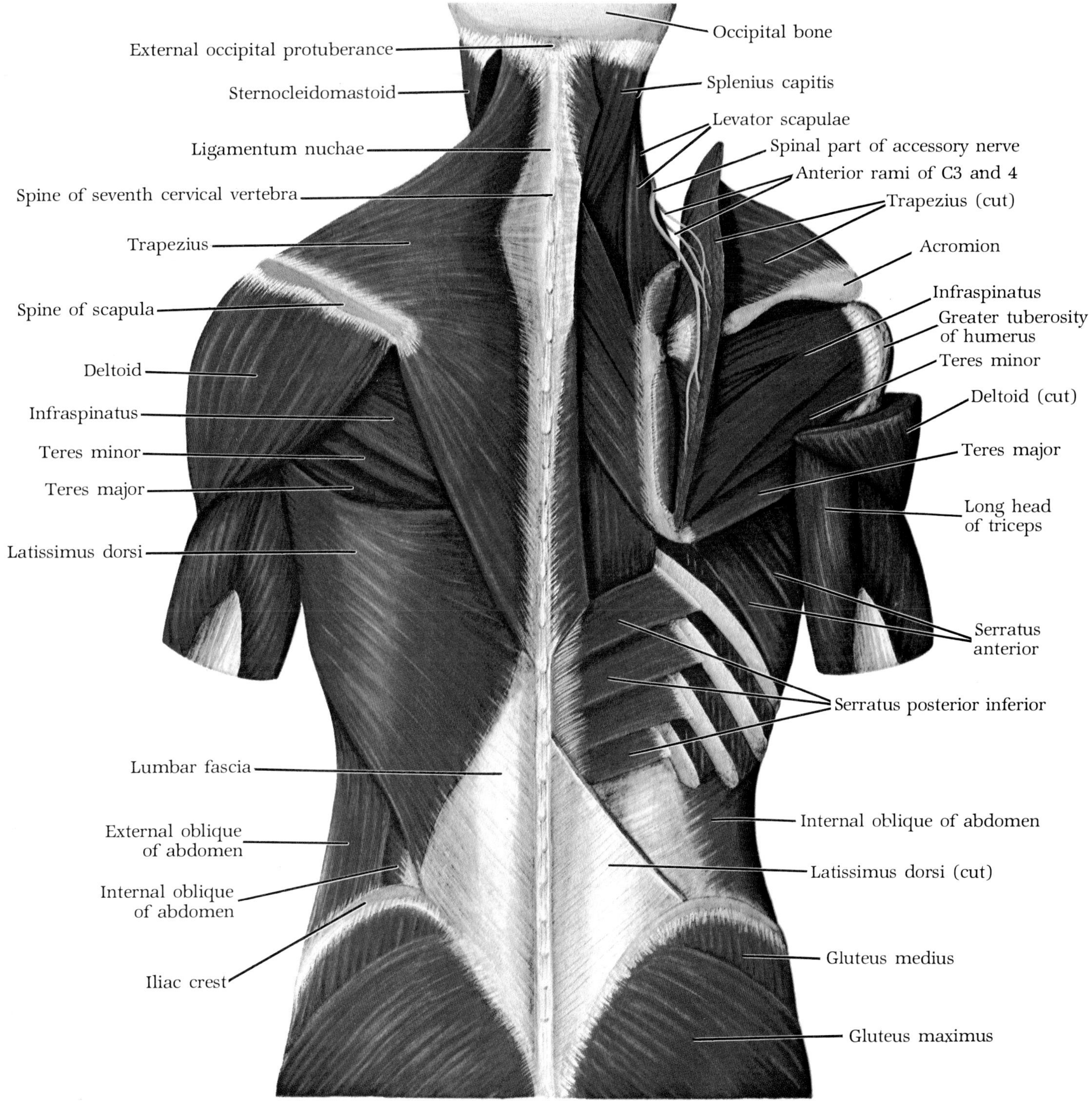
Occipital bone
External occipital protuberance
Splenius capitis
Sternocleidomastoid
Levator scapulae
Ligamentum nuchae
Spinal part of accessory nerve
Anterior rami of C3 and 4
Spine of seventh cervical vertebra
Trapezius (cut)
Trapezius
Acromion
Infraspinatus
Spine of scapula
Greater tuberosity of humerus
Deltoid
Teres minor
Deltoid (cut)
Infraspinatus
Teres minor
Teres major
Teres major
Long head of triceps
Latissimus dorsi
Serratus anterior
Serratus posterior inferior
Lumbar fascia
Internal oblique of abdomen
External oblique of abdomen
Latissimus dorsi (cut)
Internal oblique of abdomen
Gluteus medius
Iliac crest
Gluteus maximus

Figure 5-18
Superficial and deep muscles of the back. The trapezius on the right has been reflected laterally to reveal its nerve supply.

1. The curve of the lower part of the neck leading into that of the shoulder is produced by the superior margin of the trapezius. The consistency of the trapezius can be felt between the finger and thumb on physical examination of this part of the neck. It is important to compare the tone of the muscles on the two sides.

2. Because the scapula rotates around the point of attachment of the coracoid process to the clavicle by the coracoclavicular ligament, the superior and inferior fibers of the trapezius assist the serratus anterior muscle in rotating the scapula when the arm is raised above the head. The strength of these muscles can be tested by having the patient shrug his shoulders while you exert downward pressure on the superior border of the trapezius muscles.

3. The spinal part of the accessory nerve that innervates the trapezius runs inferiorly in the neck, resting on the superficial surface of the levator scapulae muscle before disappearing under the superior border of the trapezius. That portion of the accessory nerve is covered only by fascia and skin and can easily be damaged by a knife wound or surgical incision in this area. The third and fourth cervical nerves supply the trapezius muscle with sensory fibers.

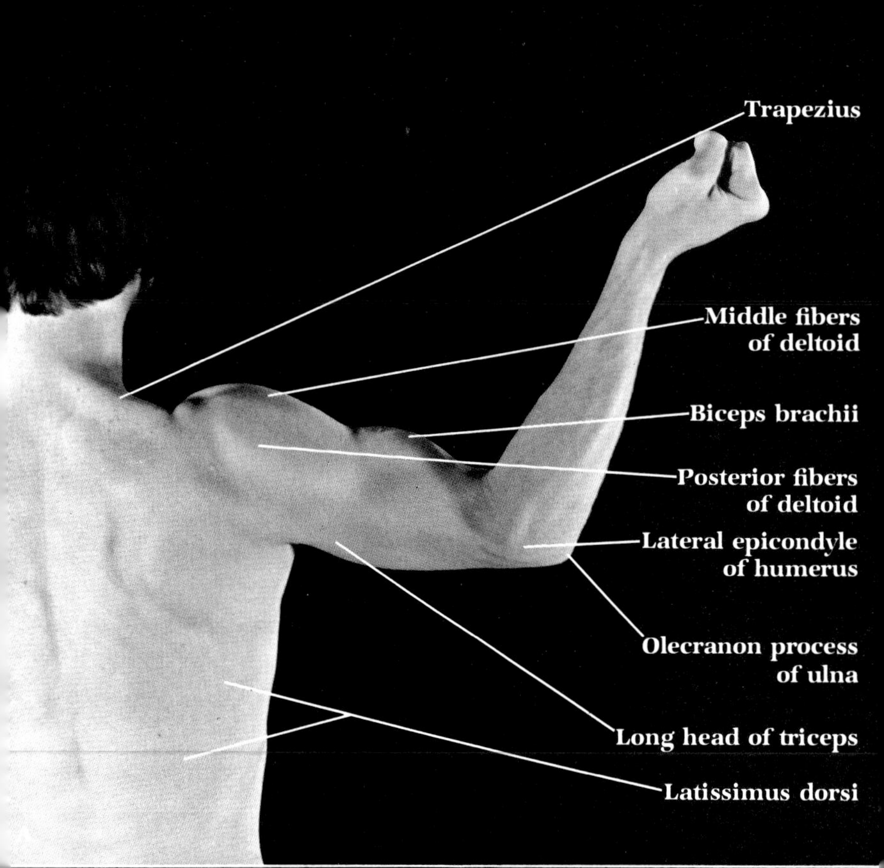
Trapezius
Middle fibers of deltoid
Biceps brachii
Posterior fibers of deltoid
Lateral epicondyle of humerus
Olecranon process of ulna
Long head of triceps
Latissimus dorsi

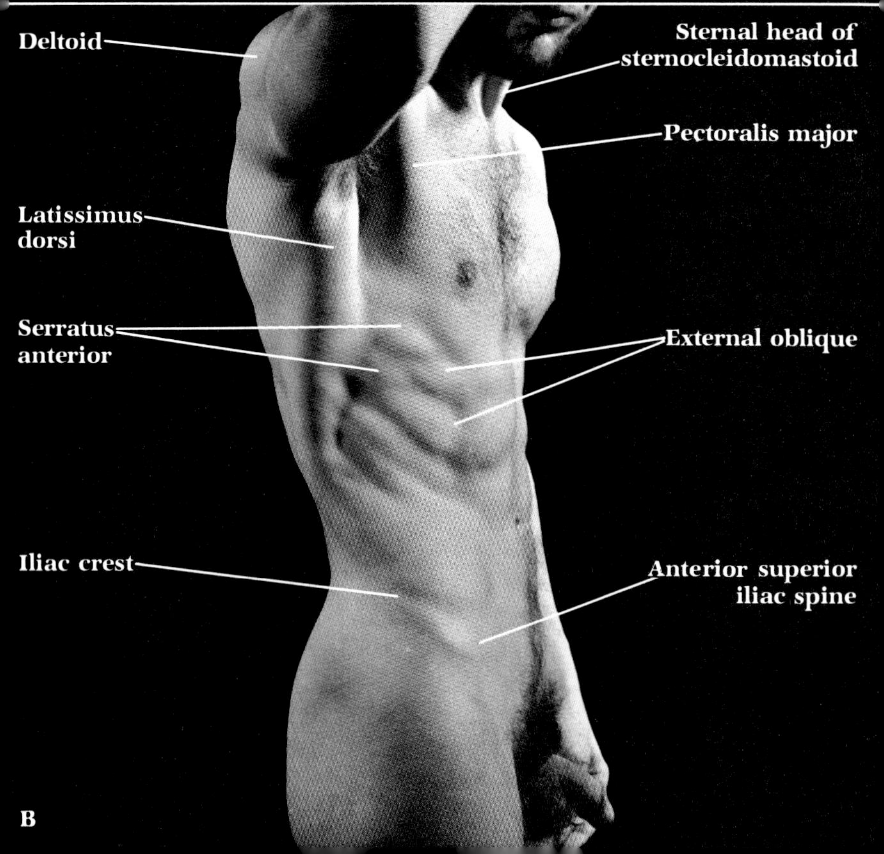
Deltoid
Sternal head of sternocleidomastoid
Pectoralis major
Latissimus dorsi
Serratus anterior
External oblique
Iliac crest
Anterior superior iliac spine

B

Figure 5-19
The upper extremity in a 27-year-old male. A. The shoulder region with the shoulder joint abducted to nearly a right angle, posterior view. B. The axilla and anterolateral view of the trunk.

1. The trapezius muscle forms the sloping contour to the neck as it passes downward from the skull to the scapula.
2. The deltoid muscle forms the smooth rounded curve of the shoulder and covers the greater tuberosity of the humerus.
3. The biceps brachii muscle forms the anterior and the triceps the posterior bulge in the arm. The biceps muscle can be made more prominent by asking the patient to flex the elbow joint against resistance. In a similar manner the triceps can be made more prominent by extending the elbow joint against resistance.
4. The latissimus dorsi muscle covers the lateral part of the thoracic cage. With the teres major it forms the posterior axillary fold. The latissimus dorsi can be made more prominent by asking the patient to extend the shoulder joint against resistance.
5. The origin of the serratus anterior muscle from the upper eight ribs can be seen on the lateral thoracic wall. The slips of origin interdigitate with those of the external oblique muscle of the anterior abdominal wall.
6. The iliac crests are palpable along their entire length. Each crest ends anteriorly at the anterior superior iliac spine.

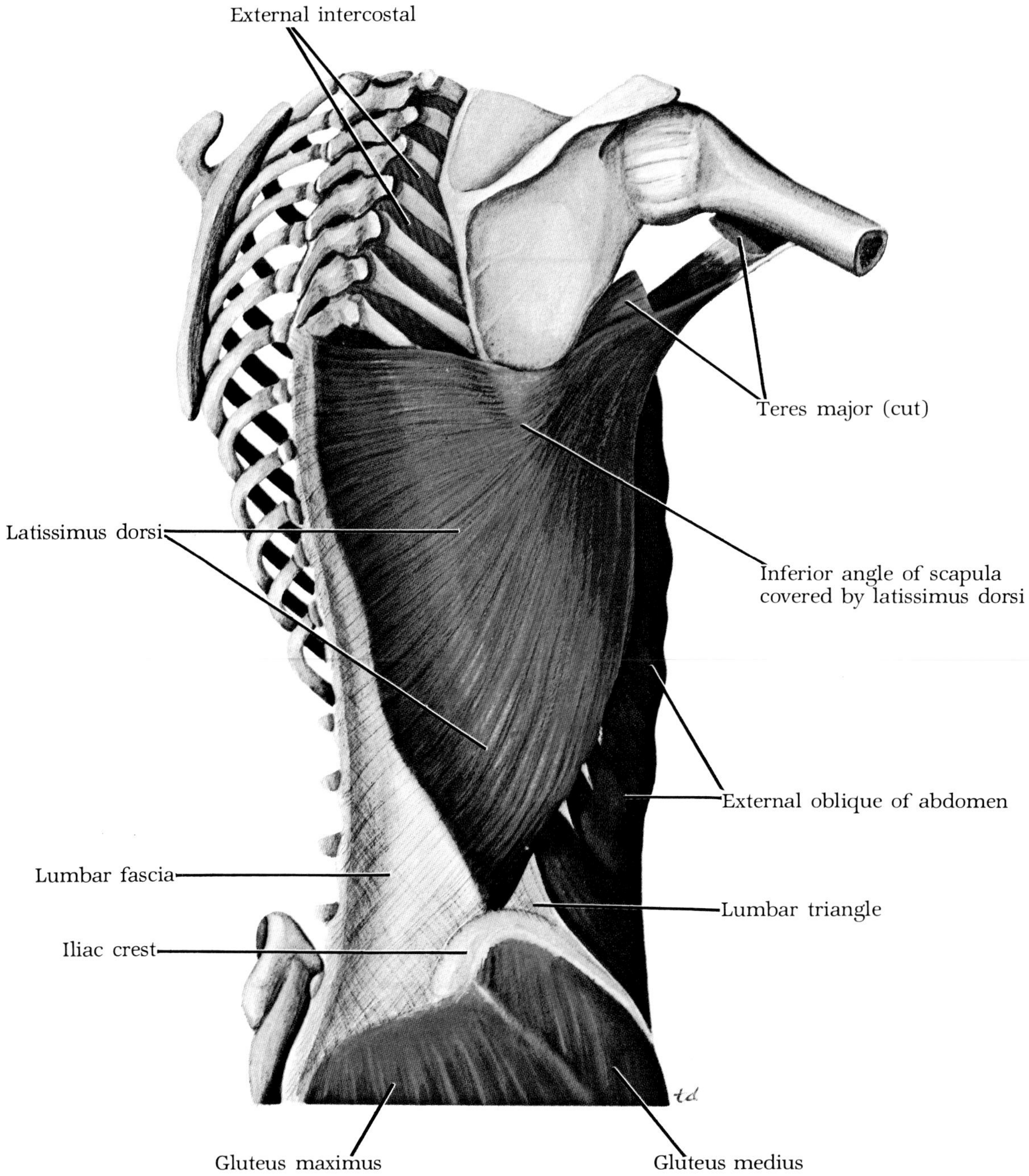
External intercostal
Teres major (cut)
Latissimus dorsi
Inferior angle of scapula covered by latissimus dorsi
External oblique of abdomen
Lumbar fascia
Lumbar triangle
Iliac crest
Gluteus maximus
Gluteus medius

Figure 5-20
Latissimus dorsi muscle, posterolateral view.

1. The latissimus dorsi is a broad, thin sheet of muscle that runs superiorly and laterally to converge on the axilla. Here it twists upon itself as it winds around the lower border of the teres major muscle to form the posterior axillary fold. Its insertion into the floor of the bicipital groove of the humerus enables it to extend, adduct, and medially rotate the arm at the shoulder joint.
2. The muscles of the two sides may be palpated by running the four fingers of each hand in a sweeping fashion over the posterior aspect of the muscle downward to the iliac crests. The lateral edge may be palpated when the patient is asked to extend and adduct the shoulder joint against resistance.
3. The latissimus dorsi muscle is rarely damaged.

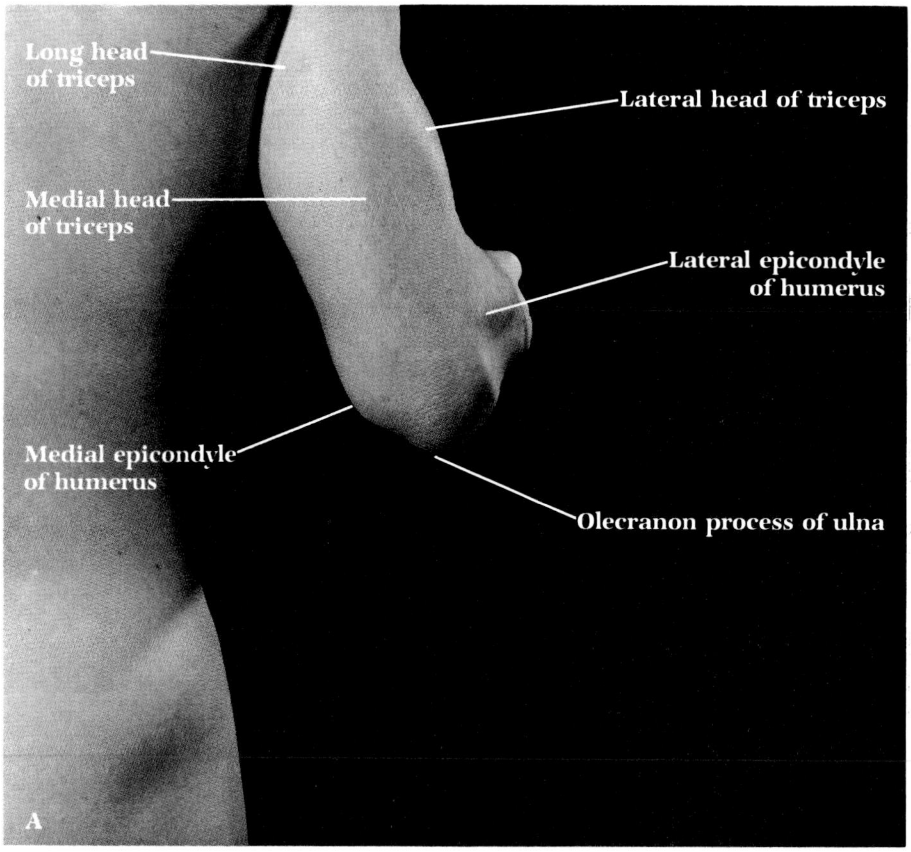
Long head
of triceps
Lateral head of triceps
Medial head
of triceps
Lateral epicondyle
of humerus
Medial epicondyle
of humerus
Olecranon process of ulna
A

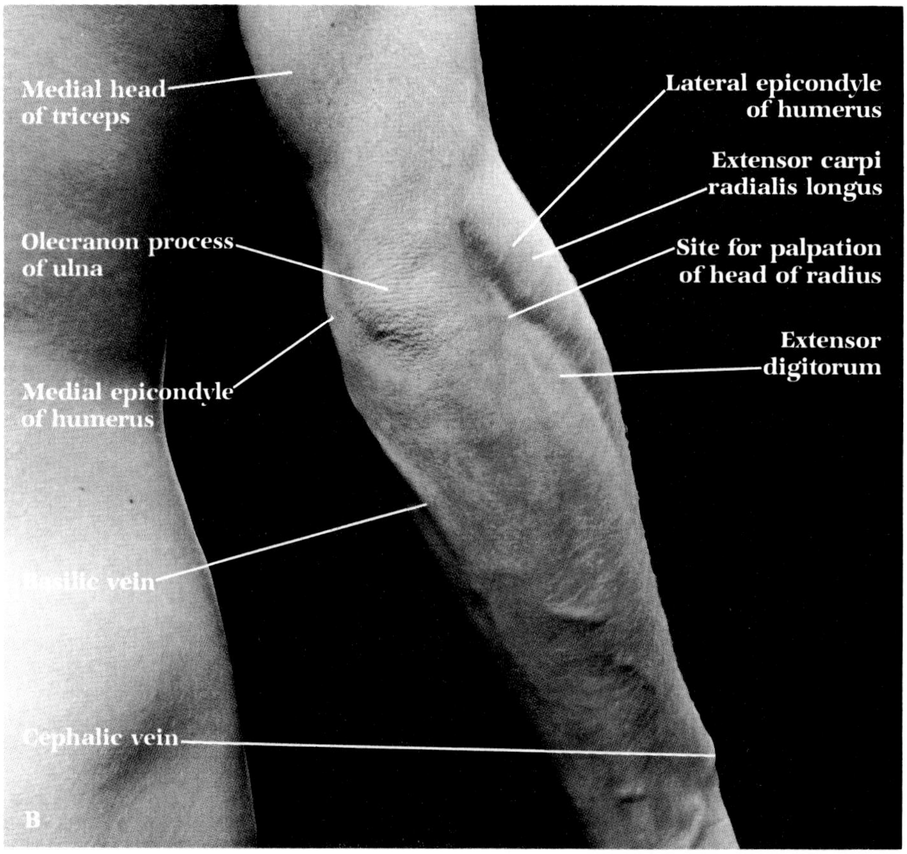
Medial head
of triceps
Lateral epicondyle
of humerus
Extensor carpi
radialis longus
Olecranon process
of ulna
Site for palpation
of head of radius
Extensor
digitorum
Medial epicondyle
of humerus
Basilic vein
Cephalic vein
B

Figure 5-21
The elbow region in a 27-year-old male, posterior view.

1. The triceps muscle forms the muscular mass on the back of the arm. The three heads form a conjoined tendon that is inserted into the olecranon process of the ulna. The muscle swelling is made more prominent by asking the patient to extend the elbow joint against resistance.
2. In the region of the elbow, the medial and lateral epicondyles of the humerus and the olecranon process of the ulna can be palpated. When the elbow joint is extended, these bony points lie on the same straight line; when the elbow is flexed, these three points form the boundaries of an equilateral triangle.
3. The head of the radius can be palpated in a depression on the posterolateral aspect of the extended elbow, distal to the lateral epicondyle. The head of the radius can be felt to rotate during pronation and supination of the forearm.

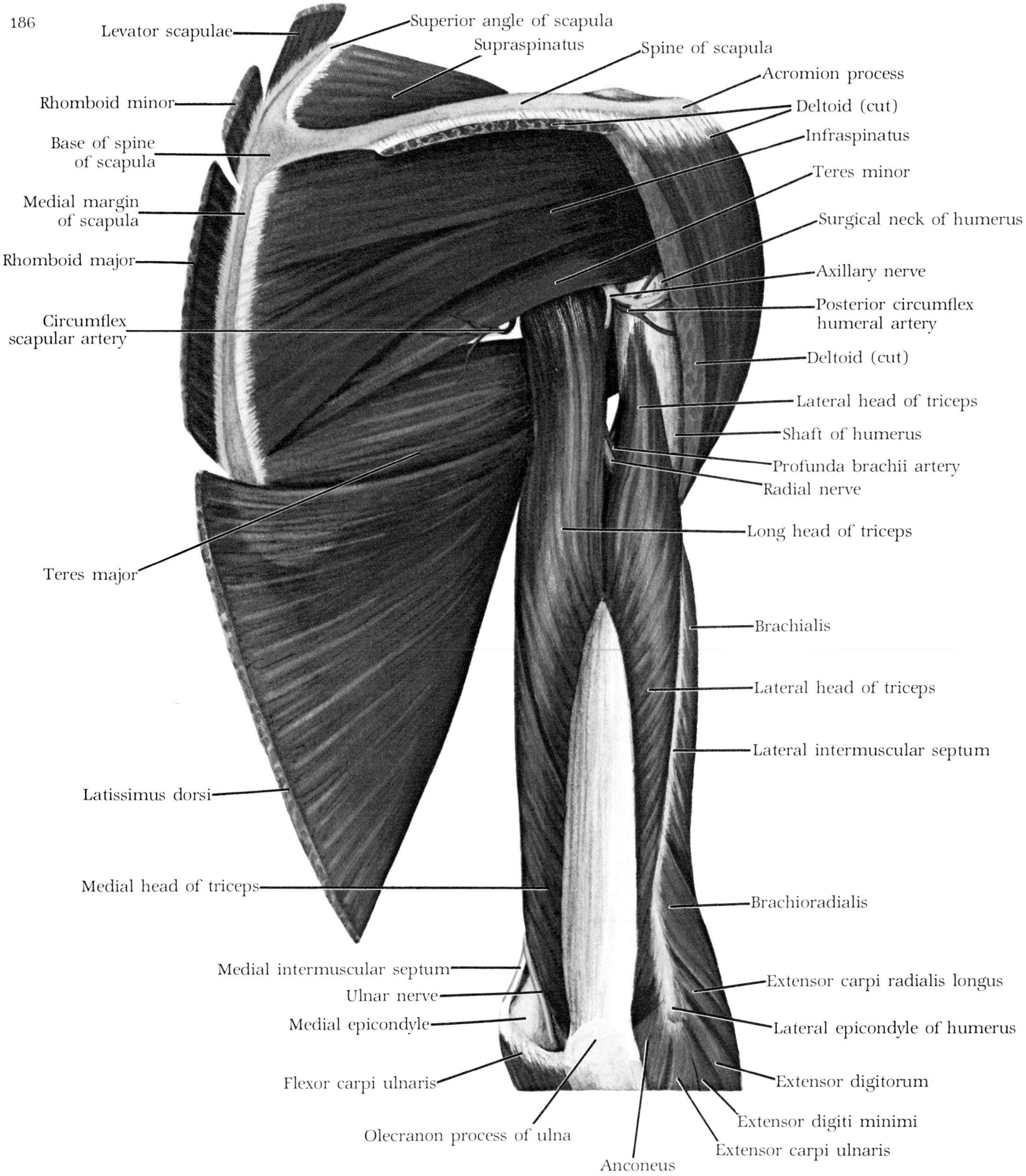
Levator scapulae
Superior angle of scapula
Supraspinatus
Spine of scapula
Acromion process
Deltoid (cut)
Infraspinatus
Teres minor
Surgical neck of humerus
Axillary nerve
Posterior circumflex humeral artery
Deltoid (cut)
Lateral head of triceps
Shaft of humerus
Profunda brachii artery
Radial nerve
Long head of triceps
Brachialis
Lateral head of triceps
Lateral intermuscular septum
Brachioradialis
Extensor carpi radialis longus
Lateral epicondyle of humerus
Extensor digitorum
Extensor digiti minimi
Extensor carpi ulnaris
Anconeus
Olecranon process of ulna
Flexor carpi ulnaris
Medial epicondyle
Ulnar nerve
Medial intermuscular septum
Medial head of triceps
Latissimus dorsi
Teres major
Circumflex scapular artery
Rhomboid major
Medial margin of scapula
Base of spine of scapula
Rhomboid minor

Figure 5-22
Muscles, nerves, and blood vessels of scapular region.

1. The tendons of the supraspinatus, infraspinatus, and teres minor muscles together with that of the subscapularis muscle form the rotator cuff that is fused to the underlying capsule of the shoulder joint. This cuff is very important in stabilizing the shoulder joint. Lesions of the cuff are a common cause of pain in the shoulder region. During the movement of abduction of the shoulder joint, the supraspinatus tendon is exposed to friction by rubbing against the acromion process. Under normal conditions the amount of friction is reduced to a minimum by the large subacromial bursa, which extends laterally beneath the deltoid muscle. Degenerative changes in the bursa are followed by degenerative changes in the underlying supraspinatus tendon; these may extend into the other tendons of the rotator cuff, resulting in a condition known as pericapsulitis.

2. The axillary nerve as it passes posteriorly through the quadrilateral space is particularly vulnerable to injury from downward displacement of the humeral head in a shoulder dislocation or fracture of the surgical neck of the humerus. Paralysis of the deltoid and teres minor muscles results, and there is a loss of skin sensation over the lower half of the deltoid muscle. The paralyzed deltoid wastes rapidly, and the underlying greater tuberosity can be readily palpated. Because the supraspinatus is the only other abductor of the shoulder, the movement of abduction is much impaired by axillary nerve injury.

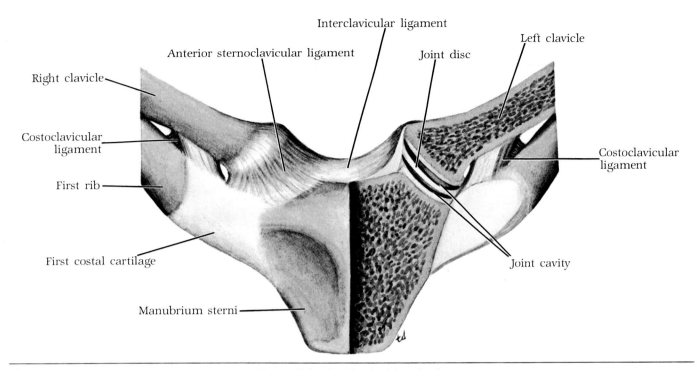

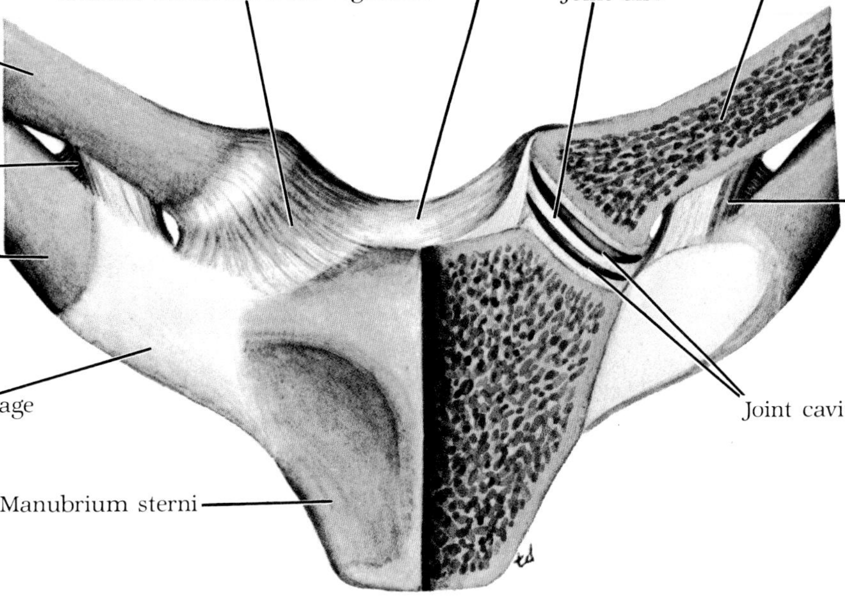
Interclavicular ligament
Left clavicle
Anterior sternoclavicular ligament
Joint disc
Right clavicle
Costoclavicular ligament
Costoclavicular ligament
First rib
First costal cartilage
Joint cavity
Manubrium sterni

Figure 5-23
Sternoclavicular joint. On the right, the clavicle and sternum have been cut and the joint opened.

1. The medial end of the clavicle is firmly held to the first costal cartilage by the very strong costoclavicular ligament. Violent forces directed along the long axis of the clavicle, if they do not result in a fracture of that bone, occasionally cause a dislocation of the sternoclavicular joint. Although the clavicle is commonly displaced anteriorly through the capsule, the dislocations occasionally are posterior. When this occurs, the medial end of the clavicle may compress the important neurovascular structures in the root of the neck. Should the costoclavicular ligament rupture completely, it is difficult to maintain the normal position of the clavicle, even after reduction has been accomplished.

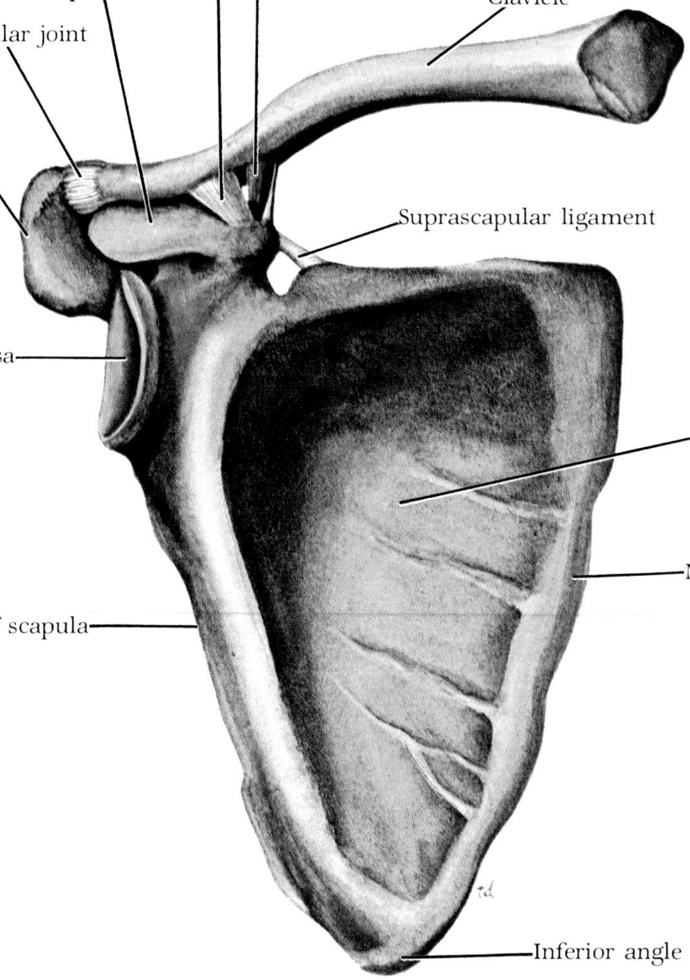

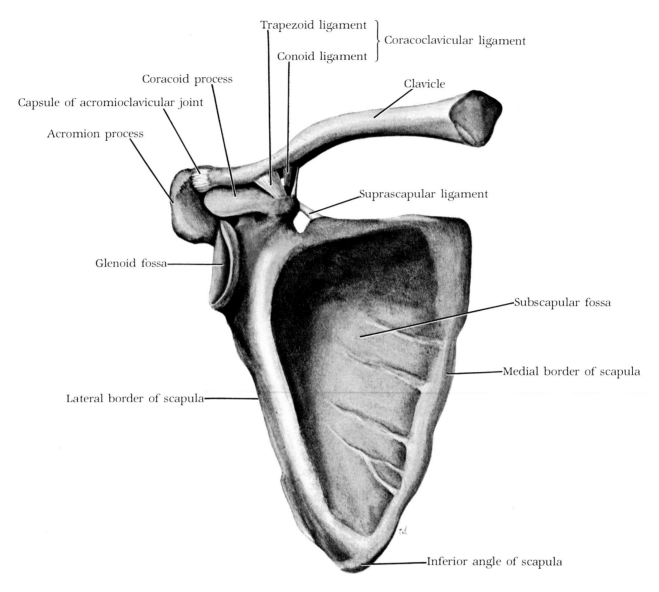

Trapezoid ligament
Conoid ligament
Coracoclavicular ligament
Coracoid process
Clavicle
Capsule of acromioclavicular joint
Acromion process
Suprascapular ligament
Glenoid fossa
Subscapular fossa
Medial border of scapula
Lateral border of scapula
Inferior angle of scapula

A

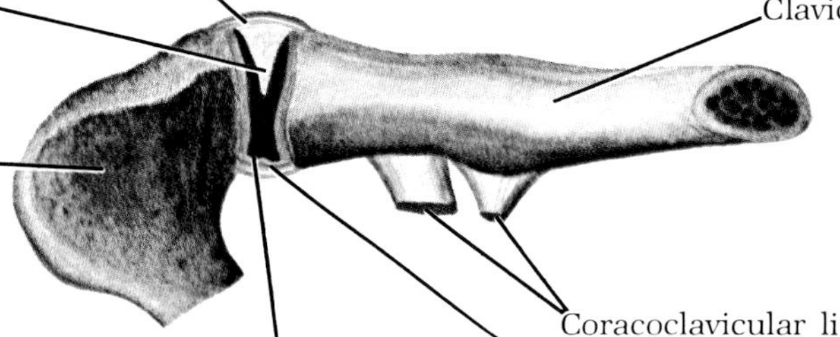

Capsule of acromioclavicular joint
Disc of acromioclavicular joint
Clavicle (cut)
Acromion process of scapula
Coracoclavicular ligament
Joint cavity
Synovial membrane

B

Figure 5-24
Acromioclavicular joints.
A. Anterior view. B. With joint opened, anterior view.

1. The plane of the articular surfaces of the acromioclavicular joint passes downward and medially, so that there is a tendency for the lateral end of the clavicle to ride up over the superior surface of the acromion. The strength of the joint depends on the very strong coracoclavicular ligament, which binds the coracoid process to the undersurface of the lateral part of the clavicle.

2. A severe blow on the point of the shoulder, as incurred during blocking or tackling in football or from any severe fall, may result in the acromion's being thrust beneath the lateral end of the clavicle, tearing the coracoclavicular ligament. The displaced lateral end of the clavicle is then easily palpable.

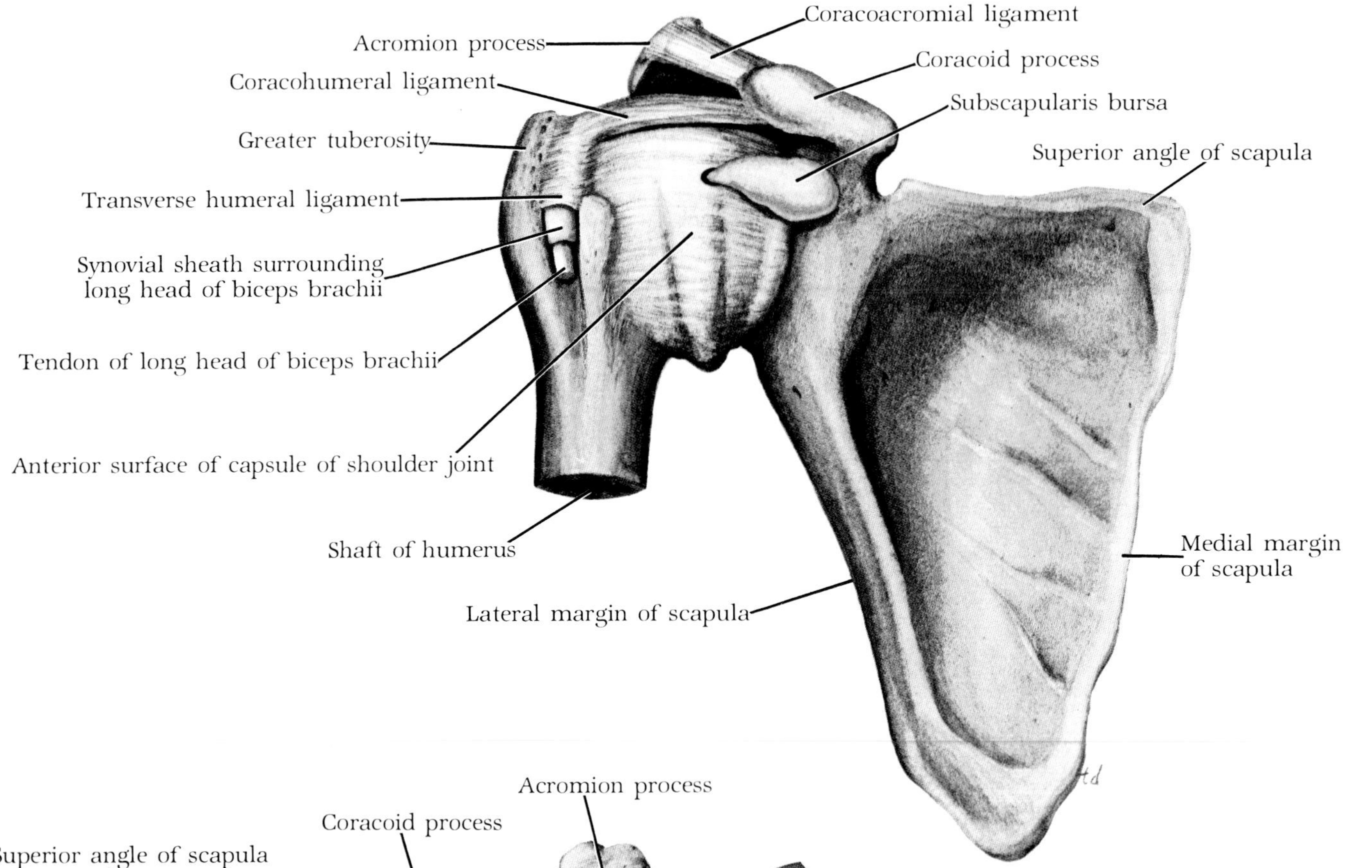
Coracoacromial ligament
Acromion process
Coracoid process
Coracohumeral ligament
Subscapularis bursa
Greater tuberosity
Superior angle of scapula
Transverse humeral ligament
Synovial sheath surrounding long head of biceps brachii
Tendon of long head of biceps brachii
Anterior surface of capsule of shoulder joint
Shaft of humerus
Medial margin of scapula
Lateral margin of scapula

A

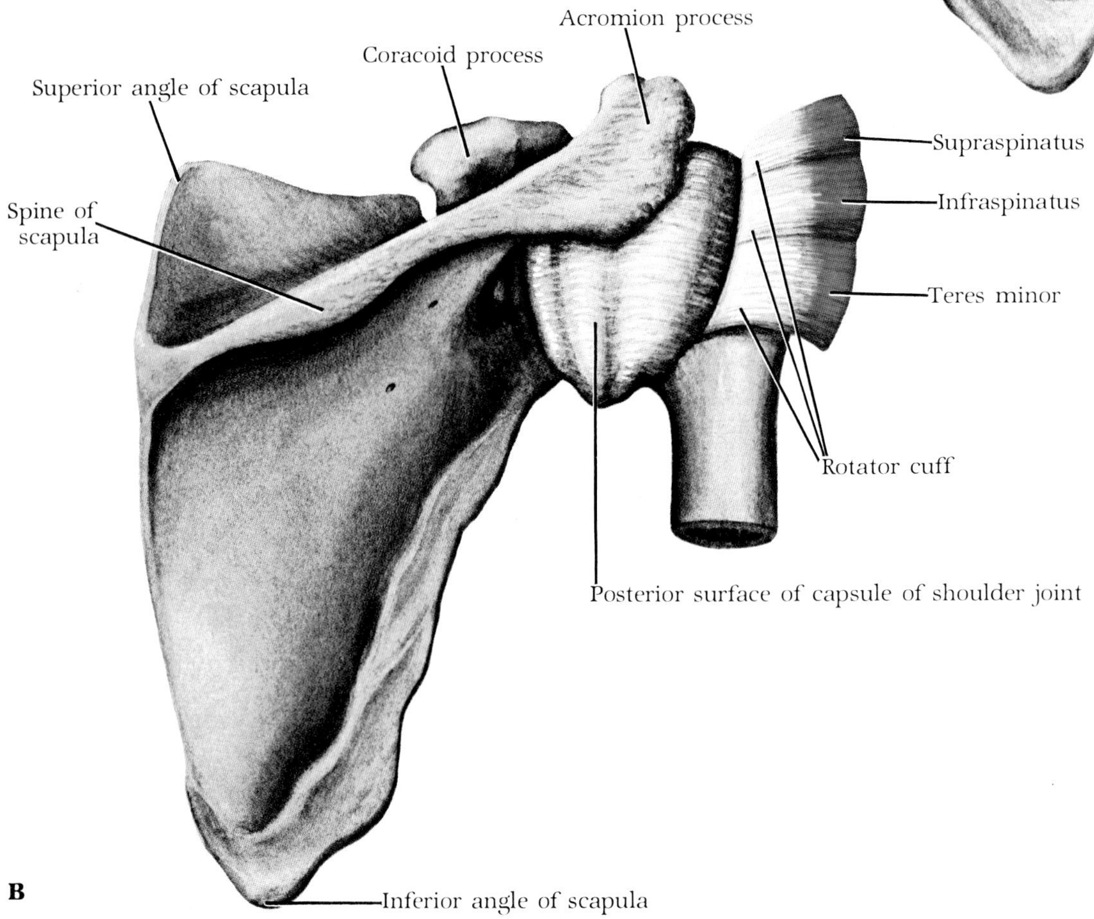
Acromion process
Coracoid process
Superior angle of scapula
Supraspinatus
Spine of scapula
Infraspinatus
Teres minor
Rotator cuff
Posterior surface of capsule of shoulder joint
Inferior angle of scapula

B

Figure 5-25
Shoulder joint. A. Anterior view.
B. Posterior view.

1. The biceps brachii muscle arises by a long head from the supraglenoid tubercle within the shoulder joint and by a short head from the tip of the coracoid process. The tendon of either head may rupture during violent movements.

2. The strength of the shoulder joint largely depends on the tone of the short muscles that surround it, namely the subscapularis anteriorly, the supraspinatus superiorly, and the infraspinatus and teres minor muscles posteriorly. The least protected and therefore the weakest part of the joint is inferiorly, where there are no short muscles.

3. The shoulder joint is the most commonly dislocated large joint. When sudden violence is applied to the humerus with the joint fully abducted, the humeral head is tilted inferiorly onto the lower weak part of the capsule, which tears, and the humeral head then comes to lie inferior to the glenoid fossa. During this movement the acromion has acted as a fulcrum. The strong flexors and adductors of the shoulder joint then usually pull the humeral head forward and upward into the subcoracoid position. Posterior dislocations are rare and are usually due to direct violence to the front of the joint.

4. The tendons of the subscapularis, supraspinatus, infraspinatus, and teres minor muscles together form the rotator cuff. They are fused to the underlying capsule of the shoulder joint and strengthen it. Degenerative changes in the rotator cuff produce the painful condition known as pericapsulitis.

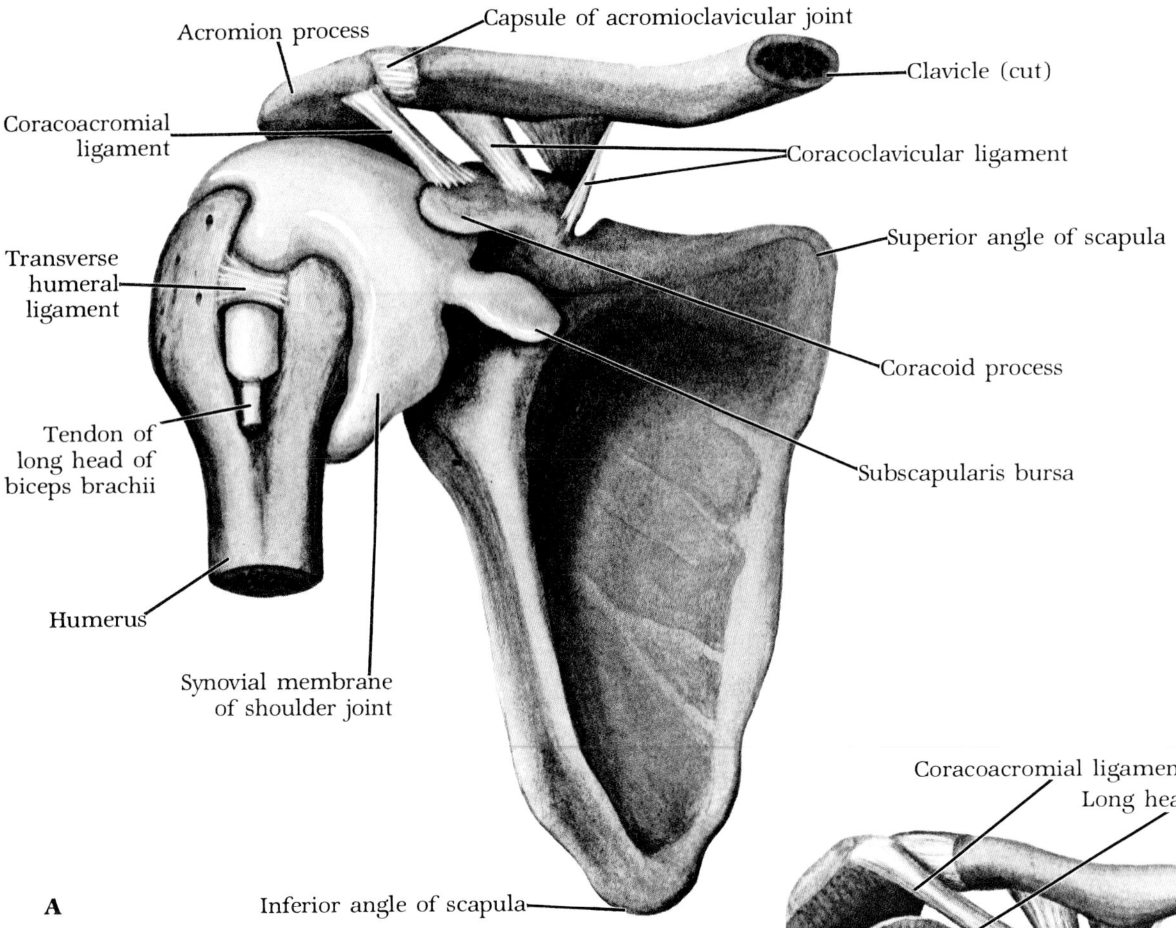
Acromion process
Capsule of acromioclavicular joint
Clavicle (cut)
Coracoacromial ligament
Coracoclavicular ligament
Superior angle of scapula
Transverse humeral ligament
Coracoid process
Tendon of long head of biceps brachii
Subscapularis bursa
Humerus
Synovial membrane of shoulder joint
Inferior angle of scapula

A

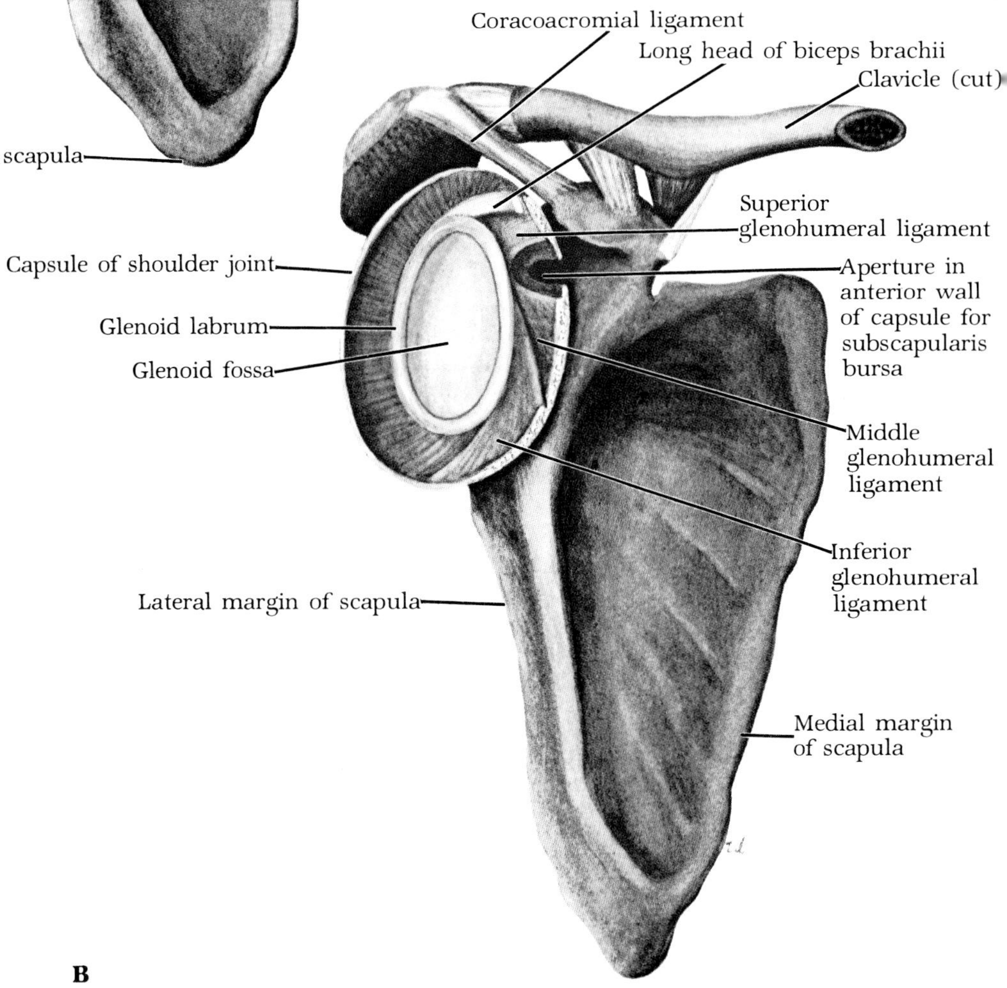
Coracoacromial ligament
Long head of biceps brachii
Clavicle (cut)
Superior glenohumeral ligament
Capsule of shoulder joint
Aperture in anterior wall of capsule for subscapularis bursa
Glenoid labrum
Glenoid fossa
Middle glenohumeral ligament
Inferior glenohumeral ligament
Lateral margin of scapula
Medial margin of scapula

B

Figure 5-26
Shoulder joint. A. Synovial membrane, anterior view. B. Glenoid fossa and capsule, lateral view. The humerus has been removed.

1. The synovial membrane of the shoulder joint is extensive. It lines the fibrous capsule and is attached to the margins of the articular surfaces of the humerus and scapula. It is prolonged around the tendon of the long head of the biceps as a synovial sheath that is held in position by the transverse humeral ligament; it is continuous with the subscapularis bursa.
2. Note the shallowness of the glenoid fossa of the scapula and the weak ligaments. It is this lack of support that makes this joint a very unstable structure.
3. The synovial membrane, capsule, and ligaments are innervated by the axillary nerve and the suprascapular nerve. The joint is sensitive to pain, pressure, excessive traction, and distention. The muscles surrounding the joint undergo reflex spasm in response to pain originating in the joint, which in turn serves to immobilize the joint and thus reduce the pain.
4. Injury to the shoulder joint is followed by pain, limitation of movement, and muscle atrophy due to disuse. It is important to appreciate that pain in the shoulder region may be caused by disease elsewhere and that the shoulder joint may be normal; for example, diseases of the spinal cord and vertebral column, and the pressure of a cervical rib can all cause shoulder pain. Irritation of the diaphragmatic pleura or peritoneum can produce referred pain via the phrenic and supraclavicular nerves.

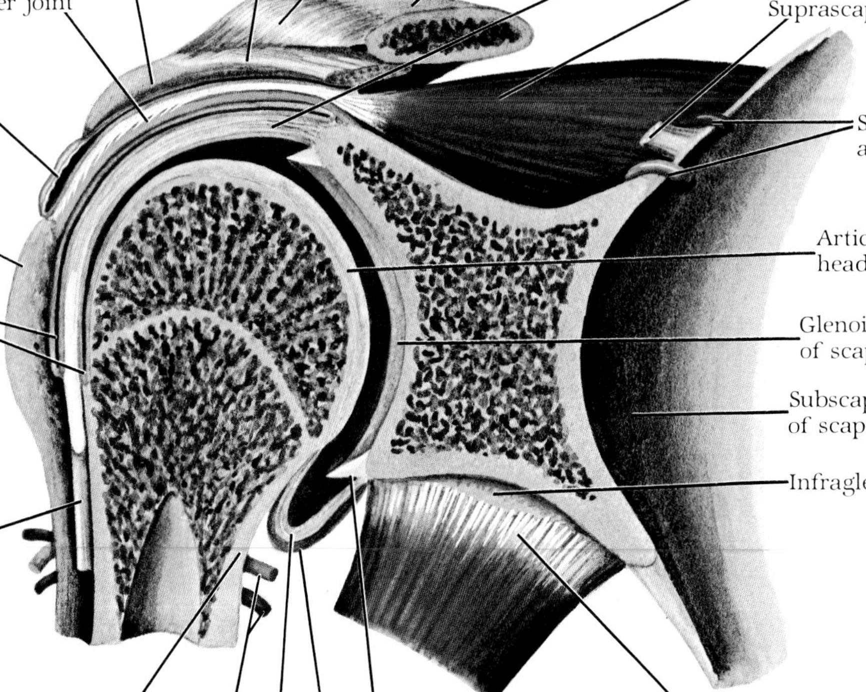
Coracoacromial ligament
Capsule of acromioclavicular joint
Clavicle (cut)
Acromion process
Tendon of long head of biceps brachii
Tendon of supraspinatus fused with capsule of shoulder joint
Supraspinatus
Suprascapular ligament (cut)
Subacromial bursa
Suprascapular artery and nerve
Greater tuberosity
Articular surface of head of humerus
Synovial sheath surrounding tendon of long head of biceps brachii
Glenoid fossa of scapula
Subscapular fossa of scapula
Infraglenoid tubercle
Tendon of long head of biceps brachii
Glenoid labrum
Long head of triceps
Slack capsule of shoulder joint
Synovial membrane
Axillary nerve and posterior circumflex humeral artery
Surgical neck of humerus

Figure 5-27
Shoulder joint; coronal section.

1. The function of the supraspinatus primarily is to keep the head of the humerus against the glenoid cavity and to assist the deltoid muscle in the movement of abduction of the arm at the shoulder joint.

2. A patient who ruptures the tendon of the supraspinatus while suffering from supraspinatus tendinitis is unable to initiate abduction of the arm. If the arm is passively assisted for the first 15 degrees of abduction, however, the deltoid can then take over and complete the movement to a right angle.

3. The tendon of the long head of the biceps brachii is intracapsular but extrasynovial; it is attached to the supraglenoid tubercle within the shoulder joint. Advanced osteoarthritic changes in the joint may lead to erosion and fraying of the tendon by osteophytic outgrowths, and rupture of the tendon can occur.

4. Note again the shallowness of the glenoid fossa and the common occurrence of shoulder joint dislocation.

5. Most dislocations of the shoulder joint start by the humerus moving inferiorly to begin with. Later, the strong flexors and adductor muscles pull the head of the humerus forward and upward into the subcoracoid position. Dislocation of the shoulder joint or a fracture dislocation in which the surgical neck of the humerus is fractured can also damage the axillary nerve as it passes through the quadrilateral space.

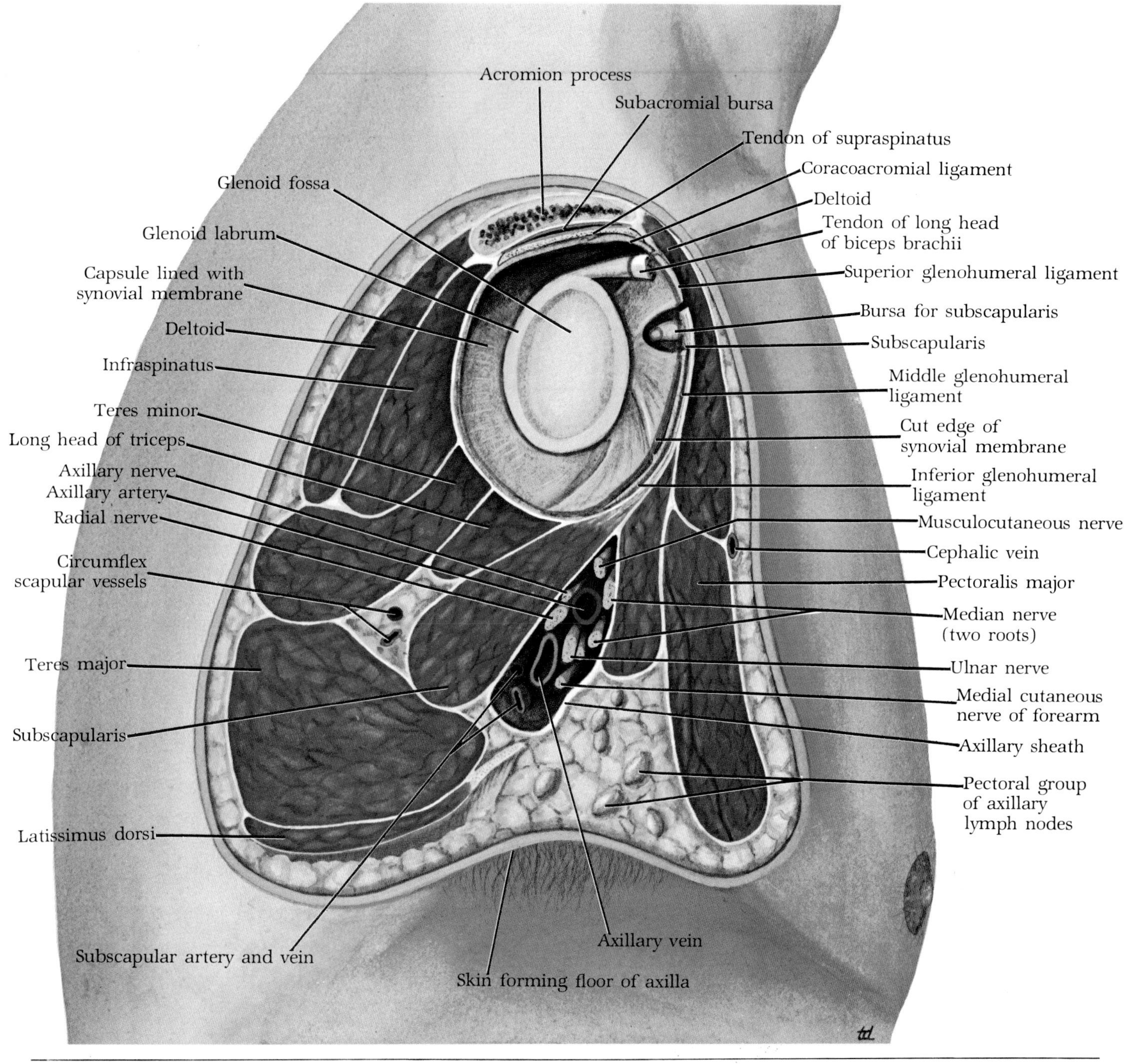
Acromion process
Subacromial bursa
Tendon of supraspinatus
Coracoacromial ligament
Deltoid
Tendon of long head of biceps brachii
Superior glenohumeral ligament
Bursa for subscapularis
Subscapularis
Middle glenohumeral ligament
Cut edge of synovial membrane
Inferior glenohumeral ligament
Musculocutaneous nerve
Cephalic vein
Pectoralis major
Median nerve (two roots)
Ulnar nerve
Medial cutaneous nerve of forearm
Axillary sheath
Pectoral group of axillary lymph nodes
Glenoid fossa
Glenoid labrum
Capsule lined with synovial membrane
Deltoid
Infraspinatus
Teres minor
Long head of triceps
Axillary nerve
Axillary artery
Radial nerve
Circumflex scapular vessels
Teres major
Subscapularis
Latissimus dorsi
Subscapular artery and vein
Axillary vein
Skin forming floor of axilla
td

Figure 5-28
Shoulder joint, showing its relations, sagittal section.

1. The tone of the short muscles that surround the shoulder joint holds the rounded head of the humerus in the shallow glenoid cavity. The absence of a deep bony cup and strong ligaments, as in the hip joint, makes the shoulder joint very mobile, but because of its lack of strength it is the most commonly dislocated large joint in the body.
2. A subglenoid displacement of the head of the humerus into the quadrilateral space can cause damage to the axillary nerve.
3. Deep pressure of the examiner's finger on the skin over the anterior part of the deltoid muscle will exert pressure on the anterior part of the rotator cuff and joint capsule. When the patient's joint is passively extended, the subacromial bursa and the supraspinatus tendon are made to move anteriorly so that they can be palpated through the deltoid muscle, just anterior to the acromion process. Deep tenderness may be due to tendinitis, tears of the tendon, or bursitis.

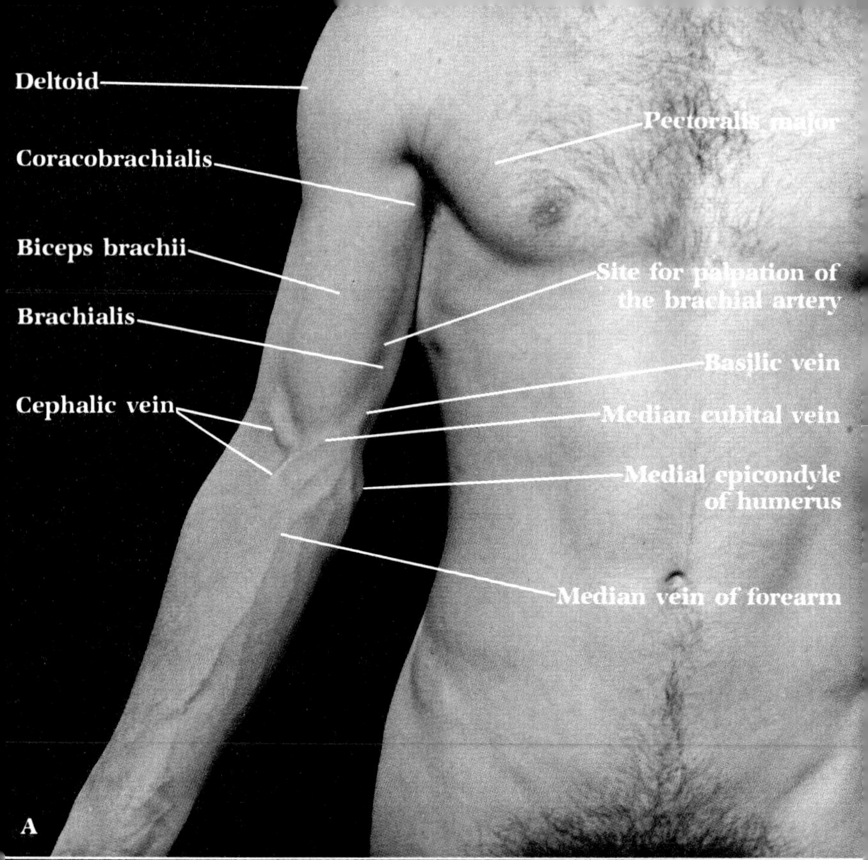
Deltoid
Coracobrachialis
Biceps brachii
Brachialis
Cephalic vein
Pectoralis major
Site for palpation of the brachial artery
Basilic vein
Median cubital vein
Medial epicondyle of humerus
Median vein of forearm
A

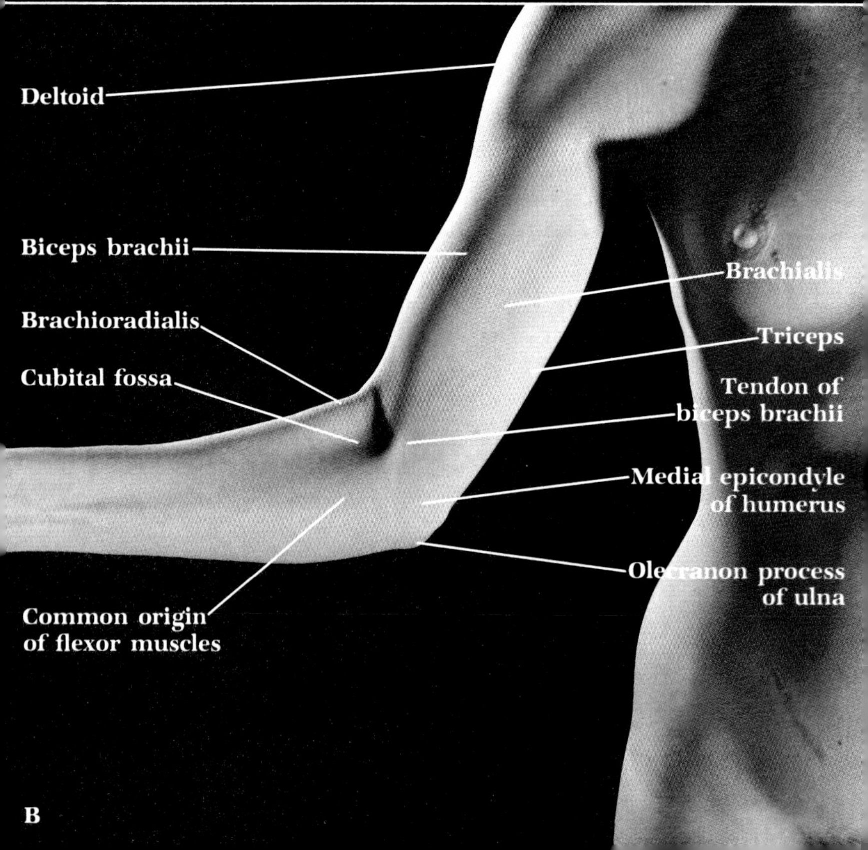
Deltoid
Biceps brachii
Brachioradialis
Cubital fossa
Common origin of flexor muscles
Brachialis
Triceps
Tendon of biceps brachii
Medial epicondyle of humerus
Olecranon process of ulna
B

Figure 5-29
Right arm and cubital fossa.
A. Anterior view in a 27-year-old male. B. Anteromedial view in a 29-year-old female.

1. The curve of the shoulder is formed by the deltoid muscle covering the underlying greater tuberosity of the humerus. If the physician places his hand over the deltoid muscle, the anterior fibers will be felt to contract when the patient flexes the shoulder joint; the middle fibers will be felt to contract when the joint is abducted; and the posterior fibers will be felt to contract when the shoulder joint is extended.
2. The anterior axillary fold is formed by the pectoralis major muscle. Adduction of the shoulder joint against resistance makes the muscle mass more prominent.
3. The biceps brachii forms a large fusiform muscle mass in the anterior part of the arm. If the patient flexes the elbow joint, or supinates the forearm at the superior and inferior radioulnar joints, against resistance, the muscle actively contracts and becomes more prominent.
4. The boundaries of the cubital fossa can be seen and felt; the brachioradialis muscle forms the lateral boundary and the pronator teres muscle forms the medial boundary. The tendon of the biceps muscle can be palpated as it passes downward into the fossa; the bicipital aponeurosis can be felt as it leaves the tendon to join the deep fascia on the medial side of the forearm.
5. The brachial artery can be felt to pulsate as it passes down the arm, overlapped by the medial border of the biceps muscle. In the cubital fossa it lies beneath the bicipital aponeurosis; at a level just below the head of the radius, it divides into its terminal branches.
6. The cephalic vein is seen on the anterior aspect of the forearm. It then ascends into the arm and runs along the lateral border of the biceps brachii muscle.
7. The basilic vein reaches the anterior aspect of the forearm just below the elbow. It then ascends into the arm and runs along the medial border of the biceps muscle. The median cubital vein (or median cephalic and median basilic veins) links the cephalic and basilic veins in the cubital fossa.

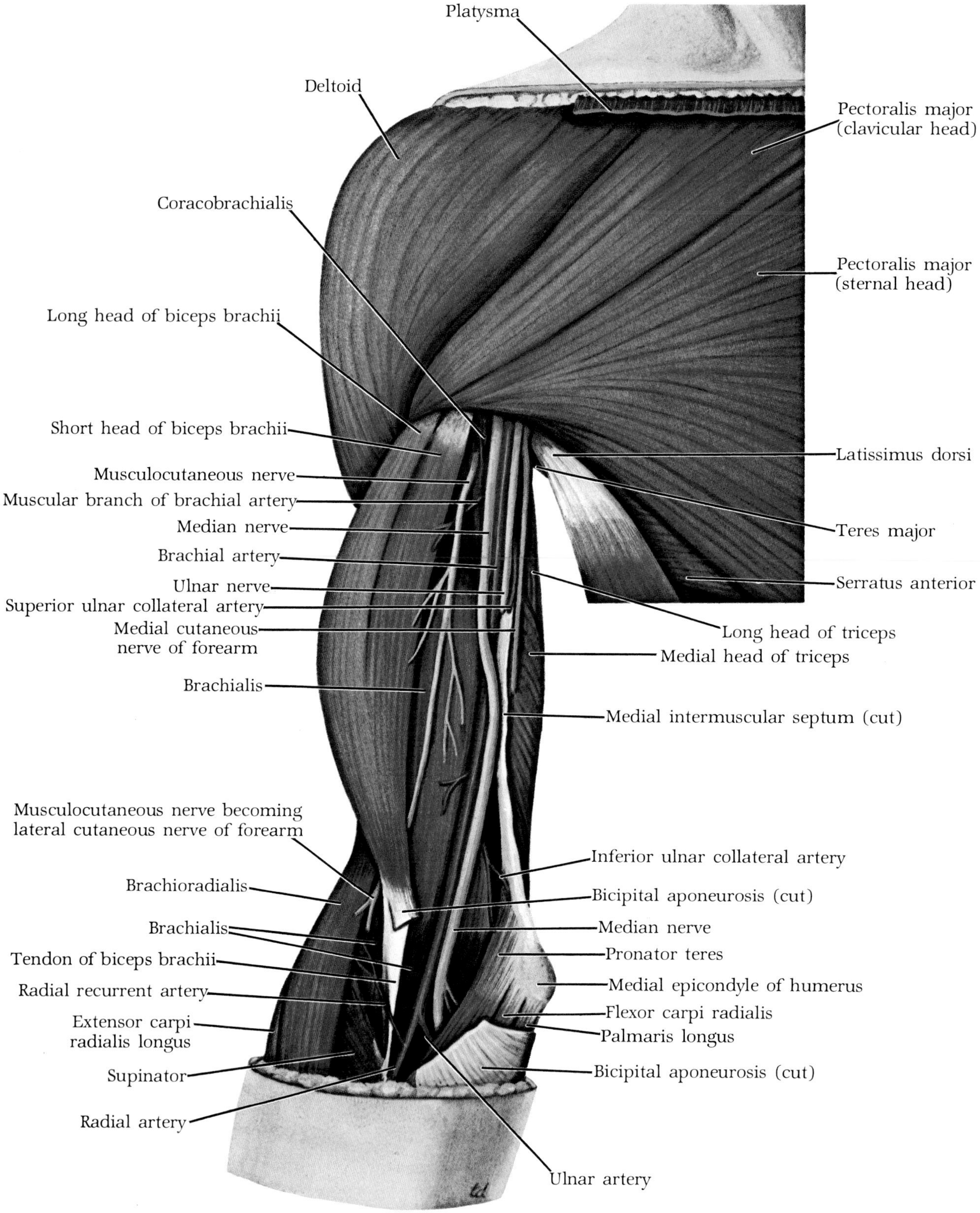
Platysma
Deltoid
Pectoralis major (clavicular head)
Coracobrachialis
Pectoralis major (sternal head)
Long head of biceps brachii
Short head of biceps brachii
Latissimus dorsi
Musculocutaneous nerve
Muscular branch of brachial artery
Teres major
Median nerve
Brachial artery
Serratus anterior
Ulnar nerve
Superior ulnar collateral artery
Long head of triceps
Medial cutaneous nerve of forearm
Medial head of triceps
Brachialis
Medial intermuscular septum (cut)
Musculocutaneous nerve becoming lateral cutaneous nerve of forearm
Inferior ulnar collateral artery
Brachioradialis
Bicipital aponeurosis (cut)
Brachialis
Median nerve
Tendon of biceps brachii
Pronator teres
Radial recurrent artery
Medial epicondyle of humerus
Extensor carpi radialis longus
Flexor carpi radialis
Palmaris longus
Supinator
Bicipital aponeurosis (cut)
Radial artery
Ulnar artery

Figure 5-30
Upper arm, anterior view. The biceps brachii muscle has been pulled slightly laterally to display the musculocutaneous nerve.

1. One or both heads of the biceps brachii muscle occasionally rupture, usually after a sudden extension of the flexed elbow joint or the lifting of heavy weights. The tendon of the long head of the biceps can be involved in degenerative disease of the shoulder joint and can rupture spontaneously.

2. Arteries of the upper limb can be damaged by penetrating wounds or may require ligation during amputation operations. If there is adequate collateral circulation around the shoulder and elbow, ligation of the axillary and brachial arteries can be performed without tissue necrosis or gangrene, provided (a) the arteries forming the collateral circulation are not diseased, and (b) the patient's general circulation is satisfactory.

3. Remember that the third part of the axillary artery can be felt in the axilla as it lies in front of the teres major muscle. The brachial artery can be palpated in the arm as it lies on the brachialis muscle and is overlapped from the lateral side by the biceps brachii muscle.

4. Volkmann's ischemic contracture is a contracture of the forearm muscles following a fracture of the lower end of the humerus. A localized segment of the brachial artery goes into spasm, reducing the arterial flow to the flexor and extensor muscles, resulting in ischemic necrosis. The arterial spasm is usually caused by an overtight plaster cast, but in some cases the fracture itself is responsible.

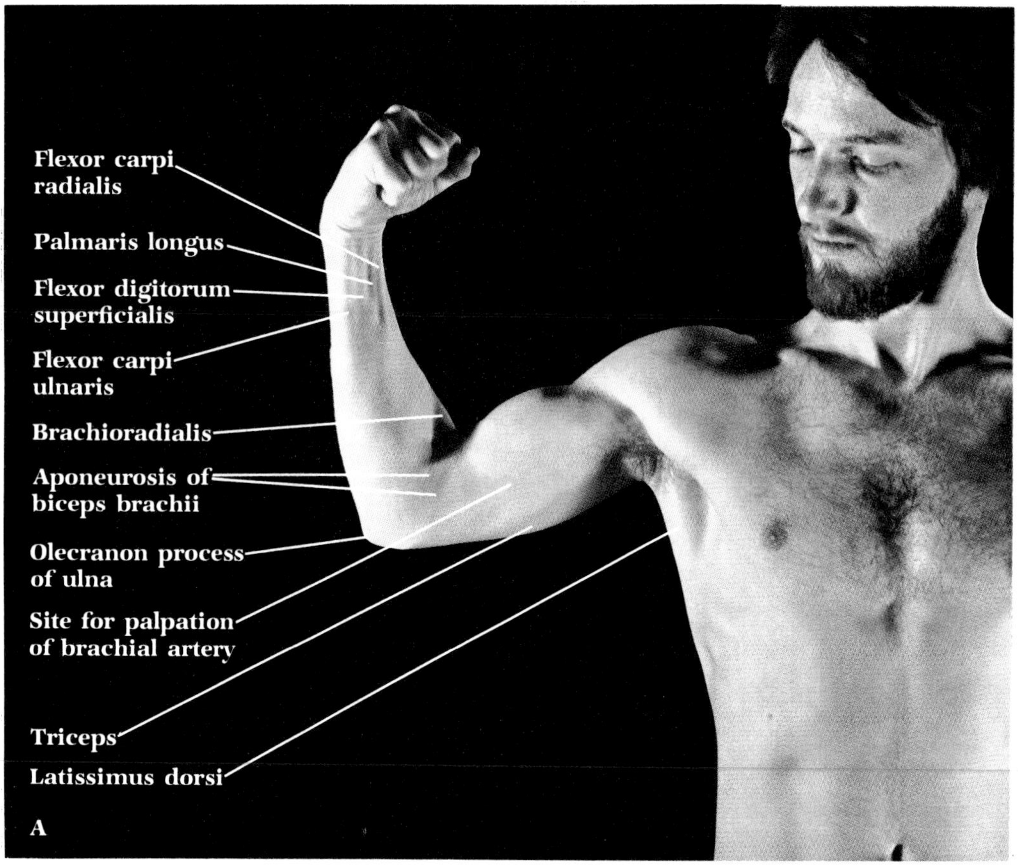
Flexor carpi radialis
Palmaris longus
Flexor digitorum superficialis
Flexor carpi ulnaris
Brachioradialis
Aponeurosis of biceps brachii
Olecranon process of ulna
Site for palpation of brachial artery
Triceps
Latissimus dorsi
A

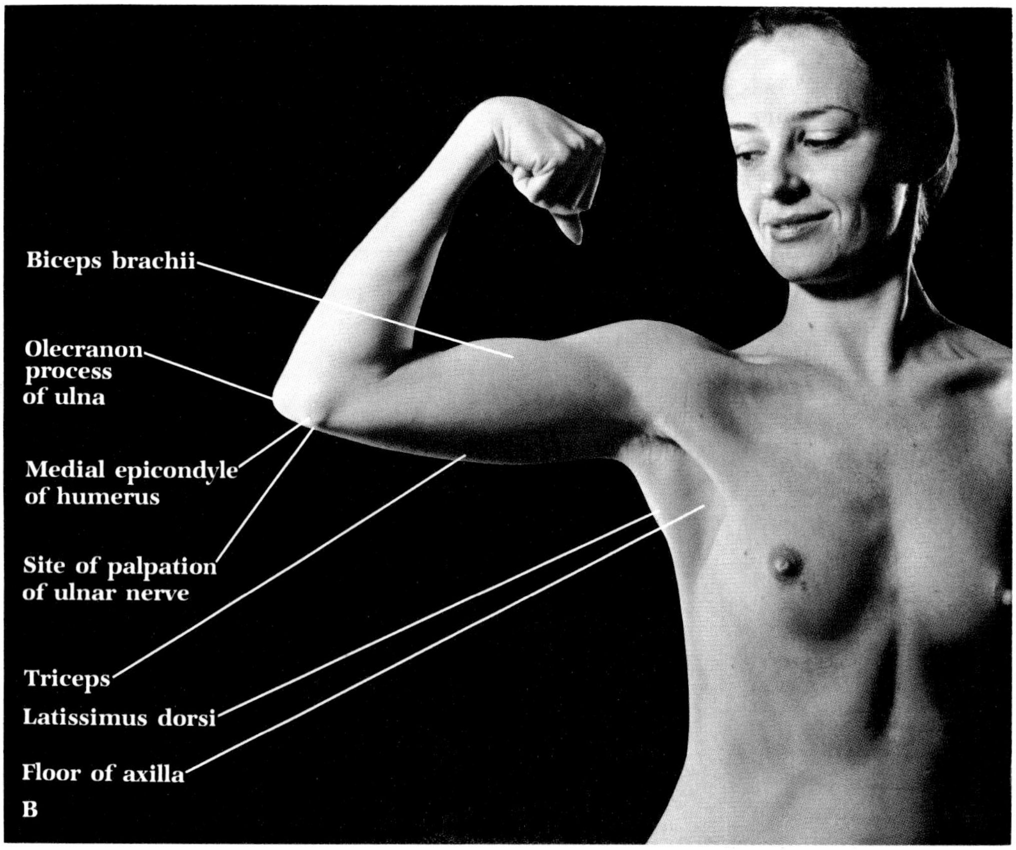
Biceps brachii
Olecranon process of ulna
Medial epicondyle of humerus
Site of palpation of ulnar nerve
Triceps
Latissimus dorsi
Floor of axilla
B

Figure 5-31
The right arm and forearm.
A. 27-year-old male. B. 29-year-old female.

1. The latissimus dorsi muscle covers the lateral part of the thoracic cage. It winds around the lower margin of the teres major muscle and is inserted into the floor of the bicipital groove of the humerus.
2. The floor of the axilla is formed by skin and fascia that connect the anterior and posterior axillary folds. As the arm is raised above the shoulder the axillary floor rises; this is due to the pull of the clavipectoral fascia that is connected above to the clavicle and below to the floor of the axilla.
3. The biceps brachii and the triceps muscles form large swellings on the anterior and posterior surfaces of the arm, respectively.
4. In the region of the elbow, the medial and lateral epicondyles of the humerus and the olecranon process of the ulna can be palpated.
5. The ulnar nerve can be palpated where it lies behind the medial epicondyle of the humerus. It feels like a rounded cord, and when it is compressed, a "pins and needles" sensation is felt along the medial part of the hand.
6. The brachial artery can be felt as it passes down the arm, overlapped by the medial border of the biceps brachii muscle.
7. The following important structures lie in front of the wrist region and should be palpated: (a) the pulsations of the radial artery anterior to the distal third of the radius and lateral to the tendon of the flexor carpi radialis muscle; (b) medial to this, the tendon of the palmaris longus muscle may be present, overlying the median nerve; (c) medial to this, a rounded group of tendons will be seen when the wrist and fingers are flexed and extended—the flexor digitorum superficialis muscle; (d) the tendon of the flexor carpi ulnaris muscle lies most medially as it goes to its insertion on the pisiform bone; (e) by careful palpation, the pulsations of the ulnar artery can be felt lateral to the tendon of the flexor carpi ulnaris; (f) the ulnar nerve lies immediately medial to the ulnar artery in this region.

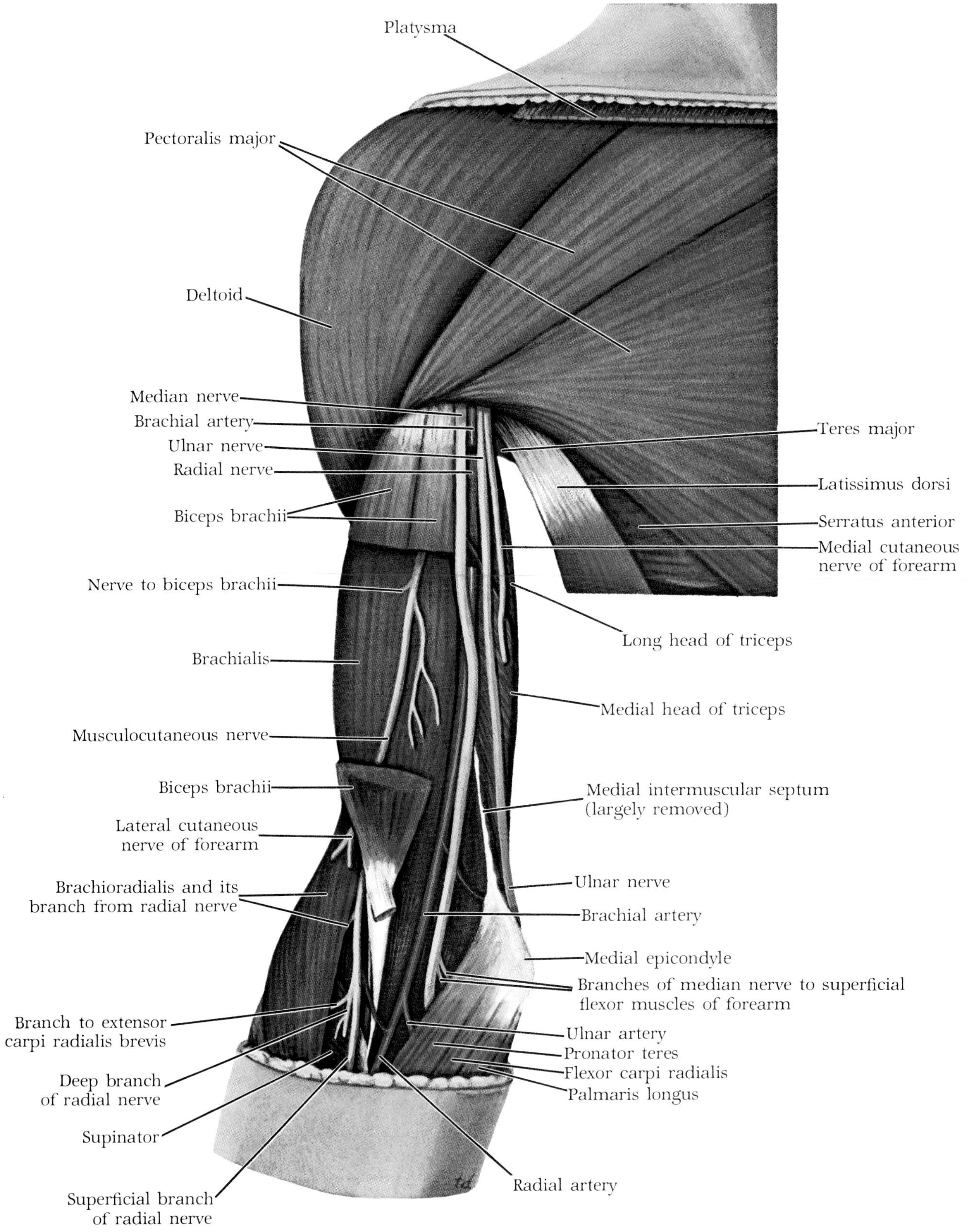
Platysma
Pectoralis major
Deltoid
Median nerve
Brachial artery
Ulnar nerve
Radial nerve
Biceps brachii
Nerve to biceps brachii
Brachialis
Musculocutaneous nerve
Biceps brachii
Lateral cutaneous nerve of forearm
Brachioradialis and its branch from radial nerve
Branch to extensor carpi radialis brevis
Deep branch of radial nerve
Supinator
Superficial branch of radial nerve
Teres major
Latissimus dorsi
Serratus anterior
Medial cutaneous nerve of forearm
Long head of triceps
Medial head of triceps
Medial intermuscular septum (largely removed)
Ulnar nerve
Brachial artery
Medial epicondyle
Branches of median nerve to superficial flexor muscles of forearm
Ulnar artery
Pronator teres
Flexor carpi radialis
Palmaris longus
Radial artery

Figure 5-32
Upper arm, anterior view. Middle portion of biceps brachii muscle has been removed to show musculocutaneous nerve lying anterior to the brachialis.

1. The musculocutaneous nerve is rarely injured because of its protected position beneath the biceps brachii muscle. If it is injured high up in the arm, the biceps and coracobrachialis are paralyzed and the brachialis muscle is weakened (the latter muscle is also supplied by the radial nerve). Flexion of the forearm at the elbow joint is then produced by the remainder of the brachialis muscle and the flexors of the forearm. When the forearm is in the prone position, the extensor carpi radialis longus and the brachioradialis muscles assist in flexion of the forearm. There is also sensory loss along the lateral side of the forearm. Wounds or cuts of the forearm can sever the lateral cutaneous nerve of the forearm, resulting in sensory loss along the lateral side of the forearm.

2. The median nerve may be injured in fractures of the lower end of the humerus. Such injury can result in the following.
Motor Loss. The pronator muscles of the forearm and the long flexor muscles of the wrist and fingers — except the flexor carpi ulnaris and the medial half of the flexor digitorum profundus — are paralyzed. Flexion of the terminal phalanx of the thumb is lost due to paralysis of the flexor pollicis longus. The muscles of the thenar eminence are paralyzed so that the thumb is then laterally rotated and adducted.
Sensory Loss. There is loss of skin sensation of the lateral half or less of the palm of the hand, the palmar aspect of the lateral three and one-half fingers, and the skin of the distal parts of the dorsal surfaces of the lateral three and one-half fingers. The area of *total* anesthesia is considerably less, however, because of the overlap of adjacent nerves.
Vasomotor Changes. The skin areas involved in sensory loss are warmer and drier than normal.

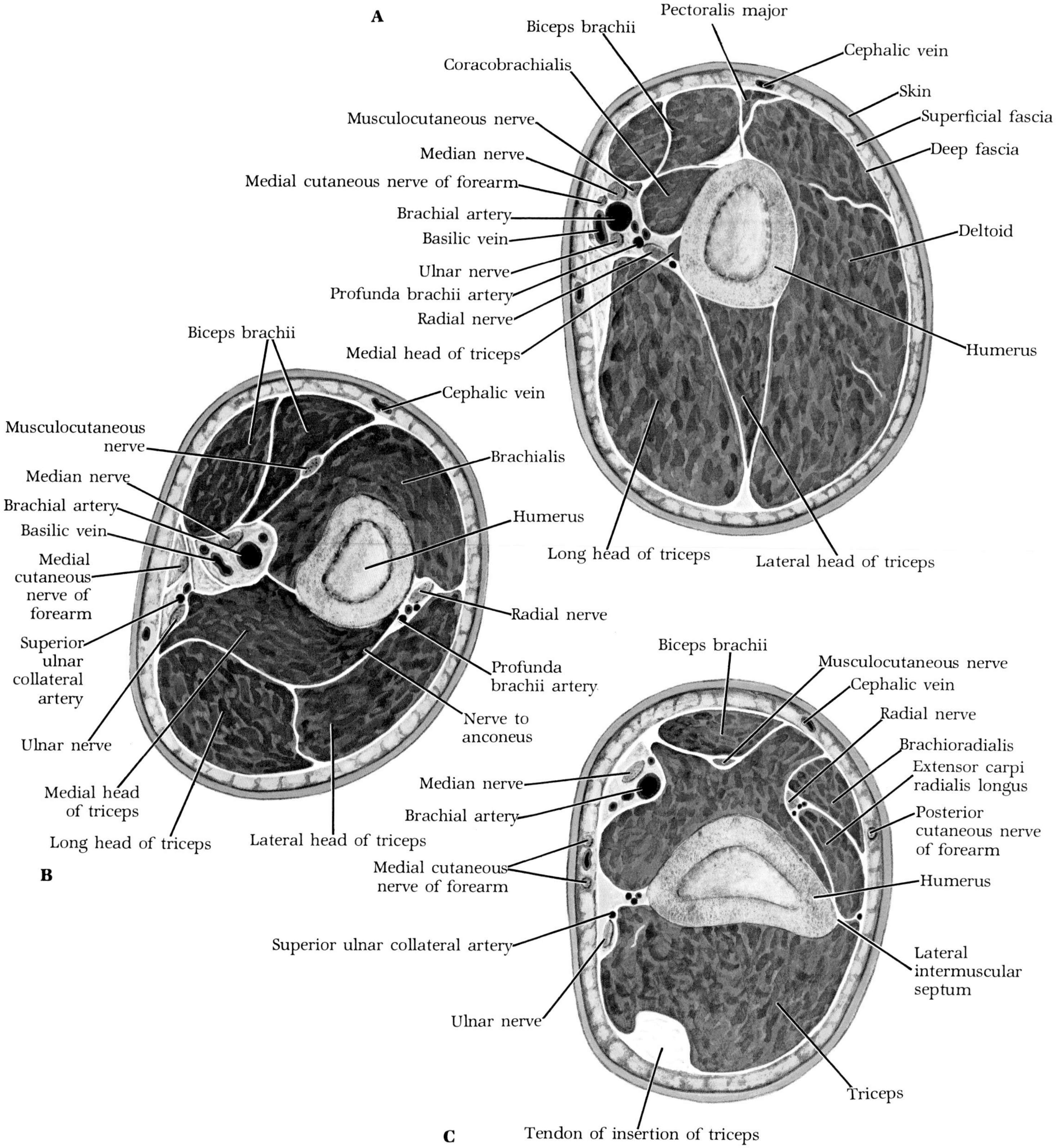
A
Pectoralis major
Biceps brachii
Cephalic vein
Coracobrachialis
Skin
Musculocutaneous nerve
Superficial fascia
Median nerve
Deep fascia
Medial cutaneous nerve of forearm
Brachial artery
Deltoid
Basilic vein
Ulnar nerve
Profunda brachii artery
Radial nerve
Humerus
Medial head of triceps
Long head of triceps
Lateral head of triceps
Biceps brachii
Cephalic vein
Musculocutaneous nerve
Brachialis
Median nerve
Brachial artery
Humerus
Basilic vein
Medial cutaneous nerve of forearm
Radial nerve
Superior ulnar collateral artery
Profunda brachii artery
Nerve to anconeus
Ulnar nerve
Medial head of triceps
Long head of triceps
Lateral head of triceps
B
Biceps brachii
Musculocutaneous nerve
Cephalic vein
Radial nerve
Brachioradialis
Median nerve
Extensor carpi radialis longus
Brachial artery
Posterior cutaneous nerve of forearm
Medial cutaneous nerve of forearm
Humerus
Superior ulnar collateral artery
Lateral intermuscular septum
Ulnar nerve
Triceps
Tendon of insertion of triceps
C

Figure 5-33
Cross sections of the upper arm. A. Superior to the level of insertion of the deltoid muscle. B. Just below level of insertion of deltoid muscle. C. Just superior to the medial epicondyle of the humerus.

1. In the proximal part of the upper arm the neurovascular bundle lies close to the shaft of the humerus. The radial nerve in particular lies in the spiral groove of the humerus deep to the triceps muscle. It is here that it is likely to be damaged in fractures of the shaft of the humerus or involved during the subsequent formation of callus.

2. The radial nerve may be injured in an unconscious patient if the back of the arm is pressed on the edge of the operating table. The prolonged application of a tourniquet to the arm of a person with a slender triceps muscle will often be followed by temporary radial palsy.

3. Injury to the radial nerve in the arm most commonly occurs in the distal part of the spiral groove, beyond the origin of the nerves to the triceps and the anconeus muscles and the origin of the cutaneous nerves. The patient is unable to extend the wrist and the fingers, and there is wrist drop. A variable small area of anesthesia is present over the root of the thumb.

4. Note that the ulnar nerve passes into the posterior compartment of the arm prior to traveling posteriorly to the medial epicondyle. It is in the latter position that the nerve is very liable to be damaged.

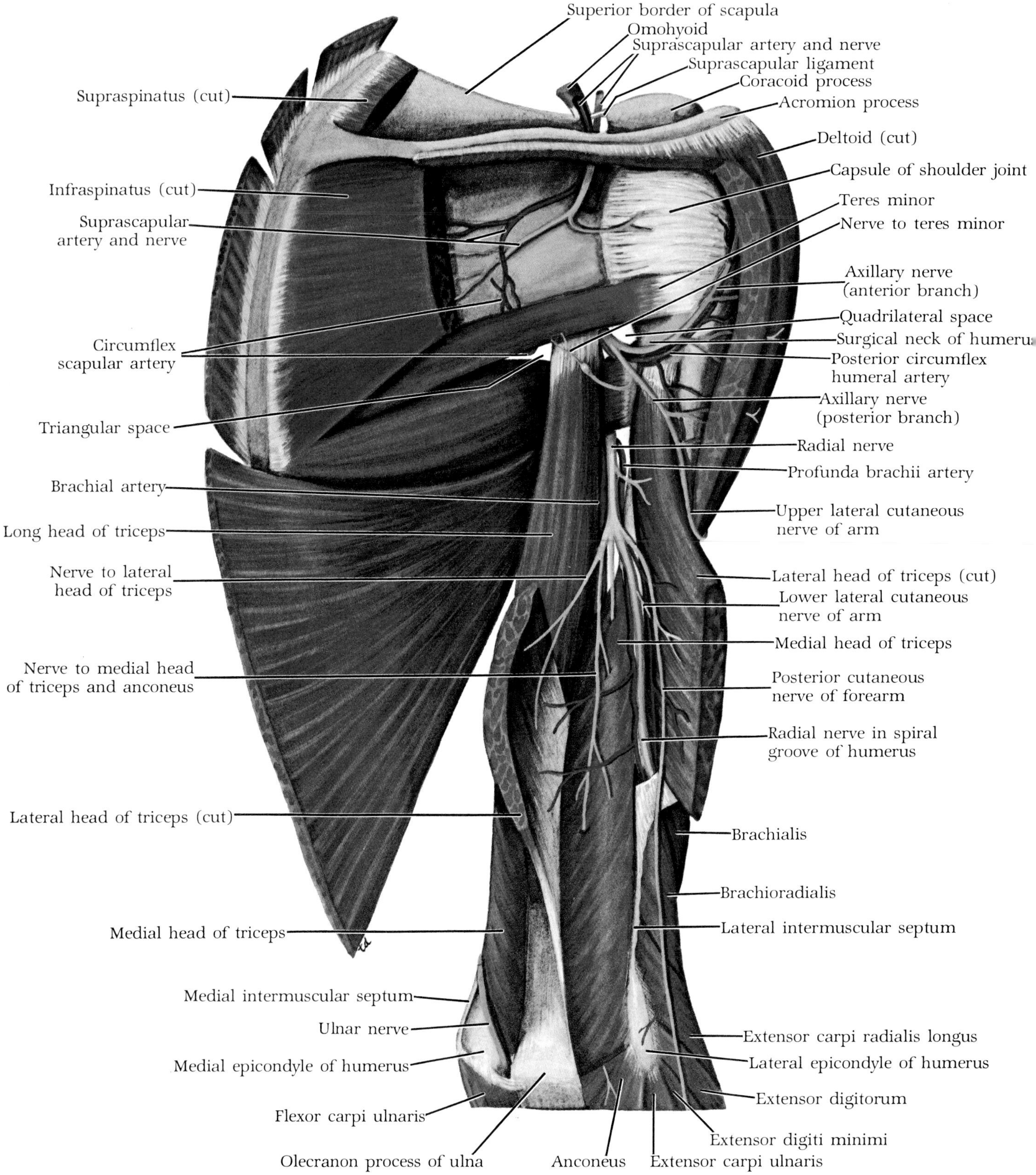
Superior border of scapula
Omohyoid
Suprascapular artery and nerve
Suprascapular ligament
Coracoid process
Acromion process
Deltoid (cut)
Capsule of shoulder joint
Teres minor
Nerve to teres minor
Axillary nerve (anterior branch)
Quadrilateral space
Surgical neck of humerus
Posterior circumflex humeral artery
Axillary nerve (posterior branch)
Radial nerve
Profunda brachii artery
Upper lateral cutaneous nerve of arm
Lateral head of triceps (cut)
Lower lateral cutaneous nerve of arm
Medial head of triceps
Posterior cutaneous nerve of forearm
Radial nerve in spiral groove of humerus
Brachialis
Brachioradialis
Lateral intermuscular septum
Extensor carpi radialis longus
Lateral epicondyle of humerus
Extensor digitorum
Extensor digiti minimi
Extensor carpi ulnaris
Anconeus
Olecranon process of ulna
Flexor carpi ulnaris
Medial epicondyle of humerus
Ulnar nerve
Medial intermuscular septum
Medial head of triceps
Lateral head of triceps (cut)
Nerve to medial head of triceps and anconeus
Nerve to lateral head of triceps
Long head of triceps
Brachial artery
Triangular space
Circumflex scapular artery
Suprascapular artery and nerve
Infraspinatus (cut)
Supraspinatus (cut)

Figure 5-34
Muscles, nerves, and blood vessels in the scapular region and the posterior compartment of the upper arm, posterior view.

1. The suprascapular nerve arises from the upper trunk of the brachial plexus and supplies the supraspinatus and the infraspinatus muscles. It is rarely injured, and then it is usually damaged with the upper trunk of the brachial plexus. Injury to the nerve produces weakness in lateral rotation of the shoulder joint.

2. The axillary nerve can be damaged in the quadrilateral space during dislocation of the shoulder joint or by fractures of the surgical neck of the humerus. The deltoid muscle is then paralyzed and rapidly atrophies, resulting in great impairment in abduction of the shoulder joint. There is also loss of cutaneous sensation over the lower half of the deltoid muscle. The paralysis of the teres minor muscle cannot be detected clinically.

3. The importance of the relationship of the radial nerve to the shaft of the humerus has been emphasized previously. Remember that the radial nerve gives off three branches in the axilla: (a) posterior cutaneous nerve of arm; (b) the nerve to the long head of the triceps; and (c) the nerve to the medial head of the triceps. Although these branches escape injury when the radial nerve is damaged in the spiral groove, the patient is unable to extend the wrist and the fingers, and there is wrist drop. A variable small area of anesthesia is present over the root of the thumb.

4. The ulnar nerve is commonly injured at the elbow where it lies posterior to the medial epicondyle of the humerus.

5. In ulnar nerve injuries, which are usually associated with fractures of the medial epicondyle, the flexor carpi ulnaris and the medial half of the flexor digitorum profundus muscles are paralyzed. The small muscles of the hand also will be paralyzed, with the exception of the muscles of the thenar eminence and the first two lumbricals, all of which are supplied by the median nerve. Loss of skin sensation will be apparent over the anterior and posterior surfaces of the medial third of the hand and the medial one and one-half fingers.

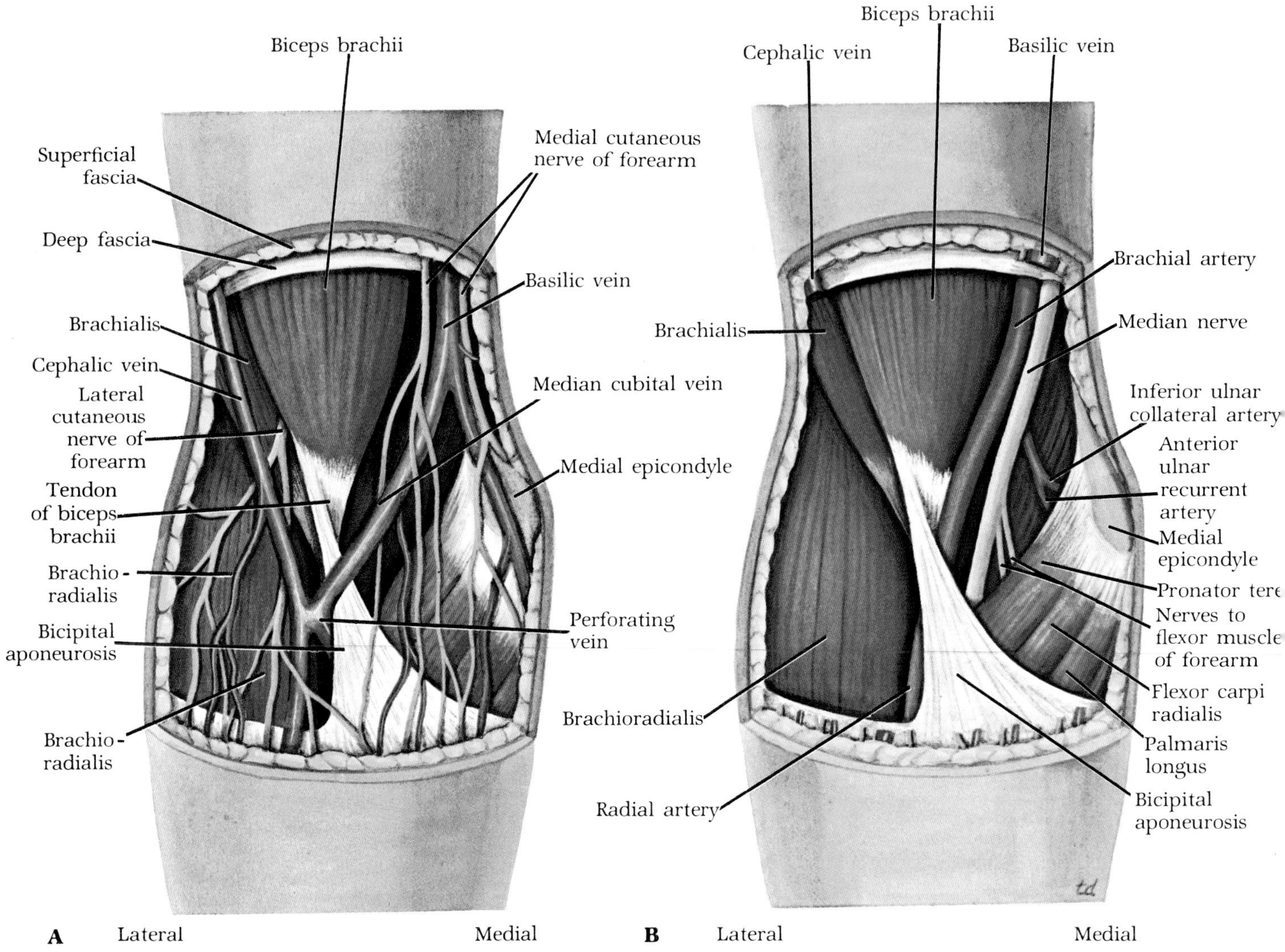
Biceps brachii
Superficial fascia
Deep fascia
Brachialis
Cephalic vein
Lateral cutaneous nerve of forearm
Tendon of biceps brachii
Brachio-radialis
Bicipital aponeurosis
Brachio-radialis
Medial cutaneous nerve of forearm
Basilic vein
Median cubital vein
Medial epicondyle
Perforating vein
Biceps brachii
Cephalic vein
Basilic vein
Brachialis
Brachioradialis
Radial artery
Brachial artery
Median nerve
Inferior ulnar collateral artery
Anterior ulnar recurrent artery
Medial epicondyle
Pronator tere
Nerves to flexor muscle of forearm
Flexor carpi radialis
Palmaris longus
Bicipital aponeurosis
A Lateral Medial
B Lateral Medial

Figure 5-35
Right cubital fossa. A. Showing subcutaneous veins and nerves. B. Showing deep structures.

1. The superficial veins are clinically very important and are used for vein puncture, blood transfusion, and cardiac catheterization. It is important that every physician know where he can give blood or obtain blood from the arm in an emergency. The cephalic and basilic veins lie in the superficial fascia. In the cubital fossa the median cubital vein is separated from the underlying brachial artery by the bicipital aponeurosis. This protects the artery from the mistaken introduction into its lumen of irritating drugs that should be injected into the vein.

2. The brachial artery can be palpated as it lies on the brachialis and is overlapped from the lateral side by the biceps brachii. Arterial blood pressure is usually estimated by placing a rubber sphygmomanometer cuff around the arm, inflating it to collapse the underlying brachial artery, and listening over the artery below the cuff with a stethoscope. As the first sound is heard in the artery — as the air is released from the cuff and the blood begins to flow — the systolic pressure is recorded. As more air escapes from the cuff, the sound becomes muffled and then disappears; it is at this point that the diastolic pressure is measured.

3. As mentioned previously the brachial artery and median nerve can be damaged in a supracondylar fracture of the humerus.

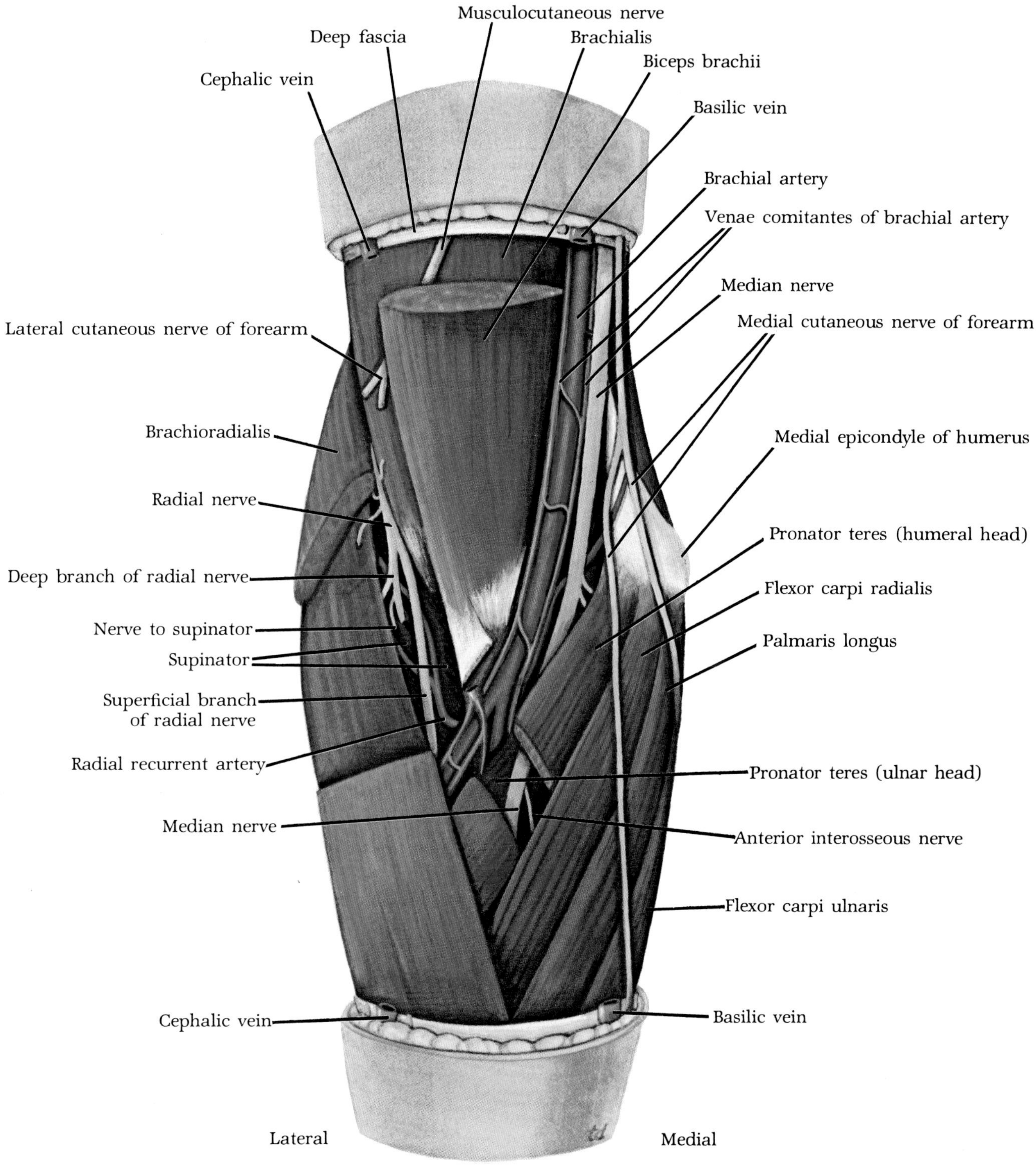
Musculocutaneous nerve
Deep fascia
Brachialis
Biceps brachii
Cephalic vein
Basilic vein
Brachial artery
Venae comitantes of brachial artery
Median nerve
Medial cutaneous nerve of forearm
Lateral cutaneous nerve of forearm
Medial epicondyle of humerus
Brachioradialis
Radial nerve
Pronator teres (humeral head)
Deep branch of radial nerve
Flexor carpi radialis
Nerve to supinator
Palmaris longus
Supinator
Superficial branch of radial nerve
Radial recurrent artery
Pronator teres (ulnar head)
Median nerve
Anterior interosseous nerve
Flexor carpi ulnaris
Cephalic vein
Basilic vein
Lateral
Medial

Figure 5-36
Right cubital fossa. Part of the brachioradialis muscle has been removed to expose the extensor carpi radialis longus muscle. A portion of the humeral head of the pronator teres muscle has also been removed to show the median nerve.

1. Physical examination of the cubital fossa reveals a muscular swelling in the proximal and lateral part of the forearm that is formed by the brachioradialis and the underlying extensor carpi radialis longus and brevis muscles. On the proximal and medial side of the forearm the pronator and flexor group of muscles form a smaller elevation, although the individual muscles are difficult to recognize since they are covered by deep fascia and the bicipital aponeurosis. The tendon of the biceps brachii can readily be felt as it passes into the cubital fossa medial to the brachioradialis muscle.
2. The median nerve and brachial artery are protected in part by the biceps brachii that overlaps them from the lateral side. Further distally the median nerve is protected by the humeral head of the pronator teres muscle and is rarely injured in this region.
3. In the interval between the brachioradialis and the brachialis muscles, the radial nerve passes anteriorly into the anterior compartment of the arm. Here it supplies the brachialis, the brachioradialis, and the extensor carpi radialis longus muscles. The deep branch of the radial nerve may be damaged in fractures of the neck of the radius or in dislocation of the proximal end of that bone.

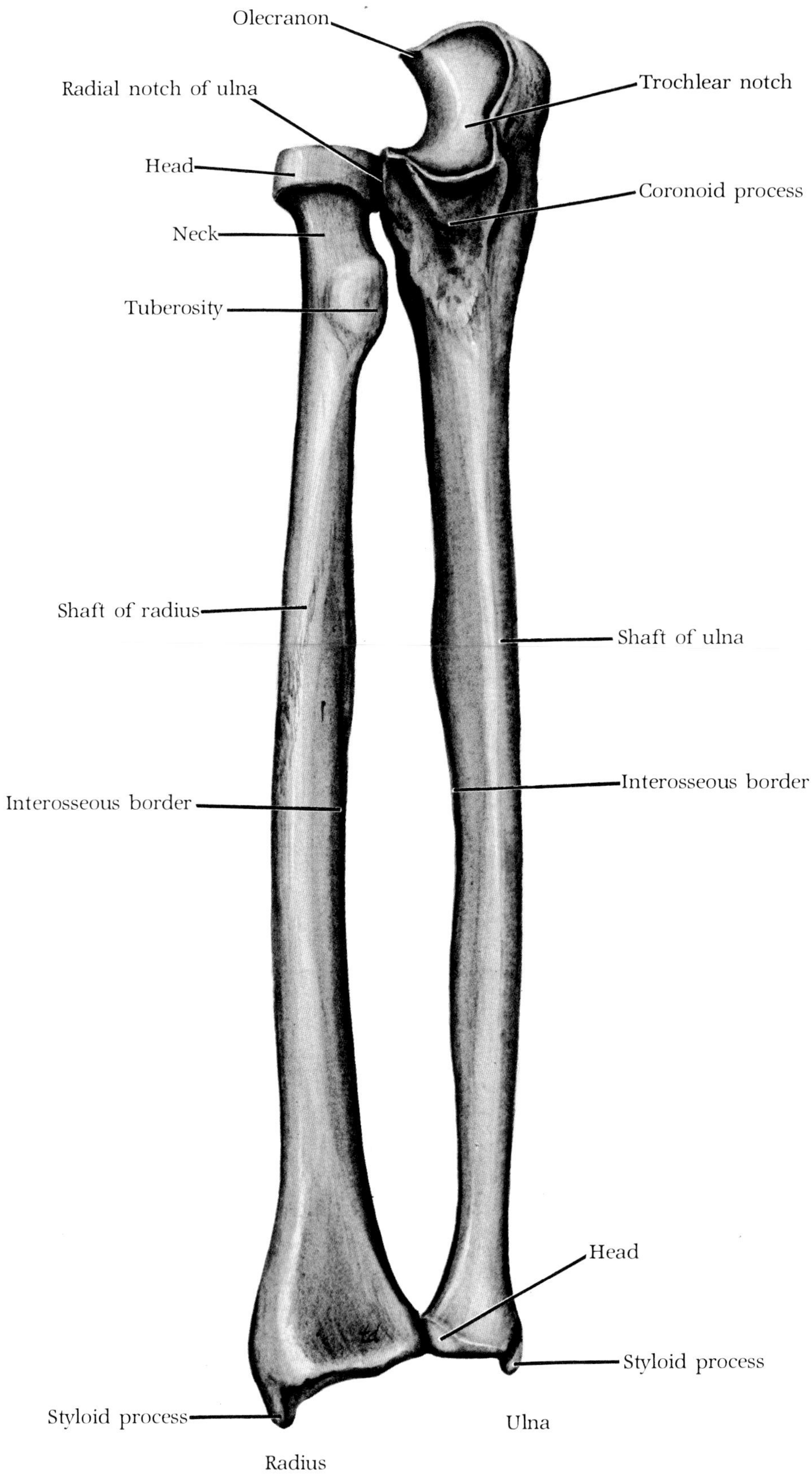
Olecranon
Radial notch of ulna
Trochlear notch
Head
Coronoid process
Neck
Tuberosity
Shaft of radius
Shaft of ulna
Interosseous border
Interosseous border
Head
Styloid process
Styloid process
Ulna
Radius

Figure 5-37
Radius and ulna, anterior view.

1. Fractures of the head of the radius can occur from falls on the outstretched hand. As the force is transmitted along the radius, the head of the radius is driven sharply against the capitulum, splitting or splintering the head.
2. Fractures of the neck of the radius occur in young children from falls on the outstretched hand.
3. Fracture of the olecranon process of the ulna can occur from falls on the outstretched hand when the elbow joint is forcibly flexed against the triceps; the triceps then avulses the olecranon.
4. Fractures of the shafts of the radius and ulna may or may not occur together. Displacement of the fragments is usually considerable and depends on the pull of the attached muscles. The proximal fragment of the radius is supinated by the supinator and the biceps brachii muscles. The distal fragment of the radius is pronated and pulled medially by the pronator quadratus muscle. The strength of the brachioradialis and extensor carpi radialis longus and brevis shorten the forearm and angulate it. In fracture of the ulna, the ulna angulates posteriorly. In order to restore the normal movements of pronation and supination, the normal anatomical relationships of the radius, ulna, and interosseous membrane must be regained.
5. Colles' fracture is a fracture of the distal end of the radius resulting from a fall on the outstretched hand. The force drives the distal fragment posteriorly and superiorly, and the distal articular surface is inclined posteriorly. Failure to restore the distal articular surface to its normal position will severely limit the range of flexion of the wrist joint.
6. Smith's fracture is a fracture of the distal end of the radius and occurs from a fall on the back of the hand. It is a reversed Colles' fracture because the distal fragment is displaced anteriorly.

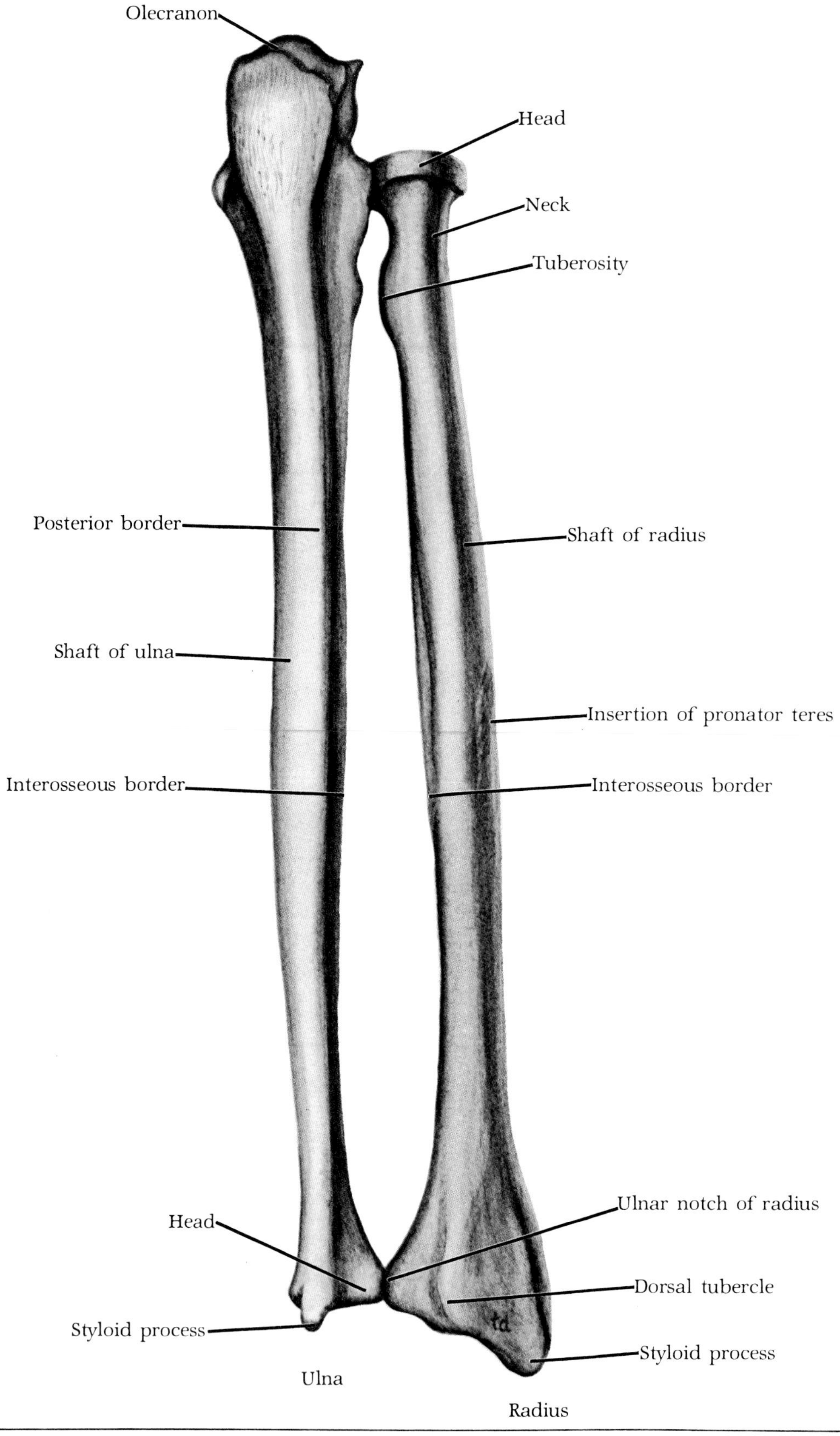
Olecranon
Head
Neck
Tuberosity
Posterior border
Shaft of radius
Shaft of ulna
Insertion of pronator teres
Interosseous border
Interosseous border
Ulnar notch of radius
Head
Dorsal tubercle
Styloid process
Styloid process
Ulna
Radius

Figure 5-38
Radius and ulna, posterior view.

1. Fractures of the forearm bones in children are commonly incomplete, especially in young children, and are referred to as greenstick fractures. Apart from angulation, displacement of the fragments rarely occurs. The treatment is manipulation to correct the deformity and at the same time complete the fracture so that the deformity does not recur in the cast.
2. Monteggia's deformity is a fracture of the proximal third of the shaft of the ulna and a dislocation of the head of the radius following a rupture of the annular ligament. The injury is caused by forced pronation or a direct blow on the posterior surface of the ulna.
3. In posterior dislocations of the elbow joint, the brachialis muscle may avulse the coronoid process of the ulna. In some dislocations the head of the radius is also fractured.

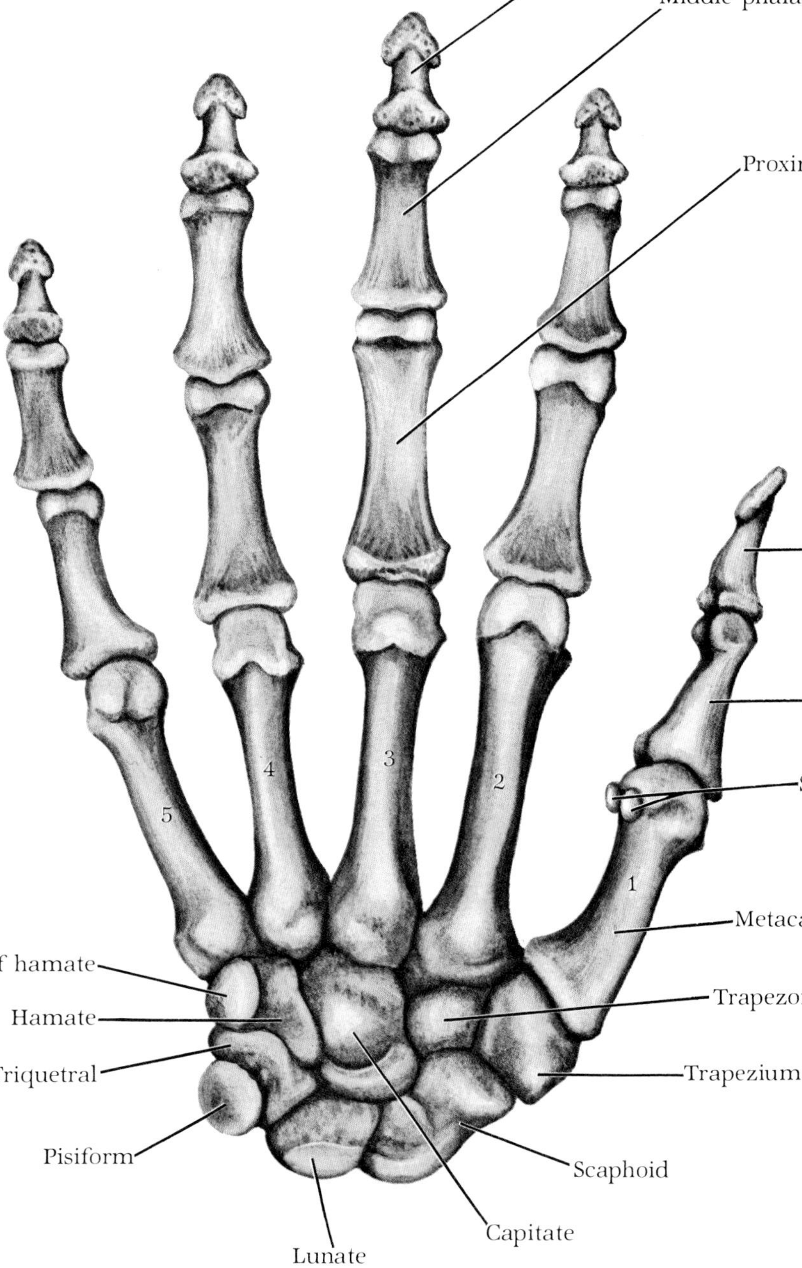
Distal phalanx
Middle phalanx
Proximal phalanx
Phalanges
Distal phalanx of thumb
Proximal phalanx of thumb
Sesamoid bones
Metacarpus
4
3
2
5
1
Metacarpal of thumb
Hook of hamate
Hamate
Carpus
Triquetral
Trapezoid
Trapezium
Pisiform
Scaphoid
Capitate
Lunate

Figure 5-39
Bones of the right hand, anterior view.

1. *Carpal Tunnel Syndrome.* The carpal tunnel, formed by the concave anterior surface of the carpal bones and closed by the flexor retinaculum, is tightly packed with the long flexor tendons of the fingers, their surrounding synovial sheaths, and the median nerve. The syndrome consists of a burning pain or a sensation of "pins and needles" along the distribution of the median nerve to the lateral three and one-half fingers and weakness of the thenar muscles. It is produced by compression of the median nerve within the tunnel.

2. Fracture of the scaphoid bone, unless treated effectively, will not unite and permanent weakness of the wrist will result. This type of fracture occurs most commonly in young adults. The fracture line usually goes through the narrowest part of the bone, which, because of its location, is bathed in synovial fluid. The blood vessels to the scaphoid enter its proximal and distal ends, although the blood supply is occasionally confined to its distal end. If the latter occurs, a fracture deprives the proximal fragment of its arterial supply, and this fragment undergoes avascular necrosis.

3. Dislocations of the lunate bone occasionally occur in young adults who fall on the outstretched hand in a way that causes hyperextension of the wrist joint. Involvement of the median nerve is common. In hyperextension of the wrist, the narrower posterior part of the lunate bone is caught between the radius and the capitate bones and squeezed anteriorly. Usually the anterior ligaments remain intact and the bone is rotated anteriorly.

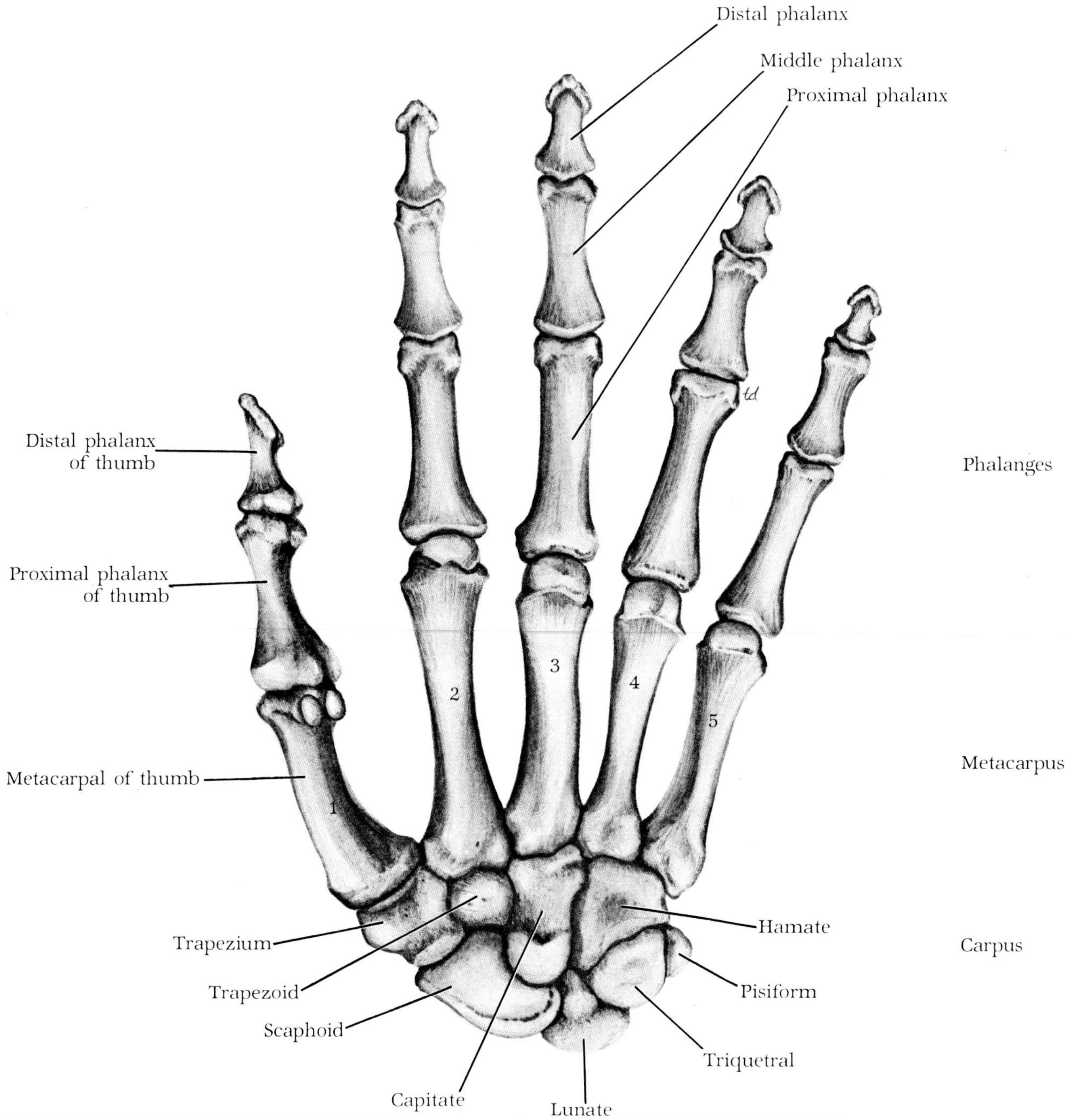
Distal phalanx
Middle phalanx
Proximal phalanx
Distal phalanx of thumb
Proximal phalanx of thumb
Metacarpal of thumb
Phalanges
Metacarpus
Carpus
1
2
3
4
5
Trapezium
Trapezoid
Scaphoid
Capitate
Lunate
Triquetral
Pisiform
Hamate

Figure 5-40
Bones of the right hand, posterior view.

1. Fractures of the metacarpal bones can occur as the result of direct violence, in which case the fracture line is transverse and the fragments are displaced anteriorly. The "indirect violence" of, for example, striking an opponent with a clenched fist commonly produces an oblique fracture of the necks of the fourth and fifth metacarpals. The distal fragment is commonly displaced proximally, thus shortening the finger posteriorly.

2. Bennett's fracture is a fracture of the base of the metacarpal of the thumb caused when violence is applied along the long axis of the thumb or the thumb is forcefully abducted. The fracture is oblique and enters the carpometacarpal joint of the thumb, causing joint instability.

3. Fractures of phalanges are common. Mallett finger occurs when the distal phalanx is violently flexed against the taut extensor tendon. The attachment of the extensor tendon to the base of the distal phalanx is then avulsed with or without a fragment of bone. This results in the last twenty degrees of extension of the terminal phalanx being lost.

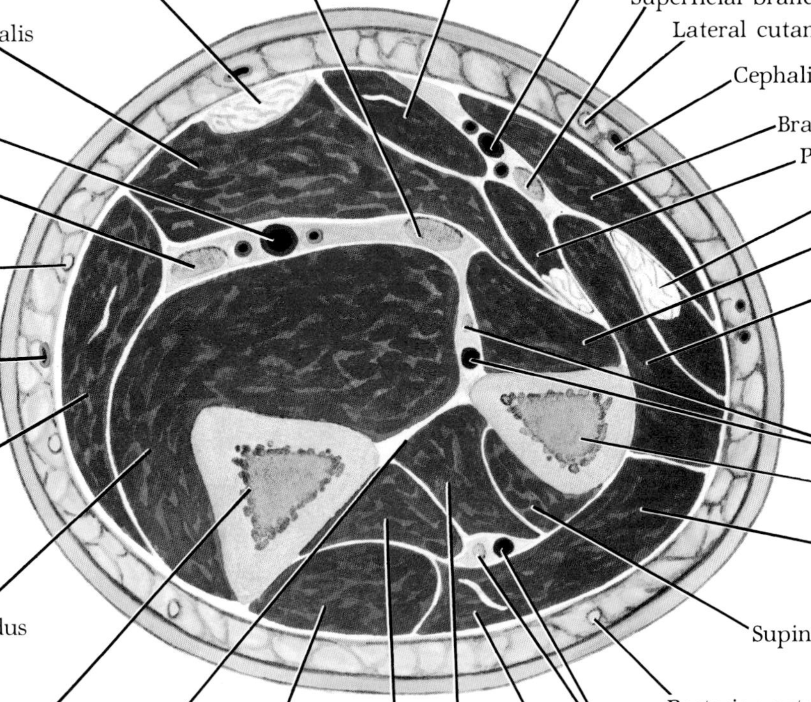

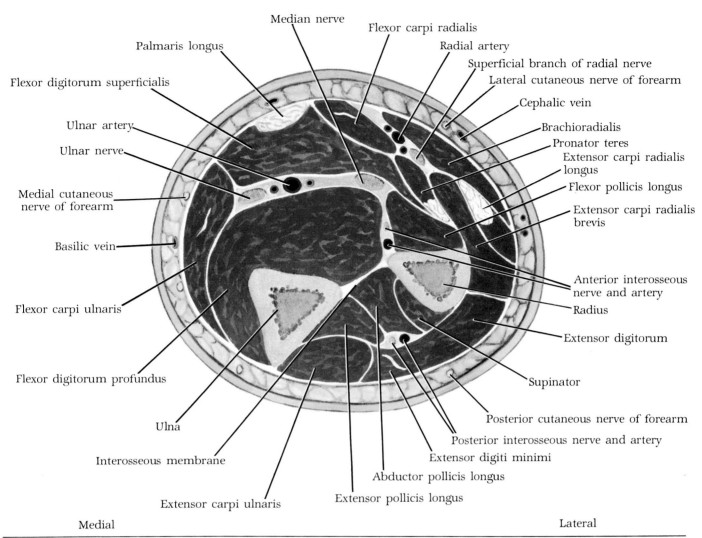

Median nerve
Flexor carpi radialis
Palmaris longus
Radial artery
Superficial branch of radial nerve
Flexor digitorum superficialis
Lateral cutaneous nerve of forearm
Cephalic vein
Ulnar artery
Brachioradialis
Ulnar nerve
Pronator teres
Extensor carpi radialis longus
Medial cutaneous nerve of forearm
Flexor pollicis longus
Extensor carpi radialis brevis
Basilic vein
Anterior interosseous nerve and artery
Radius
Flexor carpi ulnaris
Extensor digitorum
Flexor digitorum profundus
Supinator
Posterior cutaneous nerve of forearm
Ulna
Posterior interosseous nerve and artery
Interosseous membrane
Extensor digiti minimi
Abductor pollicis longus
Extensor carpi ulnaris
Extensor pollicis longus
Medial
Lateral

Figure 5-41
Cross section of forearm at level of insertion of pronator teres muscle.

1. The superficial veins of the forearm can usually be easily seen when firm pressure is applied around the upper arm. Knowledge of these veins is very important since a physician can obtain a sample of the patient's blood from them and can give blood by this route. The larger cephalic and basilic veins are fairly constant in position.

2. The strong deep fascia of the forearm surrounds the muscles and is attached to the posterior subcutaneous border of the ulna. An overtight plaster cast applied to the forearm to immobilize the fragments of the radius and ulna after their fracture can lead to Volkmann's ischemic contracture. The hemorrhage and edema at the fracture site cause a rise in pressure within the fascial compartments. The overtight plaster cast together with the strong, unyielding deep fascia prevents the compartments from expanding. At first the veins of the muscles are pressed upon, which increases the edema, and as the pressure rises, the arterial supply to the muscles and nerves is finally cut off, resulting in necrosis and replacement fibrosis: Volkmann's contracture.

3. Fractures of the forearm bones are relatively common and have been referred to previously. The various muscles attached to the bone fragments, especially the biceps brachii, supinator, pronator teres, and the pronator quadratus, will cause displacement of the bone fragments.

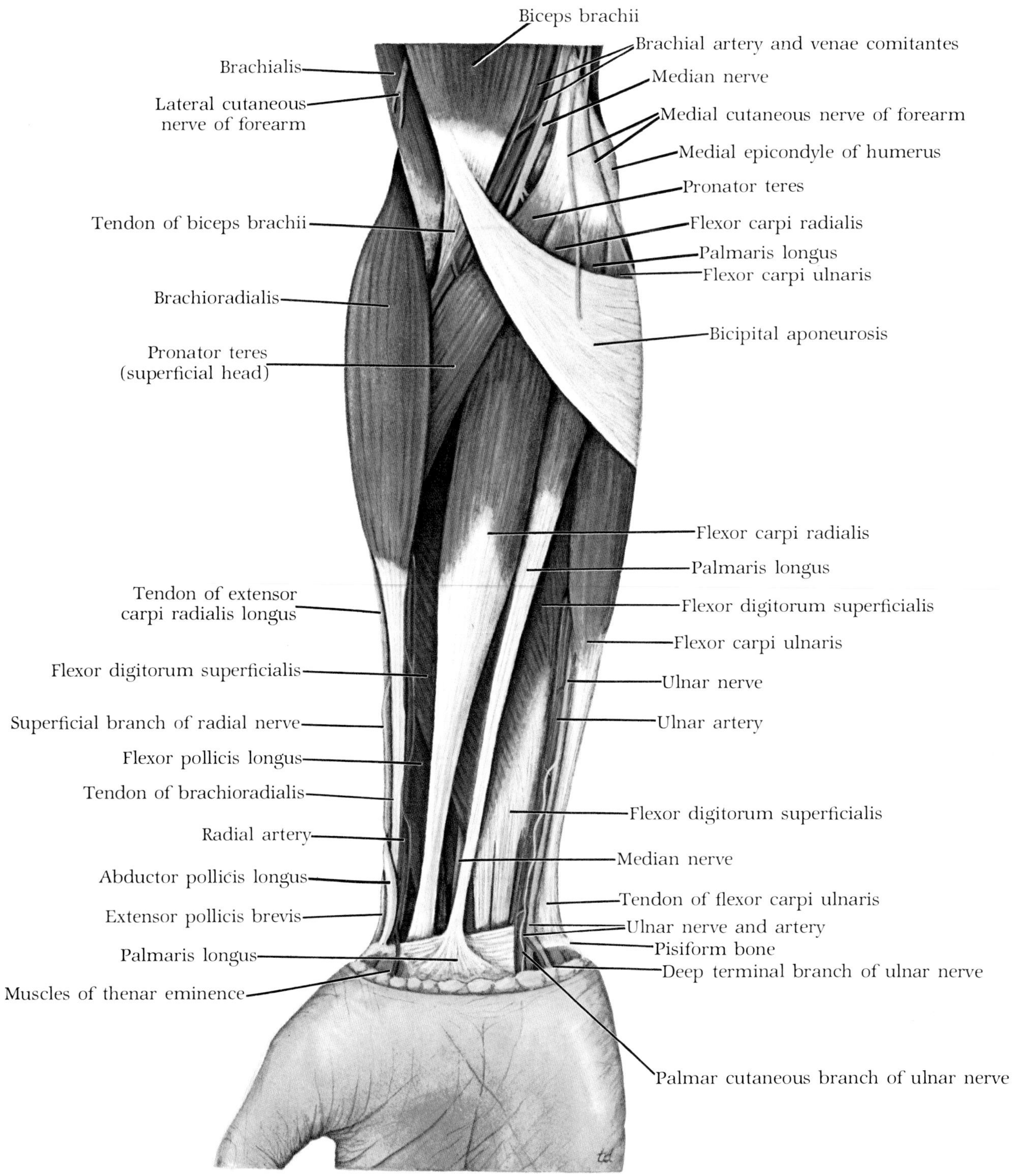
Biceps brachii
Brachial artery and venae comitantes
Brachialis
Median nerve
Lateral cutaneous nerve of forearm
Medial cutaneous nerve of forearm
Medial epicondyle of humerus
Pronator teres
Tendon of biceps brachii
Flexor carpi radialis
Palmaris longus
Flexor carpi ulnaris
Brachioradialis
Bicipital aponeurosis
Pronator teres (superficial head)
Flexor carpi radialis
Palmaris longus
Tendon of extensor carpi radialis longus
Flexor digitorum superficialis
Flexor carpi ulnaris
Flexor digitorum superficialis
Ulnar nerve
Superficial branch of radial nerve
Ulnar artery
Flexor pollicis longus
Tendon of brachioradialis
Flexor digitorum superficialis
Radial artery
Median nerve
Abductor pollicis longus
Tendon of flexor carpi ulnaris
Extensor pollicis brevis
Ulnar nerve and artery
Pisiform bone
Palmaris longus
Deep terminal branch of ulnar nerve
Muscles of thenar eminence
Palmar cutaneous branch of ulnar nerve

Figure 5-42
Forearm showing superficial structures, anterior view.

1. When the cubital fossa is palpated with the patients' elbow joint flexed against resistance, the bicipital aponeurosis can be felt as it passes downward and medially to be attached to the deep fascia of the forearm. It separates the median cubital vein from the underlying brachial artery. This separation is important, since it protects the artery from the mistaken introduction into its lumen of irritating drugs that should be injected into the vein.
2. The palmaris longus muscle may be absent on one or both sides of the forearm. Others show variation in form such as a centrally or distally placed muscle belly in the place of a proximal one.
3. When examining the muscles on the anterior aspect of the forearm, it is useful to remember that they are innervated by the median nerve and its branches except for the flexor carpi ulnaris and the medial part of the flexor digitorum profundus, both of which are innervated by the ulnar nerve. The anterolateral group of muscles, namely the brachioradialis and extensor carpi radialis longus, are supplied by branches of the radial nerve.
4. The structures that pass anterior to the flexor retinaculum are very liable to damage by cuts and penetrating wounds in the region of the wrist joint. From medial to lateral these structures are as follows: ulnar nerve, palmar cutaneous branch of ulnar nerve, ulnar artery, palmaris longus tendon, and palmar cutaneous branch of median nerve.

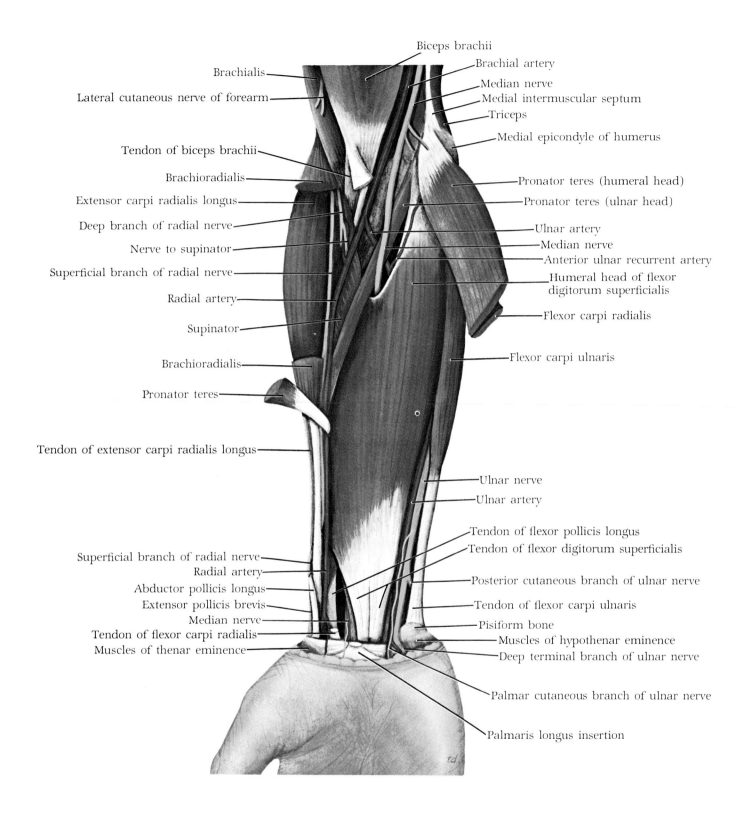

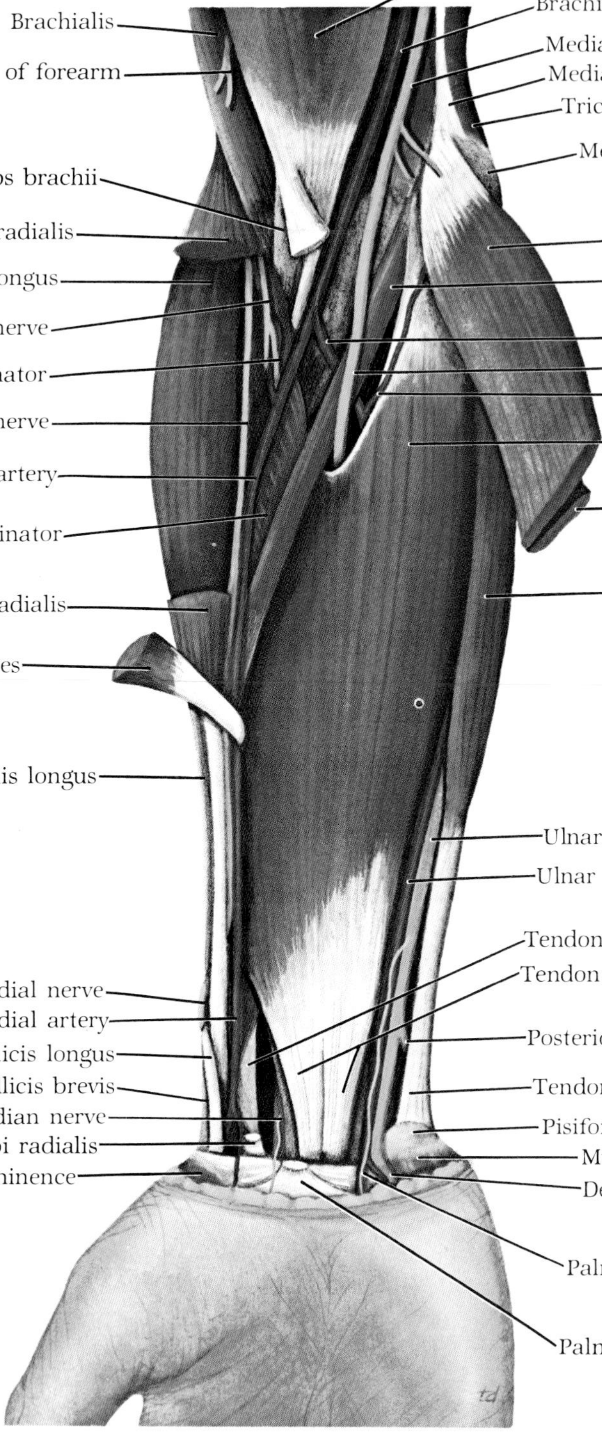
Biceps brachii
Brachialis
Brachial artery
Median nerve
Lateral cutaneous nerve of forearm
Medial intermuscular septum
Triceps
Medial epicondyle of humerus
Tendon of biceps brachii
Brachioradialis
Pronator teres (humeral head)
Extensor carpi radialis longus
Pronator teres (ulnar head)
Deep branch of radial nerve
Ulnar artery
Nerve to supinator
Median nerve
Anterior ulnar recurrent artery
Superficial branch of radial nerve
Humeral head of flexor digitorum superficialis
Radial artery
Flexor carpi radialis
Supinator
Flexor carpi ulnaris
Brachioradialis
Pronator teres
Tendon of extensor carpi radialis longus
Ulnar nerve
Ulnar artery
Tendon of flexor pollicis longus
Tendon of flexor digitorum superficialis
Superficial branch of radial nerve
Radial artery
Posterior cutaneous branch of ulnar nerve
Abductor pollicis longus
Extensor pollicis brevis
Tendon of flexor carpi ulnaris
Median nerve
Pisiform bone
Tendon of flexor carpi radialis
Muscles of hypothenar eminence
Muscles of thenar eminence
Deep terminal branch of ulnar nerve
Palmar cutaneous branch of ulnar nerve
Palmaris longus insertion

Figure 5-43
Forearm, anterior view. Most of the superficial muscles have been removed to display the flexor digitorum superficialis muscle, median nerve, superficial branch of radial nerve, and radial artery. Note that the ulnar head of the pronator teres separates the median nerve from the ulnar artery.

The median nerve is most commonly injured by stab wounds or broken glass penetrating just proximal to the flexor retinaculum. It is injured occasionally in the elbow region in a supracondylar fracture of the humerus.

1. Injuries to the Median Nerve at the Elbow
Motor. The pronator muscles of the forearm and the long flexor muscles of the wrist and fingers, with the exception of the flexor carpi ulnaris and the medial half of the flexor digitorum profundus, are paralyzed. The forearm is thereafter kept in the supine position; wrist flexion is weak and is accompanied by adduction. The latter deviation is due to paralysis of the flexor carpi radialis and the strength of the flexor carpi ulnaris and the medial half of the flexor digitorum profundus. No flexion is possible at the interphalangeal joints of the index and middle fingers, although weak flexion of the metacarpophalangeal joints of these fingers is attempted by the interossei. When the patient tries to make a fist, the index and to a lesser extent the middle finger tend to remain straight, while the ring and little fingers flex. The latter two fingers are, however, weakened by the loss of the flexor digitorum superficialis muscle.
Flexion of the terminal phalanx of the thumb is lost due to paralysis of the flexor pollicis longus. The muscles of the thenar eminence are paralyzed and wasted, resulting in flattening of the eminence. The thumb is laterally rotated and adducted.
Sensory. There is loss of skin sensation in the lateral half or less of the palm of the hand and the palmar aspect of the lateral three and one-half fingers. There is also sensory loss in the distal parts of the dorsal surfaces of the lateral three and one-half fingers. The area of total anesthesia is considerably less, due to the overlap of adjacent nerves.

2. Injuries to the Median Nerve at the Wrist
Motor. The muscles of the thenar eminence are paralyzed and wasted, resulting in flattening of the eminence. The thumb is laterally rotated and adducted, and opposition movement of the thumb is impossible. The first two lumbricals are paralyzed, which can be recognized clinically because the index and middle fingers tend to lag behind the ring and little fingers when the patient is asked to make a fist slowly.
The most serious disability of all in median nerve injuries is the loss of the ability to oppose the thumb to the other fingers and the loss of sensation over the lateral fingers. The delicate pincer action of the hand is thus no longer possible.

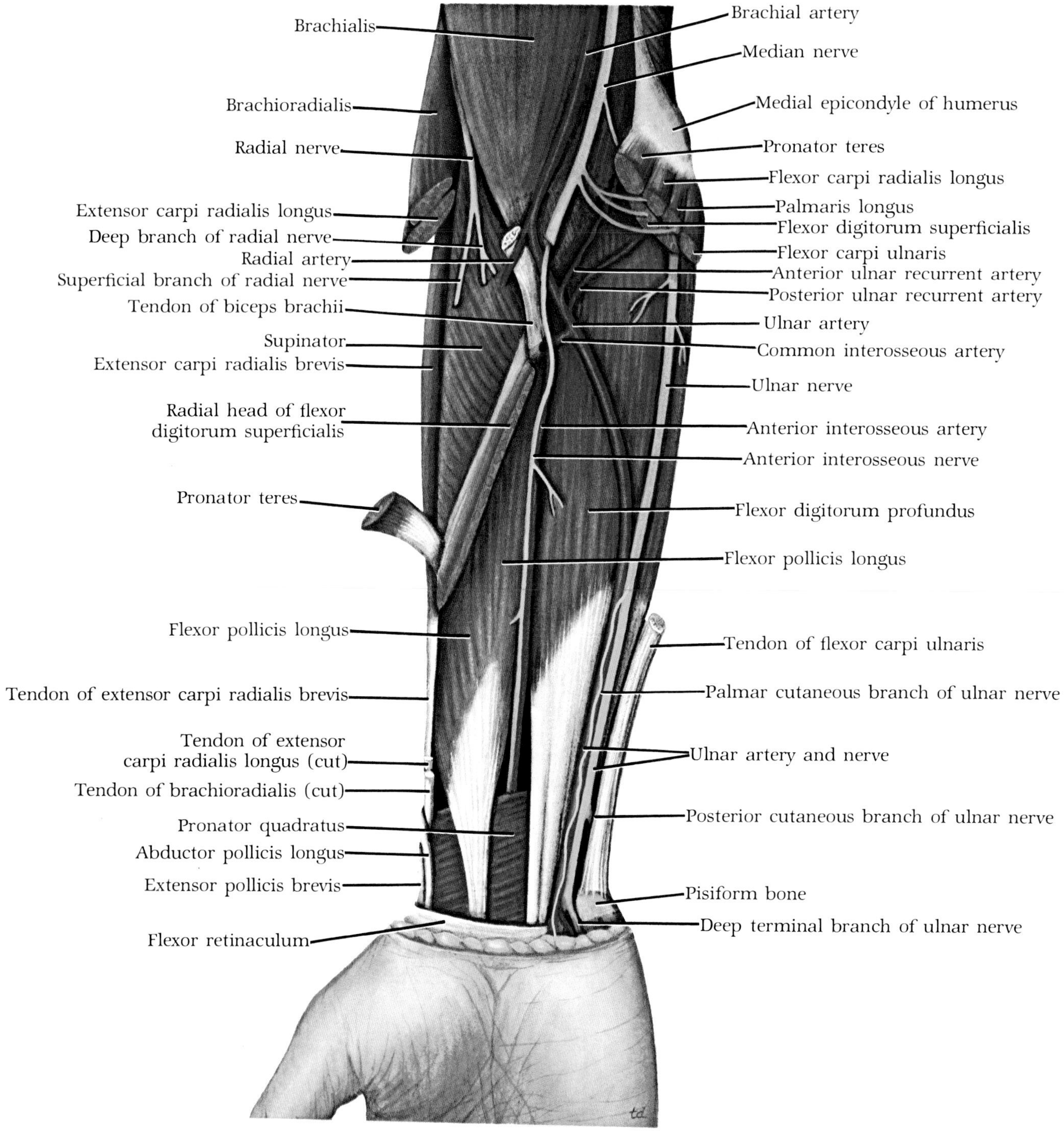
Brachialis
Brachioradialis
Radial nerve
Extensor carpi radialis longus
Deep branch of radial nerve
Radial artery
Superficial branch of radial nerve
Tendon of biceps brachii
Supinator
Extensor carpi radialis brevis
Radial head of flexor digitorum superficialis
Pronator teres
Flexor pollicis longus
Tendon of extensor carpi radialis brevis
Tendon of extensor carpi radialis longus (cut)
Tendon of brachioradialis (cut)
Pronator quadratus
Abductor pollicis longus
Extensor pollicis brevis
Flexor retinaculum
Brachial artery
Median nerve
Medial epicondyle of humerus
Pronator teres
Flexor carpi radialis longus
Palmaris longus
Flexor digitorum superficialis
Flexor carpi ulnaris
Anterior ulnar recurrent artery
Posterior ulnar recurrent artery
Ulnar artery
Common interosseous artery
Ulnar nerve
Anterior interosseous artery
Anterior interosseous nerve
Flexor digitorum profundus
Flexor pollicis longus
Tendon of flexor carpi ulnaris
Palmar cutaneous branch of ulnar nerve
Ulnar artery and nerve
Posterior cutaneous branch of ulnar nerve
Pisiform bone
Deep terminal branch of ulnar nerve
td

Figure 5-44
Deep structures of the forearm, anterior view.

The ulnar nerve is most commonly injured at the elbow, where it lies posterior to the medial epicondyle, and at the wrist, where it lies with the ulnar artery anterior to the flexor retinaculum. The injury at the elbow is usually associated with a fracture of the medial epicondyle. The superficial position of the nerve at the wrist makes it very vulnerable to damage from cuts and stab wounds.

1. Injuries to the Ulnar Nerve at the Elbow
Motor. The flexor carpi ulnaris and the medial half of the flexor digitorum profundus muscles are paralyzed. The profundus tendons to the ring and little fingers will be functionless, and the terminal phalanges of these fingers are therefore not capable of being markedly flexed. Because of paralysis of the flexor carpi ulnaris, flexion of the wrist joint results in abduction.
The small muscles of the hand will be paralyzed except the muscles of the thenar eminence and the first two lumbricals, which are supplied by the median nerve. The patient is unable to adduct and abduct the fingers and consequently is unable to grip a piece of paper placed between the fingers. It is impossible to adduct the thumb, because the adductor pollicis muscle is paralyzed. If the patient is asked to grip a piece of paper between the thumb and the index finger, he does so by strongly contracting his flexor pollicis longus and flexing the terminal phalanx (Froment's sign).
Sensory. Loss of skin sensation is detected over the anterior and posterior surfaces of the medial third of the hand and the medial one and one-half fingers.

2. Injuries to the Ulnar Nerve at the Wrist
Motor. The small muscles of the hand are paralyzed and show wasting, except for the muscles of the thenar eminence and the first two lumbricals, as described above.
Sensory. The main ulnar nerve and its palmar cutaneous branch are usually severed; the dorsal cutaneous branch, which arises from the ulnar nerve trunk about $2\frac{1}{2}$ inches (6 cm) above the pisiform bone is usually unaffected. The sensory loss will be confined to the palmar surface of the medial third of the hand and the medial one and one-half fingers and to the dorsal aspects of the middle and distal phalanges of the same fingers.
Unlike median nerve injuries, lesions of the ulnar nerve leave a relatively efficient hand. The sensation over the lateral part of the hand is intact, and the pincers action of the thumb and index finger is reasonably good, although there is some weakness due to loss of the adductor pollicis muscle.

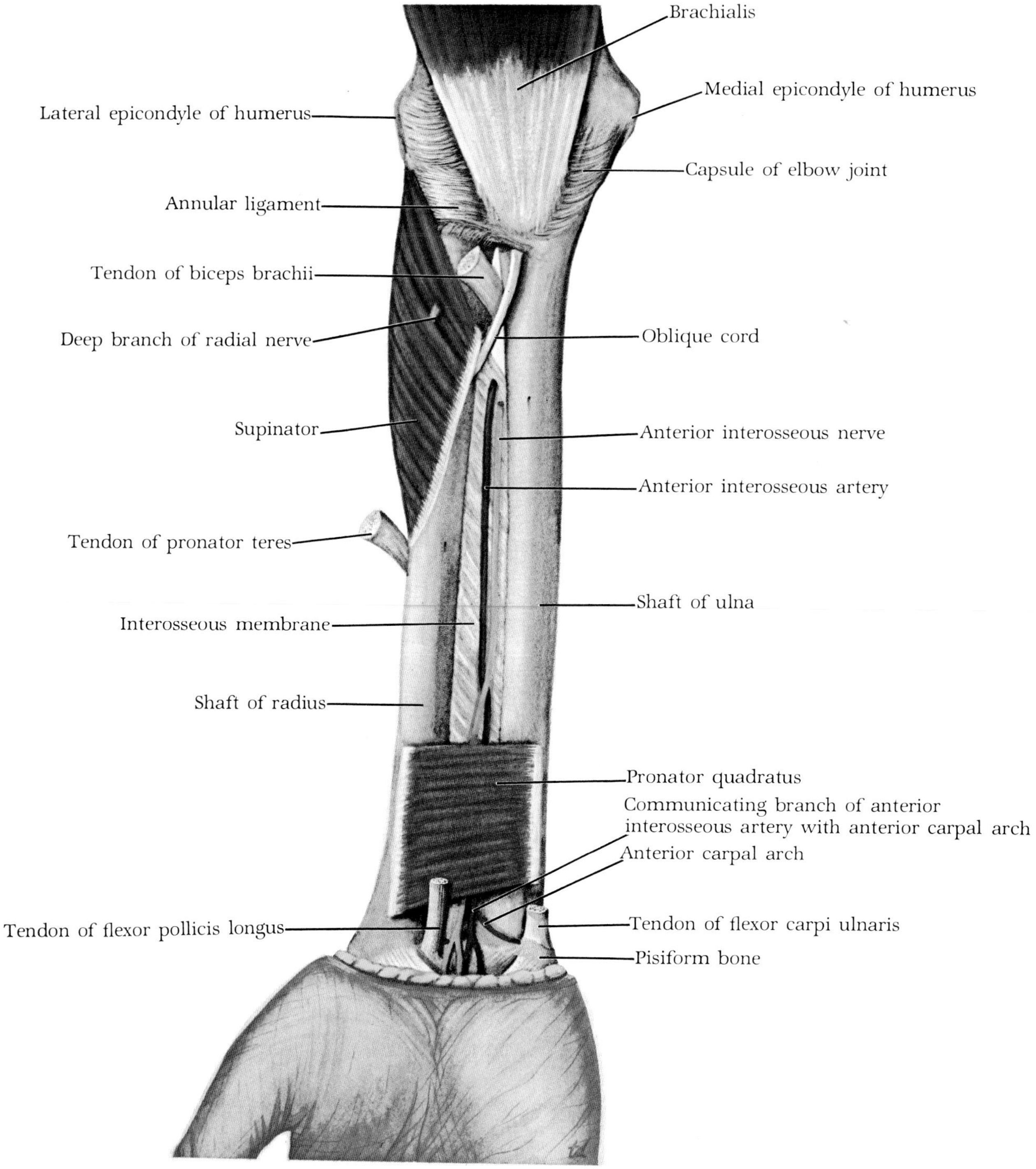
Brachialis
Medial epicondyle of humerus
Lateral epicondyle of humerus
Capsule of elbow joint
Annular ligament
Tendon of biceps brachii
Deep branch of radial nerve
Oblique cord
Supinator
Anterior interosseous nerve
Anterior interosseous artery
Tendon of pronator teres
Shaft of ulna
Interosseous membrane
Shaft of radius
Pronator quadratus
Communicating branch of anterior interosseous artery with anterior carpal arch
Anterior carpal arch
Tendon of flexor pollicis longus
Tendon of flexor carpi ulnaris
Pisiform bone

Figure 5-45
Radius and ulna, showing the supinator and pronator quadratus muscles, anterior view.

1. The deep branch of the radial nerve, as it lies in the supinator muscle, can be damaged by the surgeon when he is trying to expose the head of the radius. In this instance, the nerve supply to the supinator and the extensor carpi radialis longus muscles will be undamaged, and because the latter muscle is powerful, it will keep the wrist joint extended and wrist drop will not occur. There will be no sensory loss since this is a motor nerve.
2. The powerful supinator actions of the biceps brachii and the supinator are easily visualized in this illustration. Similarly, the pronator actions of the pronator teres and the pronator quadratus muscles can be easily understood. In fractures of the radius or ulna the displacement of the bone fragments will depend on the position of the fracture line in relation to the muscle attachments.
3. The anterior interosseous nerve, a branch of the median nerve, and the anterior interosseous artery are situated deep beneath the muscles of the forearm and consequently are rarely damaged in wounds of the forearm.

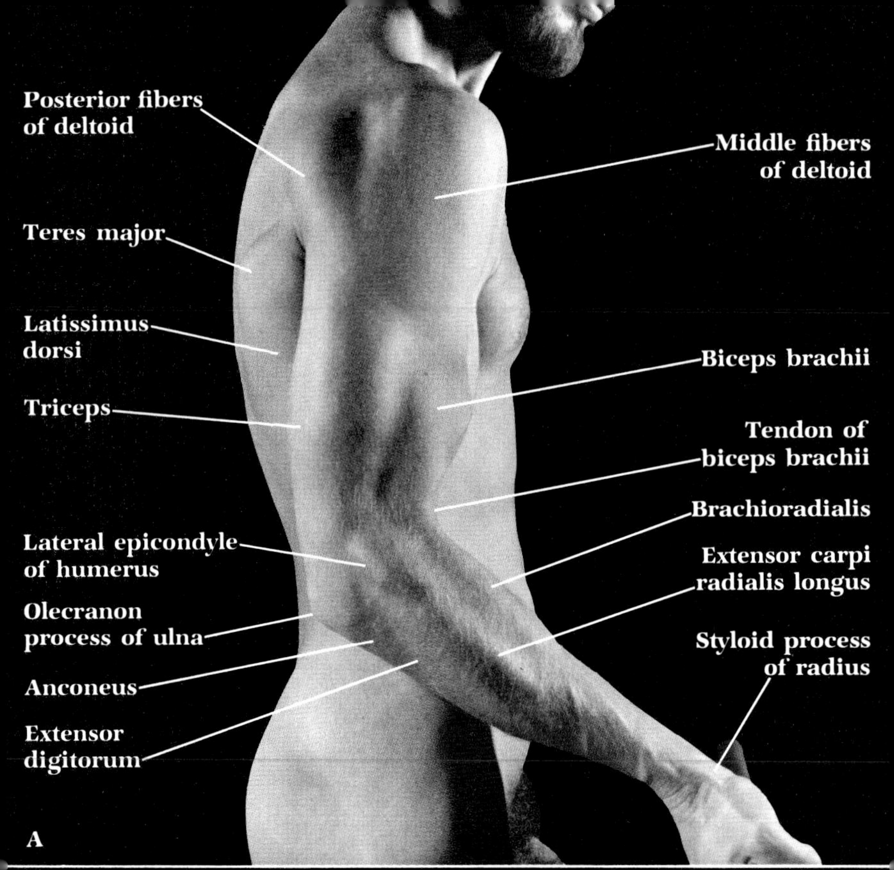
Posterior fibers of deltoid
Middle fibers of deltoid
Teres major
Latissimus dorsi
Biceps brachii
Triceps
Tendon of biceps brachii
Brachioradialis
Lateral epicondyle of humerus
Extensor carpi radialis longus
Olecranon process of ulna
Styloid process of radius
Anconeus
Extensor digitorum

A

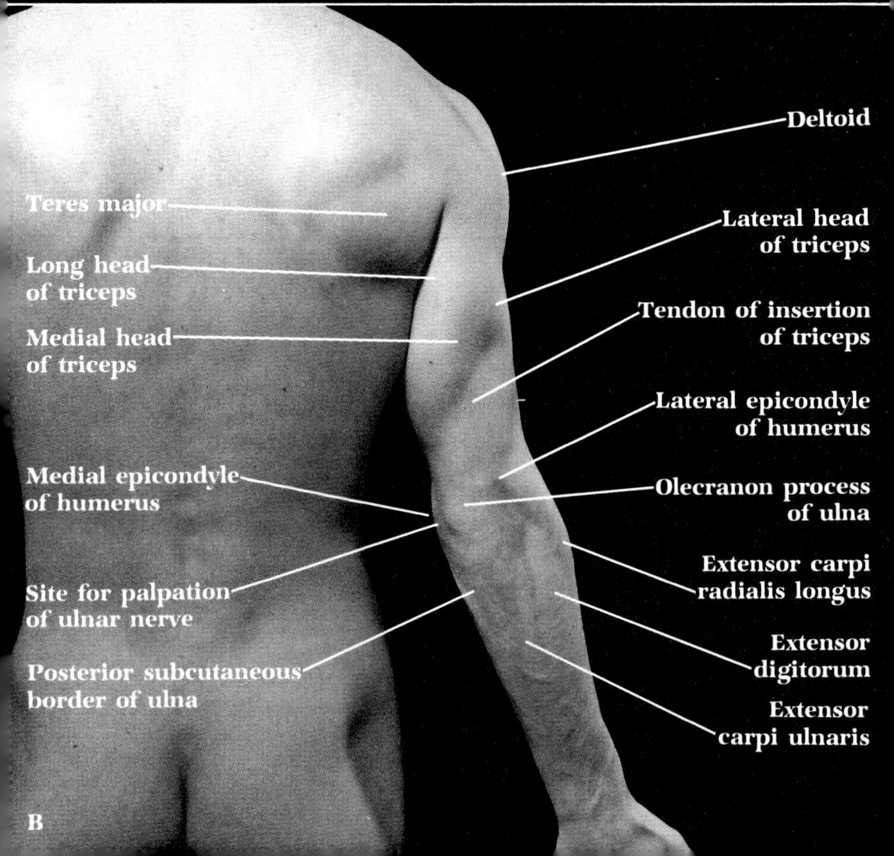
Deltoid
Teres major
Lateral head of triceps
Long head of triceps
Tendon of insertion of triceps
Medial head of triceps
Lateral epicondyle of humerus
Medial epicondyle of humerus
Olecranon process of ulna
Extensor carpi radialis longus
Site for palpation of ulnar nerve
Extensor digitorum
Posterior subcutaneous border of ulna
Extensor carpi ulnaris

B

Figure 5-46
Upper limb in a 27-year-old male. A. Lateral surface. B. Posterior surface.

1. The deltoid, biceps brachii, and the triceps muscles can easily be palpated, as noted previously.
2. In the region of the elbow, the medial and lateral epicondyles of the humerus and the olecranon process of the ulna can be palpated.
3. The head of the radius can be felt in a depression on the posterolateral aspect of the extended elbow, distal to the lateral epicondyle. The head of the radius can be felt to rotate during pronation and supination of the forearm.
4. The ulnar nerve can be palpated as a rounded cord where it lies posterior to the medial epicondyle of the humerus.
5. The posterior border of the ulna bone is subcutaneous and can be felt along its entire length.
6. At the wrist, the styloid processes of the radius and ulna can be palpated; the styloid process of the radius lies about $\frac{3}{4}$ inch (1.9 cm) distal to that of the ulna.
7. The radial artery can be felt anterior to the distal third of the radius; the tendon of the flexor carpi radialis muscle lies medial to this. Palpate the tendons of the palmaris longus, flexor digitorum superficialis, and flexor carpi ulnaris muscles as they lie anterior to the wrist region.
8. The pulsations of the ulnar artery can be felt lateral to the tendon of the flexor carpi ulnaris.
9. The "anatomical snuffbox" lies on the lateral aspect of the wrist distal to the styloid process of the radius. It is bounded medially by the tendon of the extensor pollicis longus and laterally by the tendons of the abductor pollicis longus and extensor pollicis brevis. In its floor are the styloid process of the radius, the scaphoid and trapezium bones, and the base of the first metacarpal bone.
10. The radial artery can be palpated in the anatomical snuffbox as it winds around the lateral margin of the wrist to reach the dorsum of the hand.

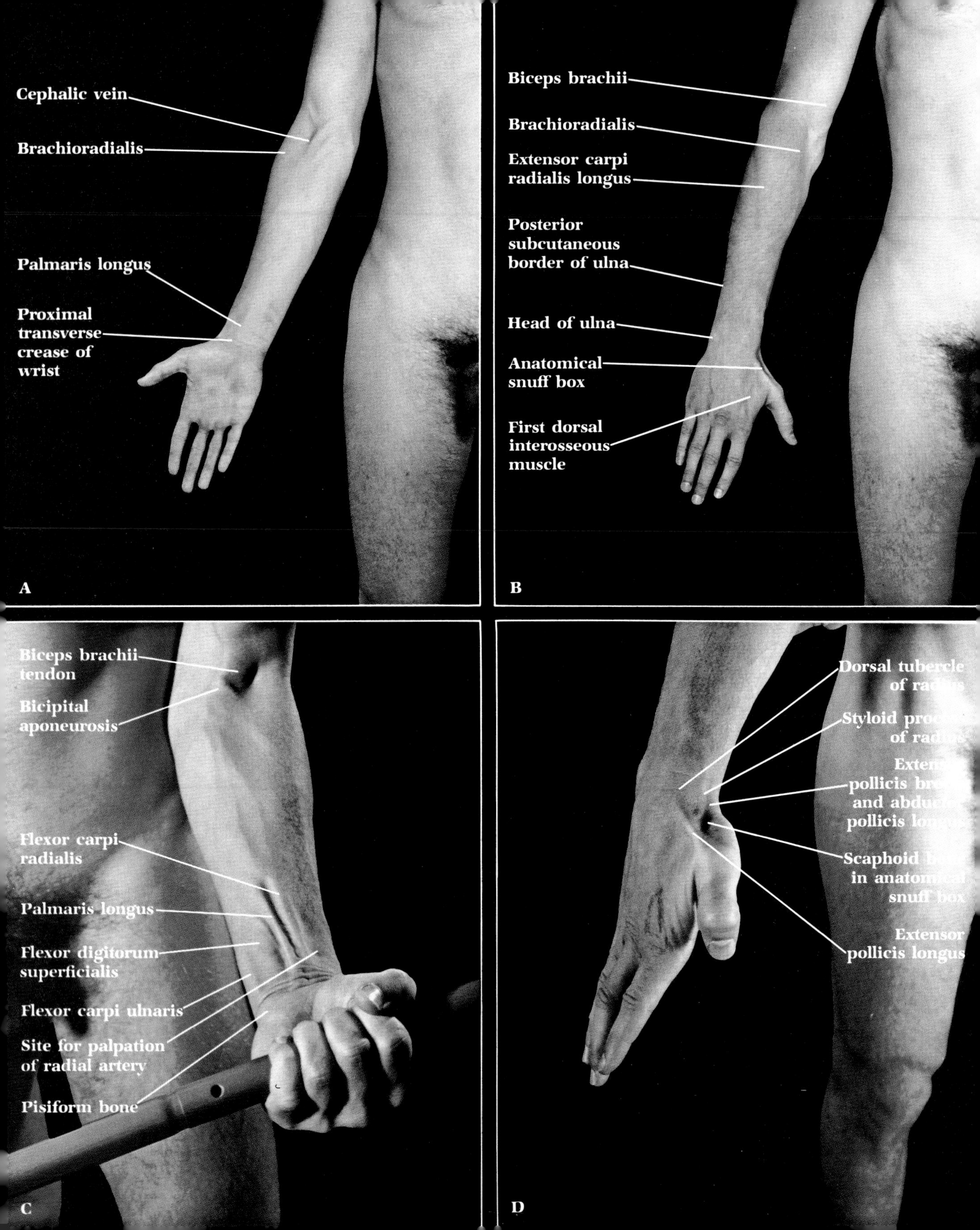

Cephalic vein
Brachioradialis
Palmaris longus
Proximal transverse crease of wrist
A
Biceps brachii
Brachioradialis
Extensor carpi radialis longus
Posterior subcutaneous border of ulna
Head of ulna
Anatomical snuff box
First dorsal interosseous muscle
B
Biceps brachii tendon
Bicipital aponeurosis
Flexor carpi radialis
Palmaris longus
Flexor digitorum superficialis
Flexor carpi ulnaris
Site for palpation of radial artery
Pisiform bone
C
Dorsal tubercle of ra
Styloid pro of ra
pollicis br and abdu pollicis lo
Scaphoid in anatomi snuff box
Extensor pollicis longus
D

Figure 5-47
Forearm and hand in a 27-year-old male. A. Right upper limb, anterior view. B. Right upper limb, forearm pronated. C. Left upper limb, anterior view. D. Right upper limb, forearm in half-prone position.

1. The posterior border of the ulna bone is subcutaneous and can be palpated along its entire length.

2. The styloid processes of the radius and ulna can be palpated at the wrist; the styloid process of the radius lies about $\frac{3}{4}$ inch (1.9 cm) distal to that of the ulna.
3. The dorsal tubercle of the radius is palpable on the posterior surface of the lower end of the radius.
4. The pisiform bone can be felt on the medial side of the anterior aspect of the wrist between the two transverse creases.
5. The hook of the hamate bone can be felt on deep palpation of the hypothenar eminence, which is a fingerbreadth distal and lateral to the pisiform bone.

6. The transverse creases seen in front of the wrist are important landmarks. The proximal transverse crease lies at the level of the wrist joint. The distal transverse crease corresponds to the proximal border of the flexor retinaculum muscle.

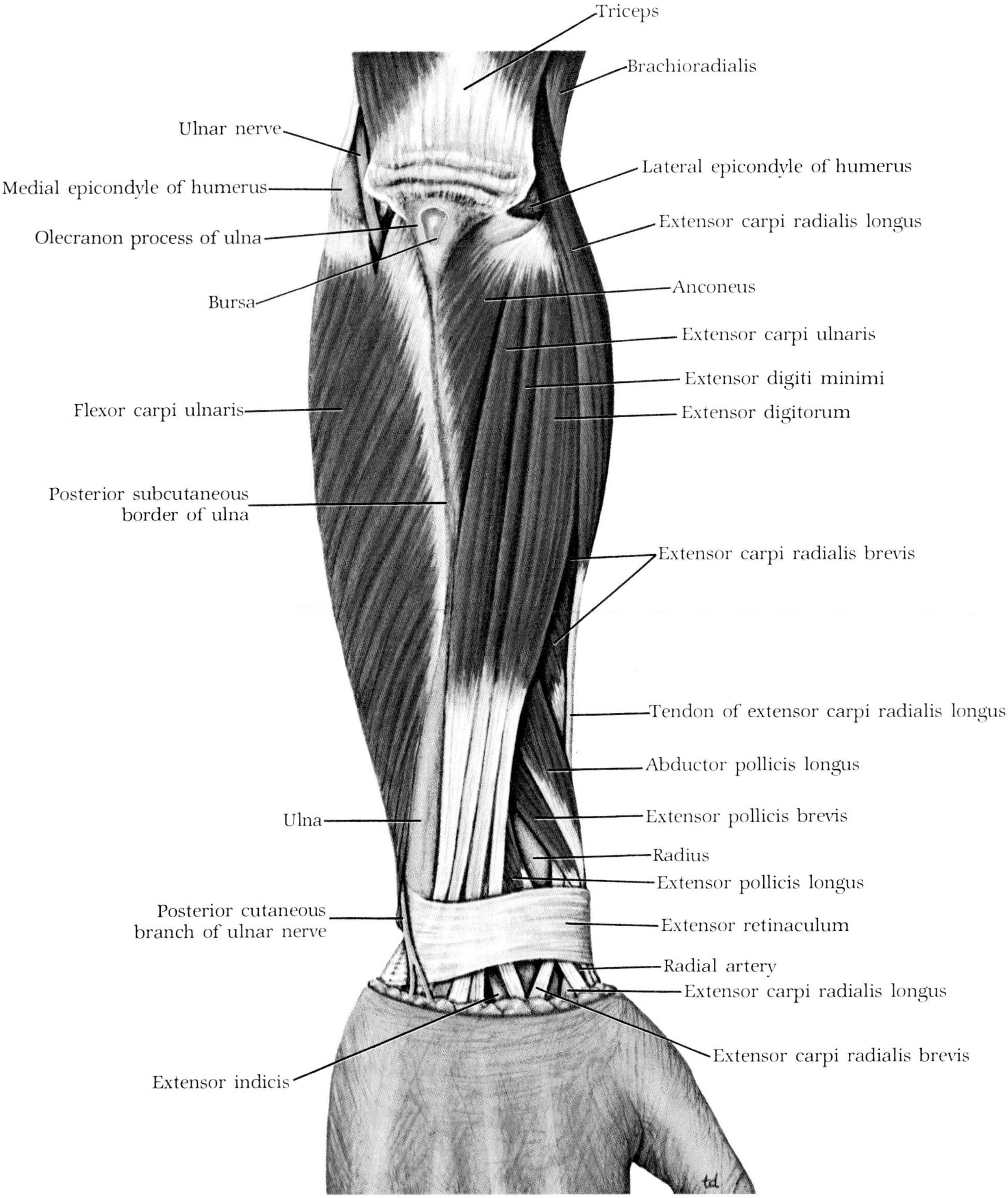
Triceps
Brachioradialis
Ulnar nerve
Medial epicondyle of humerus
Lateral epicondyle of humerus
Olecranon process of ulna
Extensor carpi radialis longus
Bursa
Anconeus
Extensor carpi ulnaris
Extensor digiti minimi
Flexor carpi ulnaris
Extensor digitorum
Posterior subcutaneous border of ulna
Extensor carpi radialis brevis
Tendon of extensor carpi radialis longus
Abductor pollicis longus
Ulna
Extensor pollicis brevis
Radius
Extensor pollicis longus
Posterior cutaneous branch of ulnar nerve
Extensor retinaculum
Radial artery
Extensor carpi radialis longus
Extensor carpi radialis brevis
Extensor indicis

Figure 5-48
Superficial muscles of the forearm, posterior view.

1. Olecranon bursitis is an inflammation of a small superficial bursa that lies between the skin and the olecranon process of the ulna. The bursa can become distended as the result of repeated trauma and produce a rounded swelling.

2. Tennis elbow, believed to be due to a partial tearing of the origin of the superficial extensor muscles from the lateral epicondyle of the humerus, is a painful condition common in tennis players, violinists, and housewives.

3. Fractures of the olecranon process are common in adults and are caused by forced flexion of the elbow joint against the contracting triceps, as when a person falls on an outstretched hand. When this occurs, insertion of the triceps tendon avulses the olecranon process.

4. The dorsal cutaneous branch of the ulnar nerve can be damaged by cuts and penetrating wounds over the distal third of the posterior surface of the ulna. The patient will experience loss of sensation on the dorsal surface of the medial third of the hand and the proximal part of the dorsal surface of the medial one and one-half fingers.

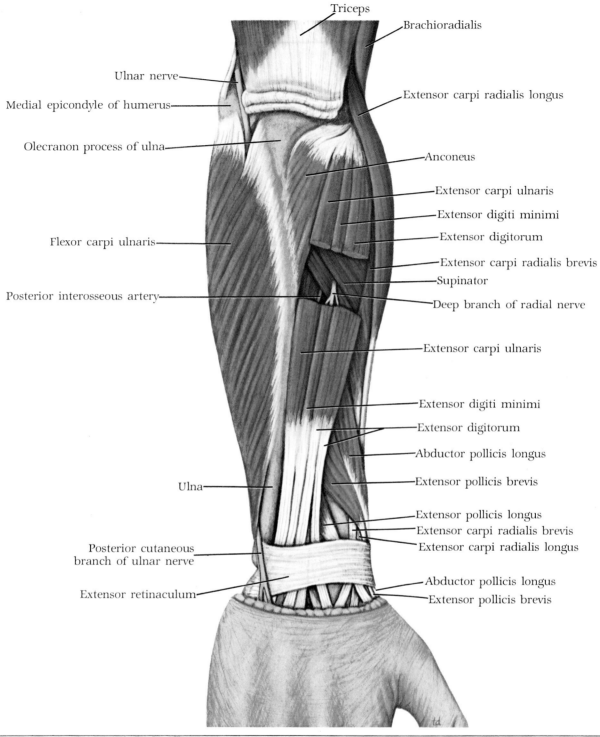

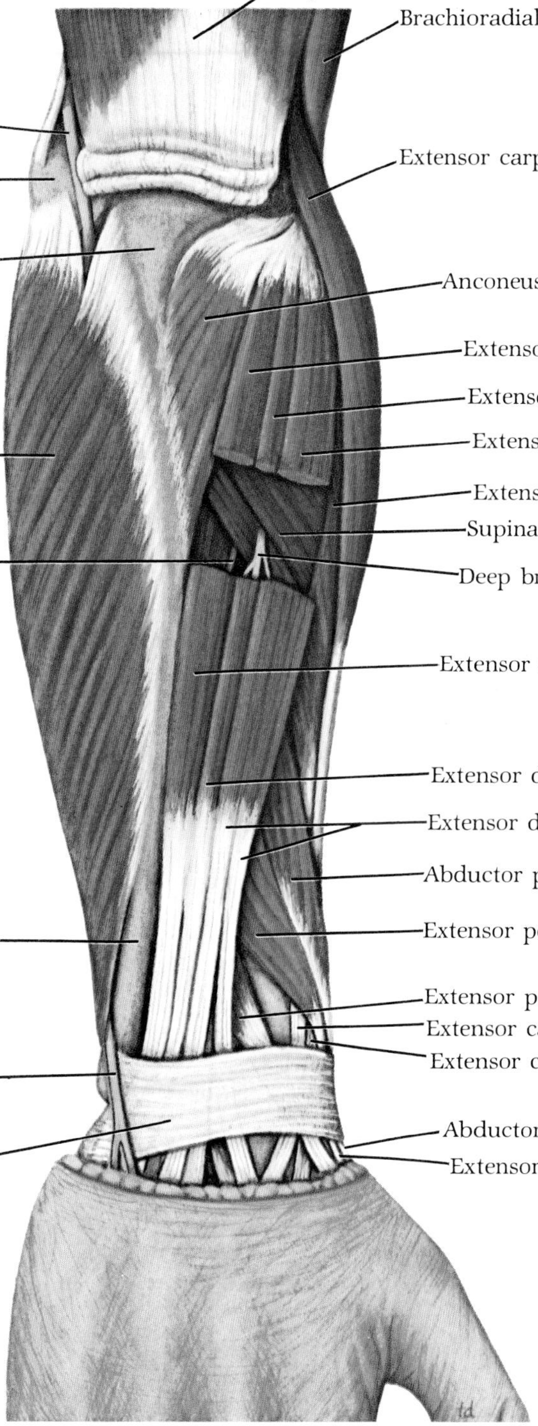
Triceps
Brachioradialis
Ulnar nerve
Medial epicondyle of humerus
Extensor carpi radialis longus
Olecranon process of ulna
Anconeus
Extensor carpi ulnaris
Extensor digiti minimi
Extensor digitorum
Flexor carpi ulnaris
Extensor carpi radialis brevis
Supinator
Posterior interosseous artery
Deep branch of radial nerve
Extensor carpi ulnaris
Extensor digiti minimi
Extensor digitorum
Abductor pollicis longus
Ulna
Extensor pollicis brevis
Extensor pollicis longus
Extensor carpi radialis brevis
Posterior cutaneous branch of ulnar nerve
Extensor carpi radialis longus
Extensor retinaculum
Abductor pollicis longus
Extensor pollicis brevis

Figure 5-49
Forearm, posterior view. Parts of the extensor digitorum, extensor digiti minimi, and extensor carpi ulnaris muscles have been removed to show deep branch of radial nerve and posterior interosseous artery. Note the presence of the ulnar nerve posterior to the medial epicondyle.

1. The ulnar nerve is very commonly injured at the elbow, where it lies posterior to the medial epicondyle of the humerus. The flexor carpi ulnaris and the medial half of the flexor digitorum profundus muscles are paralyzed. The small muscles of the hand are also paralyzed, except the muscles of the thenar eminence and the first two lumbricals, all of which are supplied by the median nerve. Loss of skin sensation is observed over the anterior and posterior surfaces of the medial third of the hand and the medial one and one-half fingers.
The injury to the nerve can be produced at the time of fracture of the medial epicondyle or can be caused later by friction of the nerve against the irregular bony surface at the site of the old fracture. Friction or tension of the nerve can also be a sequel to a gross abduction deformity resulting from a fracture of the lateral part of the distal end of the humerus. Usually the fracture has occurred in childhood, and ten to fifteen years later the patient has symptoms and signs of ulnar nerve palsy. In these cases a new bed for the ulnar nerve can be formed by transposing the nerve to the anterior surface of the medial epicondyle.
2. The deep branch of the radial nerve is well-protected in the posterior compartment of the forearm. When injury occurs, however, the patient is unable to extend the wrist and fingers, there is wrist drop, and a variable area of anesthesia is present over the root of the thumb.

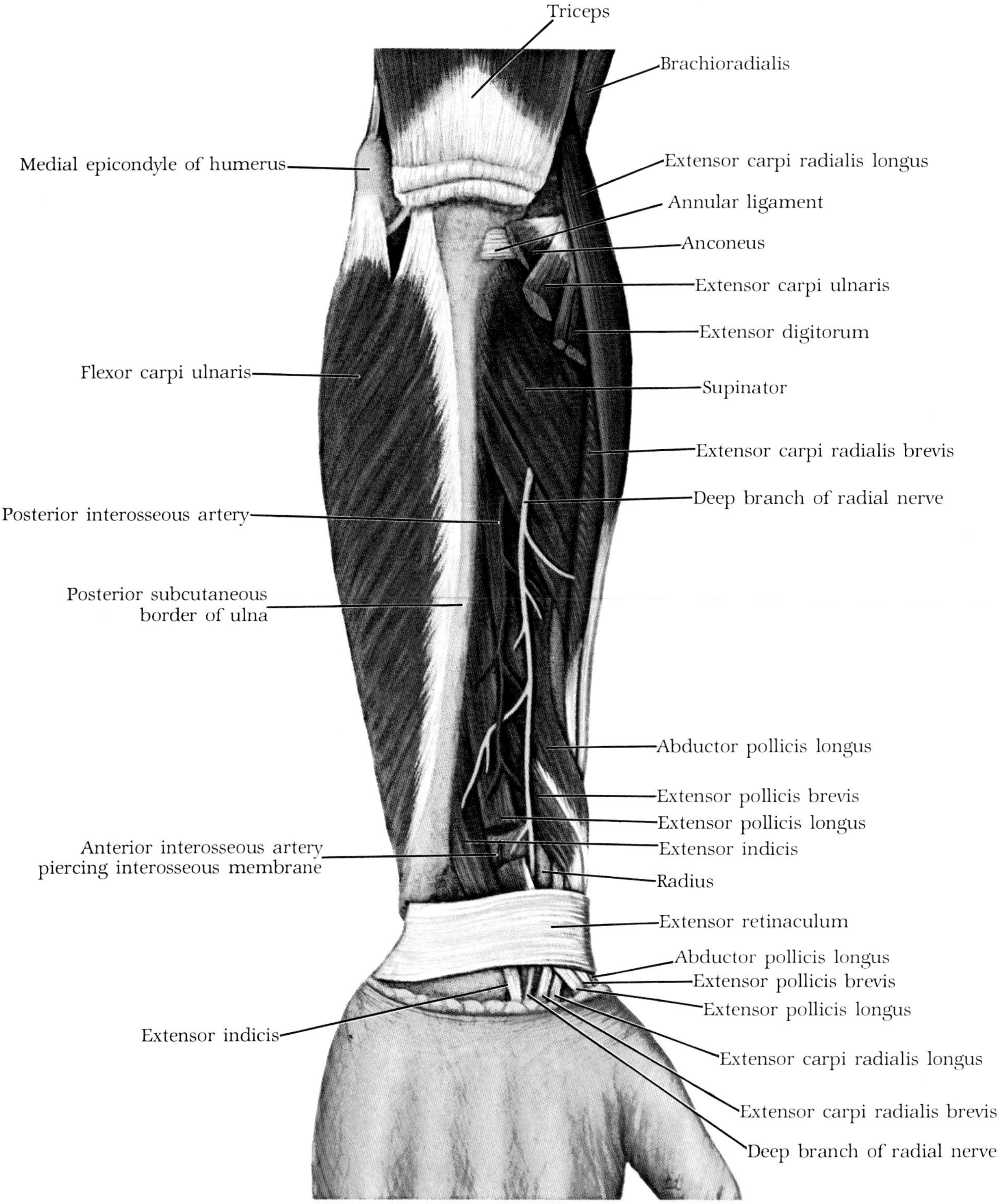
Triceps
Brachioradialis
Medial epicondyle of humerus
Extensor carpi radialis longus
Annular ligament
Anconeus
Extensor carpi ulnaris
Extensor digitorum
Flexor carpi ulnaris
Supinator
Extensor carpi radialis brevis
Deep branch of radial nerve
Posterior interosseous artery
Posterior subcutaneous border of ulna
Abductor pollicis longus
Extensor pollicis brevis
Extensor pollicis longus
Anterior interosseous artery piercing interosseous membrane
Extensor indicis
Radius
Extensor retinaculum
Abductor pollicis longus
Extensor pollicis brevis
Extensor pollicis longus
Extensor indicis
Extensor carpi radialis longus
Extensor carpi radialis brevis
Deep branch of radial nerve

Figure 5-50
Forearm, posterior view. Superficial muscles have been removed to display deep structures.

1. The close relationship of the deep branch of the radial nerve to the neck of the radius as it passes around the lateral surface of the neck of the radius within the substance of the supinator muscle should be noted. It is here that the nerve can be injured in fractures of the neck of the radius or in dislocations of the radius. The nerve can also be damaged during an operation on the neck of the radius.

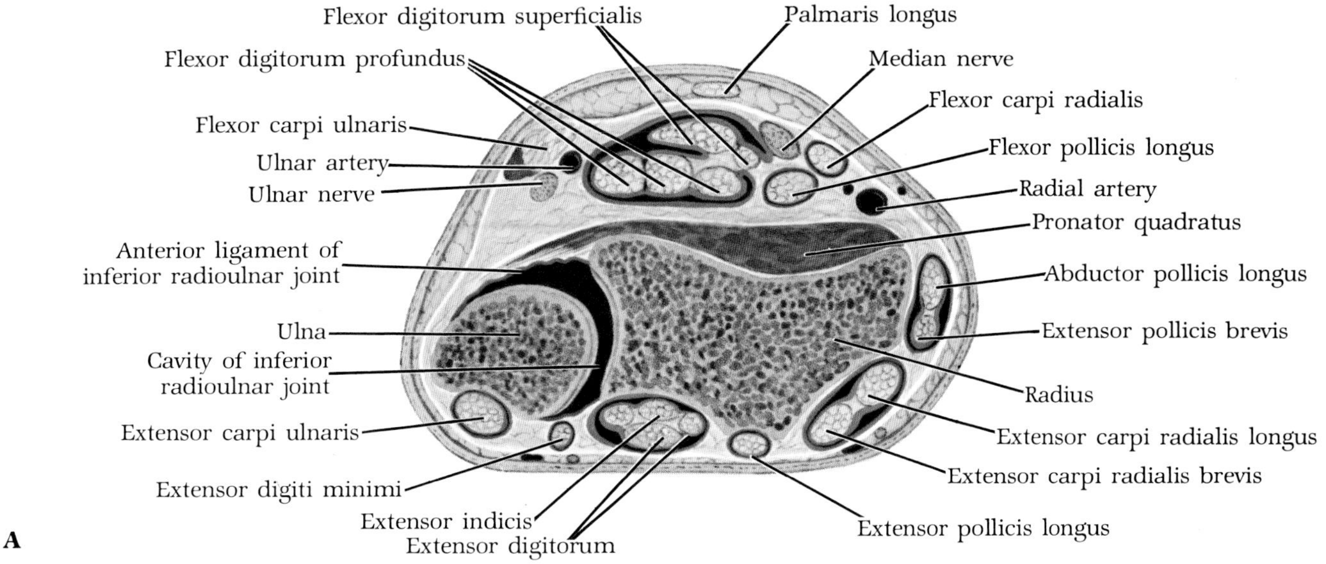
Flexor digitorum superficialis
Palmaris longus
Flexor digitorum profundus
Median nerve
Flexor carpi radialis
Flexor carpi ulnaris
Flexor pollicis longus
Ulnar artery
Radial artery
Ulnar nerve
Pronator quadratus
Anterior ligament of inferior radioulnar joint
Abductor pollicis longus
Ulna
Extensor pollicis brevis
Cavity of inferior radioulnar joint
Radius
Extensor carpi ulnaris
Extensor carpi radialis longus
Extensor carpi radialis brevis
Extensor digiti minimi
Extensor indicis
Extensor digitorum
Extensor pollicis longus
A

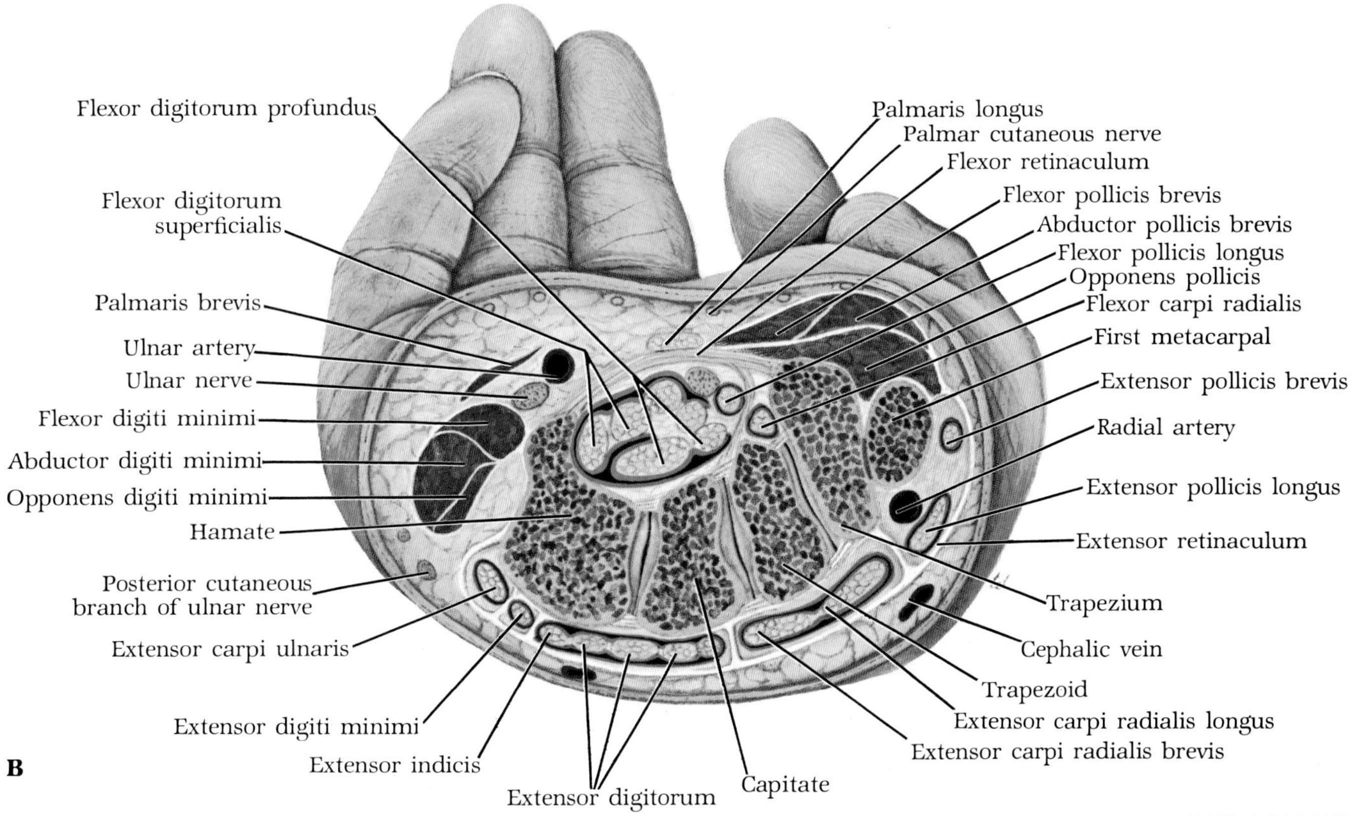
Flexor digitorum profundus
Palmaris longus
Palmar cutaneous nerve
Flexor retinaculum
Flexor digitorum superficialis
Flexor pollicis brevis
Abductor pollicis brevis
Flexor pollicis longus
Opponens pollicis
Palmaris brevis
Flexor carpi radialis
First metacarpal
Ulnar artery
Ulnar nerve
Extensor pollicis brevis
Flexor digiti minimi
Radial artery
Abductor digiti minimi
Opponens digiti minimi
Extensor pollicis longus
Hamate
Extensor retinaculum
Posterior cutaneous branch of ulnar nerve
Trapezium
Extensor carpi ulnaris
Cephalic vein
Trapezoid
Extensor digiti minimi
Extensor carpi radialis longus
Extensor carpi radialis brevis
Extensor indicis
Capitate
Extensor digitorum
B

Figure 5-51
Right forearm and hand. A. Cross section of the forearm at the level of the inferior radioulnar joint. B. Cross section of the hand showing relationship of tendons, nerves, and arteries to flexor and extensor retinacula.

1. The ulnar nerve is frequently injured at the wrist, where, with the ulnar artery, it lies anterior to the flexor retinaculum. *Motor.* The small muscles of the hand will be paralyzed and will show wasting, except for the muscles of the thenar eminence and the first two lumbricals, which are supplied by the median nerve. The patient is unable to adduct and abduct the fingers and consequently is unable to grip a piece of paper placed between the fingers. Remember that the extensor digitorum can abduct the fingers to a small extent, but only when the metacarpophalangeal joints are hyperextended.
It is impossible to adduct the thumb because the adductor pollicis muscle is paralyzed. If the patient is asked to grip a piece of paper between the thumb and the index finger, he does so by strongly contracting his flexor pollicis longus and flexing the terminal phalanx (Froment's sign).
Sensory. The main ulnar nerve and its palmar cutaneous branch are usually severed. The sensory loss is confined to the palmar surface of the medial third of the hand and the medial one and one-half fingers and to the dorsal aspects of the middle and distal phalanges of the same fingers.

2. *Carpal Tunnel Syndrome.* The carpal tunnel is formed by the concave anterior surface of the carpal bones and the flexor retinaculum. It is tightly packed with the long flexor tendons of the fingers, with their surrounding synovial sheaths, and the median nerve. The syndrome consists of burning pain or a sensation of "pins and needles" along the distribution of the median nerve to the lateral three and one-half fingers and weakness of the thenar muscles. It is produced by compression of the median nerve within the tunnel; thickening of the synovial sheaths and arthritic changes in the carpal bones are responsible in many cases. Treatment in severe cases requires division of the flexor retinaculum.

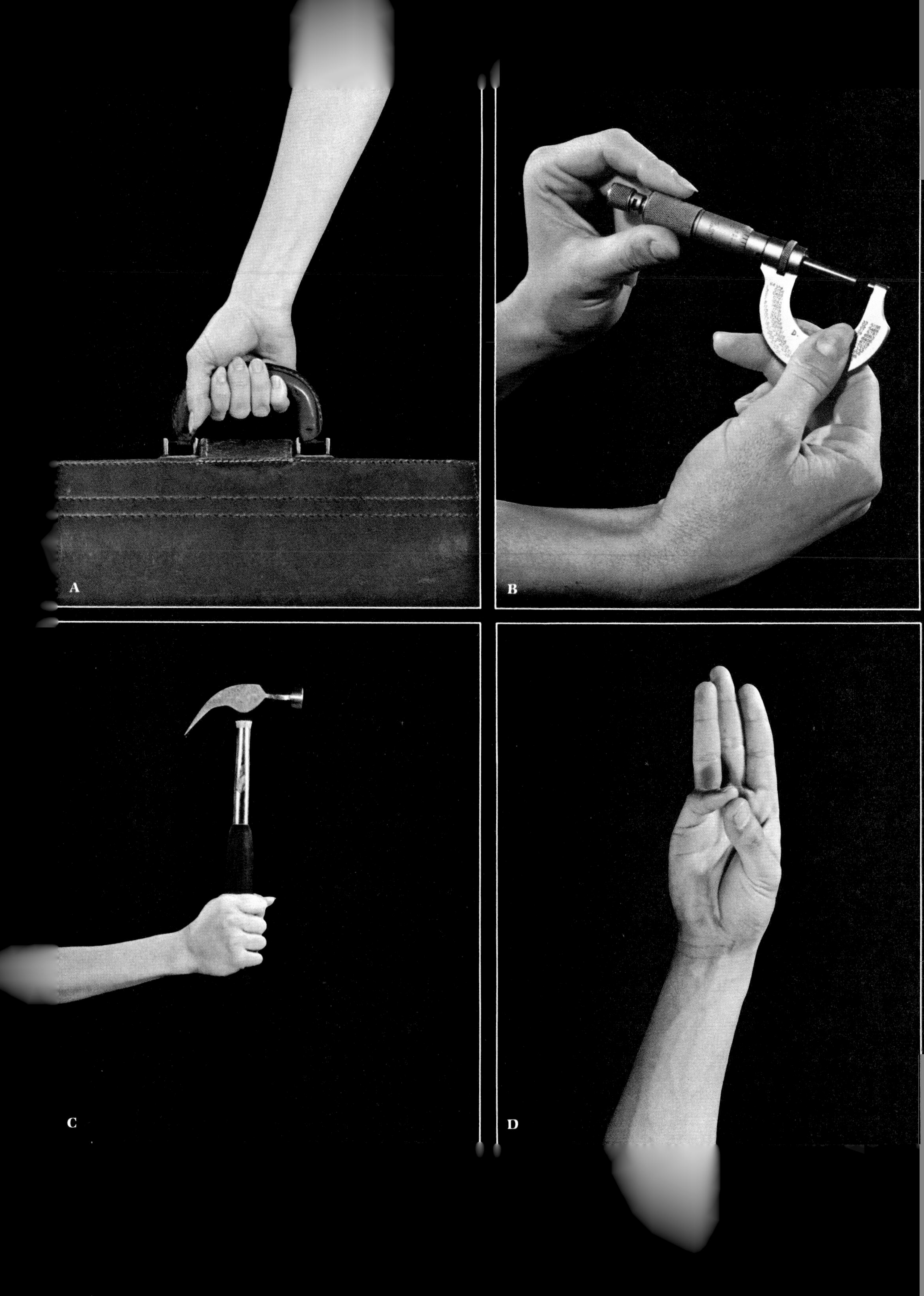
A
B
C
D

Figure 5-52
Functions of the hand. A. Static hook. B. Pinch grip. C. Power grip. D. Opposition movement of the thumb.

1. The upper limb may be regarded as a multijointed lever at the distal end of which is the important prehensile organ, the hand. Because it is the least protected part of the upper limb the incidence of hand injury is high.
2. In the static hook action of the hand the motor power is provided by the extrinsic muscles of the hand, namely, the flexor digitorum superficialis and the flexor digitorum profundus muscles. It should be noted that if the hook action is confined to the ring and little fingers, the ulnar nerve contributes to both the sensory innervation to the skin of those fingers and the nerve supply to the extrinsic muscles acting on those fingers.
3. The pinch grip involves the index finger and thumb and is used for precision movements. For the effective movement of the opposition of the thumb against the stable index finger, the median nerve supply to these two fingers must be intact; this is not only to provide innervation to the muscles of the thenar eminence, but also to give discriminative sensation to the skin areas involved in contact.
4. The power grip, the means of grasping an object, depends first on the long extensor tendons that posture the fingers ready for the grip. The fingers then flex around the object, pressing it into the palm. This action is performed mainly by the long extrinsic muscles, namely, the flexor digitorum superficialis and the flexor digitorum profundus. The index and little fingers play a very important role in this activity since they are situated on each margin of the hand.
5. The unique opposition movement of the thumb that makes the thumb functionally as important as all the remaining fingers combined is the movement of the thumb across the palm in such a manner that the anterior surface of its tip comes into contact with the anterior surface of the tip of any of the other fingers. This is accomplished by the medial rotation of the first metacarpal bone and the attached phalanges on the trapezium bone. The plane of the thumbnail comes to lie parallel with the plane of the nail of the opposed finger. The muscle producing the movement is the opponens pollicis.

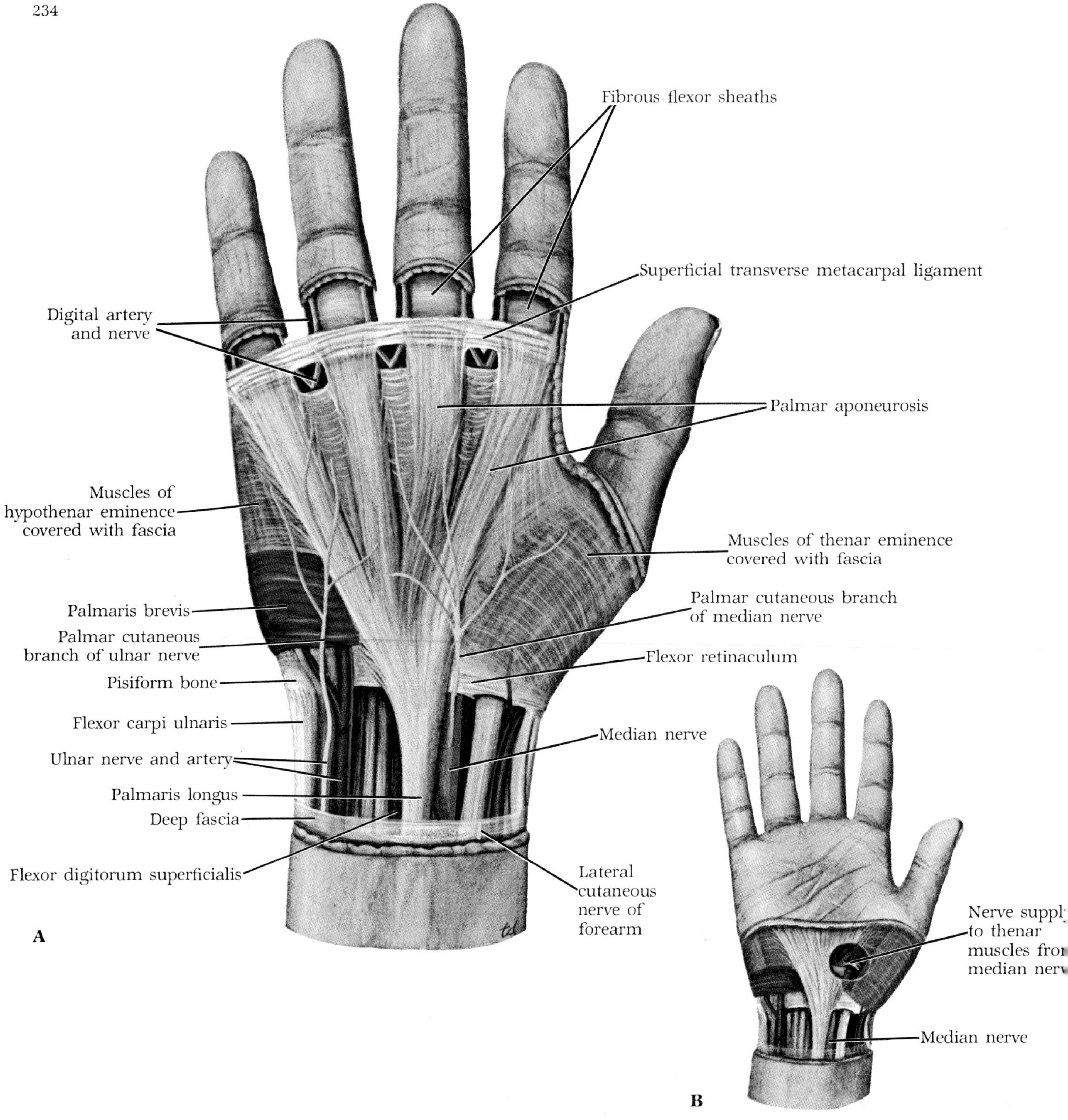
Fibrous flexor sheaths
Superficial transverse metacarpal ligament
Digital artery and nerve
Palmar aponeurosis
Muscles of hypothenar eminence covered with fascia
Muscles of thenar eminence covered with fascia
Palmaris brevis
Palmar cutaneous branch of median nerve
Palmar cutaneous branch of ulnar nerve
Flexor retinaculum
Pisiform bone
Flexor carpi ulnaris
Median nerve
Ulnar nerve and artery
Palmaris longus
Deep fascia
Flexor digitorum superficialis
Lateral cutaneous nerve of forearm
A
Nerve suppl
to thenar
muscles fro
median nerv
Median nerve
B

Figure 5-53
Palm of hand, anterior view.
A. Showing palmar aponeurosis.
B. Showing nerve supply to muscles of thenar eminence.

1. The ulnar nerve and artery are seen passing anterior to the flexor retinaculum and lateral to the pisiform bone. It is here that they are very vulnerable and can be injured by cuts or stab wounds.
2. The median nerve is seen to pass posterior to the flexor retinaculum to enter the carpal tunnel. The nerve may be damaged at the wrist by penetrating wounds.
3. Dupuytren's contracture is a localized thickening and contracture of the palmar aponeurosis. It commonly starts near the root of the ring finger and draws that finger into the palm, flexing it at the metacarpophalangeal joint. Later, the condition progresses to involve the little finger in the same manner. In long-standing cases, the pull on the fibrous flexor sheaths of these fingers results in the flexion of the proximal interphalangeal joints.
4. The nerve supply to the muscles of the thenar eminence, namely, the abductor pollicis brevis, the flexor pollicis brevis, and the opponens pollicis, arises from the median nerve about one finger's breadth distal to the tubercle of the scaphoid bone. At its origin this muscular branch of the median nerve lies beneath the lateral margin of the palmar aponeurosis. It curves proximally and comes to lie on the superficial surface of the flexor pollicis brevis, where it is exposed and likely to be injured in cuts of the hand. Injury to this nerve results in paralysis and wasting of the muscles of the thenar eminence, which then becomes flattened. The thumb is laterally rotated and adducted. Opposition movement of the thumb is impossible, and the pincers action of the hand is lost.

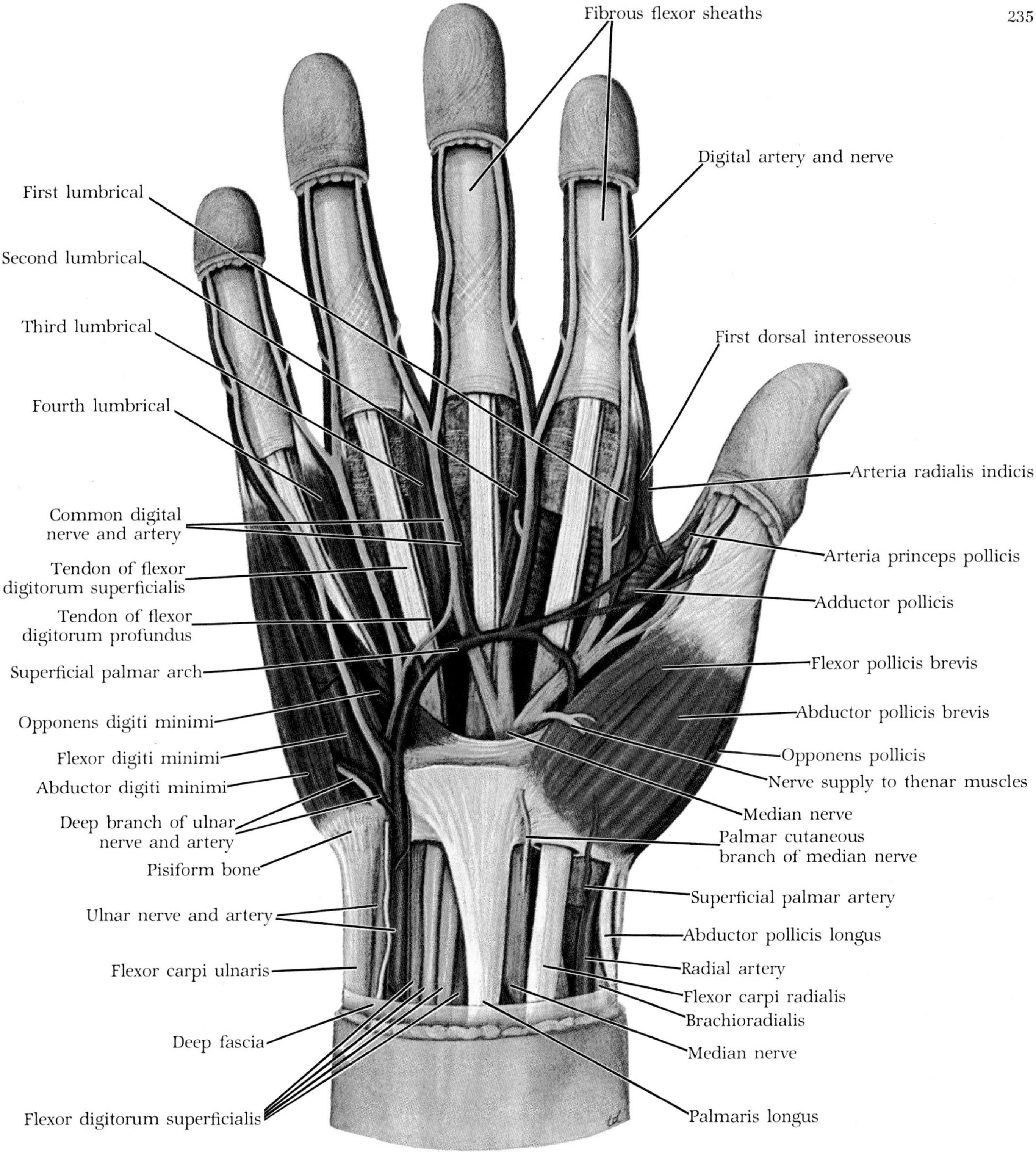
Fibrous flexor sheaths
Digital artery and nerve
First lumbrical
Second lumbrical
Third lumbrical
Fourth lumbrical
First dorsal interosseous
Arteria radialis indicis
Common digital nerve and artery
Arteria princeps pollicis
Tendon of flexor digitorum superficialis
Adductor pollicis
Tendon of flexor digitorum profundus
Superficial palmar arch
Flexor pollicis brevis
Opponens digiti minimi
Abductor pollicis brevis
Flexor digiti minimi
Opponens pollicis
Abductor digiti minimi
Nerve supply to thenar muscles
Deep branch of ulnar nerve and artery
Median nerve
Palmar cutaneous branch of median nerve
Pisiform bone
Superficial palmar artery
Ulnar nerve and artery
Abductor pollicis longus
Flexor carpi ulnaris
Radial artery
Flexor carpi radialis
Brachioradialis
Deep fascia
Median nerve
Flexor digitorum superficialis
Palmaris longus

Figure 5-54
Palm of hand, anterior view. Palmar aponeurosis has been removed to display the superficial palmar arch, median nerve, and long flexor tendons.

1. Looking at the palm of the hand without the palmar aponeurosis, one can appreciate how its toughness protects the underlying nerves, arteries, tendons, and synovial sheaths.
2. The vulnerability of the ulnar nerve and artery as they lie anterior to the flexor retinaculum and of the median nerve as it lies proximal to the flexor retinaculum must again be emphasized. Note the superficial position of the muscular branch of the median nerve as it goes to supply the muscles of the thenar eminence.
3. In testing for injury to the ulnar and median nerves, it is important to verify whether or not the sensory distribution of these nerves in the hand is intact. The palmar cutaneous branch of the ulnar nerve supplies the skin over the hypothenar eminence. Once the ulnar nerve enters the hand, the superficial branch supplies the skin of the palmar aspect of the medial one and one-half fingers, including their nail beds. The palmar cutaneous branch of the median nerve supplies the skin over the lateral half of the palm, and the digital branches of the median nerve supply the skin of the palmar aspect of the lateral three and one-half fingers; the skin of the distal parts of the dorsal surfaces of the lateral three and one-half fingers are also supplied by the digital branches.
4. The superficial palmar arterial arch is located in the central part of the palm and lies on a line drawn across the palm at the level of the distal border of the fully extended thumb. It is occasionally damaged by deep perforating wounds.

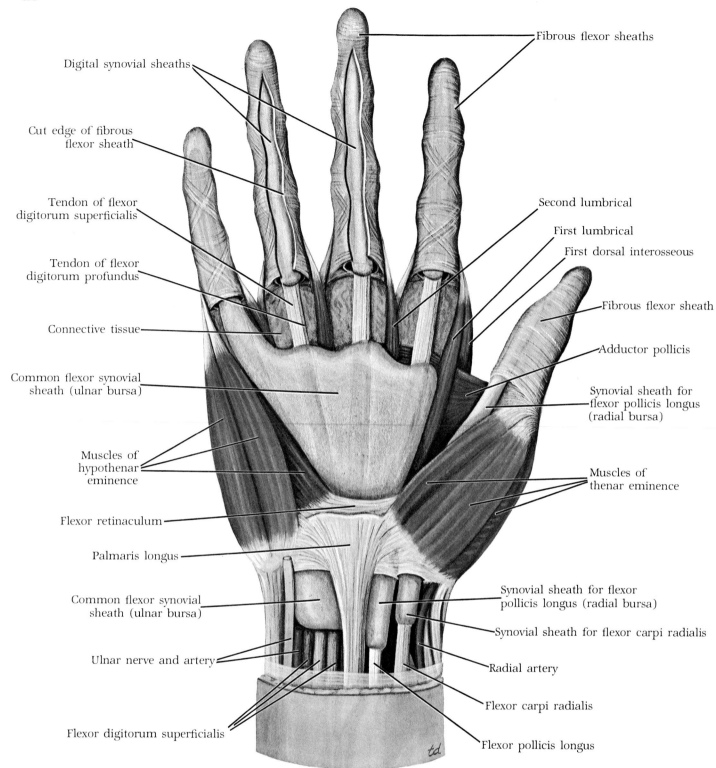

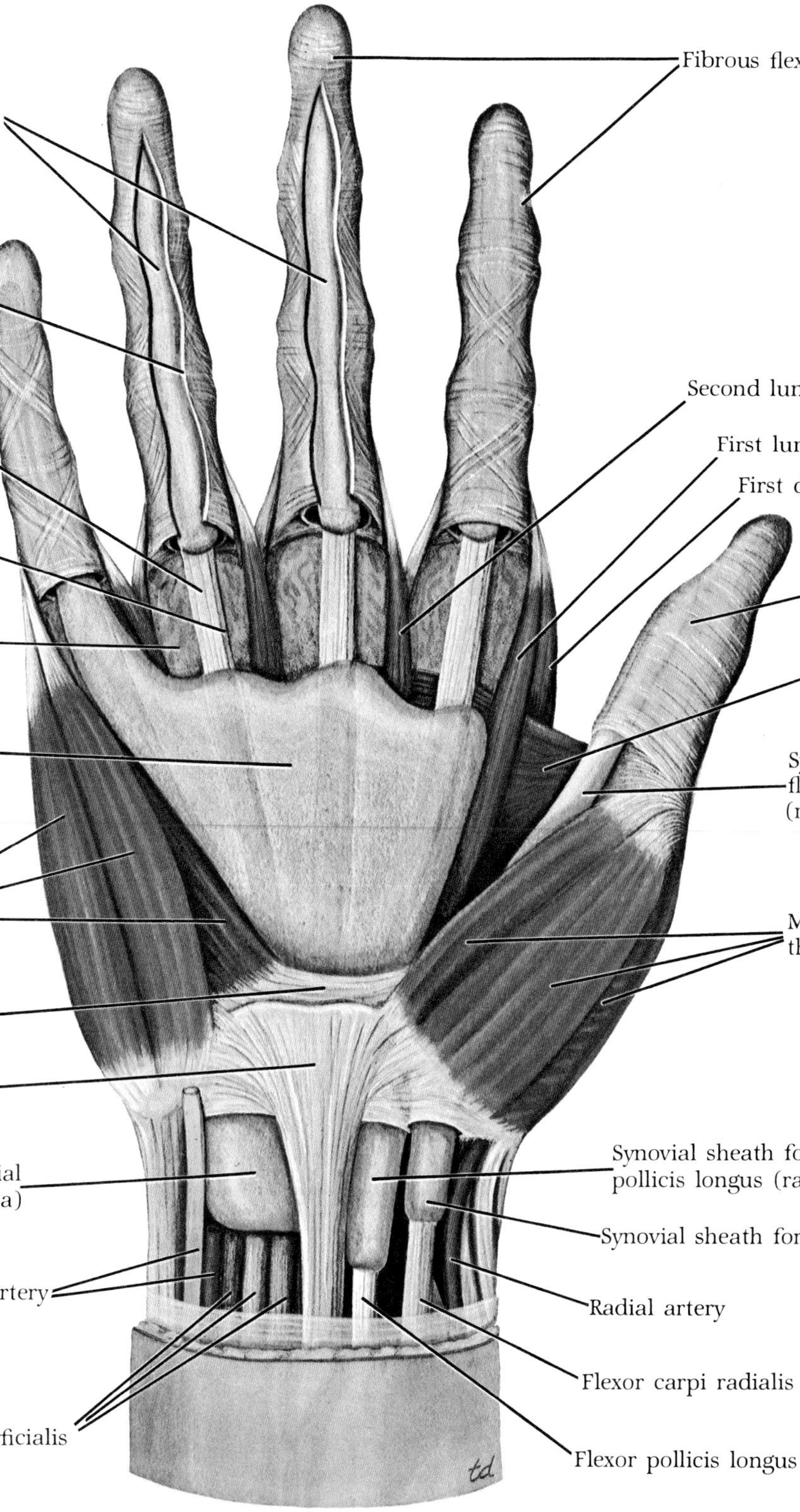
Fibrous flexor sheaths
Digital synovial sheaths
Cut edge of fibrous flexor sheath
Tendon of flexor digitorum superficialis
Tendon of flexor digitorum profundus
Connective tissue
Common flexor synovial sheath (ulnar bursa)
Muscles of hypothenar eminence
Flexor retinaculum
Palmaris longus
Common flexor synovial sheath (ulnar bursa)
Ulnar nerve and artery
Flexor digitorum superficialis
Second lumbrical
First lumbrical
First dorsal interosseous
Fibrous flexor sheath
Adductor pollicis
Synovial sheath for flexor pollicis longus (radial bursa)
Muscles of thenar eminence
Synovial sheath for flexor pollicis longus (radial bursa)
Synovial sheath for flexor carpi radialis
Radial artery
Flexor carpi radialis
Flexor pollicis longus
td

Figure 5-55
Palm of hand, showing flexor synovial sheaths, anterior view.

1. Flexion of the fingers for making a fist is produced at the distal interphalangeal joints by the flexor digitorum profundus and at the proximal interphalangeal joints by the flexor digitorum superficialis. The lumbrical and interosseous muscles are the primary flexors of the metacarpophalangeal joints. During the movement of flexion the fingers converge and adapt their position to hold a spherical object.

2. Trigger finger is a condition in which there is a sudden palpable and even audible snapping when a patient is asked to flex and extend his fingers. The condition is due to the presence of a localized swelling of one of the long flexor tendons that catches on a narrowing of the fibrous flexor sheath anterior to the metacarpophalangeal joint. It may take place either in flexion or in extension. A similar condition occurring in the thumb is called trigger thumb. The situation can be relieved surgically by incising the fibrous flexor sheath.

3. The synovial sheaths are essentially lubricating mechanisms that allow the long flexor tendons to move smoothly beneath the flexor retinaculum and the fibrous flexor sheaths of the fingers. The synovial sheath of the flexor pollicis longus communicates with the common synovial sheath of the superficialis and profundus tendons in about half the population.

4. Tenosynovitis, an infection of a synovial sheath, follows the introduction of bacteria through a small penetrating wound. In infections of the digital sheaths of the little finger and thumb, the ulnar and radial bursae quickly become involved.

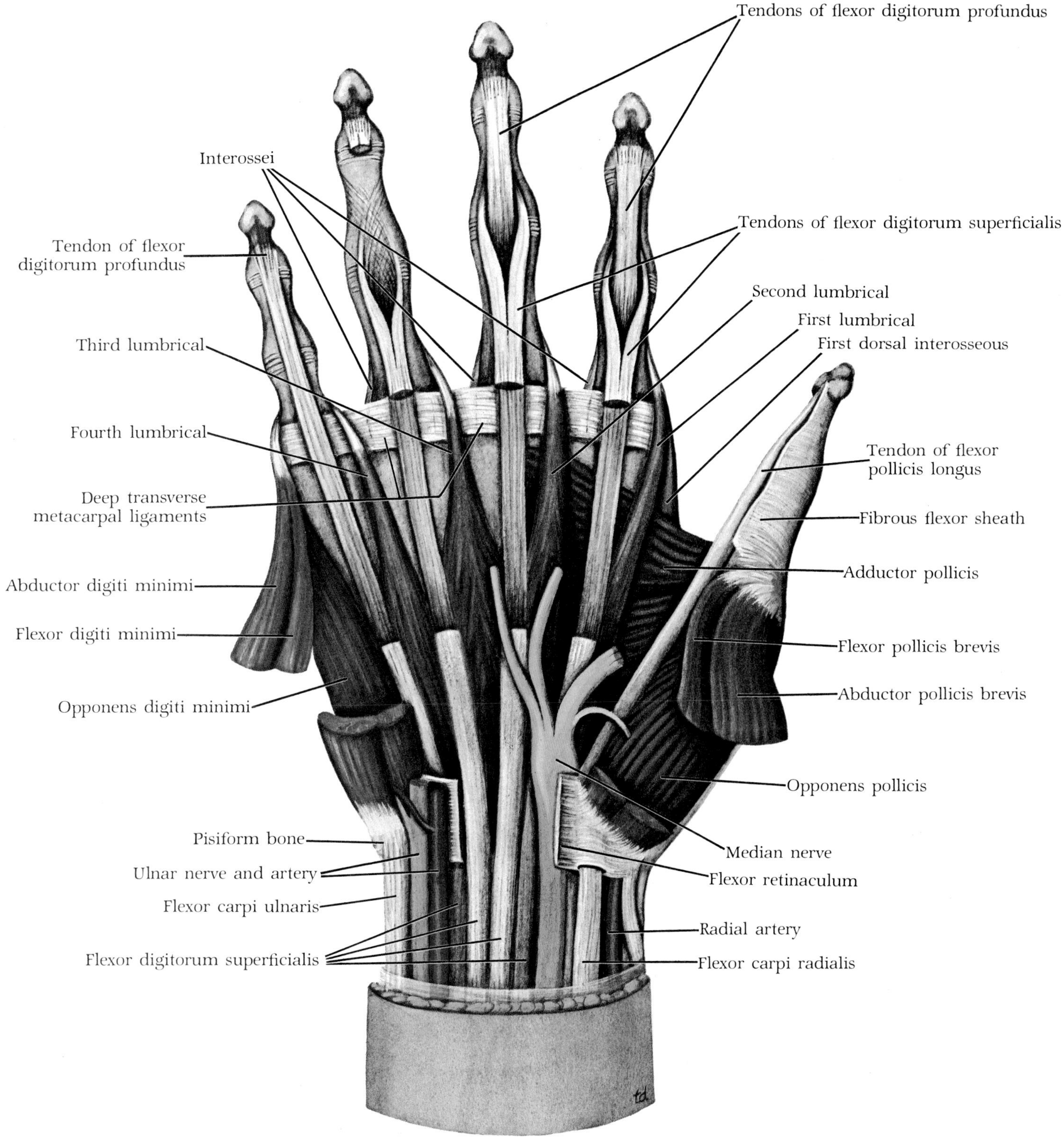
Tendons of flexor digitorum profundus
Interossei
Tendon of flexor digitorum profundus
Tendons of flexor digitorum superficialis
Second lumbrical
First lumbrical
First dorsal interosseous
Third lumbrical
Fourth lumbrical
Tendon of flexor pollicis longus
Deep transverse metacarpal ligaments
Fibrous flexor sheath
Adductor pollicis
Abductor digiti minimi
Flexor digiti minimi
Flexor pollicis brevis
Abductor pollicis brevis
Opponens digiti minimi
Opponens pollicis
Pisiform bone
Median nerve
Ulnar nerve and artery
Flexor retinaculum
Flexor carpi ulnaris
Radial artery
Flexor digitorum superficialis
Flexor carpi radialis
td

Figure 5-56
Palm of hand, anterior view. Palmar aponeurosis and greater part of flexor retinaculum have been removed to display median nerve, long flexor tendons, and the lumbrical muscles. Segments of tendons of the flexor digitorum superficialis muscle have been removed to show underlying tendons of flexor digitorum profundus muscle.

1. The rounded thenar eminence is formed by only three muscles: the abductor pollicis brevis, the flexor pollicis brevis, and the opponens pollicis; all three are supplied by a branch from the median nerve.
2. The smaller hypothenar eminence is also formed by three muscles: the abductor digiti minimi, the flexor digiti minimi, and the opponens digiti minimi. All these are supplied by the deep branch of the ulnar nerve.
3. Knowledge of the crowded carpal tunnel containing the median nerve and the long flexor tendons should again be emphasized. It is also important to remember the carpal tunnel syndrome.
4. In the flexor digitorum superficialis test, the patient's fingers are held in extension, except for the finger being tested. This isolates the flexor digitorum superficialis tendon since the tendons of flexor digitorum profundus only work together. Now ask the patient to flex the finger being examined at the proximal interphalangeal joint. If he can, the superficialis tendon is intact.
5. In the flexor digitorum profundus test, the patient's finger is held in such a way that movement at the metacarpophalangeal joint and the proximal interphalangeal joint is prevented. Because the tendons of the flexor digitorum profundus only move together, by fixing three of them, movement of the fourth prevented. Now ask the patient to flex his finger at the distal interphalangeal joint. If he can, the profundus tendon is intact.

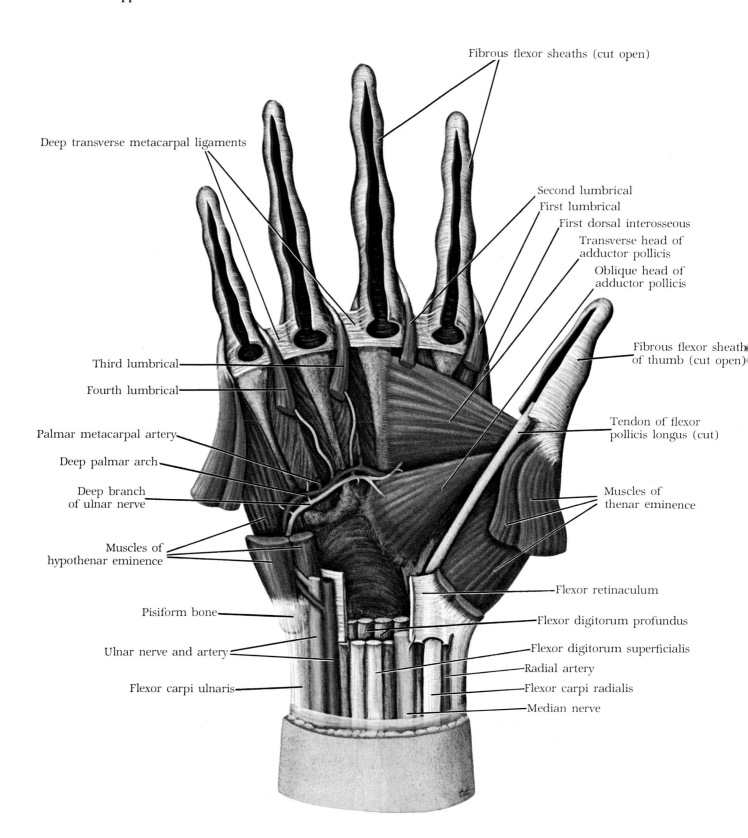

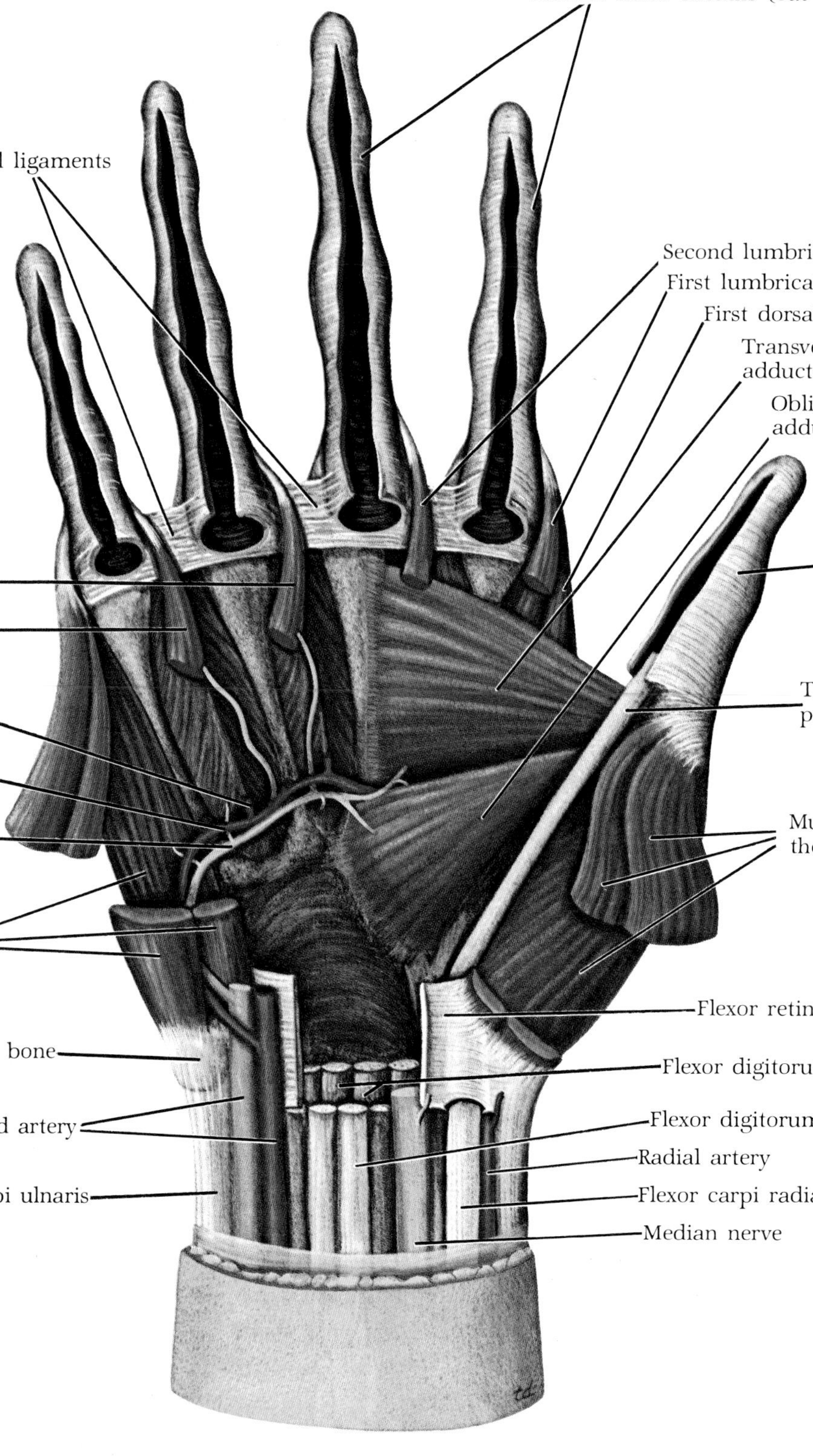
Fibrous flexor sheaths (cut open)
Deep transverse metacarpal ligaments
Second lumbrical
First lumbrical
First dorsal interosseous
Transverse head of adductor pollicis
Oblique head of adductor pollicis
Fibrous flexor sheath of thumb (cut open)
Third lumbrical
Fourth lumbrical
Palmar metacarpal artery
Tendon of flexor pollicis longus (cut)
Deep palmar arch
Deep branch of ulnar nerve
Muscles of thenar eminence
Muscles of hypothenar eminence
Flexor retinaculum
Pisiform bone
Flexor digitorum profundus
Ulnar nerve and artery
Flexor digitorum superficialis
Radial artery
Flexor carpi ulnaris
Flexor carpi radialis
Median nerve

Figure 5-57
Palm of hand, anterior view.
The long flexor tendons have been removed.

1. Nearly two-thirds of all tumors occurring in the hand are ganglia. A ganglion is a cystic swelling arising in the neighborhood of a joint or fibrous tendon sheath. It contains a clear jellylike material and is believed to be a mucinous degeneration of the fibrous connective tissue.

2. The deep branch of the ulnar nerve is well-protected in the palm and supplies the muscles of the hypothenar eminence, all the palmar and dorsal interossei, the third and fourth lumbricals, and the adductor pollicis muscle. The integrity of this nerve can be tested by asking the patient to abduct and adduct the fingers at the metacarpo-phalangeal joints and to adduct the thumb at the joint between the trapezium and the first metacarpal bone.

3. The deep branch of the ulnar nerve and the ulnar artery pass through the interval between the pisiform bone and the hook of the hamate bone. The pisohamate ligament converts this passage into a fibro-osseous tunnel. Compression of the nerve within the tunnel can produce signs and symptoms of injury.

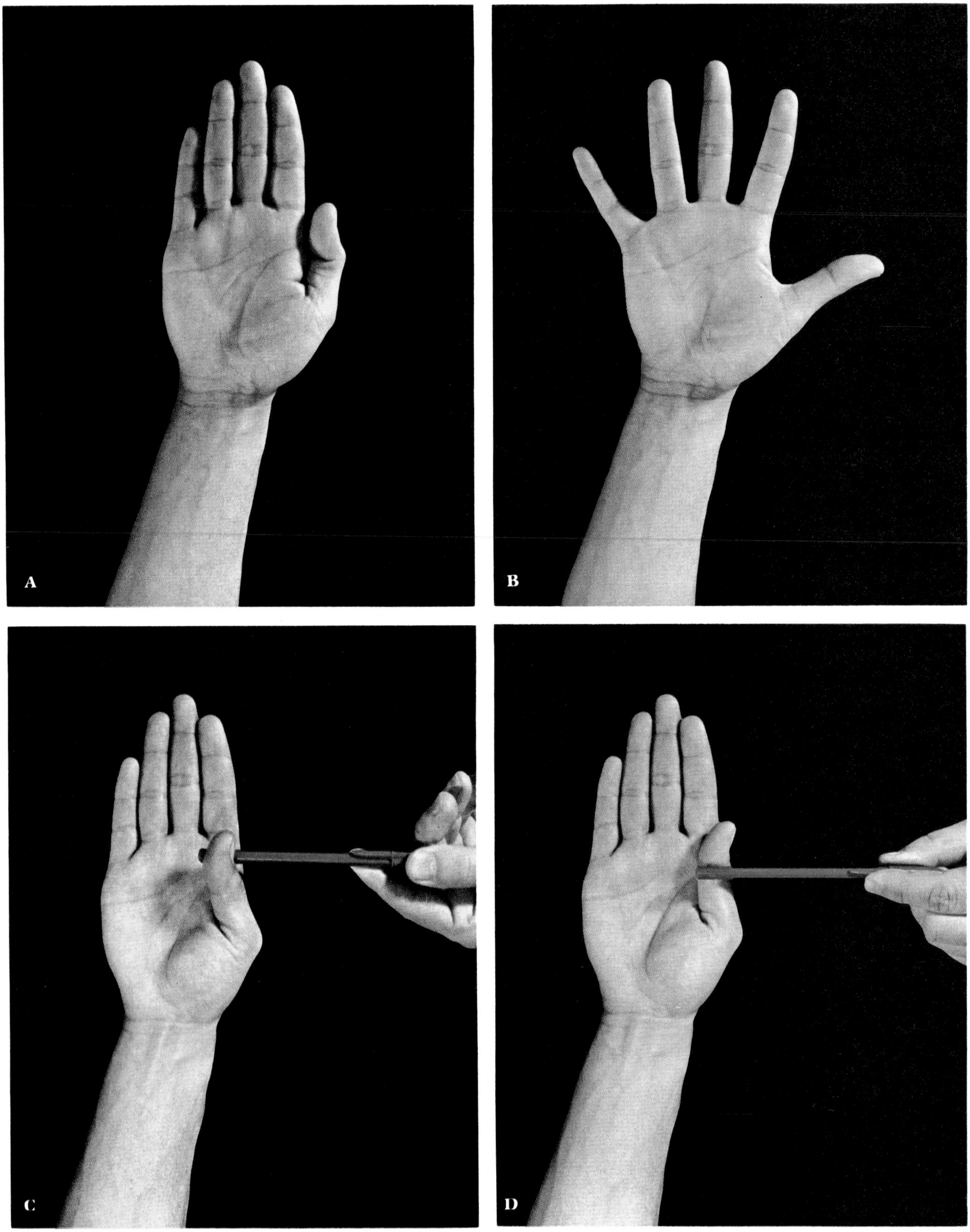
A
B
C
D

Figure 5-58
Movements of the fingers.
A. Adduction. B. Abduction.
C. Adduction of the thumb.
D. Abduction of the thumb.

1. Adduction is the movement of the fingers toward the midline of the middle finger. The movement takes place at the metacarpophalangeal joint; the muscles producing it are the palmar interossei, which are supplied by the deep branch of the ulnar nerve.

2. Abduction is the movement of the fingers (including the middle finger) away from the imaginary midline of the middle finger. The movement also takes place at the metacarpophalangeal joint; the muscles producing it are the dorsal interossei. The abductor digiti minimi muscle abducts the little finger. All these muscles are supplied by the deep branch of the ulnar nerve.

3. Adduction of the thumb is a movement in an anteroposterior plane toward the palm, the plane of the thumbnail being kept at right angles to the plane of the other fingernails. The movement takes place between the trapezium bone and the first metacarpal bone; the muscle producing it is the adductor pollicis, which is supplied by the deep branch of the ulnar nerve.

4. Abduction of the thumb is a movement in an anteroposterior plane away from the palm, the plane of the thumbnail being kept at right angles to the plane of the other fingernails. The movement takes place mainly between the trapezium bone and the first metacarpal bone; a small amount of movement takes place at the metacarpophalangeal joint. The muscles producing the movement are the abductor pollicis longus, which is supplied by the deep branch of the radial nerve, and the abductor pollicis brevis, which is supplied by the median nerve.

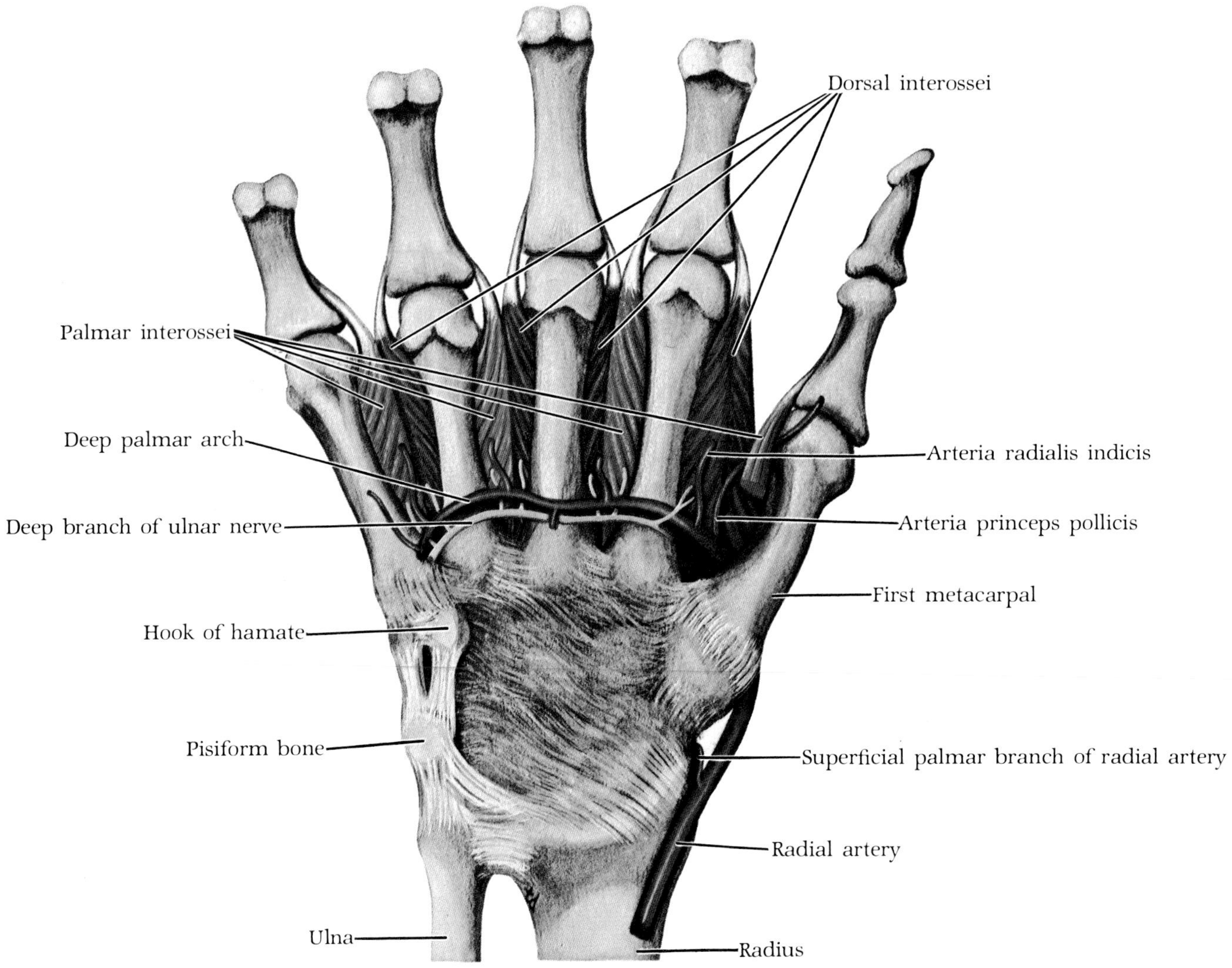
Dorsal interossei
Palmar interossei
Deep palmar arch
Deep branch of ulnar nerve
Arteria radialis indicis
Arteria princeps pollicis
First metacarpal
Hook of hamate
Pisiform bone
Superficial palmar branch of radial artery
Radial artery
Ulna
Radius

Figure 5-59
Palm of hand, showing deep palmar arch, deep terminal branch of ulnar nerve, and interossei, anterior view.

1. The deep palmar arterial arch is formed by the union of the radial artery and the deep branch of the ulnar artery. On the surface of the palm, it can be represented by a curved line drawn across the palm at the level of the proximal border of the fully extended thumb. It contributes to the arterial supply of the fingers by giving off branches that join the digital branches of the superficial palmar arch.

2. The arterial supply to the hand can be tested by first looking at the color of the nail beds. Normal nail beds are pink in color but blanch when pressure is applied to the nail; the pink color quickly returns when the pressure is released. The pulsations of the radial artery can be felt as it lies on the anterior surface of the distal third of the radius just lateral to the tendon of the flexor carpi radialis; pulsations can also be felt in the floor of the anatomical snuff box. The pulse of the ulnar artery can be felt as it crosses anterior to the flexor retinaculum, just lateral to the pisiform bone.

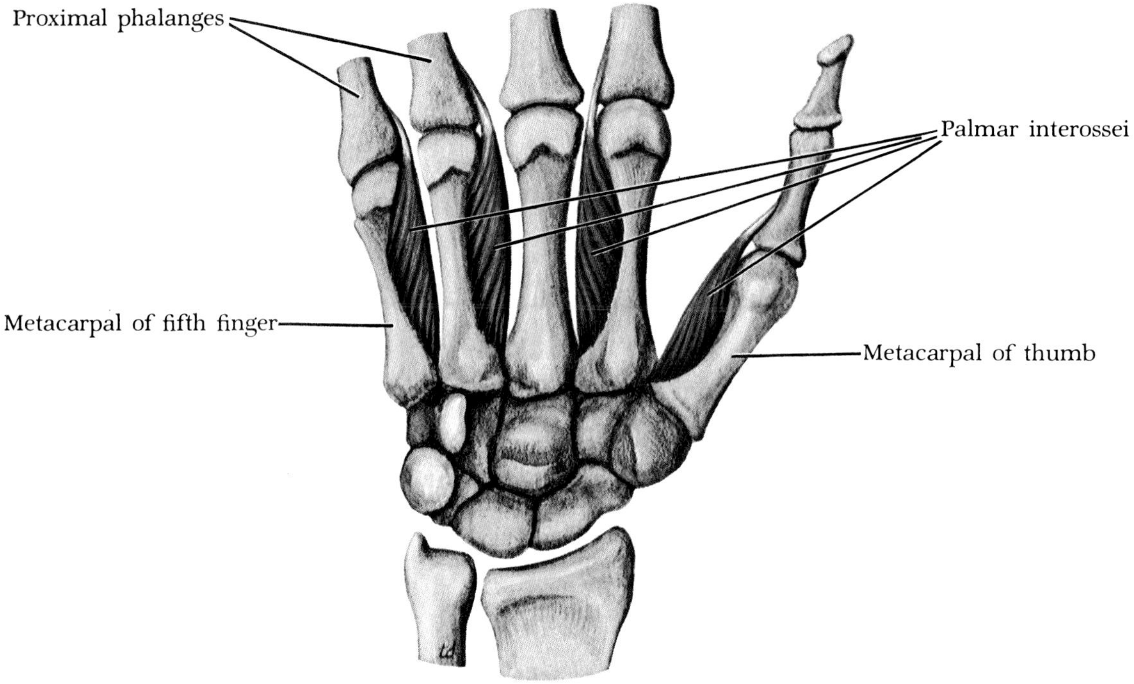

Proximal phalanges
Palmar interossei
Metacarpal of fifth finger
Metacarpal of thumb

A

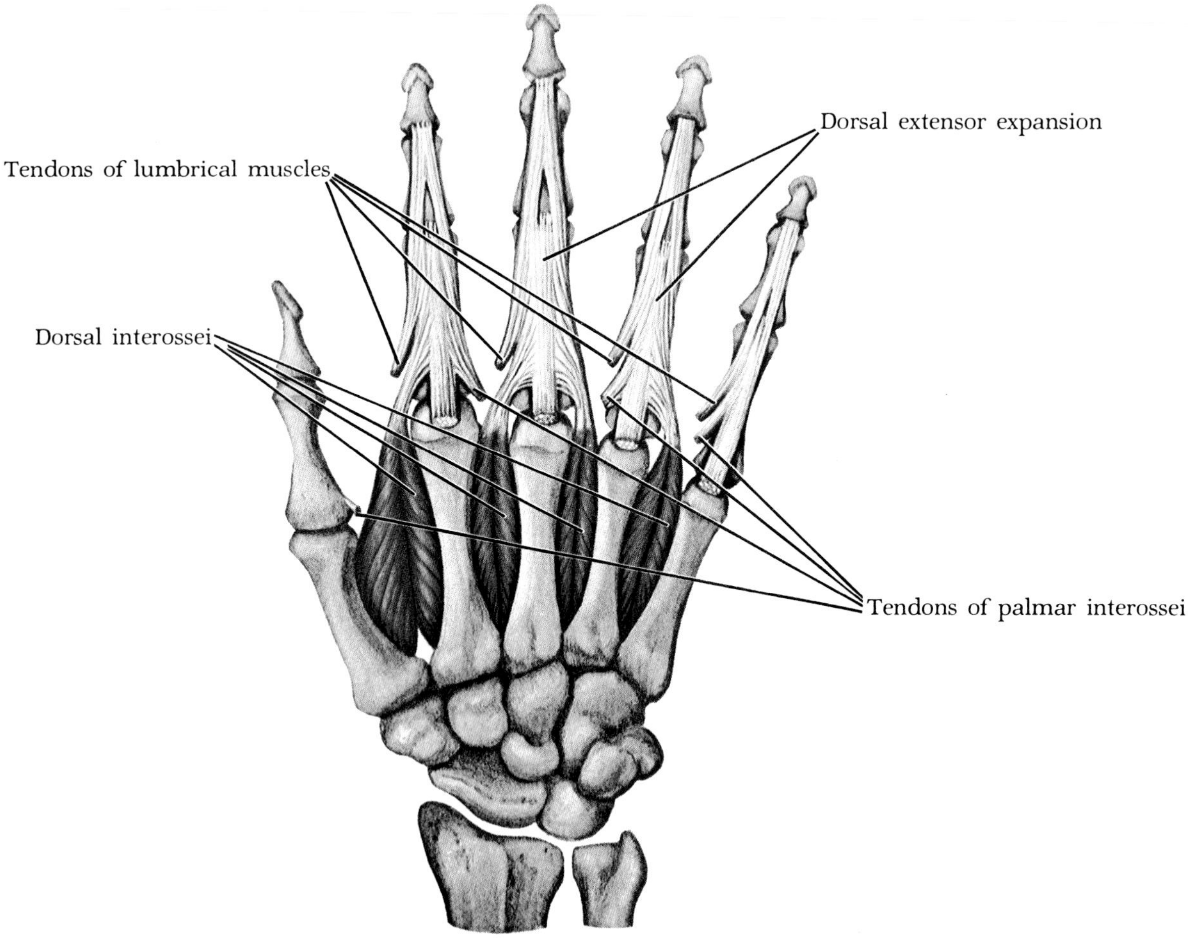

Dorsal extensor expansion
Tendons of lumbrical muscles
Dorsal interossei
Tendons of palmar interossei

B

Figure 5-60
The interossei of the right hand. A. Palmar interossei. B. Dorsal interossei.

1. The palmar interossei adduct the fingers toward the center of the third finger at the metacarpophalangeal joints, flex the metacarpophalangeal joints, and extend the interphalangeal joints. After asking the patient to spread his fingers, ask him to bring them together so that they touch. The latter part of this maneuver tests the ability of the interossei to adduct the fingers. The strength of these muscles can be tested by asking the patient to hold a piece of paper or cardboard between adjacent fingers. Compare the ability of the interossei to hold the paper while you try to withdraw it between different fingers of the same hand; then compare their performance with that of the opposite hand.

2. The dorsal interossei abduct the fingers away from the center of the third finger at the metacarpophalangeal joints, flex the metacarpophalangeal joints, and extend the interphalangeal joints. Ask the patient to spread his fingers apart. Normally he should separate his fingers in equal amounts of approximately 20 degrees during abduction. The strength of the muscles can be tested by resisting these movements.

3. Abduction and adduction of the fingers are only possible when they are in an extended position. In the flexed position of the fingers, the articular surface of the base of the proximal phalanx lies in contact with the flattened anterior surface of the head of the metacarpal bone. The two bones are held in close contact by the collateral ligaments, which in this position are taut. In the extended position of the metacarpophalangeal joint, the base of the phalanx is in contact with the rounded part of the metacarpal head, and the collateral ligaments are slack.

4. Remember that all the interosseous muscles in the hand are supplied by the deep branch of the ulnar nerve.

5. Remember also that extension of the metacarpophalangeal joints of the index, middle, ring, and little fingers is produced by the extensor digitorum; in the index finger this is aided by the extensor indicis and in the little finger by the extensor digiti minimi.

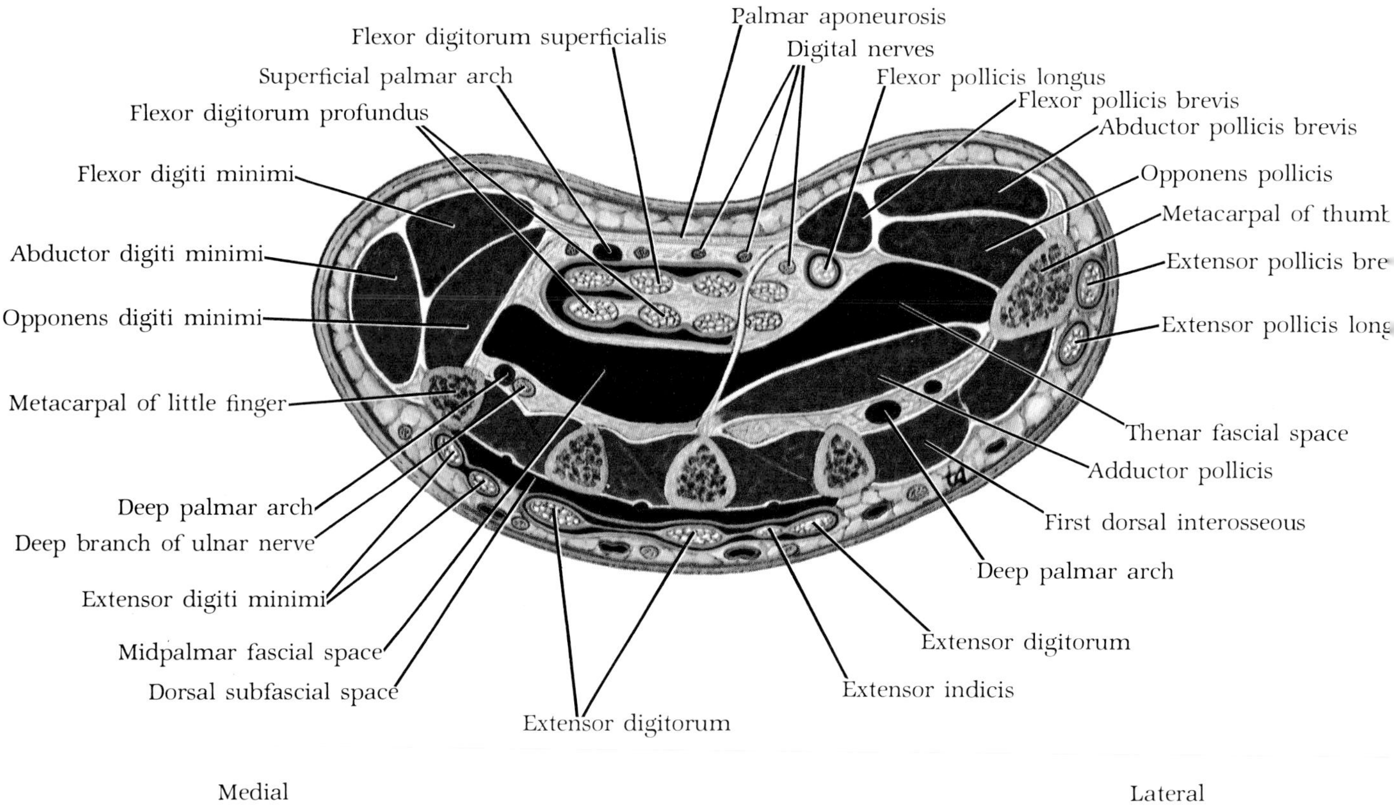
Palmar aponeurosis
Flexor digitorum superficialis
Digital nerves
Superficial palmar arch
Flexor pollicis longus
Flexor digitorum profundus
Flexor pollicis brevis
Abductor pollicis brevis
Flexor digiti minimi
Opponens pollicis
Metacarpal of thum
Abductor digiti minimi
Extensor pollicis bre
Opponens digiti minimi
Extensor pollicis long
Metacarpal of little finger
Thenar fascial space
Adductor pollicis
Deep palmar arch
First dorsal interosseous
Deep branch of ulnar nerve
Deep palmar arch
Extensor digiti minimi
Midpalmar fascial space
Extensor digitorum
Dorsal subfascial space
Extensor indicis
Extensor digitorum
Medial
Lateral

Figure 5-61
Cross section of the right hand showing the position of the midpalmar and thenar fascial spaces.

1. Normally, the fascial spaces of the palm are potential spaces filled with loose connective tissue. Their boundaries are important clinically because they limit the spread of infection. Proximally, the thenar and midpalmar spaces are closed off from the forearm by the walls of the carpal tunnel. Distally, the two spaces are continuous with the appropriate lumbrical canals.

2. The close relationship that exists between the synovial sheaths of the flexor tendons and the fascial spaces is important. The digital sheath of the index finger is related to the thenar space, while that of the ring finger is related to the midpalmar space. The sheath for the middle finger is related to both the thenar and midpalmar spaces. These relationships explain how a bacterial tenosynovitis can spread from the digital synovial sheaths to involve the palmar fascial spaces.

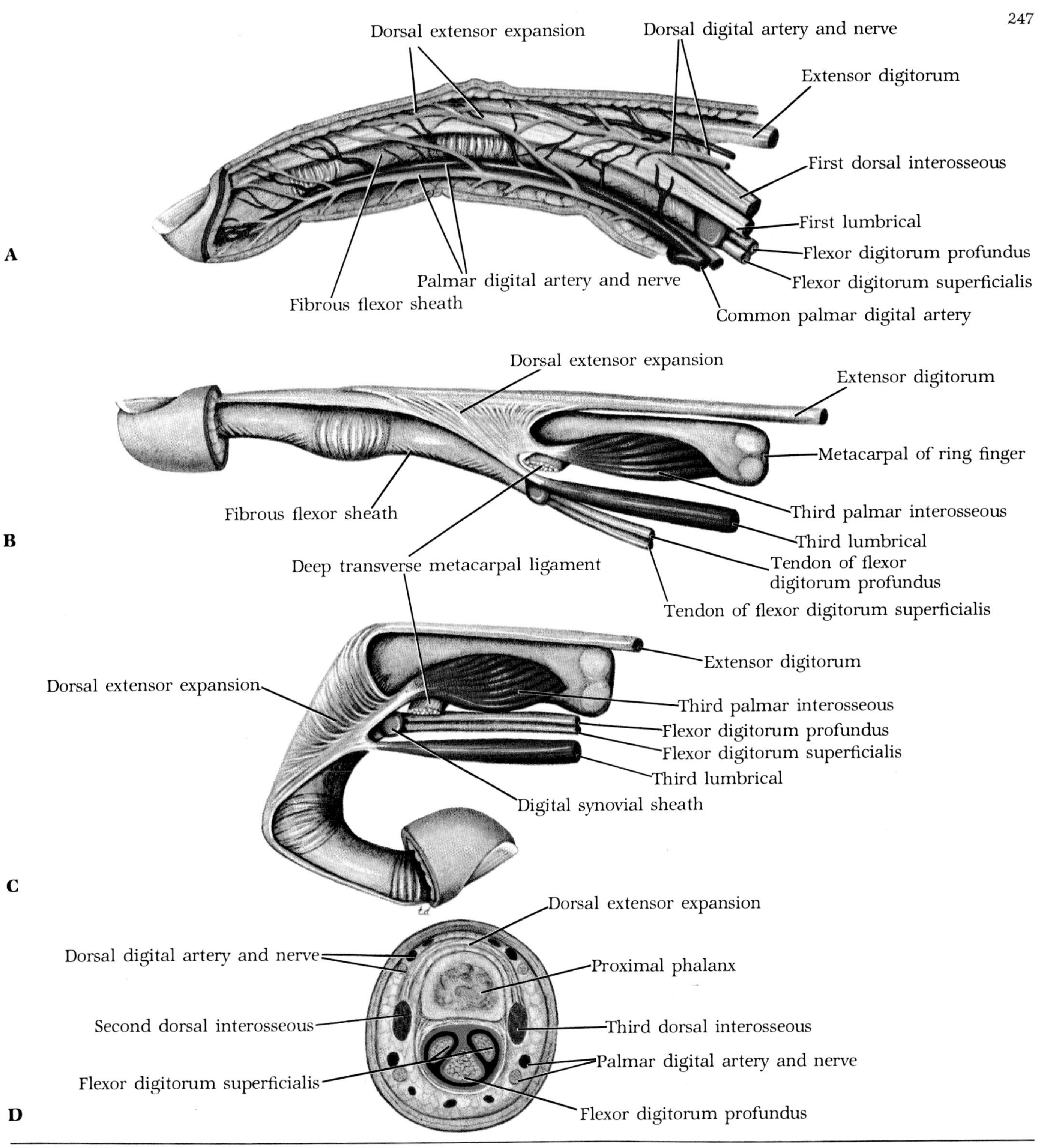
Dorsal extensor expansion
Dorsal digital artery and nerve
Extensor digitorum
First dorsal interosseous
First lumbrical
Flexor digitorum profundus
Flexor digitorum superficialis
Common palmar digital artery
Palmar digital artery and nerve
Fibrous flexor sheath
A
Dorsal extensor expansion
Extensor digitorum
Metacarpal of ring finger
Fibrous flexor sheath
Third palmar interosseous
B
Third lumbrical
Deep transverse metacarpal ligament
Tendon of flexor digitorum profundus
Tendon of flexor digitorum superficialis
Extensor digitorum
Dorsal extensor expansion
Third palmar interosseous
Flexor digitorum profundus
Flexor digitorum superficialis
Third lumbrical
Digital synovial sheath
C
Dorsal extensor expansion
Dorsal digital artery and nerve
Proximal phalanx
Second dorsal interosseous
Third dorsal interosseous
Palmar digital artery and nerve
Flexor digitorum superficialis
D
Flexor digitorum profundus

Figure 5-62
Right index finger, ring finger, and middle finger. A. Index finger, showing digital nerves and arteries, lateral view. B. Extended ring finger, showing attachments of lumbrical and interosseous muscles to dorsal extensor expansion, lateral view. C. Flexed ring finger. D. Cross-section of middle finger at level of proximal phalanx.

1. The digital arteries of the fingers are supplied by postganglionic sympathetic fibers that run in the digital nerves. The preganglionic fibers originate from cell bodies in the second to the eighth thoracic segments of the spinal cord. They synapse in the middle cervical, inferior cervical, first thoracic, or stellate ganglia. Vasospastic diseases involving digital arterioles, such as Raynaud's disease, may require a cervicodorsal preganglionic sympathectomy to prevent necrosis of the fingers. The operation is followed by arterial vasodilatation, with, as a consequence, increased blood flow to the upper limb.

2. The palmar digital nerves supply not only the skin of the palmar aspect of the fingers but the distal parts of the dorsal surfaces also.

3. Note the precise insertion of the lumbrical and interossei tendons into the dorsal extensor expansion. Note also that the fingers contain no muscle bellies and are moved only by tendons. The lumbricals and interossei flex the metacarpophalangeal joints and extend the proximal and distal interphalangeal joints.

4. The skin of the fingers is attached to bone by fibrous septa running from the skin. Without these septa the skin would move around the fingers, and it would be difficult to hold anything in the hand.

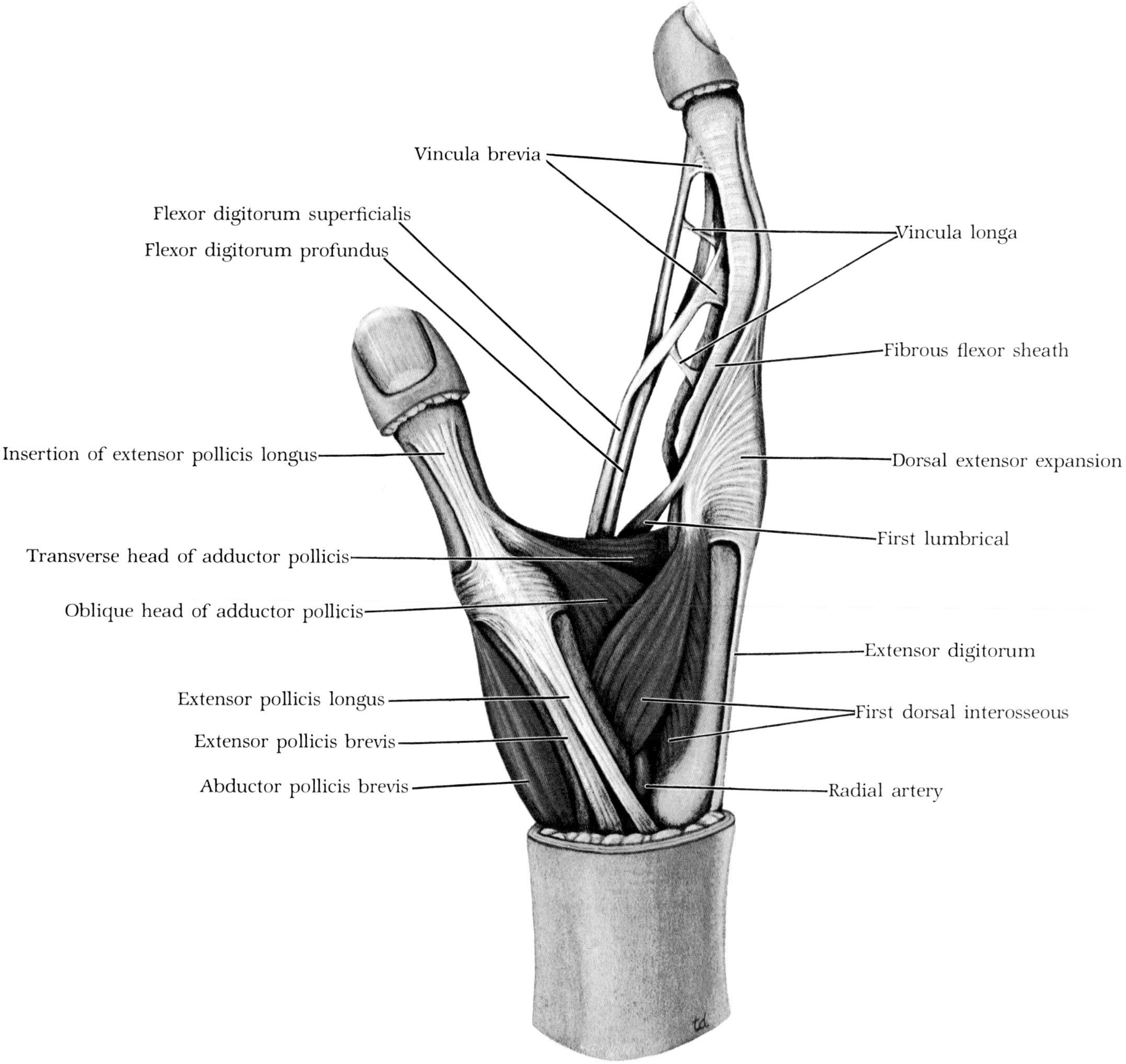
Vincula brevia
Flexor digitorum superficialis
Flexor digitorum profundus
Vincula longa
Fibrous flexor sheath
Insertion of extensor pollicis longus
Dorsal extensor expansion
First lumbrical
Transverse head of adductor pollicis
Oblique head of adductor pollicis
Extensor digitorum
Extensor pollicis longus
First dorsal interosseous
Extensor pollicis brevis
Abductor pollicis brevis
Radial artery

Figure 5-63
Right hand, showing the tendons, the vincula longa and brevia of the index finger, and the muscles in the first intermetacarpal space, lateral view.

1. The most important part of the hand is the thumb, and it is the physician's responsibility to preserve the thumb, or as much of it as possible, so that the pincer mechanism of the hand can be maintained.
2. The synovial sheaths of the flexor tendons are essentially lubricating mechanisms. The vincula longa and brevia serve as mesenteries that allow blood vessels to enter the tendons. Tenosynovitis, an infection of a synovial sheath, most commonly results from the introduction of bacteria into a sheath through a small penetrating wound, such as that made by the point of a needle.
3. Abduction of the thumb is a movement in an anteroposterior plane away from the palm. The movement takes place mainly between the trapezium and the first metacarpal bone, although a small amount of movement takes place at the metacarpophalangeal joint. The muscles producing the movement are the abductor pollicis longus (supplied by the deep branch of the radial nerve) and abductor pollicis brevis (supplied by the median nerve).
4. Adduction of the thumb is a movement in an anteroposterior plane toward the palm. The movement takes place in the same joints used in abduction. The muscle producing the movement is the adductor pollicis (supplied by the deep branch of the ulnar nerve).
5. Stenosing tenosynovitis of the abductor pollicis longus tendon and the extensor pollicis brevis tendon can occur as the result of repeated friction between these tendons and the styloid process of the radius. Edema of the tendons causes them to swell, producing more friction. Later, fibrosis of the sheath leads to stenosis. Advanced cases require surgical incision along the constricting sheath.

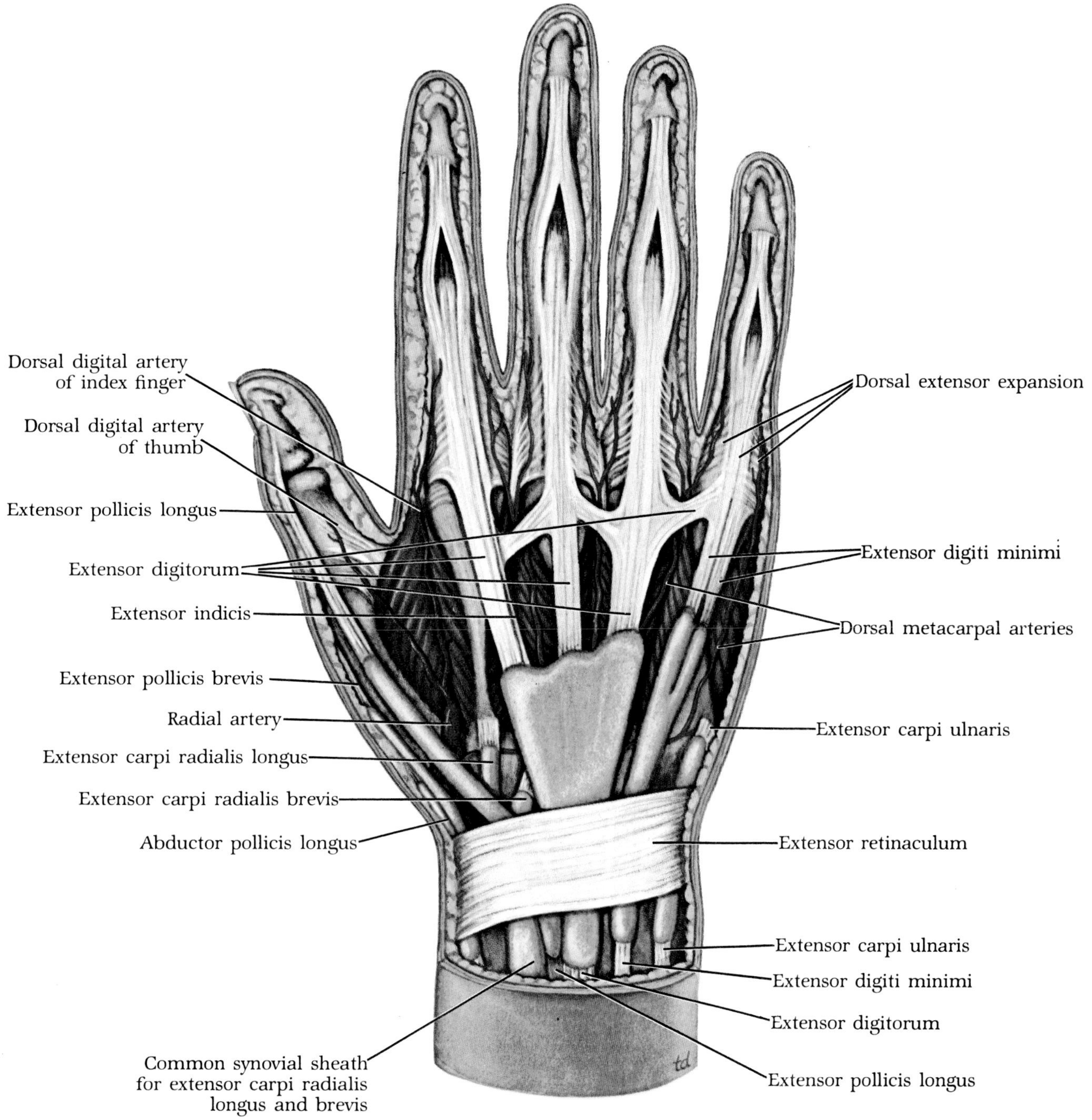
Dorsal digital artery of index finger
Dorsal digital artery of thumb
Extensor pollicis longus
Extensor digitorum
Extensor indicis
Extensor pollicis brevis
Radial artery
Extensor carpi radialis longus
Extensor carpi radialis brevis
Abductor pollicis longus
Common synovial sheath for extensor carpi radialis longus and brevis
Dorsal extensor expansion
Extensor digiti minimi
Dorsal metacarpal arteries
Extensor carpi ulnaris
Extensor retinaculum
Extensor carpi ulnaris
Extensor digiti minimi
Extensor digitorum
Extensor pollicis longus
td

Figure 5-64
Dorsal surface of hand showing long extensor tendons and their synovial sheaths.

1. Tenosynovitis can involve the tendon sheaths on the dorsal surface of the hand. Acute traumatic tenosynovitis may be related to unaccustomed movement and reveal itself by the presence of pain, swelling, and crepitus. The treatment is to rest the hand with a wrist splint and restrict the movement that causes the pain.
2. Rupture of extensor pollicis longus tendon can occur following a Colles' fracture of the distal third of the radius. Roughening of the dorsal radial tubercle produced by the fracture line can cause excessive friction on the tendon, which may then rupture. Rheumatoid arthritis can also cause rupture of this tendon.
3. Deep cuts across the dorsum of the hand can sever the tendons of the extensor digitorum. Remember that the index finger receives an additional tendon from the extensor indicis muscle, and that the tendon of the extensor digitorum to the little finger joins that of the extensor digiti minimi at the base of the little finger.
4. Avulsion of the insertion of the extensor tendon into the distal phalanx can occur if the distal phalanx is forcibly flexed when the extensor tendon is taut. The last 20 degrees of active extension is lost, resulting in a condition referred to as mallett finger.
5. In normal extension of the fingers, the extensor digitorum (and extensor indicis and digiti minimi) is the primary extensor of the metacarpophalangeal joint, and the lumbricals and interossei are the primary extensors of the proximal and distal interphalangeal joints. The fingers move in unison and extend to, or beyond, the straight position.

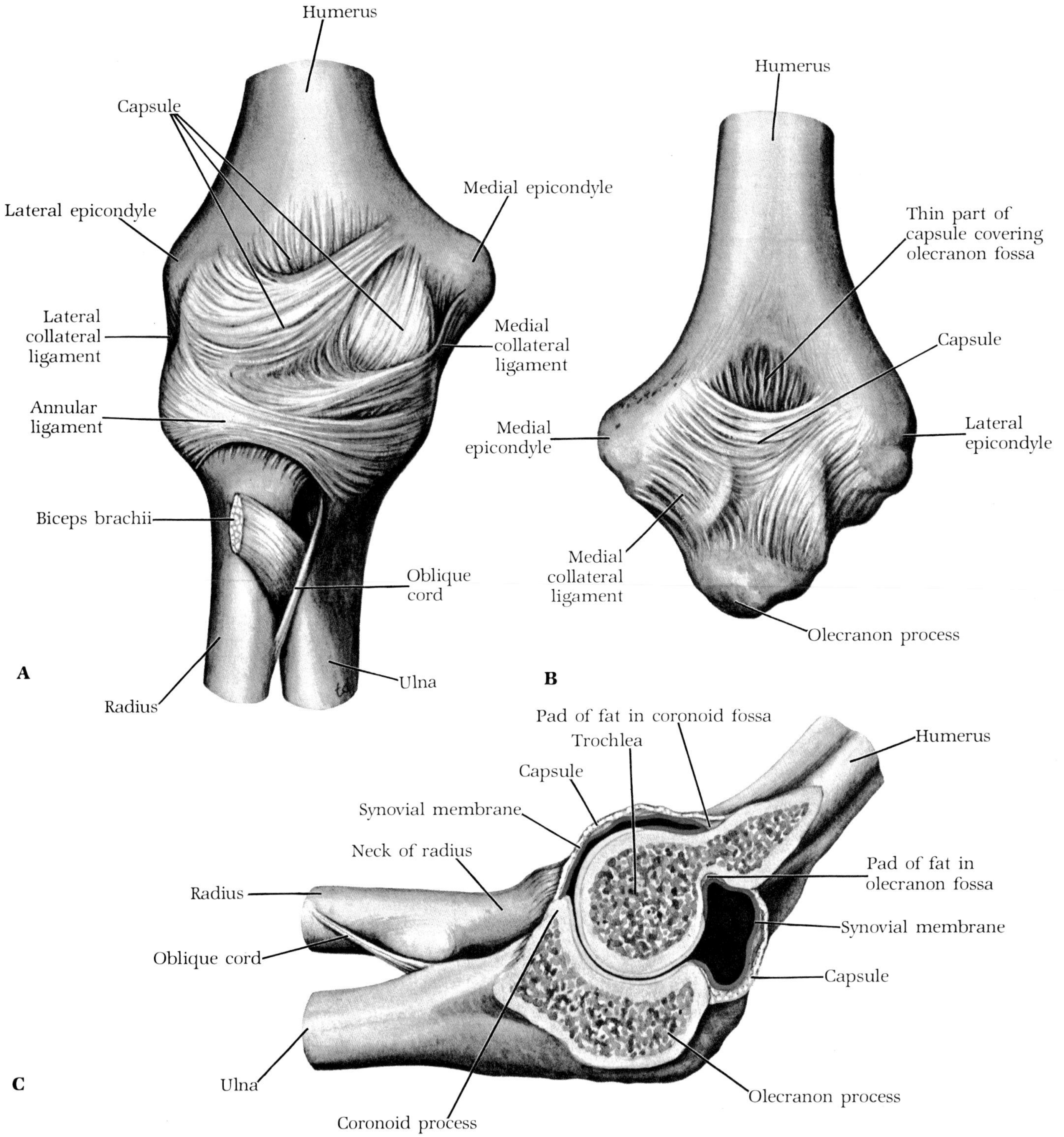
Humerus
Capsule
Lateral epicondyle
Medial epicondyle
Lateral collateral ligament
Medial collateral ligament
Annular ligament
Biceps brachii
Oblique cord
Radius
Ulna
A
Humerus
Thin part of capsule covering olecranon fossa
Capsule
Medial epicondyle
Lateral epicondyle
Medial collateral ligament
Olecranon process
B
Pad of fat in coronoid fossa
Trochlea
Humerus
Capsule
Synovial membrane
Neck of radius
Pad of fat in olecranon fossa
Radius
Synovial membrane
Oblique cord
Capsule
Ulna
Olecranon process
Coronoid process
C

Figure 5-65
Right elbow joint. A. Anterior view. B. Posterior view. C. In sagittal section.

1. The synovial cavity of the elbow joint is in direct continuity with that of the superior radioulnar joint. This means that infection of the elbow joint invariably involves the superior radioulnar joint.
2. The elbow joint is very stable because of the wrench-shaped articular surface of the olecranon and the pulley-shaped trochlea of the humerus. It also has very strong medial and lateral collateral ligaments.
3. When examining the elbow joint the physician must remember the normal relations of the bony points. In extension the medial and lateral epicondyles and the top of the olecranon process are in a straight line; in flexion, the bony points form the boundaries of an equilateral triangle.
4. Posterior dislocations of the joint are common in children because the parts of the bones that stabilize the joint are incompletely developed in childhood. Injury is caused by falling on the outstretched hand. The anterior part of the capsule is torn, and the medial and lateral ligaments are damaged.
5. In fracture-dislocations of the elbow joint, not only is there a posterior dislocation of the ulna on the humerus, but there also may be a fracture of the coronoid process, the head of the radius, or the lateral part of the distal end of the humerus. Fractures of the coronoid process are usually chip fractures caused by the pull of the brachialis muscle.

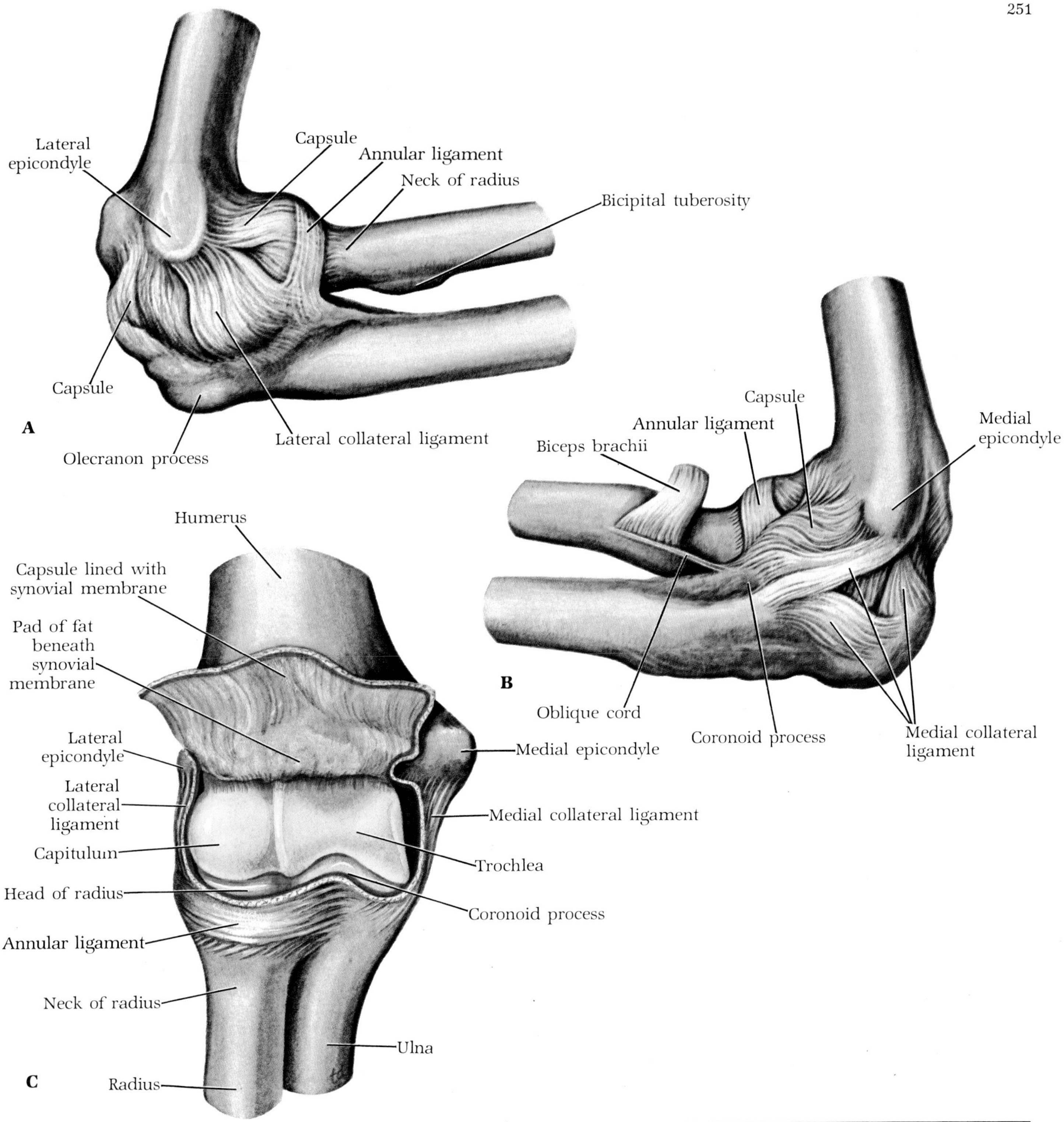
Lateral epicondyle
Capsule
Annular ligament
Neck of radius
Bicipital tuberosity
Capsule
A
Olecranon process
Lateral collateral ligament
Capsule
Annular ligament
Biceps brachii
Medial epicondyle
B
Oblique cord
Coronoid process
Medial collateral ligament
Humerus
Capsule lined with synovial membrane
Pad of fat beneath synovial membrane
Lateral epicondyle
Medial epicondyle
Lateral collateral ligament
Medial collateral ligament
Capitulum
Trochlea
Head of radius
Coronoid process
Annular ligament
Neck of radius
Ulna
C
Radius

Figure 5-66
Right elbow joint. A. Lateral view. B. Medial view. C. Anterior view of interior. The anterior part of the capsule and synovial lining has been reflected.

1. Avulsion of the epiphysis of the medial epicondyle is common in children because at that time the medial collateral ligament of the elbow joint is much stronger than the bond of union between the epiphysis and the diaphysis. The avulsion occurs when the forearm is forcibly abducted on the arm.
2. Fracture of the lateral epicondyle is also common in children. This occurs when the forearm is forcibly abducted on the arm with the elbow joint extended. Sometimes the pull of the extensor origin from the lateral epicondyle rotates the bone fragment into the joint cavity.
3. The close relationship of the ulnar nerve to the medial side of the joint posterior to the medial epicondyle often results in this nerve becoming damaged in dislocations of the joint or in fracture dislocations. The nerve lesion may occur at the time of injury or weeks, months, or even years later.
4. When examining lateral radiographs of the elbow region, it is important to remember that the lower end of the humerus is normally angulated anteriorly 45 degrees on the shaft. When examining a patient the physician should see that the medial epicondyle, in the anatomical position, is directed medially and posteriorly and faces in the same direction as the head of the humerus.
5. When the elbow joint is distended with fluid, the posterior aspect becomes swollen because of the weak lax posterior wall of the capsule. Aspiration of the joint fluid can easily be performed through the posterior aspect of the joint on either side of the olecranon process.

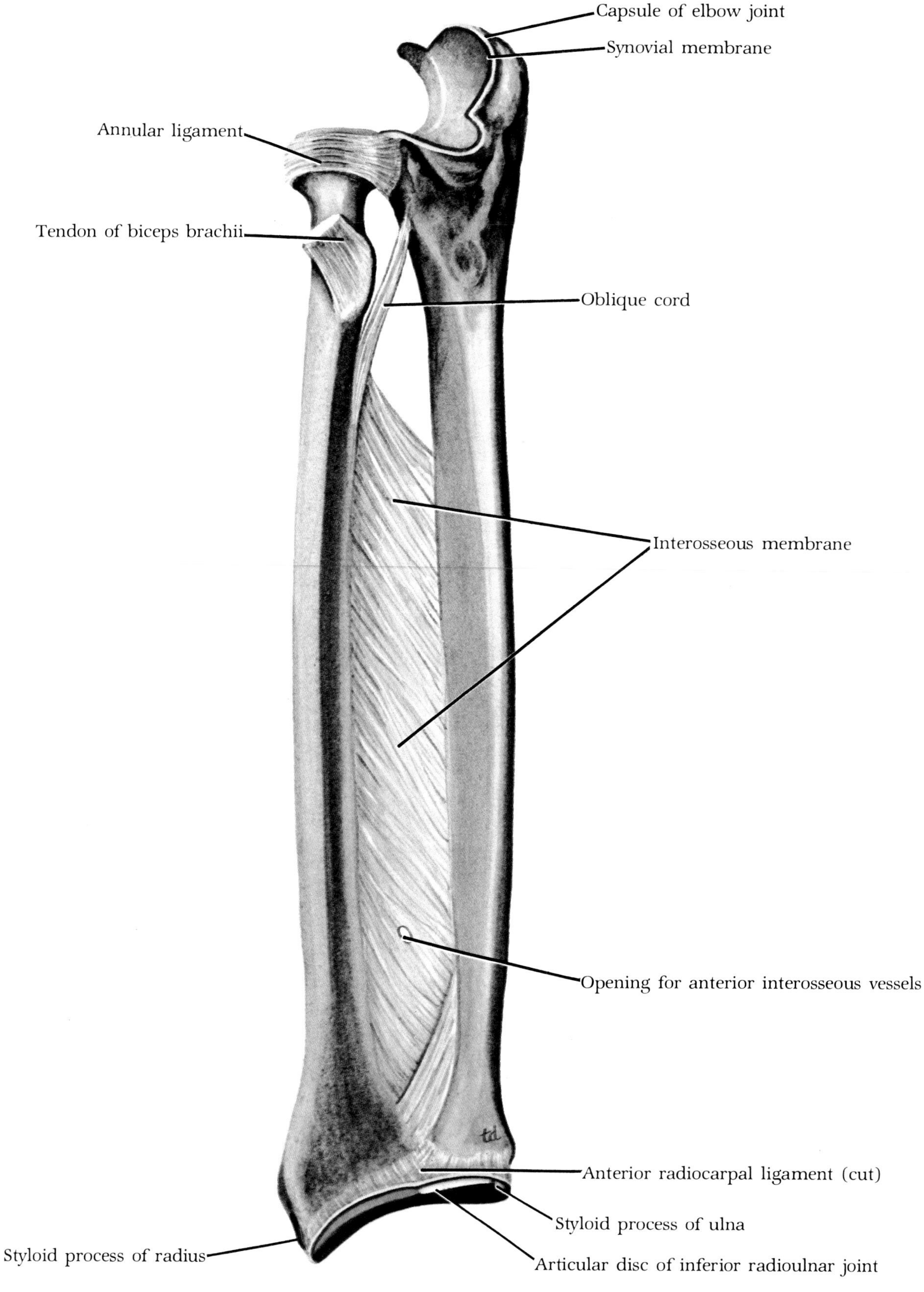
Capsule of elbow joint
Synovial membrane
Annular ligament
Tendon of biceps brachii
Oblique cord
Interosseous membrane
Opening for anterior interosseous vessels
Anterior radiocarpal ligament (cut)
Styloid process of ulna
Styloid process of radius
Articular disc of inferior radioulnar joint

Figure 5-67
Right radius and ulna, showing the attachment of the interosseous membrane, the annular ligament, and the oblique cord, anterior view.

1. Monteggia's fracture dislocation. A severe blow on the posterior surface of the forearm can cause a fracture of the proximal third of the ulna and a forward dislocation of the head of the radius on the capitulum; the annular ligament also is ruptured. Forced pronation can also cause this condition.
2. The direction of the fibers of the interosseous membrane insures that a force applied to the lower end of the radius — as incurred when falling on the outstretched hand — is transmitted from the radius to the ulna and from there to the humerus and scapula. The wrench-shaped trochlear notch of the ulna articulating with the trochlea of the humerus provides a more stable union than that between the head of the radius and the capitulum of the humerus.
3. The interosseous membrane is taut when the forearm is in the midprone position and is slack in other positions of the radioulnar joints. The forearm bones are thus most stable in the midprone position, and this is the forearm position commonly assumed when delicate movements of the fingers are performed.

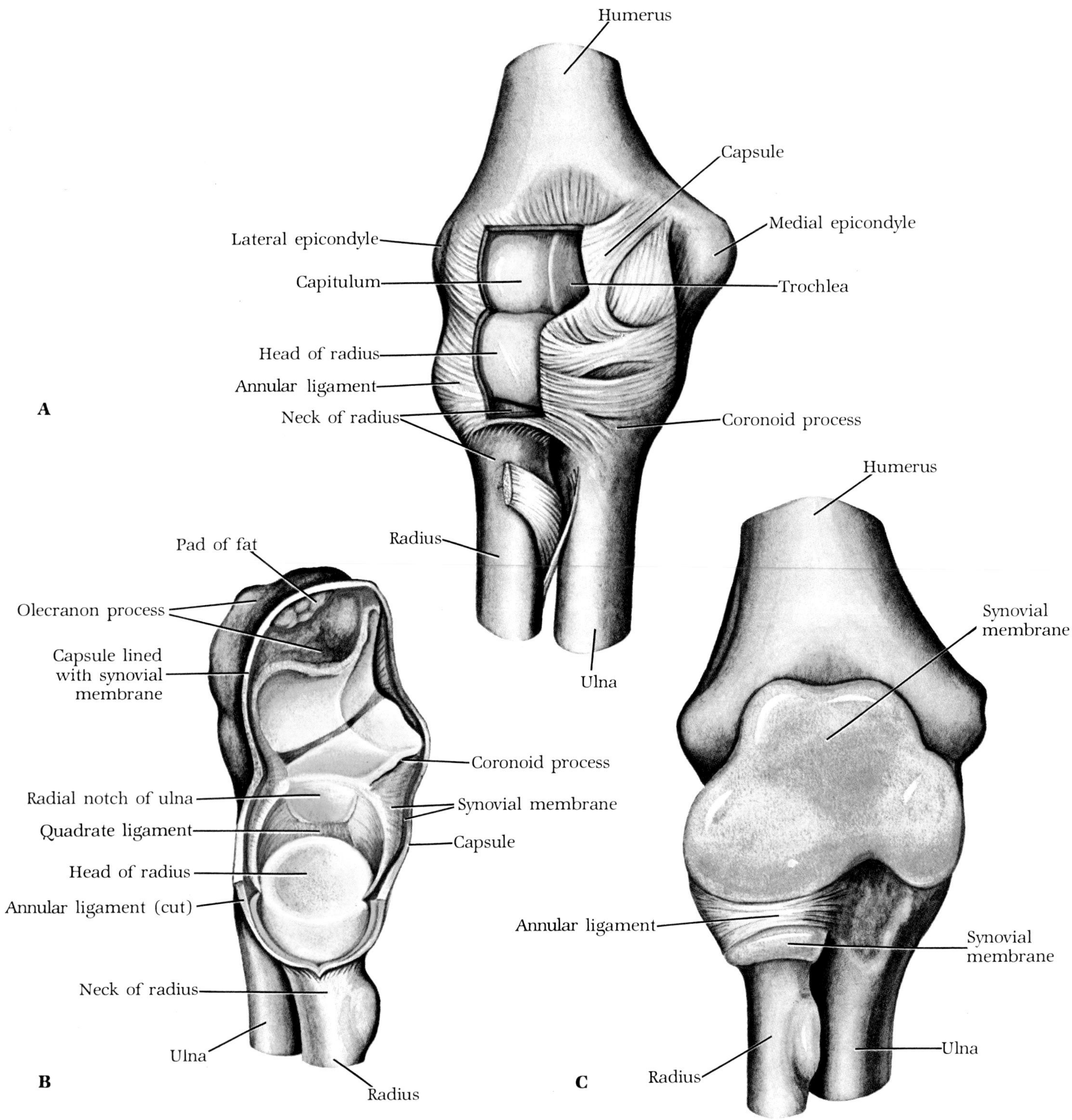
Humerus
Capsule
Lateral epicondyle
Medial epicondyle
Capitulum
Trochlea
Head of radius
Annular ligament
A
Neck of radius
Coronoid process
Radius
Ulna
Pad of fat
Olecranon process
Capsule lined with synovial membrane
Coronoid process
Radial notch of ulna
Synovial membrane
Quadrate ligament
Capsule
Head of radius
Annular ligament (cut)
Neck of radius
Ulna
B
Radius
Humerus
Synovial membrane
Annular ligament
Synovial membrane
Ulna
C
Radius

Figure 5-68
Right superior radioulnar joint.
A. Anterior view, showing interior.
B. Superior view, showing interior.
C. Synovial membrane, showing the continuity of the synovial membrane of the elbow joint and the superior radioulnar joint.

1. The synovial cavity of the superior radioulnar joint is in direct continuity with that of the elbow joint. This means that infection of the elbow joint invariably spreads to the superior radioulnar joint.
2. The strength of the superior radioulnar joint depends on the integrity of the strong anular ligament. Rupture of this ligament occurs during anterior dislocation of the head of the radius on the capitulum of the humerus.

3. In young children in whom the head of the radius is still small and undeveloped, a sudden jerk on the arm can pull the radial head inferiorly through the anular ligament. The treatment is rapid supination of the joint with the elbow in the flexed position.

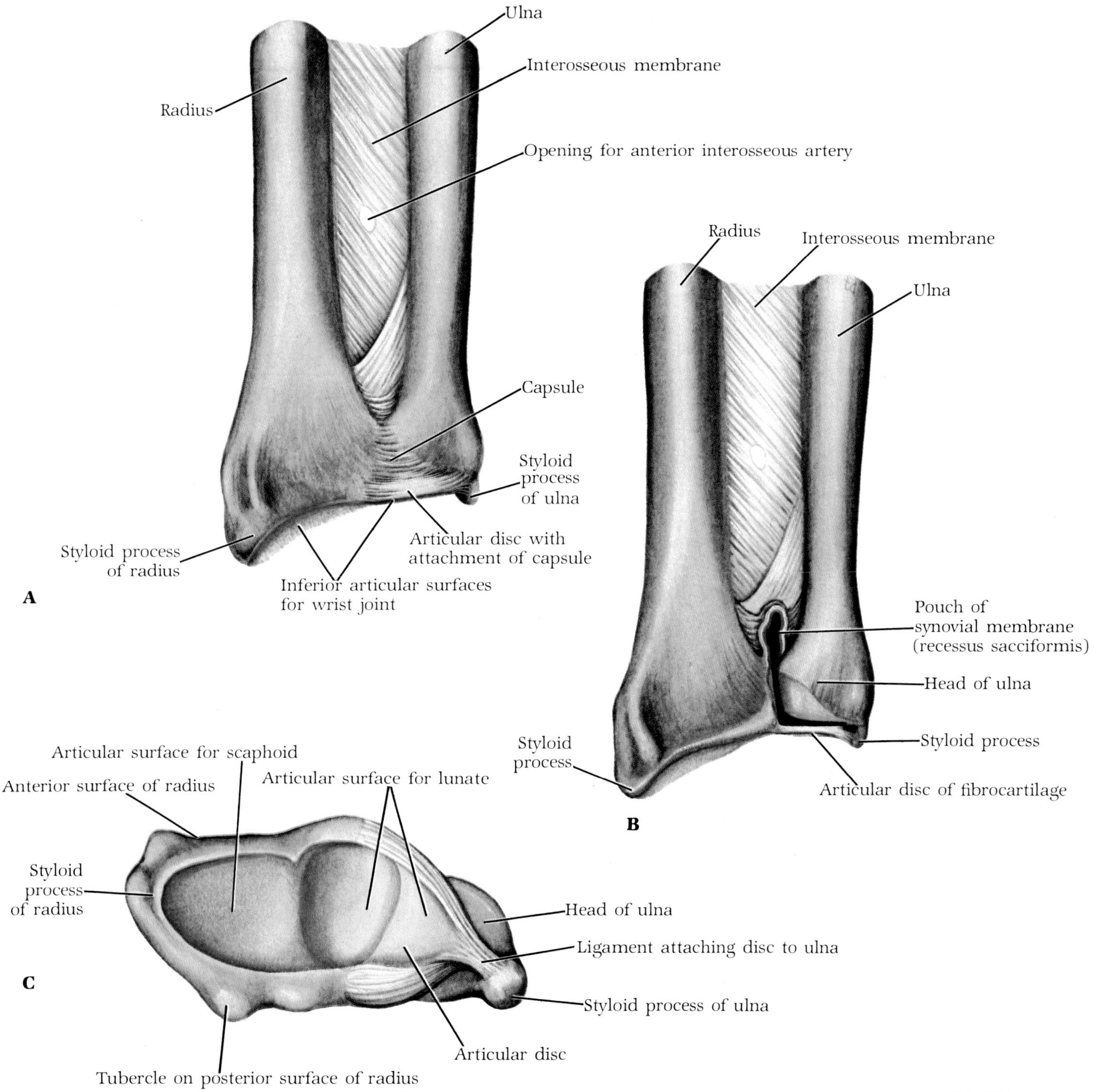
Ulna
Interosseous membrane
Radius
Opening for anterior interosseous artery
Capsule
Styloid process of ulna
Articular disc with attachment of capsule
Styloid process of radius
Inferior articular surfaces for wrist joint
A
Radius
Interosseous membrane
Ulna
Pouch of synovial membrane (recessus sacciformis)
Head of ulna
Styloid process
Styloid process
Articular disc of fibrocartilage
B
Articular surface for scaphoid
Anterior surface of radius
Articular surface for lunate
Styloid process of radius
Head of ulna
Ligament attaching disc to ulna
C
Styloid process of ulna
Articular disc
Tubercle on posterior surface of radius

Figure 5-69
Right inferior radioulnar joint.
A. Anterior view. B. Coronal section showing interior of joint.
C. Articular disc as seen from below.

1. The fibrocartilaginous triangular-shaped articular disc strongly unites the distal ends of the radius and ulna and provides stability for the rotating inferior radioulnar joint.
2. The articular disc is commonly torn in a Colles' fracture; it can also be damaged in forced extension and rotation with local tenderness and weakness of the wrist movements being the clinical finding.
3. The articular disc is sometimes incomplete, being perforated in its center. In these circumstances, the synovial cavity of the inferior radioulnar joint and that of the wrist joint are continuous.

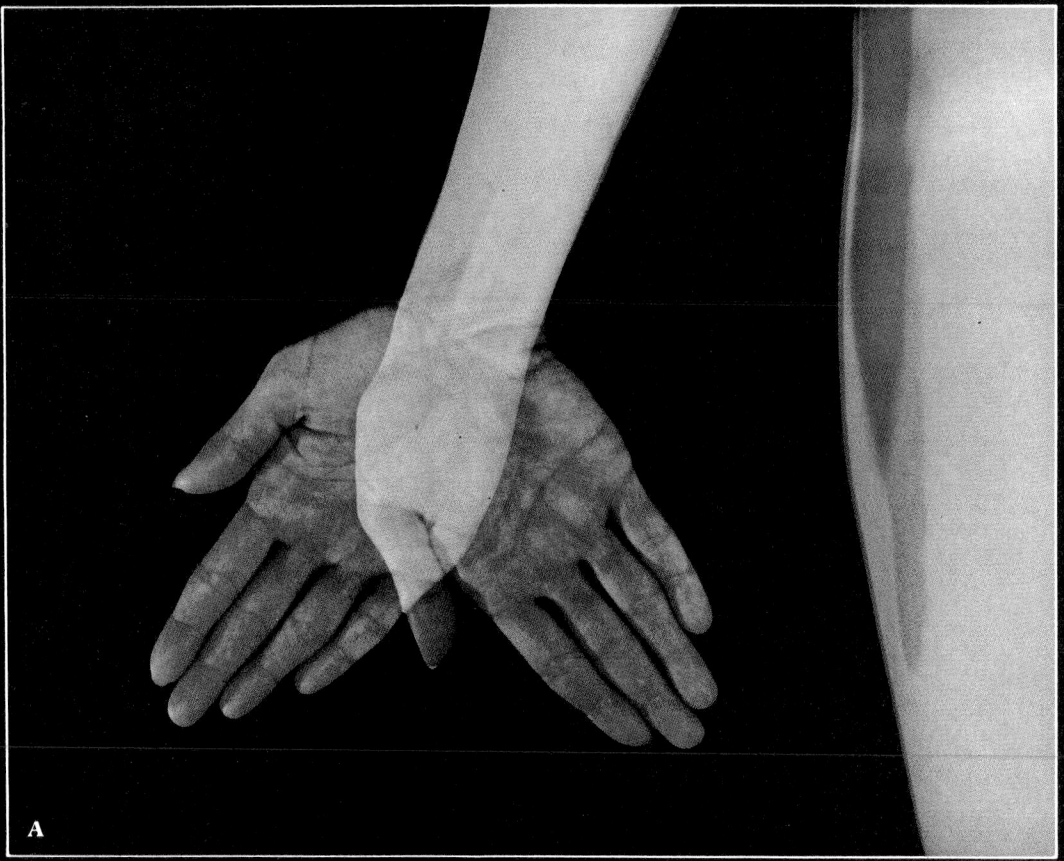
A

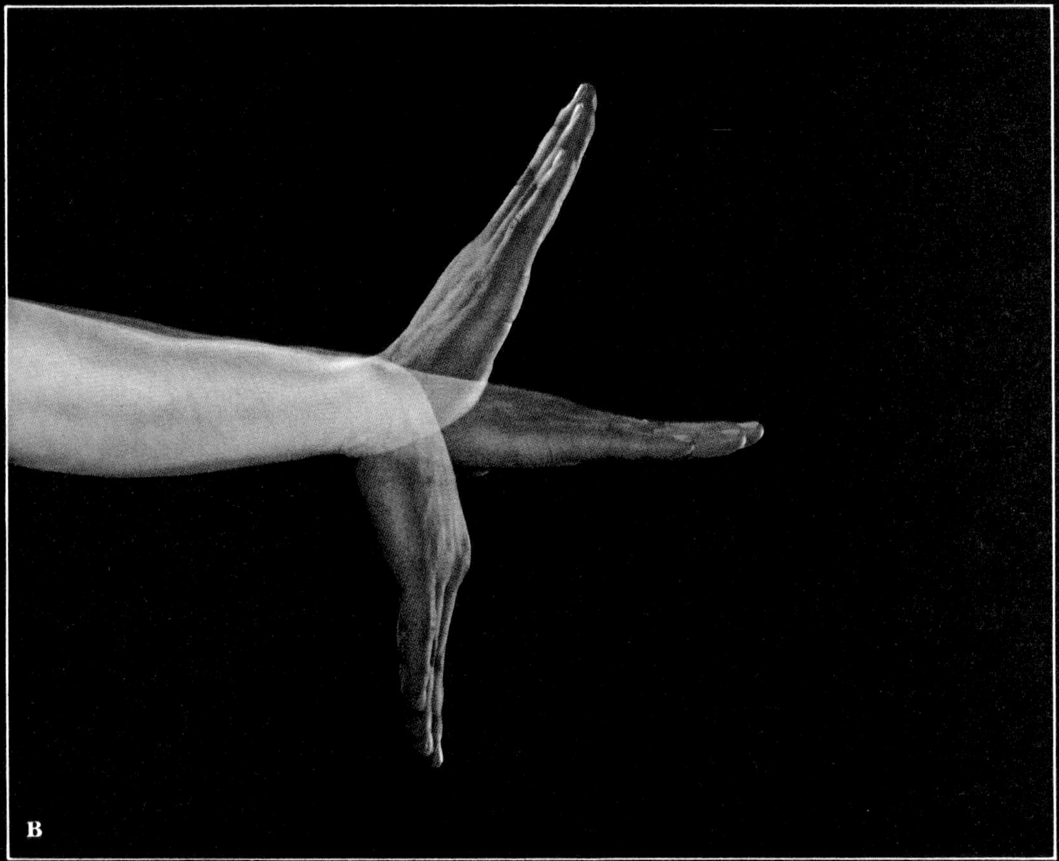
B

Figure 5-70
The wrist joint. A. Movement of abduction and adduction. B. Movement of flexion and extension.

1. The following movements are possible at the wrist joint: abduction and adduction, flexion and extension, and circumduction. Rotation is not possible because the articular surfaces are ellipsoid in shape. The lack of rotation is compensated for by the movements of pronation and supination of the forearm.
2. The range of abduction at the wrist joint is considerably less than the range of adduction due to the length of the styloid process of the radius.
3. The movement of abduction is performed by the flexor carpi radialis and the extensor carpi radialis longus and brevis muscles. These muscles are assisted by the abductor pollicis longus and extensor pollicis longus and brevis muscles.
4. The movement of adduction is performed by the flexor and extensor carpi ulnaris muscles.
5. Normal flexion allows the wrist to move about 80 degrees from the anatomical position. The normal limit for extension is less, approximately 70 degrees from the anatomical position.
6. Flexion is performed by the flexor carpi radialis, the flexor carpi ulnaris, and the palmaris longus muscles. These muscles are assisted by the flexor digitorum superficialis, the flexor digitorum profundus, and the flexor pollicis longus muscles.
7. Extension is performed by the extensor carpi radialis longus, the extensor carpi radialis brevis, and the extensor carpi ulnaris muscles. These muscles are assisted by the extensor digitorum, the extensor indicis, the extensor digiti minimi, and the extensor pollicis longus muscles.
8. It should be remembered that movements at the wrist joint are usually accompanied by movements of the intercarpal joints.

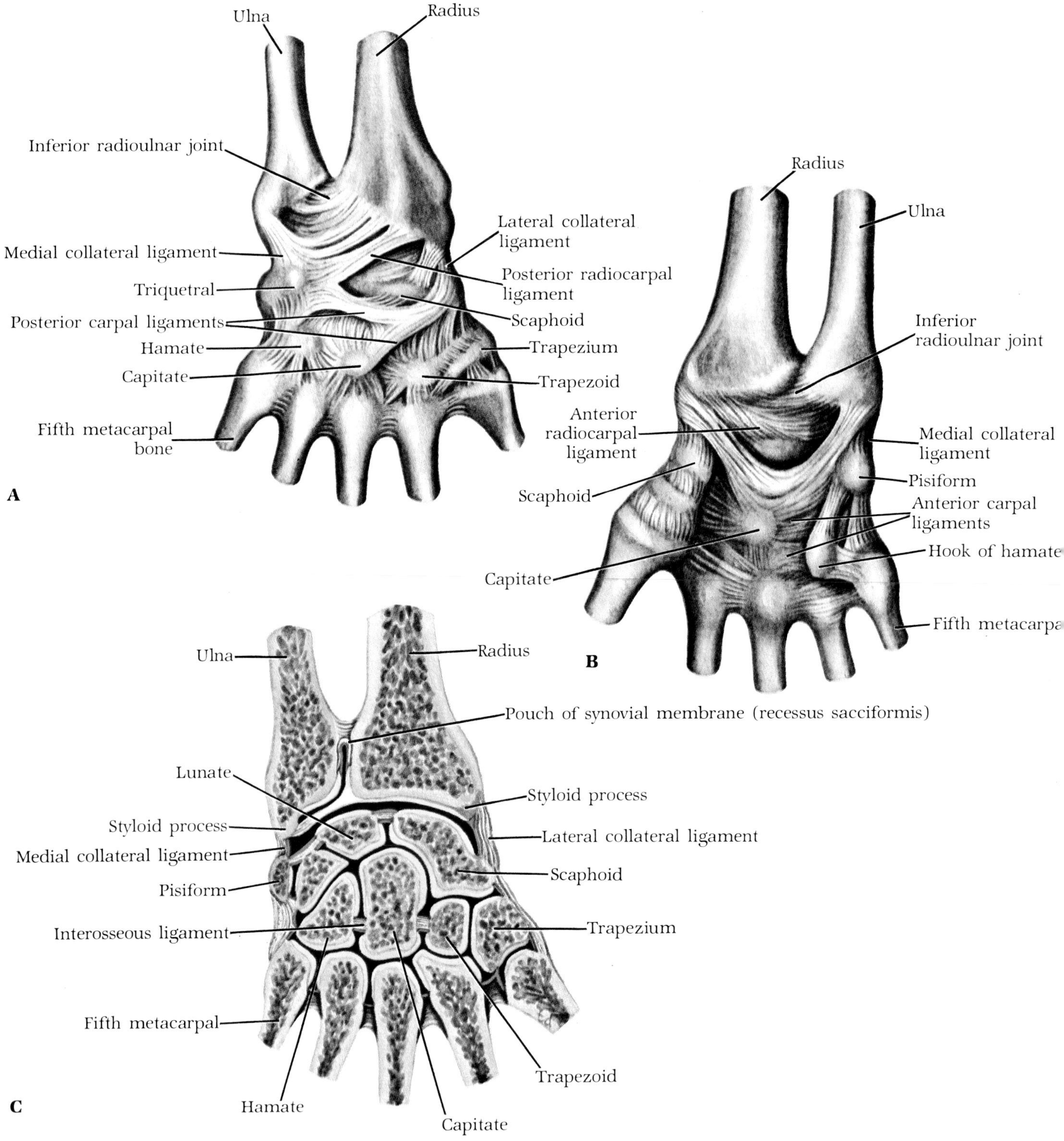
Ulna
Radius
Inferior radioulnar joint
Lateral collateral ligament
Medial collateral ligament
Posterior radiocarpal ligament
Triquetral
Scaphoid
Posterior carpal ligaments
Hamate
Trapezium
Capitate
Trapezoid
Fifth metacarpal bone
A
Radius
Ulna
Inferior radioulnar joint
Anterior radiocarpal ligament
Medial collateral ligament
Pisiform
Scaphoid
Anterior carpal ligaments
Hook of hamate
Capitate
Fifth metacarpa
B
Ulna
Radius
Pouch of synovial membrane (recessus sacciformis)
Lunate
Styloid process
Styloid process
Lateral collateral ligament
Medial collateral ligament
Scaphoid
Pisiform
Interosseous ligament
Trapezium
Fifth metacarpal
Trapezoid
C
Hamate
Capitate

Figure 5-71
Right wrist joint. A. Posterior view. B. Anterior view. C. Coronal section.

1. The wrist joint is essentially a synovial joint between the inferior end of the radius and the proximal row of carpal bones. The head of the ulna is separated from the carpal bones by the strong triangular fibrocartilaginous disc that separates the wrist joint from the inferior radioulnar joint. The wrist joint is stabilized by the very strong medial and lateral ligaments.

2. Abduction of the wrist joint is less extensive than adduction because the styloid process of the radius is longer than that of the ulna.

3. The hand can be flexed to about a right angle in flexion-extension movements but can be extended to only about 45 degrees. The range of flexion is increased by movement at the midcarpal joint.

4. A fall on the outstretched hand can tear the anterior radial carpal ligament.

5. Aspiration of the wrist joint can be performed by inserting a needle just distal to the tip of the styloid process of the ulna, between the flexor and extensor carpi ulnaris tendons.

6. When immobilizing the wrist joint (as when applying a plaster cast), it is important to place the wrist in the position of function. This is the posture adopted when the hand is about to grasp an object between the thumb and index finger and is a dorsiflexed position of about 30 degrees. Determine the exact angle by examining the opposite wrist joint in the position of function.

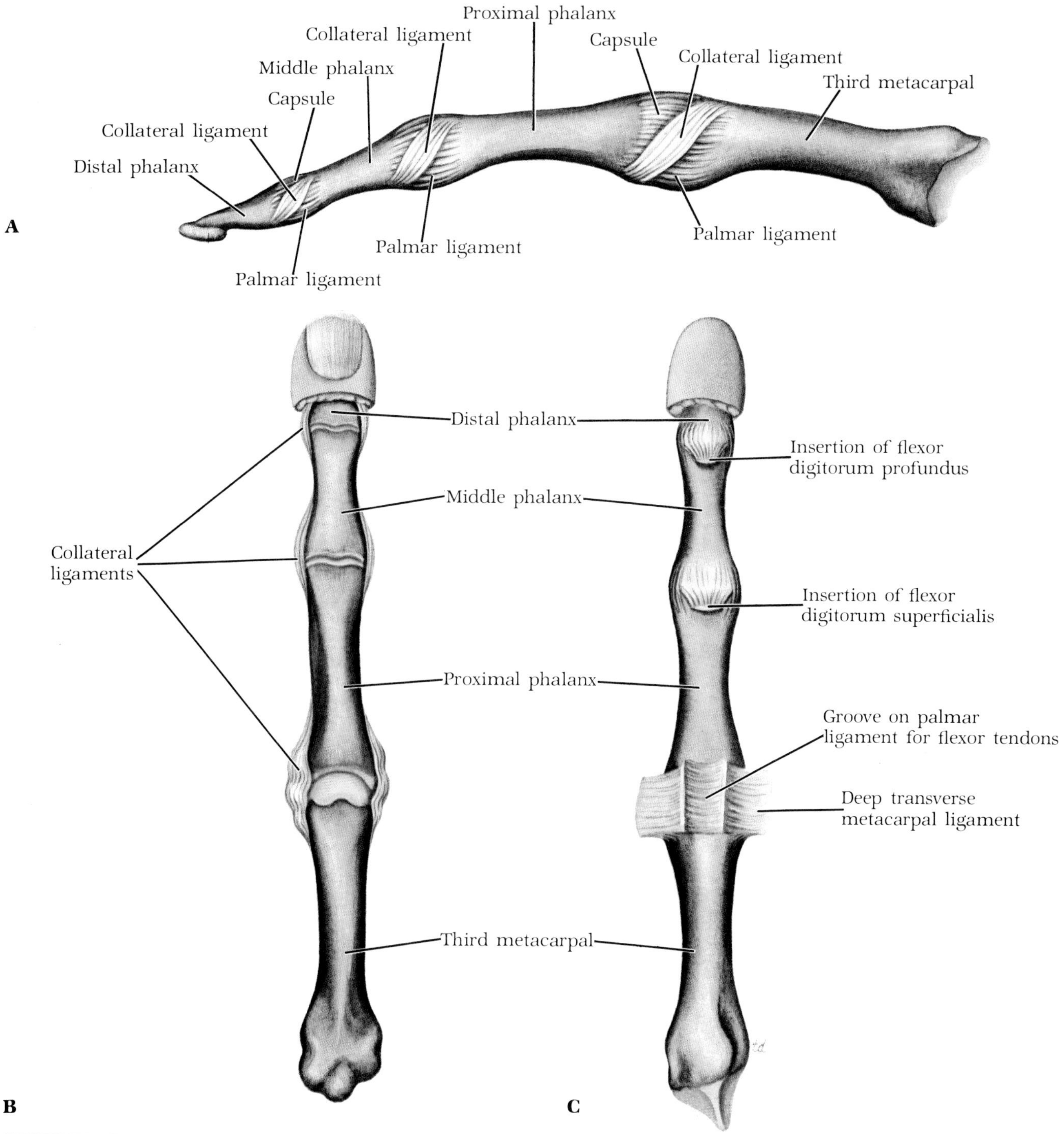
Proximal phalanx
Collateral ligament
Capsule
Collateral ligament
Middle phalanx
Third metacarpal
Capsule
Collateral ligament
Distal phalanx
A
Palmar ligament
Palmar ligament
Palmar ligament
Distal phalanx
Insertion of flexor digitorum profundus
Middle phalanx
Collateral ligaments
Insertion of flexor digitorum superficialis
Proximal phalanx
Groove on palmar ligament for flexor tendons
Deep transverse metacarpal ligament
Third metacarpal
B
C

Figure 5-72
Joints of middle (third) finger of right hand. A. Lateral view. B. Posterior view. C. Anterior view.

1. The metacarpophalangeal joints and the interphalangeal joints are similar in that they possess strong ligaments: the palmar ligaments and the collateral ligaments. On the dorsal aspect, the capsules are protected by the common extensor expansion.

2. The metacarpophalangeal joints and the interphalangeal joints are all capable of flexion and extension movements. In addition the metacarpophalangeal joints can move from side to side and do permit some passive rotation although the latter movements are only possible when the metacarpophalangeal joints are in extension. The reason for this is that due to the shape of the heads of the metacarpal bones the collateral ligaments of the metacarpophalangeal joints are taut when the joints are flexed and slack in the position of extension.

3. The muscles primarily responsible for extension of the proximal and distal interphalangeal joints are the lumbricals and interossei. The muscle responsible for extension of the metacarpophalangeal joint is the extensor digitorum (in the index and little finger it is assisted by the extensor indicis and the extensor digiti minimi, respectively). The muscle responsible for flexing the distal interphalangeal joint is the flexor digitorum profundus; the muscle primarily responsible for the flexion of the proximal interphalangeal joint is the flexor digitorum superficialis. The muscles primarily responsible for flexing the metacarpophalangeal joint are the lumbricals and interossei.

4. The characteristic ulnar deviation (adduction deformity) found at the metacarpophalangeal joints in rheumatoid arthritis is due to the destruction of the joint surfaces, capsule, and ligaments by the disease process and the oblique pull of the extensor digitorum tendons (also the extensor indicis and extensor digiti minimi for the index and little fingers).

6. The Lower Limb

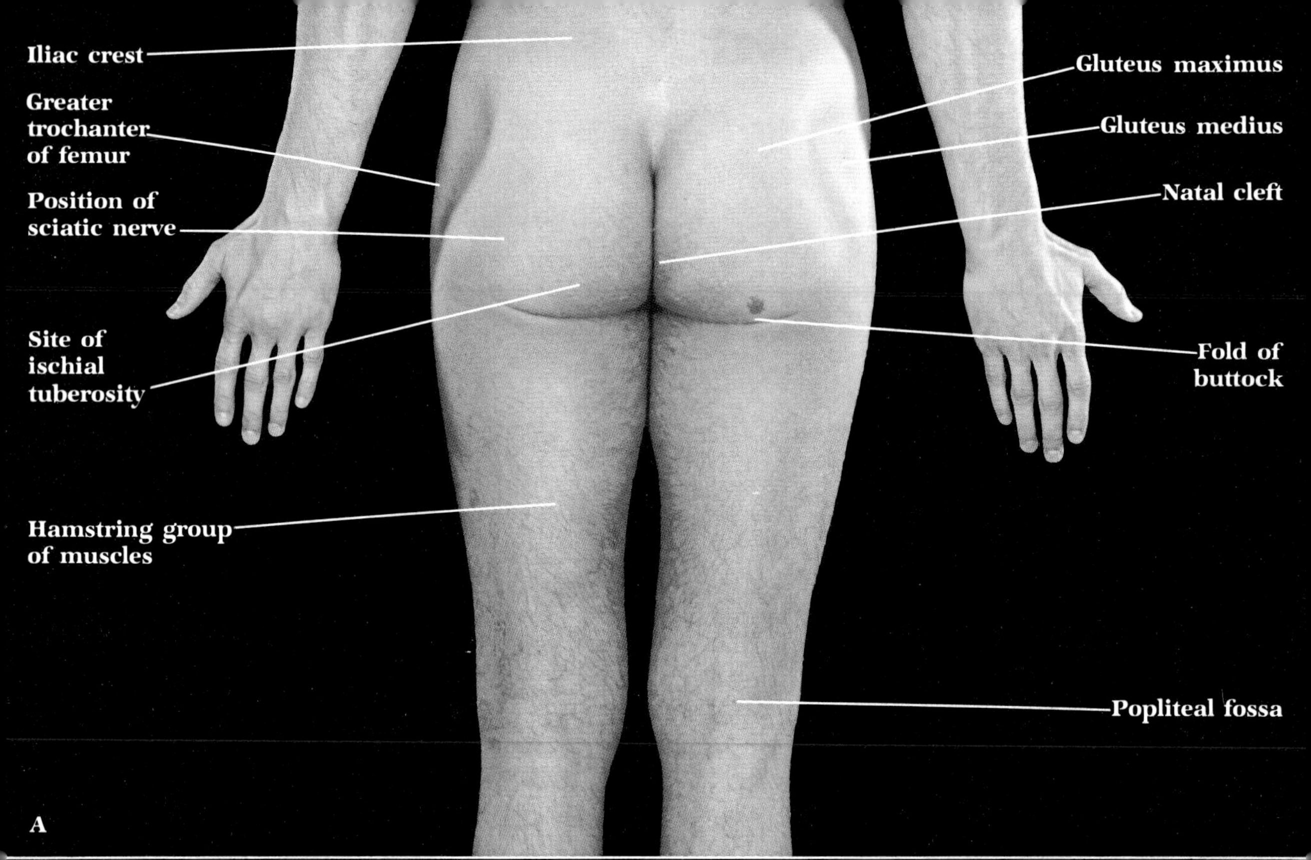

Iliac crest
Greater trochanter of femur
Position of sciatic nerve
Site of ischial tuberosity
Hamstring group of muscles
Gluteus maximus
Gluteus medius
Natal cleft
Fold of buttock
Popliteal fossa
A

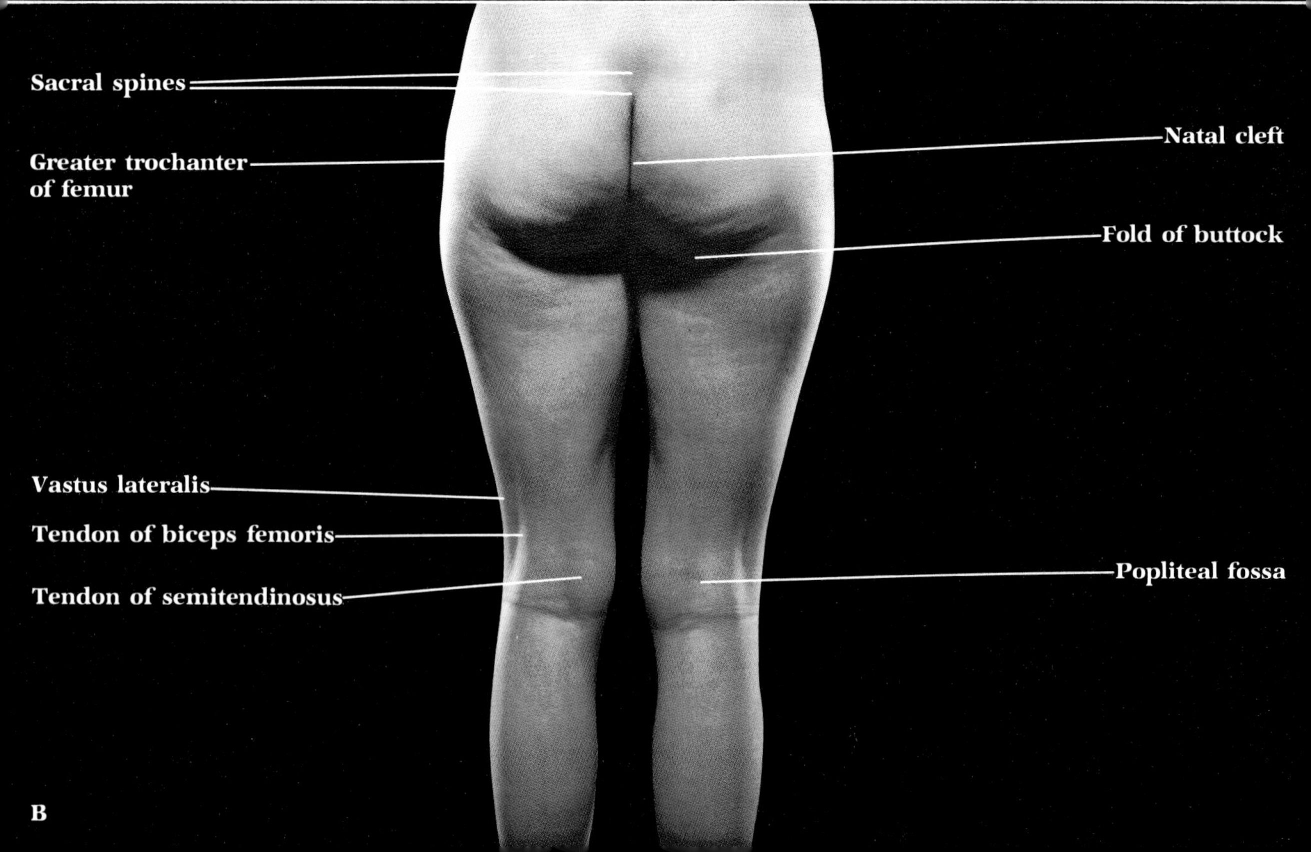

Sacral spines
Greater trochanter of femur
Natal cleft
Fold of buttock
Vastus lateralis
Tendon of biceps femoris
Tendon of semitendinosus
Popliteal fossa
B

Figure 6-1
The gluteal region and the posterior aspect of the thigh. A. 27-year-old male. B. 29-year-old female.

1. The iliac crests are easily palpable along their entire length. Each crest ends anteriorly at the anterior superior iliac spine and posteriorly at the posterior superior iliac spine, which lies beneath a skin dimple at the level of the second sacral vertebra and at the level of the middle of the sacroiliac joint.
2. The iliac tubercle is a prominence felt on the outer surface of the iliac crest about 2 inches (5 cm) posterior to the anterior superior iliac spine.
3. The ischial tuberosity can be palpated in the lower part of the buttock. With a person in the standing position, the tuberosity is covered by the gluteus maximus. When he is in a sitting position, the ischial tuberosity emerges from beneath the lower border of the gluteus maximus and supports the weight of the body.
4. The greater trochanter of the femur can be felt on the lateral surface of the thigh and moves beneath the examining finger as the hip joint is flexed and extended.
5. In a normal hip joint the upper border of the greater trochanter lies on a line connecting the anterior superior iliac spine to the ischial tuberosity.
6. The tip of the coccyx can be palpated beneath the skin in the upper part of the natal cleft.
7. The sciatic nerve in the buttock lies under cover of the gluteus maximus muscle, midway between the greater trochanter and the ischial tuberosity.
8. The biceps femoris, semitendinosus, and semimembranosus muscles, often referred to as the hamstring muscles, form the large muscle mass on the back of the thigh. They can be made more prominent by asking the patient to flex the knee joint against resistance.
9. The popliteal fossa is a diamond-shaped space situated behind the knee. It is bounded superiorly by the hamstring muscles and inferiorly by the two heads of the gastrocnemius muscle.

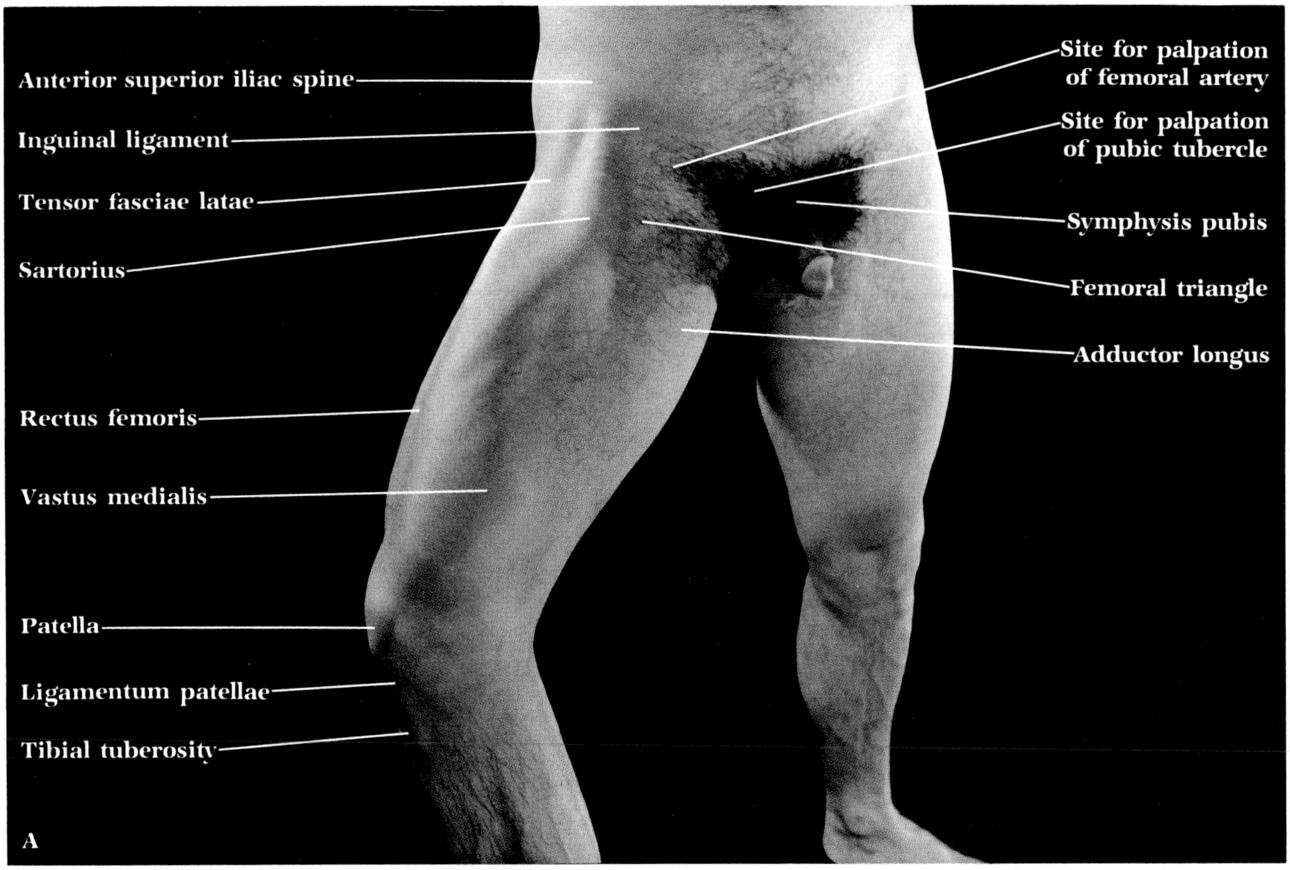
Anterior superior iliac spine
Inguinal ligament
Tensor fasciae latae
Sartorius
Rectus femoris
Vastus medialis
Patella
Ligamentum patellae
Tibial tuberosity
Site for palpation of femoral artery
Site for palpation of pubic tubercle
Symphysis pubis
Femoral triangle
Adductor longus
A

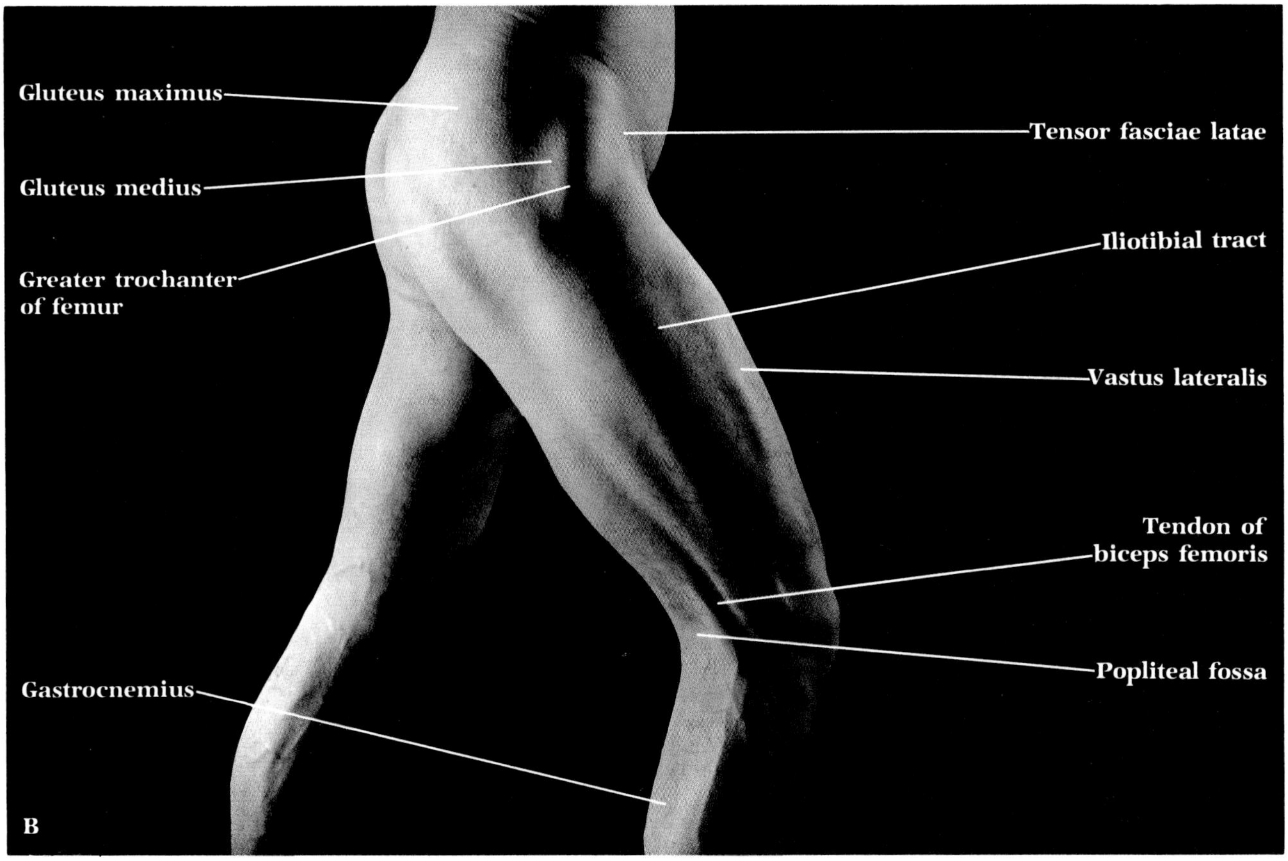
Gluteus maximus
Gluteus medius
Greater trochanter of femur
Gastrocnemius
Tensor fasciae latae
Iliotibial tract
Vastus lateralis
Tendon of biceps femoris
Popliteal fossa
B

Figure 6-2
The right lower limb of a 25-year-old male. A. The femoral and adductor regions of the thigh. B. The lateral surface of the thigh.

1. The anterior superior iliac spine can easily be felt at the anterior end of the iliac crest.
2. The pubic tubercle can be palpated in the male by invaginating the scrotum with the examining finger; the tubercle is felt on the upper border of the body of the pubis. In both sexes the tendon of the adductor longus muscle forms the medial boundary of the upper part of the thigh; if this is traced upward, it leads to the pubic tubercle.
3. The inguinal ligament may be felt along its length. It is attached laterally to the anterior superior iliac spine and medially to the pubic tubercle.
4. The femoral triangle can be seen as a depression below the fold of the groin in the upper part of the thigh. The boundaries of the triangle can be identified when the thigh is flexed, abducted, and laterally rotated. The base of the triangle is formed by the inguinal ligament, the lateral border by the sartorius muscle, and the medial border by the adductor longus muscle.
5. The saphenous opening in the fascia lata (deep fascia) of the thigh lies about 1½ inches (4 cm) below and lateral to the pubic tubercle.
6. The quadriceps femoris is a massive muscle on the anterior surface of the thigh; it is inserted into the subcutaneous patella. The ligamentum patellae connects the patella to the tibial tuberosity.
7. The superficial inguinal lymph nodes can be palpated in the superficial fascia just below and parallel to the inguinal ligament.
8. The femoral artery enters the thigh posterior to the inguinal ligament, at the midpoint of a line joining the symphysis pubis to the anterior superior iliac spine. Its pulsations are easily felt.
9. The femoral vein leaves the thigh by passing posterior to the inguinal ligament and medial to the pulsating femoral artery.
10. The inferior opening of the femoral canal lies below and lateral to the pubic tubercle.
11. The femoral nerve enters the thigh posterior to the midpoint of the inguinal ligament, i.e., lateral to the pulsating femoral artery.
12. The great saphenous vein pierces the saphenous opening in the fascia lata (deep fascia) of the thigh and joins the femoral vein 1½ inches (4 cm) below and lateral to the pubic tubercle.
13. The adductor (subsartorial) canal lies in the middle third of the thigh, immediately distal to the apex of the femoral triangle. It is an intermuscular cleft situated beneath the sartorius muscle and is bounded laterally by the vastus medialis muscle and posteriorly by the adductor longus and magnus muscles. It contains the femoral vessels and the saphenous nerve.
14. The greater trochanter of the femur can be palpated on the lateral surface of the thigh.
15. The iliotibial tract, a wide, thickened band of deep fascia that extends from the tubercle of the iliac crest to the lateral condyle of the tibia, can be felt on the lateral side of the thigh. It becomes taut when the hip and knee joints are extended.

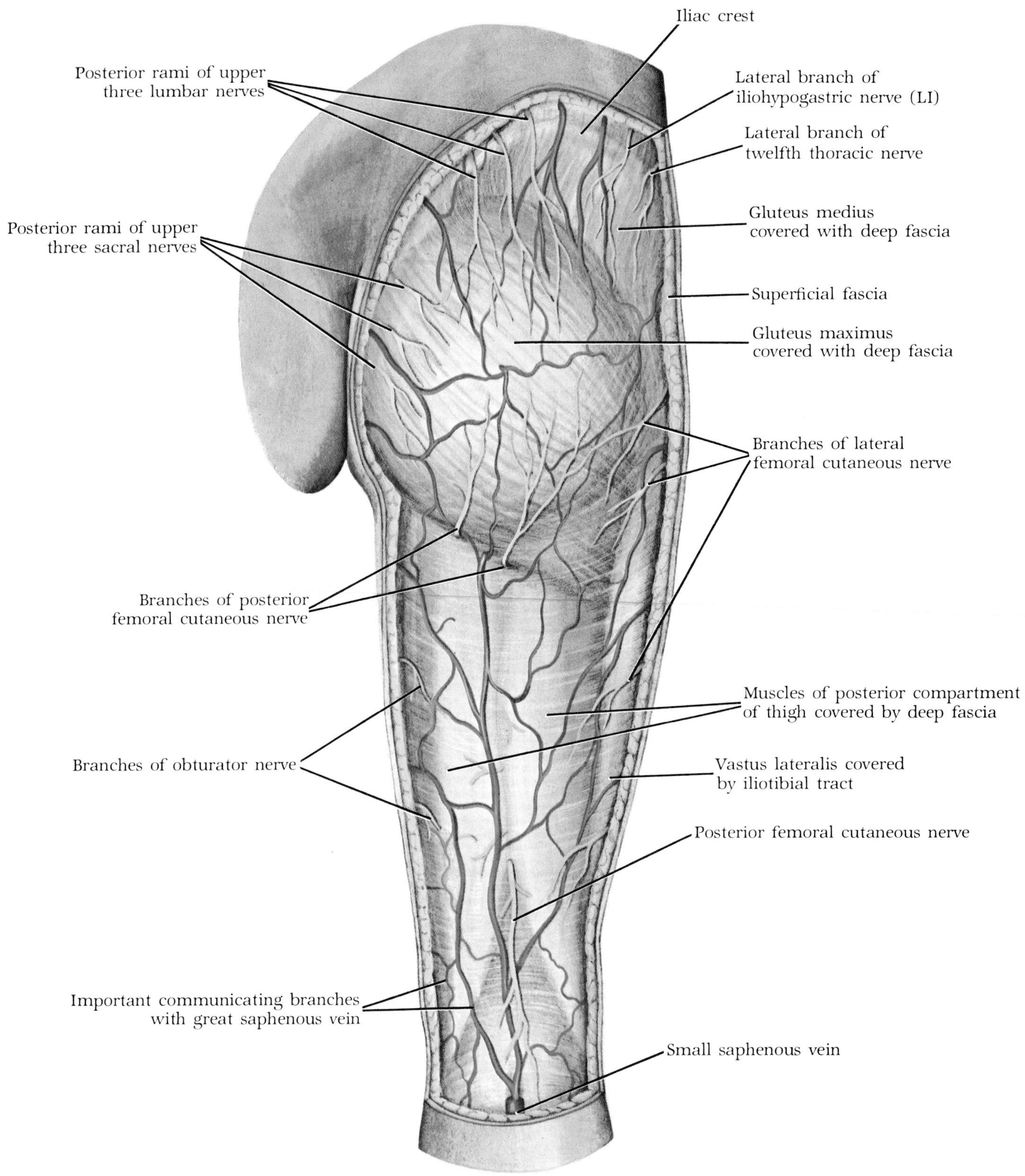
Iliac crest
Posterior rami of upper three lumbar nerves
Lateral branch of iliohypogastric nerve (LI)
Lateral branch of twelfth thoracic nerve
Gluteus medius covered with deep fascia
Posterior rami of upper three sacral nerves
Superficial fascia
Gluteus maximus covered with deep fascia
Branches of lateral femoral cutaneous nerve
Branches of posterior femoral cutaneous nerve
Muscles of posterior compartment of thigh covered by deep fascia
Branches of obturator nerve
Vastus lateralis covered by iliotibial tract
Posterior femoral cutaneous nerve
Important communicating branches with great saphenous vein
Small saphenous vein

Figure 6-3
Right gluteal region and posterior aspect of thigh, showing the superficial veins, cutaneous nerves, and deep fascia.

1. The deep fascia of the buttock is continuous below with the deep fascia (fascia lata) of the thigh. In the gluteal region it splits to enclose the gluteus maximus muscle. Superiorly it continues as a single layer that covers the outer surface of the gluteus medius and is attached to the iliac crest.
2. The iliotibial tract, a wide band of thickening of the deep fascia on the lateral surface of the thigh, is attached superiorly to the tubercle of the iliac crest and inferiorly to the lateral condyle of the tibia. It forms a sheath for the tensor fasciae latae muscle and receives the greater part of the insertion of the gluteus maximus muscle. It provides surgeons with a source of strong fibrous tissue for the fascial repair of inguinal hernias.
3. Contracture of the iliotibial tract can be associated with deformities of the leg following poliomyelitis. Surgical division of the tract in these patients is sometimes part of the treatment to correct the deformities.
4. The lymphatic drainage of the skin of the buttock is into the lateral group of the superficial inguinal nodes.

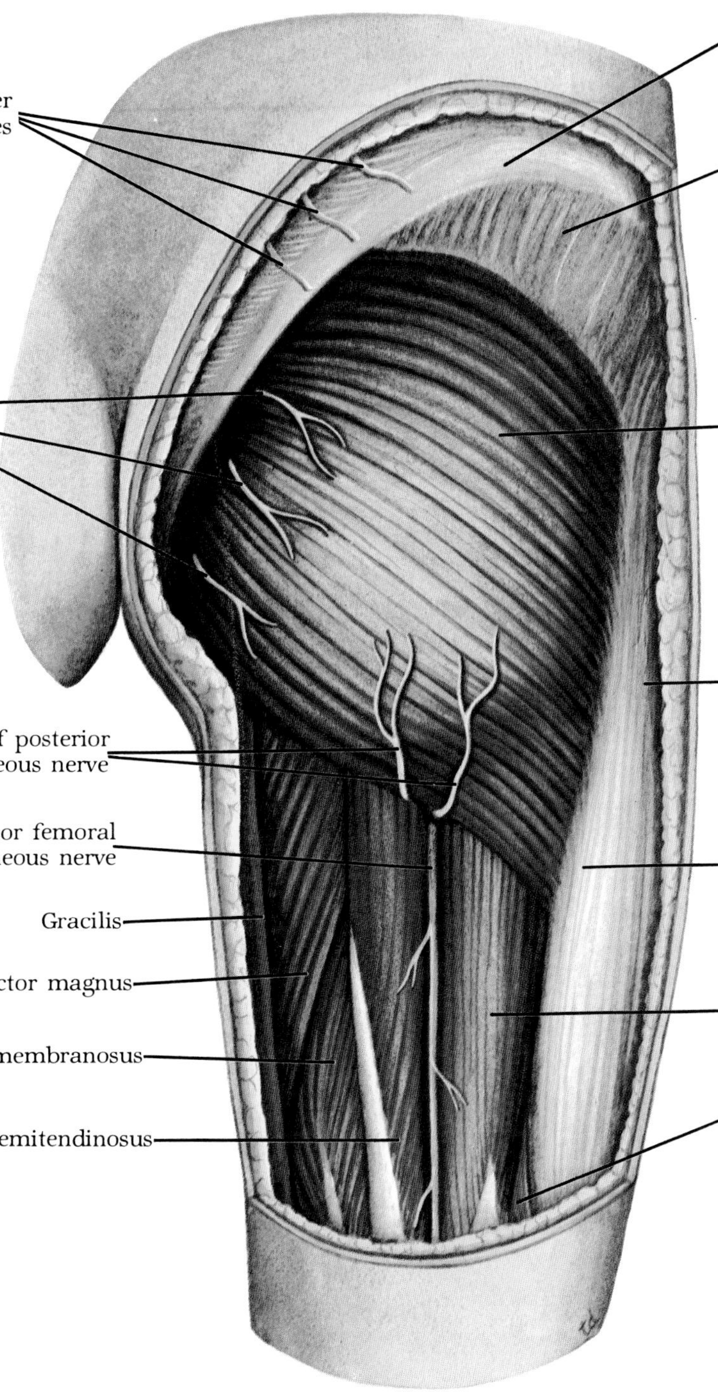

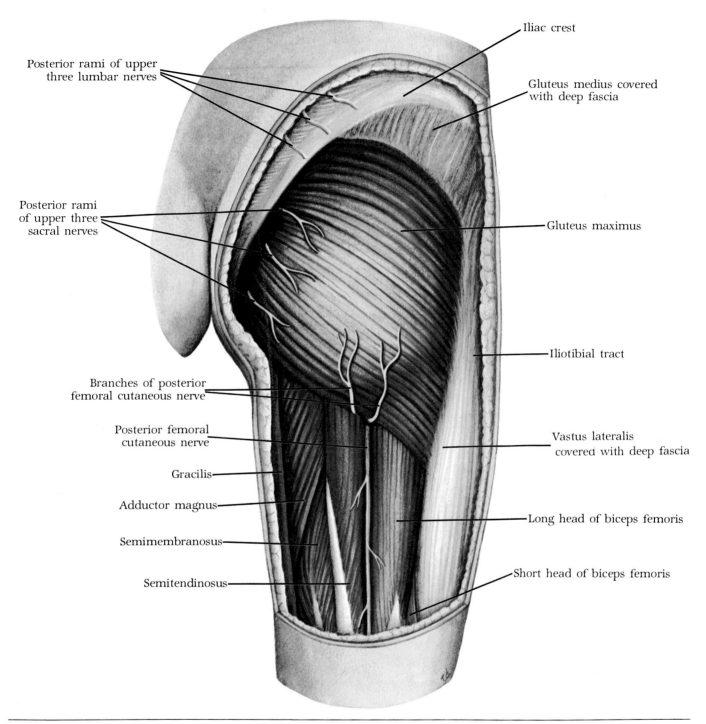

Iliac crest
Posterior rami of upper three lumbar nerves
Gluteus medius covered with deep fascia
Posterior rami of upper three sacral nerves
Gluteus maximus
Iliotibial tract
Branches of posterior femoral cutaneous nerve
Posterior femoral cutaneous nerve
Vastus lateralis covered with deep fascia
Gracilis
Adductor magnus
Long head of biceps femoris
Semimembranosus
Semitendinosus
Short head of biceps femoris

Figure 6-4
Right gluteus maximus muscle.

1. The gluteus maximus is a very large, thick muscle with coarse fasciculi that can be easily separated without damage. It extends the hip joint and is particularly used when walking up a grade or climbing stairs. It laterally rotates the hip, and it extends the knee joint by exerting traction on the iliotibial tract.

2. The great thickness of the gluteus maximus makes it an ideal muscle in which to give intramuscular injections. To avoid injury to the underlying sciatic nerve, the injection should be given well forward on the upper outer quadrant of the buttock. The majority of the nerve lesions are incomplete, and in 90 percent of injuries it is the common peroneal part of the nerve that is the most affected. In a complete section of the nerve, however, the following clinical features may be present:
Motor. The hamstring muscles are paralyzed, but weak flexion of the knee is possible due to the action of the sartorius (femoral nerve) and gracilis (obturator nerve) muscles. All muscles below the knee are paralyzed, and the weight of the foot causes it to assume the plantar-flexed position, a condition called foot drop.
Sensory. There is loss of sensation below the knee, except for that narrow area down the medial side of the lower part of the leg and along the medial border of the foot as far as the ball of the big toe that is supplied by the saphenous nerve (femoral nerve).

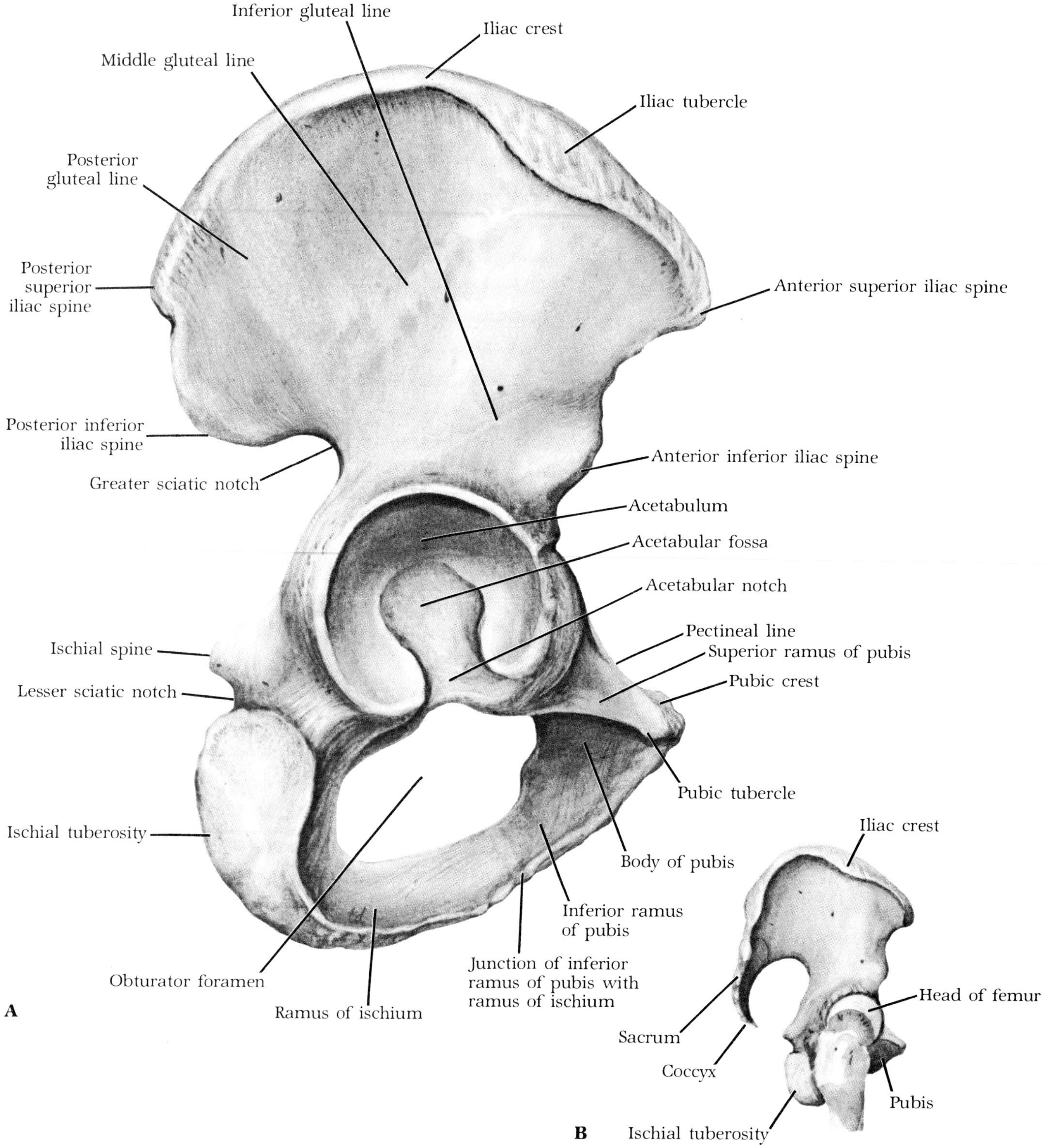
Inferior gluteal line
Iliac crest
Middle gluteal line
Iliac tubercle
Posterior gluteal line
Posterior superior iliac spine
Anterior superior iliac spine
Posterior inferior iliac spine
Greater sciatic notch
Anterior inferior iliac spine
Acetabulum
Acetabular fossa
Acetabular notch
Pectineal line
Superior ramus of pubis
Ischial spine
Pubic crest
Lesser sciatic notch
Pubic tubercle
Ischial tuberosity
Body of pubis
Inferior ramus of pubis
Junction of inferior ramus of pubis with ramus of ischium
Obturator foramen
A
Ramus of ischium
Iliac crest
Head of femur
Sacrum
Coccyx
Pubis
B
Ischial tuberosity

Figure 6-5
External surface of the right hip bone. A. The isolated bone. B. The bone, placed in the anatomical position, articulating with the sacrum and femur.

1. In the acetabulum the articular surface of the hip joint is limited to a horseshoe-shaped area and is covered with hyaline cartilage. The floor of the acetabulum is nonarticular and is called the acetabular fossa. In congenital dislocation of the hip, the upper lip of the acetabulum fails to develop adequately, and the head of the femur, having no stable platform under which it can lodge, rides up out of the acetabulum onto the gluteal surface of the ilium.

2. Fracture of the ilium (which forms part of the so-called false pelvis) occasionally occurs as the result of severe direct trauma, as in automobile and aircraft accidents. The fragments are seldom displaced, however, because of the attachment of the iliacus muscle on the inside and the gluteal muscles on the outside.

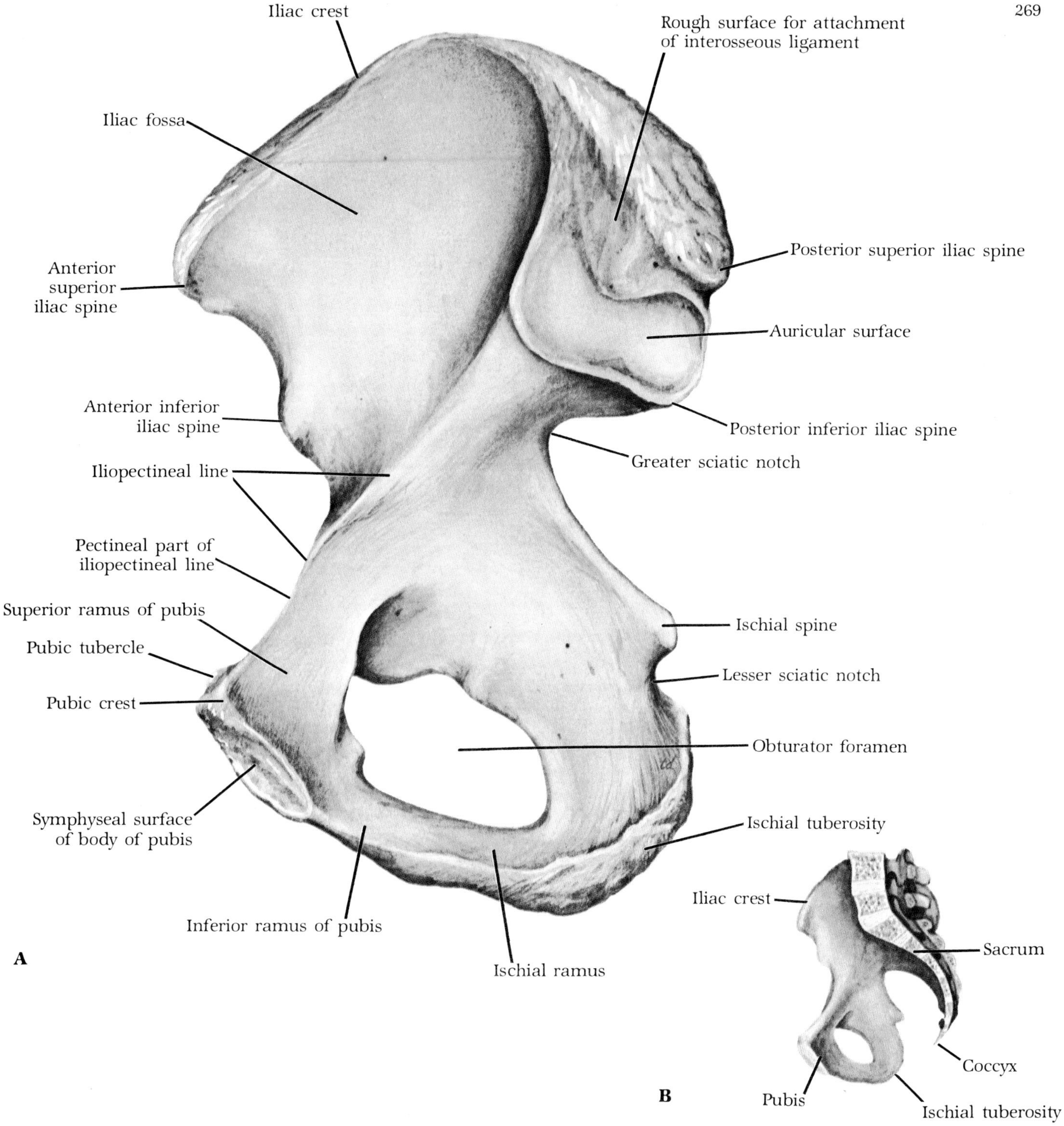
Iliac crest
Rough surface for attachment
of interosseous ligament
Iliac fossa
Posterior superior iliac spine
Anterior
superior
iliac spine
Auricular surface
Anterior inferior
iliac spine
Posterior inferior iliac spine
Greater sciatic notch
Iliopectineal line
Pectineal part of
iliopectineal line
Superior ramus of pubis
Ischial spine
Pubic tubercle
Lesser sciatic notch
Pubic crest
Obturator foramen
Symphyseal surface
of body of pubis
Ischial tuberosity
Inferior ramus of pubis
A
Ischial ramus
Iliac crest
Sacrum
Coccyx
B
Pubis
Ischial tuberosity

Figure 6-6
Internal surface of right hip bone. A. The isolated bone. B. The bone articulating with the sacrum.

1. The bony pelvis is composed of four bones: the two innominate (hip) bones, the sacrum, and the coccyx. The pelvis is divided into two parts by the pelvic brim, which is formed by the sacral promontory posteriorly, the iliopectineal lines laterally, and the symphysis pubis anteriorly. Above the brim is the false, or greater, pelvis, which flares out at its upper end and forms part of the abdominal cavity. Below the brim is the true, or lesser, pelvis.

2. The false pelvis is of little clinical importance although it does support the abdominal contents, and after the third month of pregnancy it helps to support the gravid uterus. During the early stages of labor it helps to guide the fetus into the true pelvis.

3. The true pelvis contains and protects the lower parts of the intestinal and urinary tracts and the internal organs of reproduction. Knowledge of the shape and dimensions of the female pelvis is of great importance in obstetrics since it is the bony canal through which a child passes during birth.

4. Fractures of the true pelvis are usually caused by severe "run-over" type automobile accidents. Anteroposterior compression can produce dislocation of the symphysis, or a fracture through the pubic rami accompanied by fracture of the lateral part of the sacrum. These fractures are often accompanied by tearing of the pelvic veins, resulting in severe hemorrhage and damage to the pelvic viscera.

5. Direct trauma to the greater trochanter can drive the head of the femur through the floor of the acetabulum. Bone spicules can damage the obturator nerve and pelvic viscera.

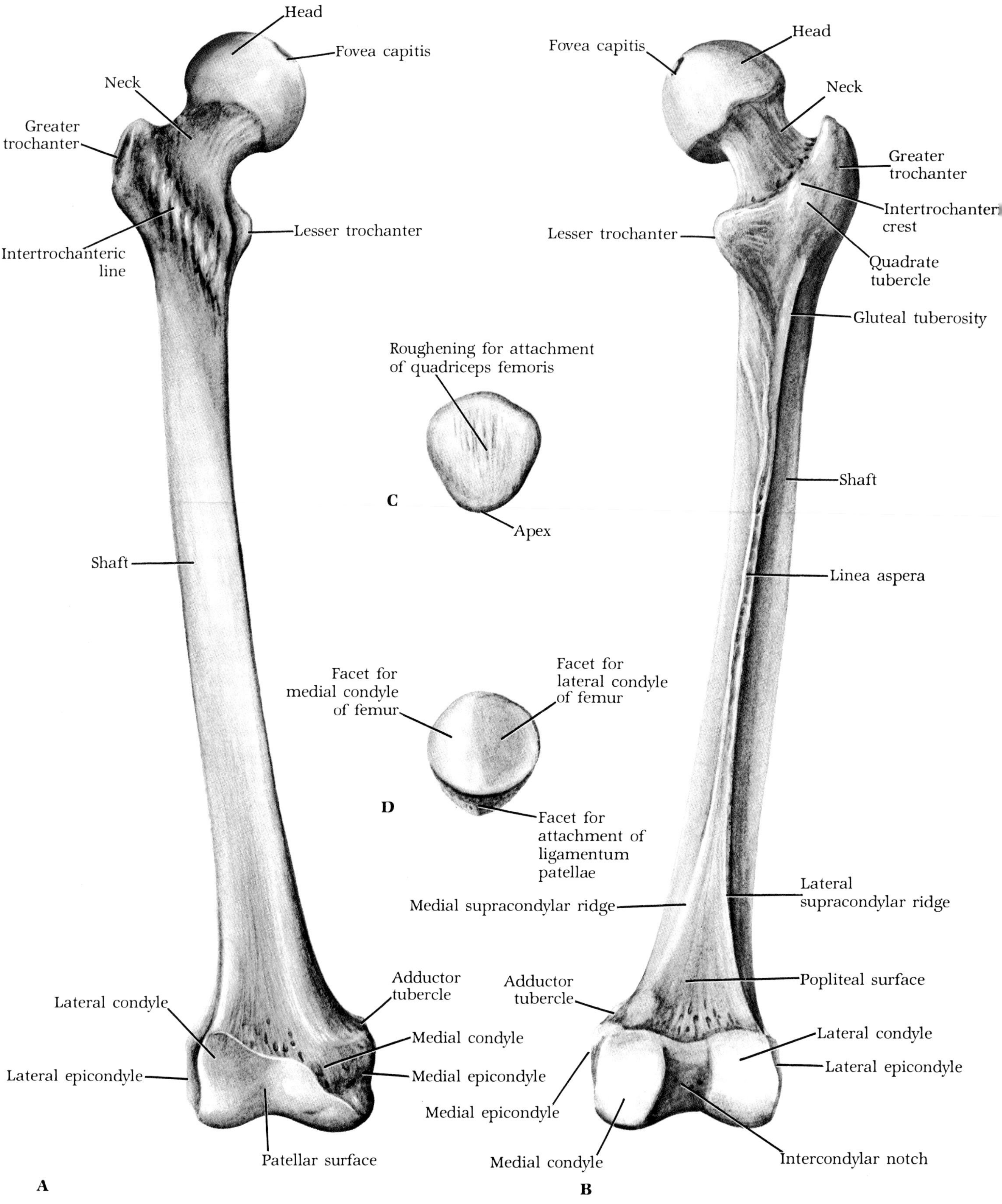
Head
Fovea capitis
Neck
Greater trochanter
Intertrochanteric line
Lesser trochanter
Shaft
Lateral condyle
Adductor tubercle
Medial condyle
Lateral epicondyle
Medial epicondyle
Patellar surface
A
Roughening for attachment of quadriceps femoris
C
Apex
Facet for medial condyle of femur
Facet for lateral condyle of femur
D
Facet for attachment of ligamentum patellae
Fovea capitis
Head
Neck
Greater trochanter
Intertrochanteric crest
Lesser trochanter
Quadrate tubercle
Gluteal tuberosity
Shaft
Linea aspera
Lateral supracondylar ridge
Medial supracondylar ridge
Adductor tubercle
Popliteal surface
Lateral condyle
Lateral epicondyle
Medial epicondyle
Medial condyle
Intercondylar notch
B

Figure 6-7
The femur and patella.
A. Right femur, anterior view.
B. Right femur, posterior view.
C. Right patella, anterior view.
D. Right patella, posterior view.

1. The neck of the femur is inclined at an angle with the shaft, the angle being about 160 degrees in the young child and about 125 degrees in the adult. The neck of the femur also projects anteriorly on the shaft; this angle can be as much as 14 degrees in the adult.
2. Coxa valga is an increase in the angle of inclination of the neck of the femur on the shaft and occurs, for example, in cases of congenital dislocation of the hip. In the latter condition, adduction of the hip joint is limited. Coxa vara is a decrease of this angle and occurs in fractures of the neck of the femur and when the femoral epiphysis slips.
3. A femoral neck fracture usually occurs in an elderly person and almost always results from a minor "trip" or stumble. Avascular necrosis of the head is a common complication. If the fragments are not impacted, there is considerable displacement. The strong muscles of the thigh pull the distal fragment upward, thus shortening the leg. The gluteus maximus, the piriformis, the obturator internus, the gemelli, and the quadratus femoris muscles rotate the distal fragment laterally.
4. Trochanteric fractures commonly occur in the young and middle-aged as the result of direct trauma. The fracture line is extracapsular, and both fragments have a profuse blood supply. If the bone fragments are not impacted, the pull of the strong thigh muscles will produce shortening and lateral rotation of the leg, as explained in 3.
5. A fracture of the upper third of the femur results in the proximal fragment being flexed by the iliopsoas muscle, abducted by the gluteus medius and minimus muscles, and laterally rotated by the gluteus maximus, the piriformis, the obturator internus, the gemelli, and the quadratus femoris muscles. The lower fragment is adducted by the adductor muscles, pulled upward by the hamstrings and quadriceps muscles, and laterally rotated by the adductors and the weight of the foot.
6. In a fracture of the middle third of the femur, the distal fragment is pulled upward by the hamstrings and the quadriceps muscles, resulting in considerable shortening. The distal fragment is also rotated backward by the pull of the two heads of the gastrocnemius muscle.
7. A fracture of the distal third of the shaft of the femur results in the same displacement of the distal fragment as seen in fractures of the middle third of the shaft. The distal fragment, however, is rotated to a greater degree by the gastrocnemius muscle and may exert pressure on the popliteal artery and thus interfere with blood flow through the leg and foot.
8. Congenital recurrent dislocation of the patella laterally is due to underdevelopment of the lateral femoral condyle.
9. The patella can be fractured by a direct blow, breaking it into a number of small fragments. Since the bone lies within the quadriceps femoris tendon, however, little separation of the fragments takes place.
10. An example of fracture of the patella by indirect violence is that caused by sudden contraction of the quadriceps, snapping the patella across the femoral condyles. Separation of the fragments usually occurs.

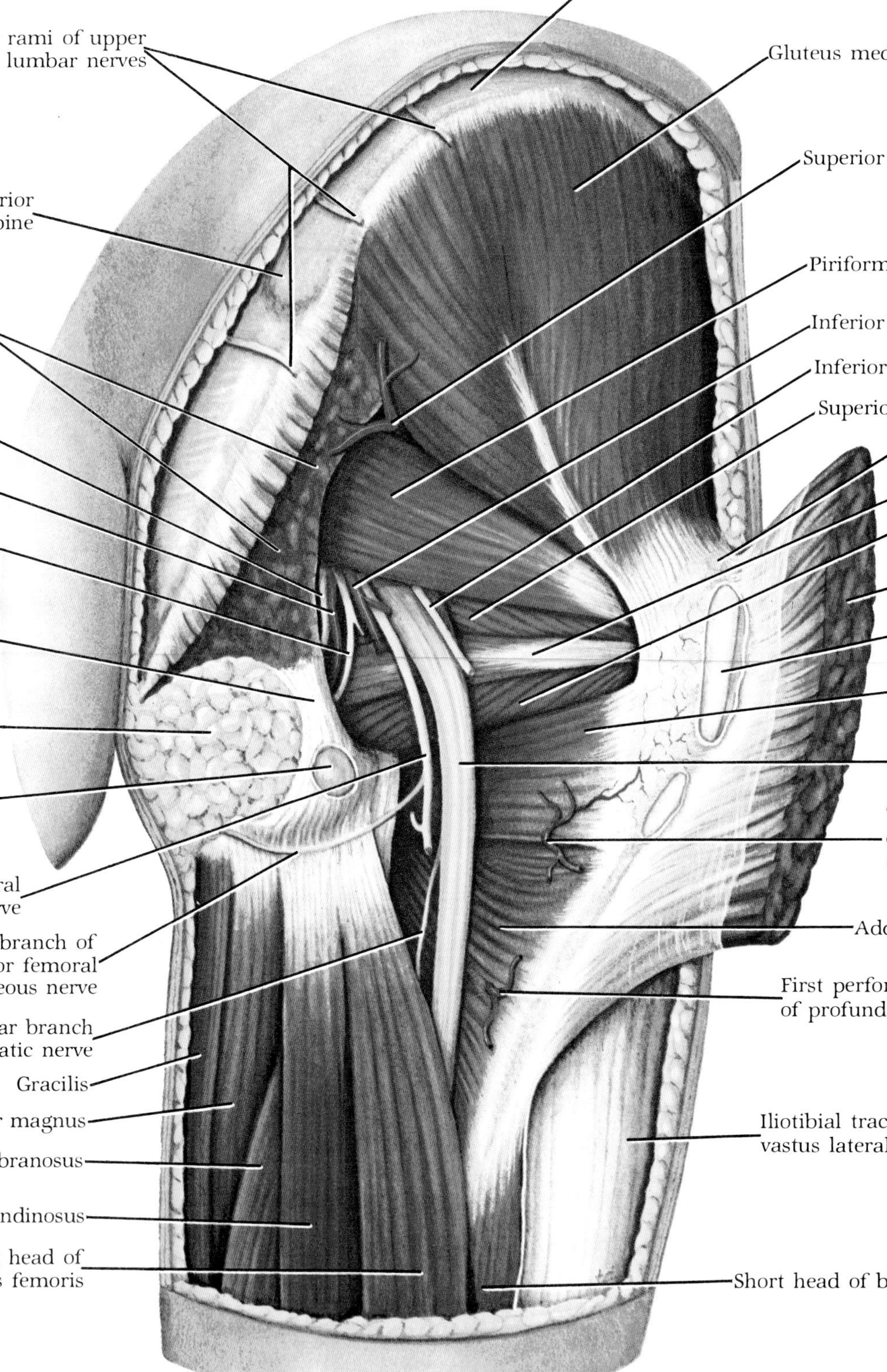
Iliac crest
Posterior rami of upper three lumbar nerves
Gluteus medius
Superior gluteal artery
Posterior superior iliac spine
Piriformis
Gluteus maximus (cut)
Inferior gluteal artery
Inferior gluteal nerve
Pudenal nerve
Superior gemellus
Greater trochanter
Internal pudendal artery
Obturator internu
Inferior gemellus
Nerve to obturator internus
Gluteus maximus (cut and reflected
Sacrotuberous ligament
Bursa over greater trochante
Quadratus femoris
Fat in ischiorectal fossa
Sciatic nerve
Bursa over ischial tuberosity
Transverse branch of medial femoral circumflex artery
Posterior femoral cutaneous nerve
Adductor magnus
Perineal branch of posterior femoral cutaneous nerve
First perforating branch of profunda femoris artery
Muscular branch of sciatic nerve
Gracilis
Adductor magnus
Iliotibial tract covering vastus lateralis
Semimembranosus
Semitendinosus
Long head of biceps femoris
Short head of biceps femoris

Figure 6-8
Structures present in right gluteal region. The greater part of gluteus maximus has been removed.

1. The gluteus medius acting with the gluteus minimus and the tensor fasciae latae strongly abducts the hip joint.
2. When walking, a person normally alternately contracts the gluteus medius and minimus muscles, first on one side and then on the other. By thus raising the pelvis first on one side and then on the other, the leg can be flexed at the hip joint and moved forward, i.e., the leg is raised clear off the ground before it is thrust forward to take the forward step.
3. Two bursae are found near the insertion of the gluteus maximus muscle: one between the tendon and the greater trochanter and the other between it and the vastus lateralis. Another bursa, the one found between the gluteus maximus and the ischial tuberosity, can become inflamed from repeated trauma, a condition that used to be called "weaver's bottom."
4. The close relationship of the sciatic nerve to the posterior surface of the hip joint makes that nerve prone to injury in a posterior dislocation of this joint.

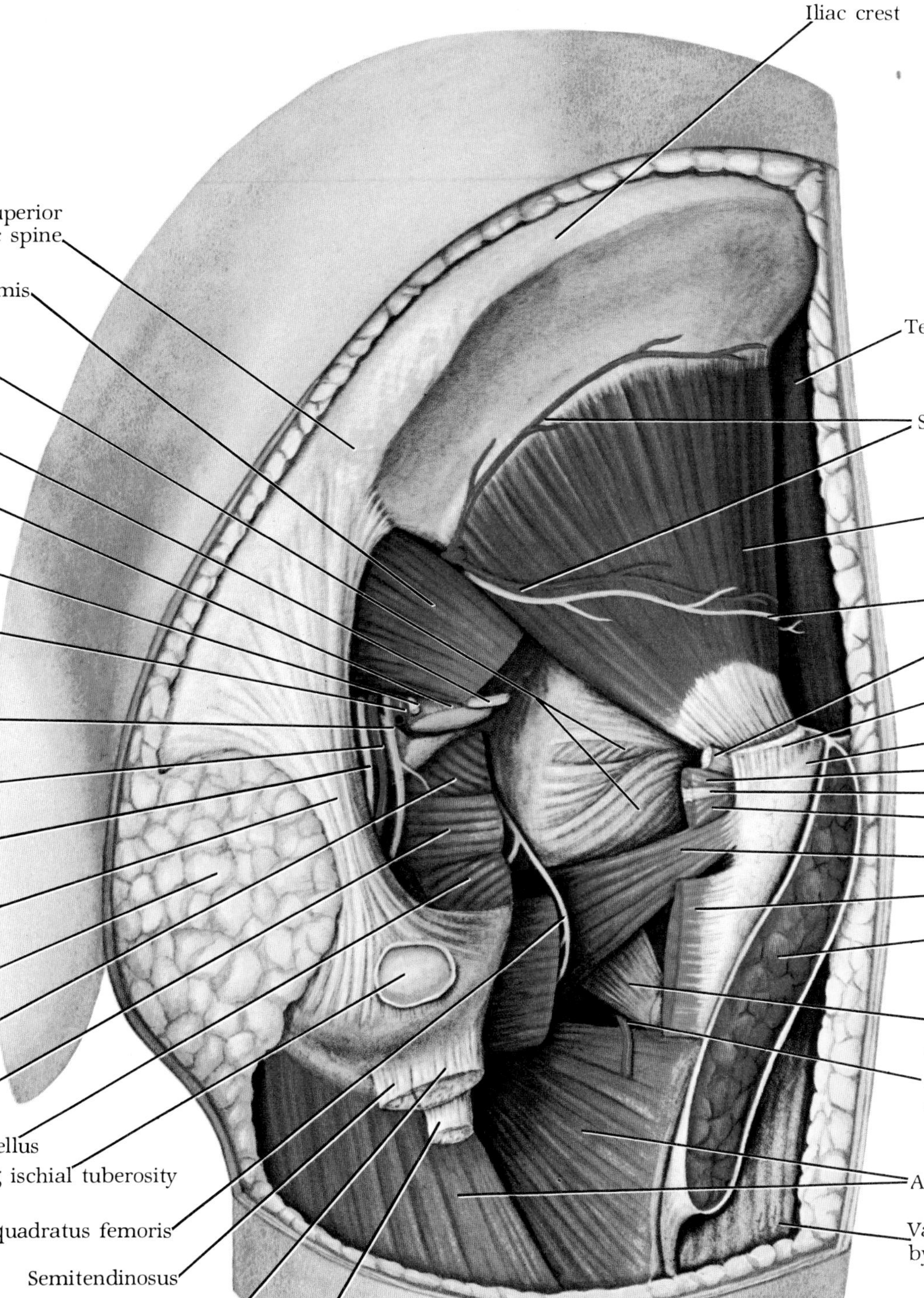

Iliac crest
Posterior superior iliac spine
Piriformis
Capsule of hip joint and ischiofemoral ligament
Inferior gluteal nerve
Sciatic nerve
Posterior femoral cutaneous nerve
Inferior gluteal artery
Nerve to obturator internus
Internal pudendal artery
Pudendal nerve
Sacro-tuberous ligament
Fat in ischiorectal fossa
Superior gemellus
Obturator internus
Inferior gemellus
Bursa covering ischial tuberosity
Nerve to quadratus femoris
Semitendinosus
Long head of biceps femoris
Semimembranosus
Tensor fasciae latae
Superior gluteal artery
Gluteus minimus
Superior gluteal nerve
Tendon of piriformis
Tendon of gluteus medius
Greater trochanter
Superior gemellus
Obturator internus
Inferior gemellus
Obturator externus
Quadratus femoris
Gluteus maximus enclosed in fascia lata
Iliopsoas tendon
Transverse branch of medial femoral circumflex artery
Adductor magnus
Vastus lateralis covered by iliotibial tract

Figure 6-9
Deep structures present in right gluteal region. The greater parts of gluteus maximus and gluteus medius have been removed.

1. Again the close relationship of the sciatic nerve to the posterior surface of the hip joint and the proneness of this nerve to injury in a posterior dislocation of this joint must be emphasized.

2. When poliomyelitis involves the lower lumbar and sacral segments of the spinal cord, the gluteus medius and gluteus minimus may be paralyzed. They are supplied by the superior gluteal nerve (L4, 5, and S1). Paralysis of these muscles seriously interferes with the ability of the patient to tilt the pelvis when walking.

3. The stability of the hip joint depends on the ball-and-socket arrangement of the articular surfaces and the very strong ligaments. In congenital dislocation of the hip, the upper lip of the acetabulum fails to develop adequately, and the head of the femur, having no stable platform under which it can lodge, rides up out of the acetabulum onto the lateral surface of the ilium beneath the gluteus minimus and gluteus medius muscles.

4. The stability of the hip joint when a person stands on one leg with the foot of the opposite side raised above the ground depends on three factors: (a) the gluteus medius and minimus must be functioning normally; (b) the head of the femur must be located normally within the acetabulum; and (c) the neck of the femur must be intact and have a normal angle with the shaft of the femur. If any one of these factors does not hold true, the pelvis will sink downward on the opposite, unsupported side. The patient is then said to exhibit a positive Trendelenburg's sign.

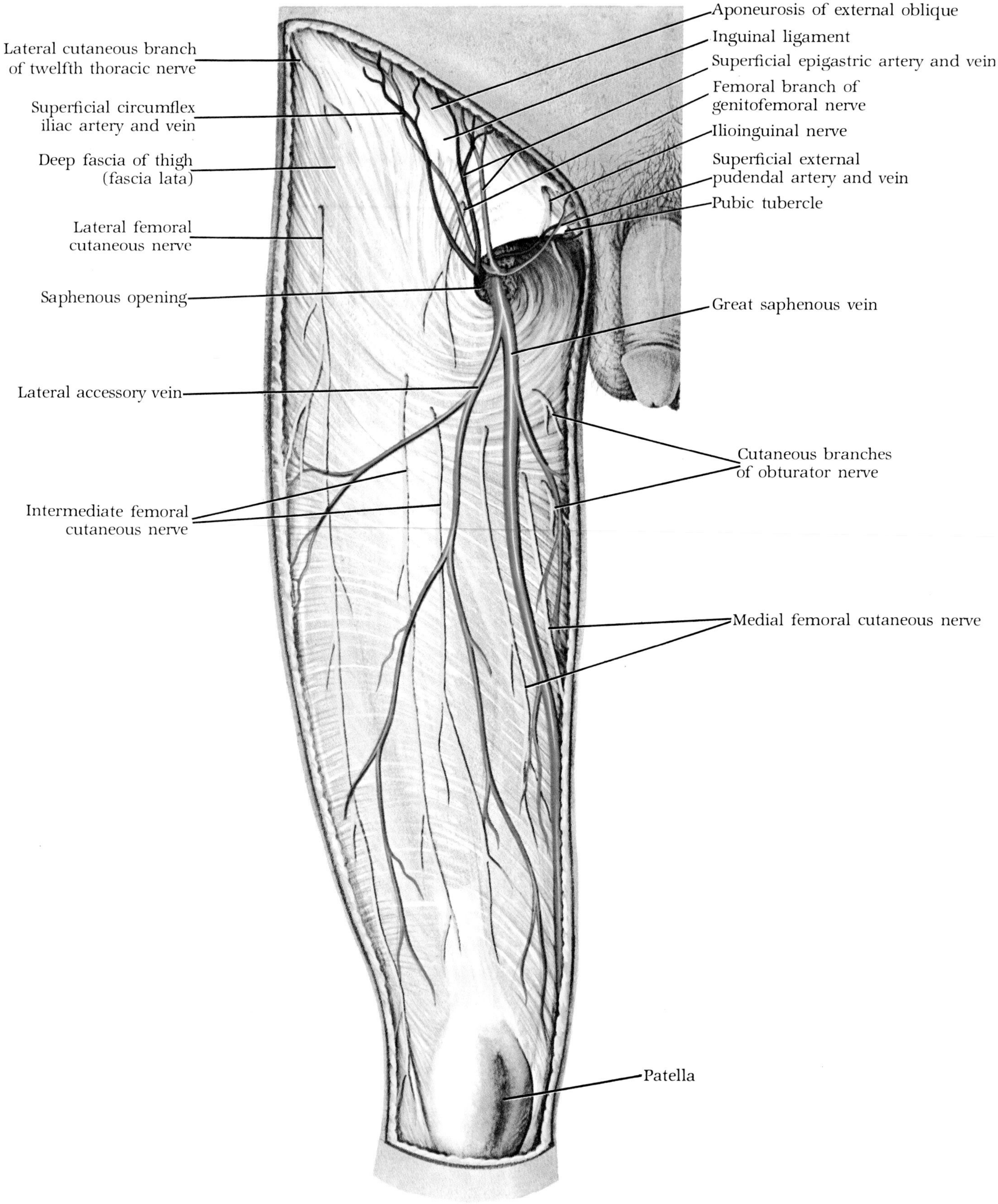
Lateral cutaneous branch of twelfth thoracic nerve
Superficial circumflex iliac artery and vein
Deep fascia of thigh (fascia lata)
Lateral femoral cutaneous nerve
Saphenous opening
Lateral accessory vein
Intermediate femoral cutaneous nerve
Aponeurosis of external oblique
Inguinal ligament
Superficial epigastric artery and vein
Femoral branch of genitofemoral nerve
Ilioinguinal nerve
Superficial external pudendal artery and vein
Pubic tubercle
Great saphenous vein
Cutaneous branches of obturator nerve
Medial femoral cutaneous nerve
Patella

Figure 6-10
Anterior surface of the right thigh. The skin and superficial fascia have been removed to display the deep fascia, superficial veins, and cutaneous nerves.

1. The superficial fascia covering the anterior abdominal wall may be divided into a superficial fatty layer (fascia of Camper) and a deep membranous layer (Scarpa's fascia). This fatty layer is continuous with the superficial fat covering the leg. The membranous layer fuses with the deep fascia of the thigh one fingersbreadth below the inguinal ligament. It is not attached to the pubis in the midline, but forms a tubular sheath for the penis (or clitoris). Below these structures it becomes continuous with Colles' fascia in the perineum. The membranous layer is important clinically, since beneath it there is a potential closed space that does not open into the thigh. Rupture of the penile urethra may be followed by extravasation of urine into the scrotum, perineum, and penis and then up into the lower part of the anterior abdominal wall deep to the membranous layer of fascia. The urine does not enter the thigh because of the attachment of this fascia to the fascia lata.

2. The great saphenous vein passes up the thigh within the superficial fascia and enters the femoral vein about $1\frac{1}{2}$ inches below and lateral to the pubic tubercle by passing through the saphenous opening in the deep fascia (fascia lata). It is a vein that commonly becomes varicosed.

3. A large area of skin on the anterior aspect of the thigh is innervated by branches from the femoral nerve (L2, 3, and 4).

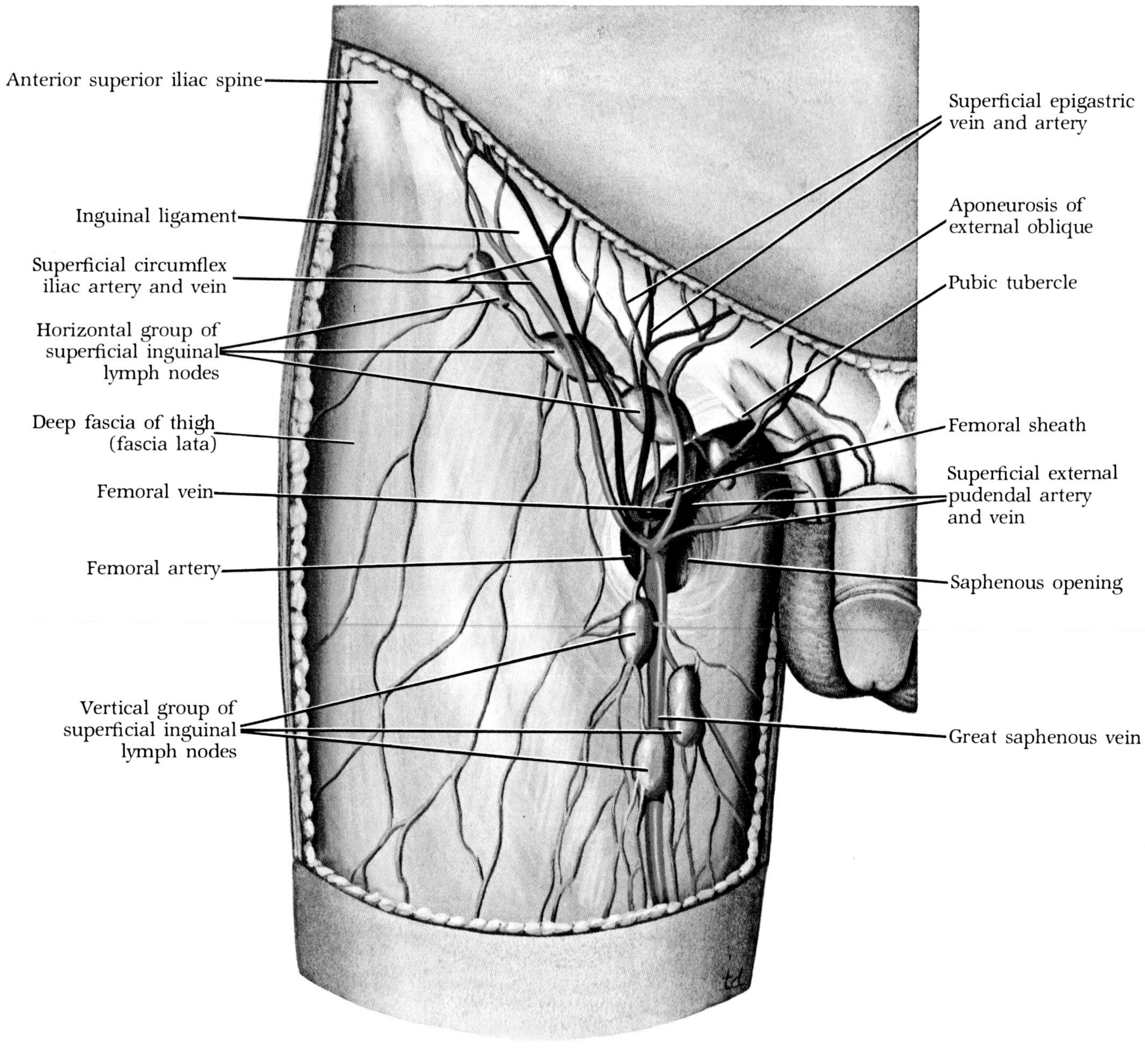
Anterior superior iliac spine
Superficial epigastric vein and artery
Inguinal ligament
Aponeurosis of external oblique
Superficial circumflex iliac artery and vein
Pubic tubercle
Horizontal group of superficial inguinal lymph nodes
Deep fascia of thigh (fascia lata)
Femoral sheath
Femoral vein
Superficial external pudendal artery and vein
Femoral artery
Saphenous opening
Vertical group of superficial inguinal lymph nodes
Great saphenous vein

Figure 6-11
Anterior surface of the right thigh. The skin and superficial fascia have been removed to display the lymph vessels, lymph nodes, and the superficial veins.

1. The saphenous opening in the deep fascia permits the great saphenous vein to enter the femoral vein and also allows the lymphatic vessels from all the superficial inguinal nodes to pass to the deep inguinal nodes. A large femoral hernial sac can emerge from the lower end of the femoral canal and protrude anteriorly through the saphenous opening.

2. The superficial inguinal lymph nodes are often chronically enlarged due to a low-grade infection from the lower half of the anal canal and the genitalia. This infection, however, may simply result from the repeated mild trauma of defecation and coitus.

3. The superficial and deep inguinal lymph nodes not only drain all the lymph from the lower limbs but also drain lymph from the skin and superficial fascia of the anterior and posterior abdominal walls below the level of the umbilicus; lymph from the external genitalia and the mucous membrane of the lower half of the anal canal also drains into these nodes. It is important to remember the long distances the lymph has to travel in some instances in order to reach the inguinal nodes. A simple infection of one of the toes can cause a painful enlargement of an inguinal lymph node.

4. The successful operative treatment of a varicosed great saphenous vein depends on the ligation and division of all the main tributaries of this vein; this will prevent a collateral venous circulation from developing. It is now common practice to also remove or strip the superficial veins. It is very important to ascertain before operative measures are undertaken that the deep veins are patent.

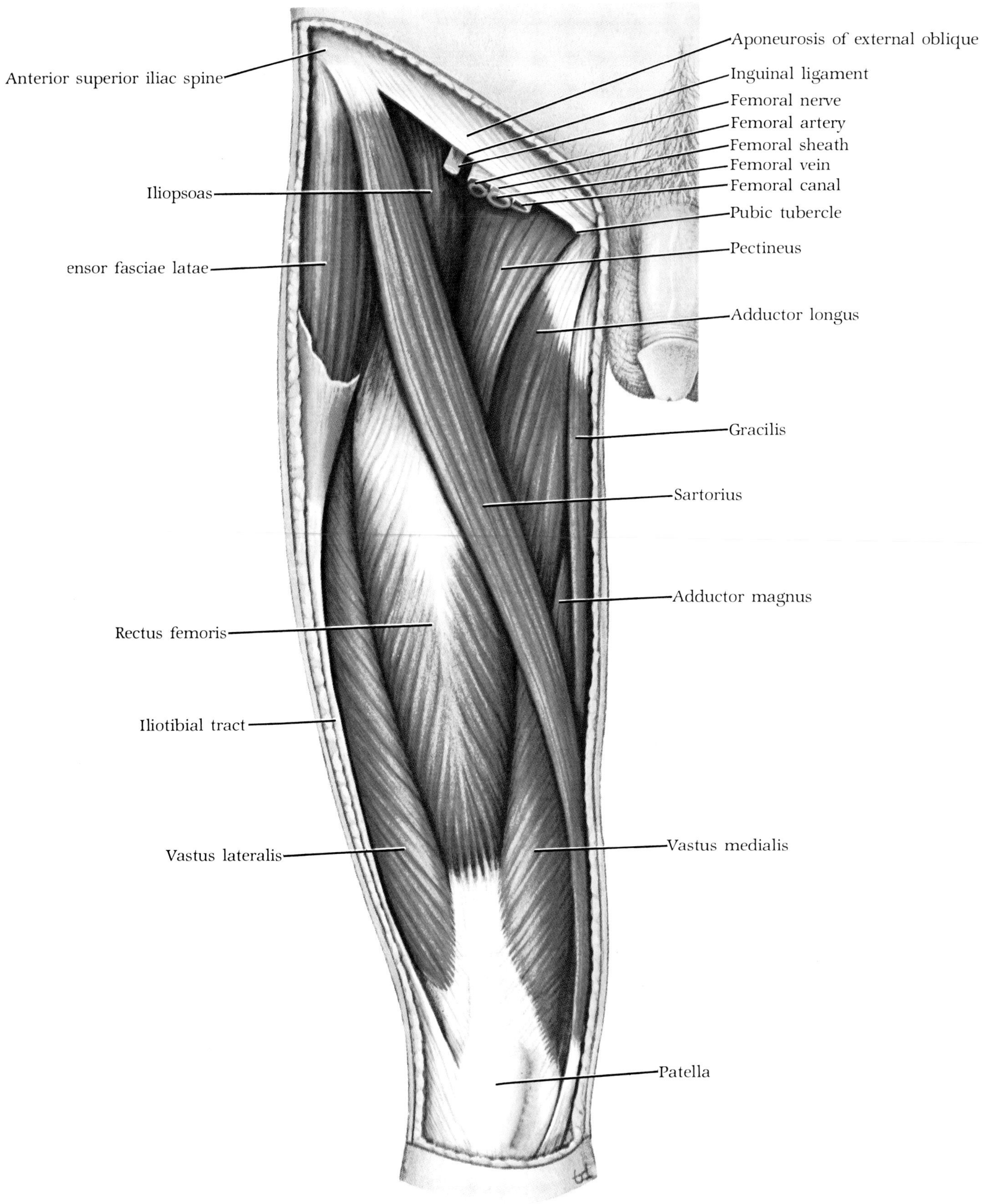
Aponeurosis of external oblique
Inguinal ligament
Femoral nerve
Femoral artery
Femoral sheath
Femoral vein
Femoral canal
Pubic tubercle
Pectineus
Adductor longus
Gracilis
Sartorius
Adductor magnus
Vastus medialis
Patella
Anterior superior iliac spine
Iliopsoas
ensor fasciae latae
Rectus femoris
Iliotibial tract
Vastus lateralis

Figure 6-12
Muscles of the femoral triangle and the anterior surface of the right thigh.

1. The position of the inguinal ligament in the groin is marked by an overlying groove. Because of the downward pull of the fascia lata of the thigh, the inguinal ligament is convex downward. The ligament is attached laterally to the anterior superior iliac spine and medially to the pubic tubercle. When examining a patient's abdomen it is often helpful to ask him or her to flex the hip joint slightly. This motion removes the downward traction of the fascia lata on the inguinal ligament and permits the greatest degree of relaxation of the anterior abdominal wall.
2. The femoral sheath is a protrusion of the fascial envelope lining the abdominal walls; it surrounds the femoral vessels and lymphatics for about an inch (2.5 cm) below the inguinal ligament. The femoral canal, the compartment for the lymphatics, occupies the medial part of the sheath. The superior opening of the femoral canal is called the femoral ring; the femoral septum, a condensation of extra peritoneal tissue, closes the femoral ring.
3. The femoral artery enters the thigh within the femoral sheath posterior to the inguinal ligament at a point midway between the anterior superior iliac spine and the symphysis pubis. It is at this point that it is easily palpated, because it can be pressed posteriorly against the pectineus and the superior ramus of the pubis.
4. The quadriceps femoris extends the knee joint, and the tone of this muscle mass greatly strengthens the knee joint. Disease of the knee joint or surgical interference with the joint results in rapid atrophy of this muscle group. Early physiotherapy and exercising of the quadriceps muscle is a vital part of the treatment of diseases of the knee joint.

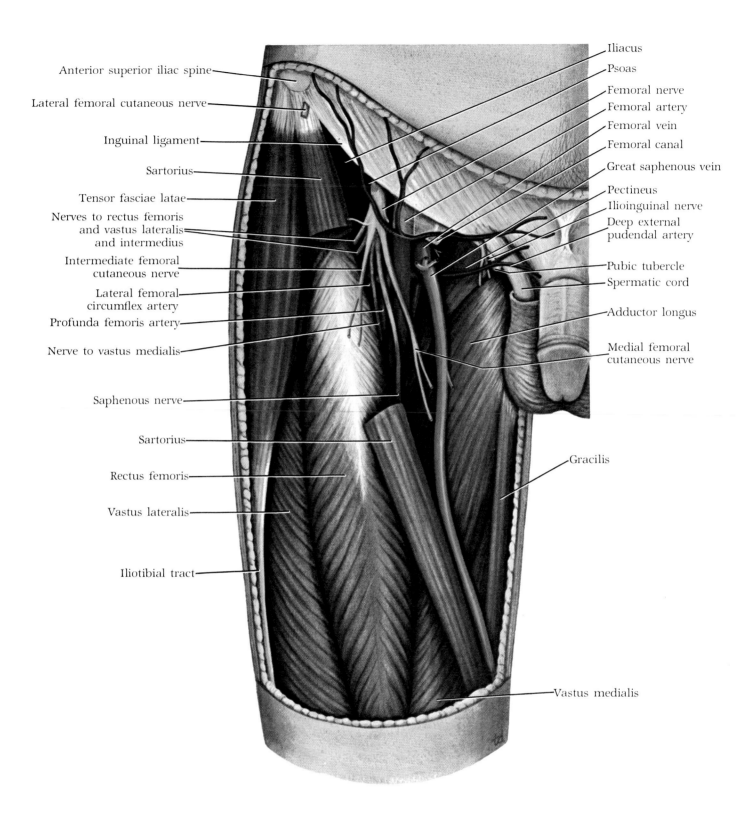

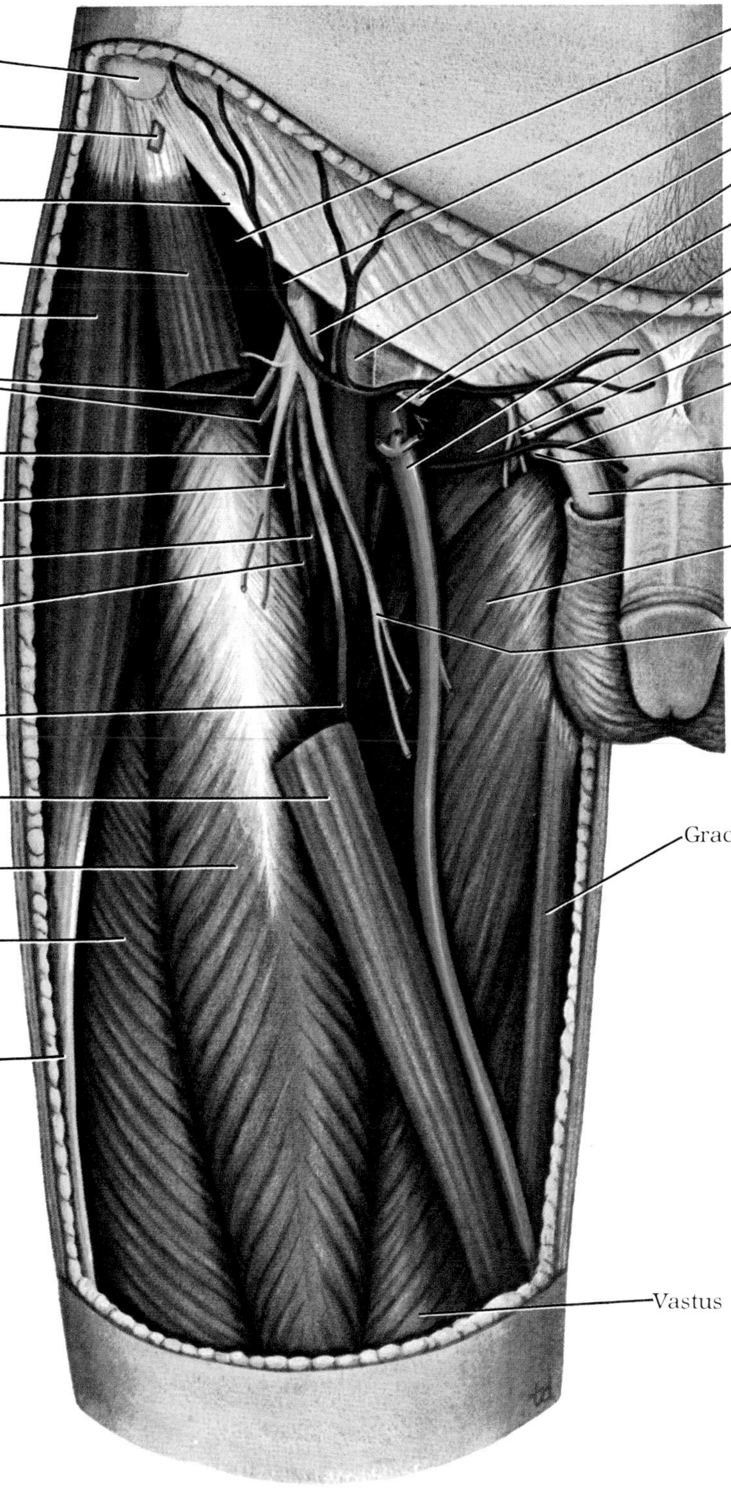
Anterior superior iliac spine
Lateral femoral cutaneous nerve
Inguinal ligament
Sartorius
Tensor fasciae latae
Nerves to rectus femoris and vastus lateralis and intermedius
Intermediate femoral cutaneous nerve
Lateral femoral circumflex artery
Profunda femoris artery
Nerve to vastus medialis
Saphenous nerve
Sartorius
Rectus femoris
Vastus lateralis
Iliotibial tract
Iliacus
Psoas
Femoral nerve
Femoral artery
Femoral vein
Femoral canal
Great saphenous vein
Pectineus
Ilioinguinal nerve
Deep external pudendal artery
Pubic tubercle
Spermatic cord
Adductor longus
Medial femoral cutaneous nerve
Gracilis
Vastus medialis

Figure 6-13
Contents of the femoral triangle and the muscles on the anterior aspect of the right thigh.

1. The femoral nerve (L2, 3, and 4) enters the thigh posterior to the inguinal ligament at a point midway between the anterior superior iliac spine and the pubic tubercle; it lies about a fingersbreadth lateral to the femoral pulse. About 2 inches (5 cm) inferior to the inguinal ligament the nerve splits up into its terminal branches.

2. The femoral nerve can be injured by stab or gunshot wounds, but a complete division of the nerve is rare. The following clinical features are present when the nerve is completely divided:
Motor. The quadriceps femoris muscle is paralyzed, and the knee cannot be extended.
Sensory. There is loss of sensation over the medial side of the lower part of the leg and along the medial border of the foot as far as the ball of the big toe, an area normally supplied by the saphenous nerve.

3. The quadriceps muscle is one of the most powerful in the body. If it is suddenly contracted with the knee partially flexed, such contraction can cause a transverse fracture of the patella. The vastus medialis is considered to be a very important part of this muscle. Its lowest fibers are almost horizontal in direction and prevent the patella from being pulled laterally during contraction of the quadriceps muscle. Following injury or disease of the knee joint the vastus medialis is the first muscle to waste and the last to recover.

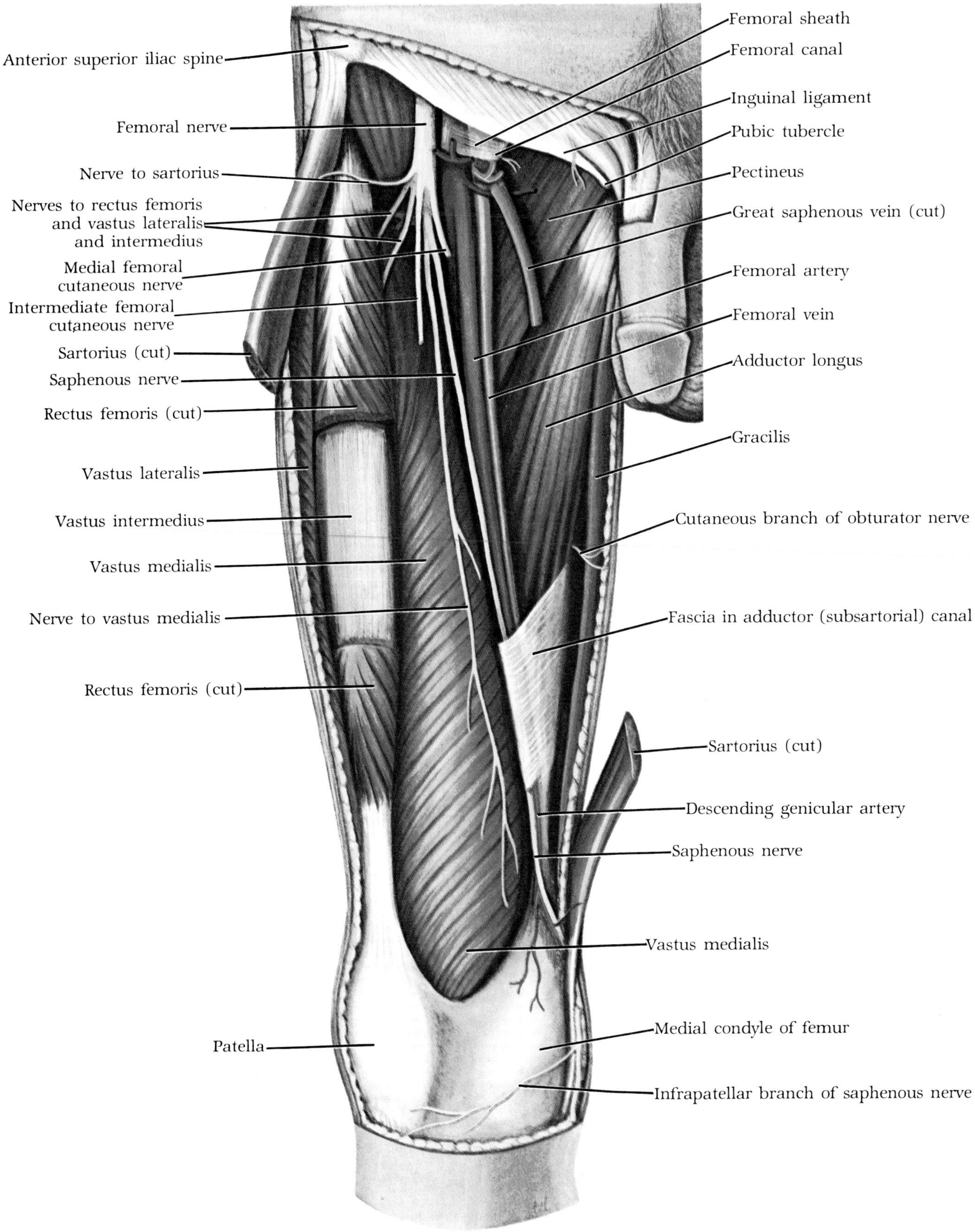
Anterior superior iliac spine
Femoral nerve
Nerve to sartorius
Nerves to rectus femoris
and vastus lateralis
and intermedius
Medial femoral
cutaneous nerve
Intermediate femoral
cutaneous nerve
Sartorius (cut)
Saphenous nerve
Rectus femoris (cut)
Vastus lateralis
Vastus intermedius
Vastus medialis
Nerve to vastus medialis
Rectus femoris (cut)
Patella
Femoral sheath
Femoral canal
Inguinal ligament
Pubic tubercle
Pectineus
Great saphenous vein (cut)
Femoral artery
Femoral vein
Adductor longus
Gracilis
Cutaneous branch of obturator nerve
Fascia in adductor (subsartorial) canal
Sartorius (cut)
Descending genicular artery
Saphenous nerve
Vastus medialis
Medial condyle of femur
Infrapatellar branch of saphenous nerve

Figure 6-14
Femoral triangle and adductor (subsartorial) canal in right lower limb.

1. A femoral hernia is more common in women than in men (possibly due to the wider pelvis and femoral canal in a woman). The hernial sac passes down the femoral canal, pushing the femoral septum before it. The neck of the sac always lies below and lateral to the pubic tubercle, which serves to distinguish this type of hernia from an inguinal hernia. The latter lies above and medial to the pubic tubercle. The neck of the sac is narrow and lies at the femoral ring. The ring is related anteriorly to the inguinal ligament; posteriorly, to the pectineal ligament and the superior ramus of the pubis; medially, to the sharp free edge of the lacunar ligament; and laterally, to the femoral vein. A femoral hernia is dangerous and should always be treated surgically.
2. Injury to the femoral artery is common. Its position is indicated by a line drawn from the midpoint of the interval between the anterior superior iliac spine and the symphysis pubis to the adductor tubercle on the femur when the hip joint is slightly flexed and laterally rotated. End-to-end anastomosis or arterial grafting (using the great saphenous vein) should be considered if the femoral artery is severely damaged. Ligation of the femoral artery below the origin of the profunda femoris artery is usually satisfactory since the collateral circulation provided by the branches of this artery is adequate to maintain the blood flow in the lower leg. Ligation of the femoral artery above the origin of the profunda femoris should only be performed prior to amputation since the collateral circulation is inadequate to prevent gangrene of the lower leg.

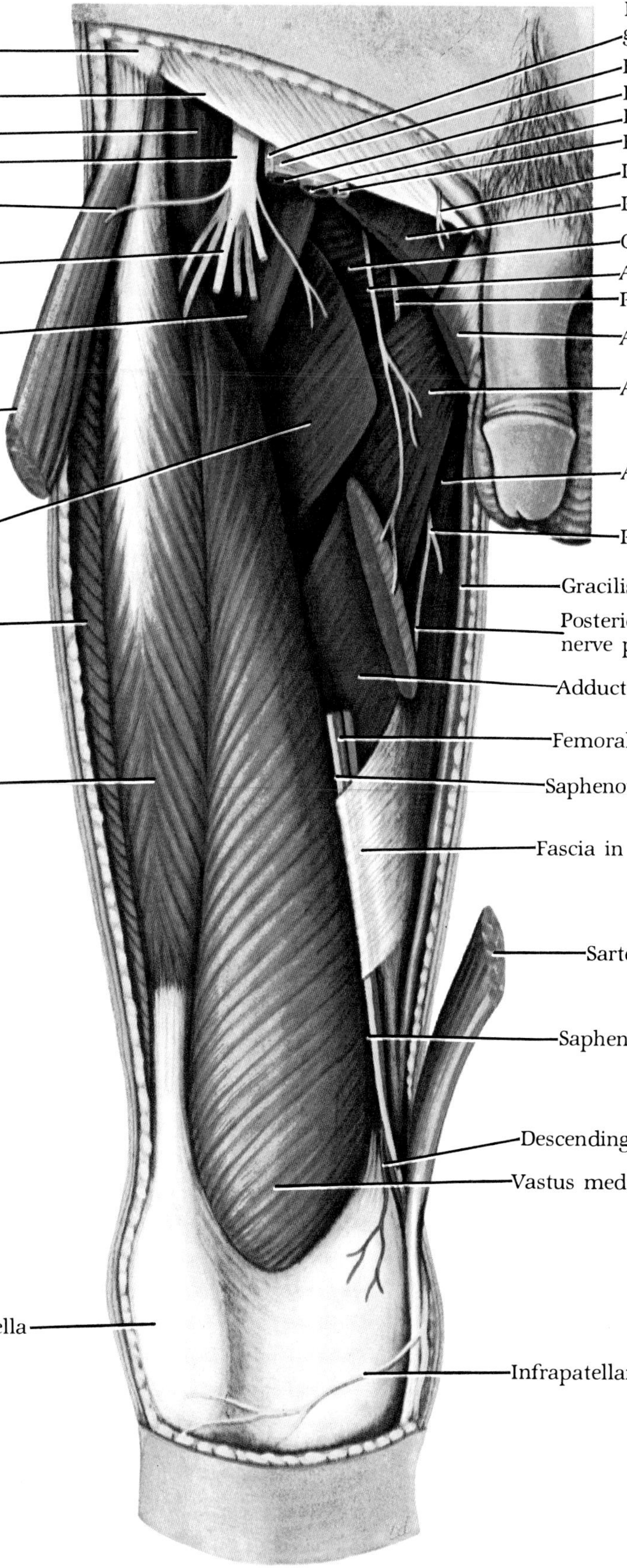
Anterior superior iliac spine
Inguinal ligament
Iliacus
Femoral nerve
Nerve to sartorius
Intermediate femoral cutaneous nerve
Psoas major
Sartorius (cut)
Pectineus (cut)
Vastus lateralis
Rectus femoris
Patella
Femoral branch of genitofemoral nerve
Femoral sheath
Femoral artery
Femoral vein
Femoral canal
Ilioinguinal nerve
Pectineus (cut)
Obturator externus
Anterior division of obturator nerve
Posterior division of obturator nerve
Adductor longus (cut)
Adductor brevis
Adductor magnus
Posterior division of obturator nerve
Gracilis
Posterior division of obturator nerve passing to knee joint
Adductor longus (cut)
Femoral vessels
Saphenous nerve
Fascia in adductor (subsartorial) canal
Sartorius (cut)
Saphenous nerve
Descending genicular artery
Vastus medialis
Infrapatellar branch of saphenous nerve

Figure 6-15
Muscles of the adductor region of the right thigh.

1. The adductor longus and adductor brevis muscles are supplied by the obturator nerve, the adductor longus being supplied by the anterior division and the adductor brevis also by the anterior division but sometimes by the posterior division. Both muscles are adductors, flexors, and lateral rotators of the hip joint.

2. The adductor magnus is really made up of two muscles. The anterior and superior groups of fibers are inserted into the linea aspera and are supplied by the posterior division of the obturator nerve; both are adductors and lateral rotators of the hip joint. The posterior fibers, especially those arising from the ischial tuberosity and inserted into the adductor tubercle, are supplied by the tibial part of the sciatic nerve and are extensors of the hip joint. The anterosuperior fibers are therefore true adductors, and the posterior vertical fibers form part of the hamstring group of muscles.

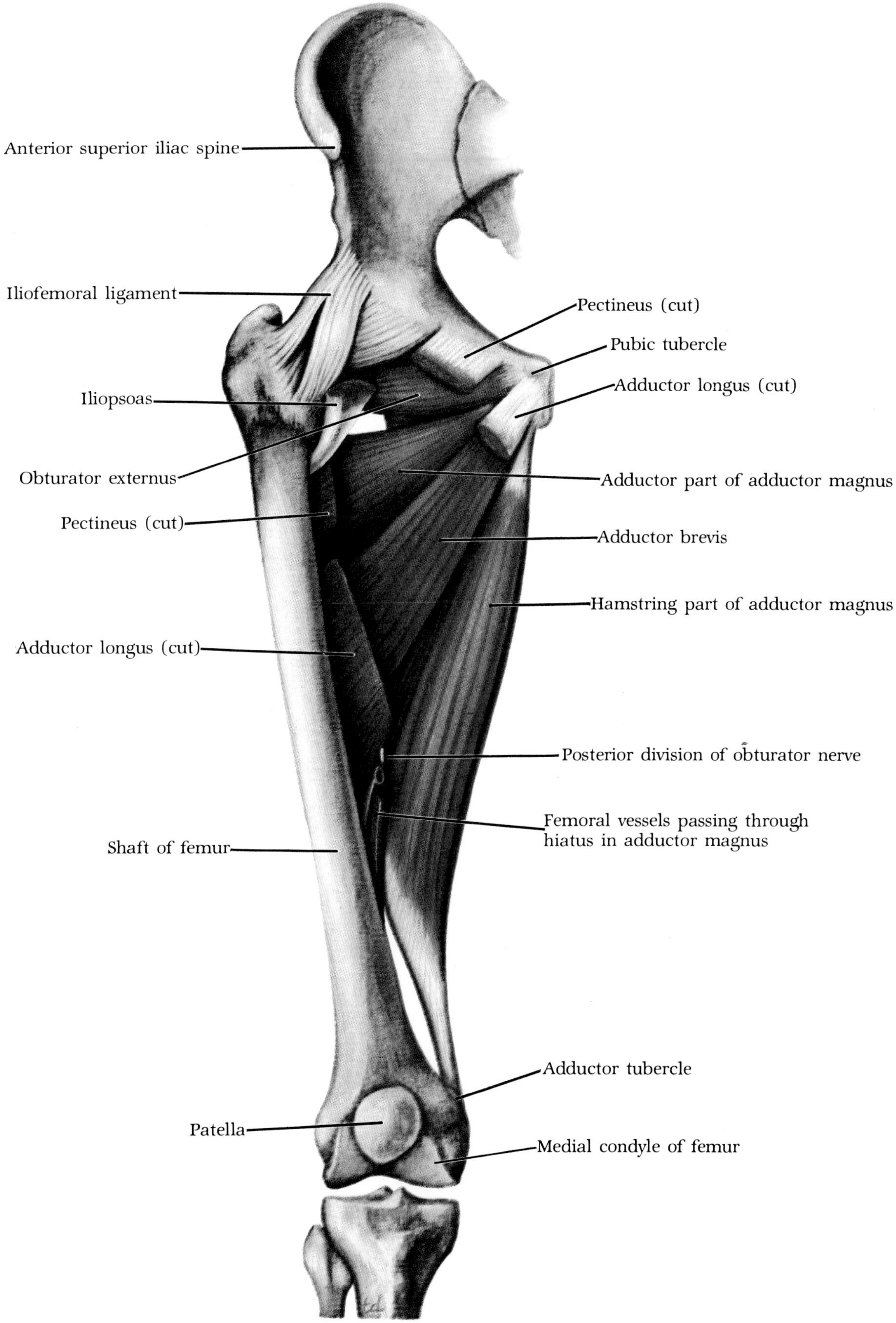
Anterior superior iliac spine
Iliofemoral ligament
Pectineus (cut)
Pubic tubercle
Adductor longus (cut)
Iliopsoas
Obturator externus
Adductor part of adductor magnus
Pectineus (cut)
Adductor brevis
Hamstring part of adductor magnus
Adductor longus (cut)
Posterior division of obturator nerve
Femoral vessels passing through hiatus in adductor magnus
Shaft of femur
Adductor tubercle
Patella
Medial condyle of femur

Figure 6-16
Muscles of the adductor region of the right thigh.

1. It is common practice in those patients with cerebral palsy who have marked spasticity of the adductor group of muscles to perform a tenotomy of the adductor longus tendon and to divide the anterior division of the obturator nerve. In addition, in some severe cases, the posterior division of the obturator nerve is crushed with pressure forceps. This operation overcomes the spasm of the adductor group of muscles but permits slow recovery of the muscles supplied by the posterior division of the obturator nerve.

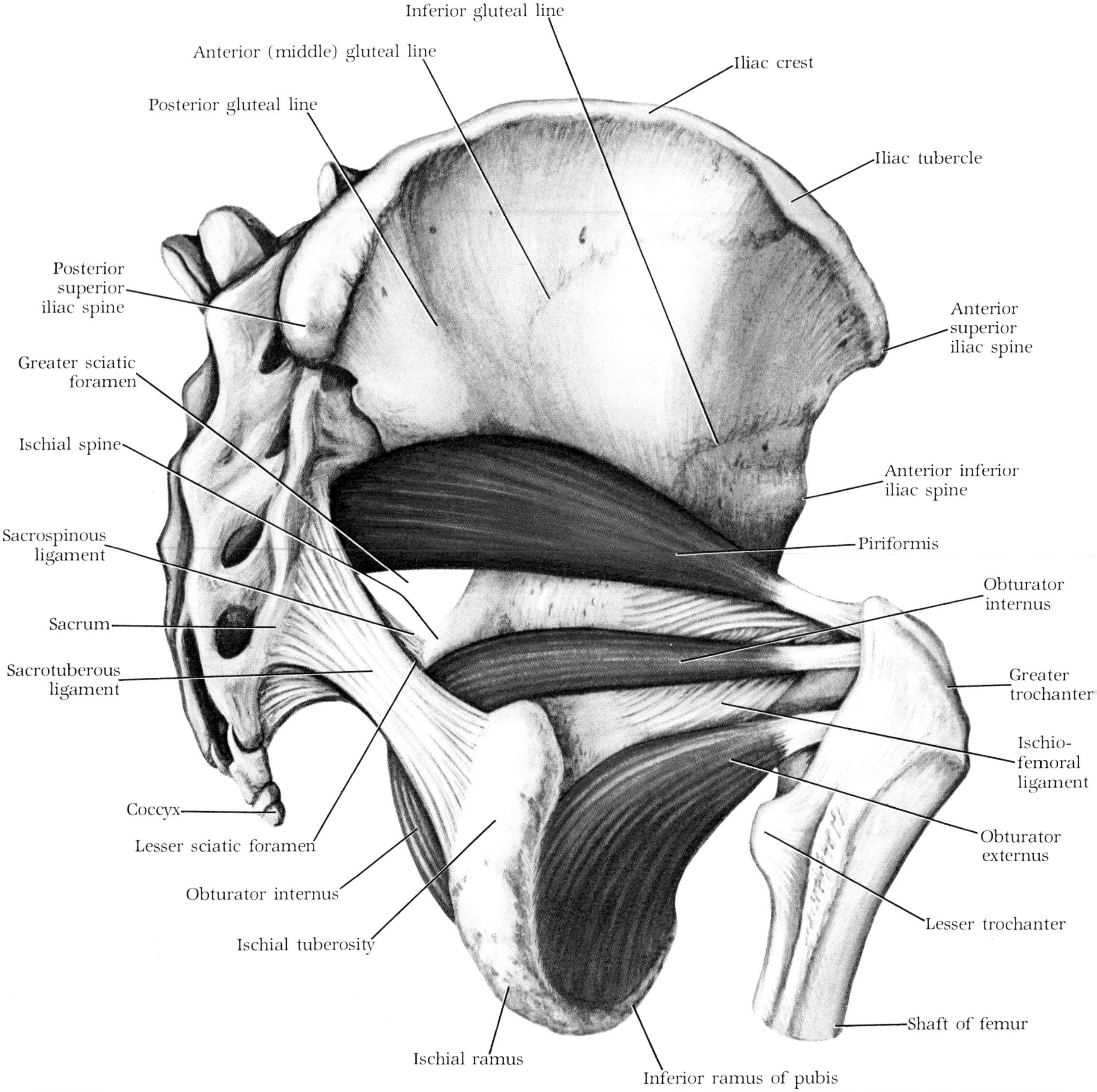
Inferior gluteal line
Anterior (middle) gluteal line
Iliac crest
Posterior gluteal line
Iliac tubercle
Posterior superior iliac spine
Anterior superior iliac spine
Greater sciatic foramen
Ischial spine
Anterior inferior iliac spine
Sacrospinous ligament
Piriformis
Obturator internus
Sacrum
Sacrotuberous ligament
Greater trochanter
Ischio-femoral ligament
Coccyx
Obturator externus
Lesser sciatic foramen
Obturator internus
Lesser trochanter
Ischial tuberosity
Shaft of femur
Ischial ramus
Inferior ramus of pubis

Figure 6-17
Right gluteal region, showing the piriformis, obturator internus, and obturator externus muscles, oblique posterior view.

1. The piriformis leaves the pelvis through the greater sciatic foramen; the obturator internus leaves the pelvis through the lesser sciatic foramen; and the obturator externus arises from the outer surface of the margins of the obturator foramen and the obturator membrane. They are all inserted into the greater trochanter of the femur and are lateral rotators of the hip joint.
2. In about 10 percent of the population the piriformis is pierced by one or both parts of the sciatic nerve.

3. To test for rotation of the hip joint have the patient lie in the supine position. Lay the examining hand on the thigh and roll the lower limb laterally and medially, using the patella and the foot as an indicator of the degree of rotation. Pain or limitation of movement can easily be detected by this method.

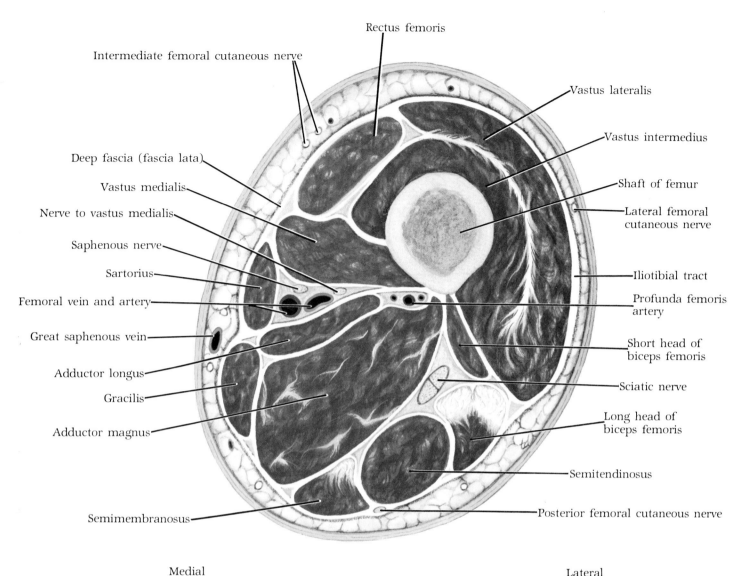

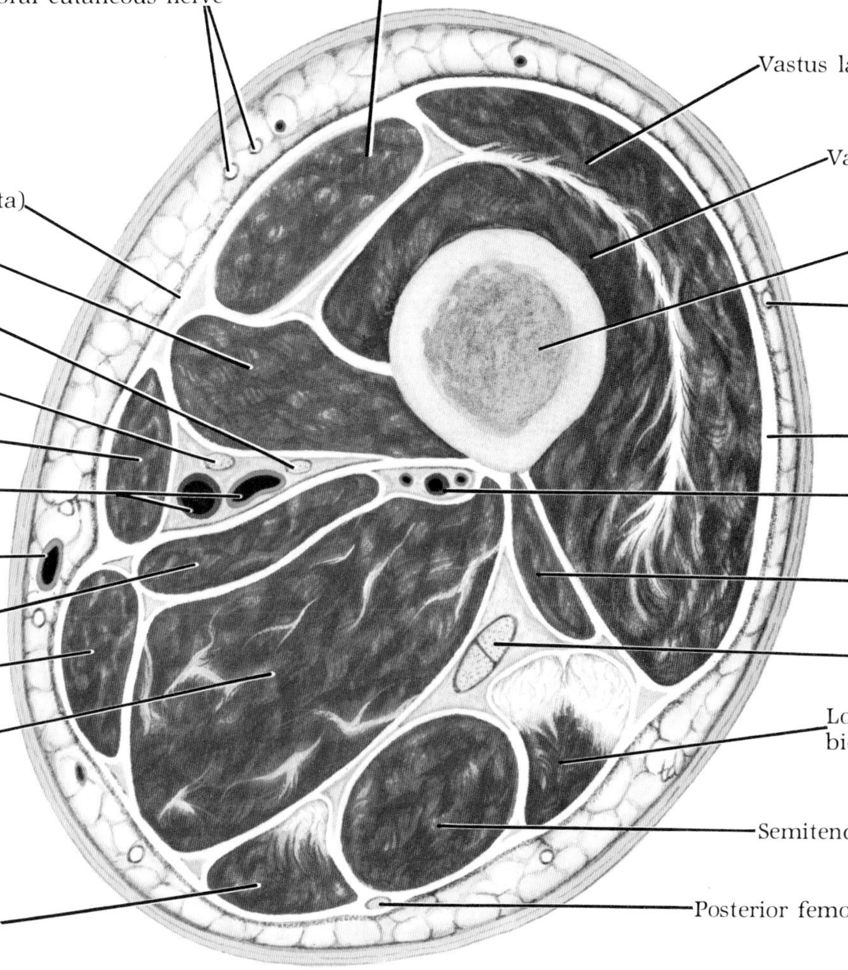
Rectus femoris
Intermediate femoral cutaneous nerve
Vastus lateralis
Vastus intermedius
Deep fascia (fascia lata)
Shaft of femur
Vastus medialis
Nerve to vastus medialis
Lateral femoral cutaneous nerve
Saphenous nerve
Sartorius
Iliotibial tract
Femoral vein and artery
Profunda femoris artery
Great saphenous vein
Short head of biceps femoris
Adductor longus
Sciatic nerve
Gracilis
Long head of biceps femoris
Adductor magnus
Semitendinosus
Semimembranosus
Posterior femoral cutaneous nerve
Medial
Lateral

Figure 6-18
Transverse section of the middle of the right thigh as seen from above. The section shows the subsartorial (adductor) canal and its contents.

1. In this region of the thigh the femoral vessels and the sciatic nerve are well protected by overlying muscles.
2. The femoral artery may be exposed surgically in the subsartorial canal for the treatment of wounds or aneurysm, and to perform a thromboendarterectomy. If it is necessary to ligate the femoral artery in this region, the circulation is reestablished through the anastomosis between the descending branch of the lateral femoral circumflex artery, the descending genicular artery, the perforating branches of the profunda artery, and the muscular branches of the popliteal artery.
3. Arterial occlusive disease of the leg is common in men. Ischemia of the muscles produces a cramplike pain with exercise that is relieved by rest, a condition known as intermittent claudication.
4. The arteries of the lower limb receive their sympathetic innervation from the lower three thoracic and upper two or three lumbar segments of the spinal cord. The preganglionic fibers pass to the lower thoracic and upper lumbar ganglia via white rami. The fibers synapse in the lumbar and sacral ganglia. The postganglionic fibers reach the femoral artery from the femoral and obturator nerves.
5. Remember that the quadriceps muscle plays a key role in stabilizing the knee joint. Of the different components of the quadriceps, the vastus medialis is the most important. Unfortunately, after injury it is the first to atrophy and the last to recover.

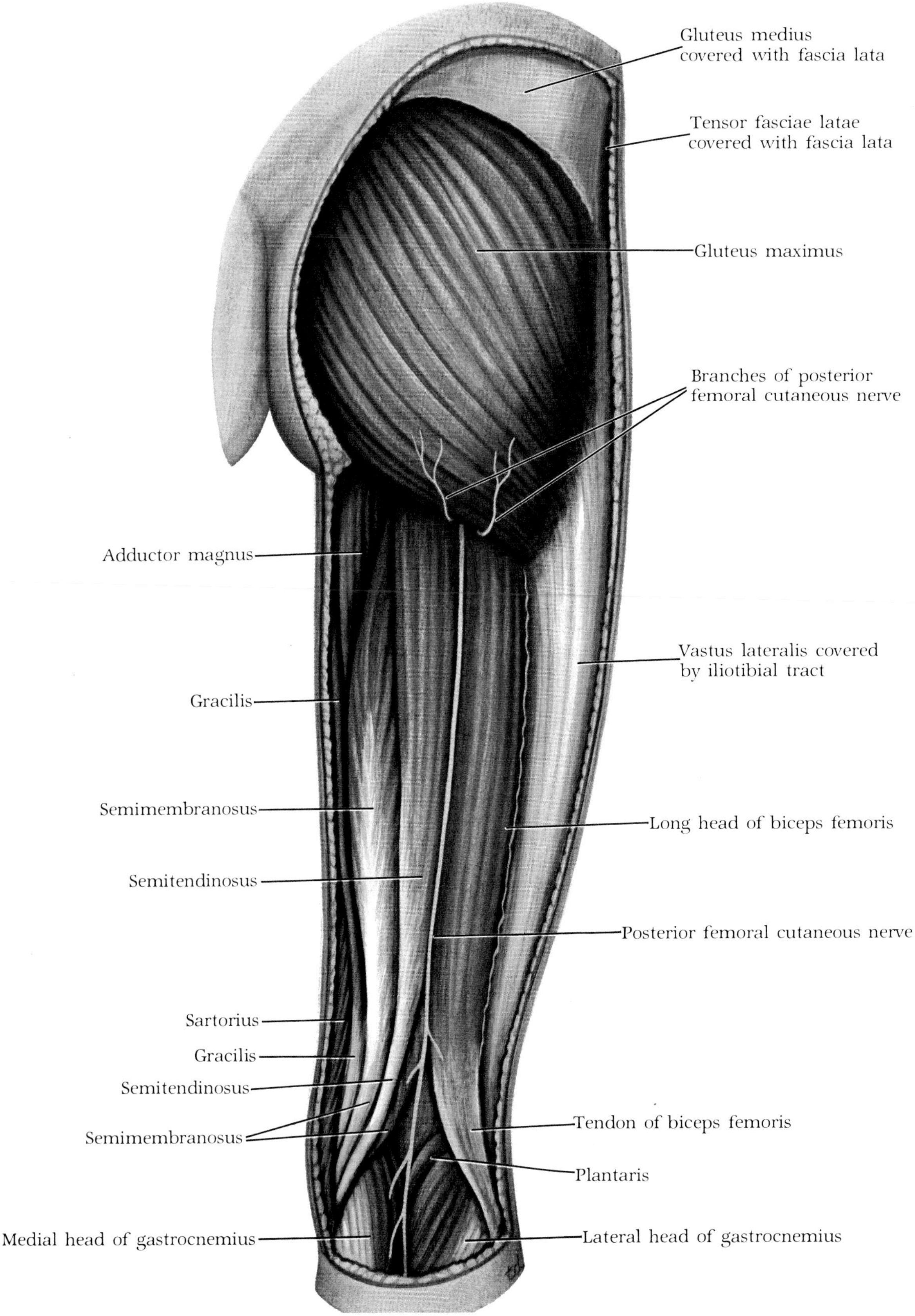
Gluteus medius
covered with fascia lata
Tensor fasciae latae
covered with fascia lata
Gluteus maximus
Branches of posterior
femoral cutaneous nerve
Adductor magnus
Vastus lateralis covered
by iliotibial tract
Gracilis
Semimembranosus
Long head of biceps femoris
Semitendinosus
Posterior femoral cutaneous nerve
Sartorius
Gracilis
Semitendinosus
Semimembranosus
Tendon of biceps femoris
Plantaris
Medial head of gastrocnemius
Lateral head of gastrocnemius

Figure 6-19
Posterior aspect of right gluteal region and thigh, showing the gluteus maximus, biceps femoris, semitendinosus and semimembranosus muscles, and the posterior femoral cutaneous nerve.

1. The sciatic nerve (L4 and 5 and S1, 2, and 3) curves laterally and downward through the gluteal region and then deep to the superficial muscles on the posterior aspect of the thigh. It is most frequently injured in the gluteal region by badly placed intramuscular injections. To avoid this injury, injections into the gluteus maximus should be given well anteriorly on the upper outer quadrant of the buttock.

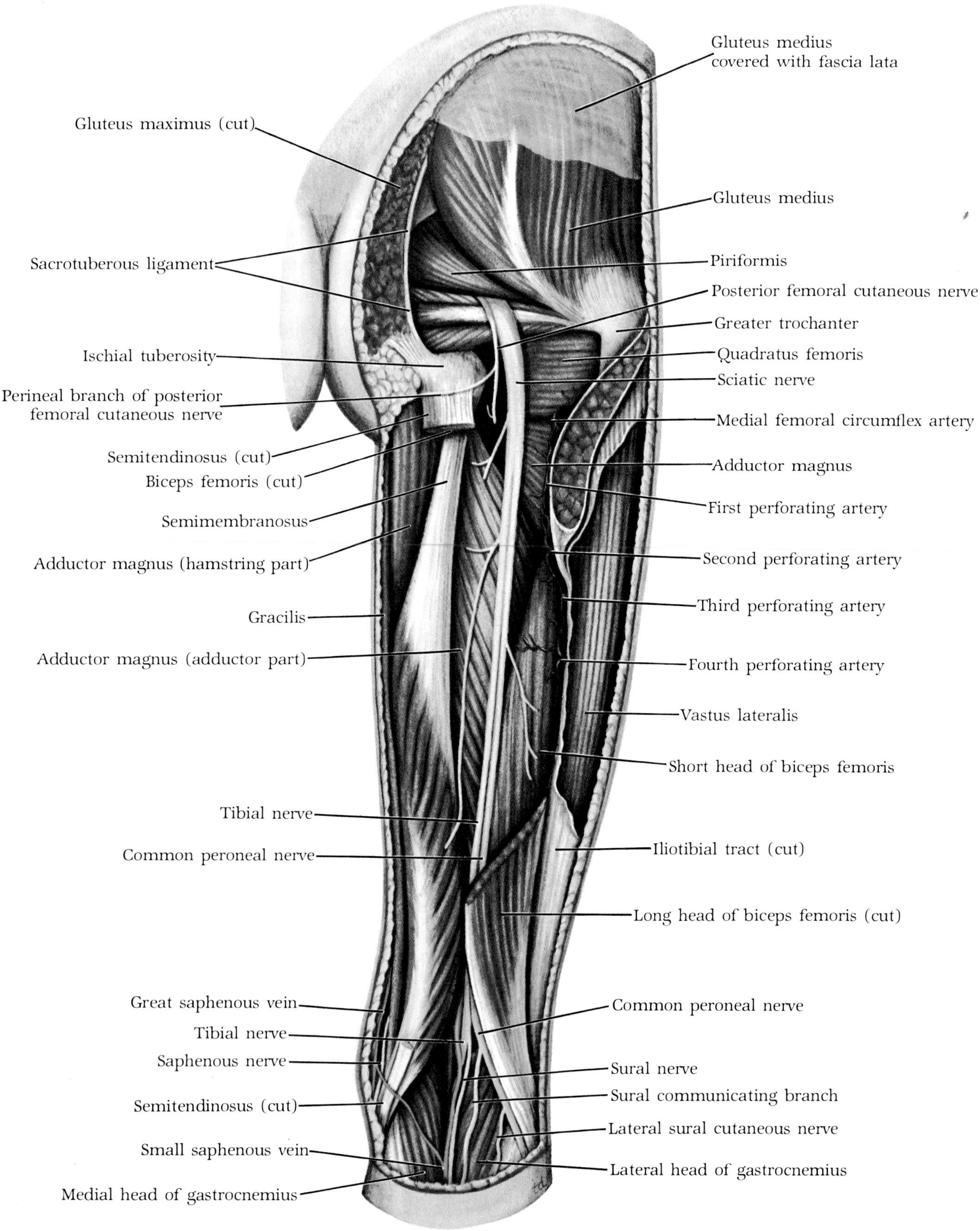
Gluteus medius
covered with fascia lata
Gluteus maximus (cut)
Gluteus medius
Sacrotuberous ligament
Piriformis
Posterior femoral cutaneous nerve
Greater trochanter
Ischial tuberosity
Quadratus femoris
Sciatic nerve
Perineal branch of posterior
femoral cutaneous nerve
Medial femoral circumflex artery
Semitendinosus (cut)
Biceps femoris (cut)
Adductor magnus
First perforating artery
Semimembranosus
Adductor magnus (hamstring part)
Second perforating artery
Third perforating artery
Gracilis
Adductor magnus (adductor part)
Fourth perforating artery
Vastus lateralis
Short head of biceps femoris
Tibial nerve
Common peroneal nerve
Iliotibial tract (cut)
Long head of biceps femoris (cut)
Great saphenous vein
Common peroneal nerve
Tibial nerve
Saphenous nerve
Sural nerve
Sural communicating branch
Semitendinosus (cut)
Lateral sural cutaneous nerve
Small saphenous vein
Lateral head of gastrocnemius
Medial head of gastrocnemius

Figure 6-20
Right thigh, posterior view. The gluteus maximus and the long head of the biceps femoris muscle have been removed to display the sciatic nerve. The semitendinosus muscle has also been removed.

1. The sciatic nerve (L4, 5 and S1, 2, and 3) curves laterally and downward through the gluteal region, situated at first midway between the posterior superior iliac spine and the ischial tuberosity, and lower down, midway between the tip of the greater trochanter and the ischial tuberosity. It then passes downward in the middle of the posterior aspect of the thigh and divides into the common peroneal and tibial nerves at a variable site above the popliteal fossa.

2. The sciatic nerve can be injured by penetrating wounds, fractures of the pelvis, or dislocations of the hip joint. It is most frequently injured by badly placed intramuscular injections in the gluteal region. The majority of the nerve lesions are incomplete, and in 90 percent of injuries, the common peroneal part of the nerve is the most affected. The following clinical features are present when there is a complete section of the nerve:
Motor. The hamstring muscles are paralyzed, but weak flexion of the knee is possible due to the action of the sartorius (femoral nerve) and gracilis (obturator nerve) muscles. All muscles below the knee are paralyzed, and the weight of the foot causes it to assume the plantar-flexed position, a condition called foot drop.
Sensory. There is loss of sensation below the knee, except for that narrow area down the medial side of the lower part of the leg and along the medial border of the foot as far as the ball of the big toe that is supplied by the saphenous nerve (femoral nerve).

3. The result of operative repair of a sciatic nerve injury is poor. It is rare for active movement to return to the small muscles of the foot, and sensory recovery is rarely complete. Loss of sensation in the sole of the foot makes the development of trophic ulcers inevitable.

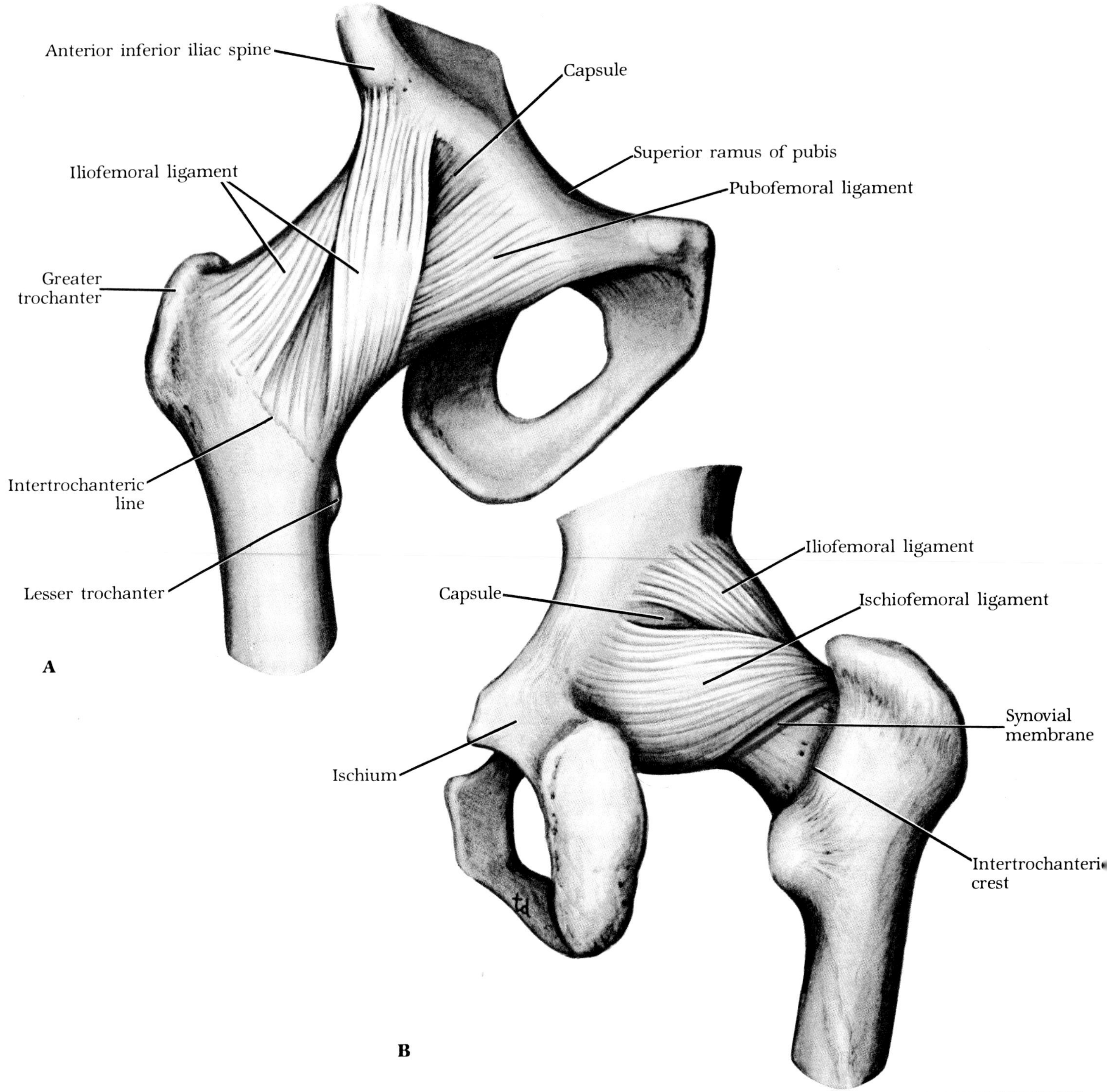
Anterior inferior iliac spine
Capsule
Superior ramus of pubis
Pubofemoral ligament
Iliofemoral ligament
Greater trochanter
Intertrochanteric line
Lesser trochanter
A
Iliofemoral ligament
Ischiofemoral ligament
Capsule
Synovial membrane
Ischium
Intertrochanteric crest
B

Figure 6-21
Right hip joint. A. Anterior view. B. Posterior view.

1. When a person stands upright in the anatomical position, the line of gravity passes posterior to the axis of movement of the hip joint, with the weight of the body thus tending to cause the hip joint to be hyperextended. It is therefore not surprising to find that the strong iliofemoral, pubofemoral, and ischiofemoral ligaments limit the movement of extension.
2. It is important to know the axial relationships of the hip joint when studying a radiograph of the hip region. For example, in an anteroposterior view the inferior margin of the neck of the femur should form a smooth continuous curve with the superior margin of the obturator foramen (Shenton's line). A line drawn at right angles to the long axis of the shaft of the femur at the junction with the neck should pass through or below the fovea centralis of the head of the femur (Skinner's line).
3. When performing an aspiration of the hip joint, the needle should be inserted into the anterior aspect of the joint 1 inch (2.5 cm) below and lateral to the midpoint between the anterior superior iliac spine and the symphysis pubis. The femoral artery should first be found and identified by palpation.

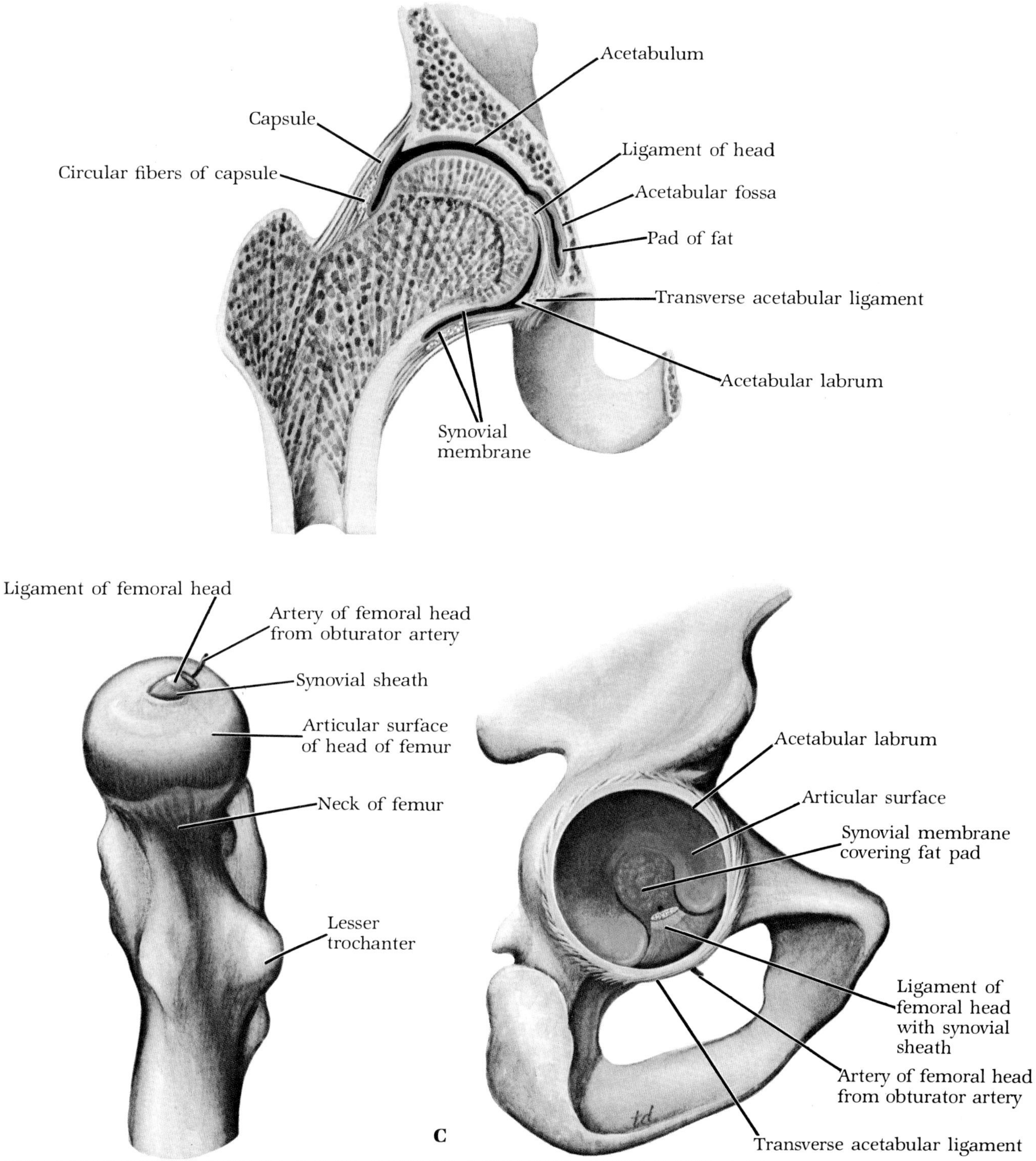
Acetabulum
Capsule
Circular fibers of capsule
Ligament of head
Acetabular fossa
Pad of fat
Transverse acetabular ligament
Acetabular labrum
Synovial membrane
A
Ligament of femoral head
Artery of femoral head from obturator artery
Synovial sheath
Articular surface of head of femur
Neck of femur
Lesser trochanter
B
Acetabular labrum
Articular surface
Synovial membrane covering fat pad
Ligament of femoral head with synovial sheath
Artery of femoral head from obturator artery
Transverse acetabular ligament
C

Figure 6-22
Right hip joint. A. Coronal section. B. Articular surface of head of femur. C. Acetabulum.

1. The stability of the hip joint depends on the ball-and-socket arrangement of the articular surfaces and the very strong ligaments.
2. In congenital dislocation of the hip, the upper lip of the acetabulum fails to develop adequately, and the head of the femur, having no stable platform under which it can lodge, rides up out of the acetabulum onto the gluteal surface of the ilium.
3. Traumatic dislocation of the hip joint is rare because of the joint's strength. When it does occur, the axial force is applied to the flexed, adducted femur. The head of the femur is displaced posteriorly through a rent in the capsule, or the posterior rim of the acetabulum is fractured. The close relationship of the sciatic nerve to the posterior surface of the joint make it prone to injury in posterior dislocations.
4. The stability of the hip joint when a person stands on one leg with the foot of the opposite side raised above the ground depends on three factors: (a) the gluteus medius and minimus muscles must be functioning normally; (b) the head of the femur must be located normally within the acetabulum; and (c) the neck of the femur must be intact and have a normal angle with the shaft of the femur. If any one of these factors does not hold true, the pelvis will sink downward on the opposite, unsupported side. The patient is then said to exhibit a positive Trendelenburg's sign.

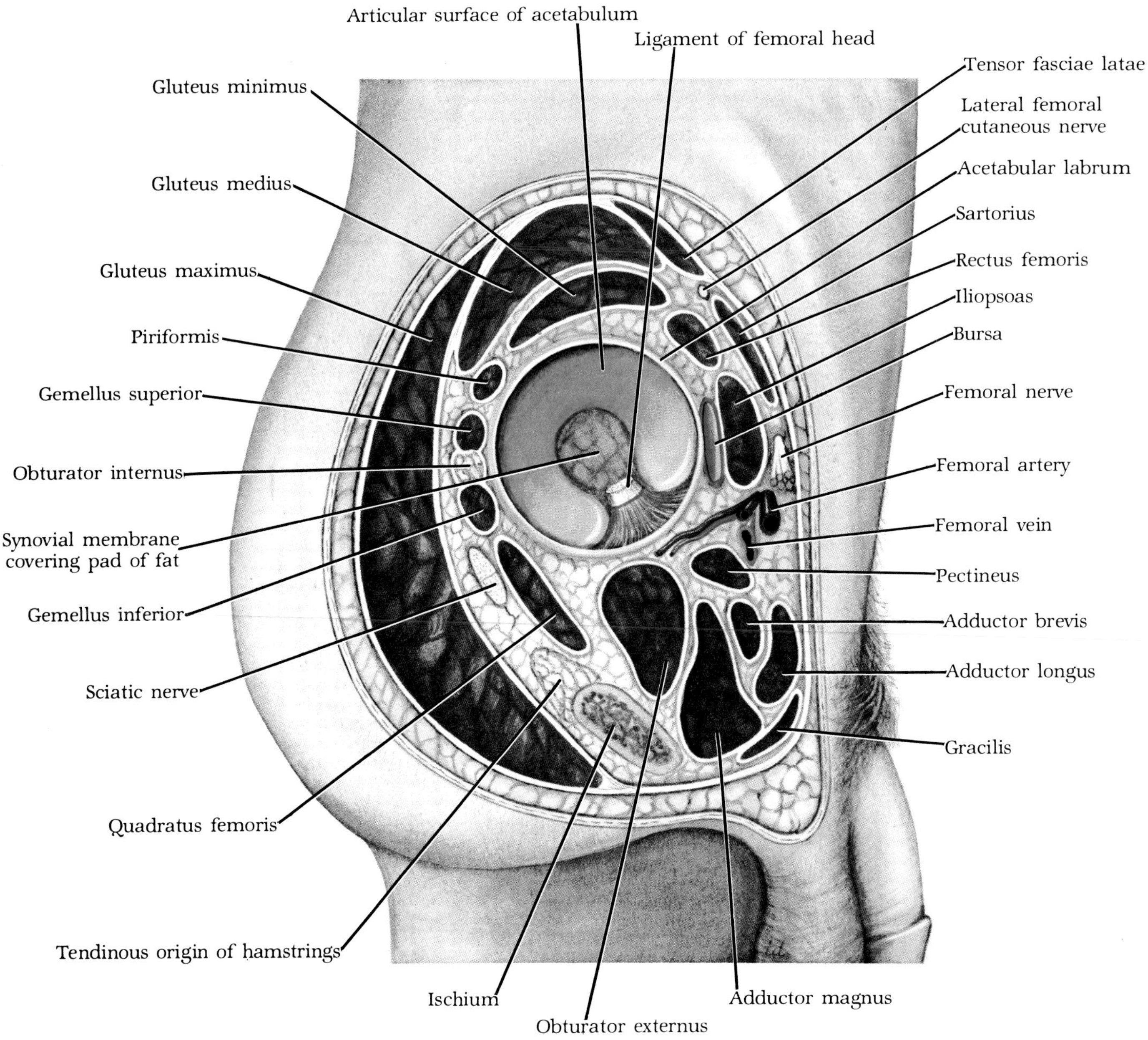
Articular surface of acetabulum
Ligament of femoral head
Tensor fasciae latae
Lateral femoral cutaneous nerve
Acetabular labrum
Sartorius
Rectus femoris
Iliopsoas
Bursa
Femoral nerve
Femoral artery
Femoral vein
Pectineus
Adductor brevis
Adductor longus
Gracilis
Gluteus minimus
Gluteus medius
Gluteus maximus
Piriformis
Gemellus superior
Obturator internus
Synovial membrane covering pad of fat
Gemellus inferior
Sciatic nerve
Quadratus femoris
Tendinous origin of hamstrings
Ischium
Obturator externus
Adductor magnus

Figure 6-23
Relations of right hip joint.

1. In order to palpate that part of the head of the femur that is not intra-acetabular, palpate the anterior aspect of the thigh just inferior to the inguinal ligament and just lateral to the pulsating femoral artery. The finger will lie directly in front of the head of the femur. Tenderness here would indicate the presence of arthritis.

2. Because of the close relationship of the sciatic nerve to the posterior surface of the hip joint, this nerve is very liable to injury in posterior dislocations of the joint.

3. Since the psoas bursa frequently communicates with the cavity of the hip joint, distention of the psoas bursa with fluid secondary to arthritis of the hip joint is a possible clinical finding. It is first seen as a tense swelling in the femoral triangle.

4. A patient with an inflamed hip joint will place the femur in the position that gives him minimum discomfort, that is, the position in which the joint cavity has the greatest capacity to contain the increased amount of synovial fluid secreted. The hip joint is partially flexed, abducted, and externally rotated.

5. Osteoarthritis is the most common disease of the hip joint in the adult and causes pain, stiffness, and deformity. The pain may be in the hip itself or referred to the knee (the obturator nerve supplies both joints). The stiffness is due to the pain and reflex spasm of the surrounding muscles. The deformity is flexion, adduction, and external rotation and is produced initially by muscle spasm and later by muscle contracture.

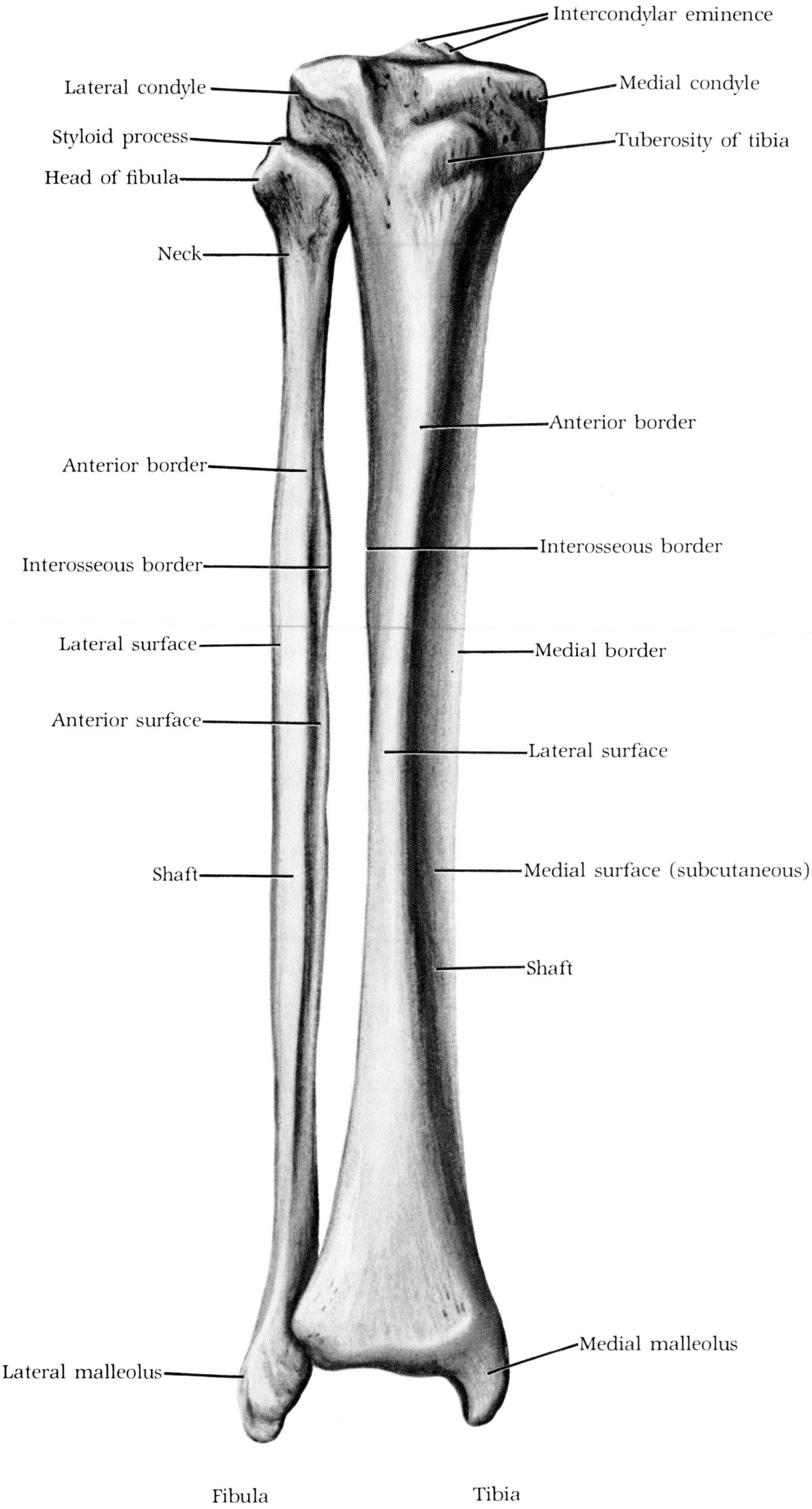
Intercondylar eminence
Lateral condyle
Medial condyle
Styloid process
Tuberosity of tibia
Head of fibula
Neck
Anterior border
Anterior border
Interosseous border
Interosseous border
Lateral surface
Medial border
Anterior surface
Lateral surface
Shaft
Medial surface (subcutaneous)
Shaft
Medial malleolus
Lateral malleolus
Fibula
Tibia

Figure 6-24
Tibia and fibula, anterior view.

1. Congenital absence of the fibula is uncommon, although there may be a partial absence in which case there is only minimal shortening of the affected leg. In patients in whom there is no fibula at all there is dysplasia of the entire leg, and the tibia is bowed and short.

2. Fractures of the tibia and fibula are common and may be caused by direct or indirect trauma. If only one bone is fractured, the other acts as a splint and there is minimal displacement. Fractures of the shaft of the tibia are often compound (open), since the entire length of the medial surface is covered only by skin and superficial fascia. Fractures of the distal third of the shaft of the tibia are prone to delayed union or nonunion. This may be due to the fact that the nutrient artery is torn at the fracture line, with a consequent reduction in blood flow to the distal fragment; it is also possible that the splint-like action of the intact fibula prevents the proximal and distal fragments from coming into apposition.

3. *Fractures of the Proximal Tibia.* Fractures of the tibial condyles (plateaus) are common in the middle-aged and elderly; they usually result from direct violence to the side of the knee joint, as for example, when a person is hit by an automobile. The tibial condyle may show a split fracture or be broken up, or the fracture line may pass between both condyles in the region of the intercondylar eminence. As the result of forced abduction of the knee joint, the medial collateral ligament can also be ruptured.

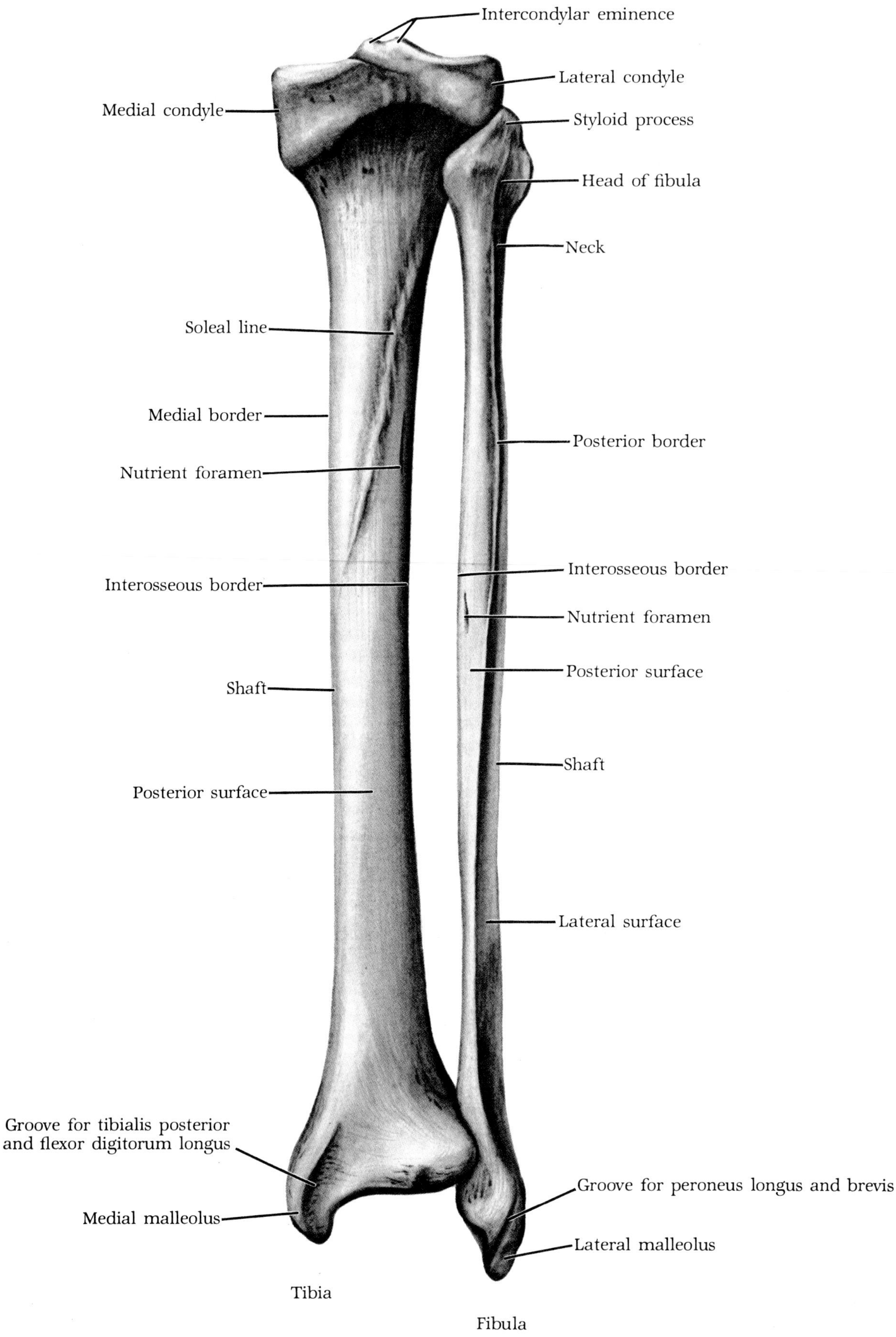
Intercondylar eminence
Lateral condyle
Medial condyle
Styloid process
Head of fibula
Neck
Soleal line
Medial border
Posterior border
Nutrient foramen
Interosseous border
Interosseous border
Nutrient foramen
Posterior surface
Shaft
Shaft
Posterior surface
Lateral surface
Groove for tibialis posterior
and flexor digitorum longus
Groove for peroneus longus and brevis
Medial malleolus
Lateral malleolus
Tibia
Fibula

Figure 6-25
Tibia and fibula, posterior view.

1. Fractures of the tibial shaft can be a result of direct violence, as when struck by an automobile, or indirect violence by torsion forces, as commonly occurs in skiing injuries. Fractures due to direct violence are transverse or comminuted, but torsion type fractures are usually spiral or oblique. Since the entire length of the medial surface of the tibia is covered only by skin and superficial fascia, compound (open) fractures are common.

2. Fractures of the distal tibia and fibula. Fracture-dislocations of the ankle joint are common, resulting from athletic endeavors especially or from a sudden misstep on an irregular surface. Forced external rotation and overeversion of the foot are the most frequent cause of fracture-dislocation. The talus is externally rotated forcibly against the lateral malleolus of the fibula. The torsion effect on the lateral malleolus causes it to fracture spirally. If the force continues, the talus moves laterally, and the medial collateral ligament of the ankle joint becomes taut and pulls off the tip of the medial malleolus. If the talus is forced to move still further, its rotary movement results in its violent contact with the posteroinferior margin of the tibia (sometimes referred to by clinicians as the posterior malleolus), which then shears off. Other less common types of fracture-dislocation are due to forced overeversion (without rotation), in which the talus presses the lateral malleolus laterally and causes it to fracture transversely. Overinversion (without rotation), in which the talus presses against the medial malleolus, will produce a vertical fracture through the base of the medial malleolus.

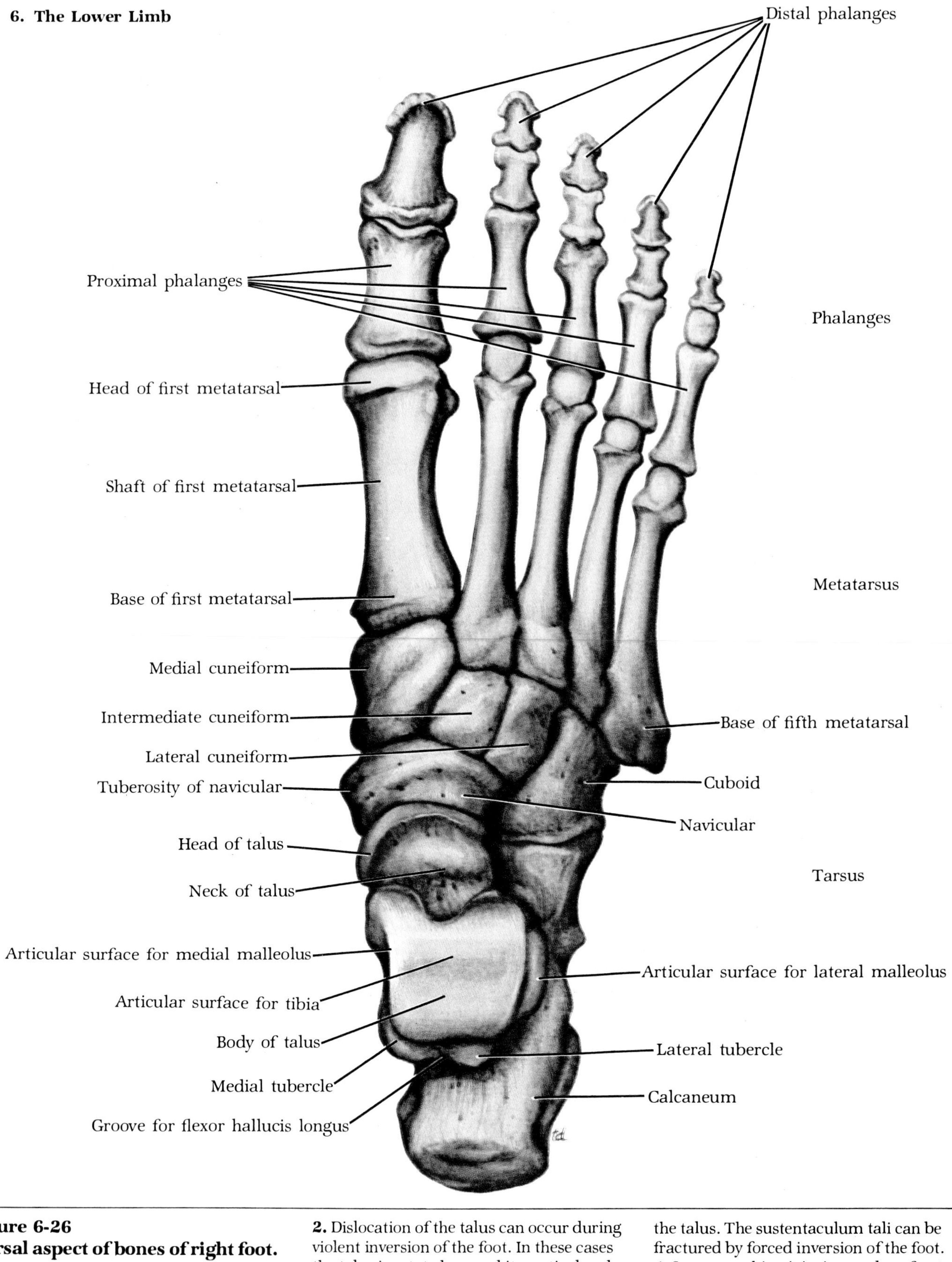
Distal phalanges
Proximal phalanges
Phalanges
Head of first metatarsal
Shaft of first metatarsal
Metatarsus
Base of first metatarsal
Medial cuneiform
Intermediate cuneiform
Base of fifth metatarsal
Lateral cuneiform
Tuberosity of navicular
Cuboid
Navicular
Head of talus
Tarsus
Neck of talus
Articular surface for medial malleolus
Articular surface for lateral malleolus
Articular surface for tibia
Body of talus
Lateral tubercle
Medial tubercle
Calcaneum
Groove for flexor hallucis longus

Figure 6-26
Dorsal aspect of bones of right foot.

1. Fractures of the talus occur at the neck or body of the talus. Neck fractures occur during violent dorsiflexion of the ankle joint when the neck is driven against the anterior edge of the distal end of the tibia. The body of the talus is fractured when jumping from a height, although the two malleoli prevent displacement.

2. Dislocation of the talus can occur during violent inversion of the foot. In these cases the talus is rotated around its vertical and longitudinal axes.

3. Compression fractures of the calcaneum result from falls from a height. The talus is driven downward into the calcaneum, crushing it in such a way that it loses vertical height and becomes wider laterally. The posterior portion of the calcaneus above the insertion of the tendo calcaneus can be fractured by the posterior displacement of the talus. The sustentaculum tali can be fractured by forced inversion of the foot.

4. Severe crushing injuries produce fractures of the remaining tarsal bones.

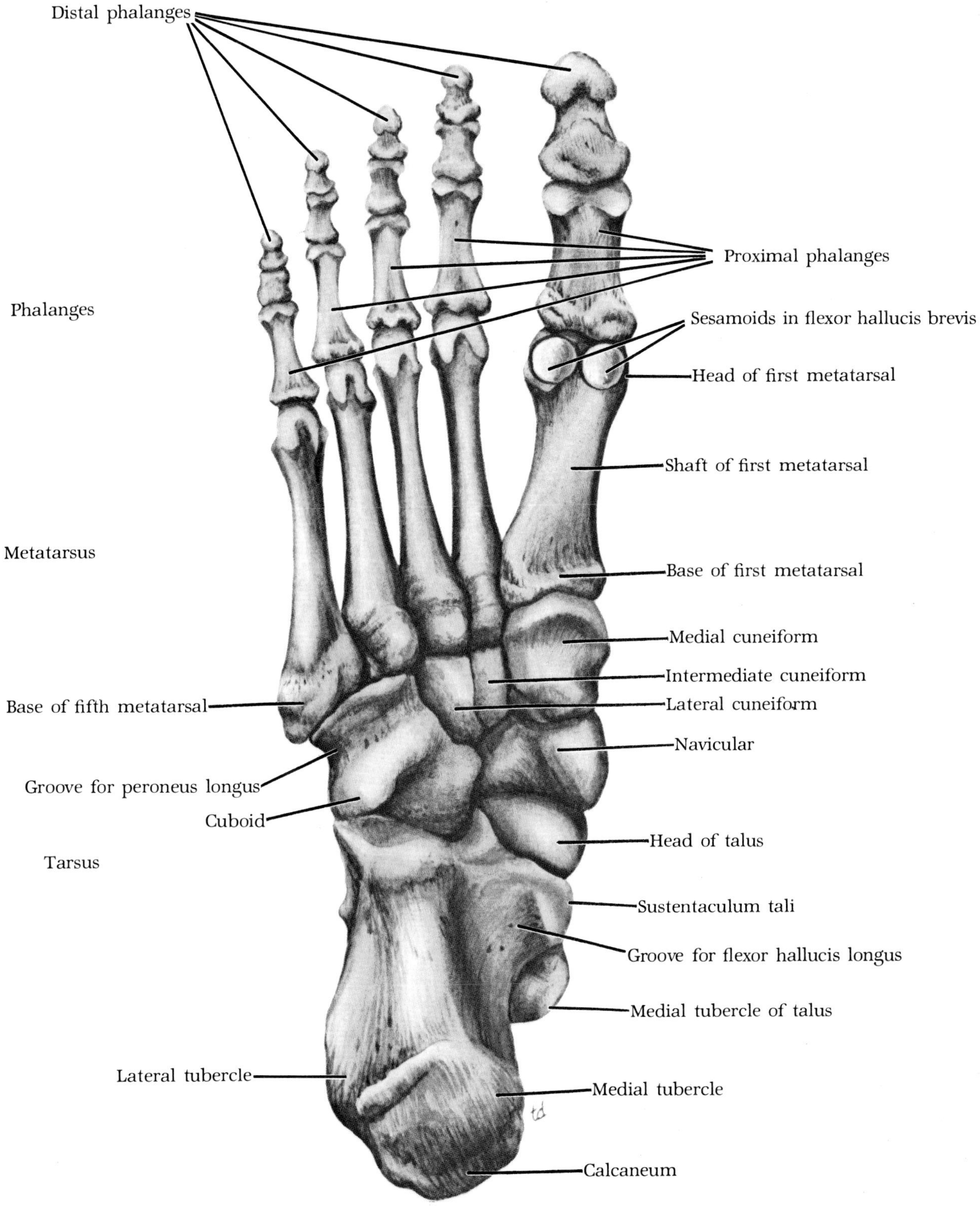
Distal phalanges
Proximal phalanges
Phalanges
Sesamoids in flexor hallucis brevis
Head of first metatarsal
Shaft of first metatarsal
Metatarsus
Base of first metatarsal
Medial cuneiform
Intermediate cuneiform
Base of fifth metatarsal
Lateral cuneiform
Navicular
Groove for peroneus longus
Cuboid
Head of talus
Tarsus
Sustentaculum tali
Groove for flexor hallucis longus
Medial tubercle of talus
Lateral tubercle
Medial tubercle
td
Calcaneum

Figure 6-27
Plantar aspect of bones of right foot.

1. *Fractures of the Metatarsal Bones.* The base of the fifth metatarsal can be fractured during forced inversion of the foot, at which time the tendon of insertion of the peroneus brevis muscle avulses the base of the metatarsal.

2. Stress fracture of a metatarsal bone occurs most frequently in the distal third of the second, third, or fourth metatarsal bones. It most commonly occurs in soldiers after long marches and can also occur in hikers or in nurses.

3. *Freiberg's Disease.* This condition affects the head of the second metatarsal bone (less often the third). There is osteoporosis of the head followed by collapse of the articular surface. Later, part of the articular surface can become detached and lie free in the joint cavity. The condition, of unknown cause, is extremely painful and can occur at any age.

4. Hallux valgus, a lateral deviation of the great toe, is a common condition. The incidence is greater in women and is associated with badly fitting shoes. It is often accompanied by a short first metatarsal bone. Once the deformity is established it is progressively worsened by the pull of the flexor hallucis longus and the extensor hallucis longus muscles. Later, osteoarthritic changes occur in the metatarsophalangeal joint, which then becomes stiff and painful; the condition is then known as hallux rigidus.

5. Gout commonly affects the metatarsophalangeal joint of the great toe. Urate crystals are deposited in the cartilage and periarticular tissues. The patient is awakened at night with an acutely painful, swollen, and tender joint.

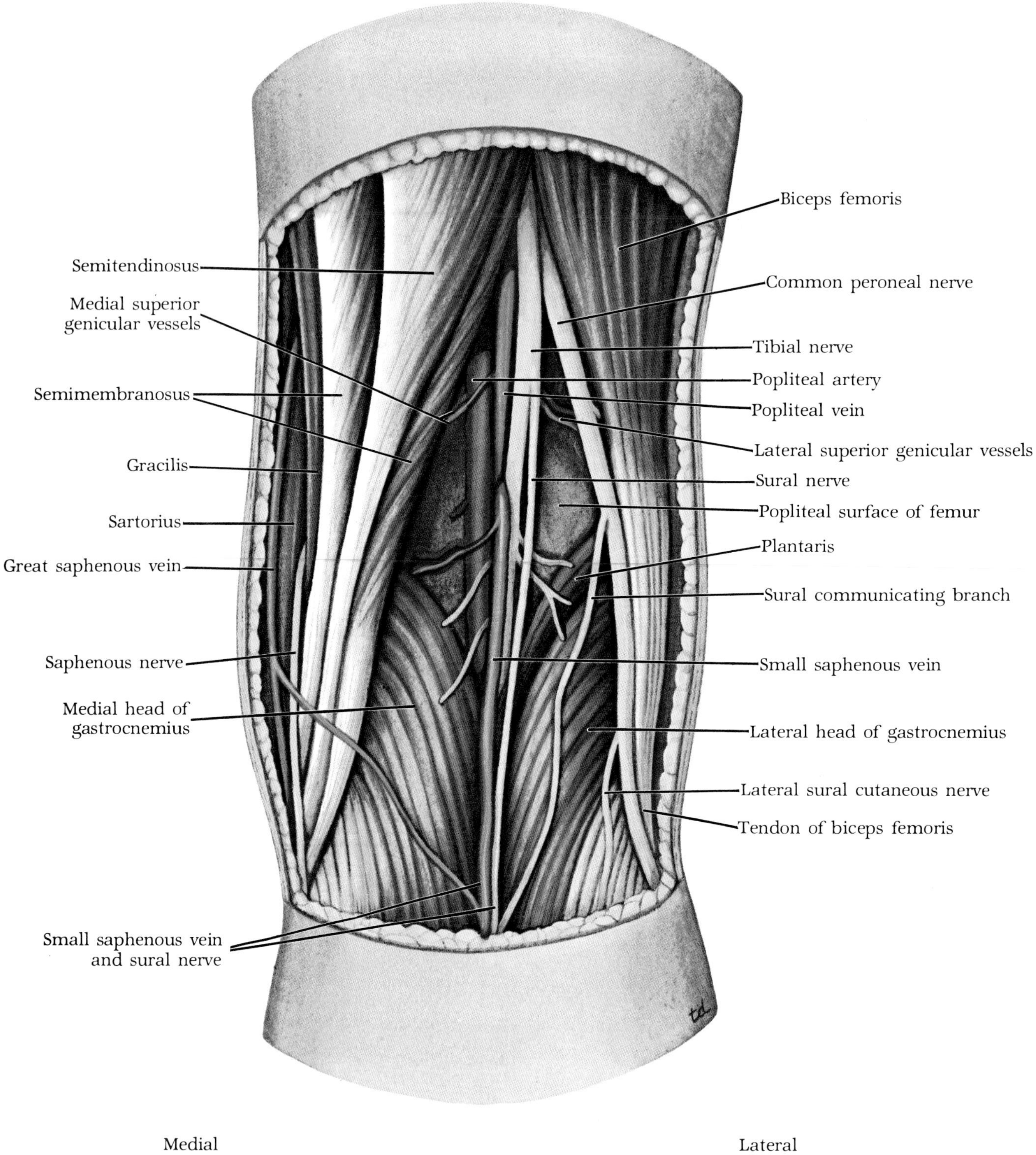
Biceps femoris
Semitendinosus
Common peroneal nerve
Medial superior genicular vessels
Tibial nerve
Popliteal artery
Semimembranosus
Popliteal vein
Lateral superior genicular vessels
Gracilis
Sural nerve
Popliteal surface of femur
Sartorius
Plantaris
Great saphenous vein
Sural communicating branch
Saphenous nerve
Small saphenous vein
Medial head of gastrocnemius
Lateral head of gastrocnemius
Lateral sural cutaneous nerve
Tendon of biceps femoris
Small saphenous vein and sural nerve
Medial
Lateral

Figure 6-28
Boundaries and contents of right popliteal fossa.

1. The popliteal artery is deeply placed in the popliteal fossa. When the examination is done with the knee extended, the deep fascia roofing over the fossa is made taut and makes the pulse difficult to detect. With the knee flexed and the patient kneeling or in the prone position, however, the fascia and neighboring muscles are relaxed, and the pulse is easily felt.
2. Aneurysms of the popliteal artery are detected by the presence of a swelling in the popliteal space that expands with each heartbeat. The most common cause of such an aneurysm is atherosclerosis.
3. A popliteal abscess due to secondary infection of the popliteal lymph nodes is deeply placed in the popliteal fossa and causes pain when the knee joint is extended. Infection of the toes or the foot causes bacteria to spread to the lymph nodes via the lymph vessels accompanying the small saphenous vein.
4. The small saphenous vein is a common site of varicosity. Although this condition is not fatal, it is responsible for considerable discomfort, pain, and suffering in innumerable persons. The condition also usually involves the great saphenous vein.

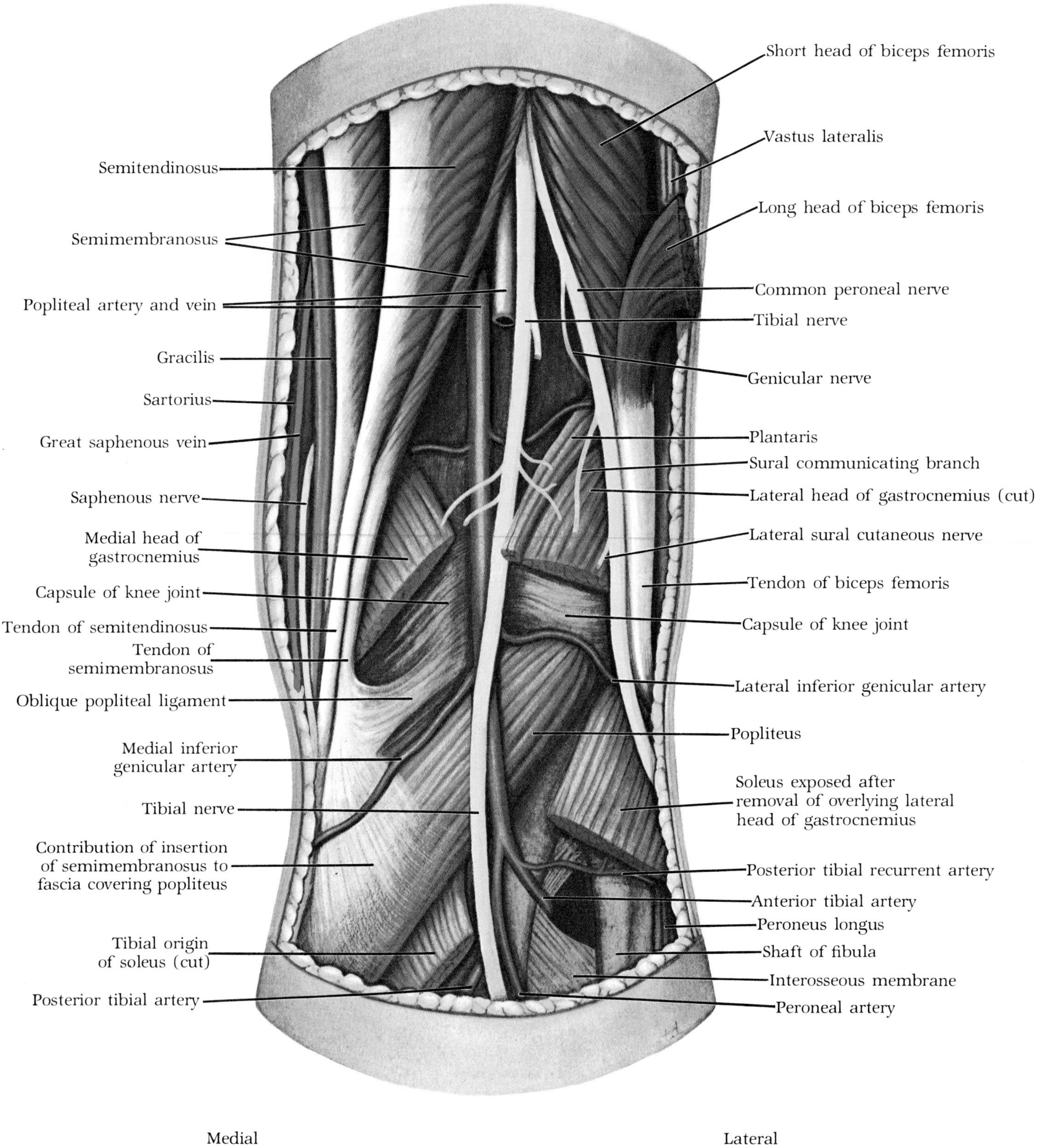
Short head of biceps femoris
Vastus lateralis
Long head of biceps femoris
Common peroneal nerve
Tibial nerve
Genicular nerve
Plantaris
Sural communicating branch
Lateral head of gastrocnemius (cut)
Lateral sural cutaneous nerve
Tendon of biceps femoris
Capsule of knee joint
Lateral inferior genicular artery
Popliteus
Soleus exposed after removal of overlying lateral head of gastrocnemius
Posterior tibial recurrent artery
Anterior tibial artery
Peroneus longus
Shaft of fibula
Interosseous membrane
Peroneal artery
Semitendinosus
Semimembranosus
Popliteal artery and vein
Gracilis
Sartorius
Great saphenous vein
Saphenous nerve
Medial head of gastrocnemius
Capsule of knee joint
Tendon of semitendinosus
Tendon of semimembranosus
Oblique popliteal ligament
Medial inferior genicular artery
Tibial nerve
Contribution of insertion of semimembranosus to fascia covering popliteus
Tibial origin of soleus (cut)
Posterior tibial artery
Medial
Lateral

Figure 6-29
Deep structures present in right popliteal fossa.

1. The semimembranosus bursa lies between the medial head of the gastrocnemius muscle and the tendon of semimembranosus. The presence of excess fluid in this bursa (bursitis) is very common. The swelling is tense when the knee joint is extended and flaccid when the joint is flexed. Compression of the swelling does not cause it to disappear because this bursa does not communicate with the synovial cavity of the knee joint.
2. "Baker's cyst" is a central swelling that occurs in the popliteal fossa in the middle-aged and elderly. It is a diverticulum of the synovial membrane of the knee joint that passes through a small hole in the posterior part of the capsule of the joint. The size of the swelling diminishes if it is compressed since the fluid within it returns to the cavity of the knee joint. The most common cause of this condition is rheumatoid arthritis.
3. The common peroneal nerve is in a very exposed position as it leaves the popliteal fossa and winds around the neck of the fibula to enter the peroneus longus muscle. Fractures of the neck of the fibula and pressure from plaster casts or splints are common causes of injury. The following clinical features of injury to this part of the peroneal nerve may occur:
Motor. The muscles of the anterior and lateral compartments of the leg are paralyzed. As a result, the opposing muscles, the plantar flexors of the ankle joint and the invertors of the subtalar and transverse tarsal joints, cause the foot to be plantar-flexed and inverted, a deformity referred to as *equinovarus.* The patient tends to walk with a high stepping gait to enable the dropped foot and toes to clear the ground.
Sensory. There is loss of sensation down the anterior and lateral sides of the leg and dorsum of the foot and toes, including the medial side of the big toe. The lateral border of the foot and the lateral side of the little toe are virtually unaffected because they are supplied by the sural nerve, mainly formed from tibial nerve. The medial border of the foot as far as the ball of the big toe, supplied by the saphenous nerve, is completely unaffected.

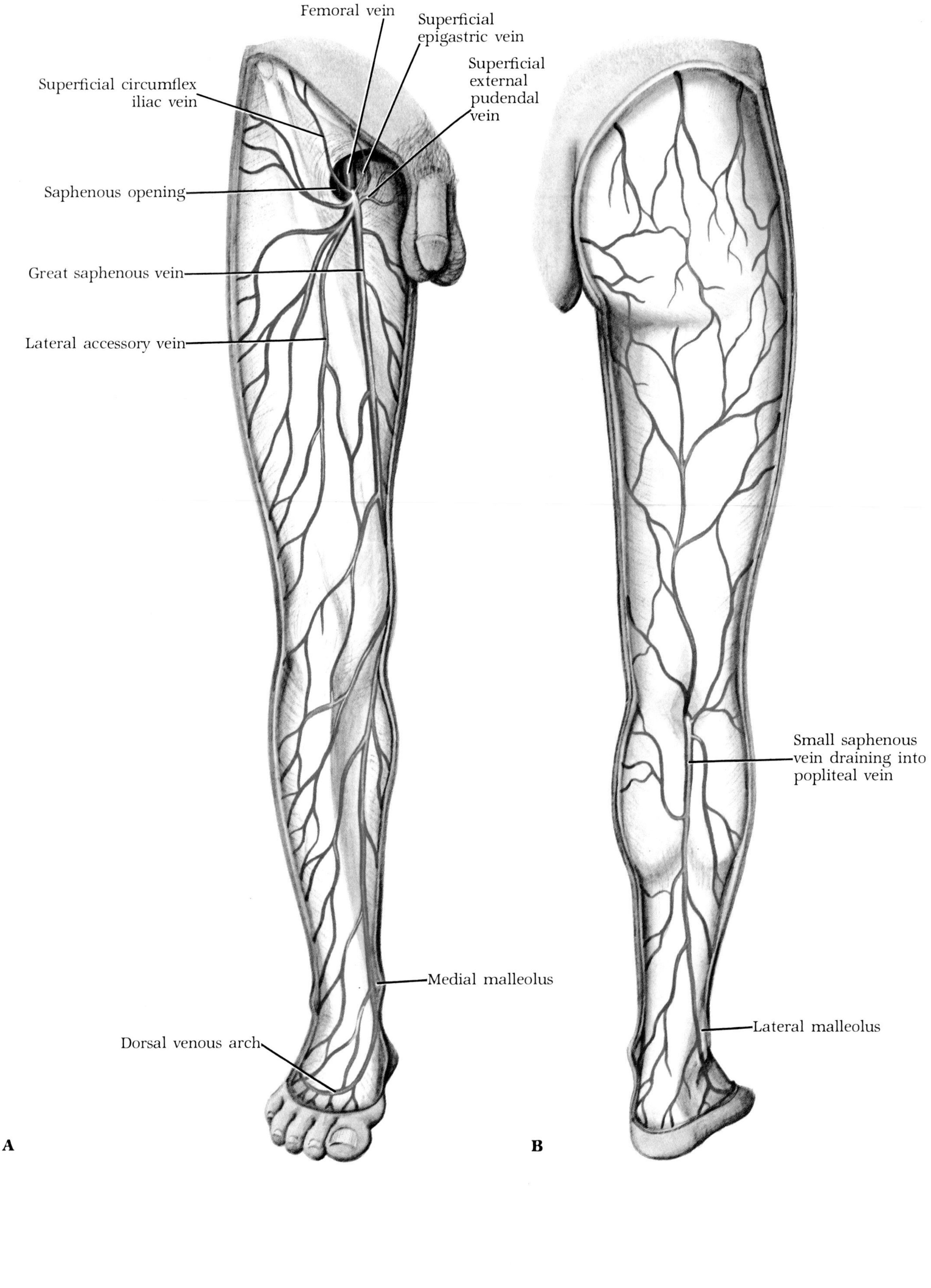
Femoral vein
Superficial epigastric vein
Superficial external pudendal vein
Superficial circumflex iliac vein
Saphenous opening
Great saphenous vein
Lateral accessory vein
Medial malleolus
Dorsal venous arch
A
Small saphenous vein draining into popliteal vein
Lateral malleolus
B

Figure 6-30
Superficial veins of the lower limb. A. Anterior view. B. Posterior view.

1. The veins of the lower limb may be divided into three groups: superficial, deep, and perforating. The great and small saphenous veins possess bicuspid valves, are superficial, and are situated beneath the skin in the superficial fascia. The constant position of the great saphenous vein in front of the medial malleolus should be remembered when patients require emergency blood transfusion. The venae comitantes to the anterior and posterior tibial, popliteal, and femoral arteries and their branches constitute the deep veins. The perforating veins, which are communicating vessels found particularly in the region of the ankle and the medial side of the lower part of the leg, run between the superficial and deep veins. These veins have valves that prevent the flow of blood from the deep to the superficial veins.
2. A varicose vein is the name given to a vein when its diameter is greater than normal and it has become elongated and tortuous. In the lower limb this condition often occurs in the tributaries of the great and small saphenous veins. The superficial dilated veins are unsightly and can cause considerable discomfort. The many causes of varicose veins include hereditary weakness of the vein walls and incompetent valves, multiple pregnancies, and thrombophlebitis of the deep veins.
3. A successful result of operative treatment of varicose veins depends on the ligation and division of all the main tributaries of the great and small saphenous veins — to prevent a collateral circulation from developing — and the ligation and division of all the perforating veins responsible for leakage of blood from the deep to the superficial veins. It is common practice to also remove or strip the superficial veins.

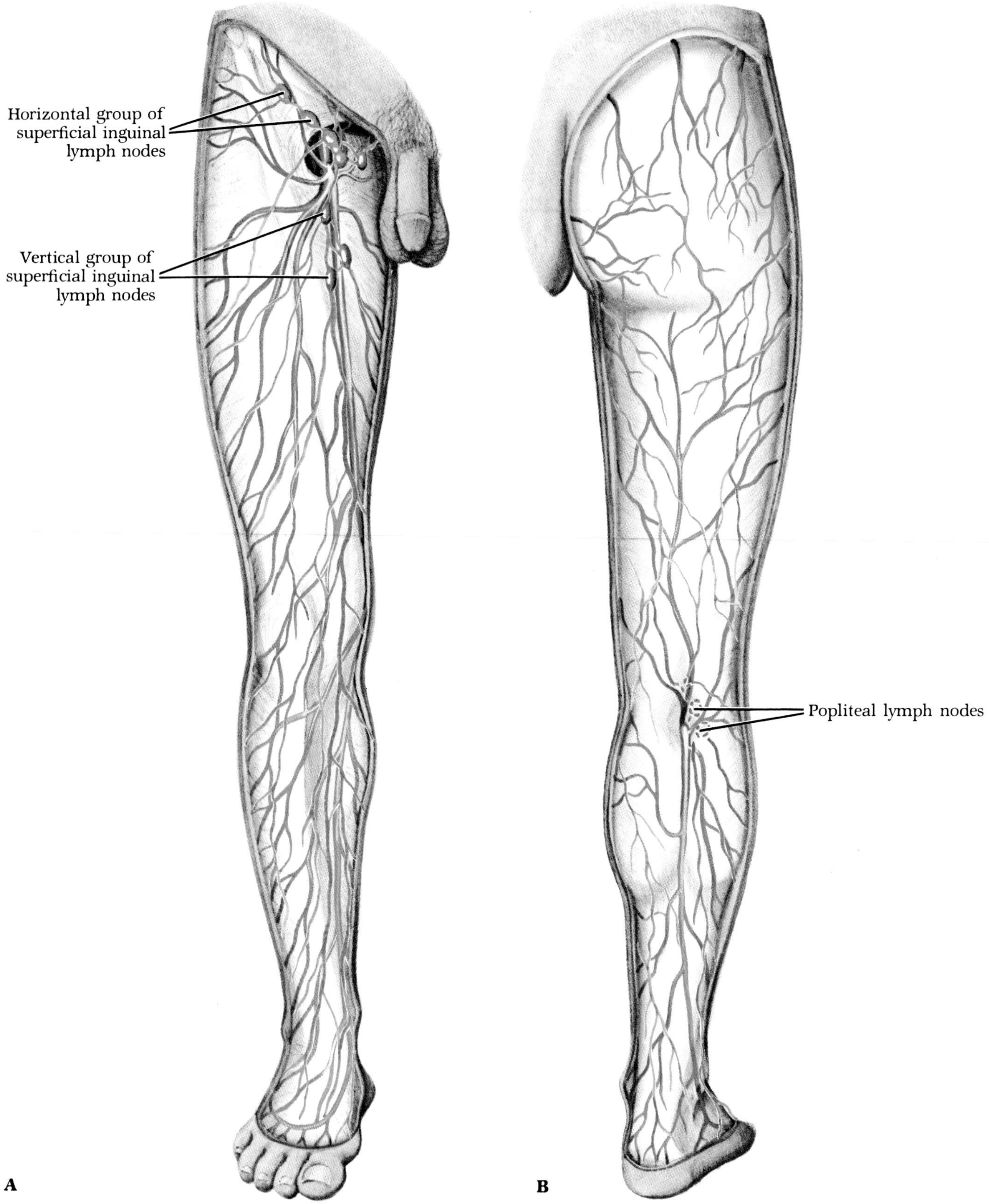
Horizontal group of superficial inguinal lymph nodes
Vertical group of superficial inguinal lymph nodes
Popliteal lymph nodes
A
B

Figure 6-31
Lymphatic drainage of the superficial tissues of the lower limb. A. Anterior view. B. Posterior view.

1. The superficial inguinal lymph nodes can be divided into a horizontal group and a vertical group. The horizontal group lies just below and parallel to the inguinal ligament; the medial members of the group receive superficial lymph vessels from the anterior abdominal wall below the level of the umbilicus and from the perineum, including vessels from the urethra, the external genitalia, and the lower half of the anal canal. The lateral members of this group receive superficial lymph vessels from the back below the level of the iliac crests. The vertical group lies along the terminal part of the great saphenous vein and receives the superficial lymph vessels from the lower limb, except those from the posterior and lateral side of the lower leg.
Efferent vessels from the superficial nodes pass through the saphenous opening and join the deep inguinal nodes that lie along the medial side of the femoral vein. The efferent lymph vessels pass through the femoral canal to join the external iliac lymph nodes.
2. The popliteal lymph nodes are situated beneath the deep fascia of the popliteal fossa. They receive superficial lymph vessels from the lateral side of the foot and leg as well as lymph from deep lymph vessels.
3. Lymphangiography is a process by which the lymphatics may be demonstrated. A radiopaque material is injected into the lymphatic vessels on the dorsum of the foot and the entire lower limb is x-rayed.
4. The superficial inguinal nodes may become enlarged due to syphilitic or malignant disease of the external genitalia. Malignant disease of the lower half of the anal canal may also result in secondary metastases in these nodes. Any infection of the superficial tissues of the abdomen below the umbilicus or in the lower limb (except the posterior and lateral aspects below the knee) commonly causes enlargement of these nodes due to spread of infection.
5. Infection or malignant disease of the superficial tissues of the back and lateral side of the leg below the knee may cause enlargement of the popliteal nodes.
6. In idiopathic hereditary lymphedema (Milroy's disease), the leg is larger than normal due to a firm edema of the tissues. The condition is both congenital and familial. The peripheral lymph vessels can be aplastic or hypoplastic.

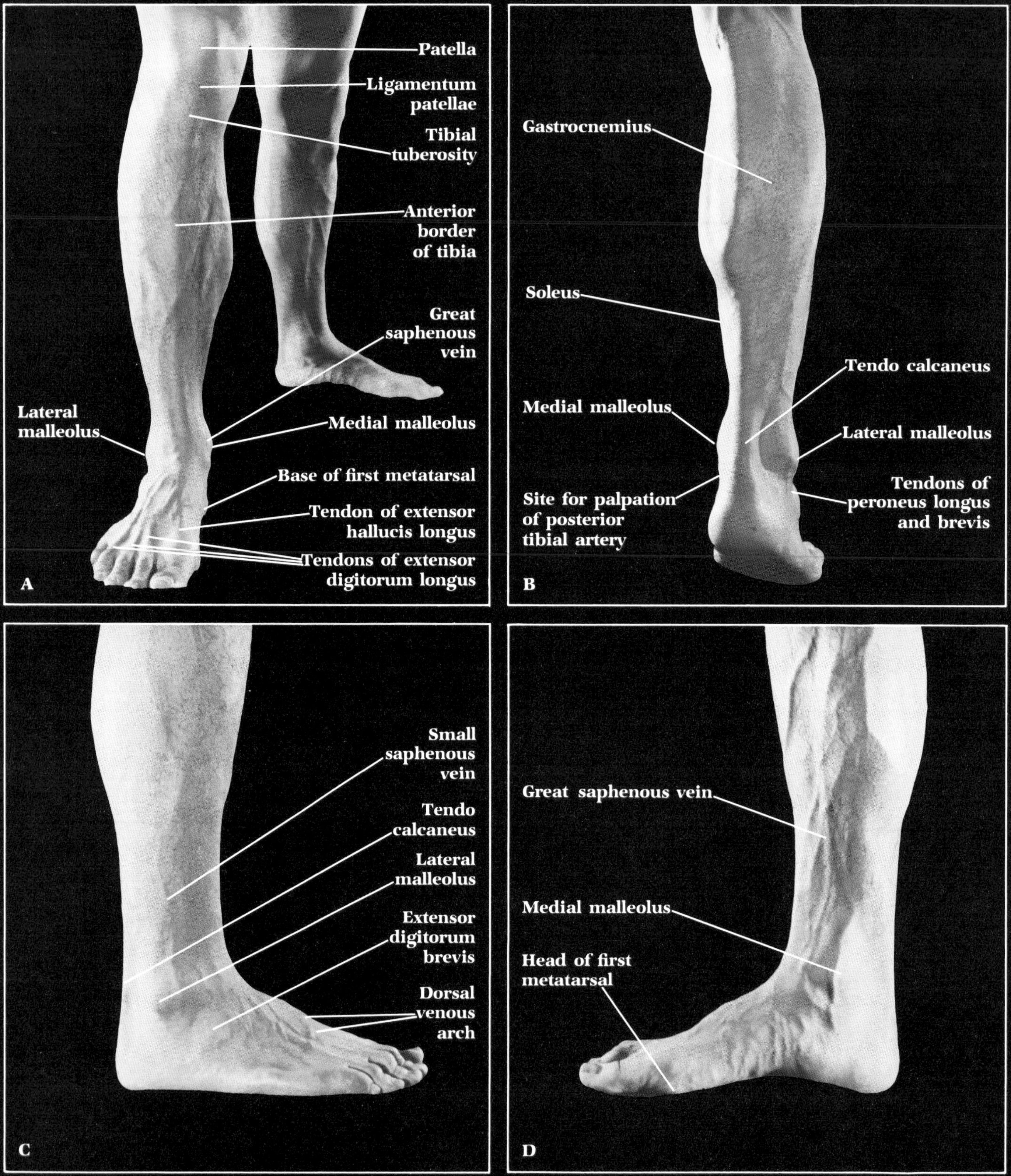
Patella
Ligamentum patellae
Tibial tuberosity
Anterior border of tibia
Great saphenous vein
Lateral malleolus
Medial malleolus
Base of first metatarsal
Tendon of extensor hallucis longus
Tendons of extensor digitorum longus
A
Gastrocnemius
Soleus
Tendo calcaneus
Medial malleolus
Lateral malleolus
Tendons of peroneus longus and brevis
Site for palpation of posterior tibial artery
B
Small saphenous vein
Tendo calcaneus
Lateral malleolus
Extensor digitorum brevis
Dorsal venous arch
C
Great saphenous vein
Medial malleolus
Head of first metatarsal
D

Figure 6-32
The right lower limb below the knee of a 25-year-old male. A. Anterior view. B. Posterior view. C. Lateral view. D. Medial view.

1. The patella and the ligamentum patellae can be easily palpated in front of the knee; the condyles of the femur and tibia can be recognized; and the joint line can be identified.
2. The adductor tubercle can be palpated on the medial aspect of the knee just above the medial condyle of the femur; the tendon of the hamstring part of the adductor magnus muscle can be felt passing to it.
3. The medial surface and anterior border of the tibia are subcutaneous and can be felt throughout their length. Note the position of the tibial tuberosity and the attachment of the ligamentum patellae to it.
4. The fibula is subcutaneous in the region of the ankle and may be followed downward to form the lateral malleolus.
5. The tip of the medial malleolus of the tibia lies about ½ inch (1.3 cm) proximal to the level of the tip of the lateral malleolus.
6. The pulsations of the posterior tibial artery can be felt halfway between the medial malleolus and the heel.
7. Posterior to the lateral malleolus are the tendons of the peroneus brevis and longus muscles.
8. The muscle mass forming the "belly" of the calf is the gastrocnemius.
9. The prominence of the heel is formed by the calcaneum bone. Above the heel is the tendo calcaneus (Achilles tendon).
10. The great saphenous vein leaves the medial part of the dorsal venous plexus (arch) of the foot and ascends *anterior* to the medial malleolus. The small saphenous vein drains the lateral part of the plexus and ascends the leg posterior to the lateral malleolus.

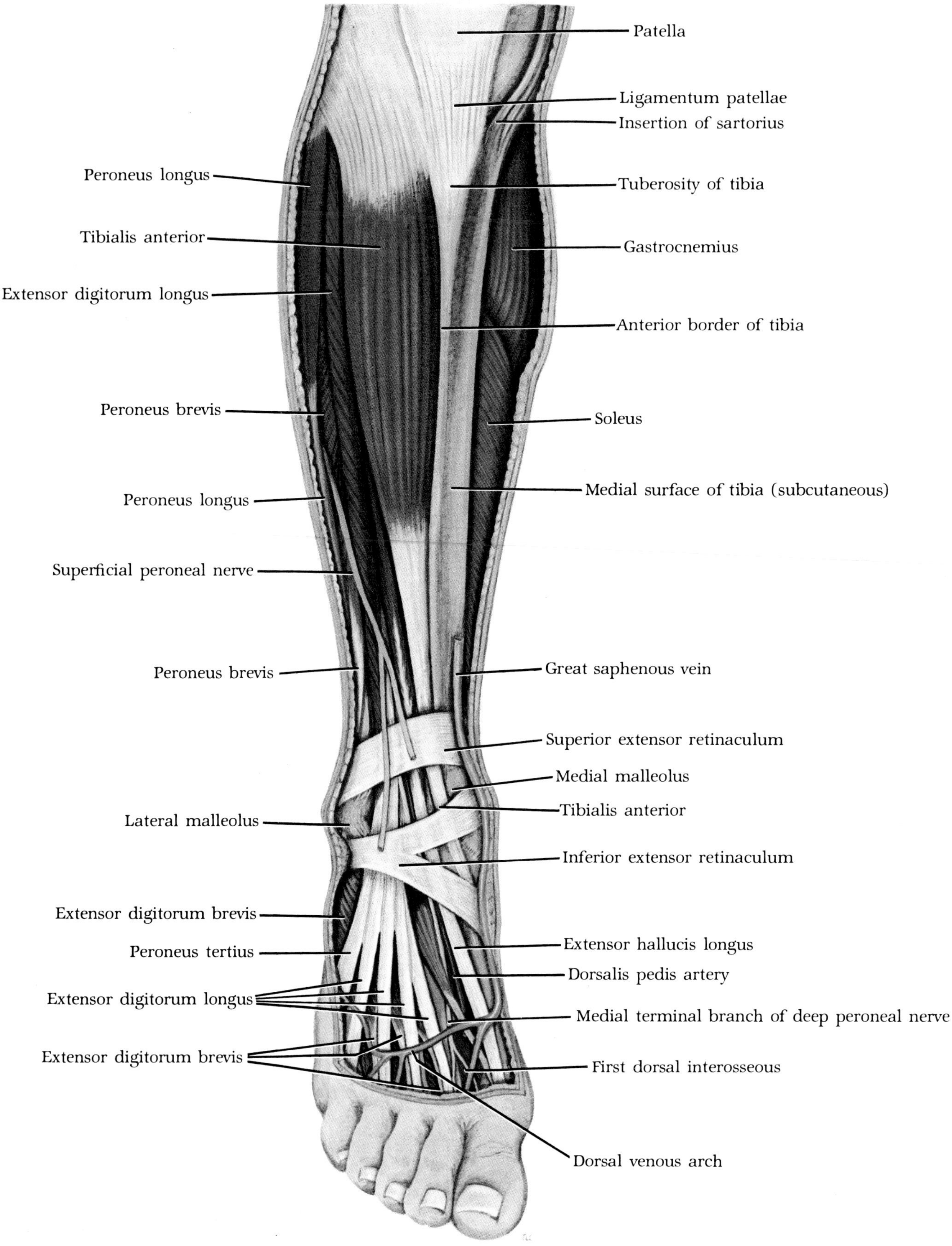
Patella
Ligamentum patellae
Insertion of sartorius
Peroneus longus
Tuberosity of tibia
Tibialis anterior
Gastrocnemius
Extensor digitorum longus
Anterior border of tibia
Peroneus brevis
Soleus
Peroneus longus
Medial surface of tibia (subcutaneous)
Superficial peroneal nerve
Peroneus brevis
Great saphenous vein
Superior extensor retinaculum
Medial malleolus
Lateral malleolus
Tibialis anterior
Inferior extensor retinaculum
Extensor digitorum brevis
Peroneus tertius
Extensor hallucis longus
Dorsalis pedis artery
Extensor digitorum longus
Medial terminal branch of deep peroneal nerve
Extensor digitorum brevis
First dorsal interosseous
Dorsal venous arch

Figure 6-33
Structures present on anterior and lateral aspects of right leg and on dorsum of foot.

1. The tibia, a long bone, is commonly involved in osteitis deformans. Essentially the disease causes thickening of the bone, softening, deformity, and, later, recalcification and hardening. There is pain in the bone and a bowing of the tibia. The skull, femur, and other bones may also be involved.

2. Tenosynovitis can affect the tendon sheaths of the peroneus longus and brevis muscles as they pass behind the lateral malleolus. Treatment is immobilization, heat, and physiotherapy.

3. Arterial disease of the leg is common. It is important that the physician be able to palpate the pulse of the dorsalis pedis artery. This artery is covered only by fascia and skin and lies between the tendons of the extensor hallucis longus and extensor digitorum longus muscles on the dorsal surface of the foot. Despite its importance it must also be kept in mind that the dorsalis pedis artery is sometimes absent, being replaced by a large perforating branch of the peroneal artery.

4. In occlusive arterial disease of the lower limb, lumbar sympathectomy, which increases the blood flow through the collateral circulation, may be advocated as a form of treatment. Preganglionic sympathectomy is performed by removing the upper three lumbar ganglia and the intervening parts of the sympathetic trunk.

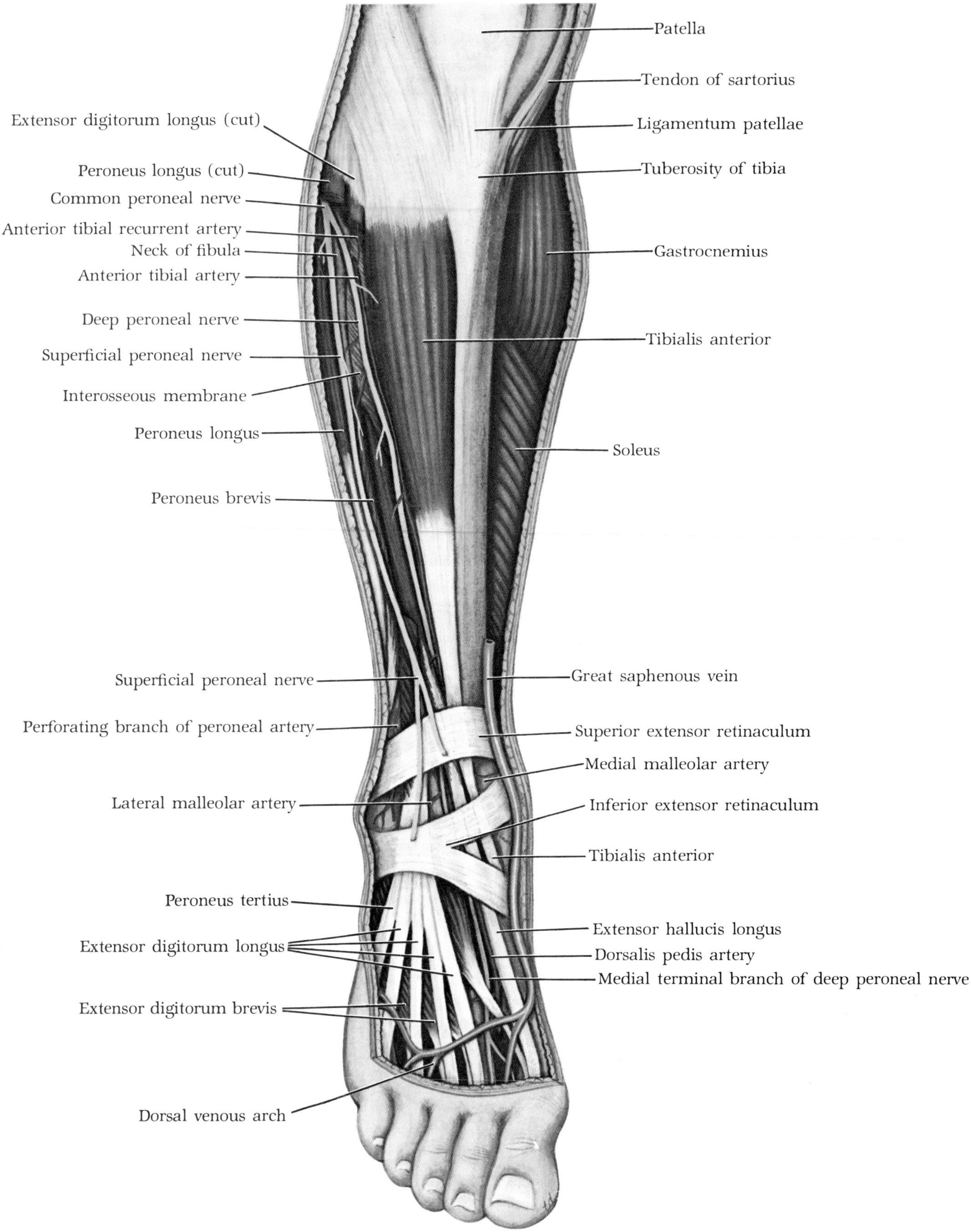

Patella
Tendon of sartorius
Extensor digitorum longus (cut)
Ligamentum patellae
Peroneus longus (cut)
Tuberosity of tibia
Common peroneal nerve
Anterior tibial recurrent artery
Neck of fibula
Gastrocnemius
Anterior tibial artery
Deep peroneal nerve
Tibialis anterior
Superficial peroneal nerve
Interosseous membrane
Peroneus longus
Soleus
Peroneus brevis
Superficial peroneal nerve
Great saphenous vein
Perforating branch of peroneal artery
Superior extensor retinaculum
Medial malleolar artery
Lateral malleolar artery
Inferior extensor retinaculum
Tibialis anterior
Peroneus tertius
Extensor hallucis longus
Extensor digitorum longus
Dorsalis pedis artery
Medial terminal branch of deep peroneal nerve
Extensor digitorum brevis
Dorsal venous arch

Figure 6-34
Structures present on anterior and lateral aspects of right leg and on dorsum of foot. Portions of the peroneus longus and the extensor digitorum longus muscles have been removed to display the common peroneal, the deep peroneal, and the superficial peroneal nerves.

1. The common peroneal nerve is in a very exposed position as it leaves the popliteal fossa and winds around the neck of the fibula to enter the peroneus longus muscle. It is at this point that it is commonly injured in fractures of the neck of the fibula and by pressure from plaster casts or splints. The following clinical features of injury are present:
Motor. The muscles of the anterior and lateral compartments of the leg are paralyzed, namely, the tibialis anterior, the extensor digitorum longus and brevis, the peroneus tertius, the extensor hallucis longus (supplied by the deep peroneal nerve), and the peroneus longus and brevis (supplied by the superficial peroneal nerve). As a result, the opposing muscles, the plantar flexors of the ankle joint and the invertors of the subtalar and transverse tarsal joints, cause the foot to be plantar-flexed and inverted, a deformity referred to as equinovarus.
Sensory. There is loss of sensation down the anterior and lateral sides of the leg and the dorsum of the foot and toes, including the medial side of the big toe. The lateral border of the foot and the lateral side of the little toe are virtually unaffected since they are supplied by the sural nerve (mainly formed from tibial nerve). The medial border of the foot as far as the ball of the big toe is completely unaffected, being supplied by the saphenous nerve. When the injury occurs distal to the site of origin of the lateral sural cutaneous nerve, the loss of sensation is confined to the area of the foot and toes.

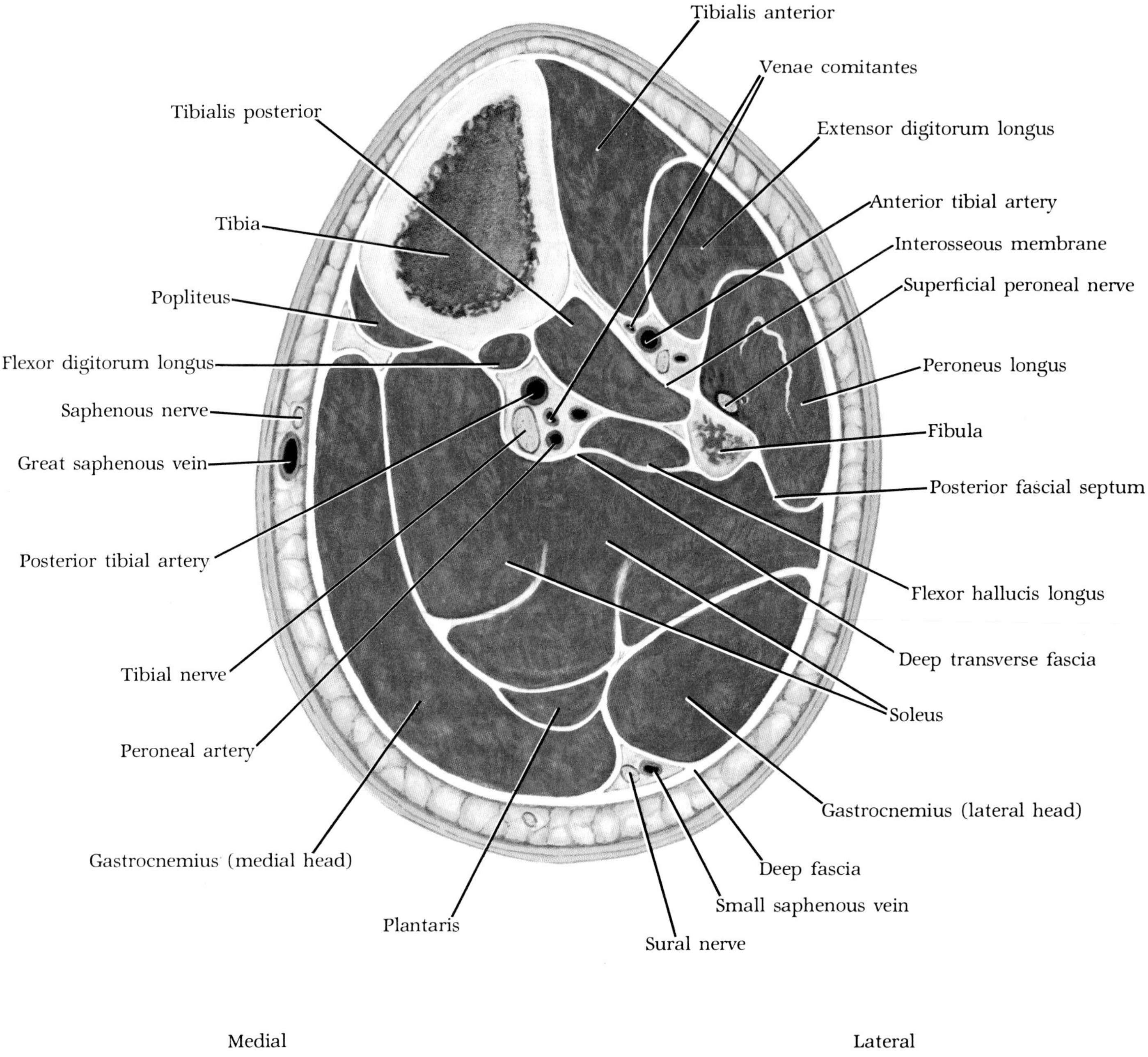
Tibialis anterior
Venae comitantes
Tibialis posterior
Extensor digitorum longus
Anterior tibial artery
Tibia
Interosseous membrane
Superficial peroneal nerve
Popliteus
Flexor digitorum longus
Peroneus longus
Saphenous nerve
Fibula
Great saphenous vein
Posterior fascial septum
Posterior tibial artery
Flexor hallucis longus
Deep transverse fascia
Tibial nerve
Soleus
Peroneal artery
Gastrocnemius (lateral head)
Gastrocnemius (medial head)
Deep fascia
Small saphenous vein
Plantaris
Sural nerve
Medial
Lateral

Figure 6-35
Transverse section through the middle of the right leg as seen from above.

1. The anterior compartment syndrome is an increase in the intracompartmental pressure that results from an increased production of tissue fluid. Marked exertion may produce this condition. The deep, aching pain in the anterior compartment of the leg that is characteristic of this syndrome can become very severe. Dorsiflexion of the foot at the ankle joint increases the severity of the pain. As the pressure rises, the venous return is diminished, thus producing a further rise in pressure. In severe cases the arterial supply is eventually cut off by compression, and the dorsalis pedis arterial pulse disappears. The tibialis anterior, the extensor digitorum longus, and the extensor hallucis longus muscles are paralyzed. Loss of sensation is limited to the area supplied by the deep peroneal nerve — the skin cleft between the first and second toes. By opening up the anterior compartment by making a longitudinal incision through the deep fascia, anoxic necrosis of the muscle can be avoided.

2. Shin splints is a common condition in athletes. The pain in the leg is not accompanied by swelling of the anterior compartment and is believed to be due to a stress fracture of the tibia.

3. Fractures of the tibia and fibula are common. Fortunately, if only one bone is fractured, the other acts as a splint and there is minimal displacement. Fractures of the shaft of the tibia are often compound, since the entire length of the medial surface is covered only by skin and superficial fascia.

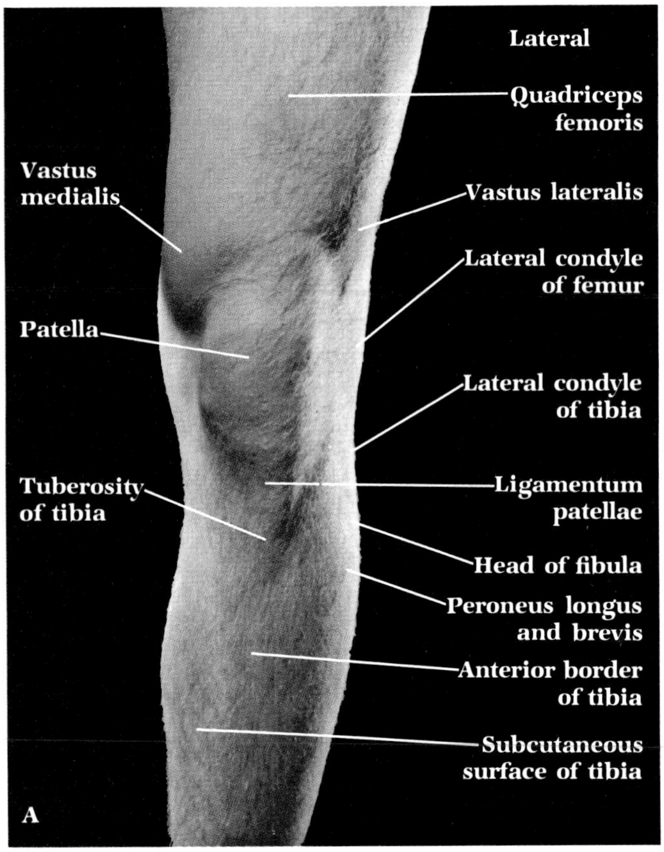
Lateral
Quadriceps femoris
Vastus medialis
Vastus lateralis
Lateral condyle of femur
Patella
Lateral condyle of tibia
Tuberosity of tibia
Ligamentum patellae
Head of fibula
Peroneus longus and brevis
Anterior border of tibia
Subcutaneous surface of tibia
A

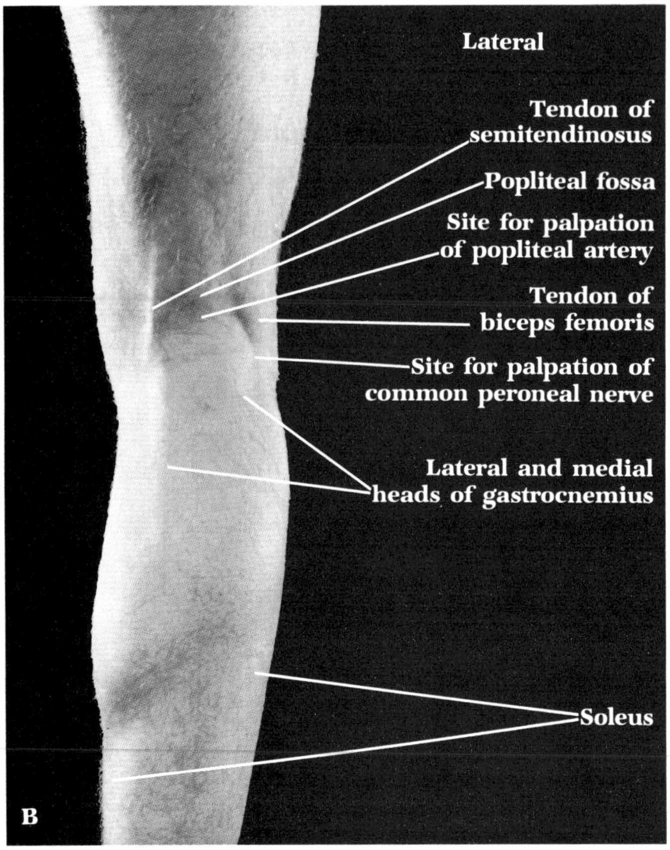
Lateral
Tendon of semitendinosus
Popliteal fossa
Site for palpation of popliteal artery
Tendon of biceps femoris
Site for palpation of common peroneal nerve
Lateral and medial heads of gastrocnemius
Soleus
B

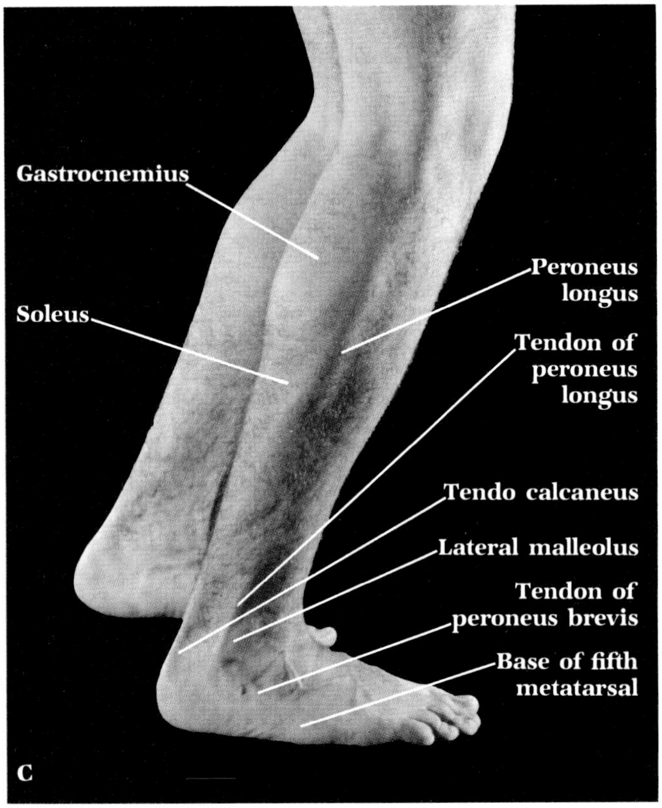
Gastrocnemius
Peroneus longus
Soleus
Tendon of peroneus longus
Tendo calcaneus
Lateral malleolus
Tendon of peroneus brevis
Base of fifth metatarsal
C

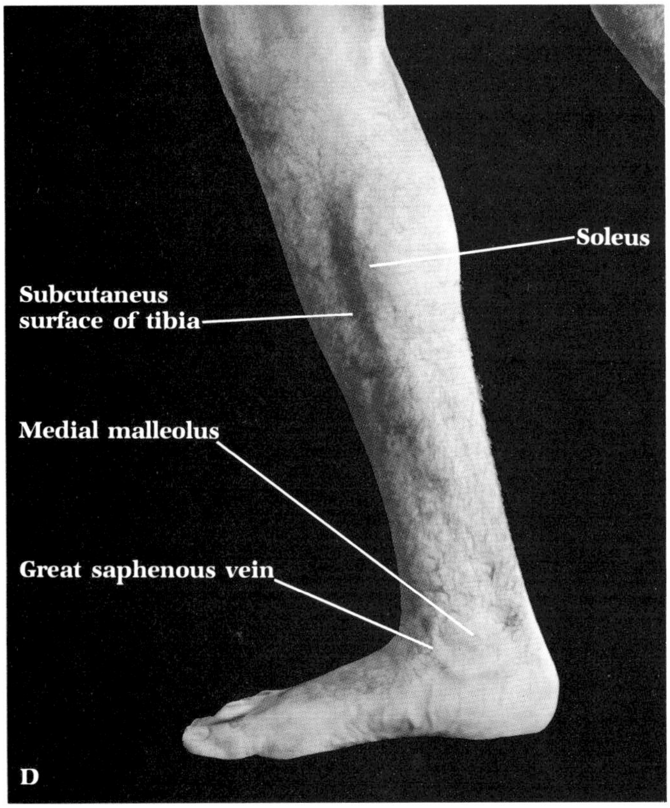
Soleus
Subcutaneus surface of tibia
Medial malleolus
Great saphenous vein
D

Figure 6-36
The regions of the knee and ankle in a 27-year-old male. A. Left knee, anterior view. B. Right knee, posterior view. C. Right ankle, lateral view. D. Right ankle, medial view.

1. The quadriceps femoris is a massive muscle situated on the anterior surface of the thigh. Three of its parts, namely, the rectus femoris, vastus medialis, and the vastus lateralis, can be palpated. The fourth part, the vastus intermedius, is deeply placed and cannot be palpated.
2. The patella and the ligamentum patellae can be easily felt in front of the knee. The condyles of the femur and tibia can be recognized on the sides of the knee, and the joint line can be identified.
3. The head of the fibula can be palpated below the lateral femoral condyle. It is situated at about the same level as the tibial tuberosity.
4. The adductor tubercle can be palpated on the medial aspect of the knee just above the medial condyle of the femur.
5. The medial and lateral collateral ligaments of the knee joint are difficult to recognize by palpation. Their anatomical location, however, may be felt with the finger pressing on the medial and lateral sides of the joint line, respectively. Damage to these ligaments will result in tenderness on palpation.
6. The popliteal fossa is a diamond-shaped space situated behind the knee. When the knee is flexed, the deep fascia that roofs over the fossa is relaxed and the boundaries are easily defined. Its upper part is bounded laterally by the tendon of the biceps femoris muscle and medially by the tendons of the semimembranosus and semitendinosus muscles. Its lower part is bounded on each side by one of the heads of the gastrocnemius muscle.
7. The popliteal artery can be felt pulsating in the depths of the popliteal space, provided that the deep fascia is fully relaxed, which is done by passively flexing the knee joint.
8. The common peroneal nerve can be palpated on the medial side of the tendon of the biceps femoris muscle, as the latter passes to its insertion on the head of the fibula. When the knee joint is partially flexed the nerve can be rolled beneath the finger.
9. The medial surface and anterior border of the tibia are subcutaneous and can be felt throughout their length.
10. The medial and lateral malleoli and the tendo calcaneus can easily be felt in the region of the ankle.
11. The tendons of the peroneus longus and brevis muscles can be felt behind the lateral malleolus.
12. On the lateral aspect of the foot the peroneal tubercle of the calcaneum can be palpated about 1 inch (2.5 cm) below and in front of the tip of the lateral malleolus. Above the tubercle the tendon of the peroneus brevis passes forward to its insertion on the prominent base of the fifth metatarsal bone. Below the tubercle the tendon of the peroneus longus passes forward to enter the groove on the underaspect of the cuboid bone.
13. The sustentaculum tali can be palpated about 1 inch (2.5 cm) below the tip of the medial malleolus.
14. Anterior to the sustentaculum tali the tuberosity of the navicular bone can be seen and palpated.

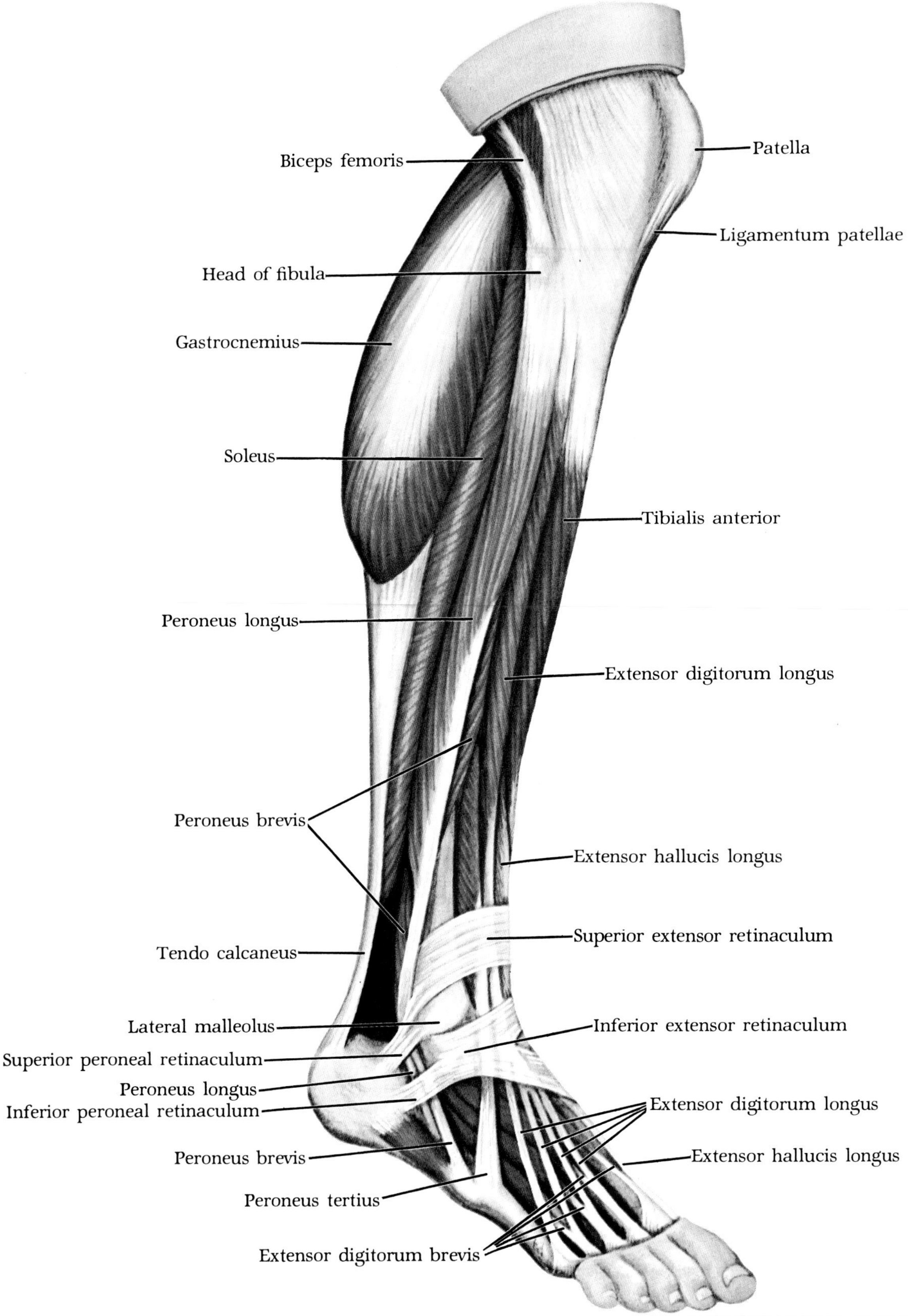
Biceps femoris
Patella
Ligamentum patellae
Head of fibula
Gastrocnemius
Soleus
Tibialis anterior
Peroneus longus
Extensor digitorum longus
Peroneus brevis
Extensor hallucis longus
Superior extensor retinaculum
Tendo calcaneus
Lateral malleolus
Inferior extensor retinaculum
Superior peroneal retinaculum
Peroneus longus
Inferior peroneal retinaculum
Extensor digitorum longus
Peroneus brevis
Extensor hallucis longus
Peroneus tertius
Extensor digitorum brevis

Figure 6-37
Muscles of the right leg and dorsum of foot as seen from the lateral side.

1. The peroneus longus and brevis muscles are supplied by the superficial peroneal nerve. They have a strong plantar flexor action at the ankle joint, and they evert the foot at the subtalar and transverse tarsal joints.
2. Tenosynovitis can affect the tendons of the peroneus longus and brevis muscles as they pass posterior to the lateral malleolus of the fibula. The treatment is immobilization, heat, and physiotherapy.
3. The tendons of peroneus longus and brevis muscles can be dislocated forward from their normal position posterior to the lateral malleolus. For this condition to occur the superior peroneal retinaculum must be torn. It usually occurs in older children and is caused by trauma.

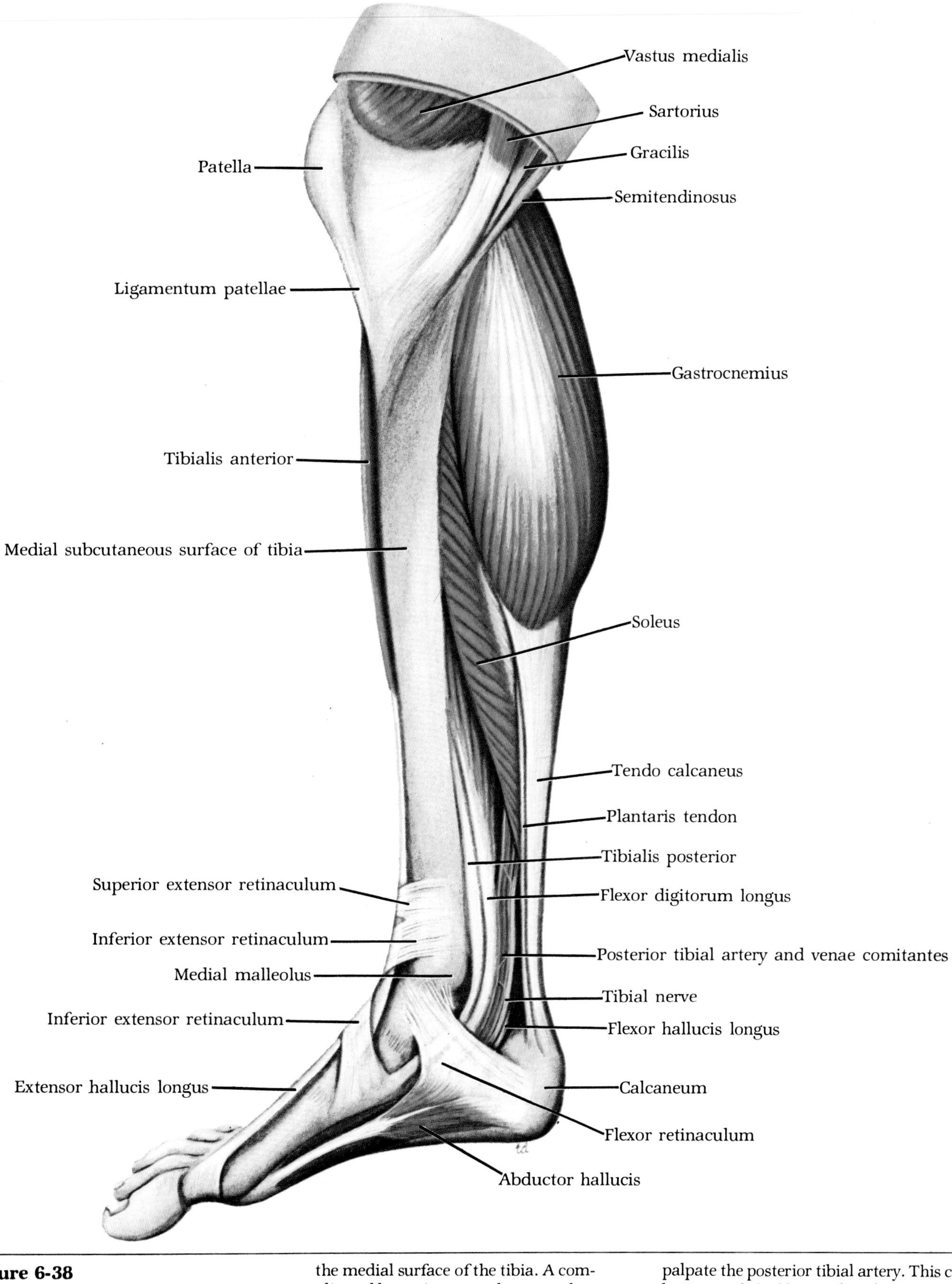
Vastus medialis
Sartorius
Gracilis
Semitendinosus
Patella
Ligamentum patellae
Gastrocnemius
Tibialis anterior
Medial subcutaneous surface of tibia
Soleus
Tendo calcaneus
Plantaris tendon
Tibialis posterior
Superior extensor retinaculum
Flexor digitorum longus
Inferior extensor retinaculum
Posterior tibial artery and venae comitantes
Medial malleolus
Tibial nerve
Inferior extensor retinaculum
Flexor hallucis longus
Extensor hallucis longus
Calcaneum
Flexor retinaculum
Abductor hallucis

Figure 6-38
Muscles of the right leg and dorsum of foot as seen from the medial side.

1. The tendons of the sartorius, gracilis, and semitendinosus muscles are the only muscles to be inserted into the upper part of the medial surface of the tibia. A complicated bursa is present between these tendons and the tibia and may become inflamed as the result of trauma to the medial side of the leg.
2. Because arterial occlusive disease is common, every physician should be able to palpate the posterior tibial artery. This can be accomplished by gently palpating the artery about halfway between the medial malleolus and the medial border of the tendo calcaneus where it lies between the tendons of the flexor digitorum longus and flexor hallucis longus muscles.

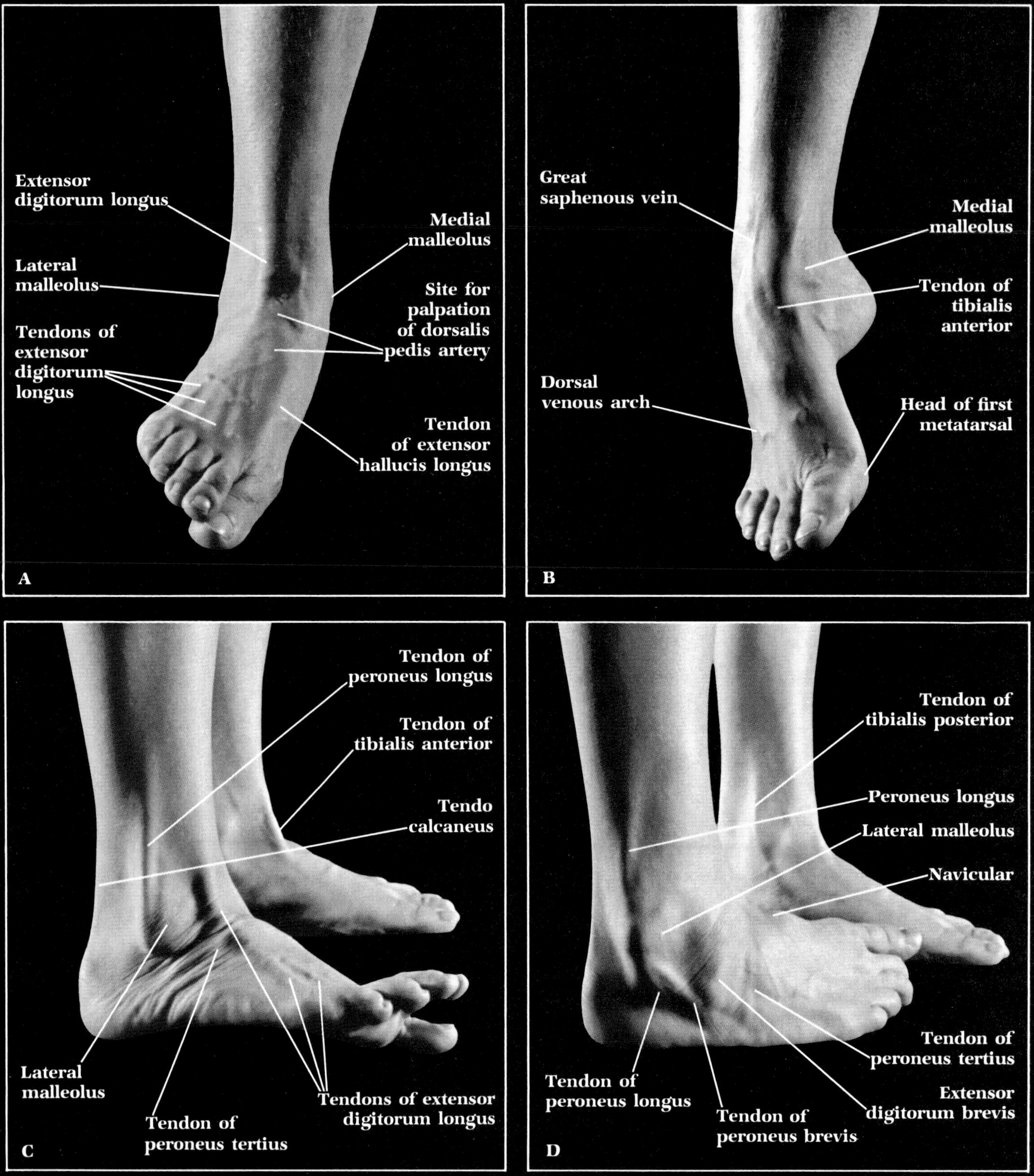
Extensor digitorum longus
Lateral malleolus
Tendons of extensor digitorum longus
Medial malleolus
Site for palpation of dorsalis pedis artery
Tendon of extensor hallucis longus
A
Great saphenous vein
Medial malleolus
Tendon of tibialis anterior
Dorsal venous arch
Head of first metatarsal
B
Tendon of peroneus longus
Tendon of tibialis anterior
Tendo calcaneus
Lateral malleolus
Tendon of peroneus tertius
Tendons of extensor digitorum longus
C
Tendon of tibialis posterior
Peroneus longus
Lateral malleolus
Navicular
Tendon of peroneus tertius
Extensor digitorum brevis
Tendon of peroneus longus
Tendon of peroneus brevis
D

Figure 6-39
The right ankle and foot. A. 27-year-old male; showing eversion, anterior view. B. 27-year-old male; showing inversion, anterior view. C. 29-year-old female; showing eversion, lateral view. D. 29-year-old female; showing inversion, lateral view.

1. The tendon of the tibialis anterior can be seen most easily on the anterior surface of the ankle joint when the foot is dorsiflexed and inverted.
2. The tendon of the extensor hallucis longus muscle lies lateral to it and can be made to stand out by extending the big toe. Lateral to the extensor hallucis longus lie the tendons of the extensor digitorum longus and peroneus tertius muscles.
3. The pulsations of the dorsalis pedis artery can be felt between the tendons of the extensor hallucis longus and extensor digitorum longus muscles, midway between the two malleoli on the front of the ankle.
4. The dorsal venous plexus (arch) can be seen on the dorsal surface of the foot proximal to the toes. The great saphenous vein leaves the medial part of the plexus and ascends the leg *anterior* to the medial malleolus. The small saphenous vein drains the lateral part of the plexus and ascends the leg posterior to the lateral malleolus.
5. The movements of the foot termed *inversion* and *eversion do not* take place at the ankle joint but occur at the subtalar and transverse tarsal joints.
6. Inversion is the movement of the foot so that the sole faces medially; this maneuver is performed by the tibialis anterior muscle, the extensor hallucis longus muscle, and the medial tendons of the extensor digitorum longus; the tibialis posterior also assists.
7. Eversion, the movement of the foot so that the sole faces in the lateral direction, is less extensive than inversion. Eversion is performed by the peroneus longus, peroneus brevis, and peroneus tertius muscles; the lateral tendons of the extensor digitorum longus also assist.

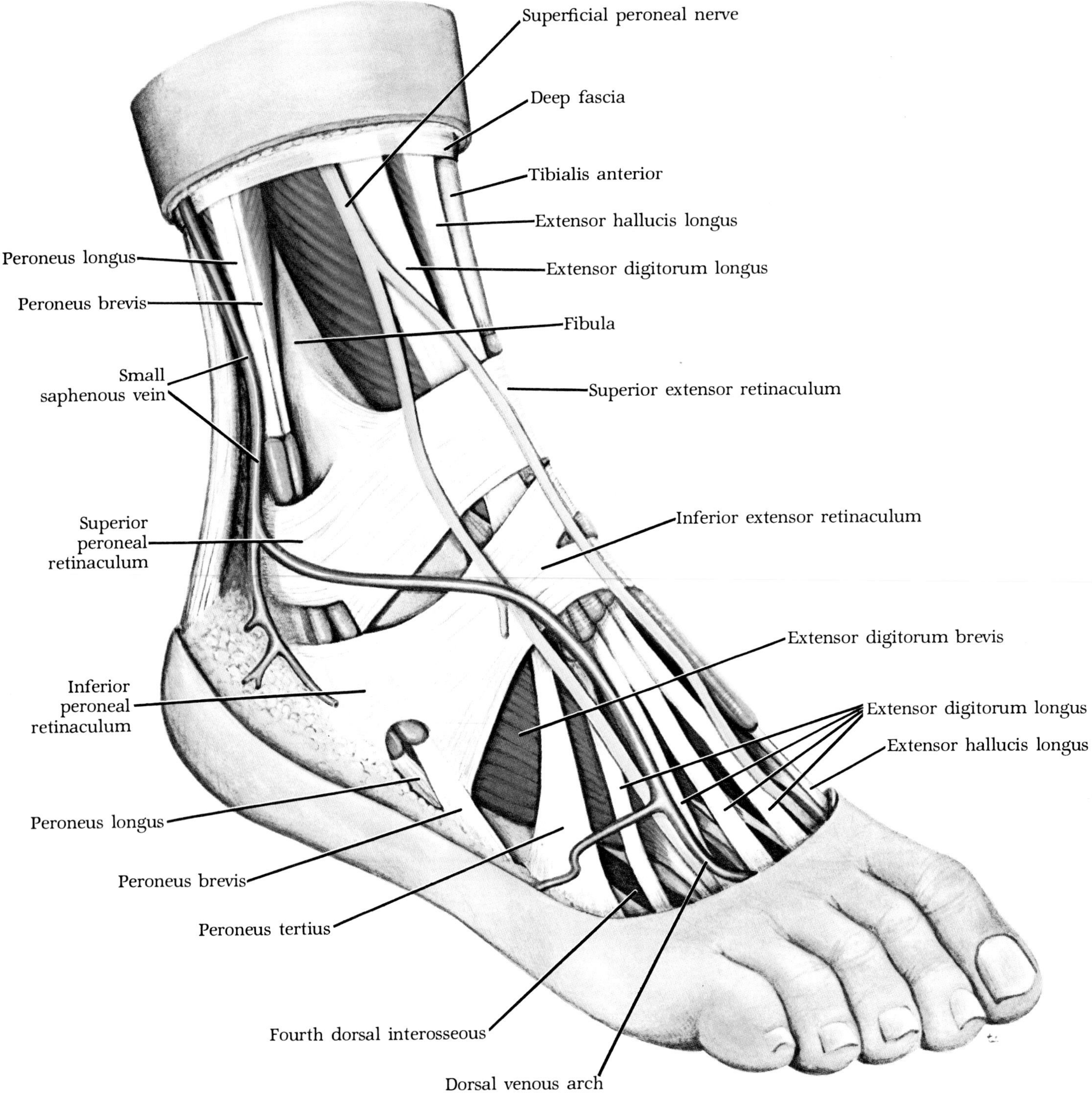
Superficial peroneal nerve
Deep fascia
Tibialis anterior
Extensor hallucis longus
Peroneus longus
Extensor digitorum longus
Peroneus brevis
Fibula
Small saphenous vein
Superior extensor retinaculum
Superior peroneal retinaculum
Inferior extensor retinaculum
Extensor digitorum brevis
Inferior peroneal retinaculum
Extensor digitorum longus
Extensor hallucis longus
Peroneus longus
Peroneus brevis
Peroneus tertius
Fourth dorsal interosseous
Dorsal venous arch

Figure 6-40
Right ankle region viewed anterolaterally.

1. Tenosynovitis can affect the tendons of the peroneus longus and brevis muscles as they pass behind the lateral malleolus of the fibula. The treatment is immobilization, heat, and physiotherapy.
2. The tendons of the peroneus longus and brevis muscles can be dislocated forward from their normal position posterior to the lateral malleolus. For this condition to occur the superior peroneal retinaculum must be torn. It usually occurs in older children and is caused by trauma in most cases.
3. A ganglion of the foot, usually occurring on the dorsum, is similar to one arising in the hand. It is a cystic swelling produced by a myxomatous degeneration of connective tissue associated with the ligaments of joints.
4. It is important to remember that the dorsal venous arch is drained laterally by the small saphenous vein that enters the leg by passing superiorly posterior to the lateral malleolus. The arch is drained medially by the great saphenous vein that passes up into the leg anterior to the medial malleolus. The constant position of the great saphenous vein in front of the medial malleolus should be remembered when patients require emergency blood transfusion.
5. The sensory nerve supply to the skin of the dorsum of the foot and toes is the superficial peroneal nerve, except for the skin cleft between the first and second toes, which is innervated by the medial terminal branch of the deep peroneal nerve. The lateral side of the foot and little toe is supplied by the sural nerve. The medial side of the foot as far as the ball of the big toe is supplied by the saphenous nerve; the medial side of the big toe is innervated by the superficial peroneal nerve.

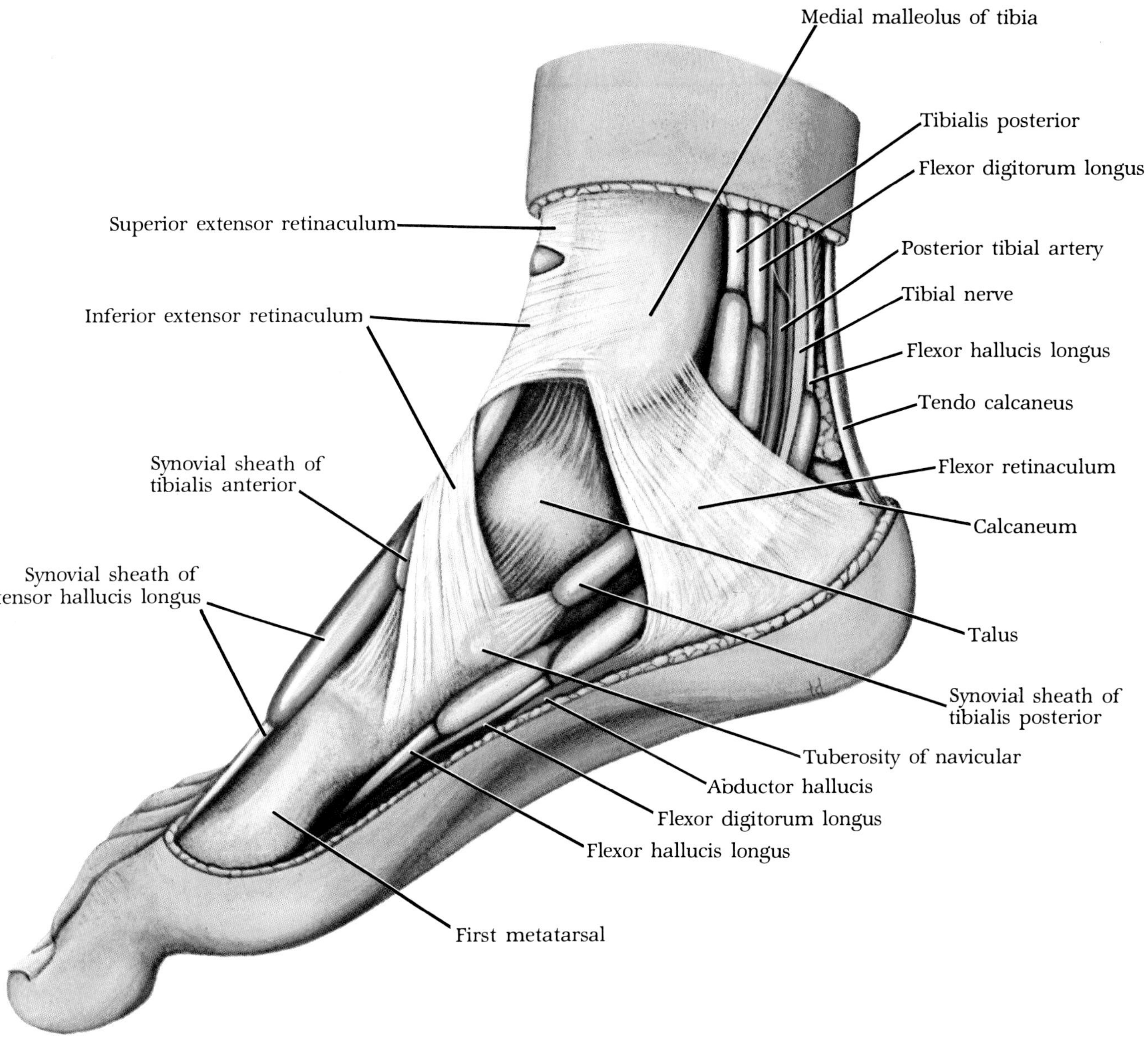
Medial malleolus of tibia
Tibialis posterior
Flexor digitorum longus
Superior extensor retinaculum
Posterior tibial artery
Tibial nerve
Inferior extensor retinaculum
Flexor hallucis longus
Tendo calcaneus
Synovial sheath of tibialis anterior
Flexor retinaculum
Calcaneum
Synovial sheath of extensor hallucis longus
Talus
Synovial sheath of tibialis posterior
Tuberosity of navicular
Abductor hallucis
Flexor digitorum longus
Flexor hallucis longus
First metatarsal

Figure 6-41
Structures passing behind the medial malleolus of the right ankle.

1. Because arterial occlusive disease of the leg is common, physicians should be able to palpate the posterior tibial artery. This is accomplished by gently palpating the artery about halfway between the medial malleolus and the medial border of the tendo calcaneus where it lies between the tendons of the flexor digitorum longus and flexor hallucis longus muscles.

2. Tenosynovitis of the flexor tendons is a rare condition. Treatment for it consists of immobilization, heat, and physiotherapy.

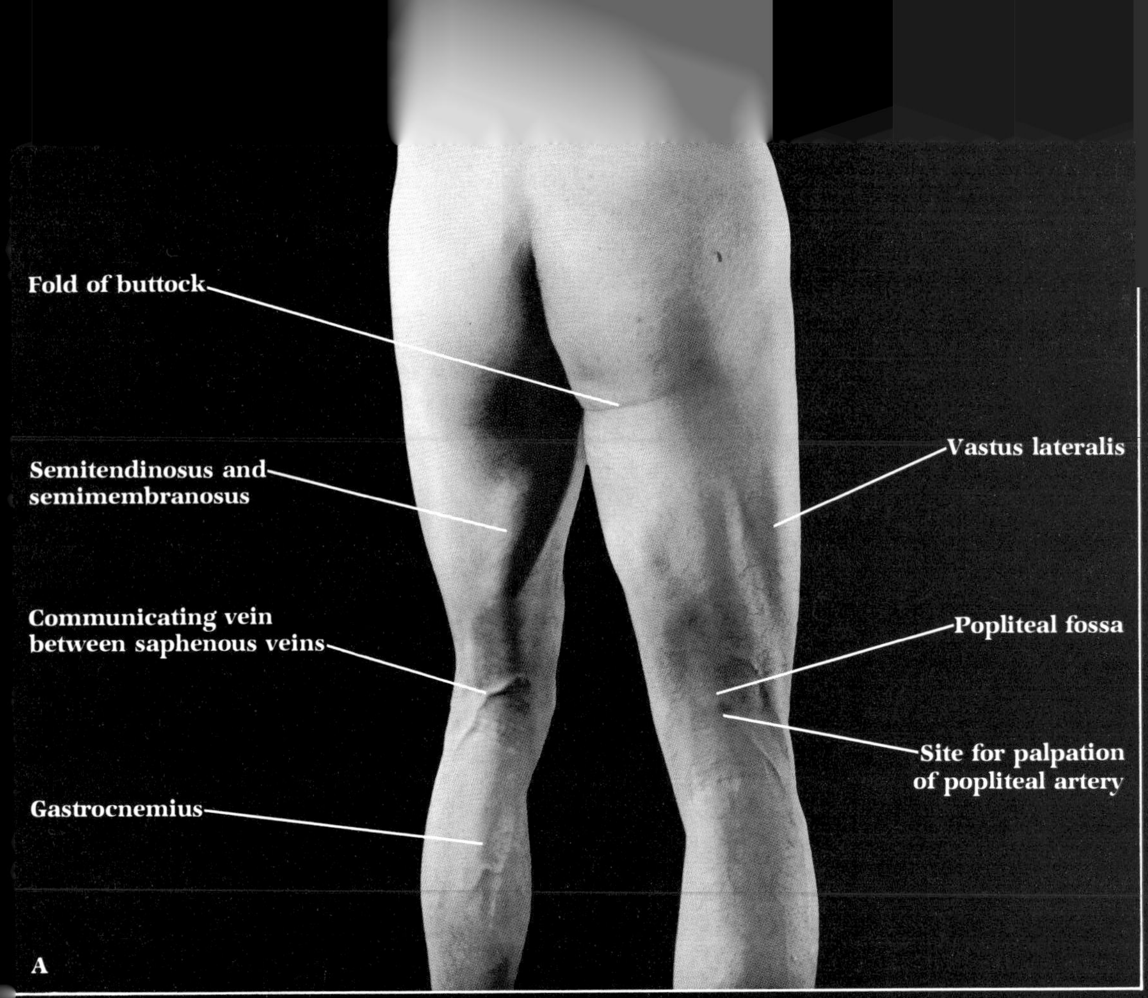
Fold of buttock
Semitendinosus and semimembranosus
Communicating vein between saphenous veins
Gastrocnemius
Vastus lateralis
Popliteal fossa
Site for palpation of popliteal artery
A

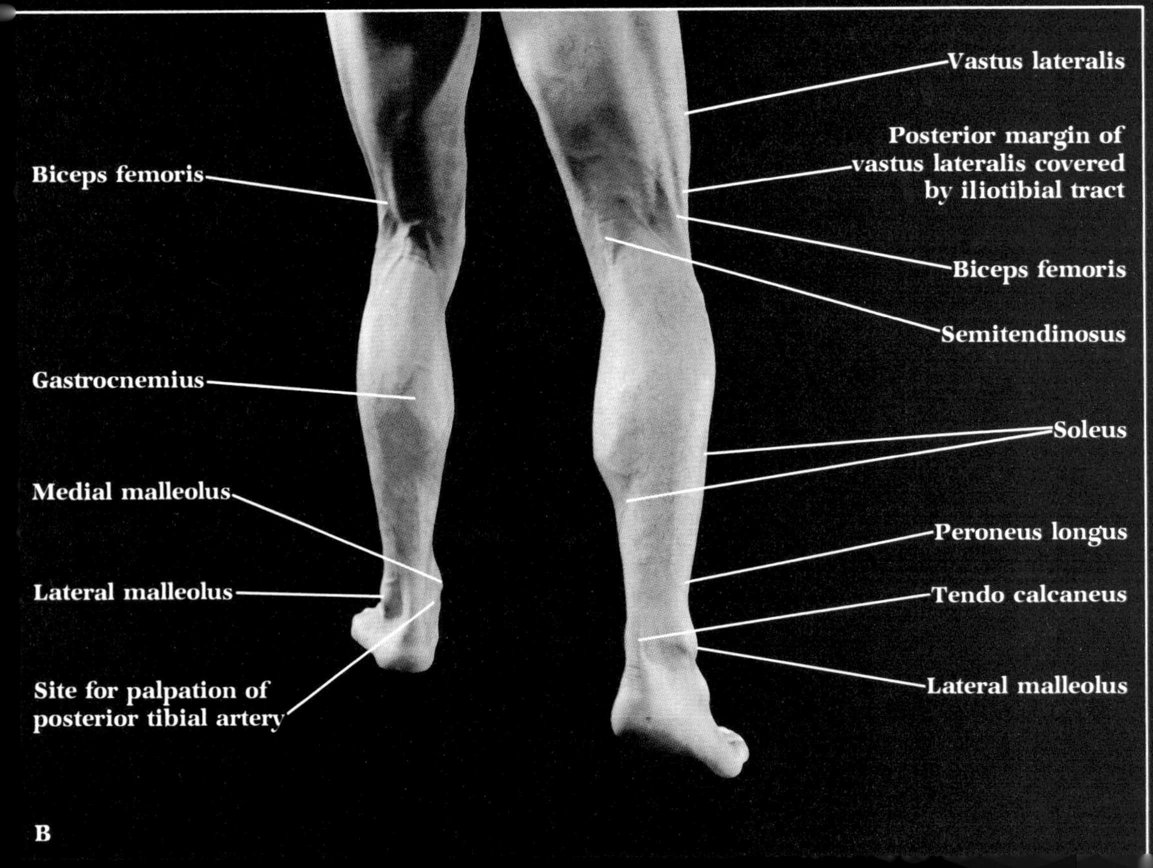
Biceps femoris
Gastrocnemius
Medial malleolus
Lateral malleolus
Site for palpation of posterior tibial artery
Vastus lateralis
Posterior margin of vastus lateralis covered by iliotibial tract
Biceps femoris
Semitendinosus
Soleus
Peroneus longus
Tendo calcaneus
Lateral malleolus
B

Figure 6-42
Posterior aspect of lower limbs in a 25-year-old male. A. Buttocks and thighs. B. Popliteal fossae, calves, and ankles.

1. The fold of the nates, or buttocks, is more prominent when a person is standing; its lower border does not correspond to the lower border of the gluteus maximus muscle.
2. The sciatic nerve in the buttock lies under cover of the gluteus maximus muscle, midway between the greater trochanter and the ischial tuberosity.
3. The biceps femoris, semitendinosus, and the semimembranosus muscles, often referred to as the hamstring muscles, form the large muscle mass on the back of the thigh. Their tendons form prominent cords on the posterior surface of the knee.
4. The popliteal fossa is a diamond-shaped space situated behind the knee. It is bounded superiorly by the hamstring muscles and inferiorly by the two heads of the gastrocnemius muscle.
5. The gastrocnemius and soleus muscles form the large muscle mass on the posterior surface of the calf.
6. In the region of the ankle the medial and lateral malleoli and the tendo calcaneus can easily be felt.
7. The popliteal artery can be palpated in the depths of the popliteal fossa, and the posterior tibial artery can be felt halfway between the medial malleolus and the heel.
8. The common peroneal nerve can be palpated on the medial side of the tendon of the biceps femoris muscle as the muscle passes to its insertion on the head of the fibula.

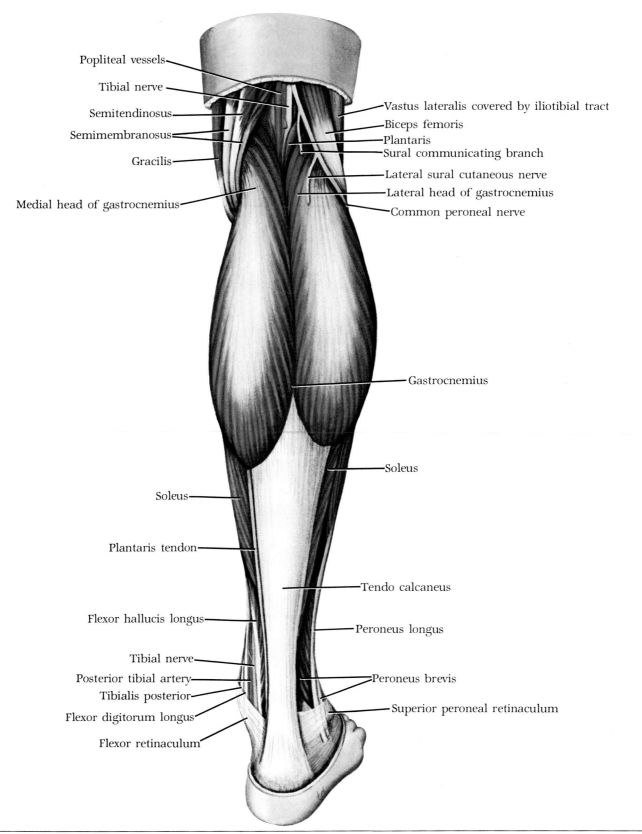

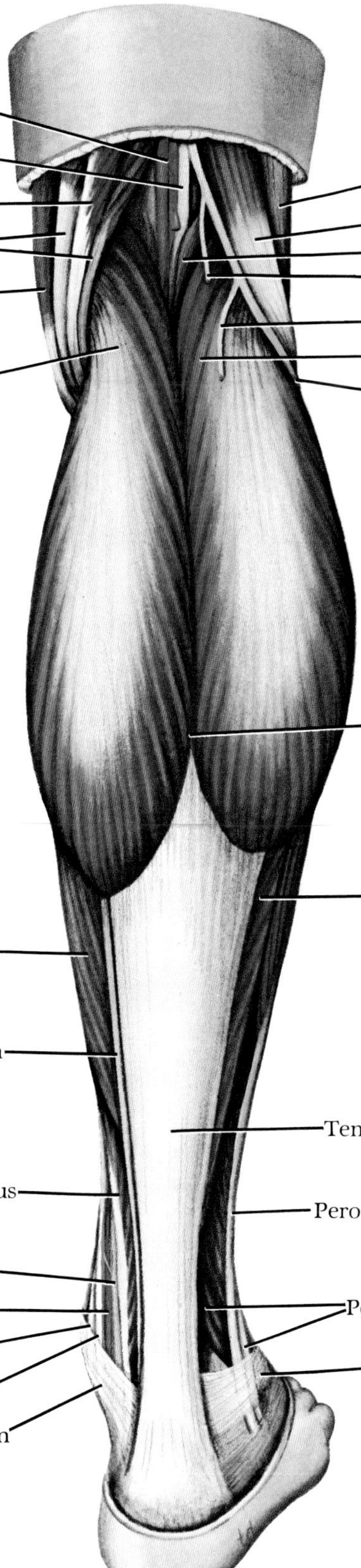
Popliteal vessels
Tibial nerve
Semitendinosus
Semimembranosus
Gracilis
Medial head of gastrocnemius
Vastus lateralis covered by iliotibial tract
Biceps femoris
Plantaris
Sural communicating branch
Lateral sural cutaneous nerve
Lateral head of gastrocnemius
Common peroneal nerve
Gastrocnemius
Soleus
Soleus
Plantaris tendon
Tendo calcaneus
Flexor hallucis longus
Peroneus longus
Tibial nerve
Posterior tibial artery
Peroneus brevis
Tibialis posterior
Superior peroneal retinaculum
Flexor digitorum longus
Flexor retinaculum

Figure 6-43
Posterior aspect of the right leg.

1. In patients with cerebral palsy and spastic contracture of the gastrocnemius and soleus muscles, the ankle joint is plantar-flexed and the foot is adducted (pes equino varus). The muscular imbalance can be greatly improved in severe cases by lengthening the tendo calcaneus, thus allowing the ankle to be dorsiflexed.

2. Rupture of the tendo calcaneus is common in middle-aged men, frequently occurring in tennis players. The rupture occurs at its narrowest part, about 2 inches (5 cm) above its insertion. There is a sudden sharp pain with immediate disability. The gastrocnemius and soleus muscles retract proximally, leaving a palpable gap in the tendon. The tendon should be sutured as soon as possible and the leg immobilized in plaster with the ankle joint plantar-flexed and the knee joint flexed.

3. Rupture of the plantaris tendon is rare, although frequently diagnosed as such a rupture it is usually due to tearing of the fibers of the soleus or partial tearing of the tendo calcaneus.

4. Bursitis of the subcutaneous adventitial bursa over the insertion of the tendo calcaneus or of the bursa between the tendo calcaneus and the calcaneum may occur as the result of poorly fitting shoes.

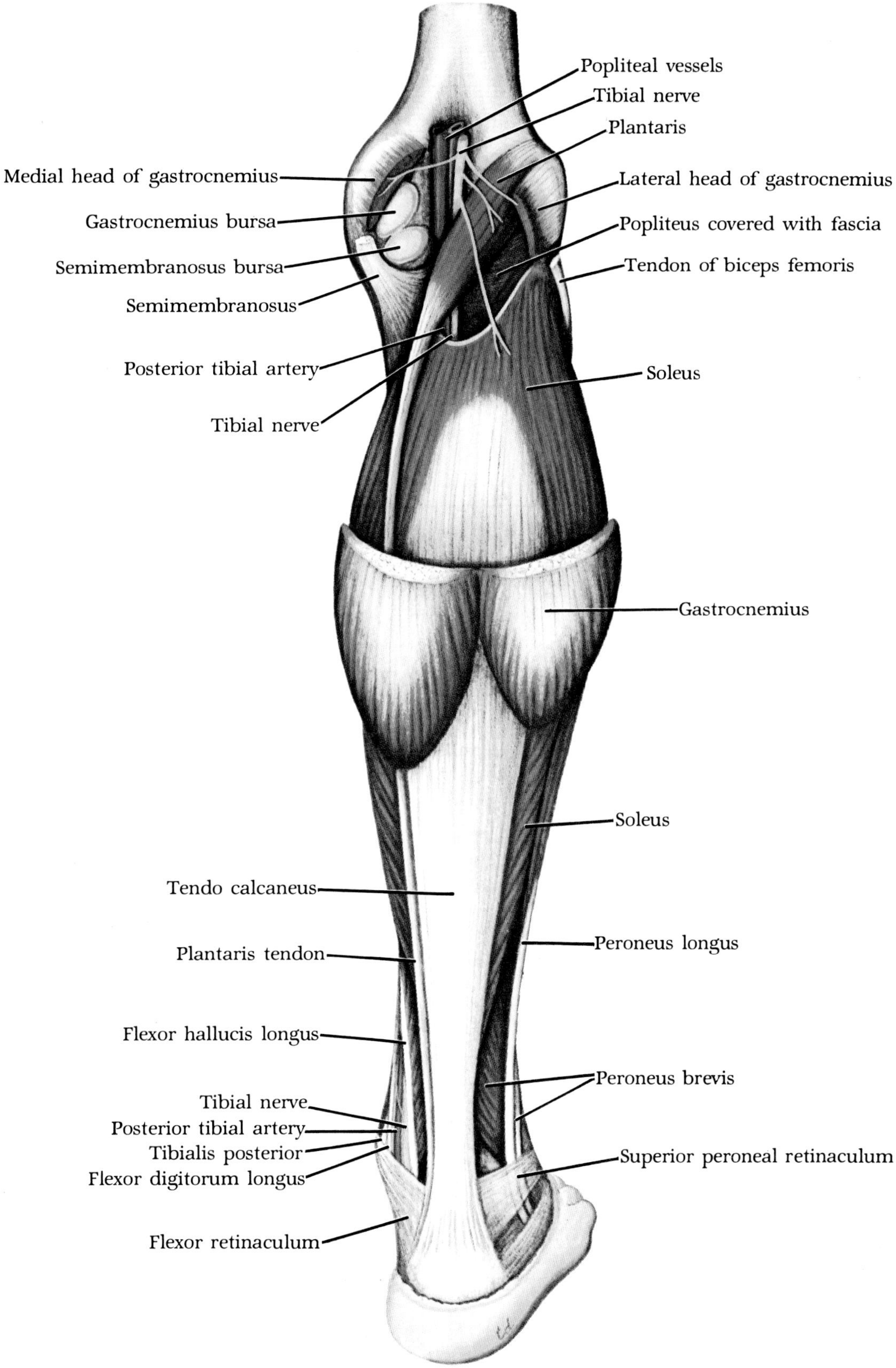
Popliteal vessels
Tibial nerve
Plantaris
Medial head of gastrocnemius
Lateral head of gastrocnemius
Gastrocnemius bursa
Popliteus covered with fascia
Semimembranosus bursa
Tendon of biceps femoris
Semimembranosus
Posterior tibial artery
Soleus
Tibial nerve
Gastrocnemius
Soleus
Tendo calcaneus
Peroneus longus
Plantaris tendon
Flexor hallucis longus
Peroneus brevis
Tibial nerve
Posterior tibial artery
Tibialis posterior
Superior peroneal retinaculum
Flexor digitorum longus
Flexor retinaculum

Figure 6-44
Posterior view of the right leg. Part of the gastrocnemius muscle has been removed.

1. The gastrocnemius and soleus muscles are the chief plantar flexors of the ankle joint. The gastrocnemius is believed to be mainly used in walking and jumping while the soleus is more concerned with stabilizing the ankle joint when a person is standing. Both muscles are supplied by the tibial nerve.

2. When present (it is often missing) the plantaris is a weak flexor of the knee joint and assists the gastrocnemius and soleus muscles in plantar-flexing the ankle joint. It represents a rudimentary muscle which, in lower animals, is inserted into the plantar aponeurosis. It can be used clinically for tendon autografts when repairing severed flexor tendons to the fingers; the tendon of the palmaris longus muscle can also be used for this purpose.

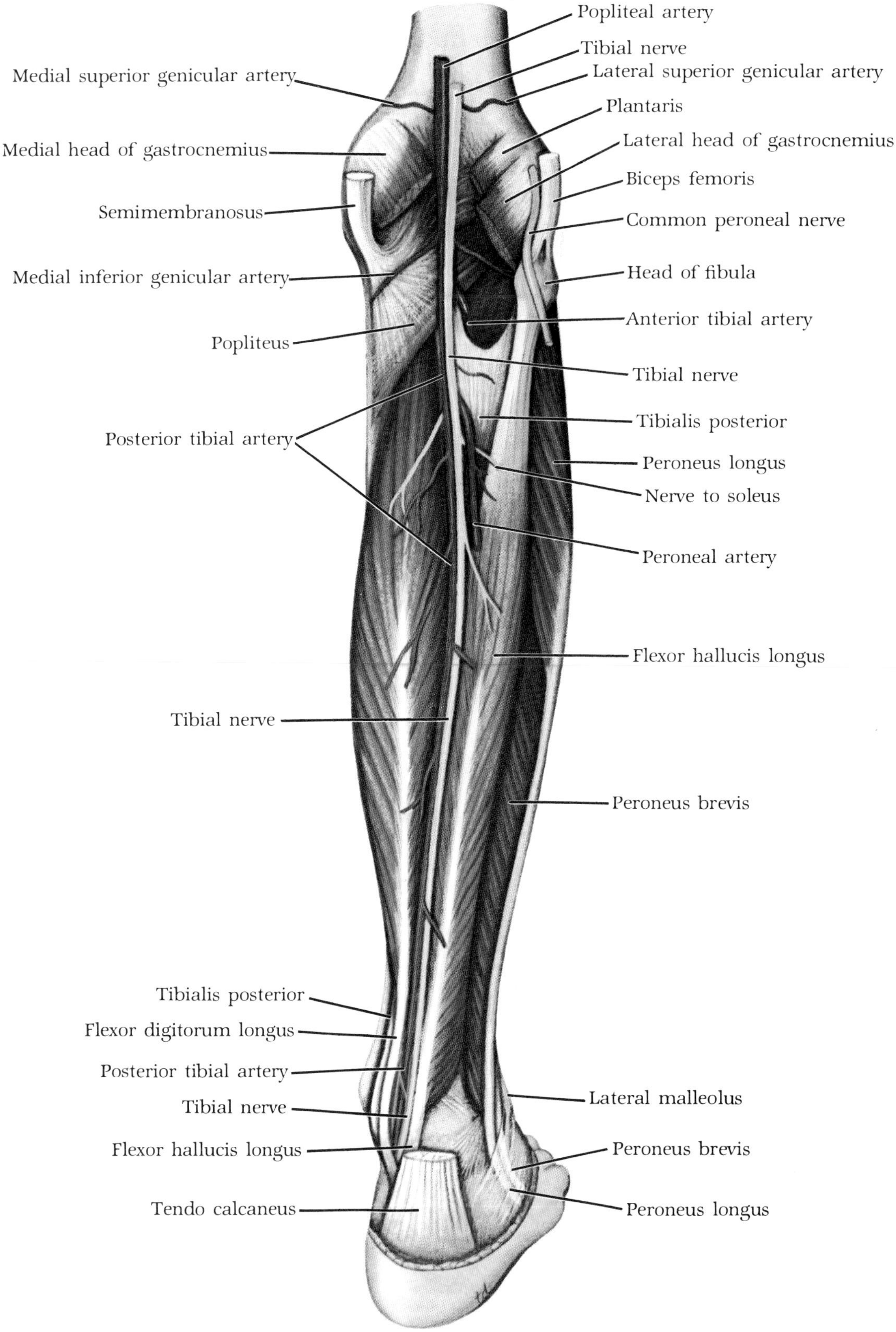
Popliteal artery
Tibial nerve
Lateral superior genicular artery
Plantaris
Lateral head of gastrocnemius
Biceps femoris
Common peroneal nerve
Head of fibula
Anterior tibial artery
Tibial nerve
Tibialis posterior
Peroneus longus
Nerve to soleus
Peroneal artery
Flexor hallucis longus
Peroneus brevis
Lateral malleolus
Peroneus brevis
Peroneus longus
Medial superior genicular artery
Medial head of gastrocnemius
Semimembranosus
Medial inferior genicular artery
Popliteus
Posterior tibial artery
Tibial nerve
Tibialis posterior
Flexor digitorum longus
Posterior tibial artery
Tibial nerve
Flexor hallucis longus
Tendo calcaneus

Figure 6-45
Deep structures present on posterior aspect of right leg.

1. The tibial nerve leaves the popliteal fossa by passing deep to the gastrocnemius and soleus muscles. Because of its deep and protected position, it is rarely injured. Complete division, however, results in the following clinical features:

Motor. All the muscles in the posterior compartment of the leg and the sole of the foot are paralyzed. The opposing muscles dorsiflex the foot at the ankle joint and evert the foot at the subtalar and transverse tarsal joints, a deformity referred to as calcaneovalgus.

Sensory. There is loss of sensation on the sole of the foot, which inevitably results in the development of trophic ulcers.

If the lesion occurs to the nerve in the lower third of the leg, the muscles in the posterior compartment of the leg will be spared.

2. Diabetic neuropathy of the tibial nerve results in hypalgesia or anesthesia of parts of the sole of the foot that is later complicated by trophic ulcers. In diabetics these ulcers often serve as a site of entry for infection.

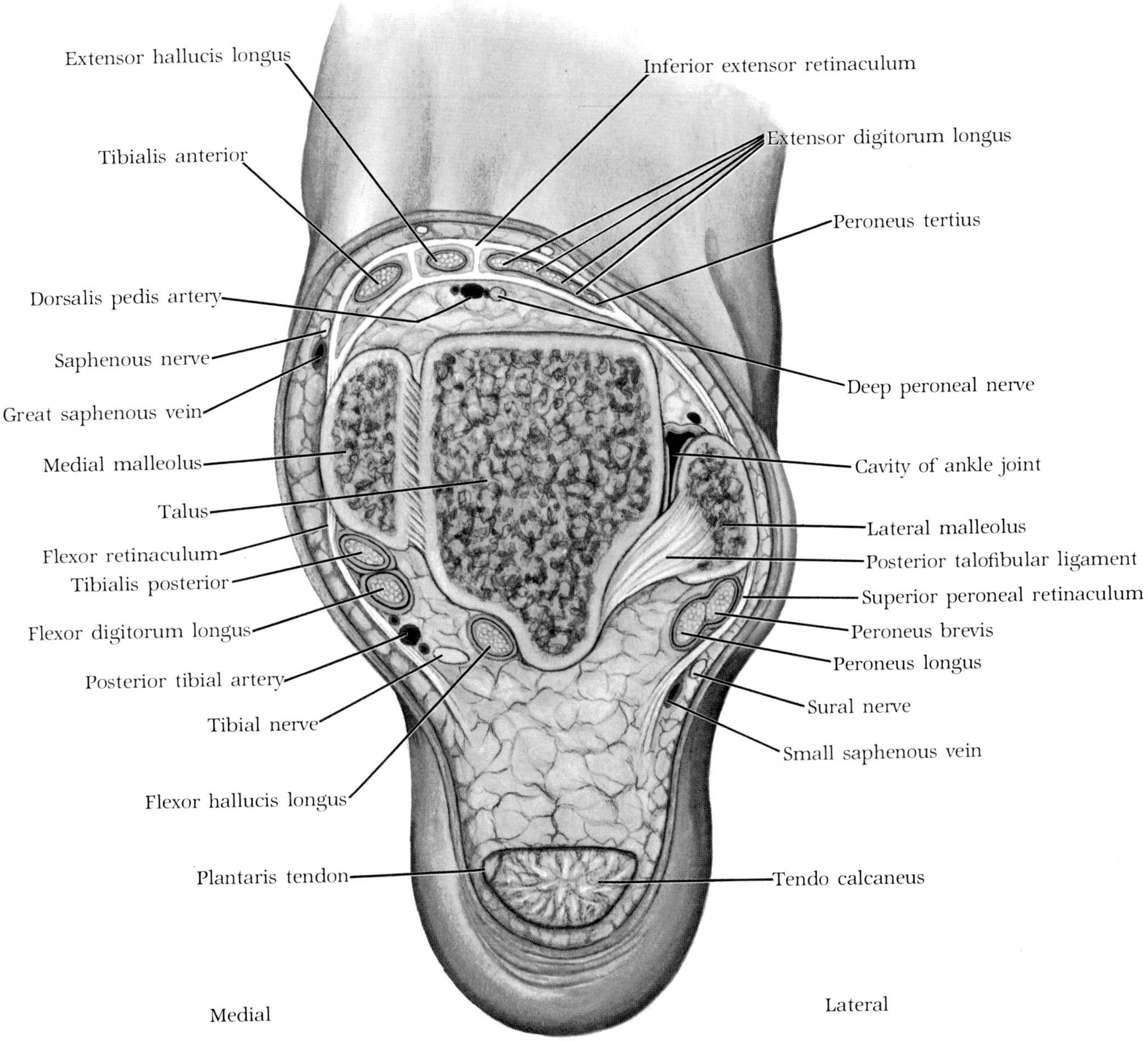
Extensor hallucis longus
Inferior extensor retinaculum
Extensor digitorum longus
Tibialis anterior
Peroneus tertius
Dorsalis pedis artery
Saphenous nerve
Deep peroneal nerve
Great saphenous vein
Medial malleolus
Cavity of ankle joint
Talus
Lateral malleolus
Posterior talofibular ligament
Flexor retinaculum
Tibialis posterior
Superior peroneal retinaculum
Flexor digitorum longus
Peroneus brevis
Peroneus longus
Posterior tibial artery
Sural nerve
Tibial nerve
Small saphenous vein
Flexor hallucis longus
Plantaris tendon
Tendo calcaneus
Medial
Lateral

Figure 6-46
Relations of right ankle joint.

1. The constant positions of the great saphenous vein anterior to the medial malleolus and the small saphenous vein posterior to the lateral malleolus must again be emphasized. This information can be life-saving for patients who require an emergency blood transfusion.

2. The positions of the dorsalis pedis artery on the dorsum of the foot and the posterior tibial artery posterior to the medial malleolus should be noted. Palpation of their pulses is important in patients with arterial disease of the lower limb.

3. As can be seen from the figure the ankle joint is very stable and dislocation is rare unless one of the malleoli is fractured.

4. Tenosynovitis can affect the sheaths of the extensor tendons and those of the peroneus longus and brevis muscles; the remaining sheaths are rarely affected. Treatment of tenosynovitis consists of immobilization, heat, and physiotherapy.

5. The tendons of the peroneus longus and brevis muscles can be dislocated forward from their normal position posterior to the lateral malleolus. For this condition to occur the superior peroneal retinaculum must be torn. It usually occurs in older children and is caused by trauma in most cases.

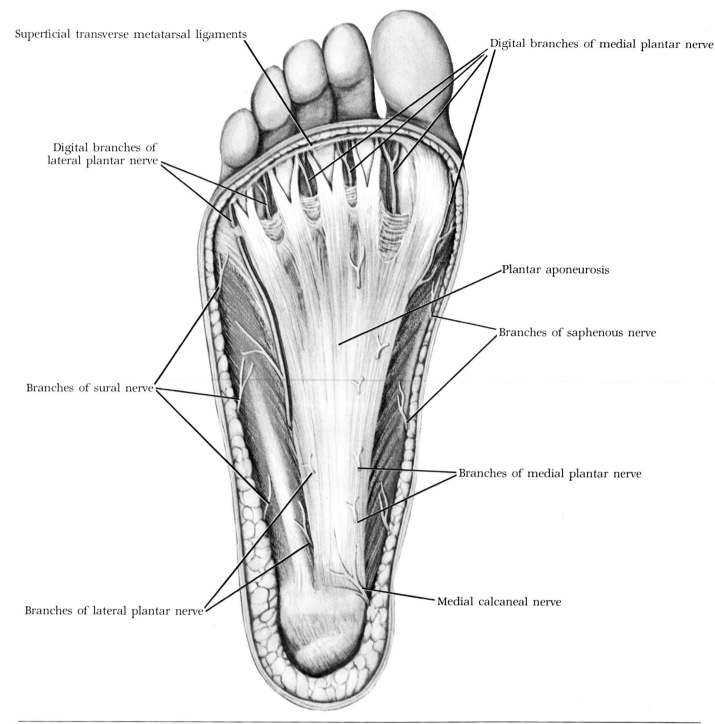

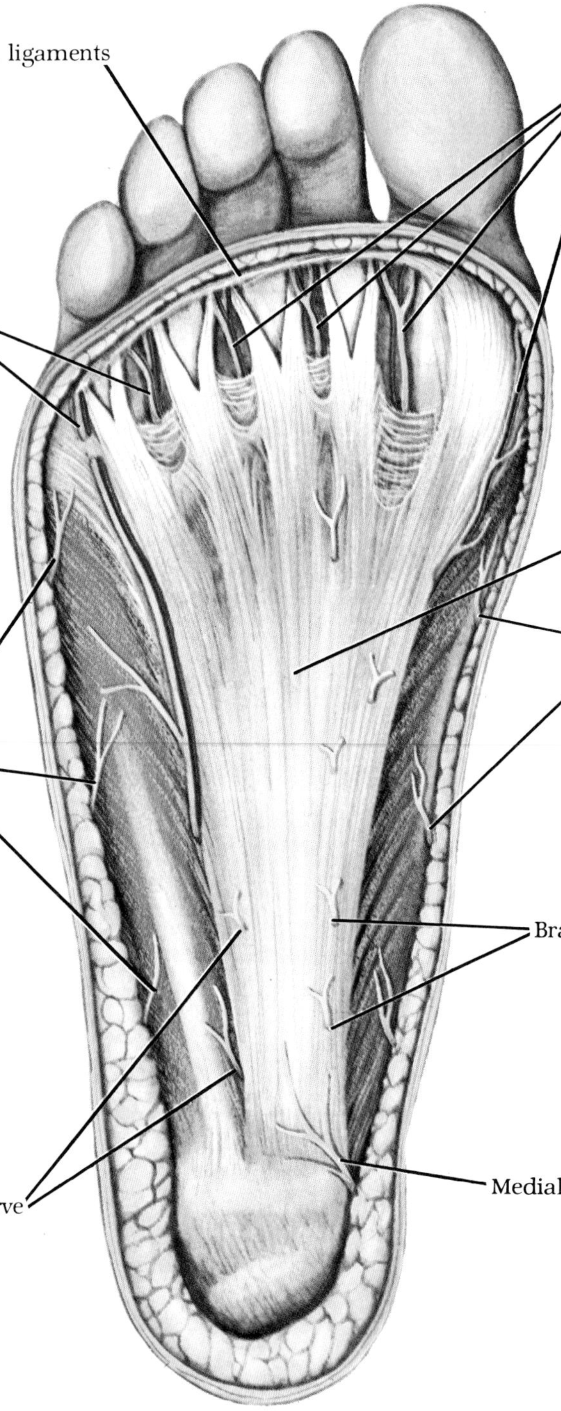
Superficial transverse metatarsal ligaments
Digital branches of medial plantar nerve
Digital branches of lateral plantar nerve
Plantar aponeurosis
Branches of saphenous nerve
Branches of sural nerve
Branches of medial plantar nerve
Medial calcaneal nerve
Branches of lateral plantar nerve

Figure 6-47
Plantar aponeurosis and cutaneous nerves of sole of right foot.

1. The sensory nerve supply to the skin of the sole of the foot is derived from the medial calcaneal branch of the tibial nerve, which innervates the medial side of the heel; branches from the medial plantar nerve, which innervate the medial two-thirds of the sole; and branches from the lateral plantar nerve, which innervate the lateral third of the sole. The saphenous nerve innervates the skin along the medial border of the foot as far as the ball of the big toe; the sural nerve innervates the skin along the entire length of the lateral border of the foot and little toe.
2. Neurotrophic ulcers of the sole of the foot are easily recognized by their punched-out appearance and location over pressure points, usually the metatarsal heads. Injury to the spinal cord, sciatic nerve, or tibial nerve can cause them in some patients. Diabetic neuropathy and tabes dorsalis can be responsible in others. The ulcers are completely anesthetic.
3. Dupuytren's contracture commonly involves the palmar aponeurosis and usually starts near the root of the ring finger, drawing that finger into the palm and flexing it at the metacarpophalangeal joint. A small number of patients also have some degree of contracture of the plantar aponeurosis, but it is usually confined to one foot and is without symptoms.
4. Plantar fasciitis, which occurs in individuals who do a lot of standing or walking, produces pain and tenderness of the sole of the foot. It is believed to be due to trauma or a pulling on the plantar aponeurosis. Repeated attacks of this condition induce ossification in the posterior attachment of the aponeurosis, forming a calcaneal spur.

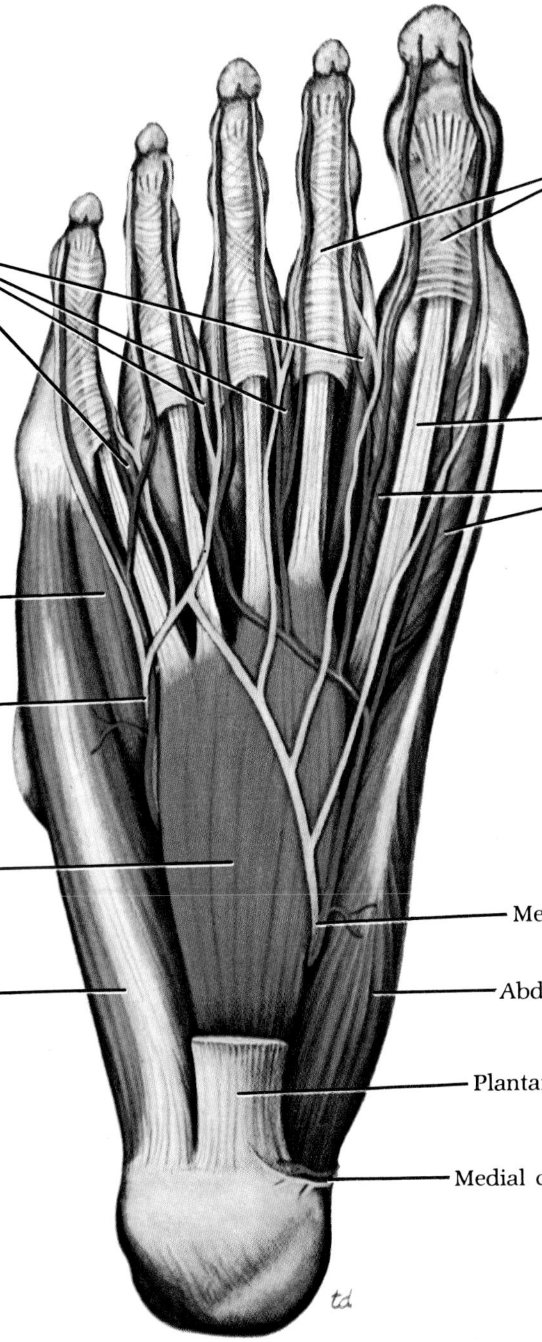

Fibrous flexor sheaths
Lumbricals
Flexor hallucis longus
Flexor hallucis brevis
Flexor digiti minimi brevis
Lateral plantar nerve and artery
Flexor digitorum brevis
Medial plantar nerve and artery
Abductor digiti minimi
Abductor hallucis
Plantar aponeurosis
Medial calcaneal nerve and artery
td

Figure 6-48
Plantar muscles of right foot, first layer. Medial and lateral plantar arteries and nerves are shown. The greater part of the plantar aponeurosis has been removed.

1. The muscles of the first layer (the abductor hallucis, flexor digitorum brevis, and abductor digiti minimi) extend from the calcaneum to the forefoot and serve as ties to brace the medial and lateral longitudinal arches, thus assisting in their support.

2. Hammer toe is a deformity commonly involving the second toe. The proximal phalanx is dorsiflexed, the middle phalanx is plantar-flexed, and the distal phalanx is dorsiflexed or plantar-flexed. The condition is often bilateral and is due to overcrowding the toes in poorly fitting shoes.
3. Fracture of the sesamoid bones of the great toe may follow sudden falls upon the feet or the dropping of a heavy weight on the forefoot.
4. Overriding toes is a congenital deformity most commonly involving the fourth and fifth toes. The fourth toe is depressed and overridden by the fifth toe. It can be corrected by applying plaster splints.

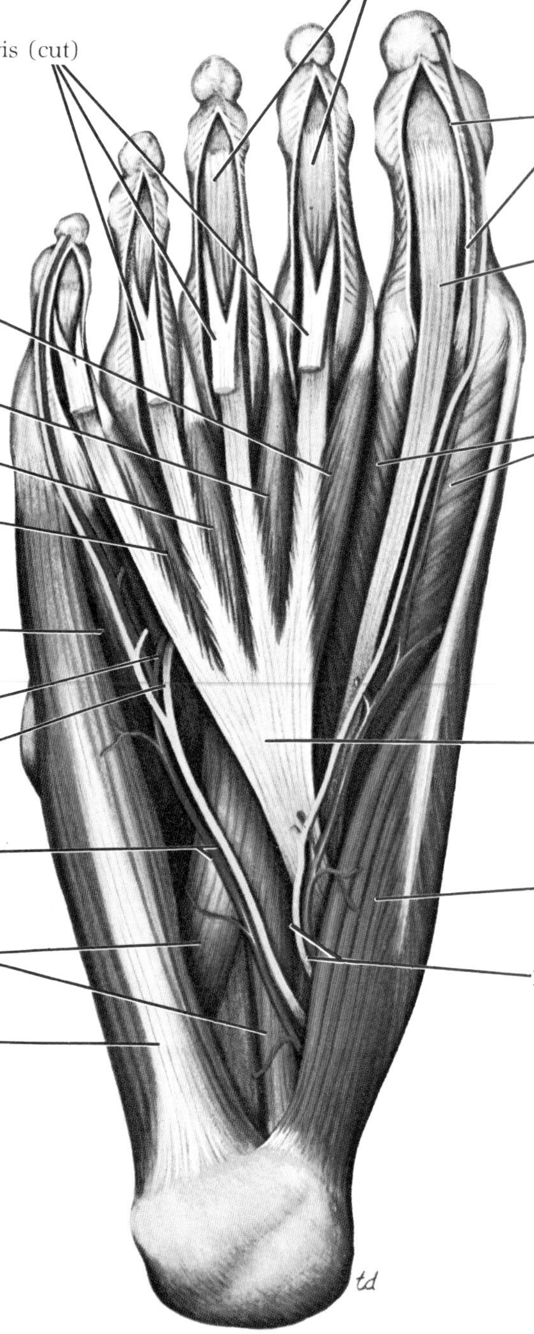
Tendons of flexor digitorum longus
Tendons of flexor digitorum brevis (cut)
Fibrous flexor sheath (cut open)
Flexor hallucis longus
First lumbrical
Second lumbrical
Third lumbrical
Fourth lumbrical
Flexor hallucis brevis
Flexor digiti minimi brevis
Plantar arch
Deep branch of
lateral plantar nerve
Flexor digitorum longus
Lateral plantar nerve and artery
Abductor hallucis
Quadratus plantae
Medial plantar nerve and artery
Abductor digiti minimi
td

Figure 6-49
Plantar muscles of right foot, second layer. Medial and lateral plantar arteries and nerves are shown. The abductor hallucis and abductor digiti minimi muscles of the first layer have been left in position.

1. The muscles of the second layer (the flexor digitorum longus, flexor hallucis longus, lumbricals, and quadratus plantae) serve as ties to brace the medial and lateral longitudinal arches, thus assisting in their support.

2. The quadratus plantae pulls the tendon of the flexor digitorum longus directly posteriorly, thus aiding this muscle to flex the lateral four toes. It can also act alone and flex these toes, using the tendon of the flexor digitorum longus when the belly of this muscle is relaxed.

3. The final thrust forward of the foot in walking is provided by the toes being strongly flexed by the long and short flexors of the foot.

4. The lumbricals prevent the lateral four toes from buckling under during walking or running by extending the toes at the interphalangeal joints when the flexor digitorum longus tendons are flexing the toes.

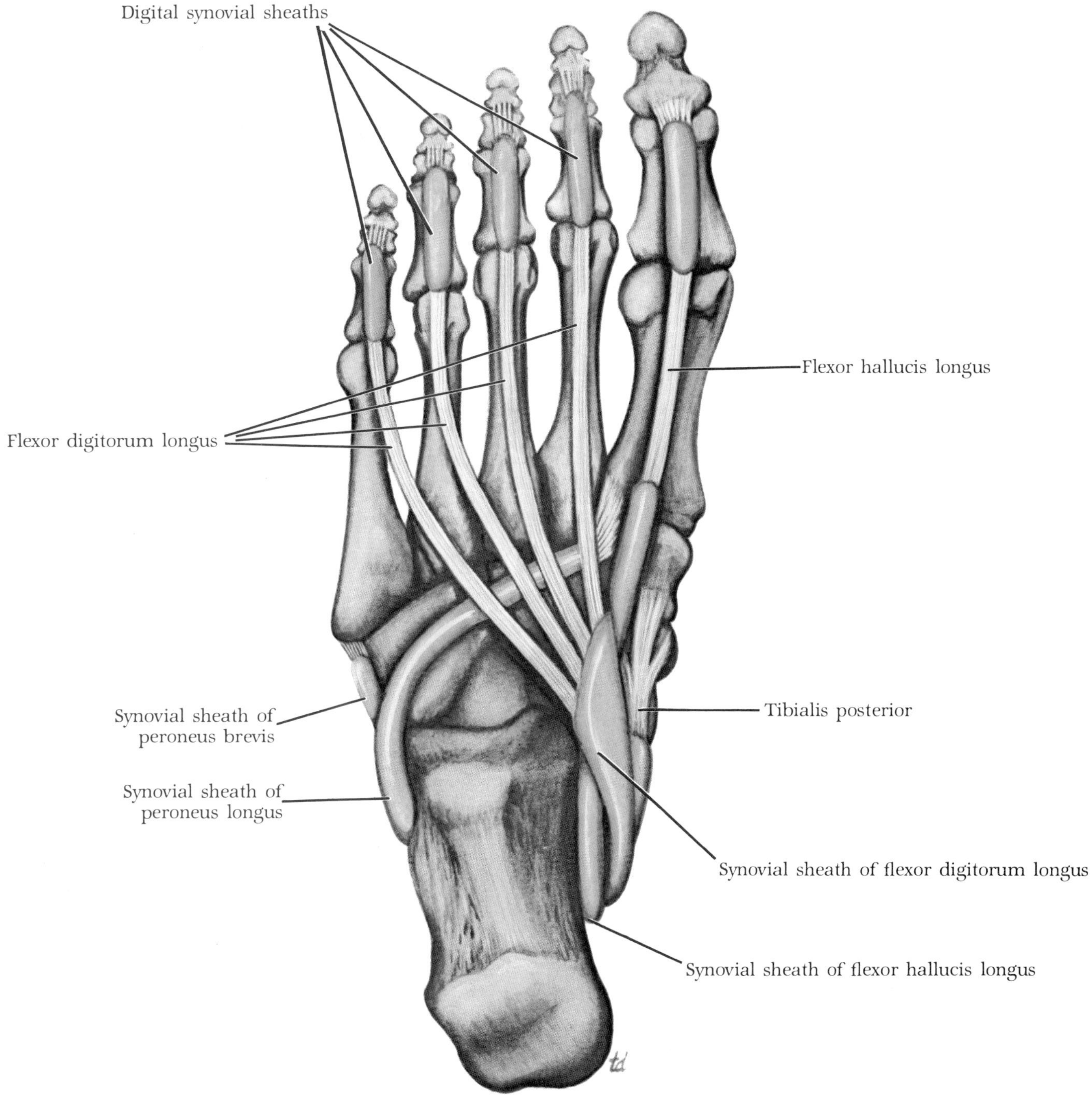
Digital synovial sheaths
Flexor hallucis longus
Flexor digitorum longus
Synovial sheath of peroneus brevis
Tibialis posterior
Synovial sheath of peroneus longus
Synovial sheath of flexor digitorum longus
Synovial sheath of flexor hallucis longus
td

Figure 6-50
Synovial sheaths of tendons present on sole of right foot.

1. The synovial sheaths in the sole of the foot are not as important as those in the hand. They rarely become injured or infected and are not continuous with the digital synovial sheaths.

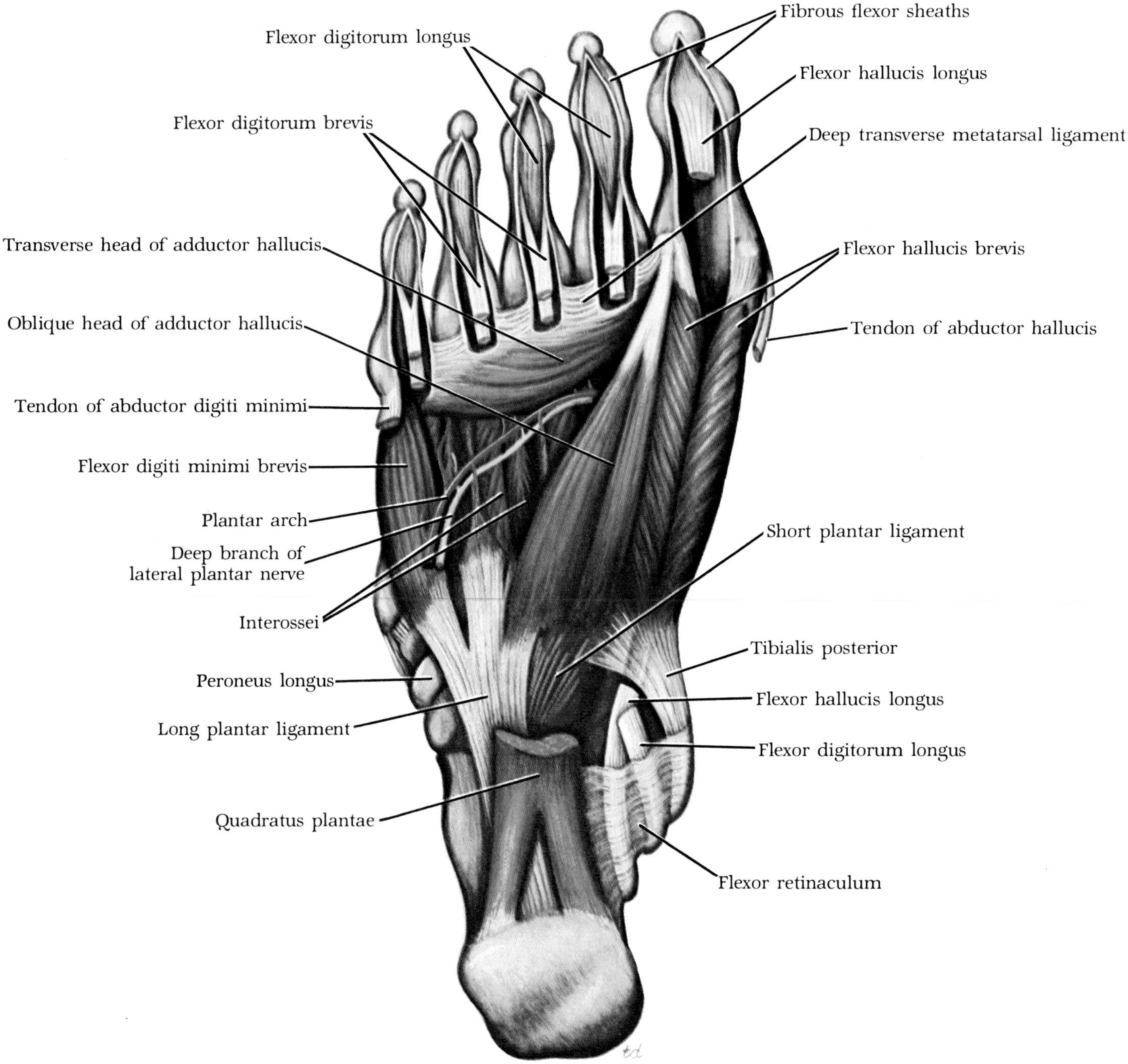
Fibrous flexor sheaths
Flexor digitorum longus
Flexor hallucis longus
Flexor digitorum brevis
Deep transverse metatarsal ligament
Transverse head of adductor hallucis
Flexor hallucis brevis
Oblique head of adductor hallucis
Tendon of abductor hallucis
Tendon of abductor digiti minimi
Flexor digiti minimi brevis
Plantar arch
Short plantar ligament
Deep branch of lateral plantar nerve
Interossei
Tibialis posterior
Peroneus longus
Flexor hallucis longus
Long plantar ligament
Flexor digitorum longus
Quadratus plantae
Flexor retinaculum

Figure 6-51
Plantar muscles of right foot, third layer. Deep branch of lateral plantar nerve and plantar arterial arch are also shown.

1. The muscles of the third layer (the flexor hallucis brevis, adductor hallucis, and flexor digiti minimi brevis) assist in supporting the medial and lateral longitudinal arches. The transverse head of the adductor hallucis muscle holds together the heads of the metatarsal bones and assists in stabilizing the forepart of the foot; it thus helps to support the transverse arch of the foot.

2. Pes planus (flatfoot) is a condition in which the medial longitudinal arch is depressed or collapsed. As a result, the forefoot is displaced laterally and everted. The head of the talus is no longer supported, and body weight forces it downward and medially between the calcaneum and the navicular bone. When the deformity has existed for some time, the plantar, calcaneonavicular, and medial ligaments of the ankle joint become permanently stretched, and the bones change shape. The muscles and tendons are also permanently stretched. There are both congenital and acquired causes of flatfoot.

3. Pes cavus (clawfoot) is a condition in which the medial longitudinal arch is unduly high. Most cases are due to muscle imbalance, in many instances a result of poliomyelitis. Bilateral clawfoot commonly accompanies lesions of the central nervous system such as Friedreich's ataxia and peroneal muscular atrophy.

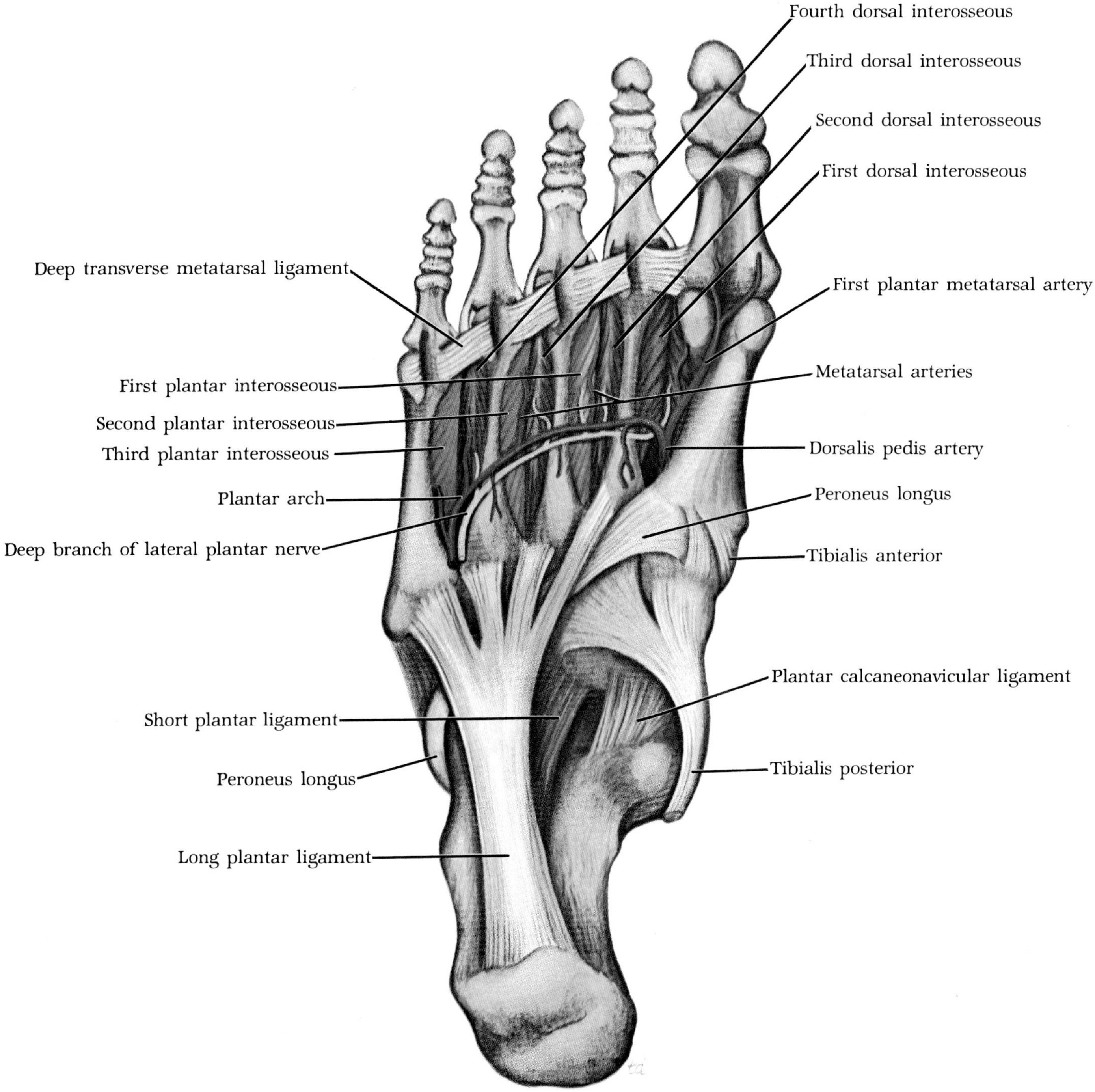
Fourth dorsal interosseous
Third dorsal interosseous
Second dorsal interosseous
First dorsal interosseous
Deep transverse metatarsal ligament
First plantar metatarsal artery
First plantar interosseous
Metatarsal arteries
Second plantar interosseous
Third plantar interosseous
Dorsalis pedis artery
Plantar arch
Peroneus longus
Deep branch of lateral plantar nerve
Tibialis anterior
Plantar calcaneonavicular ligament
Short plantar ligament
Tibialis posterior
Peroneus longus
Long plantar ligament

Figure 6-52
Plantar muscles of right foot, fourth layer. Deep branch of lateral plantar nerve and plantar arterial arch are shown. Note the insertions of the tibialis posterior and peroneus longus muscles.

1. The main function of the interossei is in walking. They flex the metatarsophalangeal joints and are thus used in the final forward propulsion of the foot; they extend the interphalangeal joints, thus preventing the toes from curling up. Since the dorsal interossei arise from two metatarsal bones, they serve to bind together the metatarsals and prevent spreading of the forefoot.
2. Morton's metatarsalgia is a severe pain of the sole of the foot which spreads to the third and fourth toes. The cause is a neuroma of the digital branch of the medial plantar nerve that supplies the skin of the cleft between the third and fourth toes. The pain occurs when the neuroma is compressed as a person walks.

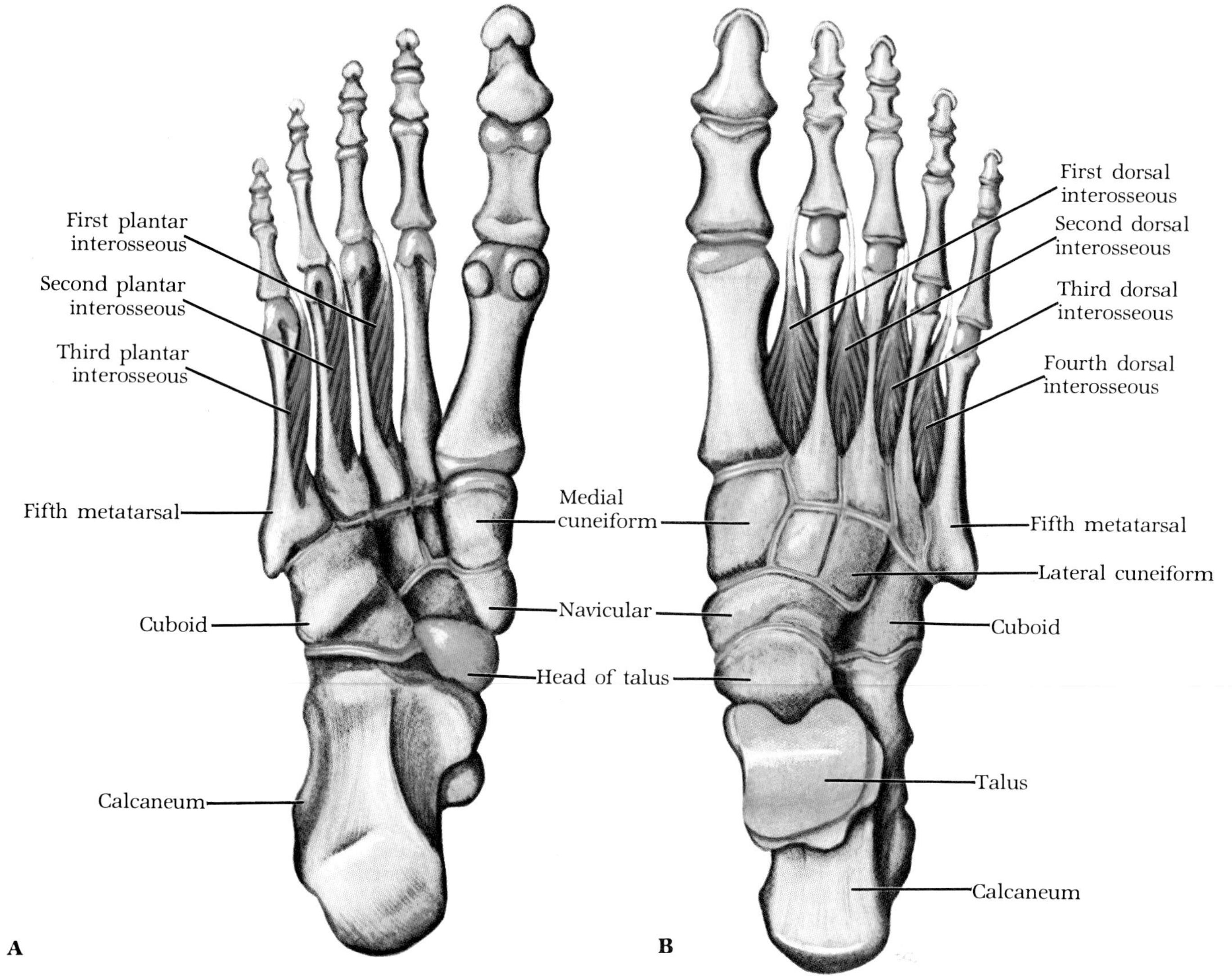
First plantar interosseous
Second plantar interosseous
Third plantar interosseous
Fifth metatarsal
Cuboid
Calcaneum
Medial cuneiform
Navicular
Head of talus
First dorsal interosseous
Second dorsal interosseous
Third dorsal interosseous
Fourth dorsal interosseous
Fifth metatarsal
Lateral cuneiform
Cuboid
Talus
Calcaneum
A
B

Figure 6-53
Right foot. A. Plantar view, showing the plantar interossei. B. Dorsal view, showing the dorsal interossei.

1. The plantar interossei adduct the toes toward the center of the second toe at the metatarsophalangeal joints. The dorsal interossei abduct the toes away from the center of the second toe at the metatarsophalangeal joints. The main function of the interossei is in walking. They flex the metatarsophalangeal joints and are thus used in the final forward propulsion of the foot; they extend the interphalangeal joints, thus preventing the toes from curling up. Since the dorsal interossei arise from two metatarsal bones, they serve to bind together the metatarsals and prevent spreading of the forefoot.
2. All the interossei are supplied by the lateral plantar nerve.
3. The arches of the foot are maintained by (a) the shape of the bones; (b) strong ligaments; and (c) muscle tone. In the standing but not moving foot the small muscles play little or no part in supporting the arches, thus forcing the bones and the ligaments to take the strain. During walking and running, however, all the small muscles of the foot become active as well as the anterior and posterior tibialis and peroneus longus muscles.

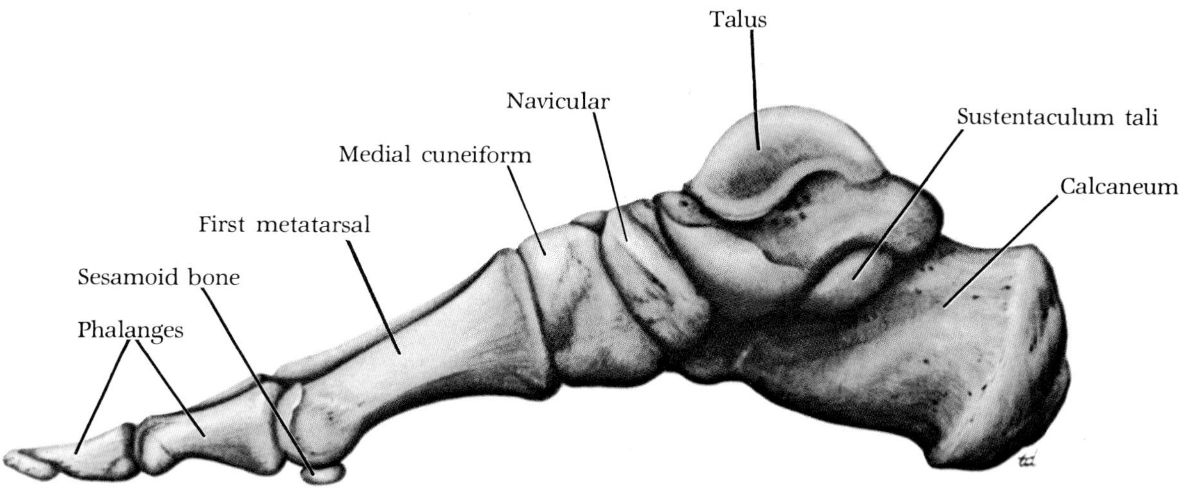
Talus
Navicular
Sustentaculum tali
Medial cuneiform
Calcaneum
First metatarsal
Sesamoid bone
Phalanges

A

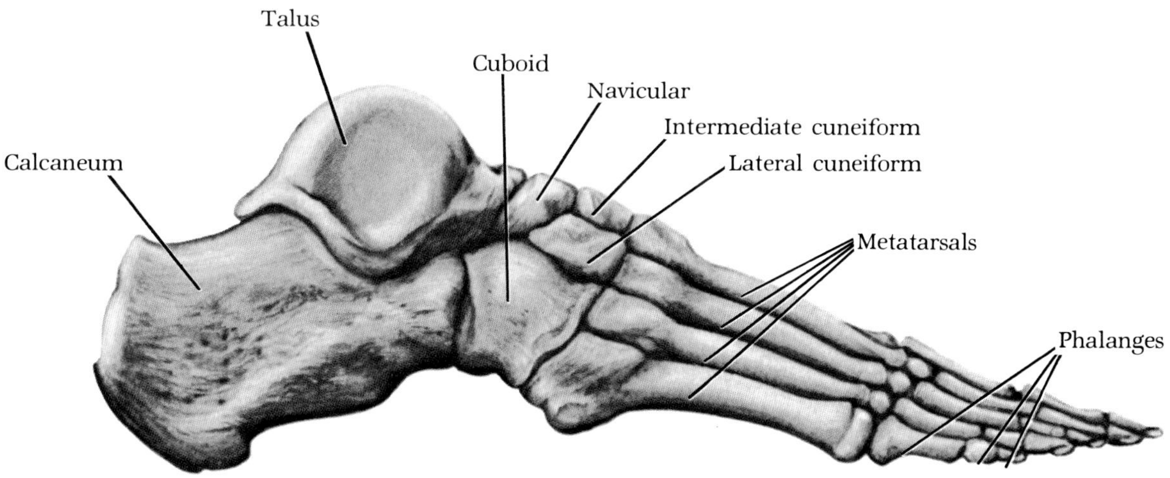
Talus
Cuboid
Navicular
Intermediate cuneiform
Calcaneum
Lateral cuneiform
Metatarsals
Phalanges

B

Figure 6-54
Bones of the right foot.
A. Medial view.
B. Lateral view.

1. The Arches of the Foot. A segmented structure can hold up weight only if it is built in the form of an arch. In the foot there are three arches present at birth.
Medial Longitudinal Arch. This is the largest and consists of the calcaneum, the talus, the navicular, the three cuneiforms, and the first three metatarsal bones.
Lateral Longitudinal Arch. This consists of the calcaneum, the cuboid, and the fourth and fifth metatarsal bones.
Transverse Arch. This arch is incomplete since its medial support is missing. It consists of the bases of the metatarsal bones, the cuboid, and the three cuneiform bones.

2. The Bearing Points of the Foot. Posteriorly there is the calcaneum and anteriorly the heads of the five metatarsal bones. The two sesamoid bones below the head of the first metatarsal raise that head away from the ground and permit the free movement of the tendon of the flexor hallucis longus muscle.
3. The arches of the foot are supported by the shape of the bones, the strong ligaments, especially those on the plantar surface of the foot, and the tone of the muscles. In the active foot the tone of the muscles is very important in supporting the arches. When the muscles are fatigued by excessive exercise (e.g., after a route march by an army recruit), by standing for long periods (e.g., in waitresses or nurses), by overweight, or by illness, the muscular support gives way, the ligaments are stretched, and pain is produced.

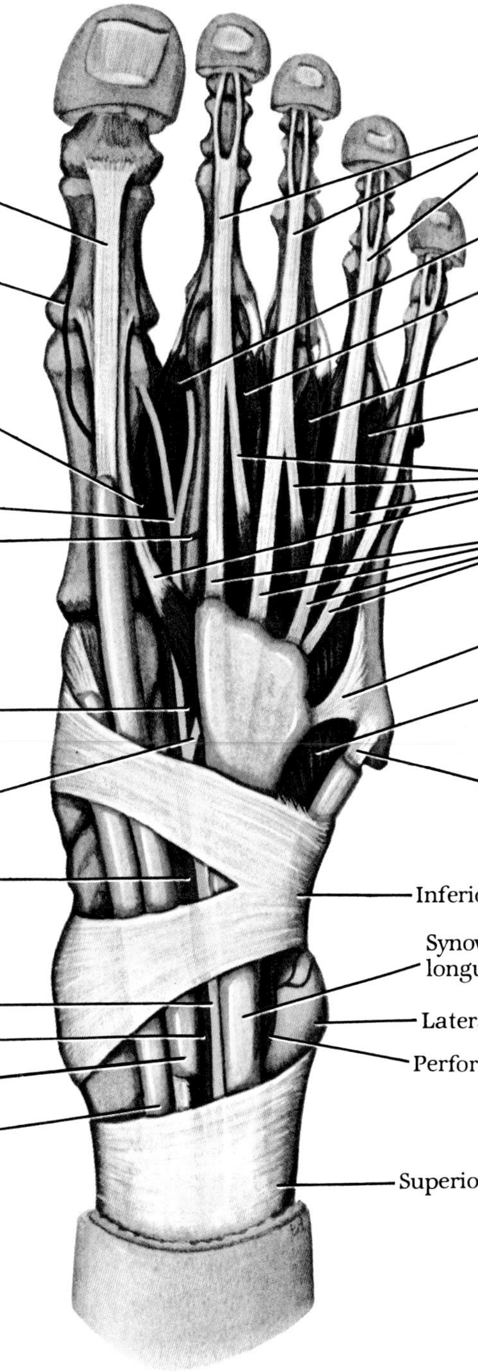
Extensor expansion
Extensor hallucis longus
First dorsal interosseous
Digital artery
Second dorsal interosseous
Third dorsal interosseous
First dorsal metatarsal artery
Fourth dorsal interosseous
Medial terminal branch of deep peroneal nerve
Extensor digitorum brevis tendons
Arcuate artery
Extensor digitorum longus tendons
Peroneus tertius
Extensor digitorum brevis
Lateral tarsal artery
Peroneus brevis
Nerve to extensor digitorum brevis
Dorsalis pedis artery
Inferior extensor retinaculum
Synovial sheath of extensor digitorum longus and peroneus tertius
Deep peroneal nerve
Lateral malleolus
Anterior tibial artery
Perforating branch of peroneal artery
Synovial sheath of extensor hallucis longus
Synovial sheath of tibialis anterior
Superior extensor retinaculum

Figure 6-55
Structures present on the dorsal aspect of the right foot.

1. A knowledge of how to palpate the pulses of the lower limb arteries is very important when seeking a diagnosis in patients with suspected arterial disease. The dorsalis pedis artery is situated between the tendons of the extensor hallucis longus and extensor digitorum longus muscles on the dorsal surface of the foot. It is covered only by skin and fascia and is easily felt. The anterior tibial artery, which lies between the two malleoli, becomes superficial just superior to the ankle joint.

2. It should be remembered that the dorsalis pedis artery is sometimes absent, being replaced by a large perforating branch of the peroneal artery.

3. A ganglion of the foot, usually occurring on the dorsum, is similar to one arising in the hand. It is a cystic swelling produced by a myxomatous degeneration of connective tissue associated with ligaments.

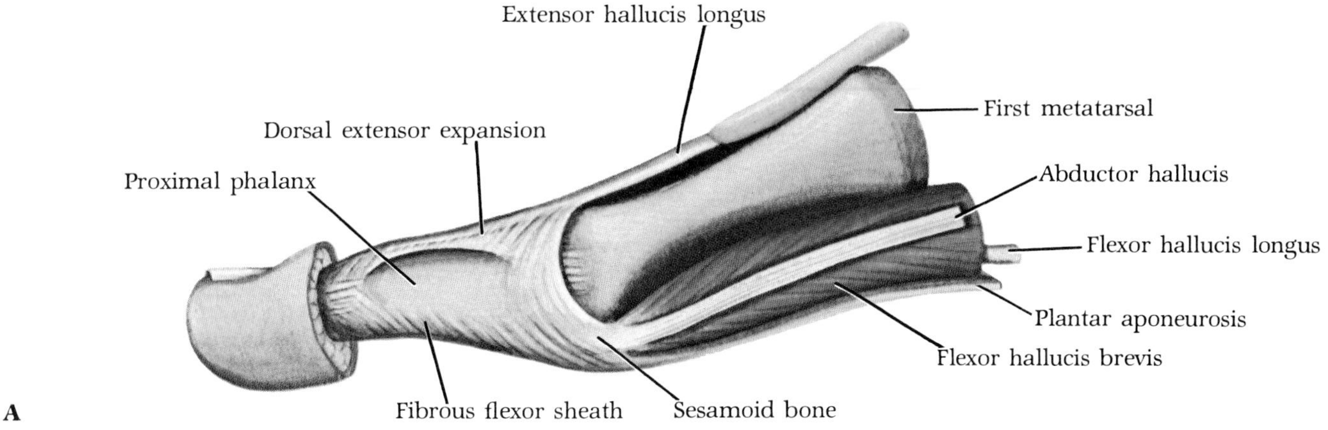
Extensor hallucis longus
First metatarsal
Dorsal extensor expansion
Abductor hallucis
Proximal phalanx
Flexor hallucis longus
Plantar aponeurosis
Flexor hallucis brevis
Fibrous flexor sheath
Sesamoid bone
A

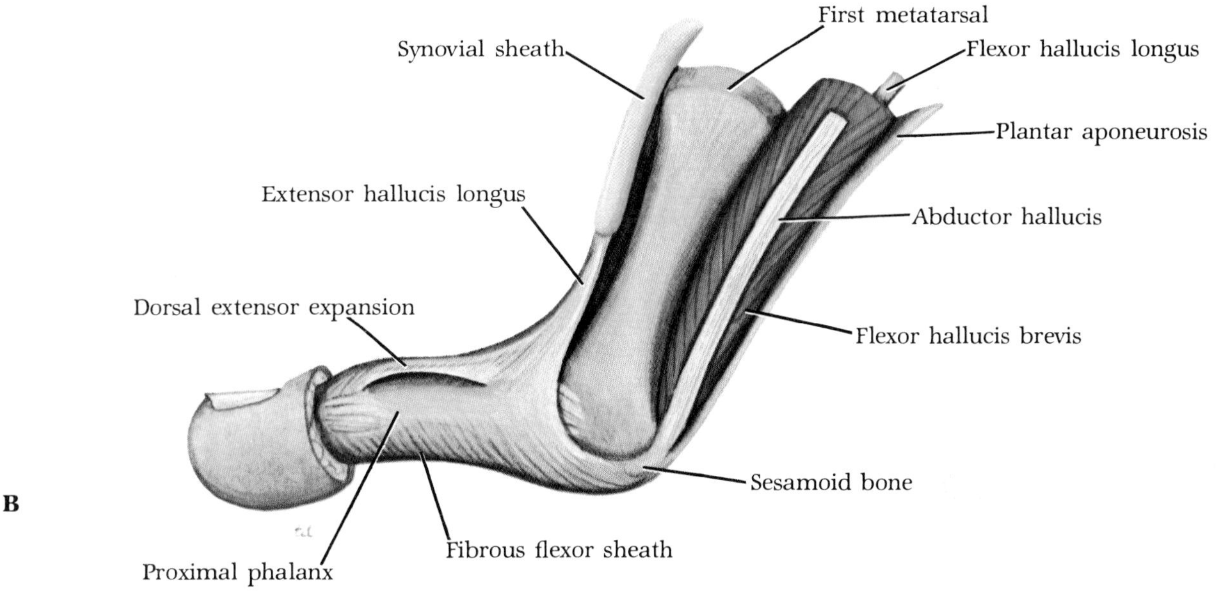
First metatarsal
Synovial sheath
Flexor hallucis longus
Plantar aponeurosis
Extensor hallucis longus
Abductor hallucis
Dorsal extensor expansion
Flexor hallucis brevis
Sesamoid bone
B
Fibrous flexor sheath
Proximal phalanx

Figure 6-56
Right big toe joints and associated tendons and fibrous sheaths.
A. Standing upright on foot.
B. Standing upright on toes.

1. When a person is standing the body weight is distributed through the foot via the heel posteriorly and the six points of contact with the ground anteriorly, namely, the two sesamoid bones under the head of the first metatarsal and the heads of the remaining four metatarsals.

2. When a person walks the body weight is thrown forward and the weight is borne successively on the lateral margin of the foot and the heads of the metatarsal bones. As the heel rises, the big toe provides the final propulsive thrust by flexion of the metatarsophalangeal joint. The flexor hallucis longus muscle and the combined actions of the flexor hallucis brevis, the abductor hallucis, and the oblique head of the adductor hallucis muscle probably play an important role in this final toe-off phase of gait. Note how the plantar aponeurosis and the "slack" in the flexor hallucis longus is taken up as the person stands upright on the toes.

3. Patients with a partially fused first metatarsophalangeal joint (hallux rigidus) or one that produces pain on movement avoid using this joint when walking. Under these circumstances the final propulsive thrust is provided by the four lateral toes, causing the patient to hold his foot obliquely.

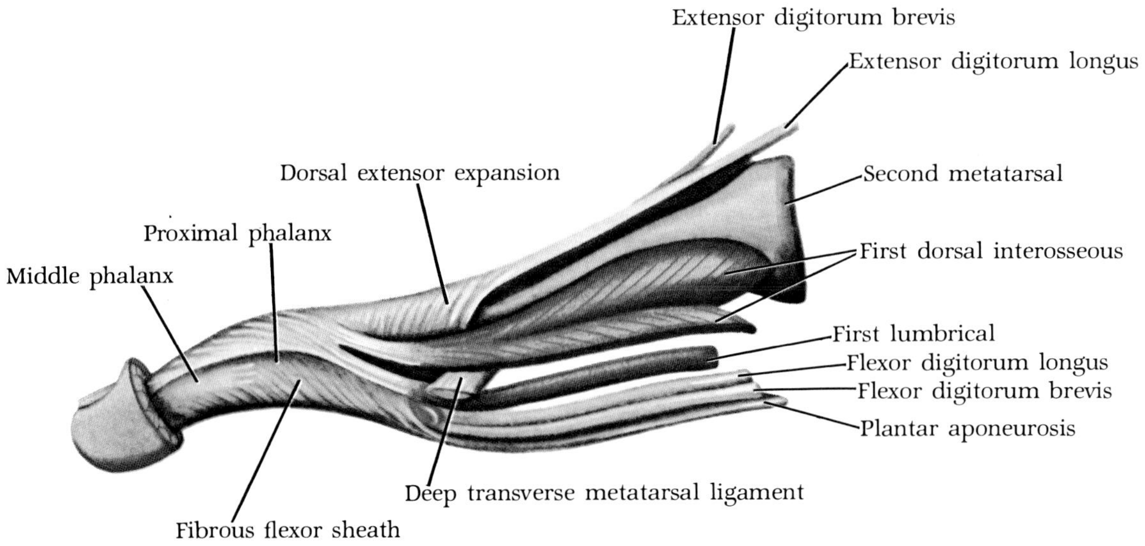
Extensor digitorum brevis
Extensor digitorum longus
Dorsal extensor expansion
Second metatarsal
Proximal phalanx
First dorsal interosseous
Middle phalanx
First lumbrical
Flexor digitorum longus
Flexor digitorum brevis
Plantar aponeurosis
Deep transverse metatarsal ligament
Fibrous flexor sheath

A

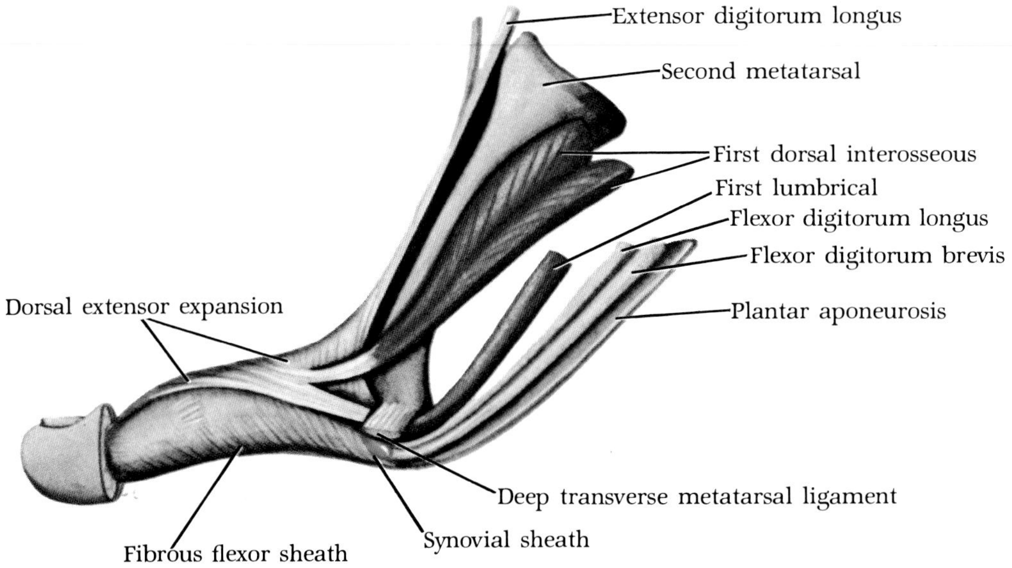
Extensor digitorum longus
Second metatarsal
First dorsal interosseous
First lumbrical
Flexor digitorum longus
Flexor digitorum brevis
Dorsal extensor expansion
Plantar aponeurosis
Deep transverse metatarsal ligament
Synovial sheath
Fibrous flexor sheath

B

Figure 6-57
Right second toe joints and associated tendons and fibrous sheaths. A. Standing upright on foot. B. Standing upright on toes.

1. When a person walks the body weight is thrown forward, and the weight is borne successively on the lateral margin of the foot and the heads of the metatarsal bones (including the two sesamoid bones under the head of the first metatarsal). As the heel rises, the toes are extended at the metatarsophalangeal joints and the plantar aponeurosis is pulled on, thus shortening the tie beams and heightening the longitudinal arches. The "slack" in the long flexor tendons is taken up, thereby increasing their efficiency. The body is then thrown forward (a) by the actions of the gastrocnemius and soleus muscles on the ankle joint, using the foot as a lever, and (b) by the toes (especially the big toe) being strongly flexed by the long and short flexors of the foot, providing the final thrust forward. The lumbrical and interossei muscles contract and keep the toes extended, so that they do not fold under because of the strong action of the flexor digitorum longus muscle.

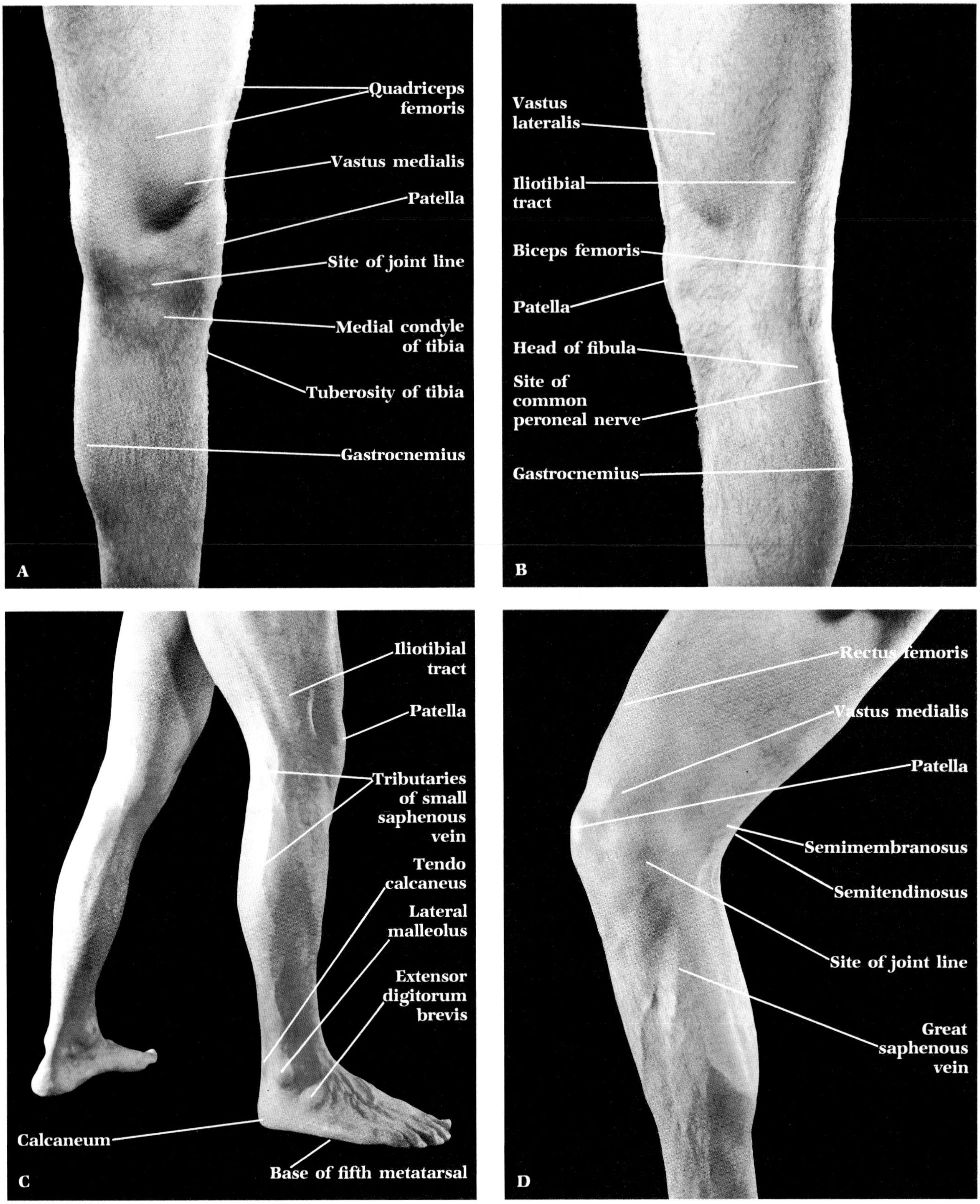
Quadriceps femoris
Vastus medialis
Patella
Site of joint line
Medial condyle of tibia
Tuberosity of tibia
Gastrocnemius
A
Vastus lateralis
Iliotibial tract
Biceps femoris
Patella
Head of fibula
Site of common peroneal nerve
Gastrocnemius
B
Iliotibial tract
Patella
Tributaries of small saphenous vein
Tendo calcaneus
Lateral malleolus
Extensor digitorum brevis
Calcaneum
Base of fifth metatarsal
C
Rectus femoris
Vastus medialis
Patella
Semimembranosus
Semitendinosus
Site of joint line
Great saphenous vein
D

Figure 6-58
The lower limbs of a 25-year-old male. A. Left leg, medial view. B. Left leg, lateral view. C. Right leg, lateral view. D. Right leg, medial view.

1. The quadriceps femoris is a most important extensor muscle for the knee joint. Its tone greatly strengthens the joint. For this reason this muscle mass must be carefully examined when disease of the knee joint is suspected.
2. Both thighs should be examined and the size, consistency, and strength of the quadriceps muscles should be tested. Reduction in size due to muscle atrophy may be tested by measuring the circumference of each thigh a fixed distance above the superior border of the patella.
3. The vastus medialis muscle mass extends further distally than that of the vastus lateralis. Remember that the vastus medialis is the first part of the quadriceps muscle to atrophy in knee joint disease and the last to recover.
4. The iliotibial tract, a wide, thickened band of deep fascia that extends from the tubercle of the iliac crest to the lateral condyle of the tibia, can be felt on the lateral side of the thigh. It becomes taut when the hip and knee joints are extended.
5. The patella and the ligamentum patellae can be felt in front of the knee. On the sides of the knee the condyles of the femur and tibia can be recognized, and the joint line can be identified.
6. The head of the fibula can be palpated below the lateral femoral condyle. It is situated at about the same level as the tibial tuberosity.
7. The anatomical location of the medial and lateral collateral ligaments of the knee joint can be felt by pressing the examining finger on the medial and lateral sides of the joint line, respectively.
8. On the dorsum of the foot the head of the talus can be palpated just in front of the malleoli.
9. The belly of the extensor digitorum brevis muscle can be seen and felt on the dorsum of the foot just below and anterior to the lateral malleolus.

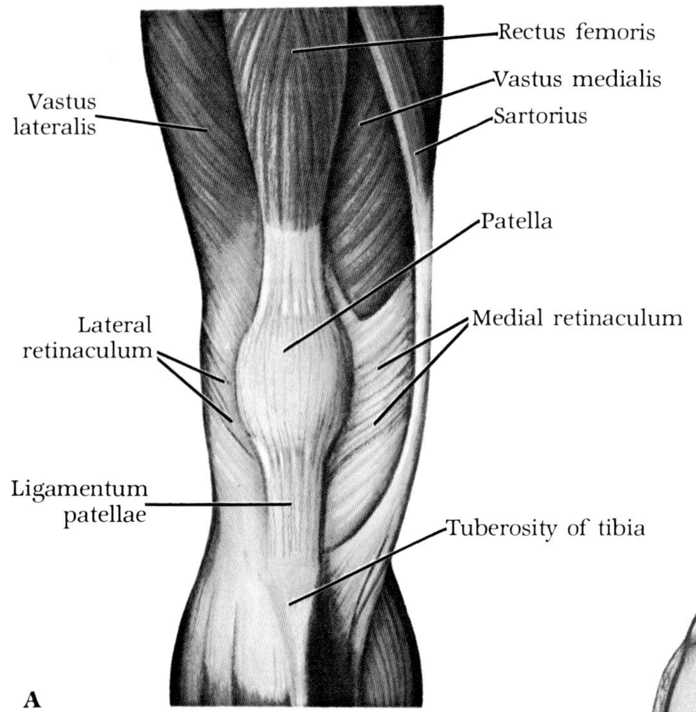
Rectus femoris
Vastus medialis
Vastus lateralis
Sartorius
Patella
Lateral retinaculum
Medial retinaculum
Ligamentum patellae
Tuberosity of tibia

A

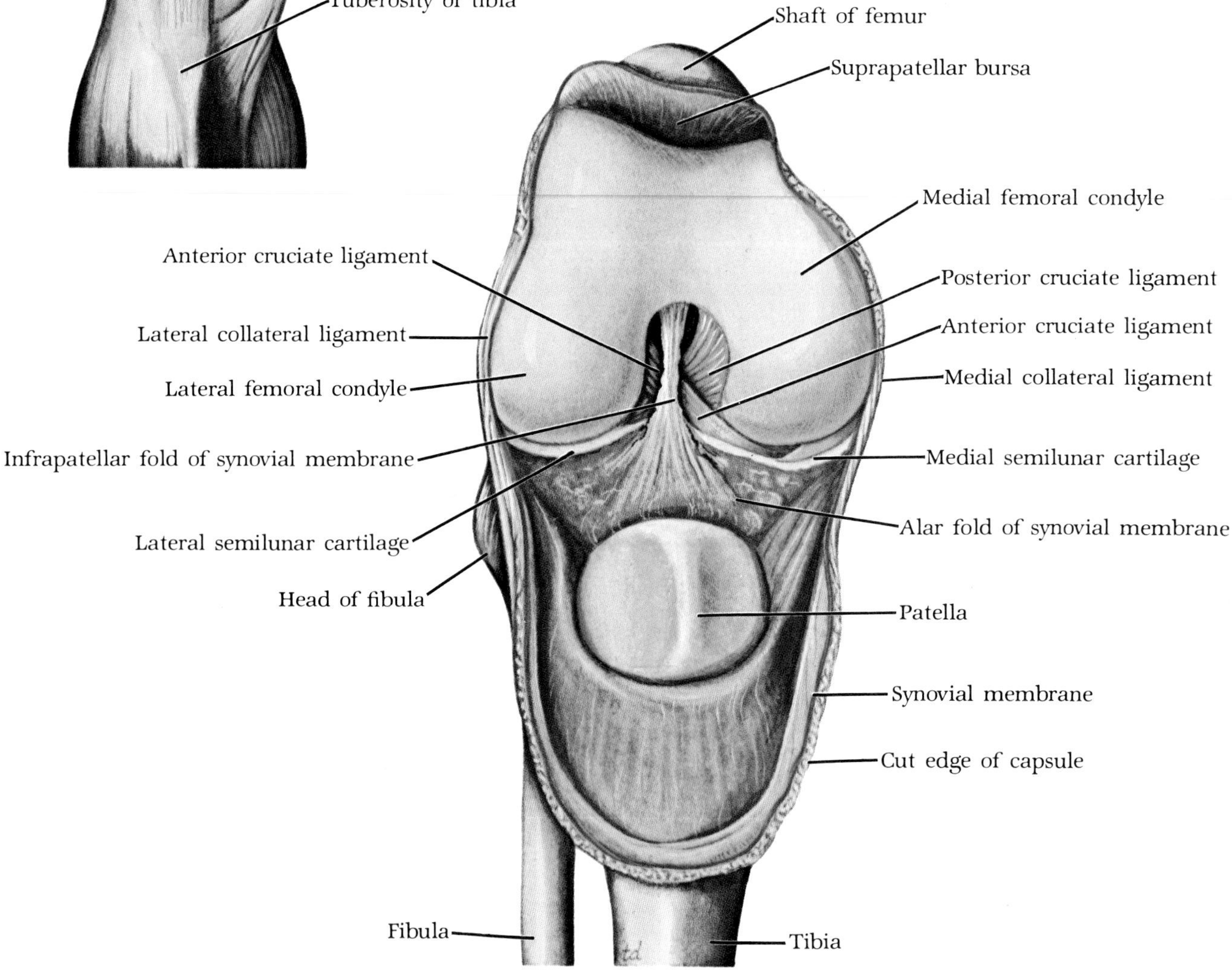
Shaft of femur
Suprapatellar bursa
Medial femoral condyle
Anterior cruciate ligament
Posterior cruciate ligament
Lateral collateral ligament
Anterior cruciate ligament
Lateral femoral condyle
Medial collateral ligament
Infrapatellar fold of synovial membrane
Medial semilunar cartilage
Alar fold of synovial membrane
Lateral semilunar cartilage
Head of fibula
Patella
Synovial membrane
Cut edge of capsule
Fibula
Tibia

B

Figure 6-59
Right knee joint. A. Anterior view, external aspect. B. Anterior view, internal aspect.

1. Four bursae are associated with the anterior aspect of the knee joint. The suprapatellar bursa lies beneath the quadriceps muscle and communicates with the joint cavity; the prepatellar bursa lies in the subcutaneous tissue between the skin and the anterior aspect of the lower half of the patella and the upper part of the ligamentum patellae; the superficial infrapatellar bursa lies in the subcutaneous tissue between the skin and the front of the lower part of the ligamentum patellae and the tibial tuberosity; and the deep infrapatellar bursa lies between the ligamentum patellae and the tibia.
2. Prepatellar bursitis, "housemaid's knee," is caused by repetitive trauma due to contact of the bursa with hard floors. Syphilis can also cause effusion into this bursa. Infrapatellar bursitis can be due to infection and causes pain and tenderness over the ligamentum patellae.
3. Effusion into the knee joint produces fullness above and on either side of the patella. The fullness is due to the filling of the suprapatellar bursa and the bulging forward of the synovial membrane and retinacula on either side of the patella. If a hand is placed on the thigh above the patella, it is possible to apply pressure to the suprapatella bursa and drive the fluid from the bursa into the main cavity of the knee. This has the effect of accentuating the bulges on either side of the patella, thus aiding in the diagnosis.
4. The patella is a sesamoid bone lying within the quadriceps tendon. The lower horizontal fibers of the vastus medialis and the large size of the lateral condyle of the femur are important in preventing lateral displacement of the patella. Congenital recurrent dislocations of the patella are due to underdevelopment of the lateral femoral condyle.

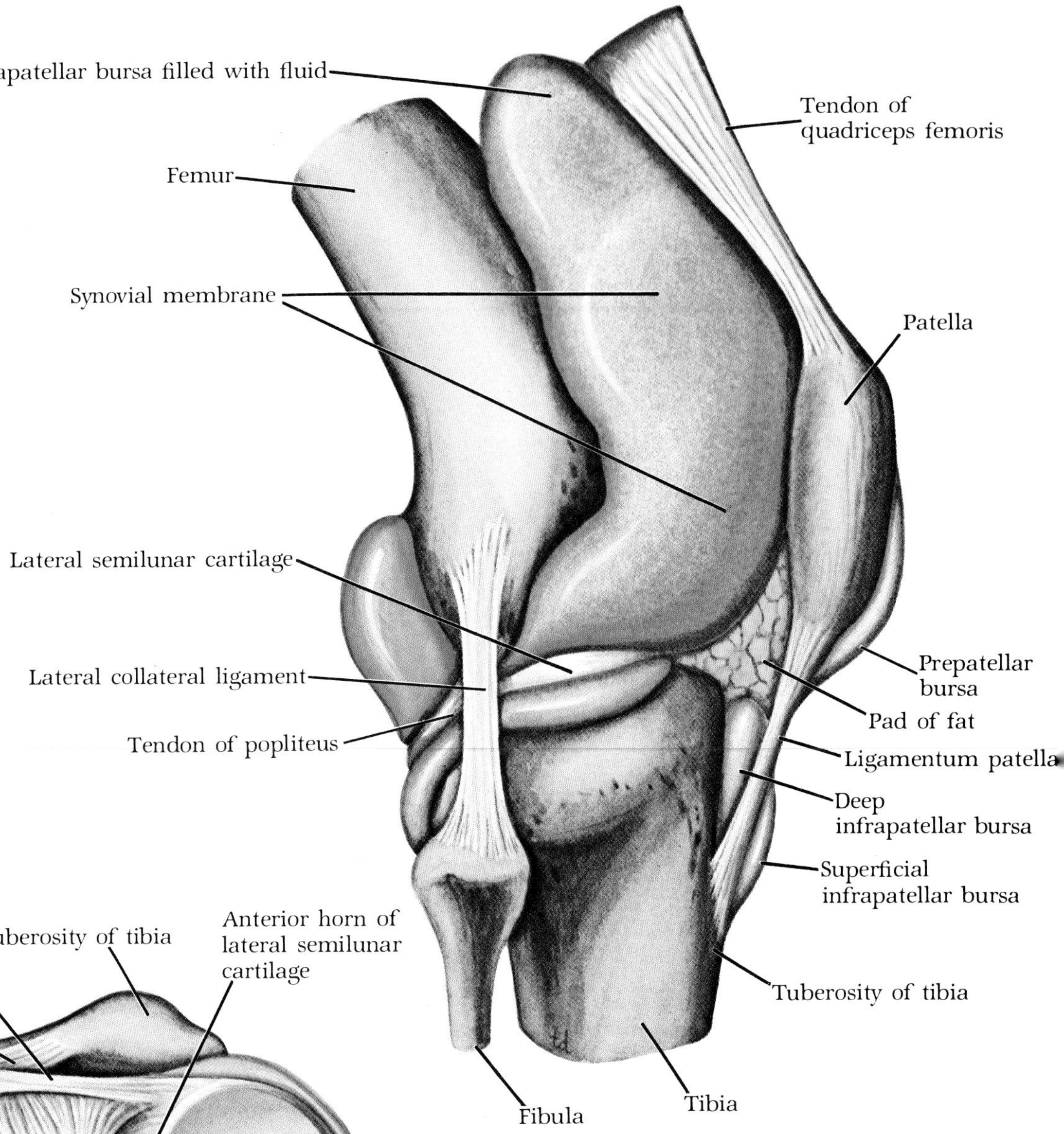
Suprapatellar bursa filled with fluid
Tendon of quadriceps femoris
Femur
Synovial membrane
Patella
Lateral semilunar cartilage
Lateral collateral ligament
Prepatellar bursa
Pad of fat
Tendon of popliteus
Ligamentum patella
Deep infrapatellar bursa
Superficial infrapatellar bursa
Tuberosity of tibia
Fibula
Tibia

A

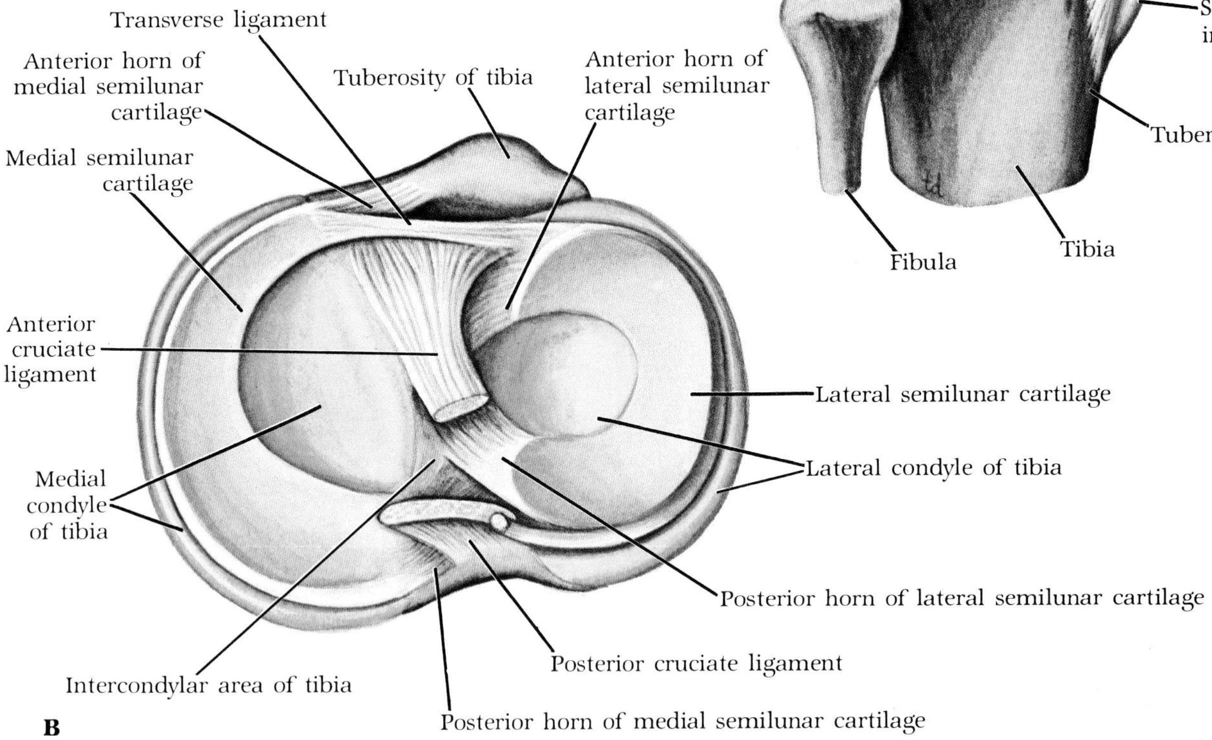
Transverse ligament
Anterior horn of medial semilunar cartilage
Tuberosity of tibia
Anterior horn of lateral semilunar cartilage
Medial semilunar cartilage
Anterior cruciate ligament
Lateral semilunar cartilage
Medial condyle of tibia
Lateral condyle of tibia
Posterior horn of lateral semilunar cartilage
Posterior cruciate ligament
Intercondylar area of tibia
Posterior horn of medial semilunar cartilage

B

Figure 6-60
Right knee joint. A. Lateral view, showing synovial membrane.
B. Condyles of tibia (superior aspect), showing semilunar cartilages and attachments of cruciate ligaments.

1. The synovial membrane of the knee joint is very extensive, and if the articular surfaces, semilunar cartilages, or ligaments of the joint are damaged, this large synovial cavity becomes distended with fluid. The communication between the suprapatellar bursa and the joint cavity results in this structure becoming distended also. The swelling of the knee extends some 3 or 4 fingersbreadth above the patella and laterally and medially beneath the aponeuroses of insertion of the vastus lateralis and medialis muscles, respectively.
2. The medial semilunar cartilage is damaged more frequently than the lateral. This is probably due to its attachment to the medial collateral ligament, which restricts its mobility. The cartilage is damaged when the knee is partially flexed and abducted; during rotation of the tibial condyles on the femoral condyles, the cartilage is ground between the apposing bony surfaces.
3. The lateral semilunar cartilage is less commonly damaged, probably due to the fact that it is not attached to the lateral collateral ligament and consequently is more mobile. Moreover, because the popliteus muscle sends a few of its fibers into the lateral cartilage, these may pull the cartilage into a more favorable position during sudden movements of the knee joint.
4. Tears or rupture of the cruciate ligaments occur when severe force is applied to the knee joint. In patients with a ruptured anterior cruciate ligament the tibia can be pulled excessively forward on the femur; with rupture of the posterior cruciate ligament, the tibia can be made to move excessively backward on the femur.

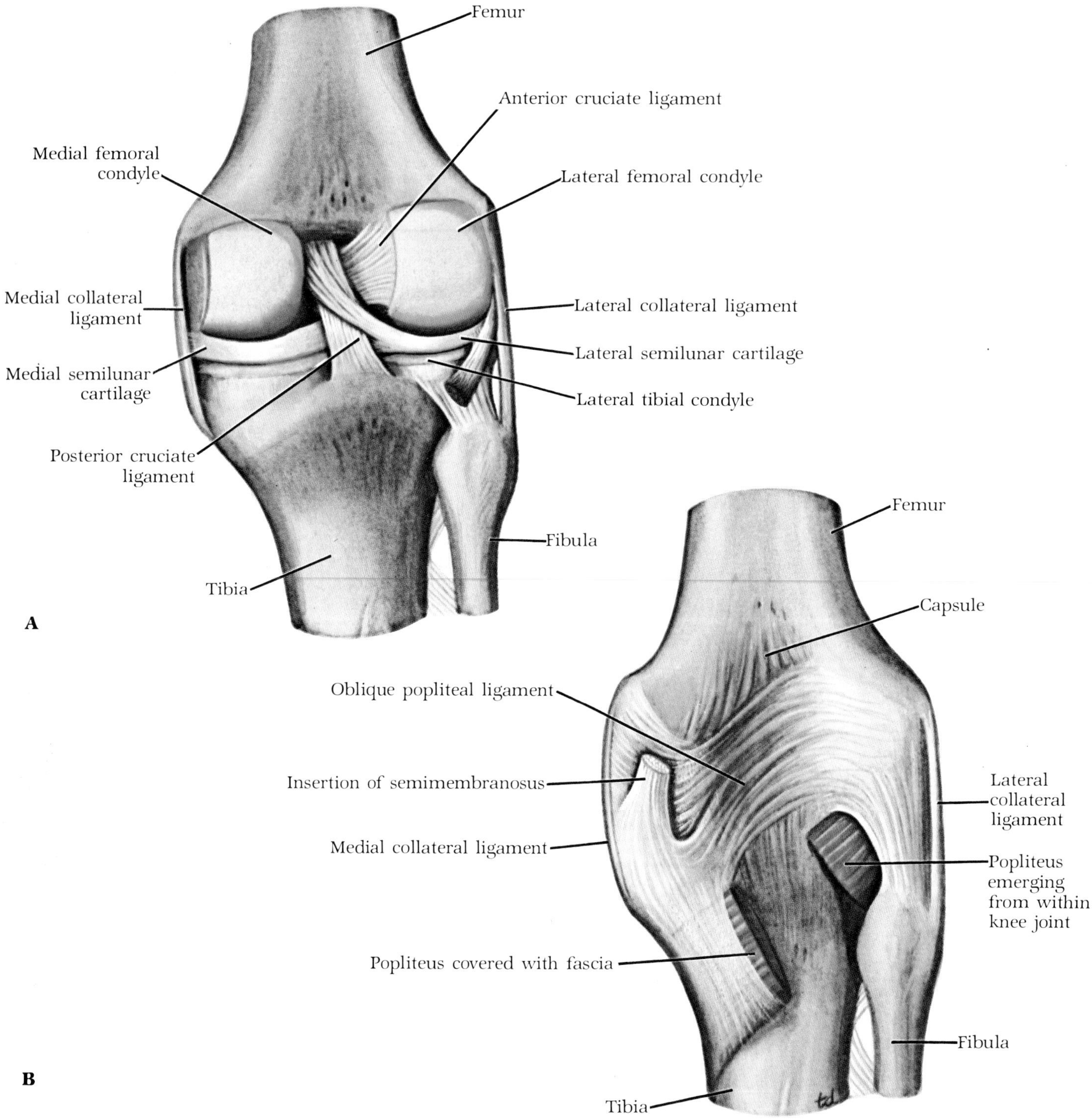
Femur
Anterior cruciate ligament
Medial femoral condyle
Lateral femoral condyle
Medial collateral ligament
Lateral collateral ligament
Lateral semilunar cartilage
Medial semilunar cartilage
Lateral tibial condyle
Posterior cruciate ligament
Fibula
Tibia
A
Femur
Capsule
Oblique popliteal ligament
Insertion of semimembranosus
Lateral collateral ligament
Medial collateral ligament
Popliteus emerging from within knee joint
Popliteus covered with fascia
Fibula
B
Tibia

Figure 6-61
Right knee joint. A. Posterior view, internal aspect. B. Posterior view, external aspect.

1. Two bursae are associated with the posterior surface of the knee joint: the popliteal bursa, which surrounds the tendon of the popliteus muscle and communicates with the joint cavity; and the semimembranosus bursa, which is related to the insertion of the semimembranosus muscle and frequently communicates with the joint cavity.
2. Popliteal bursitis produces a rounded swelling posterior to the lateral condyle of the femur. Since the bursa communicates with the joint, the condition is usually secondary to disease within the knee joint.
3. Semimembranosus bursitis is common in gamekeepers due to the excessive amount of knee flexion that occurs when walking through undergrowth. If the bursa does not communicate with the joint, it produces a tense swelling in the popliteal space when the knee joint is extended.
4. The strength of the knee joint depends on the strength of the ligaments that bind the femur to the tibia and on the tone of the muscles acting on the joint. The most important muscle group is the quadriceps femoris; if this group is well developed, it can stabilize the knee in the presence of torn ligaments.
5. The medial collateral ligament may be torn in forced abduction of the tibia on the femur; this can occur at its femoral or tibial attachments. Remember that tears of the semilunar cartilages result in local tenderness on the joint line, whereas sprains of the medial collateral ligament result in tenderness over the femoral or tibial attachments of the ligament.

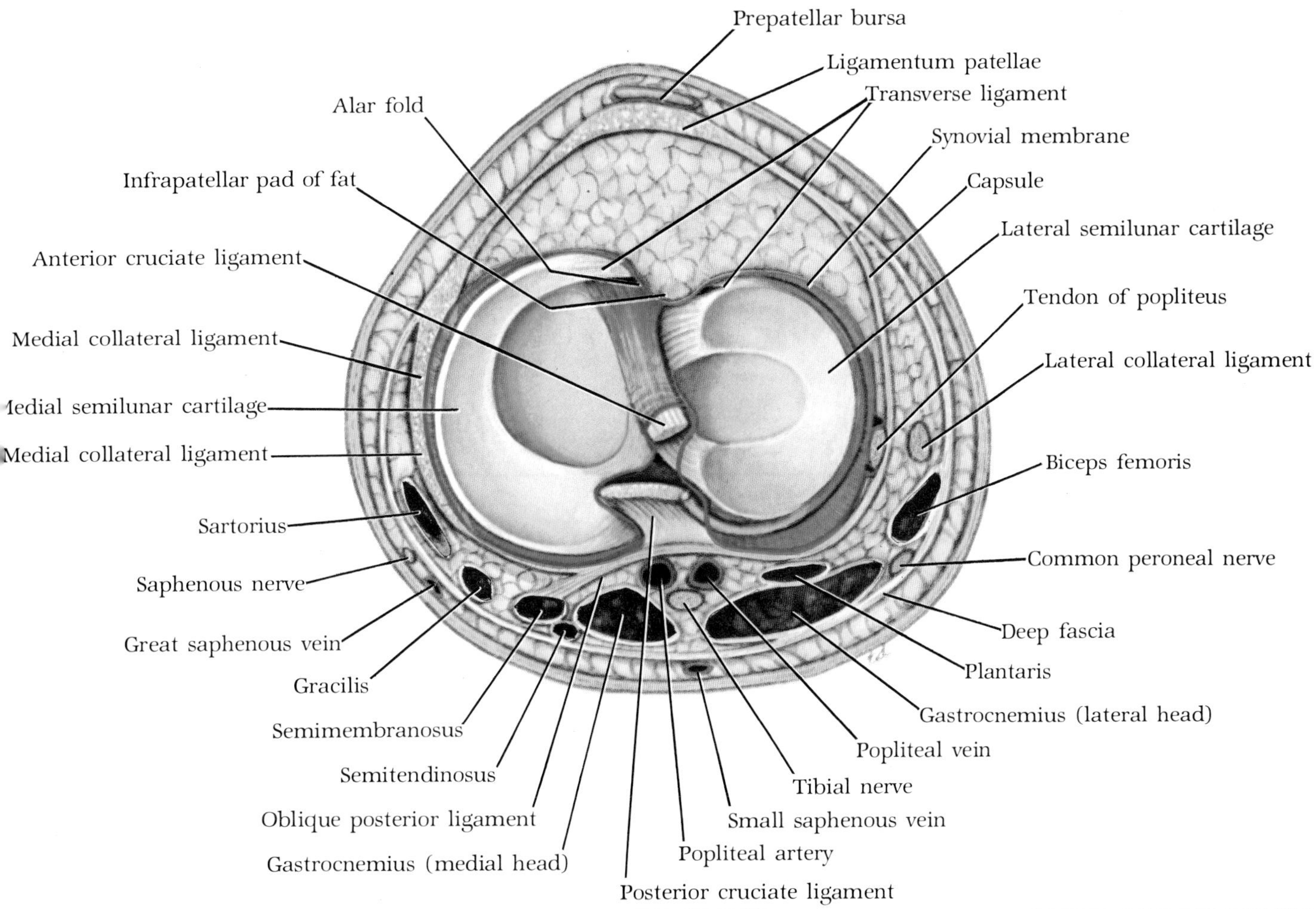
Prepatellar bursa
Ligamentum patellae
Transverse ligament
Alar fold
Synovial membrane
Infrapatellar pad of fat
Capsule
Lateral semilunar cartilage
Anterior cruciate ligament
Tendon of popliteus
Medial collateral ligament
Lateral collateral ligament
Medial semilunar cartilage
Medial collateral ligament
Biceps femoris
Sartorius
Common peroneal nerve
Saphenous nerve
Deep fascia
Great saphenous vein
Plantaris
Gracilis
Gastrocnemius (lateral head)
Semimembranosus
Popliteal vein
Semitendinosus
Tibial nerve
Oblique posterior ligament
Small saphenous vein
Gastrocnemius (medial head)
Popliteal artery
Posterior cruciate ligament

Figure 6-62
Relations of right knee joint.

1. *Wasting of the Quadriceps Femoris.* Any injury to the knee joint is quickly followed by wasting of this muscle group. The vastus medialis muscle wastes first, wastes most, and is the last to recover.
2. The popliteus muscle is particularly concerned with unlocking the knee joint during the initial phase of flexing the knee. This muscle protects the lateral semilunar cartilage by pulling it into a more favorable position during sudden flexion movements of the knee joint.
3. Cysts of the semilunar cartilages, which occasionally occur, appear to follow an injury; cysts of the lateral cartilage are more common than those in the medial cartilage.
4. *Loose Bodies.* In the knee joints of elderly patients with osteoarthritis, osteophytes may become detached and move about in the joint cavity. In young persons a fragment of articular cartilage still attached to a small piece of underlying bone may break off from the medial condyle of the femur to form a loose body; this condition is referred to as osteochondritis dissecans. Such loose bodies can become jammed between the articular surfaces, causing pain and locking the joint.
5. *Locking of the Knee Joint.* The patient sometimes states that he is unable to extend the knee but can flex it and that this phenomenon occurred suddenly. This so-called locking of the knee is due to the sudden jamming of an object between the apposing articular surfaces. The most common objects are pieces of torn semilunar cartilage or cruciate ligament or loose bodies.

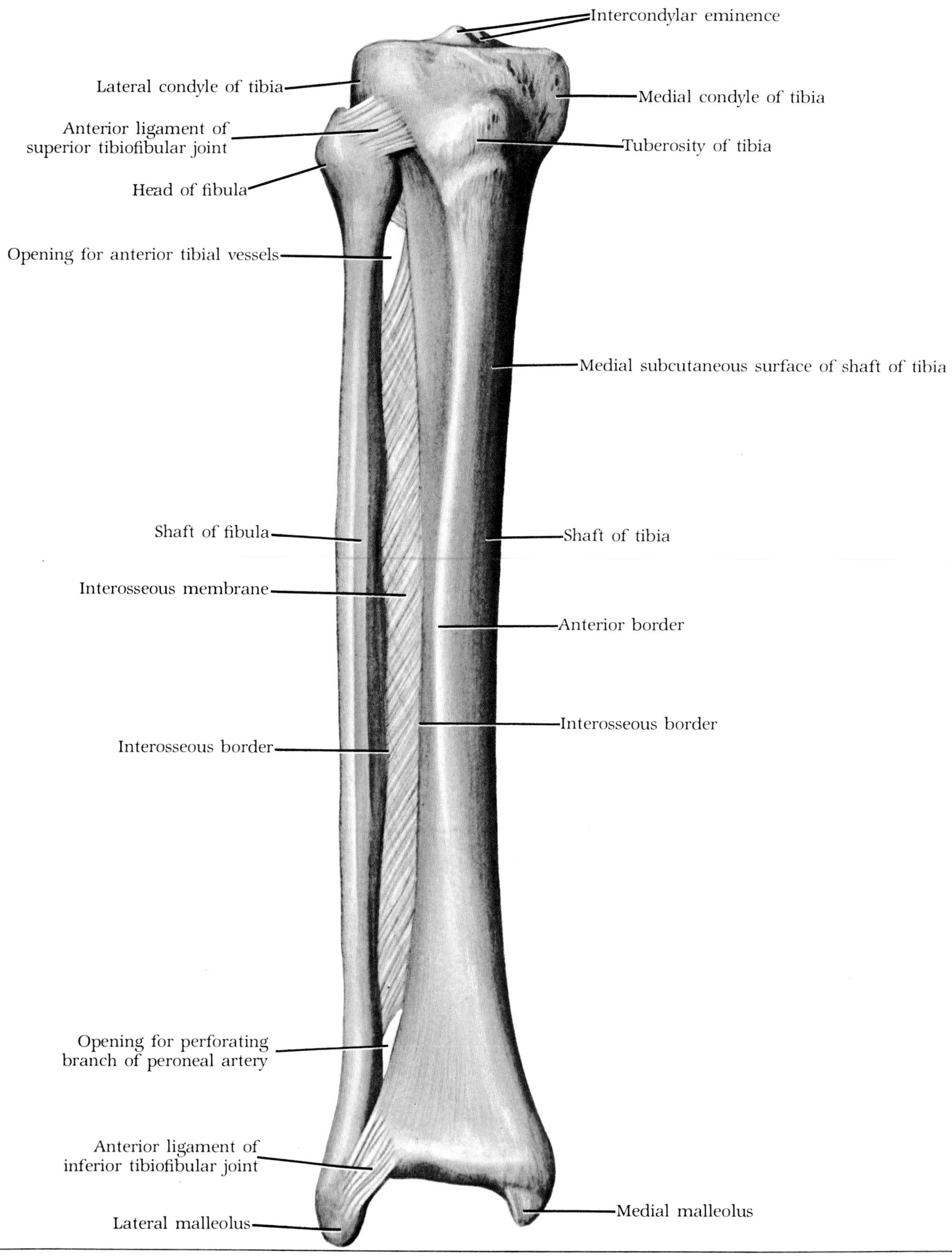
Intercondylar eminence
Lateral condyle of tibia
Medial condyle of tibia
Anterior ligament of superior tibiofibular joint
Tuberosity of tibia
Head of fibula
Opening for anterior tibial vessels
Medial subcutaneous surface of shaft of tibia
Shaft of fibula
Shaft of tibia
Interosseous membrane
Anterior border
Interosseous border
Interosseous border
Opening for perforating branch of peroneal artery
Anterior ligament of inferior tibiofibular joint
Medial malleolus
Lateral malleolus

Figure 6-63
Right tibia and fibula showing the interosseous membrane and anterior ligaments of tibiofibular joints, anterior view.

1. The stability of the ankle joint depends on the integrity of the ankle mortise, formed by the medial and lateral malleoli grasping the body of the talus, and the strong ligaments of the tibiofibular joints and the interosseous membrane.

2. Fractures of the tibia and fibula are common. Fortunately, because of the strong ligaments binding the two bones together, if one bone is fractured, the other acts as a splint and there is minimal displacement.

A

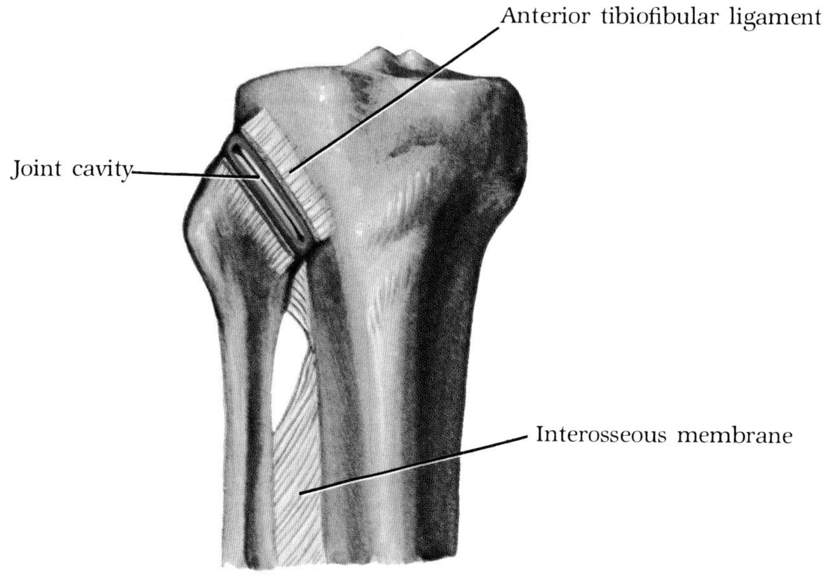
Anterior tibiofibular ligament
Joint cavity
Interosseous membrane

B

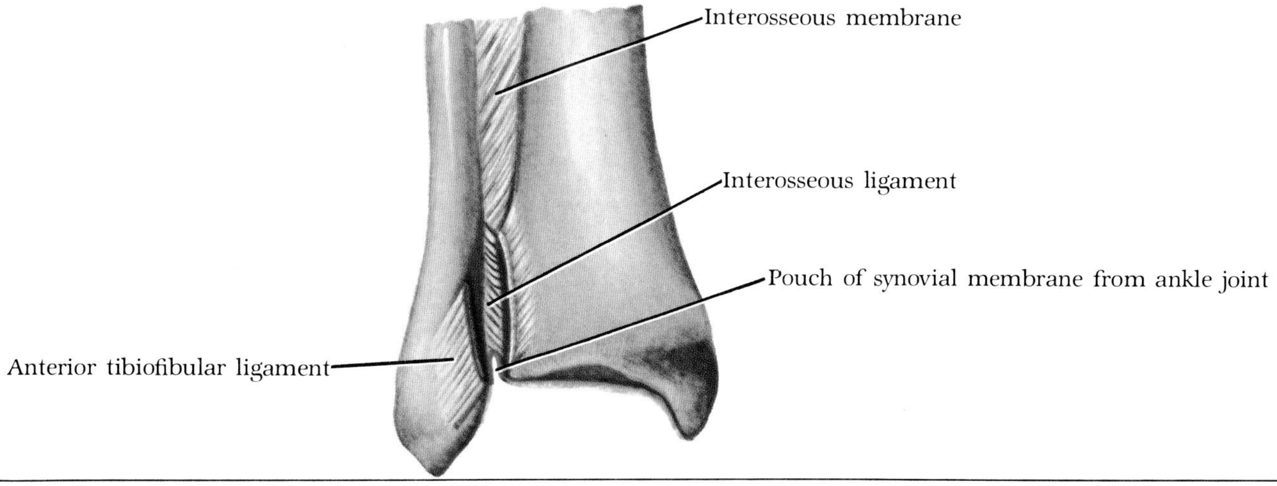
Interosseous membrane
Interosseous ligament
Pouch of synovial membrane from ankle joint
Anterior tibiofibular ligament

Figure 6-64
Right tibiofibular articulations.
A. Superior tibiofibular joint.
B. Inferior tibiofibular joint.

1. When the ankle joint is dorsiflexed the wide anterior part of the talus is forced between the medial and lateral malleoli, causing a slight amount of separation between the tibia and fibula. This movement between the bones occurs at the fibrous inferior tibiofibular joint and the synovial superior tibiofibular joint.

2. The tension produced in the various ligaments of the tibiofibular and ankle joints when the ankle is dorsiflexed results in the ankle being most stable in this position. It is not surprising, therefore, to find that all major thrusting movements of the foot, as in walking or running, take place when the ankle is dorsiflexed.

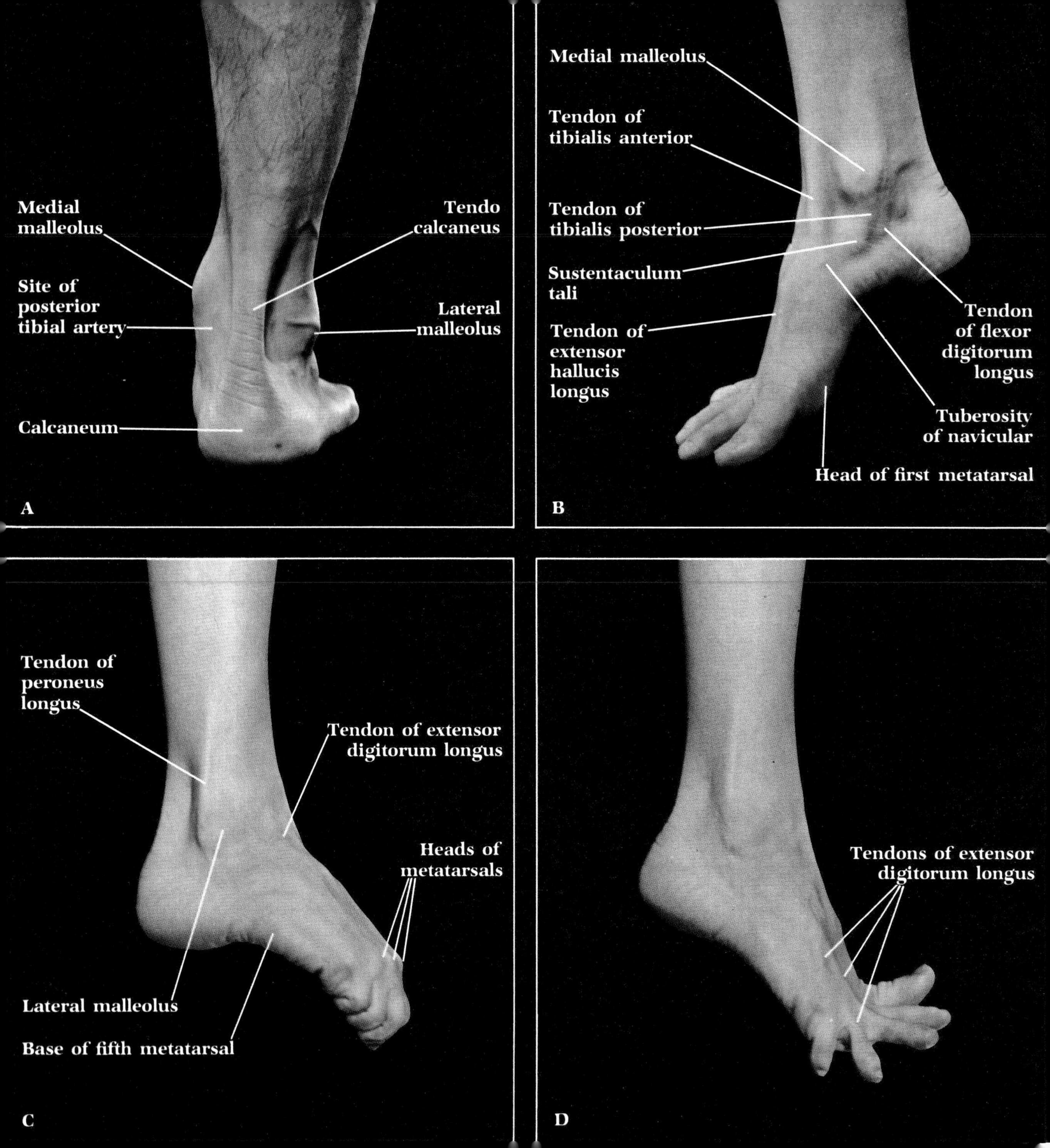
Medial malleolus
Site of posterior tibial artery
Calcaneum
Tendo calcaneus
Lateral malleolus
A
Medial malleolus
Tendon of tibialis anterior
Tendon of tibialis posterior
Sustentaculum tali
Tendon of extensor hallucis longus
Tendon of flexor digitorum longus
Tuberosity of navicular
Head of first metatarsal
B
Tendon of peroneus longus
Tendon of extensor digitorum longus
Heads of metatarsals
Lateral malleolus
Base of fifth metatarsal
C
Tendons of extensor digitorum longus
D

Figure 6-65
Right ankle. A. Of a 25-year-old male, posterior view. B. Of a 29-year-old female, medial view. C. Lateral view with toes plantar-flexed. D. Lateral view with toes extended and abducted.

1. The medial malleolus, the lateral malleolus, and the tendo calcaneus can easily be felt. The prominence of the heel is formed by the calcaneum bone.
2. In the interval between the medial malleolus and the medial surface of the calcaneum lie the following structures in the order named: (1) the tendon of the tibialis posterior muscle, (2) the tendon of the flexor digitorum longus muscle, (3) the posterior tibial vessels, (4) the posterior tibial nerve, and (5) the tendon of the flexor hallucis longus muscle. The pulsation of the posterior tibial artery can be felt halfway between the medial malleolus and the heel.
3. The tendons of the peroneus brevis and longus muscles can be palpated posterior to the lateral malleolus.
4. The sustentaculum tali can be palpated on the medial aspect of the foot about 1 inch (2.5 cm) below the tip of the medial malleolus. The tendon of the tibialis posterior muscle lies immediately above the sustentaculum tali; the tendon of the flexor hallucis longus muscle winds around its lower surface.
5. The tuberosity of the navicular bone can be palpated in front of the sustentaculum tali.
6. The prominent base of the fifth metatarsal bone can be felt on the lateral side of the foot.
7. On the dorsum of the foot note the tendons of the extensor digitorum longus muscle.

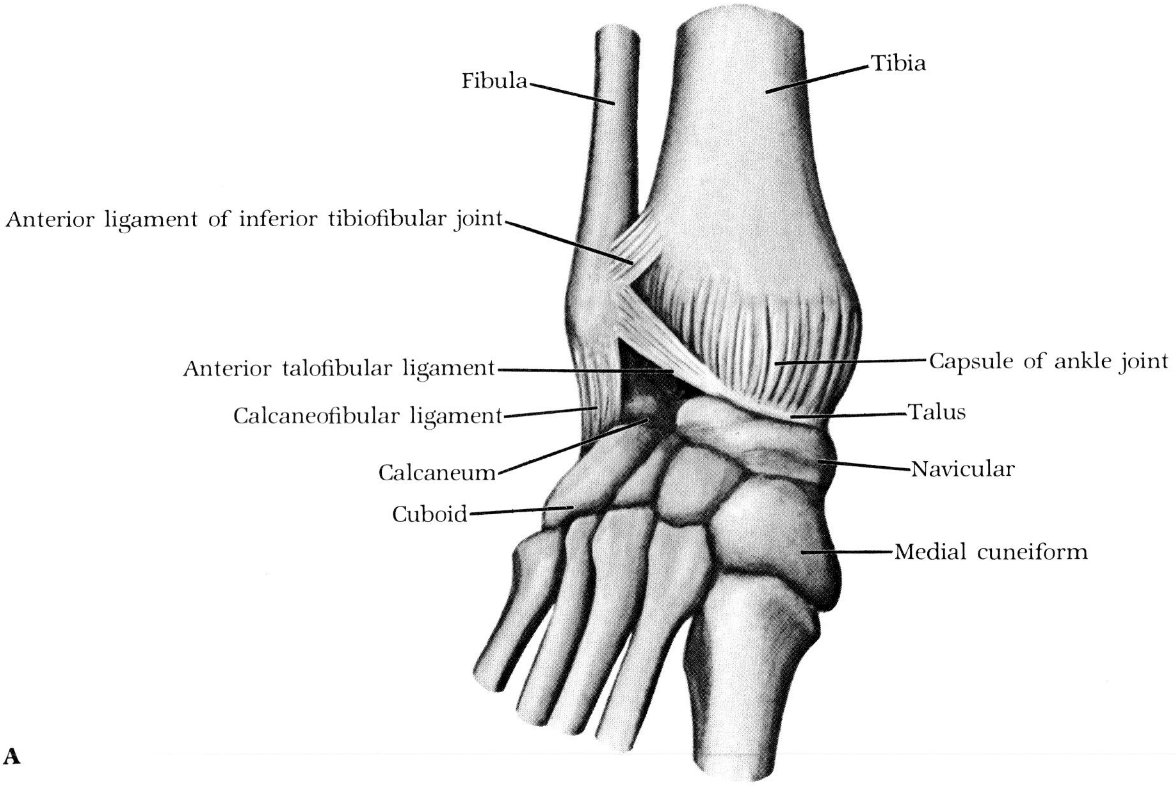
Fibula
Tibia
Anterior ligament of inferior tibiofibular joint
Anterior talofibular ligament
Capsule of ankle joint
Calcaneofibular ligament
Talus
Calcaneum
Navicular
Cuboid
Medial cuneiform

A

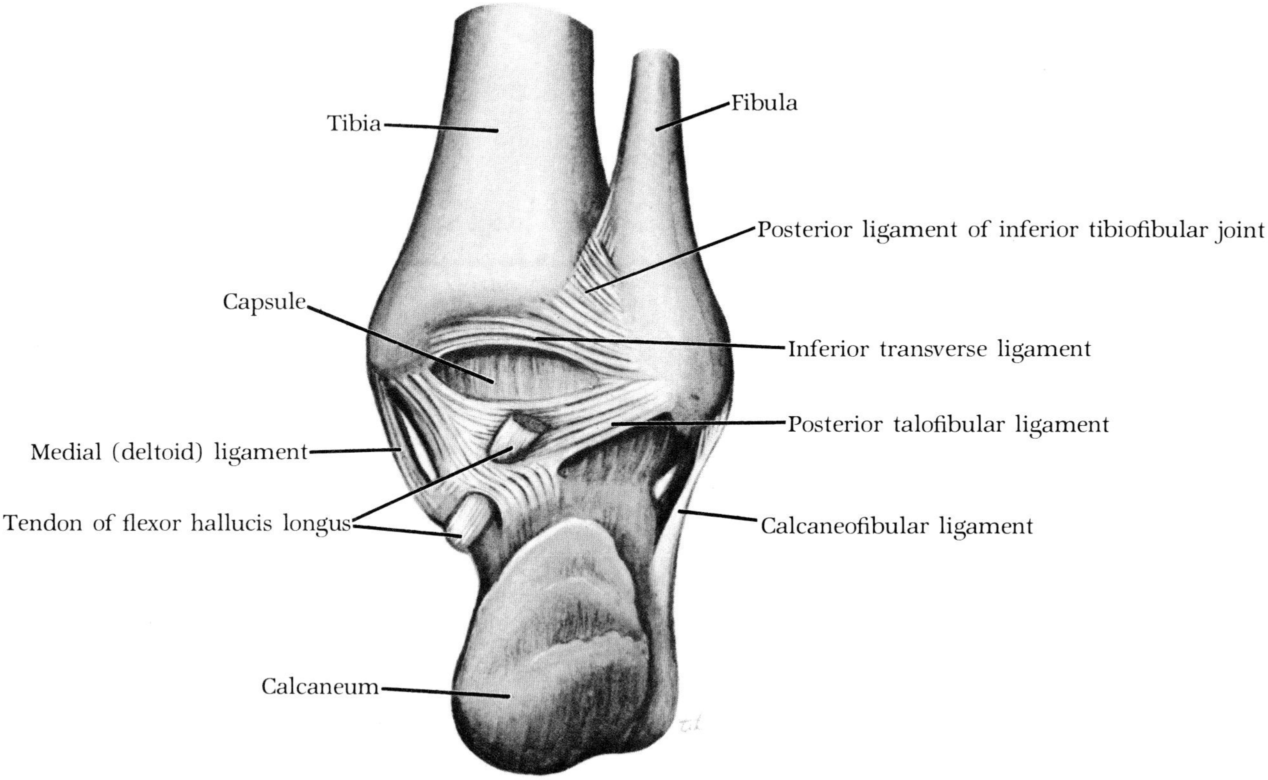
Tibia
Fibula
Posterior ligament of inferior tibiofibular joint
Capsule
Inferior transverse ligament
Posterior talofibular ligament
Medial (deltoid) ligament
Tendon of flexor hallucis longus
Calcaneofibular ligament
Calcaneum

B

Figure 6-66
Right ankle joint. A. Anterior view. B. Posterior view.

1. In the region of the ankle, the fibula is subcutaneous and may be followed inferiorly to form the lateral malleolus. The tip of the medial malleolus of the tibia lies about $\frac{1}{2}$ inch (1.3 cm) proximal to the level of the tip of the lateral malleolus. The line of the ankle joint lies on a plane 1 cm superior to the tip of the medial malleolus.
2. Fluid present in the ankle joint reveals itself by a bulging beneath the extensor tendons and anterior to the medial and lateral malleoli. There can also be bulging on either side of the tendo calcaneus on the posterior surface of the joint.
3. Two movements are possible at the ankle joint: dorsiflexion (toes pointing upward) and plantar flexion (toes pointing downward). When testing for these movements it is very important to grasp the midfoot only. Grasping the forefoot will permit the movements of inversion and eversion to take place at the tarsal joints.
4. Fracture-dislocations of the ankle joint are common. Forced external rotation and overeversion of the foot are the most frequent cause. The talus is externally rotated forcibly against the lateral malleolus of the fibula. The torsion effect on the lateral malleolus in turn causes it to fracture spirally. If the force continues, the talus moves laterally, and the medial ligament of the ankle joint becomes taut and pulls off the tip of the medial malleolus. If the talus is forced to move still further, its rotary movement results in its violent contact with the posterior inferior margin of the tibia, which then shears off. Other less common types of fracture-dislocation are those due to forced overeversion (without rotation), in which the talus presses the lateral malleolus laterally and causes it to fracture transversely, and those due to overinversion (without rotation), in which the talus presses gainst the medial malleolus; in the latter situation a vertical fracture through the base of the medial malleolus will be produced.

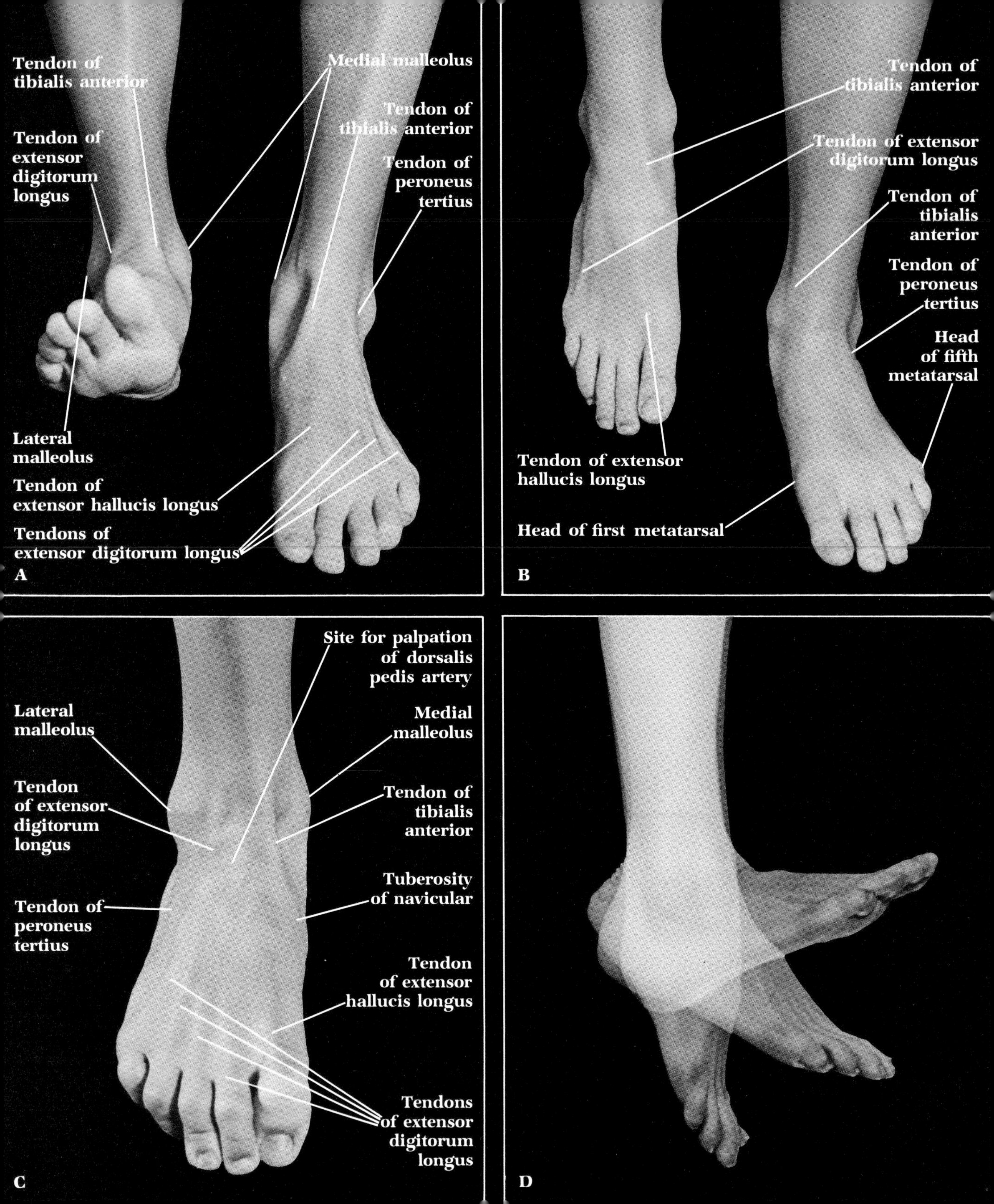
Tendon of tibialis anterior
Tendon of extensor digitorum longus
Lateral malleolus
Tendon of extensor hallucis longus
Tendons of extensor digitorum longus
Medial malleolus
Tendon of tibialis anterior
Tendon of peroneus tertius
A
Tendon of tibialis anterior
Tendon of extensor digitorum longus
Tendon of tibialis anterior
Tendon of peroneus tertius
Head of fifth metatarsal
Tendon of extensor hallucis longus
Head of first metatarsal
B
Lateral malleolus
Tendon of extensor digitorum longus
Tendon of peroneus tertius
Site for palpation of dorsalis pedis artery
Medial malleolus
Tendon of tibialis anterior
Tuberosity of navicular
Tendon of extensor hallucis longus
Tendons of extensor digitorum longus
C
D

Figure 6-67
The ankles and feet of a 29-year-old female. A, B. Anterior views. C. Anterior view, right foot. D. Movements of dorsiflexion and plantar flexion of the right ankle.

1. On the anterior surface of the ankle joint the tendon of the tibialis anterior muscle can be seen when the foot is dorsiflexed and inverted. The tendon of the extensor hallucis longus muscle lies lateral to it and can be made to stand out by extending the big toe. Lateral to the extensor hallucis longus lie the tendons of the extensor digitorum longus and peroneus tertius muscles.

2. The pulsations of the dorsalis pedis artery can be felt between the tendons of the extensor hallucis longus and extensor digitorum longus muscles, midway between the two malleoli on the front of the ankle.

3. The movements of dorsiflexion (toes pointing upward) and plantar flexion (toes pointing downward) take place between the talus and the tibia and fibula at the ankle joint. In a normal ankle joint only about 20 degrees of dorsiflexion is possible while as much as 50 degrees of plantar flexion can take place.

4. Dorsiflexion is performed by the tibialis anterior, extensor hallucis longus, extensor digitorum longus, and peroneus tertius muscles.

5. Plantar flexion is performed by the gastrocnemius, soleus, plantaris, peroneus longus, peroneus brevis, tibialis posterior, flexor digitorum longus, and flexor hallucis longus muscles.

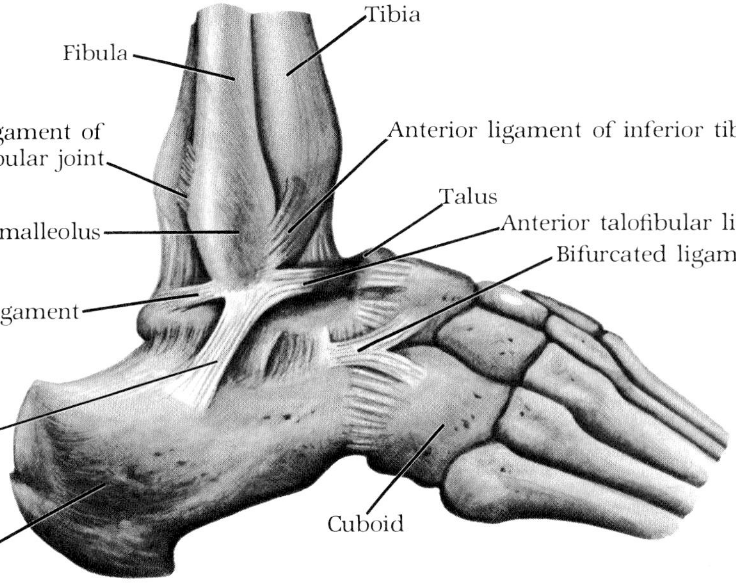
Tibia
Fibula
Posterior ligament of inferior tibiofibular joint
Anterior ligament of inferior tibiofibular joint
Talus
Lateral malleolus
Anterior talofibular ligament
Bifurcated ligament
Posterior talofibular ligament
Calcaneofibular ligament
Cuboid
Calcaneum
A

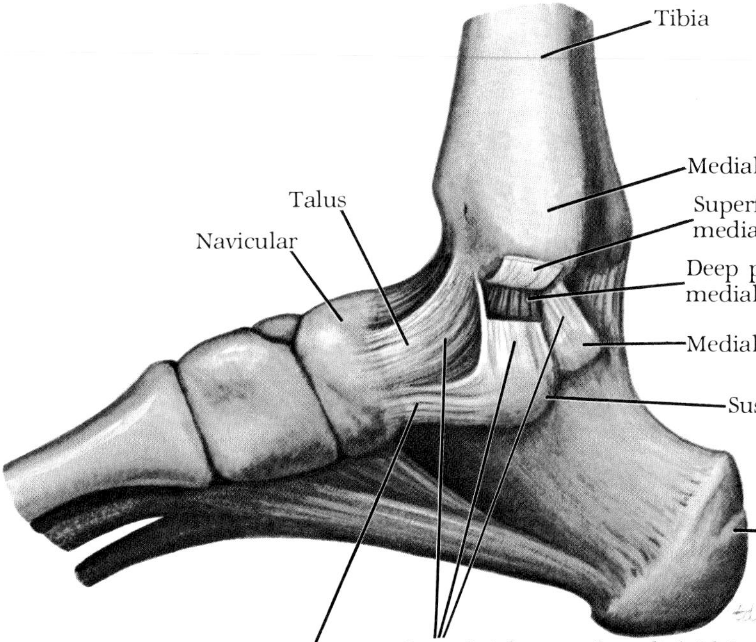
Tibia
Medial malleolus
Talus
Superficial part of medial ligament (reflected)
Navicular
Deep part of medial (deltoid) ligament
Medial tubercle of talus
Sustentaculum tali
Calcaneum
B
Superficial part of medial (deltoid) ligament
Plantar calcaneonavicular ligament

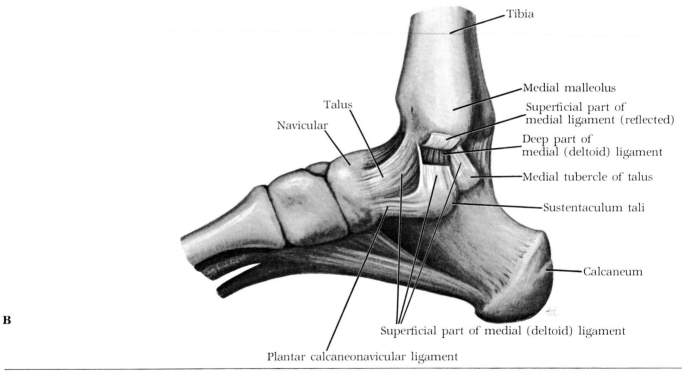

Figure 6-68
Right ankle joint. A. Lateral view. B. Medial view.

1. Swollen ankles is a very common clinical finding. Bilateral swollen ankles is usually systemic in origin and can be due to cardiac or renal insufficiency. Venous obstruction due to pressure on the inferior vena cava or lymphatic obstruction caused by malignant disease, filariasis, or congenital idiopathic hereditary lymphedema (Milroy's disease) may also be responsible for swollen ankles. One swollen ankle is usually associated with a ligamentous or bony injury.

2. Traumatic arthritis, which is commonly found in athletes, results from repeated minor trauma to the capsule and margins of the articular surfaces. Osteophytes are often present. The condition is associated with aching pain and swelling of the joint.

3. Ankle sprains are usually caused by excessive inversion of the foot. The anterior talofibular ligament and the calcaneofibular ligament are partially torn, giving rise to great pain and local swelling. A similar but less common injury can occur to the medial or deltoid ligament as the result of excessive eversion. The great strength of the medial ligament usually results in the ligament pulling off the tip of the medial malleolus.

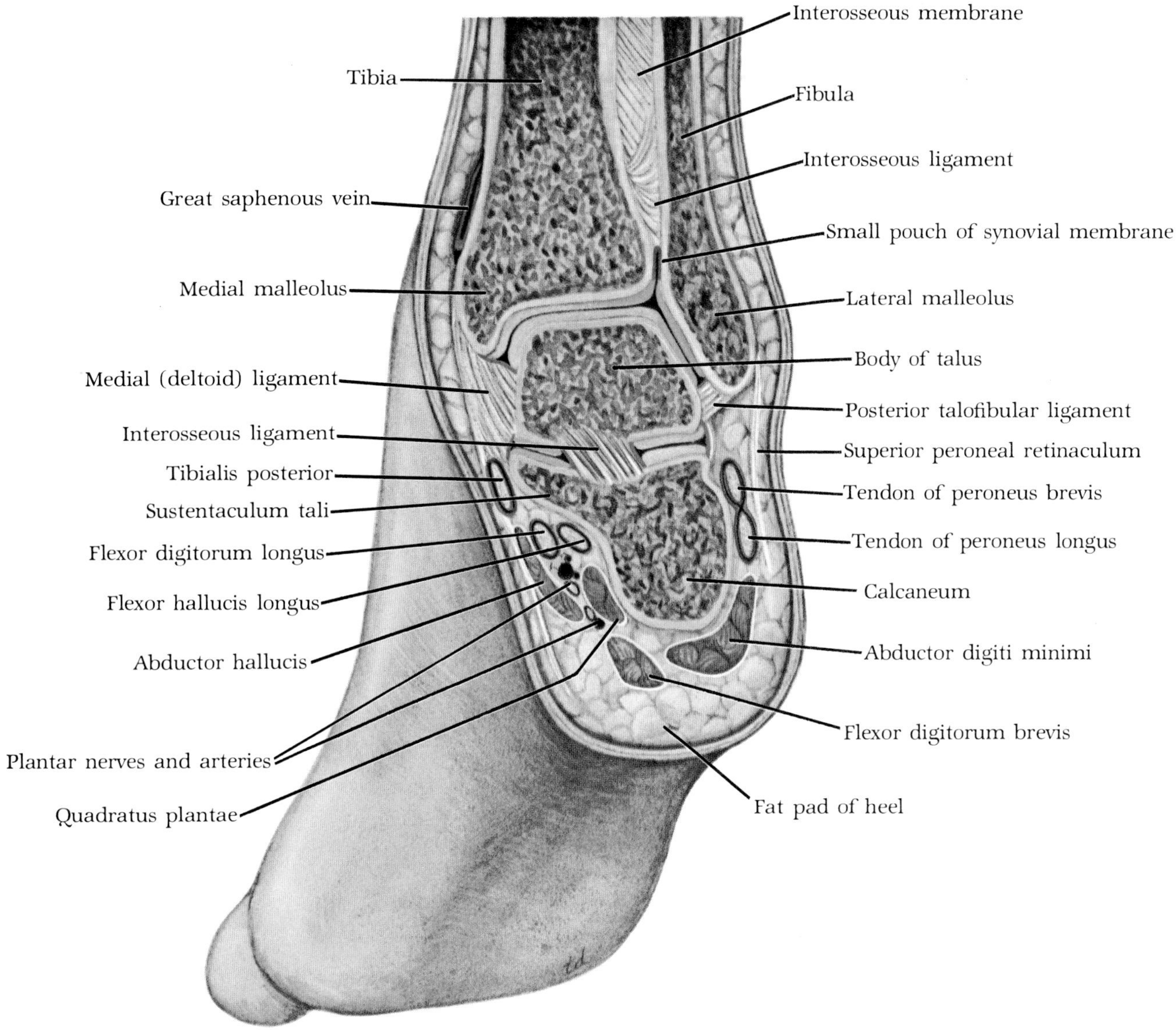

Interosseous membrane
Tibia
Fibula
Interosseous ligament
Great saphenous vein
Small pouch of synovial membrane
Medial malleolus
Lateral malleolus
Body of talus
Medial (deltoid) ligament
Posterior talofibular ligament
Interosseous ligament
Superior peroneal retinaculum
Tibialis posterior
Tendon of peroneus brevis
Sustentaculum tali
Flexor digitorum longus
Tendon of peroneus longus
Flexor hallucis longus
Calcaneum
Abductor hallucis
Abductor digiti minimi
Flexor digitorum brevis
Plantar nerves and arteries
Quadratus plantae
Fat pad of heel

Figure 6-69
Right ankle joint, vertical section.

1. The ankle joint is a hinge joint of great stability. The deep mortise formed by the lower end of the tibia and the medial and lateral malleoli securely holds the talus in position.

2. Although movements of the ankle joint are limited to dorsiflexion and plantar flexion, movements of the foot and ankle almost always involve more distal joints. In the subtalar joint the movement of inversion and eversion takes place; in the midtarsal joints forefoot adduction and abduction occurs; and in the toes the movements of flexion and extension take place.

3. If the ankle joint is to be examined, have the patient sit on the edge of the examining table with his legs hanging down. In this position the knees are flexed, which relaxes the gastrocnemius muscles. Stabilize the distal joints by grasping the heel firmly and inverting the forefoot. Now test for dorsiflexion, which should be about 20 degrees of motion, and then plantar flexion, which should be about 50 degrees of motion. Because the posterior part of the body of the talus is narrower than the anterior part, there is a slight amount of lateral mobility on plantar-flexing the ankle. This lateral mobility is absent on full dorsiflexion of the ankle joint.

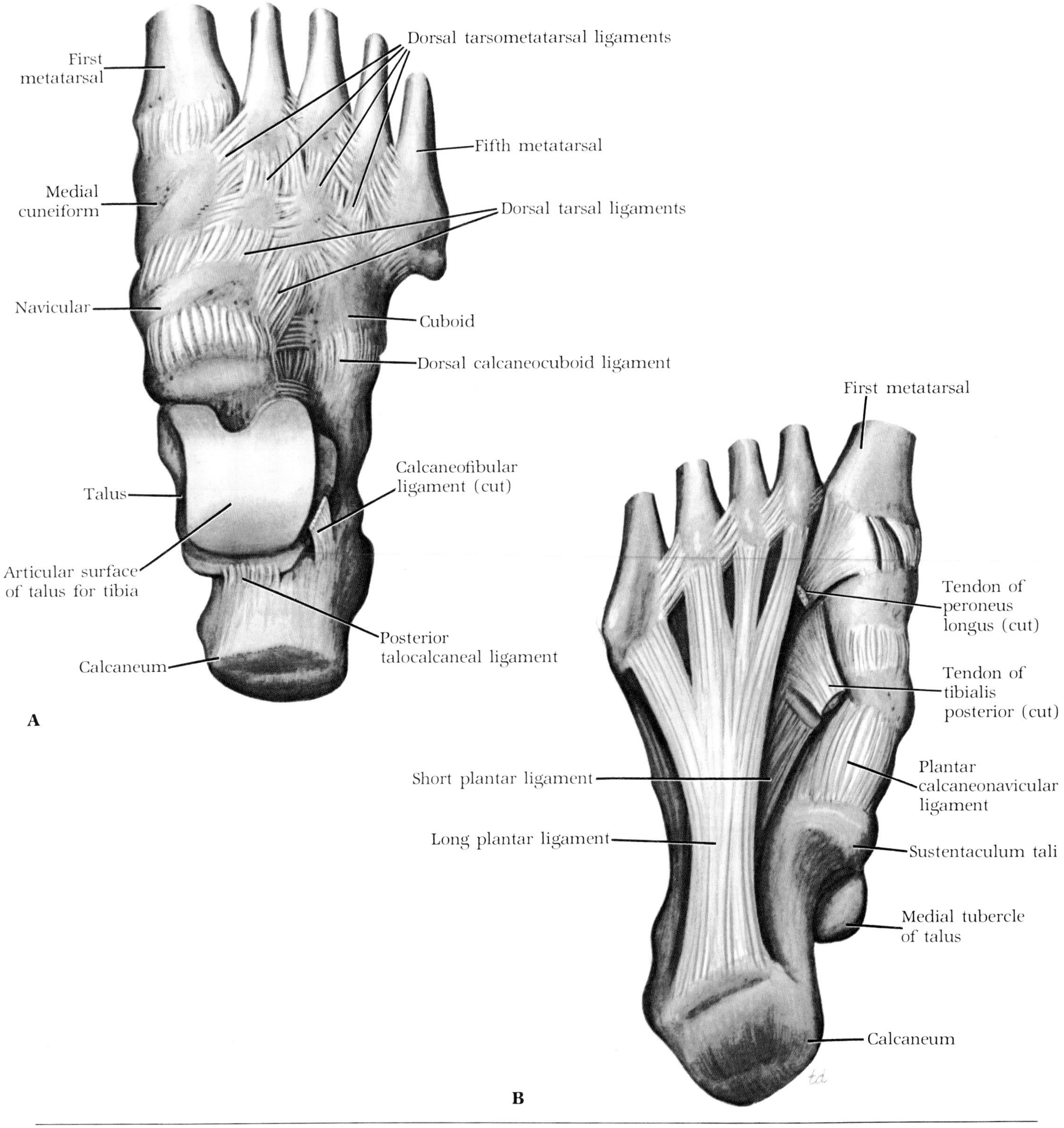

First metatarsal
Dorsal tarsometatarsal ligaments
Fifth metatarsal
Medial cuneiform
Dorsal tarsal ligaments
Navicular
Cuboid
Dorsal calcaneocuboid ligament
Talus
Calcaneofibular ligament (cut)
Articular surface of talus for tibia
Posterior talocalcaneal ligament
Calcaneum
A
First metatarsal
Tendon of peroneus longus (cut)
Tendon of tibialis posterior (cut)
Plantar calcaneonavicular ligament
Short plantar ligament
Long plantar ligament
Sustentaculum tali
Medial tubercle of talus
Calcaneum
B

Figure 6-70
Right foot. A. Dorsal surface, showing strong ligaments connecting the bones. B. Plantar surface, showing the long and short plantar ligaments.

1. The shape of the bones, the strong ligaments (especially those on the plantar surface of the foot), and the tone of the muscles play a great role in supporting the arches of the foot. The long and the short plantar ligaments are particularly important in this respect because they bind the bones together on their under aspect.

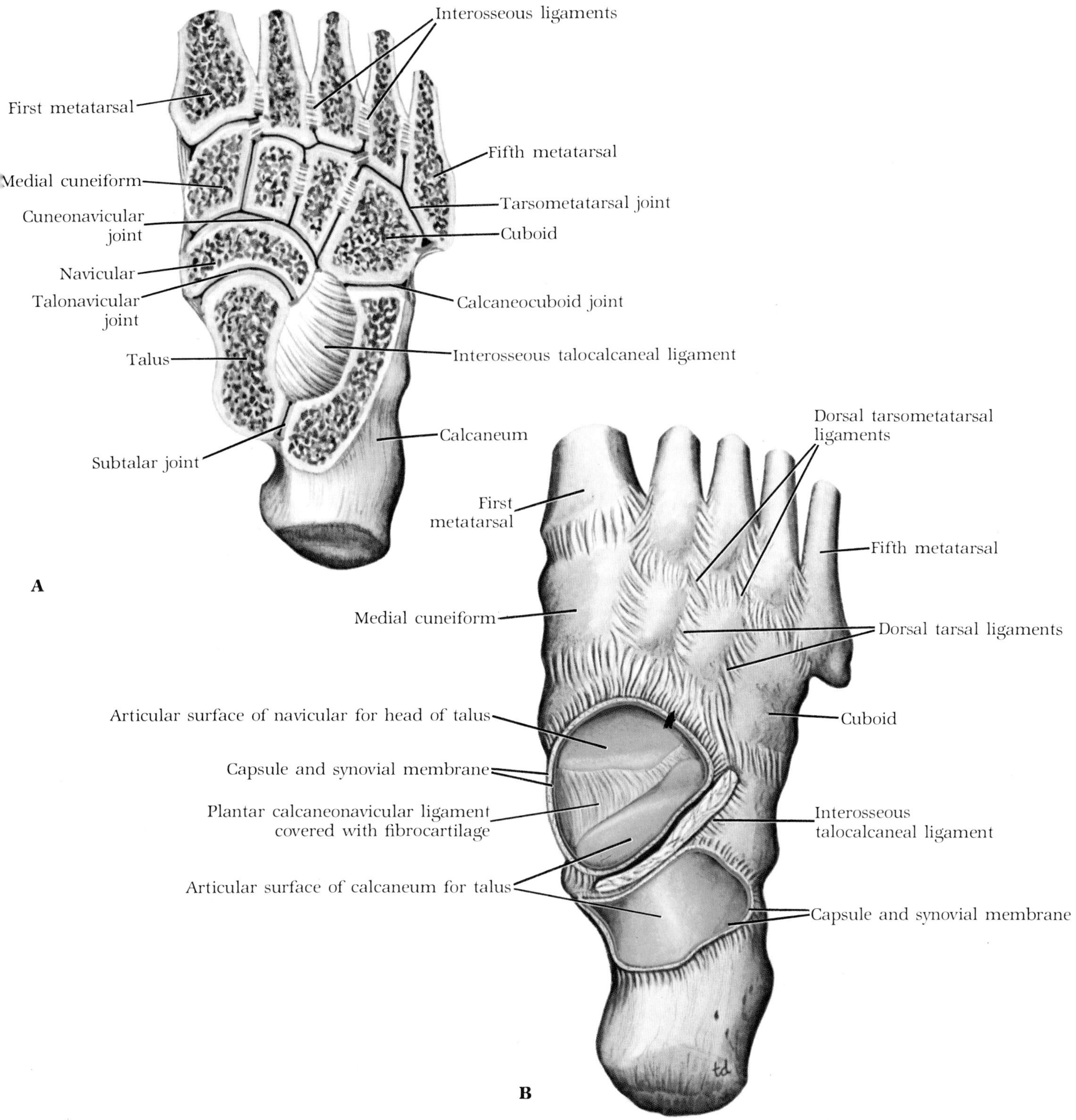

Interosseous ligaments
First metatarsal
Fifth metatarsal
Medial cuneiform
Tarsometatarsal joint
Cuneonavicular joint
Cuboid
Navicular
Talonavicular joint
Calcaneocuboid joint
Talus
Interosseous talocalcaneal ligament
Calcaneum
Subtalar joint
A
Dorsal tarsometatarsal ligaments
First metatarsal
Fifth metatarsal
Medial cuneiform
Dorsal tarsal ligaments
Articular surface of navicular for head of talus
Cuboid
Capsule and synovial membrane
Plantar calcaneonavicular ligament covered with fibrocartilage
Interosseous talocalcaneal ligament
Articular surface of calcaneum for talus
Capsule and synovial membrane
B

Figure 6-71
Right foot. A. Oblique section shows the synovial cavities of the tarsal and tarsometatarsal joints, viewed from above. B. Inferior articular surfaces of the subtalar and talocalcaneonavicular joints after removal of the talus, viewed from above.

1. The tarsometatarsal and intermetatarsal joints are synovial joints of the plane variety. The bones are connected by dorsal, plantar, and interosseous ligaments. The tarsometatarsal joint of the big toe has a separate joint cavity.

2. The subtalar, talocalcaneonavicular, and calcaneocuboid joints are of the synovial type. The talocalcaneonavicular and calcaneocuboid joints are often collectively referred to as the midtarsal or transverse tarsal joints, at which inversion and eversion movements of the foot take place. Inversion is the movement of the foot so that the sole faces medially, while eversion is the opposite movement of the foot so that the sole faces in the lateral direction. The movement of inversion is more extensive than eversion.

3. To test the movements of inversion and eversion have the patient sit on the edge of the examining table with his legs hanging down. Now grasp the lower tibia and fibula with one hand; with the other hand hold the calcaneum and invert and evert the heel. Arthritis in the subtalar joint produces pain during these movements.

7. The Head and Neck

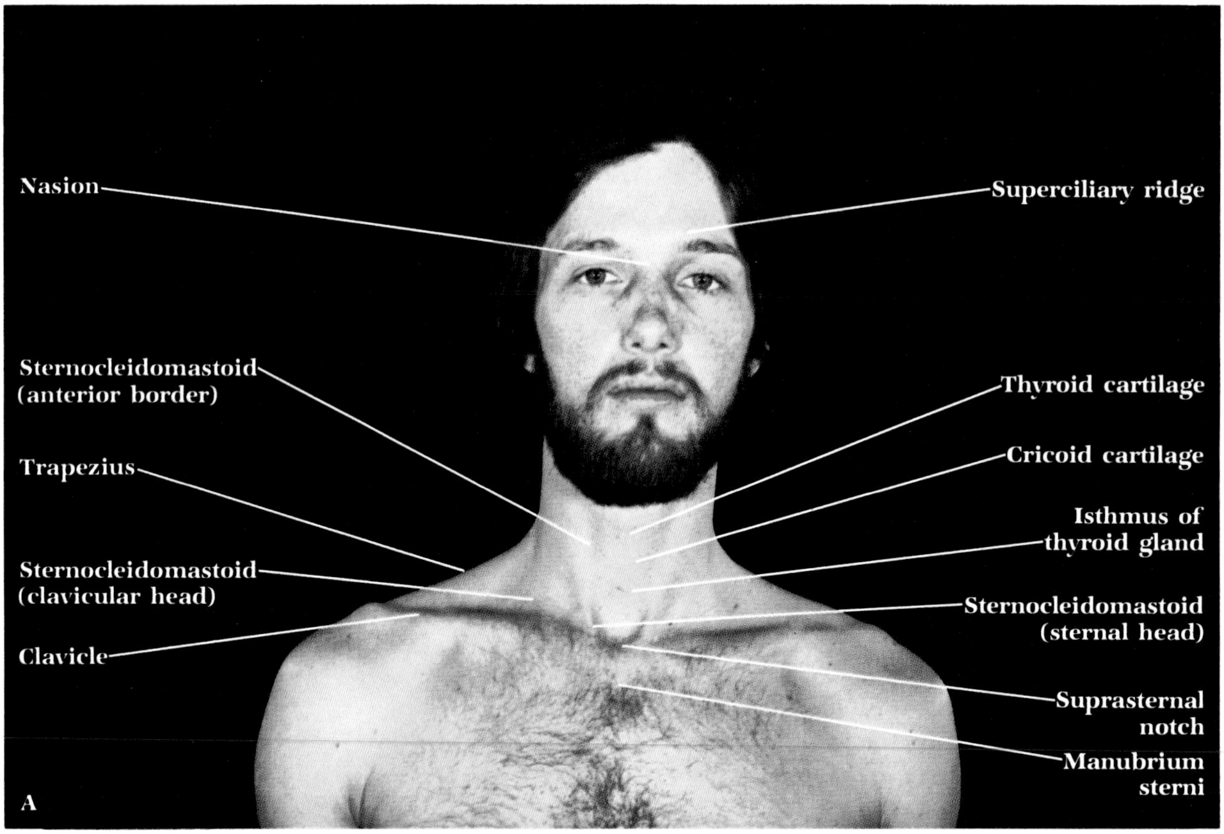
Nasion
Superciliary ridge
Sternocleidomastoid (anterior border)
Thyroid cartilage
Trapezius
Cricoid cartilage
Isthmus of thyroid gland
Sternocleidomastoid (clavicular head)
Sternocleidomastoid (sternal head)
Clavicle
Suprasternal notch
Manubrium sterni
A

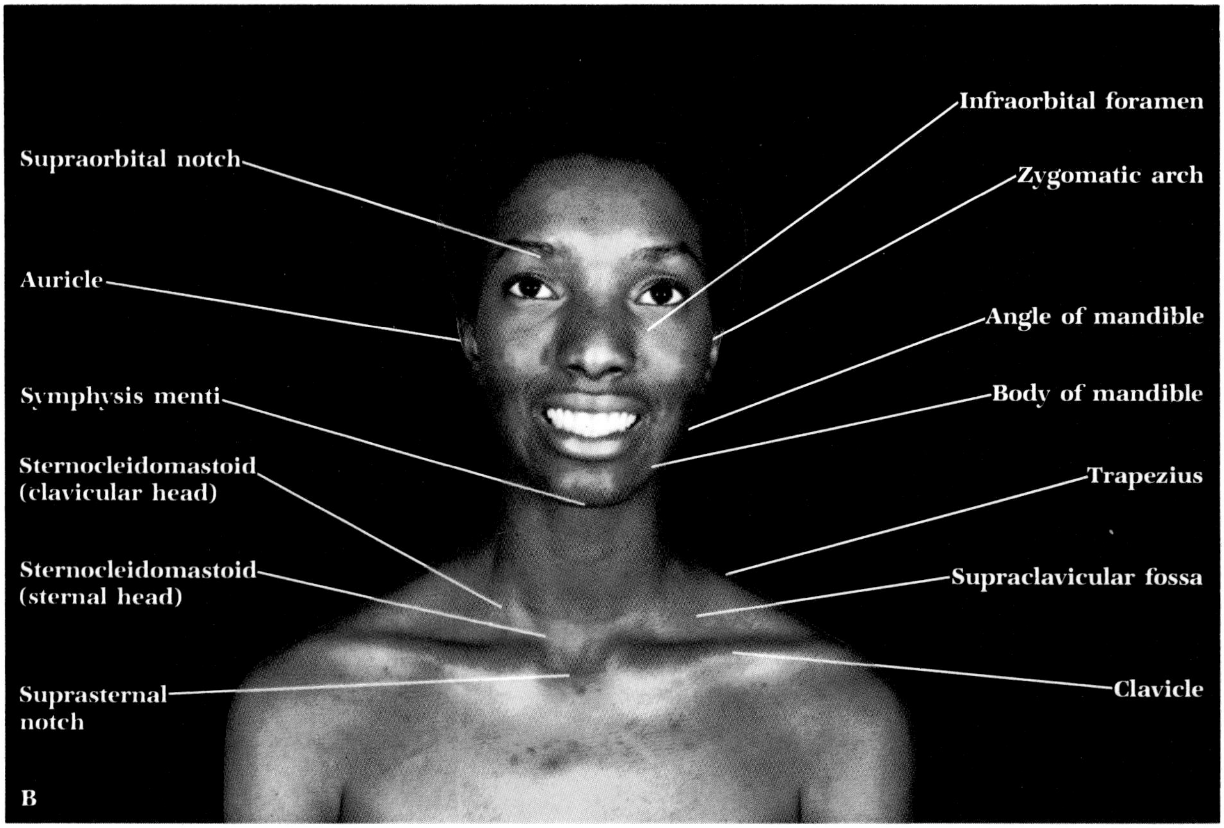
Infraorbital foramen
Supraorbital notch
Zygomatic arch
Auricle
Angle of mandible
Symphysis menti
Body of mandible
Sternocleidomastoid (clavicular head)
Trapezius
Sternocleidomastoid (sternal head)
Supraclavicular fossa
Suprasternal notch
Clavicle
B

Figure 7-1
Anterior view of head and neck. A. 27-year-old male. B. 22-year-old female.

1. The nasion is the depression in the midline at the root of the nose.
2. The superciliary ridge is a prominent ridge on the frontal bone above the upper margin of the orbit. The frontal air sinuses lie deep to this ridge on either side of the midline.
3. The zygomatic arch extends forward in front of the ear and ends in front in the zygomatic bone.
4. The body of the mandible is best examined by putting one finger inside the mouth and another on the outside. The mandible can thus be examined from the symphysis menti as far posteriorly as the angle of the mandible.
5. The upper border of the thyroid cartilage lies opposite the fourth cervical vertebra.
6. The cricoid cartilage, an important landmark in the neck, lies at the level of the sixth cervical vertebra.
7. The isthmus of the thyroid gland lies in front of the second, third, and fourth rings of the trachea.
8. The suprasternal notch can be felt between the anterior ends of the clavicles; it is the superior border of the manubrium sterni and lies opposite the lower border of the body of the second thoracic vertebra.
9. The sternocleidomastoid muscle can be palpated throughout its length.
10. The anterior border of the trapezius muscle may be felt by having the patient shrug his shoulders.
11. Each clavicle is subcutaneous throughout its entire length and can be easily palpated.

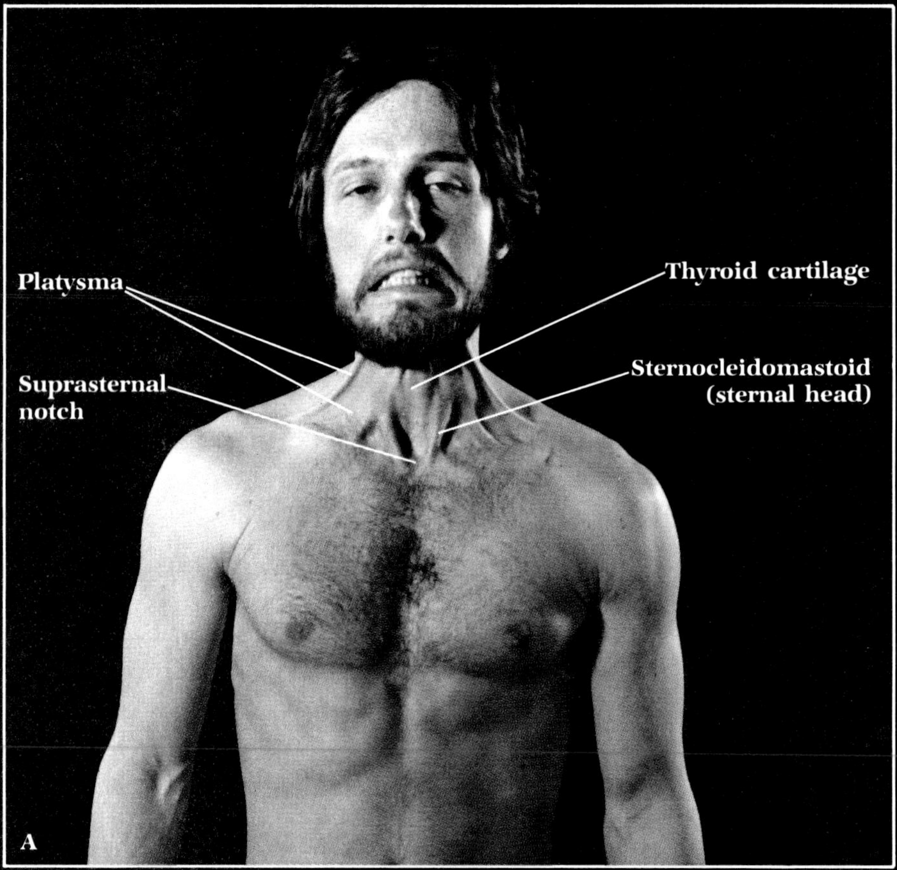
Platysma
Thyroid cartilage
Suprasternal notch
Sternocleidomastoid (sternal head)
A

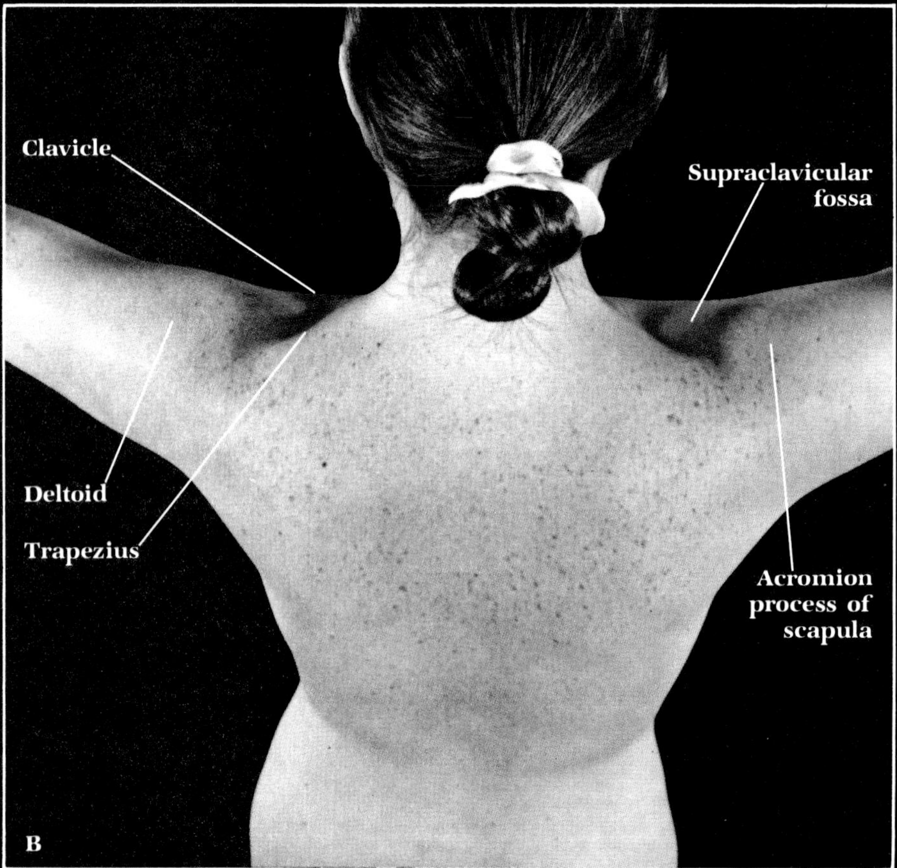
Clavicle
Supraclavicular fossa
Deltoid
Trapezius
Acromion process of scapula
B

Figure 7-2
Head and neck. A. Anterior view in a 27-year-old male with the jaws firmly clenched. B. Posterosuperior view of a 25-year-old female with the shoulder joints abducted.

1. The platysma can be seen as a sheet of muscle showing through the skin when the jaws are firmly clenched. The muscle extends down from the body of the mandible over the clavicle onto the anterior thoracic wall.
2. The suprasternal notch is the superior border of the manubrium sterni and can be felt between the anterior ends of the clavicles.
3. Each clavicle is subcutaneous throughout its length and articulates at its lateral extremity with the acromion process of the scapula. The sternoclavicular joint can be identified at the medial end of the clavicle.
4. The sternocleidomastoid muscle can be palpated throughout its length. It can be made to stand out by asking the patient to approximate his ear to the shoulder of the same side and at the same time rotate his head so that his face looks upward toward the opposite side. If the movement is carried out against resistance, the muscle will be felt to contract, and its anterior and posterior borders will be defined.
5. The anterior border of the trapezius muscle may be felt by having the patient shrug his shoulders. It can be seen to extend from the superior nuchal line of the occipital bone, downward and forward to the posterior border of the lateral third of the clavicle.
6. The supraclavicular fossa is a depression overlying the lower part of the posterior triangle of the neck and lies between the sternocleidomastoid and trapezius muscles.

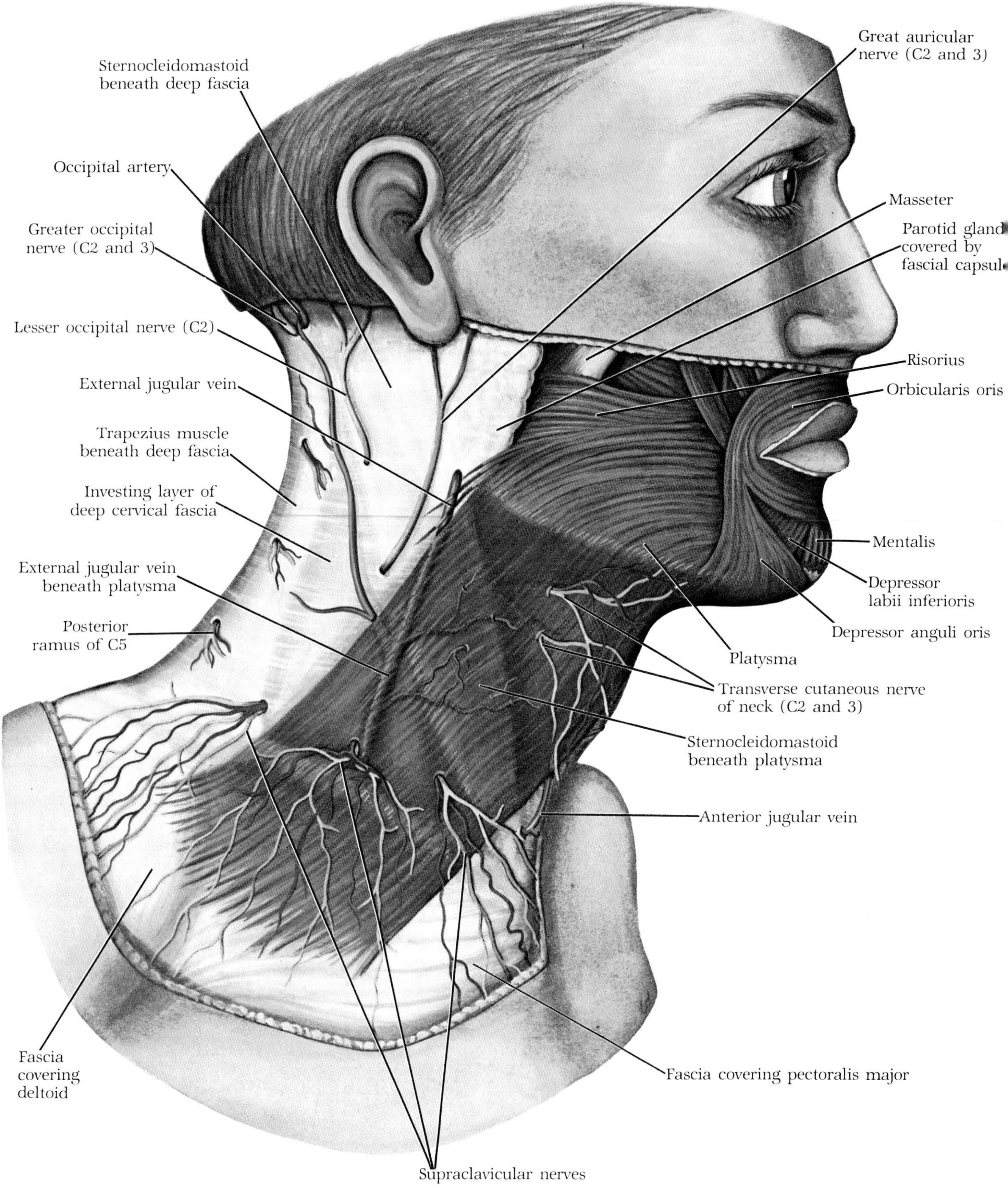
Sternocleidomastoid
beneath deep fascia
Occipital artery
Greater occipital
nerve (C2 and 3)
Lesser occipital nerve (C2)
External jugular vein
Trapezius muscle
beneath deep fascia
Investing layer of
deep cervical fascia
External jugular vein
beneath platysma
Posterior
ramus of C5
Fascia
covering
deltoid
Great auricular
nerve (C2 and 3)
Masseter
Parotid gland
covered by
fascial capsule
Risorius
Orbicularis oris
Mentalis
Depressor
labii inferioris
Depressor anguli oris
Platysma
Transverse cutaneous nerve
of neck (C2 and 3)
Sternocleidomastoid
beneath platysma
Anterior jugular vein
Fascia covering pectoralis major
Supraclavicular nerves

Figure 7-3
The platysma muscle, superficial veins, arteries, and cutaneous nerves.

1. Skin incisions in the neck produce unsightly scars unless planned correctly. When possible, incisions should be made transversely with a slight downward curve so that they lie along Langer's lines. These lines correspond to the direction of the rows of collagen in the dermis. A surgical incision made along or between these rows causes the minimum disruption of the collagen, resulting in the production of a minimum of scar tissue. On the other hand, an incision made across the rows of collagen disrupts and disturbs it, resulting in the massive production of fresh collagen and formation of a broad, ugly scar.
2. In surgical incisions in the neck, the skin and superficial fascia are incised first; the underlying platysma is divided in the same direction but at a different level. The deep fascia is then incised. When the wound is closed, the deep fascia, platysma, subcutaneous tissue, and skin are each sutured separately. A small drainage tube is essential to prevent the formation of a hematoma, with subsequent scar formation.
3. The external jugular vein serves as a useful clinical venous manometer. It runs in the superficial fascia from the angle of the mandible to just above the midpoint of the clavicle. It pierces the investing layer of deep cervical fascia about a fingerbreadth above the clavicle and drains into the subclavian vein.

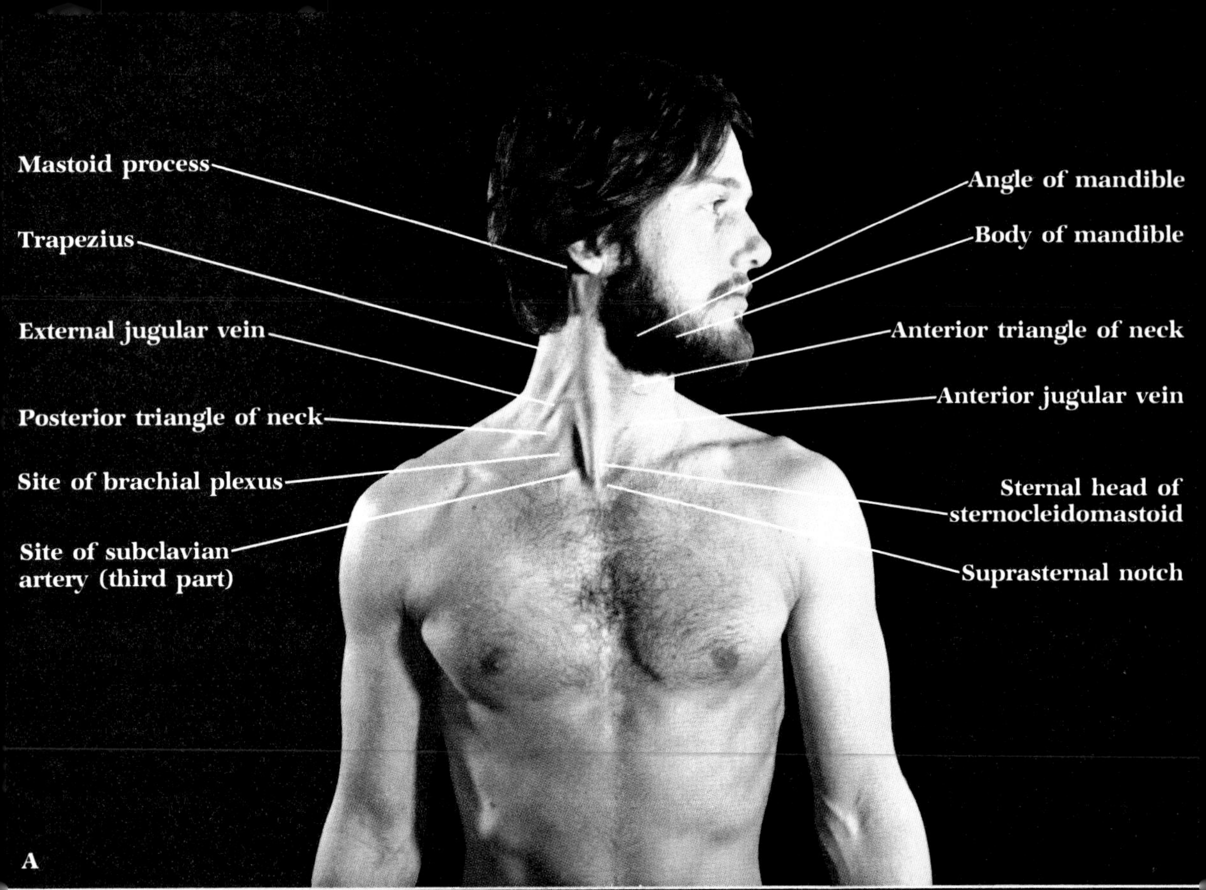
Mastoid process
Trapezius
External jugular vein
Posterior triangle of neck
Site of brachial plexus
Site of subclavian artery (third part)
Angle of mandible
Body of mandible
Anterior triangle of neck
Anterior jugular vein
Sternal head of sternocleidomastoid
Suprasternal notch
A

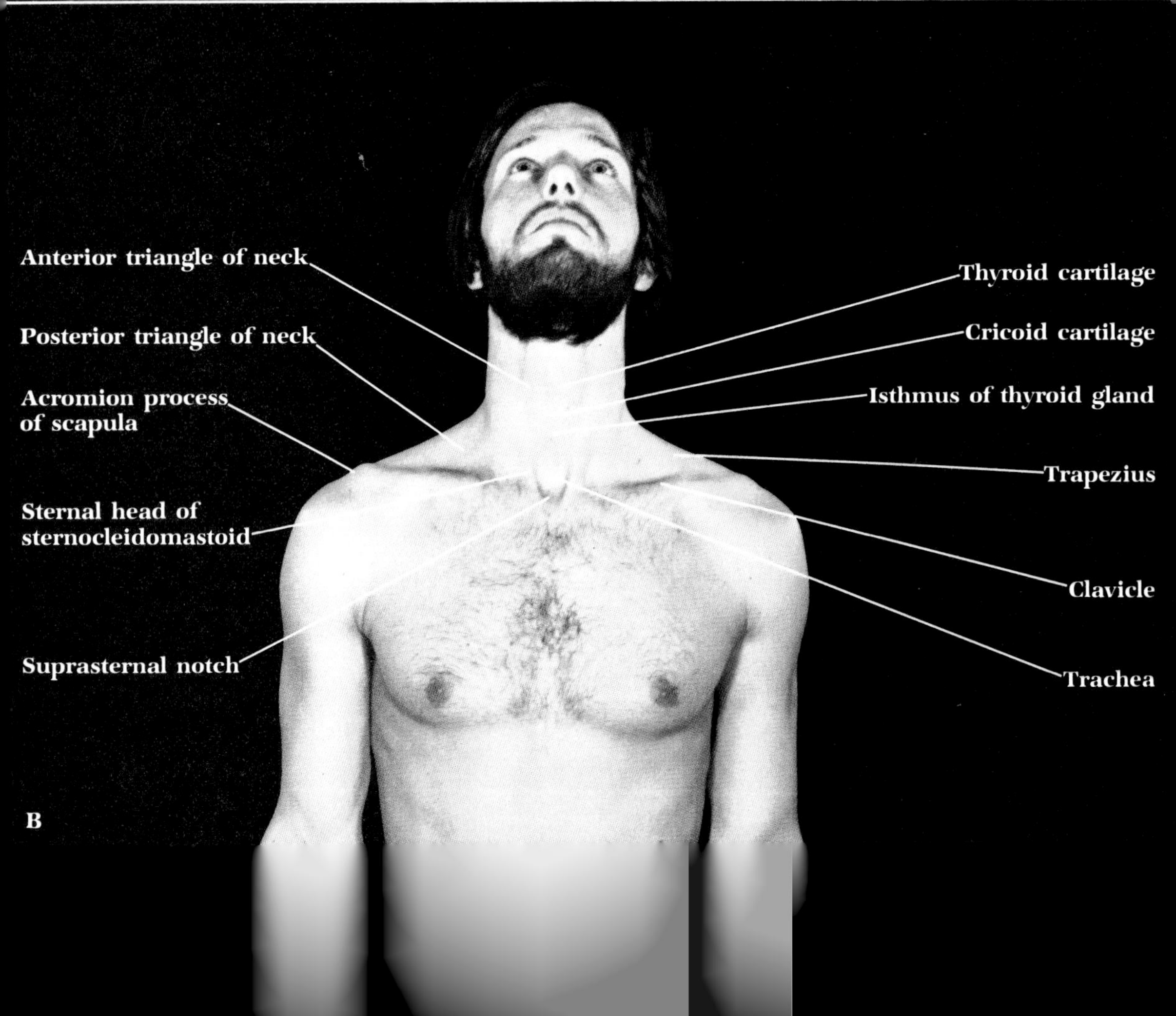
Anterior triangle of neck
Posterior triangle of neck
Acromion process of scapula
Sternal head of sternocleidomastoid
Suprasternal notch
Thyroid cartilage
Cricoid cartilage
Isthmus of thyroid gland
Trapezius
Clavicle
Trachea
B

Figure 7-4
Anterior view of the neck of a 27-year-old male. A. With the head rotated to the left. B. With the head extended.

1. The sternocleidomastoid muscle can be palpated throughout its length as it passes superiorly from the sternum and clavicle to the mastoid process. It serves to divide the neck into anterior and posterior triangles.
2. The anterior triangle of the neck is bounded by the body of the mandible, the sternocleidomastoid muscle, and the midline. The posterior triangle is bounded by the anterior border of the trapezius muscle, the sternocleidomastoid, and the clavicle.
3. The external jugular vein lies in the superficial fascia deep to the platysma. It passes downward from the region of the angle of the mandible to the middle of the clavicle. It perforates the deep fascia just above the clavicle and drains into the subclavian vein.
4. The anterior jugular vein begins just below the mandible by the union of several small veins and runs down the neck close to the midline. Just above the suprasternal notch, the veins of the two sides are united by the jugular arch.
5. The third part of the subclavian artery may be palpated on the superior surface of the first rib in the lower anterior angle of the posterior triangle.
6. The roots and trunks of the brachial plexus occupy the lower anterior angle of the posterior triangle. The upper limit of the plexus may be indicated by a line drawn from the cricoid cartilage downward to the middle of the clavicle.
7. The trachea may be palpated in the midline just above the suprasternal notch.

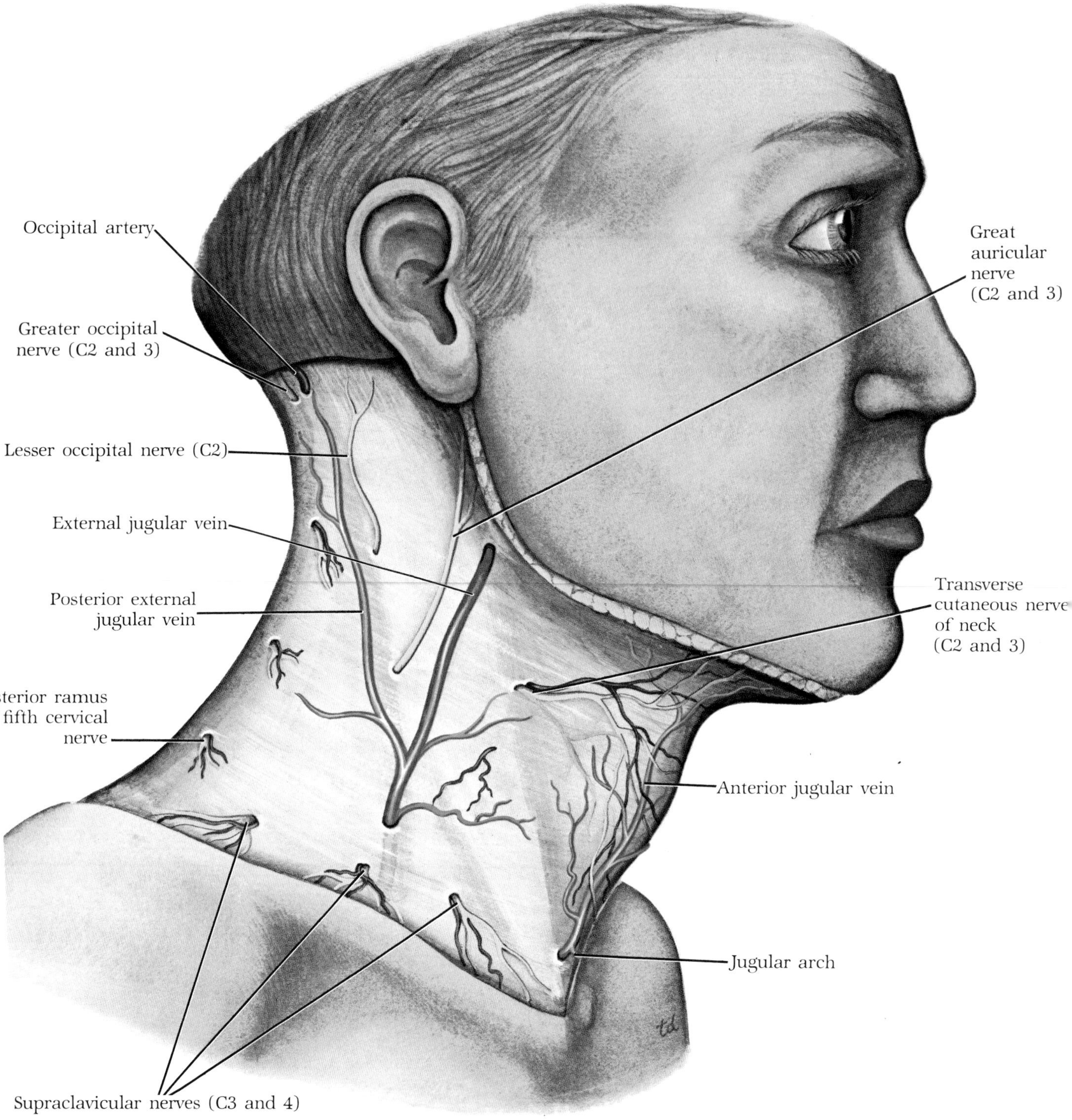
Occipital artery
Greater occipital nerve (C2 and 3)
Lesser occipital nerve (C2)
External jugular vein
Posterior external jugular vein
Posterior ramus of fifth cervical nerve
Supraclavicular nerves (C3 and 4)
Great auricular nerve (C2 and 3)
Transverse cutaneous nerve of neck (C2 and 3)
Anterior jugular vein
Jugular arch

Figure 7-5
Investing layer of deep cervical fascia with superficial veins, arteries, and cutaneous nerves; the platysma has been removed.

1. The external jugular vein is easily seen on the side of the neck as it runs downward superficial to the sternocleidomastoid muscle. Normally when the patient is lying supine with his head on pillows, the level of the blood in the external jugular veins reaches about a third of the way up the neck. As the patient sits up, the blood level falls until it is no longer visible behind the clavicle. Patients with right-sided heart failure or obstruction of the superior vena cava by a neoplasm show a raised venous pressure with higher levels of blood in the external jugular veins.

2. Local anesthesia of the skin of the anterior part of the neck is sometimes used for thyroidectomy operations, most often in patients for whom a general anesthetic is contraindicated. The transverse cutaneous nerve of the neck is blocked by the anesthetic agent as it emerges from behind the middle of the posterior border of the sternocleidomastoid muscle.

3. The deep fascia of the neck in certain areas forms distinct sheets that are easily recognizable by the surgeon at operation; these sheets are called (1) the investing, (2) the pretracheal, and (3) the prevertebral layers. Their clinical importance lies in the fact that they can determine the direction or can limit the spread of infection in the neck.

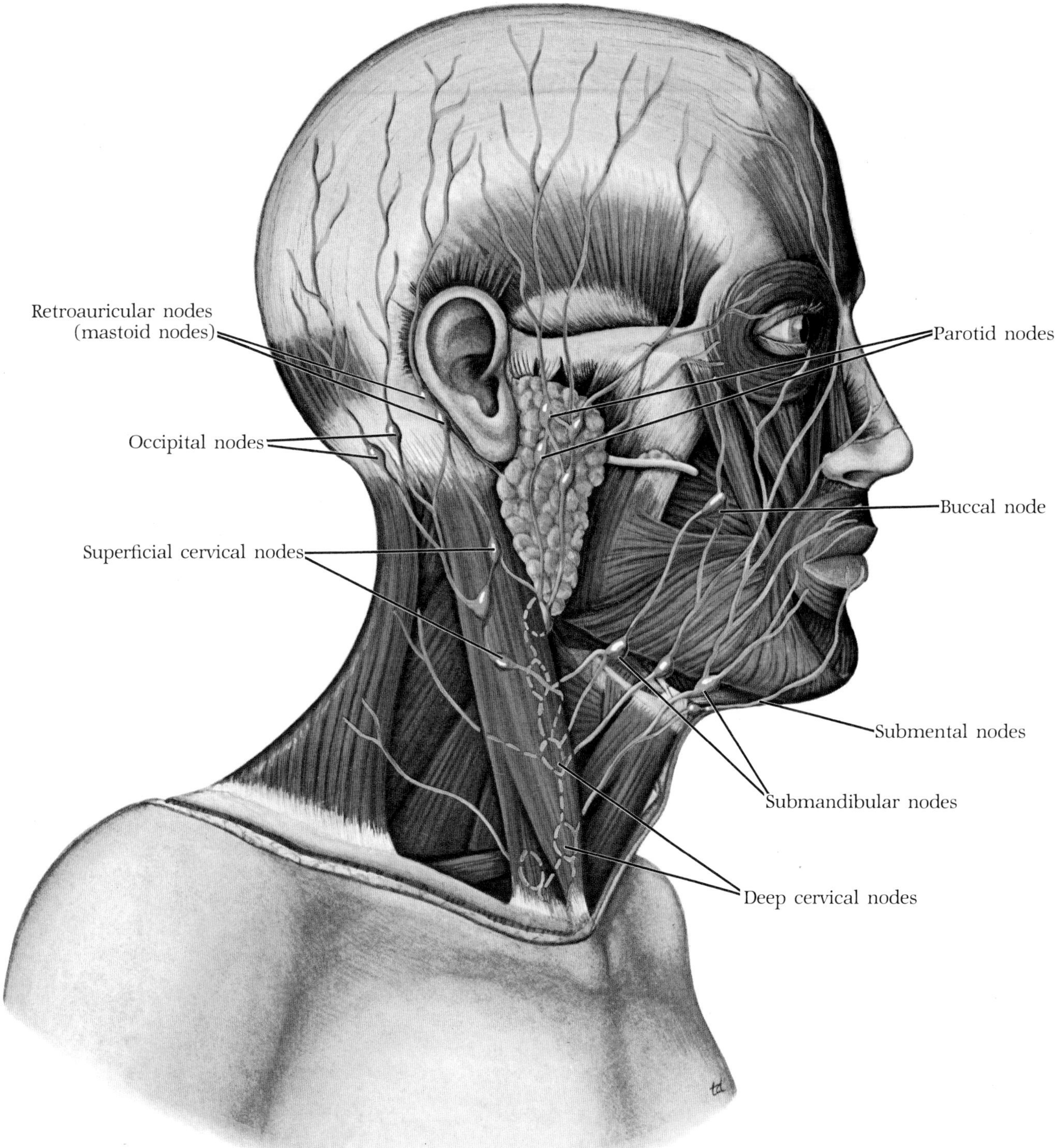
Retroauricular nodes
(mastoid nodes)
Parotid nodes
Occipital nodes
Buccal node
Superficial cervical nodes
Submental nodes
Submandibular nodes
Deep cervical nodes

Figure 7-6
Superficial lymphatic drainage of head and neck.

1. Lymph nodes in the neck should be examined from behind the patient. The examination is made easier by asking the patient to flex the neck slightly to reduce the tone of the muscles. The groups of nodes should be examined in a definite order so that none are omitted. The superficial cervical nodes extend along the external jugular vein; the occipital nodes are found at the apex of the posterior triangle.
2. Following the identification of enlarged lymph nodes, possible sites of infection or of neoplastic growth should be examined. Face, scalp, tongue, mouth, tonsil, pharynx, and so on should all be carefully looked at. If you find that the lymph nodes are enlarged and hard and you suspect malignant changes but cannot find the primary tumor, do not forget to examine the nasopharynx, piriform fossa, larynx, nasal sinuses, thyroid gland, and lungs.
3. Tuberculous infection of the deep cervical lymph nodes may result in liquefaction and destruction of one or more of the nodes. The pus is at first limited by the investing layer of the deep fascia. Later this becomes eroded at one point, and the pus passes into the less restricted superficial fascia. A dumbbell or collar-stud abscess is now present. The clinician easily perceives this superficial abscess, but he must not forget the deeply placed abscess that may have led to the superficial one.

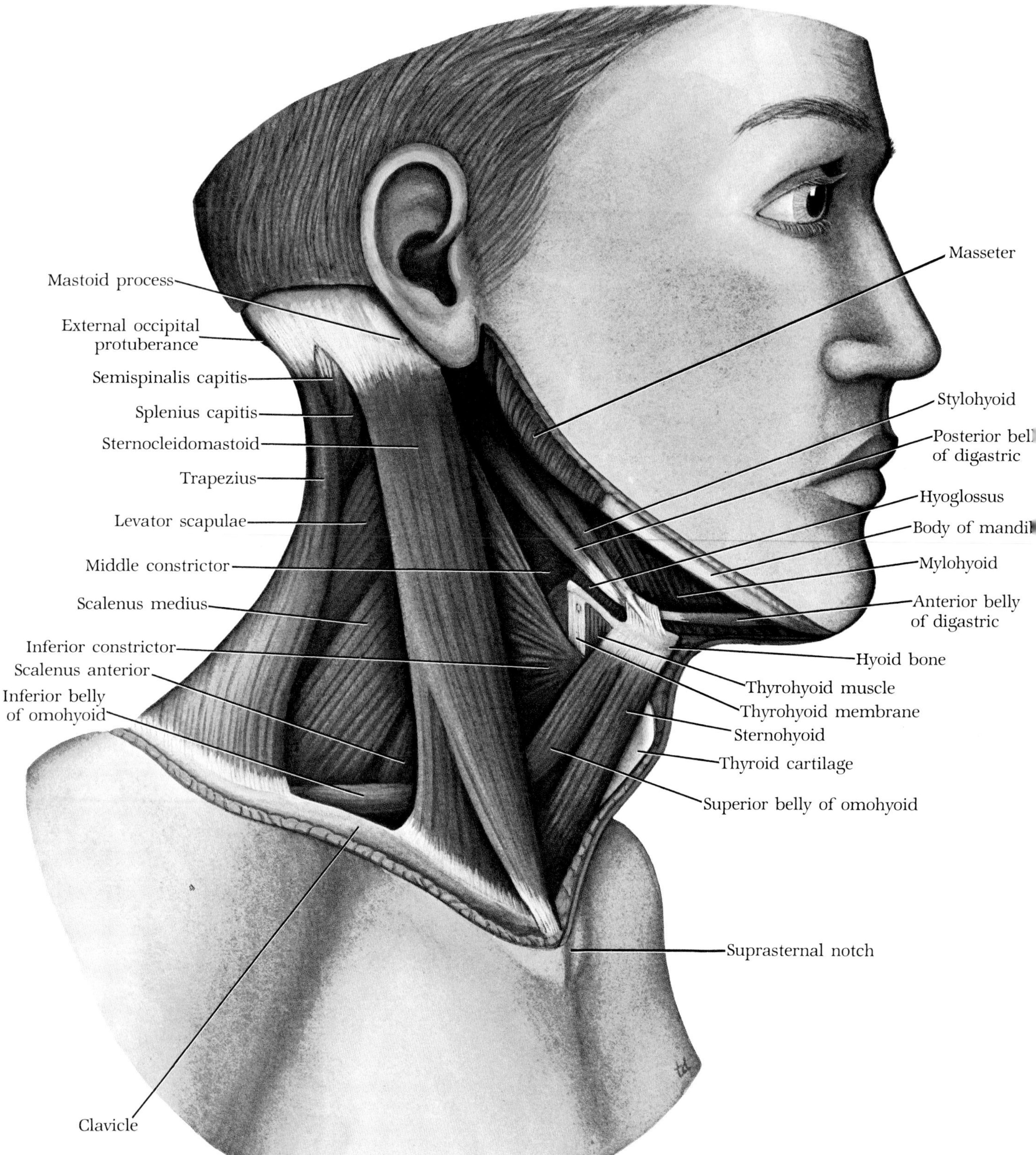

Masseter
Mastoid process
External occipital protuberance
Semispinalis capitis
Splenius capitis
Sternocleidomastoid
Trapezius
Levator scapulae
Middle constrictor
Scalenus medius
Inferior constrictor
Scalenus anterior
Inferior belly of omohyoid
Stylohyoid
Posterior bel
of digastric
Hyoglossus
Body of mandi
Mylohyoid
Anterior belly of digastric
Hyoid bone
Thyrohyoid muscle
Thyrohyoid membrane
Sternohyoid
Thyroid cartilage
Superior belly of omohyoid
Suprasternal notch
Clavicle

Figure 7-7
Muscular triangles of neck.

1. Congenital torticollis is the result of excessive stretching of the sternocleidomastoid muscle during a difficult labor. Hemorrhage that occurs into the muscle may be detected as a small rounded "tumor" during the early weeks after birth. This later becomes invaded by fibrous tissue, which in turn contracts and shortens the muscle. The mastoid process is thus pulled down toward the sterno-clavicular joint of the same side, the cervical spine is flexed, and the face looks upward to the opposite side.

2. Spasmodic torticollis is due to repeated chronic contractions of the sternocleido-mastoid and trapezius muscles and is usually psychogenic in origin. Section of the spinal part of the accessory nerve may be necessary in severe cases.

3. Branchial cyst. As the result of the downgrowth of the second pharyngeal arch in the embryo, the second, third, and fourth pharyngeal clefts are buried, forming an ectodermal-lined cavity called the cervical sinus. Normally this sinus eventually becomes obliterated, but if it does not, a branchial cyst forms. A branchial cyst gradually enlarges during childhood and in early adult life begins to protrude from beneath the anterior border of the sternocleidomastoid muscle. Surgical removal is necessary since such cysts increase in size and may become infected.

4. A branchial sinus is due to a persistence of the cervical sinus, which opens onto the skin of the neck. A branchial fistula occurs when the cervical sinus remains as a channel opening superiorly into the pharynx just posterior to the tonsil and inferiorly onto the skin along the anterior border of the sternocleidomastoid muscle. A branchial sinus or fistula should be removed in its entirety.

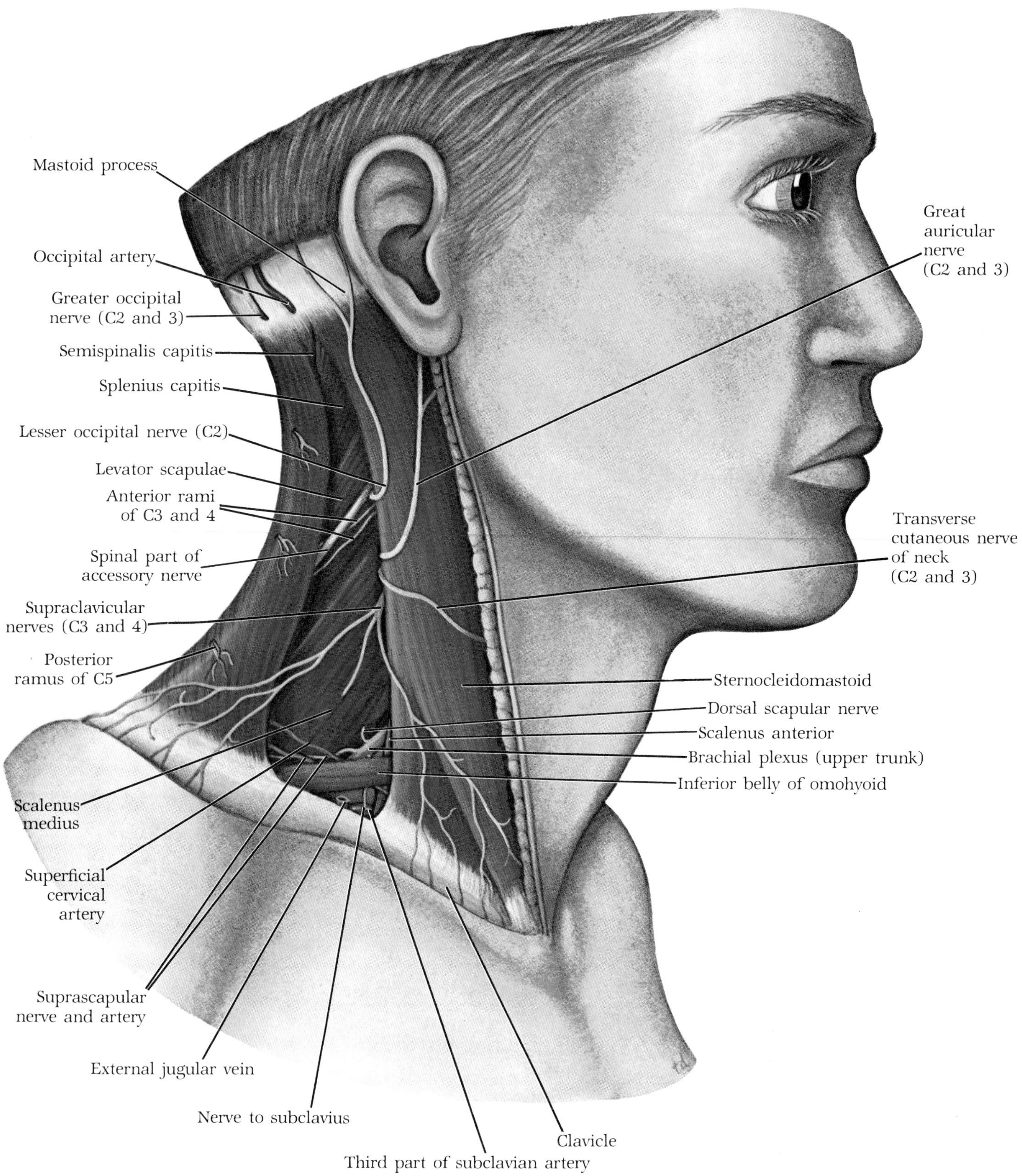

Mastoid process
Occipital artery
Greater occipital nerve (C2 and 3)
Semispinalis capitis
Splenius capitis
Lesser occipital nerve (C2)
Levator scapulae
Anterior rami of C3 and 4
Spinal part of accessory nerve
Supraclavicular nerves (C3 and 4)
Posterior ramus of C5
Scalenus medius
Superficial cervical artery
Suprascapular nerve and artery
External jugular vein
Nerve to subclavius
Third part of subclavian artery
Clavicle
Inferior belly of omohyoid
Brachial plexus (upper trunk)
Scalenus anterior
Dorsal scapular nerve
Sternocleidomastoid
Transverse cutaneous nerve of neck (C2 and 3)
Great auricular nerve (C2 and 3)

Figure 7-8
Posterior triangle of neck.

1. The spinal part of the accessory nerve crosses the posterior triangle and its course can be indicated by drawing a line from the angle of the mandible to the tip of the mastoid process and bisecting this line at right angles with a second line that passes downward across the triangle; the second line indicates the course of the nerve. This nerve may be injured during an operation or from penetrating wounds; as a result the trapezius muscle is paralyzed and will show wasting, and the shoulder will drop. The patient will experience difficulty in elevating that arm above his head.

2. The roots and trunks of the brachial plexus may be injured by stab or bullet wounds or injury may be due to traction or pressure injuries.

3. Pressure on the third part of the subclavian artery, compressing it against the upper surface of the first rib, can stop a hemorrhage resulting from laceration of the brachial or axillary arteries.

4. The brachial plexus and the subclavian artery enter the posterior triangle through a narrow, muscular, bony triangle formed by the scalenus anterior muscle, the scalenus medius muscle, and the first rib. In the presence of a cervical rib, the first thoracic nerve and the subclavian artery will be raised and angulated as they pass over the cervical rib. This may cause ischemic muscle pain in the arm and symptoms of pain in the forearm and hand, along with wasting of the small muscles of the hand.

5. The cervical dome of the pleura and the apex of the lung extend into the neck above the medial end of the clavicle. A penetrating wound above the medial part of the clavicle may involve the apex of the lung.

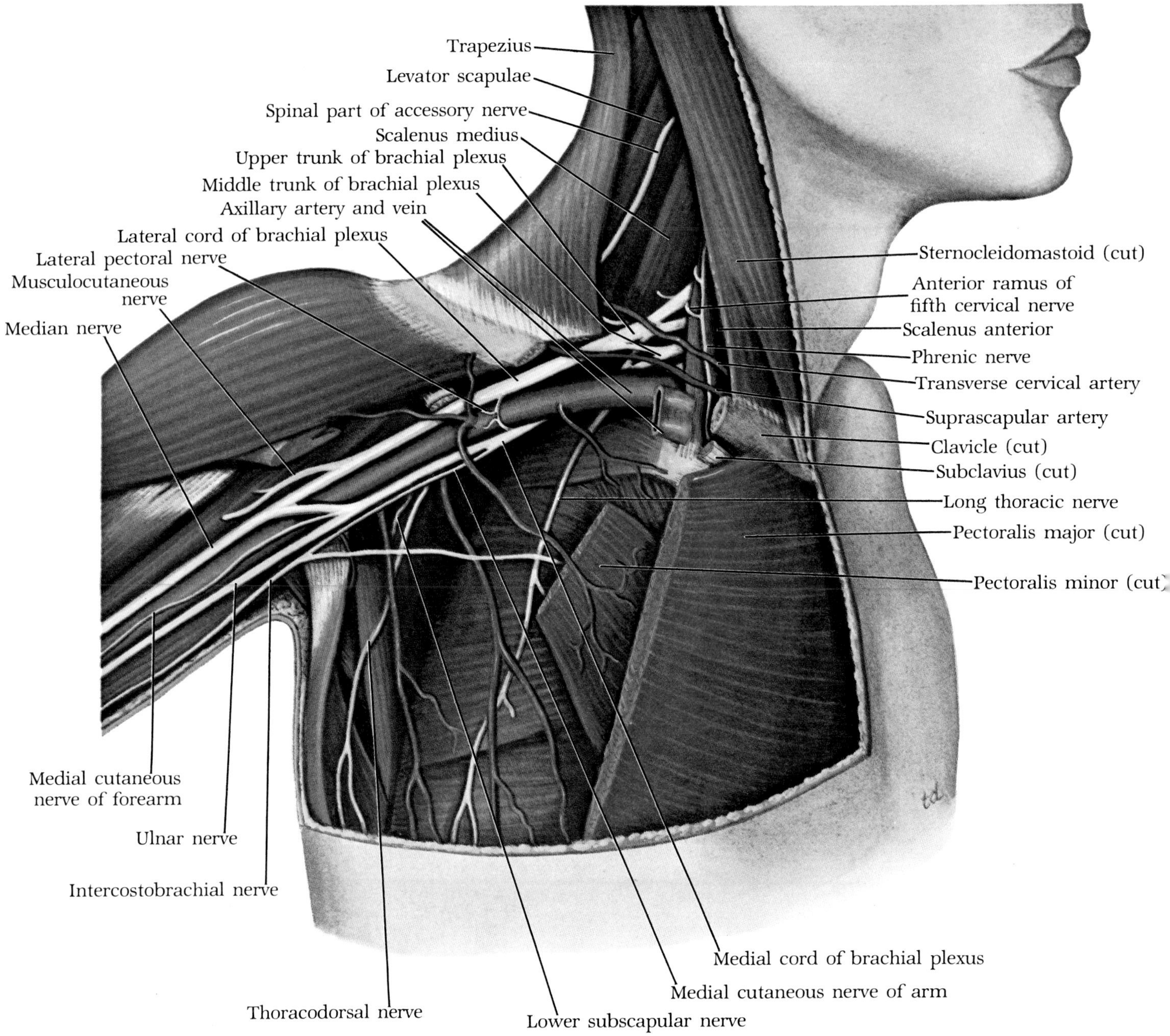
Trapezius
Levator scapulae
Spinal part of accessory nerve
Scalenus medius
Upper trunk of brachial plexus
Middle trunk of brachial plexus
Axillary artery and vein
Lateral cord of brachial plexus
Lateral pectoral nerve
Musculocutaneous nerve
Median nerve
Sternocleidomastoid (cut)
Anterior ramus of fifth cervical nerve
Scalenus anterior
Phrenic nerve
Transverse cervical artery
Suprascapular artery
Clavicle (cut)
Subclavius (cut)
Long thoracic nerve
Pectoralis major (cut)
Pectoralis minor (cut)
Medial cutaneous nerve of forearm
Ulnar nerve
Intercostobrachial nerve
Medial cord of brachial plexus
Medial cutaneous nerve of arm
Thoracodorsal nerve
Lower subscapular nerve

Figure 7-9
Posterior triangle of neck and axilla; the intermediate third of the clavicle has been removed.

1. Axillary sheath is formed as the brachial plexus and the subclavian artery emerge from between the scalenus anterior and medius muscles in the lower part of the neck. The prevertebral layer of deep cervical fascia forms a sheath that encloses these structures. The anesthesiologist can block the brachial plexus by injecting an anesthetic agent into the sheath in the axilla and milking it superiorly into the posterior triangle of the neck.

2. When malignant disease of the face, lip, tongue, and floor of the mouth has already metastasized to the regional lymph nodes, those nodes should be removed, if possible, by block dissection. The operation consists of removing, in one piece, the submental, submandibular, and deep cervical nodes; the investing layer of deep cervical fascia; the sternocleidomastoid muscle; the internal jugular vein; and all related connective tissue. The hypoglossal nerve is preserved if possible, but the submandibular salivary gland is removed.

3. Cervical lymphadenitis, an infection of the lymph nodes of the neck, is very common. The site of the original infection must be systematically looked for, especially in or on the scalp, face, nose, mouth, pharynx, and external auditory meatus.

4. Ludwig's angina is a streptococcal infection in the submandibular region. It is limited by the attachment of the investing layer of deep cervical fascia to the mandible above and the hyoid bone below. The inflammatory edema extends beneath the floor of the mouth and pushes the tongue forward and upward. Failure to treat this condition adequately results in the eventual spread of the infection downward, with edema of the vocal folds.

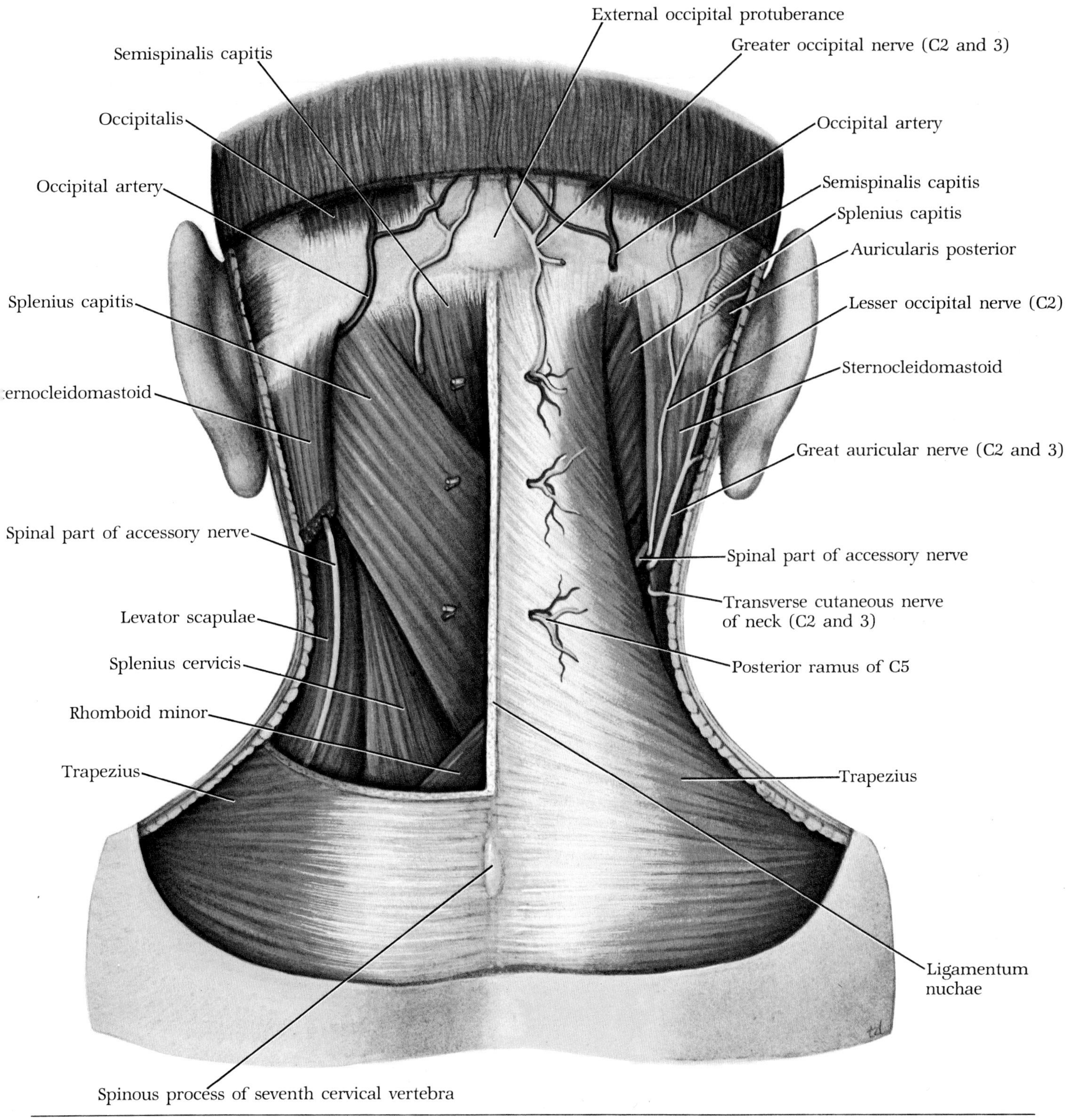
External occipital protuberance
Greater occipital nerve (C2 and 3)
Semispinalis capitis
Occipitalis
Occipital artery
Occipital artery
Semispinalis capitis
Splenius capitis
Auricularis posterior
Splenius capitis
Lesser occipital nerve (C2)
ernocleidomastoid
Sternocleidomastoid
Great auricular nerve (C2 and 3)
Spinal part of accessory nerve
Spinal part of accessory nerve
Transverse cutaneous nerve of neck (C2 and 3)
Levator scapulae
Splenius cervicis
Posterior ramus of C5
Rhomboid minor
Trapezius
Trapezius
Ligamentum nuchae
Spinous process of seventh cervical vertebra

Figure 7-10
Posterior view of the neck, showing the structures that form the floor of the posterior triangle. On the left, parts of the trapezius and sternocleidomastoid muscles have been removed.

1. The spinal part of the accessory nerve supplies the sternocleidomastoid and trapezius muscles. In order to test for a lesion of this nerve the patient should rotate his head to one side against resistance, thus bringing into action the sternocleidomastoid of the opposite side. The action of the trapezius muscles may be tested by having the patient shrug his shoulders.
2. The supraclavicular nerves supply the skin over the shoulder region. Gallbladder disease or irritation of the parietal peritoneum or parietal pleura covering the central part of the diaphragm may give referred pain to the patient's shoulder via the phrenic nerve (C3, 4, and 5) and the supraclavicular nerves (C3 and 4).
3. Tender enlargement of the occipital or mastoid lymph nodes may be due to infection of the posterior part of the scalp; the mastoid nodes are also commonly enlarged in German measles (rubella), and this condition is associated with a mild fever and a generalized rash.
4. Fractures of the skull involving the occipital bone do not become evident on the surface immediately, since the blood accumulates beneath the muscles in the posterior triangle. Later, the blood reaches the surface in the posterior triangle or near the mastoid process, causing skin discoloration or bruising.

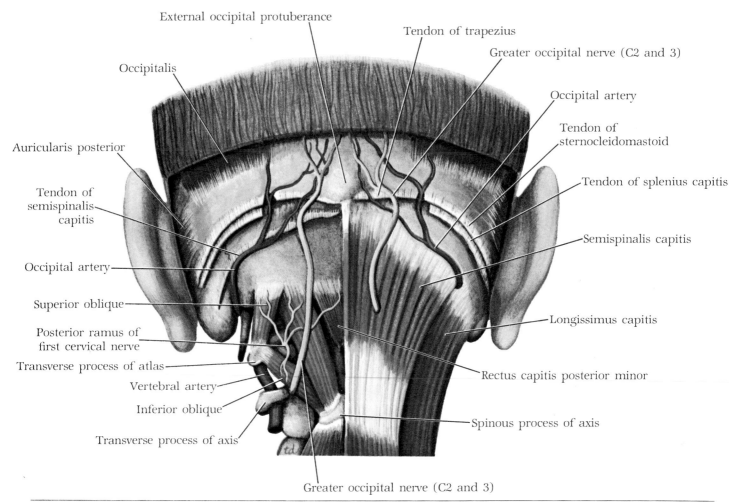

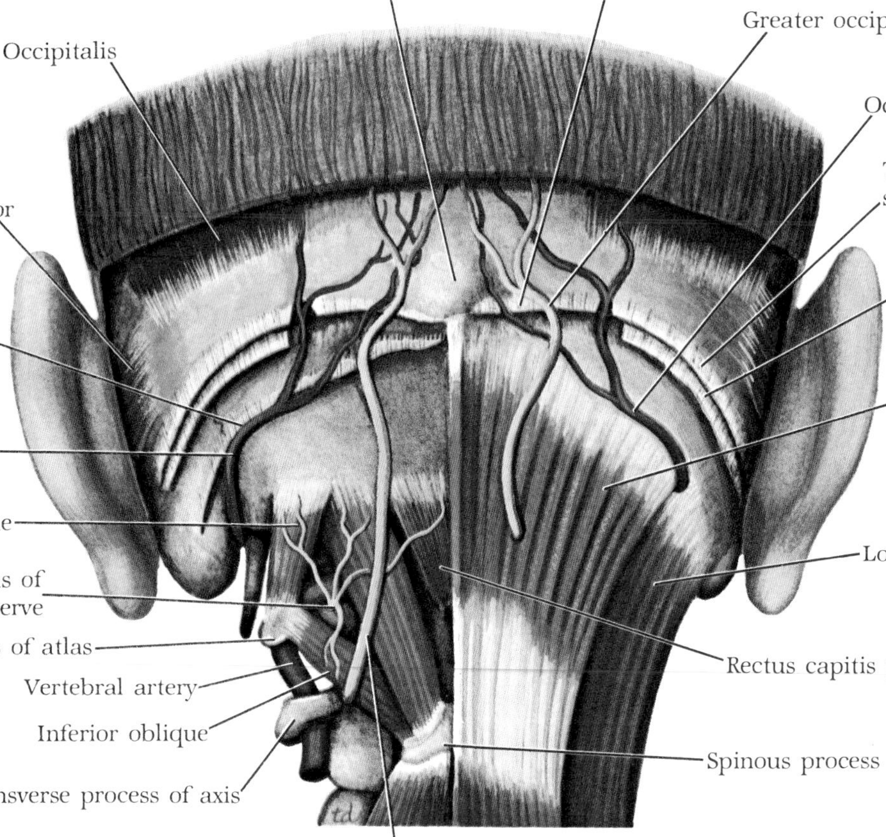
External occipital protuberance
Tendon of trapezius
Greater occipital nerve (C2 and 3)
Occipitalis
Occipital artery
Tendon of sternocleidomastoid
Auricularis posterior
Tendon of splenius capitis
Tendon of semispinalis capitis
Semispinalis capitis
Occipital artery
Superior oblique
Longissimus capitis
Posterior ramus of first cervical nerve
Transverse process of atlas
Rectus capitis posterior minor
Vertebral artery
Inferior oblique
Spinous process of axis
Transverse process of axis
Greater occipital nerve (C2 and 3)

Figure 7-11
Posterior view of the neck, showing the structures that form the boundaries of the suboccipital triangle.

1. Fracture of the atlas is commonly caused by falls on the head. The blow is transmitted from the occipital condyles to the lateral masses of the atlas, which are then forced apart, causing a fracture of the posterior or anterior arch of the atlas. The spinal cord escapes injury in about half the cases.
2. In dislocation of the atlas, if the transverse ligament that holds the odontoid process of the axis to the posterior surface of the anterior arch of the atlas is torn, the atlas will dislocate anteriorly. Needless to say, in such a case there is great danger that the spinal cord will be damaged.
3. In fracture dislocation of the atlas the odontoid process of the axis is fractured and moves forward with the anterior arch of the atlas. The spinal cord is rarely damaged.
4. Spontaneous dislocation of the atlas may occur in children following severe infection of the posterior pharyngeal wall (i.e., retropharyngeal abscess). In these circumstances there is softening and stretching of the transverse ligament.
5. The vertebral artery is commonly examined clinically by arteriography in patients with cerebrovascular disease. The angiogram may reveal the site and extent of the atherosclerotic plaques. Note the complicated course taken by the vertebral artery in this region. Having emerged from the foramen in the transverse process of the atlas it curves posteriorly behind the lateral mass of the atlas and rests on the superior border of the posterior arch of the atlas. It then travels medially to enter the vertebral canal. It ascends through the foramen magnum and pierces the dura and arachnoid mater. At the lower border of the pons it joins the vessel of the opposite side to form the basilar artery.

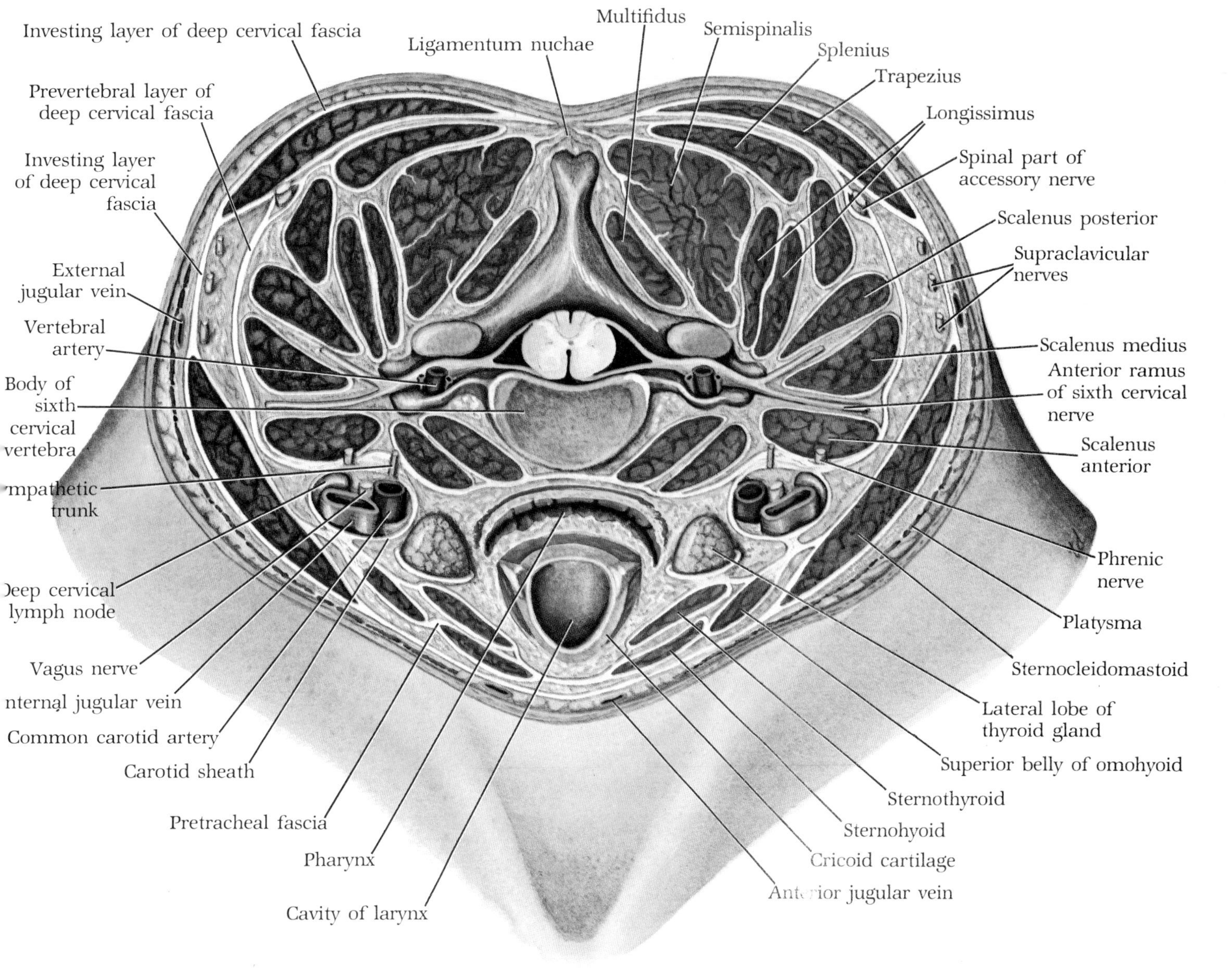
Investing layer of deep cervical fascia
Ligamentum nuchae
Multifidus
Semispinalis
Splenius
Trapezius
Longissimus
Prevertebral layer of deep cervical fascia
Investing layer of deep cervical fascia
Spinal part of accessory nerve
Scalenus posterior
Supraclavicular nerves
External jugular vein
Vertebral artery
Scalenus medius
Body of sixth cervical vertebra
Anterior ramus of sixth cervical nerve
Scalenus anterior
mpathetic trunk
Phrenic nerve
Deep cervical lymph node
Platysma
Vagus nerve
Sternocleidomastoid
nternal jugular vein
Lateral lobe of thyroid gland
Common carotid artery
Superior belly of omohyoid
Carotid sheath
Sternothyroid
Pretracheal fascia
Sternohyoid
Pharynx
Cricoid cartilage
Anterior jugular vein
Cavity of larynx

Figure 7-12
Oblique transverse section of the neck at the level of the sixth cervical vertebra.

1. The deep fascia of the neck in certain areas forms distinct sheets that are easily recognizable by the surgeon at operation; these are called (1) the investing, (2) the pretracheal, and (3) the prevertebral layers. Their clinical importance lies in the fact that they can determine the direction or can limit the spread of infection in the neck.

2. In tuberculosis of the upper cervical vertebrae the tuberculous pus is limited in front by the prevertebral layer of deep fascia. A midline swelling is formed, which bulges forward in the posterior wall of the pharynx. The pus then tracks laterally and downward behind the carotid sheath to reach the posterior triangle. Here the fascia covering the muscular floor of the triangle is weaker, and the abscess points behind the sternocleidomastoid muscle.

3. The thyroid gland is covered by a sheath derived from the pretracheal fascia. This tethers the gland to the larynx and trachea and explains why the thyroid gland follows the movements of the larynx when a person swallows. This close relationship between the trachea and the lobes of the thyroid gland commonly results in pressure on the trachea in patients with pathologic enlargement of the thyroid.

4. In the head and neck, all lymph ultimately drains into the deep cervical group of lymph nodes. Secondary carcinomatous deposits in these nodes are very common. It is important to remember that there are certain anatomical sites where the primary growth may be small and overlooked; for example, in the larynx, pharynx, esophagus, and external auditory meatus. Other possible sites of the primary growth are the bronchi, breast, and stomach. In these cases the secondary growth has spread far beyond the local lymph nodes.

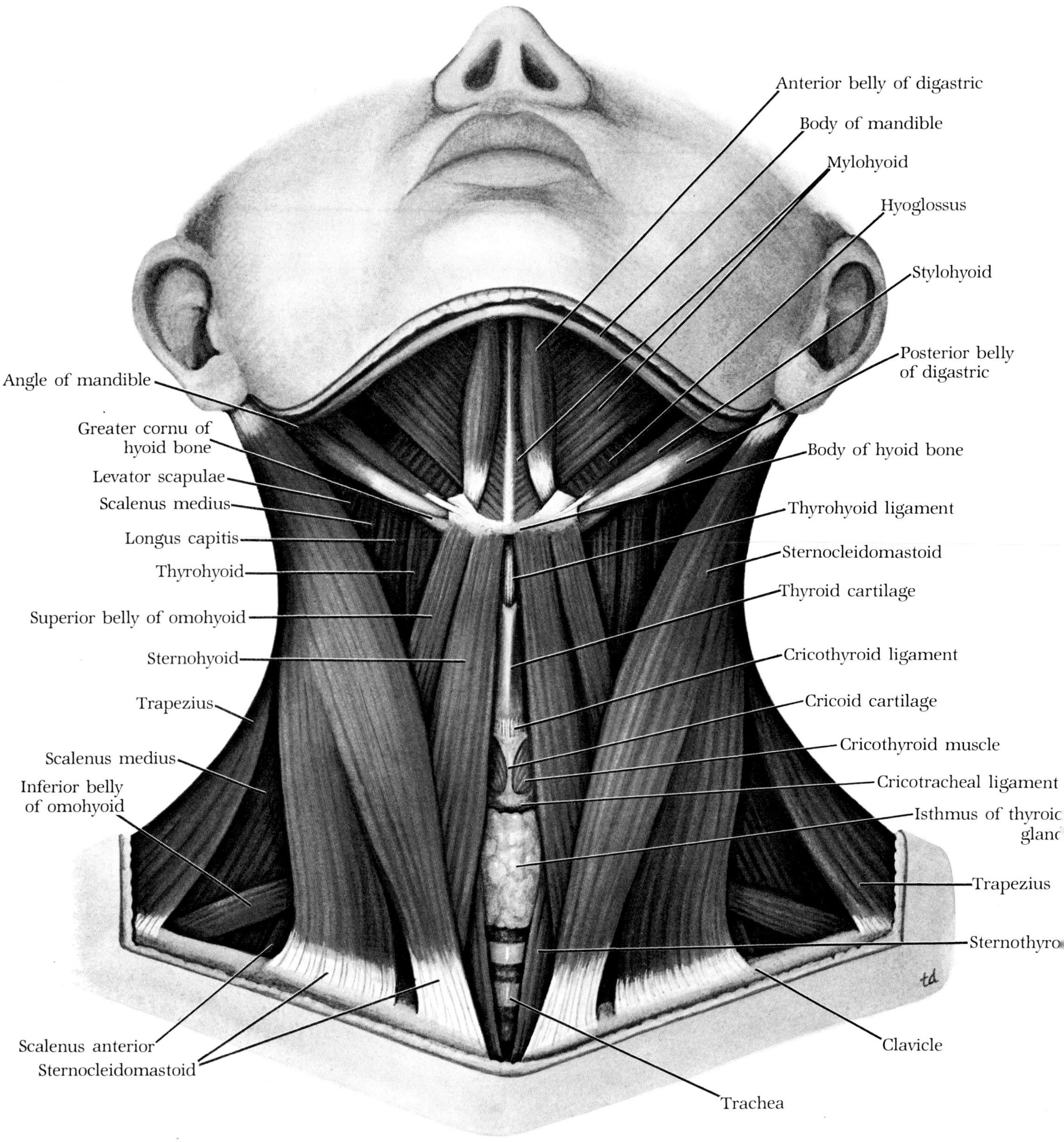
Anterior belly of digastric
Body of mandible
Mylohyoid
Hyoglossus
Stylohyoid
Posterior belly of digastric
Angle of mandible
Greater cornu of hyoid bone
Levator scapulae
Scalenus medius
Longus capitis
Thyrohyoid
Superior belly of omohyoid
Sternohyoid
Trapezius
Scalenus medius
Inferior belly of omohyoid
Body of hyoid bone
Thyrohyoid ligament
Sternocleidomastoid
Thyroid cartilage
Cricothyroid ligament
Cricoid cartilage
Cricothyroid muscle
Cricotracheal ligament
Trapezius
Clavicle
Scalenus anterior
Sternocleidomastoid
Trachea

Figure 7-13
Anterior triangle of the neck, showing the arrangement of the muscles.

1. Branchial cartilage. An irregular mass of cartilage several millimeters in diameter may be found in the subcutaneous tissues of the neck, usually just anterior to the lower third of the sternocleidomastoid muscle. It is thought that such pieces of cartilage are derived from one of the pharyngeal arches.

2. A cystic hygroma is a cystic swelling found most commonly in the lower third of the neck; it may be present at birth or it may appear during early infancy. On section, the swelling is found to consist of a large number of small cysts filled with clear lymph. It is caused by a failure of the jugular lymph sacs to join up with the main lymphatic system during embryonic development. A cystic hygroma should be excised.

3. Self-inflicted throat wounds often do not lead to death because the suicidal patient extends his neck when he makes the wound, and the carotid sheath, with its large vessels, retracts deeply under cover of the sternocleidomastoid muscle. Unaware of this fact, a suicidal patient often has to make several incisions before the great vessels of the neck are eventually sectioned. The most common sites for the wound are immediately above and below the hyoid bone.

4. A carotid body tumor arises from the carotid body situated at the bifurcation of the carotid artery. The patient usually has had it for some years before seeking medical advice. It is covered on its lateral side by the external carotid artery; it moves from side to side but not vertically.

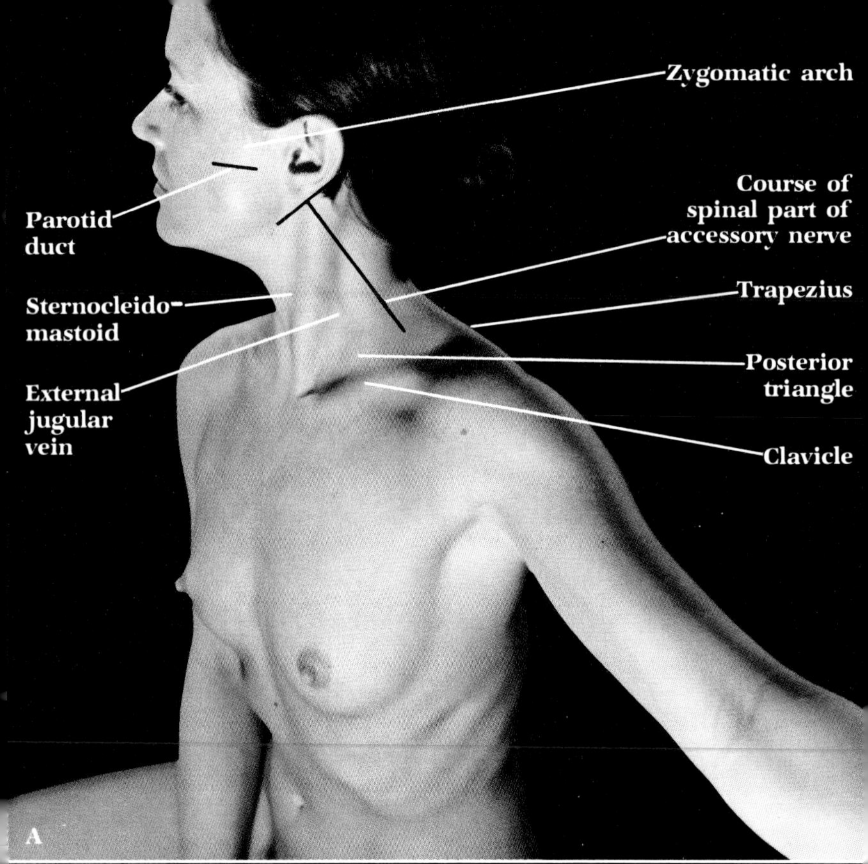
Zygomatic arch
Parotid duct
Course of spinal part of accessory nerve
Sternocleido-mastoid
Trapezius
External jugular vein
Posterior triangle
Clavicle
A

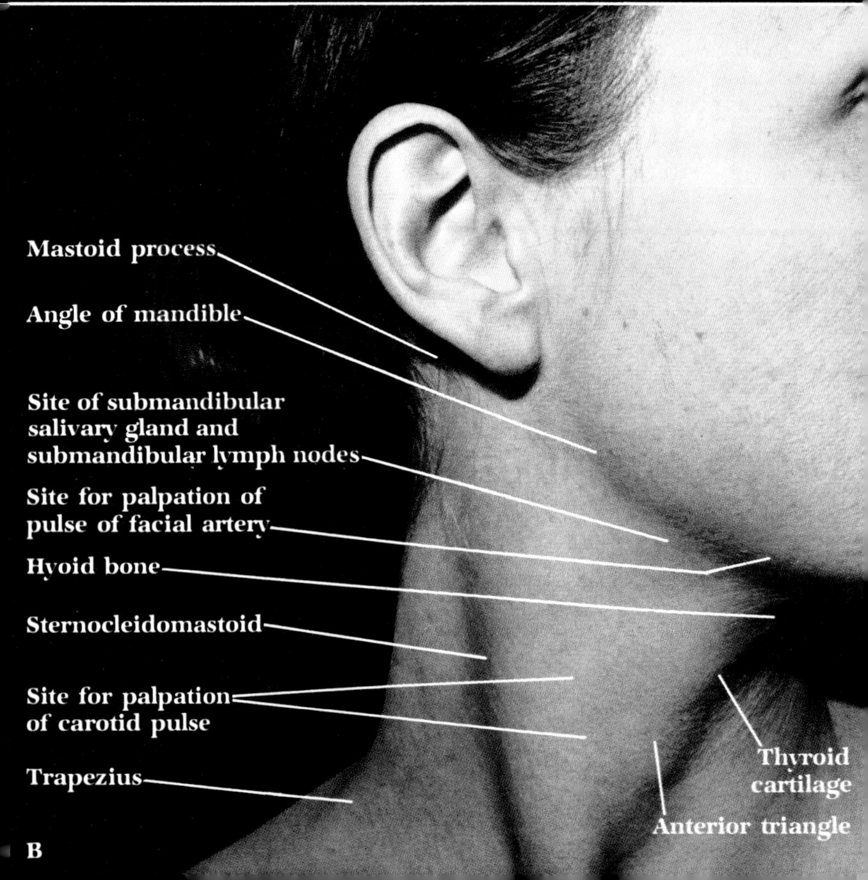
Mastoid process
Angle of mandible
Site of submandibular salivary gland and submandibular lymph nodes
Site for palpation of pulse of facial artery
Hyoid bone
Sternocleidomastoid
Site for palpation of carotid pulse
Trapezius
Thyroid cartilage
Anterior triangle
B

Figure 7-14
The neck of a 29-year-old female. A. The left posterior triangle. B. Part of the right anterior triangle.

1. The sternocleidomastoid muscle divides the neck into anterior and posterior triangles. The anterior triangle is bounded by the body of the mandible, the sternocleidomastoid, and the midline. The posterior triangle is bounded by the anterior border of the trapezius muscle, the sternocleidomastoid, and the clavicle.
2. The external jugular vein passes downward from the region of the angle of the mandible to the middle of the clavicle. It perforates the deep fascia just above the clavicle and drains into the subclavian vein.
3. The course of the spinal part of the accessory nerve may be indicated by first drawing a line from the angle of the mandible to the tip of the mastoid process. Bisect this line at right angles and extend the second line downward across the posterior triangle; the second line indicates the course of the nerve.
4. The body of the hyoid bone lies opposite the third cervical vertebra.
5. The notched upper border of the thyroid cartilage lies opposite the fourth cervical vertebra.
6. The superficial part of the submandibular salivary gland lies beneath the lower margin of the body of the mandible. The submandibular lymph nodes are situated on the surface of the submandibular salivary gland and can be palpated just below the lower border of the body of the mandible.
7. The common carotid artery bifurcates into the internal and external carotid arteries at the level of the upper border of the thyroid cartilage. The pulsations of these arteries can be felt at this level.
8. The pulsations of the facial artery can be felt as it crosses the lower margin of the body of the mandible at the anterior border of the masseter muscle.
9. The mastoid process of the temporal bone projects downward and forward from behind the ear. It is undeveloped in the newborn child but may be recognized as a bony prominence at the end of the second year.

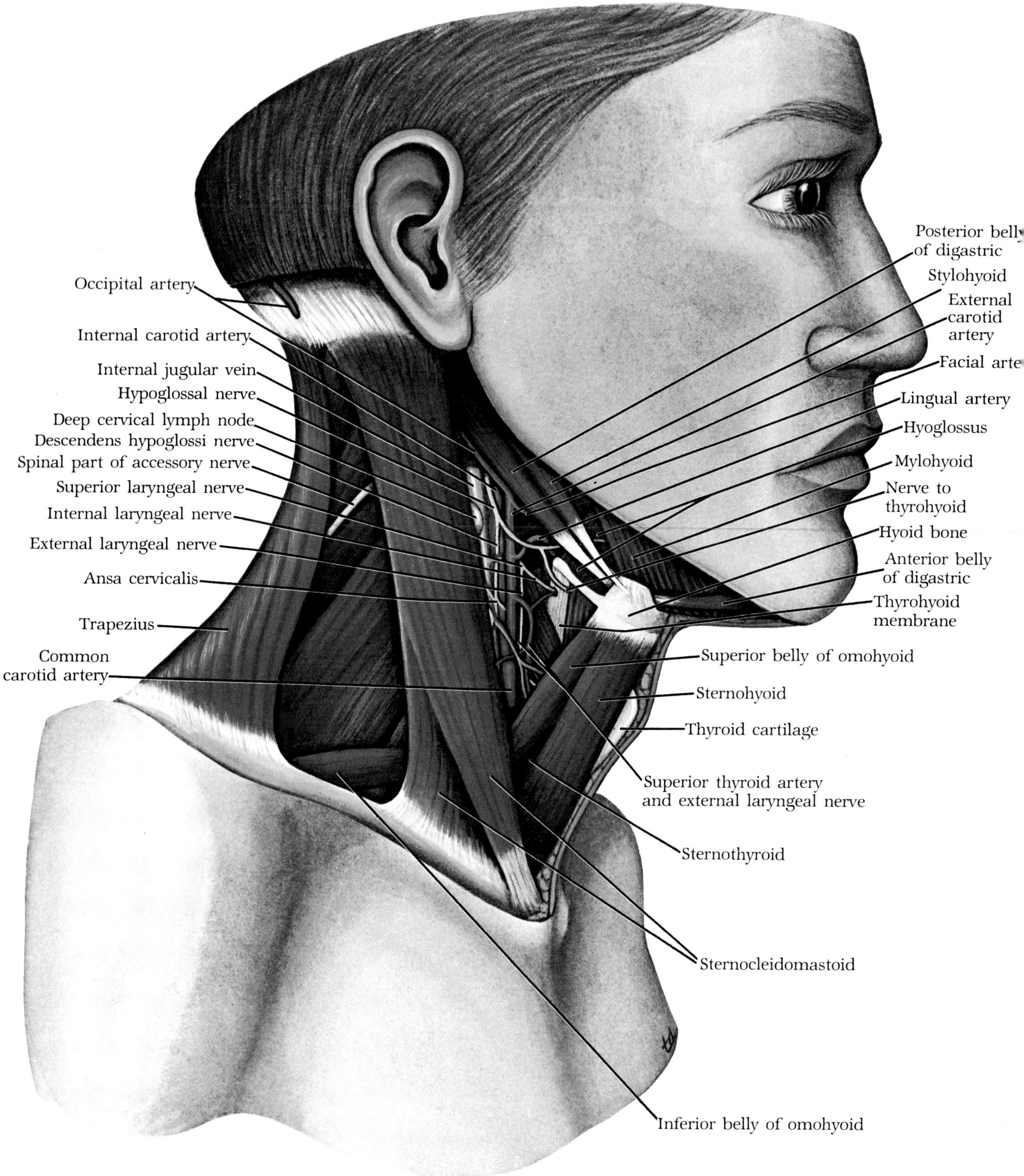
Occipital artery
Internal carotid artery
Internal jugular vein
Hypoglossal nerve
Deep cervical lymph node
Descendens hypoglossi nerve
Spinal part of accessory nerve
Superior laryngeal nerve
Internal laryngeal nerve
External laryngeal nerve
Ansa cervicalis
Trapezius
Common carotid artery
Posterior belly of digastric
Stylohyoid
External carotid artery
Facial arte
Lingual artery
Hyoglossus
Mylohyoid
Nerve to thyrohyoid
Hyoid bone
Anterior belly of digastric
Thyrohyoid membrane
Superior belly of omohyoid
Sternohyoid
Thyroid cartilage
Superior thyroid artery and external laryngeal nerve
Sternothyroid
Sternocleidomastoid
Inferior belly of omohyoid

Figure 7-15
Anterior triangle of neck, lateral view.

1. Carotid sinus hypersensitivity with syncope. Pressure on one or both carotid sinuses causes slowing of the heart rate, a fall in arterial pressure, and ipsilateral cerebral ischemia. The fainting attacks may be initiated by turning the head to one side or by wearing a tight collar. The attack usually occurs while the patient is standing; he may lose consciousness and fall, only to regain normal consciousness in a few seconds or minutes.

2. Internal carotid artery occlusion or stenosis is usually caused by an atherosclerotic plaque. When it occurs in the neck there may be visual impairment or blindness in the eye on the side of the lesion due to insufficient blood flow through the retinal artery; there also may be motor paralysis and sensory loss on the opposite side of the body due to insufficient blood flow through the middle cerebral artery.

3. The deep cervical group of lymph nodes are embedded in the carotid sheath and lie alongside the internal jugular vein. Acute and chronic infections of the head and neck often result in acute and chronic lymphadenitis. In acute lymphadenitis the nodes are enlarged and tender, and there may be an associated pyrexia. In chronic lymphadenitis, as in tuberculous lymphadenitis, the source of infection is usually the palatine tonsil of the same side. At first one lymph node is enlarged and tender and is commonly situated just below and behind the angle of the mandible. As the result of the local spread of infection other nodes later become enlarged and matted together. Abscess formation is common.

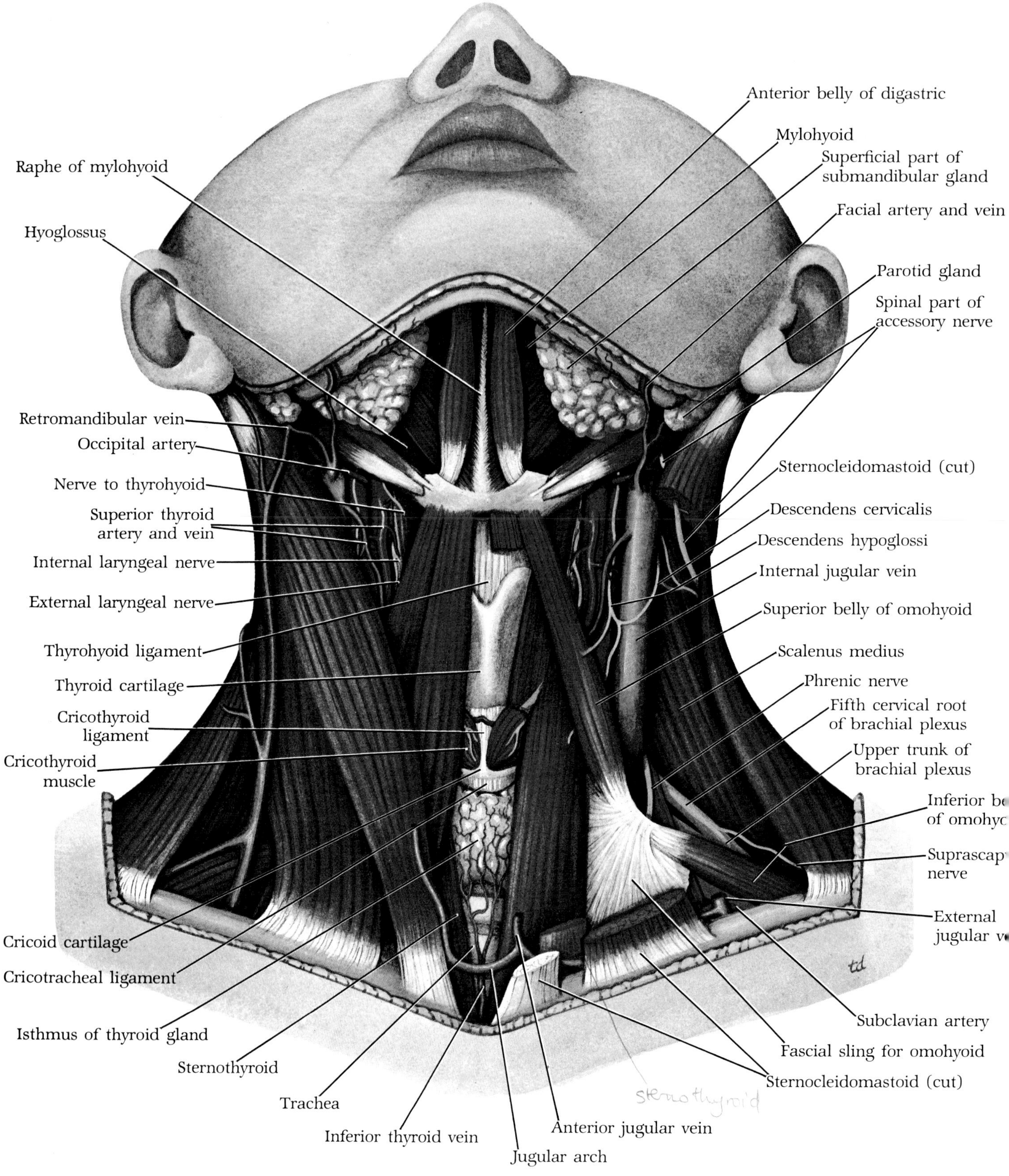

Anterior belly of digastric
Mylohyoid
Raphe of mylohyoid
Superficial part of submandibular gland
Facial artery and vein
Hyoglossus
Parotid gland
Spinal part of accessory nerve
Retromandibular vein
Occipital artery
Nerve to thyrohyoid
Sternocleidomastoid (cut)
Superior thyroid artery and vein
Descendens cervicalis
Descendens hypoglossi
Internal laryngeal nerve
Internal jugular vein
External laryngeal nerve
Superior belly of omohyoid
Thyrohyoid ligament
Scalenus medius
Thyroid cartilage
Phrenic nerve
Cricothyroid ligament
Fifth cervical root of brachial plexus
Cricothyroid muscle
Upper trunk of brachial plexus
Inferior b
of omohyo
Suprascap
nerve
External
jugular v
Cricoid cartilage
Cricotracheal ligament
Subclavian artery
Isthmus of thyroid gland
Fascial sling for omohyoid
Sternothyroid
Sternocleidomastoid (cut)
Trachea
sternothyroid
Anterior jugular vein
Inferior thyroid vein
Jugular arch

Figure 7-16
Anterior triangle of neck. The left sternocleidomastoid and sternohyoid muscles have been removed to show deeper structures.

1. During development of the thyroid gland its descent in the neck may be arrested at any point between the base of the tongue and the trachea. Lingual thyroid is the most common form of incomplete descent. The mass of tissue is found just beneath the foramen cecum in the tongue and may be of sufficient size to obstruct swallowing in the infant.
2. Persistence of the thyroglossal duct may give rise to a thyroglossal cyst or sinus. A thyroglossal cyst occurs most commonly in the region inferior to the hyoid bone and lies in the midline. As the cyst enlarges it is prone to infection, and for this reason it should be removed surgically. Since remnants of the duct often traverse the body of the hyoid bone, this may have to be excised also to prevent recurrence. Occasionally a thyroglossal cyst ruptures spontaneously or becomes infected and ruptures onto the skin surface, producing a sinus.
3. The midline structures in the neck should be readily recognized as one passes an examining finger down the neck from the chin to the suprasternal notch. These are as follows: the symphysis menti; the floor of the submental triangle with the submental lymph nodes; the body of the hyoid bone; the thyrohyoid membrane; the notched upper border of the thyroid cartilage; the cricothyroid ligament; the cricoid cartilage; the cricotracheal ligament; the first ring of the trachea; the isthmus of the thyroid gland; the fifth, sixth, and seventh rings of the trachea; and finally the suprasternal notch between the anterior ends of the clavicles.
4. Enlarged, nontender, rubbery, discrete lymph nodes of the neck would suggest Hodgkin's disease, lymphosarcoma, or reticulosarcoma. In Hodgkin's disease other lymph nodes in the axillae or inguinal regions are commonly enlarged; the spleen also may be enlarged.

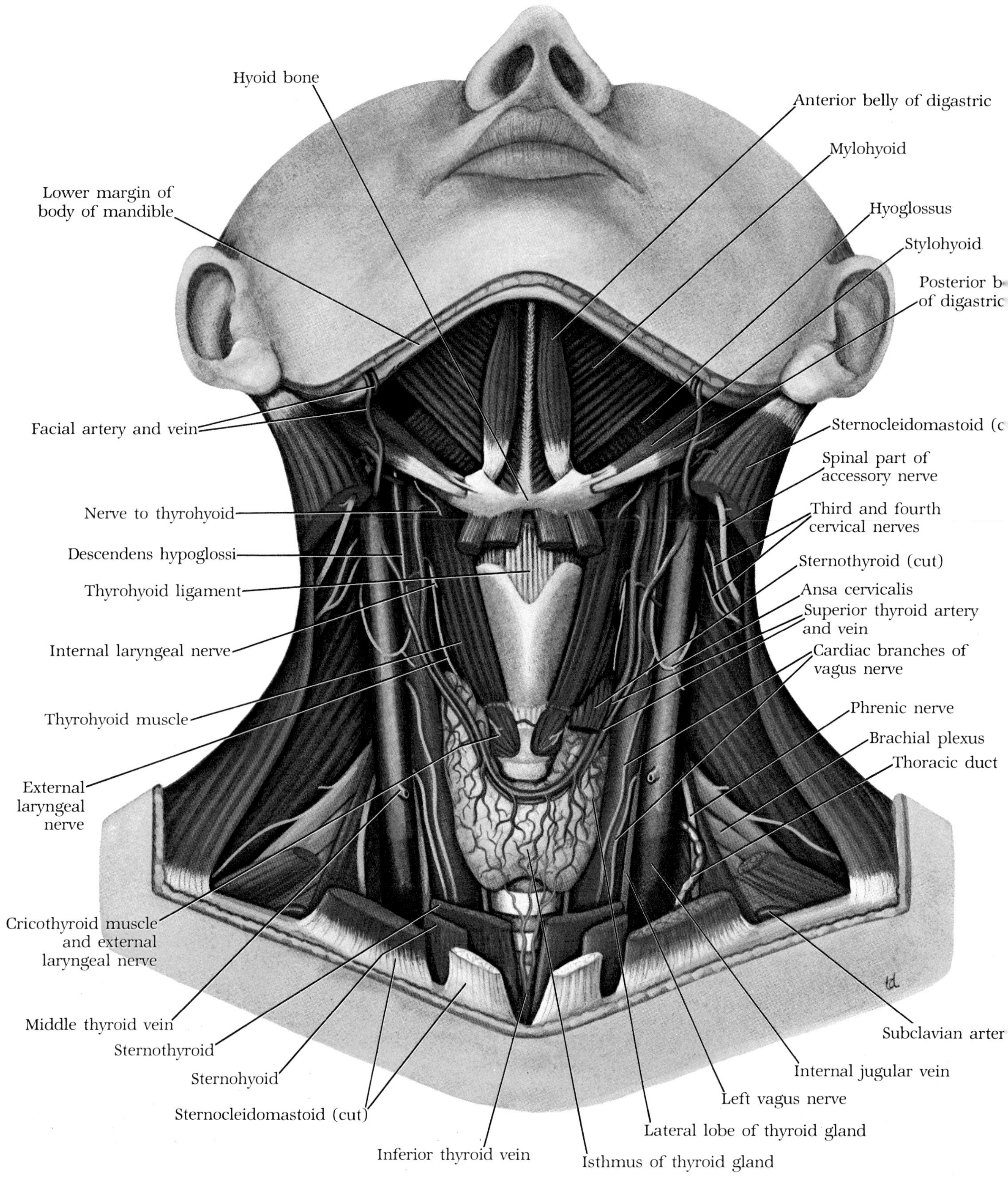
Hyoid bone
Anterior belly of digastric
Mylohyoid
Lower margin of body of mandible
Hyoglossus
Stylohyoid
Posterior b
of digastric
Facial artery and vein
Sternocleidomastoid (c
Spinal part of accessory nerve
Nerve to thyrohyoid
Third and fourth cervical nerves
Descendens hypoglossi
Sternothyroid (cut)
Thyrohyoid ligament
Ansa cervicalis
Superior thyroid artery and vein
Internal laryngeal nerve
Cardiac branches of vagus nerve
Thyrohyoid muscle
Phrenic nerve
Brachial plexus
Thoracic duct
External laryngeal nerve
Cricothyroid muscle and external laryngeal nerve
Middle thyroid vein
Subclavian arter
Sternothyroid
Internal jugular vein
Sternohyoid
Left vagus nerve
Sternocleidomastoid (cut)
Lateral lobe of thyroid gland
Inferior thyroid vein
Isthmus of thyroid gland

Figure 7-17
Anterior triangle of neck. The sternocleidomastoid, omohyoid, sternohyoid, and sternothyroid muscles have been removed to show deeper structures.

1. The thyroid gland is invested in a sheath of pretracheal fascia. This tethers the gland to the larynx and trachea and explains why the thyroid gland follows the movements of the larynx when a person swallows. The close relationship between the trachea and the lobes of the thyroid gland commonly results in pressure on the trachea in patients with pathological enlargement of the thyroid.

2. The thyroid gland is bound down to the larynx by the attachment of the sternothyroid muscles to the thyroid cartilage, and these muscles limit the upward expansion of the gland. There being no limitation to downward expansion, it is not uncommon for a pathologically enlarged thyroid gland to extend downward behind the sternum. A retrosternal goiter (any abnormal enlargement of the thyroid gland) may compress the trachea and cause dangerous dyspnea; it may also cause severe venous compression.

3. Important nerves associated with the two main arteries supplying the thyroid gland may be damaged during thyroidectomy operations. The superior thyroid artery on each side is related to the external laryngeal nerve, which supplies the cricothyroid muscle. The terminal branches of the inferior thyroid artery on each side are related to the recurrent laryngeal nerve. Damage to the external laryngeal nerve results in an inability to tense the vocal folds and thus in hoarseness. Damage to the recurrent laryngeal nerve causes paralysis of the movement of the vocal folds (for details see page 481).

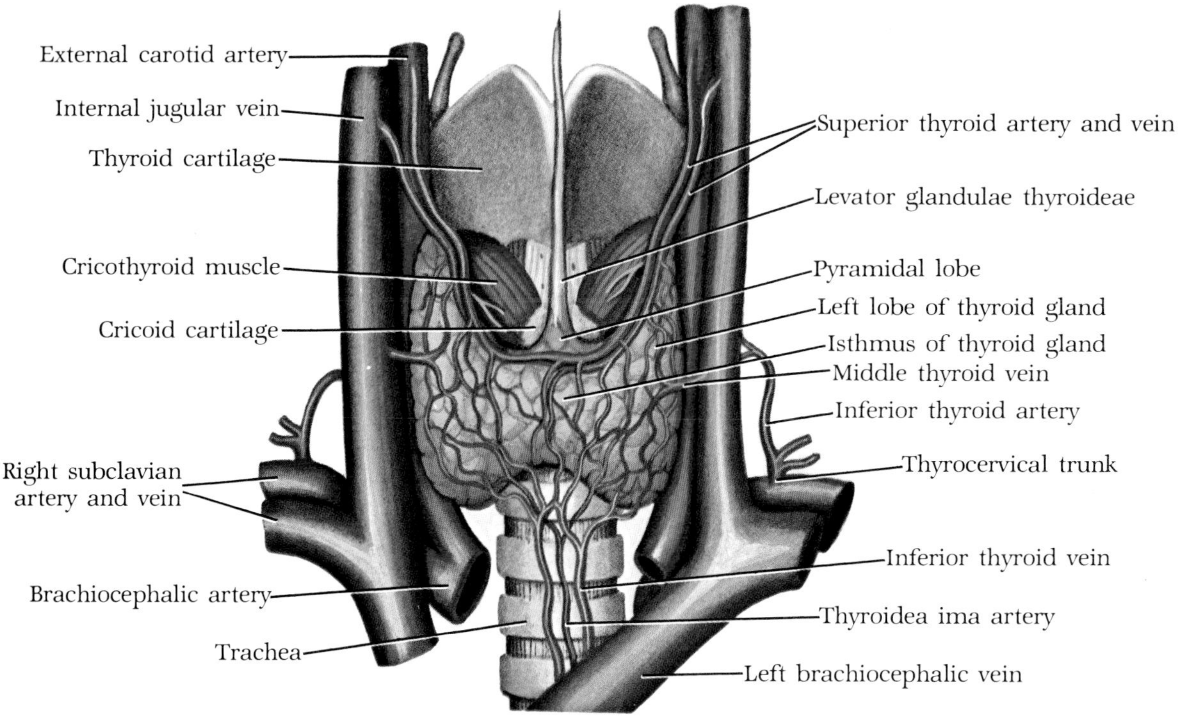
External carotid artery
Internal jugular vein
Thyroid cartilage
Cricothyroid muscle
Cricoid cartilage
Right subclavian artery and vein
Brachiocephalic artery
Trachea
Superior thyroid artery and vein
Levator glandulae thyroideae
Pyramidal lobe
Left lobe of thyroid gland
Isthmus of thyroid gland
Middle thyroid vein
Inferior thyroid artery
Thyrocervical trunk
Inferior thyroid vein
Thyroidea ima artery
Left brachiocephalic vein

A

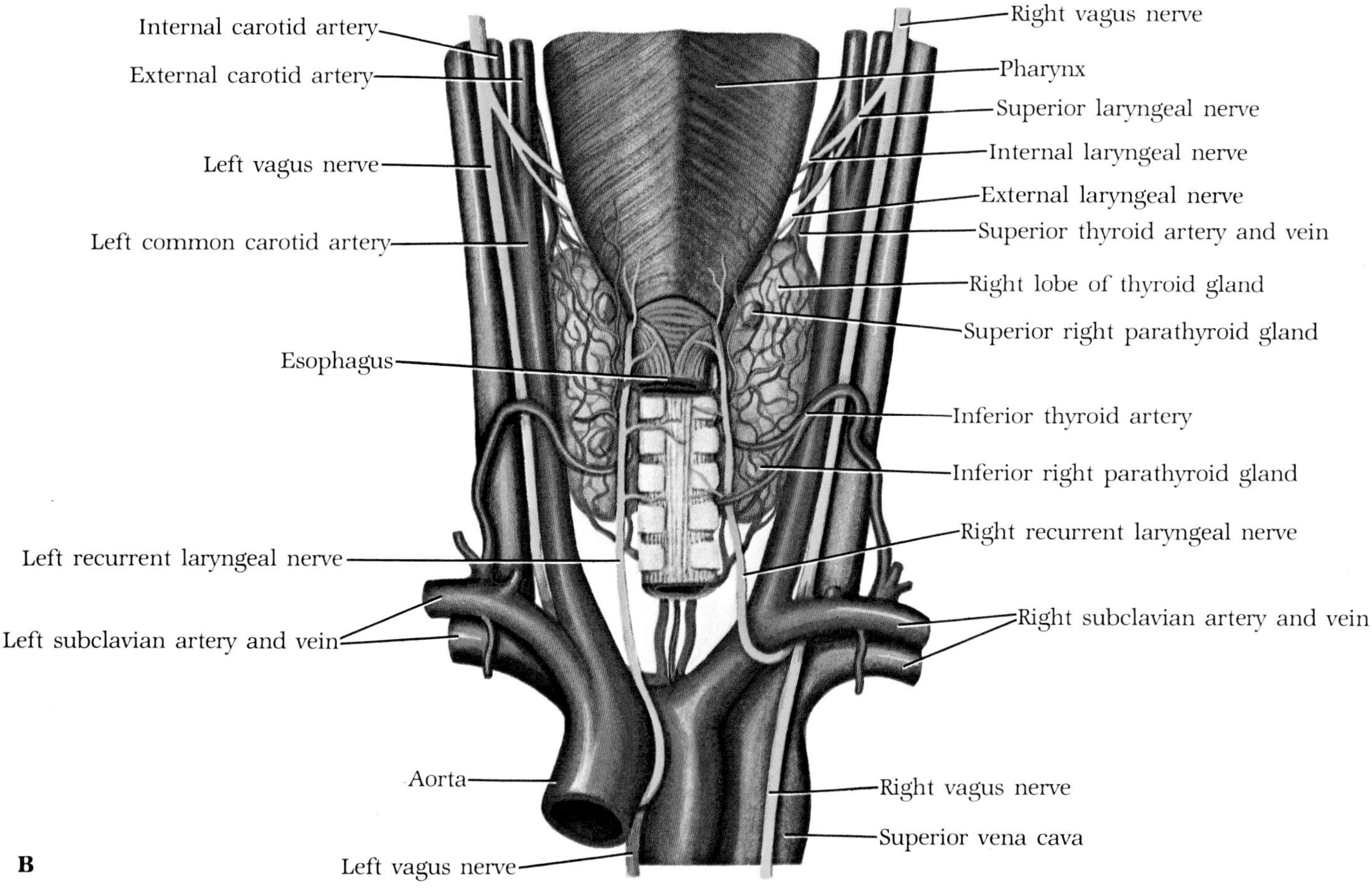
Internal carotid artery
External carotid artery
Left vagus nerve
Left common carotid artery
Esophagus
Left recurrent laryngeal nerve
Left subclavian artery and vein
Aorta
Left vagus nerve
Right vagus nerve
Pharynx
Superior laryngeal nerve
Internal laryngeal nerve
External laryngeal nerve
Superior thyroid artery and vein
Right lobe of thyroid gland
Superior right parathyroid gland
Inferior thyroid artery
Inferior right parathyroid gland
Right recurrent laryngeal nerve
Right subclavian artery and vein
Right vagus nerve
Superior vena cava

B

Figure 7-18
Thyroid gland. A. Anterior view. B. Posterior view.

1. The thyroid gland overlaps the sides of the larynx and trachea. Enlargement of the gland may compress the trachea and seriously interfere with the airway.
2. Examination of the gland is best carried out from behind the patient when his head and neck are flexed; this relaxes the infrahyoid and sternocleidomastoid muscles and permits the physician to palpate the lobes of the thyroid gland with both hands.
3. Because the thyroid gland is enclosed in a sheath of pretracheal fascia, the gland moves up and down when a person swallows.
4. The close relationship between the external laryngeal nerve and the superior thyroid artery and the recurrent laryngeal nerve and the inferior thyroid artery may result in damage to these nerves during a thyroidectomy. The injury to the nerves may be due to hemorrhage or edema, stretching or bruising, or inclusion of the nerve in an arterial ligature.
5. Hypoparathyroidism following thyroidectomy may be due to injury to the parathyroid glands' blood supply, direct injury to the glands, or actual removal of the glands. The resulting fall in the blood calcium levels may interfere with muscle contraction, although tetany is rare and then transient.
6. In a partial thyroidectomy the posterior part of each lobe of the thyroid gland is left intact so that the parathyroid glands are not disturbed.
7. Tumors of the superior parathyroid glands are easily found at operation, since these glands are constant in position. The inferior parathyroid glands may, during development, be pulled down by the thymus into the thorax. In such a case the best way to locate a tumor in the inferior glands is to follow their arterial supply from the inferior thyroid arteries.

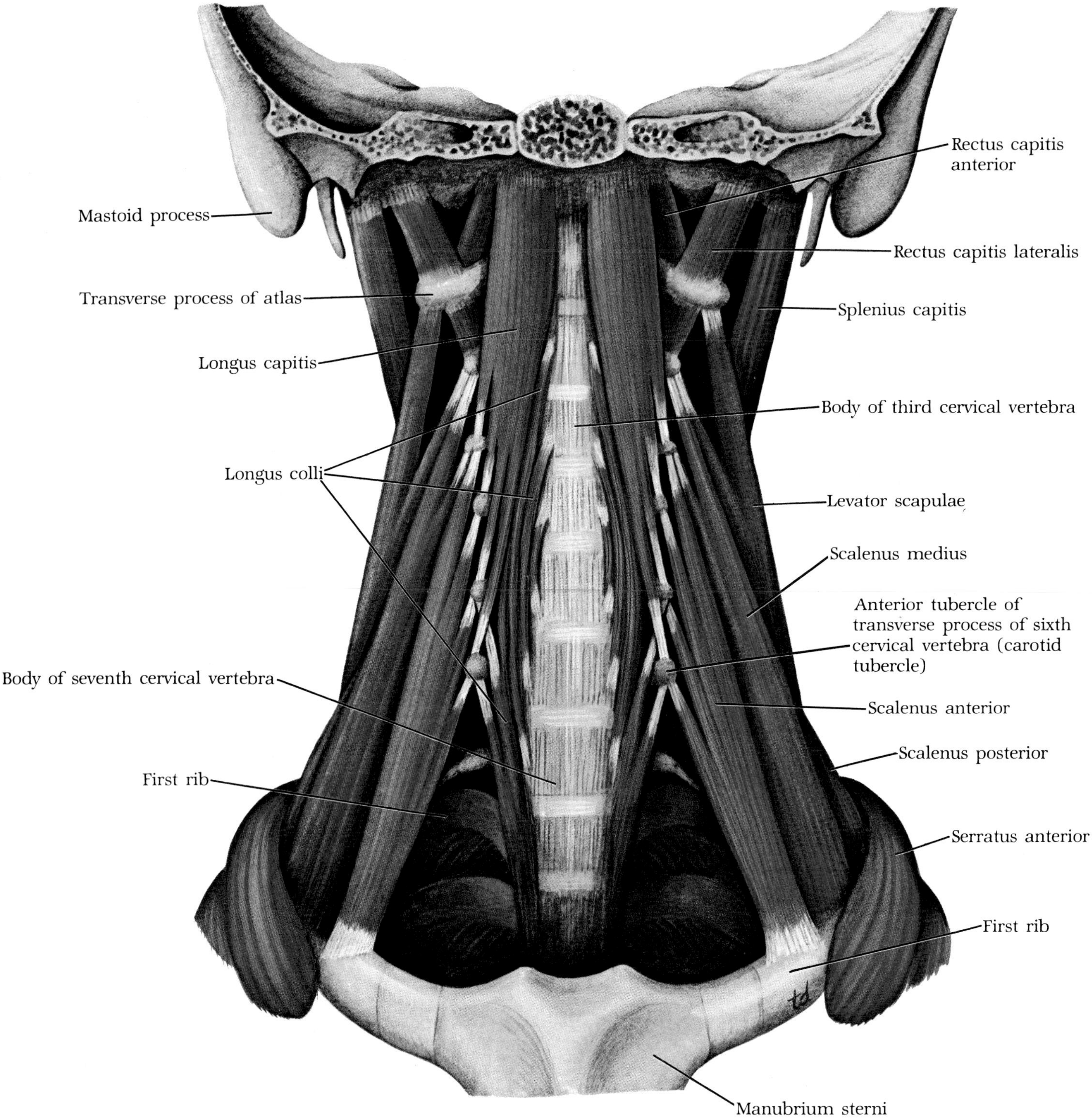
Rectus capitis anterior
Mastoid process
Rectus capitis lateralis
Transverse process of atlas
Splenius capitis
Longus capitis
Body of third cervical vertebra
Longus colli
Levator scapulae
Scalenus medius
Anterior tubercle of transverse process of sixth cervical vertebra (carotid tubercle)
Body of seventh cervical vertebra
Scalenus anterior
Scalenus posterior
First rib
Serratus anterior
First rib
Manubrium sterni

Figure 7-19
Prevertebral muscles.

1. The carotid tubercle, which is the large anterior tubercle of the transverse process of the sixth cervical vertebra, can be felt on deep palpation under the overlying muscles about a fingersbreadth lateral to the cricoid cartilage on each side. This tubercle is frequently used as a landmark for the site of injection of the stellate ganglion.
2. It is important to remember that in the midline of the neck the body of the hyoid bone, the superior border of the thyroid cartilage, and the cricoid cartilage can all be easily palpated as they lie anterior to the third, fourth, and sixth cervical vertebral bodies, respectively.
3. Tuberculous pus arising from the upper cervical vertebrae is limited anteriorly by the prevertebral layer of deep fascia. A midline swelling, which bulges forward in the posterior wall of the pharynx, is formed. The pus then tracks laterally and downward posterior to the carotid sheath to reach the posterior triangle. Here the abscess points posterior to the sternocleidomastoid muscle.
4. The following basic movements take place in the neck: (1) flexion, (2) extension, (3) rotation, and (4) lateral flexion. About 50 percent of the movements of flexion and extension of the head and neck occur at the atlanto-occipital joint and about 50 percent of rotation occurs at the atlanto-axial joint.
5. In flexion the patient should normally be able to touch his chest with his chin and in extension to look directly vertically upward. In lateral flexion he should be able to move his ear approximately halfway to his shoulder, and in rotation he should be able to move his head on both sides so that his chin is nearly in line with his shoulder.

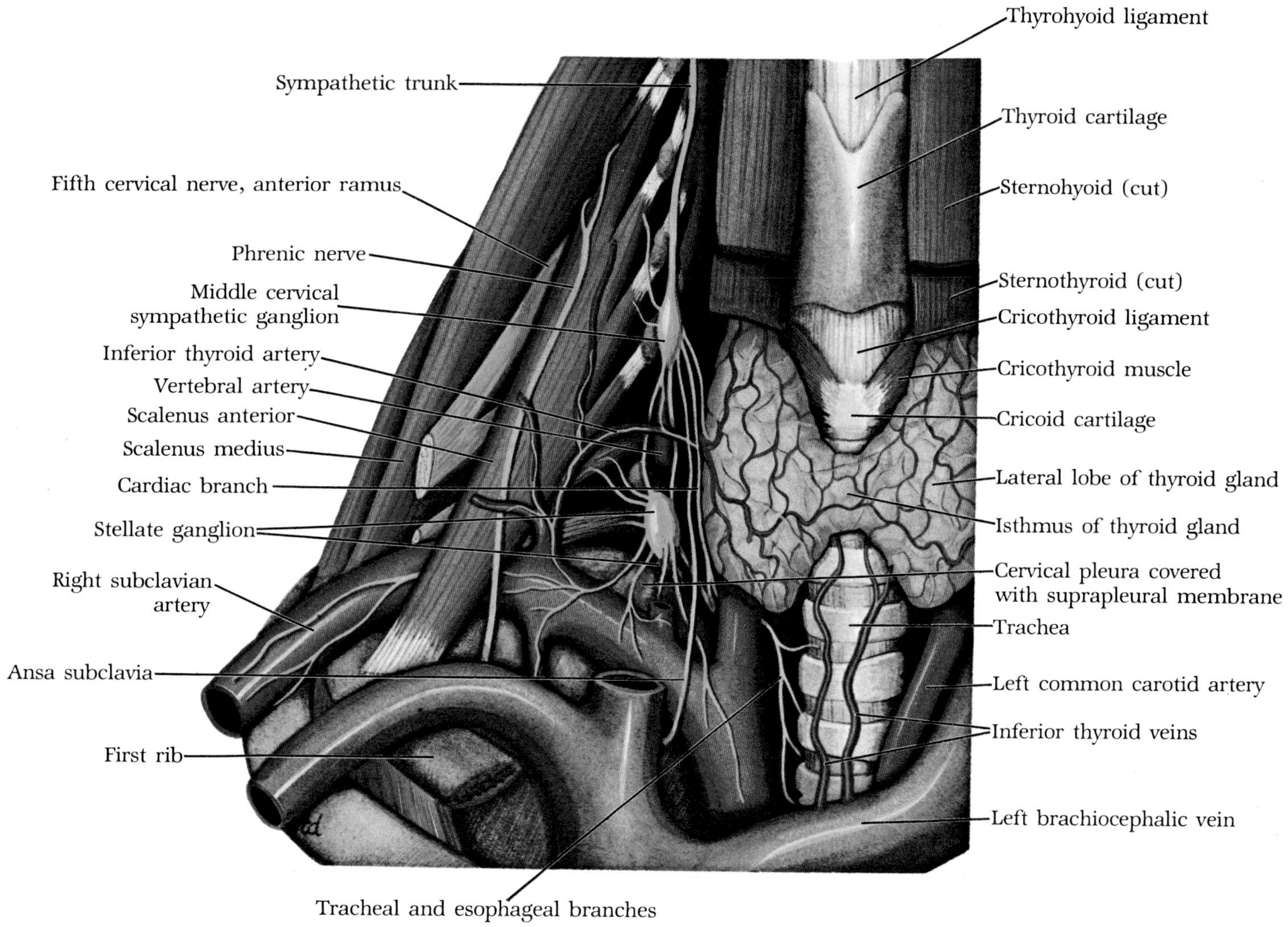
Thyrohyoid ligament
Sympathetic trunk
Thyroid cartilage
Fifth cervical nerve, anterior ramus
Sternohyoid (cut)
Phrenic nerve
Middle cervical sympathetic ganglion
Sternothyroid (cut)
Cricothyroid ligament
Inferior thyroid artery
Cricothyroid muscle
Vertebral artery
Scalenus anterior
Cricoid cartilage
Scalenus medius
Cardiac branch
Lateral lobe of thyroid gland
Stellate ganglion
Isthmus of thyroid gland
Right subclavian artery
Cervical pleura covered with suprapleural membrane
Trachea
Ansa subclavia
Left common carotid artery
Inferior thyroid veins
First rib
Left brachiocephalic vein
Tracheal and esophageal branches

Figure 7-20
Stellate ganglion.

1. There are three cervical sympathetic ganglia: superior, middle, and inferior; the inferior ganglion is often fused with the first thoracic ganglion to form the stellate ganglion.

2. The stellate ganglion is situated anterior to the head of the first rib and the transverse processes of the seventh cervical and first thoracic vertebrae. It lies posterior to the carotid sheath and anterior to the eighth cervical and first thoracic nerves.

3. Stellate ganglion block is performed by first palpating the carotid tubercle, the large anterior tubercle of the transverse process of the sixth cervical vertebra that lies about a fingerbreadth lateral to the cricoid cartilage. After raising a weal beneath the skin with a local anesthetic over this tubercle, a 5–8-cm needle is inserted directly posteriorly through the weal. At the same time the carotid sheath and the sternocleidomastoid muscle are pushed laterally and posteriorly to prevent damage to them by the needle. On hitting the carotid tubercle the needle is withdrawn about 1 cm, after which the needle should be aspirated to see if it has penetrated the vertebral artery. Fifteen ml of local anesthetic is then injected beneath the prevertebral fascia; this will effectively block the ganglion and its rami communicantes.

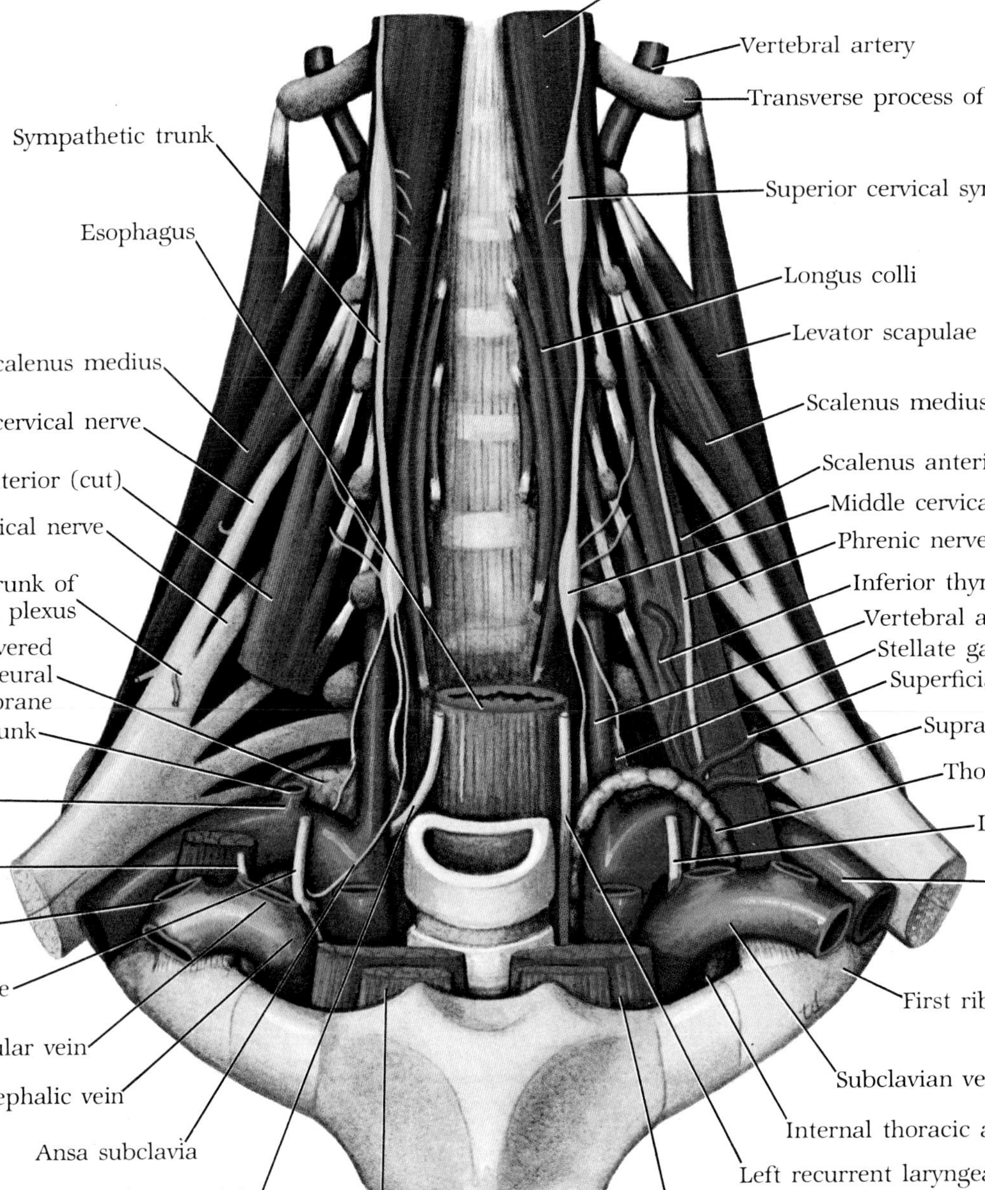

Longus capitis
Vertebral artery
Transverse process of atlas
Sympathetic trunk
Superior cervical sympathetic ganglion
Esophagus
Longus colli
Levator scapulae
Scalenus medius
Scalenus medius
Fifth cervical nerve
Scalenus anterior
Scalenus anterior (cut)
Middle cervical sympathetic ganglion
Sixth cervical nerve
Phrenic nerve
Upper trunk of brachial plexus
Inferior thyroid artery
Vertebral artery
Cervical pleura covered by suprapleural membrane
Stellate ganglion
Superficial cervical artery
Costocervical trunk
Suprascapular artery
Thoracic duct
Thyrocervical trunk
Left vagus nerve
Right phrenic nerve
Third part of subclavian artery
External jugular vein
Right vagus nerve
First rib
Internal jugular vein
Subclavian vein
Right brachiocephalic vein
Internal thoracic artery
Ansa subclavia
Left recurrent laryngeal nerve
Right recurrent laryngeal nerve
Sternothyroid
Sternohyoid

Figure 7-21
Prevertebral region and root of neck.

1. The phrenic nerve, which arises from the anterior rami of the third, fourth, and fifth cervical nerves, is of considerable clinical importance since it is the sole nerve supply to the muscle of the diaphragm. Each phrenic nerve supplies the corresponding half of the diaphragm. The nerve is often cut or crushed in the neck to paralyze the diaphragm and immobilize the lung in patients with lung tuberculosis who cannot be successfully treated with antibiotics. The paralyzed half of the diaphragm relaxes and is pushed up into the thorax by the positive abdominal pressure. It is mainly the lower lobe of the lung that is collapsed and rested by this procedure.

2. The cervical part of the sympathetic trunk possesses the superior, middle, and inferior ganglia. The inferior ganglion is most commonly fused with the first thoracic sympathetic ganglion to form the stellate ganglion. The preganglionic fibers to the upper limb leave the spinal cord in the second to the eighth thoracic nerves. On reaching the sympathetic trunk via the white rami, they ascend within the trunk and are relayed in the second thoracic, stellate, and middle cervical ganglia. Postganglionic fibers then join the roots of the brachial plexus as gray rami. Sympathectomy in the upper limb is a relatively common procedure for the treatment of arterial insufficiency. It is clear that the stellate and the second thoracic ganglia should be removed to block the sympathetic pathway to the arm completely. Unfortunately, the removal of the stellate ganglion also removes the sympathetic nerve supply to the head and neck on that side. This not only produces vasodilatation of the skin vessels but also anhidrosis, nasal congestion, and Horner's syndrome, results of which are constriction of the pupil, drooping of the upper lid, and enophthalmos. For this reason the stellate ganglion is usually left intact in sympathectomies of the upper limb.

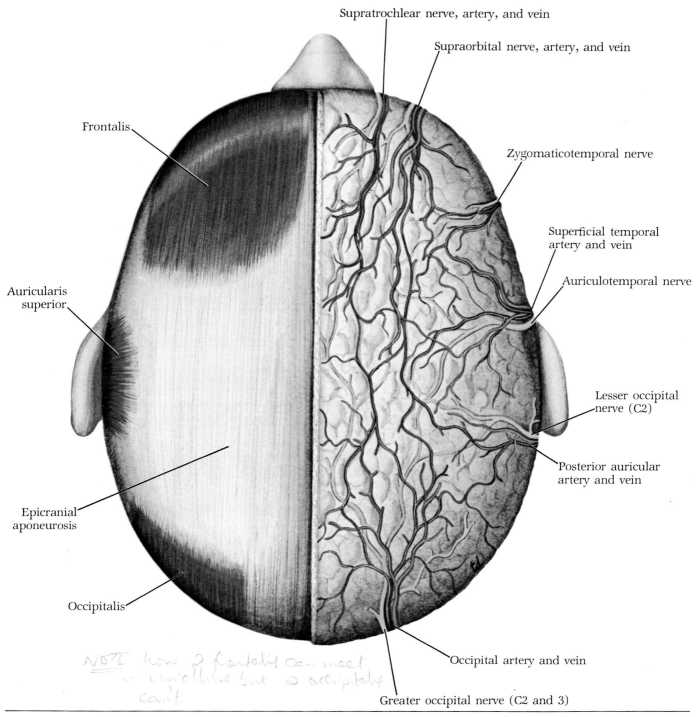

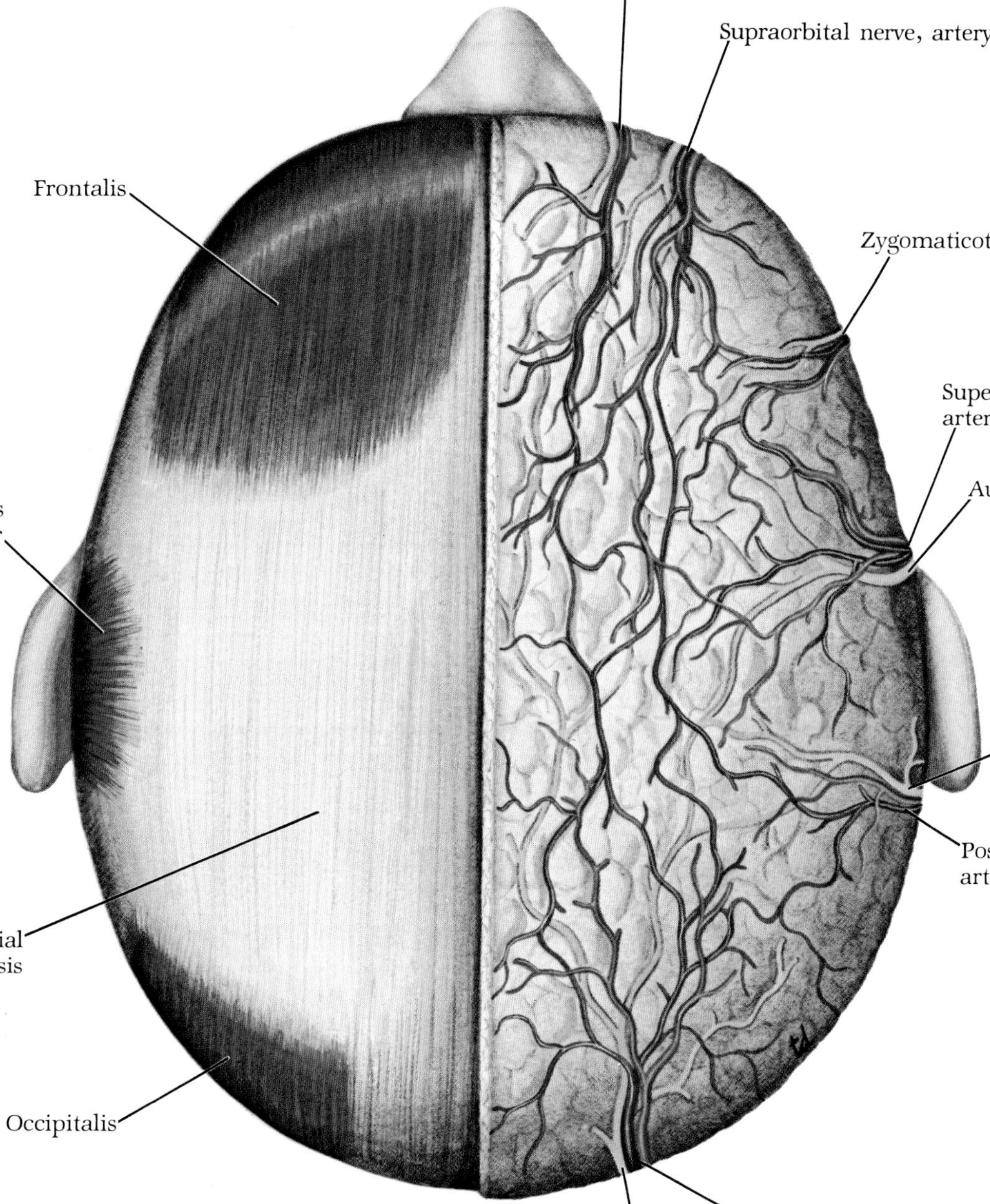
Supratrochlear nerve, artery, and vein
Supraorbital nerve, artery, and vein
Frontalis
Zygomaticotemporal nerve
Superficial temporal artery and vein
Auricularis superior
Auriculotemporal nerve
Lesser occipital nerve (C2)
Posterior auricular artery and vein
Epicranial aponeurosis
Occipitalis
Occipital artery and vein
Greater occipital nerve (C2 and 3)

Figure 7-22
The scalp, superior view. The left side shows the muscles, and the right side shows the nerve and blood supply.

1. In the scalp the skin, subcutaneous tissue, and epicranial aponeurosis are closely united with one another and are separated from the periosteum by loose areolar tissue.
2. Blood or pus may collect in the potential space beneath the epicranial aponeurosis. It tends to spread over the calvaria, being limited anteriorly by the orbital margin, posteriorly by the nuchal lines, and laterally by the temporal lines.
3. A collection of subperiosteal blood or pus is limited to one bone due to the attachment of the periosteum to the sutural ligaments.
4. Since the scalp has a profuse blood supply to nourish the hair follicles, even small lacerations of the scalp may cause a severe loss of blood. It is often difficult to stop the bleeding of a scalp wound because the arterial walls are attached to fibrous septa in the subcutaneous tissue and are unable to contract or retract to allow blood clotting to take place. Local pressure applied to the scalp is the only satisfactory method of stopping the bleeding.
5. In automobile accidents in which the head of the patient is projected forward through the windshield large areas of the scalp may be cut off the head. Because of the profuse blood supply, however, it is often possible to replace large areas of the scalp that are only hanging to the skull by a narrow pedicle. When they are sutured in place necrosis will not occur.
6. Infections of the scalp may spread by the valveless emissary veins to the skull bones, causing an osteomyelitis. Infection may spread farther by the emissary veins into the venous sinuses, producing venous sinus thrombosis.

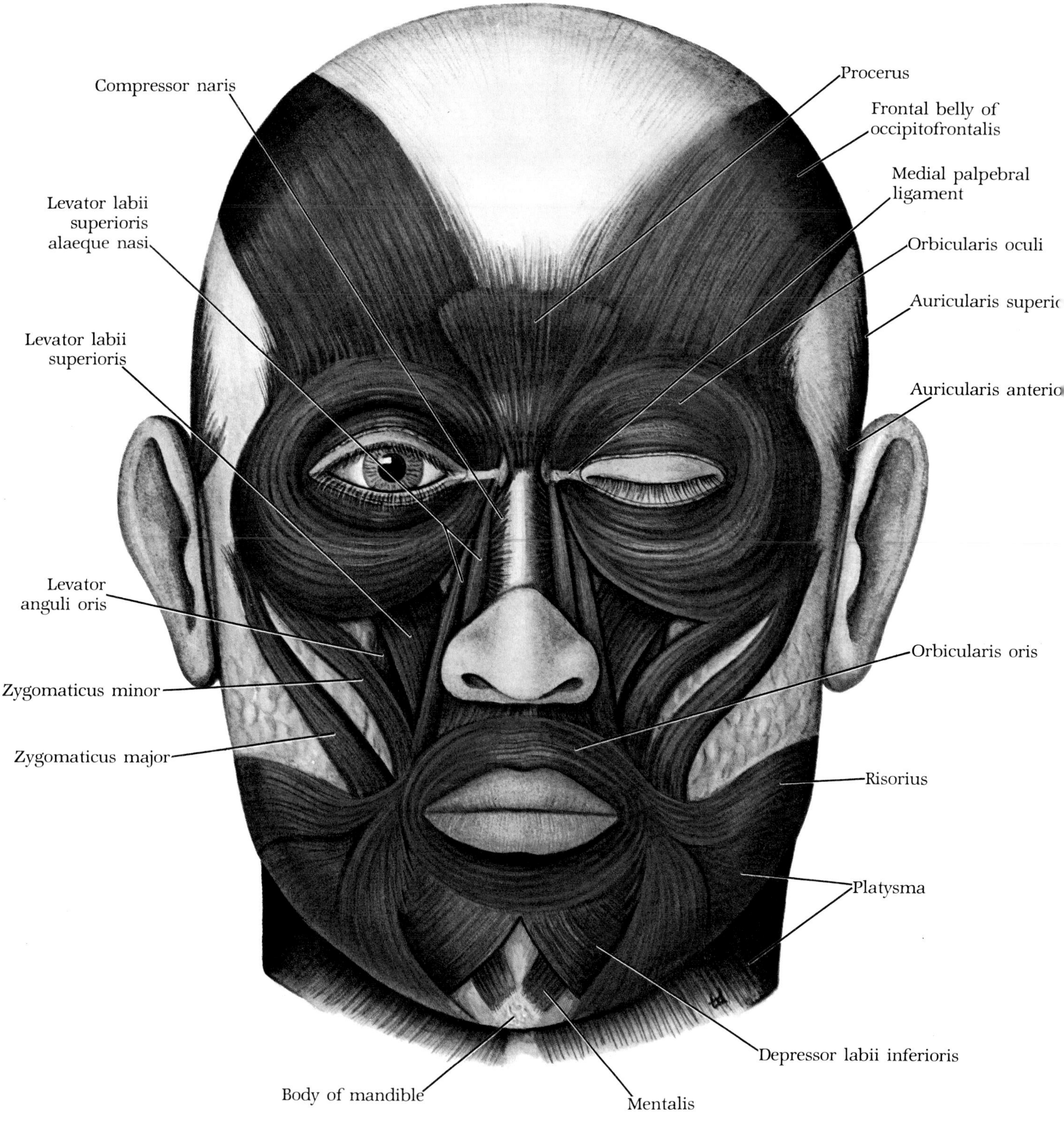
Compressor naris
Procerus
Frontal belly of occipitofrontalis
Medial palpebral ligament
Levator labii superioris alaeque nasi
Orbicularis oculi
Auricularis superic
Levator labii superioris
Auricularis anterio
Levator anguli oris
Orbicularis oris
Zygomaticus minor
Zygomaticus major
Risorius
Platysma
Depressor labii inferioris
Body of mandible
Mentalis

Figure 7-23
Muscles used for facial expression.

1. The muscles used for facial expression are derived from the mesenchyme of the second pharyngeal arch and are all supplied by the facial nerve. It is important to remember that this group of muscles also includes the buccinator, the auricular muscles, the occipitofrontalis, and the platysma muscles.

2. The facial muscles are innervated by the facial nerve. Damage to the facial nerve in the internal acoustic meatus (by a tumor), in the middle ear (by infection or operation), in the facial nerve canal (perineuritis), in the parotid gland (by a malignant tumor), or due to laceration of the face will cause distortion of the face, with the angle of the mouth pulled up on the normal side. This is essentially a lower motor neuron lesion.

3. A lesion of the corticobulbar fibers (i.e., an upper motor neuron lesion) will leave the upper part of the face normal, since the neurons supplying this part of the face receive corticobulbar fibers from both cerebral cortices.

4. The muscles used for facial expression arise from the bones of the skull and are inserted into the skin. They are embedded in superficial fascia (there is no deep fascia in the face).

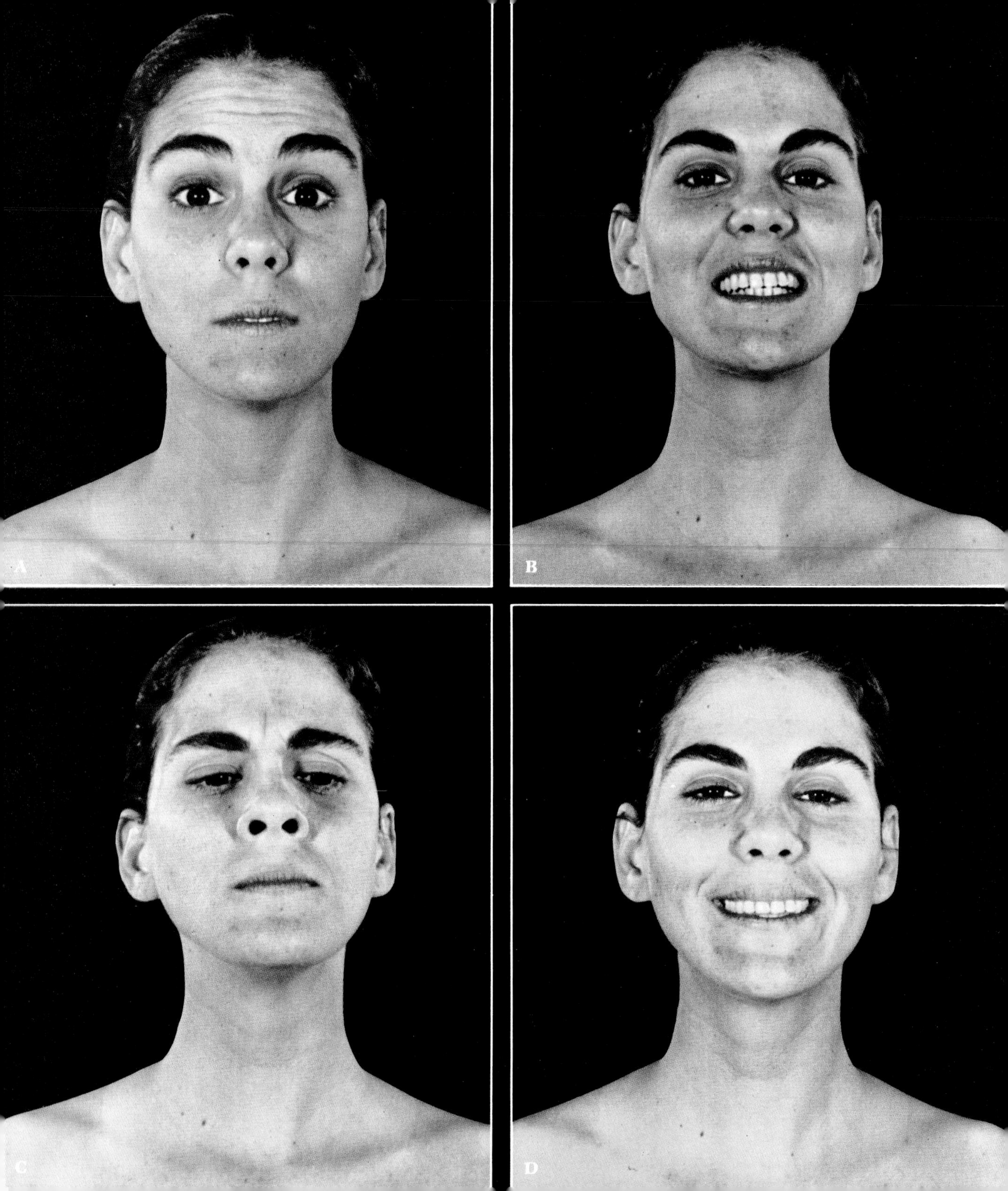
A
B
C
D

Figure 7-24
Facial expressions of a 29-year-old female. A. After contraction of the frontal belly of the occipitofrontalis muscle. B. After contraction of the dilator muscles around the mouth. C. After contraction of the corrugator supercilii muscle. D. After contraction of the risorius muscle.

1. The muscles used for facial expression are embedded in the superficial fascia; the majority arise from the bones of the skull and are inserted into the skin.
2. The orifices of the face, namely the orbit, nose, and mouth, are guarded by the eyelids, nostrils, and lips, respectively. The function of the facial muscles is to serve as sphincters or dilators of these structures and to modify the expression of the face.
3. All the muscles of the face are developed from the second pharyngeal arch and are supplied by the facial nerve.
4. The integrity of the facial nerve can be tested by asking the patient to show the teeth by separating the lips while keeping the teeth clenched, as in photograph B.

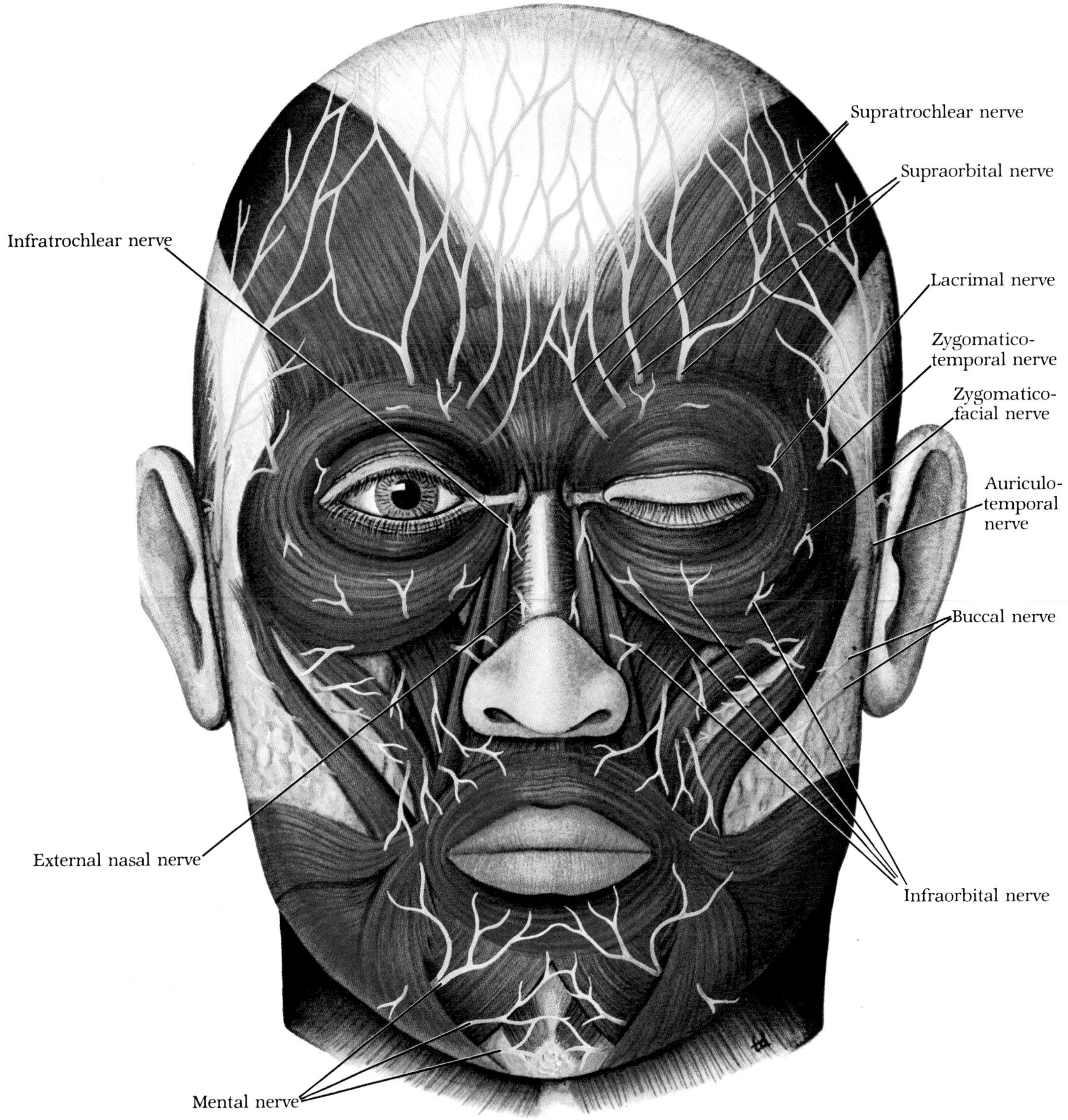
Supratrochlear nerve
Supraorbital nerve
Infratrochlear nerve
Lacrimal nerve
Zygomatico-temporal nerve
Zygomatico-facial nerve
Auriculo-temporal nerve
Buccal nerve
External nasal nerve
Infraorbital nerve
Mental nerve

Figure 7-25
Distribution of the fifth cranial nerve in the face.

1. The facial skin receives its sensory nerve supply from the three divisions of the trigeminal nerve: ophthalmic, mandibular, and maxillary. It is important to remember when testing for the integrity of these nerves that a small area of skin over the angle of the jaw is supplied by the great auricular nerve (C2 and 3).

2. Trigeminal neuralgia is a relatively common condition in which the patient experiences excruciating pain in the distribution of the mandibular or maxillary division of this nerve, although usually not in the ophthalmic division. A physician should be able to map out accurately on a patient's face the distribution of each of the divisions of the trigeminal nerve.

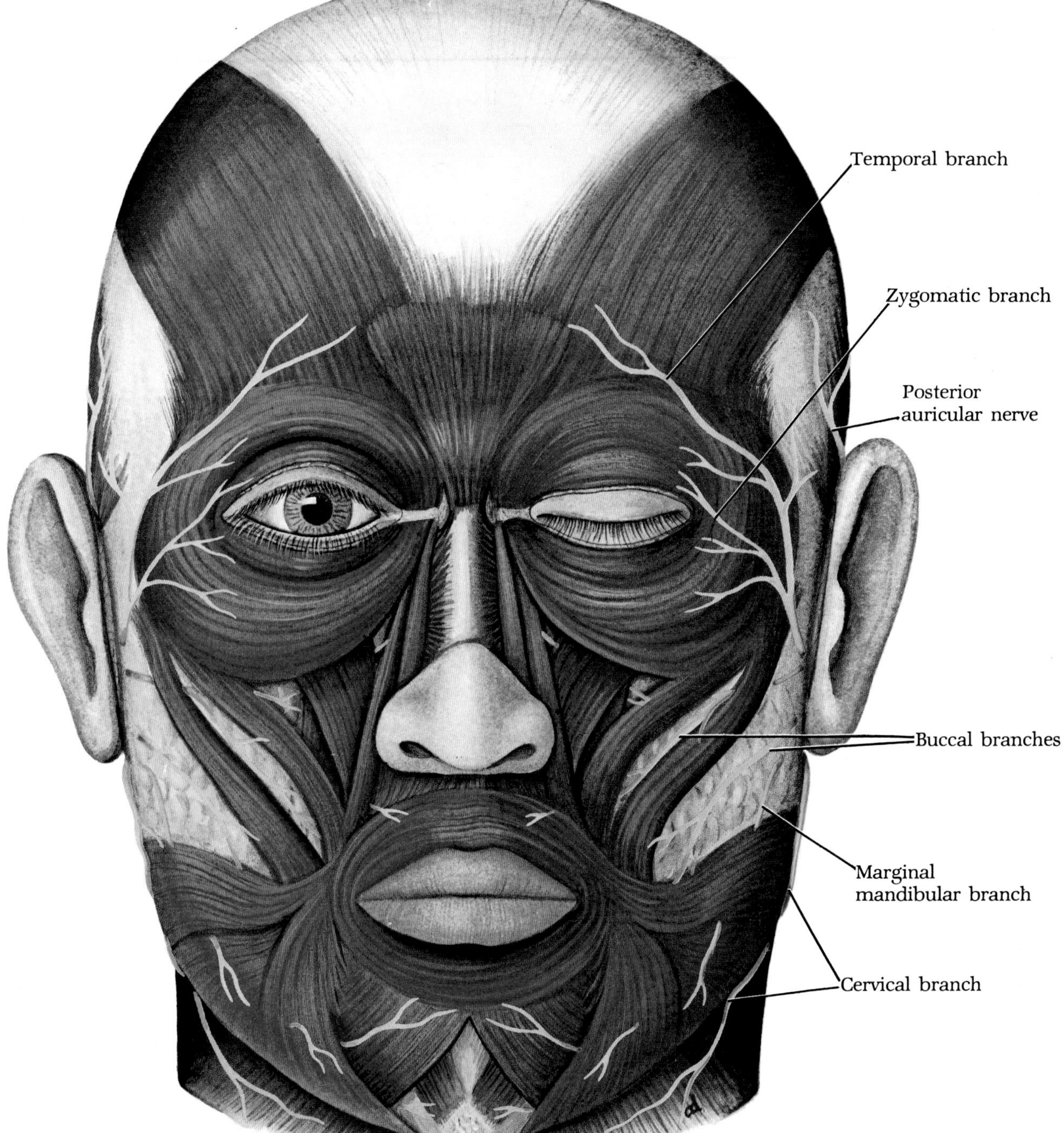
Temporal branch
Zygomatic branch
Posterior auricular nerve
Buccal branches
Marginal mandibular branch
Cervical branch

Figure 7-26
Distribution of the seventh cranial nerve in the face.

1. The facial nerve supplies the muscles used for facial expression by means of its five terminal branches that emerge from the superior, anterior, and inferior borders of the parotid gland. It is important to remember that the facial nerve *does not* supply the skin of the face.

2. In patients with facial palsy the lower eyelid will droop on the side with the palsy and the angle of the corresponding side of the mouth will sag. Tears will flow over the lower eyelid, and saliva will dribble from the corner of the mouth. To test the facial nerve, the patient is asked to show his teeth by separating the lips but keeping the teeth clenched; it may also be tested by asking the patient to close his eyes. The patient with facial palsy will be unable to expose the teeth fully on the affected side and cannot close the eye on that side.

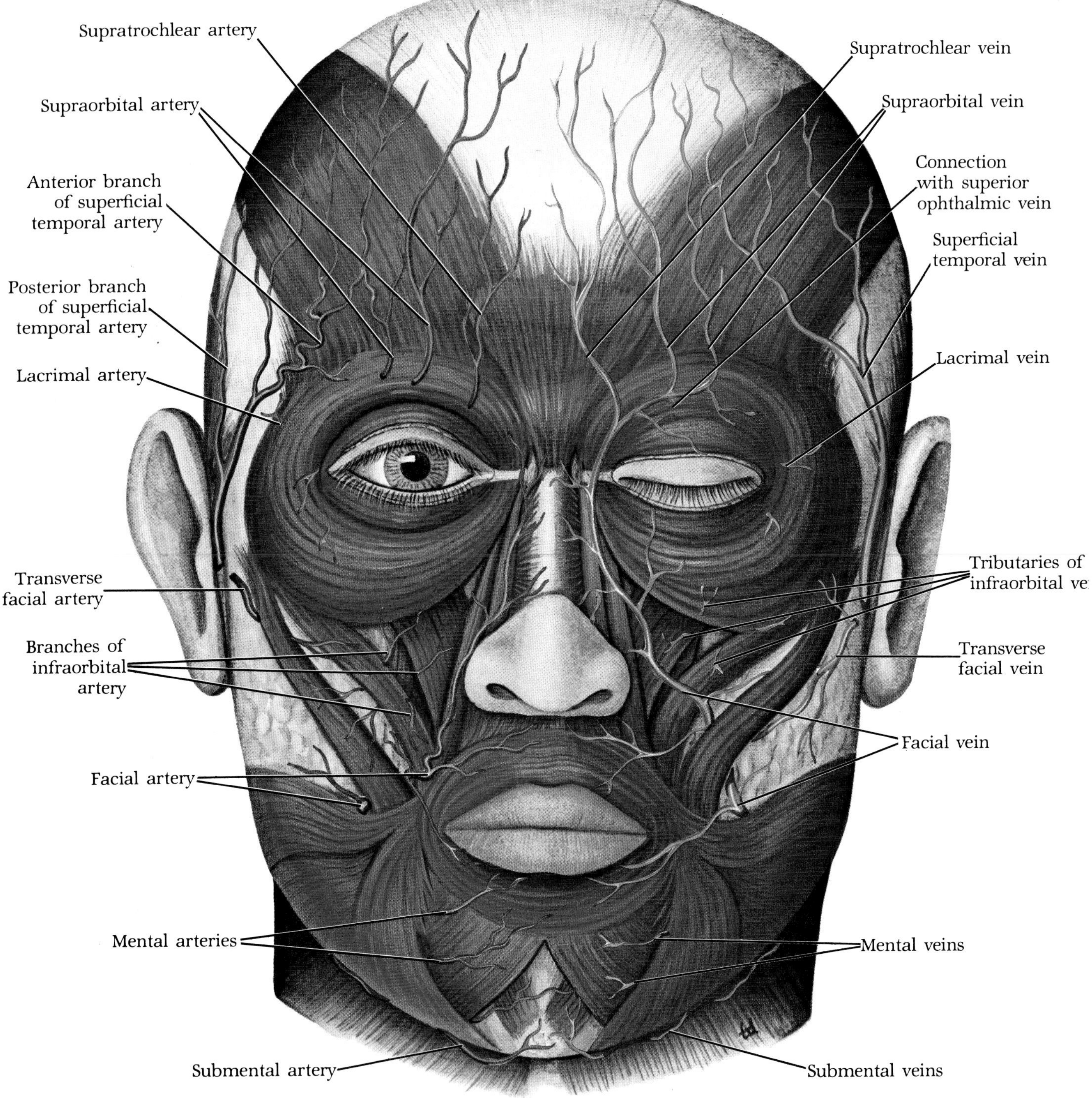

Supratrochlear artery
Supraorbital artery
Anterior branch of superficial temporal artery
Posterior branch of superficial temporal artery
Lacrimal artery
Transverse facial artery
Branches of infraorbital artery
Facial artery
Mental arteries
Submental artery
Supratrochlear vein
Supraorbital vein
Connection with superior ophthalmic vein
Superficial temporal vein
Lacrimal vein
Tributaries of infraorbital ve
Transverse facial vein
Facial vein
Mental veins
Submental veins

Figure 7-27
Blood supply of the face.

1. The blood supply to the skin of the face is profuse, so that it is rare in plastic surgery for skin flaps in this region to necrose.
2. The superficial temporal artery, as it crosses the zygomatic arch in front of the ear, and the facial artery as it winds around the lower margin of the mandible level with the anterior border of the masseter muscle, are commonly used by the anesthesiologist to take the patient's pulse during an operation.
3. The area of facial skin bounded by the nose, the eye, and the upper lip is a potentially dangerous place to have an infection. For example, a boil in this region may cause thrombosis of the facial vein, with spread of organisms through the inferior ophthalmic veins to the cavernous sinus. The resulting cavernous sinus thrombosis may be fatal unless adequately treated with antibiotics.

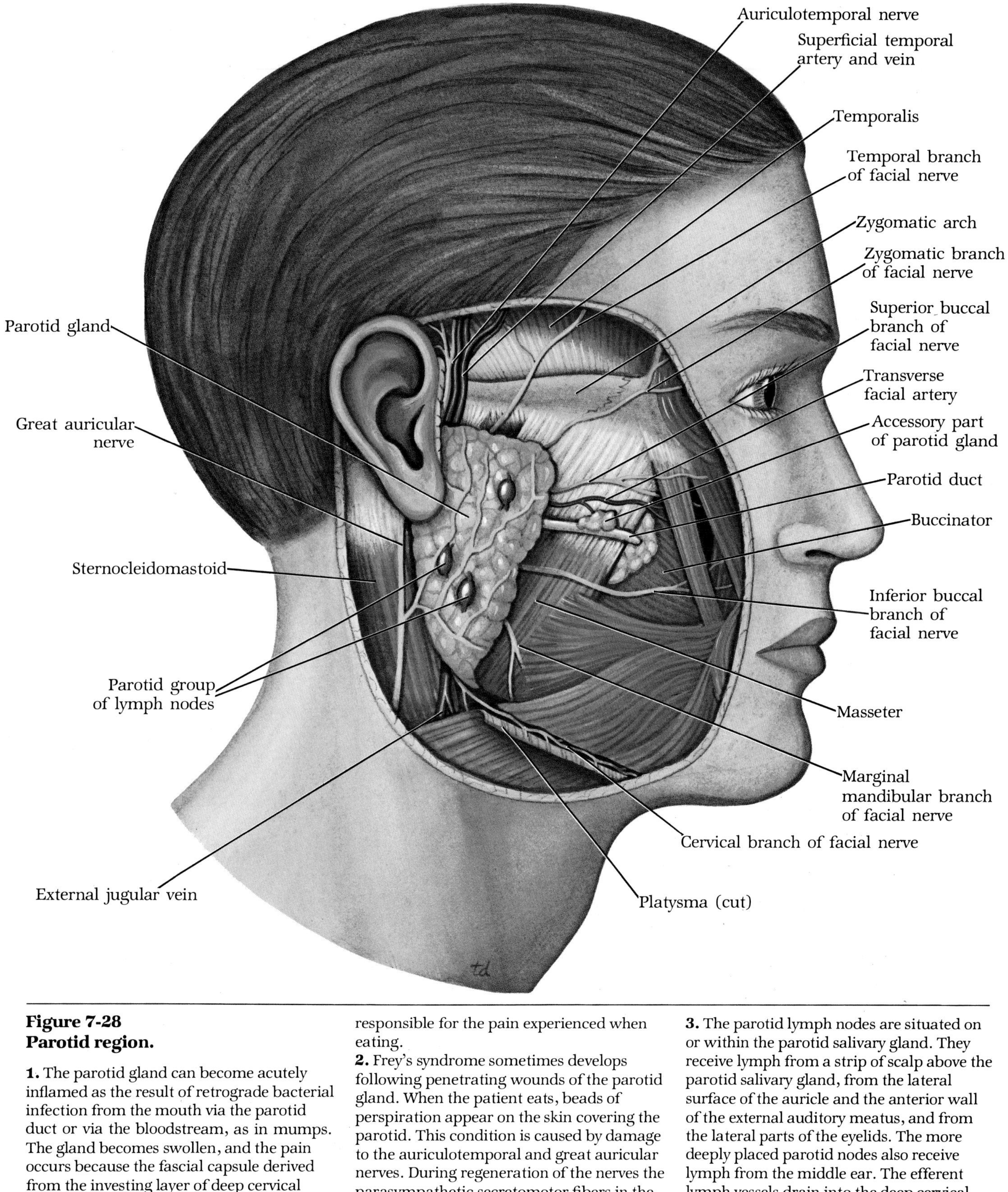
Auriculotemporal nerve
Superficial temporal artery and vein
Temporalis
Temporal branch of facial nerve
Zygomatic arch
Zygomatic branch of facial nerve
Superior buccal branch of facial nerve
Transverse facial artery
Accessory part of parotid gland
Parotid duct
Buccinator
Inferior buccal branch of facial nerve
Masseter
Marginal mandibular branch of facial nerve
Cervical branch of facial nerve
Platysma (cut)
External jugular vein
Parotid group of lymph nodes
Sternocleidomastoid
Great auricular nerve
Parotid gland
td

Figure 7-28
Parotid region.

1. The parotid gland can become acutely inflamed as the result of retrograde bacterial infection from the mouth via the parotid duct or via the bloodstream, as in mumps. The gland becomes swollen, and the pain occurs because the fascial capsule derived from the investing layer of deep cervical fascia is strong, thus limiting the swelling of the gland. In acute parotitis the swollen glenoid process, which extends medially posterior to the temporomandibular joint, is responsible for the pain experienced when eating.

2. Frey's syndrome sometimes develops following penetrating wounds of the parotid gland. When the patient eats, beads of perspiration appear on the skin covering the parotid. This condition is caused by damage to the auriculotemporal and great auricular nerves. During regeneration of the nerves the parasympathetic secretomotor fibers in the auriculotemporal nerve grow out and join the distal end of the great auricular nerve. Eventually these fibers reach the sweat glands in the facial skin. By this means, a stimulus intended for saliva production produces sweat secretion instead.

3. The parotid lymph nodes are situated on or within the parotid salivary gland. They receive lymph from a strip of scalp above the parotid salivary gland, from the lateral surface of the auricle and the anterior wall of the external auditory meatus, and from the lateral parts of the eyelids. The more deeply placed parotid nodes also receive lymph from the middle ear. The efferent lymph vessels drain into the deep cervical lymph nodes.

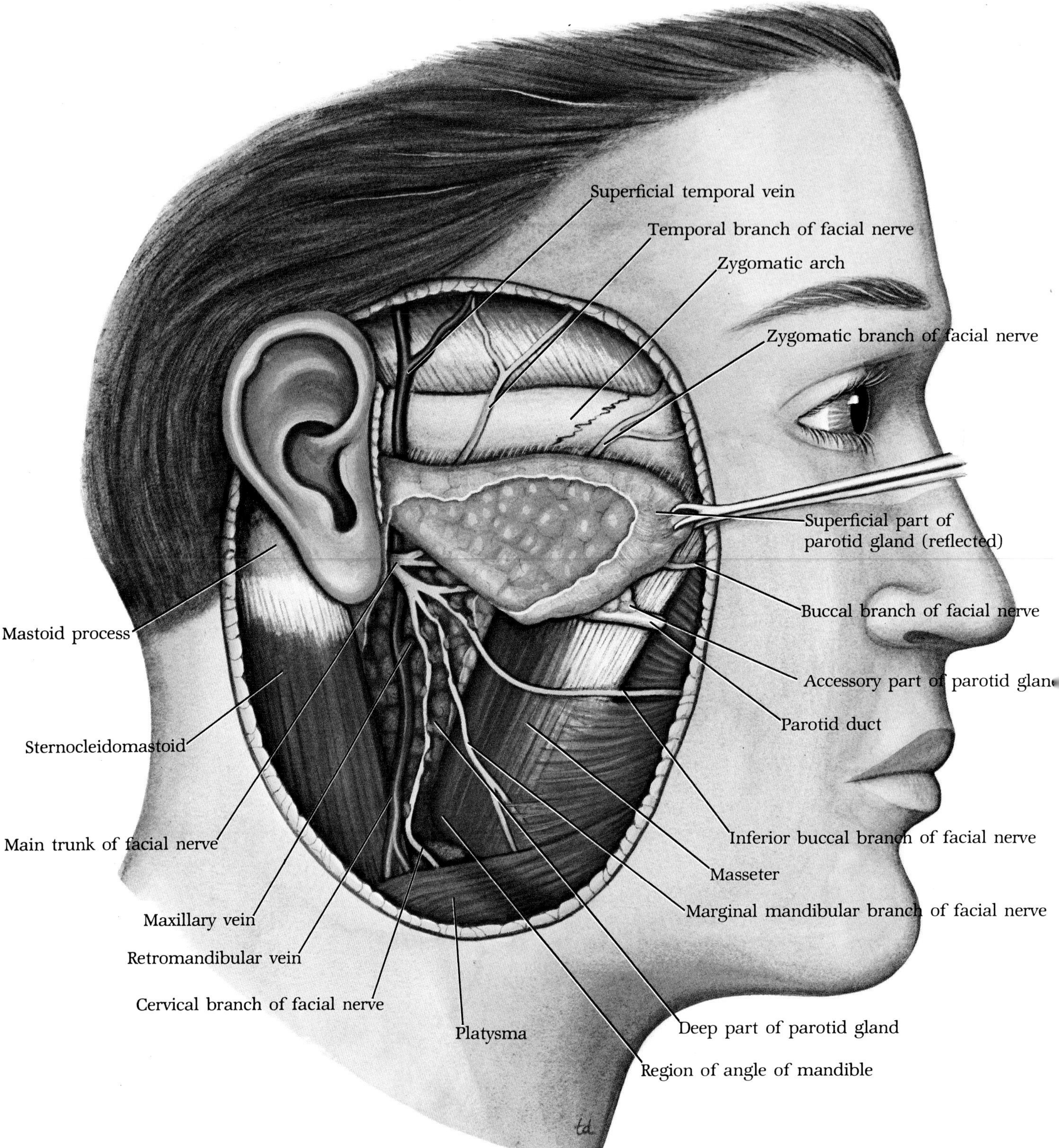
Superficial temporal vein
Temporal branch of facial nerve
Zygomatic arch
Zygomatic branch of facial nerve
Superficial part of parotid gland (reflected)
Buccal branch of facial nerve
Accessory part of parotid glan
Parotid duct
Inferior buccal branch of facial nerve
Masseter
Marginal mandibular branch of facial nerve
Deep part of parotid gland
Region of angle of mandible
Platysma
Cervical branch of facial nerve
Retromandibular vein
Maxillary vein
Main trunk of facial nerve
Sternocleidomastoid
Mastoid process

Figure 7-29
Parotid salivary gland.

1. The parotid salivary gland is wedged into the space between the mandible anteriorly and the mastoid process and the sternocleidomastoid muscle posteriorly. It consists essentially of superficial and deep parts, and the important facial nerve lies in the interval between these parts. A benign parotid neoplasm rarely, if ever, causes facial palsy. A malignant tumor of the parotid gland is usually highly invasive and quickly involves the facial nerve, causing unilateral facial paralysis.

2. The duct of the parotid gland passes forward in the face about one fingerbreadth below the zygomatic arch. At the anterior border of the masseter muscle it turns medially, pierces the buccinator muscle, and enters the vestibule of the mouth opposite the upper second molar tooth. Penetrating wounds of the face may result in damage to the duct.

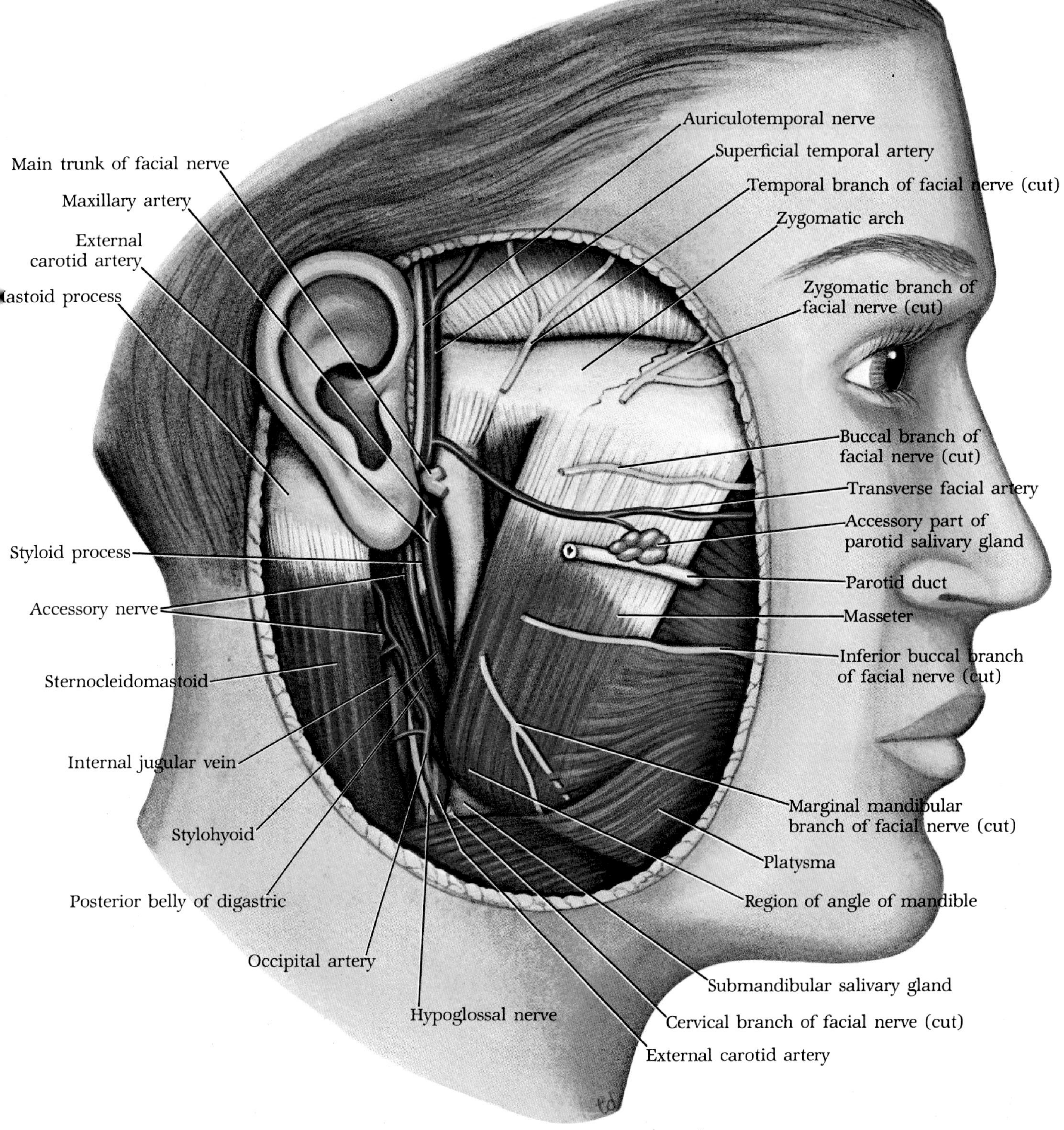
Auriculotemporal nerve
Superficial temporal artery
Temporal branch of facial nerve (cut)
Zygomatic arch
Main trunk of facial nerve
Maxillary artery
External carotid artery
astoid process
Zygomatic branch of facial nerve (cut)
Buccal branch of facial nerve (cut)
Transverse facial artery
Accessory part of parotid salivary gland
Styloid process
Parotid duct
Accessory nerve
Masseter
Inferior buccal branch of facial nerve (cut)
Sternocleidomastoid
Internal jugular vein
Marginal mandibular branch of facial nerve (cut)
Stylohyoid
Platysma
Posterior belly of digastric
Region of angle of mandible
Occipital artery
Submandibular salivary gland
Hypoglossal nerve
Cervical branch of facial nerve (cut)
External carotid artery

Figure 7-30
Parotid bed.

1. The parotid gland lies in the space between the mandible anteriorly and the mastoid process and the sternocleidomastoid muscle posteriorly. It is related medially to the posterior belly of the digastric muscle and the styloid process and its attached muscles. The gland is separated from these structures by a fascial sheath derived from the investing layer of deep cervical fascia. The sheath is strong and limits the swelling of the gland in acute parotitis, hence the severe pain.

2. Within the parotid gland are the facial nerve, the retromandibular vein, and the external carotid artery. From a surgical point of view it is very important to remember that the facial nerve and its branches within the gland all lie in one plane and are situated between the large superficial and smaller, deeper part of the gland. This fact is of great help in the surgical treatment of tumors of the parotid gland since the superficial and deep parts can be removed separately, leaving the facial nerve and its branches intact. Mixed parotid tumors most commonly occur in the superficial part of the parotid and are therefore easily removed by excising the superficial part of the gland.

3. Since advanced carcinoma of the parotid gland almost always involves the facial nerve, radical parotidectomy results in removal of the facial nerve. Later, plastic surgery procedures on the face or even hypoglossal nerve transplantation may be performed as treatment for the resulting unilateral facial paralysis.

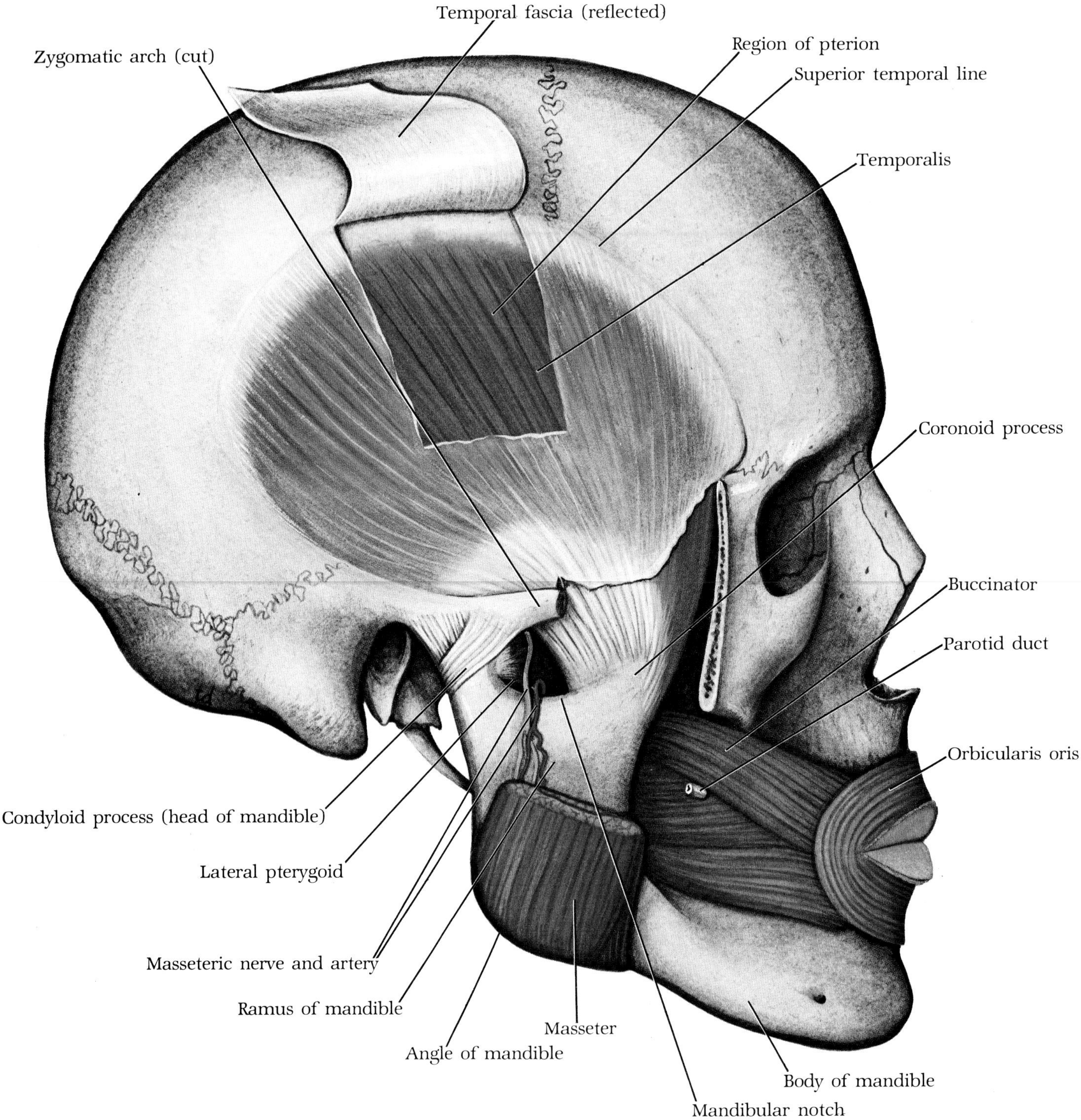
Temporal fascia (reflected)
Zygomatic arch (cut)
Region of pterion
Superior temporal line
Temporalis
Coronoid process
Buccinator
Parotid duct
Orbicularis oris
Condyloid process (head of mandible)
Lateral pterygoid
Masseteric nerve and artery
Ramus of mandible
Masseter
Angle of mandible
Body of mandible
Mandibular notch

Figure 7-31
Temporal fossa.

1. Collections of blood or pus beneath the epicranial aponeurosis of the scalp spread extensively forward to the superciliary arches above the orbital margins and posteriorly to the origin of the occipitalis muscle from the occipital bone; laterally they spread to the attachment of the aponeurosis to the superior temporal lines.

2. Fractures of the bones forming the floor of the temporal fossa may be accompanied by hemorrhage into the temporalis muscle, with the formation of a hematoma beneath the tough temporal fascia that reveals itself on the surface as a boggy swelling over the temporal fossa.

3. Extradural hemorrhage as a result of a severe blow on the side of the head is most commonly due to rupture of the middle meningeal vessels. The most frequent cause is damage to the venae comitantes, which lie lateral to the anterior division of the middle meningeal artery in a groove on the inner table of the anterior inferior part of the parietal bone (pterion). To stop the bleeding, a trephine or burr hole is made through the skull about 1½ inches (4 cm) above the midpoint of the zygomatic arch, after which the bleeding vessel can be secured.

4. The motor function of the trigeminal nerve can be tested by having the patient clench his teeth. The temporalis muscle can be palpated along with the masseter muscle and both can be felt to contract if the mandibular division of the trigeminal nerve is intact.

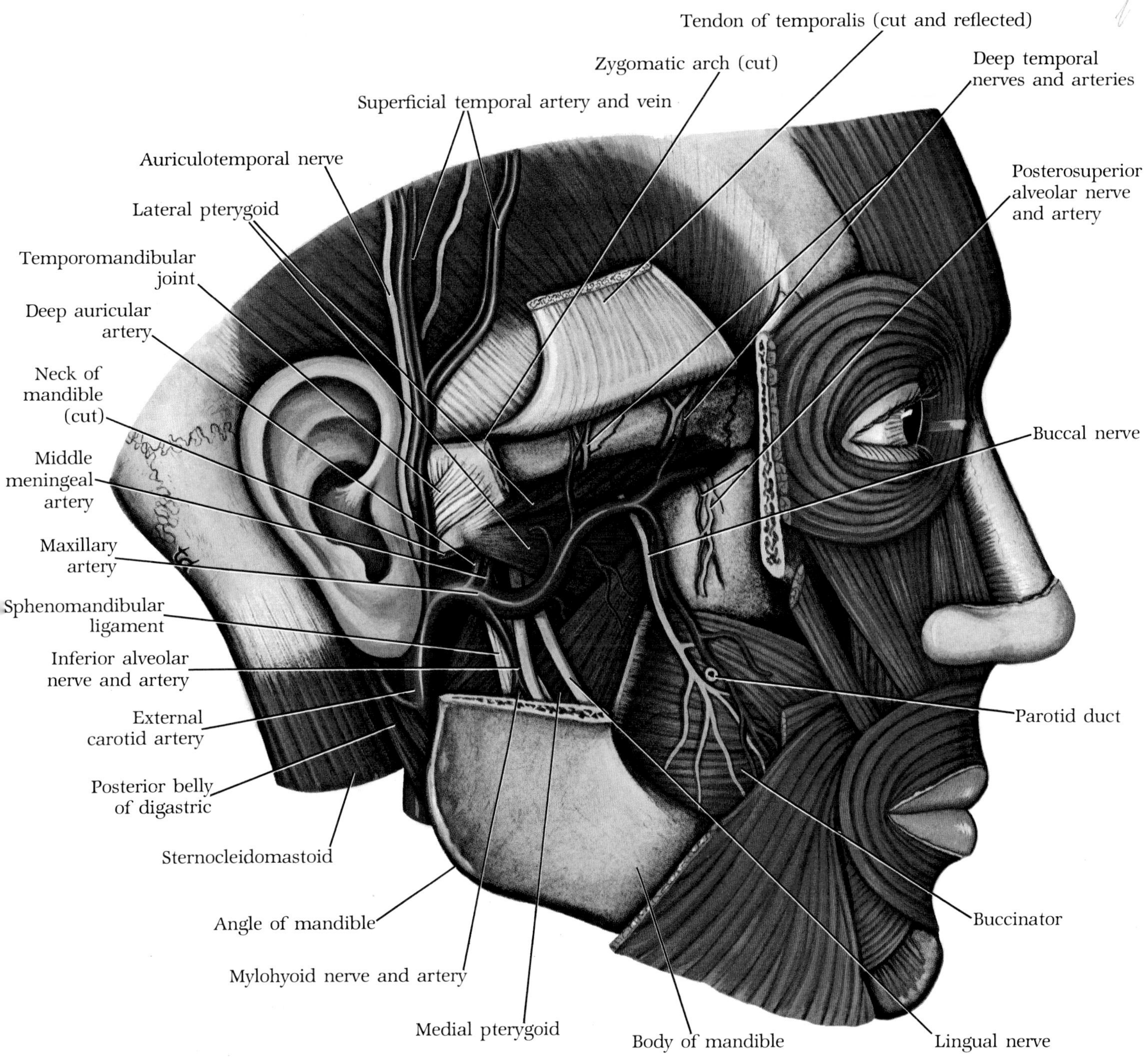
Tendon of temporalis (cut and reflected)
Zygomatic arch (cut)
Deep temporal nerves and arteries
Superficial temporal artery and vein
Auriculotemporal nerve
Posterosuperior alveolar nerve and artery
Lateral pterygoid
Temporomandibular joint
Deep auricular artery
Neck of mandible (cut)
Buccal nerve
Middle meningeal artery
Maxillary artery
Sphenomandibular ligament
Inferior alveolar nerve and artery
Parotid duct
External carotid artery
Posterior belly of digastric
Sternocleidomastoid
Angle of mandible
Buccinator
Mylohyoid nerve and artery
Medial pterygoid
Body of mandible
Lingual nerve

Figure 7-32
Infratemporal fossa. Parts of the zygomatic arch and the ramus of the mandible have been removed to display deeper structures.

1. This region of the neck is relatively free from disease, which is fortunate, because it is relatively inaccessible. Extensive malignant tumors of the maxilla may require radical resection of the maxilla, in which case the surgeon will enter this area. Hemimandibulectomy or posterior segmental resection of the mandible for malignant tumors, radioneurosis, or osteomyelitis also requires considerable knowledge of "surgical" anatomy of this region.

2. Trigeminal neuralgia is a relatively common condition of unknown cause and occurs more frequently in women. The patient experiences excruciating pain in the distribution of the mandibular or maxillary division, although not usually in the ophthalmic division. Spasms of pain are produced by cold drafts, brushing the teeth, or eating. It is distinguished by the fact that there are no physical findings. If the condition persists and the pain is not relieved by analgesics, the trigeminal sensory ganglion may be injected with alcohol that is introduced through a needle that has been inserted, under a local anesthetic, medially below the zygomatic arch. The needle is passed medially into the skull through the foramen ovale. Surgical section of the sensory root of the mandibular division of the trigeminal nerve in the middle cranial fossa is also another method of treating this condition.

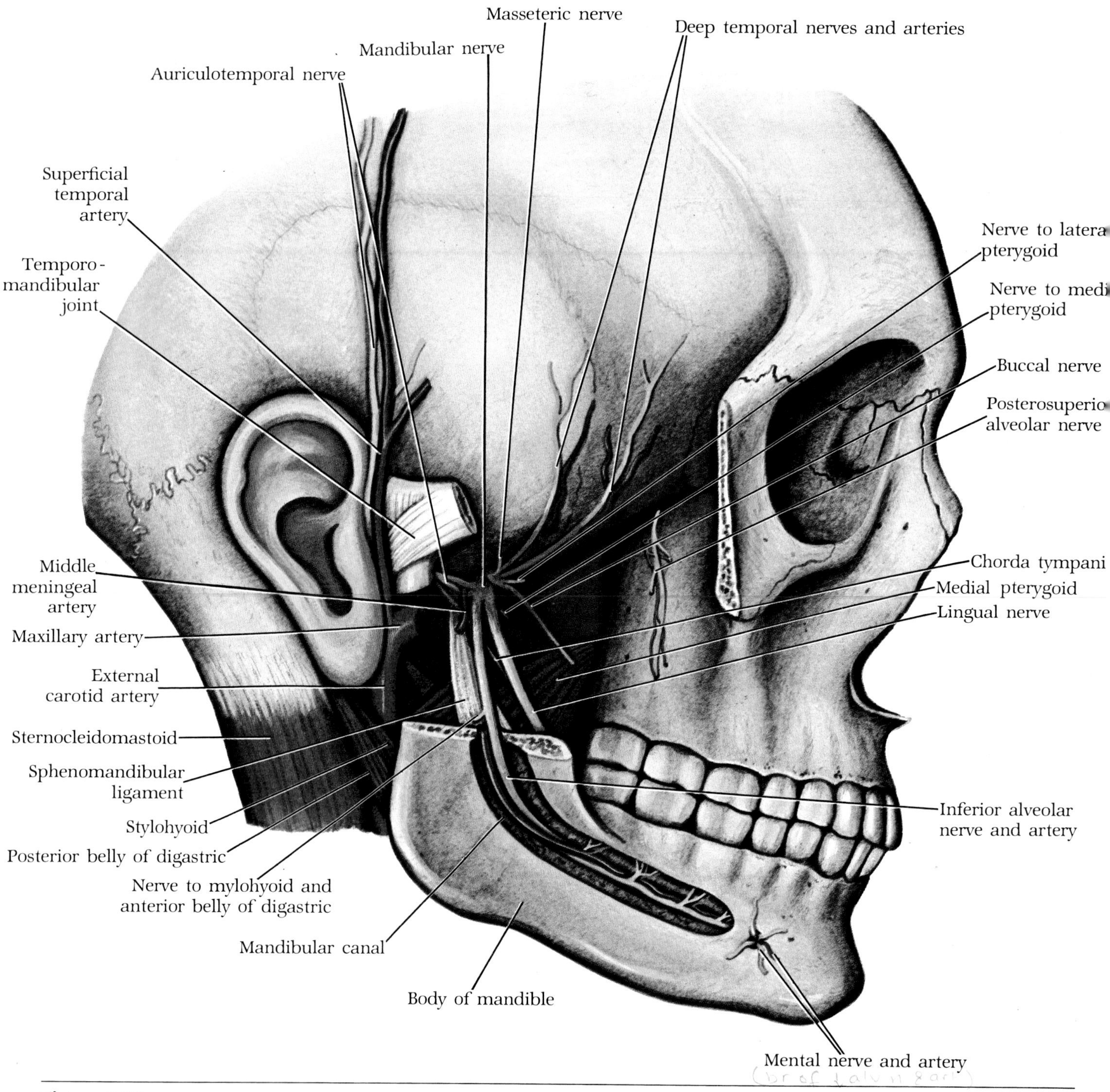
Masseteric nerve
Deep temporal nerves and arteries
Mandibular nerve
Auriculotemporal nerve
Superficial temporal artery
Temporo-mandibular joint
Nerve to latera pterygoid
Nerve to medi pterygoid
Buccal nerve
Posterosuperio alveolar nerve
Middle meningeal artery
Maxillary artery
External carotid artery
Sternocleidomastoid
Sphenomandibular ligament
Stylohyoid
Posterior belly of digastric
Nerve to mylohyoid and anterior belly of digastric
Mandibular canal
Body of mandible
Chorda tympani
Medial pterygoid
Lingual nerve
Inferior alveolar nerve and artery
Mental nerve and artery

Figure 7-33
Infratemporal fossa. Parts of the zygomatic arch and the ramus of the mandible have been removed to display the branches of the mandibular nerve and the chorda tympani. The mandibular canal has been opened to display the inferior alveolar nerve and artery.

1. The mandible is the most common bone in the face to be fractured. Because of the strong muscles of mastication the fragments are usually displaced. In fractures of the neck of the mandible the superior fragment with the head is pulled anteriorly by the lateral pterygoid muscle. In fractures of the ramus, which often occur near the angle, there is little displacement since the fragments are held together on the medial side by the attachment of the medial pterygoid muscle and laterally by the masseter muscle. In fractures of the body, the socket of the canine tooth is a site of weakness of the bone and a fracture often occurs there. Because the mucous membrane is fused to the periosteum, the fracture line is open to infection from the mouth (i.e., it is a compound fracture). In bilateral fractures of the body of the mandible, the anterior bellies of the digastric and geniohyoid muscles displace the anterior fragment of the jaw inferiorly.

2. A mandibular nerve block is used when extracting teeth from the lower jaw. With the patient's mouth open, the anterior margin of the ramus of the mandible is palpated and the pterygomandibular ligament is felt. The syringe needle is inserted through the mucuous membrane just lateral to the ligament. Two or 3 ml of an injected local anesthetic solution infiltrates the inferior alveolar nerve. It is important to remember that the central incisor teeth receive a few nerve fibers from the opposite mandibular nerve.

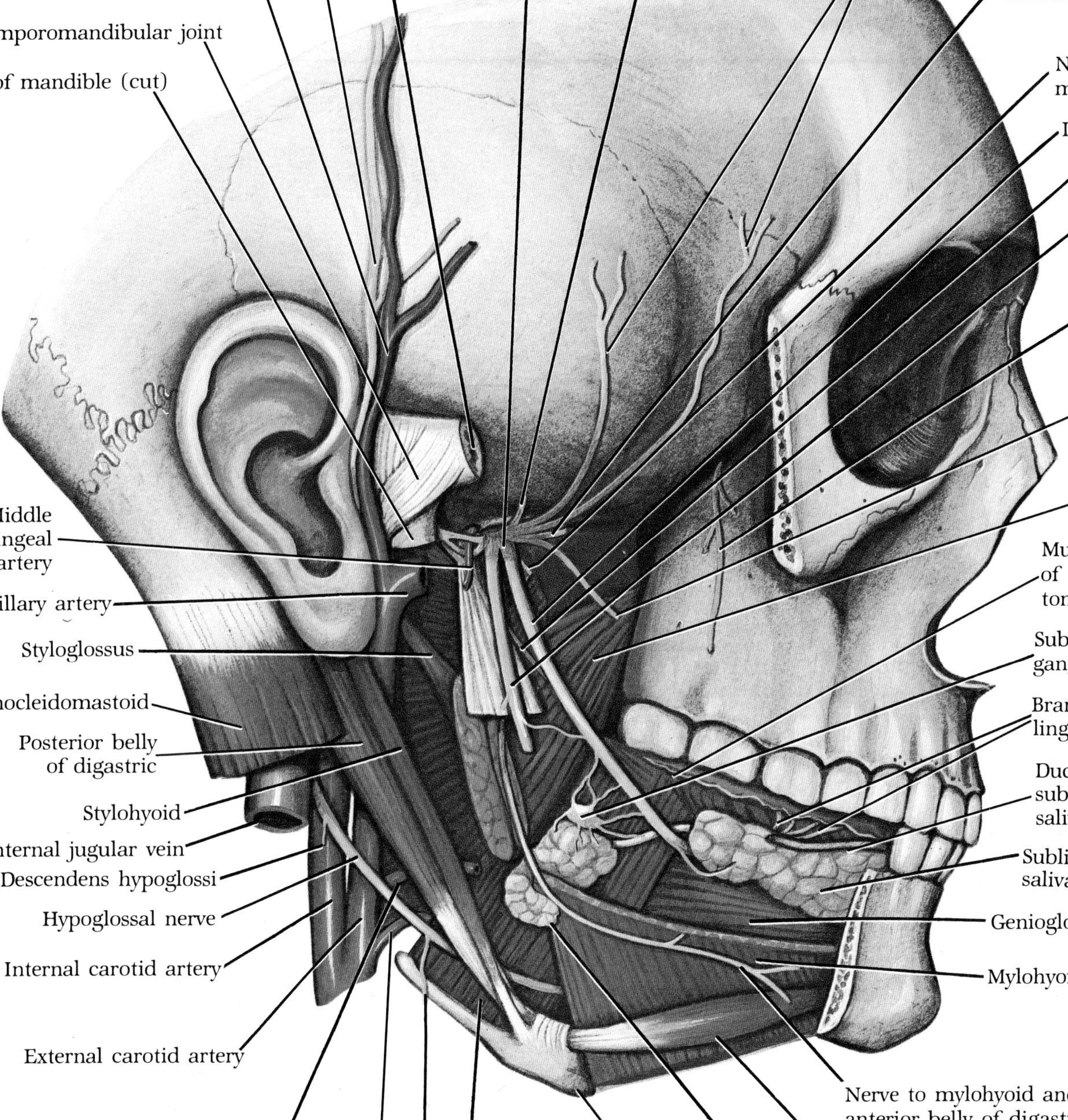

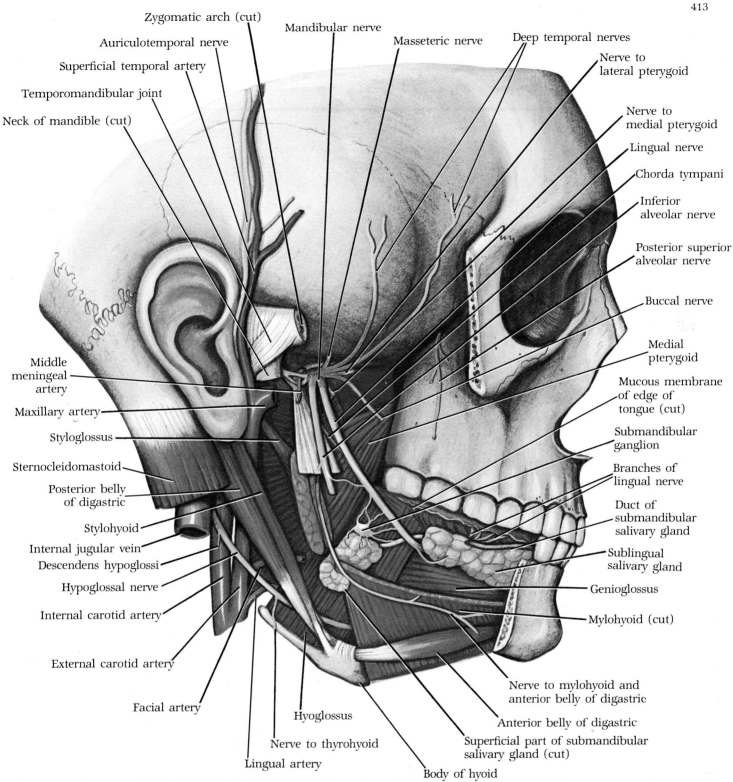

Zygomatic arch (cut)
Mandibular nerve
Masseteric nerve
Deep temporal nerves
Auriculotemporal nerve
Superficial temporal artery
Nerve to lateral pterygoid
Temporomandibular joint
Neck of mandible (cut)
Nerve to medial pterygoid
Lingual nerve
Chorda tympani
Inferior alveolar nerve
Posterior superior alveolar nerve
Buccal nerve
Medial pterygoid
Middle meningeal artery
Mucous membrane of edge of tongue (cut)
Maxillary artery
Styloglossus
Submandibular ganglion
Sternocleidomastoid
Branches of lingual nerve
Posterior belly of digastric
Duct of submandibular salivary gland
Stylohyoid
Internal jugular vein
Sublingual salivary gland
Descendens hypoglossi
Hypoglossal nerve
Genioglossus
Internal carotid artery
Mylohyoid (cut)
External carotid artery
Nerve to mylohyoid and anterior belly of digastric
Facial artery
Hyoglossus
Anterior belly of digastric
Nerve to thyrohyoid
Superficial part of submandibular salivary gland (cut)
Lingual artery
Body of hyoid

Figure 7-34
Infratemporal fossa and submandibular regions. Parts of the zygomatic arch and ramus and body of the mandible have been removed. The lateral pterygoid muscle has also been removed to display deeper structures.

1. The lingual nerve passes forward into the submandibular region from the infratemporal fossa by running beneath the origin of the superior constrictor muscle, which is attached to the posterior end of the mylohyoid line on the mandible. Here it is closely related to the last molar tooth and is liable to be damaged in cases of clumsy extraction of an impacted third molar.
2. The submandibular salivary gland is a common site of calculus formation, although the condition is rare in the other salivary glands. The presence of a tense swelling below the body of the mandible that is greatest before or during a meal and is reduced in size or absent between meals, is diagnostic of the condition. Examination of the floor of the mouth will reveal that there is no ejection of saliva from the orifice of the duct of the affected gland. The stone frequently can be palpated in the duct, which lies below the mucous membrane of the floor of the mouth.
3. The submandibular lymph nodes are commonly enlarged as the result of a pathological condition of the scalp, face, maxillary sinus, or mouth cavity. One of the most common causes of painful enlargement of these nodes is acute infections of the teeth.
4. The sublingual salivary gland opens into the mouth by numerous small ducts. Blockage of one of these ducts is believed to be the cause of cysts under the tongue.

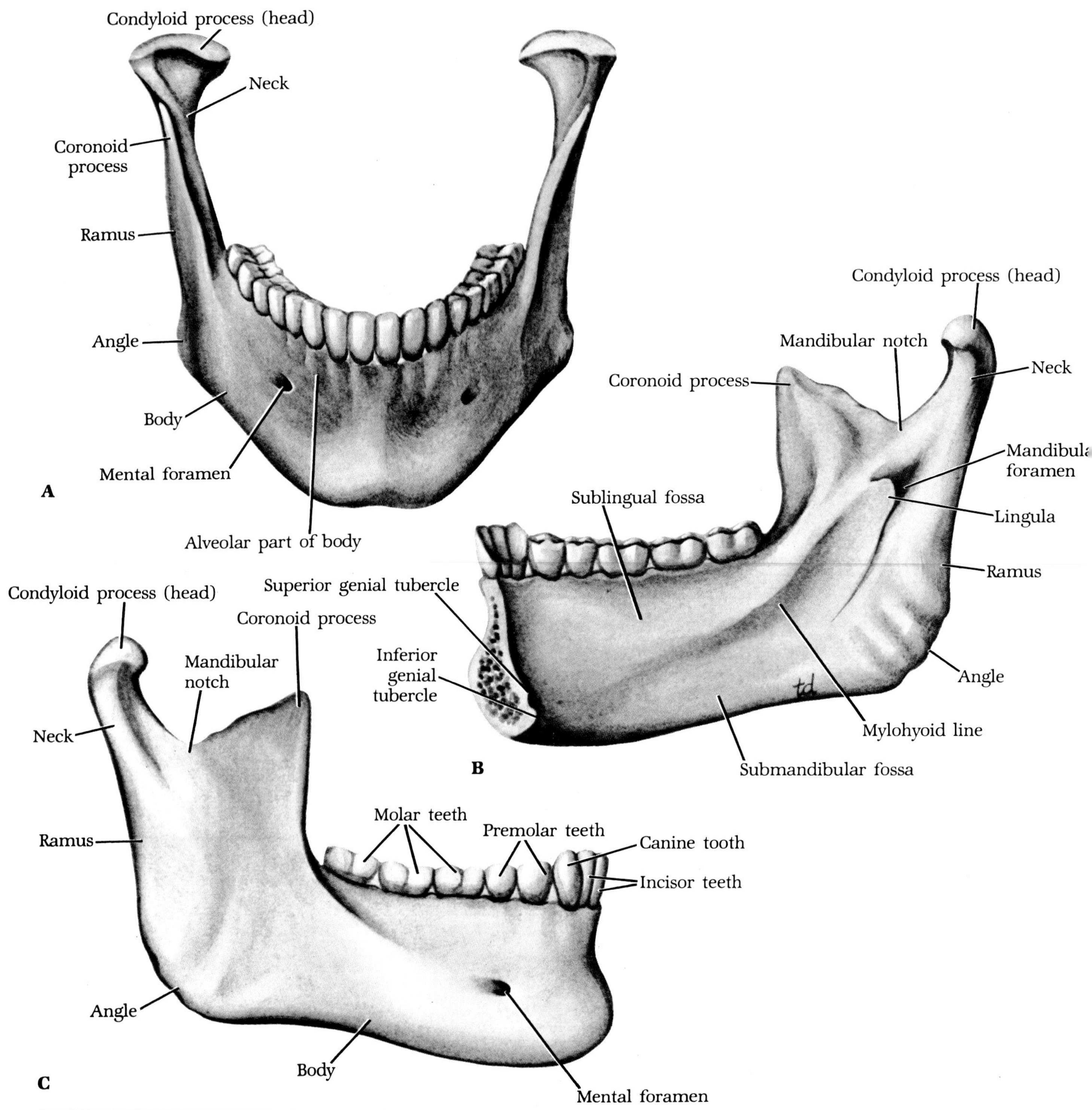
Condyloid process (head)
Neck
Coronoid process
Ramus
Angle
Body
Mental foramen
A
Alveolar part of body
Condyloid process (head)
Mandibular notch
Coronoid process
Neck
Mandibula
foramen
Lingula
Sublingual fossa
Ramus
Superior genial tubercle
Inferior genial tubercle
Angle
Mylohyoid line
B
Submandibular fossa
Condyloid process (head)
Coronoid process
Mandibular notch
Neck
Ramus
Molar teeth
Premolar teeth
Canine tooth
Incisor teeth
Angle
Body
C
Mental foramen

Figure 7-35
Mandible. A. Anterior aspect. B. Right medial aspect. C. Right lateral aspect.

1. With the patient's mouth open, the body, angle, and inferior part of the ramus of the mandible can be palpated from both inside and outside the mouth. The coronoid process can be palpated with a gloved finger inside the mouth, and its lateral surface can be felt on deep palpation outside the mouth through the relaxed masseter muscle.

2. Fractures of the mandible occur commonly in three places: the neck, the ramus, and the body. In fractures of the neck, the superior fragment with the head is pulled anteriorly by the lateral pterygoid muscle. In fractures of the ramus, which often occur near the angle, there is little displacement since the fragments are held together on the medial side by the attachment of the medial pterygoid muscle and laterally by the masseter muscle. In fractures of the body, the mucous membrane of the mouth is fused with the periosteum so that the fracture is compound. In bilateral fractures of the body of the mandible, the anterior bellies of the digastric and the geniohyoid muscles displace the anterior fragment of the jaw inferiorly.

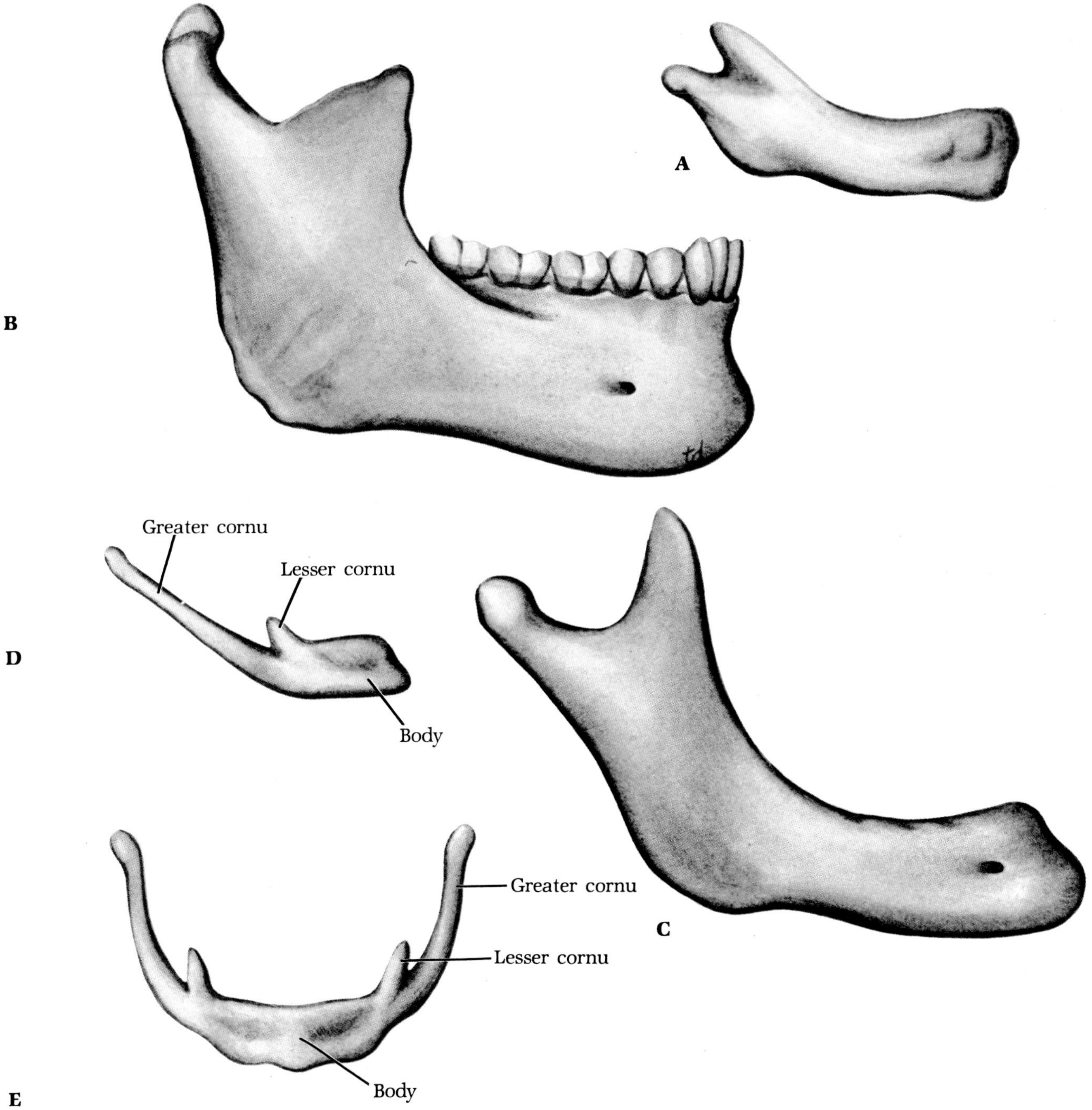

A
B
Greater cornu
Lesser cornu
D
Body
C
Greater cornu
Lesser cornu
Body
E

Figure 7-36
Mandible. A. In the infant. B. In the adult. C. In an elderly person. Hyoid bone. D. Lateral view. E. Anterior view.

1. At birth the body of the mandible is a thin shell of bone enclosing the rudimentary teeth. As the teeth erupt the body grows in height. Room for the three permanent molars is provided by the growth in height of the alveolar bone and absorption from the anterior border of the ramus. When teeth are lost, the alveolar part of the body of the mandible becomes absorbed. The angle between the inferior border of the mandible and the condylar process is greater in the newborn and the aged.

2. Swelling of the mandible may be caused by inflammation, as in alveolar abscess or osteomyelitis, by odontomes arising from epithelial or mesothelial elements of teeth, or by bone tumors, both benign or malignant. The mandible is not a common site for primary malignant neoplasms. It is, however, often involved in the late stages of carcinoma of the tongue, lip, or floor of the mouth.

3. The hyoid bone forms a base for the tongue and is suspended in position by muscles that connect it to the mandible, to the styloid process of the temporal bone, to the thyroid cartilage, to the sternum, and to the scapula. The bone is of considerable forensic importance in that it is usually found to be fractured in persons killed by strangulation.

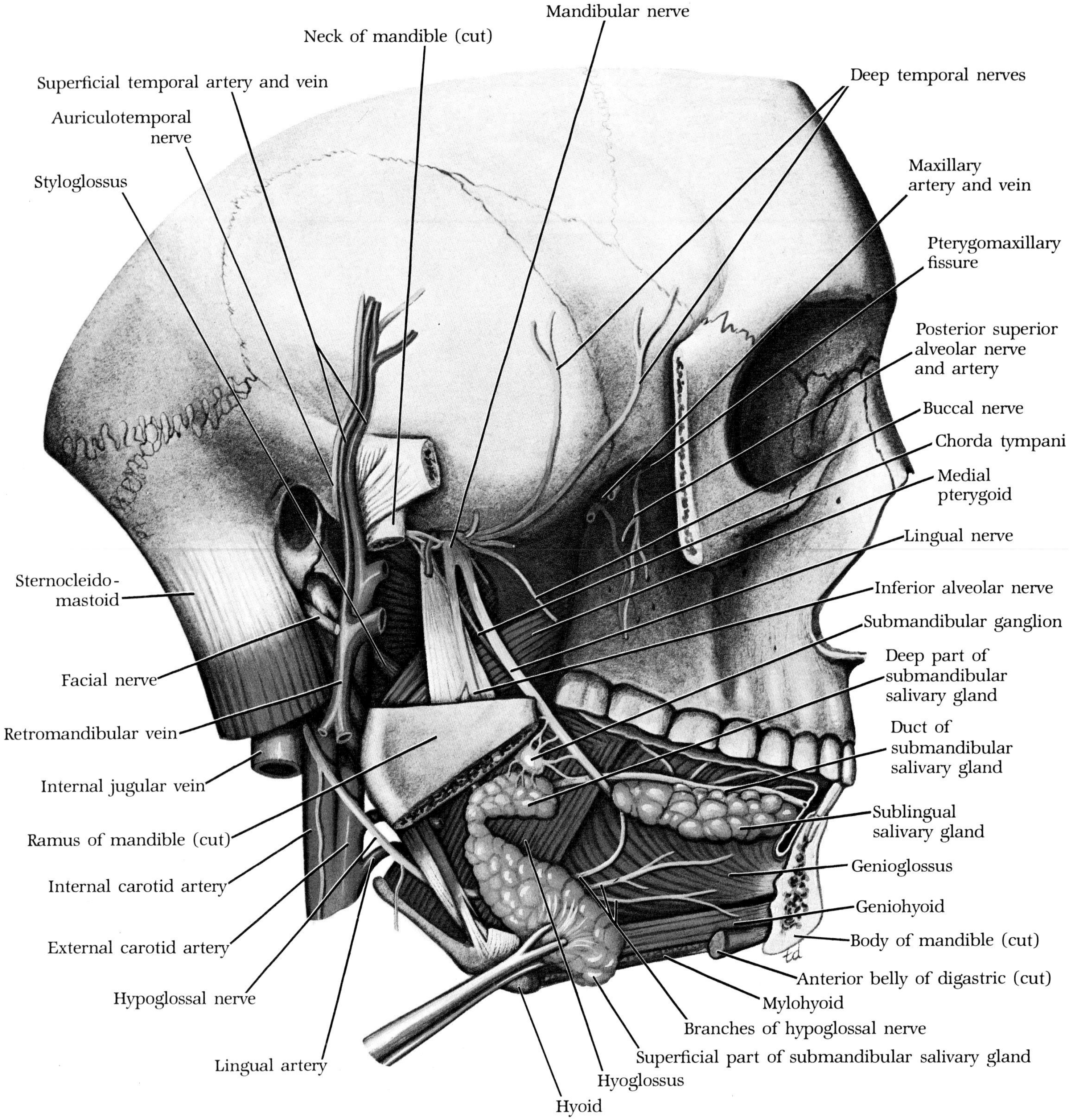
Mandibular nerve
Neck of mandible (cut)
Superficial temporal artery and vein
Auriculotemporal nerve
Styloglossus
Deep temporal nerves
Maxillary artery and vein
Pterygomaxillary fissure
Posterior superior alveolar nerve and artery
Buccal nerve
Chorda tympani
Medial pterygoid
Lingual nerve
Inferior alveolar nerve
Submandibular ganglion
Deep part of submandibular salivary gland
Duct of submandibular salivary gland
Sublingual salivary gland
Genioglossus
Geniohyoid
Body of mandible (cut)
Anterior belly of digastric (cut)
Mylohyoid
Branches of hypoglossal nerve
Superficial part of submandibular salivary gland
Hyoglossus
Hyoid
Lingual artery
Hypoglossal nerve
External carotid artery
Internal carotid artery
Ramus of mandible (cut)
Internal jugular vein
Retromandibular vein
Facial nerve
Sternocleido-mastoid
td

Figure 7-37
Infratemporal and submandibular regions. Parts of the zygomatic arch, ramus, and body of the mandible have been removed. Mylohyoid and lateral pterygoid muscles have also been removed to display deeper structures.

1. A stone in the duct of the submandibular salivary gland can often be palpated by placing a gloved finger on the floor of the mouth between the tongue and the body of the mandible. Here the duct lies just beneath the mucous membrane of the floor of the mouth.
2. The sublingual salivary gland, which lies beneath the sublingual fold of the floor of the mouth, opens into the mouth by numerous small ducts. Blockage of one or more of these ducts is believed to be the cause of cysts under the tongue.
3. Note the close relationship between the last molar tooth and the lingual nerve. The nerve may be damaged in cases of clumsy extraction of an impacted third molar.
4. The intimate relationship of the mucous membrane of the floor of the mouth, the tongue, and the body of the mandible explains how extensive carcinoma of these mouth structures readily involves the mandible by direct invasion.

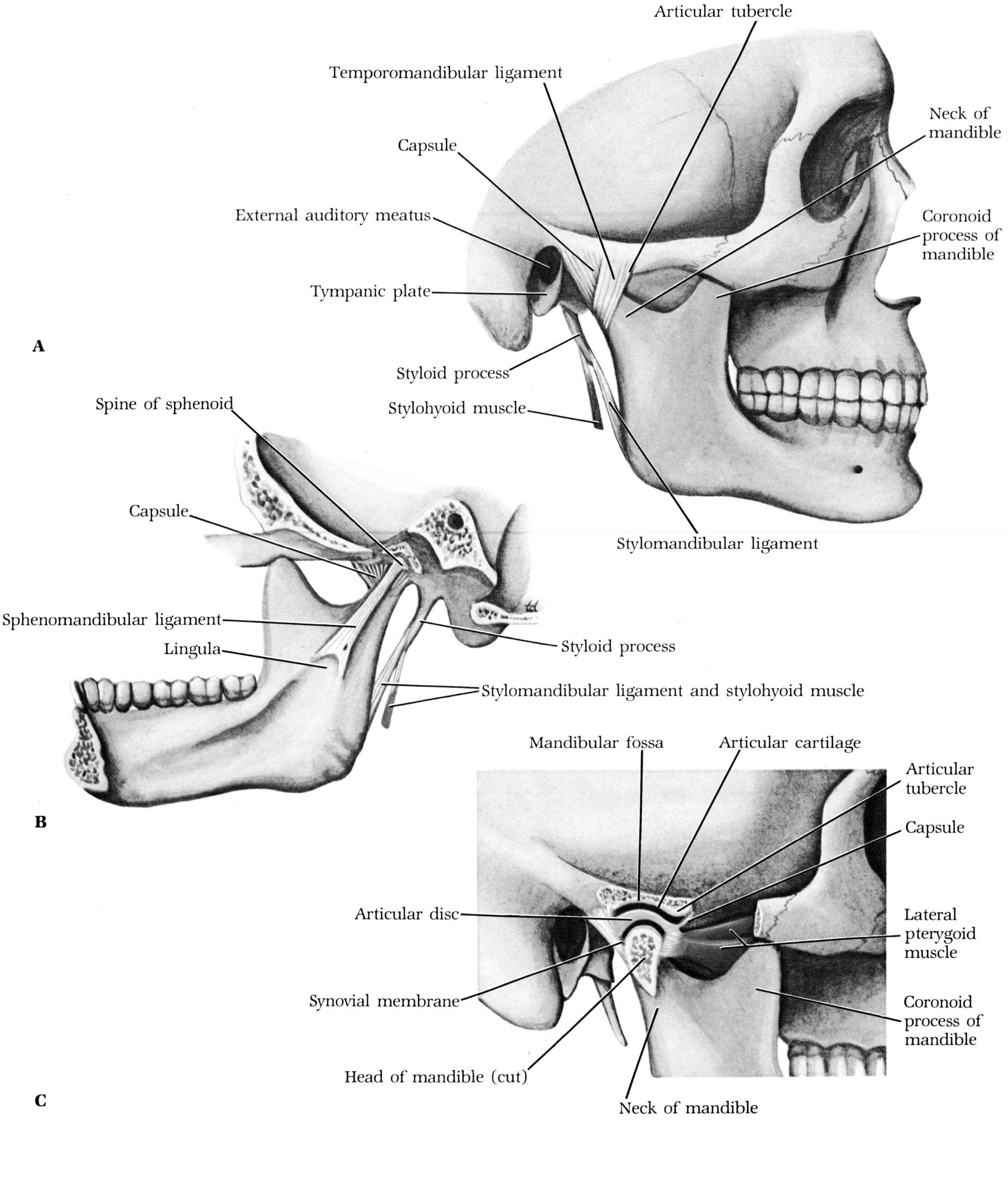
Articular tubercle
Temporomandibular ligament
Neck of mandible
Capsule
External auditory meatus
Coronoid process of mandible
Tympanic plate
A
Styloid process
Stylohyoid muscle
Spine of sphenoid
Capsule
Stylomandibular ligament
Sphenomandibular ligament
Lingula
Styloid process
Stylomandibular ligament and stylohyoid muscle
B
Mandibular fossa
Articular cartilage
Articular tubercle
Capsule
Articular disc
Lateral pterygoid muscle
Synovial membrane
Coronoid process of mandible
Head of mandible (cut)
Neck of mandible
C

Figure 7-38
Temporomandibular joint. A. Lateral aspect. B. Medial aspect. C. Sagittal section.

1. The temporomandibular joint can be easily palpated in front of the auricle. Note that as the mouth is opened, the head of the mandible rotates and moves forward below the tubercle of the zygomatic arch. While the patient opens and closes the mouth, determine if there is any crepitus or clicking. Osteoarthritis of the joint reveals itself by crepitus; a loose articular disc produces a clicking sensation, which may be audible.

2. The great strength of the lateral temporomandibular ligament prevents the head of the mandible from passing backward and fracturing the tympanic plate when a severe blow falls on the chin.

3. Dislocation of the temporomandibular joint sometimes occurs when the mandible is depressed. In this movement the head of the mandible and the articular disc both move forward until they reach the summit of the articular tubercle. In this position the joint is unstable, and a minor blow on the chin or a sudden contraction of the lateral pterygoid muscles, as in yawning, may be sufficient to pull the disc forward beyond the summit. When both sides are involved the mouth is fixed in an open position, and both heads of the mandible lie anterior to the articular tubercles. Reduction of the dislocation is easily achieved by pressing the gloved thumbs downward on the lower molar teeth and pushing the jaw backward. The downward pressure overcomes the tension of the temporalis and masseter muscles, and the backward pressure overcomes the spasm of the lateral pterygoid muscles.

4. An erupting third molar tooth in the mandible or dental infection such as an abscess may reflexly cause spasm of the masseter and temporalis muscles so that the patient is unable to open the mouth; this condition is known as trismus.

5. The tonic contraction of the masseter and temporalis muscles can also occur in tetanus.

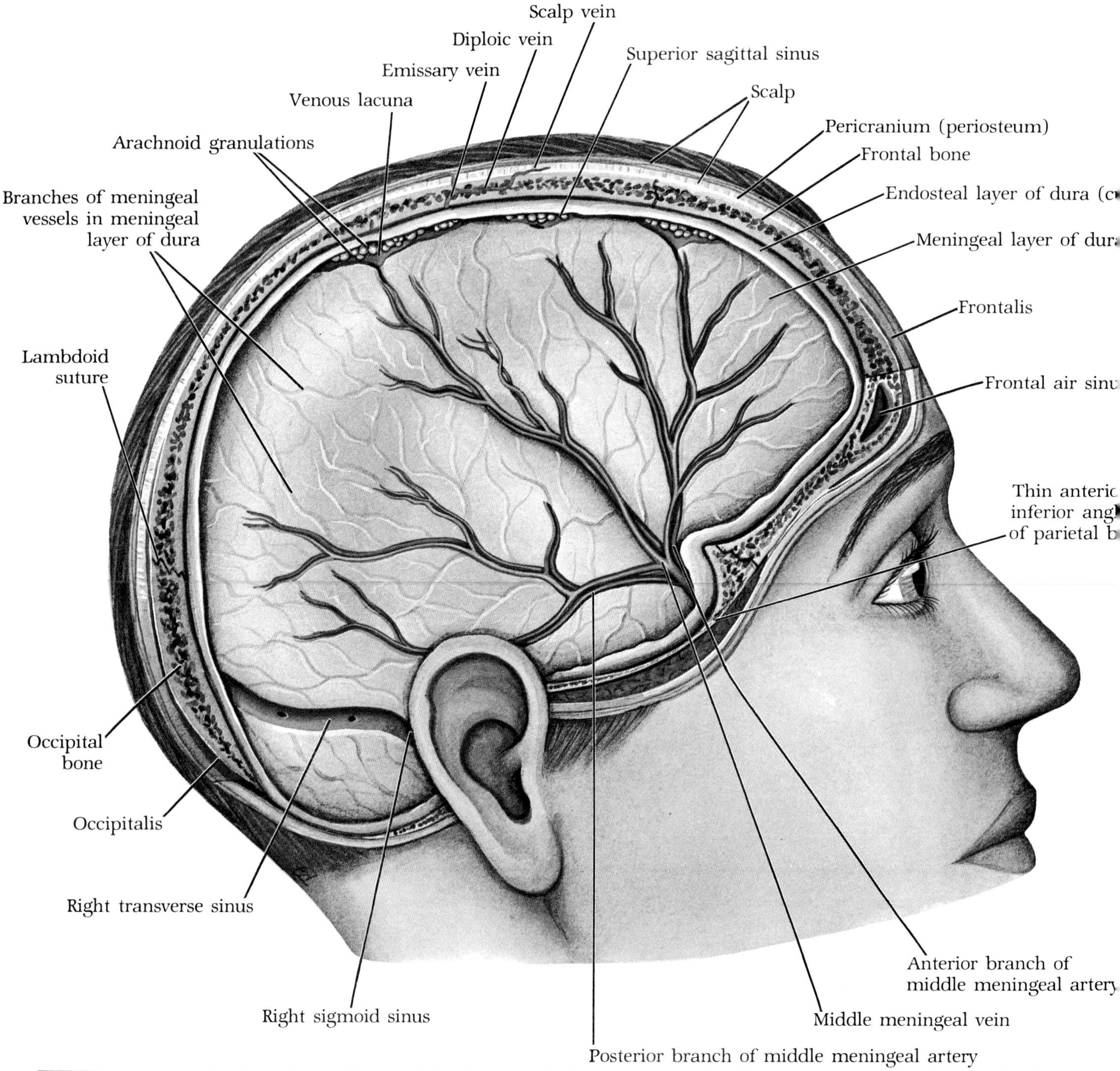
Scalp vein
Diploic vein
Emissary vein
Venous lacuna
Superior sagittal sinus
Scalp
Pericranium (periosteum)
Frontal bone
Endosteal layer of dura
Meningeal layer of
Arachnoid granulations
Branches of meningeal vessels in meningeal layer of dura
Frontalis
Lambdoid suture
Frontal air
Thin
inferior
of parietal
Occipital bone
Occipitalis
Right transverse sinus
Anterior branch of
middle meningeal
Right sigmoid sinus
Middle meningeal vein
Posterior branch of middle meningeal artery

Figure 7-39
Right side of the head with the meningeal layer of the dura exposed. Note that the skull bones and periosteal layer of dura have been removed to reveal the middle meningeal artery and vein.

1. The anteroinferior angle of the parietal bone is usually the weakest part of the lateral aspect of the skull. The presence of the anterior division of the middle meningeal artery and its vein or venae comitantes, which lie in a groove or canal on the inner table of this part of the skull, often results in their damage in fractures in this region. An extradural hematoma is formed which overlies the precentral gyrus of the cerebral hemisphere. To stop the bleeding, a trephine or burr hole is made through the skull about 1½ inches (4 cm) above the midpoint of the zygomatic arch, after which the bleeding vessel, usually a vein, can then be secured.

2. Fracture of the bones in the temporal fossa will produce bleeding into the overlying temporalis muscle, forming a hematoma beneath the tough temporal fascia. The hematoma reveals itself clinically by the appearance of a boggy swelling over the temporal fossa.

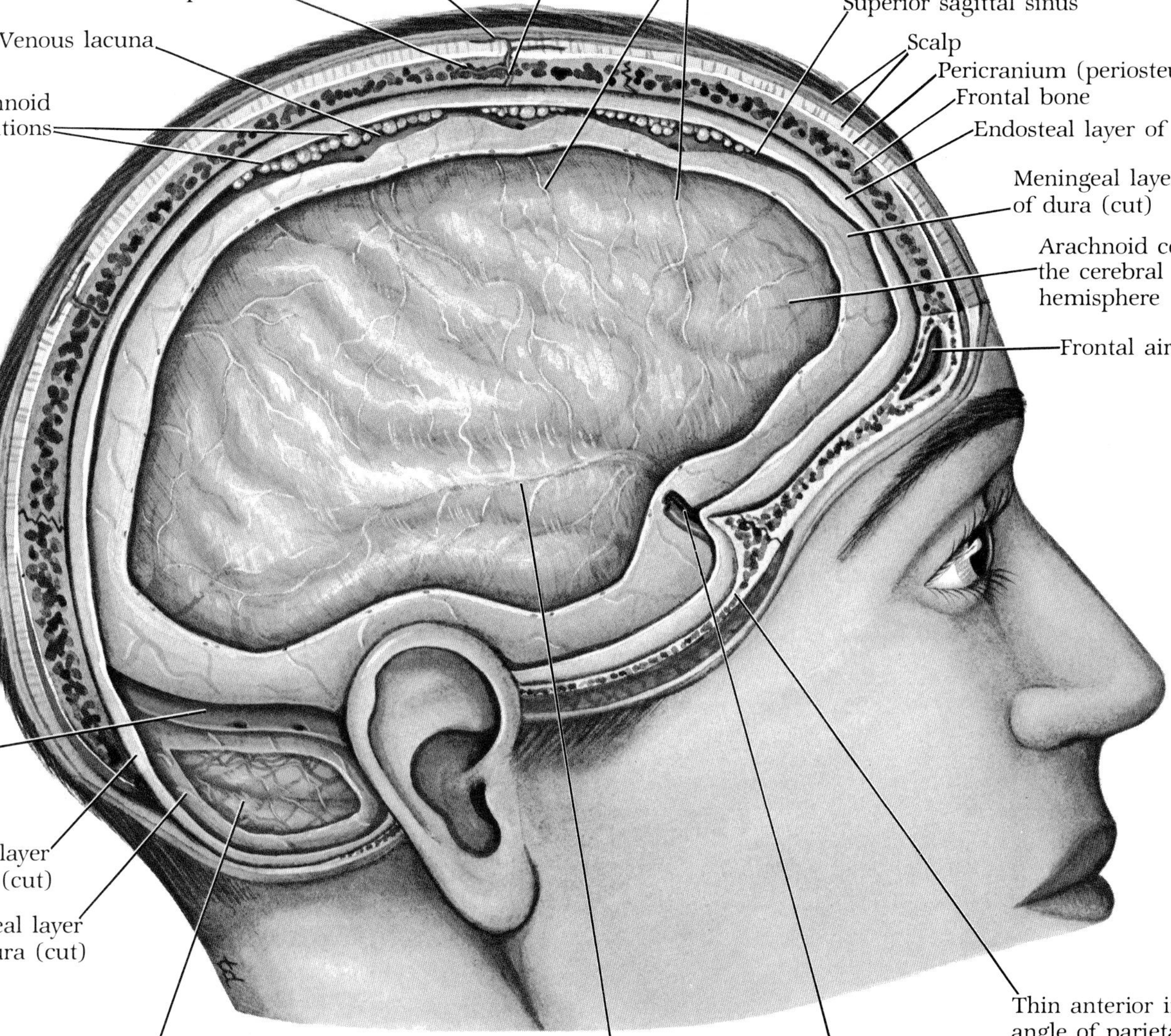
Emissary vein
Scalp vein
Superior cerebral veins
Diploic vein
Superior sagittal sinus
Venous lacuna
Scalp
Pericranium (periosteum)
Arachnoid granulations
Frontal bone
Endosteal layer of dura (cut)
Meningeal layer of dura (cut)
Arachnoid covering the cerebral hemisphere
Frontal air sinus
Right transverse sinus
Endosteal layer of dura (cut)
Meningeal layer of dura (cut)
Thin anterior inferior angle of parietal bone
Middle meningeal vessels situated between endosteal and meningeal layers of dura
Cerebellum covered with arachnoid and pia
Superficial middle cerebral vein

Figure 7-40
Right side of the head with the arachnoid mater covering the cerebrum and cerebellum exposed. The periosteal and meningeal layers of dura have been removed.

1. Note the position of the precentral and postcentral gyri relative to the skull. It is clear that a large extradural hematoma resulting from damage of the middle meningeal artery or vein may exert pressure on these gyri, causing an upper motor neuron type of paralysis and sensory loss.

2. Injuries to the brain may be produced by displacement and distortion of the neuronal tissues at the moment of impact. The brain is floating in cerebrospinal fluid in the subarachnoid space and is capable of a certain amount of anteroposterior movement, although this is limited by the attachment of the superior cerebral veins to the superior sagittal sinus. Lateral displacement of the brain is limited by the falx cerebri. The tentorium cerebelli and the falx cerebelli also restrict displacement of the brain. Blows on the side of the head produce less cerebral displacement, and the injuries to the brain consequently tend to be less severe.

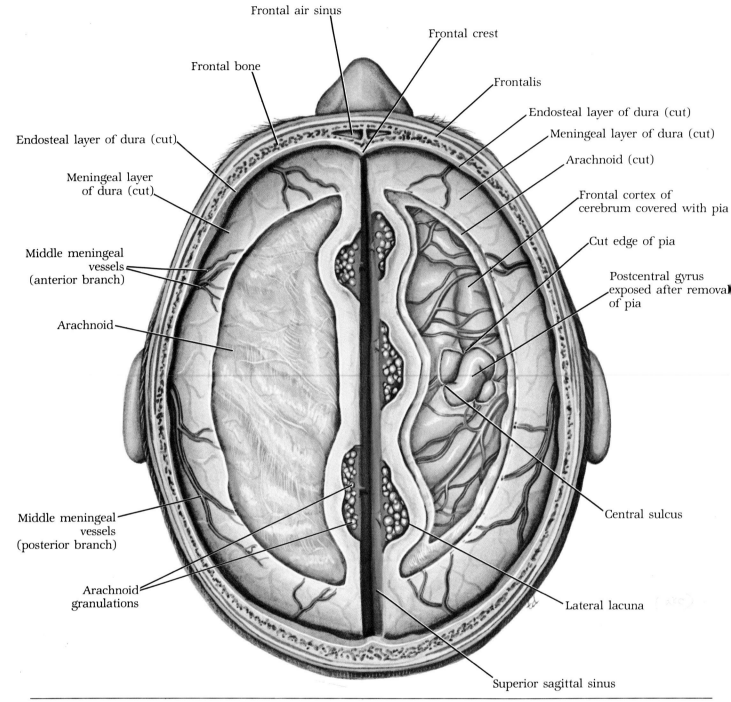

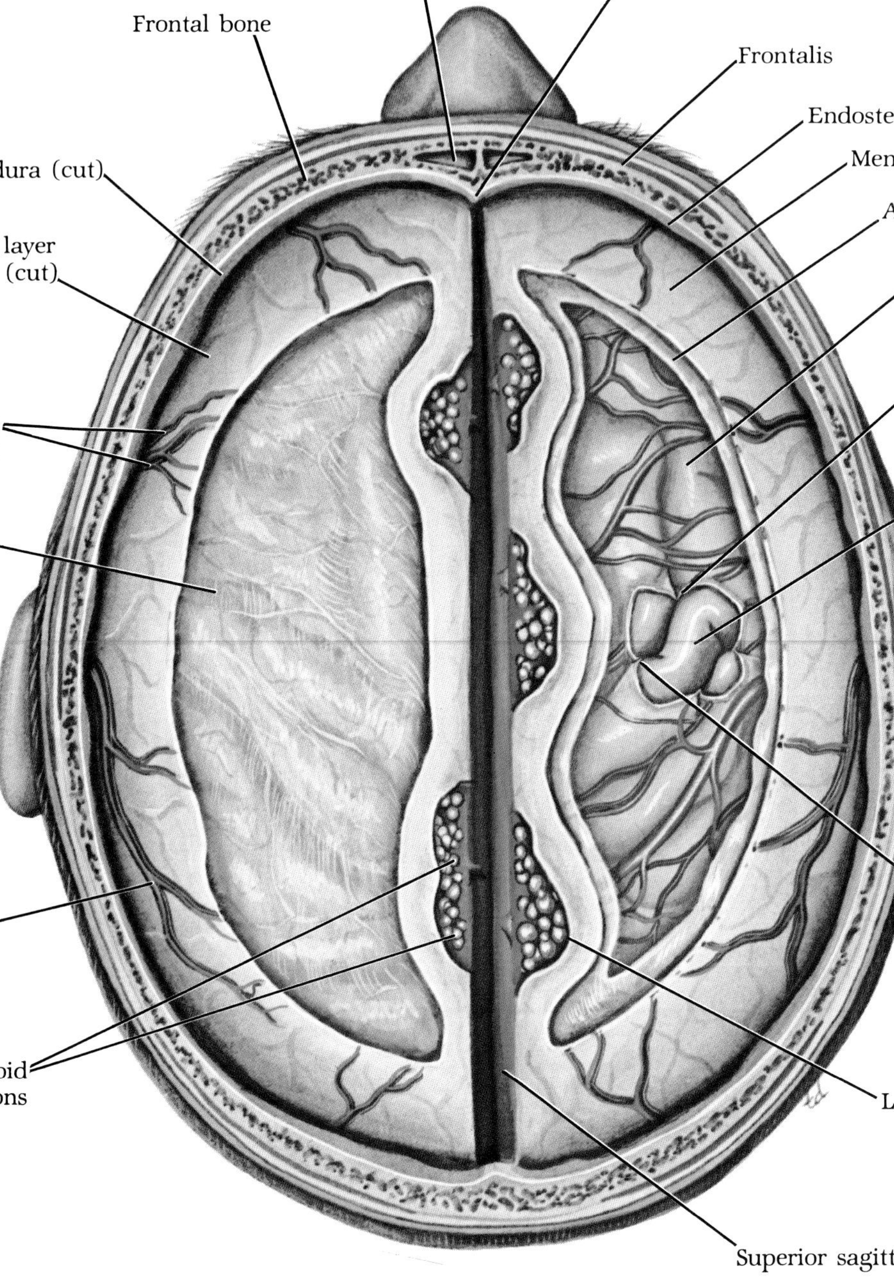
Frontal air sinus
Frontal crest
Frontal bone
Frontalis
Endosteal layer of dura (cut)
Meningeal layer of dura (cut)
Endosteal layer of dura (cut)
Arachnoid (cut)
Meningeal layer of dura (cut)
Frontal cortex of cerebrum covered with pia
Cut edge of pia
Middle meningeal vessels (anterior branch)
Postcentral gyrus exposed after removal of pia
Arachnoid
Central sulcus
Middle meningeal vessels (posterior branch)
Arachnoid granulations
Lateral lacuna
Superior sagittal sinus

Figure 7-41
Superior view of head with the calvarium removed. On the left a large portion of the meningeal layer of the dura has been removed to expose the underlying arachnoid mater. On the right large portions of the meningeal layer of dura and arachnoid mater have been removed, revealing the cerebral blood vessels and cerebral cortex covered with pia mater.

1. Note the presence of arachnoid granulations projecting into the lacunae of the superior sagittal sinus. These are sites at which the cerebrospinal fluid diffuses into the bloodstream.
2. In a baby the superior sagittal sinus may be used as a site for obtaining samples of blood. A needle is inserted through the anterior fontanelle in the midline into the sinus.
3. Subdural hemorrhage results from tearing of the superior cerebral veins at their point of entrance into the superior sagittal sinus. The cause is usually a blow on the front or back of the head, causing excessive anteroposterior displacement of the brain within the skull.
4. Meningitis is an inflammation of the pia and arachnoid mater and involves the cerebrospinal fluid present in the subarachnoid space and ventricles of the brain. It is important to realize that this space not only surrounds the brain and spinal cord but also extends around the optic nerve as far as the eyeball. Infection of the pia and arachnoid may result in the formation of pus in the subarachnoid space, which in turn may interfere with the flow of cerebrospinal fluid over the surface of the brain. Moreover, the organisms may invade the underlying brain or injure the cranial or spinal nerves or the choroid plexuses.

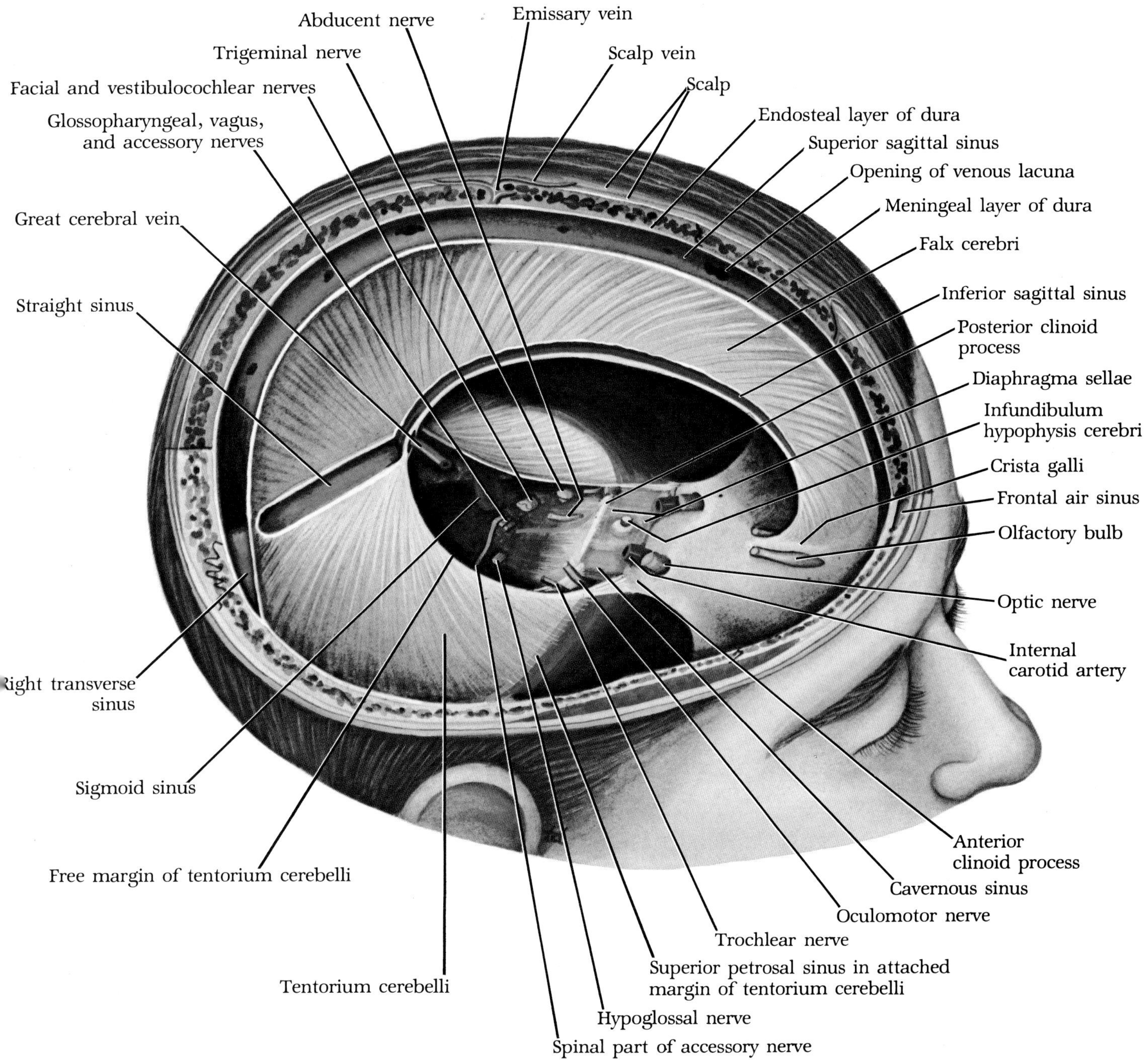
Abducent nerve
Emissary vein
Trigeminal nerve
Scalp vein
Facial and vestibulocochlear nerves
Scalp
Glossopharyngeal, vagus, and accessory nerves
Endosteal layer of dura
Superior sagittal sinus
Opening of venous lacuna
Great cerebral vein
Meningeal layer of dura
Falx cerebri
Straight sinus
Inferior sagittal sinus
Posterior clinoid process
Diaphragma sellae
Infundibulum hypophysis cerebri
Crista galli
Frontal air sinus
Olfactory bulb
Optic nerve
Internal carotid artery
Right transverse sinus
Sigmoid sinus
Anterior clinoid process
Free margin of tentorium cerebelli
Cavernous sinus
Oculomotor nerve
Trochlear nerve
Superior petrosal sinus in attached margin of tentorium cerebelli
Tentorium cerebelli
Hypoglossal nerve
Spinal part of accessory nerve

Figure 7-42
Interior of the skull, showing the dura mater and its contained venous sinuses.

1. The falx cerebri and the tentorium cerebelli serve to limit excessive movements of the brain within the skull. However, if the momentum of the brain is suddenly halted by the strong dural septa, as occurs in automobile accidents, the cerebral cortex or the cerebellum may be severely damaged. The sharp posterior border of the lesser wing of the sphenoid may also cause tearing of the cerebrum.

2. Movements of the brain relative to the skull and dural septa may seriously injure the cranial nerves that are tethered as they pass through the various foramina. Furthermore, the fragile cortical veins that drain into the dural sinuses may be torn, resulting in severe subdural and subarachnoid hemorrhage. The tortuous intracranial arteries with their strong walls are rarely damaged.

3. Infections of the scalp occasionally spread by the valveless emissary veins to the skull bones, causing an osteomyelitis. Infected blood in the diploic veins may travel by the emissary veins farther into the superior sagittal sinus, producing venous sinus thrombosis.

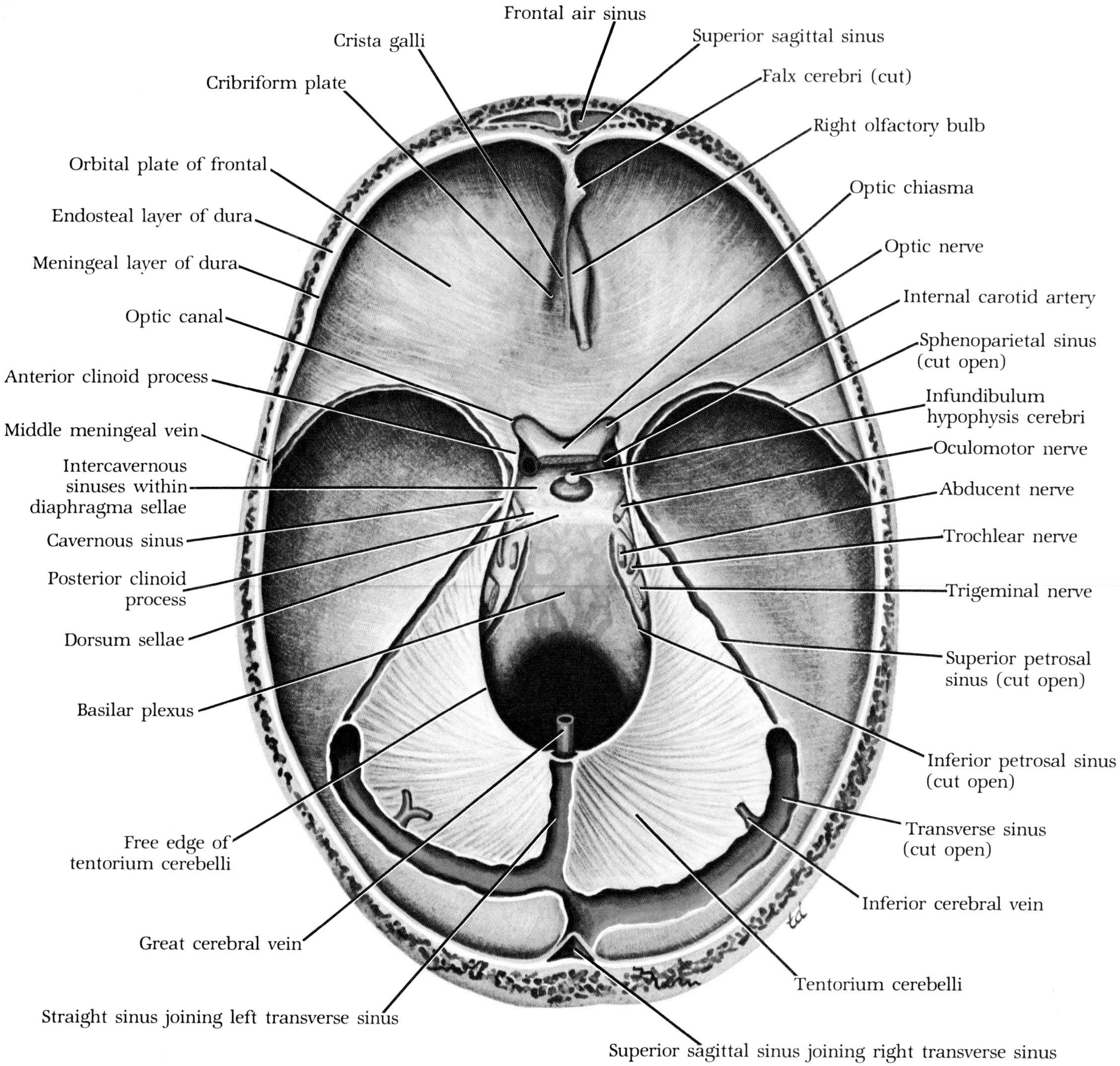
Frontal air sinus
Crista galli
Superior sagittal sinus
Cribriform plate
Falx cerebri (cut)
Right olfactory bulb
Orbital plate of frontal
Optic chiasma
Endosteal layer of dura
Optic nerve
Meningeal layer of dura
Internal carotid artery
Optic canal
Sphenoparietal sinus (cut open)
Anterior clinoid process
Infundibulum hypophysis cerebri
Middle meningeal vein
Oculomotor nerve
Intercavernous sinuses within diaphragma sellae
Abducent nerve
Cavernous sinus
Trochlear nerve
Posterior clinoid process
Trigeminal nerve
Dorsum sellae
Superior petrosal sinus (cut open)
Basilar plexus
Inferior petrosal sinus (cut open)
Transverse sinus (cut open)
Free edge of tentorium cerebelli
Inferior cerebral vein
Great cerebral vein
Tentorium cerebelli
Straight sinus joining left transverse sinus
Superior sagittal sinus joining right transverse sinus

Figure 7-43
Diaphragma sellae and tentorium cerebelli; many of the venous sinuses have been cut open.

1. Tumors of the hypophysis cerebri cause erosion and expansion of the sella turcica with thinning of its walls; this may be seen on lateral radiographs of the skull. If the tumor extends superiorly, it will exert pressure on the diaphragma sellae and the optic chiasma. The decussating fibers of the chiasma are affected first, producing bitemporal hemianopia. The surgical approach for removal of such a tumor may be transcranially, or through the sphenoidal or ethmoidal air sinuses.
2. The tentorial notch of the tentorium cerebelli serves not only for the passage of the brain stem but also for the cerebrospinal fluid in the subarachnoid space. A bacterial meningitis may cause a blocking of the space in this region, producing hydrocephalus.
3. The sharp posterior border of the lesser wing of the sphenoid and the edge of the tentorial notch may cause severe laceration of the brain tissue in violent movements of the head, as in automobile accidents.
4. The close relationship between the mastoid antrum and the sigmoid sinus may lead to thrombosis of the venous sinus in patients with otitis media with an acute mastoiditis.

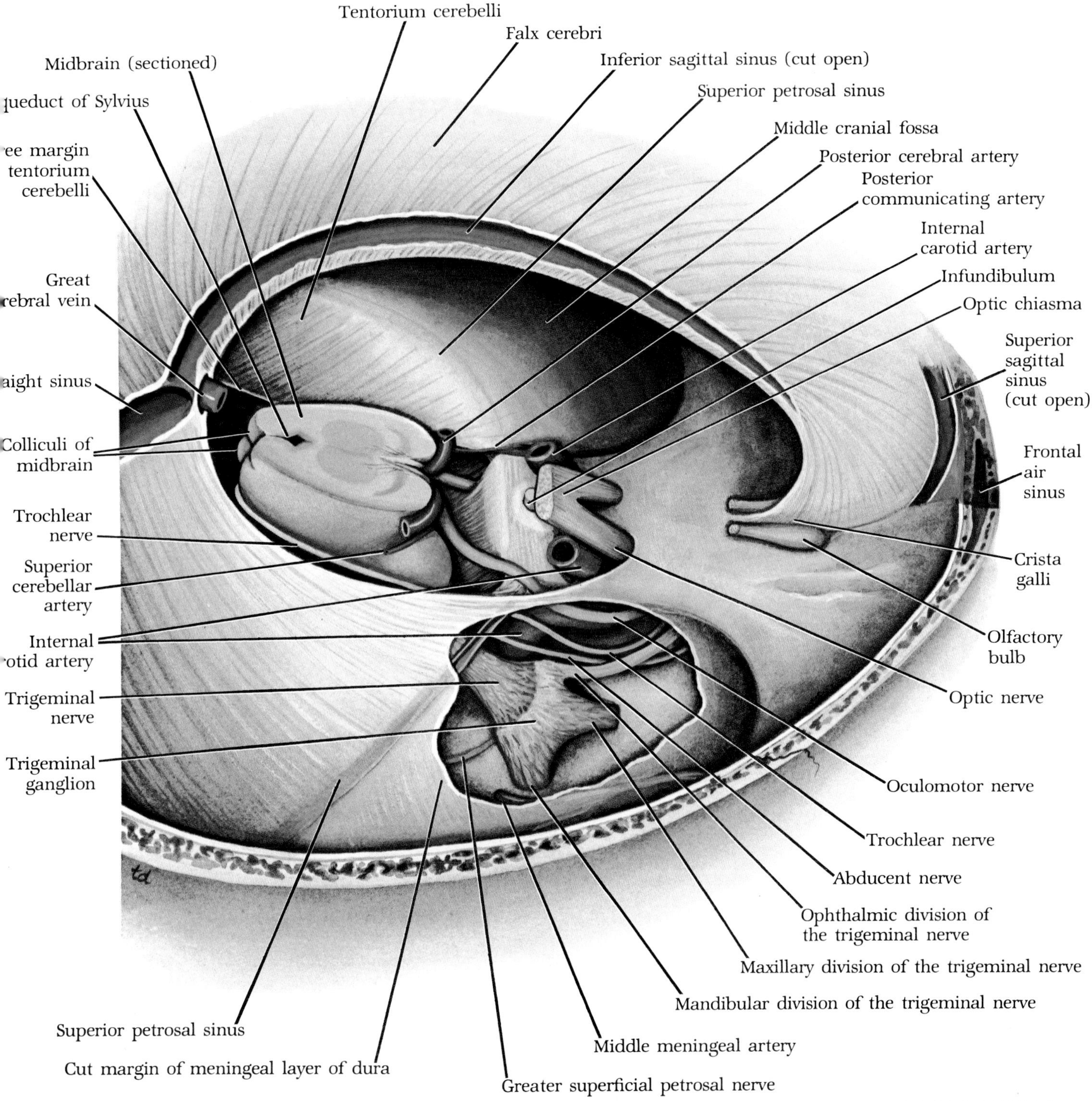
Tentorium cerebelli
Falx cerebri
Inferior sagittal sinus (cut open)
Superior petrosal sinus
Middle cranial fossa
Posterior cerebral artery
Posterior communicating artery
Internal carotid artery
Infundibulum
Optic chiasma
Superior sagittal sinus (cut open)
Frontal air sinus
Crista galli
Olfactory bulb
Optic nerve
Oculomotor nerve
Trochlear nerve
Abducent nerve
Ophthalmic division of the trigeminal nerve
Maxillary division of the trigeminal nerve
Mandibular division of the trigeminal nerve
Middle meningeal artery
Greater superficial petrosal nerve
Cut margin of meningeal layer of dura
Superior petrosal sinus
Midbrain (sectioned)
queduct of Sylvius
ee margin tentorium cerebelli
Great rebral vein
aight sinus
Colliculi of midbrain
Trochlear nerve
Superior cerebellar artery
Internal otid artery
Trigeminal nerve
Trigeminal ganglion
td

Figure 7-44
Lateral view of the interior of the skull, showing the falx cerebri, tentorium cerebelli, brain stem, and trigeminal ganglion.

1. In fractures of the anterior cranial fossa, the cribriform plate of the ethmoid bone may be damaged. This usually results in a tearing of the overlying meninges and underlying mucoperiosteum. The patient will have bleeding from the nose (epistaxis) and leakage of cerebrospinal fluid into the nose (cerebrospinal rhinorrhea).
2. In violent movements of the brain following blows to the head, the sharp edges of the falx cerebri, the tentorium cerebelli, and the lesser wing of the sphenoid may severely damage the nervous tissue.
3. Intracranial movement of the brain may result in injury to cranial nerves, since they are tethered at the point where they pass through the foramina in the skull. The oculomotor, trochlear, and especially the abducent nerve are particularly prone to injury in trauma to the head.
4. Trigeminal neuralgia, a condition characterized by the occurrence of excruciating pain in the distribution of the maxillary and mandibular divisions of the trigeminal nerve, may be treated by the needle injection of alcohol into the trigeminal ganglion through the foramen ovale. The condition may also be treated by partial surgical division of the sensory root of the trigeminal nerve.

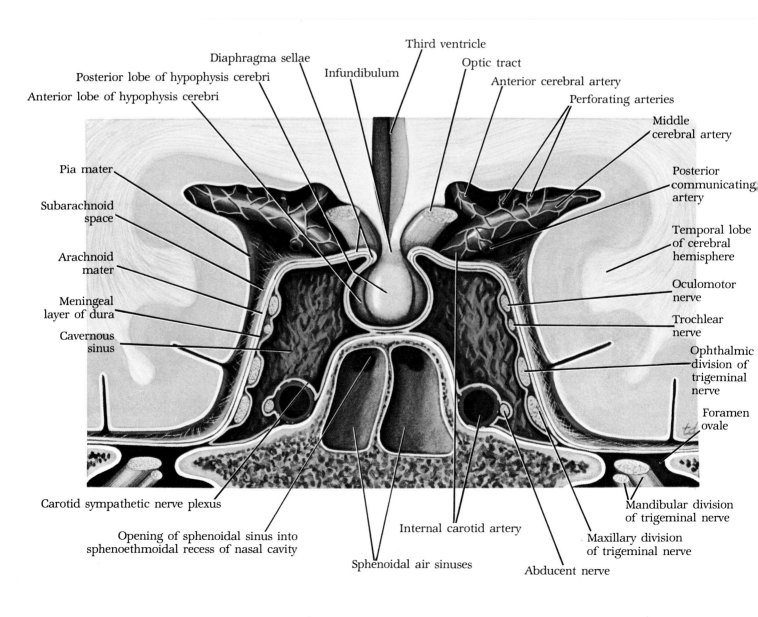

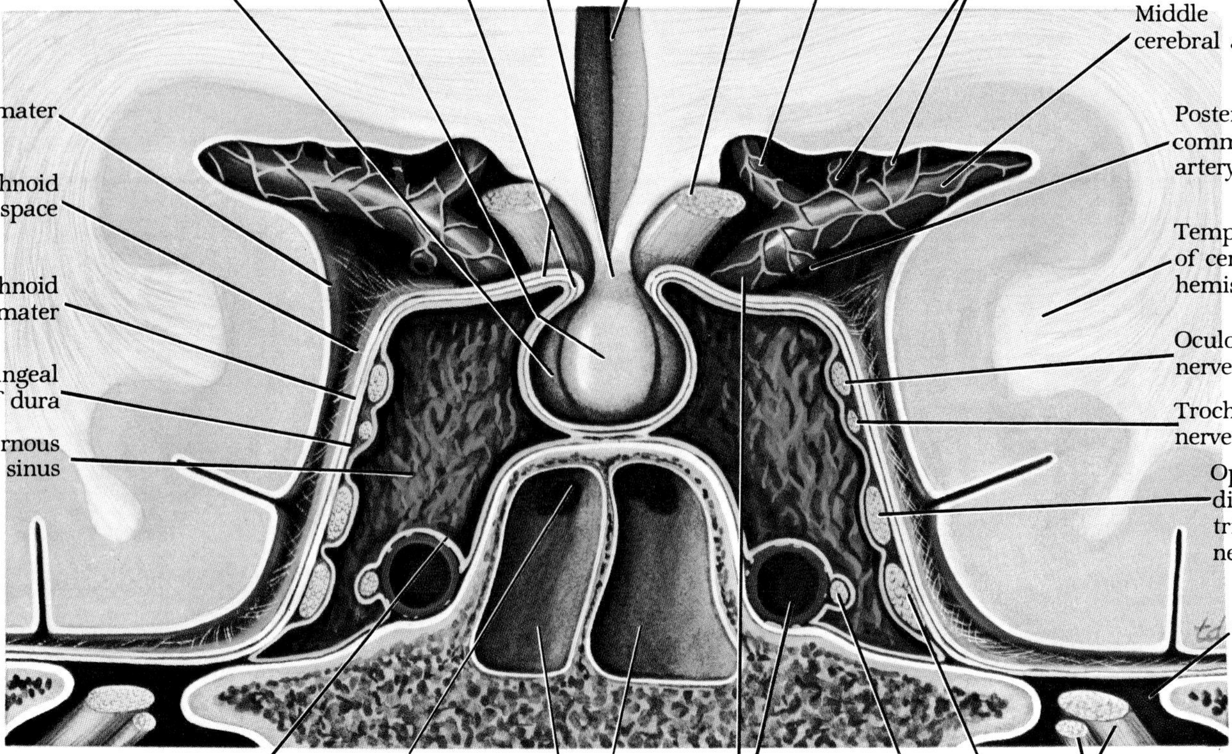
Third ventricle
Diaphragma sellae
Optic tract
Infundibulum
Posterior lobe of hypophysis cerebri
Anterior cerebral artery
Anterior lobe of hypophysis cerebri
Perforating arteries
Middle cerebral artery
Pia mater
Posterior communicating artery
Subarachnoid space
Temporal lobe of cerebral hemisphere
Arachnoid mater
Oculomotor nerve
Meningeal layer of dura
Trochlear nerve
Cavernous sinus
Ophthalmic division of trigeminal nerve
Foramen ovale
Carotid sympathetic nerve plexus
Mandibular division of trigeminal nerve
Internal carotid artery
Opening of sphenoidal sinus into sphenoethmoidal recess of nasal cavity
Maxillary division of trigeminal nerve
Sphenoidal air sinuses
Abducent nerve

Figure 7-45
Coronal section through body of sphenoid, showing hypophysis cerebri and cavernous sinuses. Note the position of the internal carotid artery and the cranial nerves.

1. Superior extension of tumors of the hypophysis cerebri will eventually exert pressure on the optic chiasma, producing bitemporal hemianopia. Inferior extension of such tumors will cause erosion of the sella turcica.
2. The facial vein and the veins of the ethmoidal air sinuses all communicate with the cavernous sinus via the ophthalmic veins. Infections of the face or ethmoidal sinusitis may spread to the orbit via the veins, producing orbital cellulitis, which may in turn spread to the cavernous sinus, producing thrombophlebitis, an extremely serious condition.
3. Damage to the wall of the internal carotid artery in severe head injuries may later result in rupture of the artery into the cavernous sinus. Roaring noises in the head associated with pulsating exophthalmos is a common finding in these cases.
4. The cranial nerves lying in relation to the cavernous sinus are usually involved in those patients with cavernous sinus thrombosis.

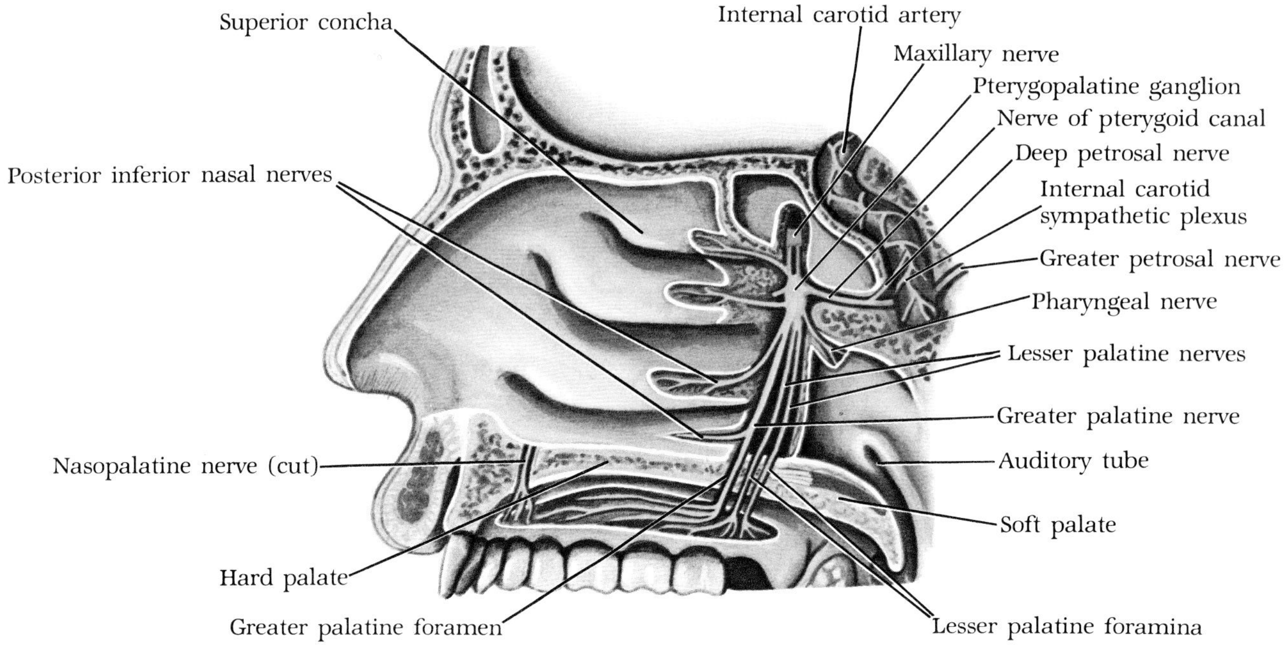
Superior concha
Internal carotid artery
Maxillary nerve
Pterygopalatine ganglion
Nerve of pterygoid canal
Deep petrosal nerve
Internal carotid sympathetic plexus
Posterior inferior nasal nerves
Greater petrosal nerve
Pharyngeal nerve
Lesser palatine nerves
Greater palatine nerve
Nasopalatine nerve (cut)
Auditory tube
Soft palate
Hard palate
Greater palatine foramen
Lesser palatine foramina

A

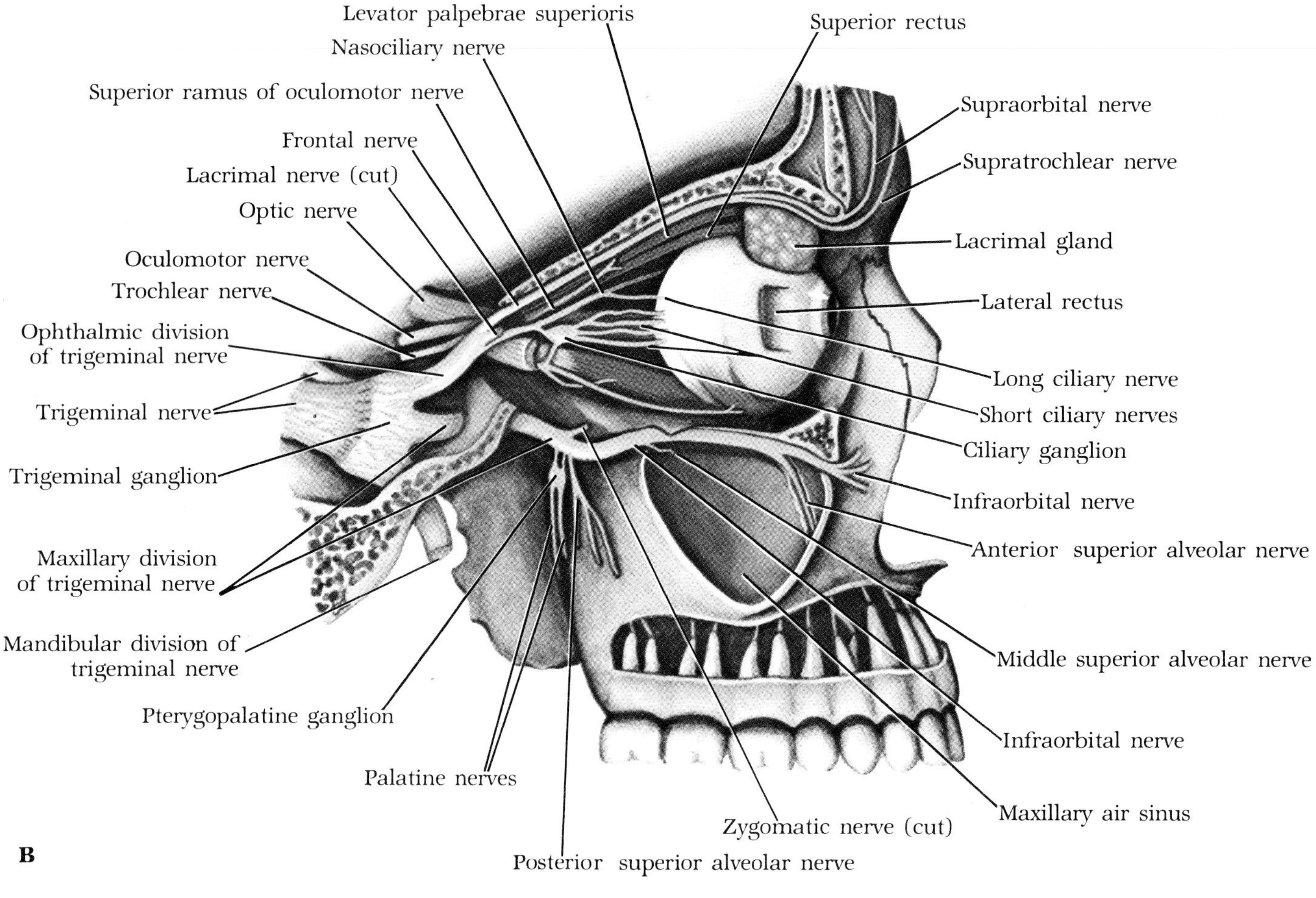
Levator palpebrae superioris
Superior rectus
Nasociliary nerve
Superior ramus of oculomotor nerve
Supraorbital nerve
Frontal nerve
Supratrochlear nerve
Lacrimal nerve (cut)
Optic nerve
Lacrimal gland
Oculomotor nerve
Trochlear nerve
Lateral rectus
Ophthalmic division of trigeminal nerve
Long ciliary nerve
Trigeminal nerve
Short ciliary nerves
Ciliary ganglion
Trigeminal ganglion
Infraorbital nerve
Anterior superior alveolar nerve
Maxillary division of trigeminal nerve
Mandibular division of trigeminal nerve
Middle superior alveolar nerve
Pterygopalatine ganglion
Infraorbital nerve
Palatine nerves
Maxillary air sinus
Zygomatic nerve (cut)
Posterior superior alveolar nerve

B

Figure 7-46
A. The lateral wall of the nose, the palate, and the pterygopalatine fossa, showing the pterygopalatine ganglion and its branches.
B. Muscles and nerves of the right orbit viewed from lateral side. Maxillary nerve and pterygopalatine ganglion are also shown.

1. The sensory nerves of the nose (apart from the olfactory nerves) consist mainly of the ophthalmic and maxillary divisions of the trigeminal nerve. Branches from the maxillary division and the posterior superior and posterior inferior nasal nerves enter the nose through the sphenopalatine foramen. The palate receives its sensory innervation from the maxillary nerve via the greater and lesser palatine nerves and the spheno-palatine nerve (nasopalatine nerve); the latter nerve reaches the palate through the incisive foramen.

2. Maxillary sinusitis gives rise to referred pain along the distribution of the infra-orbital nerve. This is due to the fact that the maxillary sinus is innervated by the maxillary and infraorbital nerves.

3. A nonpenetrating blow to the eye may result in a herniation of the orbital contents downward through a fracture in the bony orbital floor into the maxillary sinus. The infraorbital nerve may be damaged as it passes through the infraorbital canal.

4. Carious teeth are a frequent cause of facial pain. Disease of teeth in the upper jaw that are innervated by the maxillary and infraorbital nerves may cause referred pain in the distribution of the infraorbital nerve.

5. Maxillary nerve block may be performed by introducing a needle below the midpoint of the zygoma. The needle is passed medially past the anterior margin of the lateral pterygoid plate to enter the pterygomaxillary fissure. The anesthetic solution is then injected into the pterygopalatine fossa.

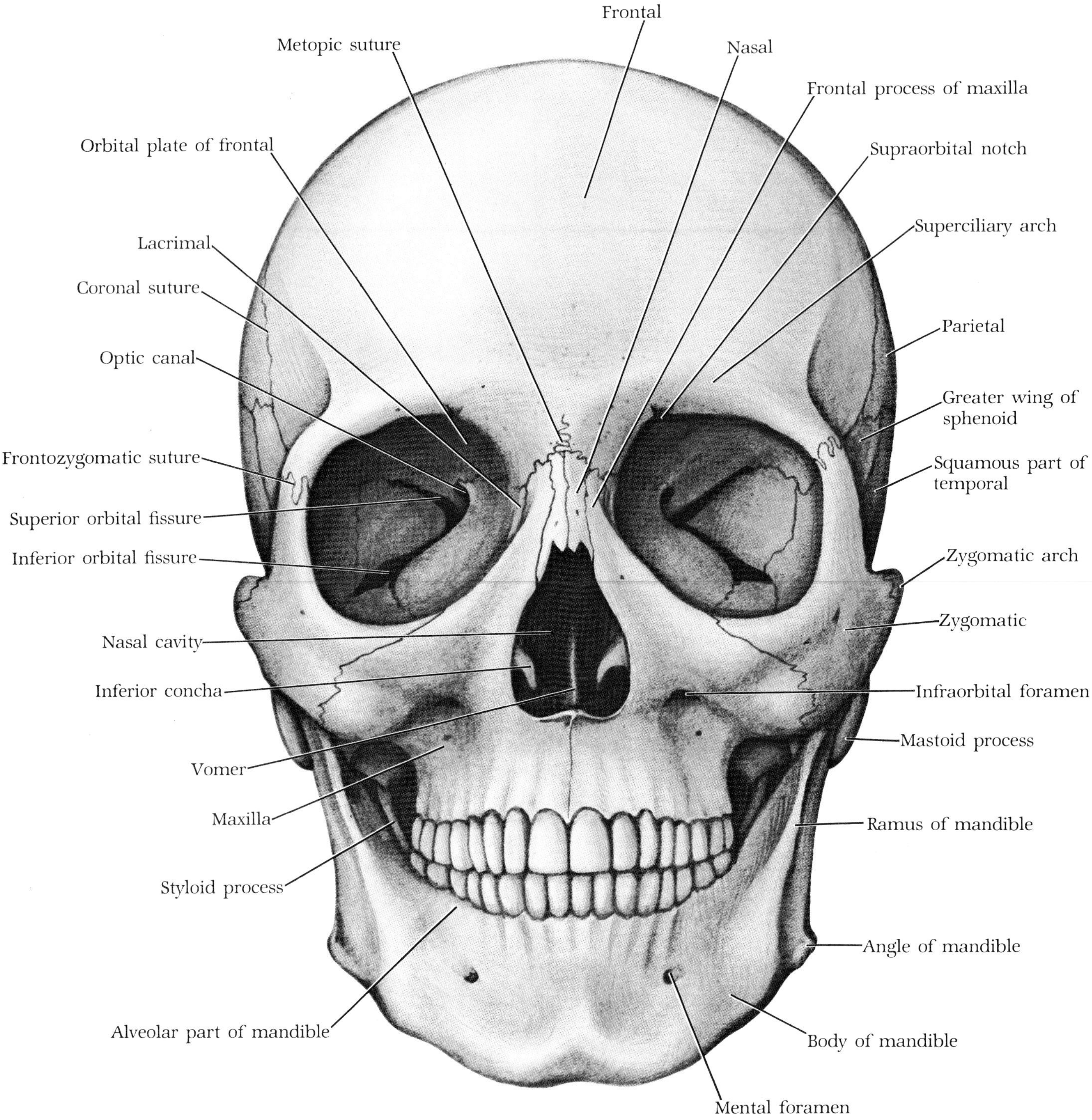
Frontal
Metopic suture
Nasal
Frontal process of maxilla
Orbital plate of frontal
Supraorbital notch
Superciliary arch
Lacrimal
Coronal suture
Parietal
Optic canal
Greater wing of sphenoid
Frontozygomatic suture
Squamous part of temporal
Superior orbital fissure
Inferior orbital fissure
Zygomatic arch
Zygomatic
Nasal cavity
Inferior concha
Infraorbital foramen
Mastoid process
Vomer
Maxilla
Ramus of mandible
Styloid process
Angle of mandible
Alveolar part of mandible
Body of mandible
Mental foramen

Figure 7-47
Skull, anterior view.

1. Fractures of the facial bones when due to a severe head injury are aften associated with damage to the brain. Clearly, treatment of the brain injury takes priority, and the facial fractures are treated when the general condition of the patient has improved.
2. Signs of fracture of facial bones include deformity, ocular disparity, or abnormal movement accompanied by crepitation and malocclusion of the teeth. Anesthesia or paresthesia of the facial skin will follow fracture of bones through which branches of the trigeminal nerve pass to the skin.
3. Fractures of the nasal bones are very common. Although the majority are simple fractures and are reduced under local anesthesia, some are associated with severe injuries to the nasal septum, which require careful treatment under general anesthesia.
4. Fractures of the maxilla commonly result from a direct anteroposterior blow to the face. Malocclusion, enophthalmos, and anesthesia of the cheek and upper lip are frequent physical findings.
5. The zygoma or the zygomatic arch may be fractured by a blow to the side of the face. Although it may occur as an isolated fracture, as from a blow from a clenched fist, it may be associated with multiple other fractures of the face, as often occurs in automobile accidents.
6. Fractures of the mandible are the most common fractures of the face and they are usually bilateral. The neck, body, angle, symphysis, and ramus are sites of fracture in decreasing orders of frequency.

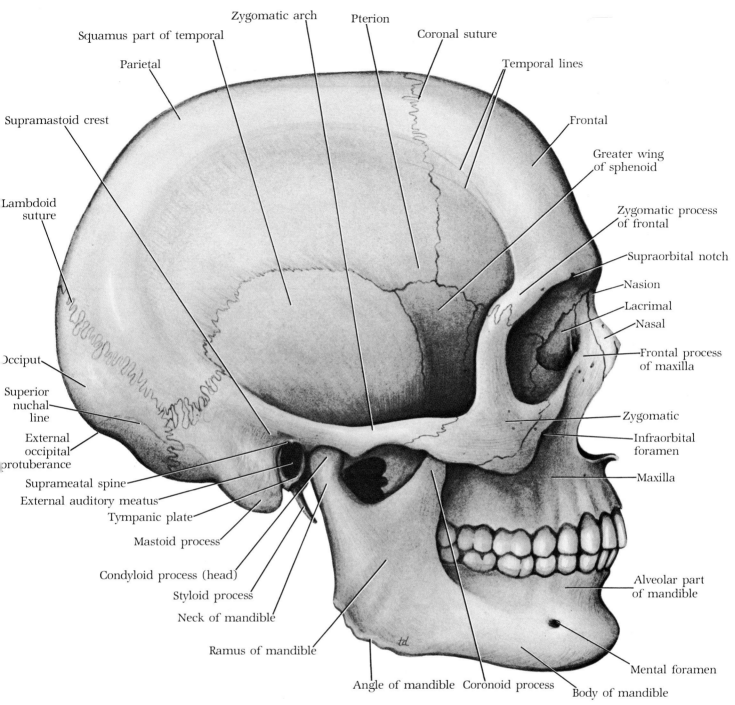

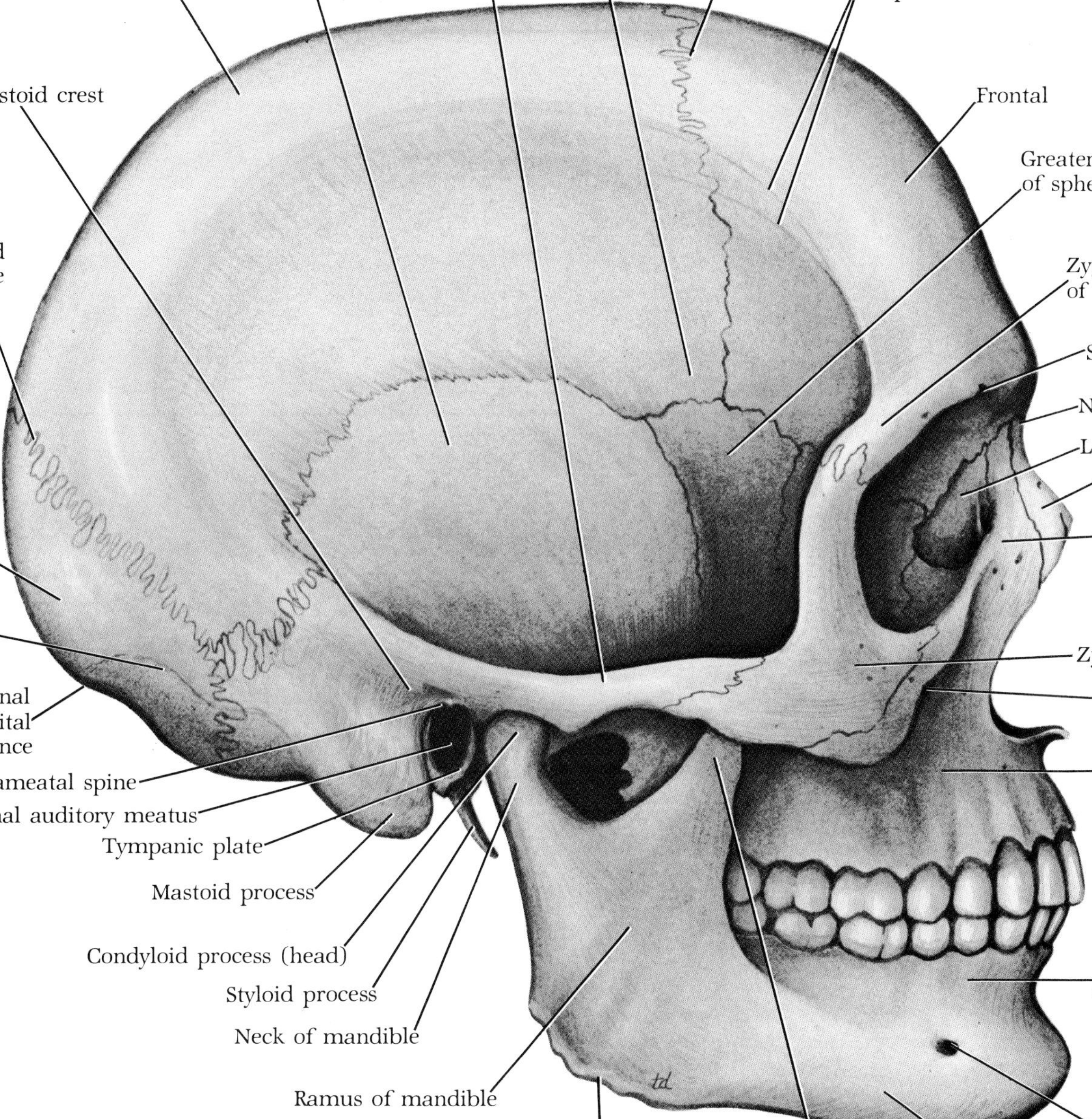
Zygomatic arch
Pterion
Squamus part of temporal
Coronal suture
Parietal
Temporal lines
Supramastoid crest
Frontal
Greater wing of sphenoid
Lambdoid suture
Zygomatic process of frontal
Supraorbital notch
Nasion
Lacrimal
Nasal
Occiput
Frontal process of maxilla
Superior nuchal line
Zygomatic
External occipital protuberance
Infraorbital foramen
Maxilla
Suprameatal spine
External auditory meatus
Tympanic plate
Mastoid process
Condyloid process (head)
Alveolar part of mandible
Styloid process
Neck of mandible
Ramus of mandible
Mental foramen
Angle of mandible
Coronoid process
Body of mandible

Figure 7-48
Skull, lateral view.

1. Fractures of the skull are very common in the adult, but much less so in the young child. In the infant, the bones are more resilient than in the adult skull, and they are separated by fibrous sutural ligaments. In the adult, the inner table of the skull is particularly brittle; moreover, the sutural ligaments begin to ossify during middle age.
2. The type of fracture that occurs in the skull will depend on the age of the patient, the severity of the blow, and the area of skull receiving the trauma. The adult skull may be likened to an eggshell in that it possesses a certain limited resilience beyond which it splinters. A severe, localized blow will produce a local indentation, often accompanied by splintering of the bone.
Blows to the vault often result in a series of linear fractures, which radiate out through the thin areas of bone. The petrous parts of the temporal bones and the occipital crests strongly reinforce the base of the skull and tend to deflect linear fractures.
3. The region of the anterior inferior angle of the parietal bone is usually the weakest part of the lateral aspect of the skull. Fractures here result from a severe blow on the side of the head. Hemorrhage occurs into the overlying temporalis muscle, causing the formation of a hematoma beneath the tough temporal fascia. This will reveal itself on the surface as a boggy swelling over the temporal fossa.
4. The presence of the anterior division of the middle meningeal artery and its venae comitantes in a groove or canal on the inner table of the anterior inferior part of the parietal bone (pterion) often results in their damage in fractures in this region. An extradural hematoma that overlies the precentral gyrus of the cerebral hemisphere is formed. To stop the bleeding a trephine or burr hole is made through the skull about 1½ inches (4 cm) above the midpoint of the zygomatic arch, after which the bleeding vessel, usually a vein, can then be secured.
5. In the young child, the skull may be likened to a table-tennis ball in that a localized blow produces a depression without splintering. This common type of circumscribed lesion is referred to as a "pond" fracture.

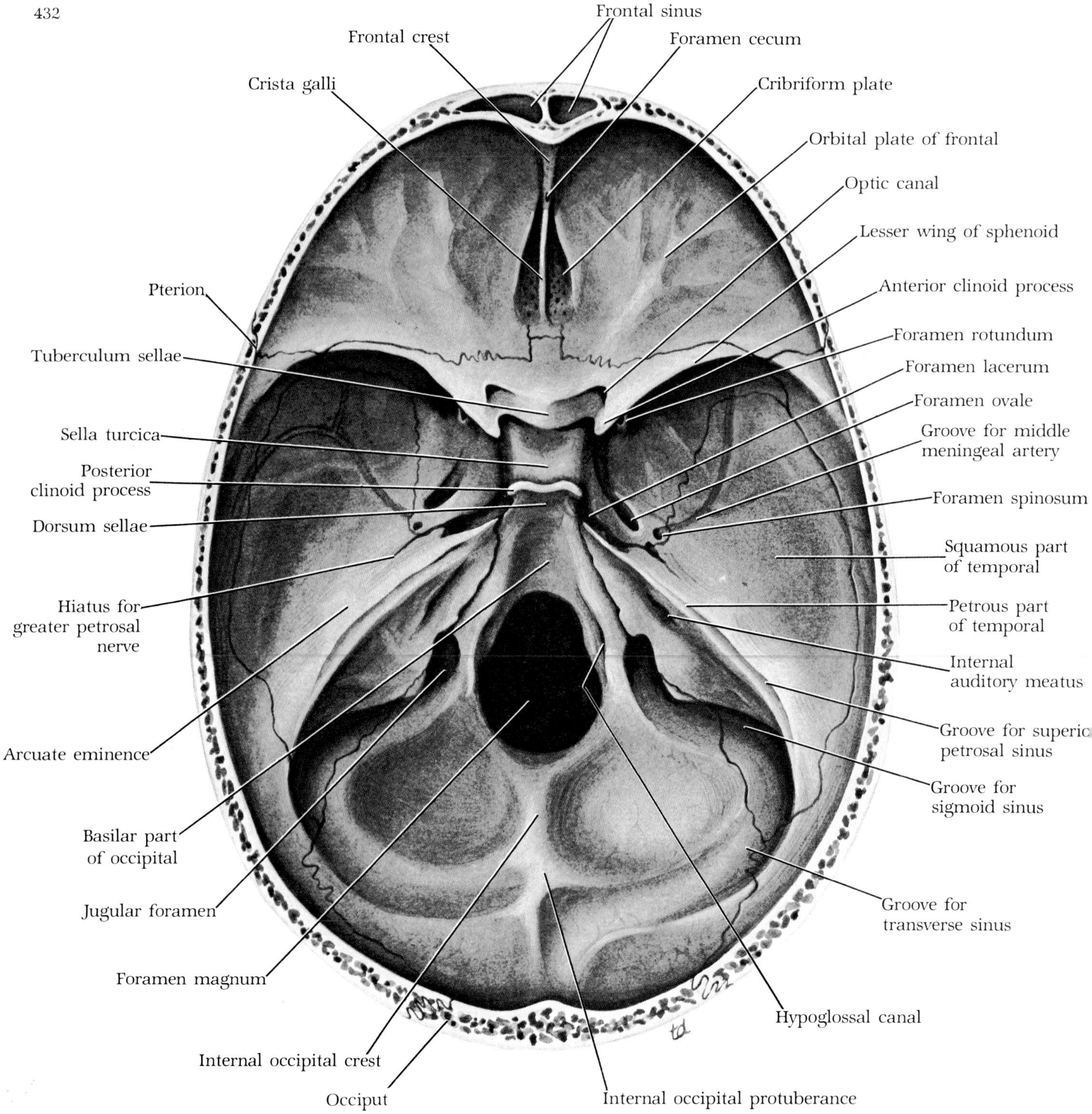
Frontal sinus
Frontal crest
Foramen cecum
Crista galli
Cribriform plate
Orbital plate of frontal
Optic canal
Lesser wing of sphenoid
Pterion
Anterior clinoid process
Foramen rotundum
Tuberculum sellae
Foramen lacerum
Foramen ovale
Sella turcica
Groove for middle meningeal artery
Posterior clinoid process
Foramen spinosum
Dorsum sellae
Squamous part of temporal
Hiatus for greater petrosal nerve
Petrous part of temporal
Internal auditory meatus
Arcuate eminence
Groove for superio
petrosal sinus
Groove for sigmoid sinus
Basilar part of occipital
Jugular foramen
Groove for transverse sinus
Foramen magnum
Hypoglossal canal
Internal occipital crest
Occiput
Internal occipital protuberance

Figure 7-49
Skull, internal view.

1. The cribriform plate of the ethmoid bone may be damaged in fractures of the anterior cranial fossa. This may result in tearing of the meninges and mucoperiosteum of the nose. The patient will have a bleeding nose (epistaxis) and leakage of cerebrospinal fluid into the nose (cerebrospinal rhinorrhea). Fractures of the orbital plate of the frontal bone will result in hemorrhage beneath the conjunctiva and into the orbital cavity, causing exophthalmos. The frontal air sinus may be involved with hemorrhage into the nose.

2. Fractures of the middle cranial fossa are common due to the presence of numerous foramina and canals in this region; the cavities of the middle ear and the sphenoidal air sinuses are particularly vulnerable. Leakage of cerebrospinal fluid and blood from the external auditory meatus is common. The seventh and eighth cranial nerves may be involved as they pass through the petrous part of the temporal bone. The third, fourth, and sixth cranial nerves may be damaged if the lateral wall of the cavernous sinus is torn. Blood and cerebrospinal fluid may leak into the sphenoidal air sinuses and then into the nose.

3. In fractures of the posterior cranial fossa, blood may escape into the nape of the neck deep to the postvertebral muscles. Some days later, it tracks between the muscles and appears in the posterior triangle, close to the mastoid process. The mucous membrane of the roof of the nasopharynx may be torn, and blood may escape there. In fractures involving the jugular foramen, the ninth, tenth, and eleventh cranial nerves may be damaged. The strong bony walls of the hypoglossal canal usually protect the hypoglossal nerve from injury.

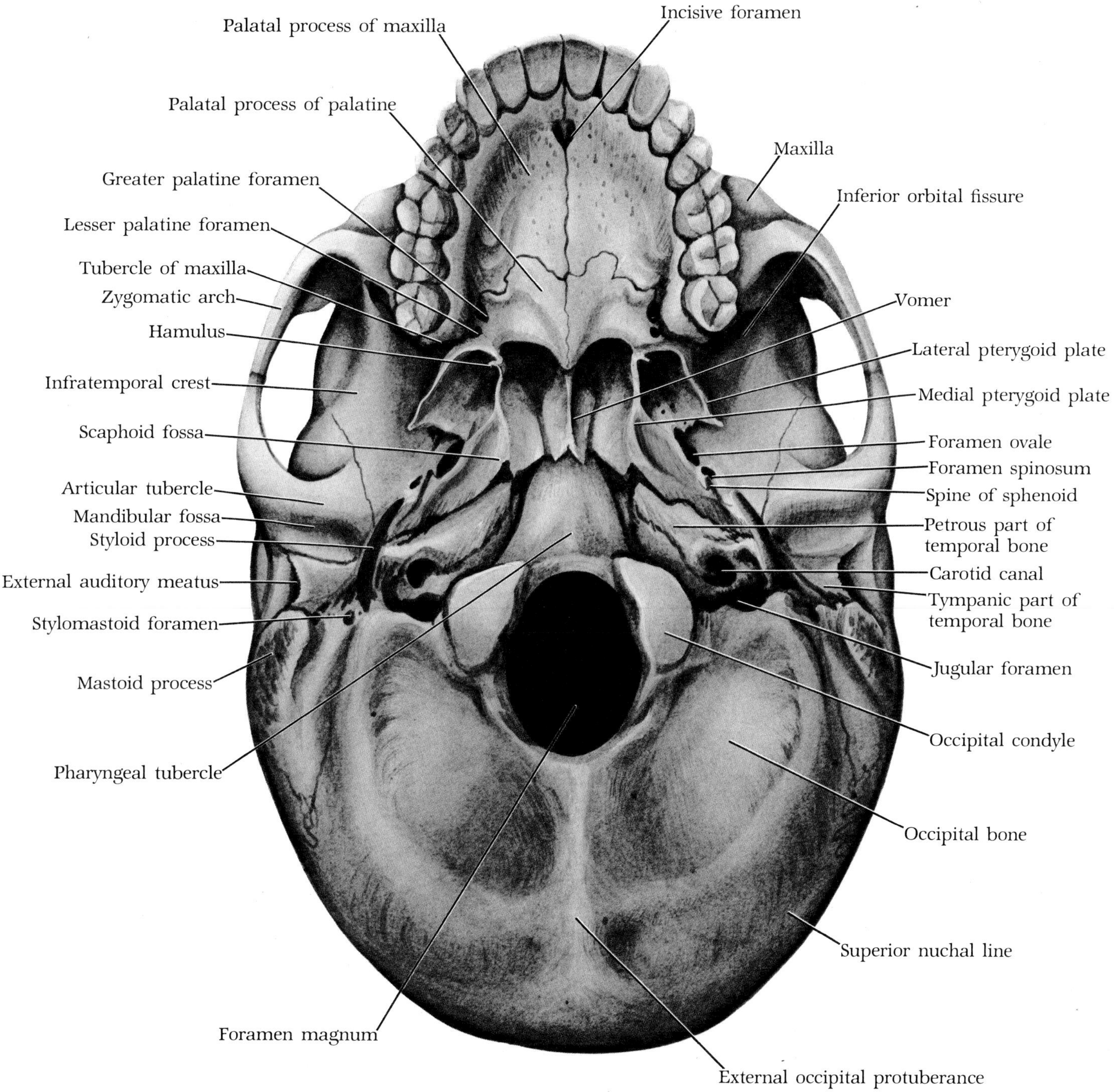
Palatal process of maxilla
Incisive foramen
Palatal process of palatine
Maxilla
Greater palatine foramen
Inferior orbital fissure
Lesser palatine foramen
Tubercle of maxilla
Zygomatic arch
Vomer
Hamulus
Lateral pterygoid plate
Infratemporal crest
Medial pterygoid plate
Scaphoid fossa
Foramen ovale
Foramen spinosum
Articular tubercle
Spine of sphenoid
Mandibular fossa
Styloid process
Petrous part of temporal bone
External auditory meatus
Carotid canal
Tympanic part of temporal bone
Stylomastoid foramen
Mastoid process
Jugular foramen
Occipital condyle
Pharyngeal tubercle
Occipital bone
Superior nuchal line
Foramen magnum
External occipital protuberance

Figure 7-50
Skull, inferior view.

1. Cleft palate. All degrees of cleft palate occur and are caused by a failure of the palatal processes of the maxilla to fuse with each other in the midline; in severe cases these processes also fail to fuse with the premaxilla. The first degree of severity is cleft uvula, and the second degree is ununited palatal processes. The third degree is ununited palatal processes and a cleft on one side of the premaxilla; this type is usually associated with unilateral cleft lip. The fourth degree of severity, which is rare, consists of ununited palatal processes and a cleft on both sides of the premaxilla; this type is usually associated with bilateral cleft lip.

2. Fractures of the maxilla occur as the result of a direct blow on the anterior aspect of the face. The upper dentition, the floor of the orbit, and the maxillary sinus may all be involved in the fracture line. The presence of malocclusion of the upper and lower teeth and involvement of the maxillary division of the trigeminal nerve with anesthesia of the upper lip and cheek may aid in the diagnosis.

3. Fractures of the middle cranial fossa are common, since this is the weakest part of the base of the skull. Anatomically this weakness is due to the presence of numerous foramina and canals in the region; the cavities of the middle ear and the sphenoidal air sinuses are particularly vulnerable.

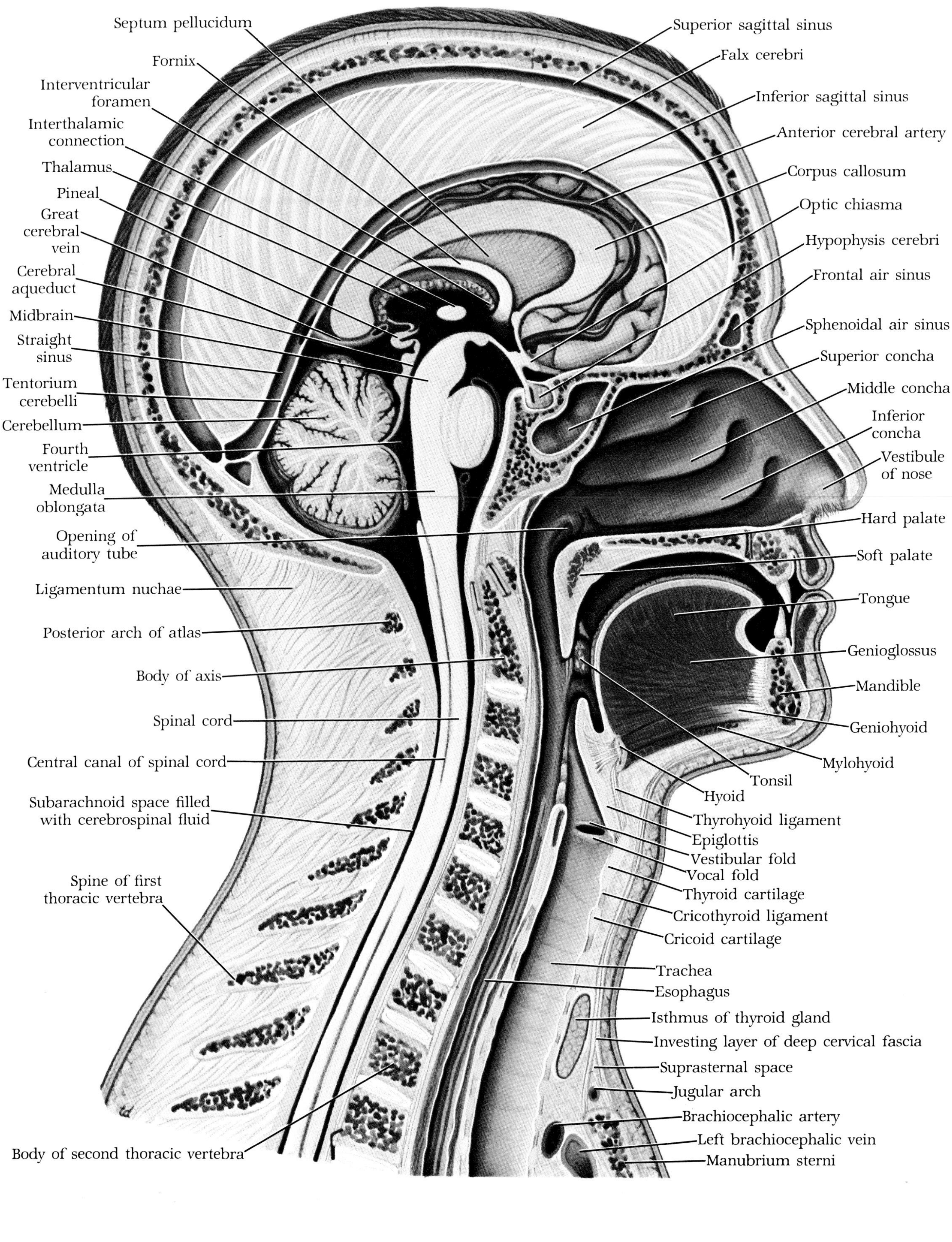

Septum pellucidum
Fornix
Interventricular foramen
Interthalamic connection
Thalamus
Pineal
Great cerebral vein
Cerebral aqueduct
Midbrain
Straight sinus
Tentorium cerebelli
Cerebellum
Fourth ventricle
Medulla oblongata
Opening of auditory tube
Ligamentum nuchae
Posterior arch of atlas
Body of axis
Spinal cord
Central canal of spinal cord
Subarachnoid space filled with cerebrospinal fluid
Spine of first thoracic vertebra
Body of second thoracic vertebra
Superior sagittal sinus
Falx cerebri
Inferior sagittal sinus
Anterior cerebral artery
Corpus callosum
Optic chiasma
Hypophysis cerebri
Frontal air sinus
Sphenoidal air sinus
Superior concha
Middle concha
Inferior concha
Vestibule of nose
Hard palate
Soft palate
Tongue
Genioglossus
Mandible
Geniohyoid
Mylohyoid
Tonsil
Hyoid
Thyrohyoid ligament
Epiglottis
Vestibular fold
Vocal fold
Thyroid cartilage
Cricothyroid ligament
Cricoid cartilage
Trachea
Esophagus
Isthmus of thyroid gland
Investing layer of deep cervical fascia
Suprasternal space
Jugular arch
Brachiocephalic artery
Left brachiocephalic vein
Manubrium sterni

Figure 7-51
Sagittal section of head and neck.

1. From this illustration a physician or dentist can orientate the various structures in the head and neck. For example, it is very useful to have this knowledge when viewing a lateral radiograph of the head and neck or when performing a tracheostomy.

2. When performing endotracheal anesthesia through a nasotracheal tube the tube is usually introduced into the larynx under direct vision using a laryngoscope. After inducing general anesthesia the laryngoscope is inserted through the mouth into the oral pharynx. The tip of the lubricated tube is then guided under direct vision between the vestibular and vocal folds. It is important that the tip of the laryngoscope be used to convert the angle formed by the mouth and the pharynx into a straight line. This is accomplished by lifting anteriorly the posterior third of the tongue and the superior edge of the epiglottis with the blade of the laryngoscope.

3. The relative positions of the nasopharyngeal tonsil and the palatine tonsils to the airway is important. Hypertrophy and chronic infection of the nasopharyngeal tonsil produces the condition called *adenoids*, which causes the patient to snore and breathe through an open mouth. The palatine tonsils are a common site of infection, producing the characteristic sore throat and pyrexia. Gross enlargement of the palatine tonsils may result in their meeting in the midline, causing obstruction to the passage of air and food.

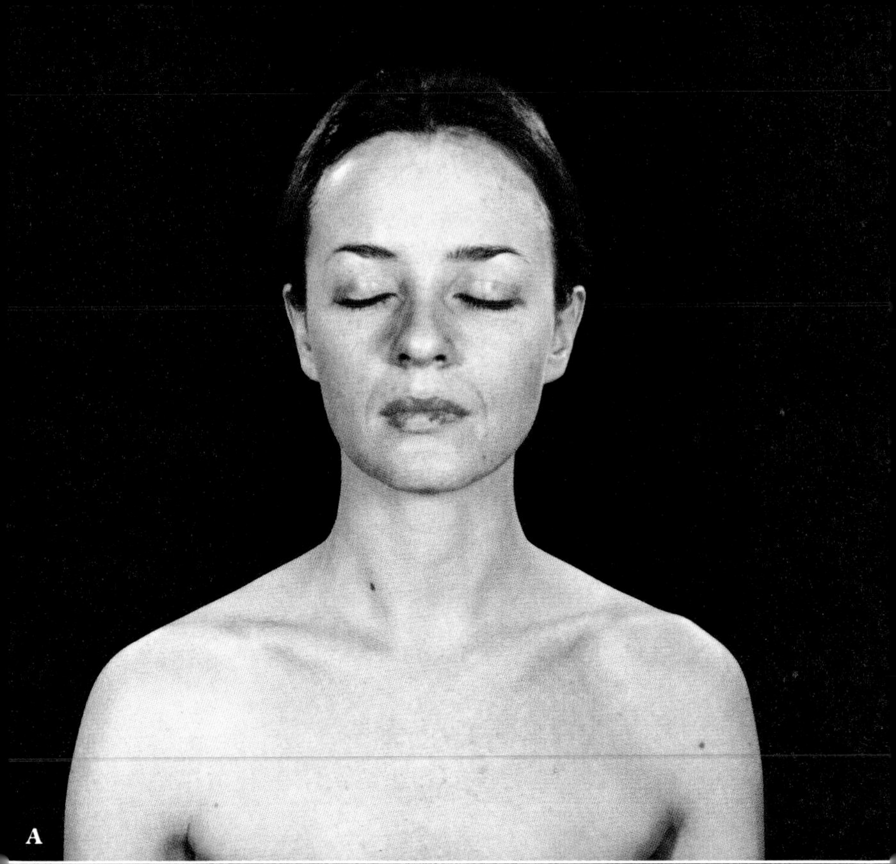

A

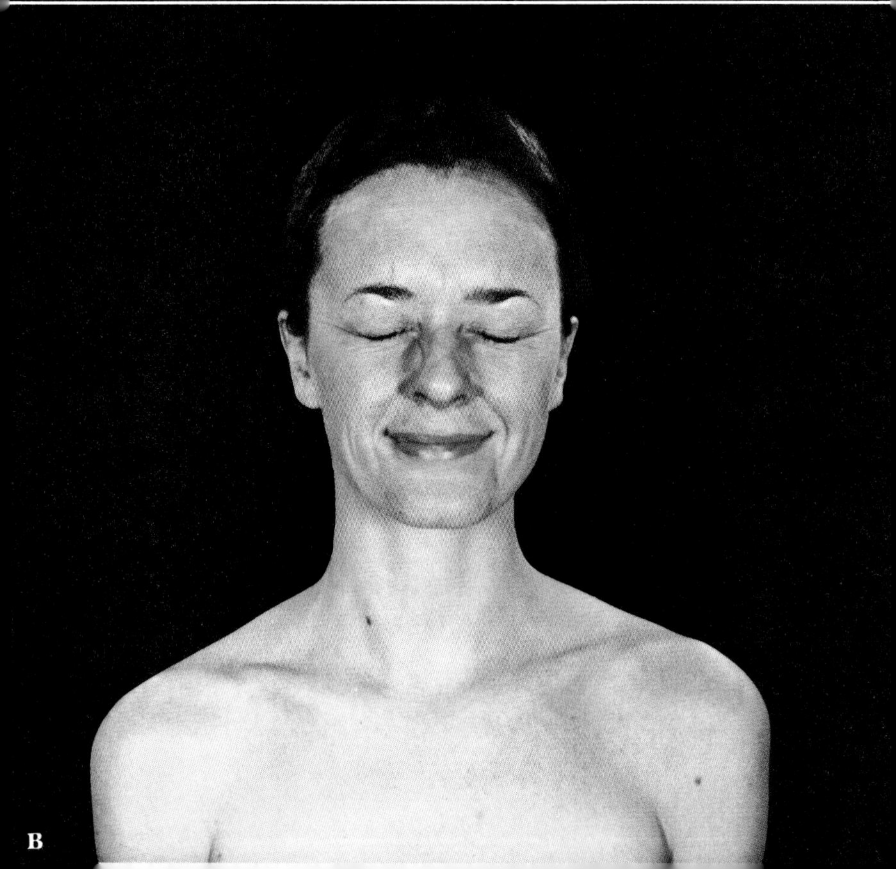

B

Figure 7-52
The act of closing the eyes in a 29-year-old female. A. Using the palpebral part of the orbicularis oculi muscle. B. Using the palpebral and orbital parts of the orbicularis oculi.

1. The palpebral part of the orbicularis oculi muscle closes the eyelids and dilates the lacrimal sac; it also directs the puncta lacrimalis into the lacus lacrimalis.
2. The orbital part of the orbicularis oculi muscle pulls on the skin of the forehead, temple, and cheek like a pursestring and draws it toward the medial angle of the orbit. The skin is thrown into prominent folds that overlap the eyelids and add further protection to the underlying eye. The movement is referred to as "screwing up the eye."
3. The orbicularis oculi muscle is supplied by the facial nerve.
4. The integrity of the facial nerve can be tested by asking the patient to firmly close the eyes. The physician then attempts to open the eyes by gently pushing up the upper lids in turn. An inability to close an eye or weakness of one of the muscles may easily be detected by this means.

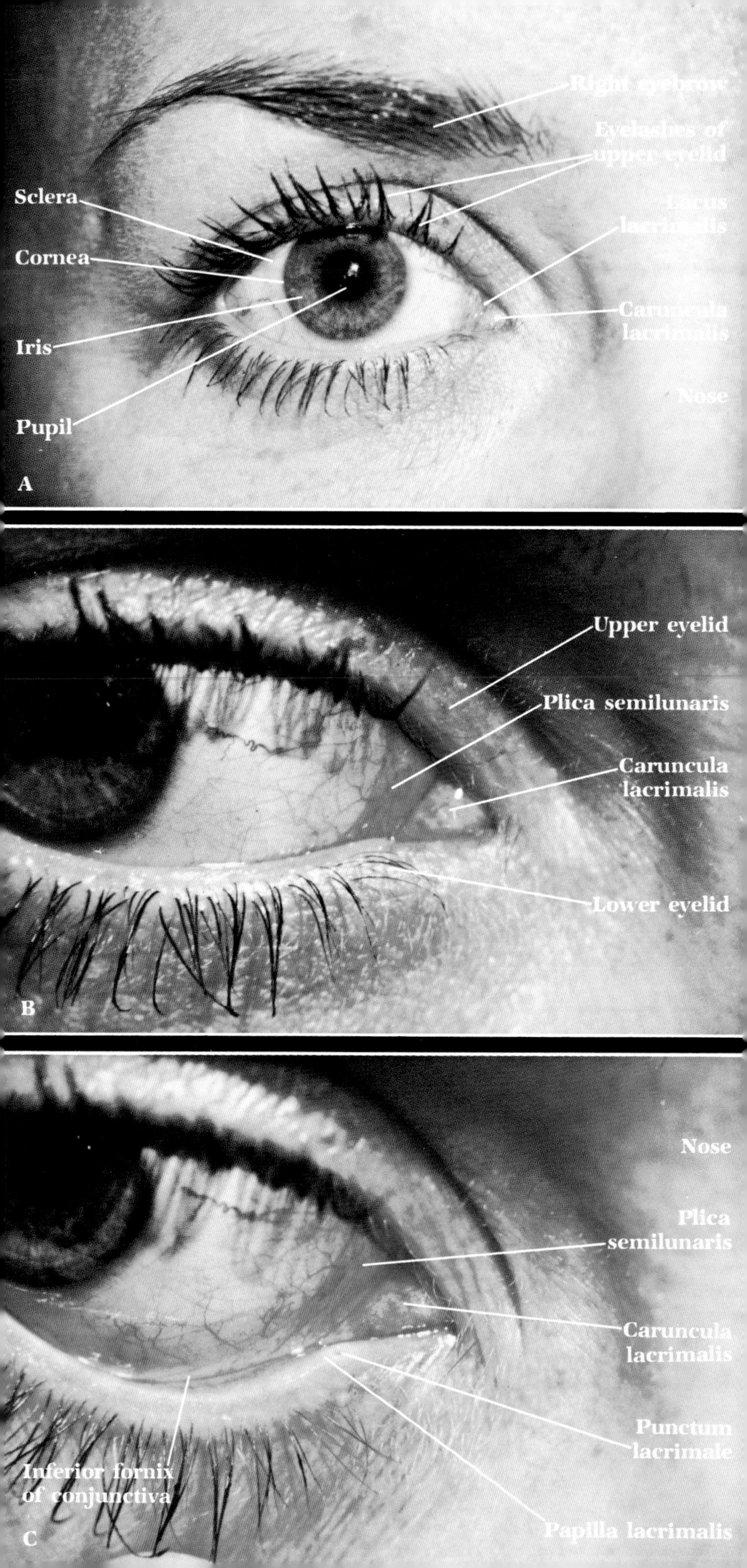
Right eyebrow
Eyelashes of upper eyelid
Lacus lacrimalis
Caruncula lacrimalis
Nose
Sclera
Cornea
Iris
Pupil
A
Upper eyelid
Plica semilunaris
Caruncula lacrimalis
Lower eyelid
B
Nose
Plica semilunaris
Caruncula lacrimalis
Punctum lacrimale
Inferior fornix of conjunctiva
Papilla lacrimalis
C

Figure 7-53
The right eye of a 29-year-old female. A. The names of the different structures. B. An enlarged view of the medial angle between the eyelids. C. The lower eyelid, pulled downward and slightly everted to reveal the punctum lacrimale.

1. The conjunctiva is a thin mucous membrane that lines the eyelids and is reflected at the superior and inferior fornices onto the anterior surface of the eyeball. The upper lateral part of the superior fornix is pierced by the ducts of the lacrimal gland. The conjunctiva thus forms a potential space, the conjunctival sac, which is open at the palpebral fissure.

2. Beneath the eyelid is a groove, the subtarsal sulcus, that runs close to and parallel with the margin of the lid. The sulcus tends to trap small foreign particles introduced into the conjunctival sac and for this reason is clinically important.

3. The *direct light reflex* can be tested by shining a light into the patient's eye: the normal pupil reflexly contracts. The nervous impulses pass from the retina via the optic nerve, optic chiasma, and optic tract. The fibers concerned with this reflex then pass to the pretectal nuclei and to the oculomotor nuclei on both sides. Parasympathetic fibers in turn pass to the ciliary ganglion, and postganglionic fibers reach the constrictor pupillae muscle via the short ciliary nerves.

4. The *consensual light reflex* is tested by shining the light in one eye and noting the contraction of the pupil in the opposite eye. This reflex is possible because the afferent pathway described in paragraph (3) travels to the parasympathetic nuclei of both oculomotor nerves.

5. The *accommodation reflex* is the contraction of the pupil that occurs when a person suddenly focuses on a near object after having focused on a distant one. The nervous impulses pass from the retina via the optic nerve, optic chiasma, optic tract, lateral geniculate body, optic radiation, and cerebral cortex of the occipital lobe of the brain. The efferent pathway passes to the parasympathetic nucleus of the oculomotor nerve. From there, the efferent impulses reach the constrictor pupillae via the oculomotor nerve, ciliary ganglion, and short ciliary nerves.

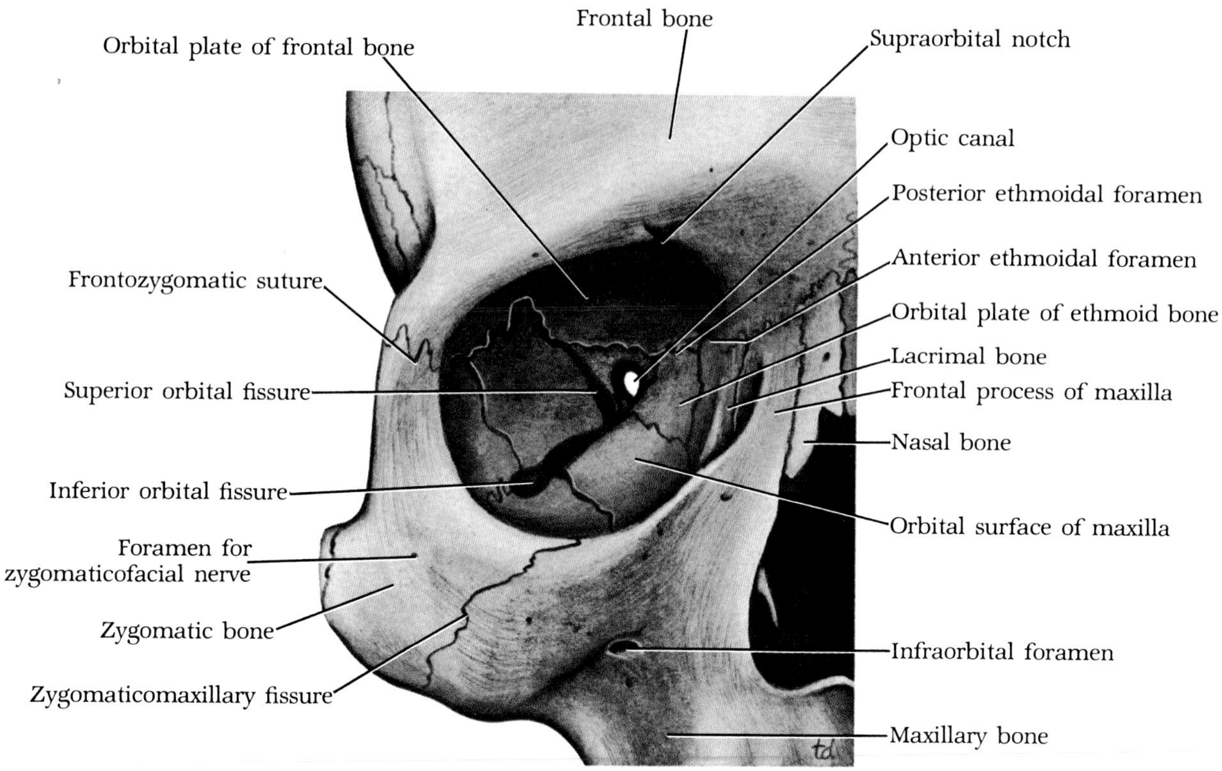

Frontal bone
Orbital plate of frontal bone
Supraorbital notch
Optic canal
Posterior ethmoidal foramen
Anterior ethmoidal foramen
Frontozygomatic suture
Orbital plate of ethmoid bone
Lacrimal bone
Superior orbital fissure
Frontal process of maxilla
Nasal bone
Inferior orbital fissure
Orbital surface of maxilla
Foramen for zygomaticofacial nerve
Zygomatic bone
Infraorbital foramen
Zygomaticomaxillary fissure
Maxillary bone

A

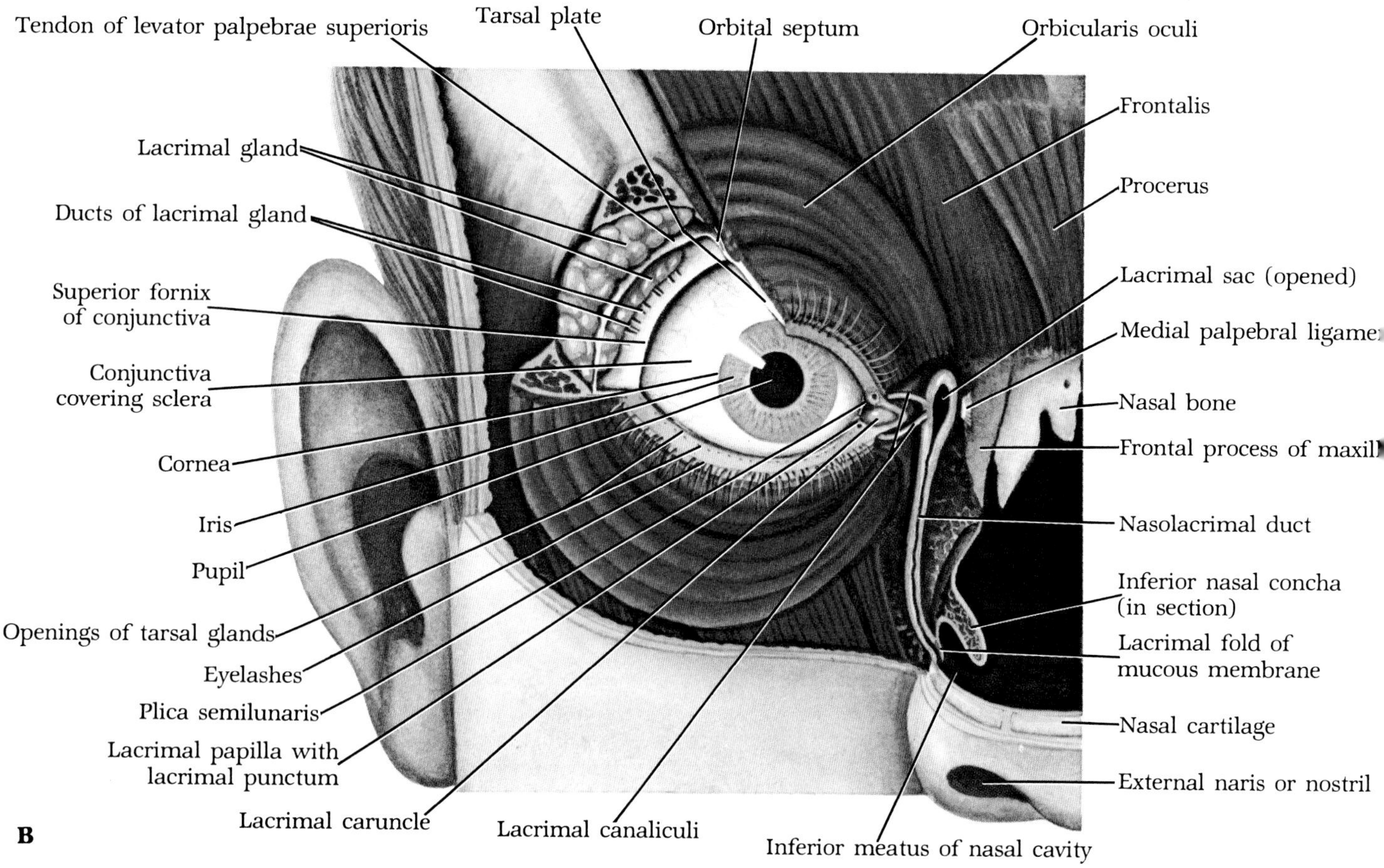

Tendon of levator palpebrae superioris
Tarsal plate
Orbital septum
Orbicularis oculi
Frontalis
Lacrimal gland
Ducts of lacrimal gland
Procerus
Superior fornix of conjunctiva
Lacrimal sac (opened)
Medial palpebral ligame
Conjunctiva covering sclera
Nasal bone
Cornea
Frontal process of maxil
Iris
Nasolacrimal duct
Pupil
Inferior nasal concha (in section)
Openings of tarsal glands
Lacrimal fold of mucous membrane
Eyelashes
Plica semilunaris
Nasal cartilage
Lacrimal papilla with lacrimal punctum
External naris or nostril
Lacrimal caruncle
Lacrimal canaliculi
Inferior meatus of nasal cavity

B

Figure 7-54
Right orbit and the eyelids.
A. Bones of the right orbit, anterior view. B. Muscles of the eyelids. The lateral part of the upper lid has been removed to reveal the lacrimal gland and its relationship to the tendon of the levator palpebrae superioris muscle and the conjunctiva.

1. Since the roof of the orbit formed by the orbital plate of the frontal bone is extremely thin, a puncture wound through the eyelids with a pointed object, e. g., a stick, may easily continue through the orbital roof into the frontal lobe of the cerebral hemisphere. The medial wall is also very thin, especially over the ethmoidal sinuses. Infections from the ethmoidal sinuses may easily spread into the orbit, causing orbital cellulitis. The floor of the orbit may be fractured from a blow to the eyeball; the orbital contents then herniate inferiorly into the maxillary air sinus.

2. The lacrimal gland is a serous gland that secretes tears that moisten, wash, and disinfect the cornea and conjunctival sac. Oversecretion can be produced with parasympathomimetic drugs, by emotion, or reflexly via the trigeminal nerve, as with conjunctivitis or mastication. Undersecretion may follow lesions of the facial nerve or the use of parasympatholytic drugs or atrophy of the gland.

3. Congenital agenesis of the lacrimal canaliculi or nasolacrimal duct produces epiphora (weeping) and, later, infection of the lacrimal sac.

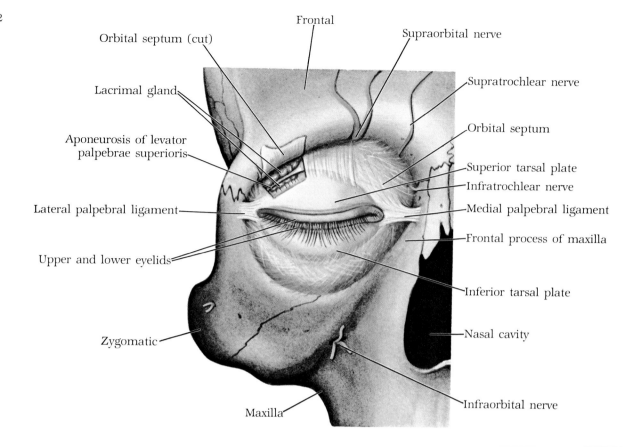

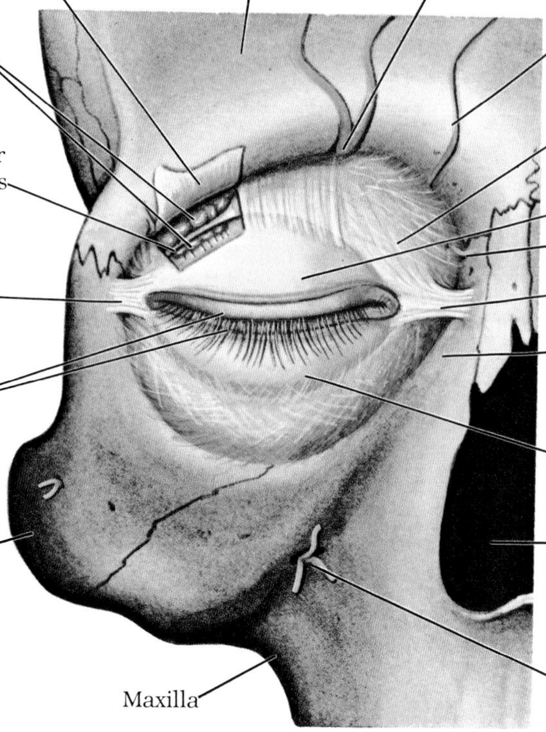
Frontal
Orbital septum (cut)
Supraorbital nerve
Lacrimal gland
Supratrochlear nerve
Orbital septum
Aponeurosis of levator palpebrae superioris
Superior tarsal plate
Infratrochlear nerve
Lateral palpebral ligament
Medial palpebral ligament
Frontal process of maxilla
Upper and lower eyelids
Inferior tarsal plate
Nasal cavity
Zygomatic
Infraorbital nerve
Maxilla

A

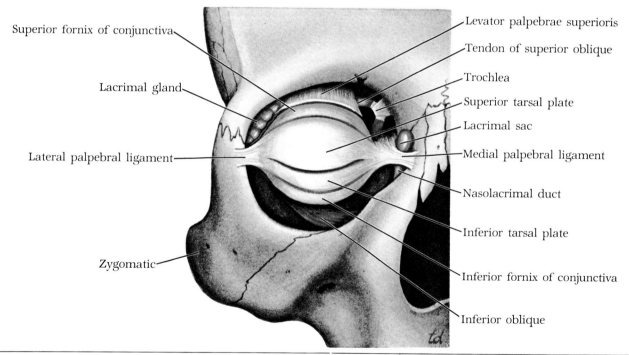

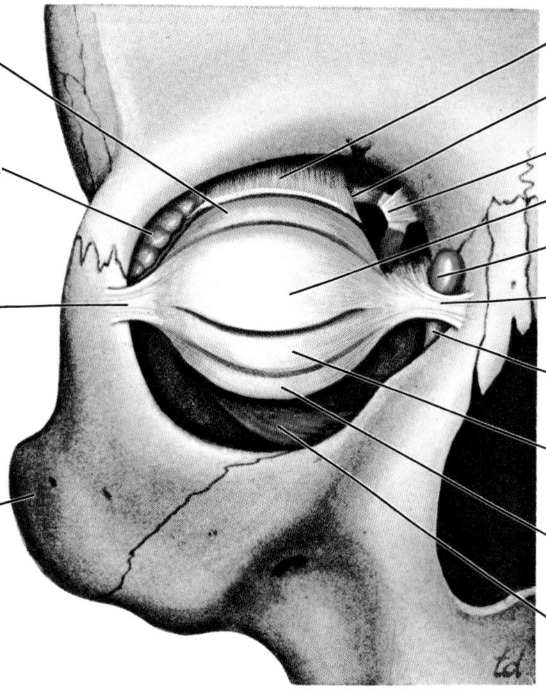
Superior fornix of conjunctiva
Levator palpebrae superioris
Tendon of superior oblique
Trochlea
Lacrimal gland
Superior tarsal plate
Lacrimal sac
Lateral palpebral ligament
Medial palpebral ligament
Nasolacrimal duct
Inferior tarsal plate
Zygomatic
Inferior fornix of conjunctiva
Inferior oblique

B

Figure 7-55
Right eye. A. The superior and inferior tarsal plates, orbital septum, and lacrimal gland. B. The superior and inferior tarsal plates, conjunctiva, and lacrimal gland, sac, and duct.

1. Acquired ptosis of the upper eyelid may be due to weakness of the striated or smooth muscular components of the levator palpebrae superioris muscle. Weakness of the striated muscle may be caused by myasthenia gravis, muscular dystrophies, or lesions of the nervous pathways to the muscle. Weakness of the smooth muscle occurs characteristically in Horner's syndrome due to a lesion of the sympathetic innervation.
2. The orbital septum effectively closes the orbital cavity anteriorly and limits the spread of blood or pus in this direction.
3. The conjunctiva is a thin mucous membrane that lines the eyelids and is reflected at the superior and inferior fornices onto the anterior surface of the eyeball. The conjunctiva thus forms a potential space, the conjunctival sac, that is open at the palpebral fissure. Since it is exposed to the outside it is liable to a large number of infections and traumatic injuries.
4. The eyelid margins are normally kept apposed to the surface of the eyeball by the tone of the palpebral part of the orbicularis oculi muscle. Senility may cause a weakness of this part of the muscle, causing the lower lid to become everted from the eyeball, a condition known as *ectropion.*

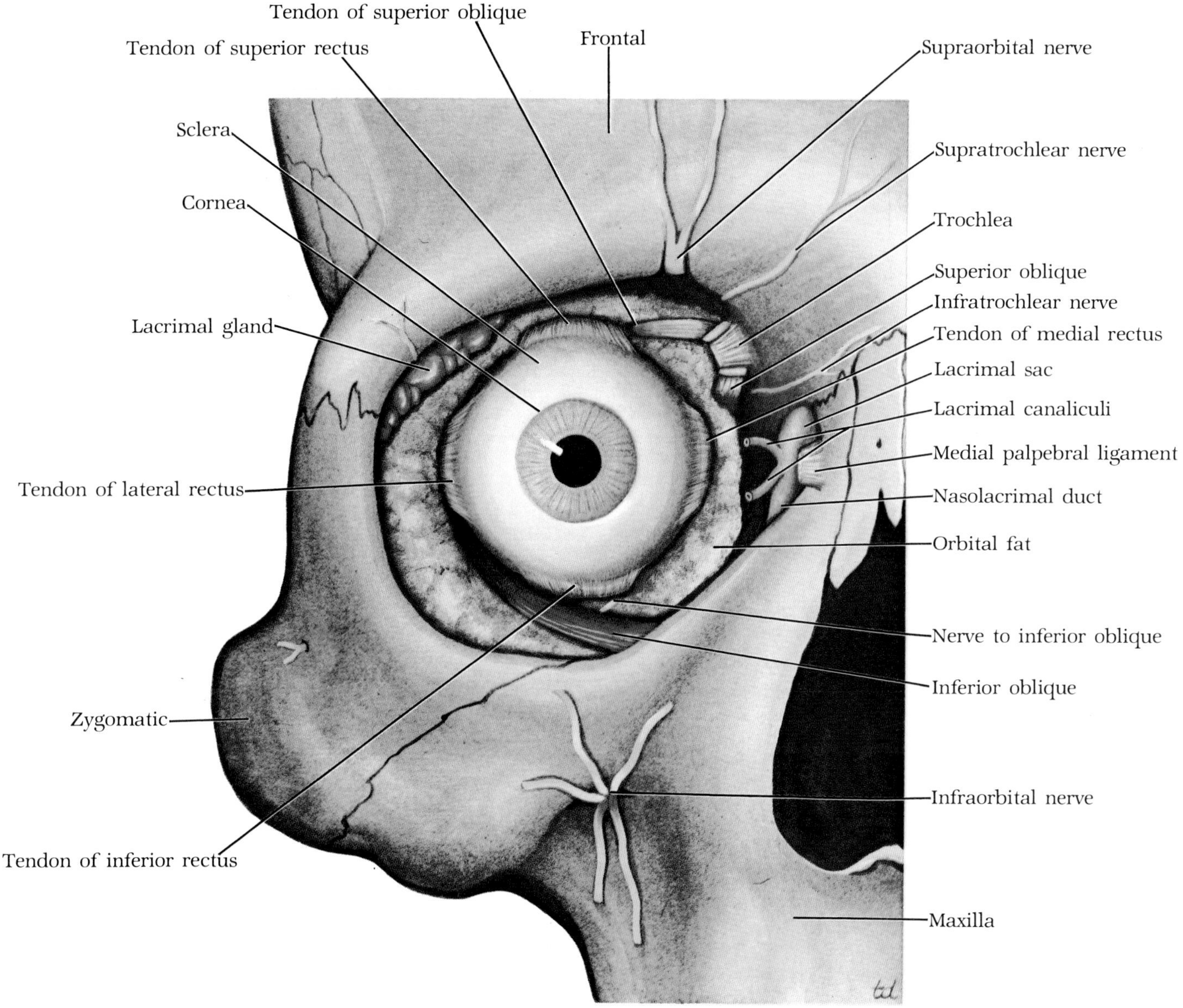
Tendon of superior oblique
Tendon of superior rectus
Frontal
Supraorbital nerve
Sclera
Supratrochlear nerve
Cornea
Trochlea
Superior oblique
Infratrochlear nerve
Lacrimal gland
Tendon of medial rectus
Lacrimal sac
Lacrimal canaliculi
Medial palpebral ligament
Tendon of lateral rectus
Nasolacrimal duct
Orbital fat
Nerve to inferior oblique
Inferior oblique
Zygomatic
Infraorbital nerve
Tendon of inferior rectus
Maxilla

Figure 7-56
Right eye showing the eyeball exposed from in front. Note the arrangement of the superior and inferior oblique muscles.

1. The superior oblique muscle as it contracts rotates the eyeball so that the cornea looks downward and laterally. It thus acts with the inferior rectus muscle to make the cornea move directly downward. It is innervated by the trochlear nerve.

2. The inferior oblique muscle as it contracts rotates the eyeball so that the cornea looks upward and laterally. It thus acts with the superior rectus muscle to make the cornea move directly upward. It is innervated by the oculomotor nerve.

3. The orbital fat fills the space between the fascial sheath of the eyeball and the bony walls of the orbit. It also fills in the intervals between the extrinsic muscles, nerves, and blood vessels.

4. The following spaces within the orbit are important clinically because they are the sites where pus or blood may accumulate: The episcleral space is between the sclera and the fascial sheath of the eyeball; the central space is between the fascial sheath of the eyeball and the muscular cone formed by the diverging recti muscles. Extravasations in this region lead to fixation of the eyeball and exophthalmos and a third space, between the muscle cone and the periosteum externally. Extravasations can extend anteriorly to involve the conjunctiva but are limited by the orbital septum.

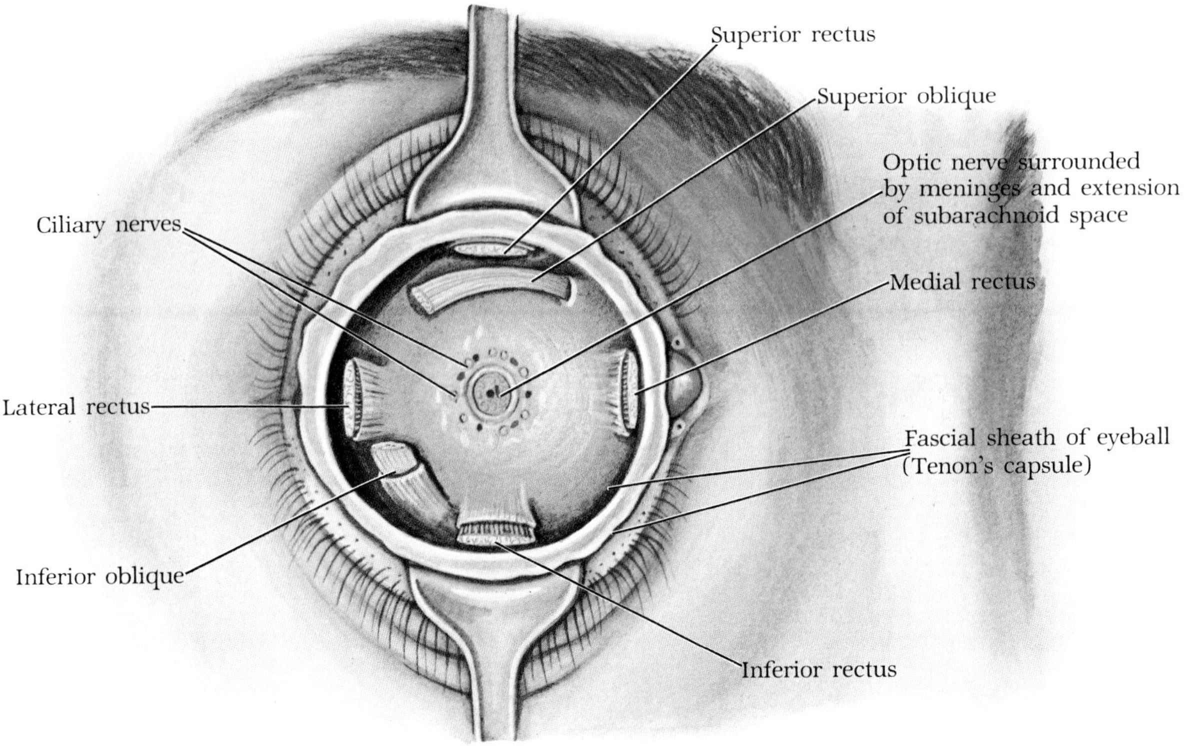
Superior rectus
Superior oblique
Optic nerve surrounded by meninges and extension of subarachnoid space
Ciliary nerves
Medial rectus
Lateral rectus
Fascial sheath of eyeball (Tenon's capsule)
Inferior oblique
Inferior rectus

A

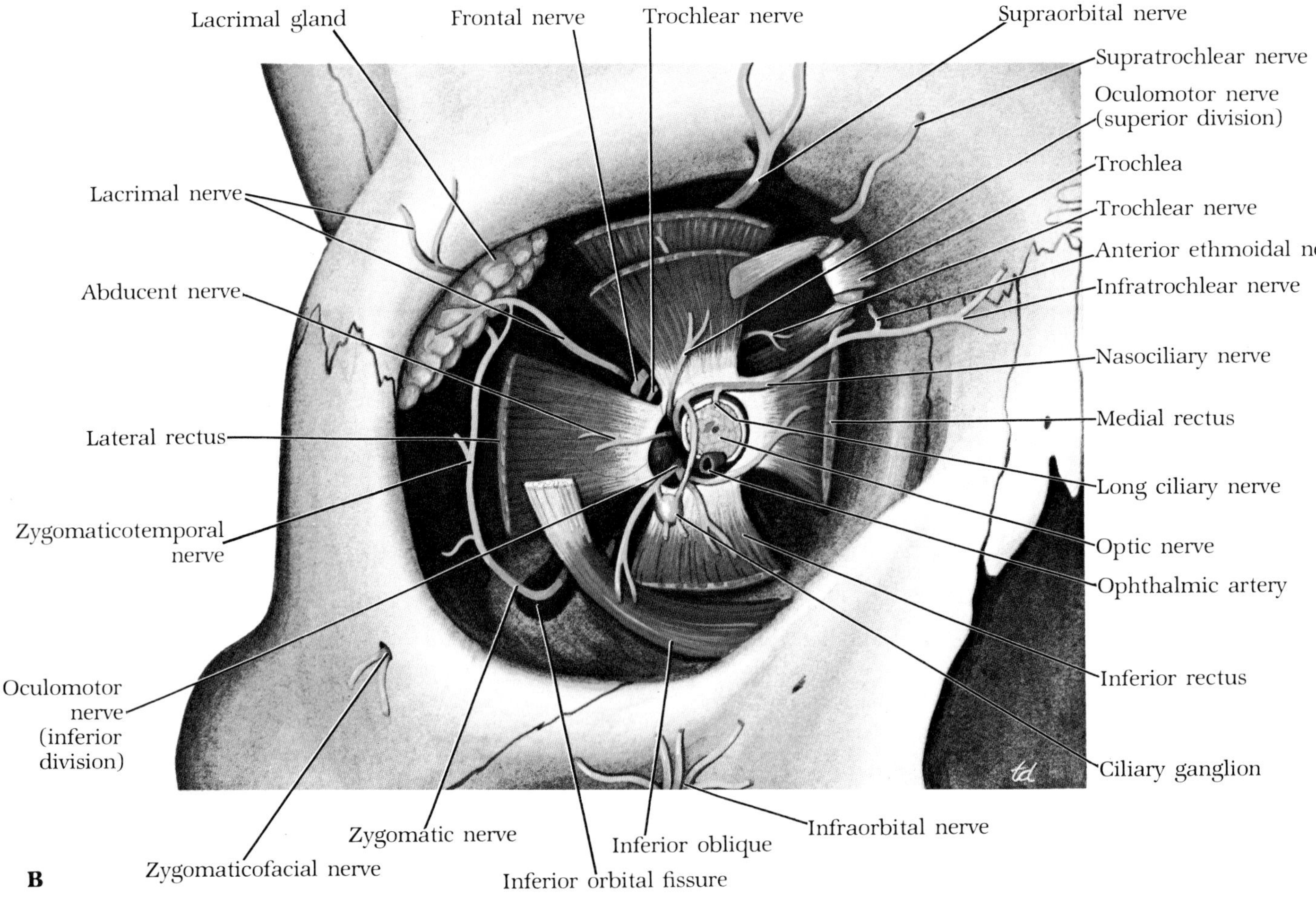
Lacrimal gland
Frontal nerve
Trochlear nerve
Supraorbital nerve
Supratrochlear nerve
Oculomotor nerve (superior division)
Trochlea
Lacrimal nerve
Trochlear nerve
Anterior ethmoidal nerve
Abducent nerve
Infratrochlear nerve
Nasociliary nerve
Medial rectus
Lateral rectus
Long ciliary nerve
Zygomaticotemporal nerve
Optic nerve
Ophthalmic artery
Inferior rectus
Oculomotor nerve (inferior division)
Ciliary ganglion
Infraorbital nerve
Zygomatic nerve
Inferior oblique
Zygomaticofacial nerve
Inferior orbital fissure

B

Figure 7-57
Right eye. A. Fascial sheath of eyeball after removal of the eyeball. B. Extrinsic muscles and their related nerves.

1. The fascial sheath of the eyeball, Tenon's capsule, consists of connective tissue that surrounds the eyeball from the corneoscleral junction anteriorly to the optic nerve posteriorly. The extrinsic muscles of the eyeball, blood vessels, and nerves pierce the sheath to reach the eyeball.

2. The fascial sheaths for the tendons of the medial and lateral recti muscles are attached to the medial and lateral walls of the orbit by triangular ligaments called the medial and lateral check ligaments. The lower part of the fascial sheath of the eyeball, which passes beneath the eyeball, connects the check ligaments and is called the suspensory ligament of the eye. By this means the eye is suspended from the medial and lateral walls of the orbit as if in a hammock. A precise knowledge of the fascial sheath of the eyeball is necessary when removing the eye and inserting orbital implants and artificial eyes.

3. The orbital muscles comprise the extrinsic voluntary muscles—the four recti and two obliques and a levator palpebrae superioris muscle—and the intrinsic involuntary muscles—the ciliary muscle and the muscle of the iris. In addition, there are unimportant small bundles of involuntary fibers scattered within the orbital cavity. However, the involuntary muscle found in relation to the insertion of the levator palpebrae superioris muscle is important. It is innervated by sympathetic nerve fibers and paralysis leads to ptosis of the upper eyelid.

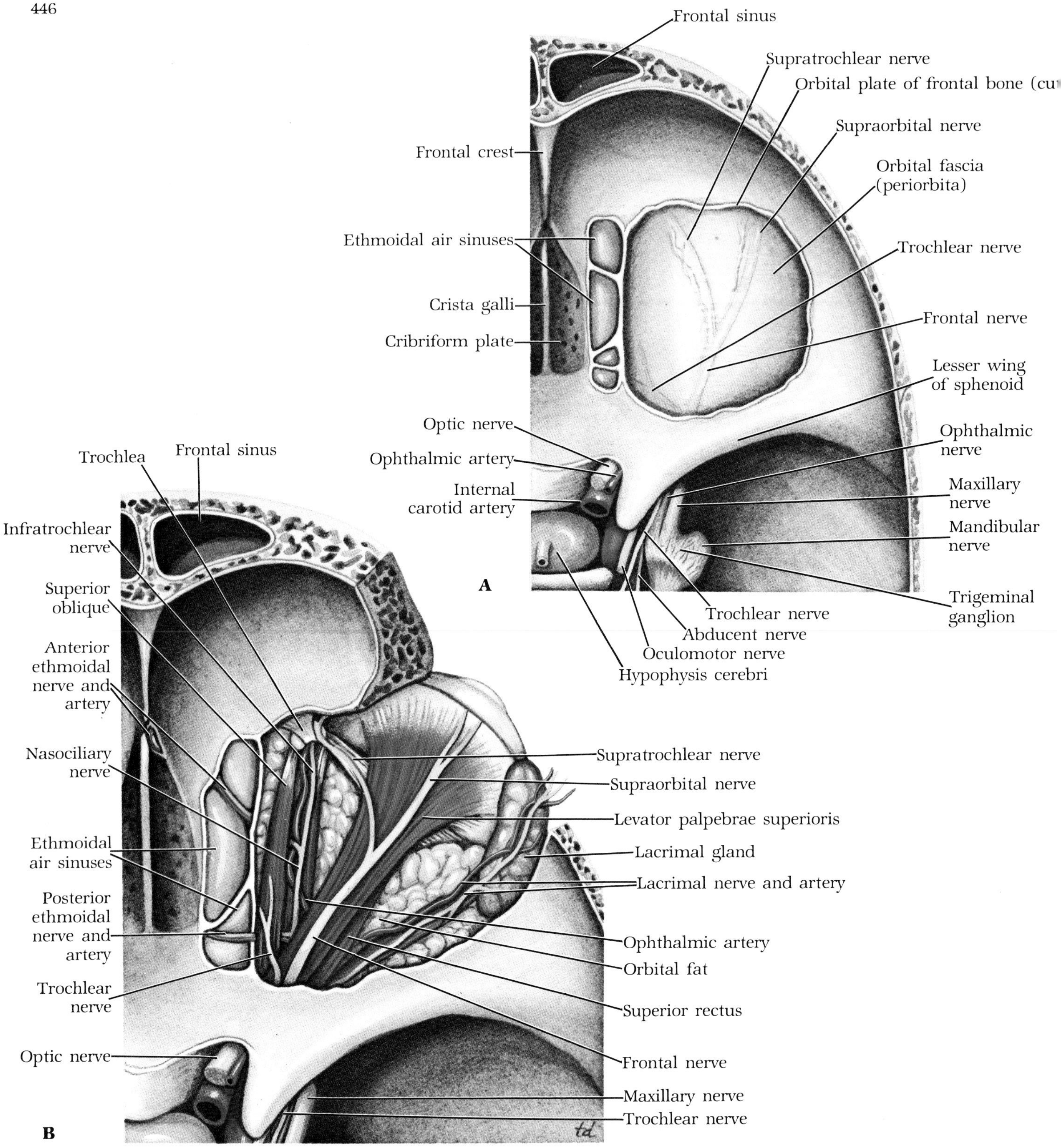
Frontal sinus
Supratrochlear nerve
Orbital plate of frontal bone (cu
Supraorbital nerve
Orbital fascia (periorbita)
Frontal crest
Ethmoidal air sinuses
Trochlear nerve
Crista galli
Frontal nerve
Cribriform plate
Lesser wing of sphenoid
Optic nerve
Ophthalmic nerve
Ophthalmic artery
Internal carotid artery
Maxillary nerve
Mandibular nerve
A
Trigeminal ganglion
Trochlear nerve
Abducent nerve
Oculomotor nerve
Hypophysis cerebri
Trochlea
Frontal sinus
Infratrochlear nerve
Superior oblique
Anterior ethmoidal nerve and artery
Nasociliary nerve
Supratrochlear nerve
Supraorbital nerve
Levator palpebrae superioris
Ethmoidal air sinuses
Lacrimal gland
Lacrimal nerve and artery
Posterior ethmoidal nerve and artery
Ophthalmic artery
Orbital fat
Trochlear nerve
Superior rectus
Optic nerve
Frontal nerve
Maxillary nerve
Trochlear nerve
B

Figure 7-58
Right orbit viewed from above. A. The orbital fascia after removal of the orbital plate of the frontal bone. B. Deeper structures after removal of the orbital fascia.

1. The orbital fascia is the periosteum that lines the bones that form the walls of the orbital cavity. It is continuous through the optic canal with the periosteal layer of dura; the meningeal layer of dura that has entered the orbit forms a sheath for the optic nerve and ends by fusing with the sclera of the eyeball. The orbital fascia is readily detached from the bones of the orbit.
2. The frontal nerve is the largest branch of the ophthalmic division of the trigeminal nerve. The supraorbital branch of the frontal nerve supplies the skin over part of the upper eyelid, the underlying conjunctiva, the frontal air sinus, and the skin over the forehead as far posteriorly as the vertex of the head. A frontal sinusitis may give rise to referred pain over the anterior part of the scalp.
3. The lacrimal gland receives its parasympathetic secretomotor nerve supply from the facial nerve via the greater petrosal nerve, the nerve of the pterygoid canal, and the pterygopalatine ganglion. The postganglionic fibers leave the ganglion and join the maxillary nerve. They then pass to the lacrimal gland via the zygomatic branch and the zygomaticotemporal nerve and the lacrimal nerve.
4. In very severe cases of exophthalmos associated with hyperthyroidism, in which the eye is in danger, the orbital plate of the frontal bone may be excised to allow the orbital contents to bulge superiorly into the anterior cranial fossa.

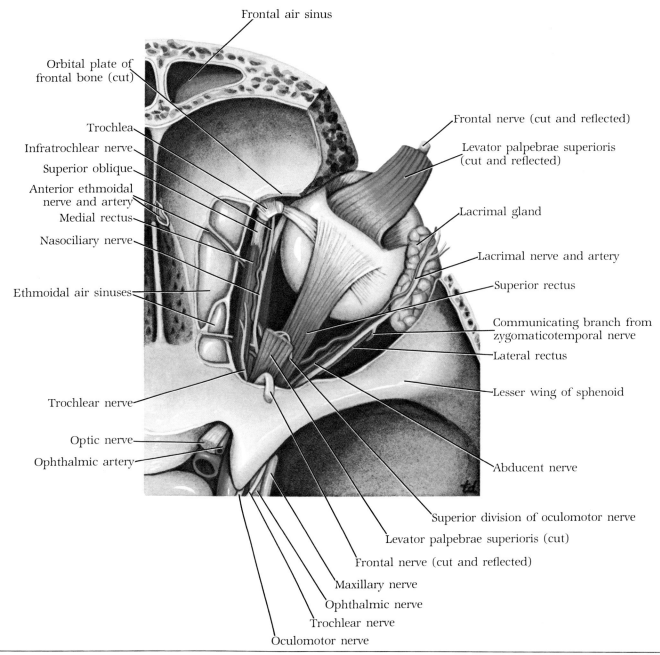

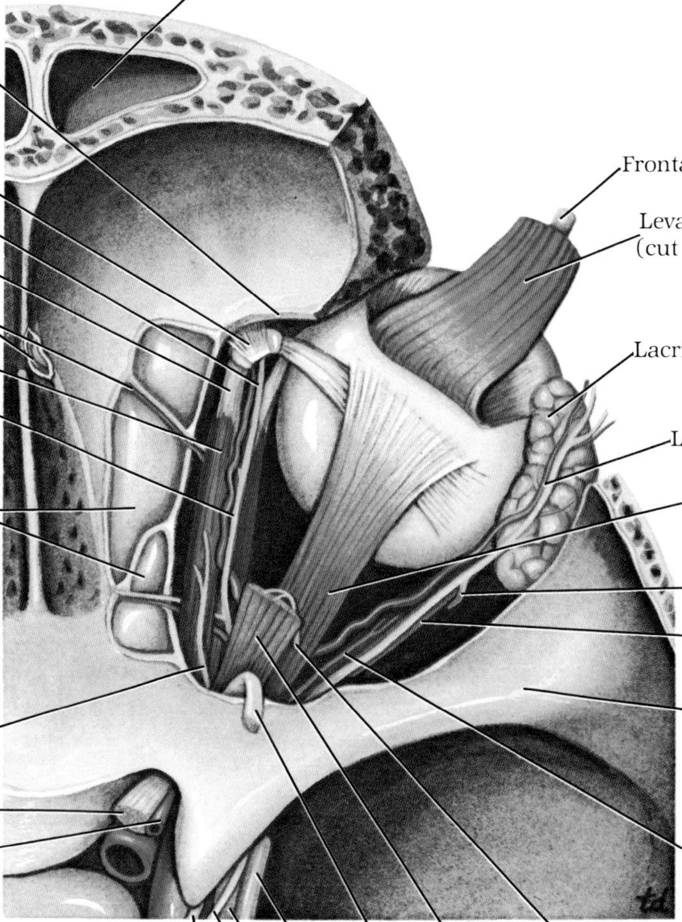
Frontal air sinus
Orbital plate of
frontal bone (cut)
Trochlea
Infratrochlear nerve
Superior oblique
Anterior ethmoidal
nerve and artery
Medial rectus
Nasociliary nerve
Ethmoidal air sinuses
Trochlear nerve
Optic nerve
Ophthalmic artery
Frontal nerve (cut and reflected)
Levator palpebrae superioris
(cut and reflected)
Lacrimal gland
Lacrimal nerve and artery
Superior rectus
Communicating branch from
zygomaticotemporal nerve
Lateral rectus
Lesser wing of sphenoid
Abducent nerve
Superior division of oculomotor nerve
Levator palpebrae superioris (cut)
Frontal nerve (cut and reflected)
Maxillary nerve
Ophthalmic nerve
Trochlear nerve
Oculomotor nerve

Figure 7-59
Right orbit viewed from above. The frontal nerve and the levator palpebrae superioris muscle have been reflected to show the superior rectus muscle.

1. The levator palpebrae superioris muscle raises the upper lid. The voluntary component of the muscle is innervated by the oculomotor nerve and the involuntary component is supplied by the sympathetic nerve. Ptosis of the upper lid may therefore follow injury to the oculomotor nerve or damage to the sympathetic outflow to the head and neck; myasthenia gravis may also cause ptosis of the upper lid.

2. Because the superior rectus muscle is inserted on the medial side of the vertical axis of the eyeball, it not only raises the cornea but rotates the cornea medially. The superior rectus is innervated by the superior division of the oculomotor nerve. Paralysis of this muscle would result in the patient's being unable to raise the cornea above the horizontal when the eye was in the abducted position (i. e., with the cornea rotated 23 degrees lateral to the median plane).

3. The superior oblique muscle, which rotates the eyeball so that the cornea looks downward and laterally, is innervated by the trochlear nerve. Paralysis of the superior oblique would result in the patient's having weakness in turning the cornea downward and laterally.

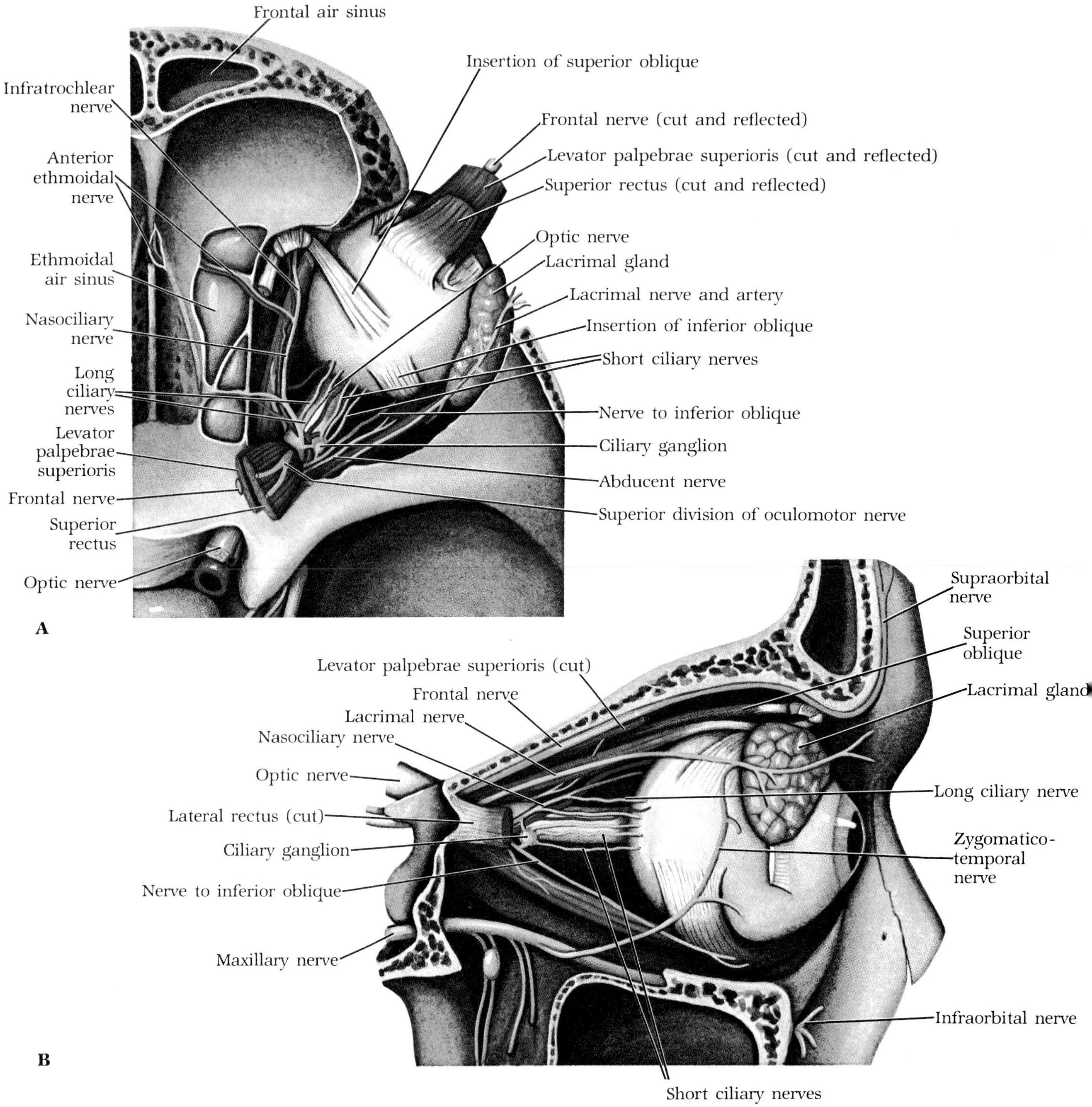

Frontal air sinus
Infratrochlear nerve
Anterior ethmoidal nerve
Ethmoidal air sinus
Nasociliary nerve
Long ciliary nerves
Levator palpebrae superioris
Frontal nerve
Superior rectus
Optic nerve
Insertion of superior oblique
Frontal nerve (cut and reflected)
Levator palpebrae superioris (cut and reflected)
Superior rectus (cut and reflected)
Optic nerve
Lacrimal gland
Lacrimal nerve and artery
Insertion of inferior oblique
Short ciliary nerves
Nerve to inferior oblique
Ciliary ganglion
Abducent nerve
Superior division of oculomotor nerve
A
Supraorbital nerve
Superior oblique
Lacrimal gland
Levator palpebrae superioris (cut)
Frontal nerve
Lacrimal nerve
Nasociliary nerve
Optic nerve
Lateral rectus (cut)
Ciliary ganglion
Nerve to inferior oblique
Maxillary nerve
Long ciliary nerve
Zygomatico-temporal nerve
Infraorbital nerve
B
Short ciliary nerves

Figure 7-60
Right orbit. A. Viewed from above after reflection of the frontal nerve, levator palpebrae superioris muscle, and the superior rectus muscles. B. Viewed from the lateral side after removal of the lateral wall of the orbit and the lateral rectus muscles.

1. The medial rectus muscle rotates the eyeball so that the cornea looks medially. It is innervated by the inferior division of the oculomotor nerve. The lateral rectus muscle rotates the eyeball so that the cornea looks laterally. It is innervated by the abducent nerve.

2. The insertions of the superior and inferior oblique muscles into the eyeball explain their actions. The superior oblique rotates the eyeball so that the cornea looks downward and laterally. The inferior oblique rotates the eyeball so that the cornea looks upward and laterally.

3. Eye movements are tested by asking the patient to follow the tip of the examiner's finger while keeping his head stationary. Small degrees of muscle weakness can be detected by the patient's experiencing double vision (diplopia) when moving his eyes.

4. The ciliary ganglion is a parasympathetic ganglion. The postganglionic nerve fibers travel to the constrictor pupillae muscle of the iris via the short ciliary nerves. The long ciliary nerves convey the sensory fibers of the eyeball and many of the sympathetic nerve fibers. The sympathetic fibers innervate the dilator pupillae muscle of the iris.

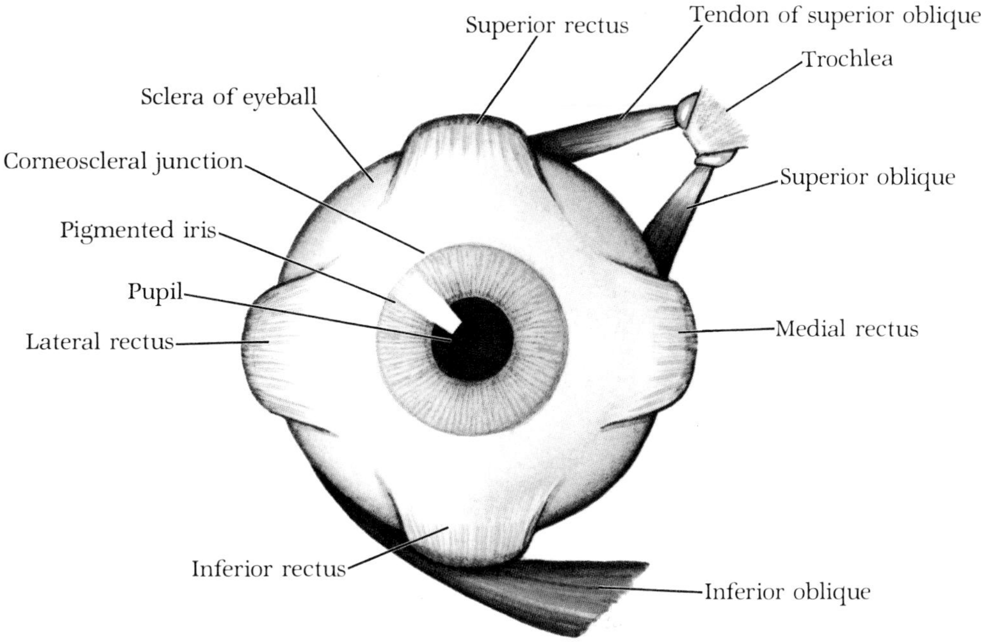
Superior rectus
Tendon of superior oblique
Trochlea
Sclera of eyeball
Corneoscleral junction
Superior oblique
Pigmented iris
Pupil
Medial rectus
Lateral rectus
Inferior rectus
Inferior oblique

A

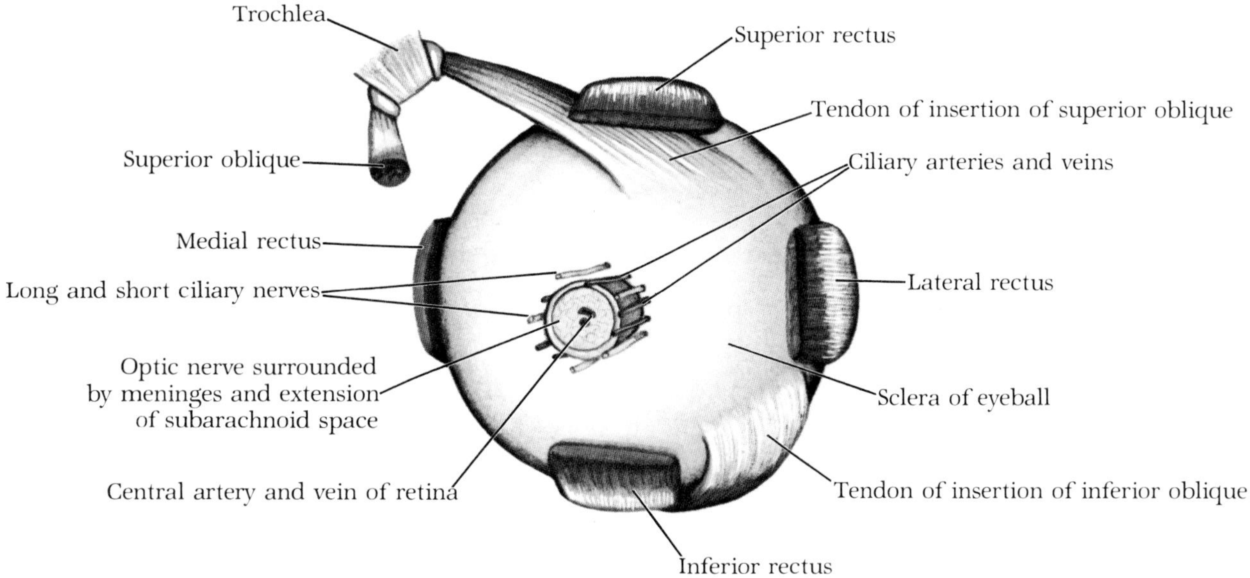
Trochlea
Superior rectus
Tendon of insertion of superior oblique
Superior oblique
Ciliary arteries and veins
Medial rectus
Long and short ciliary nerves
Lateral rectus
Optic nerve surrounded by meninges and extension of subarachnoid space
Sclera of eyeball
Central artery and vein of retina
Tendon of insertion of inferior oblique
Inferior rectus

B

Figure 7-61
Right eyeball. A. Anterior view shows the attachment of the extrinsic muscles. B. Posterior view shows the precise insertion of the superior and inferior oblique muscles.

1. Except for the movement of convergence, all the movements of the two eyes are coordinated so that the visual axes remain parallel; binocular vision is possible only as long as the visual axes remain parallel. Should the visual axes no longer be parallel, a squint (strabismus) is present.

2. The establishment of the conjugate fixation reflex by which the delicate eye movements are coordinated so that the axes of the eyes remain parallel occurs between the first five to six weeks after birth but does not become firmly established until about the eighth year. Up until then the muscular movements can become uncoordinated as a result of emotional problems.

3. The optic nerve within the orbit is surrounded by meninges and a prolongation of the intracranial subarachnoid space. A rise in cerebrospinal fluid pressure in the subarachnoid space will compress the thin walls of the retinal vein as it crosses the space to enter the optic nerve. This results in congestion of the retinal veins, edema of the retina, and bulging of the optic disc (papilledema).

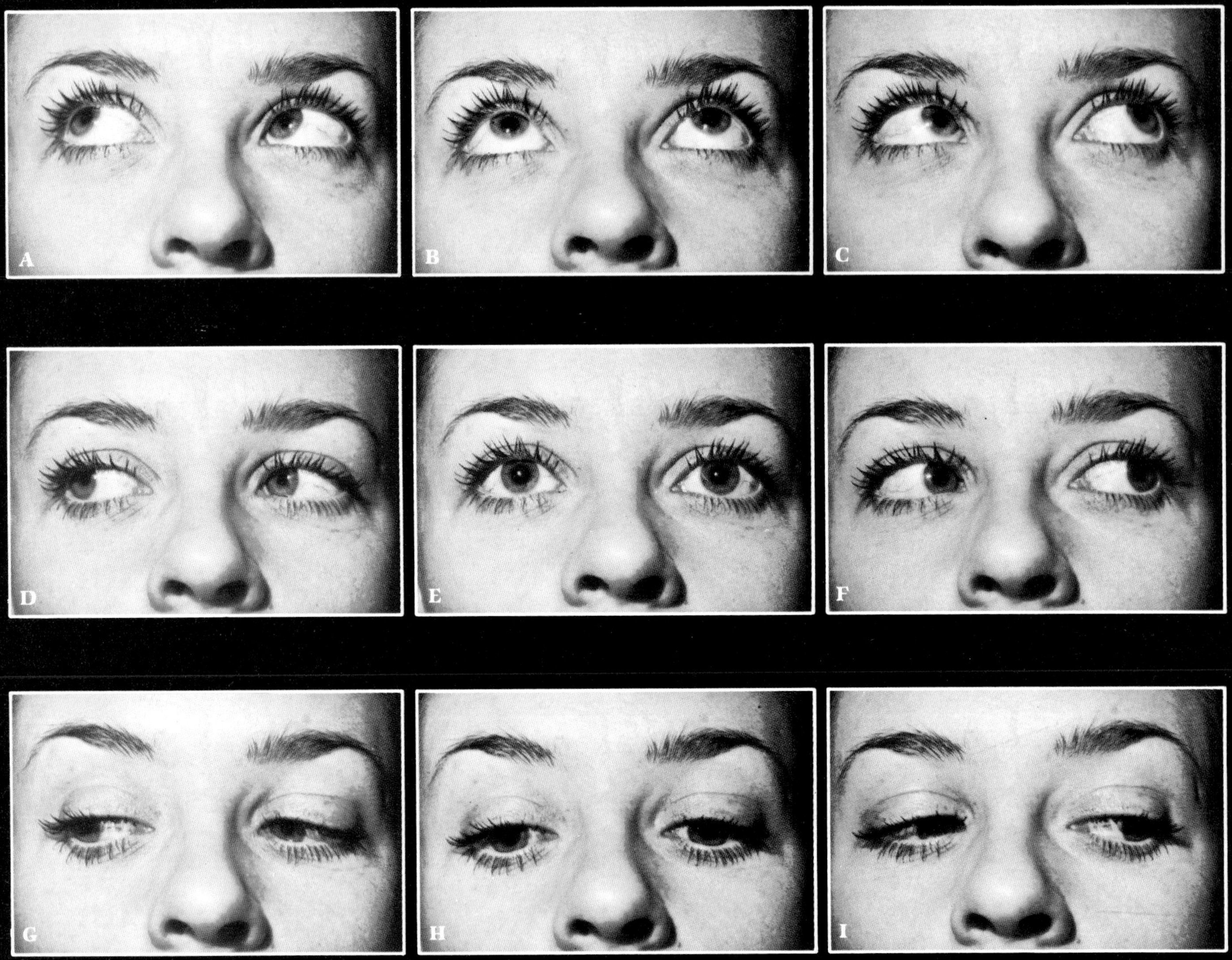
A
B
C
D
E
F
G
H
I

Figure 7-62
The cardinal positions of the right and left eyes and the actions of the recti and oblique muscles *principally responsible* for the movements of the eyes. A. Right eye, superior rectus muscle; left eye, inferior oblique muscle. B. Both eyes, superior recti and inferior oblique muscles. C. Right eye, inferior oblique muscle; left eye, superior rectus muscle. D. Right eye, lateral rectus muscle; left eye, medial rectus muscle. E. Primary position with the eyes fixed upon a distant fixation point. F. Right eye, medial rectus muscle; left eye, lateral rectus muscle. G. Right eye, inferior rectus muscle; left eye, superior oblique muscle. H. Both eyes, inferior recti and superior oblique muscles. I. Right eye, superior oblique muscle; left eye, inferior rectus muscle.

1. The cardinal positions are those in which each extraocular muscle is *mainly responsible* for the eye movement obtained.
2. Eye movements are tested by asking the patient to follow the tip of the examiner's finger while keeping his head stationary. Small degrees of muscle weakness can be detected by the patient's experiencing double vision (diplopia) when moving his eyes.
3. All the movements of the two eyes (except for the movement of convergence) are coordinated so that the visual axes remain parallel; binocular vision is possible only as long as the visual axes remain parallel.
4. The superior oblique muscle is innervated by the trochlear nerve and the lateral rectus muscle by the abducent nerve, while all the remaining extraocular muscles are innervated by the oculomotor nerve.

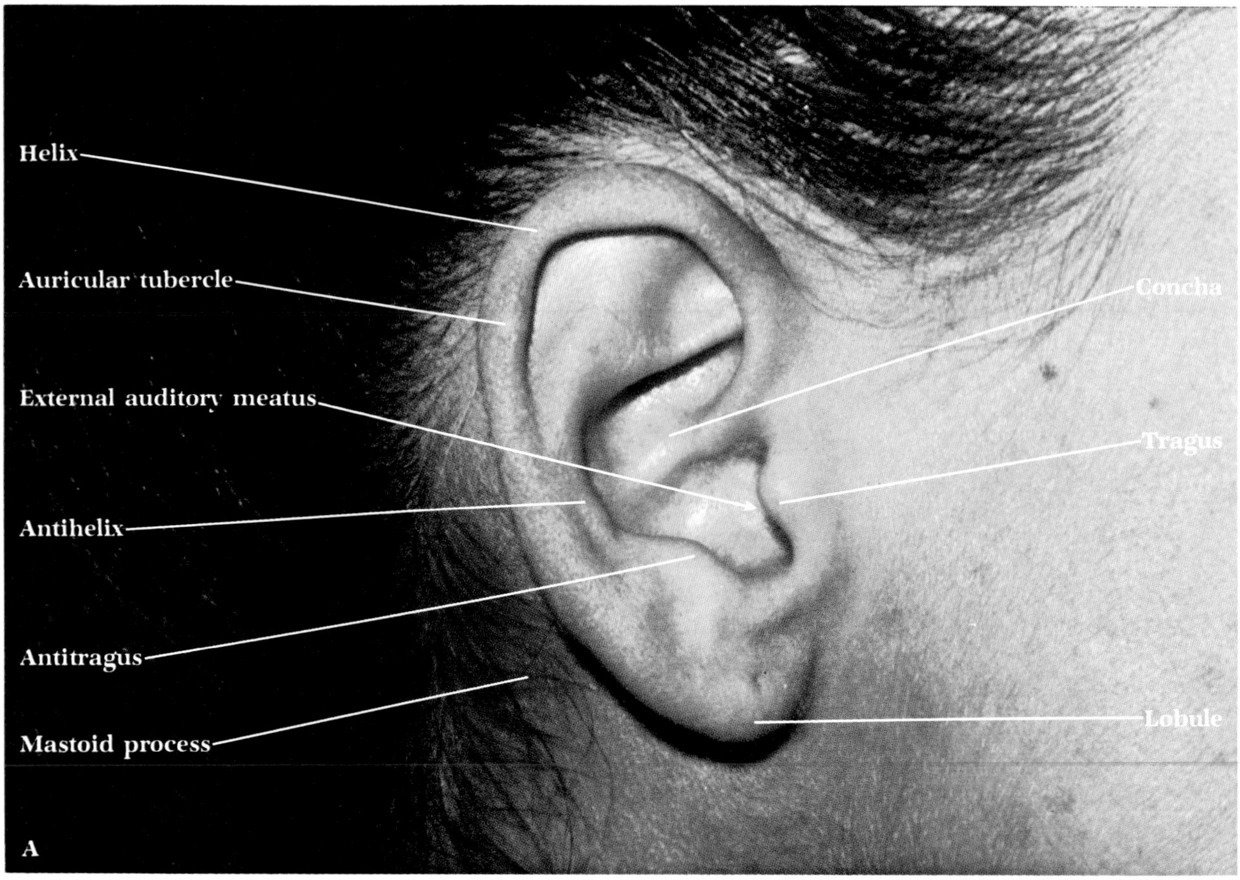

Helix
Auricular tubercle
External auditory meatus
Antihelix
Antitragus
Mastoid process
Concha
Tragus
Lobule
A

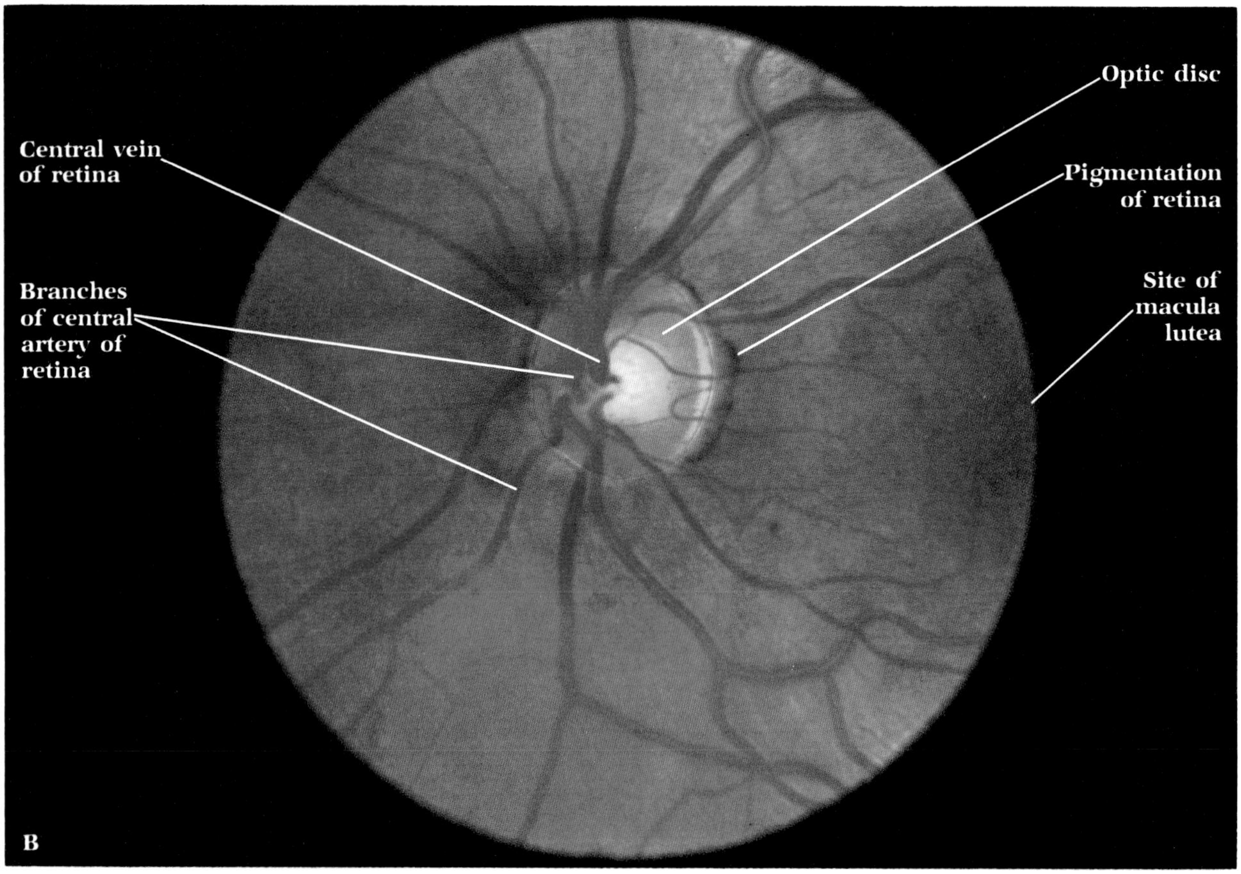

Central vein of retina
Branches of central artery of retina
Optic disc
Pigmentation of retina
Site of macula lutea
B

Figure 7-63
A. Right auricle of adult female.
B. Left fundus oculi of adult as seen through an ophthalmoscope.

1. The auricle serves to collect sound waves that are funneled into the external auditory meatus.
2. Auriscopic examination of the tympanic membrane is facilitated by first straightening the external auditory meatus by gently pulling the auricle upward and backward. Normally, the tympanic membrane is concave and pearly gray in color.
3. Embryologically, the auricle is formed from three mesenchymal proliferations derived from the first pharyngeal arch and three similar proliferations from the second pharyngeal arch, all of which fuse together. Failure to fuse produces several varieties of auricular deformity.
4. Injury to the auricle may result in an extremely painful effusion of blood under the perichondrium of the cartilaginous skeleton. If infection does not occur, the serum is absorbed after a few days but some permanent thickening usually remains.
5. The fundus oculi is red in color due to the blood circulating in the choroidal blood vessels.
6. The optic disc is pale pink in color and nearly circular in shape. The edges are usually quite sharp, and there may be some retinal pigment around the margin.
7. The center of the disc is usually pale, and the central artery and vein emerge from the middle of this area. The arteries are distinguished from the veins by being a lighter red and narrower. The vessels, especially the arteries, may show a silver streak running along the surface; this is due to reflection of light from the cylindrical surface.
8. The macula lutea is situated about 2 disc diameters lateral to the edge of the optic disc. It is usually a small circular area of dark red color.

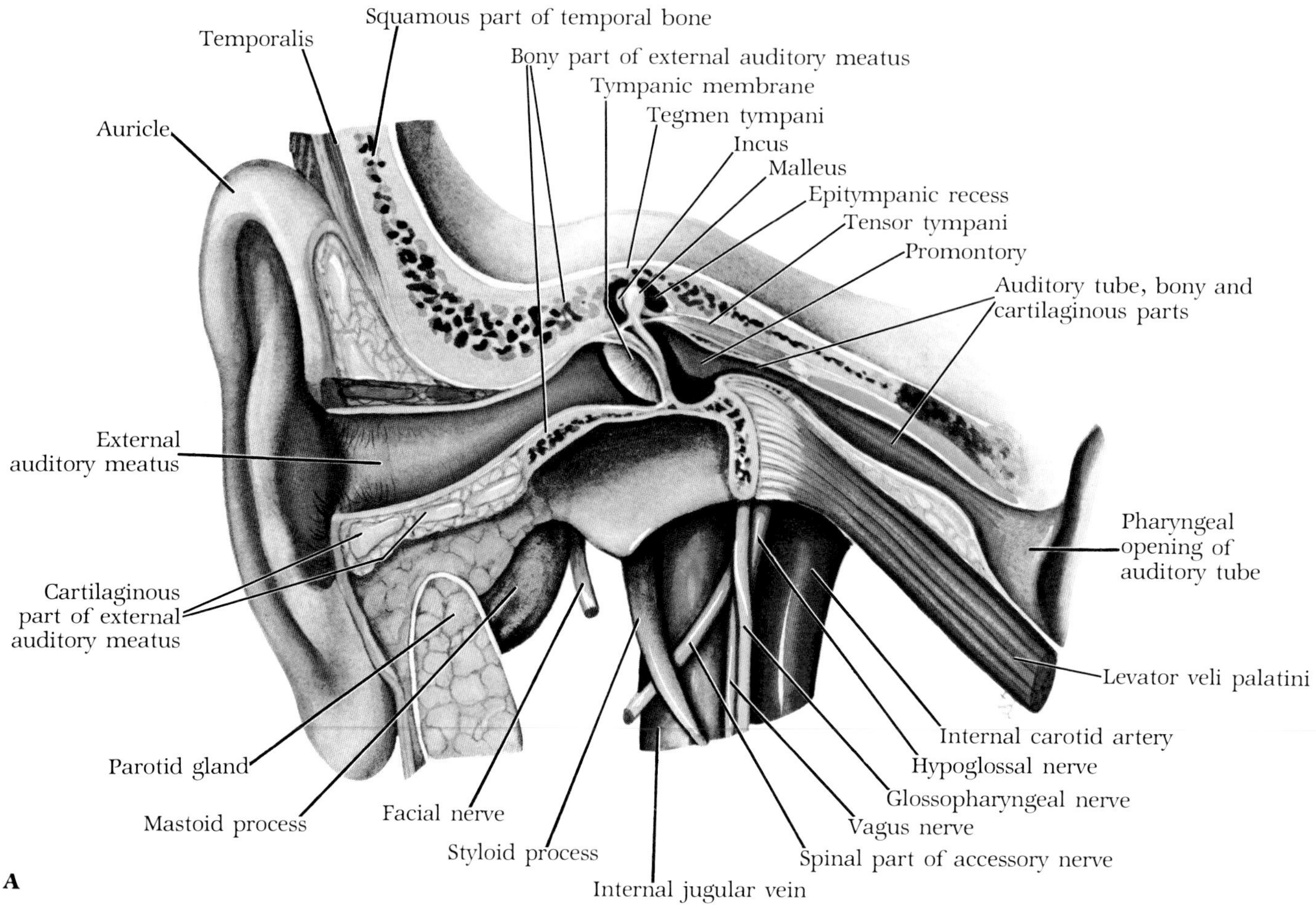
Squamous part of temporal bone
Temporalis
Bony part of external auditory meatus
Tympanic membrane
Tegmen tympani
Incus
Malleus
Auricle
Epitympanic recess
Tensor tympani
Promontory
Auditory tube, bony and cartilaginous parts
External auditory meatus
Pharyngeal opening of auditory tube
Cartilaginous part of external auditory meatus
Levator veli palatini
Internal carotid artery
Hypoglossal nerve
Parotid gland
Glossopharyngeal nerve
Mastoid process
Facial nerve
Vagus nerve
Styloid process
Spinal part of accessory nerve
Internal jugular vein

A

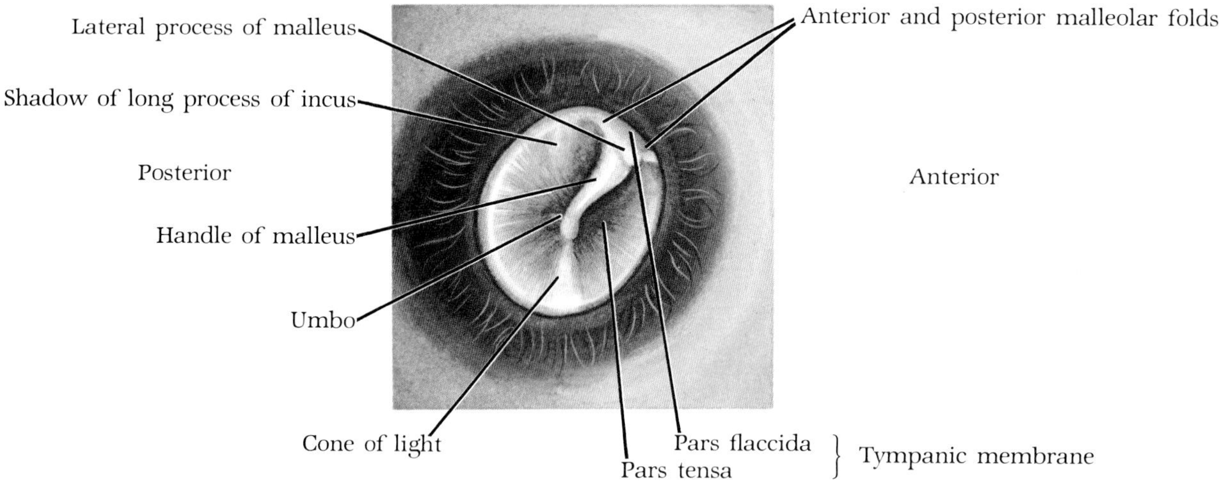
Lateral process of malleus
Anterior and posterior malleolar folds
Shadow of long process of incus
Posterior
Anterior
Handle of malleus
Umbo
Cone of light
Pars flaccida
Pars tensa
Tympanic membrane

B

Figure 7-64
Right ear. A. Auricle, external auditory meatus, tympanic cavity, and auditory tube, anterior view. B. Tympanic membrane, lateral view.

1. There should be good illumination when the ear is examined. The patient is seated sideways to the clinician, who places himself opposite the ear to be examined and directs a strong light onto it. First the skin around the ear and over the mastoid process is examined for redness or swelling; then the auricle is carefully examined.

2. Auriscopic examination of the tympanic membrane is facilitated by first straightening the external auditory meatus by gently pulling the auricle upward and backward. In infants, because of the lack of development of the bony part of the external meatus, the auricle should be pulled downward and backward. A speculum should be used in patients with well-developed vibrissae, since these interfere with the view.

3. Normally the tympanic membrane is pearly gray in color and is highly polished. For purposes of description, the membrane is divided into quadrants by imaginary lines, one of which is dropped perpendicularly from the umbo, while the other bisects this line at right angles. It is important that the physician recognize the long and short processes of the malleus, the cone-shaped light reflex (due to light reflected from the surface of the membrane), and the anterior and posterior malleolar folds.

4. A very translucent normal tympanic membrane may enable the examiner to see the long process of the incus. At right angles to this and extending posteriorly may be seen the tendon of the stapedius muscle. The pale yellow color of the center of the membrane is due to the convex promontory showing through.

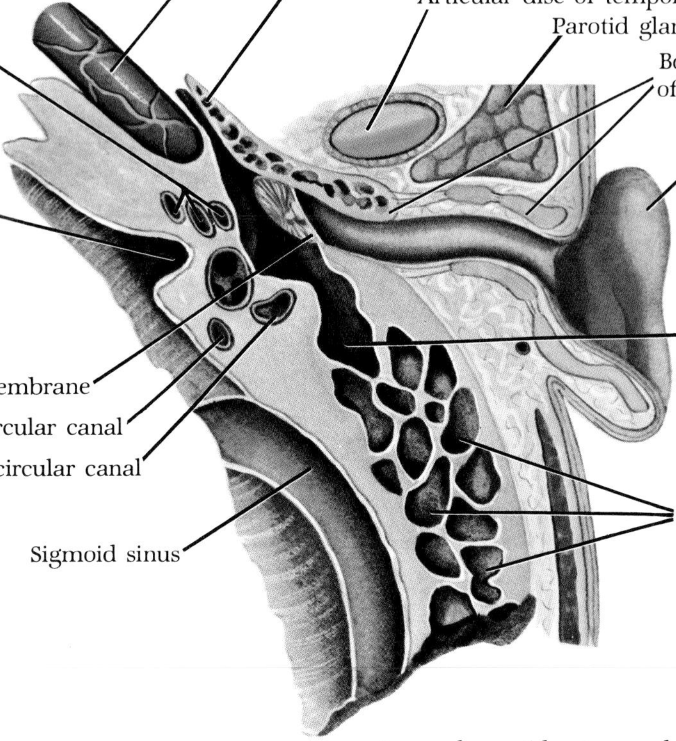
Internal carotid artery and carotid plexus
Bony part of auditory tube
Articular disc of temporomandibular joint
Parotid gland
Bony and cartilaginous parts of external auditory meatus
Auricle
Cochlea
Internal auditory meatus
Mastoid antrum
Tympanic membrane
Posterior semicircular canal
Lateral semicircular canal
Mastoid air cells
Sigmoid sinus

A

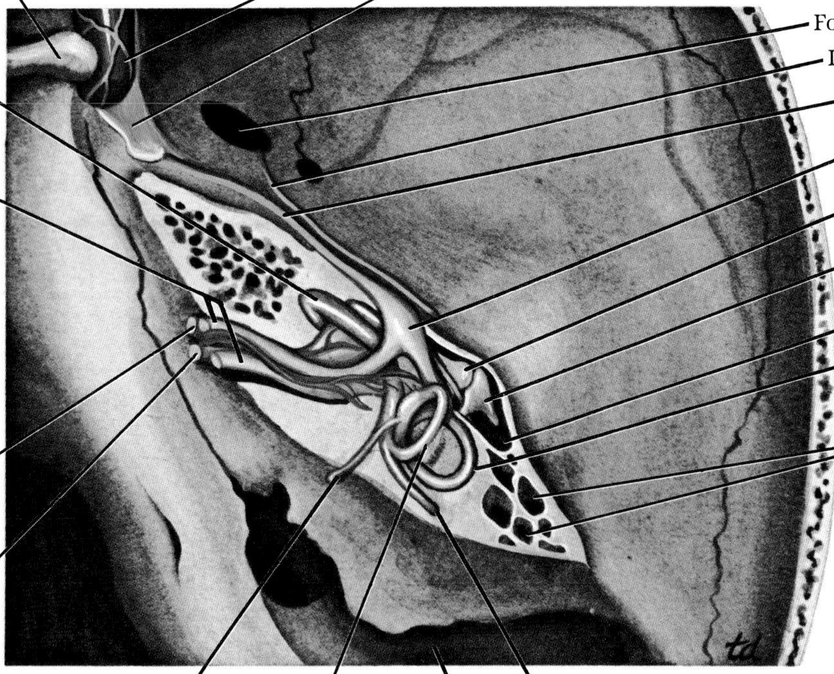
Internal carotid artery and carotid plexus
Cartilaginous plate filling in foramen lacerum
Dorsum sellae
Foramen ovale
Lesser petrosal nerve
Greater petrosal nerve
Cochlea
Geniculate ganglion of facial nerve
Facial nerve
Malleus
Incus
Mastoid antrum
Lateral semicircular canal
Mastoid air cells
Cochlear nerve
Vestibular nerve
Posterior semicircular canal
Groove for sigmoid sinus
Saccus endolymphaticus
Superior semicircular canal

B

Figure 7-65
Right ear. A. Horizontal section through the external auditory meatus, tympanic cavity, and auditory tube. B. Position of labyrinth in the skull.

1. Pathogenic organisms may gain entrance to the tympanic cavity by ascending through the auditory tube from the nasal part of the pharynx. Acute infection of the tympanic cavity (otitis media) produces bulging and redness of the tympanic membrane.
2. Inadequate treatment of otitis media may result in the spread of the infection into the mastoid antrum and the mastoid air cells (acute mastoiditis).
3. Further spread of organisms beyond the middle ear may involve the meninges and the temporal lobe of the brain that lie superiorly, producing meningitis and a cerebral abscess in the temporal lobe.
4. A spread of infection medially could cause a facial nerve palsy and labyrinthitis with vertigo.
5. Extension of the infection posteriorly beyond the mastoid antrum could cause venous thrombosis in the sigmoid sinus.
6. Neurofibromas of the vestibulocochlear nerve are relatively common. Because there are two parts to this nerve in the internal auditory meatus and the facial nerve is also present, the symptoms are referable to the vestibular, cochlear, and facial nerves. Unilateral deafness and tinnitus followed by dizziness, nystagmus, and facial paralysis are common. The paralysis is due to pressure of the expanding tumor on the facial nerve within the bony canal. With further expansion other cranial nerves become affected.

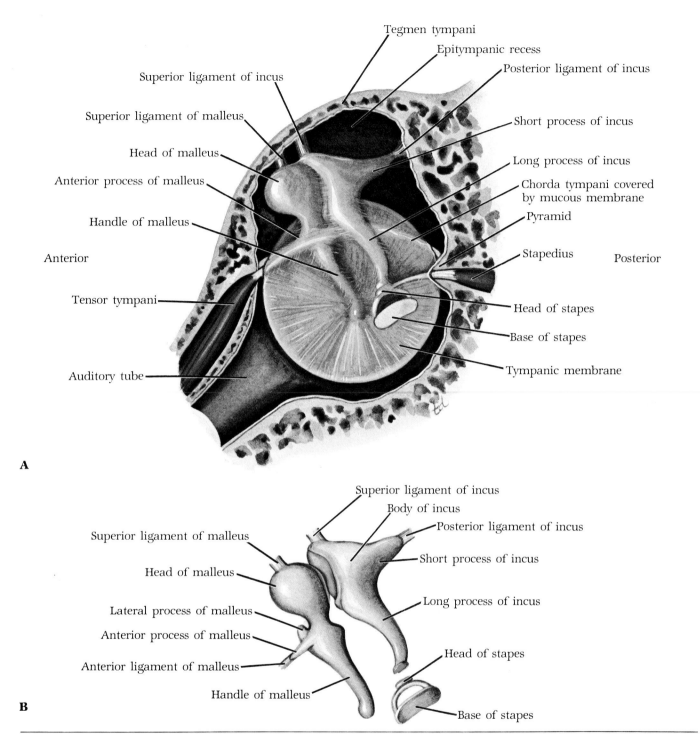

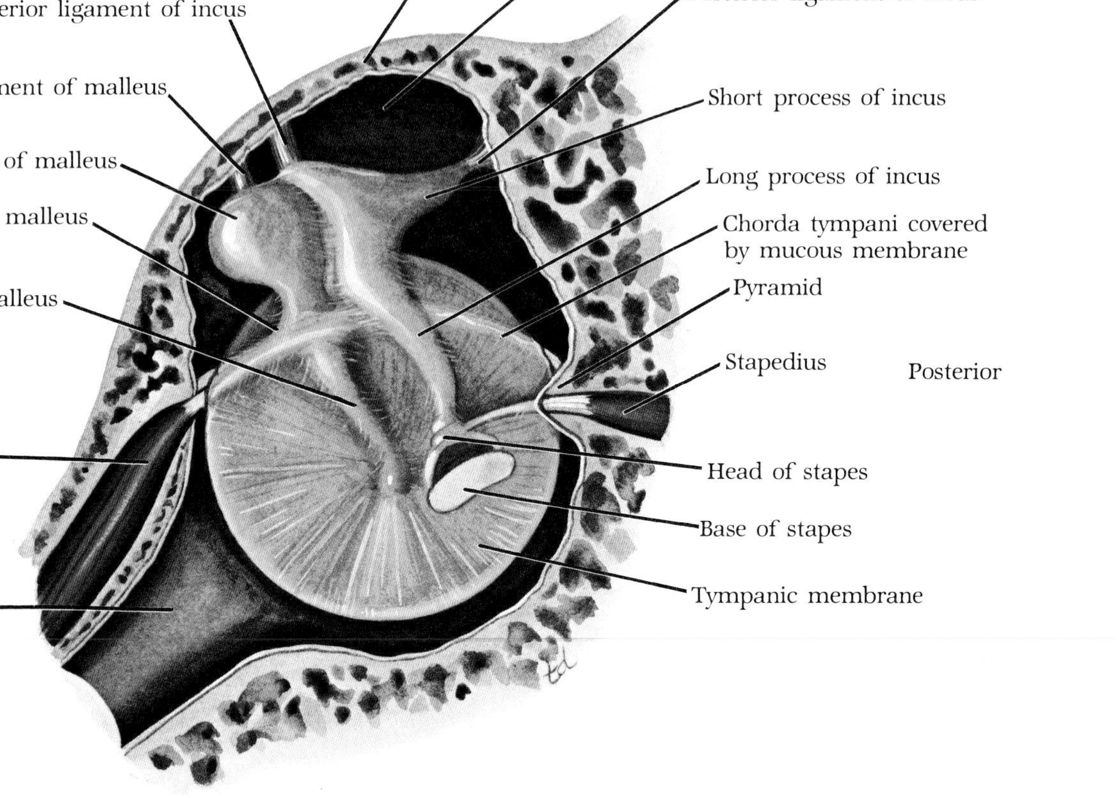
Tegmen tympani
Epitympanic recess
Superior ligament of incus
Posterior ligament of incus
Superior ligament of malleus
Short process of incus
Head of malleus
Long process of incus
Anterior process of malleus
Chorda tympani covered by mucous membrane
Pyramid
Handle of malleus
Stapedius
Anterior
Posterior
Tensor tympani
Head of stapes
Base of stapes
Tympanic membrane
Auditory tube

A

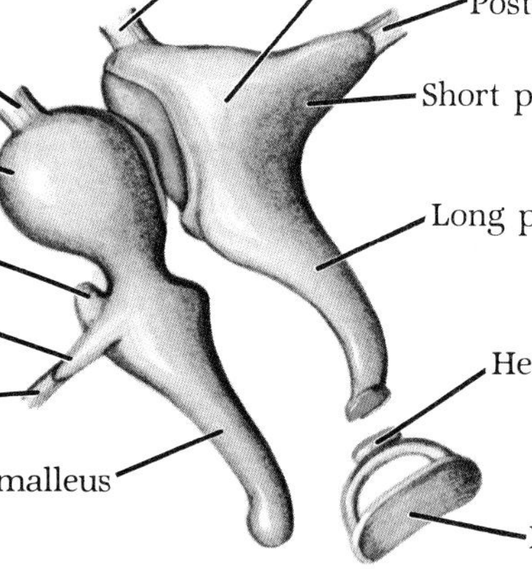
Superior ligament of incus
Body of incus
Posterior ligament of incus
Superior ligament of malleus
Short process of incus
Head of malleus
Long process of incus
Lateral process of malleus
Anterior process of malleus
Anterior ligament of malleus
Head of stapes
Handle of malleus
Base of stapes

B

Figure 7-66
Right ear. A. Tympanic membrane, malleus, incus, and stapes, medial view. B. Malleus, incus, and stapes, medial view.

1. The auricle serves to collect sound waves that are funneled into the external auditory meatus. The sound waves then strike the tympanic membrane and set it in motion. This vibratory movement is conveyed through the malleus and incus to the stapes. The arrangement of these small ossicles is such that the leverage has been increased by a ratio of 1:3 to 1. The area of the tympanic membrane is about 17 times greater than the base of the stapes, so that the effective pressure on the perilymph has been increased by a total of 22 to 1.
2. The stapedius and tensor tympani muscles diminish the sensitivity of the ear by damping down the vibrations of the ossicles. It is in this manner that the muscles reflexly prevent excessive vibration of the ossicles.
3. Impacted wax, foreign bodies in the external auditory meatus, and exostoses in the external meatus will all interfere with the conduction of sound waves to the tympanic membrane, producing symptoms of deafness. Damage to the tympanic membrane with a pointed object, e. g., a knitting needle, or rupture of the membrane by the force of a blast at the time of an explosion will cause hearing loss. Immobilization of the ossicles by an accumulation of exudate in the tympanic cavity in otitis media can also produce deafness.

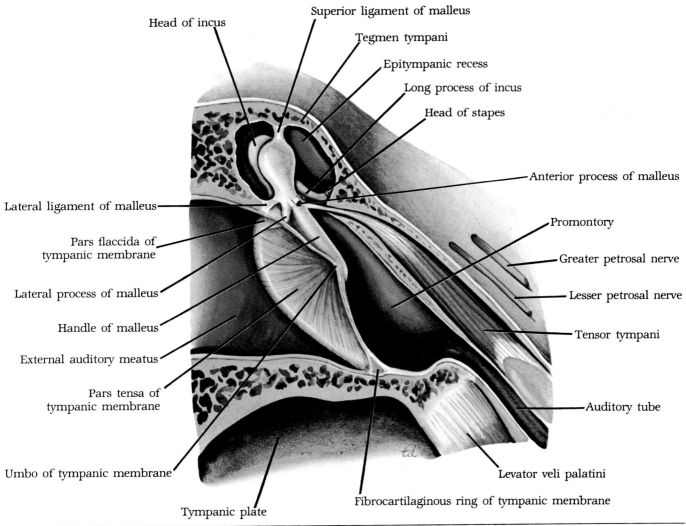

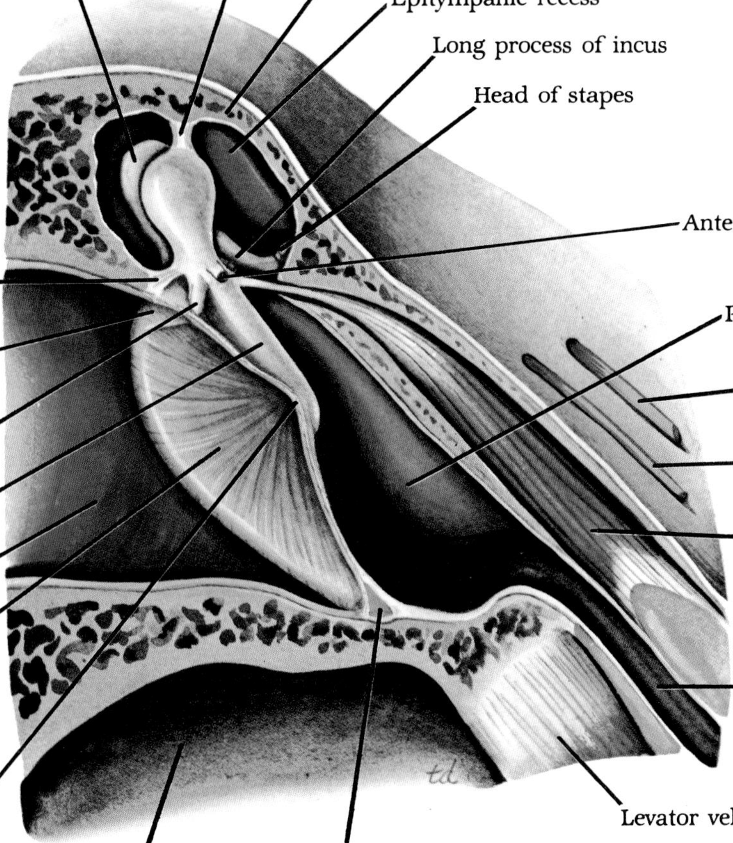
Head of incus
Superior ligament of malleus
Tegmen tympani
Epitympanic recess
Long process of incus
Head of stapes
Anterior process of malleus
Lateral ligament of malleus
Promontory
Pars flaccida of tympanic membrane
Greater petrosal nerve
Lateral process of malleus
Lesser petrosal nerve
Handle of malleus
Tensor tympani
External auditory meatus
Pars tensa of tympanic membrane
Auditory tube
Umbo of tympanic membrane
Levator veli palatini
Fibrocartilaginous ring of tympanic membrane
Tympanic plate

Figure 7-67
Right ear in coronal section, showing parts of the external auditory meatus, tympanic cavity, and auditory tube, anterior view.

1. The thin skin that lines the external auditory meatus is closely adherent to the underlying cartilage and bone, and for this reason inflammatory conditions, e. g., boils, are extremely painful in this area.

2. Sebaceous glands associated with hair follicles and ceruminous glands are present in that region of the skin lining the cartilaginous part of the external meatus. Boils in the meatus are due to infection of these hair follicles or sebaceous glands.

3. The close relationship of the pharyngeal opening of the auditory tube to the nose and pharyngeal lymphoid tissue means that infections in these areas can easily spread to the tympanic cavity via the lumen of the tube. This is especially likely to occur in young children, in whom the tube is shorter and wider than in the adult.

4. The base of the stapes is fixed to the margin of the fenestra vestibuli by the anular ligament. Movement of the base of the stapes within the fenestra vestibuli is not a piston-type motion, as was formerly believed, but is rather a rocking motion. In a common disease known as otosclerosis the base of the stapes becomes fixed by the formation of new bone. The disease is most common in women and occurs frequently in early adult life. The symptom is a deafness that becomes progressively worse.

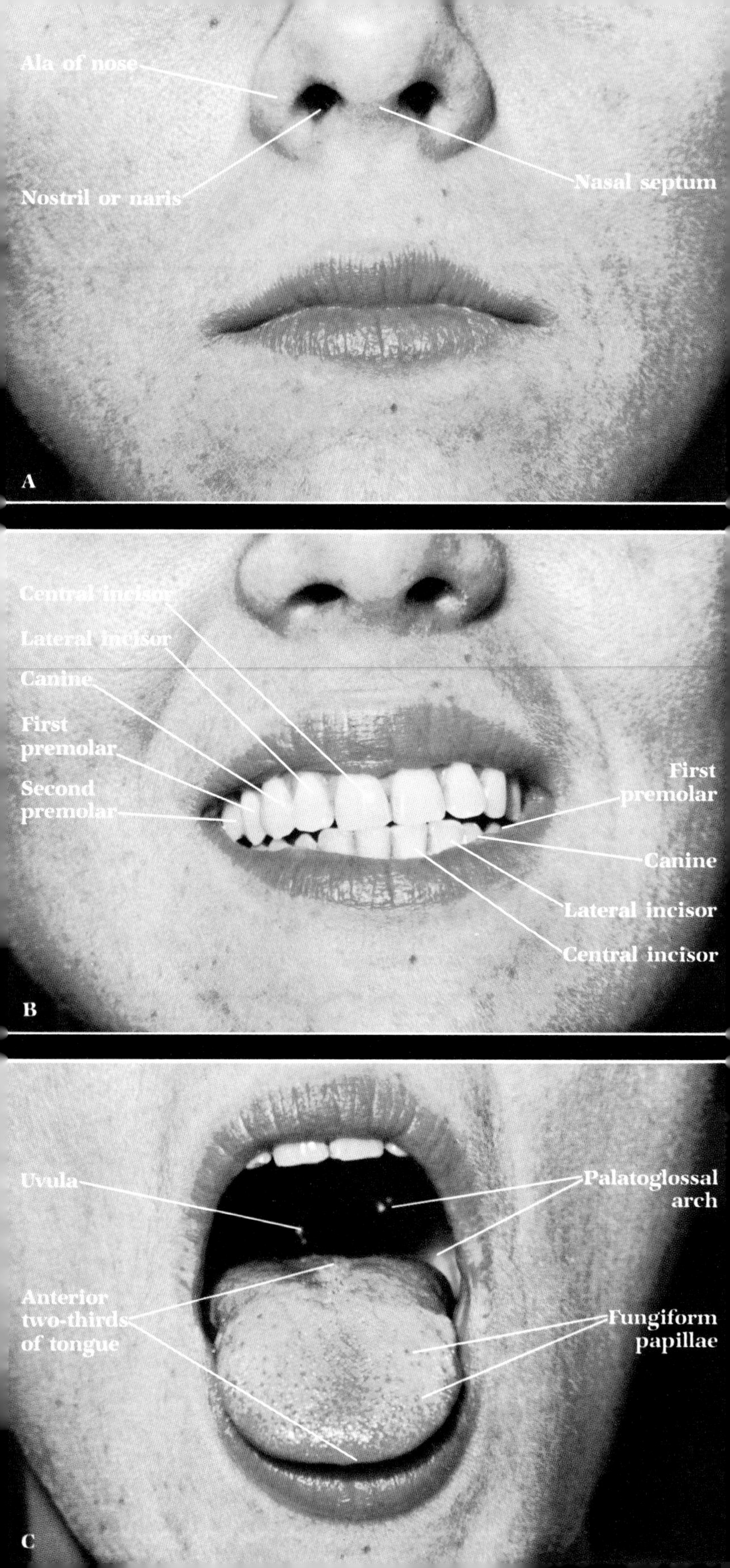
Ala of nose
Nostril or naris
Nasal septum
A
Central incisor
Lateral incisor
Canine
First premolar
Second premolar
First premolar
Canine
Lateral incisor
Central incisor
B
Uvula
Palatoglossal arch
Anterior two-thirds of tongue
Fungiform papillae
C

Figure 7-68
Perioral area of a 29-year-old female. A. External nares and the mouth closed. B. With the teeth in apposition and the lips separated. C. The mouth open and the tongue partly protruded.

1. Cleft upper lip, a congenital anomaly, may be confined to the lip or may be associated with a cleft palate. It is usually unilateral although it is sometimes bilateral. It is caused by failure of the maxillary process to fuse with the median nasal process.

2. The site and shape of a patient's nose should be noted. The nose is large in a patient with acromegaly and tends to be wider in one with hypothyroidism. In a patient with congenital syphilis the bridge of the nose is depressed.
3. Examination of the mouth begins with the lips. The normal color is vermilion, and there is vertical linear marking present in the young. With severe anemia the lips show pallor, and in patients with pernicious anemia they may have a lemon-yellow tint. Cyanosis, a blue coloration of the mucous membrane or skin, is due to hypoxia or the presence of abnormal hemoglobins in the capillaries.
4. Fissuring of the lips at the angles of the mouth can be caused by *Candida albicans*, iron-deficiency anemia, and vitamin B_{12} deficiency. A primary syphilitic chancre usually occurs on the upper lip; radiating white scars seen in the corner of the mouth are usually due to old syphilitic infections. Carcinomatous ulcers of the lips occur.
5. A patient's teeth should be examined and their condition noted. Abnormally formed teeth occur in congenital syphilis.
6. The gums should be bright pink in color, firm, and adherent to the necks of the teeth. Inflamed gums bleed easily, are pale in anemia, and show a blue line with lead poisoning.
7. The tongue is normally moist and pinkish in color. Its upper surface is velvety in appearance due to the many epithelial papillae.

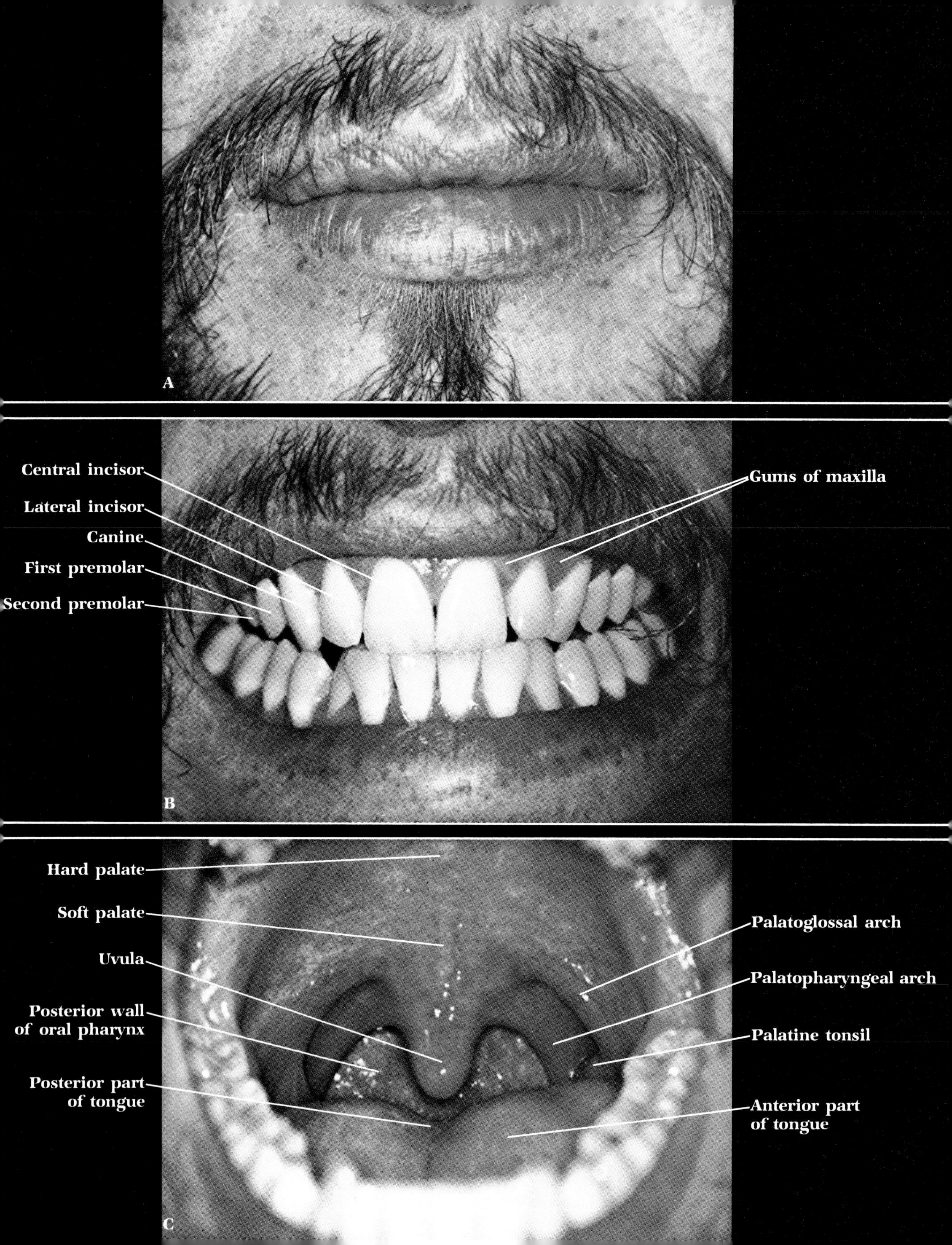
A
Central incisor
Lateral incisor
Canine
First premolar
Second premolar
Gums of maxilla
B
Hard palate
Soft palate
Uvula
Posterior wall
of oral pharynx
Posterior part
of tongue
Palatoglossal arch
Palatopharyngeal arch
Palatine tonsil
Anterior part
of tongue
C

Figure 7-69
Area of the mouth of a 27-year-old male. A. With the mouth closed. B. With the teeth in apposition and the lips separated. C. With the mouth open to reveal the oral part of the pharynx.

1. Examination of the mouth begins with the lips. Note the color and the presence of crusts, ulcers, or fissures. Palpate the lips between finger and thumb for the presence of indurated areas.
2. The patient is asked to separate his lips, keeping his upper and lower teeth in apposition. Inspect the teeth. Are they present in the correct numbers? Which teeth are missing? Are any teeth broken, loose, or obviously decayed? Inspect the gums for color and firmness.
3. The patient is now asked to open his mouth wide, keeping his tongue at first within the mouth. With a flashlight and spatula examine the inner surfaces of the cheek. Note the color of the mucous membrane; normally it is pink and shiny. Now extend the patient's head and examine the hard and soft palate.
4. Examine the tongue. Normally it is moist and pinkish in color. Its upper surface is velvety in appearance due to the many epithelial papillae. The tip and sides of the tongue should be smooth and free from ulcers.
5. Ask the patient to put out his tongue. Note the size, mobility, and consistency of the organ. Does the organ protrude straight out in the midline? Deviation to one or the other side may be due to a lesion of the hypoglossal nerve or fixation of part of the tongue by a malignant neoplasm.
6. Examination of the back of the mouth and pharynx reveals the posterior part of the tongue inferiorly, and the palatoglossal, the palatopharyngeal arches, and the palatine tonsil laterally. The soft palate moves upward when the patient says "ah" and the uvula moves backward in the midline.
7. Always conclude an examination of the mouth area by palpation of the submental and submandibular lymph nodes. Inflammatory and malignant diseases rapidly spread to involve these nodes.

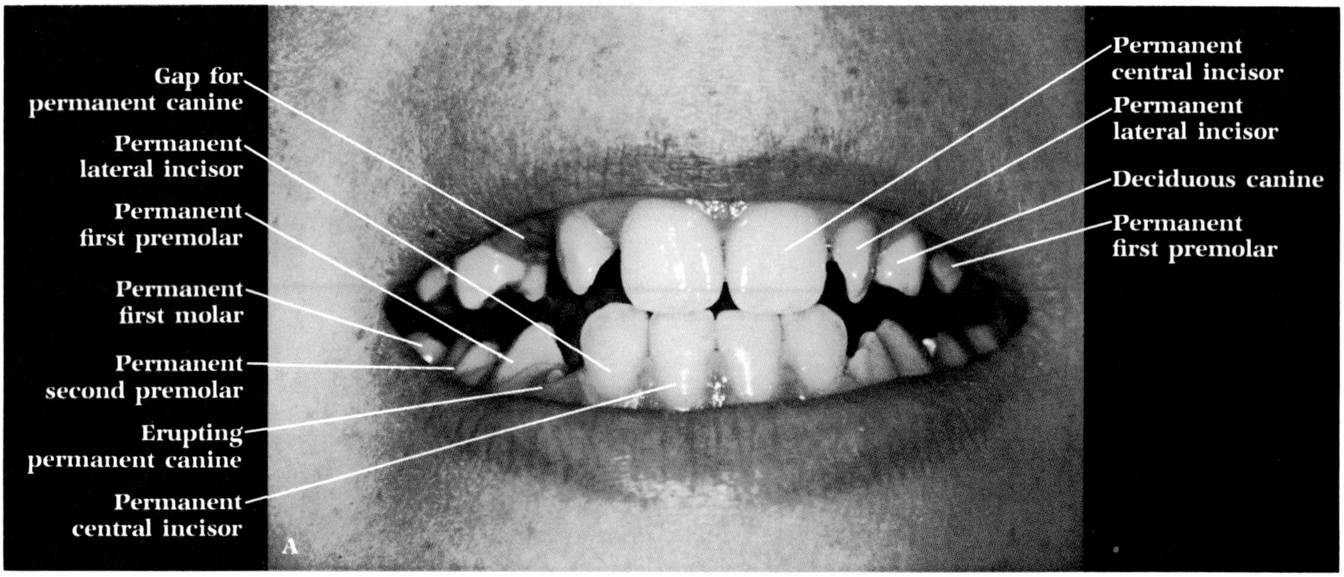
Gap for permanent canine
Permanent lateral incisor
Permanent first premolar
Permanent first molar
Permanent second premolar
Erupting permanent canine
Permanent central incisor
Permanent central incisor
Permanent lateral incisor
Deciduous canine
Permanent first premolar
A

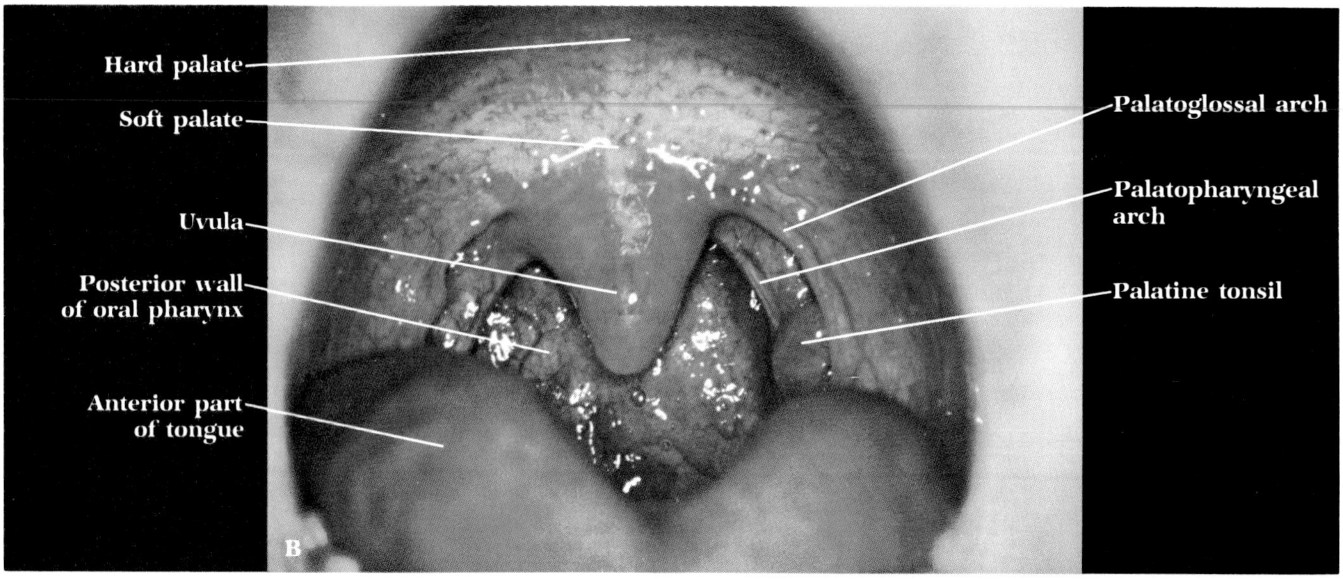
Hard palate
Soft palate
Uvula
Posterior wall of oral pharynx
Anterior part of tongue
Palatoglossal arch
Palatopharyngeal arch
Palatine tonsil
B

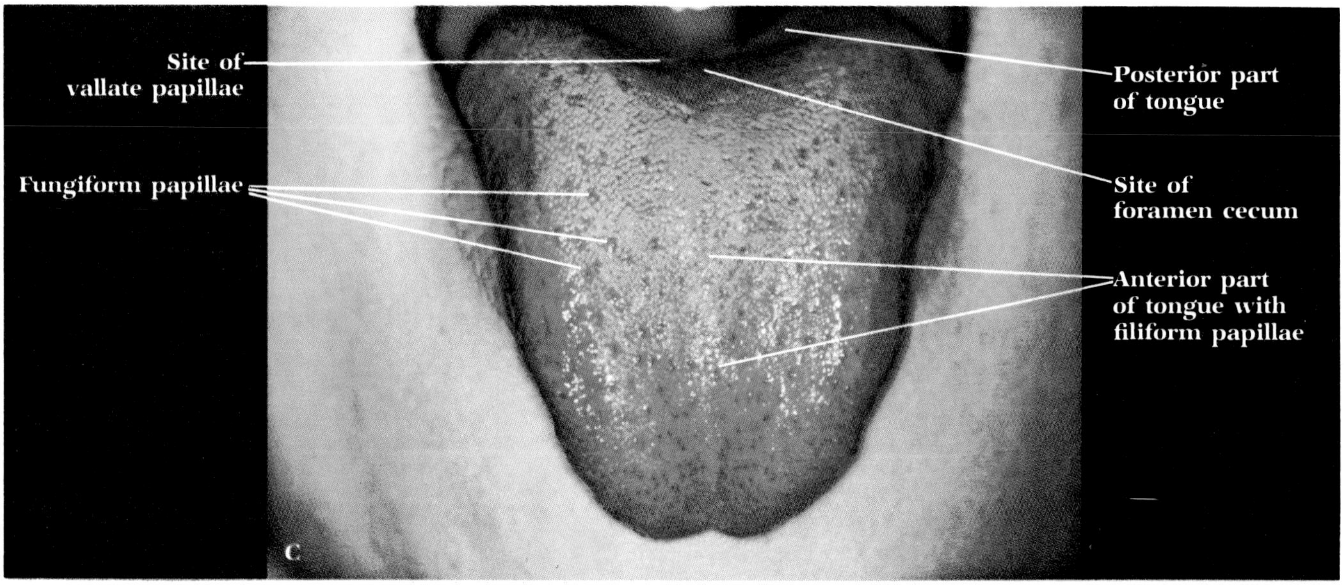
Site of vallate papillae
Fungiform papillae
Posterior part of tongue
Site of foramen cecum
Anterior part of tongue with filiform papillae
C

Figure 7-70
Lips, mouth, and tongue of a 10-year-old boy. A. With the teeth in apposition and the lips separated. B. With the mouth open, revealing the oral part of the pharynx. C. With the tongue protruding.

1. Each person has two sets of teeth, which make their appearance at different times of life. The first set, called the *deciduous teeth*, is temporary; the second set is called the *permanent teeth.* All physicians should have a general knowledge of the times of eruption of the teeth in the two sets.

2. The deciduous teeth are 20 in number: four incisors, two canines, and four molars in each jaw. They begin to erupt at about the sixth month and have all erupted by the end of the second year. The teeth of the lower jaw usually appear before those of the upper jaw.

3. The permanent teeth are 32 in number: four incisors, two canines, four premolars, and six molars in each jaw. They begin to erupt at the sixth year. The last tooth to erupt is the third molar, and this may take place anywhere between the seventeenth and thirtieth years. The teeth of the lower jaw usually appear before those of the upper jaw.

4. Examination of the back of the mouth and pharynx is sometimes difficult, especially with children. Ask the patient to open his mouth and keep the tongue relaxed and in the floor of the mouth. Use the tongue depressor gently and press downward on the anterior two-thirds of the tongue. Keep the tongue depressor away from the posterior third of the tongue, since this part receives its sensory nerve supply from the glossopharyngeal nerve. Once this nerve is stimulated, the gag reflex will be initiated and the patient will not be able to cooperate.

5. The palatine tonsils reach their maximum normal size in early childhood. After puberty they gradually atrophy, along with other lymphoid tissue in the body.

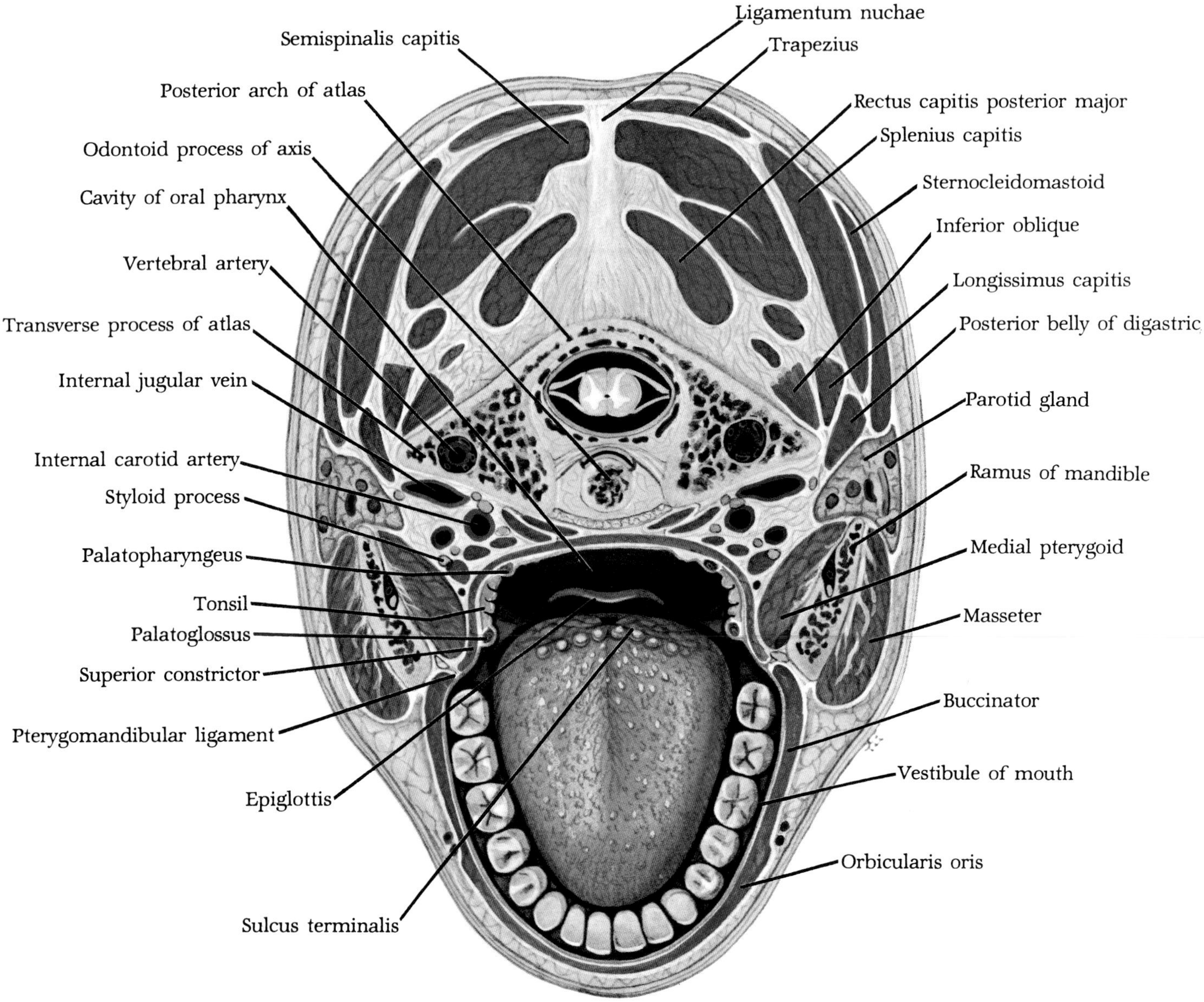
Ligamentum nuchae
Semispinalis capitis
Trapezius
Posterior arch of atlas
Rectus capitis posterior major
Splenius capitis
Odontoid process of axis
Sternocleidomastoid
Cavity of oral pharynx
Inferior oblique
Vertebral artery
Longissimus capitis
Transverse process of atlas
Posterior belly of digastric
Internal jugular vein
Parotid gland
Internal carotid artery
Ramus of mandible
Styloid process
Palatopharyngeus
Medial pterygoid
Tonsil
Masseter
Palatoglossus
Superior constrictor
Buccinator
Pterygomandibular ligament
Vestibule of mouth
Epiglottis
Orbicularis oris
Sulcus terminalis

Figure 7-71
Horizontal section through the neck at the level of the atlas, showing the mouth and oral part of the pharynx.

1. Acute retropharyngeal abscess is due to infection of the prevertebral lymph nodes and is common in young children. The sites of entry of the organisms are the tonsils and pharynx. In this condition the posterior pharyngeal wall bulges forward into the cavity of the pharynx. Chronic retropharyngeal abscess may be caused by a tuberculous infection spreading from a cervical vertebra.

2. Acute tonsillitis with sore throat and pyrexia is a common disease. The infection usually spreads rapidly to involve the "tonsillar lymph node," which is a member of the deep cervical group of nodes situated close to the angle of the mandible. Chronic tonsillitis may result from repeated attacks of acute tonsillitis and is usually accompanied by hypertrophy; tonsillectomy may be required as treatment of this condition.

3. The palatine tonsil has a profuse blood supply from the descending palatine, ascending palatine, ascending pharyngeal, and lingual, and, most important of all, the facial artery. Serious venous bleeding may occur following a tonsillectomy operation.

4. Peritonsillar abscess (quinsy) is due to spread of infection from the palatine tonsil to the loose connective tissue outside the capsule of the tonsil.

5. Note the close relationship that exists between the lingual nerve and the lower third molar tooth. A clumsy extraction of an impacted lower third molar could damage this nerve.

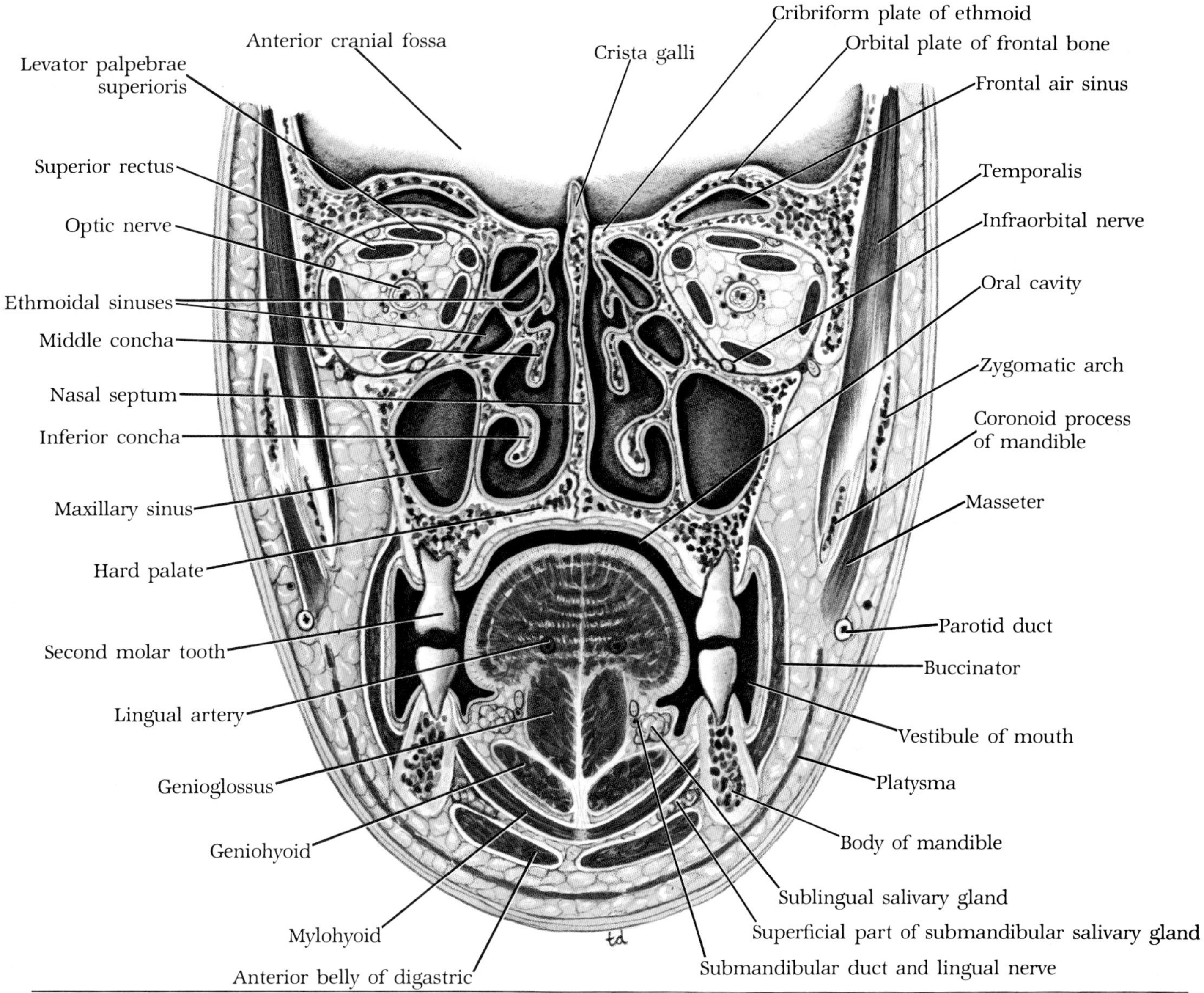
Cribriform plate of ethmoid
Anterior cranial fossa
Crista galli
Orbital plate of frontal bone
Levator palpebrae superioris
Frontal air sinus
Superior rectus
Temporalis
Optic nerve
Infraorbital nerve
Ethmoidal sinuses
Oral cavity
Middle concha
Zygomatic arch
Nasal septum
Coronoid process of mandible
Inferior concha
Masseter
Maxillary sinus
Hard palate
Parotid duct
Second molar tooth
Buccinator
Lingual artery
Vestibule of mouth
Genioglossus
Platysma
Geniohyoid
Body of mandible
Sublingual salivary gland
Mylohyoid
Superficial part of submandibular salivary gland
Anterior belly of digastric
Submandibular duct and lingual nerve
td

Figure 7-72
Coronal section through the head, showing the nose, paranasal sinuses, and mouth cavity.

1. Infection of the paranasal sinuses usually follows an upper respiratory infection. The infection spreads from the nasal cavity and most commonly first involves the maxillary sinus. Maxillary sinusitis may also occur secondary to a peridontal abscess.

2. The close relationship of the ethmoidal sinuses to the orbit should be noted. An acute ethmoidal sinusitis may spread to the orbit, causing orbital cellulitis and thrombosis of the orbital veins.

3. The frontal, ethmoidal, and maxillary sinuses may be palpated clinically for areas of tenderness. The frontal sinus may be examined by pressing the finger upward beneath the medial end of the superior orbital margin. It is here that the floor of the frontal sinus is closest to the surface. The ethmoidal sinuses may be palpated by pressing the finger medially against the medial wall of the orbit. The maxillary sinus may be examined for tenderness by pressing the finger against the anterior wall of the maxilla below the inferior orbital margin; pressure over the infraorbital nerve may reveal increased sensitivity.

4. The mucous membrane of the tongue may atrophy in certain deficiency diseases, as for example, in vitamin B complex deficiency, hypochromic anemia, and pernicious anemia.

5. The close relationship of the submandibular duct to the floor of the mouth may enable one to palpate a calculus within the duct in cases of periodic swelling of the submandibular salivary gland.

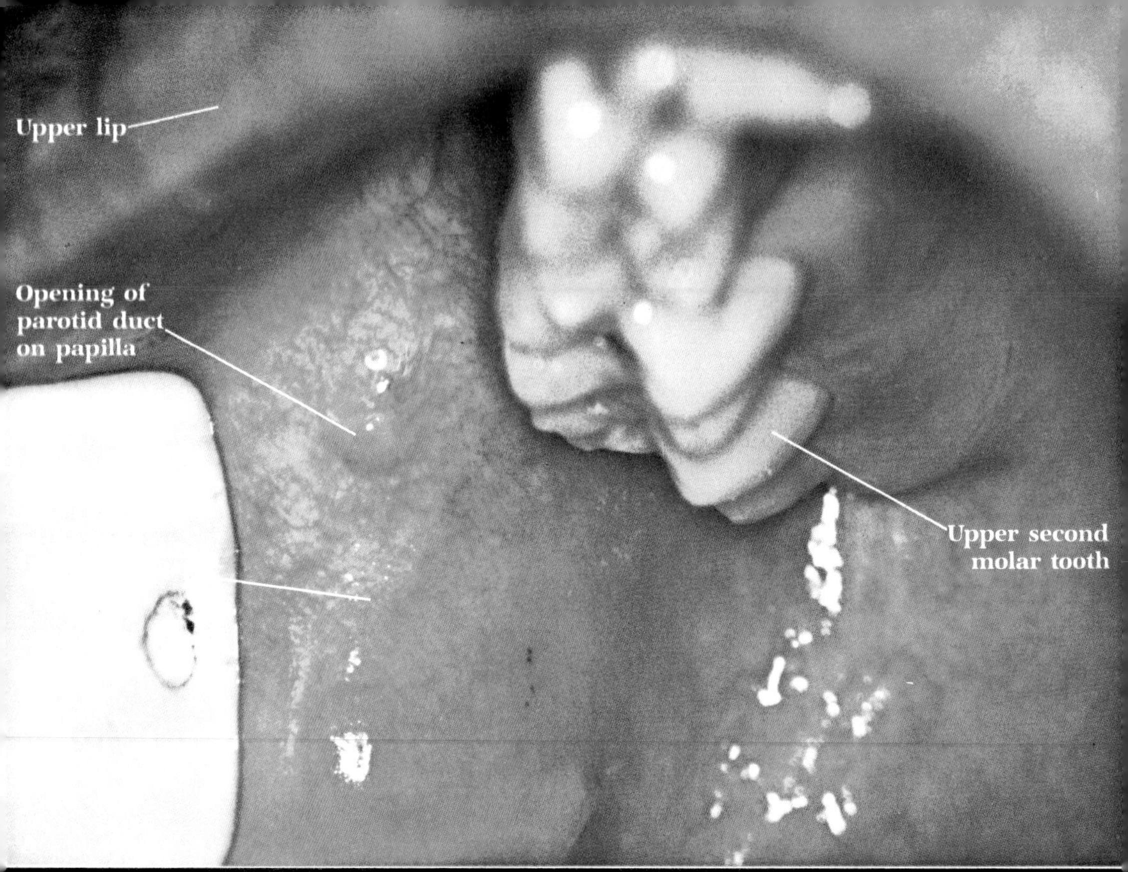
Upper lip
Opening of parotid duct on papilla
Upper second molar tooth

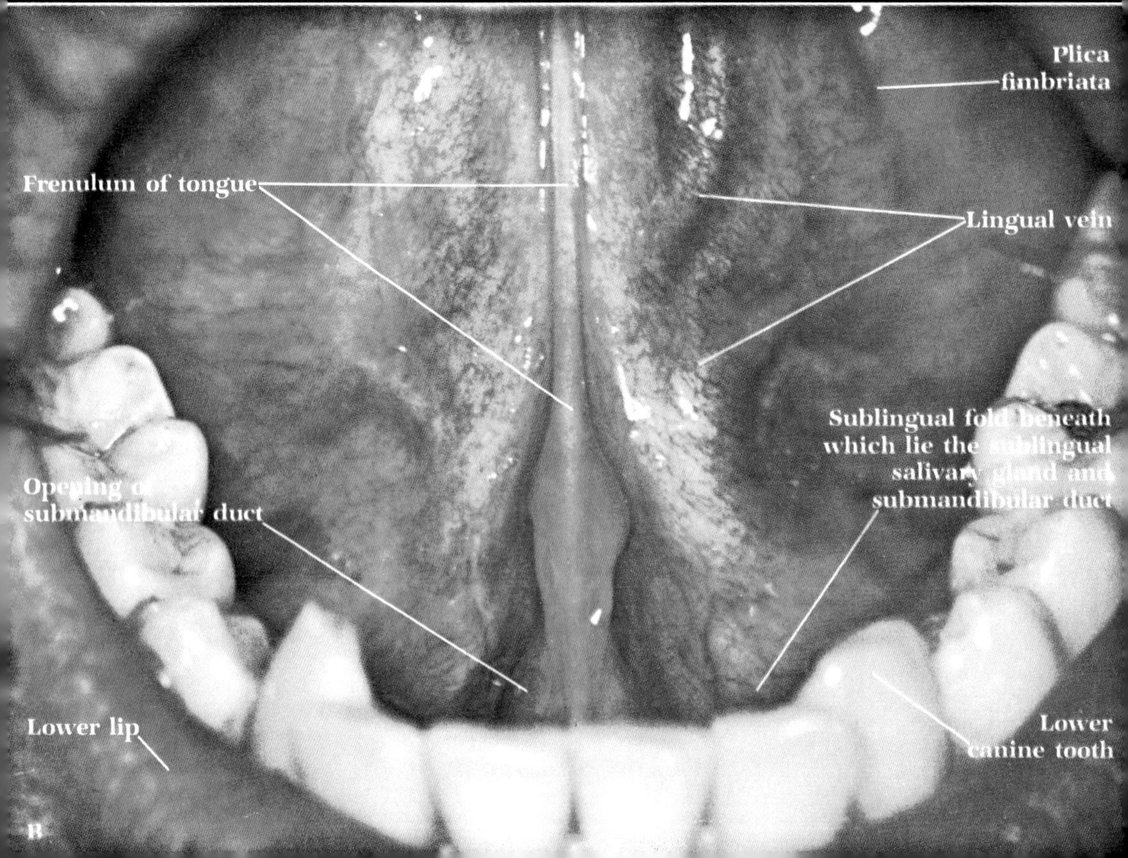
Plica fimbriata
Frenulum of tongue
Lingual vein
Sublingual fold beneath which lie the sublingual salivary gland and submandibular duct
Opening of submandibular duct
Lower lip
Lower canine tooth
B

Figure 7-73
The interior of the mouth of a 27-year-old white male. A. The mucous membrane of the right cheek. B. The inferior surface of the tongue and the adjacent part of the floor of the mouth.

1. The parotid duct runs forward from the parotid gland one fingerbreadth below the zygomatic arch. At the anterior border of the masseter muscle, it turns medially and opens into the mouth opposite the upper second molar tooth. The orifice of the parotid duct may be visualized by gently pushing the cheek laterally with a spatula and viewing the mucous membrane with a flashlight.

2. In patients with acute parotitis there may be some reddening of the mucous membrane around the orifice of the duct, and pressure on the inflamed gland may cause purulent saliva to exude from the duct.

3. The color of the mucous membrane of the cheek is normally pinkish-red. In Addison's disease, however, brownish areas of pigmentation may be present.

4. On the second day of measles, before the generalized skin rash appears, small bluish white spots—Koplik's spots—may be seen on the mucous membrane of the cheek.

5. The deep part of the submandibular gland, the submandibular duct, and the sublingual gland may be palpated through the mucous membrane covering the floor of the mouth in the interval between the tongue and the mandible. The submandibular duct opens into the mouth on the side of the frenulum of the tongue. The close relationship of the submandibular duct to the floor of the mouth may enable one to palpate a calculus in the duct in cases of periodic swelling of the submandibular salivary gland.

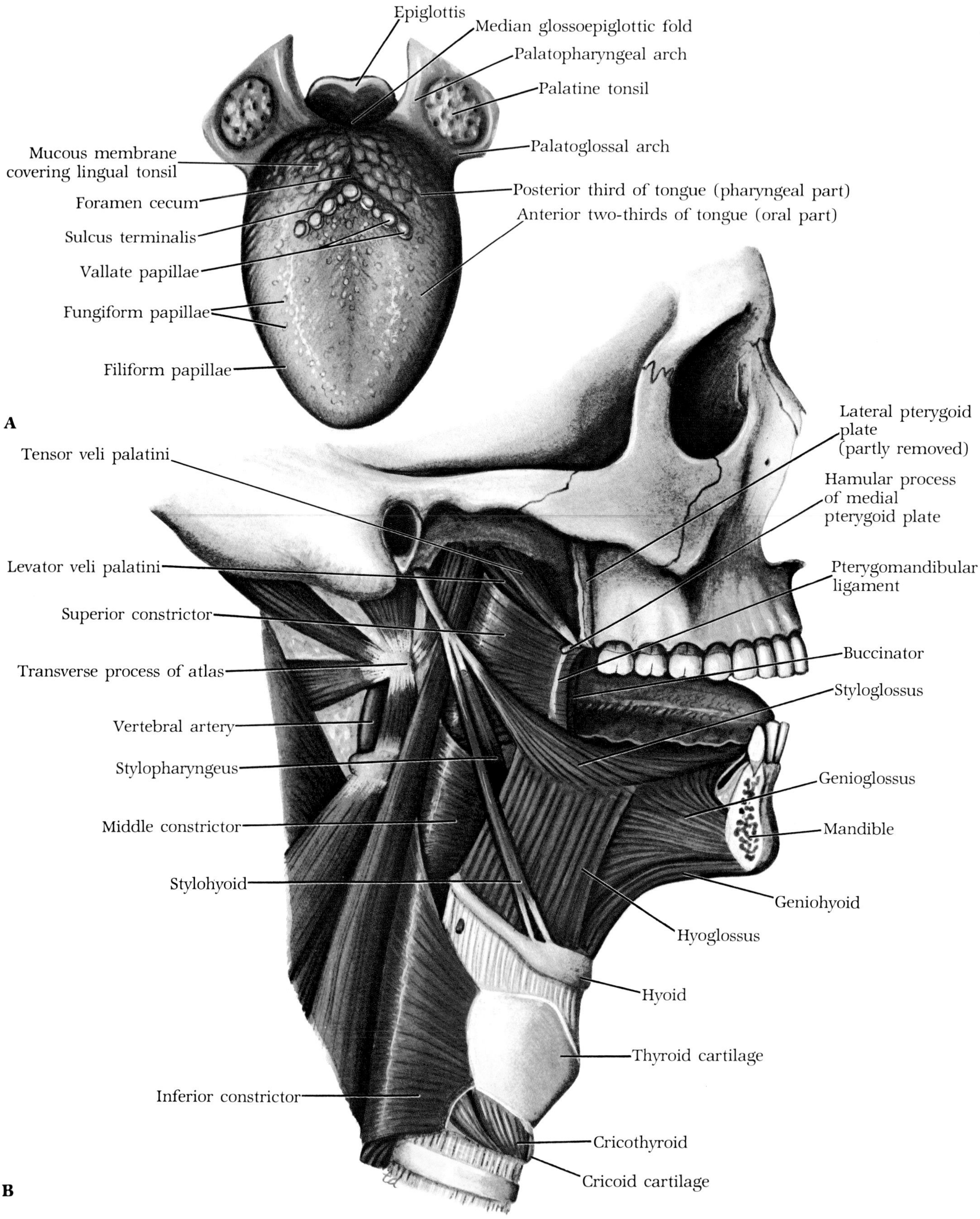
Epiglottis
Median glossoepiglottic fold
Palatopharyngeal arch
Palatine tonsil
Mucous membrane covering lingual tonsil
Palatoglossal arch
Foramen cecum
Posterior third of tongue (pharyngeal part)
Anterior two-thirds of tongue (oral part)
Sulcus terminalis
Vallate papillae
Fungiform papillae
Filiform papillae
A
Lateral pterygoid plate (partly removed)
Tensor veli palatini
Hamular process of medial pterygoid plate
Levator veli palatini
Pterygomandibular ligament
Superior constrictor
Buccinator
Transverse process of atlas
Styloglossus
Vertebral artery
Stylopharyngeus
Genioglossus
Middle constrictor
Mandible
Stylohyoid
Geniohyoid
Hyoglossus
Hyoid
Thyroid cartilage
Inferior constrictor
Cricothyroid
Cricoid cartilage
B

Figure 7-74
A. Dorsum of the tongue. B. Right side of neck, showing the muscular attachments of the tongue to the mandible, hyoid, and styloid process.

1. A wound of the tongue is often caused by a blow on the chin when the tongue is partly protruded from the mouth. It may also occur when a patient accidentally bites his tongue while eating, during recovery from an anesthetic, or during an epileptic attack. Bleeding is halted by grasping the tongue between finger and thumb posterior to the laceration, thus occluding the branches of the lingual artery.
2. The tongue is readily examined clinically. It may become furred in infections of the oral cavity, from tobacco smoking, or due to general dehydration. It also becomes dry in dehydration. Atrophy of the mucous membrane occurs in vitamin B complex deficiency, hypochromic anemia, and pernicious anemia.
3. The tongue may become enlarged (macroglossia) in such conditions as edema, as in wasp sting, or muscular hypertrophy, as in cretins; or due to tumors such as lymphangioma, hemangioma, or lingual thyroid.
4. Deviation of a protruded tongue may be caused by a lesion of the hypoglossal nerve or by extensive infiltration by a carcinoma on one side; in either case the tongue will be deviated toward the side of the lesion.
5. Carcinoma of the tongue, although less common than previously, still occurs. It tends to spread rapidly to regional lymph nodes. The tip of the tongue is drained by the submental nodes, the middle third of the tongue by the submandibular nodes, and the posterior third by the deep cervical nodes.

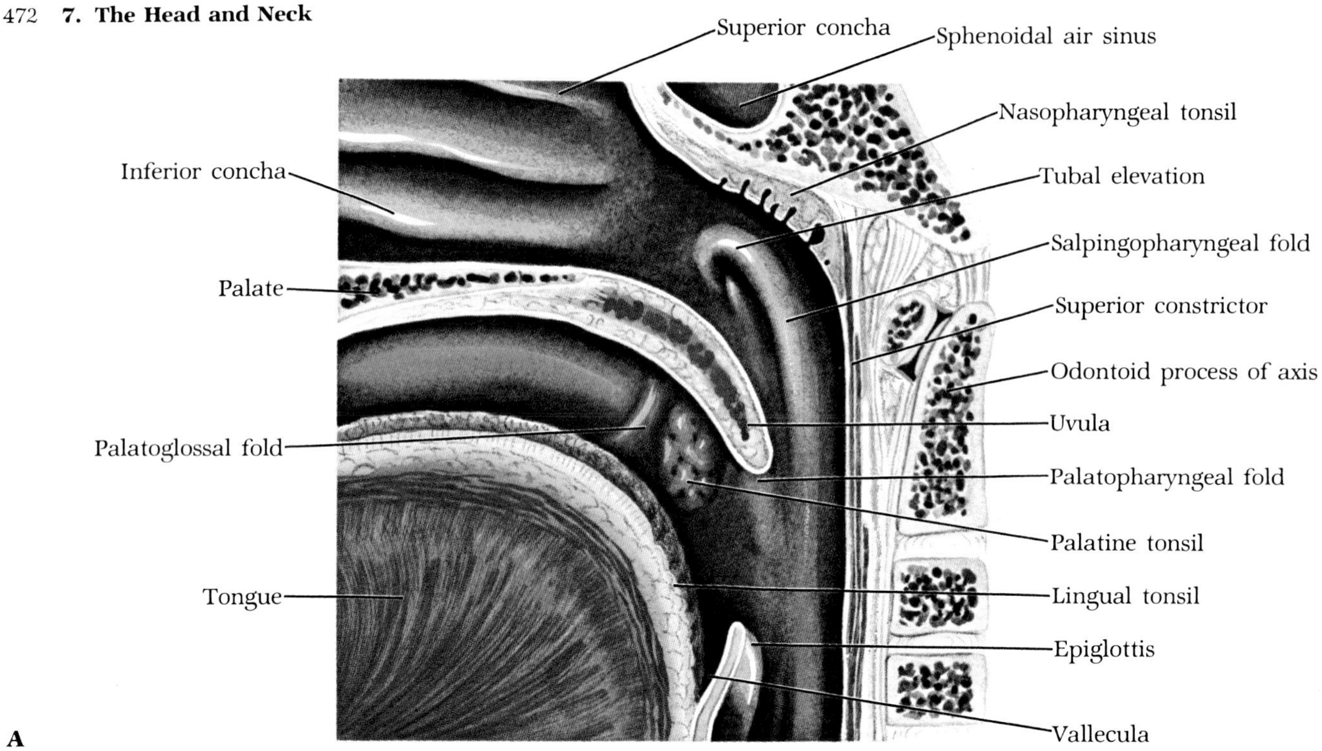
Superior concha
Sphenoidal air sinus
Nasopharyngeal tonsil
Inferior concha
Tubal elevation
Salpingopharyngeal fold
Palate
Superior constrictor
Odontoid process of axis
Uvula
Palatoglossal fold
Palatopharyngeal fold
Palatine tonsil
Tongue
Lingual tonsil
Epiglottis
Vallecula
A

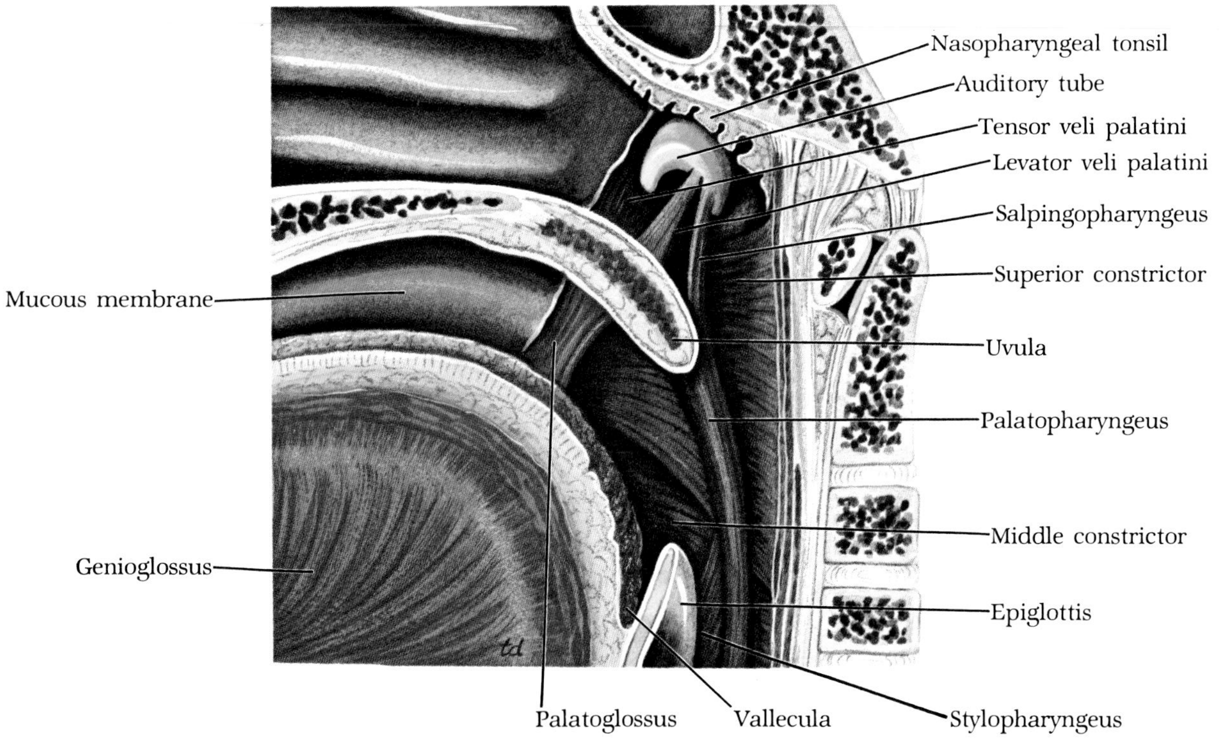
Nasopharyngeal tonsil
Auditory tube
Tensor veli palatini
Levator veli palatini
Salpingopharyngeus
Superior constrictor
Mucous membrane
Uvula
Palatopharyngeus
Middle constrictor
Genioglossus
Epiglottis
B
Palatoglossus
Vallecula
Stylopharyngeus

Figure 7-75
A. Junction of nose with nasal part of pharynx and mouth with oral part of pharynx. Note position of tonsil and opening of auditory tube. B. Muscles of soft palate and upper part of pharynx.

1. Cleft palate. Cleft palate is commonly associated with cleft upper lip. All degrees of cleft palate occur and all are caused by failure of the palatal processes of the maxilla to fuse with each other in the midline, and in severe cases, the failure of these processes also to fuse with the premaxilla. The first degree of severity is cleft uvula, the second degree is ununited palatal processes, and the third degree is ununited palatal processes and a cleft on one side of the premaxilla. The fourth degree of severity, which is rare, consists of ununited palatal processes and a cleft on both sides of the premaxilla.

2. In order for the physician to examine the palate, the patient should open his mouth to the maximum extent and extend his head backward. Movement of the soft palate is tested when the patient says "ah"; normally the soft palate rises and the uvula moves backward in the midline. Loss of movement of half the soft palate or a deviation of the uvula to one side would suggest a lesion of the vagus nerve or a neoplastic infiltration of the soft palate.

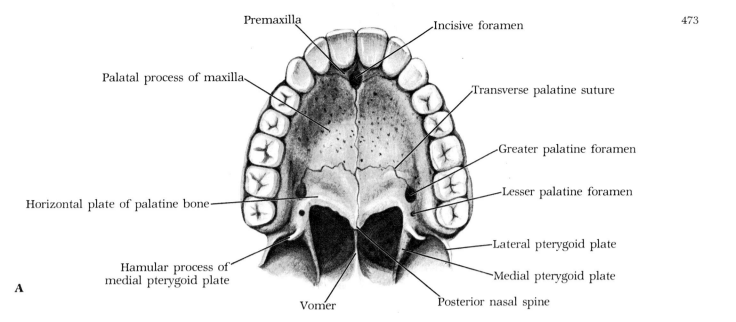

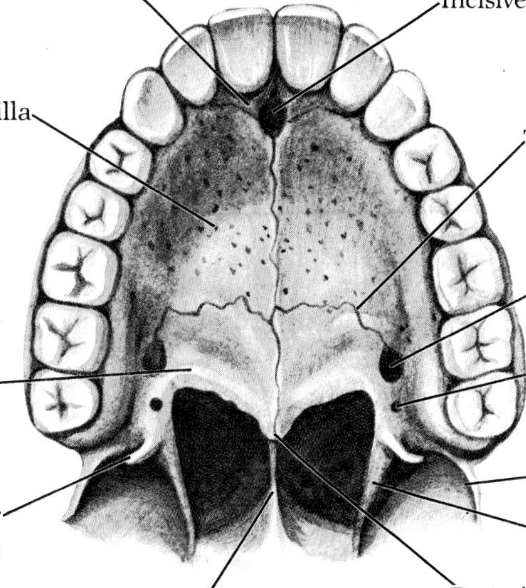
Premaxilla
Incisive foramen
Palatal process of maxilla
Transverse palatine suture
Greater palatine foramen
Lesser palatine foramen
Horizontal plate of palatine bone
Lateral pterygoid plate
Hamular process of
medial pterygoid plate
Medial pterygoid plate
Vomer
Posterior nasal spine

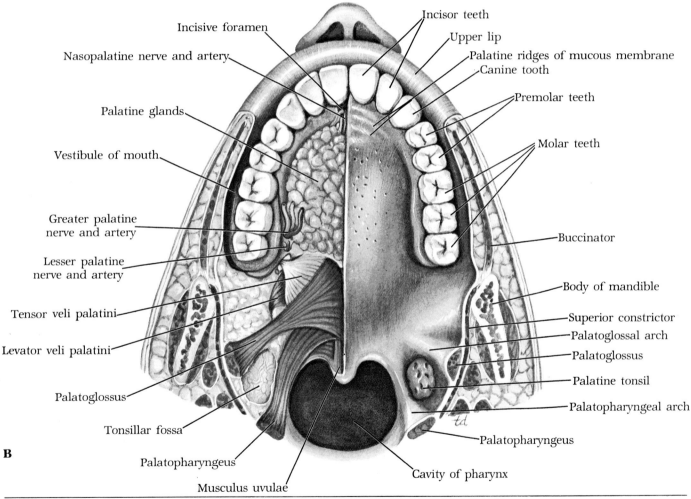

A

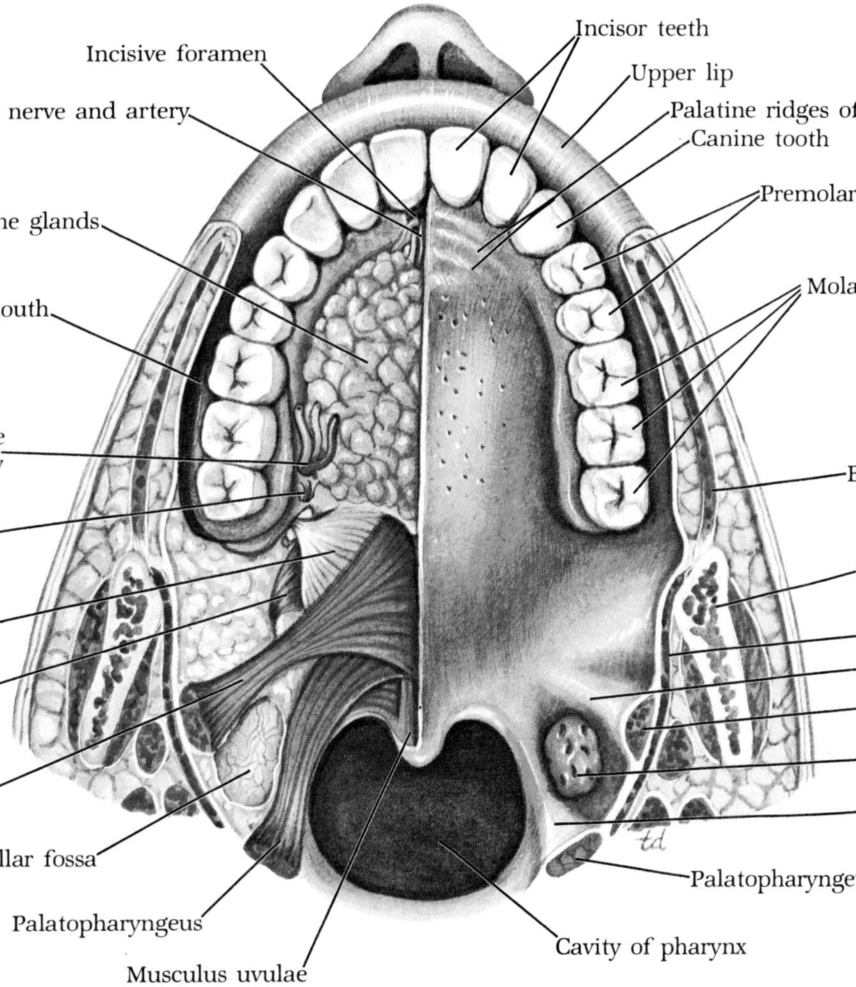
Incisor teeth
Incisive foramen
Upper lip
Nasopalatine nerve and artery
Palatine ridges of mucous membrane
Canine tooth
Premolar teeth
Palatine glands
Molar teeth
Vestibule of mouth
Greater palatine
nerve and artery
Buccinator
Lesser palatine
nerve and artery
Body of mandible
Tensor veli palatini
Superior constrictor
Palatoglossal arch
Levator veli palatini
Palatoglossus
Palatine tonsil
Palatoglossus
Palatopharyngeal arch
Tonsillar fossa
Palatopharyngeus
Palatopharyngeus
Cavity of pharynx
Musculus uvulae

B

Figure 7-76
Palate. A. Hard palate, inferior view. B. Horizontal section through the head and neck showing the inferior surface of the palate. On the left the mucous membrane has been removed to reveal the underlying structures.

1. During the process of swallowing, the nasal part of the pharynx is shut off from the oral part of the pharynx. This is accomplished when the soft palate is elevated by the contraction of the levator veli palatini muscle on each side. The posterior pharyngeal wall is also pulled forward by the contraction of the upper fibers of the superior constrictor muscle. At the same time the palatopharyngeal arches move medially as the result of the contraction of the palatopharyngeus muscles.
2. A swelling of the palate observed clinically may be due to a tumor of the mucous membrane or underlying bone. It could also be produced by extension inferiorly of a malignant tumor of the maxillary sinus. Tumors of isolated salivary glands in the palatal mucosa may also produce palatal swellings. Chronic infections such as syphilitic gummas rarely cause palatal swellings.
3. Holes in the hard palate can be due to an incomplete repair of a cleft palate, erosion of the hard palate by a badly fitting denture, or, rarely, necrosis of a syphilitic gumma.

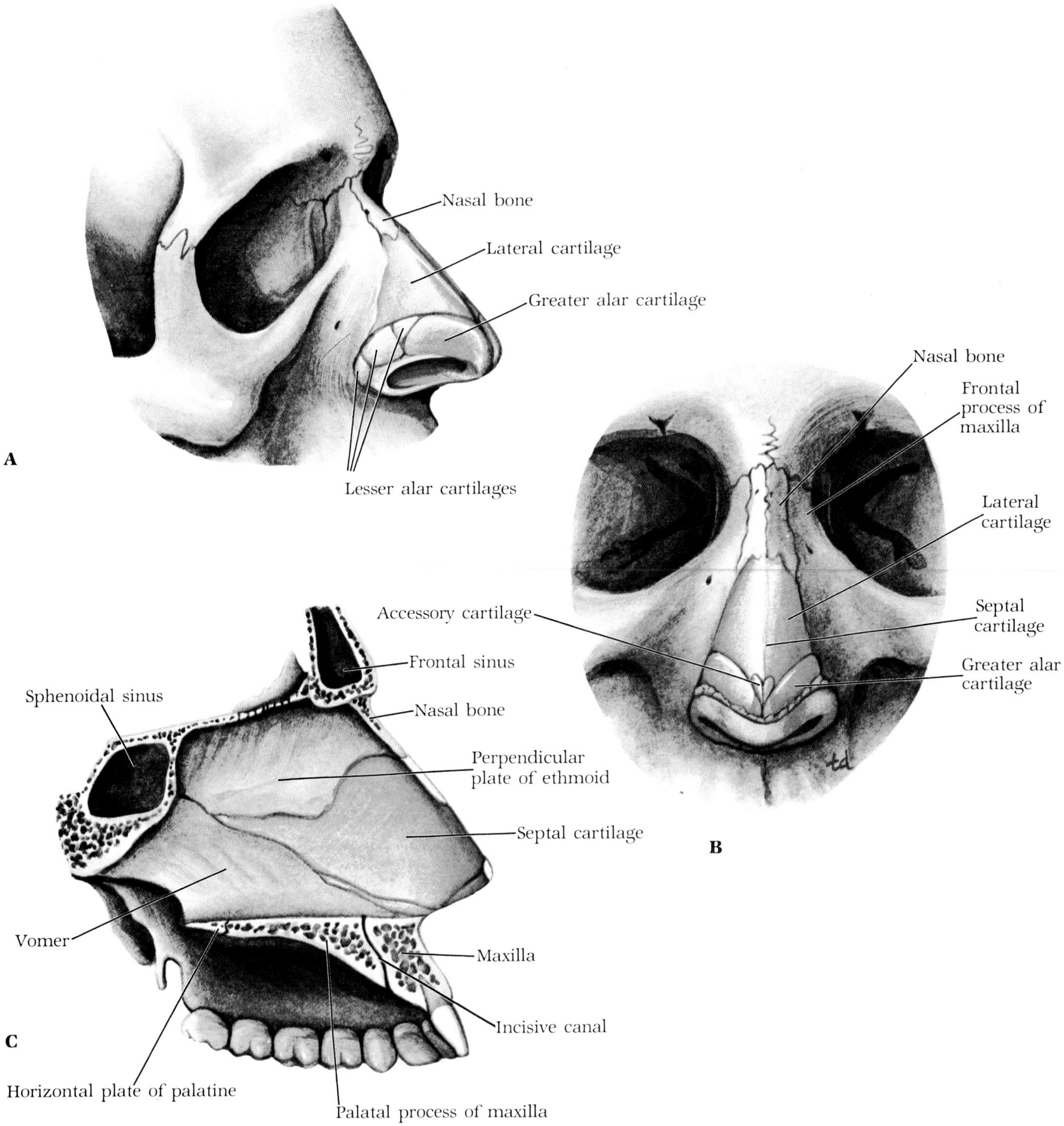
Nasal bone
Lateral cartilage
Greater alar cartilage
Lesser alar cartilages
A
Nasal bone
Frontal process of maxilla
Lateral cartilage
Septal cartilage
Greater alar cartilage
Accessory cartilage
B
Frontal sinus
Sphenoidal sinus
Nasal bone
Perpendicular plate of ethmoid
Septal cartilage
Vomer
Maxilla
Incisive canal
C
Horizontal plate of palatine
Palatal process of maxilla

Figure 7-77
External nose and nasal septum. A. Lateral view of bony and cartilaginous skeleton of external nose. B. Anterior view of bony and cartilaginous skeleton of external nose. C. Bony and cartilaginous skeleton of nasal septum.

1. Fractures of the nasal bones are very common. Although the majority are simple fractures and can be reduced under local anesthesia, some are associated with severe injuries to the nasal septum, which require careful treatment under general anesthesia.
2. Foreign bodies in the nose are frequently found in children. Peas, beans, balloons, and small toys are often found impacted high up in the nasal cavity between the septum and the conchae. These should be carefully removed under local or general anesthesia, and special care should be exercised to avoid damage to the cribriform plate of the ethmoid or the walls of the ethmoidal air sinuses.
3. Examination of the nasal cavity can be carried out by inserting a speculum through the external nares or by means of a mirror in the pharynx. If the latter procedure is used, the choanae and the posterior border of the septum can be visualized.
4. The nasal septum is rarely situated in the midline. A severely deviated septum may interfere with drainage of the nose and the paranasal sinuses.

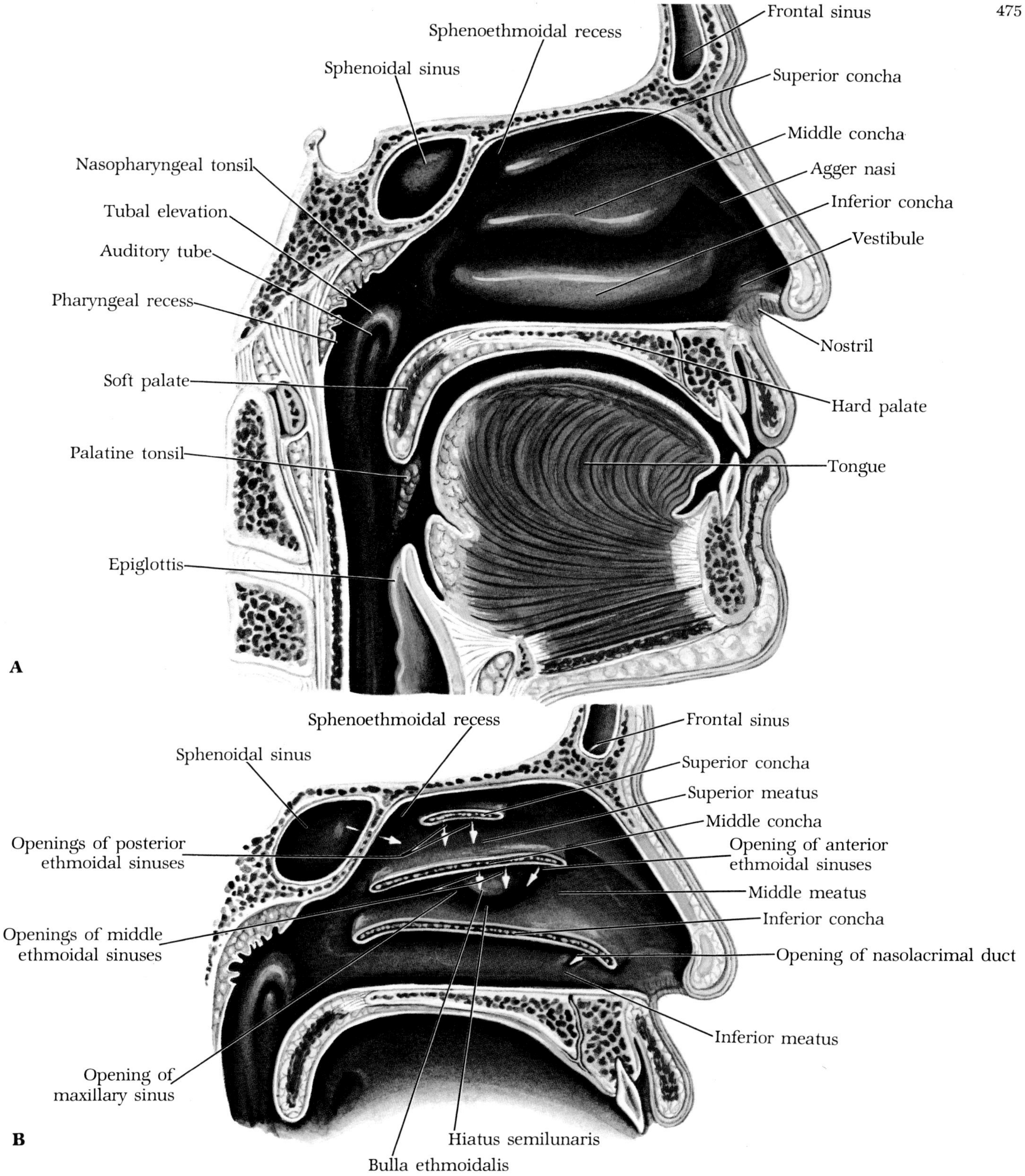
Frontal sinus
Sphenoethmoidal recess
Sphenoidal sinus
Superior concha
Middle concha
Nasopharyngeal tonsil
Agger nasi
Inferior concha
Tubal elevation
Vestibule
Auditory tube
Pharyngeal recess
Nostril
Soft palate
Hard palate
Palatine tonsil
Tongue
Epiglottis
A
Sphenoethmoidal recess
Frontal sinus
Sphenoidal sinus
Superior concha
Superior meatus
Middle concha
Openings of posterior ethmoidal sinuses
Opening of anterior ethmoidal sinuses
Middle meatus
Inferior concha
Openings of middle ethmoidal sinuses
Opening of nasolacrimal duct
Inferior meatus
Opening of maxillary sinus
B
Hiatus semilunaris
Bulla ethmoidalis

Figure 7-78
Nose, mouth, and pharynx.
A. Sagittal section through nose, mouth, and pharynx. B. Lateral wall of nose and nasal part of pharynx.

1. Bleeding from the nose (epistaxis) is a common complaint and may occur as the result of an upper respiratory infection, trauma from nose-picking or a blow to the nose, blood diseases, and hypertension. The arterial supply to the nose is profuse, being supplied by branches of the internal and external carotid arteries. The majority of epistaxes originate on the nasal septum just above the vestibule, a site of anastomosis between the sphenopalatine artery and the septal branch of the superior labial branch of the facial artery.

2. Infection of the nasal cavity can spread in a number of directions. The paranasal sinuses are especially prone to infection. Organisms may spread via the nasal part of the pharynx and the auditory tube to the middle ear. It is also possible for organisms to ascend to the meninges of the anterior cranial fossa, along the sheaths of the olfactory nerves, through the cribriform plate, and produce meningitis.

3. The paranasal sinuses are cavities found in the interior of the maxilla, frontal, sphenoid, and ethmoid bones; they communicate with the nasal cavity through relatively small apertures. They are lined with mucoperiosteum and filled with air. Infection of the paranasal sinuses is a common complication of nasal infections.

4. Nasal polyps occur in adults in the region of the middle turbinate. They may be regarded as edematous protuberances of mucous membrane and often originate in the ethmoidal sinuses; they cause nasal obstruction and discharge.

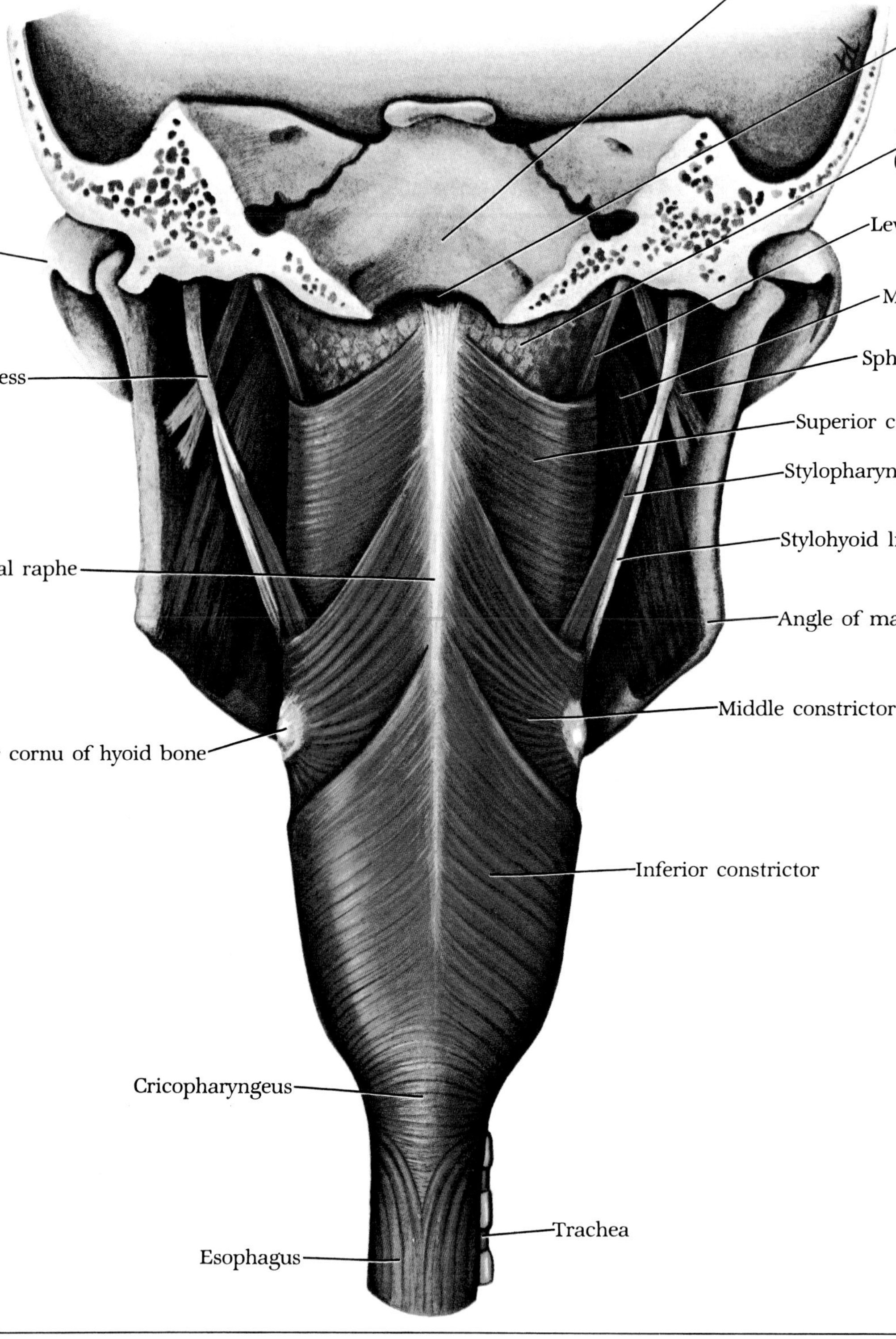
Basilar part of occipital bone
Pharyngeal tubercle
Pharyngobasilar fascia (fibrous layer of pharynx)
Levator veli palatini
Medial pterygoid
Sphenomandibular ligament
Superior constrictor
Stylopharyngeus
Stylohyoid ligament
Angle of mandible
Middle constrictor
Inferior constrictor
Zygomatic arch
Styloid process
Pharyngeal raphe
Tip of greater cornu of hyoid bone
Cricopharyngeus
Trachea
Esophagus

Figure 7-79
Pharynx, posterior view, showing the three constrictor muscles and the stylopharyngeus muscle.

1. Pharyngeal pouch. In the lower part of the posterior surface of the inferior constrictor muscle there is a potential gap between the upper oblique and the lower horizontal fibers (cricopharyngeus muscle). This area is marked by a dimple in the lining mucous membrane. It is believed that the function of the cricopharyngeus is to prevent the entry of air into the esophagus. Should the cricopharyngeus fail to relax during swallowing, the internal pharyngeal pressure may rise and force the mucosa and submucosa of the dimple posteriorly, thus producing a diverticulum. Once the diverticulum has been formed, it may gradually enlarge and fill with food with each meal. Unable to expand posteriorly because of the vertebral column, it turns downward, usually on the left side. The presence of this pouch filled with food causes difficulty in swallowing (dysphagia).

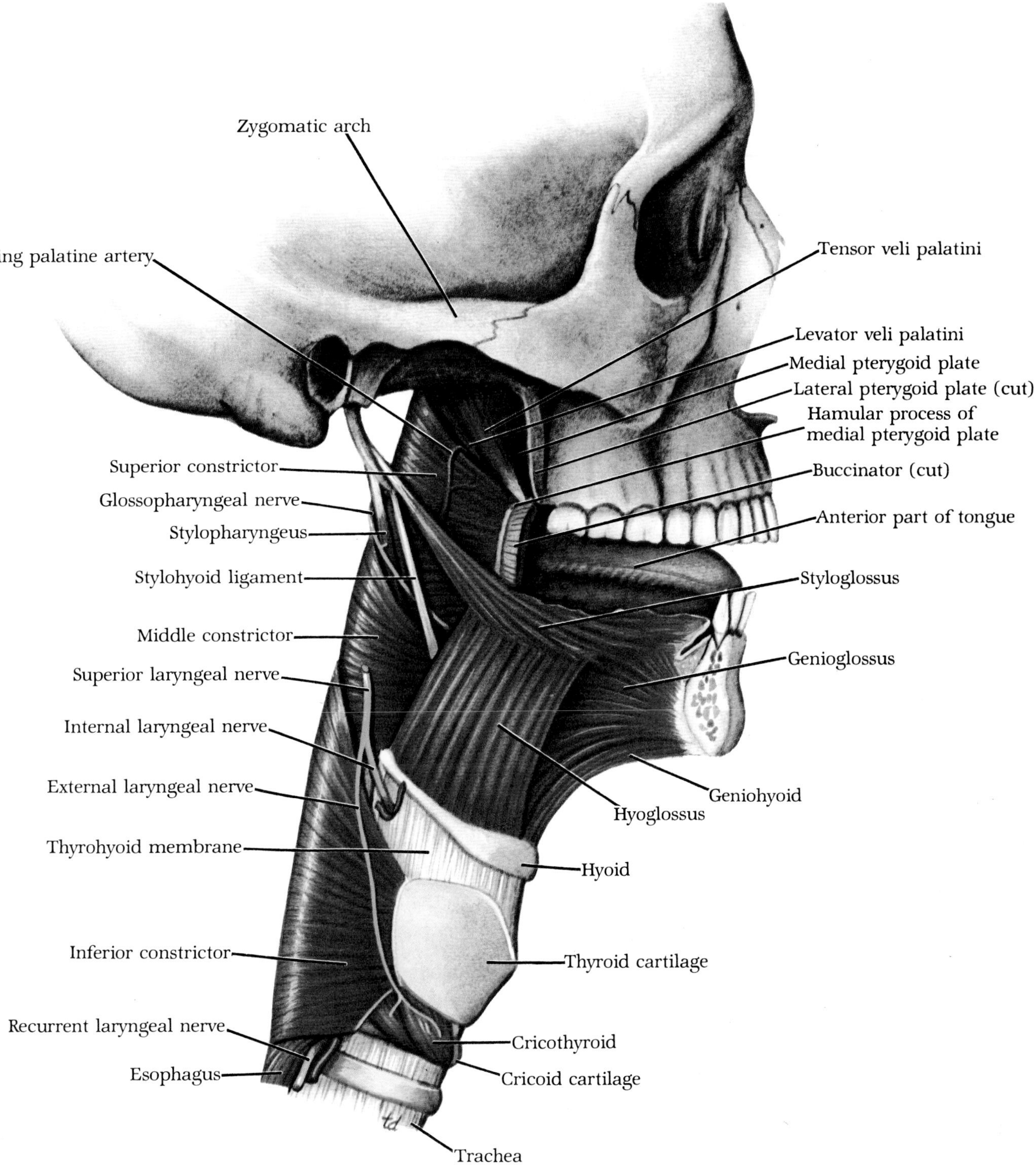
Zygomatic arch
Ascending palatine artery
Tensor veli palatini
Levator veli palatini
Medial pterygoid plate
Lateral pterygoid plate (cut)
Hamular process of medial pterygoid plate
Superior constrictor
Buccinator (cut)
Glossopharyngeal nerve
Anterior part of tongue
Stylopharyngeus
Stylohyoid ligament
Styloglossus
Middle constrictor
Genioglossus
Superior laryngeal nerve
Internal laryngeal nerve
External laryngeal nerve
Geniohyoid
Hyoglossus
Thyrohyoid membrane
Hyoid
Inferior constrictor
Thyroid cartilage
Recurrent laryngeal nerve
Cricothyroid
Esophagus
Cricoid cartilage
Trachea

Figure 7-80
Pharynx, lateral view, showing the three constrictor muscles.

1. Mechanism of swallowing. After food enters the mouth, it is usually broken down by the grinding action of the teeth and is mixed with saliva. The food is repeatedly passed between the opposing teeth as the result of the movements of the tongue and the action of the buccinator muscles of the cheeks. The thoroughly mixed food is now formed into a bolus on the dorsum of the tongue and pushed upward and backward against the undersurface of the hard palate. This is brought about by the contraction of the styloglossus muscles on both sides, which pull the root of the tongue upward and backward. The contraction of the palatoglossus muscles now squeezes the bolus backward into the oral part of the pharynx. The process of swallowing is an involuntary act from this point onward.
The nasal part of the pharynx is now shut off from the oral part of the pharynx by the elevation of the soft palate, the pulling forward of the posterior pharyngeal wall by the upper fibers of the superior constrictor muscle, and the contraction of the palatopharyngeus muscles.
The larynx and laryngeal part of the pharynx are now pulled upward by the contraction of the stylopharyngeus, salpingopharyngeus, thyrohyoid, and palatopharyngeus muscles. The main part of the larynx is thus elevated to the posterior surface of the epiglottis, and the entrance into the larynx is closed.
The bolus moves downward over the epiglottis, the closed entrance into the larynx, and reaches the lower part of the pharynx as the result of the successive contraction of the superior, middle, and inferior constrictor muscles. Some of the food slides down through the piriform fossae. Finally, the lower fibers of the inferior constrictor muscle (cricopharyngeus) relax, and the bolus enters the esophagus.

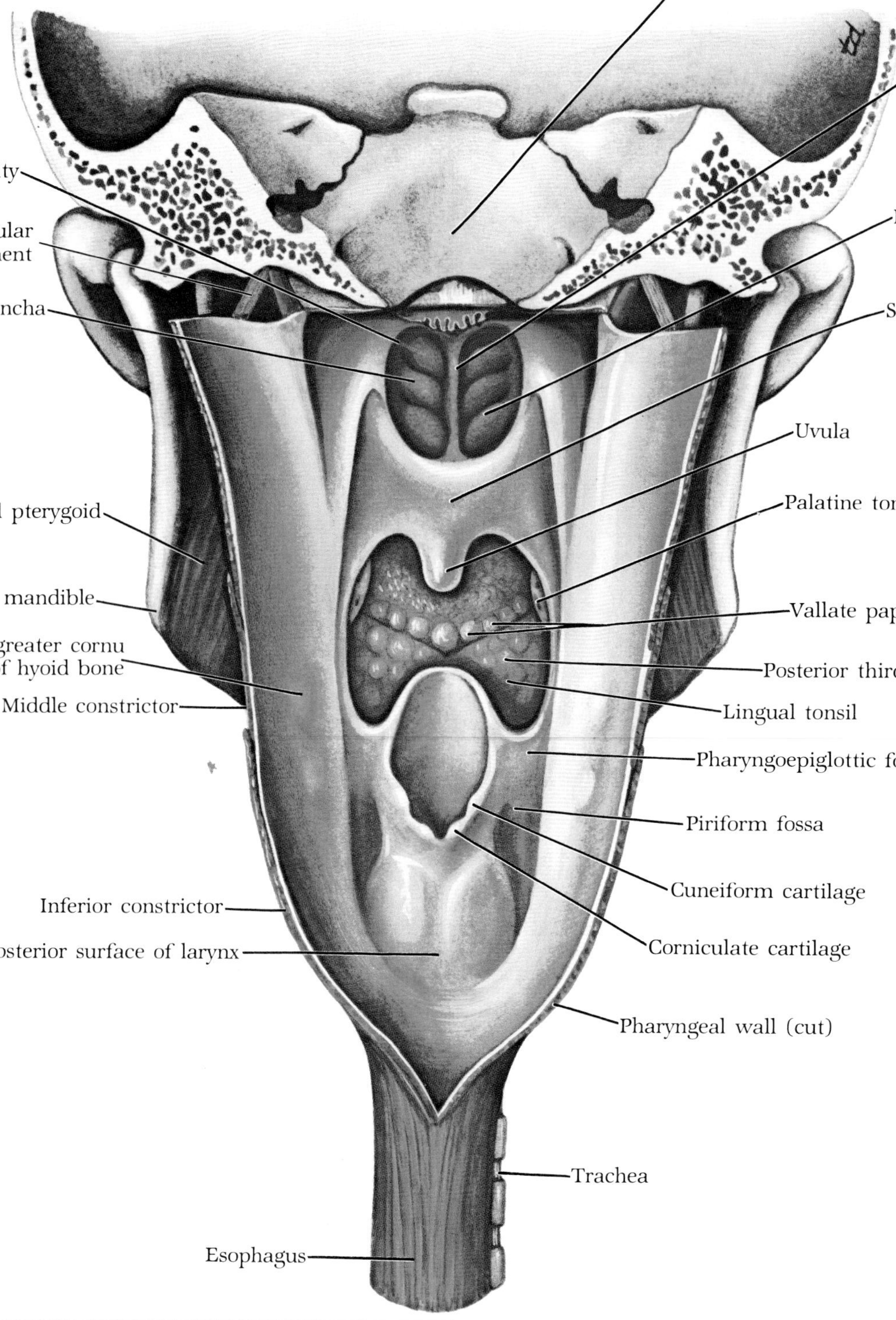
Basilar part of occipital bone
Nasal septum
Nasal cavity
Inferior concha
Sphenomandibular ligament
Middle concha
Soft palate
Uvula
Medial pterygoid
Palatine tonsil
Angle of mandible
Vallate papillae
Tip of greater cornu of hyoid bone
Posterior third of tongue
Middle constrictor
Lingual tonsil
Pharyngoepiglottic fold
Piriform fossa
Cuneiform cartilage
Inferior constrictor
Corniculate cartilage
Posterior surface of larynx
Pharyngeal wall (cut)
Trachea
Esophagus

Figure 7-81
Interior of pharynx as seen on posterior view. The posterior wall has been cut in the midline, and the right and left halves have been turned laterally.

1. At the junction of the mouth with the oral part of the pharynx and the junction of the nose with the nasal part of the pharynx, there are collections of lymphoid tissue of considerable clinical importance. The palatine, tubal, nasopharyngeal, and lingual tonsils all form part of a ring of lymphoid tissue that may be invaded by bacteria. Infection may spread to the deep cervical lymph nodes, causing tender swellings in the neck.
2. Excessive hypertrophy of the nasopharyngeal tonsils, usually associated with infection, causes the organs to be enlarged, after which they are known as adenoids. The posterior nasal openings become blocked, and this causes the patient to breathe through the mouth or snore loudly at night.
3. The close relationship between infected adenoids and the auditory tube may be the cause of deafness and recurrent otitis media. Adenoidectomy is the treatment of choice in cases of infected hypertrophied adenoids.
4. The nasal part of the pharynx can be viewed clinically by checking a mirror passed through the mouth.
5. The piriform fossa is a common lodging site for sharp ingested foreign bodies such as fishbones. The presence of such a foreign body immediately causes the patient to gag violently. Once the object has become jammed, it is difficult for the patient to remove it without a physician's assistance.

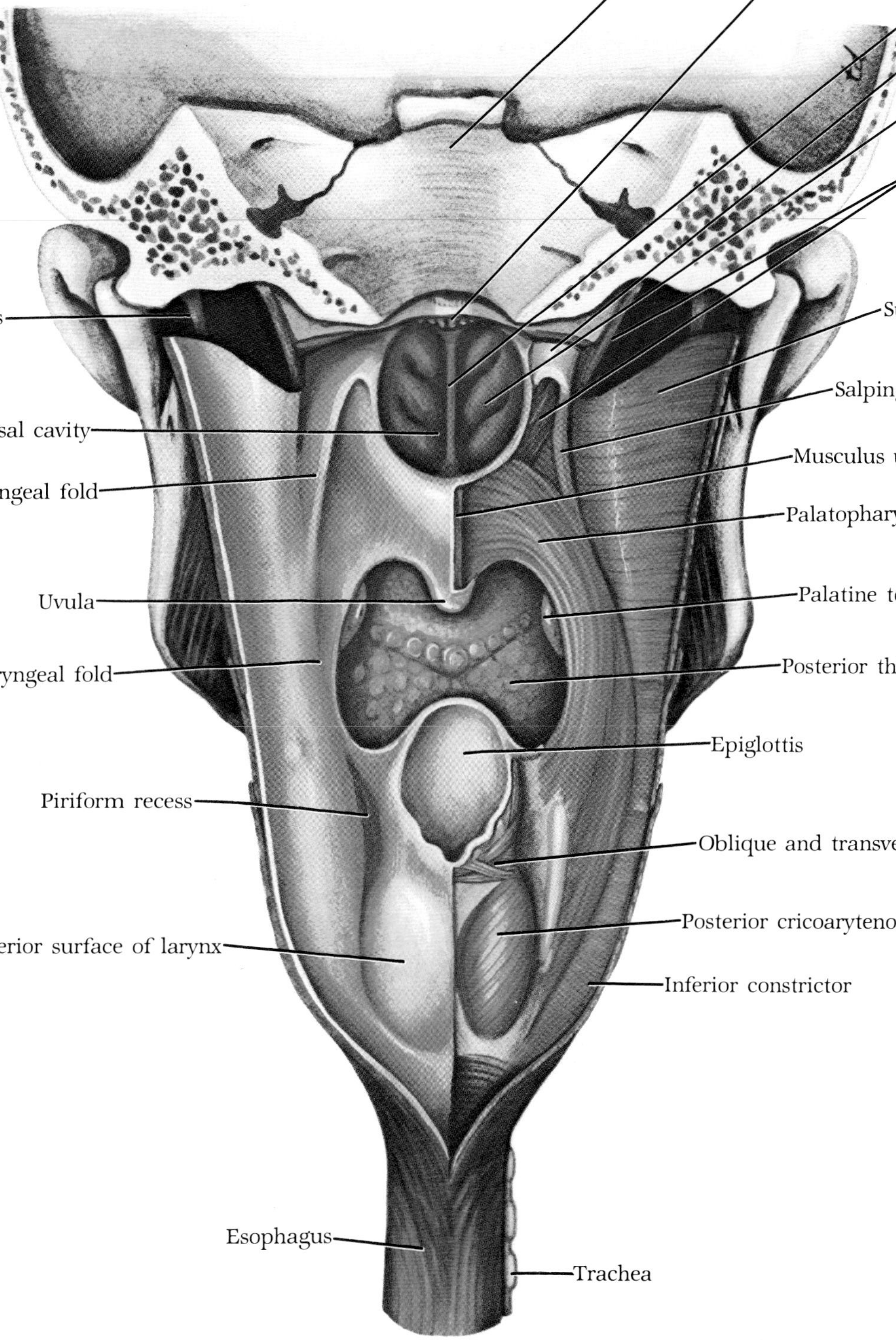
Dorsum sellae
Nasopharyngeal tonsil
Nasal septum
Auditory tube
Middle concha
Levator veli palatini
Styloid process
Superior constrictor
Salpingopharyngeus
Nasal cavity
Musculus uvulae
Salpingopharyngeal fold
Palatopharyngeus
Uvula
Palatine tonsil
Palatopharyngeal fold
Posterior third of tongue
Epiglottis
Piriform recess
Oblique and transverse arytenoid muscle
Posterior cricoarytenoid
Posterior surface of larynx
Inferior constrictor
Esophagus
Trachea

Figure 7-82
Interior of pharynx as seen on posterior view. The posterior wall has been cut in the midline, and the right and left halves have been turned laterally. In addition, the mucous membrane has been removed from the right half to display the underlying muscles.

1. Benign and malignant tumors occur in the pharynx. Angiofibromas occur most commonly in the nasopharynx. Although an angiofibroma is a benign tumor it is capable of expanding into the nose and paranasal sinuses. Carcinoma of the pharynx is very common in Japan and China. It commonly occurs in the lateral wall of the nasopharynx or on the lateral side of the posterior third of the tongue. In the laryngeal part of the pharynx it commonly occurs in the aryepiglottic fold, the piriform fossa, the lateral pharyngeal wall, or posterior to the cricoid cartilage.

2. Malignant disease of the pharynx is often symptomless to begin with. A slight sore throat, difficulty in swallowing (dysphagia), or a swelling in the neck due to spread of the malignancy to involve the deep cervical lymph nodes, may be the only symptoms or signs.

3. Hoarseness of the voice associated with malignant disease of the pharynx indicates that the tumor has already invaded the larynx and involved the vocal folds.

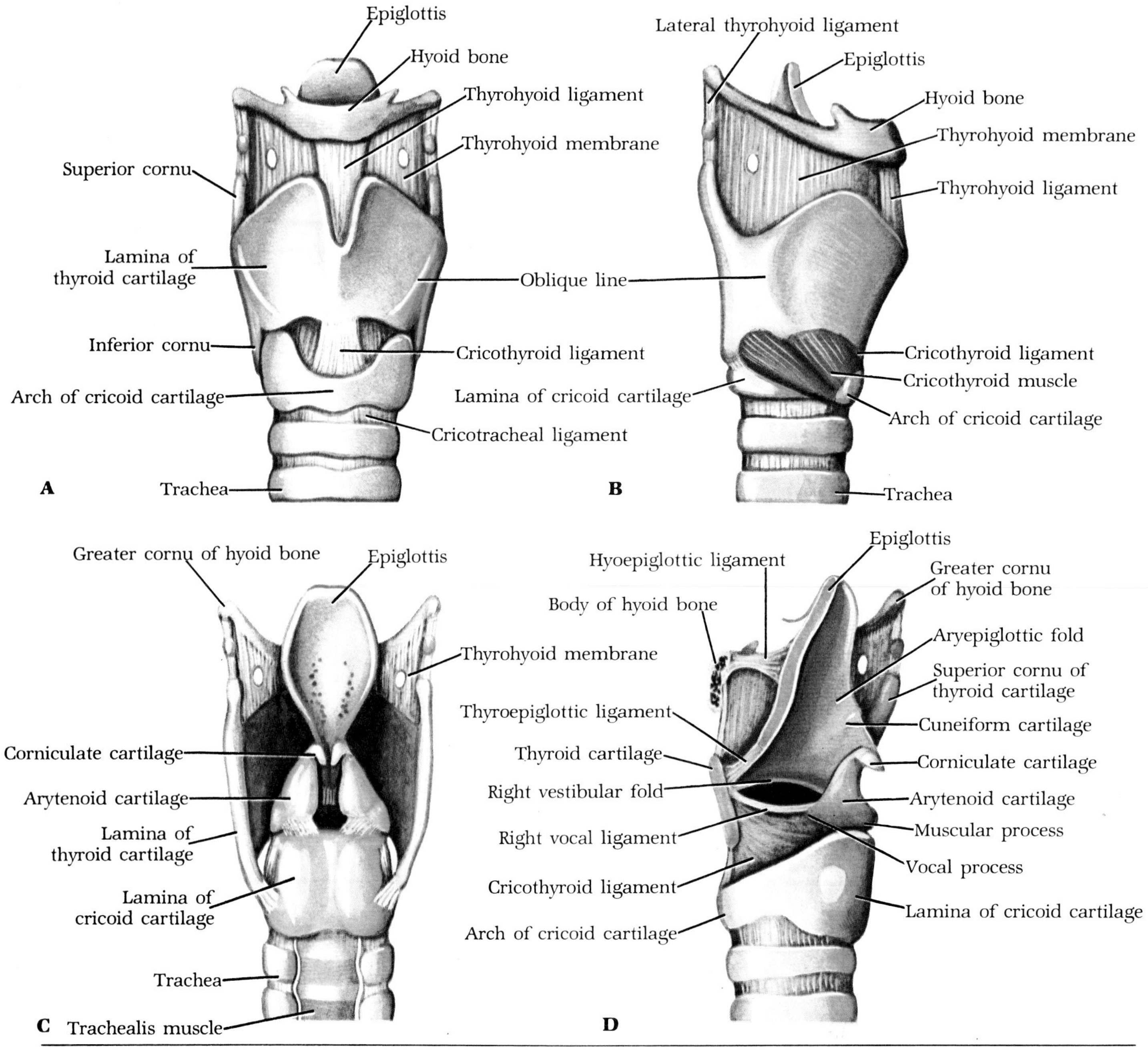

Epiglottis
Hyoid bone
Thyrohyoid ligament
Thyrohyoid membrane
Superior cornu
Lamina of thyroid cartilage
Oblique line
Inferior cornu
Cricothyroid ligament
Arch of cricoid cartilage
Cricotracheal ligament
A
Trachea
Lateral thyrohyoid ligament
Epiglottis
Hyoid bone
Thyrohyoid membrane
Thyrohyoid ligament
Cricothyroid ligament
Cricothyroid muscle
Lamina of cricoid cartilage
Arch of cricoid cartilage
B
Trachea
Greater cornu of hyoid bone
Epiglottis
Thyrohyoid membrane
Corniculate cartilage
Arytenoid cartilage
Lamina of thyroid cartilage
Lamina of cricoid cartilage
Trachea
C
Trachealis muscle
Epiglottis
Hyoepiglottic ligament
Greater cornu of hyoid bone
Body of hyoid bone
Aryepiglottic fold
Superior cornu of thyroid cartilage
Thyroepiglottic ligament
Cuneiform cartilage
Thyroid cartilage
Corniculate cartilage
Right vestibular fold
Arytenoid cartilage
Muscular process
Right vocal ligament
Vocal process
Cricothyroid ligament
Lamina of cricoid cartilage
Arch of cricoid cartilage
D

Figure 7-83
The larynx and its ligaments. A. Anterior view. B. Lateral view. C. Posterior view. D. Left lamina of thyroid cartilage has been removed to display interior of larynx.

1. The muscles of the larynx are innervated by the recurrent laryngeal nerves, with the exception of the cricothyroid muscle, which is supplied by the external laryngeal nerve. Both these nerves are in a vulnerable position during operations on the thyroid gland because of the close relationship between them and the arteries of the gland.

2. The left recurrent laryngeal nerve may be involved in a bronchial or esophageal carcinoma or in secondary metastatic deposits in the mediastinal lymph nodes. The right and left recurrent laryngeal nerves may be damaged in malignant involvement of the deep cervical lymph nodes.

3. Section of the external laryngeal nerve produces weakness of the voice due to the inability to tense the vocal fold. This is because the cricothyroid muscle is paralyzed.

4. A unilateral complete section of the recurrent laryngeal nerve results in the vocal fold on the affected side assuming the position midway between abduction and adduction. It lies just lateral to the midline. Speech is not greatly affected, however, since the other vocal fold compensates to some extent and moves toward the affected vocal fold.

5. A unilateral partial section of the recurrent laryngeal nerve causes the affected vocal fold to assume the adducted midline position. In this condition the abductor muscles are paralyzed to a greater degree than the adductor muscles.

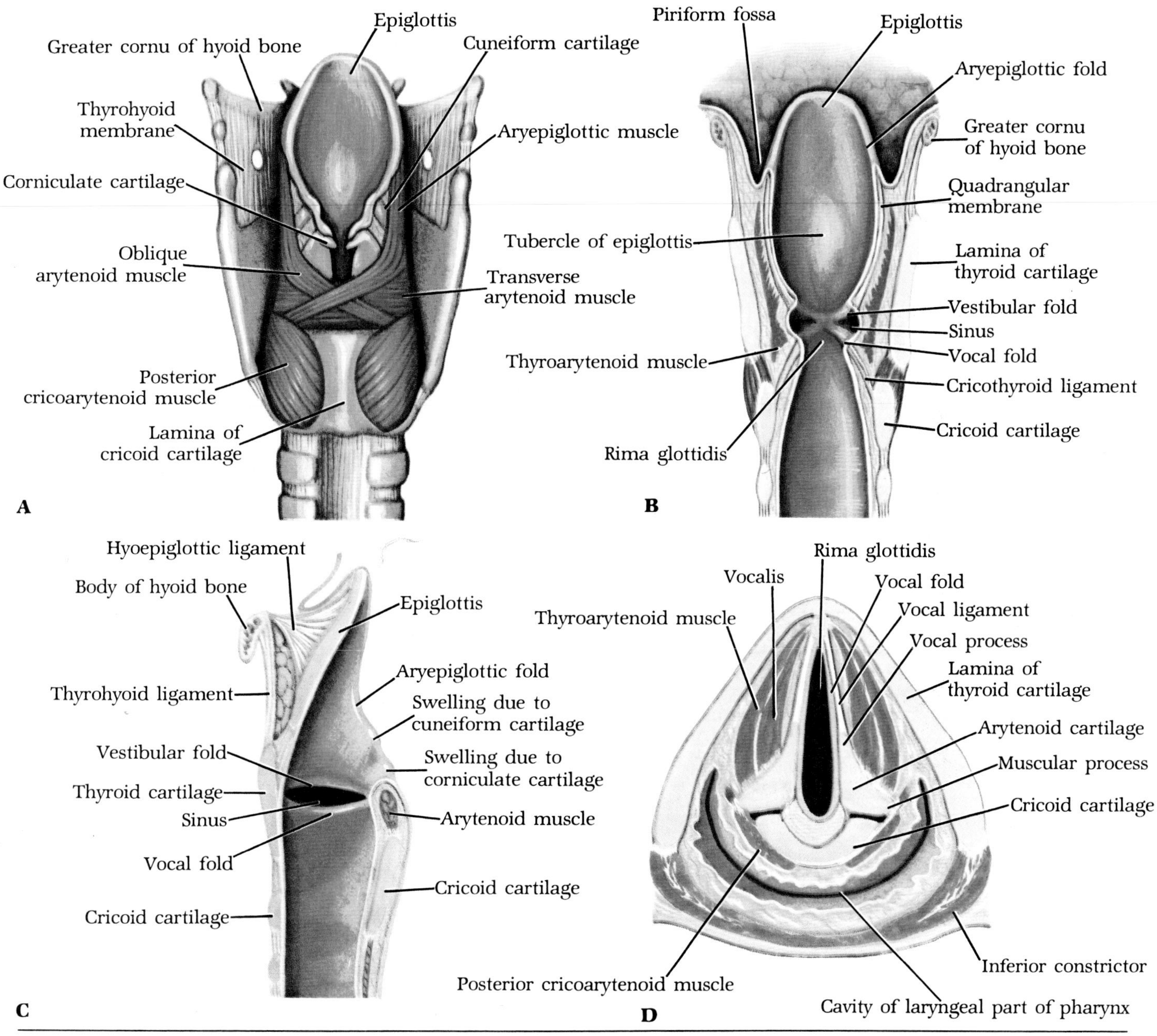
Greater cornu of hyoid bone
Epiglottis
Cuneiform cartilage
Thyrohyoid membrane
Aryepiglottic muscle
Corniculate cartilage
Oblique arytenoid muscle
Transverse arytenoid muscle
Posterior cricoarytenoid muscle
Lamina of cricoid cartilage
A
Piriform fossa
Epiglottis
Aryepiglottic fold
Greater cornu of hyoid bone
Quadrangular membrane
Tubercle of epiglottis
Lamina of thyroid cartilage
Vestibular fold
Sinus
Vocal fold
Thyroarytenoid muscle
Cricothyroid ligament
Cricoid cartilage
Rima glottidis
B
Hyoepiglottic ligament
Body of hyoid bone
Epiglottis
Aryepiglottic fold
Thyrohyoid ligament
Swelling due to cuneiform cartilage
Vestibular fold
Swelling due to corniculate cartilage
Thyroid cartilage
Sinus
Arytenoid muscle
Vocal fold
Cricoid cartilage
Cricoid cartilage
C
Rima glottidis
Vocalis
Vocal fold
Thyroarytenoid muscle
Vocal ligament
Vocal process
Lamina of thyroid cartilage
Arytenoid cartilage
Muscular process
Cricoid cartilage
Inferior constrictor
Posterior cricoarytenoid muscle
Cavity of laryngeal part of pharynx
D

Figure 7-84
Larynx. A. Muscles of larynx, posterior view. B. Coronal section through larynx. C. Sagittal section through larynx. D. Horizontal section through larynx at level of rima glottidis.

1. Bilateral partial section of the recurrent laryngeal nerve results in bilateral paralysis of the abductor muscles and in the vocal folds' moving together. Acute breathlessness (dyspnea) and stridor follow, and a tracheotomy is necessary.

2. Inhalation of foreign bodies is common in children. Plastic bags, balloons, and other objects may become impacted within the cavity of the larynx. If the foreign body cannot be removed or dislodged by inverting the child, a tracheotomy should be performed.

3. Edema of the mucous membrane of the larynx above the level of the vocal folds may encroach on the airway, and in severe cases a tracheotomy may have to be performed. Some common causes of acute edema of the larynx are infection (streptococcal and diphtherial infections), angioneurotic edema, injury in automobile accidents, or corrosive fluids or gases.

4. The interior of the larynx may be inspected indirectly through a laryngeal mirror passed through the open mouth into the oral pharynx. A more satisfactory method is the direct method, in which a laryngoscope is used.

5. Tumors of the larynx. Malignant tumors are more common than benign tumors. The majority of malignant tumors occur in men between 40 and 60 years of age and most commonly originate on the vocal folds.

8. The Back

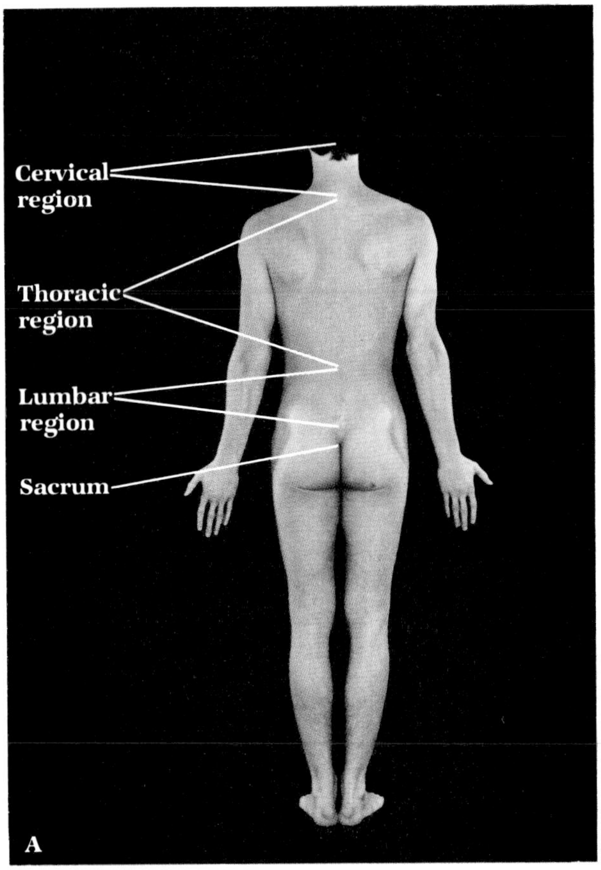
Cervical region
Thoracic region
Lumbar region
Sacrum
A

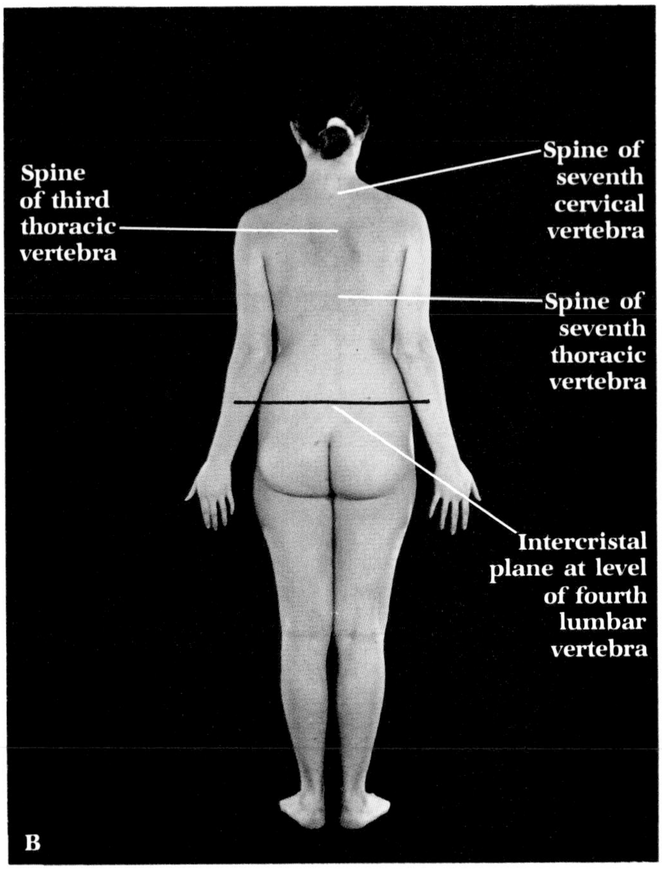
Spine of third thoracic vertebra
Spine of seventh cervical vertebra
Spine of seventh thoracic vertebra
Intercristal plane at level of fourth lumbar vertebra
B

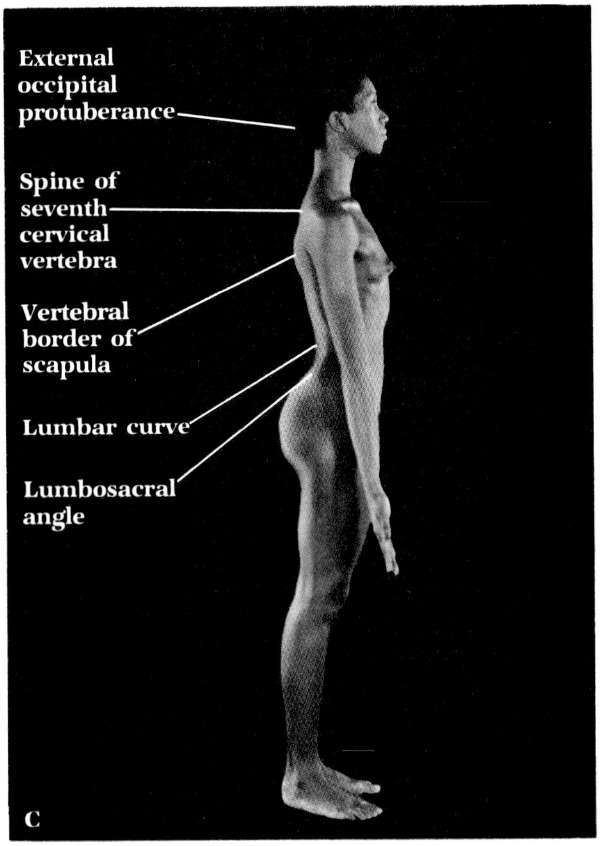
External occipital protuberance
Spine of seventh cervical vertebra
Vertebral border of scapula
Lumbar curve
Lumbosacral angle
C

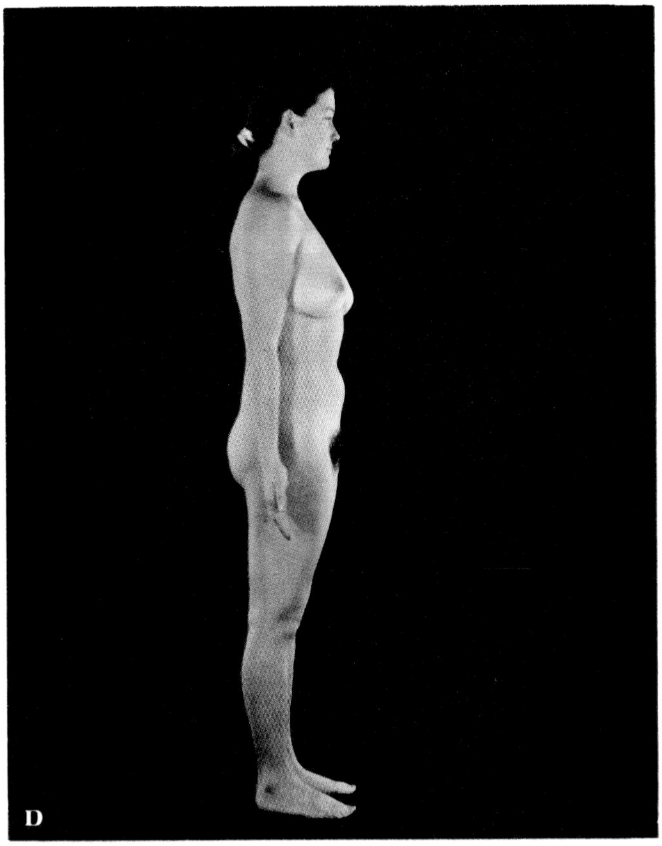
D

Figure 8-1
The back. A. 27-year-old male, posterior view. B. 25-year-old female, posterior view. C. 22-year-old female, lateral view. D. 25-year-old female, lateral view.

1. When one is inspecting the back the patient should be examined from head to foot, and his arms should hang loosely at his sides. All clothes and shoes should be removed.
2. Observe the posterior aspect of the back as a whole and compare the two sides with reference to an imaginary line passing downward from the external occipital protuberance to the natal cleft.
3. The external occipital protuberance lies at the junction of the head and neck. Place the index finger on the skin in the midline. Note that it can be drawn downward from the protuberance in the nuchal groove. The first spinous process to be felt is that of the seventh cervical vertebra (the first to sixth cervical spines are covered by the ligamentum nuchae).
4. The spines of all the thoracic and lumbar vertebrae can be individually palpated and are most easily felt when the trunk is flexed. The most prominent spine is that of the first thoracic vertebra.
5. The median sacral crest (fused spines of sacrum) can be felt in the upper part of the natal cleft.
6. The sacral hiatus lies about 2 inches (5 cm) above the tip of the coccyx beneath the skin of the natal cleft.
7. The inferior surface and tip of the coccyx can be palpated in the natal cleft about 1 inch (2.5 cm) behind the anus.
8. By inspecting the lateral contour of the back the curves of the vertebral column can be examined. Normally the posterior surface is concave in the cervical region, convex in the thoracic region, and concave in the lumbar region. The sacrum and coccyx are convex posteriorly. The lumbar region meets the sacrum at a sharp angle, the lumbosacral angle.
9. The spinal cord in the adult extends down to the level of the lower border of the spine of the first lumbar vertebra. In the child it may extend to the third or fourth lumbar spine.
10. The intercristal plane (horizontal black line in photograph B) passes across the highest points on the iliac crests and lies on the level of the body of the fourth lumbar spine. It is at this level in the adult that a lumbar puncture needle may be safely inserted into the subarachnoid space without damaging the spinal cord.

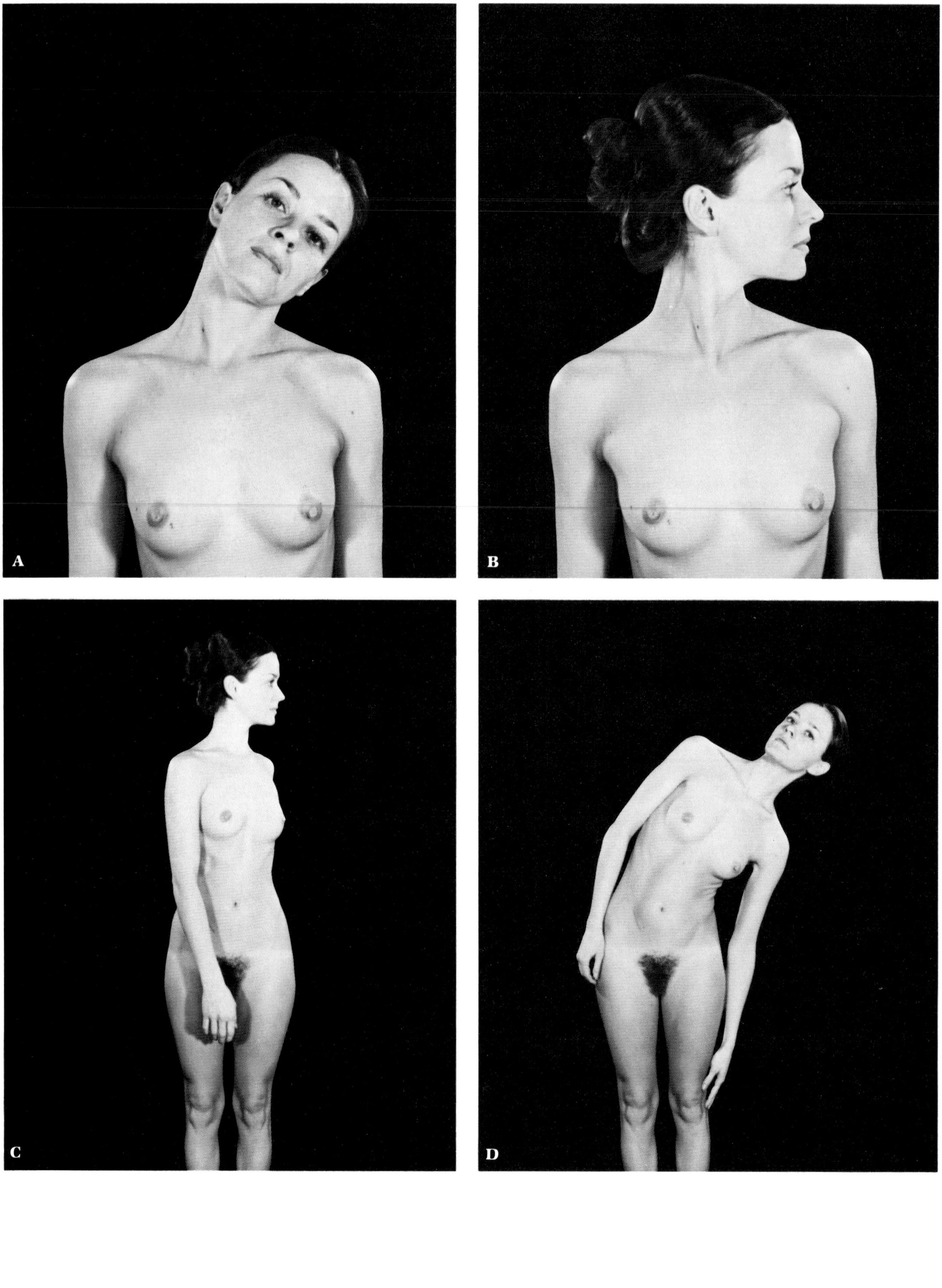
A
B
C
D

Figure 8-2
Movements of the vertebral column in a 29-year-old female. A. Lateral flexion of the cervical vertebrae. B. Lateral rotation of the cervical vertebrae. C. Lateral rotation of the thoracic and lumbar vertebrae. D. Lateral flexion of the thoracic and lumbar vertebrae.

1. When examining a patient's back the normal range of movement of the different parts of the vertebral column should be tested.

2. In the cervical region, flexion, extension, lateral rotation, and lateral flexion are possible. Remember that about half of the movement we refer to as flexion is carried out at the atlanto-occipital joints. In flexion the patient should be able to touch his chest with his chin, and in extension he should be able to look directly upward. In lateral rotation the patient should be able to place his chin nearly in line with his shoulder. Half of lateral rotation occurs between the atlas and the axis. In lateral flexion the head can normally be tilted 45 degrees to each shoulder. It is important that the shoulder is not raised when this movement is being tested.

3. In the thoracic region the movements are limited by the presence of the ribs and sternum. Rotation is possible because the articular processes lie on an arc of a circle whose center is within the vertebral body. When testing for rotation make sure that the patient does not rotate the pelvis.

4. In the lumbar region, flexion, extension, lateral rotation, and lateral flexion are possible. Flexion and extension are fairly free. Lateral rotation is, however, limited by the interlocking of the articular processes. Lateral flexion in the thoracic and lumbar regions is tested by asking the patient to slide, in turn, each hand down the lateral side of the thigh.

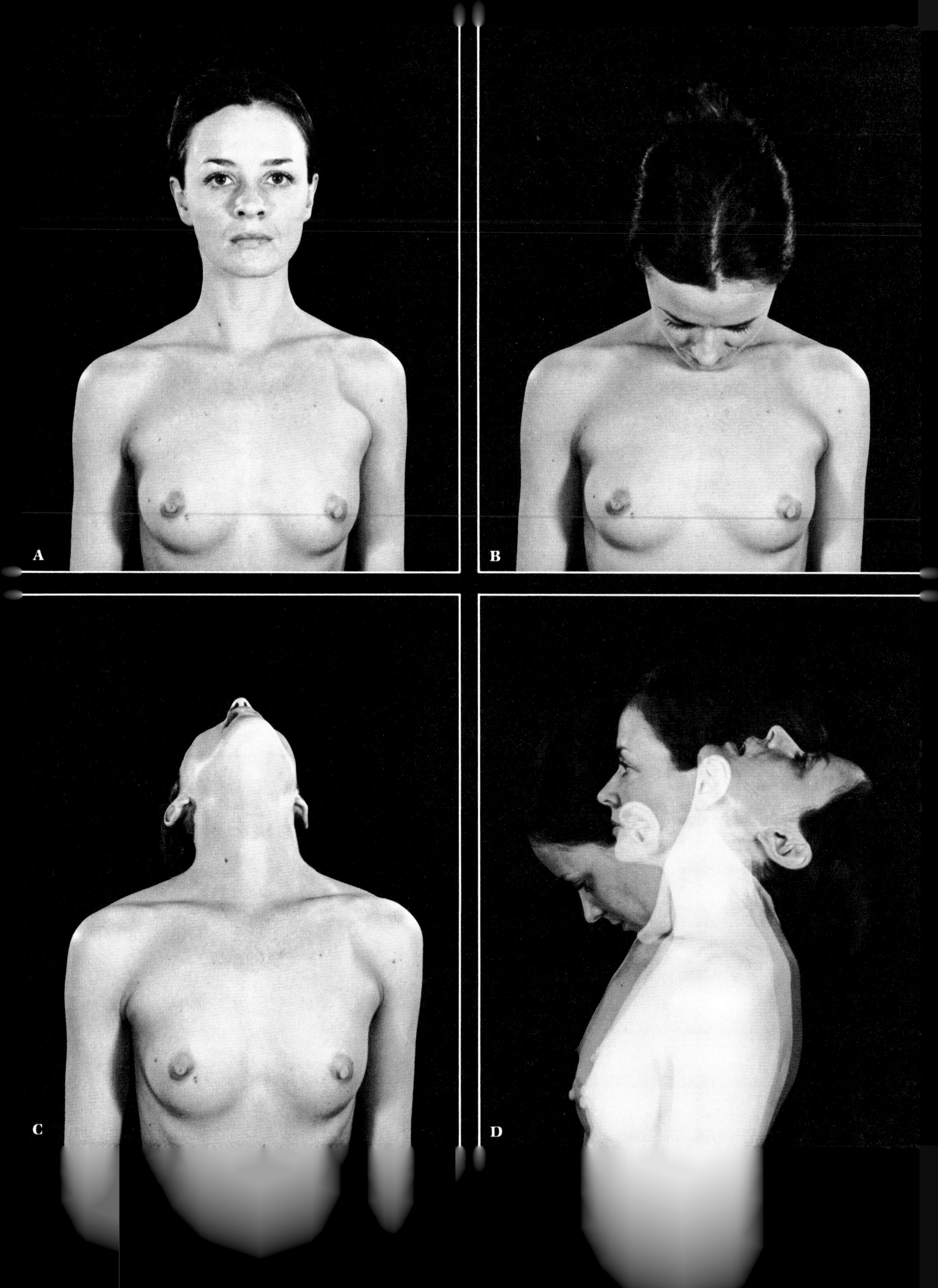
A
B
C
D

Figure 8-3
Movements of the cervical region of the vertebral column in a 29-year-old female. A. The anatomical position. B. Flexion. C. Extension. D. Movement from full flexion to full extension.

1. Although the vertebrae are held in position relative to one another by strong ligaments that severely limit the degree of movement possible between adjacent vertebrae, the summation of all these movements gives the vertebral column as a whole a remarkable degree of mobility.

2. In the cervical region, the intervertebral discs are relatively thick so that the movements of flexion, extension, lateral rotation, and lateral flexion are fairly free.

3. It is important to remember that in the cervical region about half of flexion is carried out at the atlanto-occipital joints and about half of lateral rotation occurs between the atlas and the axis.

4. Flexion is produced by the longus cervicis, the scalenus anterior, and the sternocleidomastoid muscles. Extension is produced by the postvertebral muscles. Lateral flexion is produced by the scalenus anterior and medius and the trapezius and sternocleidomastoid muscles. Rotation is produced by the sternocleidomastoid on one side and the splenius on the other side.

5. In photograph D note that the trunk moves to compensate for the altered position of the head so that the patient can keep her balance.

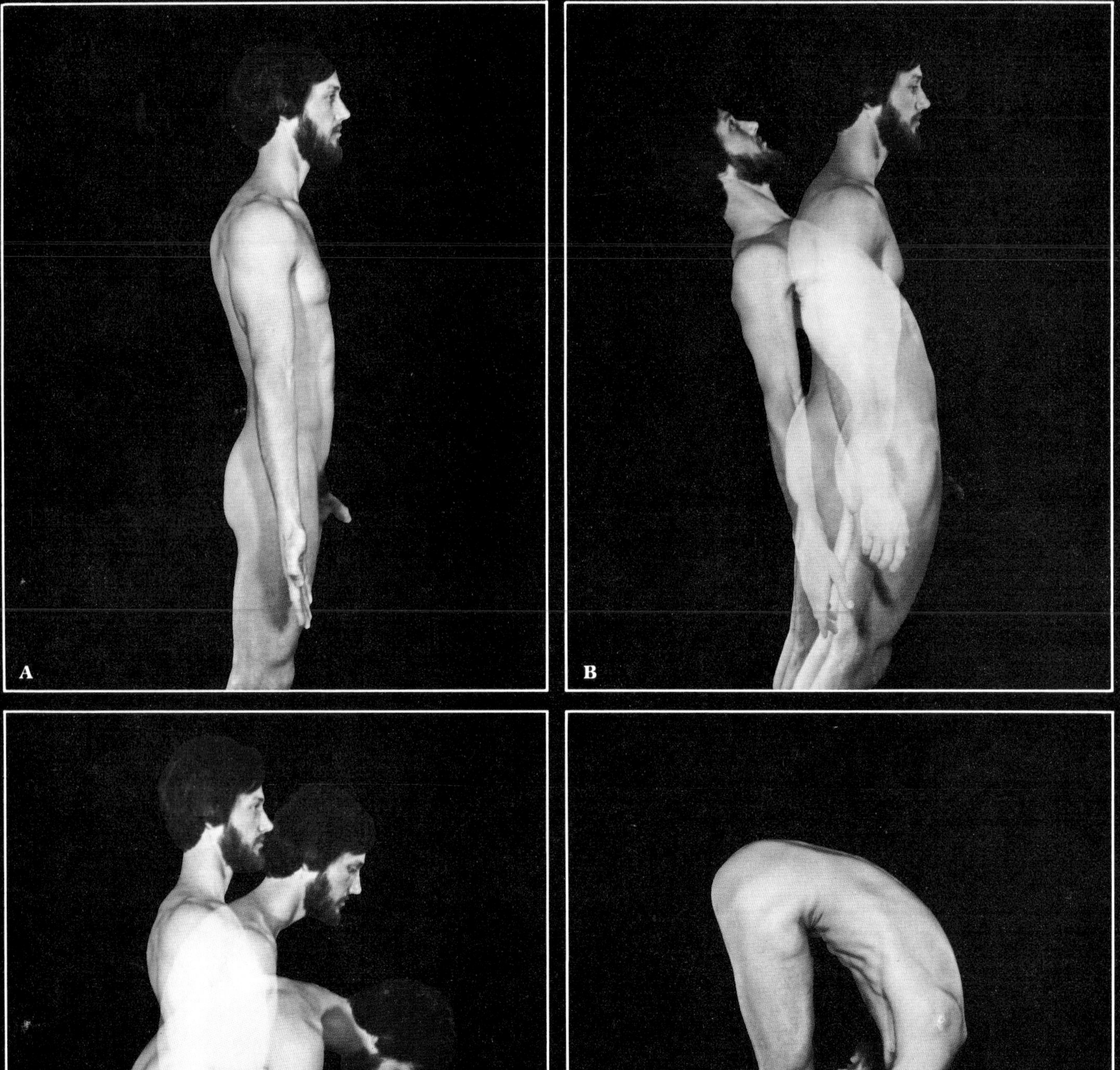

A
B
C
D

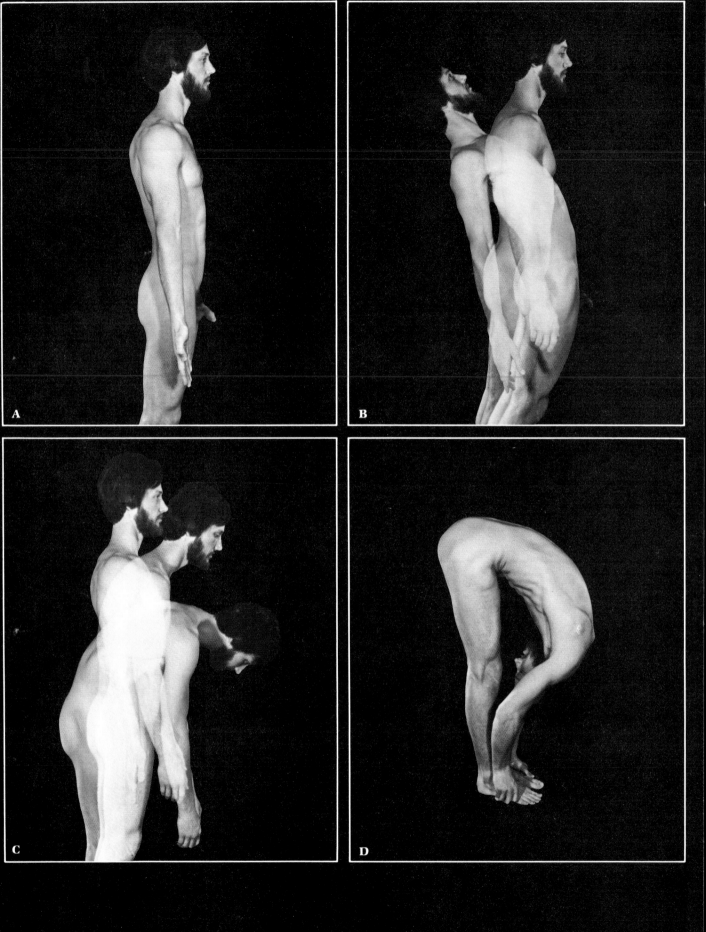

Figure 8-4
Movements of the vertebral column in a 27-year-old male, lateral view. A. Anatomical position. B. Extension. C. Flexion. D. Flexion of the vertebral column and hip joints.

1. During the movements of flexion and extension of the vertebral column the vertebrae rock forward and backward on one another; the range of movement in each region of the column is largely dependent on the thickness of the intervertebral discs (with the greatest thickness in the cervical and lumbar regions) and the shape and direction of the articular processes. In the thoracic region, the ribs, costal cartilages, and sternum severely restrict the range of movement between the upper thoracic vertebrae.
2. It is important to remember that much of the movement we refer to as flexion is carried out at the atlanto-occipital joints and the hip joints.
3. When being tested for flexion the patient is asked to touch his toes with the knee joints extended. Note the point at which the pelvis begins to rotate, i.e., when the flexion movement of the hip joints begins to contribute to the overall movement of flexion. The degree of flexion is measured by the distance of the tips of the fingers from the ground.
4. Extension is measured by asking the patient to bend backward. The physician should place a supporting hand on the patient's back so that he does not lose his balance and fall backward. Note in photographs B and C how the body maintains its balance in movements of the vertebral column by compensatory movements of the hip joints.

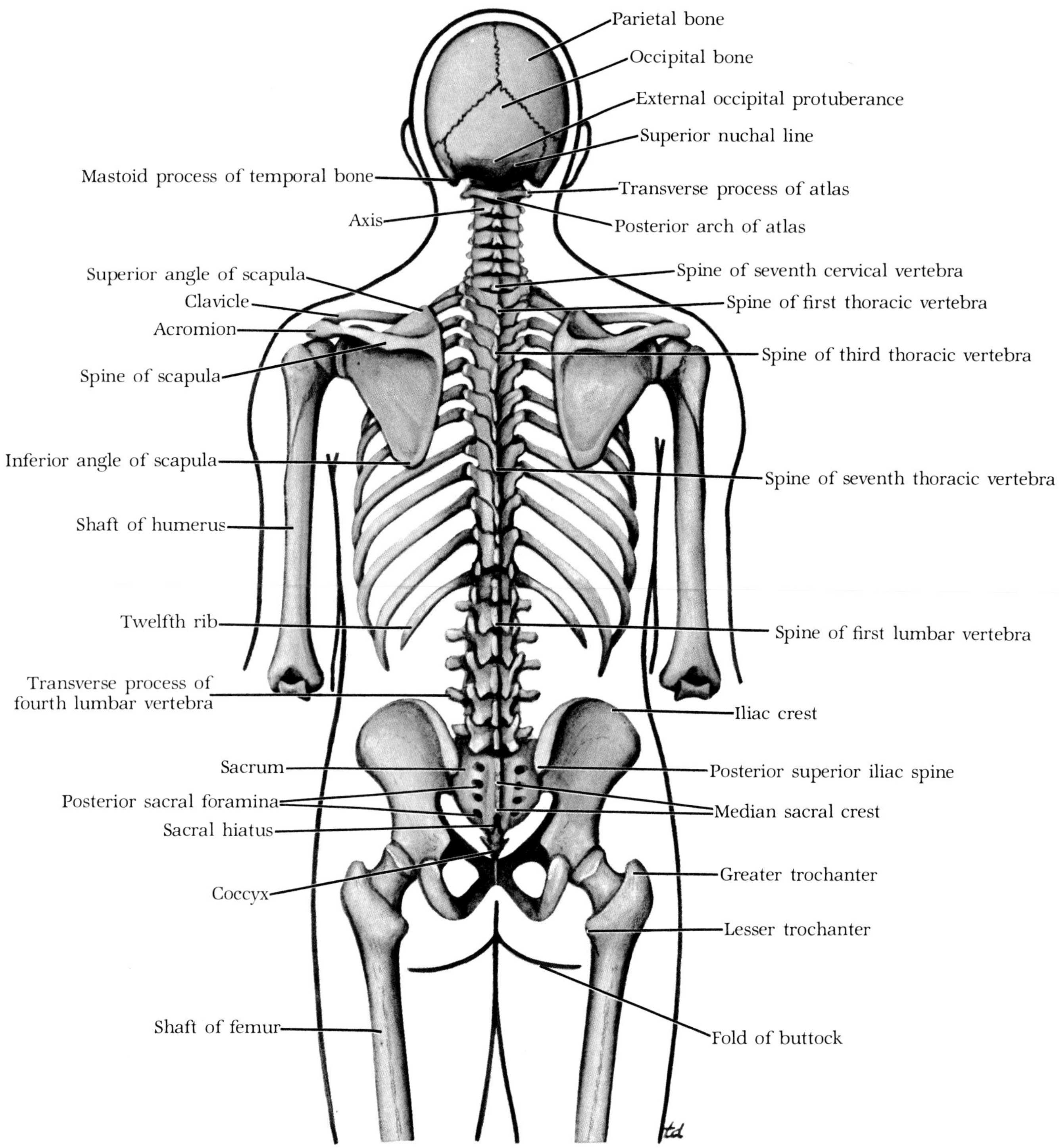
Parietal bone
Occipital bone
External occipital protuberance
Superior nuchal line
Mastoid process of temporal bone
Transverse process of atlas
Axis
Posterior arch of atlas
Superior angle of scapula
Spine of seventh cervical vertebra
Clavicle
Spine of first thoracic vertebra
Acromion
Spine of third thoracic vertebra
Spine of scapula
Inferior angle of scapula
Spine of seventh thoracic vertebra
Shaft of humerus
Twelfth rib
Spine of first lumbar vertebra
Transverse process of fourth lumbar vertebra
Iliac crest
Sacrum
Posterior superior iliac spine
Posterior sacral foramina
Median sacral crest
Sacral hiatus
Greater trochanter
Coccyx
Lesser trochanter
Shaft of femur
Fold of buttock

Figure 8-5
Bones of the back, posterior view.

1. The vertebral column consists of the vertebrae, a series of small bones poised one upon the other; their position is maintained primarily by the normal tone of the muscles, especially the deep back muscles, which are acting on the vertebrae. Paralysis of the muscles, as in poliomyelitis, may lead to gross deformity, while nerve irritation may lead to pain, muscle spasm, and postural changes.
2. The vertebral column not only supports the weight of the head and trunk, which it transmits to the hip bones and lower limbs, but it also protects the delicate spinal cord. Should this bony and ligamentous structure be severely damaged, however, careless movement of the vertebrae may result in their being the instrument for severing the spinal cord.
3. Scoliosis is the term used to describe a lateral deviation of the vertebral column. It is most commonly found in the thoracic region and may be due to muscular or vertebral defects. Paralysis of muscles due to poliomyelitis may cause severe scoliosis, as can the presence of a congenital hemivertebra. Very often scoliosis is compensatory, being due to a short leg or hip disease.

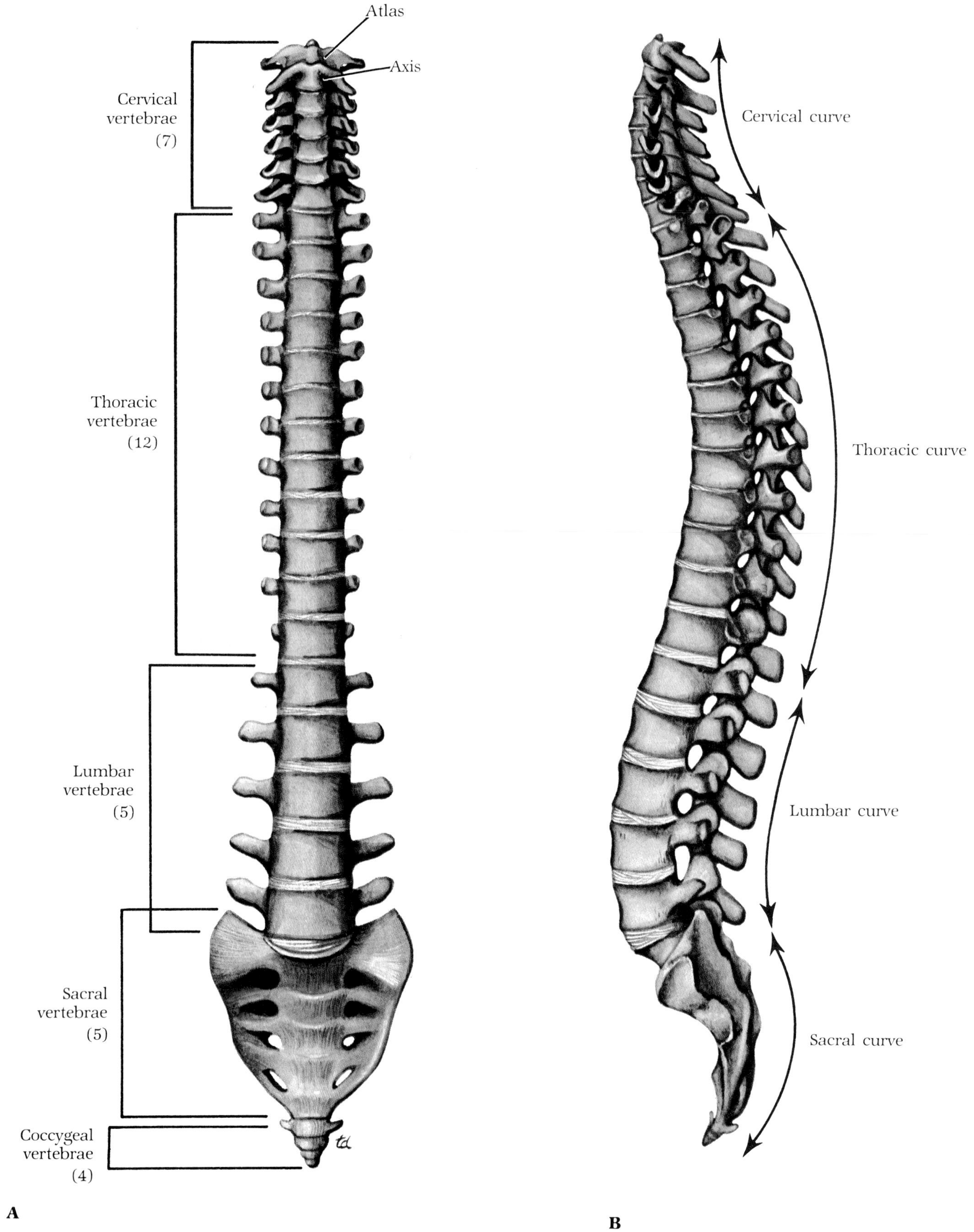
Atlas
Axis
Cervical vertebrae (7)
Thoracic vertebrae (12)
Lumbar vertebrae (5)
Sacral vertebrae (5)
Coccygeal vertebrae (4)
A
Cervical curve
Thoracic curve
Lumbar curve
Sacral curve
B

Figure 8-6
Vertebral column. A. Anterior view. B. Lateral view.

1. The intervertebral discs take up one-quarter of the length of the vertebral column. They are thickest in the cervical and lumbar regions, where the movements of the vertebral column are greatest. They may be regarded as semielastic discs lying between the rigid bodies of adjacent vertebrae. Their physical characteristics allow them to serve as shock absorbers when the load on the vertebral column is suddenly increased, as when jumping from a height. Their elasticity allows the rigid vertebrae to move one upon the other, although unfortunately this resilience is gradually lost with age.
2. Kyphosis is the term used to describe an exaggeration in the sagittal curvature present in the thoracic part of the vertebral column. It may be due to muscular weakness or structural changes in the vertebral bodies or intervertebral discs. In sickly adolescents poor muscle tone produces a gently curved kyphosis of the upper thoracic region. Crush fractures or tuberculous destruction of the vertebral bodies leads to acute angular kyphosis. Degeneration of the intervertebral discs in the aged leads to senile kyphosis.
3. Lordosis is the term used to describe an exaggeration in the sagittal curvature present in the lumbar region. It may be due to an increase in the weight of the abdominal contents produced by, for example, a gravid uterus or a large ovarian tumor, or it may be due to a disease of the vertebral column such as spondylolisthesis. The possibility that it is postural compensation for a kyphosis in the thoracic region or a disease of the hip joint (congenital dislocation) must not be overlooked.

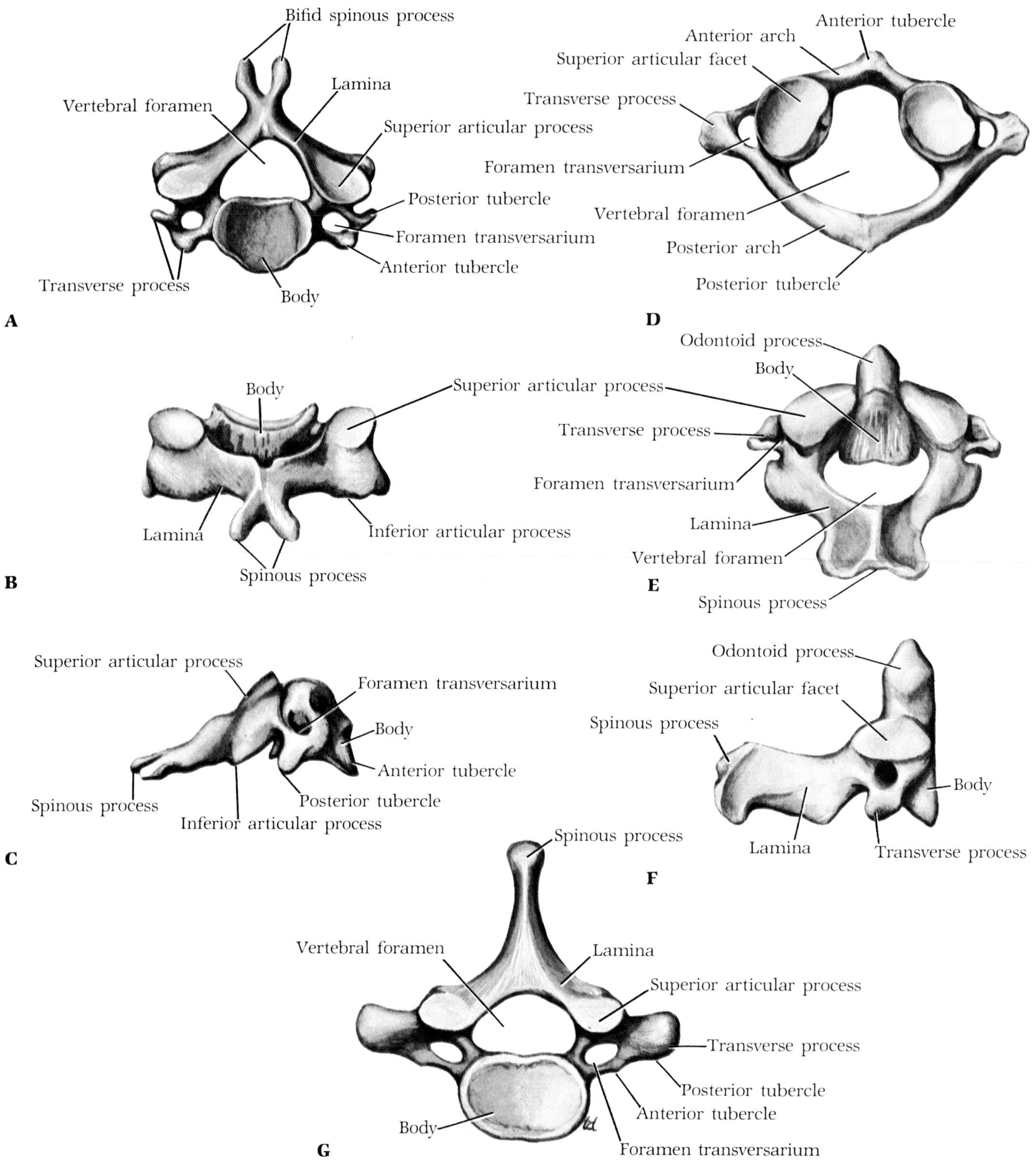
Bifid spinous process
Lamina
Vertebral foramen
Superior articular process
Posterior tubercle
Foramen transversarium
Anterior tubercle
Transverse process
Body
A
Anterior tubercle
Anterior arch
Superior articular facet
Transverse process
Foramen transversarium
Vertebral foramen
Posterior arch
Posterior tubercle
D
Body
Superior articular process
Lamina
Inferior articular process
Spinous process
B
Odontoid process
Body
Superior articular process
Transverse process
Foramen transversarium
Lamina
Vertebral foramen
Spinous process
E
Superior articular process
Foramen transversarium
Body
Anterior tubercle
Posterior tubercle
Spinous process
Inferior articular process
C
Odontoid process
Superior articular facet
Spinous process
Body
Lamina
Transverse process
F
Spinous process
Vertebral foramen
Lamina
Superior articular process
Transverse process
Posterior tubercle
Anterior tubercle
Body
Foramen transversarium
G

Figure 8-7
Cervical vertebrae. A, B, C. Typical cervical vertebra (fourth). D. Atlas. E, F. Axis. G. Seventh cervical vertebra.

1. Dislocation without fracture of the cervical part of the vertebral column occurs commonly between the fourth and fifth or fifth and sixth cervical vertebrae where mobility is greatest. In a unilateral dislocation the inferior articular process of one vertebra is forced forward over the anterior margin of the superior articular process of the vertebra below. The spinal nerve on the same side is usually nipped in the intervertebral foramen, producing severe pain. Although the large size of the vertebral canal allows the spinal cord to escape damage in most cases, bilateral cervical dislocations are almost always associated with severe injury to the spinal cord.
2. Flexion-compression fractures of the cervical vertebral bodies commonly occur in the lower three cervical vertebrae and follow acute forced flexion of the neck.
3. Vertical compression forces applied to the vertebral column may result in a "bursting" fracture of the vertebral body; the posterior fragment of the body may be displaced backward and compress the spinal cord.
4. Fractures of the spinous processes or transverse processes of the cervical vertebrae, which are rare, may be caused by avulsion or a direct blow.
5. The atlas may be fractured by direct downward pressure, as when a patient falls on the head. The occipital condyles are driven downward, forcing the superior articular facets of the atlas apart.
6. Severe acute flexion, extension, or rotation of the head may result in a fracture of the odontoid process of the axis.

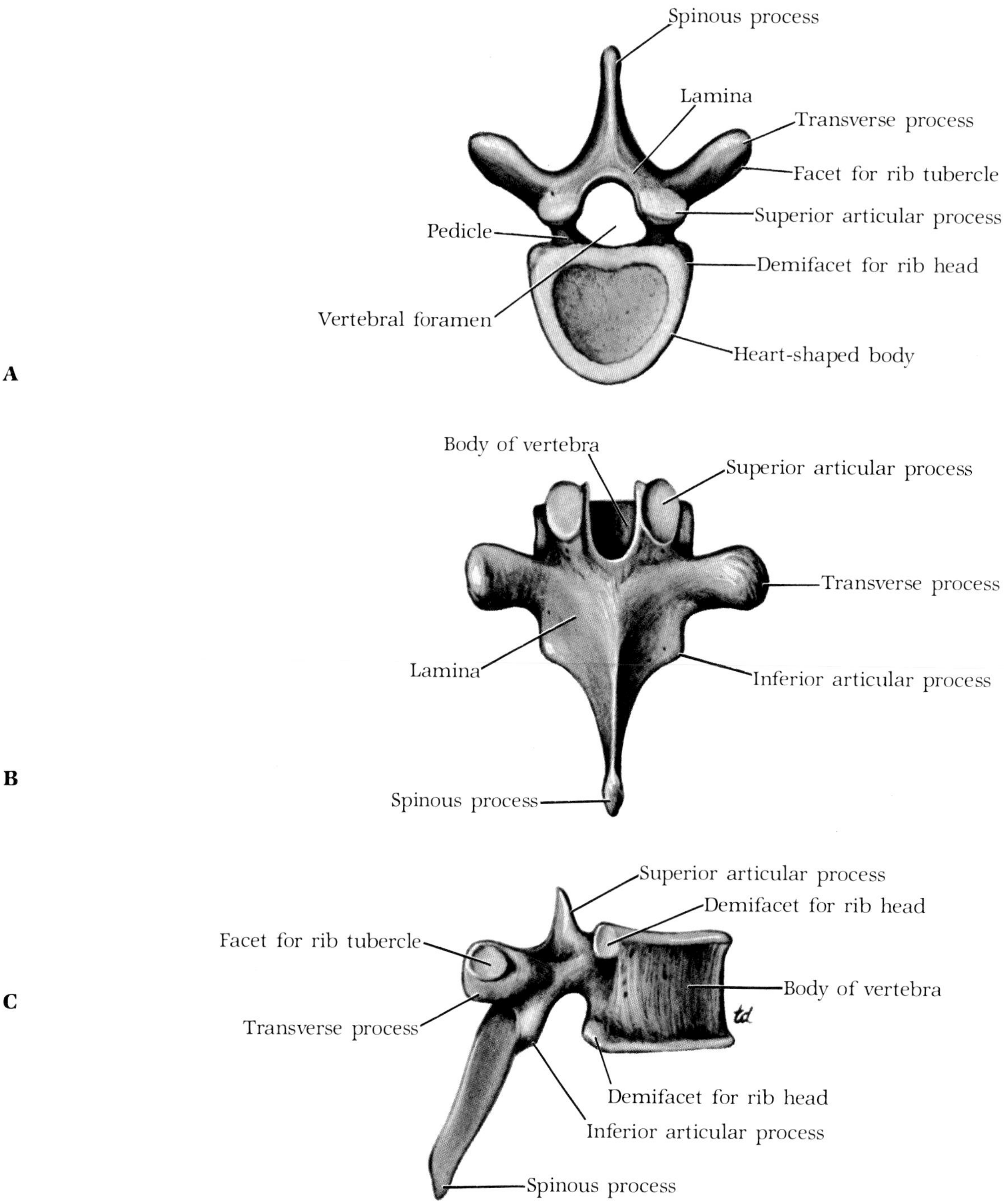
A
Spinous process
Lamina
Transverse process
Facet for rib tubercle
Superior articular process
Pedicle
Demifacet for rib head
Vertebral foramen
Heart-shaped body
B
Body of vertebra
Superior articular process
Transverse process
Lamina
Inferior articular process
Spinous process
C
Superior articular process
Demifacet for rib head
Facet for rib tubercle
Body of vertebra
Transverse process
td
Demifacet for rib head
Inferior articular process
Spinous process

Figure 8-8
Typical thoracic vertebra (sixth). A. Superior view. B. Posterior view. C. Lateral view.

1. Because the rib cage provides stability, fractures and dislocations of the upper and middle thoracic vertebrae are uncommon.
2. Compression fractures of the thoracic vertebral bodies are usually caused by an excessive flexion-compression type of injury and occur at the sites of maximum mobility, i.e., the lower thoracic region. With this type of fracture the body of the vertebra is crushed although the strong posterior longitudinal ligament remains intact. Because the vertebral arches remain unbroken and the intervertebral ligaments are intact, vertebral displacement and spinal cord injury do not occur.
3. Fracture-dislocations are caused by an excessive flexion-compression type of injury. Because the articular processes are fractured and the ligaments torn, the vertebrae involved are unstable and the spinal cord is usually severely damaged or severed, with accompanying paraplegia. Note that the vertebral foramen is relatively small in the thoracic region; consequently, the spinal cord is easily damaged in this area.
4. Fractures of the spinous processes, transverse processes, or laminae are caused by direct injury or, in rare cases, by severe muscular activity.

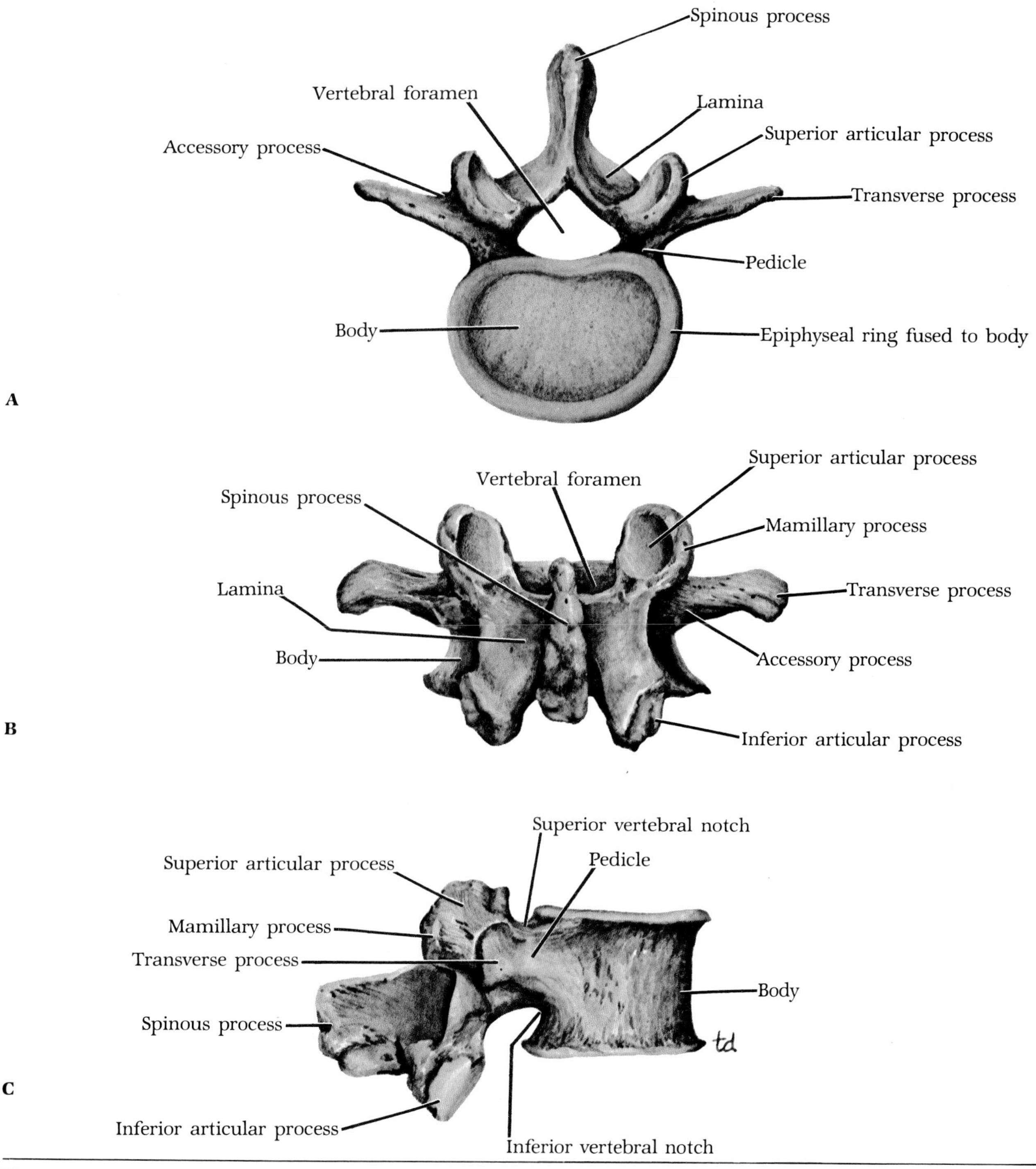
Spinous process
Vertebral foramen
Lamina
Accessory process
Superior articular process
Transverse process
Pedicle
Body
Epiphyseal ring fused to body
A
Superior articular process
Vertebral foramen
Spinous process
Mamillary process
Lamina
Transverse process
Body
Accessory process
Inferior articular process
B
Superior vertebral notch
Pedicle
Superior articular process
Mamillary process
Transverse process
Body
Spinous process
td
C
Inferior articular process
Inferior vertebral notch

Figure 8-9
Typical lumbar vertebra (third). A. Superior view. B. Posterior view. C. Lateral view.

1. Each intervertebral foramen is bounded above and below by the pedicles of adjacent vertebrae, anteriorly by the lower part of the vertebral body and by the intervertebral disc, and posteriorly by the articular processes and the joint between them. The spinal nerve thus is very vulnerable and may be pressed on or irritated by disease of the surrounding structures. Herniation of the intervertebral disc, fractures of the vertebral bodies, and osteoarthritis involving the joints of the articular processes or the joints between the vertebral bodies, may all result in pressure, stretching, or edema of the emerging spinal nerve.

2. Fractures of the spinous processes, transverse processes, or laminae are caused by direct injury or, in rare cases, by severe muscular activity.

3. Compression fractures of the vertebral bodies are usually caused by an excessive flexion-compression type of injury.

4. Fracture-dislocations are also caused by an excessive flexion-compression type of injury. Because the articular processes are fractured and the ligaments torn, the vertebrae are unstable and the spinal cord or cauda equina is usually severely damaged.

5. In spondylolisthesis, the body of a lower lumbar vertebra, usually the fifth, moves forward on the body of the vertebra below and carries with it the whole of the upper portion of the vertebral column. The condition is due to a congenital defect in the pedicles of the migrating vertebra.

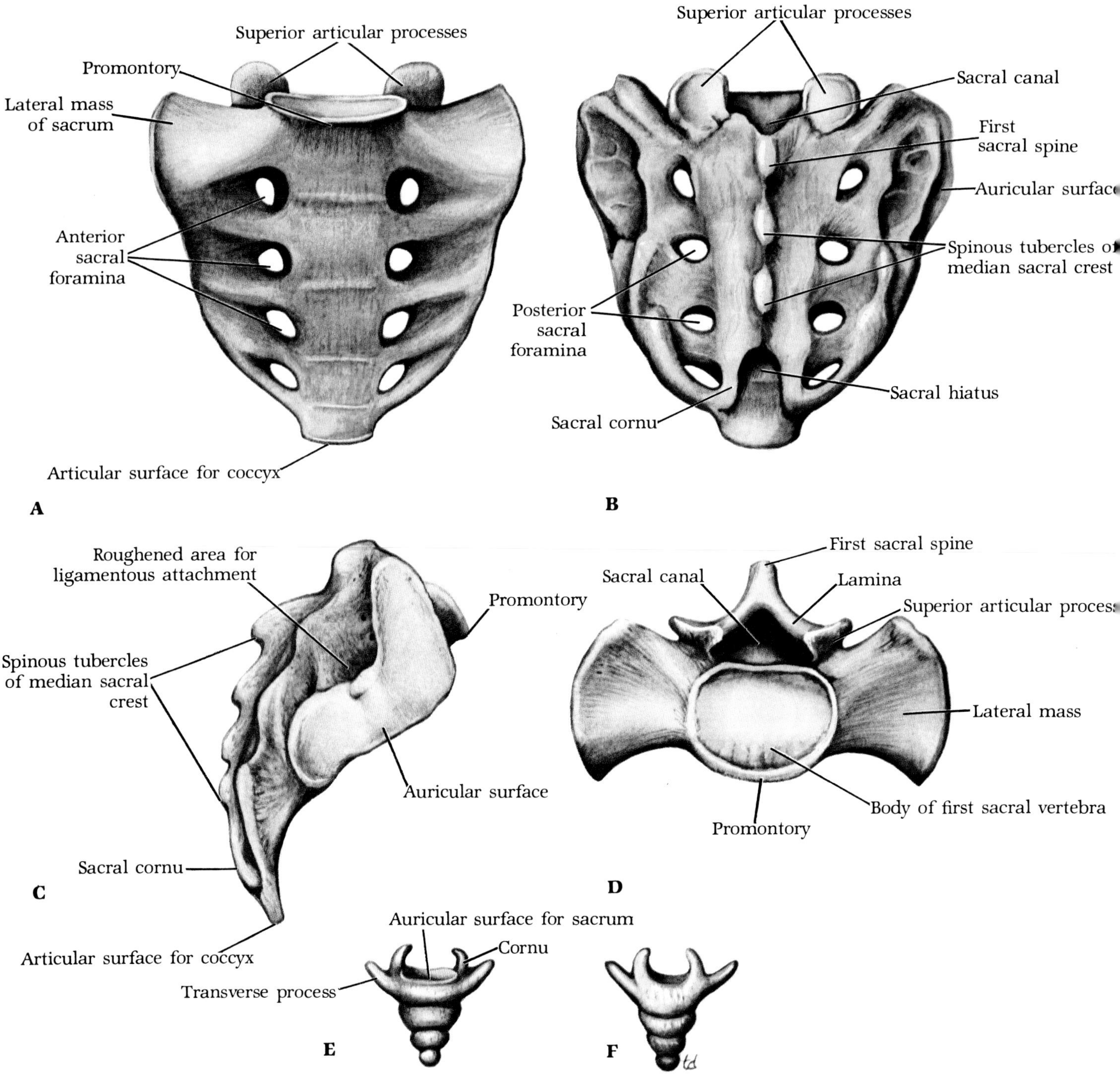
Superior articular processes
Promontory
Lateral mass of sacrum
Anterior sacral foramina
Articular surface for coccyx
A
Superior articular processes
Sacral canal
First sacral spine
Auricular surface
Spinous tubercles of median sacral crest
Posterior sacral foramina
Sacral hiatus
Sacral cornu
B
Roughened area for ligamentous attachment
Promontory
Spinous tubercles of median sacral crest
Auricular surface
Sacral cornu
C
Articular surface for coccyx
First sacral spine
Sacral canal
Lamina
Superior articular process
Lateral mass
Body of first sacral vertebra
Promontory
D
Auricular surface for sacrum
Cornu
Transverse process
E
F

Figure 8-10
Sacrum in the male. A. Anterior view. B. Posterior view. C. Lateral view. D. Superior view. Coccyx. E. Anterior view. F. Posterior view.

1. The first sacral vertebra may be partly or completely separated from the second sacral vertebra. Occasionally, radiography of the vertebral column has shown the fifth lumbar vertebra to be fused with the first sacral vertebra.

2. *Caudal Analgesia.* An anesthetic solution can be injected into the sacral canal through the sacral hiatus. The solution then acts on the spinal nerves of the second, third, fourth, and fifth sacral and the coccygeal segments of the cord as they emerge from the dura mater. The needle must be confined to the lower part of the sacral canal, since the meninges extend down as far as the lower border of the second sacral vertebra. Caudal analgesia is used obstetrically during parturition to block pain fibers from the cervix part of the uterus and to anesthetize the perineum.

3. Fractures of the sacrum may be associated with severe pelvic injuries. The fracture line usually involves the lateral mass of the sacrum and the anterior sacral foramina. Isolated fractures of the sacrum are rare.

4. The coccyx may be injured when a person falls on a hard surface while in the sitting position. There may be subperiosteal bruising, fracture, or fracture-dislocation of the sacrococcygeal joint. The injury is associated with considerable pain, which may continue for many months.

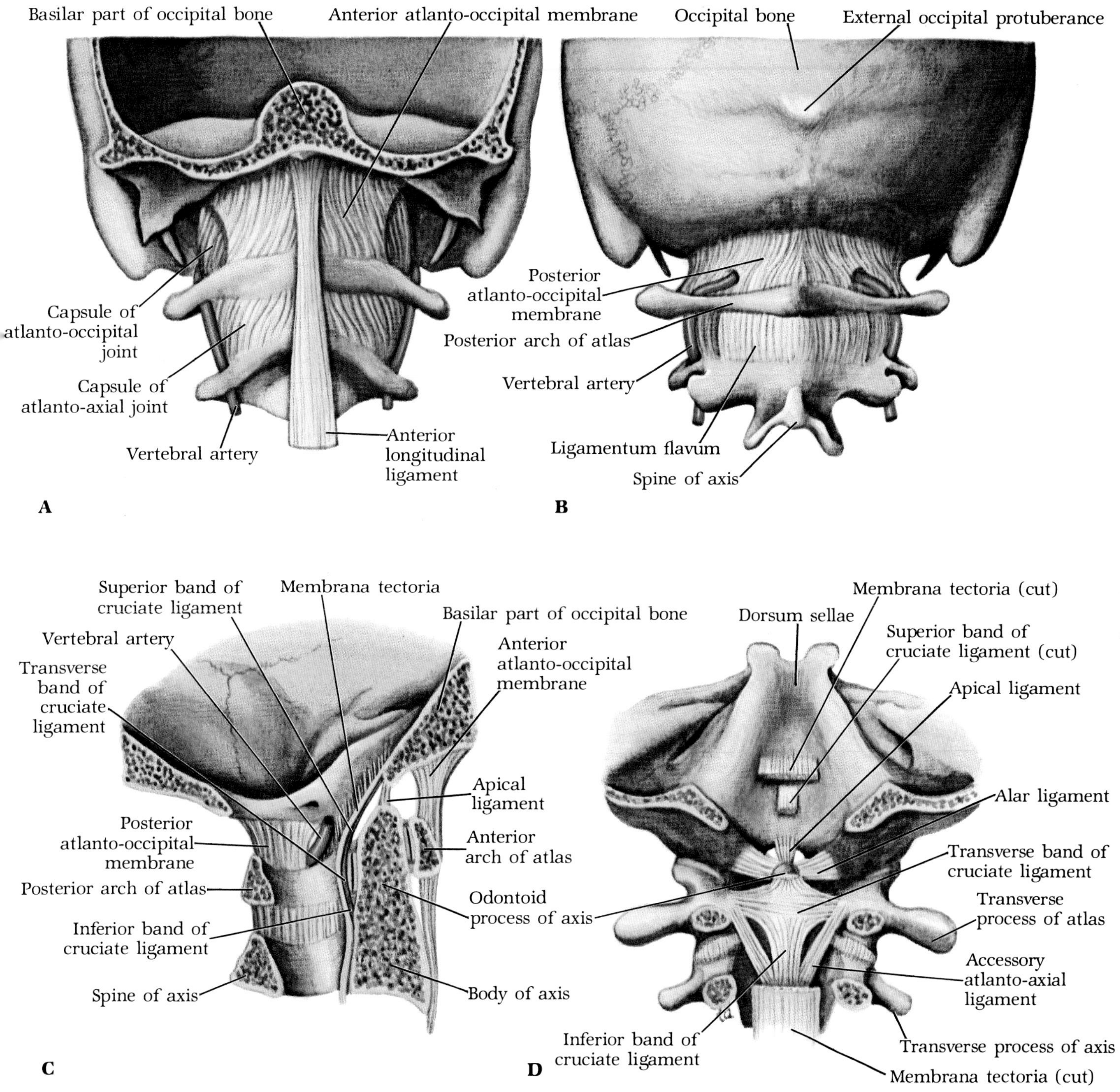
Basilar part of occipital bone
Anterior atlanto-occipital membrane
Occipital bone
External occipital protuberance
Capsule of atlanto-occipital joint
Capsule of atlanto-axial joint
Vertebral artery
Anterior longitudinal ligament
Posterior atlanto-occipital membrane
Posterior arch of atlas
Vertebral artery
Ligamentum flavum
Spine of axis
A
B
Superior band of cruciate ligament
Membrana tectoria
Basilar part of occipital bone
Vertebral artery
Anterior atlanto-occipital membrane
Transverse band of cruciate ligament
Apical ligament
Posterior atlanto-occipital membrane
Anterior arch of atlas
Posterior arch of atlas
Odontoid process of axis
Inferior band of cruciate ligament
Spine of axis
Body of axis
C
Membrana tectoria (cut)
Dorsum sellae
Superior band of cruciate ligament (cut)
Apical ligament
Alar ligament
Transverse band of cruciate ligament
Transverse process of atlas
Accessory atlanto-axial ligament
Inferior band of cruciate ligament
Transverse process of axis
Membrana tectoria (cut)
D

Figure 8-11
Atlanto-occipital joints. A. Anterior view. B. Posterior view. Atlanto-axial joints. C. Sagittal section. D. Posterior view.

1. *Fracture of the Atlas.* This is commonly caused by falls on the head. The blow is transmitted from the occipital condyles to the lateral masses of the atlas, which are then forced apart, causing a fracture of the posterior or anterior arch of the atlas. The spinal cord escapes injury in about half the cases.

2. *Dislocation of the Atlas.* The stability of the atlas largely depends on the integrity of the transverse ligament that holds the odontoid process of the axis to the posterior surface of the anterior arch of the atlas. If this ligament is torn, the atlas dislocates anteriorly. In such a situation there is great danger that the spinal cord will be damaged.

3. *Fracture-Dislocation of the Atlas.* If the fracture occurs at the base of the odontoid process, the atlas will dislocate anteriorly. In this instance the odontoid process moves forward with the anterior arch of the atlas, so that there is less likelihood that the spinal cord will be damaged. Note that in young children the atlas normally has a greater degree of forward movement on the axis in flexion than it does in the adult.

4. Spontaneous dislocation of the atlas may occur in children after a severe infection of the posterior pharyngeal wall (i.e., a retropharyngeal abscess). In these circumstances there is softening and stretching of the transverse ligament.

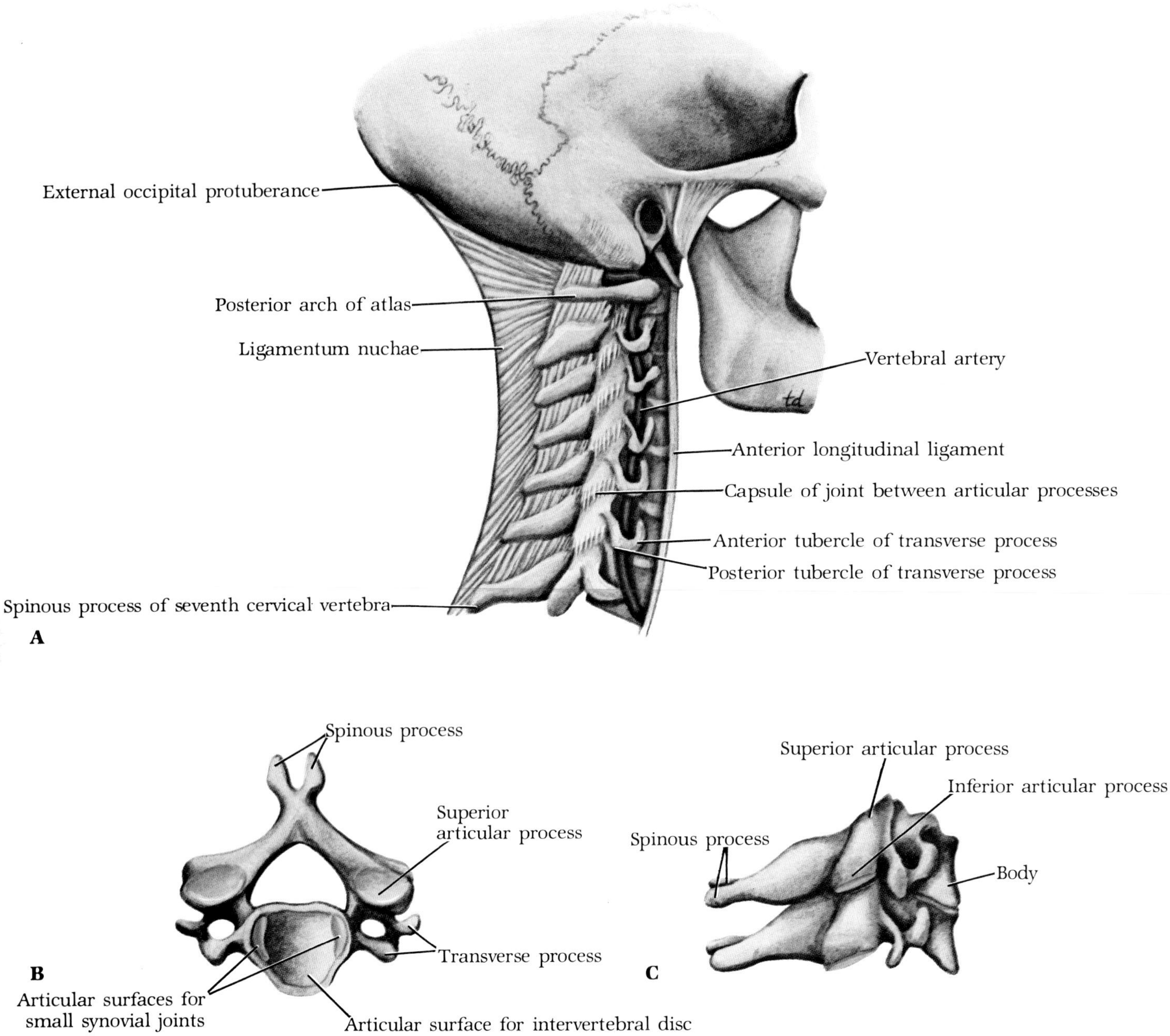
External occipital protuberance
Posterior arch of atlas
Ligamentum nuchae
Vertebral artery
Anterior longitudinal ligament
Capsule of joint between articular processes
Anterior tubercle of transverse process
Posterior tubercle of transverse process
Spinous process of seventh cervical vertebra
A
Spinous process
Superior articular process
Transverse process
B
Articular surfaces for small synovial joints
Articular surface for intervertebral disc
Superior articular process
Inferior articular process
Spinous process
Body
C

Figure 8-12
Joints between cervical vertebrae. A. Lateral view. B, C. Articular surfaces.

1. With the exception of the first two cervical vertebrae, the remainder of the mobile vertebrae of the spine articulate with each other by means of cartilaginous joints between their bodies and by synovial joints between their articular processes. In the cervical region there are, in addition, small synovial joints present at the sides of the intervertebral discs between the upper and lower surfaces of the bodies of the vertebrae.
2. The cartilaginous joint between the vertebral bodies is extremely strong because of the firm attachment of the anulus fibrosus to the vertebral bodies, which is reinforced by the attachment of the anterior and posterior longitudinal ligaments.
3. The synovial joints between the articular processes are strengthened by the ligamentum flava and the ligamentum nuchae (homologous with the interspinous and supraspinous ligaments in other vertebral regions).
4. *Dislocations.* These commonly occur between the fourth and fifth or fifth and sixth cervical vertebrae where mobility is greatest. In a unilateral dislocation the inferior articular process of one vertebra is forced forward over the anterior margin of the superior articular process of the vertebra below. The spinal nerve on the same side is usually nipped in the intervertebral foramen, producing severe pain. Fortunately the large size of the vertebral canal allows the spinal cord to escape damage in most cases.
Bilateral cervical dislocations are almost always associated with severe injury to the spinal cord. Death occurs immediately if the upper cervical vertebrae are involved, since the respiratory muscles, including the diaphragm, are paralyzed.
5. *Flexion Fractures.* If forced flexion of the cervical vertebrae occurs and the posterior ligaments remain intact, the vertebral bodies are compressed, producing a wedge-shaped deformity of the anterior parts of the bodies. Because the articular processes are intact and if the posterior ligaments are intact, the injury is stable.
6. *Extension Fractures.* These commonly occur in automobile accidents when the neck is hyperextended. The anterior longitudinal ligament is stretched or ruptured, permitting the vertebral bodies to separate. In severe cases there may be extensive damage to the spinal cord.

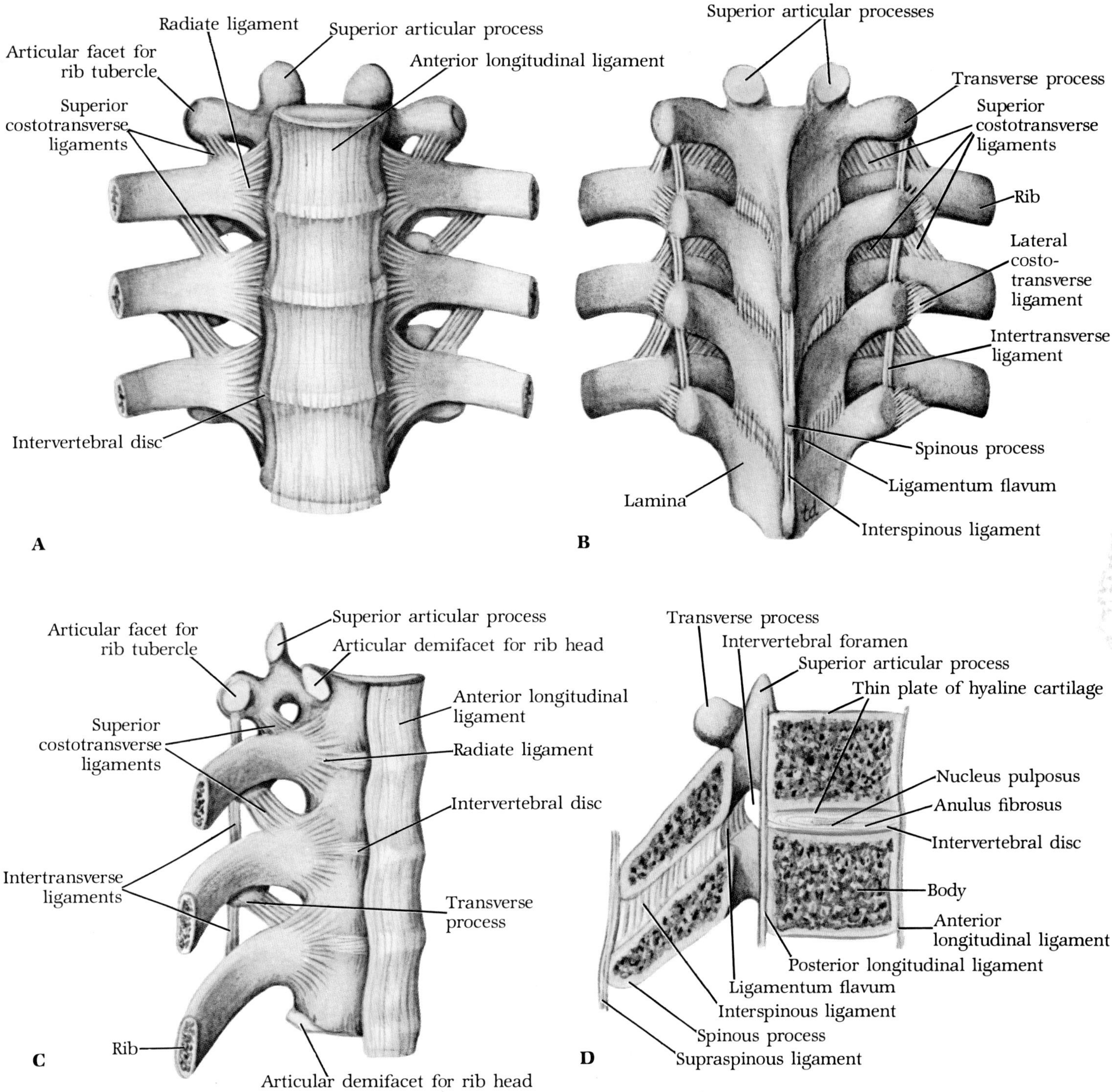
Radiate ligament
Superior articular process
Articular facet for rib tubercle
Anterior longitudinal ligament
Superior costotransverse ligaments
Intervertebral disc
A
Superior articular processes
Transverse process
Superior costotransverse ligaments
Rib
Lateral costo-transverse ligament
Intertransverse ligament
Spinous process
Ligamentum flavum
Lamina
Interspinous ligament
B
Superior articular process
Articular facet for rib tubercle
Articular demifacet for rib head
Anterior longitudinal ligament
Superior costotransverse ligaments
Radiate ligament
Intervertebral disc
Intertransverse ligaments
Transverse process
Rib
C
Articular demifacet for rib head
Transverse process
Intervertebral foramen
Superior articular process
Thin plate of hyaline cartilage
Nucleus pulposus
Anulus fibrosus
Intervertebral disc
Body
Anterior longitudinal ligament
Posterior longitudinal ligament
Ligamentum flavum
Interspinous ligament
Spinous process
Supraspinous ligament
D

Figure 8-13
Joints between thoracic vertebrae. A. Anterior view. B. Posterior view. C. Lateral view. D. Sagittal section.

1. In the thoracic region the intervertebral joints are strengthened by the thoracic cage. Fractures of the upper thoracic vertebral bodies are usually accompanied by buckling of the sternum.

2. *Flexion Fractures.* Compression fractures of the vertebral bodies are usually caused by an excessive flexion-compression type of injury and usually occur in the lower thoracic region. The body of the vertebra is crushed and becomes wedge-shaped, although the strong posterior longitudinal ligament remains intact. Because the vertebral arches remain unbroken and the intervertebral ligaments are intact, vertebral displacement and spinal cord injury do not occur.

3. *Fracture-Dislocations.* These are caused by an excessive flexion-compression type of injury and take place at the site of maximum mobility, i.e., the junction of the thoracic with the lumbar region of the vertebral column. Because the articular processes are fractured and the ligaments torn, the vertebrae are unstable and the spinal cord is usually severely damaged or severed, with accompanying paraplegia. Note that the size of the vertebral canal in the thoracic region is small compared to the size of the spinal cord; thus even small vertebral displacements are often accompanied by considerable damage to the spinal cord.

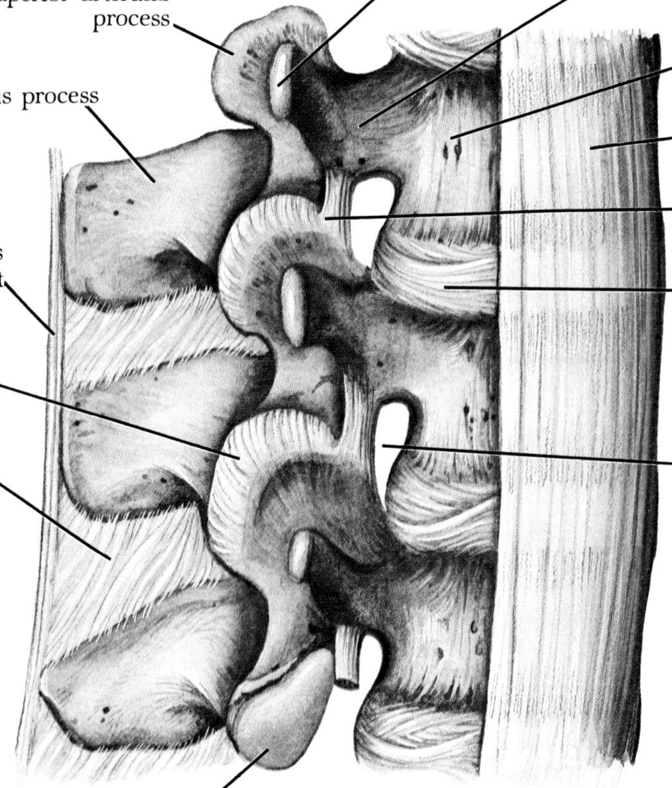
Superior articular process
Transverse process
Pedicle
Body of vertebra
Spinous process
Anterior longitudinal ligament
Ligamentum flavum
Supraspinous ligament
Intervertebral disc
Capsule of joint
Interspinous ligament
Intervertebral foramen
A
Inferior articular process

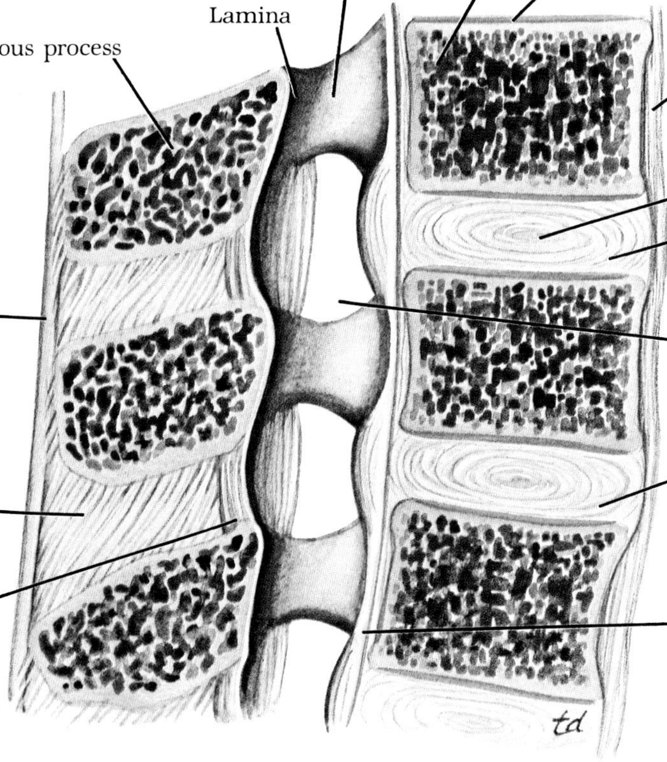
Pedicle
Body
Thin plate of hyaline cartilage
Lamina
Spinous process
Anterior longitudinal ligament
Nucleus pulposus
Anulus fibrosus
Supraspinous ligament
Intervertebral foramen
Intervertebral disc
Interspinous ligament
Ligamentum flavum
Posterior longitudinal ligament
td
B

Figure 8-14
Joints between lumbar vertebrae. A. Lateral view. B. Sagittal section.

1. Flexion-compression fractures and fracture-dislocations are common in the lumbar region. They may result from a fall from a height onto the heels or buttocks or from a heavy object falling on the back when a person is in a crouched position.

2. In fracture-dislocations of the lumbar region, two anatomical facts aid the patient. First, the spinal cord in the adult extends down only as far as the level of the lower border of the first lumbar vertebra. Second, the large size of the vertebral foramen in this region gives the roots of the cauda equina ample room. Nerve injury may therefore be minimal in this region.

3. In the lumbar region the intervertebral discs are thickest, the joint surfaces are largest, and the range of movement between adjacent vertebrae is greatest. These conditions particularly prevail in the lower lumbar region, and it is here that herniation of intervertebral discs and osteoarthritis are commonly found.

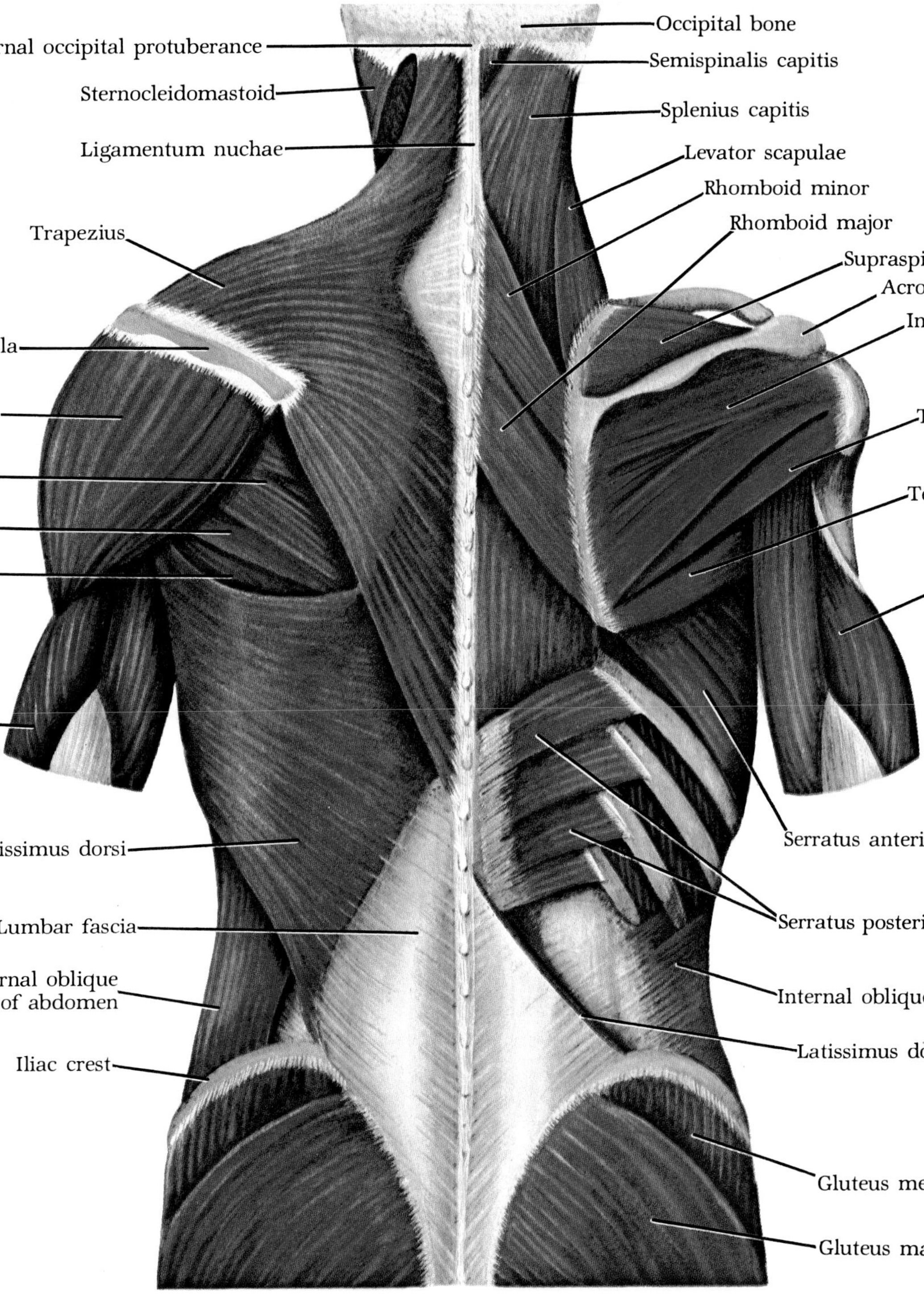

Occipital bone
External occipital protuberance
Semispinalis capitis
Sternocleidomastoid
Splenius capitis
Ligamentum nuchae
Levator scapulae
Rhomboid minor
Rhomboid major
Trapezius
Supraspinatus
Acromion process
Infraspinatus
Spine of scapula
Deltoid
Teres minor
Infraspinatus
Teres minor
Teres major
Teres major
Triceps
Triceps
Serratus anterior
Latissimus dorsi
Serratus posterior inferior
Lumbar fascia
External oblique of abdomen
Internal oblique of abdomen
Latissimus dorsi (cut)
Iliac crest
Gluteus medius
Gluteus maximus

Figure 8-15
Superficial layer of muscles of the back.

1. The two scapulae afford protection against trauma to the upper part of the rib cage. The tone of the muscles attached to them maintains the normal postural position of the scapulae.
2. Paralysis of the trapezius muscle would result in a dropped shoulder. Paralysis of the levator scapulae and rhomboid muscles would produce a sagging of the medial border of the scapula, and the patient would experience difficulty in bracing the shoulders backward. Paralysis of the serratus anterior muscle would result in winged scapula.
3. The so-called auscultatory triangle lies between the superior border of the latissimus dorsi, the lateral border of the trapezius, and the medial border of the rhomboid major muscles. Because there is a deficiency of the superficial back muscles in this area, a stethoscope placed here will be closer to the lungs, and the breath sounds may be heard more clearly than elsewhere.
4. The lumbar triangle is bounded laterally by the posterior border of the external oblique muscle, medially by the lateral border of the latissimus dorsi muscle, and inferiorly by the iliac crest; the floor of the triangle is formed by the internal oblique muscle. Infection of the vertebral column with abscess formation may result in pus tracking laterally and inferiorly and emerging from beneath the latissimus dorsi in the lumbar triangle.
5. Again note that the curve of the shoulder is produced by the greater tuberosity of the humerus displacing the deltoid laterally.

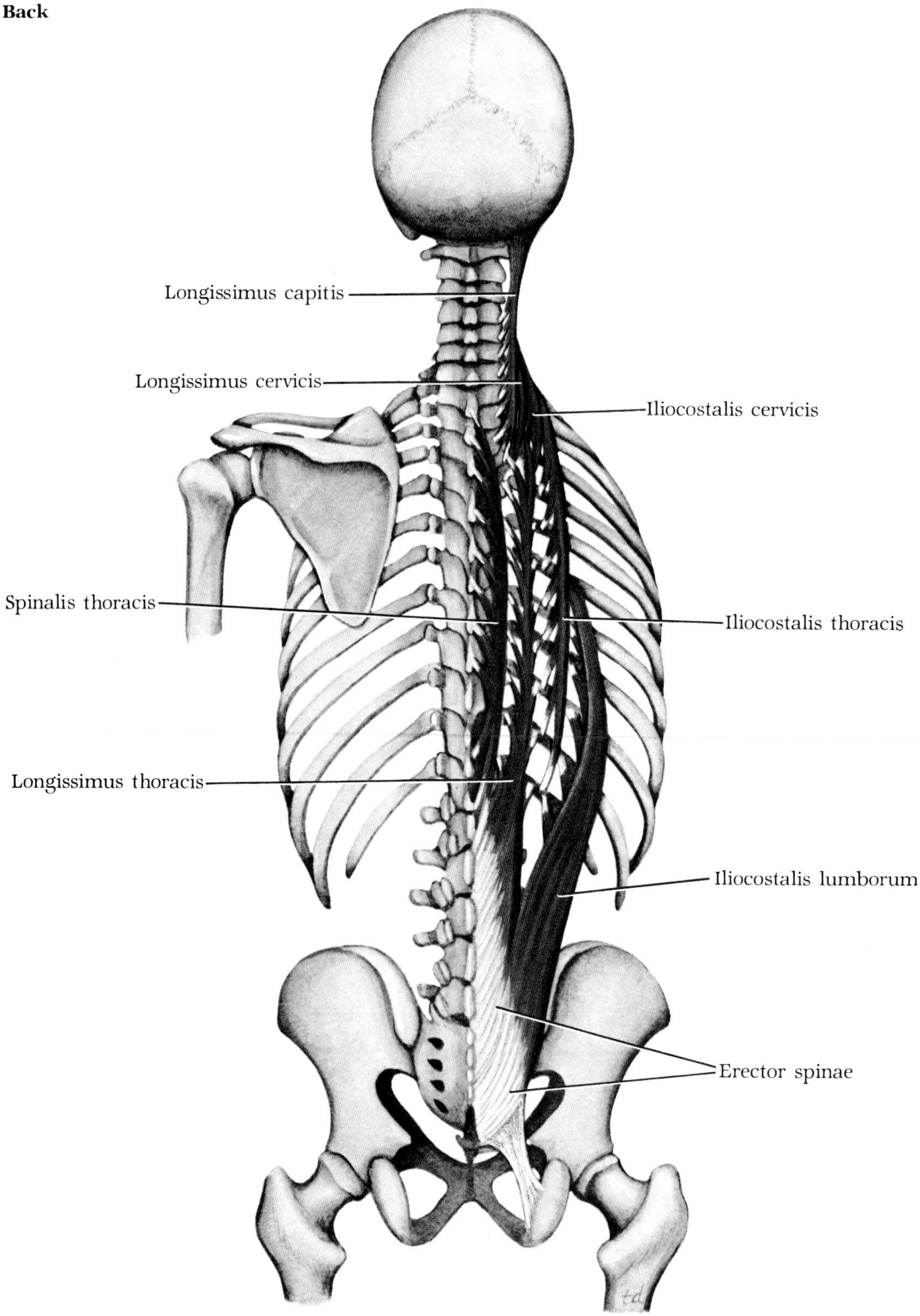
Longissimus capitis
Longissimus cervicis
Iliocostalis cervicis
Spinalis thoracis
Iliocostalis thoracis
Longissimus thoracis
Iliocostalis lumborum
Erector spinae

Figure 8-16
Intermediate layer of muscles of the back.

1. The postvertebral muscles are well developed in man because when a person is in the standing position the line of gravity falls anterior to the vertebral column. The postural tone of these muscles is the major factor responsible for the maintenance of the normal curves of the vertebral column.
2. The muscles of the back form a broad, thick column of muscle tissue that occupies the hollow on each side of the spinous processes. They extend from the sacrum to the skull and lie beneath the thoracolumbar fascia.
3. The muscle mass is composed of many separate muscles of varying length. Each individual muscle may be regarded as a string that, when pulled on, causes one of several vertebrae to be extended or rotated on the vertebra below. Since the origins and insertions of the different groups of muscles overlap, entire regions of the vertebral column can be made to move smoothly.
4. To examine these muscles have the patient disrobe completely. With the patient standing upright ask him to extend his head to relax the overlying fascia. Having positioned yourself behind the patient, run your fingers down the postvertebral muscles, examining them for increased tone or tenderness. Look for scoliosis or other vertebral deformities produced by a paralysis of the muscles, as in poliomyelitis, or muscle spasm caused by irritation of motor nerves. If the postvertebral muscles are in spasm on both sides, they stand out as hard ridges and may completely hide the normal curvature of the lumbar part of the vertebral column. Because the postvertebral muscles are segmentally innervated, small areas of atrophy may be detected if the segmental nerves are injured.

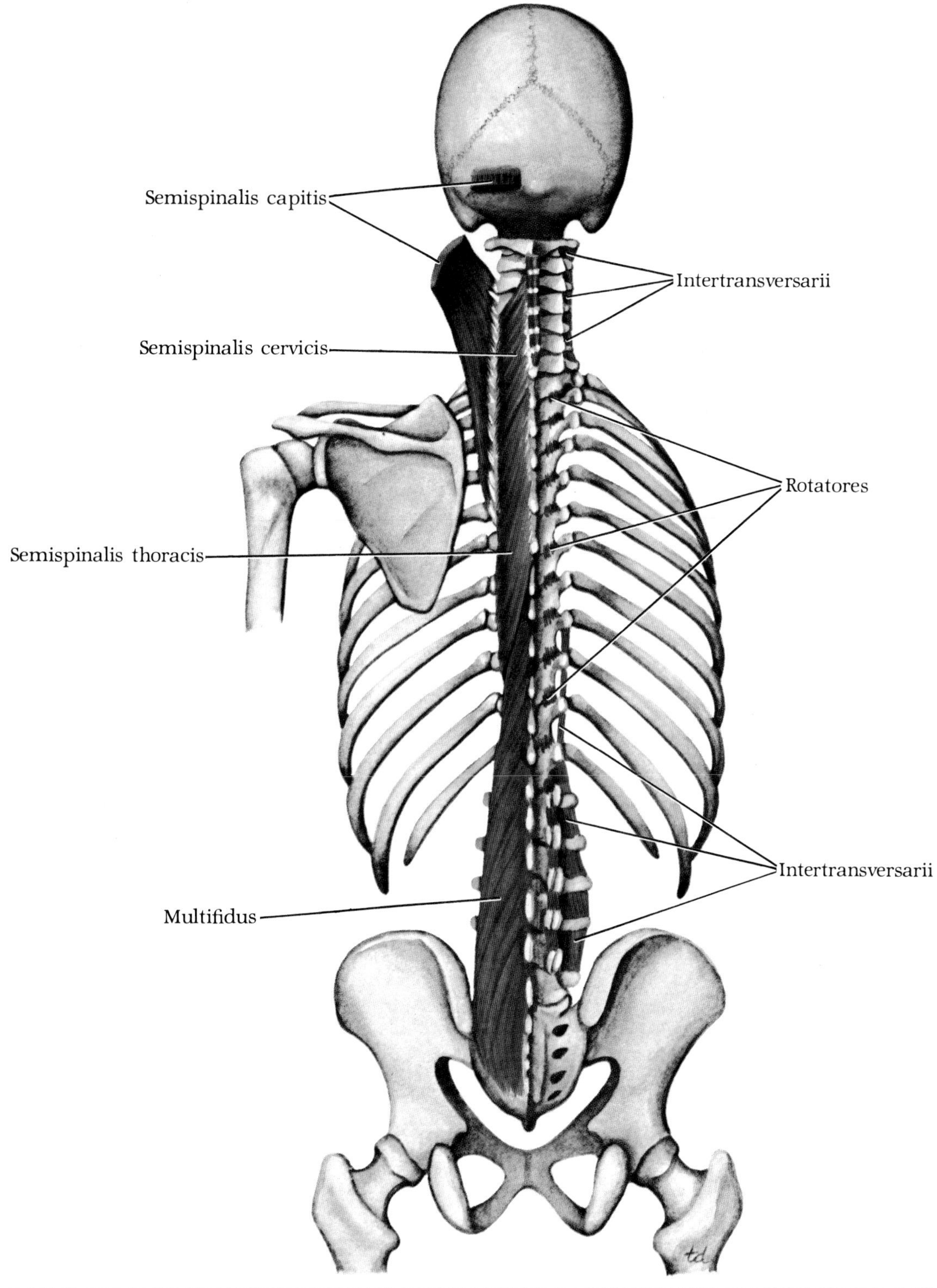
Semispinalis capitis
Intertransversarii
Semispinalis cervicis
Rotatores
Semispinalis thoracis
Intertransversarii
Multifidus

Figure 8-17
Deep layer of muscles of the back.

1. The spinous processes and transverse processes of the vertebrae serve as levers that facilitate muscle actions. The muscles of longest length lie superficially and run vertically from the sacrum to the rib angles, the transverse process, and the upper vertebral spinous processes. The muscles of intermediate length run obliquely from transverse processes to spinous processes. The shortest and deepest muscle fibers run between spinous processes and between transverse processes of adjacent vertebrae.

2. The normal range of movement of the different parts of the vertebral column should be tested during examination of a patient's back. Remember that much of the movement we refer to as flexion is carried out at the atlanto-occipital joints and the hip joints. To determine flexion the patient is asked to touch his toes with the knee joints extended. The degree of flexion is measured by the distance of the tips of the fingers from the ground. Extension is measured by asking the patient to bend backward. Lateral flexion is demonstrated by asking the patient to slide each hand down the lateral side of his thigh. In the test for movement of rotation, it is important that the patient's pelvis be fixed while he rotates the upper part of his trunk. In the cervical region fifty percent of rotation occurs between the atlas and the axis, while the remainder occurs between the other five cervical vertebrae.

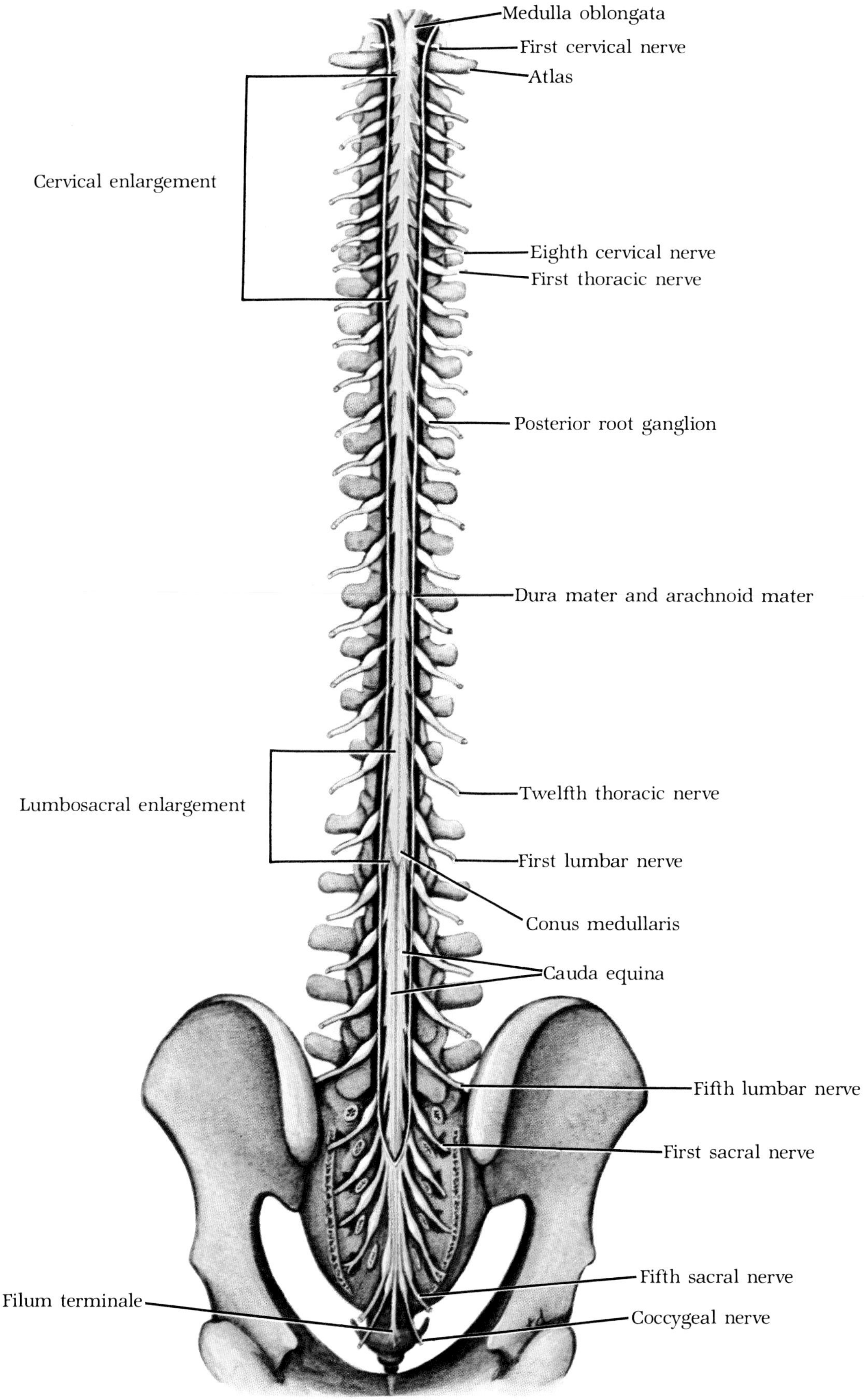
Medulla oblongata
First cervical nerve
Atlas
Cervical enlargement
Eighth cervical nerve
First thoracic nerve
Posterior root ganglion
Dura mater and arachnoid mater
Twelfth thoracic nerve
Lumbosacral enlargement
First lumbar nerve
Conus medullaris
Cauda equina
Fifth lumbar nerve
First sacral nerve
Fifth sacral nerve
Filum terminale
Coccygeal nerve

Figure 8-18
Spinal cord lying within vertebral canal, posterior view.

1. The spinal cord begins at the foramen magnum where it is continuous with the medulla oblongata of the brain. In the adult it terminates at the level of the lower border of the first lumbar vertebra. In the young child it is relatively longer and usually ends at the upper border of the third lumbar vertebra.

2. The spinal nerve roots pass from the spinal cord to the level of their respective intervertebral foramina; here they unite to form a spinal nerve.

3. Because of the disproportionate growth in length of the vertebral column during early development, as compared with that of the spinal cord, the length of the nerve roots increases progressively from above downward. In the upper cervical region the spinal nerve roots are short and run almost horizontally, but the roots of the lumbar and sacral nerves below the level of the termination of the cord form a vertical leach of nerves around the filum terminale called the cauda equina.

4. The nerve tracts of the spinal cord cannot regenerate after they have been severed. Injury to the nerve roots, on the other hand, is not incompatible with complete recovery since regeneration occurs as in any other peripheral nerve.

5. Fracture-dislocations of the vertebral column occur most commonly at the thoracolumbar junction. Here the lumbar, sacral, and coccygeal segments of the spinal cord lie at a level between the tenth thoracic and the first lumbar vertebrae. Unfortunately it is these segments of the spinal cord that are responsible for receiving sensation from the lower limbs, controlling movements of the lower limbs, and controlling bladder, rectal, and genital functions.

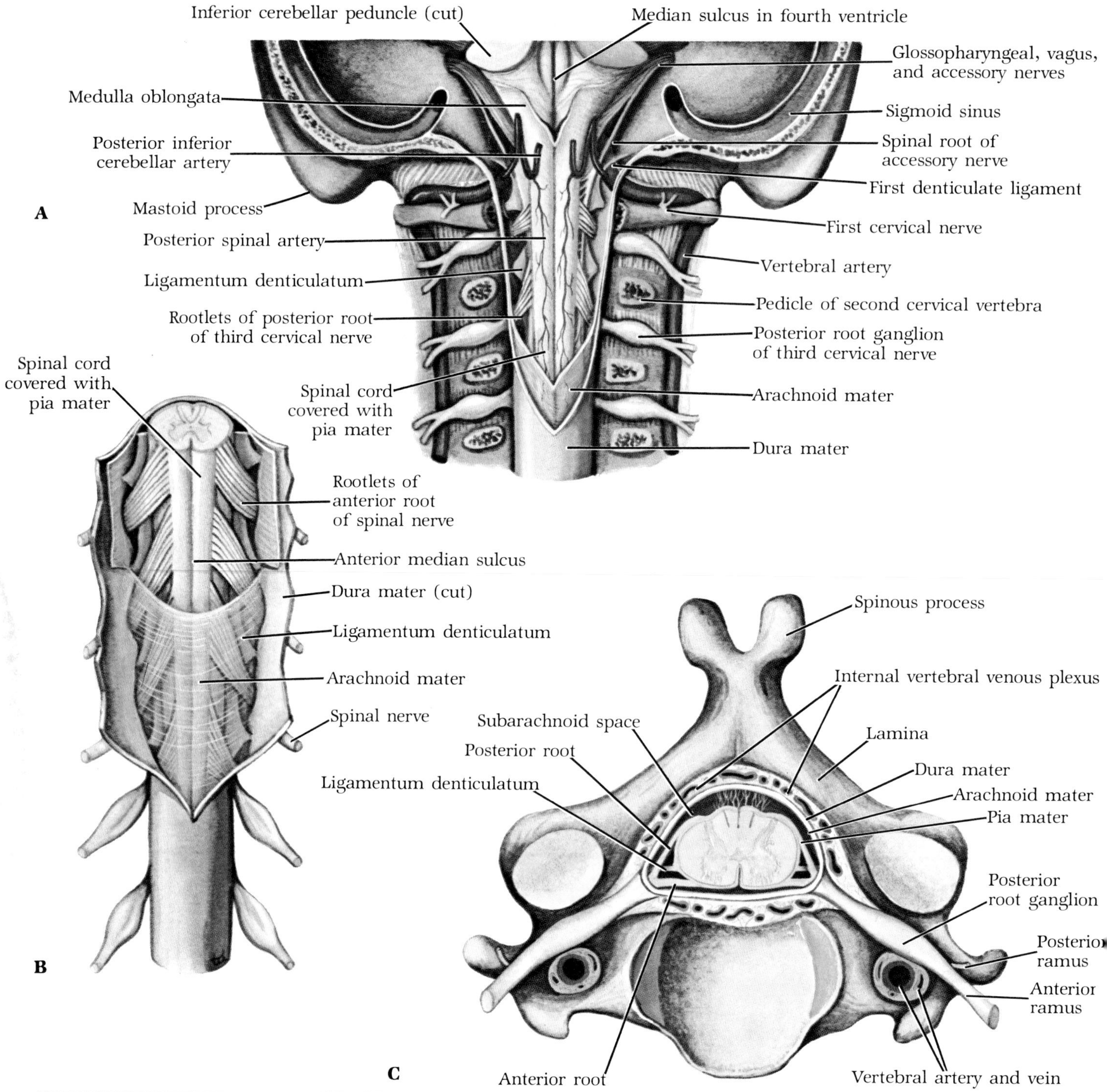
Inferior cerebellar peduncle (cut)
Median sulcus in fourth ventricle
Glossopharyngeal, vagus, and accessory nerves
Medulla oblongata
Sigmoid sinus
Posterior inferior cerebellar artery
Spinal root of accessory nerve
First denticulate ligament
A
Mastoid process
First cervical nerve
Posterior spinal artery
Vertebral artery
Ligamentum denticulatum
Pedicle of second cervical vertebra
Rootlets of posterior root of third cervical nerve
Posterior root ganglion of third cervical nerve
Spinal cord covered with pia mater
Spinal cord covered with pia mater
Arachnoid mater
Dura mater
Rootlets of anterior root of spinal nerve
Anterior median sulcus
Dura mater (cut)
Ligamentum denticulatum
Arachnoid mater
Spinal nerve
B
Spinous process
Internal vertebral venous plexus
Subarachnoid space
Lamina
Posterior root
Dura mater
Ligamentum denticulatum
Arachnoid mater
Pia mater
Posterior root ganglion
Posterio ramus
Anterior ramus
C
Anterior root
Vertebral artery and vein

Figure 8-19
Spinal cord, veins, and meninges. A. Medulla oblongata, upper cervical spinal cord, and meninges, posterior view. B. Spinal cord and meninges, anterior view. C. Cervical spinal cord and meninges, transverse section.

1. In patients suspected of having an intracranial tumor where there is a marked rise in intracranial pressure, lumbar puncture should never be performed. The sudden withdrawal of cerebrospinal fluid from the lumbar subarachnoid space may cause the medulla oblongata and part of the cerebellum to prolapse inferiorly through the foramen magnum, producing fatal results.

2. Prolapse of an intervertebral disc is common in the lower cervical region. A central prolapse may cause compression of the spinal cord with signs and symptoms of injury to the nerve pathways within the cord. A lateral prolapse may cause nerve root compression with pain in the arm.

3. The most common tumor of the vertebral column is a secondary carcinoma, the primary tumor being frequently located in the breast, thyroid, lung, kidney, or prostate. The spinal cord may be compressed as the result of the collapse of the vertebral body or direct extension of the tumor into the vertebral canal.

4. In the cervical region, dislocation or fracture-dislocation is common, but the large size of the vertebral canal often protects the spinal cord from severe injury. However, when there is considerable displacement, the cord is sectioned and death occurs immediately. Respiration ceases if the lesion occurs above the segmental origin of the phrenic nerves.

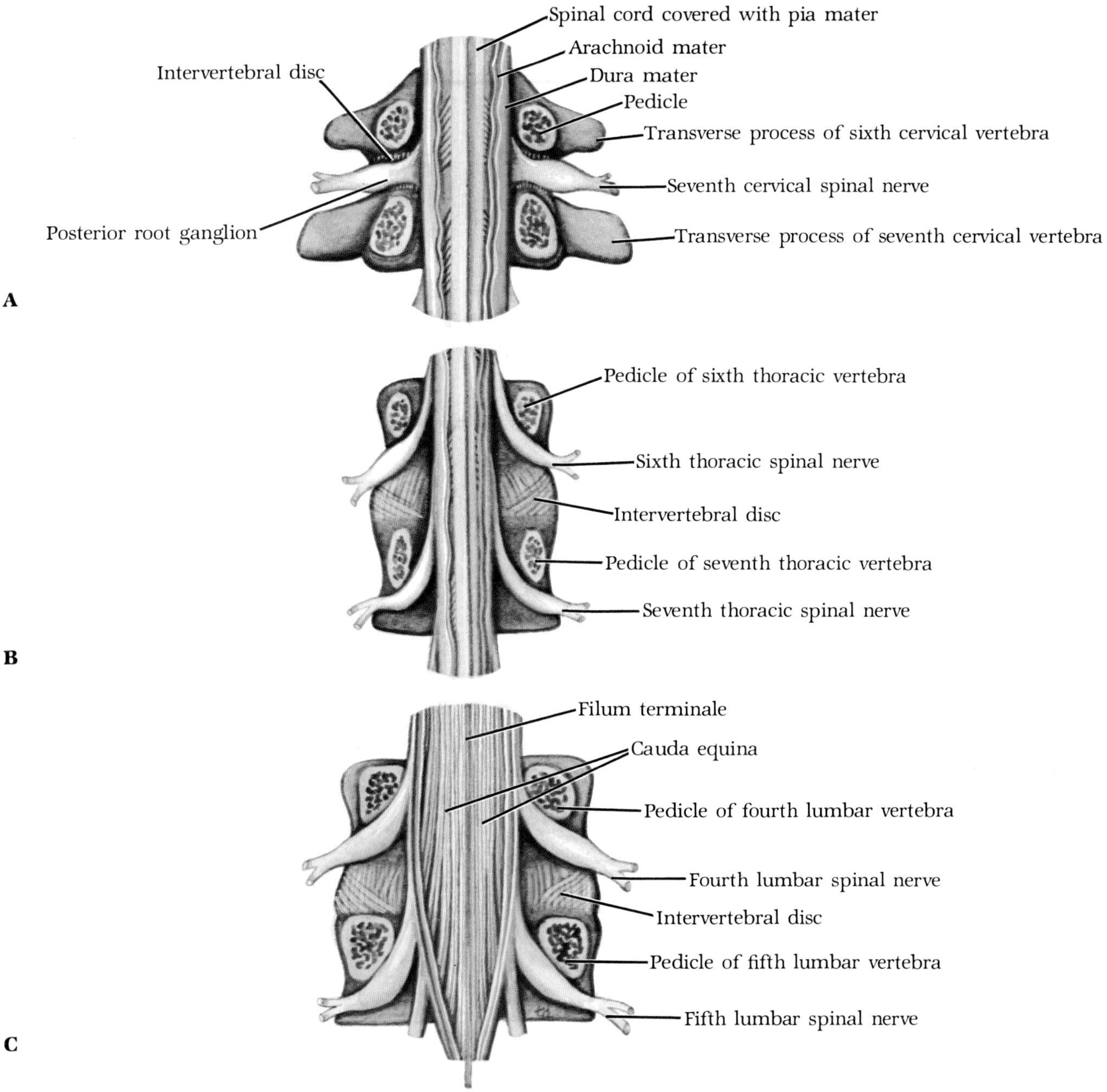
Spinal cord covered with pia mater
Arachnoid mater
Intervertebral disc
Dura mater
Pedicle
Transverse process of sixth cervical vertebra
Seventh cervical spinal nerve
Posterior root ganglion
Transverse process of seventh cervical vertebra
A
Pedicle of sixth thoracic vertebra
Sixth thoracic spinal nerve
Intervertebral disc
Pedicle of seventh thoracic vertebra
Seventh thoracic spinal nerve
B
Filum terminale
Cauda equina
Pedicle of fourth lumbar vertebra
Fourth lumbar spinal nerve
Intervertebral disc
Pedicle of fifth lumbar vertebra
Fifth lumbar spinal nerve
C

Figure 8-20
Emerging spinal nerves and their relationship to the intervertebral discs. A. In the cervical region. B. In the thoracic region. C. In the lumbar region.

1. *Cervical Disc Herniation.* In the cervical region the spinal nerve roots are short and run almost horizontally to reach their respective intervertebral foramina. Since there are eight spinal nerves in this region and only seven vertebrae, and since spinal nerves one through seven emerge above the corresponding vertebrae, the eighth cervical nerve is left to emerge between the seventh cervical vertebra and the first thoracic vertebra. The discs most liable to herniate are those between the fifth and sixth or sixth and seventh vertebrae. Lateral protrusions cause pressure on a spinal root. Thus, for example, protrusion of the disc between the fifth and sixth cervical vertebrae compresses the C6 nerve root. Pain is felt in the lower part of the back of the neck and shoulder and along the arm in the distribution of the nerve root involved.
Central protrusions may press on the spinal cord and the anterior spinal artery and involve the nervous pathways within the cord.

2. *Thoracic Disc Herniation.* This is rare.

3. *Lumbar Disc Herniation.* In the lumbar region the roots of the lumbar, sacral, and coccygeal nerves are long and form the cauda equina. They run almost vertically downward, lying posterior to a number of intervertebral discs. A lateral herniation may press on one or two roots and often involves the nerve root going to the intervertebral foramen just below. A large central herniation may compress the whole cauda equina, producing paraplegia. Since the sensory roots most commonly pressed upon are the fifth lumbar and the first sacral, pain is usually felt down the back and lateral side of the leg, radiating to the sole of the foot. In severe cases there may be paresthesia or actual sensory loss. Pressure on the motor roots causes muscle weakness. Involvement of the fifth lumbar motor root produces weakness of dorsiflexion of the ankle, while pressure on the first sacral motor root causes weakness of plantar flexion, and the ankle jerk may be diminished or absent. A large, centrally placed protrusion may give rise to bilateral pain and muscle weakness in both legs. Acute retention of urine may also occur. Lumbar disc herniations are more common than cervical disc herniations.

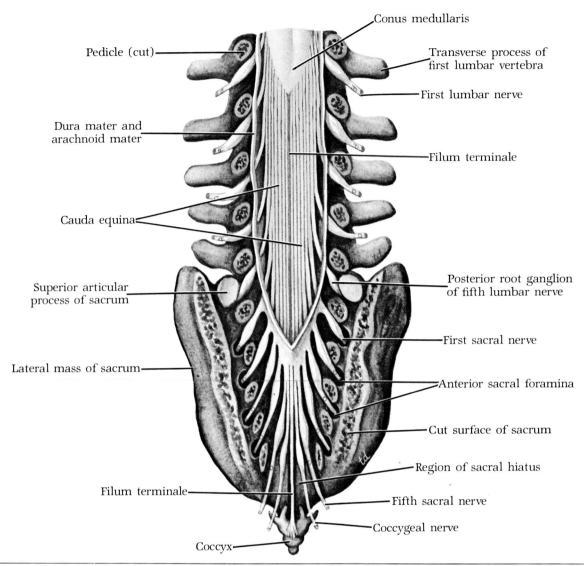

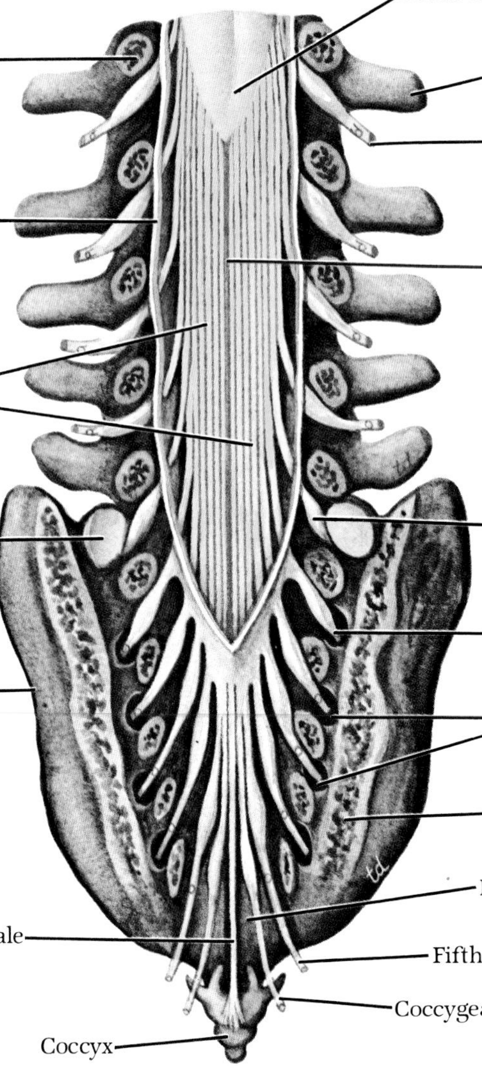
Conus medullaris
Pedicle (cut)
Transverse process of first lumbar vertebra
First lumbar nerve
Dura mater and arachnoid mater
Filum terminale
Cauda equina
Posterior root ganglion of fifth lumbar nerve
Superior articular process of sacrum
First sacral nerve
Lateral mass of sacrum
Anterior sacral foramina
Cut surface of sacrum
Region of sacral hiatus
Filum terminale
Fifth sacral nerve
Coccygeal nerve
Coccyx

Figure 8-21
Sacrum, posterior view. The laminae have been removed to show sacral nerve roots lying within the sacral canal.

1. The subarachnoid space extends inferiorly as far as the lower border of the second sacral vertebra. The lower lumbar and upper sacral parts of the vertebral canal are thus occupied by the subarachnoid space, which contains the lumbar and sacral nerve roots (the cauda equina) and the filum terminale. The lower sacral part of the vertebral canal is occupied by the lower sacral nerve roots, the roots of the coccygeal nerve, and the filum terminale; note how the nerve roots emerge from the subarachnoid space and are enclosed in a dural sheath as they pass to the intervertebral foramina (modified in the sacrum). Here they unite to form the spinal nerves, and the anterior and posterior rami exit through the anterior and posterior sacral foramina, respectively.

2. *Caudal Anesthesia.* An anesthetic solution can be injected into the sacral canal through the sacral hiatus. The solution then passes upward in the loose connective tissue and bathes the spinal nerves as they emerge from the dural sheath. Obstetricians use this method of nerve block to relieve the pains of the second stage of labor. Its advantage is that, administered by this method, the anesthetic does not affect the infant.

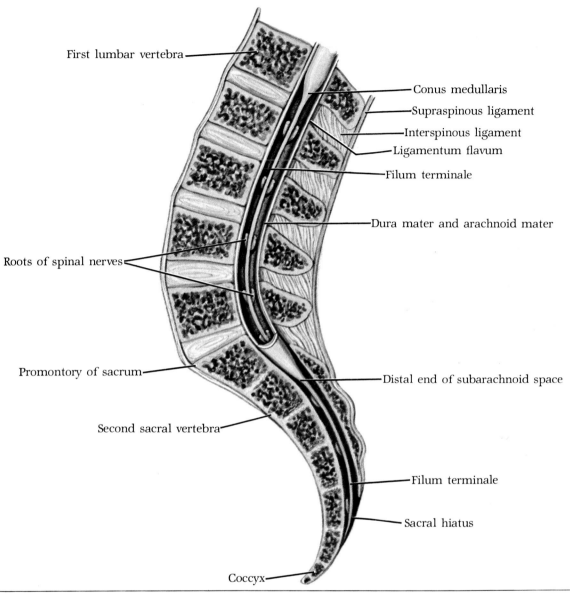

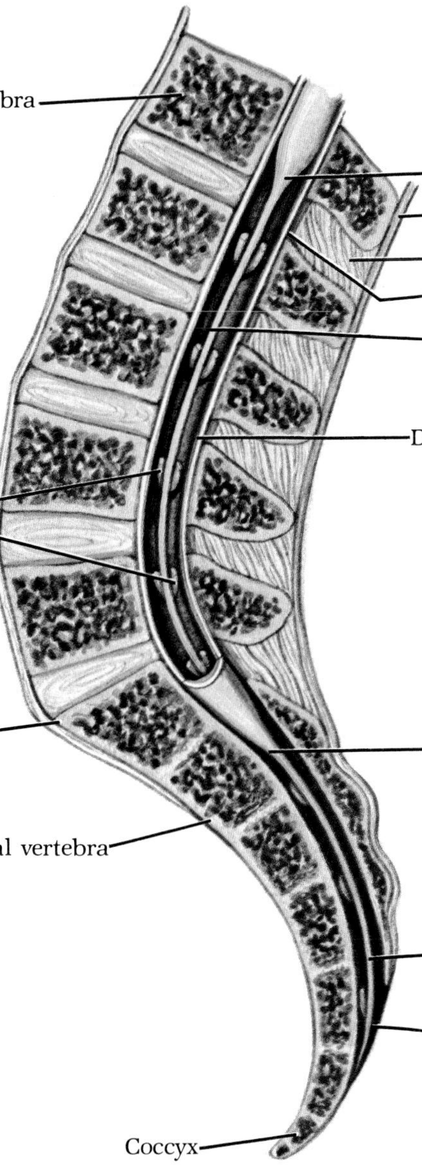
First lumbar vertebra
Conus medullaris
Supraspinous ligament
Interspinous ligament
Ligamentum flavum
Filum terminale
Dura mater and arachnoid mater
Roots of spinal nerves
Promontory of sacrum
Distal end of subarachnoid space
Second sacral vertebra
Filum terminale
Sacral hiatus
Coccyx

Figure 8-22
Median sagittal section through lumbosacral region, showing conus medullaris, filum terminale, and lower limit of subarachnoid space.

1. The cauda equina is a vertical leach of lumbar and sacral nerve roots around the filum terminale below the level of the termination of the spinal cord.
2. Lumbar puncture may be performed to withdraw a sample of cerebrospinal fluid for examination or to record its pressure. It is also used for injecting drugs into the subarachnoid space to combat infection or induce anesthesia. Since in the adult the spinal cord terminates inferiorly at the level of the lower border of the first lumbar vertebra (in the infant it may reach as low as the third lumbar vertebra) and the subarachnoid space extends inferiorly as far as the lower border of the second sacral vertebra, a needle can be introduced into the lower part of the subarachnoid space without damage to the spinal cord; the nerve roots are usually pushed to one side without damage. With the patient lying on his side with the vertebral column well flexed, the space between adjoining laminae in the lumbar region is opened to a maximum. An imaginary line joining the highest points on the iliac crests passes over the spinous process of the fourth lumbar vertebra. With careful aseptic technique and under local anesthesia, the lumbar puncture needle, fitted with a stylet, is passed into the vertebral canal above or below the fourth lumbar spine. The following structures will be penetrated: (a) skin, (b) superficial fascia, (c) supraspinous ligament, (d) interspinous ligament, (e) ligamentum flavum, (f) areolar tissue containing the internal vertebral venous plexus, (g) dura mater, and (h) arachnoid mater. With the patient in the recumbent position, the normal cerebrospinal fluid pressure is about 120 mm of water.

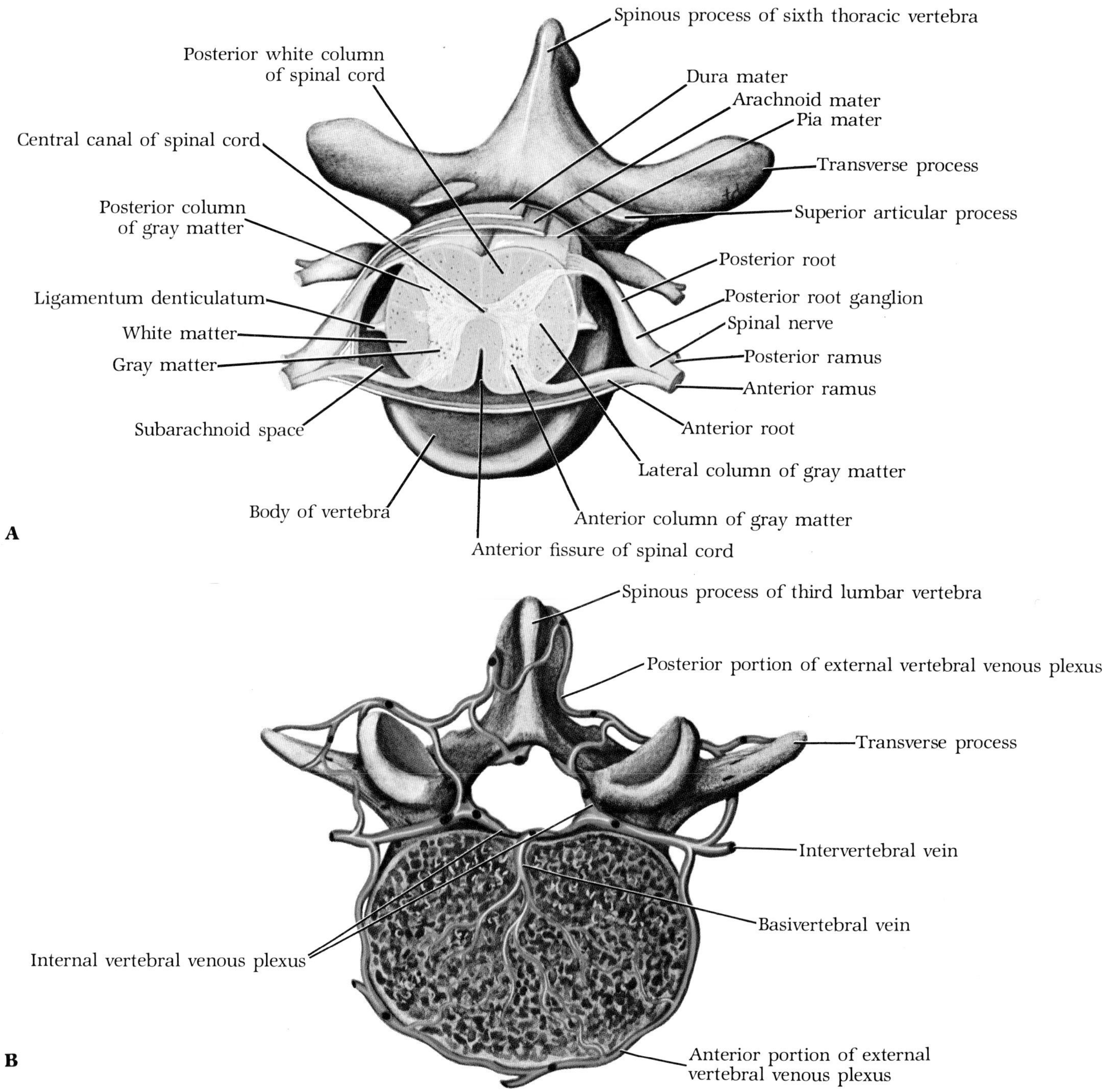
Spinous process of sixth thoracic vertebra
Posterior white column of spinal cord
Dura mater
Arachnoid mater
Pia mater
Central canal of spinal cord
Transverse process
Posterior column of gray matter
Superior articular process
Posterior root
Ligamentum denticulatum
Posterior root ganglion
Spinal nerve
White matter
Gray matter
Posterior ramus
Anterior ramus
Subarachnoid space
Anterior root
Lateral column of gray matter
Body of vertebra
Anterior column of gray matter
Anterior fissure of spinal cord
A
Spinous process of third lumbar vertebra
Posterior portion of external vertebral venous plexus
Transverse process
Intervertebral vein
Basivertebral vein
Internal vertebral venous plexus
Anterior portion of external vertebral venous plexus
B

Figure 8-23
Spinal cord and venous drainage of vertebra. A. Spinal cord and meninges lying within vertebral foramen of a thoracic vertebra. B. Venous drainage of a lumbar vertebra showing the internal and external vertebral venous plexuses.

1. Along the whole length of the spinal cord thirty-one pairs of spinal nerves are attached by the anterior, or motor roots, and the posterior, or sensory roots. Each root is attached to the cord by a series of rootlets, which extend the whole length of the corresponding segment of the cord. Each posterior nerve root possesses a posterior root ganglion, the cells of which give rise to peripheral and central nerve fibers.

2. In fracture-dislocations of the thoracic region, displacement is often considerable, and the small size of the vertebral canal results in severe injury to the spinal cord.

3. When moving a patient with spinal cord injury, it is imperative to keep the vertebral column in a neutral position, i.e., neither flexed nor extended. Never, never lift the patient by his feet and shoulders, because the movements of flexion may severely crush or sever the spinal cord. The stretcher should be placed on the ground alongside the patient, and he should be gently lifted with support beneath the head, cervical, thoracic, lumbar, pelvic, and leg regions by many assistants.

4. The longitudinal thin-walled, valveless vertebral venous plexus is important clinically since it provides a route by which carcinoma of the prostate may metastasize to the vertebral column and the cranial cavity. The internal vertebral plexus communicates above with the intracranial venous sinuses and segmentally with the veins of the thorax, abdomen, and pelvis. The internal venous plexus is not subject to direct external pressures when the intra-abdominal or pelvic pressures rise. However, coughing or sneezing, by raising the intrapelvic pressure, would aid in the spread of malignant cells from the pelvis to the cranium via the intervertebral veins and the internal vertebral venous plexus.

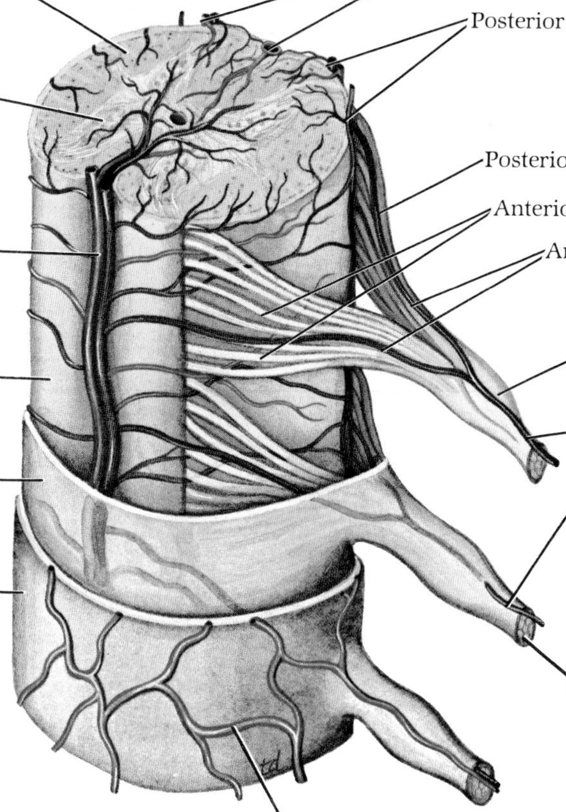

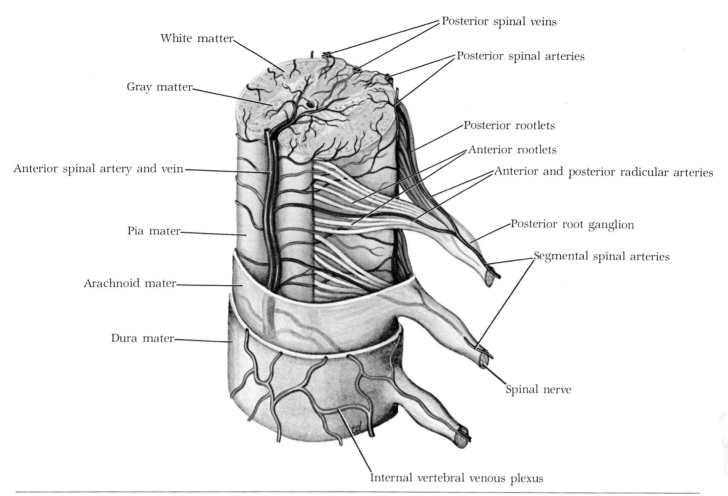

Posterior spinal veins
White matter
Posterior spinal arteries
Gray matter
Posterior rootlets
Anterior rootlets
Anterior spinal artery and vein
Anterior and posterior radicular arteries
Posterior root ganglion
Pia mater
Segmental spinal arteries
Arachnoid mater
Dura mater
Spinal nerve
Internal vertebral venous plexus

Figure 8-24
Blood supply and venous drainage of the spinal cord.

1. The spinal cord is supplied by two posterior spinal arteries and one anterior spinal artery. The two posterior arteries are branches of the vertebral or posterior inferior cerebellar arteries, and the anterior spinal artery is formed from the union of a branch from each vertebral artery. The spinal arteries receive important additional tributaries from the segmental arteries, which are branches of the vertebral, posterior intercostal, and lumbar arteries; these arteries enter the vertebral canal through the intervertebral foramina. One segmental artery in the lower thoracic or upper lumbar region is particularly important, since it forms the main supply of blood to the distal portion of the anterior spinal artery; when it is realized that the anterior spinal artery supplies the greater part of the spinal cord, except the posterior gray horns and the posterior white columns (supplied by the posterior spinal arteries), it will be appreciated that occlusion of this segmental artery will deprive most of the lower end of the spinal cord of its blood supply. Orthopedic and neurosurgeons should treat this segmental artery with great respect.

Index